£4.60

Papers in Plant Physiology

Consulting Editors

Papers in Plant Physiology

Edited by

WILLIAM S. HILLMAN

Brookhaven National Laboratory

HOLT, RINEHART AND WINSTON, INC.

New York • Chicago • San Francisco • Atlanta • Dallas • Montreal • Toronto • London • Sydney

Library of Congress Catalog Card Number: 72-110096
SBN: 03-081196-1
Printed in the United States of America
9 8 7 6 5 4 3 2 1

Preface

Purpose and Character of the Collection

This book is intended to provide a view of modern plant physiology through an acquaintance with some of the most interesting research papers in the recent literature. To the reader previously unacquainted with this general area, but with some basic knowledge of biology, these papers should give an idea of the concepts, procedures, and materials currently used in studying the activities of the vascular plants. Because of this relatively restricted scope, work with other organisms is included only when it bears directly on processes in the higher plants. In addition, questions of completely general biological significance (such as respiration) receive relatively less attention than topics that are peculiarly significant to plants. Even with such restrictions, the field is still very large for a single volume.

Since most of the papers collected here have appeared within the last ten years, and many are much more recent, some of the problems that a historical treatment of plant physiology might define as central receive relatively little attention. Even within this time limitation, however, the inclusion of a paper here does not necessarily imply that it is highly original or fundamental in its area. Although some of the selections may indeed be regarded as "classical," these have been included only when they seem particularly useful to the reader, and not because of their historical position—it is an unfortunate fact that the most creative investigators do not necessarily write the most lucid reports. Many relatively trivial criteria were also used in selecting the papers. In general, articles with fewer pages seemed preferable to those with many, yet thorough, well-documented discussions were preferred to perfunctory ones. Naturally the subjective element in such selections remains very large; it is hardly exaggerating to say that an equally appropriate collection could be made without including any of the papers chosen here. Above all, a collection made to illustrate current questions, approaches, and procedures should not be mistaken for either a connected set of case-histories or a source of authoritative "facts."

The arbitrary but pragmatic criterion of language—since all the papers are in English—necessarily misrepresents the literature as a whole. Though a surprisingly large fraction of research in non-English-speaking countries is now published in English, much is not. Thus, even allowing for the fact that during the period covered, a disproportionately large share of the world's research was being done in the United States, the restriction to the English language inevitably imposed an additional distortion in the same direction. In this connection, however, note that a substantial proportion of the work even in American journals is by scientists trained or working abroad.

As to organization, the papers have simply been grouped into six sections of approximately equal length. The section titles are self-explanatory except perhaps for that of Section VI, which is intended as short-hand for a concern with structure and organization, and with environmental effects on them. Like the selection of particular papers, both the organization into sections and the placement of papers within them is quite subjective, and not to be taken too seriously.

A Note on Reading Scientific Papers

Since for some readers this book may offer the first experience with the kinds of article published in scientific journals, some remarks on the typical structure of such articles may be helpful. Except for the "letter" or "short communication," whose brevity makes explicit organization unnecessary, most papers today consist of several parts: an introduction, a section on materials and methods, an experimental or results section, a discussion, and a summary or abstract. These usually occur in the order indicated, except that the abstract may be placed either at the beginning or the end, and that two or more of the others (usually either the methods and results or the results and discussion) may be combined. But even though the form is fairly standard, individual writers use the various parts quite differently. This is particularly true for what may be most important to readers of this volume, the explanation of the significance of the work. This is usually dealt with in the introduction and discussion sections, but often one of these is very brief, placing the main task on the other. More rarely, most of the useful discussion is actually with the results. In addition, the helpfulness of the abstract varies immensely; often enough, abstracts make sense only to those already thoroughly familiar with the specialty.

While these practices may at first be both foreign and annoying to the inexperienced reader, there are of course good reasons for most of them, and they are neither perverse nor deliberately obscure. On the contrary, it is extremely useful when one is pressed for time to be able, for instance, to scan a recent issue of a journal by reading only the abstracts; to avoid introductory and background material in a specialty with which one is very familiar; or to find the details of a method without reading the whole article. For many readers of this collection, however, it may be best to read an article starting with the introduction and discussion (and abstracts) first, in order to get some idea of why the work was done, and only then to proceed to the results and the methods sections. This practice will almost surely not do for all the papers, but for most it will probably be better than proceeding straight through. Finally, the bibliographies are of course to be regarded not only as sources of evidence for various statements or assumptions in the paper, but also as useful starting points for going back into the earlier literature. In addition, they can often indicate the names of important contributors to the subject in question and the journals in which they publish.

Plant Physiology and Plant Physiologists, or, Who Writes the Papers and How Good Are They?

The scope of plant physiology is extremely broad, running from near-ecological concern with the interactions of gross environmental factors all the way to molecular level studies of the consequences of minor changes in protein conformation. Yet the number of plant physiologists, no matter how defined, is not very great, and many of them work strictly on applied problems. The consequence of this is that some fundamental topics are studied by a relative handful of people; this, in turn, means that the results of technical errors, poor experimental plans, or plain sloppy thinking can persist unchallenged for many years in a way that is hardly imaginable in more narrowly defined and intensely competitive fields. Hence it is often impossible to be sure of the quality of a paper even some years after its publication. The reader should keep this constantly in mind, for since it is a characteristic of plant physiology as a whole that some of the finest and some of the most fatuous work can occur side by side in the same journal, the present collection is unlikely to escape the problem entirely. The chief difficulty, of course, is that there is no wholly effective detection mechanism for bad science—each of us carries his own, but would be a fool to believe it infallible.

Although investigations relevant to plant physiology appear in probably hundreds of periodicals, the vast proportion of the significant work appears in fewer than a dozen. That more than one-third of the papers selected here come from *Plant Physiology* is most probably a joint result of the language restriction and the bias in favor of short papers (which are much less common, for example, in British journals), but it may also reflect the unavoidable prejudices of a member of its editorial board. Depending on the journal, the process through which a paper must go before appearing varies widely. Some journals use a rigorous refereeing procedure in which the paper is checked by two or more experts in its field and accepted, modified, or rejected according to their recommendations. This procedure is not at all foolproof, though it usually helps. Other, equally respectable journals have a much less formal procedure and leave things to the discretion of one or more editors. A few journals—the prestigious *Proceedings of the National Academy of Sciences* (U.S.), for example—accept any paper recommended by any member of the sponsoring body. In all types of journals, the author with a substantial reputation may be accorded different (not always easier—sometimes more critical) treatment than a newcomer to the field. In short, the mere publication of a paper is no guarantee of its quality; nor is its appearance in this collection. In reading here, as elsewhere, the reader should keep his critical faculties alert. After all, the fun of science has at least two parts—arriving at new insights, and showing how wrong the other fellow was; papers selected to demonstrate the first are still fair game for the second.

In this connection, early in the planning of this work it had seemed desirable to introduce each group of papers with a brief but thorough summary of the background and present status of the problem covered. However, it soon became obvious that this was impossibly overambitious, meaning nothing less than an outline text in plant physiology—a text, moreover, which because of its brevity would in the end be more misleading than valid. Thus the reader will find no reviews or references here besides those in the papers themselves. In general, the latter do a reasonably good job of setting the context at the time of writing. Evaluating the work in terms of more recent developments, of course, poses another problem, a problem for which the only general solution is for the reader to approach the literature of reviews and monographs as well as original experimental papers. Probably the best entry into the review literature is the *Annual Review of Plant Physiology* (Annual Reviews, Inc., Palo Alto, California). Since 1950, the yearly volumes have provided a guide to the changing views and data in almost any area of plant physiology; the reader should make it a habit to consult both recent and past issues, with their cumulative indices, on any subject under consideration.

I would like to acknowledge the support of Brookhaven National Laboratory, under contract to the U. S. Atomic Energy Commission, during the compilation of this book. For valuable suggestions and criticisms on proposed selections I am much indebted to my friends and colleagues Winslow R. Briggs, Geoffrey Hind, Frank Loewus, John M. Olson, William K. Purves, and Bruce B. Stowe.

Upton, L.I., N.Y.
July, 1969 *W. S. H.*

Credits

Ahmed, S. and H. J. Evans, "Cobalt: a micronutrient for the growth of soybean plants under symbiotic conditions," *Soil Science* **90:** 205-210. Copyright 1960, The Williams & Wilkins Company, Baltimore, Maryland. Reprinted by permission.

Borthwick, H. A. and R. J. Downs, "Roles of active phytochrome in control of flowering *Xanthium pennsylvanicum*," *Botan. Gaz.* **125:** 227-231 (1964), The University of Chicago Press.

Mancinelli, A. L., H. J. Borthwick, and S. B. Hendricks, "Phytochrome action in tomato-seed germination," *Botan. Gaz.* **127:** 1-5 (1966), The University of Chicago Press.

Wagner, E. and H. Mohr, "Kinetic studies to interpret 'high energy phenomena' of photomorphogenesis on the basis of phytochrome," *Photochem. Photobiol.* **5:** 397-406. Copyright 1966, Pergamon Press. Reprinted by permission.

Butler, W. L., H. W. Siegelman, and C. O. Miller, "Denaturation of phytochrome," *Biochemistry* **3:** 851-857. Copyright 1964, The American Chemical Society. Reprinted by permission.

Wright, S. T. C., "Growth and cellular differentiation in the wheat coleoptile *(Triticum vulgare)* II. Factors influencing the growth response to gibberellic acid, kinetin and indolyl-3-acetic acid," *Journal of Experimental Botany,* Vol. 17, pp. 165-176 (1966). By permission of the Clarendon Press, Oxford.

Weinberger, P. and M. Measures, "The effect of two audible sound frequencies on the germination and growth of a spring and winter wheat," *Canadian Journal of Botany* **46:** 1151-1158 (1968). Reproduced by permission of the National Research Council of Canada.

Contents

1 Water, Minerals, Ion Uptake

2 Translocation, Metabolism, Biosynthesis

3 Photosynthesis

6 Cell, Organism, Environment

1
Water, Minerals, Ion Uptake

THE INFLUENCE OF SEVERAL TRANSPIRATION SUPPRESSANTS ON TRANSPIRATION, PHOTOSYNTHESIS, AND WATER-USE EFFICIENCY OF COTTON LEAVES

By R. O. SLATYER* and J. F. BIERHUIZEN*†

[*Manuscript received August* 2, 1963]

Summary

The influence of transpiration suppressants on transpiration and apparent photosynthesis of cotton leaves was investigated under moderate and high light intensity ($7 \cdot 5$ and $22 \cdot 5 \times 10^4$ erg cm^{-2} sec^{-1}) and low windspeed ($1 \cdot 5$ and $3 \cdot 1$ cm sec^{-1}) conditions. The substances used were Adol 52 (a formulation of cetyl alcohol), OED Green (a formulation of oxyethylene docosanol), S-600 (a plastic transplanting spray), and phenyl mercuric acetate.

All substances caused significant reductions in transpiration which were associated with marked increases in the diffusive resistance of the leaf to water vapour transfer due primarily to stomatal closure. All substances also caused reductions in photosynthesis which appeared to be due in part to increased resistance to CO_2 diffusion, and in part to metabolic inhibition of the photosynthetic process.

Phenyl mercuric acetate, at concentrations of 10^{-4} and 10^{-5}M, was the only substance which caused proportionately greater reductions in transpiration than in photosynthesis. This was achieved by stomatal closure without apparent effect on the photosynthetic apparatus. This caused a significant reduction in the transpiration ratio.

It is emphasized that while substances such as those tested can induce substantial reductions in transpiration, for maximum effectiveness the applied material should either develop a low permeability film over the entire leaf, or induce stomatal closure, without metabolically interfering with photosynthesis.

I. INTRODUCTION

For a number of years attention has been given to the possibility of suppressing transpiration from plants by reducing the permeability of the leaf surface to water vapour movement. One obvious means of achieving this is to manipulate the stomatal apparatus and many examples of phenomena or substances known to induce stomatal closure are reported in the literature (Heath 1959). An alternative procedure is to impose a low permeability film over the entire leaf, as has been the aim in nursery practice during planting out of seedlings (Miller, Nielson, and Bandemer 1937; Miller *et al.* 1950).

Interest in this problem has been stimulated recently by experiments concerned with suppression of evaporation from exposed water surfaces (Mansfield 1955, 1958) and, as a result, a considerable number of investigations have been conducted, using not only various formulations of cetyl alcohol (Roberts 1961) but also a range of plastic films (Gale 1961; Gale and Poljakoff-Mayber 1964), wax emulsions (Miller

* Division of Land Research and Regional Survey, CSIRO, Canberra.

† Present address: Institute of Land and Water Management Research, Wageningen, Netherlands.

et al. 1950; Allen 1955), and dilute concentrations of metabolic inhibitors (Zelitch 1961; Smith and Buchholtz 1962; Stoddart and Miller 1962; Zelitch and Waggoner 1962*a*, 1962*b*). Most of these materials have been observed to reduce transpiration to various degrees, but the various modes of action are still not properly evaluated, even though there is good evidence that the inhibitor group acts primarily through inducing stomatal closure (Heath 1959; Zelitch and Waggoner 1962*a*, 1962*b*).

From the physiological viewpoint, it is important not only to understand the nature of transpiration suppression but also to investigate the effects of compounds, such as those given above, on photosynthesis and plant growth. Although transpiration suppression, if achieved without serious injury to treated plants, could in itself be of considerable practical value, the benefit would be greatly enhanced if suppression could be obtained without a proportional reduction in photosynthesis.

Since both transpiration and photosynthesis involve gaseous diffusion across the leaf–air interface it seems probable, from a superficial examination, that an increase in the resistance across this zone would affect both processes to a similar degree. However, more detailed examination reveals an important difference in the two diffusion pathways.

Transpiration involves the evaporation of water from sites primarily located in the walls of internal mesophyll cells, and its diffusion through two resistances, connected in series, to the free air outside the leaf. The first of these resistances is encountered by the diffusing vapour as it moves up to, and through, the stomatal pore, and in this paper it will be termed leaf resistance r_l. (Some evaporation takes place from the epidermal cell walls and the vapour diffuses directly through the leaf cuticle. The amount is very small, however, and this pathway will be neglected in this analysis.) The vapour then diffuses across the external boundary layer, which sheathes the leaf, finally reaching the free air beyond where convective transfer processes dominate gas transfer. This external resistance in the boundary layer will be termed r_a.

Transpiration can therefore be represented by an expression basically similar to Fick's law where the total flux, T, in terms such as g (or cm^3) cm^{-2} sec^{-1} can be written as

$$T = \Delta c/(r_a+r_l), \tag{1}$$

where Δc is the difference in water vapour concentration at the evaporating sites and in the free air [in g (or cm^3) cm^{-3}] and (r_a+r_l) is the total resistance to flow in sec cm^{-1}.

Photosynthesis also involves these internal and external resistances since CO_2, in entering the plant, must pass through the same pathway in the reverse direction. These resistances for CO_2 transfer, r'_a and r'_l, are related to r_a and r_l by the relationship

$$(r'_a+r'_l)/(r_a+r_l) = D/D', \tag{2}$$

where D and D' are, respectively, the diffusion coefficients of water vapour and CO_2 in air. Their approximate values are 0·24 and 0·14 cm^2 sec^{-1} so that the ratio of D/D' is about 1·7. However, the CO_2 must diffuse, not only to the mesophyll cell walls, but through the walls to the photosynthetic sites in the chloroplasts, and this liquid phase transfer involves an additional resistance, r'_m, connected in series to r'_a

and r'_l. Consequently, an expression for photosynthesis, P, can be written in a form similar to that for transpiration in equation (1) as follows

$$P = \Delta c'/(r'_a + r'_l + r'_m), \tag{3}$$

where $\Delta c'$ is the concentration difference between the photosynthesizing sites in the chloroplasts and the free air, and $(r'_a + r'_l + r'_m)$ is the total resistance to CO_2 transfer, all in the same units as used previously.

Few values for r'_m are available, but most evidence suggests that it is generally several times as high as $(r_a + r_l)$ (Gaastra 1959, 1962). Because of this, and because r_a is usually relatively low under even moderate windspeed conditions (0·5–2·0 sec cm^{-1}) it is apparent that factors which influence r_l may cause $(r_a + r_l)$ to increase several fold and hence cause a marked reduction in transpiration. The effect on $(r'_a + r'_l + r'_m)$, however, and hence on photosynthesis, could be comparatively small so that a significant improvement in efficiency of water use could be obtained.

Because of the potential significance of this phenomenon, it was decided to conduct a series of experiments to investigate the direct effects of various transpiration suppressants on both transpiration and photosynthesis first, and then to interpret these effects in terms of the various diffusion resistances known to occur in the water vapour and CO_2 pathways. The experiments were conducted on the leaves of young cotton plants in conjunction with comprehensive studies of transpiration (Slatyer and Bierhuizen 1964*a*) and photosynthesis (Bierhuizen and Slatyer 1964*a*) from which a considerable amount of basic material concerning diffusive resistances had been obtained. The experiments were preceded by preliminary investigations on whole plants (Slatyer and Bierhuizen 1964*b*).

II. Experimental

The experiments were conducted with a large number of cotton plants (cv. Pope) grown in earthenware pots in a heated, naturally lit, greenhouse. During the experimental period environmental conditions were fairly constant, maximum day temperature ranging from 35–40°C and daylength from 14–15 hr. At night minimum temperature was kept at 25°C with artificial heating. The plants were grown at a high level of nutrition and with frequent watering to minimize water stress. They were used for experimental purposes when they had reached a height of 35–40 cm. Measurements were made on one leaf which was selected for size and uniformity by using a standard leaf template of 100 cm^2. In this way the leaf material actually used for each experiment was very similar in geometry and seldom varied more than $\pm 5\%$ in leaf area. Even so leaf area was measured, by tracing the leaf outline and planimetering the enclosed area, and proportional adjustments made. In this paper leaf area therefore means the area of the leaf outline and *not* the total area of the upper and lower surfaces (i.e. $2 \times$outline area) as in some investigations.

When the plants had reached the desired growth stage, those required for experimental purposes were selected for uniformity and the transpiration suppressant treatments imposed. The following four substances, all of which had been used by other investigators, were used and concentrations were selected following preliminary experiments which indicated lower limits below which no significant effects could be observed and upper limits above which serious injury was caused during the test period:

(1) *Adol 52*: This material was obtained from Archer Daniels Inc., U.S.A. The stock material consisted of 95% cetyl alcohol plus about 1·5% C_{14}-alcohols and 1·5% C_{18}-alcohols. The solid material was made up into a stable water emulsion and the actual concentrations used (weight of solids/volume of emulsion) were 1·25, 2·5, 12·5, and 25%.

(2) *OED Green*: This material was supplied through the courtesy of Nikken Chemicals Ltd., Japan. It consisted of a 55 : 45 mixture of $C_{22}H_{45}O(C_2H_4O)H$ and $C_{18}H_{37}O(C_2H_4O)H$ and was made up into a 10% water emulsion. Actual concentrations used (weight of solids/volume of emulsion) were 0·5, 1·0, 2·0, and 5·0%.

(3) *S-600*: This material was obtained from Synchemicals Ltd., London. (It is sold as VL. 600 by the B. F. Goodrich Co. in the U.S.A.) No specific details of its composition or concentration were available, but it is a stable colloidal water dispersion of a modified vinyl resin (Witty 1953). Actual concentrations of the stock emulsion used were 5, 10, 50, and 100% (v/v) dilutions.

(4) *Phenyl mercuric acetate*: This compound was dissolved in 70% alcohol and stirred into water at concentrations of 10^{-5}, 5×10^{-5}, 10^{-4}, 10^{-3}, and 10^{-2}M.

To all solutions and emulsions a spreader (Tween 20) was added at a dilution of 1 : 2000. The control plants were treated with distilled water plus spreader.

The treatments were imposed between 4 and 5 p.m. on day 0. Each plant was inverted and dipped in the appropriate treatment solution, removed, and allowed to drain until dripping ceased. (Spraying was used initially but it was found that the amount of material applied could vary by a factor of more than 2 and it was therefore discontinued.) Two series of experiments were then conducted, one being short and one long term. The short-term experiments involved two sets of observations on a group of plants receiving all treatments on days 1 and 3 after dipping. Measurements of transpiration, photosynthesis, and leaf and air temperature were made on the same leaf of each plant at two light intensities, 7·5 and $22\cdot5\times10^4$ erg cm^{-2} sec^{-1} (corresponding to 2000 and 6000 f.c.), and at two windspeeds, 1·5 and 3·1 cm sec^{-1}. Other conditions, held constant, were bulk air temperature 35°C, bulk air relative vapour pressure 0·6, bulk CO_2 concentration 0·03%.

In the long-term experiments observations were made on days 2, 4, 8, 12, 18, and 25 after dipping. In this case only six treatments were employed, apart from the control. These were S-600: 10 and 50%; OED: 1·0 and 5·0%; and phenyl mercuric acetate: 5×10^{-5} and 10^{-4}M. The same measurements were made, again using the same leaf on each occasion, but at one light intensity ($22\cdot5\times10^4$ erg cm^{-2} sec^{-1}) and one windspeed (1·5 cm sec^{-1}) only. The same bulk conditions of temperature, humidity, and CO_2 concentrations were employed.

The experiments were conducted in conjunction with a large series of related investigations on transpiration and photosynthesis of cotton plants (Slatyer and

Bierhuizen 1964*a*; Bierhuizen and Slatyer 1964*a*, 1964*b*). Full details of the experimental techniques used for transpiration and photosynthesis measurements can be found in these papers, together with details of the calculations used to determine external and internal resistances to water vapour and CO_2 transfer. All photosynthesis data refer to apparent photosynthesis and therefore include respiration effects.

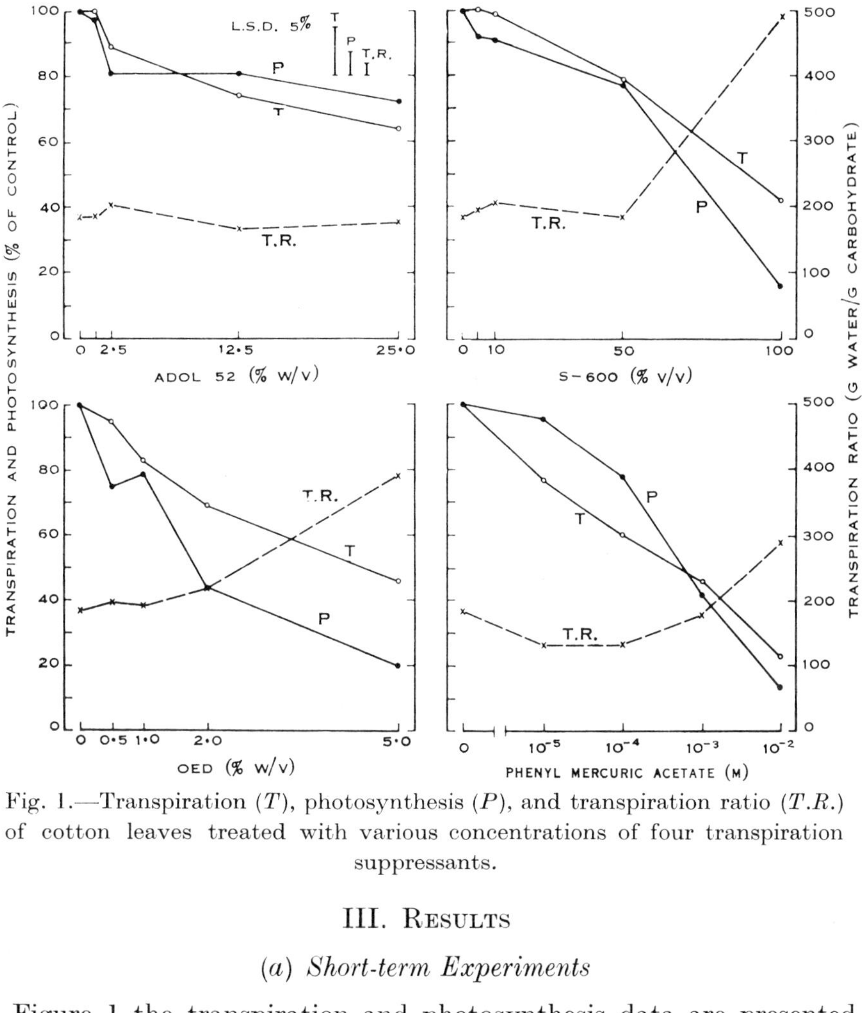

Fig. 1.—Transpiration (*T*), photosynthesis (*P*), and transpiration ratio (*T.R.*) of cotton leaves treated with various concentrations of four transpiration suppressants.

III. Results

(*a*) *Short-term Experiments*

In Figure 1 the transpiration and photosynthesis data are presented for all concentrations of the four transpiration suppressants investigated. The primary data were collected in terms of weight (in grams) of water and volume (in cm^3) of CO_2 per unit leaf area per unit time, but for ease of comparison are plotted as percentages of the untreated controls. Also included is the "transpiration ratio" calculated by dividing the transpiration rate by the rate of carbohydrate production, assuming that 44 g CO_2 is associated with 30 g carbohydrate in the plant. Differences required for significance at the 5% level were calculated from the variation found amongst the four nil-treatment controls. Converted to percentages, as used in Figure 1, differences in transpiration, to be significant at this level, had to exceed 15% and in photosynthesis had to exceed 7·5%. Differences in transpiration ratio had to exceed 20.

Because of space limitations and the fact that the other treatments provided similar results, the data given represent only the high light intensity ($22 \cdot 5 \times 10^4$ erg cm^{-2} sec^{-1}), high windspeed ($3 \cdot 1$ cm sec^{-1}) treatment and are the means of the observations made 1 and 3 days after treatment imposition. Under these conditions, the photosynthetic apparatus was light-saturated (Bierhuizen and Slatyer 1964*a*) so that photosynthesis was limited by the rate of diffusion of CO_2 to the chloroplasts. Also, stomatal resistance was minimal. The effects of any factors which influenced r_l, r'_l, or r'_m would therefore be expected to be most apparent.

Figure 1 shows that with Adol 52 photosynthesis and transpiration both decreased to a similar extent with increasing concentration, the transpiration ratio remaining virtually the same. With S-600, the trend was similar at low concentrations but at the highest concentration transpiration declined rapidly. However, the decrease in photosynthesis was even more pronounced, resulting in a marked increase in transpiration ratio. The same response pattern was observed with OED.

The effect of phenyl mercuric acetate on transpiration was also generally similar, but photosynthesis was less affected at concentrations below 10^{-3}M. This resulted in a significant decrease in the transpiration ratio at these concentrations.

The total and internal resistances to water vapour transfer, (r_a+r_l) and r_l, and to CO_2 transfer $(r'_a+r'_l+r'_m)$ and r'_m were calculated from the data of Figure 1 and are given in Table 1. Actual values of r_l were obtained from equation (1) using an r_a value of $1 \cdot 6$ sec cm^{-1} which was obtained by Slatyer and Bierhuizen (1964*a*) for a windspeed of $3 \cdot 1$ cm sec^{-1} using identical environmental conditions and the same plant material.

Estimates of $(r'_a+r'_l+r'_m)$ were obtained using equation (3) and assuming that $\Delta c'$ was equal to $0 \cdot 03\%$ CO_2 so that $(r'_a+r'_l+r'_m) = \Delta c'/P$. Then, r'_m can be found by subtraction, using equation (2) to calculate $(r'_a+r'_l)$. It should be emphasized that the validity of this approach to provide estimates of r'_m, which are in fact estimates of liquid-phase resistance to diffusion of CO_2, depends on the chloroplast concentration of CO_2 being effectively zero and on the validity of equation (2). The first assumption appears to be reasonable in the control treatments which were light-saturated at the CO_2 level used, but when the photosynthetic reaction is partially inhibited by the transpiration suppressants it is probable that the concentration of CO_2 at the chloroplasts is significantly greater than zero. If this is so r'_m is overestimated but it still gives an indication of the effective resistance which can be used to explain observed reductions in rate of photosynthesis. The second assumption, involving equation (2), will introduce errors if the "film"-type compounds are differently permeable to CO_2 and water vapour. However, at the concentrations used, transpiration suppression did not approach the stage associated with stomatal closure and it seems probable that the effective pore size of the films was relatively large and differential permeability effects relatively small.

Figure 2 is of value in explaining the results, and the specific effects of the various transpiration suppressants, of Figure 1. It can be seen from Figure 2 that Adol 52 has only a small effect on r_l and r'_m and the absence of marked changes in the ratio of the total resistances $(r'_a+r'_l+r'_m):(r_a+r_l)$ is a reflection of the relative constancy of the transpiration ratio. In the case of S-600 a similar pattern is observed except at the highest concentration where a steep increase in r_l is associated with a steeper increase in r'_m, and a consequent increase in the transpiration ratio. In OED both r_l and r'_m increase through the range of concentrations. With phenyl mercuric acetate the explanation of the previously observed reduction in the transpiration ratio is seen to be due to r'_m remaining at a steady value at concentrations up to 10^{-4}M, while a progressive increase occurred in r_l. Subsequently r'_m increased more rapidly than r_l, the final value being of a similar order to that observed with OED.

TABLE 1

SUMMARY OF RESISTANCES IN COTTON LEAVES TO DIFFUSION OF WATER VAPOUR AND CARBON DIOXIDE

Treatment	Concentration	Resistances to Water Vapour Movement		Resistances to Carbon Dioxide Movement	
		(r_a+r_l)	r_l	$(r'_a+r'_l+r'_m)$	r'_m
Adol 52	Nil	2·3	0·7	11·3	7·4
	5% (w/v)	2·4	0·8	11·6	7·5
	10% (w/v)	2·6	1·0	11·8	7·4
	50% (w/v)	2·4	0·8	12·4	8·3
	100% (w/v)	3·1	1·5	15·6	10·3
S-600	Nil	2·3	0·7	11·4	7·5
	5% (v/v)	2·4	0·8	11·5	7·4
	10% (v/v)	2·3	0·7	12·8	7·9
	50% (v/v)	2·6	1·0	12·9	8·5
	100% (v/v)	12·1	10·5	59·6	39·6
OED	Nil	2·6	1·0	12·1	7·6
	5% (w/v)	2·9	1·3	14·9	10·0
	10% (w/v)	3·4	1·8	15·3	9·5
	20% (w/v)	5·5	3·9	21·0	11·4
	50% (w/v)	9·6	8·0	50·4	33·8
Phenyl mercuric acetate	Nil	2·2	0·6	10·0	6·3
	10^{-5}M	3·8	2·2	12·3	5·8
	10^{-4}M	5·3	3·7	14·6	5·6
	10^{-3}M	9·7	8·1	22·6	6·1
	10^{-2}M	21·5	19·9	70·8	34·2

As mentioned above, it is difficult to obtain, from estimates of r'_m, a definite indication as to whether the observed inhibitions of photosynthesis were due to a real increase in r'_m and so act physically or are only apparent, not directly affecting CO_2 diffusion, and hence act metabolically. Accordingly, additional data on r'_m are presented in Figure 3 from the experiments at a light intensity of $7 \cdot 5 \times 10^4$ erg cm^{-2} sec^{-1} in which the photosynthetic apparatus was not light-saturated; in other words, when the rate of diffusion of CO_2 to the chloroplasts was presumably not limiting the rate of photosynthesis. On this assumption it may be proposed that, should the transpiration suppressants act by directly increasing r'_m, this increase should be delayed and reduced at the lower light intensity where CO_2 diffusion was not directly limiting. If, on the other hand, the mode of action is by metabolic inhibition, it is more probable that the reduction in photosynthesis, and hence the proportional change in r'_m, could be expected to be similar at both low and high light intensities.

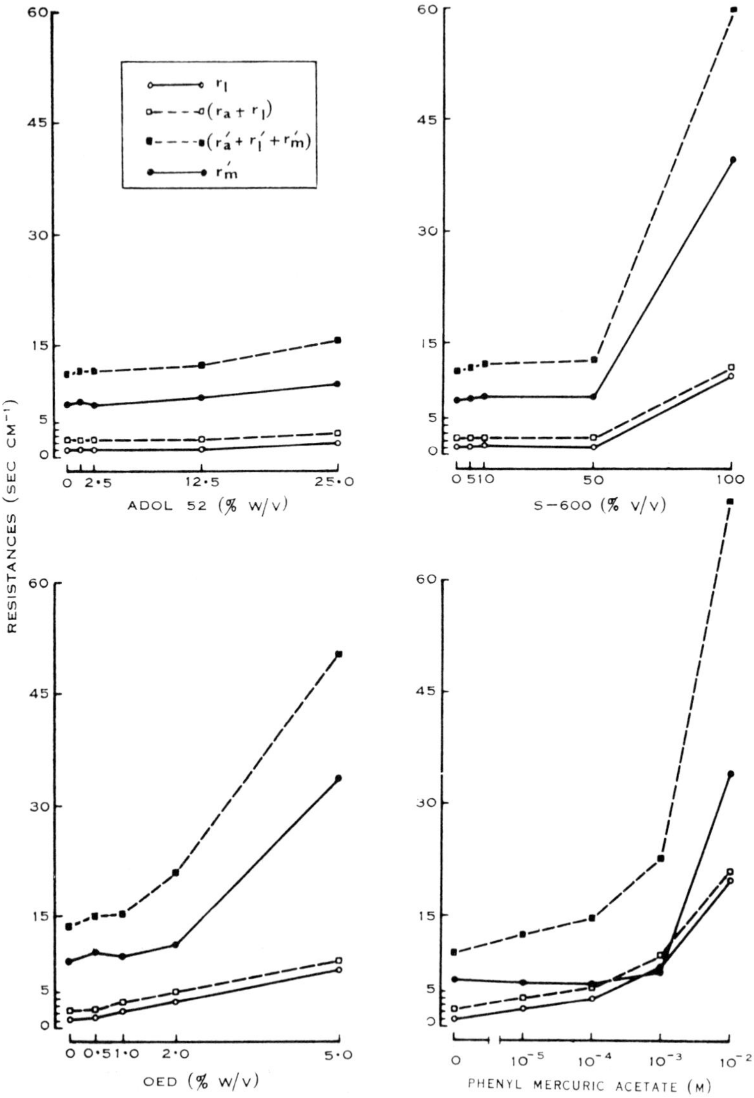

Fig. 2.—Total (r_a+r_l) and internal (r_l) diffusion resistances to water vapour transport and total $(r'_a+r'_l+r'_m)$ and internal (r'_m) diffusion resistances to CO_2 transport in cotton leaves treated with various concentrations of four transpiration suppressants.

From Figure 3 it can be seen that the ratio of $r'_{m(7\cdot5)}$ to $r'_{m(22\cdot5)}$ is relatively constant in all treatments. Although it was not possible to calculate meaningful values for least significant difference, all values lay within a range of ratios of 1·8–2·1 so that any direct effects of the treatments on r'_m appear to be small.

(b) Long-term Experiments

In Figure 4 data for transpiration, photosynthesis, and transpiration ratio are plotted against time for the long-term experiments which continued for 25 days following treatment imposition. The experiments were conducted at a windspeed of $1 \cdot 5$ cm sec^{-1} and a light intensity of $22 \cdot 5 \times 10^4$ erg cm^{-2} sec^{-1}. As in Figure 1, differences required for significances at the 5% level were calculated from the variation found among nil-treatment control plants. To be significant at this level differences in transpiration had to exceed $0 \cdot 15$ g 100 cm^{-2} hr^{-1}; in photosynthesis $1 \cdot 2$ mg 100 cm^{-2} hr^{-1}; and in transpiration ratio 14.

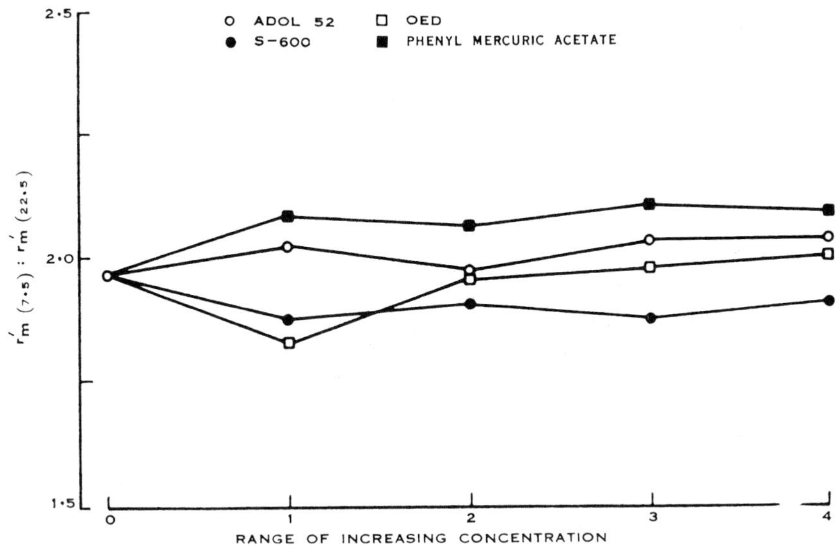

Fig. 3.—Influence of concentration of transpiration suppressants on the ratio of r'_m at $7 \cdot 5 \times 10^4$ erg cm^{-2} sec^{-1} $[r'_{m(7 \cdot 5)}]$ to the value at $22 \cdot 5 \times 10^4$ erg cm^{-2} sec^{-1} $[r'_{m(22 \cdot 5)}]$. The numbers 0–4 refer to increasing concentrations of transpiration suppressants in the ranges used for Figures 1 and 2.

The initial responses are very similar to those reported earlier but it is most interesting to observe the subsequent trends in the treatment and control plants. In the control, both transpiration and photosynthesis declined throughout the experimental period, the final values of transpiration being about one-half, and those of photosynthesis less than 40%, of the original. In consequence the transpiration ratio increased significantly. By comparison transpiration of all of the high-concentration treatments remained fairly steady during the experimental period and photosynthesis only declined slightly, resulting in a relatively small increase in transpiration ratio. In the low-concentration treatments the general response was intermediate, photosynthesis declining at about the same rate as the control, but transpiration less rapidly.

The actual values of the transpiration ratio are also of interest. The high-concentration OED treatment was significantly higher than the control throughout the experimental period and at both phenyl mercuric acetate concentrations were significantly lower, the differences being a similar order to those observed in Figure 1.

In Figure 5, values of the diffusive resistances r_l and r'_m are plotted from the data of Figure 4. It is at once apparent that the marked reduction in transpiration rate

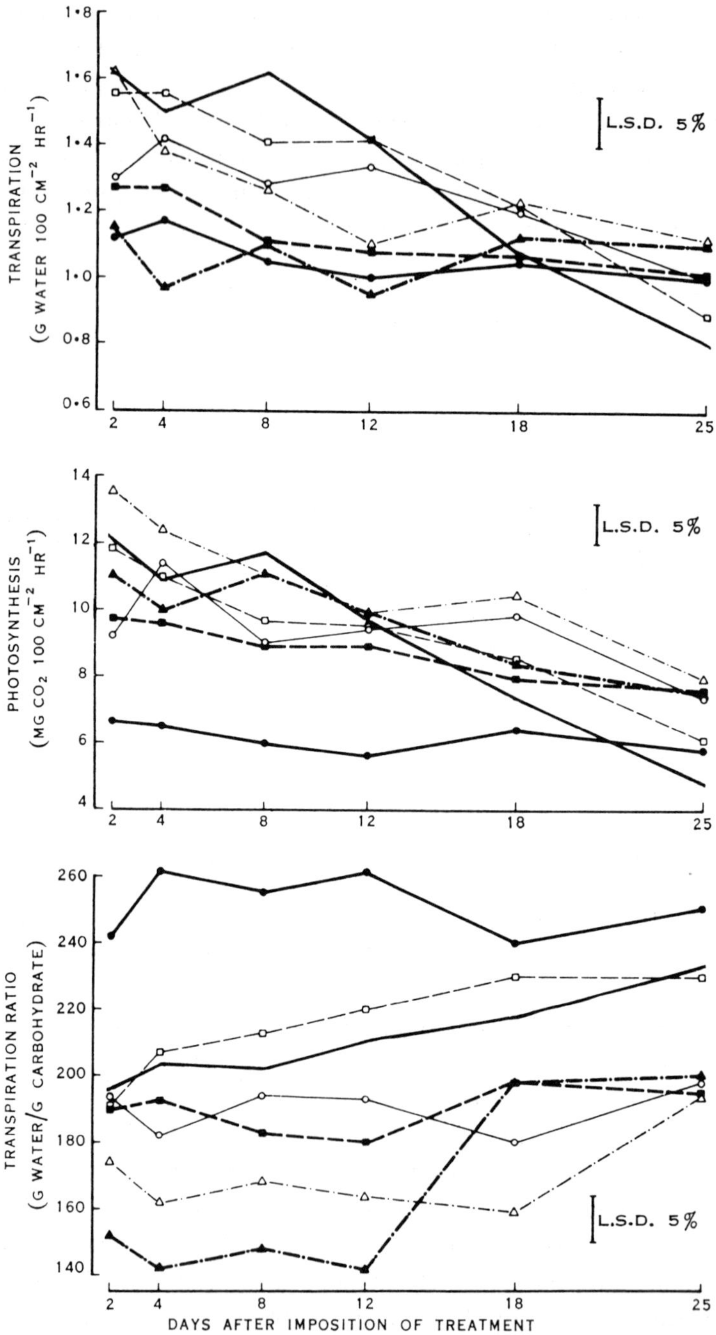

Fig. 4.—Effect of three transpiration suppressants on transpiration, photosynthesis, and transpiration ratio of cotton leaves over a 25-day period. —— Nil (control); ○ OED, 1%; ● OED, 2%; □ S-600, 10%; ■ S-600, 50%; △ phenyl mercuric acetate, 5×10^{-5}M; ▲ phenyl mercuric acetate, 10^{-4}M.

of the control, and the related increase in the transpiration ratio, is due to a steep increase in r_l in the latter part of the experimental period, r'_m increasing to a relatively small degree.

In the initial part of the period there was little change in r_l, the high-concentration treatments giving, in general, values 2–3 times as high as the control and the low-concentration treatments, 1–2 times. Subsequently r_l increased in most treatments

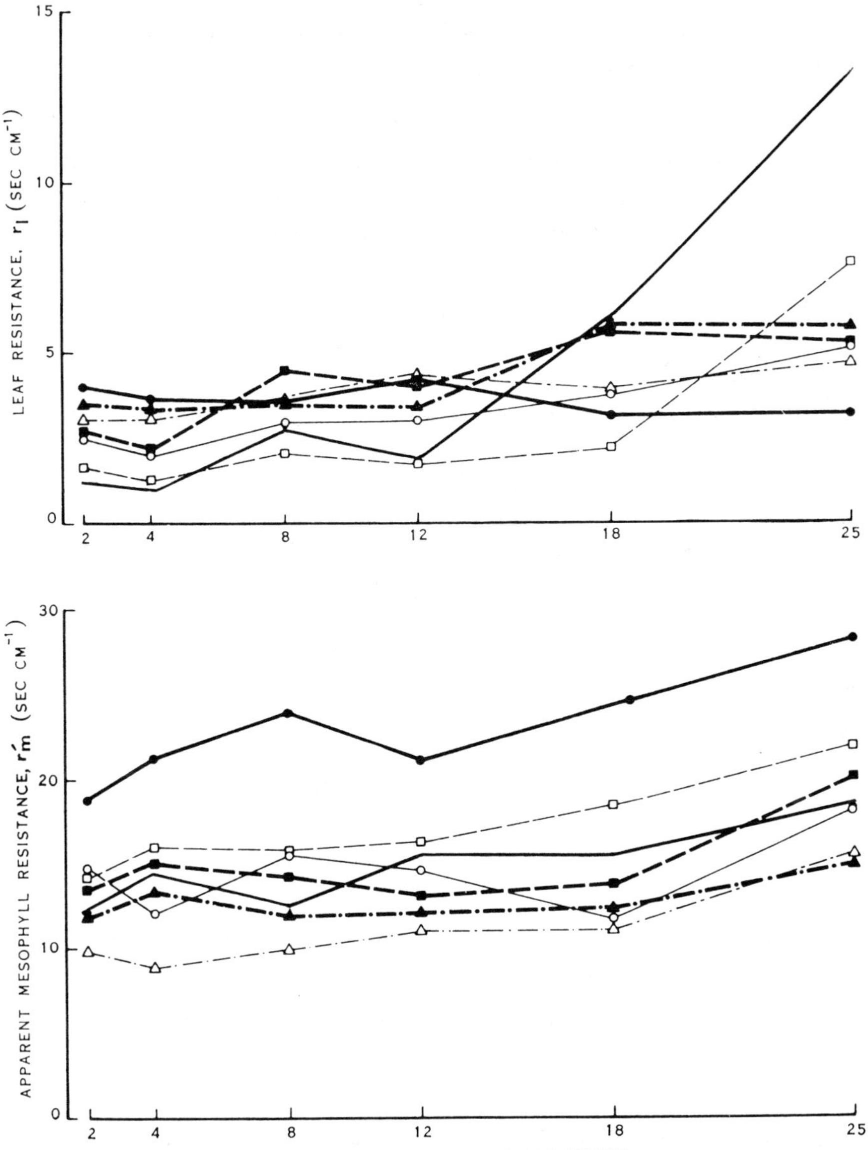

Fig. 5.—Effect of three transpiration suppressants on specific diffusive resistances to water vapour movement (r_l) and CO_2 movement (r'_m) in cotton leaves over a 25-day period. —— Nil (control); ○ OED, 1%; ● OED, 2%; □ S-600, 10%; ■ S-600, 50%; △ phenyl mercuric acetate, 5×10^{-5}M; ▲ phenyl mercuric acetate, 10^{-4}M.

but in no case as rapidly as in the control. By comparison, treatment values of r'_m showed a similar trend to those of the control over the whole period, and the actual values of r'_m did not differ markedly from those of the control except in the 20% OED treatment.

(c) *Effect of Transpiration Suppressants on Leaf Temperature*

Reduction in transpiration inevitably results in a greater proportion of the incident radiation which reaches the leaf being dissipated by processes other than latent heat transfer. Since the heat storage capacity of the leaf is small, and the proportion of energy used in photosynthesis is also small, the primary source of energy dissipation is by transfer of sensible heat from leaf to air.

While it is not the purpose of this paper to consider the nature of the sensible heat transfer process, it is of interest to examine the effect of transpiration suppression on leaf temperature, since most plants have fairly specific temperature thresholds and injury could be expected to occur if these were exceeded.

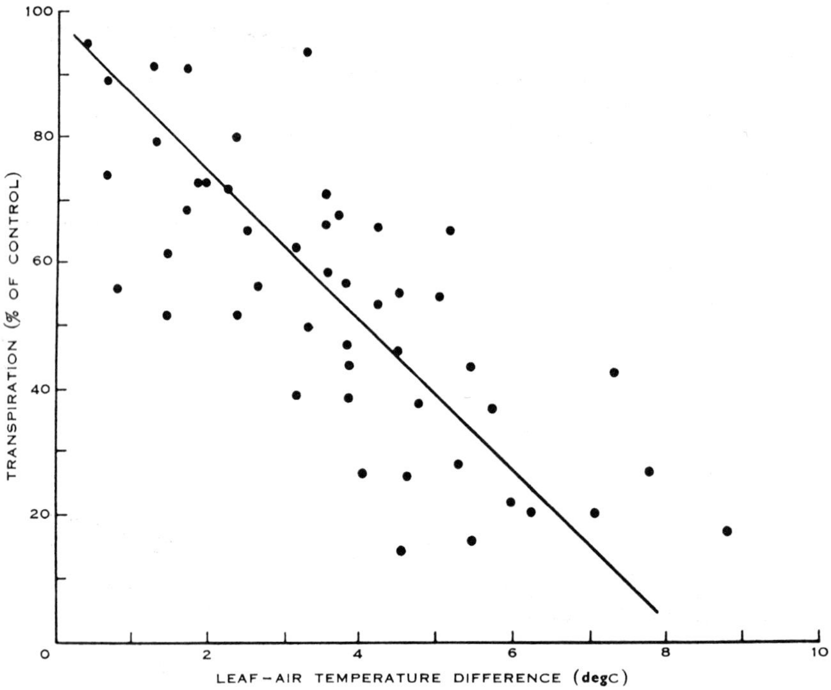

Fig. 6 —Effect of transpiration suppression on leaf–air temperature difference.

Accordingly, in Figure 6, the difference between leaf temperature and bulk air temperature, measured at the leaf chamber air inlet, is plotted against the percentage reduction in transpiration for all treatments used in both the short-term and long-term experiments.

The data, although providing a degree of scatter, give an approximate estimate of the order of leaf temperature change expected. With the windspeeds (and hence r_a values) used it can be seen that a reduction of 50% in transpiration is associated with an increase of leaf temperature over air temperature of about 4 degC, and that complete inhibition would be expected to give a difference of 8–9 degC.

IV. Discussion

Two main conclusions appear to stand out from the results, firstly that all the transpiration suppressants used appeared to cause significant reductions in transpiration; secondly that, except in the case of low concentrations of phenyl mercuric acetate, no treatments significantly improved water-use efficiency by reducing the transpiration ratio. These results are in good agreement with those obtained for whole plants by Slatyer and Bierhuizen (1964*b*).

However, it is appropriate to comment that the reductions in transpiration in the case of Adol 52, OED, and S-600 were achieved only at rather higher concentrations than those used elsewhere. Witty (1953), for example, recommends the use of S-600, as an antidesiccant, at equivalent concentrations of 10% and the manufacturers of OED claim significant reductions in transpiration at equivalent concentrations of less than 1%. By comparison results with phenyl mercuric acetate provided good agreement with the experiments of Zelitch and Waggoner (1962*a*, 1962*b*) at the concentrations used by these workers.

The mode of action of the suppressants on transpiration through an effect on r_l (Fig. 2) is almost certainly due to stomatal resistance, since cuticular resistance has been shown to be so high (32 sec cm^{-1}, cf. Slatyer and Bierhuizen 1964*a*). An increased cuticle resistance would only be of significance when the stomatal resistance reached a similar order of magnitude. The apparent degree of stomatal closure can be seen from Figure 2 to be only partial, even in the highest concentration of phenyl mercuric acetate where r_l reached a value of about 20 sec cm^{-1}. With S-600 and OED the values did not exceed 10 sec cm^{-1}. Even so this was still 10 times as high as the value for open stomata. It is also possible that some reduction in transpiration could be achieved by increasing the albedo of the leaf. However, net radiation, measured above the highest S-600 treatment (which was the most effective for this purpose) gave values only 5% lower than the control. There was no significant difference in the net radiation received by the other treatments, compared with the control, even at the highest concentrations used.

For maximum effectiveness a transpiration suppressant should operate on r_l alone so that r'_m remains unchanged and transpiration is reduced proportionally more than photosynthesis. In the control treatments of the present experiments the values of r_l averaged 1·0–1·5 sec cm^{-1} and r'_m 8–12 sec cm^{-1} so that with r_a at 1·5 sec cm^{-1}, trebling of r_l could be expected to cause a 50% reduction in transpiration and, if r'_m remained unchanged, only a 20% reduction in photosynthesis. In the case of the 10^{-4}M phenyl mercuric acetate treatments this readjustment was almost achieved and the transpiration ratio was reduced by 25% indicating stomatal aperture alone was affected. In all the other treatments, however, increases in r_l were linked with increases in r'_m and in those cases photosynthesis was reduced, proportionally, more than transpiration, causing an increase in the transpiration ratio.

Since phenyl mercuric acetate tends to act as a metabolic inhibitor, it was anticipated that at the low concentrations stomatal closure would be induced and that at the high concentrations there would be direct inhibition of photosynthesis. The evidence presented supports this view. By comparison the primary mode of action of the other substances was expected to be through the formation of a relatively impermeable film over the entire leaf. In the case of OED and Adol 52 it was also thought that a monomolecular film may have spread over the internal substomatal surfaces.

Both of these phenomena may have occurred but the fact remained that the apparent value of r'_m also was affected and, rather than a direct increase being caused by film formation over the mesophyll surfaces, the primary effect appeared to be on metabolic inhibition of photosynthesis itself. However, it should be mentioned again that the procedure for calculating r'_m involves possible errors as outlined previously. Conclusions based on assumed values of r'_m should therefore be interpreted cautiously.

It is unlikely that other factors, such as increased leaf temperature, would have metabolically suppressed photosynthesis, since the highest temperatures measured did not exceed 42°C and cotton is frequently grown in areas with daily maximum temperatures exceeding 45°C without injury (Eaton 1955). In consequence it appears that these compounds produced undesirable side effects and it would seem that, when more other substances are being examined, emphasis should be placed either on the ability to develop and maintain relatively impermeable films on plant leaves or to induce stomatal closure, without seriously inhibiting the metabolic effectiveness of the photosynthetic apparatus.

The long-term experiments provided the interesting result that effective reductions in transpiration could be maintained for 25 days, even though reductions in photosynthesis were also maintained, and the transpiration ratio did not change greatly. Also of interest was the progressive decline, particularly in transpiration, of the control at a more rapid rate than in the treated plants so that at the end of the test period lowest transpiration was observed in the control. The most probable explanation for this phenomenon is that in the control, total growth of the plants was unimpeded and a substantial amount of new leaf growth occurred during the test period. Associated with this was probably a considerable degree of solute translocation to the new tissue with consequent senescence of the test leaves. By comparison, in the treated plants, total growth was suppressed along with photosynthesis and less new leaf material developed. In consequence less translocation would be expected and the test leaves could be expected to remain relatively more active than in the controls. Good supporting evidence for these conclusions, obtained with whole plants, has been obtained by Slatyer and Bierhuizen (1964*b*).

The observed increases in leaf temperature due to transpiration suppression, although substantial, were almost certainly without deleterious effects on cotton growth. However, it is probable that growth of some species could be impaired by increases of 5–10 degC, should leaf temperatures exceed physiological threshold levels.

The work of Raschke (1956, 1960) and Takechi and Haseba (1962) enable some estimates of expected temperature change to be made, assuming different degrees of transpiration suppression. Taking an extreme case of 80% suppression under radiation levels of $1 \cdot 0$ cal cm^{-2} min^{-1}, it can be assumed that about $0 \cdot 8$ cal cm^{-2} min^{-1} will have to be dissipated by sensible heat transfer processes. Assuming also that the heat transfer coefficient is 0·06 cal cm^{-2} min^{-1} $degC^{-1}$ for both leaf surfaces (appropriate for a windspeed of about 1 m sec^{-1} and a leaf width of about 10 cm) the leaf–air temperature difference would need to be approximately 13 degC to dissipate this amount of energy. Under most outdoor conditions effective windspeed would exceed 1 m sec^{-1} and radiation levels of the order quoted are unusual, but this estimate provides an example of the care which should be exercised if heat-sensitive plants are under investigation.

In conclusion it may be emphasized that, although only one transpiration suppressant produced a significant improvement in water-use efficiency, all substances tested caused marked reductions in transpiration rate, without apparent long-term ill effects on the plants used. While considerable scope still exists, therefore, for a compound acting more specifically on r_l and, hence causing a real and pronounced improvement in water-use efficiency, the present results indicate that transpiration suppression is already possible and may have far-reaching practical implications if carried out in an appropriate biological and economic situation and with an understanding of the physical and biological phenomenon involved.

V. Acknowledgments

It is desired to acknowledge the assistance of Miss J. L. G. Sheaffe and Miss D. Coutts who conducted many of the measurements and computations involved in this investigation.

VI. References

Allen, R. M. (1955).—Foliar treatments improve survival of long leaf pine plantings. *J. For.* **53**: 724–7.

Bierhuizen, J. F., and Slatyer, R. O. (1964*a*).—Photosynthesis of cotton leaves under a range of environmental conditions in relation to internal and external diffusive resistances. *Aust. J. Biol. Sci.* **17**: (in press).

Bierhuizen, J. F., and Slatyer, R. O. (1964*b*).—An apparatus for the continuous and simultaneous measurement of photosynthesis and transpiration under controlled environmental conditions. CSIRO Aust. Div. Land Res. Reg. Surv. Tech. Pap. No. 24.

Eaton, F. M. (1955).—Physiology of the cotton plant. *Annu. Rev. Pl. Physiol.* **6**: 299–328.

Gaastra, P. (1959).—Photosynthesis of crop plants as influenced by light, carbon dioxide, temperature, and stomatal diffusion resistance. *Meded. Landbouwhogesch., Wageningen* **59**: 1–68.

Gaastra, P. (1962).—Photosynthesis of leaves and field crops. *Neth. J. Agric. Sci.* **10**: 311–24.

Gale, J. (1961).—Studies on plant antitranspirants. *Physiol. Plant.* **14**: 777–86.

Gale, J., and Poljakoff-Mayber, A. (1964).—Antitranspirants as a research tool for the study of the effects of water stress on plant behaviour. Proc. Symp. Eco-Physiol. UNESCO Arid Zone Res. (In press.)

Heath, O. V. S. (1959).—The water relations of stomatal cells and the mechanisms of stomatal movement. In "Plant Physiology". (Ed. F. C. Steward.) Vol. 2. pp. 193–250. (Academic Press, Inc.: New York.)

Mansfield, W. W. (1955).—Influence of monolayers on the natural rate of evaporation of water. *Nature* **175**: 247–9.

Mansfield, W. W. (1958).—The influence of monolayers on evaporation from water storages. I. The potential performance of monolayers of cetyl alcohol. *Aust. J. Appl. Sci.* **9**: 245–54.

Miller, E. J., Nielson, J. A., and Bandemer, S. L. (1937).—Wax emulsions for spraying nursery stock and other plant materials. Mich. Agric. Exp. Sta. Spec. Bull. No. 282.

Miller, E. J., Gardner, V. R., Petering, H. G., Comar, C. L., and Neal, A. L. (1950).—Studies on the development, preparation, properties, and applications of wax emulsions for coating nursery stock and other materials. Tech. Bull. Mich. State Coll. Agric. Exp. Sta. No. 218.

Raschke, K. (1956).—Über die physikalischen Beziehungen zwischen Wärmeübergangszahl, Strahlungsaustausch, Temperatur und Transpiration eines Blattes. *Planta* **48**: 200–38.

Raschke, K. (1960).—Heat transfer between the plant and the environment. *Annu. Rev. Pl. Physiol.* **11**: 111–26.

Roberts, W. J. (1961).—Reducing transpiration from plants. *J. Geophys. Res.* **66**: 3309–12.

Slatyer, R. O., and Bierhuizen, J. F. (1964*a*).—Transpiration from cotton leaves under a range of environmental conditions in relation to internal and external diffusive resistances. *Aust. J. Biol. Sci.* **17**: 115–30.

Slatyer, R. O., and Bierhuizen, J. F. (1964*b*).—The effect of several foliar sprays on transpiration and water use efficiency of cotton plants. *Agric. Meteorol.* **1**: (in press).

Smith, D., and Buchholtz, K. P. (1962).—Transpiration rate reduction in plants with atrazine. *Science* **136**: 263–4.

Stoddart, E. M., and Miller, P. M. (1962).—Chemical control of water loss in growing plants. *Science* **137**: 224–5.

Takechi, O., and Haseba, T. (1962).—Studies on the leaf temperature (2). On the heat transfer from the leaf of citrus in laminar flow. *J. Agric. Meteorol., Tokyo* **18**: 92–7.

Witty, P. M. (1953).—The uses of a water-dispersible plastic as an aid in park maintenance. Proc. 4th Ann. Mtg. Florida Parks Inst. pp. 1–7.

Zelitch, I. (1961).—Biochemical control of stomatal opening in leaves. *Proc. Nat. Acad. Sci., Wash.* **47**: 1423–33.

Zelitch, I., and Waggoner, P. E. (1962*a*).—Effect of chemical control of stomata on transpiration of intact plants. *Proc. Nat. Acad. Sci., Wash.* **48**: 1297–9.

Zelitch, I., and Waggoner, P. E. (1962*b*).—Effect of chemical control of stomata on transpiration and photosynthesis. *Proc. Nat. Acad. Sci., Wash.* **48**: 1101–8.

Reprinted from CROP SCIENCE
Vol. 1:26–28, 1961.

A Method of Studying Stresses Occurring in Plant Tissue During Freezing[1]

Charles Robert Olien[2]

THE immediate injuries resulting from exposure of cereal plants to low temperature are due to the formation of ice crystals within the tissue. Different types of stress can occur, depending on the location and rate of ice formation. The stress resulting from the formation of ice crystals within the cells causes their death because of mechanical disruption of the protoplasm. Hardy plants can survive the formation of some extracellular ice, death occurring only at lower temperatures due to external pressure or severe cell contraction (2, 4, 5). Continual evaluation of the distribution of water in plant tissue during freezing, with respect to both extra- and intracellular location of both the liquid and solid forms, would provide a means of determining the principal type of stress affecting any given plant tissue. The following method provides a means of continually evaluating changes in the content of liquid extracellular water during freezing and thawing of a region of a plant.

DEVELOPMENT OF THE METHOD

Extracellular water is considered to be the continuous liquid system in plant tissue through which substances in solution can diffuse freely and includes the film of moisture surrounding cells as well as water less closely associated with living protoplasts such as in xylem vessels. Electrolytes do not diffuse freely into and out of living protoplasts, their water content being isolated from the principal path of migration. The fact that cellular organization of living tissue constitutes an effective diffusion barrier is demonstrated by the high resistance of normal tissue to a weak direct electric current (about 5×10^6 ohms per centimeter along the longitudinal axis of leaves of cereal seedlings) and by the decrease in resistance to less than 5% of this value after the tissue has been killed.

The diffusion barrier is not broken by low voltage (in the range of 0 to 3 volts per centimeter), since within this range the resistance of the tissue is independent of the voltage applied. When slightly higher voltages are used, the resistance decreases reversibly with increasing voltage, indicating a weakening of the barrier. The diffusion barrier can be irreversibly destroyed by applying a sufficiently high voltage. Within the low voltage range, the tissue is unaffected by the amount of current up to a total of about 0.8 coulombs (the equivalent of 10 microampere days) (6).

The method of obtaining freezing patterns was developed from previous studies of electrophoresis of foreign ions and a stem rust toxin through the extracellular water of wheat leaves (7). The effect of environmental factors on the resistance of the electrophoretic unit used for studying migration provided the basis for evaluating changes in the liquid extracellular moisture content of plant tissue during freezing. The same type of unit was used for both problems and consisted of two electrical contacts and the plant tissue extending between them. Other units, in which a film of various solutions on paper or glass replaced the plant tissue, were used in developing the method of studying stress caused by freezing. The contact was made through a paste of finely divided charcoal moistened with an isotonic solution which is applied to about two square centimeters of the surface of the tissue. Dry graphite or platinum provides the best connection between the paste and the wires of the electrical circuit. The resistance of the contact is much less than that of the tissue and does not change appreciably with slight changes in the moisture content of the paste. However, the resistance will be high if the paste is very wet due to separation of the charcoal particles or if it becomes too dry due to poor contact with the tissue.

When a slight D.C. voltage is applied, a direct current is carried by ions which migrate through the continuous liquid system extending between the two electrical contacts. A direct rather than an alternating current is used, so that isolated water (intracellular) is excluded from the conducting system. The path of migration of ions through the tissue was traced by introducing foreign charged substances from the contacts. In leaves most of the current was found to be carried through the film of moisture on the mesophyll cells; but in roots a large part of the current seems to be carried through the xylem vessels. The current value does not drift while conditions are held constant, since the large surface of the charcoal contacts and the high resistance of the tissue minimize polarization. The current (I) in amperes is defined as the number of coulombs which flow per second. The number of coulombs involved is equal to the number of chemical equivalents of electrolyte (n) multiplied by the Faraday constant (F = 96,500 coulombs per equivalent). The average velocity of the ions ($\bar{u}$) divided by the distance between the contacts (l) gives the average proportion of the distance between the contacts each ion travels per second. Therefore, $I_i = \bar{u}_i l^{-1} n_i F$ for ions of one type, and the total current $(I) = \sum_{i=1}^{n} I_i$.

The derivation for determining the velocity of a charged particle through the liquid of an electrophoretic unit is given by Briggs (1) and by Glasstone (3), in which they followed the work of Helmholtz and Perrin. The force driving the particle has been shown to be proportional to the product of the voltage (E) applied across the unit and the electrostatic charge density (q) of the particle. Frictional force opposes the migration of a particle and is proportional to the product of the viscosity of the solution (N_s) and the velocity gradient $\left(\frac{u}{t}\right)$ in the layer surrounding the particle where (t) represents the thickness of the region of shear. These opposing forces come to equilibrium soon after applying the voltage, so that $u = Eq(aN_s)^{-1}$, where a is the proportionality constant and includes the factor t.

Viscosity is the only property of the medium which must be considered in evaluating the force restricting migration when the pathway between the two electrodes consists of a volume of liquid. When the volume is reduced to a film as in the extracellular liquid in leaf tissue, the resistive property of the medium is considered as effective viscosity (N) which may be higher than the viscosity of the bulk solution (N_s) due to orienting forces at the interface. Therefore, $u = Eq(aN)^{-1}$ or $u = Eqa^{-1} f$, where f stands for effective fluidity, the reciprocal of N. The average velocity of ions of one type then equals $Eqa^{-1}\bar{f}$, where $\bar{f}$ is the average effective fluidity. Therefore, $I_i = Eq_iF\ (a_il)^{-1}\ n_i\ \bar{f}$, and since $Eq_i F(a_il)^{-1}$ is approximately a constant (b_i) for any given electrophoretic unit operating at a constant voltage, $I_i = b_i n_i \bar{f}$ and the total current is proportional to the average effective fluidity of the medium when only the temperature and moisture content of the film are varied.

The current was found to be inversely proportional to the viscosity of the solution in bulk when only temperature was varied. The current is directly proportional to the film thickness or extracellular moisture content (M) when only the moisture content is varied within the limits of a film (data presented in figure 1). Therefore $\bar{f} = dMN_s^{-1}$, where d is the proportionality constant, and $I_i = c_i n_i MN_s^{-1}$, where c is the product of b and d. The total current $(I) = MN_s^{-1} \sum_{i=1}^{n} n_i c_i$ and $I = nMN_s^{-1} \sum_{i=1}^{n} \alpha_i c_i$, where α_i equals the proportion of the electrolyte made up of ions of a given type and where n equals the total number of equivalents of ions of all types. $\sum_{i=1}^{n} \alpha_i c_i$ will remain constant if there is

[1] Cooperative investigation, Crops Research Division, ARS, USDA, and Michigan State University. Journal Article 2570 from the Michigan Agr. Exp. Sta.

[2] Physiologist, Crops Research Division, ARS, USDA.

no change in the proportion of different ions during an experiment. Therefore, $I = KnMN_s^{-1}$ and if the subscript o is used to designate values of these factors at a chosen reference temperature, x to designate the values at any given temperature, and r to designate the relative value at any given temperature, then $I_o N_{s_o} = Kn_o M_o$, $I_x N_{s_x} = Kn_x M_x$, and $I_r N_{s_r} = n_r M_r$. Since n_r equals unity if there is no injury, the relative liquid extracellular moisture content equals $I_r N_{s_r}$.

I_r is equal to the reciprocal of the relative resistance of the tissue. In order to evaluate I_r accurately, the resistances should be determined by measuring I in microamperes at several different low voltages and extrapolating to zero electrical stress, since at very low voltage only the continuous liquid system through which ions can diffuse freely is involved. Care should be taken not to include contact resistance with resistance of the tissue. The contact resistance can be evaluated by making two contacts close together at each end of the part of the plant being studied, measuring the resistance, and comparing with the resistance of the tissue being studied. The removal of liquid water by evaporation or by the formation of ice crystals causes the conducting film of moisture in the tissue to diminish, and changes in liquid extracellular water are evaluated during freezing and thawing.

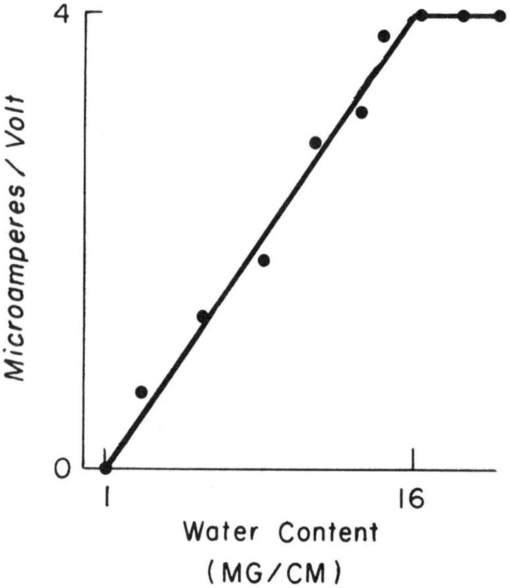

Figure 1—Microamperes of current per volt passing through a 1-cm.-wide strip of filter paper moistened with a 400-ppm NaCl solution plotted against the water content of the strip as it dries at 70° F.

RESULTS

The method is being used to study winter cereals having various degrees of hardiness induced by controlling environmental conditions during plant development. The plants are tested during a gradual change in temperature which causes stress of the type that could occur naturally. Various regions of intact plants have been studied independently without interfering with the freezing processes or causing injury other than that due to freezing. The regions studied usually consist of five centimeters of tissue between the electrical contacts. Portions of single roots and leaves in various stages of maturity as well as the mesocotyl and crown have been studied extensively during the past three years. The freezing patterns presented in figures 2 through 4 are based on data from single experiments with Dicktoo barley but are typical of the patterns obtained from studies of 6 varieties of barley as well as several varieties of wheat and rye.

Three distinct patterns have been found for the relationship between extracellular liquid water content and temperature during freezing of regions of plants in different stages of hardiness. The temperature at which death of the tissue occurred was estimated by evaluating injury in a series of tissues of the same type exposed to successively lower temperatures. Notes were taken on survival and injury, and deterioration of the tissue was studied microscopically. Leaves of hardened plants characteristically have a pattern of the type presented in figure 2. Most of the extracellular water is closely associated with the cells in this tissue as a film surrounding the living protoplasts, and the formation of ice is accompanied by cellular dehydration so that the extracellular moisture diminishes gradually. Injury accompanies severe cell contraction. The pattern in figure 3 is characteristic of nonhardy leaves frozen without dehydrating. Instead of decreasing, the content of extracellular liquid water suddenly increases when freezing begins. The

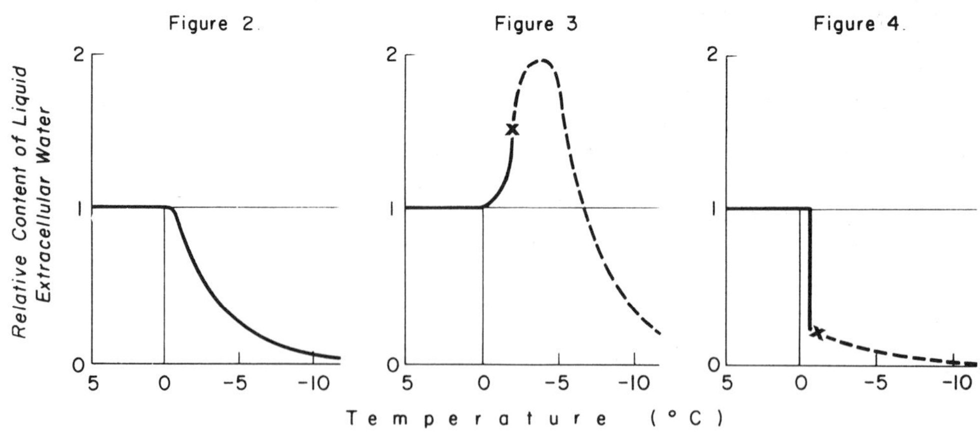

Figures 2, 3, 4—Freezing patterns in which the relative content of liquid extracellular water was plotted against temperature during gradual freezing of plant tissue. X indicates point at which severe injury occurred. Fig. 2. From leaves of hardened plants and roots of hardened dehydrated plants. Fig. 3. From leaves of nonhardened plants. Fig. 4. From roots of hydrated plants.

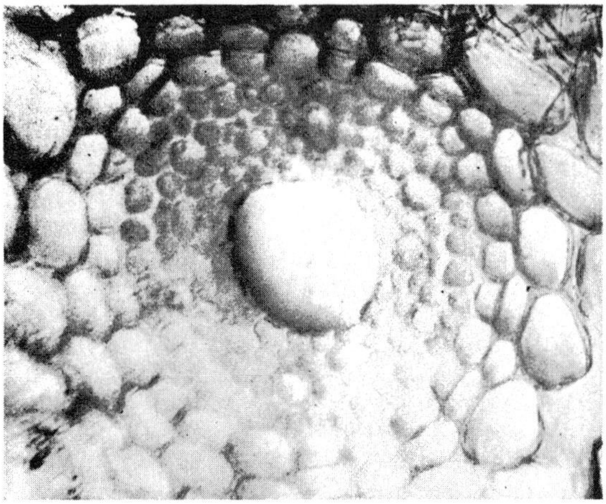

Figure 5—Cross section of a root from a hardened winter barley plant. X200.

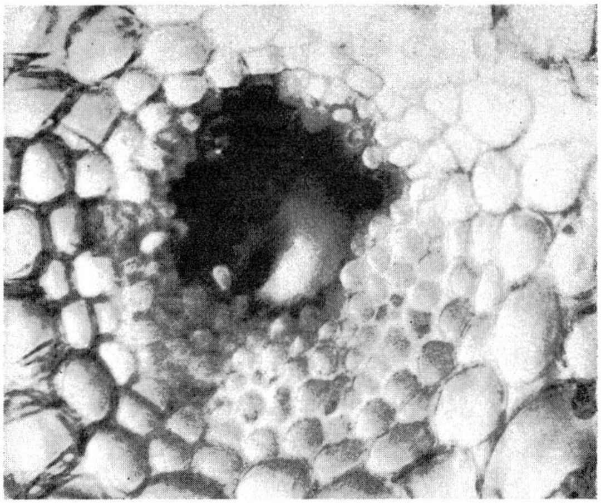

Figure 6—Cross section of a root from a hardened winter barley plant which was gradually frozen (to —4° C.) and thawed and then kept in a plant growth chamber at 75° F. for 4 days before sectioning. X200.

large increase results from disruption of the diffusion barrier probably due to intracellular ice formation causing cell rupture. Severe cell contraction is not probable due to the vapor pressure of water at this temperature. Cellular damage can be observed microscopically as a result of this treatment. These electrophoretic patterns seem to agree with what would be expected based on Levitt's (4) review of cellular behavior during freezing.

The third pattern (figure 4) is characteristic of normally hydrated roots. The content of extracellular liquid water abruptly decreases to less than 25% of its original value when freezing is initiated and is not accompanied by appreciable cellular dehydration. The sudden freezing of the water in the vascular tissue was suspected, since it is not closely associated with the living protoplasts. Severe injury was found in the vicinity of the xylem vessels when the tissue was examined microscopically. A cross section of a normal root with a single xylem vessel is shown in figure 5. The size of the vessel and the thick walls of the cells surrounding it indicate that considerable disruption might result from expansion of water in the vessel during freezing. Figure 6 shows a cross section of a similar root a few days after freezing and thawing. The injury surrounding the xylem vessel extends on one side into the tissue containing phloem elements, and several of the protoxylem vessels have been injured. The entire wall of the xylem vessel has been destroyed. When hydrated plants are exposed to temperatures below that required for initiating ice formation, more of the tissue within the endodermis is destroyed. Apparently the core of ice increases in size during cellular dehydration, extending the disruption of the tissue. Vascular injury from freezing results in the development of a brown core of deteriorating tissue within the roots. This zone of deterioration extends beyond the roots into the base of the crown. When injury is slight, the deteriorating vascular elements can be seen in the lower portion of sectioned crowns. When injury is severe, all the tissue in the lower part of the crown and in a conical zone extending up into the crown deteriorates. Moderate injury inhibits the formation of a new root system, and severe injury prevents the formation of new roots.

The upper part of the crown and the leaf tissue, having an electrophoretic pattern shown in figure 2, can survive lower temperatures than the base of the crown and the roots of normally hydrated, hardened plants can survive. Such plants can be exposed to temperatures low enough to severely damage the base of the crown without injury to the upper part of the plant. The plants appear normal during the first week after returning them to an environment suitable for growth. Some elongation of young leaves occurs. But the plants die during the second or third week, since the root system is not regenerated.

SUMMARY

At least three distinct patterns of distribution of extracellular water occur during freezing of different types of tissue from winter cereals. Each pattern is associated with a particular type of stress. The method permits studying various regions of a plant independently and evaluating the type and extent of stress occurring in each region without interfering with the process or injuring the plant. Separating components of hardiness by identifying the nature of the limiting stress associated with each phase of hardening may be important in understanding hardiness and selecting crops for particular locations.

REFERENCES

1. Briggs, David R. The determination of the ζ-potential on cellulose.—A method. J. of Phys. Chem. 32:641–675. 1928.
2. Dexter, S. T. The evaluation of crop plants for winter hardiness. Advances in Agronomy 8:203–239. Academic Press Inc., New York. 1956.
3. Glasstone, Samuel. Textbook of Physical Chemistry. 2nd Ed. D. van Nostrand Co., Inc., New York. Chap. 14. 1946.
4. Levitt, J. The Hardiness of Plants. Academic Press Inc., New York. Part 1. 1956.
5. Luyet, B. J., and Gehenio, P. M. The mechanism of injury and death by low temperature. Biodymanica 3:33–97. 1940.
6. Olien, C. R. Studies on necrosis in the infection of Khapli emmer by the stem rust fungus, *Puccinia graminis* var. *tritici*. Dissertation Abstracts 17:1171. 1957.
7. ————. Electrophoretic displacement of the necrotic area from the region of mycelial development in Khapli emmer infected with race 56 of *Puccinia graminis* var. *tritici*. (Abstract). Phytopathology 47:26. 1957.

How Mangroves Desalinate Seawater

By

P. F. Scholander

Scripps Institution of Oceanography University of California, San Diego
La Jolla, California

(Received April 21, 1967)

Abstract

Water relations of a number of mangroves and other halophytes were studied. All mangroves had high negative pressure in the xylem (30 to 60 atm) which more than balances the osmotic pressure of the seawater (25 atm). Freezing a short stem section of an attached twig always resulted in wilting caused by cavitation of the tensile sap. Semipermeable membranes in the leaves remained functioning in spite of being iced or heated (50°C), or being poisoned by dinitrophenol and carbon monoxide (13 atm). Chloroform and ether ruptured the membranes.

When some 40 atm pressure was applied to the root system of a decapitated seedling potted in seawater, a flow of nearly fresh water ensued. This separation kept on for hours in spite of chilling and poisoning with dinitrophenol or 20 atm carbon monoxide. A reversal of the gradient made roots lose water to the seawater. One may conclude, therefore, that mangroves separate fresh water from the sea by a simple non-metabolic ultrafiltration process, in a system where high tension obtains in the xylem and where turgor of the living cells is produced by an active maintenance of high solute pressure.

Introduction

Walter and Steiner (10), showed that the freezing point depression in juices of crushed roots and leaves of mangroves usually exceeded that of the seawater. In 1962 (6, 8) it was found that the xylem sap in many mangroves was close to fresh water; *i.e.,* with a salt concentration less than 0.03 per cent. In salt-secreting species, however, it could be 10 times this. Finally, in 1964, the hydrostatic pressure of the xylem sap was found to be negative by at least 25 atm, and at zero turgor pressure the freezing point depression in the parenchyma cells of the leaves closely balanced the negative pressure in the xylem. This strongly suggested that the mangroves derive their fresh water from the sea by a simple ultrafiltration. A single experiment on a pressurized root supported this (7).

On a recent expedition to the Great Barrier Reef in Australia on the laboratory vessel "Alpha Helix", studies of water relations were extended to a variety of mangroves and other halophytes. It was our special aim to inquire into the nature of the processes responsible for the hydrostatic-osmotic equilibrium between the xylem sap and leaf cells, and also the nature of the process by which the roots separate fresh water from seawater.

Materials

Most plants were obtained in the vicinity of the anchorage of the "Alpha Helix" at Flinders Islands, and some on the adjacent coast of Queensland. The following species were used: *Acrostichum aureum, Aegialitis annulata, Aegiceras corniculatum, Avicennia marina, Bruguiera gymnorrhiza, Canavalia obtusifolia, Ceriops Tagal, Enhalus acoroides, Euphorbia atotu, Excoecaria agallocha, Ipomoea pes-caprae, Lumnizera littorea, Messerschmidtia argentea, Osbornia octodonta, Pandanus* sp., *Pisonia grandis, Poaceae* sp., *Rhizophora mucronata,* and mistletoe, *Salicornia australis, Scaevola sericea, Sonneratia alba, Suaeda* sp., *Suriana maritima, Tecticornia cinerea, Thespesia populnea.*

Results

1. *Sap tension in xylem of mangroves and some other halophytes*

In Figure 1 is given the hydrostatic sap tension in twigs taken from a variety of intertidal mangroves and plants from sandy beaches and salt flats, as determined directly by the pressure-bomb technique (6, 9). All mangroves had a sap tension in excess of 24–26 atm, which corresponds to the osmotic pressure of the seawater. In rainy weather, and at night, the sap tension approached this lower limit, which in fact is the partial solvent pressure in the sea (4, 5), and the highest tensions were found on hot, sunny days.

Many plants growing further up on the sandy beach had only moderate tensions of 10 to 20 atm; *i.e.,* within the range of inland species. A striking example is the beach morning-glory, *Ipomoea,* which no doubt derives its water from low-salinity residues left in the sand from rain and dew. Tensions even higher than those in mangroves were seen occasionally in *Salicornia* or *Tecticornia* growing on muddy salt flats. In three *Rhizophora* a mistletoe had a sap tension exceeding that of the host by 1–2 atmospheres.

One might wonder if totally submerged marine vascular plants would show gross osmotic-hydrostatic separations; thus, placing the leaves of an *Enhalus* in the bomb with the firm basal part protruding, the balancing pressure was found to be ambient and the squeezed-out juices had a freezing point depression equal to that of the seawater. *Enhalus* has a submerged plum-sized fruit which is eaten by the Aboriginies and is slightly salty to taste. However, both rind and pulp were isotonic to seawater.

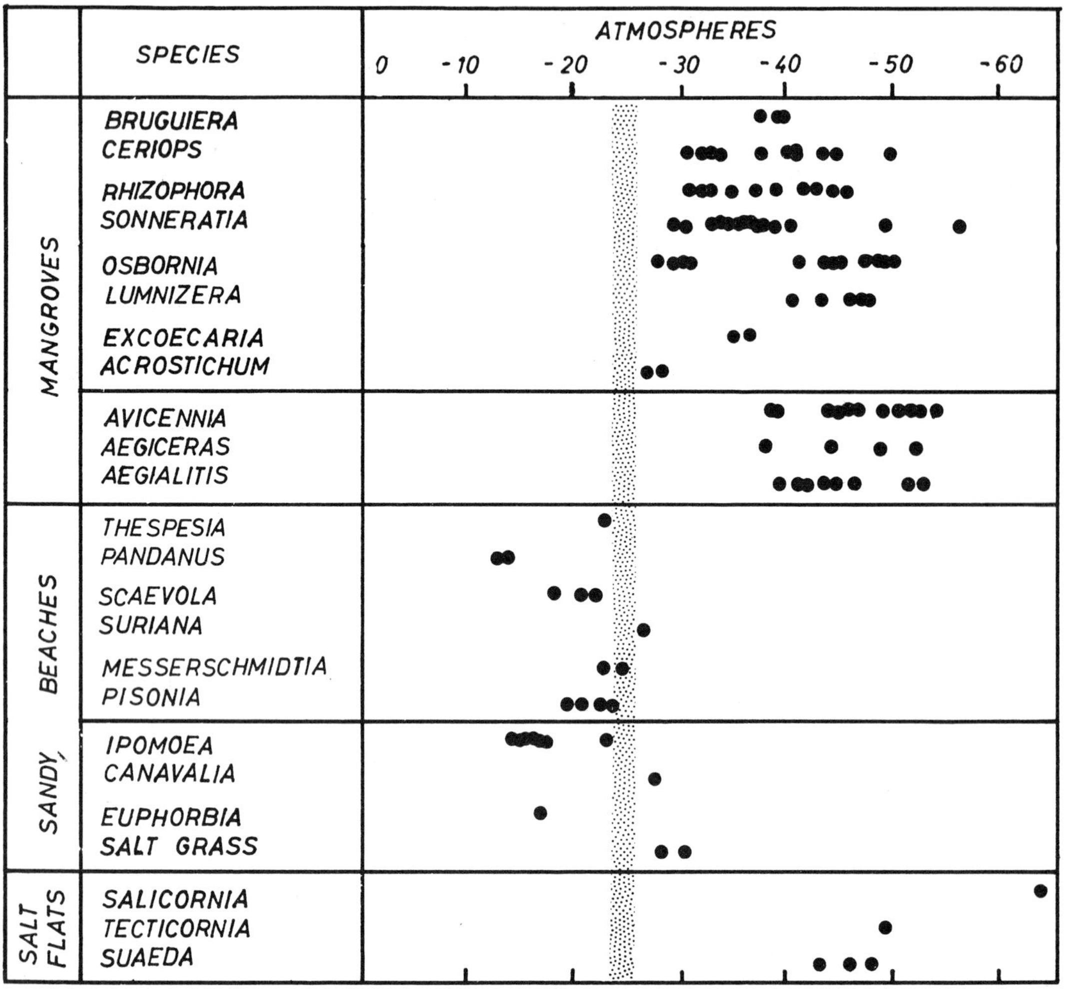

Figure 1. *Xylem sap tension in a number of mangroves and other halophytes.* Shaded area is the negative solvent pressure of sea water.

2. *Demonstration of tensile sap in xylem by freezing*

When water freezes to ice, dissolved gases are expelled (1), and when a spinning Briggs capillary is frozen in the middle, it cavitates immediately (9). Therefore, if a narrow section of a tensile twig were to be frozen through *in situ,* the liquid should cavitate and recede in either direction, and when thawed, the twig should remain vapor-locked and should wilt. This was tested on a series of mangroves and other plants, namely, *Ceriops, Rhizophora, Sonneratia, Osbornia, Lumnizera, Excoecaria, Avicennia, Aegialitis, Thespesia* and an unidentified salt grass (Poaceae).

In all of these species a 1–5 mm thick twig was frozen through *in situ* between two little pieces of dry ice. In *Aegialitis* the distal leaves wilted in two hours, and in all other species they wilted and died in the course of one to several days. When an *Aegialitis* twig was ring barked and the bare xylem

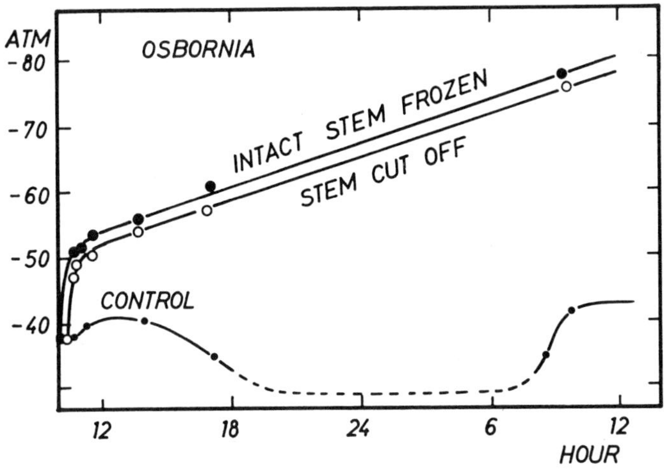

Figure 2. *Rise in the xylem sap tension as the result of cutting or ring-freezing the stem.*

covered up to prevent drying, it kept green and fresh during the next several weeks.

In one experiment (Figure 2), a branch of *Osbornia* with a dozen little twigs was cut off and tied back in place. A similar neighboring branch was ring-frozen *in situ* for a few minutes and thawed. A third branch served as control. In all three, the sap tension was followed with the bomb technique by cutting off one little twiglet at a time throughout 24 hours. It will be seen from Figure 2 that after an initial steep tension increase because of loss of turgor, dehydration progressed at a similar rate whether the stem was cut off or ring-frozen; hence, the latter had its water supply completely severed. The control twig showed the usual diurnal fluctuation in tension. It may therefore be concluded that the wilting observed after ring-freezing is caused by rupture of tensile water.

3. *Osmotic balance in leaves*

Leafy shoots were mounted in the pressure bomb and the balancing pressure was followed stepwise as water was removed. When the inverse of the pressure is plotted against the total volume of water removed (7), one gets a linear relation from the point of zero turgor, provided the leaf cells behave like perfect osmometers (Figures 3 and 4). The linear part extrapolates to the total amount (= 100 %) of water which is being concentrated, *i.e.,* the intracellular water. A steep initial drop of the curve reflects rapid loss of turgor, and a steep drop at the end signifies that the vacuole and other free fluid spaces have collapsed, leaving the packed cell structures and colloids to resist further removal of water. This packing took place usually before half of the water had been lost.

The extruded sap has generally only a few hundredths percent of salts. The first and last sample, however, may be more concentrated. The first may be contaminated by natural xylem sap, for salt-secreting species like *Avicennia* and *Aegialitis* have 0.2–0.5 % sodium chloride in the sap. The last rise may be caused by the very high pressure difference (60–90 atm) which conceivably may rupture the membrane.

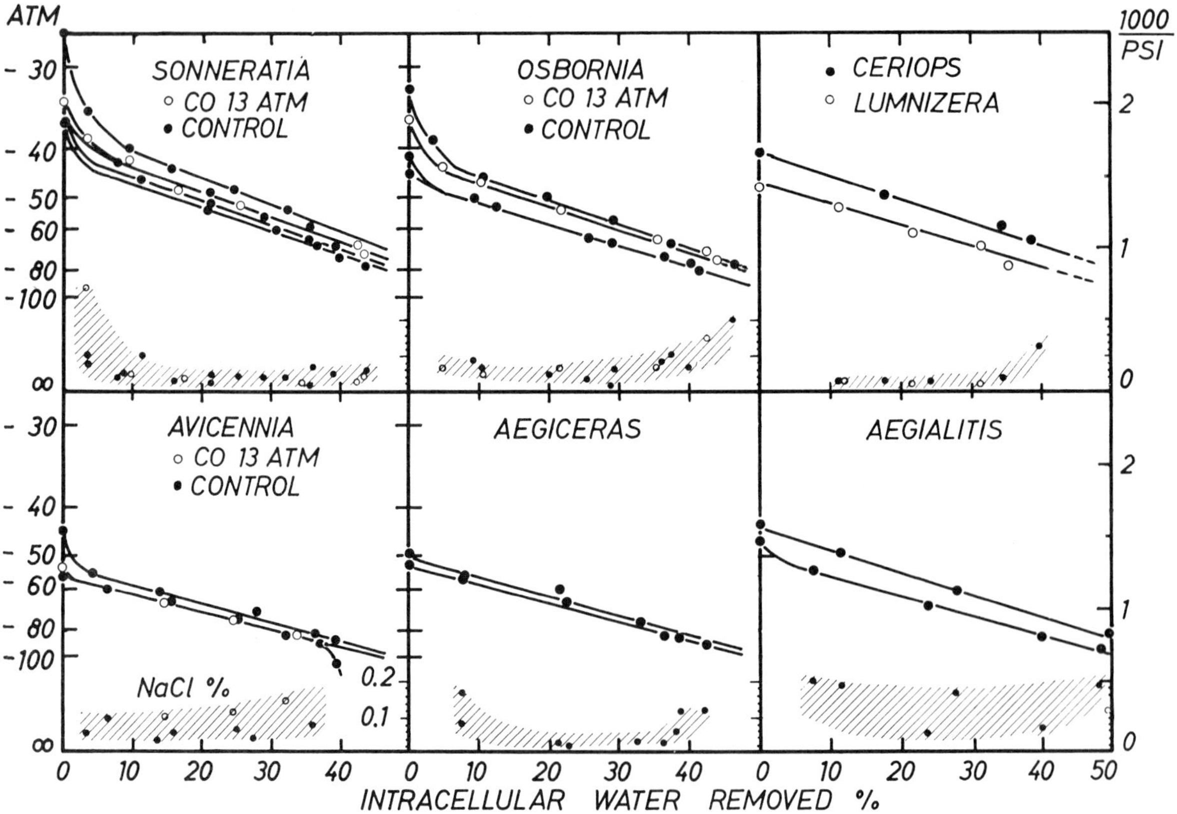

Figure 3. *Pressure volume curves in mangroves with and without poison.* Shaded area percentage of sodium chloride (determined as chloride) in the extrusion sap. One curve for each experimental series. Curves extrapolate to 100 % at infinite pressure.

4. *Stress experiments on leaves*

The near-perfect osmotic hydrostatic relation obtaining in the leaves has been challenged by various stresses involving the effects of temperature and poisons.

a. *The effect of temperature.* — Leafy shoots were put into an iced pressure bomb. Zero degree had little or no deterioration in the salt separation, but resulted in a slight drop in the balancing pressure. S. L. Miller (University of California, personal communication) demonstrated that this drop is proportional to the absolute temperature; *i.e.,* it is simply the temperature effect on osmotic pressure. Chilling, however, tended to slow down the rate of water extrusion, sometimes very markedly (*Ceriops, Sonneratia, Osbornia, Lumnizera* and *Ipomoea*). In other experiments the bomb was placed in heated water. Usually, no detrimental effect was found on the balancing pressure or salt separation until 50°C. Even at 57°C some leaves were resistant (*Sonneratia* and *Osbornia*).

b. *The effect of dinitrophenol.* — Leafy shoots were cut off under water and transferred to vials with a 0.1 % solution of dinitrophenol. A measured amount of the solution was transpired up into the leaves, followed by fresh

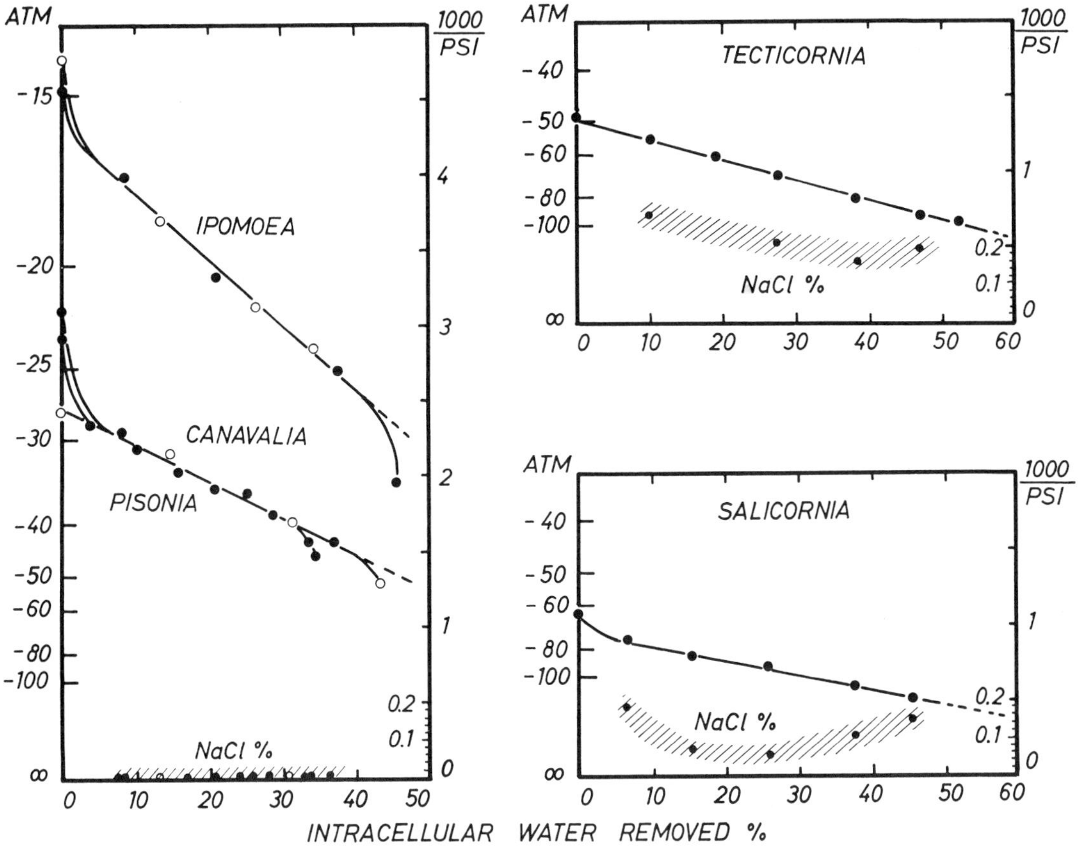

Figure 4. *Pressure volume curves in mangroves.* Shaded area percentage of sodium chloride (determined as chloride) in the extrusion sap. Curves extrapolate to 100 % at infinite pressure.

water. The shoot was transferred to the pressure bomb and sap was squeezed back out again by excess pressure. Even at the last stages of collapse of the cells (70 atm), the sap retained its yellow stain. Little or no effect on the balancing pressure or the salt separation could be detected, although the poison must have at least contacted the cell membrane and may even have penetrated into the cells. Species tested were *Osbornia* and *Sonneratia*.

c. *The effect of carbon monoxide and other gases.* — With the twig mounted in the bomb, this was charged with 13.5 atm of carbon monoxide, followed by nitrogen, and a pressure volume curve was taken. The open circles in *Sonneratia, Osbornia* and *Avicennia* (Figure 3) denote the carbon monoxide experiments. This enormous dose of monoxide had no deleterious effect on the salt separation and balancing pressure.

Other experiments were run, using pure nitrogen, after the air had been washed out of the bomb so that the system was virtually anaerobic. This had no influence on balancing pressure and salt separation (*Avicennia, Ipomoea*). Neither had a 24-hours pretreatment, and subsequent run, with 1 atm nitrous oxide plus nitrogen any effect (*Sonneratia*).

d. *The effect of chloroform and ether.* — After 10 minutes' control run

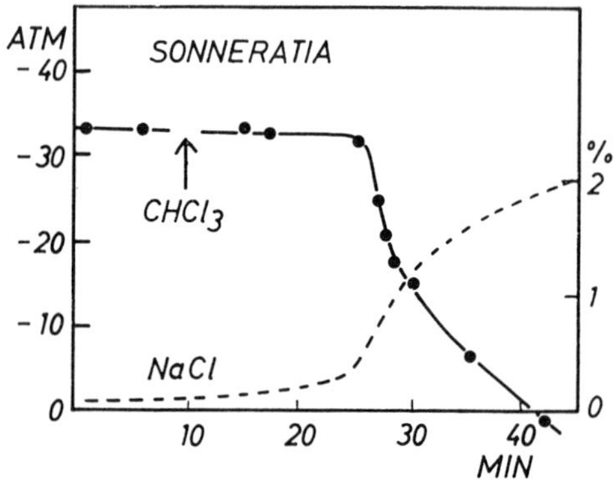

Figure 5. *Effect of chloroform on sap tension and concentration on the twig in situ.*

with a *Sonneratia* twig, the bomb was opened and a cotton pad soaked in chloroform was inserted. After 12 minutes the balancing pressure dropped precipitously and ultimately fell below ambient; *i.e.,* the stem exuded liquid (Figure 5). From a normal, barely detectable value of chloride, the concentration increased to over 2 ⁰/₀, and discoloration and high viscosity showed that the cell membranes were indeed ruptured. Saturation with ether produced similar structural damage but required longer time.

A plastic bag tied around a twig *in situ* was carefully injected with chloroform, forming a pool at the bottom. It usually required 10–12 minutes before total damage set in; *i.e.,* until the twig would lose its balancing pressure and the xylem became flooded with salts and viscid materials. However, a neighboring twig handled in the same way might recover completely, or one might see that in one twig only some of the leaves would be damaged. The chloroform tended toward an "all or none" effect. Similar results were obtained on *Lumnizera* and *Osbornia.*

5. *Ultrafiltration through the roots of mangroves*

Having established that the sap tension in the xylem was always great enough to be able to sustain a simple ultrafiltration of the seawater, this possibility was explored by subjecting the isolated root system to seawater of elevated hydrostatic pressure, keeping the xylem sap at ambient pressure (cp. 7, 11).

Seedlings with 4 to 8 leaves were potted in shell-sand which filled a standard 13 × 16 cm coffee can provided with drain holes. The plants, kept in shade, were watered with seawater or half seawater, and a few plants were also successfully grown in half seawater without shell sand. After several weeks of healing of the damaged root system, the potted plant was put into a close fitting pressure bomb and the stem cut off 2 cm above the sand. Stripping down the bark, the xylem stump was fitted through the rubber gland and the lid screwed on. The gland was tightened up and nitrogen pressure admitted until sap kept coming.

Many of these preparations leaked straight seawater at any pressure, but some, especially *Avicennia* and *Ceriops,* kept on delivering nearly fresh water. In Figures 6 A, the root of an *Avicennia* plant with 10 cm^3 water con-

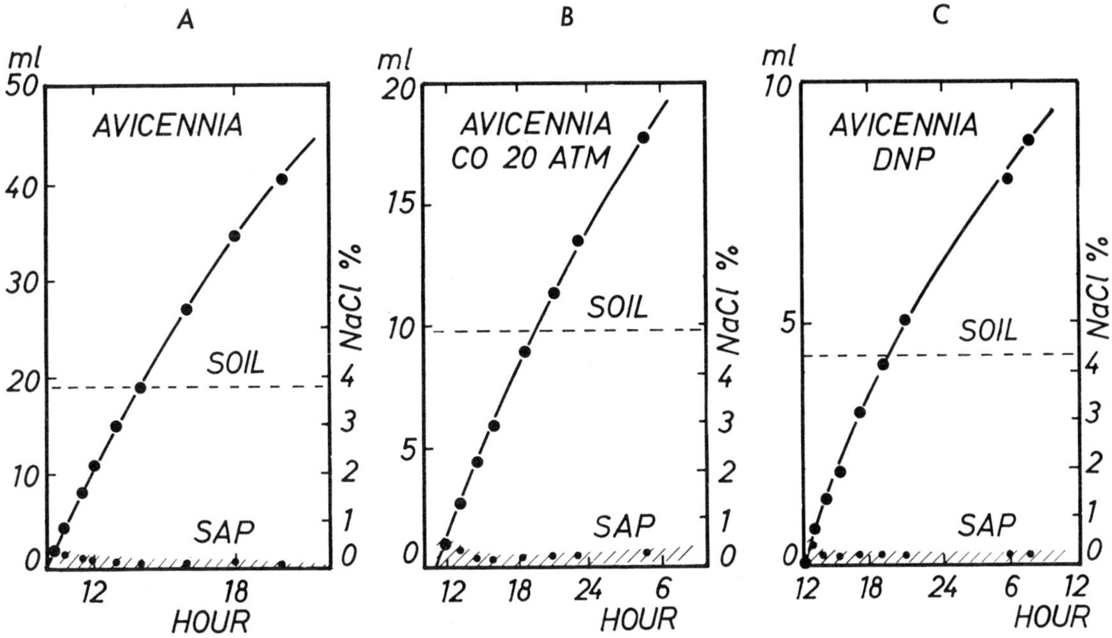

Figure 6. *Reverse osmosis in the roots of Avicennia with and without poison.* Shaded area sodium chloride concentration in sap. Broken line NaCl concentration in the sand at end of experiment.

tent was pressed with 40–45 atm for 10 hours, and yielded 40 cm³ of 0.2 ⁰/₀ sodium chloride. The water sample obtained from the sand by centrifugation at the end of the experiment had a salinity of 3.8 ⁰/₀. The roots, therefore, rejected some 95 ⁰/₀ of the salts, which corresponds to the *in situ* concentration of xylem sap in this species.

In Figure 7 B is given a similar result with a *Ceriops* root of 3 cm³ capacity which yielded 6 cm³ of 0.2 ⁰/₀ sodium chloride from a 3.7 ⁰/₀ substrate; *i.e.*, it rejected some 95 ⁰/₀ of the salts, which is a little less efficient than in a normal *Ceriops*.

6. *Stress experiments on roots*

As in the case of leaves, one might ask whether the salt rejection is a simple physical process. Various stresses were therefore applied to the system.

a. *Effect of poisons.* — Figure 6 B shows an experiment where 20 atm of carbon monoxide was introduced into the bomb prior to nitrogen. It will be seen that the root of 7 cm³ water content produced some 20 cm³ of fluid in 24 hours, and that the salt rejection was excellent, being close to 95 ⁰/₀. In another experiment, (Figure 6), dinitrophenol was added to the sea water until it drained out yellow through the sand, but there was no apparent effect by the poison, although this must at least have contacted the membrane. *Osbornia* behaved similarly.

b. *Effect of chilling.* — In other experiments, roots were chilled by plac-

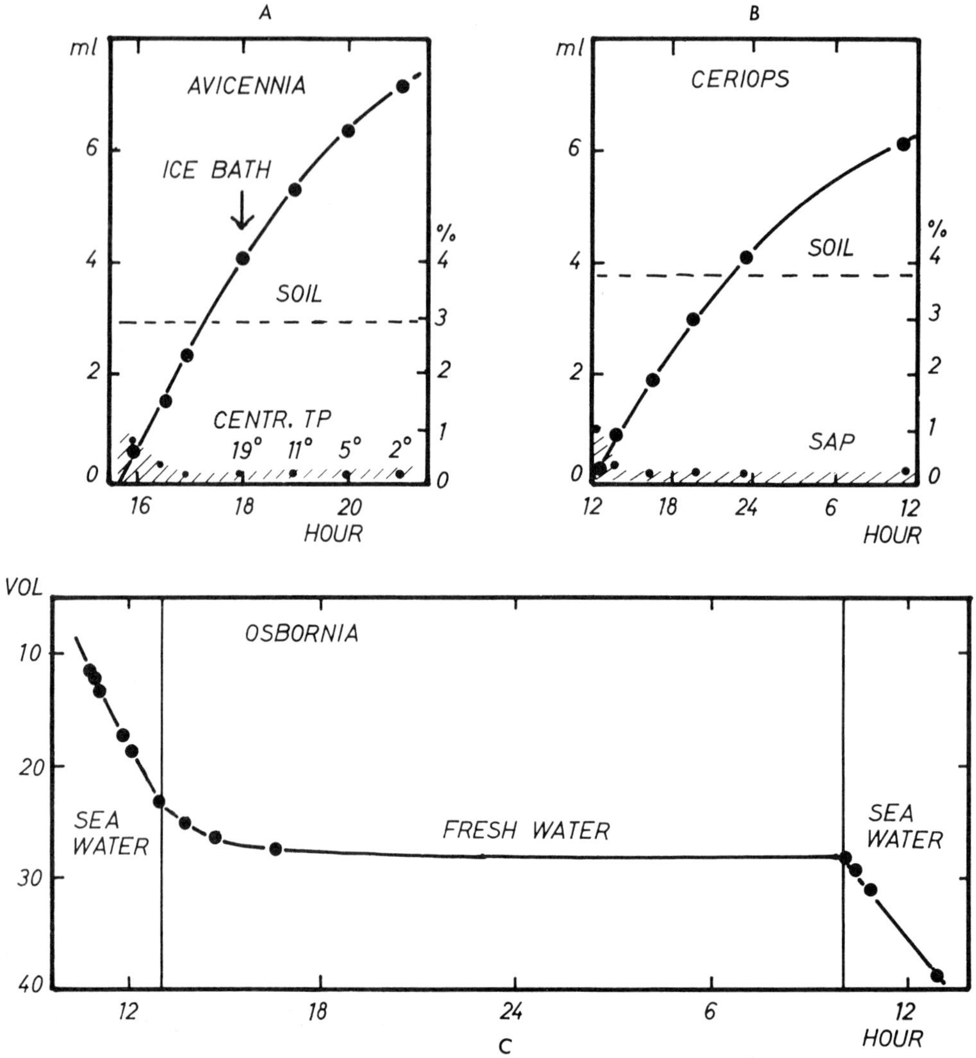

Figure 7. Upper panels: *Reverse osmosis in the roots of chilled Avicennia* (A) *and normal Ceriops* (B). Shaded area sodium chloride concentration in sap. Broken line concentration in shell sand. Lower panel: *Reversal of flow in root of Osbornia* (C) *whose xylem was conneceted to a capillary with fresh water, and whose roots were immersed* (1) *in sea water,* (2) *in fresh water, and* (3) *in sea water.*

ing the bomb in ice water (Figure 7 A). Controls showed that the temperature dropped to 2.2°C in the center of the sand and close to zero at the periphery. This slowed down the delivery of water, but it did not affect the separation of fresh water from the sea water, which remained 95 %.

c. *Reversal of flow.* — If the process which separates fresh water from seawater were a simple ultrafiltration, one should be able to reverse it by reversing the gradient. Such an experiment will be seen in figure 7 C. A young *Osbornia* plant was growing successfully in sea water without sand. The stem was cut off under fresh water and the root was joined to a capillary filled with same. Sea water around the roots immediately started a reverse xylem flow, which stopped when the substrate was changed to fresh water.

Discussion

Many mangroves luxuriate where there is some influx of fresh water, but others, like *Osbornia, Rhizophora, Sonneratia* and *Avicennia,* do very well without. Pure seawater as a source of fresh water is unique in that it has an extremely low activity (water potential) which is practically constant and always known. More precisely, one may state that the solute pressure against the free surface has lowered the partial solvent pressure by 25 atm. A non-transpiring mangrove at equilibrium is indeed pervaded by the negative partial solvent pressure (–25 atm) of the seawater throughout living cells and xylem sap. The negative sap pressure is ultimately contingent upon a high solute (= osmotic) pressure in the living cells which usually exceeds that of the seawater by 10–30 atm. This provides turgor in the leaves even when the sap tension has reached values of 40–50 atm during the height of transpiration.

Our bomb experiments on leaves demonstrated that a nearly perfect osmotic-hydrostatic relation obtained between the cells and the xylem sap. It was further shown that this balance was maintained by a purely mechanical ultrafiltration process, for neither temperature nor metabolic poisons interfered. Only structural damage to the membrane by chloroform or ether destroyed the separation. Similarly, it was shown that when differential pressure was applied to the root system of a decapitated mangrove seedling, nearly pure fresh water would exude from the xylem when the applied pressure exceeded the osmotic pressure of the seawater. The salt separation remained unaffected by strong metabolic poisons and chilling; so also in the roots the water was separated by a simple ultrafiltration.

It is interesting to consider the relative amounts of water and ions that flow, experimentally or naturally, through the cell membrane. Thus, when excess pressure is applied to the leaves in the bomb, they deliver a solution which frequently has less than 10 mg salt per 100 cm^3 water; *i.e.,* with a water/salt ratio of 10,000 to 1. In the root pressure experiments this ratio was 500 to 1. In natural xylem sap of non-salt-secreting mangroves, it may be 5–10,000 to 1, and in ordinary plants it may be as high as 100,000 to 1 (1, 6). During normal transpiration, therefore, many thousand times more water than salt flows through the membrane. This in itself suggests that the salt secretion is confined to a very small part of the membrane. It is highly significant that the salt gland cells of *Aegialitis* show a complex organelle system rather than a simple membrane (1). One may therefore suspect that the salt "pumps" in ordinary cells might appear as discrete organelle loci which very sparsely interrupt the vast and undifferentiated plasmalemma whose main function is simply to retain solutes while letting water through.

The finding that the retention of salt within the cells was not damaged by metabolic poisons or low temperature suggests, at first glance, that the pumps could not reverse by overload. Another possibility, however, cannot be excluded, namely, that the solute concentration is maintained by a very slow steady state pumping. If the pumps failed, the back flux might also be very slow and could have eluded our detection.

One may conclude, therefore, that xylem sap formation by the roots in mangroves and other halophytes involves essentially an ultrafiltration of the seawater combined with an ion transport. For most components the latter proceeds along a downhill gradient. The negative xylem pressure is ultimately produced by a high salt concentration in the living cells which gives a solute pressure which usually exceeds that of the seawater. This excess maintains turgor in the cells, even when transpiration may raise the sap tension to 40–60 atm.

This research was one of the projects of Operation Billabong, carried out aboard the newly commissioned Scripps Institution of Oceanography research vessel, the R/V ALPHA HELIX, in North Queensland's waters, Australia, in mid-1966, supported by National Science Foundation grant GB-4259. Additional support was furnished by research grant GM-10521 from the U.S. Department of Health, Education and Welfare, Public Health Service.

We are grateful to the many Australian colleagues who made possible the Australian-American expedition, in particular, Sir Hugh Ennor, Vice Chancellor, University of Canberra; Sir Fred Schonell, Vice Chancellor, University of Queensland; Dr. Ralph Slatyer, Division of Land Research, CSIRO; and Mr. G. G. T. Harrison, Chief Inspector of Fisheries, Queensland.

I am indebted to Mr. Vince Vlasoff and his staff, and to Captain James L. Faughn and his crew, for logistics and other assistance on shore camp and ship. Dr. L. J. Brass, Flecker Herbarium at Cairns, Dr. S. T. Blake, Botanic Museum, Brisbane, Dr. R. D. Hooglund and Dr. F. Jones of CSIRO, Canberra, Australia, rendered invaluable taxonomic and horticultural assistance; and I am grateful to Miss Peggy McNally and Miss Dee Bradstreet for expert technical aid. Much of this work was inspired by discussions with the many colleagues on the expedition.

References

1. Atkinson, M. R., Findlay, G. P., Hope, A. B., Pitman, M. G., Saddler, H. D. W. & West, K. R.: Salt regulation in the mangroves Rhizophora mucronata Lam. and Aegialitis annulata R. Br. — Aust. J. Biol. Sci. 20: 589–599. 1967.
2. Hemmingsen, E.: Permeation of gases through ice. — Tellus 11: 355–359. 1959.
3. Herzfeld, K. F.: Thermodynamische und kinetische Betrachtungen über das Zustandekommen der Dampfdruckerniedrigung von Lösungen. — Physikalische Zeitschrift 38(2): 58–64. 1937.
4. Scholander, P. F.: The role of solvent pressure in osmotic systems. — Proc. Nat. Acad. Sci. 55: 1407–1414. 1966.
5. — Osmotic mechanism and negative pressure. — Science 156: 67–69. 1967.
6. — Bradstreet, E. D., Hammel, H. T. & Hemmingsen, E. A.: Sap concentrations in halophytes and some other plants. — Plant Physiol. 41: 529–532. 1966.
7. — Hammel, H. T., Bradstreet, E. D. & Hemmingsen, E. A.: Sap pressure in vascular plants. — Science 148: 339–346. 1965.
8. — — Hemmingsen, E. & Garey, W.: Salt balance in mangroves. — Plant Physiol. 37: 722–729. 1962.
9. — Hemmingsen, E. & Garey, W.: Cohesive lift of sap in the rattan vine. — Science 134: 1835–1838. 1961.
10. Walter, H. & Steiner, M.: Die Ökologie der ost-afrikanischen Mangroven. — Z. Bot. 30: 65–193. 1936.
11. Weatherly, P. E.: The pathway of water movement across the root cortex and leaf mesophyll of transpiring plants. — *In* The Water Relations of Plants (A. J. Rutter and F. H. Whitehead, ed.), pp. 85–100. Blackwell Scientific Publications, London. 1963.

Reprinted from PLANT PHYSIOLOGY, Vol. 41, No. 4, April, 1966, pages 617-622

Changes During Recovery from Sodium Deficiency in Atriplex

P. F. Brownell[1] and Margaret E. Jackman
Department of Botany, University of Adelaide, South Australia

Received September 15, 1965.

Summary. Although the concentration of sodium in leaves of *Atriplex* plants increased rapidly after receiving sodium, no growth response was detectable for about 6 days. It was found that respiration rate increased to its maximum within 3 days. Chlorophyll content also increased from an early stage, whereas the concentrations of sugars and starch did not increase, and ratios of soluble to total nitrogen did not decrease until later.

The respiratory response appears to be specific to sodium as different salts of sodium caused similar responses, and no other univalent cation substituted for sodium. In addition, both growth response and respiration rate tended towards their maxima with the same concentration of applied sodium. The rate of anaerobic CO_2 production increased when sodium was fed to leaves, suggesting that the effect of sodium is in the glycolytic sequence.

The first decisive evidence that sodium is an essential element for plant growth was obtained by Allen and Arnon (1), who showed that small amounts of sodium are required by the blue-green alga, *Anabaena cylindrica*. Subsequently, it was found that sodium is essential for the angiosperm, *Atriplex vesicaria* Heward ex Benth (4, 5), and more recently, a similar requirement has been shown in certain other species of *Atriplex* (unpublished experiments).

Preliminary experiments indicated that there is a delay of about 6 days following the supply of sodium to sodium-deficient *Atriplex* plants before a significant increase in dry weight occurs. A metabolic response to sodium detected early in this period may represent a primary step in recovery, and hence contribute to an understanding of the function of sodium in the nutrition of *Atriplex*. Therefore, a study was made of the changes in leaf respiration rate, chlorophyll content, soluble and insoluble nitrogen fractions, and sugar and starch contents, following the addition of sodium to sodium-deficient plants.

Materials and Methods

Plant Culture. The species were *Atriplex nummularia* Lindl, and *A. inflata* F.v.M., and seeds were obtained from Griffith, New South Wales, and Renmark, South Australia respectively. *A. nummularia* was used except where otherwise indicated. The procedures for germination and growth of plants under conditions of low sodium have been described previously (4). The capacity of culture vessels was 2 liters. The age of plants in all experiments ranged from 34 days in summer to 46 days in winter. Except in 1 experiment in which a short-term response was sought, sodium was fed by the addition of sodium sulfate to the culture solution in amounts varying between 0.1 and 0.6 meq/liter. The concentration used is included with the appropriate experimental results.

Sodium Content. Sodium determinations in plant material were carried out on boiling water extracts (2) using an atomic absorption instrument (3).

Leaf Respiration. Respiration rate of leaf tissues was determined manometrically by measuring O_2 uptake of intact leaves (0.2-1 g fr wt) placed on moist, convoluted filter paper in the dark at 27°. Changes in rate of uptake over a period of 1 to 2 hours were small (fig 4).

For later experiments (see below) alternative methods of introducing sodium to the tissue were tested, the aim being to supply sodium to all cells in the tissue as rapidly as possible. The O_2 uptake of leaf disks (7 mm in diameter) of sodium-fed plants (0.6 meq/liter sodium sulfate 20 days previously) was measured under the following conditions: A) disks placed on moist, convoluted filter paper, 590 μl/g fresh weight per hour; B) vacuum infiltrated with water and placed on moist convoluted filter paper, 265 μl/g fresh weight per hour; C) submerged in water, 360 μl/g fresh weight per hour; and D) vacuum infiltrated and submerged, 282 μl/g fresh weight per hour. Thus, submerging or vacuum infiltrating the tissues markedly reduced the rate of O_2 uptake compared with that of the untreated tissue. This inhibitory effect of water on O_2 uptake is similar to that found by Yocum and Hackett (14) work-

[1] Present address: Department of Agricultural Biochemistry and Soil Science, Waite Agricultural Research Institute, Private Bag No. 1, Glen Osmond, South Australia.

ing with aroid spadix, and Ohmura and Howell (12) working with corn, bean and barley tissues. In addition the rate of O_2 uptake by submerged tissues could be increased to that of the untreated tissues by replacing air in the flask with O_2, although the rates of O_2 uptake by untreated tissues in air and O_2 were similar. Even though this limitation to gas exchange can apparently be overcome, an increase in O_2 tension could conceivably induce other complicating effects. For this reason the sodium was introduced through cut stems. The cut ends of sodium deficient shoots were placed in sodium sulfate solution (2.2 meq/liter) for a period (1 and one-half hours) determined by observing the progress of a dye (eosin) into a similar shoot under the same conditions for transpiration. At the same time control shoots were placed in distilled water. Immediately following this treatment, the leaves were cut from the stems and the rate of O_2 uptake measured.

A similar method was used to feed sucrose, but in this experiment cut petioles of detached leaves were immersed in 0.1 M sucrose for 1 and one-half hours before O_2 uptake was measured. Sodium was added to the culture solution 43 hours before harvesting.

Anaerobiosis. CO_2 evolution by leaves was measured for at least 1 hour in Warburg vessels containing O_2-free nitrogen. After the anaerobic measurements, the leaves were returned to air, and the rate of O_2 uptake determined.

Chlorophyll. Samples of leaf material (0.3–6 g fr wt) were ground in cold 80 % (v/v) acetone and filtered. The OD of the solutions were measured at 645 and 663 mμ using a Shimadzu Type QR-50 spectrophotometer after adjusting the volume to give readings between 0.1 and 0.6. The concentrations of chlorophylls a and b and total chlorophyll were then calculated from the extinction coefficients reported by MacKinney (11).

Soluble and Insoluble Nitrogen Fractions. Fresh leaf material (0.5–3 g) was ground in ice cold 80 % (v/v) ethanol, centrifuged at 1350 × g for 4 minutes, and the nitrogen contents of both supernatant fraction and residue determined by the method of McKenzie and Wallace (10).

Sugars and Starch. Harvests were made at noon. Sugars were extracted from leaves (0.5–2 g fr wt) with boiling 80 % (v/v) ethanol, the starch solubilized from the residue with dilute perchloric acid, and both were then estimated using the anthrone method (9).

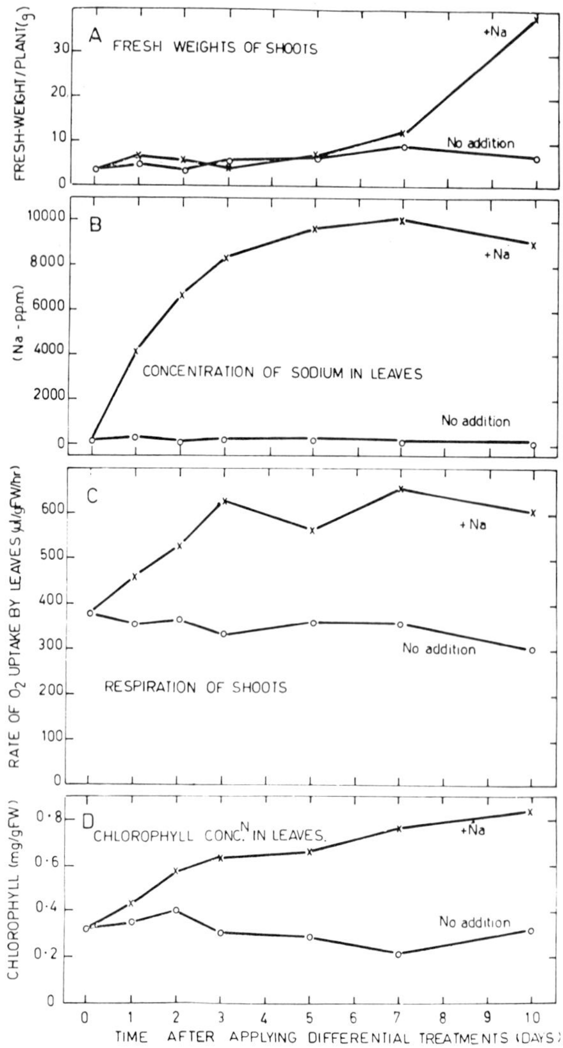

FIG. 1. Changes in fresh weight of shoots (A), sodium concentration (B), rate of O_2 uptake by leaves (C), and chlorophyll concentration (D) following the addition of sodium sulfate (0.6 meq/liter) to 32-day-old sodium-deficient plants. All harvests were made at 9 AM. Each point is derived from 2 cultures of 8 plants, of which 6 plants were used for chlorophyll determinations, and the remainder for measurements of sodium concentration and respiration rate. Total fresh weights of shoots of all 16 plants were recorded.

Results and Discussion

Changes During Recovery from Sodium Deficiency. The sequence of changes that follow the feeding of sodium to sodium-deficient *Atriplex* plants is shown in figure 1. For at least 5 days, growth of plants which received sodium (expressed as fr wt of shoots) was similar to that of plants to which no sodium was added (fig 1 A). After 7 days however, the growth of sodium-fed plants began to increase rapidly compared with that of the untreated controls. This delay in growth response occurred despite a rapid uptake of sodium into the shoots from the time of application (fig 1 B). By the fifth day the plants had taken up most of the sodium supplied in the culture solution, and the concentration in the leaves thus remained steady until after the seventh day when it showed a slight decrease presumably due to the onset of rapid growth.

The respiration rate per unit fresh weight of shoots (fig 1 C) increased rapidly (for about 3 days)

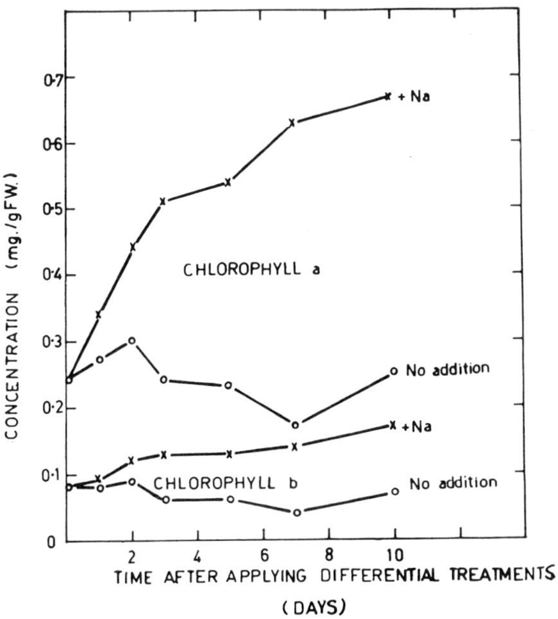

FIG. 2. Changes in concentrations of chlorophylls a and b following the addition of sodium sulfate to sodium-deficient plants. Data obtained from the same experiment as figure 1.

after receiving sodium, reaching a rate about twice that of the sodium-deficient plants. Thereafter there was little further change in rate.

The concentration of chlorophyll in the leaves increased rapidly to almost double the initial concentration by the third day after the addition of sodium (fig 1 D), and continued to increase throughout the 10 days of the experiment. During this period the chlorophyll a concentration in sodium-fed plants increased to a value nearly 3 times the initial value, while chlorophyll b concentration doubled (fig 2). These results were obtained during the summer, and the increase in chlorophyll concentration was much more rapid than in the winter months when growth was slow.

From this experiment it appears that even though fresh weight responses do not occur until some time after the addition of sodium to sodium-deficient *Atriplex* plants, both respiration rate and chlorophyll concentration increase at an early stage.

Marked differences in ratios of 80 % ethanol-soluble to total nitrogen were apparent between sodium-fed and deficient shoots by 20 days after supplying sodium but no difference was detectable 2 days after the addition of sodium (table I), although by this time there was a definite respiratory increase. Hence, the apparent long-term effect of sodium on nitrogen fractions is likely to be indirect, and only one of many changes expected to take place in the later stages of recovery.

The concentrations of both sugars and starch in leaves of sodium-deficient plants were low (fig 3). During recovery, however, these levels gradually rose to several times those found in deficient plants.

Further Respiration Experiments. Since it was one of the first detected, the respiratory response was

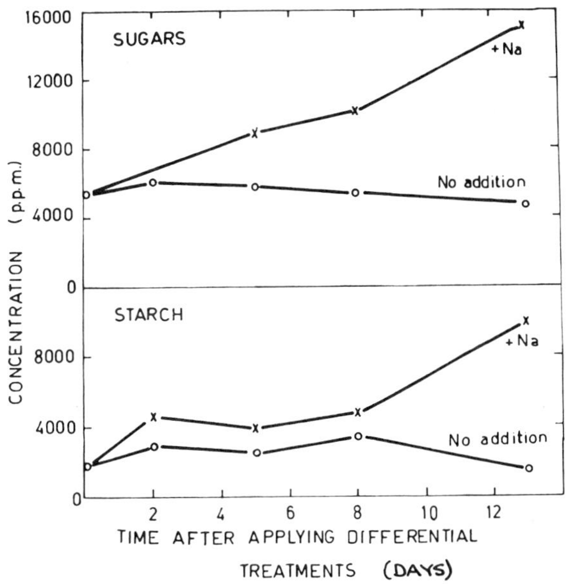

FIG. 3. Changes in concentrations of sugars and starch following the addition of sodium sulfate (0.6 meq/liter) to 45-day-old sodium-deficient plants. Points are the means of determinations on 6 samples each containing the leaves of 2 plants.

Table I. *Concentrations of Soluble and Insoluble Nitrogen in Leaves of* Atriplex

	*Atriplex nummularia** 2 days after receiving sodium				*Atriplex inflata*** 20 days after receiving sodium			
	Fr wt per plant (g)	Nitrogen content (% fr wt) Ethanol extract	Nitrogen content (% fr wt) Residue	Soluble N / Total N	Fr wt per plant (g)	Nitrogen content (% fr wt) Ethanol extract	Nitrogen content (% fr wt) Residue	Soluble N / Total N
No sodium added	1.33	0.29	0.42	0.41	0.11	0.32	0.39	0.45
Sodium added***	1.76	0.25	0.39	0.39	0.50	0.17	0.46	0.27

* Data are the means of determinations on 5 individual plants.
** Data are the means of determinations on 4 samples of 2 plants when no sodium was added and 1 plant when sodium was fed.
*** 0.60 and 0.10 meq/liter sodium sulfate to *Atriplex nummularia* and *Atriplex inflata*, respectively.

studied more fully by examining the effects of different salts of sodium, other univalent cations, varying concentrations of sodium, feeding sodium to cut shoots, feeding sucrose, and the effect of sodium on rate of anaerobic CO_2 production.

Whether supplied as chloride or as sulfate, equivalent amounts of sodium brought about similar increases in rate of O_2 uptake (fig 4). It appeared, therefore that these increases depended on the sodium ion, and were not due to excess cation absorption (6, 7), as, with the latter phenomenon, different rates of O_2 uptake could be expected as a result of the different mobilities of chloride and sulfate ions.

The rates of O_2 uptake by leaves of sodium-deficient plants 48 hours after receiving (in the culture solutions) sulfates of the univalent cations (0.10 meq/liter) were as follows: lithium, 368 µl/g fresh weight per hour; sodium, 640 µl/g fresh weight per hour; potassium, 364 µl/g fresh weight per hour; rubidium, 354 µl/g fresh weight per hour; and caesium, 345 µl/g fresh weight per hour. There was no difference from the control (370 µl/g fr wt per hr) except for the marked increase in the case of sodium. It seems improbable that this is salt-stimulated or anionic respiration (8, 13) for 2 reasons; first, the salt-stimulated response generally occurs only in tissues initially containing low concentrations of ions, whereas the concentrations of ions in leaf tissues of *Atriplex* are high in comparison with the low concentrations

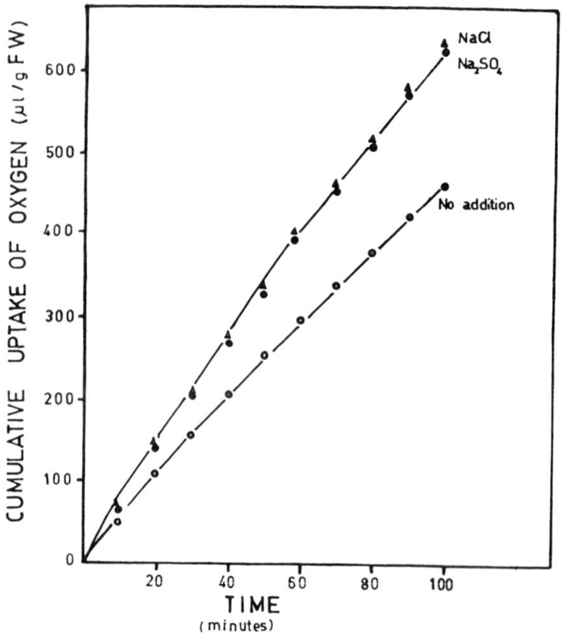

FIG. 4. Effect of adding sodium as either chloride or sulfate on the rate of O_2 uptake by sodium-deficient plants. Sodium chloride or sodium sulfate (0.22 meq/liter) was added to the culture solution 48 hours before harvesting. Points are the mean rates of O_2 uptake of 4 Warburg flasks each containing the leaves of 2 plants.

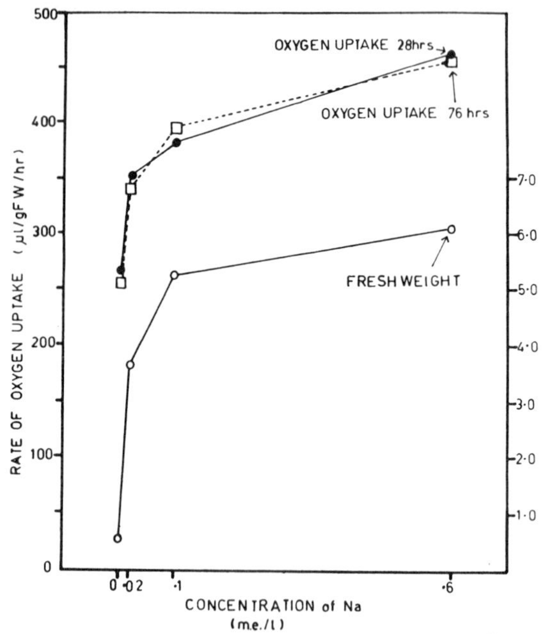

FIG. 5. Effect of concentration of sodium in culture solution on fresh weight and rate of O_2 uptake. Fresh weight data are means of 5 cultures of 4 plants each, sodium being supplied 16 days after germination and the plants harvested 32 days later. The rates of O_2 uptake were determined 28 and 76 hours after supplying sodium to 46-day-old plants. Points are the mean rates of uptake of 3 Warburg flasks each containing the leaves of 3 plants.

of sodium required to elicit a response; second, salt-stimulated respiration may be brought about by various ions, unlike the respiratory response in *Atriplex* where there is apparently a specific requirement for sodium.

Figure 5 shows that both respiration rate (O_2 uptake) and growth (fr wt) tend towards a maximum value with similar concentrations of sodium in the culture solution.

From the results of this work, it appears that lack of sodium limits the rate of respiration of *Atriplex* leaf tissues. If this leads to a reduction in phosphorylation, growth could conceivably be restricted by the decreased amounts of high energy phosphate compounds available for cellular work and syntheses.

When sodium was introduced to the tissue through cut stems, the rate of O_2 uptake by the leaves was 20 % greater than that of the untreated leaves (table II) and analyses showed that there had been an increase in the sodium concentration. This respiration response, only 1 and one-half hours after the application of sodium, was detected in a considerably shorter time than in other experiments when sodium was added to the culture solution. It is difficult to determine precisely the time required for this response to sodium, first because it is not known when sodium arrives at the actual site of action, and second because

Table II. *Effect of Feeding Sodium to Cut Shoots on Rate of O_2 Uptake and Sodium Concentration of Leaves*

	O_2 uptake** (μl per g fr wt per hr)	Sodium Conc*** (meq Na per kg dry wt)
No sodium added	211	12.1
Added sodium*	256	18.5

* Cut shoots were placed in sodium sulfate (2.2 meq/liter) for 1 and one-half hours in the light.
** Values are the mean rates of uptake of 6 Warburg flasks each containing the leaves of 2 to 3 plants.
*** Values are the means of 2 determinations on the leaves from 6 Warburg flasks.

it would not be expected to reach all cells in a tissue at the same time.

Since the concentrations of sugars and starch were low in leaves of sodium-deficient plants (fig 3), it seemed possible that lack of respiratory substrates could limit their respiration rate. Hence, cut leaves were supplied with sucrose and their respiration rates measured. The results (table III) showed that addition of sucrose increased the respiration rates of leaves from both sodium-deficient and sodium-fed plants suggesting that the small amounts of endogenous substrate were limiting the rate of respiration.

Table III. *Effect of Feeding Sucrose to Cut Leaves on Rate of O_2 Uptake*

Data are the mean rates of O_2 uptake of 3 Warburg flasks each containing the leaves of 2 to 3 plants.

	O_2 uptake (μl/g fr wt/hr) Without sucrose	With sucrose**
No sodium added	244	322
Added sodium*	302	400

* 0.6 meq/liter sodium sulfate added to culture solution 43 hours before harvesting.
** Cut petioles were immersed in 0.1 M sucrose for 1 and one-half hours in the light.

The supply of sodium to the culture solution 43 hours before harvesting also caused an increase in the respiration rate of leaves which received no additional sucrose. This could have been due to either a direct effect of sodium on the respiratory system or alternatively to an indirect effect, for example by increasing the amount of substrate derived from photosynthesis during the period after addition of sodium. When sucrose was fed in an attempt to remove the substrate limitation, there was still a response to sodium, suggesting that its effect is a direct one on some part of the respiration process. If however, the substrate limitation was not entirely removed by the exogenous sucrose, the apparent effect of sodium on respiration rate could be through its effect on photosynthesis. In all experiments, sodium was supplied in the light when conditions were favourable for high transpiration.

In an initial attempt to locate the part of the respiratory system requiring sodium, rates of CO_2 output of sodium-deficient and sodium-fed plants were compared. The results obtained (table IV) showed that the addition of sodium to sodium-deficient plants stimulated the rate of anaerobic CO_2 production and that the rate of O_2 uptake when leaves were returned to air was unaffected by the anaerobic treatment. Hence, the increase in rate of CO_2 production under anaerobic conditions, suggests that sodium acts during the glycolytic stages of respiration. However, the possibility of an earlier effect of sodium on some other system leading to this respiratory response should not be overlooked.

Table IV. *Effect of Sodium on Rate of Anaerobic CO_2 Production of Sodium-deficient* Atriplex nummularia *Leaves*

Data are the mean rates of gas exchange of 16 Warburg flasks each containing the leaves of 2 to 3 plants.

	Anaerobic phase CO_2 Output	Aerobic phase O_2 Uptake
No sodium added	85	316
Added sodium*	105	426

* 0.1 meq/liter sodium sulfate was supplied to the culture solution 24 hours before harvesting.

Acknowledgment

We thank Miss Helen Cameron for her assistance in the respiration determinations and for the preparation of figures and Mr. Spencer Knowles for his help in the analysis of starch and sugars.

Literature Cited

1. Allen, M. B. and D. I. Arnon. 1955. Studies on nitrogen-fixing blue-green algae. II. The sodium requirement of *Anabaena cylindrica*. Physiol. Plantarum 8: 653–60.
2. Black, R. F. 1956. Effect of sodium chloride in water cultures on the ion uptake and growth of *Atriplex hastata*. Australian J. Biol. Sci. 9: 65–80.
3. Box, G. F. and A. Walsh. 1959. Si-Ro-Spec. C.S.I.R.O. Ind. Res. News 17.
4. Brownell, P. F. 1965. Sodium as an essential micronutrient element for a higher plant (*Atriplex vesicaria*). Plant Physiol. 40: 460–68.
5. Brownell, P. F. and J. G. Wood. 1957. Sodium as an essential micronutrient element for *Atriplex vesicaria* Heward. Nature 179: 635–36.
6. Jacobson, L. 1955. Carbon dioxide fixation and ion absorption in barley roots. Plant Physiol. 30: 264–69.

7. JACOBSON, L. AND L. ORDIN. 1954. Organic acid metabolism and ion absorption in roots. Plant Physiol. 29: 70–75.
8. LUNDEGARDH, H. 1955. Mechanisms of absorption, transport, accumulation, and secretion of ions. Ann. Rev. Plant Physiol. 6: 1–24.
9. MCCREADY, R. M., J. GUGGOLZ, V. SILVIERA, AND H. S. OWENS. 1950. Determination of starch and amylose in vegetables. Anal. Chem. 22: 1156–58.
10. MCKENZIE, H. A. AND H. S. WALLACE. 1955. The Kjeldahl determination of nitrogen. A critical study of digestion conditions-temperature, catalyst and oxidizing agent. Australian J. Chem. 7: 55–70.
11. MACKINNEY, G. 1941. Absorption of light by chlorophyll solutions. J. Biol. Chem. 140: 315–22.
12. OHMURA, T. AND R. E. HOWELL. 1960. Inhibitory effect of water on oxygen consumption by plant materials. Plant Physiol. 35: 184–88.
13. ROBERTSON, R. N. AND J. S. TURNER. 1945. Studies in the metabolism of plant cells. 3. The effects of cyanide on the accumulation of potassium chloride and on respiration; the nature of salt respiration. Australian J. Exptl. Biol. 23: 63–79.
14. YOCUM, C. S. AND D. P. HACKETT. 1957. Participation of cytochromes in the respiration of the aroid spadix. Plant Physiol. 32: 186–91.

Reprinted from SOIL SCIENCE
Vol. 90, No. 3, September, 1960
Printed in U.S.A.

COBALT: A MICRONUTRIENT ELEMENT FOR THE GROWTH OF SOYBEAN PLANTS UNDER SYMBIOTIC CONDITIONS

SHAUKAT AHMED AND HAROLD J. EVANS

North Carolina Agricultural Experiment Station[1]

Received for publication July 15, 1960

Evidence presented in a previous report (1) indicated that soybean plants (*Glycine Max* Merr. variety Roanoke) inoculated with *Rhizobium japonicum* and grown in purified nutrient solutions lacking combined nitrogen were distinctly benefited from the addition of traces of Co. The magnitude of the response resulting from the addition of 1 part per billion of the element was 52 per cent. Plants lacking Co exhibited mild symptoms of nitrogen deficiency and contained less total nitrogen than those receiving it. Prior to that report Bolle-Jones and Mallikarjuneswara (3) demonstrated that small amounts of Co significantly increased the growth and dry weights of both the rubber and tomato plants, but no visual symptoms of Co deficiency were observed and the chlorophyll content of leaves did not vary appreciably. Indispensability of the element for the growth of these species was not, therefore, claimed.*

In view of the established Co requirement for sheep (2, 6), certain blue-green algae (5), and other microorganisms, and of the encouraging results obtained in initial experiments by the authors and by Bolle-Jones and Mallikarjuneswara (3), a systematic attempt was made to refine cultural and purification methods in order to critically examine the possible Co requirement of leguminous plants. As a working hypothesis the authors proposed that the Co requirement for legumes grown in symbiosis with *Rhizobia* may be greater than that for the plant per se. Experiments completed to date, therefore, have been limited to symbiotically grown soybean plants. The results of experiments summarized in this paper provide strong evidence that Co is an essential element for the growth of inoculated soybean plants in the absence of combined nitrogen. The possibility of a Co requirement for higher plants supplied with adequate combined nitrogen is being investigated. Full details of these investigations will be presented at a later date.

PURIFICATION AND CULTURAL METHODS

The methods used for purification of mineral salts and water and the cultural techniques were similar to those previously described (1). Deionized water for all preparations was obtained by passing distilled water through an ion exchange column (IR 120 and IRA 400 saturated with H^+ and OH^- respectively[2]) 32 inches in length. The efficiency of removal of Co by the column was confirmed by use of Co60. All polyethylene vessels used for the preparation or storage of nutrient or stock solutions were washed with hot detergent solution, rinsed with distilled or tap water, thoroughly washed with 3 *N* HCl and finally rinsed with deionized water.

The basic culture solution contained the following essential macronutrient elements (me. per liter) and the essential micronutrient elements (in ppm.): K^+, 5; Mg^{++}, 4; Ca^{++}, 13; PO_4^{+++}, 3; SO_4^{++}, 19; Mn^{++}, 0.15; Cu^{++}, 0.01; Zn^+, 0.05;

[1] Paper No. 1194 from the Department of Botany and Bacteriology of the North Carolina Agricultural Experiment Station. This investigation was supported entirely by the North Carolina Agricultural Experiment Station and by a grant (G 7580) to Harold J. Evans from the National Science Foundation. The senior author, a Ph.D. candidate at N. C. State College, is sponsored by the United States ICA Program and the Atomic Energy Commission of Pakistan, Karachi. The authors express their appreciation for the competent technical assistance of Thea Schostag and Oliver Whittington.

* While this paper was in press, the authors noted a brief paper by H. M. Reisenaur "Cobalt in nitrogen fixation by a legume," [*Nature* (1960) 186: 375–376]. Reisenaur reports that small amounts of Co resulted in a 66 per cent increase in weight and a marked increase in the N content of symbiotically grown alfalfa plants.

[2] Obtained from Rohm and Haas Co., Philadelphia.

BO_3^{---}, 0.1; MoO_4^{--}, 0.02; Cl^-, 0.2; and Fe^{+++}, 1.0 every week. In experiment I the macronutrient salts were supplied as K_2SO_4, $MgSO_4 \cdot 7 H_2O$, $CaSO_4 \cdot 2 H_2O$ and $Ca(H_2PO_4)_2 \cdot H_2O$; and the micronutrient elements as $MnSO_4 \cdot 4 H_2O$, $CuSO_4 \cdot 4 H_2O$, $ZnSO_4 \cdot 7 H_2O$, H_3BO_3, $Na_2MoO_4 \cdot 2 H_2O$, NaCl, and ferric etheylenediamine di-o-hydroxyphenyl acetate (FeEDDHA). In experiments II and III PO_4^{---} at 3 me. per liter was supplied as a mixture of 1 part KH_2PO_4 and 5 parts K_2HPO_4. The concentration of K_2SO_4 and $CaSO_4 \cdot 2 H_2O$ were adjusted in order to maintain the same ionic concentrations as those used in experiment I. Also in experiments II and III, Fe at a cencentration of 1 ppm. was supplied each week as ferric citrate instead of FeEDDHA.

All the macronutrient salts and $MnSo_4 \cdot 4 H_2O$ were purified by the 1-nitroso-2-naphthol procedure described by Bolle-Jones and Mallikarjuneswara (3). Excess reagent was removed from the stock solution with charcoal which had been washed previously with redistilled HC1 (3 *N*). All other essential micronutrient element salts, except Fe, were purified by twice crystallizing from deionized water. Co and other elements (except Ni) added to the basic nutrient solution were prepared from spectrographically standardized material[3]; Ni was purified by twice crystallizing $NiSO_4 \cdot 4 H_2O$. V was added as V_2O_5 (dissolved in redistilled HCl); Ga as Ga $(No_3)_3 \cdot 8 H_2O$; Al as $AlCl_3 \cdot 6 H_2O$; and Co as the metal dissolved in redistilled HCl.

For the preparation of purified ferric citrate, 18 g. of iron carbonyl[4] were dissolved in 350 ml. of redistilled HCl (6 *N*) after which sufficient H_2O_2 was added to convert Fe^{++} to Fe^{+++}. The $FeCl_3$ solution was thoroughly shaken 3 times with 100-ml. portions of redistilled ether saturated with redistilled HCl (4). The portions of ether were combined and were extracted 3 times with 100-ml. portions of redistilled HCl saturated with ether. The $FeCl_3$ in the ether was transferred to the aqueous phase by shaking with 5 successive 100-ml. portions of deionized water. The $FeCl_3$ solution was freed from ether by boiling and then analyzed for its Fe content (7). The $FeCl_3$ solution was converted to ferric citrate by boiling with an amount of purified sodium citrate equivalent to the quantity of Fe. The sodium citrate solution was purified by the 1-nitroso-2-naphthol procedure (3).

In order to obtain adequate nitrogen fixation and growth of inoculated soybean plants in water cultures it was essential to maintain the pH of the culture solution near neutrality. This was accomplished by adjusting the pH of culture solutions at 3-day intervals with purified $Ca(OH)_2$, prepared by purifying a solution of calcium acetate by the 1-nitroso-2-naphthol procedure. Calcium acetate was crystallized by evaporation and concentration of the hot solution, and the salt, after drying, was placed in a silica dish with cover and then ignited for 8 hours at 950°C. in a muffle furnace. The CaO was stored in a desiccator over solid KOH.

Soybean seeds (variety Roanoke) were germinated between folds of acid-washed and thoroughly rinsed filter paper that was kept moist with 0.1 strength purified nutrient solution. The germinating seedlings used in experiments I and II were inoculated with *Rhizobium japonicum* cells from fresh nodules. These were washed by suspension and centrifugation in 0.1 strength nutrient solution. In experiment III seedlings were inoculated directly with a pure culture of *Rhizobium japonicum.*[5]

When the radicals of seedling had grown to a length of about 3 inches, 5 seedlings were transferred to each of the 12-liter polyethylene culture vessels containing the appropriate nutrient solution. Seedlings were supported in the polyethylene vessel covers by acid-washed Dacron wool, and each vessel was continuously aerated through Dacron wool filters. For the first 15 days after transplanting, one-half-strength nutrient solutions were utilized, after which the solutions were brought to full strength. Within 4 days after transplanting into the culture vessels the cotyledons were removed from seedlings in order to reduce the source of cobalt from the seed. One week after transplanting, seedlings in each vessel were thinned to 4 uniform plants. Culture solutions were changed at 3-week intervals throughout the growth period of each experiment.

All experiments were conducted in growth chambers[6] with internal dimensions (in feet) of 5 by 8 by 8. After experiment I had been completed, all metal surfaces in the growth chambers

[3] Obtained from Johnson Matthey and Co., Ltd., London.

[4] Obtained from General Aniline Works of New York.

[5] Purchased from the American Type Culture collection of Washington, D. C.

[6] Purchased from Fleming-Pedlar Ltd., Winnipeg, Canada.

TABLE 1

A summary of experimental results showing the effect of cobalt on the dry weight of soybean shoots

Expt. No.	Shoot Growth: Type	Shoot Growth: Period	Results of Various Treatments: Control	1 ppb. Ni, Al, Ga	1 ppb. V	1 ppb. Co as Vitamin B_{12}	0.1 ppb. Co	1.0 ppb. Co	50 ppb. Co
		weeks	*% of control†*						
I	1st axillary	3	100	—	—	123**	—	150**	138**
	2nd axillary	3	100	—	—	133*	—	152**	148**
II	Initial	5	100	106	110	—	169**	188**	—
III	Initial	5	100	89	92	—	147*	127	—
	1st axillary	3	100	117	103	—	427**	386**	—

† The data for all control treatments with the exception of that for the 1st axillary growth in experiment III are based on means of 8 replicate cultures. All other data are based on means of 4 replicate cultures.

* Significant at 5%

** Significant at 1%

were cleaned and coated with a clear lacquer[7] and the air circulation systems fitted with fiber glass filters that were wrapped with 4 layers of cheesecloth. The light intensity in the chambers was maintained at 2000 foot-candles at a distance of 2 feet above the plants. The temperature varied between 79 and 81°F. and the day length in the experiment I was 14 hours; in experiments II and III the day length was increased to 15 hours. All experiments were designed as randomized blocks with 4 replications of each treatment except the control treatments, which included 8 replications. Additional controls were used in experiment III to demonstrate that the deficiency symptoms could be cured by addition of Co and not by other elements added to the nutrient solution.

In experiments I and III all plants were cut back just above the third node above the cotyledons. Axillary growth of shoots harvested after the initial growth, or after previous axillary growth, was obtained by cutting shoots at the third node above the previous cut. Experiment II was terminated after the initial harvest because a fungus infection of roots occurred at that time.

RESULTS AND DISCUSSION

Three different experiments have been conducted to determine the influence of Co on the growth and chemical composition of soybean plants in the absence of combined nitrogen. In all experiments, cultures were supplied with the basic nutrient solution, and the additional elements indicated in table 1 constituted the different treatments. The effect of Co and other elements tested on the relative yield of dry matter in all experiments is summarized in table 1. The actual mean yields for shoots from control cultures in the 3 experiments in the order presented in table 1 are, respectively: 6.1, 16.6, 1.9, 2.0, and 3.4 g. dry matter per culture. In experiment I increases in dry matter of first and second axillary shoots resulting from the addition of Co ranged between 23 and 52 per cent, and, as indicated, these differences are statistically significant. Vitamin B_{12} (at a rate of 1 ppb. Co) was included as one treatment in this experiment and, from the results, it is apparent that this addition resulted in a significant yield increase. Since the culture solutions were not aseptic, this response does not provide evidence that the vitamin molecule per se was utilized. It is very possible that Co in the vitamin could have been released by microbial activity. The dry weights of roots in experiment I were not significantly affected by the different treatments.

In experiments II and III the addition of Co at 0.1 or 1.0 ppb. resulted in a striking increase in the dry weight of shoots, but Ni, Al, and Ga, as a combination addition at the rate of 1 ppb. each, or the addition of V at 1 ppb., failed to significantly influence the yield. In experiment II 1 ppb.

[7] Styreseal No. 60, Kyanize Paints Inc., Boston.

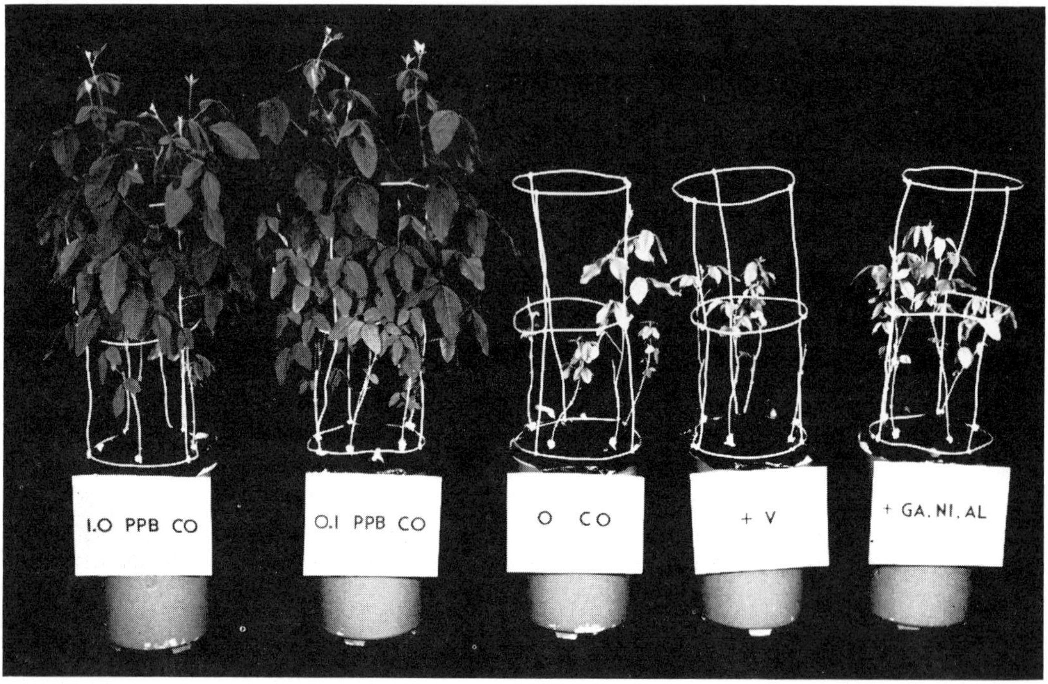

FIG. 1. The second axillary growth of one replication of soybean cultures in experiment III. At the time of the photograph, plants had been growing for 4 weeks after the previous cutting. In the two cultures on the right of the photograph, V at 1 ppb. and Ga, Ni, and Al, each at 1 ppb., were added instead of Co.

Co resulted in a greater response than 0.1 ppb. of the element; this difference was not apparent, however, in experiment III. From the overall results it would appear that 1 ppb. of Co is sufficient for maximum response. The most striking effect of Co on the growth of soybean shoots is illustrated by the data for the yields of the first axillary shoots (table 1) in experiment III. The mean yields of shoots (in g. dry weight per culture) for 0 Co; 1 ppb. Ni, Al, and Ga; 1 ppb. V; 0.1 ppb. Co; and 1 ppb. Co treatments were, respectively: 3.4, 4.0, 3.5, 14.6, and 13.2. The L.S.D. for comparison of these means was 3.0 at 5 per cent and 4.6 at 1 per cent. Co levels of 0.1 and 1.0 ppb. resulted in 3.27- and 2.86-fold increases in dry weight respectively. The second growth of axillary shoots in experiment III exhibited the greatest response to Co and these differences are illustrated by a photograph (fig. 1)[8] of one replication of each culture in the experiment. All plants from cultures receiving Co were dark green and vigorous in appearance, whereas those lacking Co, or those treated with V, or a combination Ga, Ni, and Al instead of Co, exhibited pronounced symptoms of nitrogen deficiency and severe growth retardation. At the time the photograph was made (4 weeks after harvesting the first axillary growth), 4 plants in cultures lacking Co had died from the extreme deficiency. The third growth of axillary shoots in experiment III are now being subjected to conditions that will initiate flowering and seed production. Yield data for shoots or roots, therefore, are not available at this time. From the dry weight of roots produced in experiments I and II, and from the appearance of roots produced in experiment III, it seems that Co has a much greater and more consistent effect on the development of shoots than roots. The roots, however, were not subjected to the cutting-back technique applied to shoots.

After the first axillary shoots were harvested in experiment III, 4 Co-deficient cultures were used to determine whether or not the deficiency symptoms could be cured by addition of the element. Co, at the rate of 5 ppb., was added to 2 cultures

[8] *(Added in proof)* The second axillary growth (fig. 1) was harvested 7 weeks after the previous cutting and the mean yields of shoots in g. dry matter per culture were: 1 ppb. Co, 43; 0.1 ppb. Co, 42; 1 ppb. V, 7; 1 ppb. Ni, Ga, Al, 6; 0 Co, 4. The yields of roots in g. dry matter per culture were: 1 ppb. Co, 13; 0.1 ppb. Co, 15; 1 ppb. V, 5; 1 ppb. Ni, Al, Ga, 5; 0 Co, 4.

while 2 other Co-deficient cultures were kept as controls. After 10 days plants in the pots treated with Co showed definite evidence of recovery. After 21 days the new leaves of Co-treated plants were dark green, and a random sample of the third trifoliate leaves from the top of the plants contained a mean of 3.6 mg. chlorophyll per g. fresh weight and 4.8 per cent total N on a dry matter basis. Comparable samples of leaves from Co-deficient plants contained a mean of 1.6 mg. chlorophyll per g. fresh weight and 4.1 per cent total N. The chlorophyll content of the third trifoliate leaves from plants receiving 0.1 ppb. Co from the beginning of the experiment was 4.4 mg. per g. fresh weight, and the N content, on a dry weight basis, was 5.4 per cent. This evidence, together with the apparent striking increase in growth resulting from the Co addition, is conclusive evidence that the symptoms were cured.

In another treatment, Ru, Cr, W, Pd, Ag, Sn, Bi, and U, at a concentration of 1 ppb. of each element, was added to Co-deficient plants. These elements were chosen because they are reported to form complexes with 1-nitroso-2-naphthol. Although the experiment is not yet completed, none of these elements alleviated the symptoms within 20 days.

The plant material from the various experiments outlined in table 1 were analyzed for certain chemical constituents. In experiment I, dry shoots from the first and second axillary growth of control cultures contained means of 2.16 and 2.53 per cent N, respectively. In comparison, the N contents of samples of shoots from cultures treated with either 1 or 50 ppb. Co were increased by amounts ranging between 12 and 26 per cent. In experiment II addition of Co failed to significantly increase the N content of shoots; in experiment III, however, initial shoots from cultures with Co contained 15 to 18 per cent more N than samples of shoots from control cultures. In the same experiment the first axillary shoots from cultures treated with Co exhibited an even greater increase in N content, ranging between 41 and 44 per cent more than the N content of shoots from control cultures. Also in experiment III, leaves from the first axillary shoots of cultures receiving the following treatments: 0 Co (control); 1 ppb. each of Ni, Al, and Ga; 1 ppb. of V; 0.1 ppb. Co; and 1.0 ppb. Co, contained, respectively, means of 2.04, 1.97, 2.19, 2.29, 3.50, and 3.30 mg. chlorophyll per g. of fresh weight. Chlorophyll analyses of leaves from other experiments have not been carried out.

Analyses of plant material from experiment II indicated that dry shoots from control cultures contained 0.04 ppm. or less of Co. The mean Co content of dry shoots from cultures treated with 1 ppb. Co was less than 0.1 ppm., and that of shoots from cultures treated with 50 ppb. Co ranged between 3.5 and 1.0 ppm. The Co content of roots from experiment I contained the following mean Co contents on a dry weight basis: control, 0.05 ppm.; 1 ppb. Co (as B_{12}), 0.1; 1 ppb. Co, 0.5; 50 ppb. Co, 4.6. The Co analyses of plants from the other experiments have not been completed. Since Co is known to be a constituent of the vitamin B_{12} molecule, nodules from plants receiving the various treatments in experiment I have been assayed for this vitamin by the *Ochromonas malhamensis* method. Results have shown that nodules from control cultures contained a mean of 0.13 millimicrograms of B_{12} per g. of fresh nodules. The B_{12} content of nodules, in the same units, from cultures treated with 1 ppb. Co (as B_{12}), 1 ppb. Co, and 50 ppb. Co were 7.8, 0.65, and 519, respectively. Vitamin B_{12} analyses of nodules from other experiments have not been completed.

SUMMARY

The effect of small amounts of Co on the growth of soybean plants, under symbiotic conditions, were investigated in three different experiments. Plants were grown under controlled conditions in highly purified nutrient solutions lacking combined N and were forced, therefore, to depend upon symbiotic nitrogen fixation. Plants in culture solution without added Co were severely retarded in growth and displayed a chlorosis of leaves similar to that caused by N deficiency. Four of 16 Co-deficient plants died as a result of the severe symptoms. The addition of 0.1 to 1.0 parts per billion (ppb.) of Co to nutrient solutions throughout the growth period in all experiments resulted in normal vigorous plant growth. Also, the addition of small amounts of Co to severely deficient plants cured the symptoms on all new growth within a period of 10 to 21 days. No evidence has been obtained that V, Ni, Ga, Al, or a series of 8 elements presumed to have been removed by the purification procedure, would substitute for Co as a necessary mineral for the symbiotic growth of soybean plants.

In addition to the growth response from traces

of Co, analyses indicated that the percentage total N in shoots and the chlorophyll content of leaves of deficient plants were markedly decreased. Nodules from plants grown in culture solutions supplied with Co contained much greater concentrations of vitamin B_{12} than those from plants grown in Co-deficient solutions.

It is concluded that the results of experiments summarized in this paper provide strong evidence that Co is an essential element for the growth of soybean plants under symbiotic conditions. The possible requirement of the element for non-legumes or for legumes supplied with combined N remains an open question and is under investigation.

REFERENCES

(1) Ahmed, S., and Evans, H. J. 1959 Effect of cobalt on the growth of soybeans in the absence of supplied nitrogen. *Biochem. and Biophys. Research Communications* 1: 271–275.

(2) Anderson, J. P., and Andrews, E. D. 1952 Response to vitamin B_{12} of grazing cobalt deficient lambs. *Nature* 170: 807.

(3) Bolle-Jones, E. W., and Mallikarjuneswara, V. R. 1957 Cobalt: effects on the growth and composition of Hevea. *J. Rubber Research Inst. Malaya* 15: 128–140.

(4) Hewitt, E. J. 1952 Sand and water culture methods used in the study of plant nutrition. *Tech. Comm. No. 22 of the commonwealth Bureau of Horticulture and Plantation Crops East Malling, Maidstone, Kent,* p. 195.

(5) Holm-Hansen, O., Gerloff, G. C., and Skoog, F. 1954 Cobalt as an essential element for blue-green algae. *Physiol. Plant* 7: 665–675.

(6) Marston, H. R., and Lee, H. J. 1952 Cobalt in the nutrition of ruminants. Response of cobalt-deficient sheep to massive doses of vitamin B_{12}. *Nature* 170: 791.

(7) Sandell, E. B. 1944 *Colorimetric Determination of Traces of Metals,* p. 271. Interscience Publishers, Inc., New York.

Monovalent Cations and Growth Regulation. I. Growth Responses in Cucumber Hypocotyl Segments[1]

William K. Purves

Department of Biological Sciences, University of California, Santa Barbara

Revised manuscript received June 10, 1965.

Summary. The elongation of etiolated cucumber (*Cucumis sativus* L.) hypocotyl segments was stimulated by KCl and a number of other potassium salts at a concentration of 0.02 N K^+. The effect of KCl was not evident until 14 hours after the beginning of treatment.

NaCl, LiCl, and RbCl enhanced elongation of the segments, and their dosage-response curves were similar to that for KCl. At supraoptimal concentrations ($\geqslant$ 0.05 M), LiCl inhibited segment growth. CsCl was inhibitory at all concentrations tested and was thus the only alkali metal cation to lack growth-promoting activity in this system. NH_4Cl also promoted elongation, but not as effectively as did the alkali cations.

$CaCl_2$, $CoCl_2$, and $MgCl_2$ did not enhance growth when tested at the same concentrations as the monovalent cations. $CoCl_2$ promoted strongly at 10^{-3} M, and $NiCl_2$ was slightly active at 10^{-4} M.

It is suggested that the nutritional requirement for K in higher plants results from a specific involvement in certain enzyme systems and from a relatively nonspecific role related to the elongation response described here.

Potassium and certain other alkali metal cations have been reported to be active in the *Avena* coleoptile section test (4, 9, 15) and the split pea curvature test (13). These cations also promote shoot growth of intact *Avena* seedlings (3). Since preliminary studies showed that potassium markedly promoted the elongation of etiolated cucumber hypocotyl segments, the role of alkali metal cations was investigated in this biological system. The present paper will describe the elongation responses of cucumber hypocotyl segments to lithium, sodium, potassium, rubidium, and cesium ions, as well as to ammonium ions and the divalent cations of calcium, cobalt, magnesium, and nickel.

Materials and Methods

Seeds of *Cucumis sativus* L. cv. National Pickling (Burpee) were soaked for 2 hours in distilled water and sown in vermiculite (Terralite, Zonolite Company) saturated with distilled water. The plant material was kept in complete darkness except for a period of roughly 20 minutes during which time hypocotyls were sectioned under diffuse green light. The light source was a green fluorescent tube wrapped with 3 layers of green and 3 layers of amber du Pont cellophane. Seedlings were grown at 22 ± 1° for 5 to 6 days. At this age, the hypocotyls were 5 to 8 cm in length, and the hypocotyl hook was in the form of an acute angle (fig 1A).

Five-mm hypocotyl segments were obtained by removing the cotyledons and epicotyl from each seedling and cutting the hypocotyls 5 mm below the concave surface of the hypocotyl hooks (fig 1A, B). Sets of 5 segments were incubated for 20 hours at 27° in Syracuse dishes containing a circle of filter

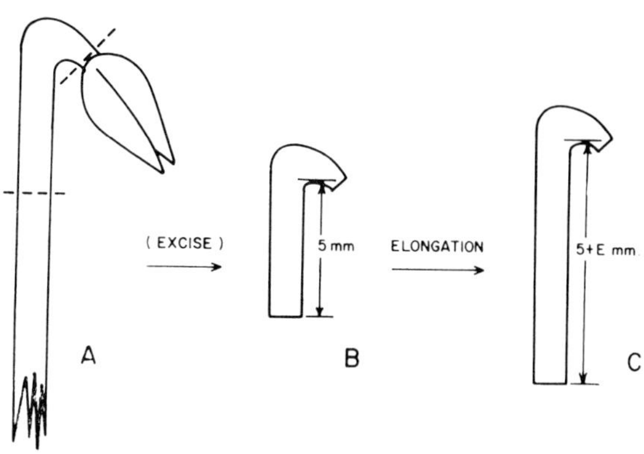

FIG. 1. Harvesting and measuring of cucumber hypocotyl segments. 1A, portion of seedling showing hypocotyl hook and cotyledons, as well as positions of first and second cuts. 1B, freshly harvested segment, showing method of measurement. 1C, segment after growth period, having elongated E mm.

[1] Supported by National Science Foundation grant G21758.

paper and 2 ml of test solution. Thus, the segments were bathed in test medium but prevented from submerging and suffering an O_2 deficit. Each treatment (test medium) was run in duplicate.

The incubation media consisted of distilled, deionized water containing 10^{-4} M gibberellin A_3 plus test additives. (Gibberellin was slightly synergistic with the cations and was included in all experiments reported here; however, the active cations were also effective in its absence.) At the end of the incubation period, segments were measured to the nearest 0.1 mm from the concave surface of the hook to the end of the straight portion of the segment (fig 1C). All measurements were made under a dissecting microscope with a calibrated ocular micrometer. Although absolute growth measurements varied from experiment to experiment, qualitative agreement between experiments was excellent (i.e., if treatment A gave greater elongation than treatment B in 1 experiment, it also did so in repetitions of the experiment).

Salts used in the experiments were twice recrystallized from reagent grade stocks. Gibberellin A_3 was obtained from Abbott and Company (lot no. 746-2059).

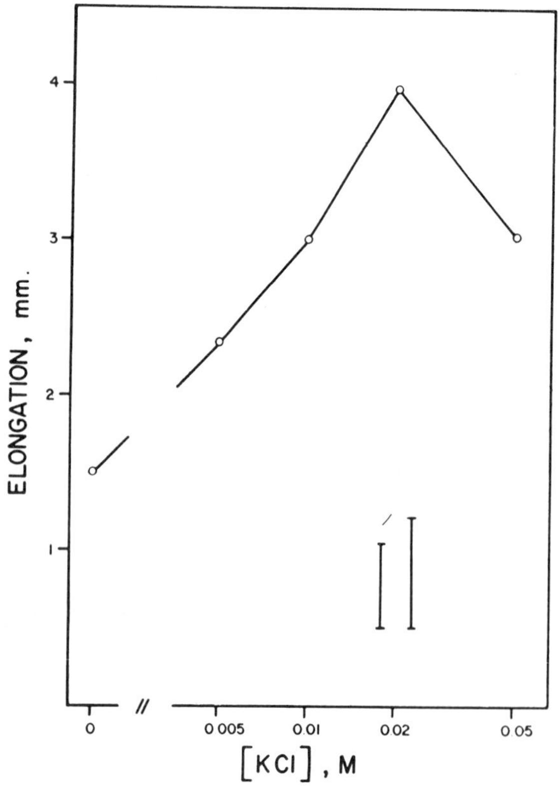

FIG. 2. Dosage response curve for KCl. Three pooled experiments. Each point represents the mean of 30 segments. Least significant differences at 1 % and 5 % probability levels are shown.

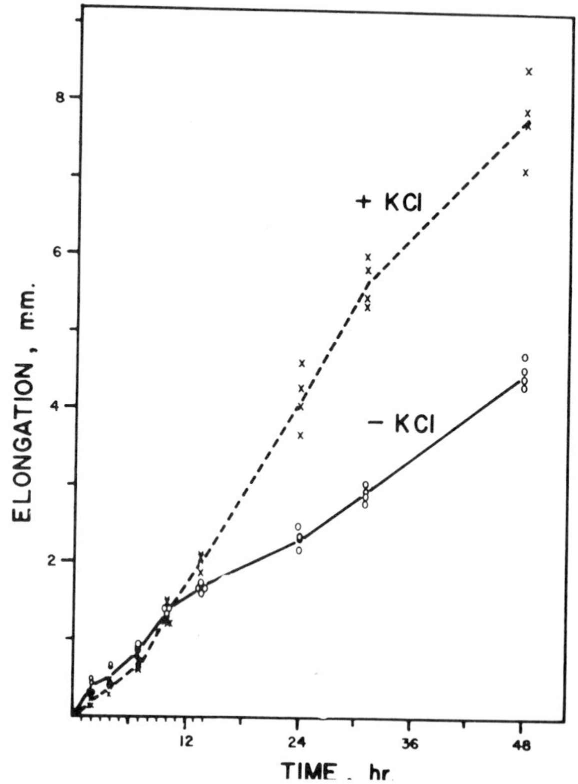

FIG. 3. Time course of elongation in presence and absence of 0.02 M KCl. Two experiments; results from duplicate dishes are shown. Each point represents the mean of 5 segments.

Results

Elongation in Response to KCl. The elongation of hypocotyl segments in response to varying dosages of KCl is shown in figure 2. A maximum promotion was obtained with a 0.02 M KCl dosage, and the 0.05 M dosage was clearly supraoptimal. A series of additional experiments employing smaller concentration increments (e.g., 0.02, 0.03, 0.04, 0.05 M) gave variable results for optimal dosage. In some cases, 0.02 M was the optimal concentration, while in others 0.04 M KCl was definitely more effective.

The time course of the KCl effect at the 0.02 M dosage level is shown in figure 3. Elongation over controls was not evident until 14 hours after initiation of the KCl treatment. It would appear that the divergence of the control and KCl curves is associated with both a decrease in control elongation rate and an increase in elongation rate of the KCl-treated segments.

Elongation in Response to Other K Salts. To determine whether the KCl effect was due to the cation or the anion, the 5 salts KNO_3, KI, KBr, K_2SO_4, and K_2HPO_4 were tested for activity at the 0.02 N K^+ dosage level. All 5 salts were found to be active (table I). In addition, studies with different concentrations of the 5 salts gave dosage-response curves

Table I. *Effects of K Salts on Elongation of Cucumber Hypocotyl Segments*

Treatment*	Elongation, mm**
Control	1.95
KCl	2.87
KBr	3.05
KI	2.26
KNO_3	2.76
K_2HPO_4	2.73
K_2SO_4	2.37

* All salts were tested at 0.02 N K^+.
** Mean of 20 segments.

similar to each other and to the KCl curve. These data strongly suggest that the cation, K^+, is the source of activity.

Elongation in Response to Other Monovalent Cations. Dosage-response curves for NaCl, LiCl, RbCl, CsCl, and NH_4Cl are shown in figure 4. This figure is a composite of several experiments; therefore, the control values vary from salt to salt. In each experiment, a treatment with 0.02 M KCl was included as a reference point. The effect of Na (fig 4A) was similar to that of K. At the optimal concentration, 0.05 M, NaCl promoted segment elongation as effectively as did 0.02 M KCl. Like K, Na has been reported to enhance the growth of *Avena* coleoptile sections (4, 9, 15) and of intact *Avena* seedlings (3).

LiCl increased elongation at the concentration range 0.005 to 0.02 M (fig 4B). However, inhibition was observed at higher concentrations. Segments treated with 0.02 M LiCl elongated more than those treated with 0.02 M KCl.

The Rb response (fig 4C) was similar to those induced by K and Na. However, it would appear that the optimal concentration of RbCl was higher than that for K (i.e., between 0.05 and 0.1 M, rather than 0.02 M). The elongation in response to 0.05 M RbCl was somewhat less than that induced by 0.02 M KCl. Promotive effects of both Li and Rb on *Avena* section growth have been reported (15). However, both Rb and Li inhibited coleoptile growth at high concentrations, while only Li caused inhibition within the range of concentrations tested in the cucumber system reported here.

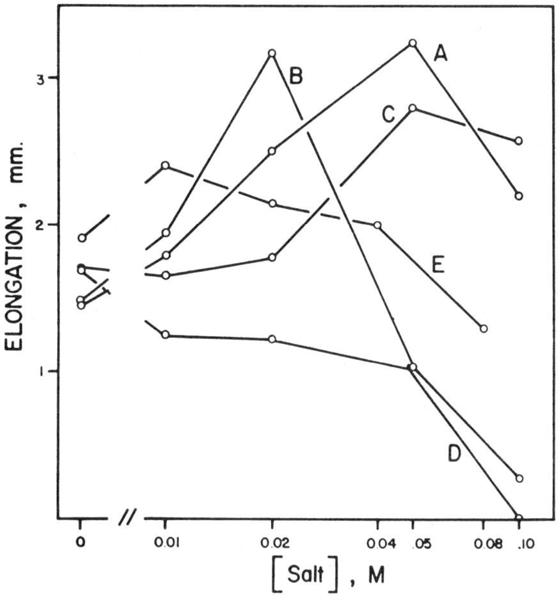

FIG. 4. Dosage response curves for alkali metal chlorides and NH_4Cl. Results of 13 experiments. Each point represents the mean of 30 to 40 segments. Curve A, NaCl; B, LiCl; C, RbCl; D, CsCl; E, NH_4Cl.

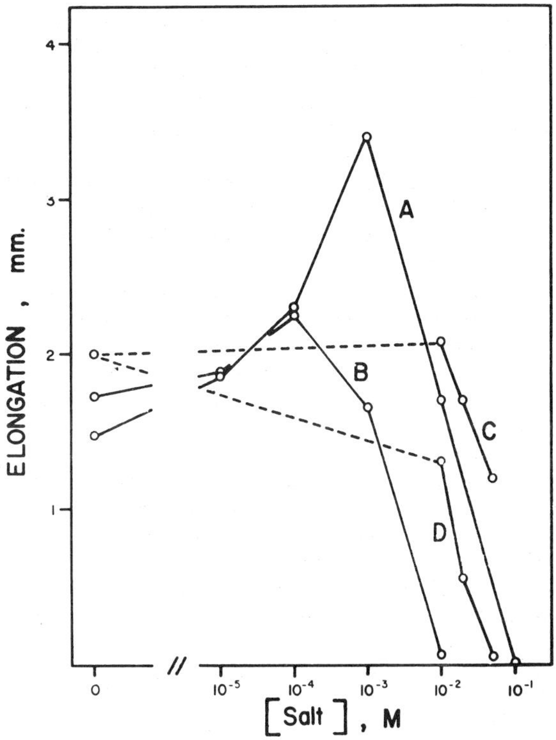

FIG. 5. Dosage response curves for chlorides of divalent metals. Results of 7 experiments. Each point represents the mean of 20 to 30 segments. Curve A, $CoCl_2$; B, $NiCl_2$; C, $MgCl_2$; D, $CaCl_2$.

CsCl inhibited elongation at all concentrations shown in figure 4D, with inhibition increasing with dosage. Concentrations of CsCl below 0.005 M were neither promotive nor inhibitory.

Like the alkali metal cations, NH_4^+ increased elongation at an optimal concentration of 0.02 M NH_4Cl (fig 4E). This response, however, was less than that to 0.02 M KCl. The NH_4^+ effect appears to be the first example of an NH_4-induced promotion of elongation in a system responsive to the alkali metal cations.

Elongation in Response to Divalent Cations. When $CaCl_2$, $CoCl_2$, $MgCl_2$, and $NiCl_2$ were tested at the same concentrations (0.01–0.1 M) as were effective with the monovalent cations, only inhibition was observed (fig 5). When tested at concentrations below 0.005 M, $CaCl_2$ and $MgCl_2$ were without effect. Since Co and Ni are known to promote elongation in certain systems (6, 14), the effects of low dosages of $CoCl_2$ and $NiCl_2$ on cucumber segments were examined. $CoCl_2$ proved strongly active at an optimal concentration of 10^{-3} M (fig 5A), while $NiCl_2$ was only slightly active at its optimum of 10^{-4} M (fig 5B).

Discussion

While other systems, such as *Avena* coleoptile sections, respond to some of the alkali metal cations, cucumber hypocotyl segments are distinguished by the magnitude and rapidity of their responses to these ions. Thimann and Schneider (15) reported K-induced increases in *Avena* section elongation. About 50 % promotion over control elongation was observed after 24 hours and about 80 % after 63 hours of growth. Cooil (4) reported a maximum K response of 85 % by *Avena* sections after 45 hours of growth. In the present study, cucumber hypocotyl segments showed a K-induced promotion of 167 % over controls after only 20 hours of growth (fig 2).

Although only inhibitory effects of Cs were observed (fig 4D), it seems reasonable to suspect that Cs would take part in the same growth-promoting mechanism as the other alkali metal cations. Cs would not be set apart on the basis of ionic size, since the hydrated Cs^+ and Rb^+ ions have virtually identical radii (5). Cs is known to mimic the action of K in certain metabolic reactions of rat liver mitochondria (10) and lactobacilli (2, 8). In any event, the hypothetical stimulation predicted for Cs is masked in the cucumber hypocotyl by an apparent toxicity, even at low levels. Toxic effects of CsCl on the germination and growth of wheat have been reported by Serrano (12).

The inhibitory effect of Li on cucumber hypocotyl elongation (fig 4B) is similar to that on *Avena* coleoptile sections reported by Thimann and Schneider (15). Inhibitory effects of Li on seed germination (12) and on geotropic bending (11) have also been reported, and various inhibitory and formative effects of Li on animal development are well known (7). The nature of the inhibition in cucumber hypocotyls is under further investigation.

In cucumber segments, there are no major differences in positive responses to the alkali metal cations other than Cs. On the other hand, the specific requirement of intact higher plants for K as an essential element is absolute. Partial substitutions of Na and Rb for K as an essential element have been reported in a few cases (1). Where K acts as an enzyme cofactor, Na and Li generally inhibit enzyme activity, while K and often Rb promote (5). A comparison of the enzymic data with the data on cucumber segment elongation gives rise to the following suggestions. The specificity for K as an essential element results from the requirement for K as a cofactor for specific enzymes. The bulk requirement may, however, be related to the mechanisms active in growth promotion in the cucumber system described here. The relative nonspecificity of the cucumber system would, then, appear to explain the partial replacement of K by other alkali metal cations in the mineral nutrition of intact plants.

Acknowledgment

The technical assistance of Mr. Richard D. Glenn is gratefully acknowledged.

Literature Cited

1. Baumeister, W. 1958. Hauptnährstoffe. In: Encycl. Plant Physiol. W. Ruhland, ed. 4: 482–557.
2. Clark, J. A. and R. A. MacLeod. 1954. The effect of some inorganic ions on a bacterial apyrase. J. Biol. Chem. 211: 541–47.
3. Cooil, B. J. 1951. The influence of various sodium and potassium salts upon the growth of young *Avena* seedlings. Plant Physiol. 26: 822–31.
4. Cooil, B. J. 1952. Relationships of certain nutrients, metabolites and inhibitors to growth in the *Avena* coleoptile. Plant Physiol. 27: 49–69.
5. Mahler, H. R. 1961. Interrelationships with enzymes. In: Mineral Metabolism. C. L. Comar and F. Bronner, eds. Academic Press, New York, 1B: 743–879.
6. Miller, C. O. 1954. The influence of cobalt and sugars upon the elongation of etiolated pea stem segments. Plant Physiol. 29: 79–82.
7. Needham, J. 1942. Biochemistry and Morphogenesis. Cambridge University Press, London.
8. Nossal, P. M. 1951. The decarboxylation of malic acid by *Lactobacillus arabinosus*. Biochem. J. 49: 407–13.
9. Ordin, L., T. H. Applewhite, and J. Bonner. 1956. Auxin-induced water uptake by *Avena* coleoptile sections. Plant Physiol. 31: 44–53.
10. Pressman, B. C. and H. A. Lardy. 1952. Influence of potassium and other alkali cations on respiration of mitochondria. J. Biol. Chem. 197: 547–56.
11. Schrank, A. R. 1953. Effect of inorganic ions and their conductances on geotropic curvature of the *Avena* coleoptile. Plant Physiol. 28: 99–104.
12. Serrano, M. 1954. Los elementos Li, Na, K, Rb y Cs en la germinacion y crecimiento de plantulas de trigo. Farmacognosia 14: 27–51.
13. Spear, I. and K. V. Thimann. 1949. The effect of onion juice on the growth response to auxin. Plant Physiol. 24: 587–600.
14. Thimann, K. V. 1956. Studies on the growth and inhibition of isolated plant parts. V. The effects of cobalt and other metals. Am. J. Botany 43: 241–50.
15. Thimann, K. V. and C. L. Schneider. 1938. The role of salts, hydrogen-ion concentration and agar in the response of the *Avena* coleoptile to auxins. Am. J. Botany 25: 270–80.

Iron Translocation
I. Plant Culture, Exudate Sampling, Iron-Citrate Analysis

Lee O. Tiffin

United States Department of Agriculture, Agricultural Research Service, Soil and Water Conservation Research Division, Mineral Nutrition Laboratory, Plant Industry Station, Beltsville, Maryland

Received September 7, 1965.

Summary. Plant culture, exudate sampling, and analytical methods designed to ascertain the form of iron translocated are presented.

Restoration of iron to sunflower plants precultured at different Fe levels resulted in exudate iron concentrations ranging from 0.2 to 31 $\times 10^{-5}$ M. Citrate was from 3 to 89 $\times 10^{-5}$ M. Iron and citrate were highest in exudates from iron-deficient plants. Citrate/Fe ratios were between 1 and 3 for exudates of deficient plants. Exudate from normal plants gave a citrate/Fe ratio of 15.

Malate, iron, and a fraction of the citrate in stem exudates migrated electrophoretically to similar positions in acetate buffer. Extracts of narrow bands from the iron-containing areas gave curves suggesting that citrate bound the iron. Citrate that was not combined with iron migrated in a slower band. The effect of iron on citrate migration was confirmed in several related experiments.

The stability of Fe-citrate was demonstrated electrophoretically in malate buffer. Citrate retained iron against malate.

Data given in this paper indicate that citrate binds iron in sunflower exudate. The data suggest that citrate carries iron in intact plants.

It is generally assumed that metals supplied to plant roots may be intercepted in stem exudates to provide information about their translocation (2). In the present study, iron was supplied to sunflower roots, and exudates were used to sample translocated iron. The quantities of iron involved in other phases (root absorption and accumulation) have been considered mostly in relation to translocation.

Restoration of iron to plants precultured at different iron levels results in wide variations of this metal in the leaves (8, 14) and stem exudates (3, 14, 15). When iron is supplied as a synthetic chelate, the roots of sunflower (13) and soybean (14) absorb the metal but leave most of the chelating agent in the nutrient medium.

In the present study, sunflower plants precultured at different iron levels were given similar treatments at decapitation. Assays were carried out to determine quantities of citrate and iron in the exudates. The primary objective was to determine the form of iron translocated. A brief report of this work is published (11).

In later experiments (12) the iron treatments were reversed. Plants were precultured similarly but given different iron levels at decapitation. The purpose was to determine the effect of iron supplied in the exudation period on the citrate and iron content of the exudates. Attention has also been given to the changes in concentration of citrate and iron that occur during exudation.

Materials and Methods

Nutrient Solution. The standard nutrient was a modified Steinberg solution (10, 14). Elements were approximately one-fifth the Steinberg concentration, except for lower phosphorus. The nutrient contained (mg/liter): 51 Ca, 6.6 Mg, 59 N, 3 P, 44 K, 5 S, 0.130 Mn, 0.070 B, 0.040 Zn, 0.010 Cu, 0.010 Mo, and 0.056 Fe. The iron was supplied as ferric ethylenediamine di(*o*-hydroxyphenylacetate) (Fe EDDHA).

After germination, 8 liters of this nutrient were used in all phases of plant culture except when specified otherwise. After growth periods of 10 days, the nutrient usually was changed. Solutions were replaced more often when fast-growing species or large numbers of plants were grown in the 8-liter volumes.

General Culture and Sampling. Plants were grown in a controlled light chamber under 8 hours dark, 16 hours at 1500 ft-c, and 22 ± 2°. Harvests of plants (i.e. decapitations) were made after plants had received 6 to 8 hours of light. The exudate was collected at 22° in the dark. The general cultural and exudate sampling procedures are given in figure 1. Sunflower plants, *Helianthus annuus* L. var. Greystripe, were used in experiments reported here. Details concerning plant age and treatments are given in the legend of table I.

Stem Elongation. Hypocotyls of plants often are too short to withstand the bending necessary for

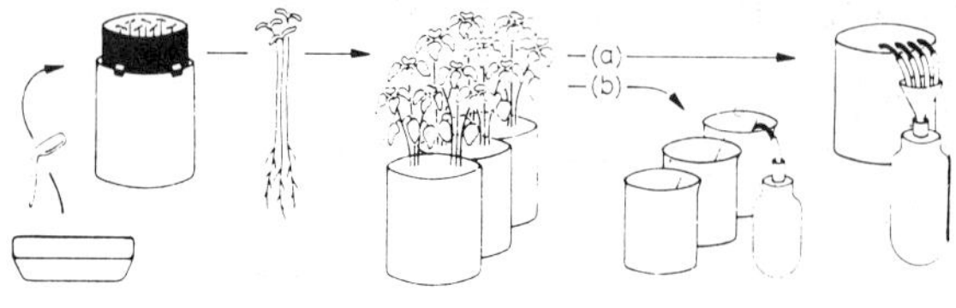

FIG. 1. Diagram of plant culture and exudate sampling methods. Plants are carried through 3 periods: I. Seeds are germinated 2 to 5 days on wet muslin in covered glass trays. The muslin, supported on a stainless frame, extends into water about 1 cm below the seeds. Seedlings are transferred and grown 2 to 8 days under partial shading in an enclosed frame to elongate the stems. Plants with stems 10 to 15 cm long are grouped for transfer. II. This period extends from one to several weeks. In the earlier part plants are cultured to obtain healthy root and top growth; in the latter part they are conditioned for specific tests in the exudation period. III. Plants are decapitated and exudate is delivered by plastic tubing into vials held in an icepack. Two methods are suggested: A) Exudate is pooled from several groups of plants (precultured similarly) to obtain a large volume of exudate or smaller volumes at time intervals, and B) exudate is collected from single groups of plants.

exudate to drip from the cut stems. Seedlings are elongated (fig 1) to prevent stem injury and facilitate exudate collection.

Electrophoresis Apparatus and Buffers. An apparatus was constructed to accommodate large papers and volumes of buffer. The system included two 46 × 46 cm plexiglass sheets[1] to hold the treated papers, a plate glass support of the same dimensions, and 2 electrode compartments of 4-liter capacity. The use of two 20 × 57 cm sheets of Whatman No. 3 paper in the apparatus permitted fractionation of up to 12 ml of exudate.

The buffer used most was sodium acetate (2 liters per compartment) pH 5.4 (6). Other buffers were citrate and malate, both at 10^{-2} M and pH 5.4 (by NaOH). Generally, buffer was used at pH near that of the plant exudates. Exudate pH ranged between 5.2 and 5.6.

Electrophoresis, Isotope Detection, Elution. The red chelate, FeEDDHA, was used as the electrophoretic marker. This chelate moved compactly and was stable in all buffers tested. About 2 μl of a 0.1 M solution was applied per spot. After loading of exudate and marker, the papers were clamped between the plastic sheets and subjected to electrophoresis at 450 v for 2.3 hours, or until FeEDDHA migrated a prescribed distance. The papers were dried on the plastic support in a forced draft oven at 20°. The FeEDDHA spots were marked, prior to drying, to indicate solute migration in the drying period. Radioiron and ^{14}C-labeled acids were located after electrophoresis by X-ray film. The distribution patterns in the figures were obtained by direct photoprinting of the X-ray film positives. Bands were cut from the electropapers and eluted with water to recover exudate compounds.

Analysis of Solutions. Radioiron in the nutrient solutions, raw exudates, and extracts of electropapers was assayed by counting dried samples in a proportional counter. Specific activity of nutrient iron was used to estimate iron quantitatively in stem exudate. The values obtained were compared wtih total iron determined by orthophenanthroline (9). Citric acid was determined by the pentabromoacetone method (5).

The L-malate assay described by Hohorst (7) was modified slightly and used to determine L-malic acid in extracts from electrophoresis papers. The reaction mixture (3 ml) contained 1.5 ml glycine buffer pH 9.5, 0.5 ml of 5 × 10^{-3} M β-DPN, 0.2 ml (200 units) of L-malic dehydrogenase (Sigma Chemical Company), and standard L-malic acid or sample and water, if necessary to obtain volume. OD change at 340 mμ and 25° was measured by a DU spectrophotometer in a 10 minute period, beginning approximately 3 minutes after addition of standard or sample. The standard curve was obtained with zero to 2 μmoles of L-malic acid.

Experiments and Results

Effects of Preculture. The effects of preculture on the iron and citrate levels in sunflower exudate are given in table I. The 4 plant groups were treated alike for 25 days and then were transferred to conditioning nutrients. After growth in these nutrients for 6 days, groups 1 and 2 were chlorotic, and groups 3 and 4 were green. The chlorotic plants absorbed and translocated about the same amounts of iron. The citrate concentrations and also the citrate/iron ratios for groups 1 and 2 were similar. Plants in group 3 were under iron deficiency stress (although green), and thus absorbed iron in quantity similar to that absorbed by the chlorotic plants. But the concentration of citrate was only about one-third and the citrate/iron ratio about one-half that of chlorotic plants. The exudate from group 4 was quite different. Compared to the other exudates, the citrate and iron were low and the citrate/Fe relatively high.

Electrophoresis of Stem Exudates. Electrophoresis of 6-ml volumes of exudates 1, 2, and 3 (table I) was carried out in acetate buffer, and the dried papers were exposed to X-ray film (fig 2). The components from bands A, B, and C for each exudate are shown in table II. Extracts for D bands were carried through all analyses, but no components were found. This indicates a distinct break in the migrating fronts of iron, citrate, and malate. Table II shows the following distribution: Most of the iron was in band C; all of the malate was in band C; sufficient citrate to bind all the iron (1:1) was in band C; and most of the total citrate (without iron) was in band B.

Results in table II indicate that malic acid in the exudates applied at the origin ranged between 0.2 and 0.9 mM. If citric acid had not been present in band C it would be reasonable to conclude that malate bound the iron. A point in support of citrate binding

Table I. *Effect of Preculture on Fe and Citrate Levels in Sunflower Exudate*

Plant age (days) and treatments: 0, germinated seed; 3, seedlings into standard nutrient and shaded to elongate stems; 6, groups into full light and standard nutrient; 13 and 20, renewed nutrient; 25, groups into conditioning nutrients; 31, transferred each group (3 plants) to 1.5 'it(rs of absorption nutrient (1 μc $^{59}Fe/\mu mole$ Fe), decapitated, and collected exudate 18 hours. The conditioning and absorption nutrients are standard nutrients with iron changes specified.

Plant group	Conditioning nutrient FeEDDHA	Absorption nutrient FeEDDHA	Post** absorption nutrient FeEDDHA	Iron in exudate	Citrate in exudate	Ratio: citrate / iron
	Molar	$M \times 10^{-5}$	$M \times 10^{-5}$	$M \times 10^{-5}$	$M \times 10^{-5}$	
1*	10^{-8}	1	0.08	31	89	2.9
2*	10^{-7}	1	0.04	28	80	2.9
3	10^{-6}	1	0.15	20	32	1.6
4	10^{-5}	1	0.84	0.2	3	15.0

* Plants were chlorotic.
** After containing plants 18 hours.

Table II. *Sunflower Exudate Components in Wide Electrophoretic Bands*

Band widths are shown in figure 2.

Exudate number	Extract component	Quantities in band extracts (mμmoles)* A	B	C
1	Iron	1	2	130
	Citrate	140	540	150
	Malate	0	0	860
2	Iron	0	1	101
	Citrate	80	410	130
	Malate	0	0	470
3	Iron	0	2	49
	Citrate	40	80	60
	Malate	0	0	280

* Values given indicate the migration of components originally in 1 ml of exudate applied at the origin.

is suggested by the distributions of this acid in bands B and C for exudates 1 and 3. Comparing the ratios, it is obvious that 540/150 does not equal 80/60. A similar inequality can be noted for exudates 2 and 3. These comparisons indicate that a disproportionate amount of the citrate of exudate 3 moved into band C. This distribution suggests that iron was bound to citrate and thus regulated the amount of citrate that migrated in the faster band. The effect of iron on citrate migration was confirmed in later experiments in which different ratios of iron:citrate were run in acetate buffer.

Figure 3 shows the electrophoretic migration of ^{59}Fe in an experiment similar to that given in figure 2. Conditions were the same except that the X-ray film was exposed 3 hours. The short exposure was used to locate the most concentrated ^{59}Fe band. The purpose was to establish over a narrow width the iron distribution curve and compare it with the curves for citrate and malate. The narrow band c was cut out to obtain the peak for ^{59}Fe activity, and the other bands were taken on each side as shown in the figure. No components were detected in extracts of band e. The quantities of iron, citrate, and malate in the other extracts are given in table III. The results show the malate peak in band b and considerable overlapping of malate in band c. The peaks of citrate and iron were in band c.

Taken together, tables II and III show 2 peaks for citrate. The first and largest is in band B, table

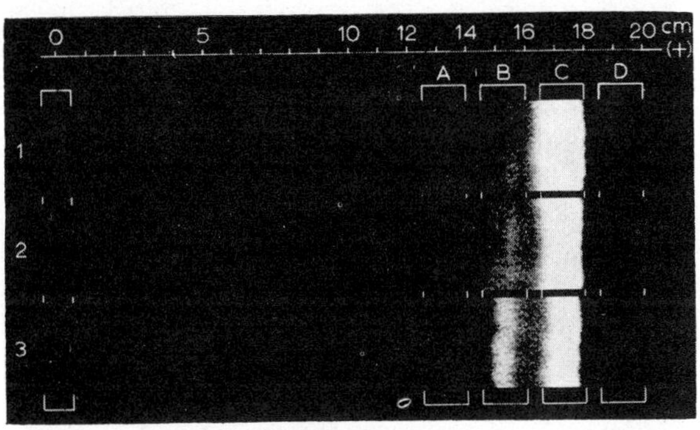

FIG. 2. Electrophoretic distribution of ^{59}Fe in sunflower stem exudate. X-ray film exposure was 30 hours. The 3 exudates are characterized in table I. Brackets at zero cm indicate areas of exudate loading. Bands A to D show areas eluted (elution data, table II). Electrophoresis conditions: Whatman No. 3 paper, 0.05 M acetate buffer, pH 5.4, 450 v, 2.3 hours.

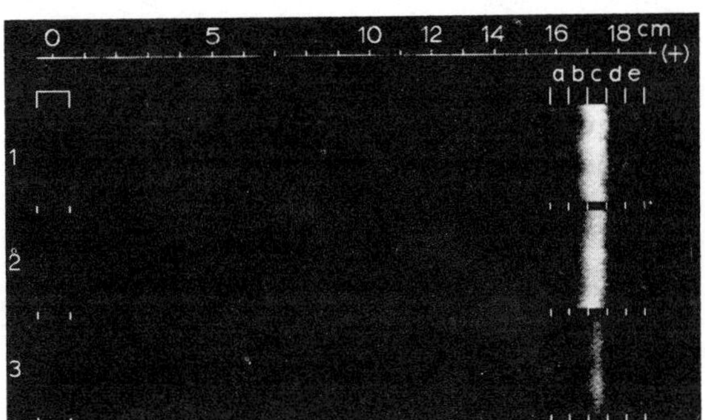

FIG. 3. Electrophoretic distribution of ^{59}Fe in sunflower stem exudate. X-ray film was exposed 3 hours. The 3 exudates are characterized in table I. Bands a to e show areas eluted (elution data, table III). Electrophoresis conditions: Whatman No. 3 paper, 0.05 M acetate buffer, pH 5.4, 450 v, 2.3 hours.

Table III. *Sunflower Exudate Components in Narrow Electrophoretic Bands*

Band widths are shown in figure 3.

Exudate number	Extract component	Quantities in band extracts (mμmoles)* a	b	c	d
1	Iron	10	53	136	12
	Citrate	15	33	85	28
	Malate	216	507	291	0
2	Iron	8	39	121	12
	Citrate	19	30	77	29
	Malate	40	240	144	0
3	Iron	5	23	60	2
	Citrate	11	15	29	0
	Malate	0	240	140	0

* Values given indicate the migration of components originally in 1 ml of exudate applied at the origin.

II. This peak centers at about 15 cm in figure 2. The second peak is smaller and is in band c, table III; its center is at about 17 cm, figure 3. The overall pattern indicates that total citrate migrates as 2 fractions. The faster one appears to be associated with iron. This association is inferred from the symmetry of the iron and citrate curves in table III. It is necessary to recognize, however that citrate in table III is not equimolar with iron. To be consistent with table II, more citrate should have been found in bands b and c.

Several points are noted in summary of the exudate electrophoresis. First, the data do not show unambiguously that citrate carried iron. The malate overlapped the citrate and iron in the acetate system. However, there were indications of citrate-iron complexing. They are as follows: 1) Sufficient citrate to bind Fe (1:1) was in band C for all exudates (table II). 2) Proportionally more of the citrate of exudate 3 was in band C than was noted for the other exudates (table II); this suggested an iron function in citrate distribution. 3) Tables II and III, together, show 2 citrate fractions: one without iron, and the other in the area of concentrated iron. Implicit in this distribution was the suggestion that some factor was causing part of the citrate to migrate at a faster rate. From the symmetry of the citrate and iron curves, the factor appeared to be iron.

Experiments, based on the above findings, were designed to answer the following questions: Does Fe-citrate migrate faster than citrate? Will a strong competitor take Fe from Fe-citrate? Does malate compete strongly with citrate for Fe?

Electrophoresis of Organic Acids and Fe-Acid Complexes. Electrophoresis of ^{14}C citrate and $Fe^{14}C$ citrate was carried out in acetate and citrate buffers. Figure 4 shows distributions of citrate in acetate buffer. Except for slight streaking toward the anode, the ^{14}C citrate migrated compactly to 1 position. The increases of iron (paths 2, 3, 4) caused increasing quantities of citrate to migrate as the faster Fe-citrate fraction. It is evident from this experiment that citrate can be separated into 2 bands and that the relative amounts in each band can be varied appreciably by iron.

The electrophoresis in citrate buffer (not shown) also demonstrated an effect of iron on the migration of citrate. But there was a striking difference in the ^{14}C distribution. On path 4 in the acetate buffer (fig 4), most of the $Fe^{14}C$ citrate migrated about 15 cm. In contrast, the electrophoresis of this sample in citrate buffer gave a heavy ^{14}C streak from 14.5 to 17 cm. The interpretation is that during electrophoresis an exchange of Fe was taking place between labeled citrate and citrate buffer. As the labeled molecules lost Fe to the buffer they migrated more slowly. This would account for the streaking. This result illustrates the effect of a strong competitor. It also recalls the results in tables II and III where malate overlapped citrate and iron. The crucial question is whether malate successfully competed for iron.

The following experiment was designed to test malate competition. Malate buffer was the supporting electrolyte. It was assumed that citrate would migrate at a slower rate if it lost iron to malate. Figure 5 shows that malic acid and its iron complexes (paths 1, 2, 3) migrated to similar positions. The free citrate (path 4) moved behind malate, but the iron compounds of citrate (paths 5, 6) moved well

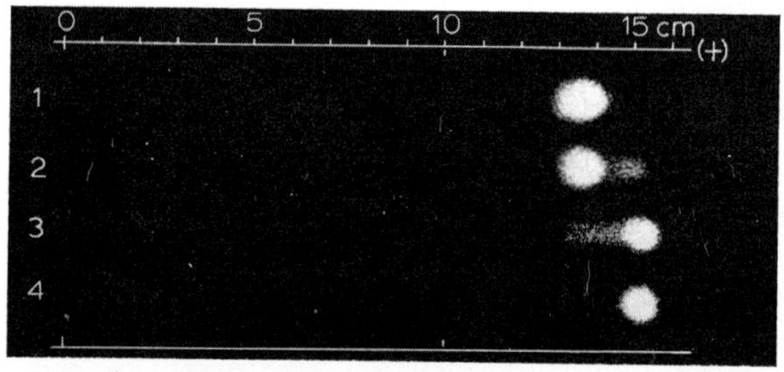

FIG. 4. Radiograph showing the effect of iron on the electromigration of ^{14}C citrate in acetate buffer. ^{14}C citrate (10^{-2} M) was separated at citrate/iron ratios of 1 : 0, 1 : 0.1, 1 : 0.5, 1 : 1 on paths 1 to 4 respectively.

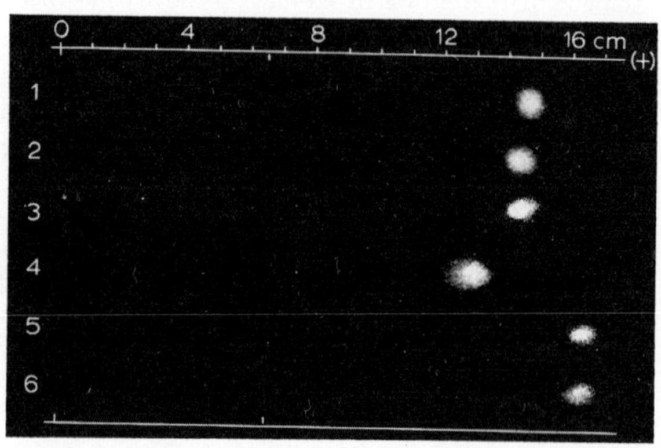

FIG. 5. Radiograph showing the effect of iron on the electromigration of organic acids in malate buffer. The following were separated on paths 1 to 6 respectively: ^{14}C-malate, $Fe^{14}C$ malate, ^{59}Fe malate, ^{14}C citrate, $Fe^{14}C$ citrate, ^{59}Fe citrate. Iron, organic acids, and malate buffer were 10^{-2} M.

ahead of all other compounds. It is noted that in malate buffer, as in other buffers, the Fe-citrate migrated faster than the iron-free acid. But the most significant fact was that the malate saturating the paper did not pull iron from citrate.

It is believed that the results given in this section help explain the exudate data in tables II and III and confirm that citrate was the agent principally involved in the iron transport.

Discussion

Responses of Normal and Chlorotic Plants. Exudate from green sunflower (group 4, table I) contained low concentrations of citrate and iron. Citrate was present at about 3 % of the concentration found in exudates of chlorotic plants, and iron was less than 1 %. A controlled uptake of iron is evident from the fact that most of the iron supplied to the roots was excluded. The chlorotic plants absorbed large quantities of iron and translocated it at concentrations considerably above the external medium.

In contrast to values of 2 or 3 it is noted that normal plants gave a citrate/iron ratio of 15. Although more work needs to be done, it is probable that values well above unity, i.e., citrate concentrations in considerable excess of iron in the exudate, are characteristic of plants grown with adequate iron.

If the combining quantities of citrate/iron are 1:1 in the exudate of normal plants, the ratio of citrate/iron = 15/1 suggests that 1/15 of the citrate is involved in transport of iron and that 14/15 could function in the solubility and transport of other metals.

Further Studies, Translocation Concepts. All exudates in table I gave citrate in molar excess of iron. By using various isotopes and electrophoresis it should be possible to determine whether citric acid binds other trace metals in stem exudates.

The citrate/iron ratios of 15 and 1.6 are far out of proportion. The question arises as to the possibility of getting ratios below unity. The following paper (12) gives evidence that iron can flood into roots and be translocated at concentrations greater than those of citrate.

It will be necessary to revaluate the soybean work (15) which emphasized iron binding by malate. It has been reported (3) that soybean exudates contain relatively large amounts of citric acid. Preliminary runs of soybean exudate in acetate buffer show 2 iron-containing bands with relatively little iron in the faster band. This distribution is quite different from that of sunflower, and will require further study.

With methods presently available it is not possible to demonstrate the binding of iron by citrate or other molecules in the xylem vessels. There is evidence, nevertheless, which supports the concept of acid-iron translocation in vivo. It is summarized as follows: A) Iron precipitates in aqueous media at physiological pH when solubilizing agents are not present. B) Iron binding in solutions of organic acids can be readily demonstrated. C) Stabilities of organic acid chelates of iron are relatively high (1,4). D) Soluble iron and organic acids are usually present in stem exudates of plants. E) Citrate-iron complexing is indicated by exudate electrophoresis.

The role of citric acid is most often considered in relation to cellular metabolism. But released from metabolic sites and moving in the xylem, this acid would be expected to perform other functions. Iron binding is suggested as one of those functions. Further study should indicate whether citric acid participates in the translocation of other trace metals.

[1] Plexiglass pattern P-4, Rohm and Haas Company, Philadelphia 5, Pennsylvania. Trade names and company names are included for the benefit of the reader, and do not imply endorsement by the United States Department of Agriculture.

Acknowledgment

The author thanks Dr. John C. Brown and Dr. Sterling B. Hendricks of this laboratory for their interest and helpful discussions in the course of this work.

Literature Cited

1. Bjerrum, J., G. Schwarzenbach, and L. G. Sillen. 1957. Stability Constants, Part I. The Chemical Society, London.
2. Bollard, E. G. 1960. Transport in the xylem. Ann. Rev. Plant Physiol. 11: 141–66.
3. Brown, J. C. and L. O. Tiffin. 1965. Iron stress as related to the iron and citrate occurring in stem exudate. Plant Physiol. 40: 395–400.
4. Chaberek, S. and A. E. Martell. 1959. Organic Sequestering Agents. Wiley and Sons, New York.
5. Dickman, S. R. and A. A. Cloutier. 1951. Factors affecting the activity of aconitase. J. Biol. Chem. 188: 379–88.
6. Gomori, G. 1955. Preparation of buffers for use in enzyme studies. In: Methods in Enzymology, Vol. I. Colowick, S. P. and N. O. Kaplan, eds. Academic Press, New York. p 138–46.
7. Hohorst, H.-J. 1963. L-(−)-malate determination with malic dehydrogenase and DPN. In: Methods of Enzymatic Analysis. H. U. Bergmeyer, ed. Academic Press, New York. p 328–32.
8. Jacobson, L. and J. J. Oertli. 1956. The relation between iron and chlorophyll contents in chlorotic sunflower leaves. Plant Physiol. 31: 199–204.
9. Saywell, L. G. and B. B. Cunningham. 1937. Determination of iron, colorimetric *o*-phenanthroline method. In. Eng. Chem., Anal. Edition 9: 67–69.
10. Steinberg, R. A. 1953. Symptoms of molybdenum deficiency in tobacco. Plant Physiol. 28: 319–22.
11. Tiffin, L. O. 1965. Translocation of iron by citrate in plant exudates. Plant Physiol. 40: xii.
12. Tiffin, L. O. 1966. Iron Translocation. II. Citrate/iron ratios in plant stem exudates. Plant Physiol. 41: 515–18.
13. Tiffin, L. O. and J. C. Brown. 1959. Absorption of iron from iron chelate by sunflower roots. Science 130: 274–75.
14. Tiffin, L. O. and J. C. Brown. 1961. Selective absorption of iron from iron chelates by soybean plants. Plant Physiol. 36: 710–14.
15. Tiffin, L. O. and J. C. Brown. 1961. Iron chelates in soybean exudate. Science 135: 311–13.

Reprinted from PLANT PHYSIOLOGY, Vol. 41, No. 3, March, 1966, pages 515–518.
(PRINTED IN U.S.A.)

Iron Translocation
II. Citrate/Iron Ratios in Plant Stem Exudates

Lee O. Tiffin
United States Department of Agriculture,
Soil and Water Conservation Research Division, Mineral Nutrition Laboratory,
Plant Industry Station, Beltsville, Maryland

Received September 7, 1965.

Summary. Stem exudates of sunflower, soybean, cucumber, and tomato were analyzed for citrate and iron content. Generally, the lowest iron treatments given at decapitation were associated with lowest citrate levels in the exudate and intermediate treatments with the highest citrate levels. Citrate reached 6.2×10^{-4} M in sunflower exudate, 7.3×10^{-4} M in soybean, 9.4×10^{-4} M in cucumber, and 1.8×10^{-4} M in tomato. Some of these values represent a doubling or tripling of citrate when compared to the lowest ones obtained. High Fe depressed the citrate content of exudate in some cases.

As iron was raised in the nutrient, the increases of iron in the exudate were proportionately greater than those of citrate. The highest Fe treatments led to exaggerated uptake of iron by sunflower. In some cases iron was translocated in molar excess of citrate. A time-response experiment with tomato showed a rapid flooding of iron into the roots but a relatively slow release of iron into xylem exudate.

Similar electrophoretic patterns of iron were obtained despite changes in citrate and iron with time. Exudates from sunflower, cucumber, and tomato gave iron concentrations in the position of iron-citrate.

In the previous paper (2) it was shown that preculture had pronounced effects on the iron and citrate content of sunflower exudate. But despite these effects, the analysis of exudate from both green and chlorotic plants indicated that iron was carried by citrate.

The effect of iron supplied in the exudation period is considered in the present paper. The plant groups in each experiment were precultured alike but were given different iron treatments at decapitation. Assays were carried out to determine citrate/iron ratios in the exudates.

Another consideration has been the length of the exudation period. The primary concern was whether citrate and iron change drastically as exudation proceeds. In experiments with tomato, the exudates were collected at several intervals to determine citrate and iron variability with time.

Materials and Methods

The experimental plants were soybean, *Glycine max* L., Merr., var. Hawkeye; sunflower, *Helianthus annuus* L. var. Greystripe; cucumber, *Curcurbita sativas* L. var. Burpee's Sunnybrook; and tomato, *Lycopersicon esculentum* Mill., var. Marglobe.

Exudate sampling and analytical procedures given previously (2) were employed here. Eight liters of standard nutrient (2) were used in the extended growth period after germination and up to the time of decapitation. This nutrient contained Fe as 10^{-6} M FeEDDHA. The absorption nutrient is the standard nutrient with Fe concentration varied as shown in the tables.

Experiments and Results

Sunflower Experiments. Table I shows the distribution of iron and citrate in 20-hour exudation periods of sunflower. The nutrient iron probably was limiting in the first 2 treatments of experiment 1 and the first treatment of experiment 2. This was suggested by the relatively higher percentages of iron left in the nutrient after 20 hours. Associated with the low iron supply were the low exudate iron levels, low citrate levels, and high citrate/iron ratios.

In general, intermediate iron treatments were associated with highest citrate production. The results show clearly that up to certain limits of nutrient iron it is possible to triple citrate concentration in the exudate. The highest treatments depressed the citrate content of the exudates and gave citrate/iron ratios near unity or considerably lower. Other than the quantitative tests (table I), the only attempt to characterize these exudates was by electrophoresis. The radiograph (fig 1) shows 4 samples from the second experiment (table I). The streaking of radioiron on the paper was associated with the low citrate/

iron ratios. Probably a less reactive support than paper will be necessary to resolve the binding patterns of samples that contain excess iron. The iron spots closest to the anode are in the position of iron citrate.

Soybean Experiment. Results from soybean (table II) were similar to those reported for sunflower. Nutrient iron probably was limiting up to the fourth level (5 × 10^{-6} M). This is suggested by the quantities of iron left in the nutrient. Plant group 4 left only 10 % of the iron after 20 hours. A fairly consistent relationship was noted in the ability of the roots to translocate iron at concentrations considerably above those of the nutrient. The ratio of exudate Fe/nutrient Fe was 20 for group 1. The other groups gave ratios from 29 ot 36. Exudate with the highest concentration of iron still contained citrate in molar excess.

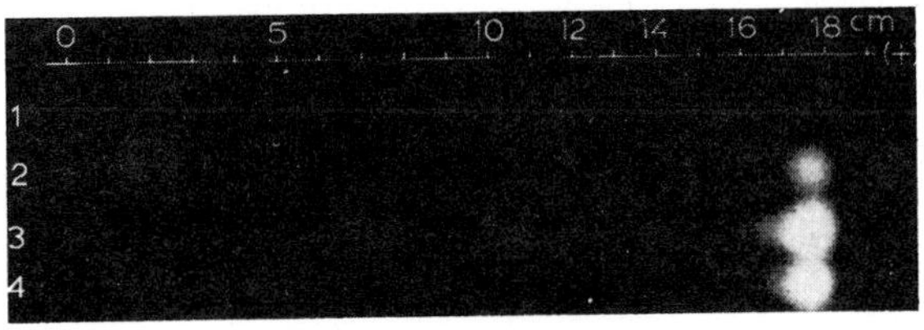

FIG. 1. Electrophoretic distribution of ^{59}Fe in sunflower stem exudates. The 4 exudates are characterized in table I, experiment 2. Electrophoretic conditions: Whatman No. 3 paper, 0.05 M acetate buffer, pH 5.4, 450 v, 2.5 hours.

Cucumber Experiment. Results from cucumber are not given in detail because of similarities to sunflower and soybean. The plants were given iron treatments from 1 to 50 × 10^{-6} M FeEDDHA. Iron in the exudate ranged from 6.7 to 235 × 10^{-6} M. Citrate was from 5.1 to 9.4 × 10^{-4} M. The citrate/iron ratios were from 76 down to 4.

Table I. *Effect of Variable Fe Supply on the Distribution of Fe and Citrate in 20-Hour Exudation Periods of Sunflower*

Plant age (days) and treatment (expt 1): 0, germination; 3, seedlings into standard nutrient and shaded; 6, groups into full light and standard nutrient; 13 and 20, renewed nutrient; 25, transferred each group (3 plants) to 1 liter absorption nutrient (1 μc ^{59}Fe/μmole Fe) and decapitated. Preculture in both experiments was identical, but plants in experiment 2 were decapitated the twenty-sixth day and received 0.5 μc ^{59}Fe/μmole Fe.

Plant group	Absorption nutrient FeEDDHA	Fe left in nutrient	Nutrient Fe into exudate	Exudate volume	Fe in exudate	Citrate in exudate	Ratio: citrate / Fe
	M × 10^{-6}	%	%	ml	M × 10^{-6}	M × 10^{-4}	
Expt 1							
1	1.0	33	24	29	8	2.0	25.0
2	2.5	35	47	36	33	3.6	10.9
3	5.0	7	62	49	63	1.9	3.0
4	7.5	12	54	18	225	6.0	2.7
5	10.0	15	49	26	190	6.0	3.1
6	25.0	33	35	21	425	5.0	1.2
7	50.0	60	21	22	465	4.6	1.0
Expt 2							
1	1	39	12	21	6	2.0	33.3
2	10	9	52	23	220	3.0	1.3
3	50	41	38	18	1050	6.2	0.6
4	100	51	18	21	870	3.6	0.4

Table II. *Effect of Variable Fe Supply on the Distribution of Fe and Citrate in 20-Hour Exudation Periods of Soybean*

Plant age (days) and treatment: 0, germination; 3, seedlings into standard nutrient and shaded; 7, groups into full light and standard nutrient; 18, transferred each group (10 plants) to 1 liter absorption nutrient (5 μc ^{59}Fe/μmole Fe) and decapitated.

Plant group	Absorption nutrient FeEDDHA	Fe left in nutrient	Nutrient Fe into exudate	Exudate volume	Fe in exudate	Citrate in exudate	Ratio: citrate / Fe
	M × 10^{-6}	%	%	ml	M × 10^{-6}	M × 10^{-4}	
1	0.5	33	14	7.0	10	4.0	40
2	1.0	18	28	9.3	30	3.8	12.7
3	2.5	13	41	12.8	80	5.0	6.3
4	5.0	10	48	13.4	180	5.7	3.2
5	7.5	28	35	11.0	240	7.3	3.0
6	10.0	32	35	12.3	290	7.1	2.5

Table III. *Distributions of Fe and Citrate in an Exudation Period of Tomato*

Plant age (days) and treatment: 0, germination; 7, seedlings into standard nutrient and shaded; 17, plants into full light and standard nutrient; 25 and 31, renewed nutrient; 36, transferred 2 plants into 1.9 liters absorption nutrient (2.5×10^{-6} M FeEDDHA, 4 μc ^{59}Fe/μmole Fe) and decapitated.

Sampling time	Absorption nutrient Fe	Nutrient Fe into exudate*	Nutrient Fe into exudate (cumulative)	Fe in exudate	Citrate in exudate	Ratio: citrate / Fe
hr	%	%	%	$M \times 10^{-6}$	$M \times 10^{-6}$	
0	100					
3	19	10.1	10.1	20.0	120	6.0
5	7	12.3	22.4	24.3	140	5.8
8	3	12.4	34.8	24.5	110	4.5
12	1	8.7	43.5	17.3	20	1.2
17	1	4.8	48.3	9.5	6	0.6
22.5	1	3.2	51.5	6.3	7	1.1

* Exudate volume for each collection was 24 ml.

Table IV. *Distributions of Citrate and Fe in Exudates of Tomato Plants Treated with Different Nutrient Fe Levels*

Plant age (days) and treatment: 0, germination; 6, seedlings into standard nutrient and shaded; 12, plants into full light and standard nutrient; 22 and 29, renewed nutrient; 35, transferred 8 plants to 8 liters absorption nutrient (5 μc ^{59}Fe/μmole Fe, group 1; 1 μc ^{59}Fe/μmole Fe, group 2) and decapitated.

	Group 1 Treatment: 2×10^{-6} M FeEDDHA				Group 2 Treatment: 10×10^{-6} M FeEDDHA			
Exudate sampling time	Exudate volume	Exudate citrate	Exudate Fe	Ratio: citrate / Fe	Exudate volume	Exudate citrate	Exudate Fe	Ratio: citrate / Fe
hr	ml	$M \times 10^{-5}$	$M \times 10^{-5}$		ml	$M \times 10^{-5}$	$M \times 10^{-5}$	
3	47	10.8	1.5	7.2	41	11.0	4.8	2.3
6	55	11.1	4.5	2.4	50	13.5	12.5	1.1
9	45	10.9	6.6	1.6	39	13.2	18.8	0.70
12	51	9.0	4.6	1.9	41	14.0	18.4	0.76
15	49	6.0	3.1	1.9	40	14.8	17.3	0.85
18	42	4.8	2.2	2.1	40	15.0	14.7	1.0
21	46	5.0	1.1	4.5	38	16.3	11.8	1.4
24	38	4.8	0.9	5.3	31	14.5	9.7	1.5
27	24	2.8	0.9	3.1	21	9.6	9.5	1.0

Tomato Experiments. Table III shows results from a tomato experiment in which exudate was collected at 6 intervals. The plants lowered nutrient iron rapidly. Less than 10 % remained at 5 hours. This pattern indicates flooding of root iron pools. Of the total nutrient iron (zero time) 81 % had moved into the roots at 3 hours, 10 % had moved into the exudate and, by difference, 71 % was left in the roots. In contrast to the rapid uptake by the roots, the release into the xylem was relatively slow.

The concentration of iron in the exudate held at about 24×10^{-6} M (ca. 10-fold over the original nutrient) for 8 hours and then decreased to about one-fourth that value. About one-half the total iron supplied to the roots was in the exudate at the end of the experiment.

Citrate reached a peak in the second collection, but in the final collection was about one-twentieth its highest concentration. Citrate concentrations declined more rapidly than iron. This was reflected in the drop in citrate/iron values from 6 to approximately 1.

These results suggested that a higher initial supply of iron would have extended the high citrate curve for a longer time. This was confirmed by treating plants with different levels of iron (table IV). The citrate from plants treated with 10^{-5} M Fe continued to increase up to the twenty-first hour. The citrate from plants given 2×10^{-6} M Fe began to decrease after the nin th hour. Compared quantitatively, the low-Fe plants produced 30.4 μmoles of citrate, and the high-Fe plants produced 46.7 μmoles. Electrophoresis of the exudates gave patterns very similar to those given by sunflower in figure 1. All 18 exudates gave a fast-moving spot in the position of iron citrate.

Discussion

Citrate/Fe Ratios Above and Below Unity. As discussed previously (2), green sunflower plants, precultured on 10^{-5} M Fe and treated with the same level for exudate collection, produced exudate with a citrate/Fe ratio of 15. The iron in the exudate of the plants was 2×10^{-6} M. A ratio in which citrate is in considerable excess is believed to be characteristic of exudates from normal plants.

In contrast, the experiments with sunflower and other species reported here involved preculture with 10^{-6} M Fe. The plants were green but under iron deficiency stress as shown by their response when iron was restored at decapitation. In general, the magnitude of the citrate/Fe ratio in exudates depended on iron treatment. The sunflower data in table I show the ratios obtained under a wide range of treatments. On low Fe the citrate/Fe ratio was 74, but on high Fe the ratio was 0.4. Thus manipulation of iron supply can invert a ratio to give iron in excess of citrate.

One implication of these findings is the restriction placed on the translocation concept. It is not possible to generalize under all conditions that iron is carried by citrate. On the other hand, an excess of iron over citrate in stem exudate does not change the concept that citrate is the agent normally involved in iron transport. Apparently the control of iron influx by iron-deficient plants is not the same as that of normal plants. Restoration of high iron to deficient plants therefore results in overloading and, consequently, a translocation of iron in abnormal amounts.

If the combining quantities of citrate and iron are 1:1, then an excess of iron in the exudate would require that considerable iron be carried by other ligand(s). If true, this could have important implications for iron translocation and possibly for the translocation of several other trace metals. It would be necessary of course to prove 1) that citrate carries other metals and 2) that iron can displace them when it reaches certain levels in the exudate.

Changes in Exudate with Time. The iron and citrate curves (tables III, IV) indicate continuous change as exudation proceeds. Nutrient iron accumulated rapidly by tomato (table III) was released relatively slowly into the xylem exudate. Citrate increased until about the fifth hour, but fell sharply after the eighth hour. By providing additional iron, the decline in citrate could be delayed a considerable time, although not indefinitely (table IV). Apparently a certain minimum iron content in root pools is required to maintain a high rate of citrate release.

The tomato exudates (table IV) reveal a sharp decrease in citrate at about 21 hours; exudate volume also declined. The total citrate released by group 1 in the last 3 hours was 0.67 μmole compared to 1.82 μmoles in the previous 3 hours. For group 2 these values were 2.0 and 4.5 μmoles of citrate. Because iron concentrations were still high in the final collections, the citrate decline does not appear to be associated with iron shortage.

Effect of Iron on Citrate Release. It has been shown that the citrate content of exudates can be increased by providing additional iron to plant roots. This was true for all the plants studied. This effect is clearly shown in the exudates of tomato (table IV) where high-Fe plants gave increases of citrate for about 21 hours. The citrate in the exudates of low-Fe plants began to decrease at about 9 hours.

Although a release of additional citrate into the xylem can be induced by iron treatment, the source of the acid is not known. The difference in the quantities of citrate released by the tomato groups (table IV) could represent a difference in citrate production and release from respiratory pools, or possibly a difference in its release from remote compartments (1).

Acknowledgment

The author thanks Dr. John C. Brown and Dr. Sterling B. Hendricks of this laboratory for their interest and helpful discussions during the course of this work.

Literature Cited

1. MacLennan, D. H., H. Beevers, and J. L. Harley. 1963. Compartmentation of acids in plant tissues. Biochem. J. 89: 316–27.
2. Tiffin, L. O. 1966. Iron translocation. I. Plant culture, exudate sampling, iron-citrate analysis. Plant Physiol. 41: 510–14.

Reprinted from PLANT PHYSIOLOGY, Vol. 39 No. 5, September, 1964, pages 862–868.

(PRINTED IN U.S.A.)

Control by Iron of Chlorophyll Formation and Growth in Euglena gracilis [1, 2]

C. A. Price and E. F. Carell [3]

Plant Biology Department, Rutgers—The State University, New Brunswick, New Jersey

Iron has long been recognized as an essential element for growth and chlorophyll formation in higher plants (2) and, more recently, for photosynthetic bacteria (7, 8, 9).

Despite a wide variety of studies on iron deficiency chlorosis (1, 18), there is no specific information on the site of action of iron in chlorophyll formation. The purpose of our study was to identify this site or sites.

The use of higher plants for physiological and biochemical studies presents certain disadvantages. For reasons outlined elsewhere (15), the alga *Euglena gracilis* combines certain favorable characteristics of higher plants and microorganisms for investigations involving trace metals. Since, in addition, the physiology of chlorophyll production in *Euglena* at least superficially resembles that in higher plants (20) and an iron requirement for growth has been established (13), we thought that *Euglena* might be especially suitable for investigating the role of iron in chlorophyll formation.

In the present paper we report on some of the physiological variables concerned in the control of chlorophyll synthesis and growth by iron. We shall show that the iron-dependent chlorophyll-forming system of *Euglena* provides a remarkable test system for the study of the means by which an inorganic ion controls a metabolic process.

Methods and Materials

Euglena gracilis (Klebs), *Z* strain (Hutner), was originally obtained from Dr. S. Hutner.

Growth Conditions. The culture medium was as described previously (4, 15). The culture flasks were shaken at 25° in a Model G-25 Rotary Shaker (New Brunswick Scientific Company), modified so that the cultures could be illuminated through a large window in the top of the shaker. Two 22-w daylight fluorescent lamps were suspended to provide 25 or 50 ft-c illumination at the level of the culture media. Growth was monitored turbidimetrically in corrected Klett units (K_c) (15).

All subsequent incubations were carried out at 25°.

Purification of Medium Components, Glassware, etc. Potassium phosphate, magnesium sulfate, calcium nitrate, malic acid, and the surfaces of culture flasks and other containers were purified as described elsewhere (15). Deionized water, hydro-
fied as follows: reagent grade glutamic acid was
chloric acid, and ammonium hydroxide were prepared
according to Thiers (19). Glutamic acid was puri-
made M in metal-free hydrochloric acid and 1.5 M

---

[1] Revised manuscript received Feb. 6, 1964.

[2] From the doctoral thesis of E. F. Karali (now Carell). Supported in part by Grant G 9815 from the National Science Foundation to Dr. W. R. Robbins. This is a paper of the Journal Series of the New Jersey Agricultural Experiment Station, Rutgers—The State University, Plant Biology Department, New Brunswick.

[3] Present address: Department of Botany and Bacteriology, North Carolina State College of the University of North Carolina, Raleigh, North Carolina.

in potassium thiocyanate. The solution was extracted with either ethyl acetate or amyl alcohol (16) several times until the organic phase was colorless. The aqueous solution was titrated to pH 3.0 with metal-free ammonium hydroxide and stored overnight in the refrigerator. The resulting crystals were collected on a large sintered-glass funnel previously soaked in 50 % nitric acid, and washed with 95 % ethanol until the filtrate was free of thiocyanate (negative to iron solution). The glutamic acid was air dried, weighed, suspended in water, dissolved and titrated to pH 6 with ammonium hydroxide, and diluted to a M concentration.

Iron Determination. A thiocyanate method (16) scaled down to 10 ml final volume, was used. The standard deviation of replicate samples was less than 0.02 μg with 0.2 μg amounts of iron. A sample in which iron was not detected was unlikely therefore to contain more than 0.06 μg. Solutions of the inorganic components of the medium were analyzed directly for iron. Organic samples or cell material were dry-ashed in platinum dishes (19) prior to iron determination. At the suggestion of Dr. R. E. Thiers, a small amount of organic matter (filter paper) was added to platinum dishes when standard solutions of inorganic iron were to be ashed; otherwise recovery was low.

Protein Nitrogen. Samples of cell suspension containing 0.2 to 1 mg of total nitrogen were diluted to about 5 ml volume with 6 % trichloracetic acid, heated in a boiling water bath for 15 minutes, cooled, and centrifuged. The supernatant fractions were discarded and the sediments washed 3 times with 6 % trichloroacetic acid and transferred to Kjeldahl flasks calibrated to 50 ml. The sample was digested, cooled, and brought to 50 ml with distilled water. The acid digestion was then followed by nesslerization (10, 15).

Chlorophyll. One ml of cell suspension was mixed with 4 ml of acetone in centrifuge tubes cooled in ice water. After one hour in the dark the samples were remixed and centrifuged, and the extinction of the clear supernatant solution determined at 663 mμ in a Beckman DU spectrophotometer. (The residue from a single extraction was colorless and further extractions did not increase the yield.) The concentration and amounts of chlorophyll were then calculated from the extinction coefficient for chlorophyll, as reported by MacKinney (11).

Chromatography of chlorophyll was carried out by a slight modification of the glass fiber method of Hager (3). Cells extracted in boiling acetone were mixed with an equal volume of ethyl acetate and washed twice with 10 % (w/v) sodium chloride and once with water (6). The organic phase was dried with a minimum amount of sodium sulfate, evaporated to dryness at reduced temperature and pressure, and dissolved in a minimum volume of acetone. The solution of pigments was applied to sheets of glass fiber (S & S No. 26) prepared as described by Hager (3). The amount of pigment applied was less than 30 (E_{660} optical density units) $\cdot \mu l \cdot cm^{-1}$. The chromatograms were run descending in petroleum ether (30° to 60°): kerosene: absolute ethanol (50: 48: 2) under pressure for about 1 hour.

Hager's glass fiber method has the advantage over chromatography on cellulose paper of many-fold greater capacity and superior resolution. Under our conditions chlorophylls a and b ran at R_F of about 0.8. The exact values varied from one run to another. Chlorophyllides, obtained by extracting spinach or light-grown *Euglena* with cold acetone, ran at R_F values of about 0.2.

Results

Growth as a Function of Iron Level. The components of the growth medium were analyzed for iron and those that contributed more than 0.2 μg iron per liter of final medium were purified. Unpurified glutamic acid from several sources contained sufficient iron to satisfy completely the requirements of the alga. By means of our extraction method, we succeeded in purifying glutamate sufficiently to limit growth to about 10 % of the iron-supplemented controls (table 1).

In order to establish the iron requirement for growth we studied the relationship between iron concentration in the growth medium and the kinetics of growth of the organism. Ferrous sulfate was added to the purified medium to make a final concentration of added iron between 0 and 3.6×10^{-5} M; 50 ml volumes were inoculated in 250 ml nephelo-culture flasks and the cultures shaken at 25° under laboratory light. Growth was measured continuously until the cells ceased to grow. The time course of growth of some cultures are shown in figure 1. Over most of the growth period the cultures containing 5×10^{-7} M iron or more showed exponential kinetics; at lower concentrations of iron growth kinetics were at no point exponential.

Preliminary Chlorophyll Studies. Euglena were grown under high light intensity with and without 3×10^{-5} M added iron and the growth and chlorophyll contents determined. The culture with added iron reached 740 K_c and without added iron, 78 K_c. In addition to the limitations of growth in the absence of added iron, the chlorophyll content was disproportionately low. The chlorophyll content with added iron was 38 μg per mg protein nitrogen. Without added iron the concentration of chlorophyll was 14 μg per mg protein nitrogen or 37 % of the controls. In different experiments the iron contents of the chlorotic cells were 0.23 to 0.5 μg iron per mg protein nitrogen compared to more than 20 μg per mg protein nitrogen in control cells. Thus, *Euglena* reacted to low nutrient iron levels in a manner characteristic of plants generally.

There were 2 possible reasons for the low level of chlorophyll in the iron-deficient cells: decreased synthesis of chlorophyll or increased destruction. Clearly, if we were to study the role of iron in

Table I. *Iron Contents of Medium Components before and after Purification*

The iron contents of medium components were determined in unpurified and, where necessary, purified preparations. For purification procedures, see Methods. The contents are listed as μg detected and as μg calculated per liter of final medium.

Component	Unpurified			Purified	
	Amount analyzed (mg)	Iron found (μg)	μg iron per liter medium	Iron found (μg)	μg iron per liter medium
Potassium phosphate	306	0.06	0.06		
Magnesium sulfate	200	0.45	0.45	0.06	0.06
Calcium nitrate	200	ND*			
Manganese sulfate	1.7	0.09	0.09		
Copper sulfate**	...	...	...		
Zinc sulfate	2.5	ND*	ND*		
Metal-free HCl	12,000			0.10/ml	0.25
Thiamine hydrochloride	120	0.32	0.16		
Cyanocobalamin	0.02	ND*	ND*		
Ethanol	6,000	ND*	ND*		
Malic acid	540	9.10	4.55	0.12	0.06
Ammonium glutamate	3,000	50.00	50.00		
Ammonium glutamate	3,000			2.5	2.5
Sum of Components			55.31		3.18***

* ND, not detected.

** Not analyzed because copper interferes with iron assay. If contaminated with as much as 1% iron, copper sulfate would still only contribute about one μg of iron to the medium.

*** Sum includes iron in those components which did not require purification; estimate does not include iron in deionized water or in copper sulfate.

chlorophyll synthesis, we had to exclude interference due to destruction. Iron-sufficient and iron-deficient cells were grown under 500 ft-c, harvested, and made up in concentrated cell suspensions containing about equal concentrations of chlorophyll. The cells were then shaken in a water bath shaker at 1500 ft-c and samples removed at intervals for chlorophyll determination. We expected that if differential bleaching of chlorophyll was occurring, the iron-deficient cells would lose chlorophyll under the high light intensity, while the iron-sufficient cells would lose less or not be affected at all. As shown in table II this did not occur. Both suspensions increased their chlorophyll concentration, the iron-sufficient controls more rapidly than the deficient cells. While still higher light intensities might have caused bleaching, the low levels of chlorophyll observed in iron-deficient cultures occurred at light intensities of less than 500 ft-c. Thus, simple bleaching could not account for the earlier results.

Chlorophyll Formation in Resting Cells. The rates of chlorophyll synthesis of cells exposed to high light intensity and also the experiments of Huzisige et al. (5) suggested to us that we might be able to induce *Euglena* to synthesize large amounts of chlorophyll by growing the cells under low light and then incubating a washed cell suspension under higher light intensities.

We tested this by growing cells at 25 ft-c, washing and suspending the cells in buffer, and then shaking them under 1500 ft-c (fig 2). No, or in this case, little chlorophyll synthesis occurred in the iron-deficient cells, whereas in the iron-sufficient controls, synthesis occurred rapidly after an initial lag period and was, for an interval, nearly linear with time.

This experiment provided a convenient means for comparison of the rates of chlorophyll synthesis; the cells could grow at normal rates under low light, but the chlorophyll content would be low. By then subjecting a suspension to high light, the cells would be induced to form easily measurable amounts of chlorophyll within a few hours. Since this procedure appeared to constitute a general test system, we sought to characterize it further.

Iron-deficient cells were suspended in different medium components and the rates of chlorophyll formation measured. The rate of chlorophyll synthesis was promoted by iron; the effect of iron was not increased by ethanol, glutamate or other components of the medium (table III and fig 2).

The promotion by iron of chlorophyll synthesis in iron-deficient cell suspensions indicated to us that the iron lesion was partially reversible. While the addition of iron to the deficient cells increased the rate of chlorophyll synthesis, it did not alter the increased length of the lag period that occurs in iron deficiency. Under severe iron-deficiency conditions, no chlorophyll was detected over a 20-hour interval, even with addition of iron.

An optimum concentration for the promotion of chlorophyll formation by iron was found at about 10^{-5} M added iron (fig 3); significant promotion of chlorophyll synthesis was obtained with as little as 3×10^{-8} M iron added to the deficient suspension.

Promotion of chlorophyll synthesis by added iron was found to be independent of the pH of the induction medium from pH 3.5 to 8.

Identity of Newly Formed Chlorophyll. We observed that 80% acetone extracts of cells exposed to light under the standard conditions absorbed with

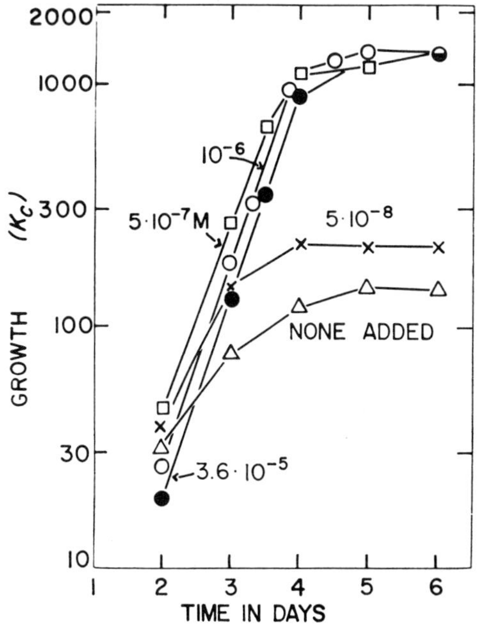

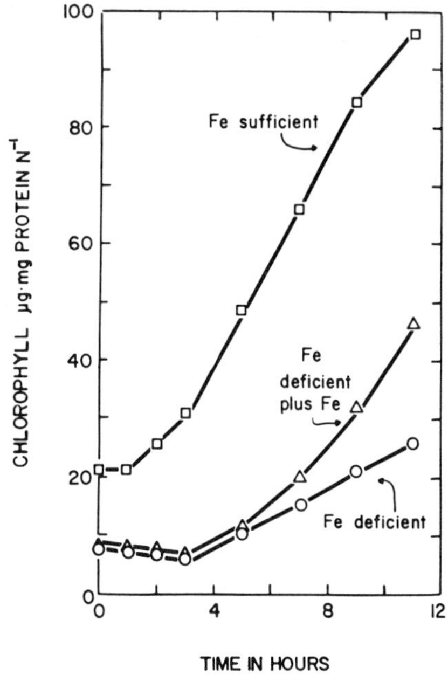

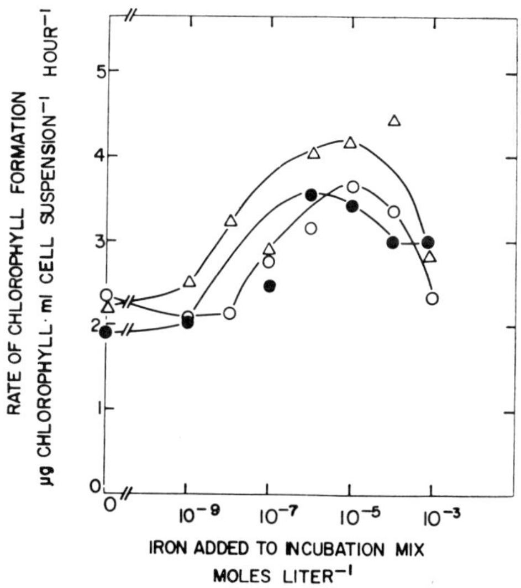

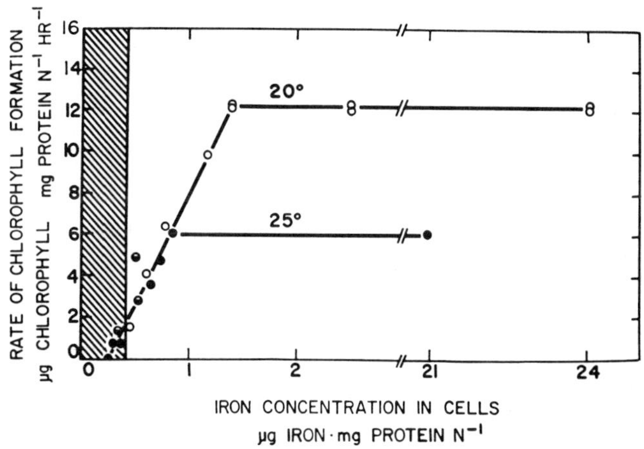

FIG. 1 (*upper left*). Time course of growth of *Euglena gracilis* with different amounts of added iron. Growth monitored in nephelo-culture flasks turbidimetrically in the light at 25°; iron (as ferrous sulfate) added to final concentrations as follows: none, △; 5×10^{-8}, X; 5×10^{-7}, □; 10^{-6}, ○; 3.6×10^{-5}, ●. Ordinate units (K_c) are corrected Klett units (15).

FIG. 2 (*upper right*). Time course of chlorophyll synthesis. Cells grown under low (50 ft-c) light intensity with 3×10^{-5} M iron (□) or low iron (1.8×10^{-7} M) (○, △). After harvest the cells were resuspended in 10^{-3} M phosphate buffer pH 6, with (△) or without (○) 3×10^{-5} M added iron and incubated under high light intensity; samples taken at various intervals for chlorophyll analysis.

FIG. 3 (*lower left*). Rates of chlorophyll synthesis with addition to the cell suspension of various concentrations of iron. Cells grown with 7×10^{-8} M added ferrous sulfate under 50 ft-c light intensity, harvested and resuspended in 10^{-3} M phosphate buffer, pH 6, at various concentrations of ferrous sulfate as shown; rate of chlorophyll synthesis estimated over following intervals: 8 to 11 hours (○), 11 to 13 hours (△), and 13 to 15 hours (●); suspensions contained 0.47 mg protein nitrogen and 0.14 µg (0.0025 µmole) cellular iron per ml of incubation mixture.

FIG. 4 (*lower right*). Rates of chlorophyll synthesis in *Euglena* as a function of the iron content of the cells. The details of the calculations are in the text; the shaded area represents a region where growth is inhibited. The different symbols represent results from different experiments. In one experiment the cells had been grown at 20° (○), in the remainder at 25°. All rates of chlorophyll formation were determined during incubation at 25°.

Table II. *Time Course of Chlorophyll Levels in Cells Exposed to High Light Intensity*

Euglena was grown under 450 ft-c, harvested, washed and resuspended in water at concentrations such that suspensions had nearly equal concentrations of chlorophyll. Then the cells were exposed to 1500 ft-c and samples were taken for chlorophyll analysis at various times. The iron-sufficient cells had 0.87 mg protein-N/ml; iron-deficient cells had 1.8 mg protein-N/ml.

Concentration of iron added to medium	Chlorophyll concentration (μg chlorophyll/ml cell suspension)		
	Zero time	27 hrs	63 hrs
3×10^{-5} M	34	75	97
None added	26	45	53

a peak in the red at 663 mμ characteristic of chlorophyll a; this was true for iron-sufficient cells, iron-deficient cells, and iron-deficient cells treated with iron. We also examined such extracts by glass fiber chromatography. Judging from the R_F and absorption spectrum, the principal green pigment formed was probably the chlorophyllide of chlorophyll a. Chlorophyll a, the chlorophyllide of chlorophyll b, phenophytin a and a fourth unidentified green pigment were also formed in a 10-hour incubation, but in much smaller amounts (about 0.1 as much as the principal pigment). Increases in β-carotene were also observed.

Protein Synthesis. We wished to determine if the observed increases in chlorophyll were accompanied by any net increase in cell protein. Samples of different treatments were taken at zero time and after 9 hours, and analyzed for cell protein. No significant change in cell protein was detected under these conditions (table IV). Thus, chlorophyll formation was disengaged from net protein synthesis.

The observation that no net synthesis of protein occurs when chlorophyll is synthesized in our system does not exclude the possibility that protein turnover occurs. However, we find chlorophyll synthesis is also insensitive to dinitrophenol; in fact, the rate is actually increased about 30 % (table V).

Chlorophyll Formation as a Function of Iron Content. We reasoned that if iron controls a single step in the formation of chlorophyll, we might find a simple relation between the concentration of iron in the cells and the rate of chlorophyll synthesis. We examined, therefore, the rate of chlorophyll formation in cells grown with 10^{-5} M and 10^{-6} M added iron. The choice of the lower concentration of added iron was based on the finding that at 10^{-6} M there was normal growth (cf fig 1) but decreased chlorophyll formation, and that reversal of chlorosis and maximum rates of chlorophyll synthesis were obtained with about 10^{-6} M iron added to the cell suspension.

Cells grown with 10^{-5} M and 10^{-6} M added iron were harvested at different stages of growth in order to obtain cells with varying levels of cell iron. The cells were then suspended in buffer and induced to

Table III. *Rates of Chlorophyll Synthesis with Various Additions to Euglena Cell Suspensions*

Euglena was grown under low light intensity either with (iron-sufficient) or without (iron-deficient) the addition of 3×10^{-5}M iron, then harvested, washed, resuspended in water, and shaken at 25° under 1500 ft-c of light. Chlorophyll synthesis was measured as μg chlorophyll • hour^{-1} • mg protein-N^{-1}.

Composition of suspending medium	Maximum rate of of chlorophyll synthesis	
	Iron-sufficient	Iron-deficient
Water	4.83	0.94
Water	6.5	0.56
Water	6.25	0.51
Water	6.0	...
Complete growth medium less iron	...	0.7
Complete growth medium + iron	...	4.0
2×10^{-3} M glutamate + iron	...	3.0
2×10^{-3} M glutamate + 0.05 M ethanol + iron	...	2.3
10^{-3} M glycylglycine buffer + 0.05 M ethanol + iron	...	3.3

Table IV. *Protein Content During Interval of Induced Chlorophyll Synthesis*

Cell iron content μg • mg protein-N^{-1}	Time	Protein nitrogen (mg/ml cell suspension)	Protein nitrogen % increase
0.27	0	0.358	...
	9 hours	0.345	− 3.6
1	0	0.224	...
	9	0.226	+ 0.7
1*	0	0.224	...
	9	0.226	+ 0.7
21	0	0.268	...
	9	0.270	+ 0.6

* Incubated with 3×10^{-5} M iron; analyzed for iron prior to incubation.

Table V. *Rates of Chlorophyll Synthesis in the Presence of Varying Concentration of Dinitrophenol*

Cells suspended in 10^{-3} M phosphate buffer, pH 6.8, and under conditions as described in table III.

Concentration of dinitrophenol (M)	Rate of chlorophyll synthesis μg • (ml of incubation mixture)$^{-1}$ • hour^{-1}
0	2.44
0	2.42
10^{-6}	2.44
3×10^{-6}	2.71
10^{-5}	2.76
3×10^{-5}	3.21

form chlorophyll. Cells harvested early during growth, contained enough iron to form chlorophyll at normal rates. Cells harvested at more advanced stages of growth, contained less cell iron and synthesized chlorophyll at reduced rates.

Figure 4 contains combined data from different experiments of this sort. The rising part of the curve is the least mean square regression. A clear linear relationship was found between low concentrations of cell iron and the rates of chlorophyll synthesis. Beyond a critical value of cell iron there was no increase in rates of chlorophyll formation as cell iron increased.

The linear part of the curve passes through 2 regions, one (fig 4, shaded area) where growth is limited by iron and the other where growth is normal.

Discussion

Euglena gracilis has a number of promising characteristics as a test system for the study of the means by which iron controls chlorophyll synthesis and growth: under our specified conditions chlorophyll synthesis is rapid, highly repeatable and absolutely dependent on the iron content of the cells. Growth is also dependent on iron, but chlorophyll synthesis can occur independent of growth. Thus *Euglena* appears to represent the kind of organism recommended by Pirson (15) for the solution of problems in mineral nutrition.

While we have not obtained direct information by which the site or sites of iron action can be identified, the data does permit certain generalities about iron in *Euglena.*

From the observation that decreased chlorophyll synthesis occurred in cells which had been growing at normal rates (fig 1, 3), it is clear that the iron required for growth is associated with sites that are distinct from the iron required for growth. Another observation, that the x-intercept is positive in the chlorophyll-iron curve, shows that there exists a pool of iron in the cell which is absolutely unavailable for chlorophyll synthesis. Oertli and Jacobson (12) published a composite graph for a variety of higher plants, qualitatively similar to our figure 4, and also showing iron unavailable for chlorophyll synthesis.

In view of our finding that 2,4-dinitrophenol fails to inhibit chlorophyll formation (table V), it seems unlikely that oxidative phosphorylation could be the energy source for this synthetic process. It may be that chlorophyll formation is dependent on photophosphorylation. This possibility is supported by Carell's (unpublished) finding that inhibitors of photophosphorylation do inhibit chlorophyll formation.

We shall report later (Carell and Price, in press) on the employment of this *Euglena* system in studying the relation of porphyrin metabolism to chlorophyll synthesis.

Summary

When *Euglena gracilis* are grown first under low light intensity and then shaken under high light intensity in a nonnutrient buffer, chlorophyll is synthesized rapidly with high repeatability, and in absolute dependence on the iron content of the cells.

Chlorophyll formation in iron-deficient cells is promoted by iron but not by other components of the growth medium. The iron deficiency "lesion" is partially and specifically reversible.

A graph of chlorophyll synthesis as a function of the iron content of the cells shows the rate of chlorophyll synthesis rising linearly from zero to a saturating value of iron, beyond which the rate remains constant over a 20-fold range of iron contents.

Chlorophyll synthesis may be as little as one-third of control rates at iron contents where growth remains normal.

Chlorophyll synthesis may occur without net synthesis of protein and is not inhibited by 2,4-dinitrophenol.

The form of the newly found chlorophyll appears to be principally chlorophyllide a.

Acknowledgments

We thank Dr. W. R. Robbins and Dr. H. E. Clark for their criticisms and suggestions.

Literature Cited

1. Brown, J. C. 1956. Iron chlorosis. Ann. Rev. Plant Physiol. 7: 171–89.
2. Gris, E. 1844. Nouvelles expériences sur l'action des composés ferrugineux solubles, appliqués a la végétation, et specialement au traitement de la chlorose et de la débilité des plantes. Compt. Rend. Acad. Sciences (Paris) 19: 1118–19.
3. Hager, A. 1959. Die Trennung der Chloroplastenfarbstoffe an Glasfaserpapier. Papierchromatographie in der Botanik. H. F. Linskens, ed, Springer-Verlag, Berlin. p 226–27.
4. Hutner, S. H., M. K. Bach, and G. I. M. Ross. 1956. A sugar-containing basal medium for vitamin B-12-Assay with *Euglena;* application to body fluids. J. Protozool. 3: 101–12.
5. Huzisige, H., T. Terada, M. Nishimura, and T. Uemura. 1957. Effect of amino acids and streptomycin on the chlorophyll formation in *Euglena.* Biol. J. Okayama Univ. 3: 209–22.
6. Jeffrey, S. W. 1961. Paper-chromatographic separation of chlorophylls and carotenoids from marine algae. Biochem. J. 80: 336–42.
7. Larsen, H. 1952. On the culture and general physiology of the green sulfur bacteria. J. Bacteriol. 64: 187–96.
8. Lascelles, J. 1956. The synthesis of porphyrin and bacteriochlorophyll by cell suspension of *Rhodopseudomonas spheroides.* Biochem. J. 62: 78–93.
9. Lascelles, J. 1959. Adaptation to form bacteriochlorophyll in *Rhodopseudomonas spheroides:* changes in activity of enzymes concerned in pyrrole synthesis. Biochem. J. 72: 508–18.

10. LATIES, G. G. 1953. The physical environment and oxidative and phosphorylative capacities of higher plant mitochondria. Plant Phsyiol. 28: 557–75.
11. MACKINNEY, G. 1941. Absorption of light by chlorophyll solutions. J. Biol. Chem. 140: 315–22.
12. OERTLI, J. J. AND JACOBSON, L. 1960. Some quantitative considerations in iron nutrition of higher plants. Plant Physiol. 35: 683–88.
13. ONDRATSHEK, K. 1941. Über den Mineralsalzbedarf heterotropher Flagellaten. Arch. Mikrobiol. 12: 241–53.
14. PIRSON, A. 1955. Functional aspects in mineral nutrition of green plants. Ann. Rev. Plant Physiol. 6: 71–144.
15. PRICE, C. A. AND VALLEE, B. L. 1962. *Euglena gracilis,* a test organism for study of zinc. Plant Physiol. 37: 428–33.
16. SANDELL, E. B. 1944. Colorimetric Determination of Traces of Metals. Interscience Publishers, Inc., New York.
17. SISTROM, W. R. 1962. Observations on the relationship between the formation of photopigments and the synthesis of protein in *Rhodopseudomonas spheroides.* J. Gen. Microbiol. 28: 599–605.
18. SOMERS, I. I. AND J. W. SHIVE. 1942. The iron and manganese relations in plant metabolism. Plant Physiol. 17: 582–602.
19. THIERS, R. E. 1955. Separation, concentration, and contamination. In: Trace Analysis, J. H. Yoe, and H. J. Koch, Jr., eds. New York Acad. Med. 957: 637–66.
20. ZUMSTEIN, H. 1900. Zur Morphologie and Physiologie der *Euglena gracilis,* Klebs. Jahrb. Wiss. Botan. 34: 140–98.

Sodium Absorption by Barley Roots: Role of the Dual Mechanisms of Alkali Cation Transport[1]

D. W. Rains[2] and Emanuel Epstein

Department of Soils and Plant Nutrition, University of California, Davis, California 95616

Received September 21, 1966.

Summary. Radioactively labeled Na^+ absorbed by barley roots was sequestered in an intracellular compartment or compartments ("inner" spaces) in which it was only very slowly exchangeable with exogenous Na^+. Absorption of this fraction proceeded at a constant rate for at least 1 hour.

When the rate of Na^+ absorption was examined over the range of concentrations, 0.005 to 50 mM, the isotherm depicting the relation showed dual kinetics as follows. Over the range, 0.005 to 0.2 mM, a single Michaelis-Menten term describes the relation between the concentration of Na^+ and the rate of its absorption. The mechanism of Na^+ absorption operating over this range of concentrations, mechanism 1 of alkali cation transport, is severely inhibited in the presence of Ca^{2+} and virtually rendered inoperative for Na^+ transport by the combined presence of Ca^{2+} and K^+. The mechanism is equally effective in Na^+ transport whether Cl^- or F^- is the anion, but is somewhat inhibited when the anion is SO_4^{2-}.

Over the high range of concentrations, 0.5 to 50 mM Na^+, a second, low-affinity mechanism of Na^+ absorption comes into play. In the presence of Ca^{2+} and K^+, this mechanism 2 is the only one to transport Na^+ effectively, since Na^+ absorption via mechanism 1 is virtually abolished under these conditions.

Anaerobic conditions, low temperature, and the uncoupler, 2,4-dinitrophenol, inhibit Na^+ absorption both at low and high Na^+ concentrations.

Sodium is the predominant cation in the world's oceans and the preponderant soluble cation in many of the soils of those very large land areas lying in the arid and semi-arid regions of the earth. However, its absorption by plants and plant tissues has been studied much less than that of potassium, the only alkali metal known to be universally essential for all higher plants (7). In the present work, concepts and findings gained primarily from experiments on the absorption of potassium (and rubidium) have been applied to the study of sodium absorption by excised barley roots.

The previous conclusions referred to are 2-fold. 1) Calcium ions in the medium play a crucial, indeed an indispensable, role in selective ion transport by plant tissues. This is so for absorption of alkali cations (4, 5, 14, 16, 18, 19, 22, 23), halides (1, 3, 4, 16, 19, 23), and for other ions as well (25). The role of calcium in preventing or reversing damage to the cellular transport mechanisms caused by sodium is particularly striking (4, 5). The effects of calcium deprivation are profound; they involve damage to normal respiratory patterns and other aspects of metabolism in addition to ion transport (12, 15). Neglect of this factor is doubtless responsible for many contradictory and perplexing statements in the literature. 2) The absorption of potassium and many other inorganic ions by plant tissues is effected by dual mechanisms which differ drastically, often in diametrically opposed fashion, in their properties and responses to the external ionic environment (8). This dual pattern characterizes the absorption of so many ions, by various tissues of so many plant species, as to lead to the claim that it may be well-nigh universal in mature tissues of higher plants (8).

Available evidence indicates that dual transport mechanisms operate in the absorption of sodium as they do in the case of other ions (2, 4, 5, 9, 13, 17, 20). However, with the exception of the experiments from this laboratory (4, 5, 9, 20), these investigations were done without regard to the essential role of calcium in membrane transport and without inclusion of calcium in the experimental solutions. It therefore became necessary to conduct a systematic series of experiments on the kinetics of sodium absorption, its selectivity, and the role of the associated anions and other factors, similar to the experiments in which the dual pattern of potassium transport has been delineated (9, 10).

---

[1] Supported by grants from the National Science Foundation and the Office of Saline Water, United States Department of the Interior.

[2] Present address: Kearney Foundation of Soil Science, University of California, Davis, California 95616.

Material and Methods

The method for growing roots of barley, *Hordeum vulgare,* was as described before (5). Briefly, seeds are germinated for 24 hours in aerated water, in the dark. Seedlings are then grown 5 days over an aerated 0.2 mM solution of $CaSO_4$, in the dark. On the day of the experiment, the roots are thoroughly rinsed in water, excised, repeatedly rinsed again, and suspended in 4 liters of water.

For the experiments proper, the technique described earlier (11) was used. Samples of roots weighing 1.00 g (fr wt) were suspended in the aerated experimental solution, usually at 30°, which contained Na^+ in radioactively labeled form. Sodium was labeled with ^{22}Na. The pH, initially 5.7, varied by no more than 0.2 unit during the experimental runs. The absorption period, usually of 20 minutes except when time was a variable, was discontinued by procedure B of Epstein et al. (11), i.e., the fraction of the labeled Na^+ ions associated with the tissue in diffusible and readily exchangeable form was removed by immersion of the sample in a solution containing unlabeled Na^+ at 5 mM, as well as 0.5 mM Ca^{2+}. Reference 11 should be consulted for details and discussion of the technique.

Results

In the technique used the tissue is transferred, after the period of absorption of labeled substrate ion, to an unlabeled solution of the same ion for removal of that fraction of labeled ions superficially associated with the cell wall. The implication is that another fraction is transported by cellular processes into "inner" spaces of the cells where they are not readily subject to exchange. That this is so for Na^+ was shown in an experiment in which intact seedlings absorbed labeled Na^+ for 4 hours from a 0.1 mM solution also containing 0.5 mM Ca^{2+}. At the end of this period the roots had attained an accumulation ratio of 150 to 1. The roots were excised, briefly rinsed, and 1.00 g samples suspended in unlabeled solution of identical composition. In 2 hours the roots lost only 5 % of the labeled Na^+ they contained. Repetition of the experiment with 10 mM labeled Na^+ during the absorption period and 10 mM unlabeled Na^+ during the desorption period gave the same result; only 5 % of the labeled Na^+ was lost in 2 hours.

Figure 1 brings out 2 points. First, at 0.1 mM NaCl, absorption of the non-exchangeable ("inner" space) fraction of Na^+ proceeds at a steady rate for an appreciable period, and second, Ca^{2+} at 0.5 mM causes a marked reduction in the rate of Na^+ absorption. Repetition of this experiment at 10 mM NaCl gave similar results, except that the rate

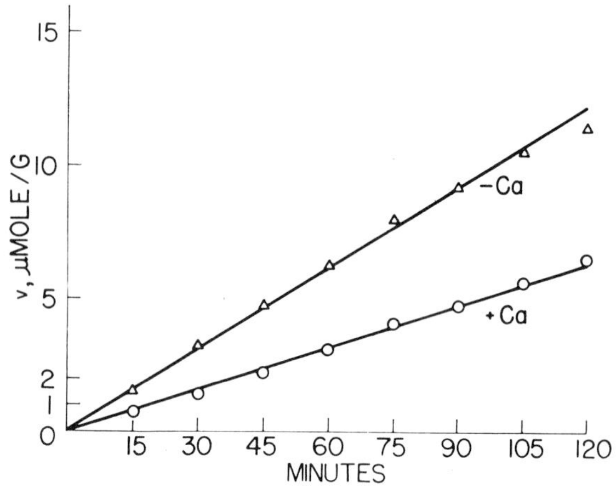

FIG. 1. Absorption of Na^+ as a function of time in the presence and absence of Ca^{2+}. NaCl : 0.1 mM; $CaCl_2$ when added: 0.5 mM.

of Na^+ absorption remained constant for only 1 hour, and the inhibitory effect of Ca^{2+} was smaller.

The effect of Ca^{2+} and of the Na^+ concentration on the rate of Na^+ absorption was examined in the experiment shown in figure 2. Both in the absence and the presence of Ca^{2+} at 0.5 mM, the absorption isotherm strictly conforms to the Michaelis-Menten relationship. The maximal rates calculated for non-limiting concentrations of Na^+ differed by only about 10 % in the 2 runs, but the apparent affinity of the transport mechanism for Na^+ was drastically lowered by the presence of Ca^{2+}, the Michaelis constant being 0.06 mM in the absence of Ca^{2+} and 0.32 mM at 0.5 mM Ca^{2+}. When in addition to 0.5 mM Ca^{2+}, K^+ was present at 1.0 mM, absorption of Na^+ over the range of concentrations shown

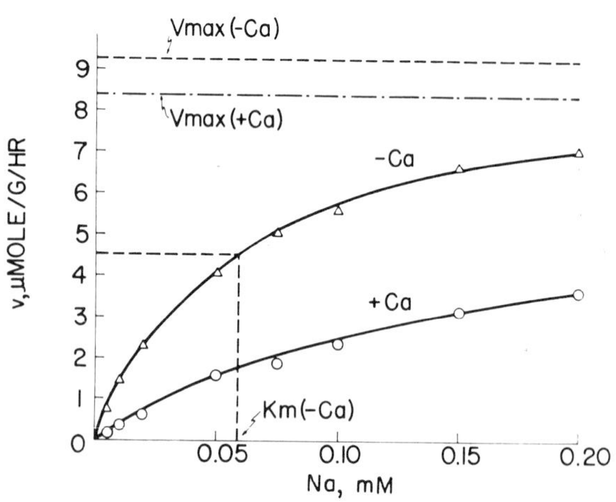

FIG. 2. Rate, v, of absorption of Na^+ as a function of NaCl concentration in the presence and absence of $CaCl_2 \cdot Na^+$: 0.005 mM to 0.2 mM; Ca^{2+} when added: 0.5 mM. Circles and triangles represent experimental data; the lines are plots of the Michaelis-Menten equation. −Ca: K_m, 0.06 mM; V_{max}, 9.2 μmoles/g/hr. +Ca: K_m, 0.32 mM; V_{max}, 8.4 μmoles/g/hr.

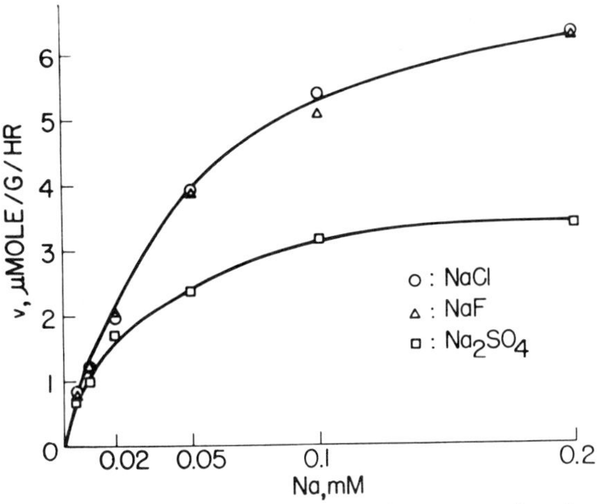

FIG. 3. Rate, v, of absorption of Na^+ as a function of Na^+ concentration and concomitant anion. No Ca^{2+} added.

in figure 2, i.e., up to 0.2 mM, was virtually abolished (see below, and fig 6).

The effect of the anion on the rate of Na^+ absorption over this same range of concentrations is shown in figure 3. With SO_4^{2-} as the anion, the rate of Na^+ absorption was lower than with Cl^-. On the other hand, substitution of F^- for Cl^- as the anion was without effect on the rate of Na^+ absorption.

In figure 4 are shown the results of an experiment in which the rate of Na^+ absorption was determined as a function of the Na^+ concentration over a very wide range of concentrations of NaCl, from 0.005 mM to 50 mM. No Ca^{2+} was added. Up to 0.2 mM Na^+, the isotherm depicting the relation between the concentration of Na^+ and its rate of absorption shows saturation kinetics, virtually leveling off at 0.2 mM (also cf. the $-Ca^{2+}$ run in fig 2). However, at higher concentrations the rate of Na^+ absorption greatly exceeds the rate calculated on the basis of the relation applying over the low range of concentrations (the dashed line). Similar dual kinetics are observed when

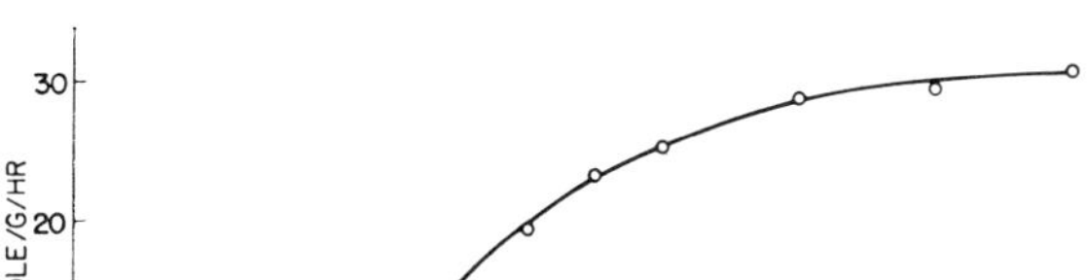

FIG. 4. Rate, v, of absorption of Na^+ as a function of NaCl concentration (0.005–50 mM). No Ca^{2+} added. Concentration scale broken between 0.2 and 0.5 mM. Dashed line represents calculated maximal velocity for the high-affinity mechanism 1 (7.2 μmoles/g/hr).

the anion is SO_4^{2-} (fig 5), but the actual rates of Na^+ absorption are lower in this case than with Cl^- as the anion.

The same wide range of concentrations of Na_2SO_4 (0.005–50 mM Na^+) was explored in the experiment shown in figure 6, but in contrast to the previous 2 experiments the present one was done in the presence of both Ca^{2+} (0.5 mM) and K^+ (1.0 mM). Under these conditions, Na^+ absorption over the low range of concentrations (below 0.2 mM) was virtually eliminated. Results of a similar ex-

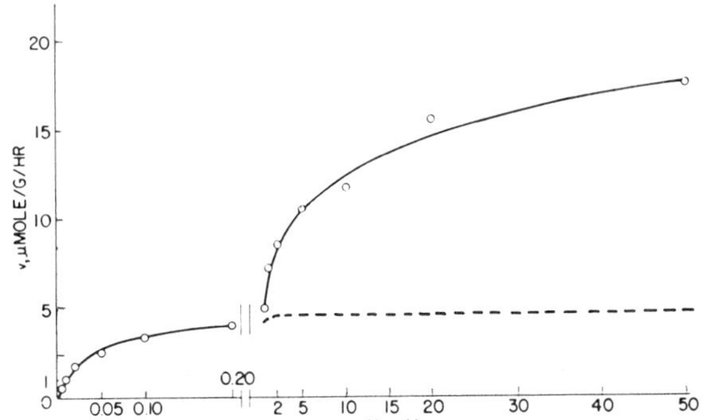

FIG. 5. Rate, v, of absorption of Na^+ from the SO_4^{2-} salt as a function of Na_2SO_4 concentration. No Ca^{2+} added. Na^+ : 0.005 to 50 mM. Concentration scale broken between 0.2 and 0.5 mM. Dashed line represents calculated maximal velocity for the high-affinity mechanism 1 (4.8 μmoles/g/hr).

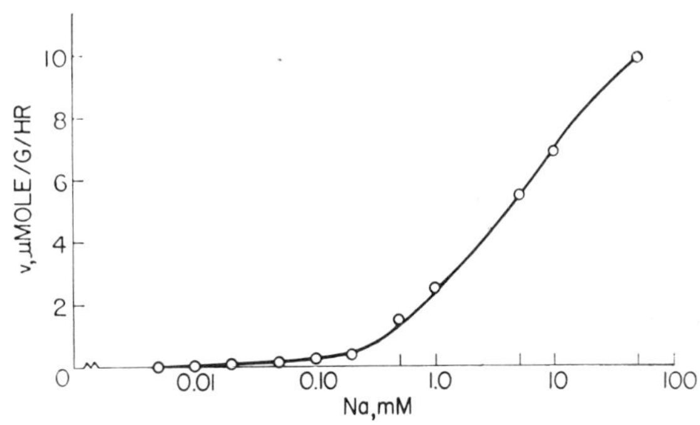

FIG. 6. Rate, v, of absorption of Na^+, plotted arithmetically, as a function of Na_2SO_4 concentration. Na^+ concentration: 0.005 to 50 mM, plotted logarithmically. K^+ as K_2SO_4 : 1.0 mM; Ca^{2+} as $CaSO_4$: 0.5 mM.

periment with Cl^- instead of SO_4^{2-} as the anion have been shown earlier (20). With SO_4^{2-} as the anion, rates of Na^+ absorption were lower than with Cl^- as the anion, as observed before (cf. figs 4 and 5).

Table I shows the results of various inhibitory treatments on the rate of Na^+ absorption at several concentrations of Na^+, both with and without Ca^{2+} and K^+.

Table I. *Effect of Various Metabolic Inhibitors on Na^+ Absorption by Barley Roots*

Treatment	Concentration of ions present	% of Control
N_2	0.1 mM Na^+	50
N_2	10 mM Na^+ + 1 mM K^+ + 0.5 mM Ca^{2+}	35
0.1 mM DNP	0.1 mM Na^+	16
0.1 mM DNP	5 mM Na^+ + 1 mM K^+ + 0.5 mM Ca^{2+}	18
0.1 mM Azide	0.1 mM Na^+	34
0.1 mM Azide	5 mM Na^+ + 1 mM K^+ + 0.5 mM Ca^{2+}	52
		Q_{10}
Q_{10} (20–30°)	0.1 mM Na^+	1.8
Q_{10} (20–30°)	10 mM Na^+ + 1 mM K^+ + 0.5 mM Ca^{2+}	2.3

Discussion

The work reported here leads to the following principal conclusions. The absorption of Na^+ by barley root tissue is mediated by 2 distinct mechanisms. Of these 2 the first, which operates at low concentrations, is severely reduced in effectiveness for Na^+ transport when Ca^{2+}, and especially, when both Ca^{2+} and K^+ are present. Under these conditions, which are the normal, physiological ones, absorption of Na^+ proceeds almost wholly via the second, low-affinity mechanism which does not come into play except at fairly high concentrations of Na^+. On the basis of considerations given below it is concluded that the same dual carrier mechanisms which effect the absorption of K^+ (6, 8, 10) also mediate the transport of Na^+, selection between these closely related alkali cations being accomplished through the differential affinities of these mechanisms for the 2 elements.

Mechanism 1. Over the low range of concentrations, up to 0.2 mM, the absorption of K^+ has been shown to be mediated by a high-affinity mechanism, mechanism 1, with a Michaelis constant of about 0.02 mM (8, 10). Absorption of Na^+ over this same range of concentrations is shown in figures 2, 3 and the left side of figures 4 and 5 to exhibit similar kinetics. In addition to this similarity, the following evidence shows that absorption of Na^+ at these concentrations is via the same mechanism which transports K^+ at low K^+ concentrations, i.e., mechanism 1 of alkali cation transport.

Absorption of Na^+ over this low concentration range is virtually abolished in the presence of Ca^{2+} at 0.5 mM and K^+ at 1.0 mM (fig 6, and fig 1 of ref 20), indicating a common mechanism with much higher affinity for K^+ than for Na^+, so that under these conditions it is virtually preempted by K^+ to the exclusion of Na^+ (20). If this conclusion is correct, it follows that absorption of K^+ by this mechanism should be highly indifferent to Na^+, and this is so (5, 6, 10). Mechanism 1 can be rendered somewhat effective for Na^+ absorption only by imposition of the physiologically abnormal conditions of the absence of both Ca^{2+} and K^+.

Like K^+ absorption via mechanism 1 of alkali cation transport (6, 10), Na^+ absorption via this mechanism does not depend on the presence of an absorbable anion. The results were the same whether the anion was Cl^-, which is readily absorbed (3), or F^-, which is not (24). With SO_4^{2-} as the anion, there was this difference between K^+ and Na^+ absorption by this mechanism: K^+ absorption was almost totally unaffected by substitution of SO_4^{2-} for Cl^- as the anion (6, 10; also cf. 17, 23), whereas Na^+ absorption was appreciably slowed down thereby, at least over the 0.05 to 0.2 mM range of Na^+ concentrations (fig 3).

Mechanism 2. At concentrations above 0.2 mM Na^+, a second mechanism of Na^+ absorption comes into play (fig 4 and 5), as is the case in K^+ absorption (6, 10). Since in this range of concentrations, Na^+ severely and competitively inhibits the absorption of K^+ (10, 21), absorption of Na^+ at these concentrations is considered to be mediated by the same mechanism which effects the absorption of K^+ at these concentrations, i.e., mechanism 2 of alkali cation transport (10, 21). But while mechanism 1 is highly selective for K^+ and can be rendered somewhat effective for Na^+ transport only as discussed above, mechanism 2 shows appreciably higher affinity for Na^+ than for K^+ (10, 21). Further common characteristics of Na^+ and K^+ absorption at high concentrations will be discussed in the succeeding paper (21).

With Ca^{2+} present at 0.5 mM and K^+ at 1.0 mM, mechanism 1 is unavailable for Na^+ transport, and absorption of Na^+ then does not occur except at those high concentrations where the low-affinity mechanism 2 comes into play (fig 6 and refs 9, 20, and 21). If the findings of the present research are of general validity, and this seems likely (8), the work thus furnishes a rationale for the almost universal prevalence of K^+ over Na^+ in plant cells.

No definite explanation can as yet be offered for the interaction by which Ca^{2+} renders mechanism 1 highly selective for K^+ vis-a-vis Na^+ ions. The most attractive hypothesis is that the configuration of the active site of the carrier is such in the presence of Ca^{2+} as to make for a preferential "fit" for K^+ (and Rb^+) ions, whereas in the absence of Ca^{2+} the geometry is loosened to the extent where other ions, sodium included, are also admitted. Because of the far-reaching effects which omission of Ca^{2+} has on various other phases of metabolism (12, 15) additional functions of Ca^{2+} are almost certainly involved.

Interpretation of the effects of inhibitory conditions (table I) is rendered difficult because absorption of K^+, added to deny mechanism 1 to Na^+, is itself inhibited by these agents. It is in any event clear that absorption of Na^+ is severely inhibited by the various antimetabolic conditions that were imposed.

Literature Cited

1. ARISZ, W. H. 1964. Influx and efflux of electrolytes. Acta Botan. Neerl. 13: 1–58.
2. BANGE, G. G. J. 1959. Interactions in the potassium and sodium absorption by intact maize seedlings. Plant Soil 11: 17–29.
3. ELZAM, O. E. AND E. EPSTEIN. 1965. Absorption of chloride by barley roots: kinetics and selectivity. Plant Physiol. 40: 620–24.
4. ELZAM, O. E. 1966. Absorption of sodium, potassium and chloride by two species of *Agropyron* differing in salt tolerance. Thesis (Ph.D.), University of California, Davis.
5. EPSTEIN, E. 1961. The essential role of calcium in selective cation transport by plant cells. Plant Physiol. 36: 437–44.
6. EPSTEIN, E. 1965. Salinity and the Pattern of Selective Ion Transport in Plants. United States Department of the Interior, Office of Saline Water, Saline Water Res. Develop. Progr. Rep. 161.
7. EPSTEIN, E. 1965. Mineral metabolism. In: Plant Biochemistry. J. Bonner and J. E. Varner, eds. Academic Press, New York. 438–66
8. EPSTEIN, E. 1966. Dual pattern of ion absorption by plant cells and by plants. Nature 212: 1324–27.
9. EPSTEIN, E. AND D. W. RAINS. 1965. Carrier-mediated cation transport in barley roots: kinetic evidence for a spectrum of active sites. Proc. Natl. Acad. Sci. 53: 1320–24.
10. EPSTEIN, E., D. W. RAINS, AND O. E. ELZAM. 1963. Resolution of dual mechanisms of potassium absorption by barley roots. Proc. Natl. Acad. Sci. 49: 684–92.
11. EPSTEIN, E., W. E. SCHMID, AND D. W. RAINS. 1963. Significance and technique of short-term experiments on solute absorption by plant tissue. Plant Cell Physiol. 4: 79–84.
12. FOOTE, B. D. AND J. B. HANSON. 1964. Ion uptake by soybean root tissue depleted of calcium by ethylenediaminetetraacetic acid. Plant Physiol. 39: 450–60.
13. FRIED, M. AND J. C. NOGGLE. 1958. Multiple site uptake of individual cations by roots as affected by hydrogen ion. Plant Physiol. 33: 139–44.
14. HANDLEY, R., A. METWALLY, AND R. OVERSTREET. 1965. Divalent cations and the permeability to Na of the root meristem of *Zea mays*. Plant Soil 22: 200–06.
15. HANSON, J. B. 1960. Impairment of respiration, ion accumulation, and ion retention in root tissue treated with ribonuclease and ethylenediamine tetraacetic acid. Plant Physiol. 35: 372–79.
16. HOOYMANS, J. J. M. 1964. The role of calcium in the absorption of anions and cations by excised barley roots. Acta Botan. Neerl. 13: 507–40.
17. JACKSON, P. C. AND H. R. ADAMS. 1963. Cation-anion balance during potassium and sodium absorption by barley roots. J. Gen. Physiol. 46: 369–86.
18. JACOBSON, L., R. J. HANNAPEL, D. P. MOORE, AND M. SCHAEDLE. 1961. Influence of calcium on selectivity of ion absorption process. Plant Physiol. 36: 58–61.
19. MARSCHNER, H. AND K. MENGEL. 1966. Der Einfluss von Ca- und H-Ionen bei unterschiedlichen Stoffwechselbedingungen auf die Membranpermeabilität junger Gerstenwurzeln. Z. Pflanzenern. Düng. Bodenk. 112: 39–49.
20. RAINS, D. W. AND E. EPSTEIN. 1966. Transport of sodium in plant tissue. Science 148: 1611.
21. RAINS, D. W. AND E. EPSTEIN. 1966. Sodium absorption by barley roots: its mediation by mechanism 2 of alkali cation transport. Plant Physiol. 42: 319–23.
22. RAINS, D. W., W. E. SCHMID, AND E. EPSTEIN. 1964. Absorption of cations by roots. Effects of hydrogen ions and essential role of calcium. Plant Physiol. 39: 274–78.
23. TORII, K. AND G. G. LATIES. 1966. Dual mechanisms of ion uptake in relation to vacuolation in corn roots. Plant Physiol. 41: 863–70.
24. VENKATESWARLU, P., W. D. ARMSTRONG, AND L. SINGER. 1965. Absorption of fluoride and chloride by barley roots. Plant Physiol. 40: 255–61.
25. WALLACE, A., E. FROLICH, AND O. R. LUNT. 1966. Calcium requirements of higher plants. Nature 209: 634.

Reprinted from PLANT PHYSIOLOGY, Vol. 42, No. 3, March, 1967, pages 319–323.
(PRINTED IN U.S.A.)

Sodium Absorption by Barley Roots: Its Mediation by Mechanism 2 of Alkali Cation Transport[1]

D. W. Rains[2] and Emanuel Epstein
Department of Soils and Plant Nutrition, University of California, Davis, California 95616

Received September 21, 1966.

Summary. When barley roots absorb Na^+ at concentrations ranging from 1 to 50 mM, in the presence of low concentrations of Ca^{2+} and K^+, absorption of Na^+ is mediated by carrier mechanism 2 of alkali cation transport, mechanism 1 being unavailable for Na^+ transport under these conditions. The absorption isotherm depicting the rate of Na^+ absorption as a function of the external Na^+ concentration, over the 1 to 50 mM range of concentrations, shows several inflections. This stepwise response occurs whether Cl^- or SO_4^{2-} is the counterion, but actual rates of Na^+ absorption are lower in the latter case.

When the concentration of Na^+ is 50 mM, and the concentration of either K^+ or Ca^{2+} is increased from nil to 50 mM, the rate of absorption of Na^+ is diminished not as a smooth function of increasing concentrations of the interfering ions but stepwise. Similarly, when the concentration of K^+ is 50 mM, and the concentration of either Na^+ or Ca^{2+} is increased from nil to 50 mM, the rate of absorption of K^+ is diminished not as a smooth function of increasing concentrations of the interfering ions but stepwise.

Together, this evidence supports the previous conclusion to the effect that mechanism 2 of alkali cation transport possesses a spectrum of carrier sites with different ionic affinities.

When both K^+ and Na^+ are presented at equivalent concentrations over the 1 to 50 mM range, mechanism 2 transports Na^+ almost exclusively, and mechanism 1 K^+ almost exclusively. These findings support previous conclusions to the effect that the active sites of mechanism 2 have higher affinity for Na^+ than for K^+, whereas the reverse is true for mechanism 1.

Two mechanisms effect the absorption of alkali cations by plant roots and other plant tissues (see ref 5 for a comprehensive review). The 2 differ drastically in many features relating to ionic affinities and selectivities, effects of counterions, and kinetics. One of the outstanding differences between them is that the high-affinity mechanism, mechanism 1 of alkali cation transport, which is operative even at very low concentrations, is highly selective for K^+ as against Na^+, whereas the low-affinity mechanism 2, which only comes into play at concentrations of 0.5 to 1 mM and above, transports Na^+ as well as K^+ (see 11 for references). In the preceding paper (11), Na^+ transport in barley roots has been examined over a wide range of concentrations, in terms of the activities of both mechanisms 1 and 2. In the present paper, the transport of Na^+ and K^+ by mechanism 2 is scrutinized in detail.

[1] Supported by grants from the National Science Foundation and the Office of Saline Water, United States Department of the Interior.

[2] Present address: Kearney Foundation of Soil Science, University of California, Davis, California 95616.

Materials and Methods

Excised roots of barley, *Hordeum vulgare*, 'Arivat', were the experimental material. The methods throughout were those described in the preceding paper. Seedlings were grown as described earlier (3), and the technique of conducting the absorption experiments with excised roots was as outlined before (8). Sodium was labeled with ^{22}Na. Potassium was labeled with ^{86}Rb in the experiments shown in figures 3 and 5, and with ^{42}K in the experiment of figure 6.

Results

It has been shown before that mechanism 2 of alkali cation transport in barley roots shows

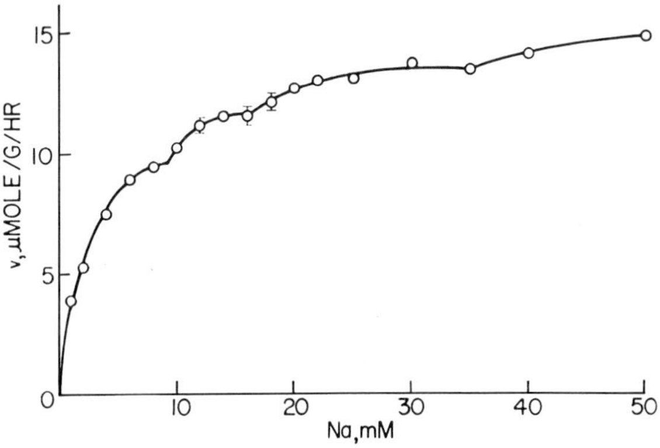

FIG. 1. Rate, v, of absorption of Na^+ as a function of the concentration of Na_2SO_4. Sodium as Na_2SO_4: 1 to 50 mM; $CaSO_4$; 0.5 mM; K^+ as K_2SO_4: 1 mM. pH 5.7 initially, changing little during the absorption period. Temp 30°. All treatments replicated; circles represent the means of 2 values indicated by the short horizontal lines. Horizontal lines not drawn where the distance between them would have been equal to or less than the diameter of the circle.

a certain heterogeneity (6). When the rate of absorption of K^+ or Na^+ is examined over the high range of concentrations where mechanism 2 contributes to absorption, i.e., from 0.5 or 1.0 to 50 mM, the absorption isotherm depicting the rate of absorption as a function of the external concentration has several inflections (6). In those experiments K^+ and Na^+ were furnished as the Cl^- salts. Figure 1 shows that the kinetics of Na^+ absorption over the high range of concentrations exhibit the same complex pattern with SO_4^{2-} as the anion. The rate of Na^+ absorption rises not as a smooth function of increasing Na_2SO_4 concentrations in the solution but stepwise. The inflections in the isotherm occur at similar if not identical concentrations, whether Cl^- (6) or SO_4^{2-} (fig 1) is the counter-ion, viz., at about 10, 18 to 20, and 30 to 35 mM Na^+.

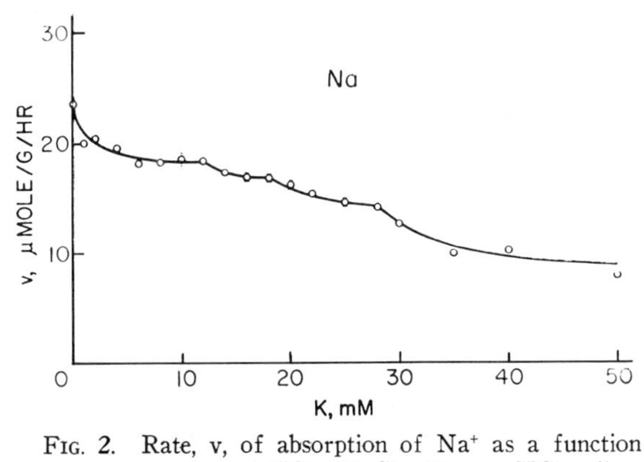

FIG. 2. Rate, v, of absorption of Na^+ as a function of the concentration of KCl. NaCl: 50 mM; KCl: nil to 50 mM; $CaCl_2$: 0.5 mM. Other conditions and conventions as for figure 1.

Figures 2 to 5 present the results of experiments in which this heterogeneity was demonstrated in a novel way. In each of these experiments, the concentration of the ion whose absorption was to be measured (the substrate ion) was 50 mM, while the concentration of another cation (the interfering ion) varied from nil to 50 mM. The rate of absorption of the substrate ion was examined as a function of the concentration of the interfering ion.

In the experiment shown in figure 2, the substrate ion was Na^+, present at 50 mM. When the concentration of K^+, the interfering ion, was varied, the rate of Na^+ absorption decreased not as a smooth function of increasing concentrations of KCl but stepwise. The reverse of this experiment is shown in figure 3 in which K^+ was the substrate ion, present at 50 mM, and the concentration of Na^+ varied from nil to 50 mM.

Figures 4 and 5 show the results of experiments similarly designed, in which the effect of Ca^{2+}

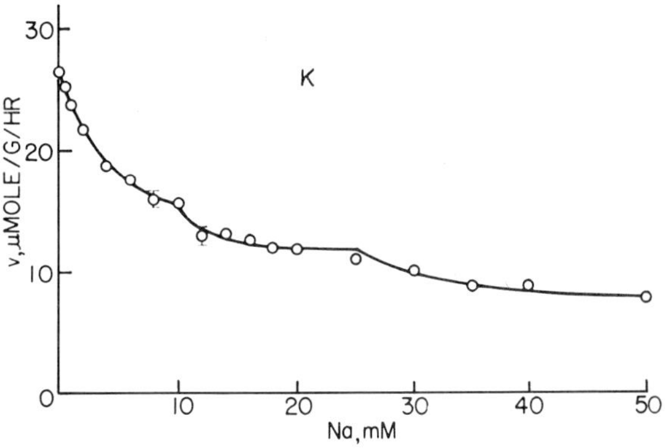

FIG. 3. Rate, v, of absorption of K^+ as a function of the concentration of NaCl. KCl: 50 mM; NaCl: nil to 50 mM; $CaCl_2$: 0.5 mM. Other conditions and conventions as for figure 1.

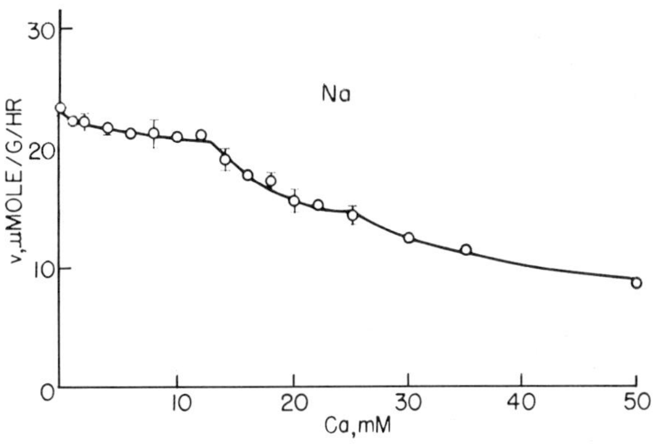

FIG. 4. Rate, v, of absorption of Na^+ as a function of the concentration of $CaCl_2$. NaCl: 50 mM; $CaCl_2$: nil to 50 mM; KCl: 1 mM. Other conditions and conventions as for figure 1.

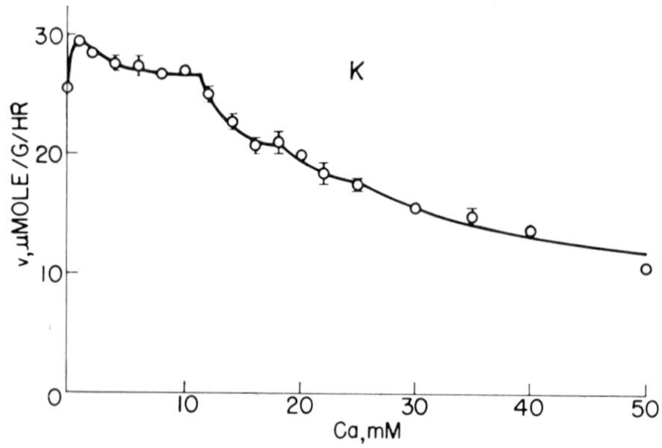

FIG. 5. Rate, v, of absorption of K^+ as a function of the concentration of $CaCl_2$. KCl: 50 mM; $CaCl_2$: nil to 50 mM. Other conditions and conventions as for figure 1.

on the rate of absorption of Na^+ and K^+, respectively, was examined.

In the experiments reported so far, the absorption of either Na^+ or K^+ was measured. Figure 6 presents the results of an experiment in which the concentrations of both Na^+ and K^+ varied from 1 to 50 mM. Any given value on the abscissa denotes the concentration of each ion. On the ordinate are plotted the rates of absorption of Na^+, K^+, and the sum of these rates. These values were obtained by first counting the total radioactivity, then letting the 12.4 hour ^{42}K activity decay (5 days) and finally counting the ^{22}Na activity alone. The rates of K^+ absorption were obtained by subtracting the values for ^{22}Na absorption from the total.

Figure 6 shows that under these conditions, with Na^+ and K^+ present in equivalent concentrations, the rate of absorption of K^+ barely rises

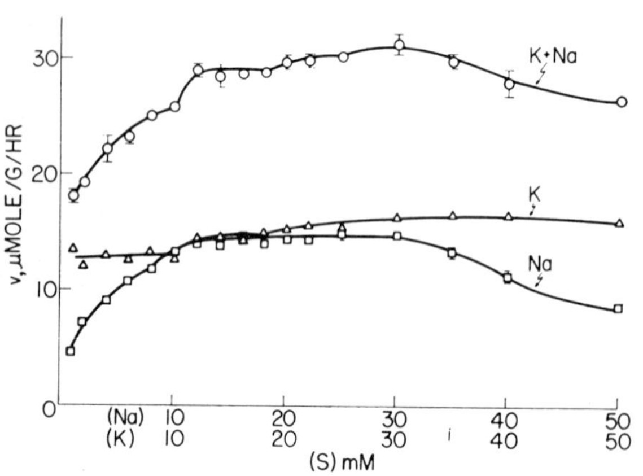

FIG. 6. Rates, v, of absorption of Na^+, K^+, and $Na^+ + K^+$ as a function of their concentrations. Concentrations of K^+ and Na^+: 1 to 50 mM; $CaCl_2$: 0.5 mM. Other conditions and conventions as for figure 1.

over this range (1–50 mM), and the total rate of absorption of $K^+ + Na^+$ (top curve) mainly reflects the absorption characteristics of Na^+ (bottom curve).

Discussion

Earlier findings (6) were to the effect that mechanism 2 of alkali cation transport, which becomes operative at concentrations of about 0.5 or 1.0 mM and above, is heterogeneous. When the concentration of K^+ or Na^+ was varied from 1 to 50 mM, the isotherm describing the rate of absorption as a function of the external concentration of the substrate ion showed a number of inflections. The interpretation offered was that carrier mechanism 2 possesses a spectrum of active sites which differ somewhat, though not greatly, in their affinity for a given ion. At concentrations in the range of a few millimoles per liter, the mechanism 2 site with the highest affinity for the substrate ion mainly effects the transport of the ion by this mechanism. At progressively higher concentrations, additional carrier sites of mechanism 2 come into play until, at 50 mM, a total of 4 such sites participate in the transport of the ion via this mechanism (6).

If this interpretation is sound, 2 consequences are predictable. First, the role of the counterion should be secondary. That this is indeed so is shown in figure 1. The absorption of Na^+ over the high concentration range (1–50 mM) shows the same pattern when SO_4^{2-} is the anion as when the counterion is Cl^- (fig 2 of ref 6). The pattern of the process of absorption is thus mainly a function of the concentration of Na^+.

Second, if the inflections in the absorption isotherm over the range of mechanism 2 (fig 1) denote the progressive occupation of additional carrier sites by the substrate ion in response to increasing external concentrations, then at a high external concentration of the substrate ion, increasing the concentration of a competing ion should result in a stepwise reduction of the rate of absorption of the substrate ion, each step or inflection denoting the displacement of substrate ion from 1 of the carrier sites.

Figures 2 to 5 show that this prediction, based on the hypothesis of a spectrum of active sites in carrier mechanism 2, is borne out by the evidence. Increasing concentrations of K^+ cause a stepwise diminution of the rate of Na^+ absorption, when Na^+ is present at 50 mM (fig 2), and conversely, increasing concentrations of Na^+ reduce the rate of absorption of K^+, present at 50 mM, in stepwise fashion (fig 3). Similar findings are shown with Ca^{2+} as the interfering ion for absorption of Na^+ (fig 4) and K^+ (fig 5).

The addition of the lowest concentration of Ca^{2+} used (1 mM) caused a slight increase in the rate of K^+ absorption over that of the control lacking Ca^{2+} (fig 5). However, it would probably be more proper to say that in the complete absence of Ca^{2+}, the rate of K^+ absorption was somewhat less than in the presence of a low concentration of Ca^{2+}. As documented elsewhere (see 11 for references) the presence of Ca^{2+} at a low concentration is essential for unimpaired membrane transport of ions, and this condition, rather than the absence of Ca^{2+}, should be looked upon as the physiological control. Omission of Ca^{2+} quickly results in unphysiological responses denoting damage to membrane and transport processes (3, 11). In the present case, reversal by 1.0 mM Ca^{2+} of the negative effect of its omission outweighed its inhibitory effect, resulting in the slight net increase apparent in figure 5 at 1 mM.

It has been shown that barley roots absorb Cl^- present at concentrations ranging from 1 to 50 mM via a mechanism, halide carrier mechanism 2, which shows a heterogeneity similar to that of mechanism 2 of alkali cation transport (2). These findings have since been confirmed for roots of 2 species of wheatgrass, Agropyron (1), and for corn roots (9, 12). It might thus be surmised that the transport of the anion is the primary act, and that the kinetics of the transport of alkali cations shown here and in references 1, 6, and 9 merely parallel those of chloride transport. However, like the absorption isotherm of the experiment shown in figure 1, in which SO_4^{2-} was the anion, the competition experiments shown in figures 2 to 5 reemphasize a point made earlier (7). It is that the absorption of alkali cations at high concentrations is effected by a cation carrying mechanism in the operation of which the anion plays only a subsidiary role. In each of these experiments, the interfering cation was added as the chloride. Despite the progressive increase in Cl^- concentration, the effects of the added salts on the rate of absorption of the substrate cations were negative. Furthermore, there were marked differences in the fine structure of these isotherms which were functions of the cations involved. These results therefore reflect cationic affinities for the active sites of carrier mechanism 2.

However, one point concerning the role of the anion requires comment. The transport of K^+-Rb^+ at high concentrations is quite markedly influenced by the identity of the anion (4, 7, 9). Specifically, substitution of SO_4^{2-} for Cl^- as the anion severely inhibits the operation of mechanism 2 with these substrate cations, their rates of absorption rising only slightly above the maximal rate due to the operation of mechanism 1 (fig 8 of ref 4, fig 2 of ref 7, fig 10 of ref 9). Sodium on the other hand, is transported readily by mechanism 2 whether the anion is Cl^- or SO_4^{2-} (figs 4 to 6 of ref 11, fig 1 of ref 10), although the rate of absorption of Na^+ is consistently lower with SO_4^{2-} than with Cl^- as the anion. That the rate of Na^+ absorption via mechanism 2 is depressed so much less severely than is that of K^+-Rb^+ by this mechanism is probably related to the considerably higher affinity of mechanism 2 for Na^+ (7, 11 and see below), and possibly to other factors also.

The results of the experiment shown in figure 6 lead to the conclusion that with both Na^+ and K^+ present at equivalent concentrations, over the high range (1–50 mM), mechanism 2 transports Na^+ almost exclusively, i.e., its active sites have a higher affinity for Na^+ than for K^+. The argument is as follows.

Over the range 1 to 10 mM the rate of K^+ absorption in this experiment averaged 12.8 μmole/g/hr, close to the maximal velocity of K^+ absorption via mechanism 1 in these roots (11.9 μmole/g/hr in ref 7, and 11.8 in ref 6). Since the rate of absorption of K^+ (or Rb^+) via this mechanism is not appreciably affected by Na^+ (3, 4, 7) it is concluded that this absorption of K^+ is almost entirely through the operation of mechanism 1. Over the entire concentration range explored, the rate of K^+ absorption rose by only 4.5 μmole/g/hr above this value. This small additional uptake must indeed be considered due to the operation of mechanism 2, inasmuch as the rate of K^+ absorption via mechanism 1 is essentially at the theoretical maximum at 1 mM K^+ (7).

The absorption of Na^+ via mechanism 1 of alkali cation transport (see figs 2, 3 and the left side of figs 4 and 5 of ref 11) is virtually abolished in the presence of Ca^{2+} and K^+ (see fig 1 of ref 10 and fig 6 of ref 11). The absorption of Na^+ shown in figure 6 of the present paper is therefore due almost entirely to the operation of mechanism 2 of alkali cation transport. The stepwise nature of the Na^+ absorption isotherm in this figure supports this conclusion. Thus, when both Na^+ and K^+ ions are present at equivalent concentrations in the high range where mechanism 2 contributes to absorption, absorption of K^+ is almost entirely effected by mechanism 1 and absorption of Na^+ almost entirely by mechanism 2. This evidence for preferential Na^+ transport by mechanism 2 confirms the previous conclusion (7) to the effect that mechanism 2 has considerably higher affinity for Na^+ than for K^+.

The dual pattern of ion transport discussed in this and the preceding paper (11) in terms of the absorption of Na^+ has been shown to apply widely, for various ions, for many species of plants, and for different types of tissue (5). Specifically, the complex fine structure of the absorption isotherm characteristic of absorption via the type 2 mechanisms, discovered in experiments with barley roots (2, 6), has recently been confirmed in work with roots of *Agropyron elongatum* and *A. intermedium* (1) and of *Zea mays* (9, 12). It is therefore likely that the pattern which has emerged is quite general, possibly universal in higher plants. It appears to offer a rationale for phenomena of specificity and mutual ion effects ("antagonisms"), and in particular, for both the preferential absorption of K^+ and the frequently observed interference by other ions, Na^+ included, with the absorption of K^+. The former phenomenon, of highly selective absorption, is due to the operation of the type 1 mechanisms, whereas the latter, viz. fairly indiscriminate competition, is a characteristic of the type 2 mechanisms. That this generalization holds for other ions as well is brought out in a recent review (5).

Acknowledgment

We are grateful for the cooperation of the Director and staff of the University of California Lawrence Radiation Laboratory at Livermore, California, who furnished ^{42}K for this investigation.

Literature Cited

1. ELZAM, O. E. 1966. Absorption of sodium, potassium, and chloride by two species of *Agropyron* differing in salt tolerance. Thesis (Ph.D.), University of California, Davis.
2. ELZAM, O. E., D. W. RAINS, AND E. EPSTEIN. 1964. Ion transport kinetics in plant tissue: complexity of the chloride absorption isotherm. Biochem. Biophys. Res. Commun. 15: 273–76.
3. EPSTEIN, E. 1961. The essential role of calcium in selective cation transport by plant cells. Plant Physiol. 36: 437–44.
4. EPSTEIN, E. 1965. Salinity and the Pattern of Selective Ion Transport in Plants. United States Department of the Interior, Office of Saline Water, Res. Develop. Progr. Rep. 161.
5. EPSTEIN, E. 1966. Dual pattern of ion absorption by plant cells and by plants. Nature 212: 1324–27.
6. EPSTEIN, E. AND D. W. RAINS. 1965. Carrier-mediated cation transport in barley roots: kinetic evidence for a spectrum of active sites. Proc. Natl. Acad. Sci. 53: 1320–24.
7. EPSTEIN, E., D. W. RAINS, AND O. E. ELZAM. 1963. Resolution of dual mechanisms of potassium absorption by barley roots. Proc. Natl. Acad. Sci. 49: 684–92.
8. EPSTEIN, E., W. E. SCHMID, AND D. W. RAINS. 1963. Significance and technique of short-term experiments on solute absorption by plant tissue. Plant Cell Physiol. 4: 79–84.
9. LÜTTGE, U. AND G. G. LATIES. 1966. Dual mechanisms of ion absorption in relation to long distance transport in plants. Plant Physiol. 41: 1531–39.
10. RAINS, D. W. AND E. EPSTEIN. 1965. Transport of sodium in plant tissue. Science 148: 1611.
11. RAINS, D. W. AND E. EPSTEIN. 1967. Sodium absorption by barley roots: the role of the dual mechanisms of alkali cation transport. Plant Physiol. 42: 314–18.
12. TORII, K. AND G. G. LATIES. 1966. Dual mechanisms of ion uptake in relation to vacuolation in corn roots. Plant Physiol. 41: 863–70.

Reprinted from PLANT PHYSIOLOGY, Vol. 42, No. 1, January, 1967, pages 37–46.
(PRINTED IN U.S.A.)

Mineral Ion Contents and Cell Transmembrane Electropotentials of Pea and Oat Seedling Tissue[1]

N. Higinbotham, Bud Etherton,[2] and R. J. Foster[3]
Department of Botany, Washington State University, Pullman, Washington 99163

Received September 6, 1966.

Summary. The relationships of concentration gradients to electropotential gradients resulting from passive diffusion processes, after equilibration, are described by the Nernst equation. The primary criterion for the hypothesis that any given ion is actively transported is to establish that it is not diffusing passively. A test was made of how closely the Nernst equation describes the electrochemical equilibrium in seedling tissues. Segments of roots and epicotyl internodes of pea (*Pisum sativum* var. Alaska) and of roots and coleoptiles of oat (*Avena sativa* var. Victory) seedlings were immersed and shaken in defined nutrient solutions containing eight major nutrients (K^+, Na^+, Ca^{2+}, Mg^{2+}, Cl^-, NO_3^-, $H_2PO_4^-$ and SO_4^{2-}) at 1-fold and 10-fold concentrations. The tissue content of each ion was assayed at 0, 8, 24, and 48 hours. A near-equilibrium condition was approached by roots for most ions; however, the segments of shoot tissue generally continued to show a net accumulation of some ions, mainly K^+ and NO_3^-. Only K^+ approached a reasonable fit to the Nernst equation and this was true for the 1-fold concentration but not the 10-fold. The data suggest that for Na^+, Mg^{2+}, and Ca^{2+} the electrochemical gradient is from the external solution to the cell interior; thus passive diffusion should be in an inward direction. Consequently, some mechanism must exist in plant tissue either to exclude these cations or to extrude them (e.g., by an active efflux pump). For each of the anions the electrochemical gradient is from the tissue to the solution; thus an active influx pump for anions seems required. Root segments approach ionic equilibrium with the solution concentration in which the seedlings were grown. Segments of shoot tissue, however, are far removed from such equilibration. Thus in the intact seedling the extracellular (wall space) fluid must be very different from that of the nutrient solution bathing the segments; it would appear that the root is the site of regulation of ion uptake in the intact plant although other correlative mechanisms may be involved.

The significance of cell electropotentials, PDs, as an important force having definite relationships to ionic gradients in plant cells has been discussed recently by Dainty (4) and by Briggs, Hope, and Robertson (3). The existence of cell transmembrane PDs in higher plants has been established (6, 7, 10, 11, 12, 20) and evidence has been obtained showing that, under certain conditions at least, the PD is closely related in magnitude to the transmembrane concentration gradient of K^+ (cell interior to exterior) (11) as in muscle fibers (13) and Chara cells (15). Whether or not this is true for K^+ or other major nutrient ions in cells of higher plants at ionic equilibrium (or homeostasis) is the subject of this study.

An ionic species which is free to diffuse through the cell membrane, that is, one which is not actively transported in a process requiring metabolic energy, when at equilibrium should show a definite relationship between the activity and the electrical gradients. For cells in a solution containing K^+ and Cl^- the relationship of concentration and potential, if distributed by diffusion only, would be expected at equilibrium to conform to the Nernst equation, e.g.:

$$E_{io} = \frac{RT}{zF} \ln \frac{[K^+_o]}{[K^+_i]} = \frac{RT}{zF} \ln \frac{[Cl^-_i]}{[Cl^-_o]} \qquad \text{I}$$

[1] This work was supported in part by National Science Foundation Grant G9560 and by funds provided by the State of Washington Initiative Measure 171.

[2] Present address: Biology Department, Vassar College, Poughkeepsie, New York.

[3] Department of Agricultural Chemistry and Department of Chemistry.

where E_{io} is the electropotential between inside and outside the cell; R, the gas constant in joules degree $^{-1}$ equiv. $^{-1}$; T, absolute temperature; z, the valence of the ion; F, the faraday in coulombs/equivalent; and the brackets indicating the concentration of the ion concerned in the inner or outer phases. Similar equations can be written for each of the several ions. For each ion the following derivation is useful (21, 22):

$$\frac{\varnothing_i}{\varnothing_o} = \frac{[C_o]}{[C_i]\ \exp(zFE_{io}/RT)} \qquad \text{II}$$

where C represents the concentration inside or outside as indicated by the subscript, and E_{io}, the transmembrane PD. Thus the external concentration $[C_o]$ is in electrochemical equilibrium with the internal concentration when (at 25°, and in mv):

$$[C_i] = [C_o]\ \exp. \pm [z\ (0.039)\ E_{io}] \qquad \text{III}$$

where the plus or minus sign of the exponent refers to the cation or anion, respectively.

It was the purpose of this study to grow and treat seedling tissues under defined conditions and to ascertain experimentally the electrochemical activities between the tissue and external solution of K^+, Na^+, Ca^{2+}, Mg^{2+}, Cl^-, NO_3^-, SO_4^{2-}, and $H_2PO_4^-$. The electropotential gradients between the vacuole and the external medium under these conditions have been reported previously (6). Evidence suggests that there is little or no difference in PD between the cytoplasm and vacuole (7).

Materials and Methods

The plants used were peas, *Pisum sativum* var. Alaska, and oats, *Avena sativa* var. Victory. Seeds were soaked 2 hours prior to planting. For segments of pea epicotyl and oat coleoptile seeds were planted in vermiculite well rinsed with nutrient solution (as specified below) and placed in an incubator in the dark at 25°. Peas were grown for 7 days in darkness; oats were grown for 90 to 96 hours in the dark but were given a 2-hour exposure to red light in the 72- to 76-hour interval. For experiments using root segments, seeds were germinated on glass beads in nutrient solution then transferred to a cheese-cloth net with roots immersed in aerated nutrient solution at a controlled temperature of 25°.

At the times indicated, tissue segments were excised and placed in deionized water. Samples of about 1 g each were blotted and weighed, one being taken for the 0-time interval, the others being divided into 0.2 g samples and placed in 50 ml of nutrient solution in 250 ml Erlenmeyer flasks and shaken for periods as given for each experiment. At the proper time, samples were removed, rinsed in deionized water, blotted, weighed, and frozen for subsequent chemical assay procedures. The usual proportion of tissue weight to solution volume was thus 4 g/liter: all samples were duplicated.

Root segments 1 to 2 cm long were excised from the terminal 2.5 cm portion after discarding the terminal 0.5 cm. With oat coleoptiles and the third internode of pea epicotyls the uppermost cm portion was discarded and the next 2 cm segments were used. The hollow coleoptile segments were filled with solution prior to immersion.

The nutrient solution had the same salts in the same proportion as that used previously and referred to as 1X nutrient (11). Its composition in mmoles per liter was: KCl, 1.0; $Ca(NO_3)_2 \cdot 4H_2O$, 1.0; $MgSO_4 \cdot 7H_2O$, 0.25; $NaH_2PO_4 \cdot H_2O$, 0.904, plus Na_2HPO_4, 0.048. The pH value was 5.5 to 5.8. In some experiments this solution was used at 10-fold concentration and, as such, is referred to as 10X nutrient.

Chemical Assay Methods. Tissue samples harvested for assay were placed in about 10 ml of deionized water and frozen until assays could be made. For assay these samples were thawed, boiled in deionized water for 15 minutes and the resulting hot solution filtered (using Whatman no. 1 filter paper) into a 25 ml volumetric flask. The residue was boiled a second time in 5 ml H_2O and this solution was added to the first extract. This is referred to as the hot water extract. The combined extracts were brought to volume and aliquots subsequently assayed for K^+, Na^+, Ca^{2+}, Mg^{2+}, Cl^-, NO_3^-, SO_4^{2-}, and $H_2PO_4^-$. These assays are assumed, arbitrarily, to approximate the minimal activities of each of the ions within the tissue. The authors recognize that this assumption cannot be completely correct since several compartments for ions may exist in intact tissue (3, 19) and measurement of ionic partitioning between these requires special methods. Breaking down the membrane barriers may appreciably alter the partitioning. However, it seemed to us that assays of the hot water extractible fraction would provide us with a reasonable estimate of the minimal average ionic activity in the tissue and this combined with the acid digest of the tissue residue, described below, would provide estimates of total ion content as an upper limit of average electrochemical activity within the protoplasm.

The tissue residue, representing the hot water insoluble materials, was removed from the filter paper and subsequently solubilized in 10 ml of concentrated HNO_3. The residue was allowed to stand for 1 day in cold HNO_3 then warmed without boiling 2 to 3 times driving off HNO_3 but not allowing the sample to dry completely. The acid digest was then transferred to a 25 ml volumetric flask and brought to volume for assay of K^+, Na^+, Ca^{2+}, Mg^{2+}, SO_4^{2-}, and $H_2PO_4^-$.

Potassium and Sodium. Potassium and sodium were determined by flame photometry using a Beckman DU spectrophotometer with photomultiplier and flame attachments. The solution used for standardizing was made up in 0.1 N HCl and was composed as follows in mM: KCl, 8.0; NaCl, 2.0; $CaNO_3$, 1.0; and $MgSO_4$, 1.0. Dilutions as required were also in 0.1 N HCl solution. Assays were made using the 766.5 mμ wavelength for potassium and the 589.3 mμ wavelength for sodium.

Calcium and Magnesium. Calcium and magnesium were assayed by flame photometry using additions of NH_4EDTA according to the method of Greweling (8). The standard solution used was as follows in mM: 1.25 $MgSO_4 \cdot 7H_2O$; 1.25 $CaNO_3$; 12.5 KCl; and 25 EDTA to which NH_4OH had been added as specified in the procedure. Dilutions made for producing standard curves were also made in the NH_4EDTA solution. The wavelength used for Ca^{2+} was 422 mμ and, for Mg^{2+}, 285.2 mμ. Aliquots of the water extract were brought to the proper mixture by addition of an equal volume of 50 mM NH_4EDTA. Aliquots of the acid extract were gently heated to dryness after addition of one-third volume of 30 % H_2O_2 and then the residue dissolved in 25 mM NH_4EDTA.

Chloride. Chloride was determined electrolytically using a chloride titrator.

Nitrate. Nitrate was assayed colorimetrically by the method described by Morris and Gonzales-Mas (18).

Phosphate. For phosphate assays the colorimetric method of Griswold et al. was used (9). The values are reported as $H_2PO_4^-$ although phosphate also occurs as HPO_4^{2-} in the tissue, the proportion depending upon pH.

Sulfate. Sulfate was assayed by a method using colorimetry in the ultraviolet range (wavelength 350 mμ) and chloranilic acid as indicator (1, 17). Aliquots of solution containing water extracts or acid digests were passed through a column of Dowex 50 (X 8, 50 to 100 mesh) to remove cations. To the effluent was added 2 drops of concentrated HNO_3 and about one-half volume of 30 % H_2O_2. This solution was evaporated to dryness at 100°. The residue was taken up into solution with 1 ml of the indicator suspension brought to volume and assayed using a Beckman DU spectrophotometer.

Results

Hot Water Extractible Ions of Pea Root Tissue; Cations. Pea root and epicotyl segments were assayed for ion content after the following 3 regimes: 1) seedlings were grown in nutrient solution of 1× concentration and segments treated for periods up to 48 hours in 1× solution: 2) seedlings were grown in 10× solution and segments treated in 1× solution and 3) seedlings were grown in 10× solution and segments treated in 10× solution.

The results showed quite clearly that in root segments the concentrations of the 4 major cations are essentially constant with respect to time over the 48-hour period in regimes 1 (fig 1), and 3 (fig 2). In regime 2 involving a change from 10× nutrient solution during growth to 1× nutrient during segment immersion there was some loss of K^+ and Na^+, but equilibration was approached. Since the results of regime 1 closely resembled those for regime 2 in all cases data for the latter are not reported here.

Pea Roots; Anions. In contrast to the pattern of cation uptake some of the anions showed an appreciable amount of net accumulation initially in 10× solution (fig 4) and in the 1× solutions (fig 3); however, no great net change of the anions as a whole took place. Also, there appeared to be a reasonably close approach in 10× solution to a steady state after the first 8 hours for tissue content of SO_4^{2-} and NO_3^-. However, the root segments continued to gain $H_2PO_4^-$ and lose Cl^- over the 48-hour period. A surprisingly high accumulation of SO_4^{2-} occurred.

Pea Epicotyl; Cations. In segments of pea epicotyl under regimes 1 and 3 only K^+ was accumulated in appreciable amounts; with the other cations, Na^+, Ca^{2+}, and Mg^{2+} there was a small net uptake (figs 5, 6). There was relatively little difference in the initial K^+ content whether the seedlings were grown in 1× or 10× solutions. However, segments treated in 10× solution gained much more K^+ than those in 1× solution.

Pea Epicotyl; Anions. Under both regimes 1 and 3 there was a net increase of anions during the 48-hour period (figs 7, 8). The order of these during the first 24 hours was: $NO_3^- > H_2PO_4^- > SO_4^{2-} > Cl^-$. In both regimes NO_3^- showed a loss during the 24- to 48-hour interval but in view of the experimental error this may not be significant.

It is apparent that stem segments were not in equilibrium with the concentrations of any of the 8 ions in the external medium, although there was a wide range in the rates and amounts of net uptake. For several ions, however, the rates of net change in tissue content were negligible and the data allow for some evaluations in terms of electrochemical activities. These are dealt with in a later section.

Oat Root; Cations. As in the case of pea roots, oat root segments in regimes 1 and 3 (figs 9, 10) showed little net change in content of most cations. In 1× solution there was some loss of K^+ but in 10× solution there was a slight gain. However, the fact that roots grown in 1× solution had a higher initial K^+ content than those grown in 10× solution is unexpected and this would appear to be either a result of biological variability or experimental error.

Oat Root; Anions. In regimes 1 and 3 little change in content occurred except for NO_3^-. This ion showed a loss in 1× solution corresponding approximately to the loss found in K^+ (fig 9,

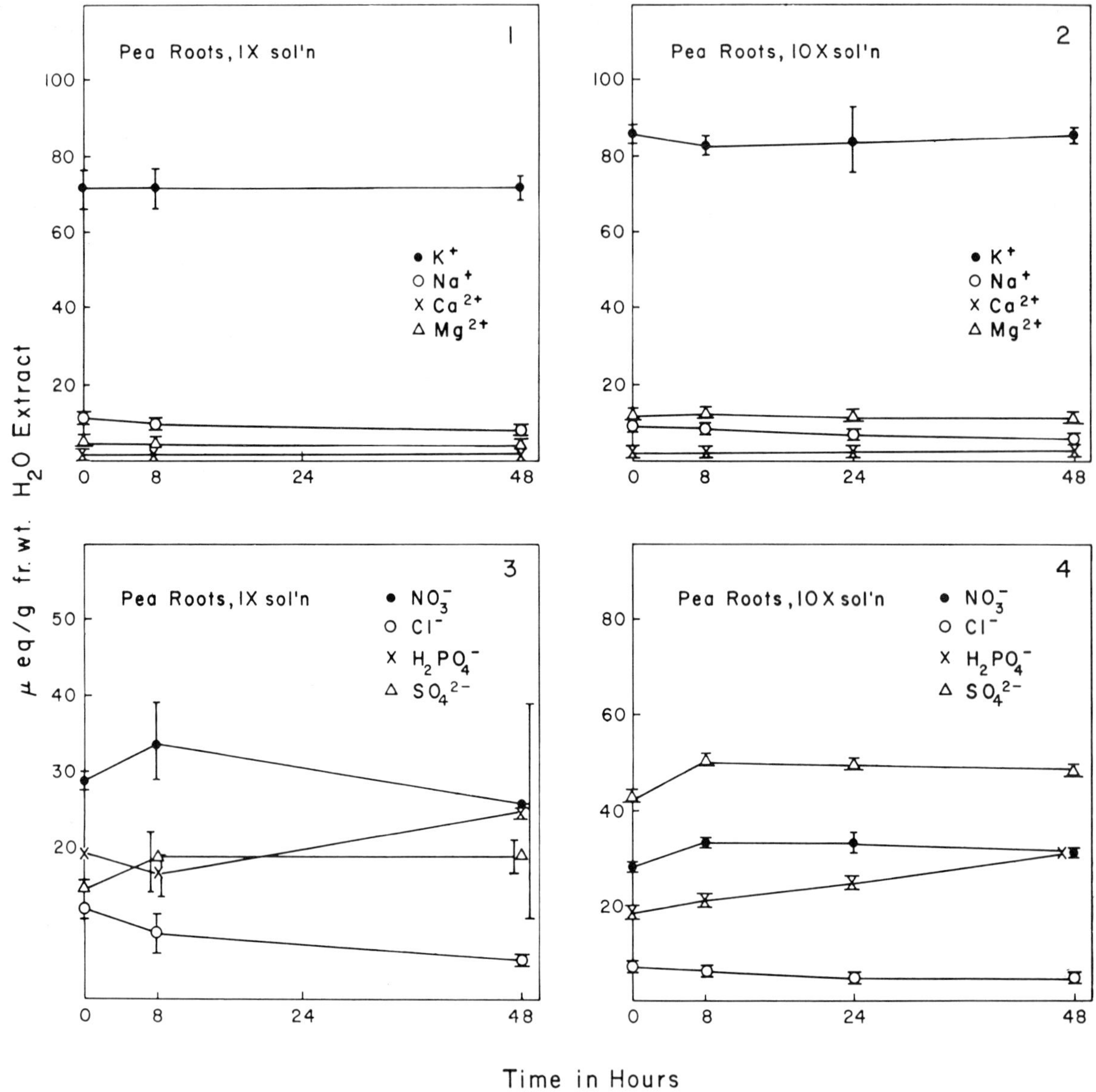

FIGS. 1 to 4. The time course of ion content of pea root segments of cations and anions in solutions as indicated in each figure. The vertical lines at each point show the mean ± SD.

Table I. *Ion Accumulation Ratios at 48 Hours*

The analyses were obtained with hot water soluble extract.

Tissue regime*		Cations				Anions			
		K^+	Na^+	Mg^{2+}	Ca^{2+}	NO_3^-	Cl^-	SO_4^{2-}	$H_2PO_4^-$
Pea roots	1	72	8.0	6	0.80	13	5.3	38	25
	3	8.6	0.6	4.7	0.16	1.6	0.5	9.6	3.2
Pea epicotyls	1	59	1.9	5.8	1.3	14	4.4	22	19
	3	9.5	0.34	1.0	0.17	1.3	0.7	3.5	3.7
Oat roots	1	57	4.7	33	1.6	23	3.5	9.0	19
	3	6.5	0.3	4.2	0.14	1.5	0.5	2.2	1.2
Oat coleoptiles	1	63	5.2	24	1.2	12	21	9.6	15
	3	10.1	2.7	4.1	0.6	4.6	5.3	2.0	2.6

* For an explanation of these treatments see the text (paragraph 1 under Results).

same tissue as in the data of fig 11). Tissues grown and treated in 10× solution showed a similar loss of NO_3^- (fig 12) in this case not matched by a loss of K^+ (fig 10, data from the same tissue as in fig 12). As in the case of the cation content there is higher content of certain ions, e.g., NO_3^- and $H_2PO_4^-$, in the 1× solution than in 10× solution.

Oat Coleoptile; Cations. The net changes in cation content of oat coleoptiles in regimes 1 and 3 are shown in figures 13 and 14. As in pea epicotyl segments the coleoptile tissue was not in a near-equilibrium state with the external solution and a marked gain of K^+ occurred in 48 hours. Appreciable increases in other cations also took place; however, the final content of Na^+, Mg^{2+}, and Ca^{2+} was still markedly below that found for K^+.

Oat Coleoptile; Anions. Appreciable net gains of anions occurred in regimes 1 and 3 during the 48-hour period (figs 15, 16). Only SO_4^{2-} appeared to approach equilibrium.

Ion Accumulation Ratios and Comparisons of Ions and of Tissues. In table I a summary is presented of the ion accumulation ratios (in μeq g^{-1} fr wt of tissue/μeq ml^{-1} of external solution)

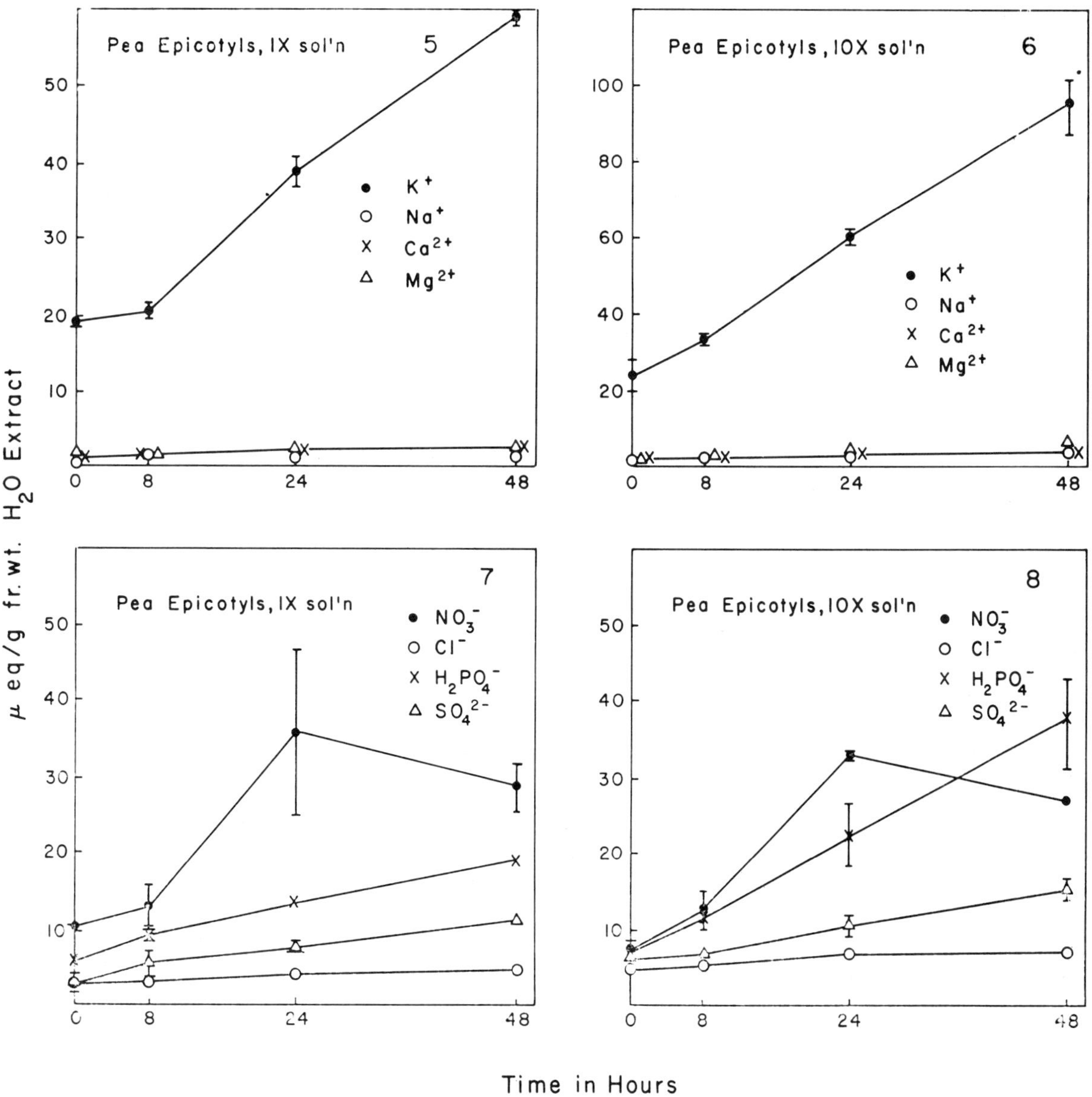

FIGS. 5 to 8. The time course of ion content of pea epicotyl segments of cations and anions in solutions as indicated in each figure. The vertical lines at each point show the mean ± SD. Points lacking vertical bars had smaller SDs than could be shown (excepting in fig 8 in which the 48-hour point for NO_3^- is for 1 sample).

Table II. *Assay of Ions in the Acid-digested Residue of Tissue from Hot Water Extractions*
Values are reported from the 48-hour period only.

		Cations μeq/g fresh wt				Anions μeq/g fresh wt	
Tissue regime*		K^+	Na^+	Mg^{2+}	Ca^{2+}	SO_4^{2-}	$H_2PO_4^-$
Pea roots	1	6.0 ± 0.0	1.8 ± 0.2	4.7 ± 0.4	4.7 ± 0.6	3.8 ± 2.4	0.6 ± 0.0
	3	4.7 ± 0.9	3.4 ± 0.9	6.6 ± 0.9	4.5 ± 0.3	3.9 ± 0.7	8.8 ± 1.3
Pea epicotyls	1	3.4 ± 0.6	1.4 ± 0.0	4.0 ± 0.3	7.1 ± 0.5	0.8 ± 1.0	1.0 ± 0.5
	3	1.9 ± 0.3	3.0 ± 1.2	3.1 ± 0.1	8.2 ± 0.3	...	...
Oat roots	1	6.4 ± 0.5	1.9 ± 0.4	3.0 ± 0.5	1.3 ± 0.0	4.2 ± 1.3	0.4 ± 0.1
	3	5.8 ± 1.6	2.0 ± 0.4	5.3 ± 0.3	2.1 ± 0.1	1.6 ± 0.2	6.3 ± 1.5
Oat coleoptiles	1	17.5 ± 1.3	3.4 ± 1.1	5.4 ± 1.6	5.0 ± 0.1	1.6 ± 0.1	29.0 ± 0.6
	3	16.1 ± 0.2	7.6 ± 0.7	9.3 ± 0.4	10.7 ± 0.2	5.0 ± 1.2	23.1 ± 0.3

* As in table I.

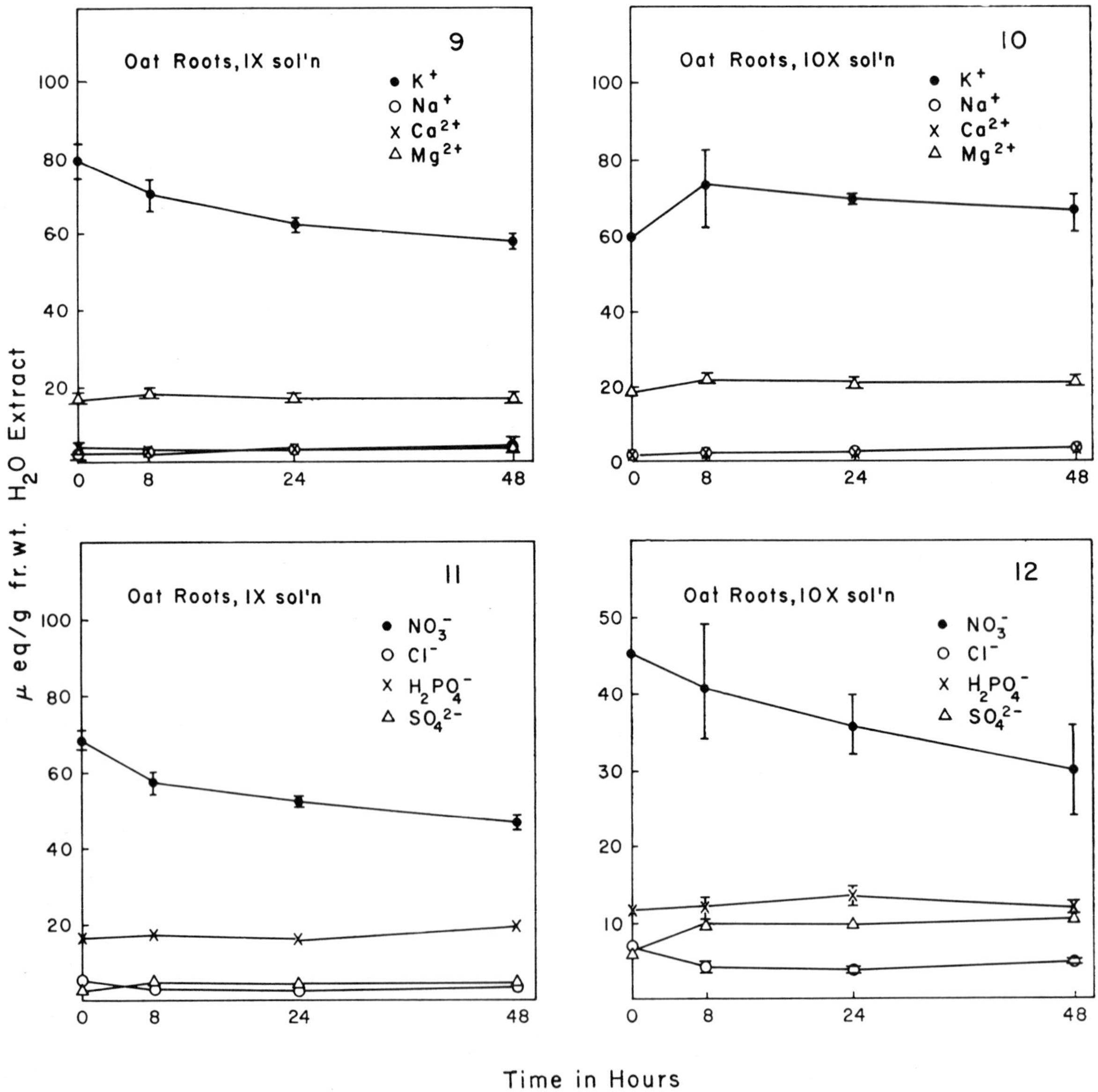

FIGS. 9 to 12. The time course of ion content of oat root segments of cations and anions in solutions as indicated in each figure. The vertical lines at each point show the mean ±SD.

of the hot water extractible ions. In all cases the cations were in the following order $K^+ > Mg^{2+} > Na^+ > Ca^{2+}$, with K^+ accumulation greatly exceeding all the others. The anions, however, show no definite pattern except that NO_3^- accumulation exceeded that for Cl^-. In general, the order was the same as that of the initial seedling content.

Assays of HNO_3 Digest of Residue from Hot Water Extracted Tissue. The assays of ions from the acid digest of the 48-hour samples are given in table II. These values are similar to those of the shorter time periods (not reported). In the case of K^+, SO_4^{2-}, and $H_2PO_4^-$ the fraction remaining in the residue was small compared to that of the hot water extractible fraction. However, in other cases, notably Na^+, Mg^{2+}, and Ca^{2+}, the bound fraction is quite large, sometimes exceeding the hot water extracted portion (compare figs 1–16).

Relationship of Ion Concentration Gradients to Cell Electropotentials. Equation I defines the diffusional relationships of ion concentrations and electropotentials at equilibrium. Cell electropotential measurements have been made for each of the tissues under conditions essentially identical to those obtaining for the ion accumulation study. These have been reported previously in part (6, 11) and additional measurements made since tend to confirm the reliability of these values within an error of about 10 %. (Of course with multicellular tissue it is impossible to measure electropotentials and accumulation ratios with the same cell as in the case of the giant-celled algae.) Having these values we may now estimate the electrochemical activity gradient between the cells and the ambient solutions. It is quite apparent from figures 1 to 16 that the tissue contents are not static; nevertheless, as a first approximation it is meaningful to compare the ion content found by assay with that predicted from measured cell electropotentials using equation II (and assuming that an

Table III. *Concentration of Ions (μeq/g Water Content) of Pea and Oat Seedling Tissues*

The ion concentration values are those found experimentally and as predicted from experimentally measured cell potentials in 1X solution at 24 hours. The H_2O content was calculated as 95 % of fresh weight. Actual values ranged from 94.2 % to 95.7 %. The values for ion content are those for the H_2O extraction. The predicted diffusion equilibrium values calculated from equation II assuming equilibration exists) using measurements of cell electropotentials (E_{in}) from Etherton (6), namely, −110, −119, −84, and −105 mv (interior negative) for pea roots, pea epicotyls, oat roots and oat coleoptiles, respectively.

	Pea				Oat			
	Roots		Epicotyls		Roots		Coleoptiles	
Ion	Exptl.*	Predicted	Exptl.*	Predicted	Exptl.*	Predicted	Exptl.*	Predicted
K^+	75	74	41	104	66	27	42	60
Na^+	8	74	2	104	3	27	4	60
Mg^{2+}	3	2700	3	5400	17	350	7	1800
Ca^{2+}	2	10,800	3	21,600	3	1400	2	7200
NO_3^-	28	0.0272	38	0.0192	56	0.0756	18	0.0332
Cl^-	7	0.0136	4	0.0096	3	0.0378	15	0.0166
$H_2PO_4^-$	21	0.0136	14	0.0096	17	0.0378	9	0.0166
SO_4^{2-}	19	0.000094	8	0.000048	4	0.00071	5	0.000138

* Values interpolated from time course curves. Experimental values at 24 hours in a separate experiment, not reported, are not greatly different.

Table IV. *Concentrations of Ions (μeq/g Water Content) of Pea and Oat Seedling Tissues*

The ion concentration values are those found experimentally and as predicted from measured cell potentials in 10X solution at 24 hours. H_2O content calculated as 95 % of fresh weight. Actual values ranged from 94.2 % to 95.7 %. The values for ion content are those for the H_2O extraction. The predicted diffusion equilibrium values calculated from equation III using measurements of cell electropotential ($=E_{in}$) from Etherton (6), namely, −108, −119, −71, and −102 mv for pea roots, pea epicotyls, oat roots and oat coleoptiles, respectively.

	Pea				Oat			
	Roots		Epicotyls		Roots		Coleoptiles	
Ion	Exptl.	Predicted	Exptl.	Predicted	Exptl.	Predicted	Exptl.	Predicted
K^+	89	680	64	1040	73	159	79	535
Na^+	7	680	3	1040	3	159	24	535
Mg^{2+}	12	22 750	3	54,000	22	1275	17	14,250
Ca^{2+}	3	91,000	2	216 000	3	5100	10	59 000
NO_3^-	35	0.294	35	0.192	38	1.250	61	0.372
Cl^-	6	0.147	7	0.096	4	0.625	42	0.186
$H_2PO_4^-$	26	0.147	24	0.096	14	0.625	19	0.186
SO_4^{2-}	52	0.00109	11	0.00049	11	0.0197	15	0.00175

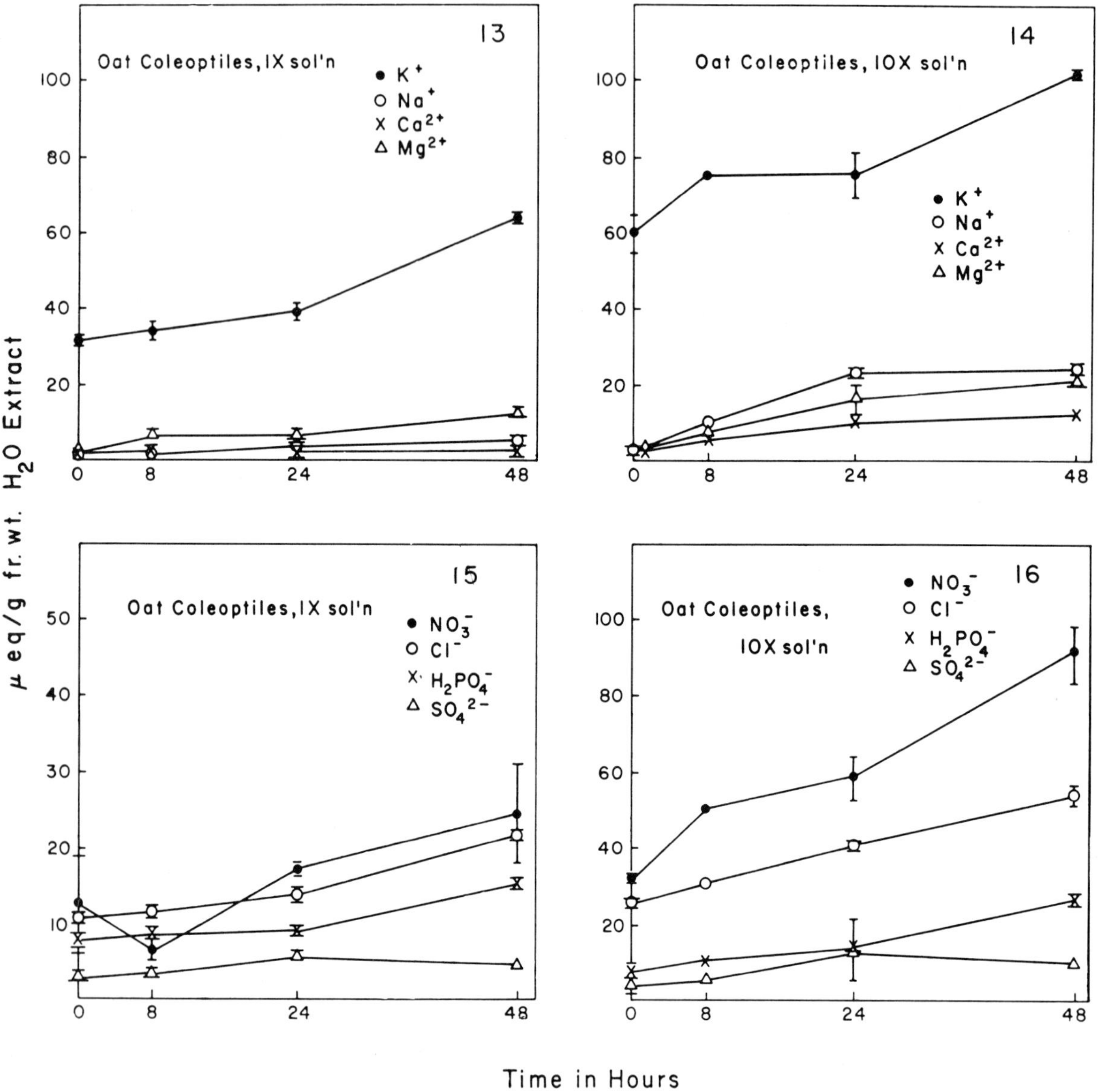

FIGS. 13 to 16. The time course of ion content of oat coleoptile segments of cations and anions in solutions as indicated in each figure. The vertical lines at each point show the mean ± SD.

Table V. *Estimates of K^+ Flux Rates and Permeability Coefficients, P_K, over the 24 to 48 Hour Period*

Tissue	$[K_o^+]$*	$[K_i^+]$**	E_{io}	$Ø_i/Ø_o$	$Ø_K$***	$Ø_o$	$Ø_i$	$P_K \times 10^5$
Pea roots	1	72	−110	1.04	...	...	...	...
	10	84	−108	8.03	...	...	...	...
Pea epicotyls	1	39	−119	2.69	2.32	1.37	3.69	7.83
	10	61	−119	17.20	3.93	0.24	4.17	0.89
Oat roots	1	...	− 84	...	...	...	...	...
	10	70	− 71	2.26	−0.58	...	...	...
Oat coleoptiles	1	40	−105	1.50	2.66	5.32	7.98	19.0
	10	75	−102	7.14	3.01	0.49	3.50	0.86

* $[K_o^+]$ is the external potassium ion concentration times 10^6 eq ml^{-1}.
** $[K_i^+]$ is the potassium ion concentration in the tissue times 10^6 eq (g fr wt)$^{-1}$ at $t = 24$ hours. This is approximately the time at which E (PD) was measured.
*** $Ø_K$ is the net influx in units of 10^{-10} eq g^{-1} sec^{-1}. It applies to the 24 to 48 hour interval.

equilibrium exists). These values (table III) with 1× nutrient solutions suggest that among the 8 ions studied only K^+ approaches the relationship predicted for diffusion. Of the ions other than K^+ only Na^+ is within 1 order of magnitude of the tissue concentration predicted and some ions, e.g., Ca^{2+} and SO_4^{2-}, may be several thousand-fold removed.

At a 10× concentration of nutrient solution, the electrochemical activities of none of the ions are close to the predicted diffusional equilibrium values (table IV). Again K^+ approaches electrochemical equilibrium more closely than any other ion. In shoot tissues, at least, net accumulation of K^+ is proceeding during immersion; thus, in accord with equation II, the flux ratio would have to be known before it could be concluded that active transport of K^+ is involved.

Calculations of Flux Values. Equation II permits the calculation of diffusional flux ratios for tissue samples not in equilibrium. Since: $\phi = \phi_i - \phi_o$ (where ϕ is the net flux, ϕ_i, the influx, and ϕ_o the efflux) each rate may be estimated. Also, since for K^+ (4):

$$\phi = -P_K \frac{EF/RT\ [(K_o^+) - (K_i^+)\ (Ki^+)e^{EF/RT}]}{1-e^{EF/RT}} \qquad \text{IV}$$

the permeability coefficient of K^+ may be calculated. These estimates are summarized in table V. Small experimental errors lead to quite large errors in the estimate unless ϕ_i is relatively large and the ϕ_i/ϕ_o ratio is appreciably removed from unity; for this reason several calculations have been omitted.

The flux ratios show quite clearly that a discrepancy exists between diffusion theory and the experimental measurements. For example, in pea roots at 1× the flux ratio approaches 1 as predicted; however, at 10× the flux ratio, predicted from cell potential and the concentration gradient, is 8.0 despite the fact that the data showed no significant net uptake. We feel that the data showing no net uptake are reliable and that either the measured potential value is in large error, or K^+ is not diffusing independently. While the potential measurement may be subject to some error the deviation required, −55 mv as compared to the measured potential of −71 mv, is too great to seem likely. Larger deviations in PD would be required in the other tissues. Consequently, the suggestion of some other mechanism, such as an anion pump, appears necessary to explain the various points of evidence in a more logical way.

The flux rates, calculated for those cases in which the data are suitable, give values within a reasonable order of magnitude. With cells of 35 × 200 μ the cell surface area per gram of tissue is about 1000 cm^2, thus the flux rates are in the range of 0.04 to 2.5 × 10^{-12} eq cm^{-2} sec^{-1} as in the case of similar measurements of other plants (4, 19). It is not reasonable, however, as these estimates indicate, to expect the ϕ_o to decrease with increasing internal concentration.

Discussion

The overall conclusion from this study using the Nernst equation as a criterion is that none of the 8 major ions is in electrochemical equilibrium in pea or oat seedling tissues although K^+ approaches it; thus none appear to move passively. However, estimates of the electrochemical gradients provided by this study suggest that for Na^+, Mg^{2+}, and Ca^{2+} the energy gradient is from the external solution to cell interior. Thus passive diffusion would be from outside to inside and any active transport of these ions, if it occurs, would represent an efflux pumping process. This finding seems to be in distinct contrast to the view that in higher plants any ion which is accumulated (against the chemical gradient) by metabolically active tissue is energetically transported inward.

Each of the anions was accumulated against the energy gradient. The disparity from electrochemical equilibrium was much greater at 1× than at 10× concentrations. It may be inferred that metabolic energy is required for accumulation of each anion. However, as noted by Briggs, Hope, and Robertson (3), the energy of transport may be necessary for only 1 co-ion of a pair, the other co-ion may then diffuse obligatorily because of the requirement for electroneutrality. In view of the fact that cation inward movement is downhill in the electrochemical gradient, we are left with the conclusion that only the anion accumulation process needs obviously to be the active influx transport. Furthermore, evidence that inhibitors such as dinitrophenol may rapidly depress the cell PD suggests that cell potentials are dependent upon metabolism and that ion transport may likely be electrogenic (7, 10, 14), and thus generate the cell PD. In this case cation (particularly K^+) diffusion would occur as a result of the PD rather than the PD being a consequence of such diffusion. However, once accumulation has occurred, blocking the metabolic pump would prevent the pump PD but then would permit a diffusion potential to occur in accord with the Nernst equation.

The argument may be raised that cell electropotentials are not closely related to ion movement through cell membranes. If ions were transferred in a combined form, e.g., with a co-ion or organic carrier complex, and, thus, in an uncharged state, then they would neither be influenced by the electropotential gradient nor would diffusion of such carrier complexes generate a diffusion potential. Some evidence that K^+ and Cl^- may move inwardly in an electroneutral transport system in beet root tissue has been reported (20). Also the conclusion in the past of other workers from research based on the carrier hypothesis (5) seems to lead to a model of ion uptake completely mediated by electrically neutral carriers. However, on theoretical grounds, it appears very unlikely that any differentially permeable membrane is perfect in the sense implied, i.e., with no ions passing in a charged state. Rather, the assumptions of some leakage in the bulk phase seems more reasonable, e.g., through pores, and, in this case, ions should influence or be influenced by the electropotential field. Furthermore, there has appeared a considerable body of evidence in support of the view that ion movement is related in some way to cell PDs (2, 3, 4, 6, 7, 10, 11, 12, 13, 14, 16, 20). An excellent example of inwardly directed Cl^- passive diffusion in potato disks, quantitatively predictable from electrochemical theory, has been reported by Laties, McDonald, and Dainty (16). In *Nitella translucens* Walker and Hope (23) have clearly demonstrated a marked effect of cell potential on flux rates of K^+, Na^+, and Cl^-.

An interesting aspect of ion regulatory activity of intact plants was revealed in this study. Root tissues approached equilibration in the seedling growth solution; thus excised segments showed a relatively constant ion content in the 48-hour immersion period. In contrast the initial mineral content of top portions was quite independent of the concentrations in the external growth solution. A relatively rapid net uptake by excised epicotyl or coleoptile segments occurred during immersion in 1× and 10× solutions. It seems apparent that the mineral concentrations of the extracellular (wall space) fluid bathing cells of the tops is either drastically different (generally lower) in concentration or that excision strongly modifies ion absorption characteristics. Unfortunately there appears to be little or no information on free space ion content of shoot tissues. The site of the regulatory mechanism would appear to reside in the root system. Further evidence for this conclusion is the fact that in 10× solution the ion content of roots was about the same as in 1× solution.

Acknowledgments

The authors acknowledge the aid of James K. Sugg and Wayne S. Pierce in some of the chemical assays. We are also indebted to Dr. L. B. Kirschner and Dr. Alan R. Koch for suggestions and reading the manuscript.

Literature Cited

1. Bertolacini, R. J. and J. E. Barney. 1958. Ultraviolet spectrophotometric determination of sulfate, chloride and fluoride with chloranilic acid. Anal. Chem. 30: 202–05
2. Briggs, G. E. 1963. Rate of uptake of salts by plant cells in relation to an anion pump. J. Exptl. Botany 14: 191–97.
3. Briggs, G. E., A. B. Hope, and R. N. Robertson. 1961. Electrolytes and plant cells. Blackwell Scientific Publications, Oxford.
4. Dainty, J. 1962. Ion transport and electrical potentials in plant cells. Ann. Rev. Plant Physiol. 13: 379–402.
5. Epstein, E. 1960. Spaces, barriers, and ion carriers: ion absorption by plants. Am. J. Botan. 47: 393–99.
6. Etherton, B. 1963. Relationship of cell transmembrane electropotential to potassium and sodium accumulation ratios in oat and pea seedlings. Plant Physiol. 38: 581–85.
7. Etherton, B. and N. Higinbotham. 1960. Transmembrane potential measurements of cells of higher plants as related to salt uptake. Science 131: 409–10.
8. Grewfling, T. 1961. Flame photometry of calcium and magnesium in ashed plant materials; resolution of anion interferences with EDTA Chem. Anal. 50: 40–43.
9. Griswold, B. L., F. L. Humoller, and A. R. McIntyre. 1951. Inorganic phosphates and phosphate esters in tissue extracts. Anal. Chem. 23: 192–94.
10. Higinbotham, N. 1964. Electropotentials and ion transport in cells of seed plants Abstr. of Tenth Intern'l Botan. Congress, p 169–70.
11. Higinbotham, N., B. Etherton, and R. J. Foster. 1964. Effect of external K, NH_4, Na, Ca, Mg and H ions on the cell transmembrane electropotential of *Avena* coleoptile. Plant Physiol. 39: 196–203.
12. Higinbotham, N., A. B. Hope, and G. P. Findlay. 1964. Electrical resistance of cell membranes of *Avena* coleoptiles. Science 143: 1448–49.
13. Hodgkins, A. L. and P. Horowicz. 1959. The influence of potassium and chloride on the membrane potential of single muscle fibres. J. Physiol. 148: 127–60.
14. Hope, A. B. 1965. Ionic relations of cells of *Chara australis*. X. Effects of bicarbonate ions on electrical properties. Australian J. Biol. Sci. 18: 789–801.
15. Hope, A. B. and N. A. Walker. 1961. Ionic relations of cells of *Chara australis* R. Br. IV. Membrane potential differences and resistances. Australian J. Biol. Sci. 14: 26–44.
16. Laties, G. G., I. R. McDonald, and J. Dainty. 1964. Influence of the counter-ion on the absorption isotherm for chloride at low temperature. Plant Physiol. 39: 254–62.
17. Lloyd, A. G. 1959. Studies on sulphatases: The use of barium chloranilate in the determination of enzymatically liberated sulfate. Biochem. J. 72: 133–36.
18. Morris, M. P. and A. Gonzales-Mas. 1958. Simple colorimetric method for the determination of nitrate in forage crops. J. Agric. Food Chem. 6: 456–57.
19. Pitman, M. G. 1963. The determination of the salt relations of the cytoplasmic phase in cells of beetroot tissue. Australian J. Biol. Sci. 16: 647–68.
20. Poole, R. J. 1966. The influence of the intracellular potential on potassium uptake by beetroot tissue. J. Gen. Physiol. 49: 551–63.
21. Teorell, T. 1949. Membrane electrophoresis in relation to bio-electrical polarization effects. Arch. Sci. Physiol. 3: 205–19.
22. Ussing, H. H. 1949. The distinction by means of tracers between active transport and diffusion. Acta Physiol. Scand. 19: 43–56.
23. Walker, N. A. and A. B. Hope. 1965. The effect of membrane potential on ionic fluxes in *Nitella translucens*. Abstracts. Australian Acad. Sci. Conference on the physiology of giant algal cells.

2
Translocation, Metabolism, Biosynthesis

Reprinted from PLANT PHYSIOLOGY, Vol. 43, No. 6, June, 1968, pages 877–882.
(PRINTED IN U.S.A.)

Bidirectional Translocation of Sugars in Sieve Tubes of Squash Plants[1]

P. Trip and P. R. Gorham
Division of Biosciences, National Research Council, Ottawa, Canada

Received December 28, 1967.

Abstract. **Two streams of sugars moving in opposite directions in the petiole of a half-grown leaf were demonstrated by feeding tritiated glucose to a fully grown leaf of a squash plant (*Cucurbita melopepo* Bailey) and $^{14}CO_2$ to the half-grown one. Autoradiographic evidence indicates that the movement of both streams occurred within the same sieve tubes. The data do not fit the mass flow theory of translocation which requires unidirectional flow of sugar solution in the lumen of the sieve tube.**

The term bidirectional translocation describes the simultaneous passage of 1 or more substances upward and downward within the same section of a stem (1). Several studies (1, 2, 4, 10) have demonstrated bidirectional translocation in stems and at least 2 (1, 4) have demonstrated that bidirectional movement can occur in the same phloem bundles. The technique used by Biddulph and Cory (1) in which ^{14}C and ^{32}P were used as tracers was not suitable for showing whether or not the same sieve tubes were involved in the simultaneous upward and downward movements of tracers. Using fluorescein and ^{14}C, Eschrich (4) was able to show in 12 experiments involving 68 aphids that both upward and downward-moving tracers appeared in the honeydew from about half of the aphids, while one or the other of the tracers appeared in the honeydew of the remaining half. He concluded that there existed either a bidirectional movement in a single sieve tube, or the tracers moved side by side in a homodromous "loop-path" involving sieve-tube anastomoses. However, the 2 to 8 hours between tracer application and the appearance of labeled honeydew were sufficient for radial diffusion of tracers (9, 11) from neighboring sieve tubes to occur. This could account for double labeling in some of the cells tapped by aphids. Therefore, the critical question, whether or not bidirectional movement occurs within the same sieve tube, has not been conclusively answered (12). The present work describes the autoradiographic detection of ^{14}C- and tritium (T)-labeled translocates which were moving in opposite directions in the same sieve tubes of squash plants.

[1] Issued as National Research Council of Canada No. 10193.

Materials and Methods

Squash plants (*Cucurbita melopepo* Bailey, var. *torticollis*) were germinated and grown under controlled conditions which were the same as those described earlier (10) except that the incandescent illumination was not water-filtered.

In a series of 4 identical experiments, 1 mc of glucose-6-T (specific activity = 450 μc/mmole) was introduced into 3-week-old plants via a flap formed by a side vein of a mature leaf. The glucose-6-T was dissolved in 50 μliter of water which was taken up at a steady rate of about 1 μliter/minute. Autoradiographs of the flap tissue showed that the label entered the leaf via the xylem and then was transferred via the mesophyll to the phloem. After 160 minutes, 100 μc of $^{14}CO_2$, 70 % ^{14}C labeled, was offered for assimilation in a lucite chamber (10) to a half-grown leaf of the same plant (fig 1). After another 20 minutes the plant was frozen and sawed into pieces for freeze-drying at −40° (9). Lyophilized tissue was vacuum-embedded with paraffin, sectioned (10 μ) and mounted on Kodak NTB-10 plates, exposed, developed, and mounted for microscopic examination as previously described (6). Gross distribution of tritium and ^{14}C was assayed by counting 5 mm pieces and 10 μ sections of petiole in a liquid scintillation spectrometer (9).

Results and Discussion

Typical distributions of ^{14}C and tritium (T) in the petioles of the fed leaves are shown in figure 1. The gradient of ^{14}C was the reverse of that of the tritium indicating that bidirectional movement was

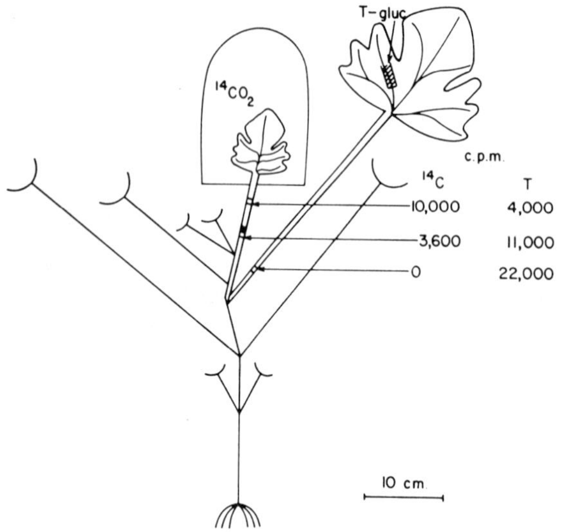

FIG. 1. Diagram of a squash plant indicating method of supply and gross distribution of tracer. T-glucose was supplied for 3 hours and $^{14}CO_2$ for 20 minutes. Autoradiographs were made from the blackened section of the petiole of the half-grown leaf.

occurring. In the tissues represented by the blackened region of figure 1, the advancing fronts of upward and downward movement had just passed one another and there was not enough time for radial movement of tracers from the sieve tubes to have occurred (6, 9). Paper chromatography of 80 % ethanol extracts of these tissues indicated that most of the radioactivity supplied as glucose-6-T to the leaf was present in the form of the transport sugars stachyose, raffinose and sucrose, and only a minor proportion as T-labeled glucose, fructose, and several unknowns. Similarly, the radioactivity from the $^{14}CO_2$ assimilation was present in the petiole in these same transport sugars.

Figure 2A shows an autograph obtained from the section of petiole indicated by blackening in figure 1 in which bidirectional movement was occurring. All bundles contained the diffuse patches typical of ^{14}C autographs in the outer phloem caps (detail shown in fig 2C). In addition, bundles in the right half of figure 2A contained intensely black, less diffuse patches typical of T-autographs superimposed on the ^{14}C-autographs (detail shown in fig 2E). The right half of figure 2A corresponds with the side of the plant on which the T-fed leaf was attached.

To differentiate between T-exposed and ^{14}C-exposed grains, the T radiation and part of the ^{14}C radiation were screened out. This was done by dipping NTB-10 plates in warm aqueous 20 % gelatin so that half of each plate was covered with a gelatin layer approximately 10 μ thick. After cooling and drying, serial sections were mounted on both halves of the plate. ^{14}C standards of known emission were applied to screened and unscreened plates and exposed for the same amount of time as the sections. Grain counts per 25 μ^2 were made using an oil immersion lens and a 5 μ ocular grid micrometer. The ratio of screened to unscreened counts was 1:1.5. Screened sections adjacent to the sections pictured in figures 2C and 2E are shown in figures 2B and 2D respectively. In 2D the screening of T radiation causes the autograph to look like a typical ^{14}C-autograph such as figure 2C. In 2B the removal of one-third of the ^{14}C radiation results in a lighter autograph than 2C. The autographs of cross sections were thus useful in indicating which bundles were most likely to be doubly labeled. Longitudinal sections discussed later were taken from these bundles.

In a separate experiment, 1 mc of glucose-6-T was administered alone for 3 hours. Figure 2F shows a radial section obtained from tissue in which tritium is the only tracer present. Such sections cut through a narrow zone of phloem so that it was usual to find only 1 or 2 streaks of grains side by side. In figure 2F, sieve tubes 1 and 3 contained tracer while 2 did not. The largest sieve tubes, such as number 3, were 30 μ in diameter [although in *Cucurbita* spp., sieve tubes may reach a diameter of 60 μ (3)]. The width of the streaks of silver grains varied from 6 to 30 μ. This variability in width is attributed to a combination of displacement of the protoplasm during lyophilizing and embedding and the improbability of cutting sections in planes that are continuously parallel with the long axes of the sieve tubes. This autoradiograph may be compared with an earlier one (ref 9, fig 3F) in which a single sieve tube contained label and the sieve

FIG. 2. Autoradiographs of transverse and radial sections of young petioles of squash containing labeled translocate. A) Sections taken from blackened area indicated in figure 1. All bundles contain tracer. Bundles on the right are more heavily labeled: Exposure 42 hours. × 15. C, E, B and D Detail of individual bundles. E contains patches of grains characteristic of T exposure superimposed on the ^{14}C exposed area. These patches are lacking in C. B and D are sections serial to C and E respectively from which all T and one-third ^{14}C have been screened out. The ^{14}C patch at D resembles a normal ^{14}C patch as in C. Exposure 42 hours. × 90. F) Radial section through phloem from separate experiment in which only T was offered for 3 hours. Sieve tubes 1 and 3 contain label while 2 does not. Exposure 2 days. × 215. G) Radial section along white line indicated in E. The indistinct borders of the streaks are ascribed to the presence of ^{14}C. Exposure 2 days. × 215. H) Similar to G but cut at a greater angle from the vertical plane of the tubes and with less exposure (16 hr) to facilitate grain counting. × 215.

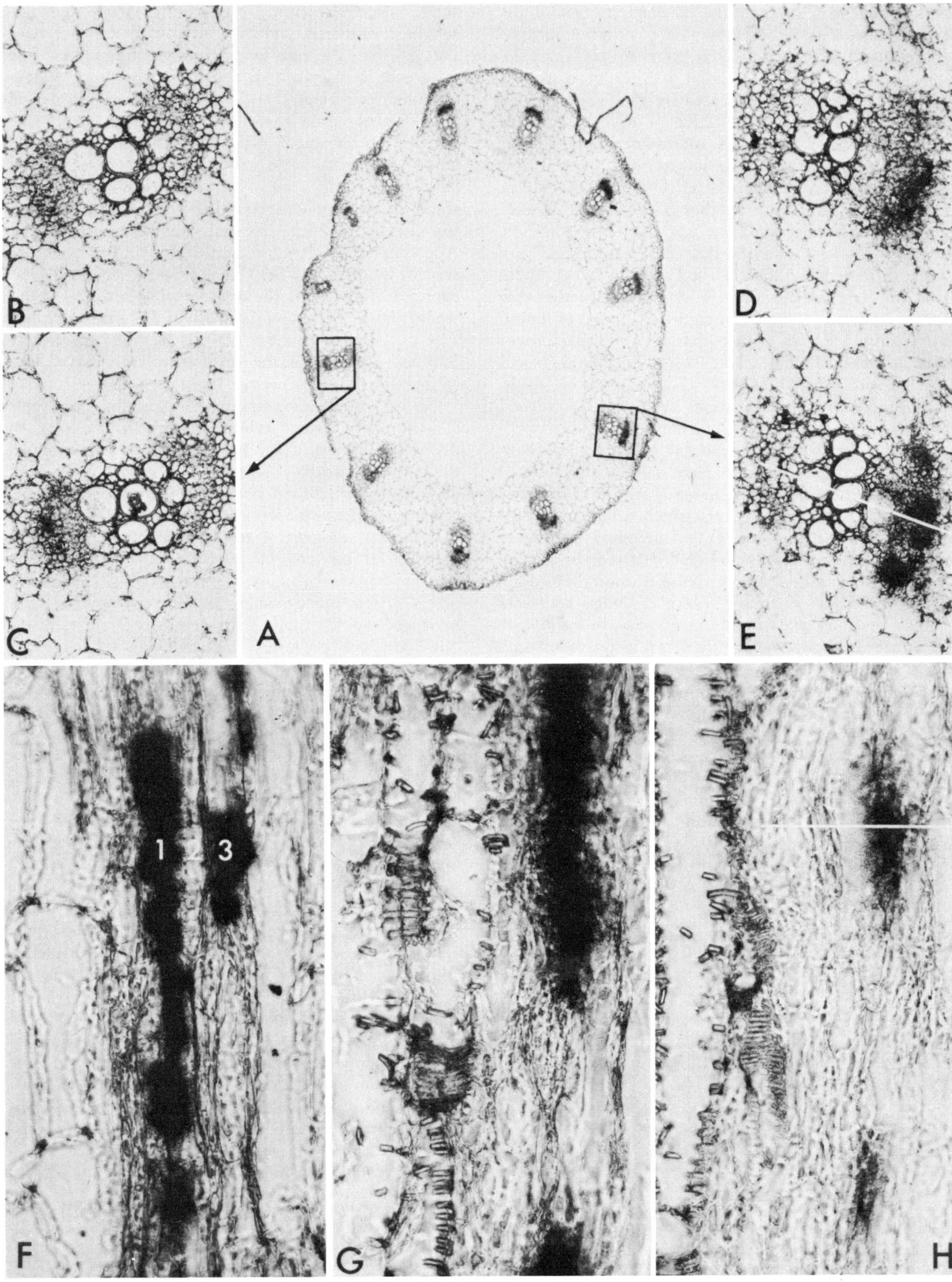

Insert

plate was clearly visible. In this earlier work it was shown that, in time, the label moved out of the sieve tubes and therefore these autoradiographs represented substances which were in transit.

Figure 2G shows a radial section of a doubly labeled bundle cut as indicated by the white line in figure 2E. It is representative of several hundred such sections. Parts of 2 autoradiographic streaks are shown. A study of serial sections showed that the gap between the streaks is the result of cutting out of one and into another sieve tube. Radioactivity was only apparent in single streaks one below the other. The borders of these streaks are more diffuse than those in figure 2F in which there was no ^{14}C present. The question of whether the autoradiographic streaks in figure 2G were produced by tritium, ^{14}C, or a combination of both isotopes was approached in 2 ways: (1) Scintillation spectrometry of serial sections; (2) Comparative autoradiography of adjacent serial sections on screened and unscreened NTB-10 emulsions.

Three serial sections, each 10 μ thick, adjoining the ones used for figure 2G were individually assayed in the scintillation spectrometer. All 3 had a tritium: ^{14}C ratio of 3.6:1.0, which was similar to the ratio in a 5 mm segment of petiole immediately above. It was therefore concluded that both tracers were present in the sections used for autoradiography.

Grain counts per 25 μ^2 were made on transects across the long axes of the streaks. The solid line in figure 3a shows the profile of grain counts caused by tritium for a transect across the lower third of the long streak shown in figure 2F. Distances on the abscissa are plotted from a point on one side where the grain count was at background across the radioactive streak until background was reached again on the other side. The dotted line represents the grain counts for the adjacent screened section. It indicates that the screening stopped the tritium radiation completely.

Figure 3b shows the profile of grain counts obtained from a transect across the middle section of the upper streak of figure 2H (white line). The solid line representing the counts from the unscreened emulsion, shows a peak with a wider base (60 μ) than the one for tritium in figure 3a. The difference is ascribed to the presence of ^{14}C whose stronger beta emission produces silver grains on NTB plates up to 15 μ from the source (6). The dotted line, representing the counts from the screened emulsion, has a peak width of about 35 μ, and a base that is still 50 μ wide. Because the mean standard deviation of counting was 2 grains/25 μ^2, the dips in the peak are not considered significant. Applying the standard ratio for unscreened to screened counts (1.5:1) to the peak screened ^{14}C count of figure 3b, the peak unscreened ^{14}C count would have been approximately 10 grains/25 μ^2. The unscreened count actually observed was nearly 3 times greater, so the difference must have been caused by the presence of tritium.

When autoradiographing longitudinal sections 10 μ thick, of sieve tubes that are 20 to 30 μ in diameter, there is a distinct probability that some of the autoradiographic streaks will refer, at least in part, to superimposed portions of 2 adjoining sieve tubes as well as to sections of individual sieve tubes. Since autoradiographs tend to obscure cytological detail, it is not possible to be sure whether a particular streak represents the tracers in 1 sieve tube or 2. It is conceivable, therefore, for a streak produced by ^{14}C and T to refer to 2 sieve tubes, one of which contained ^{14}C and the other T. A streak might also refer to an individual sieve tube (or parts of 2 adjacent sieve tubes) containing both tracers. In the first case there would be a pronounced difference in the ratio of tritium: ^{14}C peak counts for transects made at either end of the streak as compared with the middle. In the second case, there would be no such difference.

Since no difference in ratio of tritium: ^{14}C peak counts was observed between transects at the ends or in the middle, it is concluded that upward moving tritium and downward moving ^{14}C were present in the same sieve tubes.

Bidirectional movement of sugars in the same sieve tube has been observed in these experiments. The mass flow theory of translocation, involving the unidirectional movement of a solution through the lumen of a sieve tube, is no longer adequate to explain these observations. While a mass flow of solution may account for the movement in one direction, in some way the sieve tube protoplast must control and direct the translocation process to permit concurrent movement in opposite directions. Transcellular strands in sieve tubes (5, 8) provide a possible mechanism for separating upward and downward movements in the same cell.

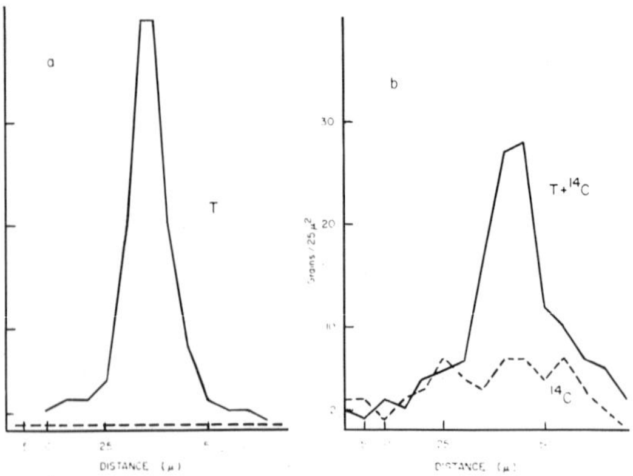

FIG. 3. Grain counts along transects across streaks of grains as pictured in figure 2H. a) Tritium only. b) Tritium and ^{14}C. Solid line: Unscreened emulsion. Dotted line: Screened emulsion.

Acknowledgments

We thank Mr. R. Whitehead for photographic work and Miss B. Sinnott for technical assistance.

Literature Cited

1. Biddulph, O. and R. Cory. 1960. Demonstration of two translocation mechanisms in studies of bidirectional movement. Plant Physiol. 35: 689–95.
2. Biddulph, O. and R. Cory. 1965. Translocation of ^{14}C metabolites in the phloem of the bean plant. Plant Physiol. 40: 119–29.
3. Esau, K. 1953. Plant Anatomy. John Wiley and Sons, New York. p 647 and 650.
4. Eschrich, W. 1967. Bidirektionelle Translokation in Siebröhren. Planta 73: 37–49.
5. Evert, R. F., L. Murmanis, and I. B. Sachs. 1966. Another review of the ultrastructure of *Cucurbita* phloem. Ann. Botany 30: 120, 563–85.
6. Perkins, H. J., C. D. Nelson, and P. R. Gorham. 1959. A tissue-autoradiographic study of the translocation of ^{14}C-labeled sugars in the stems of young soybean plants. Can. J. Botany 37: 871–77.
7. Schumacher, W. 1948. Zur Frage nach den Stoffbewegungen im Pflanzenkörper. Naturwissenschaften 35: 176–79.
8. Thaine, R. 1962. A translocation hypothesis based on the structure of plant cytoplasm. J. Exptl. Botany 13: 152–60.
9. Trip, P. and P. R. Gorham. 1967. Autoradiographic study of the pathway of translocation. Can. J. Botany 45: 1567–73.
10. Webb, J. A. and P. R. Gorham. 1964. Translocation of photosynthetically assimilated ^{14}C in straight-necked squash. Plant Physiol. 39: 663–72.
11. Webb, J. A. and P. R. Gorham. 1965. Radial movement of ^{14}C translocates from squash phloem. Can. J. Botany 43: 97–103.
12. Zimmerman, M. H. 1960. Transport in the phloem. Ann. Rev. Plant Physiol. 11: p 179.

Reprinted from PLANT PHYSIOLOGY, Vol. 42, No. 2, February, 1967, pages 191–194.
(PRINTED IN U.S.A.)

Phloem Water Relations and Translocation[1]

Merrill R. Kaufmann and Paul J. Kramer
Department of Botany, Duke University, Durham, North Carolina 27706

Received October 10, 1966.

Summary. Satisfactory measurements of phloem water potential of trees can be obtained with the Richards and Ogata psychrometer and the vapor equilibration techniques, although corrections for loss of dry weight and for heating by respiration are required for the vapor equilibrium values. The psychrometer technique is the more satisfactory of the 2 because it requires less time for equilibration, less tissue, and less handling of tissue. Phloem water potential of a yellow-poplar tree followed a diurnal pattern quite similar to that of leaves, except that the values were higher (less negative) and changed less than in the leaves.

The psychrometer technique permits a different approach to the study of translocation in trees. Measurements of water potential of phloem discs followed by freezing of samples and determination of osmotic potential allows estimation of turgor pressure in various parts of trees as the difference between osmotic potential and total water potential. This technique was used in evaluating gradients in water potential, osmotic potential, and turgor pressure in red maple trees. The expected gradients in osmotic potential were observed in the phloem, osmotic potential of the cell sap increasing (sap becoming more dilute) down the trunk. However, values of water potential were such that a gradient in turgor pressure apparently did not exist at a time when rate of translocation was expected to be high. These results do not support the mass flow theory of translocation favored by many workers.

Numerous measurements have been made of the water content of tree trunks and bark, but no measurements of the water potential of the phloem are known to have been published. Bier (4) and Baranyay (1) measured the relative turgidity of the bark, but those measurements did not permit calculation of the water potential. Information concerning the water potential, osmotic potential, and turgor pressure of leaves and phloem would be useful in determining whether conditions in trees are favorable for the operation of the mass flow theory of translocation proposed by Munch (13) and supported by many other investigators, most recently by Zimmermann (20–23), Biddulph and Cory (3), and Ford and Peel (6).

This paper reports the results of a study of phloem and leaf water potential in yellow-poplar (*Liriodendron tulipifera* L.) and red maple (*Acer rubrum* L.) by use of the Richards and Ogata (17) thermocouple psychrometer and the vapor equilibration techniques (18).

The terminology of water relations used in this paper was recently described by Kramer, Knipling, and Miller (10), following suggestions by Slatyer and Taylor (19). Since the potential of pure water is set at zero and addition of solutes reduces both the water potential and osmotic potential below that of pure water, water and osmotic potentials are represented by negative numbers. A solution having an osmotic potential of −5 bars has a higher osmotic potential and is more dilute than a solution having an osmotic potential of −10 bars.

Comparison of the Psychrometer and Vapor Equilibration Techniques

Methods. Discs of phloem tissue with intact outer bark, 1.3 cm in diameter, were removed with a sharp cork borer from the boles of yellow-poplar trees 6 to 8 cm in diameter. Water potentials of samples consisting of 2 discs were measured in thermocouple psychrometers having small chambers, and the values were corrected for heat of respiration (2). Eight to 12 hours were required for equilibration. Samples of 3 discs were weighed and placed in vapor equilibration chambers over sodium chloride solutions ranging in water potential from zero (pure water) to −15 bars. The chambers were kept in a water bath at 27° for 22 hours before the final weighing. Thermocouples were imbedded in discs in 2 of the chambers with reference junctions in the solution to measure disc temperature relative to solution temperature.

Results and Discussion. In 2 of 5 determinations by the vapor equilibration method discs lost weight over pure water. Positive water potentials

[1] Work supported by National Science Foundation Grant No. GB-3934.

Table I. *Comparison of the Psychrometer and Vapor Equilibration Techniques for Measurement of Water Potential in Phloem Tissue of Yellow-Poplar Trees*

	Water potential	
Psychrometer	Uncorrected Vapor Equilibration	Corrected Vapor Equilibration*
−5.3	+6.3**	−4.7
−5.6	+2.5**	−8.1
−9.2	−2.5	−9.0
−11.9	−5.2	−14.5
−16.6	−11.2	−20.3

* Corrected for loss of dry weight and respiration heating.
** Estimated by extrapolation.

for these 2 determinations were estimated by extrapolation. The results obtained by the 2 methods are compared in table I, where it can be seen that all water potentials obtained by vapor equilibration (Uncorrected) are higher than those obtained with the psychrometers on similar samples. These discrepancies result from loss of dry weight and from the fact that respiration made the tissue 0.049 to 0.103° warmer than the solution with which it was being equilibrated.

A correction for dry weight loss was made by a method similar to the one used by Kreeb and Onal (11). A correction for the increased vapor pressure resulting from the temperature difference was calculated by the following equation:

$$\triangle \Psi = \frac{RT \ ln(e/e_o)}{\overline{V}_w}$$

where $\triangle \Psi$ is the change in water potential in bars, R is the gas constant, T is the absolute temperature, e and e_o are sample and saturation vapor pressures, and $\overline{V}_w$ is the molal volume of water. Warming of the discs by respiration produced errors of 4.1 to 8.6 bars. After these corrections, the vapor equilibration values still differed from the values determined with the psychrometer (table I). Apparently other unidentified factors are involved.

Determination of Diurnal Patters of Phloem and Leaf Water Potential

Methods. While water potential of the phloem can be measured approximately with either of the methods used for leaf tissue, the psychrometer method was used for subsequent experiments because it requires less time to obtain a value, less tissue, and less handling of tissue. Duplicate samples of phloem and leaves were collected at 6-hour intervals for 48 hours from a yellow-poplar tree 5 m tall, growing at the edge of a forest stand. Phloem samples were collected at a height of 1 m, and leaf samples at 2.5 m. Moisture changes of the outer bark were minimized by protecting the sampling area with aluminum foil, and holes resulting from removal of phloem discs were closed with glazing compound. Soil water potential was higher than −1 bar.

Results and Discussion. Distinct and similar diurnal patterns of change in water potential of leaves and phloem were observed on 2 successive days (fig 1), although the range is much smaller in the phloem than in the leaves. The daytime decrease in water potential of the phloem must result from loss of water rather than from increase in amount of osmotically active solute because both Huber et al. (9) and Zimmermann (21) reported that a wave of carbohydrates moves down the stem at night. Such an increase in soluble carbohydrate would decrease the osmotic potential of the phloem sap and decrease the water potential at night, causing a pattern opposite to the one observed here. Perhaps the small range in water potential observed in the phloem results from the opposing effects of changes in water content and in amount of osmotically active solutes.

The loss of water from the phloem probably is chiefly by lateral movement to the xylem rather than by longitudinal movement to the leaves; transpiration through bark is quite small (8). Exchange of both water and solutes between xylem and phloem is relatively rapid, and water probably moves from the phloem to the xylem during the day when transpiration lowers the water potential in the xylem, and in the reverse direction at night. It would be very interesting to compare the water potential of xylem and phloem, but it is not yet possible to make direct measurements of the water potential of the xylem of a tree trunk. One can

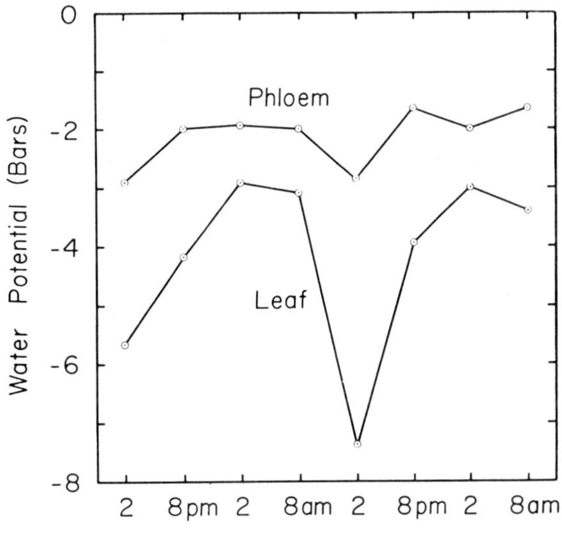

FIG. 1. Diurnal pattern of phloem and leaf water potential of a yellow-poplar tree.

reasonably assume, however, that it follows the same daily course as the water potential of the leaves, but perhaps the range of variation is intermediate between that for leaves and for phloem.

Gradients in Water Potentials, Osmotic Potential, and Turgor Pressure

Methods. The ability to measure phloem water potential makes possible a new approach to the study of gradients in turgor pressure and osmotic potential in trees. Traditionally these gradients have been studied using exudation from cut sieve tubes (7, 13) and recently from aphid stylets (12). Existence of such gradients is an important feature of the mass flow theory of translocation supported by many workers. Duplicate leaf samples were collected at a height of 6 m from red maple trees about 12 m tall. Triplicate phloem samples were collected at heights of 5 m and 1 m above the ground, and duplicate soil samples were taken at depths of 10 to 15 cm approximately 2 m from the trees. Samples were collected from 2 trees during late July and early August on 3 clear days following rains. Conditions were such that production and translocation of photosynthate were expected to be high. Collections were made between 7 and 10 PM because gradients in concentration of solutes are large during this part of the day (9, 21). After determination of leaf and phloem water potentials, samples were frozen for several hours, and then osmotic potentials of the tissue were measured. Turgor pressure was calculated as the difference between water potential and osmotic potential.

Results and Discussion. A gradient of decreasing water potential was observed from the soil through the phloem to the leaves (table II), with one exception. Osmotic potential was the lowest (cell sap most concentrated) in the leaves and increased (sap became more dilute) down the trunk. Similar gradients in osmotic potential in the phloem have been observed by other workers (9, 16, 20), although Phillis and Mason (15) found lower sugar concentrations in sap expressed from leaf mesophyll cells than in sap from petiolar phloem. The gradient in osmotic potential was most pronounced from leaves to phloem and less pronounced from upper to lower phloem. The steeper gradient observed in the upper part of the trees may result from a higher concentration of salt in the leaves, since our measurements yield total osmotic potentials resulting from both sugars and salts. Our leaf samples contained few large veins (see 15). A large difference in turgor pressure was observed between the leaves and the phloem, but a turgor pressure gradient in the phloem was *not* observed.

General Discussion and Conclusions

Because of the great distances over which it must occur, considerable effort has been directed toward the study of translocation in trees. Evidence has been accumulated supporting 2 opposing theories of the mechanism of translocation: the mass flow theory first proposed by Munch (13), and the theory of metabolic transport. Conflicting lines of evidence have led several workers (5, 20) to suggest that translocation may involve several mechanisms. Significantly, Biddulph and Cory (3) have apparently demonstrated that metabolic transfer and transport resembling mass flow can occur simultaneously in the same plant.

In view of our incomplete understanding of the translocation system, however, more knowledge is needed of the relationships among gradients of water potential, osmotic potential, and turgor pressure in trees. Of particular interest are gradients in turgor pressure. According to the mass flow theory, the accumulation of osmotically active substances (sugars) in the phloem of leaves in conjunction with removal of these substances in and near the root system causes a higher hydrostatic pressure in the upper part of the conducting network, resulting in a massive downward flow of water and dissolved materials. However, the development of a turgor pressure gradient depends not on the osmotic potential alone (20, 21), but rather on the relationship between osmotic poten-

Table II. *Gradients in Water Potential, Osmotic Potential, and Turgor Pressure in Red Maple Trees Sampled at 7 to 10* PM

Sample	(Leaf 6 m)	Phloem (5 m)	Phloem (1 m)	Soil
Water potential				
Tree A	−4.4	−5.4	−3.4	−0.7
Tree A	−4.3	−3.7	−2.3	−0.1
Tree B	−5.3	−4.4	−2.9	−0.1
Osmotic potential				
Tree A	−14.6	−9.3	−7.6	...
Tree A	−14.5	−8.9	−7.5	...
Tree B	−13.4	−10.0	−8.3	...
Turgor pressure				
Tree A	+10.2	+3.9	+4.2	...
Tree A	+10.2	+5.2	+5.2	...
Tree B	+8.2	+5.6	+5.4	...

tial and the total water potential. This can be seen in the water potential equation,

$$\Psi_w = \Psi_s + \Psi_p$$

where Ψ_w is the total water potential, Ψ_s the solute or osmotic potential, and Ψ_p the turgor pressure. It is quite possible, as demonstrated by our data (see table II), that a gradient can exist in osmotic potential without a gradient in turgor pressure, because of a difference in the total water potential through the conducting system. A gradient in turgor pressure can occur only when the gradient of water potential is less steep than the gradient in osmotic potential. It might be asked if the removal of discs of phloem from trees releases the pressure in the sieve tubes and invalidates the method we used. However, no exudation was observed from any of the incisions; hence there appears to have been no release of pressure. The turgor pressures reported in this paper are necessarily an average of all cells in the phloem and not merely those of the sieve tubes. On the other hand, it is not likely that appreciable differences occur in water potential, osmotic potential, or turgor pressure between the sieve tubes and other cells in the phloem. The osmotic potentials (−7.5 to −10.0 bars) and the gradients in osmotic potential (0.4 bar per meter) we observed for phloem samples containing several cell types are similar to those reported for exudate from sieve tubes (14; also see 22).

These measurements failed to demonstrate the existence of the gradient of decreasing turgor pressure from the upper to lower regions of the phloem which must exist if mass flow of solutes is to occur.

Literature Cited

1. Baranyay, J. A. 1964. Bark moisture relationships in western hemlock branches infected by dwarf mistletoe. Can. J. Botany 42: 1313–19.
2. Barrs, H. D. 1965. Comparison of water potentials in leaves as measured by two types of thermocouple psychrometer. Australian J. Biol. Sci. 18: 36–52.
3. Biddulph, O. and R. Cory. 1965. Translocation of ^{14}C metabolites in the phloem of the bean plant. Plant Physiol. 40: 119–29.
4. Bier, J. E. 1959. The relation of bark moisture to the development of canker diseases caused by native, facultative parasites. III. Cephalosporium canker on western hemlock. Can. J. Botany 37: 1140–42.
5. Esau, K., H. B. Currier, and V. I. Cheadle. 1957. Physiology of phloem. Ann. Rev. Plant Physiol. 8: 349–74.
6. Ford, J. and A. J. Peel. 1966. The contributory length of sieve tubes in isolated segments of willow, and the effect on it of low temperatures. J. Exptl. Botany 17: 522–33.
7. Hartig, Th. 1860. Beitrag zur physiologischen Forstbotanik. Allgem. Forstz. und Jagdztg. 36: 257–61.
8. Huber, B. 1956. Die Transpiration von Sprossachsen und anderen nicht foliosen Organen. In: Encyclopedia of Plant Physiol. W. Ruhland, ed. 3: 427–35.
9. Huber, B., E. Schmidt, and H. Jahnel. 1937. Untersuchen über den Assimilatstrom. I. Tharandt. forstl. Jahrb. 88: 1017–50.
10. Kramer, P. J., E. B. Knipling, and L. N. Miller. 1966. Terminology of cell-water relations. Science 153: 889–90.
11. Kreeb, K. and M. Onal. 1961. Über die gravimetrische Methode zur Bestimmung der Saugsspannung und das Problem des negativen Turgors. II. Mitteilung die Berücksichtigung von Atmungsverlusten während der Messungen. Planta 56: 409–15.
12. Mittler, T. E. 1958. Sieve-tube sap via aphid stylets. In: The Physiology of Forest Trees, Chapter 19. K. V. Thimann, ed. The Ronald Press Company, New York.
13. Münch, E. 1930. Die Stoffbewegungen in der Pflanze. Gustav Fischer Verlagsbuchhandlung. Jena, Germany.
14. Pfeiffer, M. 1937. Die Verteilung der osmotischen Werte im Baum im Hinblick auf die Münchsche Druckstromtheorie. Flora 132: 1–47.
15. Phillis, E. and T. G. Mason. 1933. Studies on the transport of carbohydrates in the cotton plant. III. The polar distribution of sugar in the foliage leaf. Ann. Botany 47: 585–634.
16. Pitman, G. B. 1966. Studies on the pheromone of Ips confusus (Le Conte). III. The influence of host material on pheromone production. Contrib. Boyce Thompson Inst. 23: 147–58.
17. Richards, L. A. and G. Ogata. 1958. Thermocouple for vapor pressure measurement in biological and soil systems at high humidity. Science 128: 1089–90.
18. Slatyer, R. O. 1958. The measurement of diffusion pressure deficit in plants by a method of vapor equilibration. Australian J. Biol. Sci. 11: 349–65.
19. Slatyer, R. O. and S. A. Taylor. 1960. Terminology in plant- and soil-water relations. Nature 187: 922–24.
20. Zimmermann, M. H. 1957. Translocation of organic substances in trees. II. On the translocation mechanism in the phloem of white ash (*Fraxinus americana* L.). Plant Physiol. 32: 399–404.
21. Zimmermann, M. H. 1958. Translocation of organic substances in the phloem of trees. In: The Physiology of Forest Trees, Chapter 18. K. V. Thimann, ed. The Ronald Press Company, New York.
22. Zimmermann, M. H. 1960. Transport in the phloem. Ann. Rev. Plant Physiol. 11: 167–90.
23. Zimmermann, M. H. 1961. Movement of organic substances in trees. Science 133: 73–79.

Measurement of Turgor Pressure and Its Gradient in the Phloem of Oak

H. T. Hammel[1]

Physiological Research Laboratory, Scripps Institution of Oceanography, University of California, La Jolla, California 92037

Received January 10, 1968.

Abstract. A direct method is described for measuring the pressure in secondary phloem sieve tubes of oak trees. One end of a 26-gauge stainless steel tube was shaped such that when it penetrated the outer bark and transected a few sieve elements, it was stopped by the xylem so that small openings in the end allowed phloem sap to enter the tube. The other end of the stainless tube (phloem needle) was joined to a long glass capillary sealed at its other end to form a manometer for measuring phloem sap pressure. A method for measuring the average osmotic and turgor pressures in cells of leaves is also described. Phloem turgor pressures varied greatly in a series of phloem punctures around the trunk at 1.5 and at 6.3 meters. The variation in turgor pressure was always greater than the variation in osmotic pressure. In a series of turgor pressures arranged in descending order, the values in a sequence for the upper level was usually a little (0–3 atm) larger than the values for the lower level. These results may suggest that translocation of assimilate is favored by a small turgor pressure gradient, but they do more to emphasize the complications in measuring gradients in an elastic low resistance distribution system composed of contiguous longitudinal conduits. The results also imply that the sieve tubes are inflated with assimilate fluid under high pressure which can readily move longitudinally and with less pressure drop than would be necessary if the sieve tubes were rigid.

During summer and autumn, the osmotic pressure of phloem sap has been shown to be highest near the leaves and to decrease toward the roots in white ash (9) and red maple (3). In many species, the phloem sap must be under pressure, since it exudes from the transected sieve tubes (2, 10). During transpiration, the water potential in xylem sap and in the extracellular sap is lowered by the hydrostatic tension in this sap due to the evaporation of water from the free surfaces within the leaf (6). The xylem sap tension is greatest (hydrostatic pressure is most negative and water potential is lowest) in the leaves and decreases toward the roots, affording the pressure gradient to overcome gravity and the resistance to flow through the xylem. If the osmotic pressure within the phloem and mesophyll in the leaf is less (more negative) than the negative hydrostatic pressure in the xylem and extracellular sap, then the water potential in the cells may be increased by a positive hydrostatic pressure (turgor pressure) to effect equal water potential in xylem sap and phloem or leaf cells. The relationship between the osmotic pressure and the turgor pressure in the phloem sap and the hydrostatic tension in the xylem sap is less clear in the limbs and trunk of woody plants where the secondary phloem and xylem are separated by the cambium. Evaluation of turgor pressures along the length of the phloem system by indirect measurements is a formidable task; and the thesis that phloem transport proceeds as a mass flow from the leaves toward the roots, propelled by a turgor pressure gradient, has not been verified to everyone's satisfaction. Nevertheless, there is a rootward movement of sap through the secondary phloem under some conditions (10).

This report describes 2 methods which may provide valid determinations of turgor pressures within woody plants. One method purports to be a direct determination of phloem sap pressure in the trunk and limbs of trees and the other is a semi-direct method for obtaining the average turgor pressure in the cells of the leaf. More experience with measurements extending through an entire season may be required to yield a full measure of confidence in the methodology.

Materials and Methods

All observations were made on 1 tree during October and early November. A mature red oak (*Quercus rubrum,* 10 meters tall with a bole diameter of 36 cm at 1 meter above ground) was used to measure pressure gradients in the vascular system.

Direct Measurement of Turgor Pressure in Secondary Phloem. A double bevel was ground on 1 end of a 3.5 cm length of stainless steel tubing

[1] This manuscript was prepared while the author was a Fellow of the John B. Pierce Foundation Laboratory, New Haven, Connecticut.

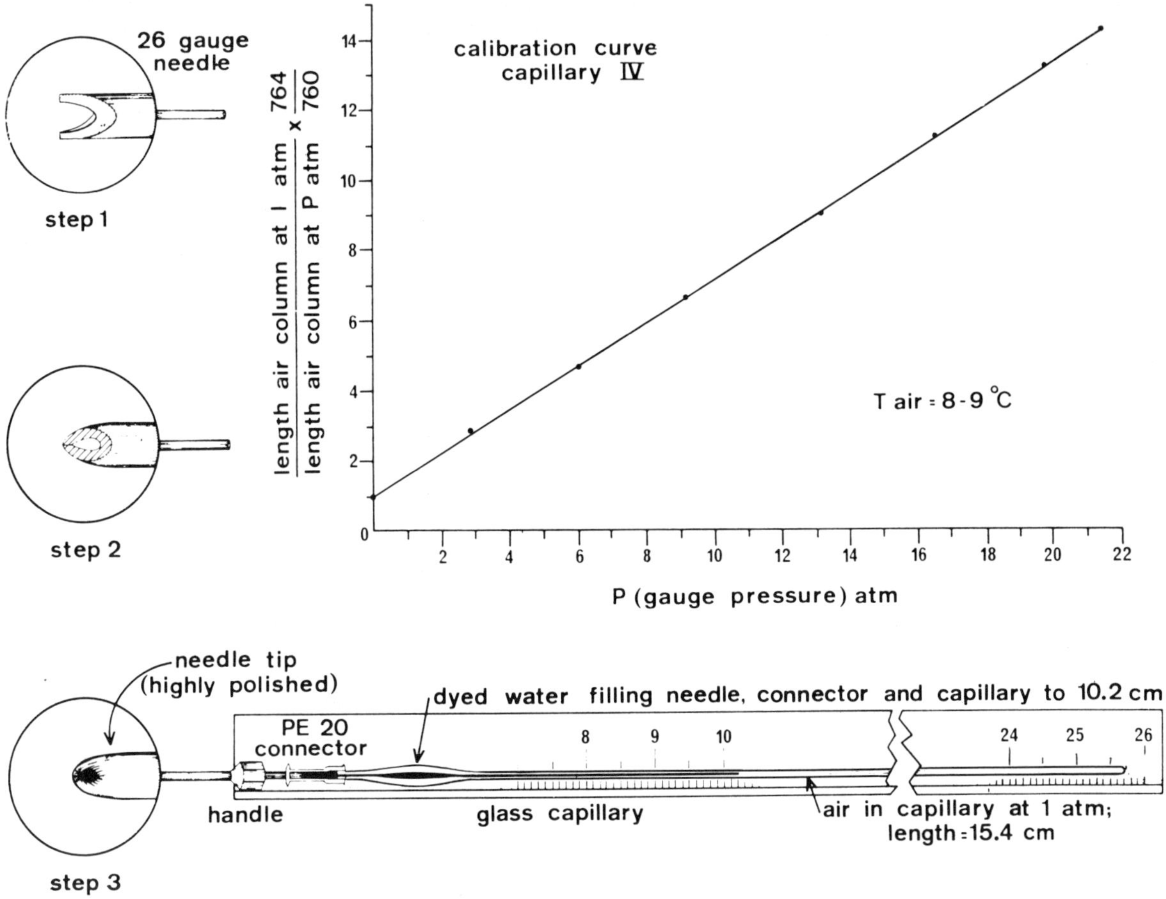

FIG. 1. Construction and calibration of phloem needle and capillary for direct measurement of turgor pressure in secondary phloem of trees.

(0.47 mm OD and 0.25 mm ID) as illustrated in step 1, figure 1. The 2 tapered points were then squeezed together as shown in step 2, and then highly polished as in step 3. A handle for holding the needle, a brass piece with a hole along its axis, was slipped over the needle and soft soldered in place with about 1 cm of the needle extending beyond the handle. The other end of the needle was joined to a glass capillary with a short length of PE-20 Polyethylene tubing. The glass capillary was drawn from stock glass with 5.5 mm OD and 0.5 mm ID. The portion of the capillary containing the air column had an inside diameter of approximately 25 μ so that the volume of a 15 cm long air column was about 0.08 mm^3. The needle, connector and capillary were filled with an indicator fluid (distilled water containing methylene blue) as follows: A 0.5 ml syringe containing the dyed water and fitted with a 26 gauge hypodermic needle with a short length of PE-20 Polyethylene tubing slipped over its tip was joined to the phloem needle. The dyed water was then flushed through the phloem needle, its connector, through the glass capillary and out the open end of the capillary. After blotting the dyed water from the tip of the capillary, the water was withdrawn into the capillary until the air column following it was 10 to 20 cm in length. The end of the capillary was then sealed in a flame.

The capillary was calibrated by joining the phloem needle with an adjustable source of pressure with a master test gauge to read the pressure to $\pm$ 0.1 atm. The ratio of the length of the air column at ambient pressure to the length of the air column for gauge pressure P $\times$ Barometric Pressure in mm Hg $\div$ 760 mm Hg was plotted as a function of the gauge pressure P. A calibration curve for a typical capillary is illustrated in figure 1.

The phloem sap pressure was measured by pressing the phloem needle through the bark of the trunk until it stopped abruptly. Occasionally nothing would happen. Most often the air column was immediately compressed by an outrush of dyed water into the end of the capillary. Occasionally the phloem needle could be twisted slightly in its hole resulting in an immediate compression; and occasionally the air column would be slowly compressed, with or without twisting the phloem needle. Slow compression of the air column seldom gave as high a compression ratio as rapid compression and these slow determinations were not recorded. Occasionally there was rapid compression followed by an immediate receding of the dyed water column. These determinations were not recorded as they were usually accompanied by the appearance of sap at the bark surface around the needle indicating leakage. The criteria used for accepting a determination were rapid compression of the air column and a steady reading of the length of the compressed air column.

The design of the needle tip seemed to insure that the needle would stop at the appropriate depth at each penetration. The smooth blunt end seemed to stop at the wood surface, the interface between the cambium and the xylem, after crushing the functional phloem sieve tubes lying just exterior to the cambium. The phloem needle was held with its 2 openings directed up and down during penetration, so that when the phloem sieve tubes were ruptured, their contents could drain into the needle. Although the volume of phloem exudate required to make a pressure determination was less than 0.1 mm^3, this volume was both many times the volume of a single cell and only a small fraction of the available volume. If the needle were withdrawn, 10 to 20 mm^3 of sap would exude through the needle hole, draining contents from many cells above and below the penetration and, also, to a less extent from lateral cells. The highly polished surface of the needle, the small openings on the 2 sloping surfaces at the end opening into the larger lumen of the needle combined to minimize plugging of the needle. Two separate penetrations of the bark never plugged the needle; 3 to 4 penetrations would often not plug the needle. After the second or third penetration, the needle was routinely disconnected under water from the PE-20 Polyethylene connector and back flushed with water from a syringe and then reconnected under water to the PE-20 connector. The high polish on the end of the needle and diminishing dimensions toward the tip combined to minimize leakage between the needle surface and the wall of the hole. At the lower levels of the trunk where the bark was up to 1 cm thick, a one-half cm pilot hole was made in the bark with a straight sewing needle with diameter less than the diameter of the phloem needle. This pilot hole seemed to reduce the tendency to plug the phloem needle without increasing the risk of leakage. At the upper levels of the trunk, the pilot hole was not used. It was always necessary to read the length of the compressed air column with the aid of a 10x power spectacle loupe.

Measurement of Osmotic Pressure of Secondary Phloem Sap. Following a series of penetrations to determine the secondary phloem turgor pressure, 3 separate penetrations were made with a sewing needle, and the exudate from each was immediately collected in a micro-pipette. The other end of the pipette was drawn to a taper and sealed in a flame and the phloem sap was centrifuged for 5 minutes to separate all cell organelles from the supernatant cell fluid. A melting point determination was made on all 3 samples in a sealed, square, insulated bath containing the sample pipette and a standard thermometer. The bath temperature could be held constant within $\pm$ 0.01° for many hours and rapidly changed to another temperature at will. The sample was continuously stirred with a fine glass stirring rod while the last tiny ice crystal just melted after about 1 hour of observation. The melting point was thus determined to $\pm$ 0.01°. The method was an adaptation of a simpler method described by Scholander *et al.* (7). The melting point was converted to osmotic pressure at the temperature of the sap when taken from the tree, assuming that the major osmotic activity was due to sucrose (8) which has a freezing point depression of 2.05° per mol for a 1 molal solution.

Semi-Direct Determination of Intracellular Turgor and Osmotic Pressures and Extracellular or Xylem Sap Pressure in Leaves. After measurements of the phloem pressures in the trunk at a lower and an upper level which required 1 to 2 hours to complete, a twig of 5 to 10 leaves was cut from the upper level and another from the lower level. The twigs were immediately enclosed in a plastic bag and held there for 10 to 15 minutes until they could be transferred to a pressure chamber maintained at the outdoor temperature.

To measure the xylem sap tension, the pressure in the chamber was increased until the sap returned to the cut end of the twig outside the chamber (5). If an over-pressure was applied to the twig, then water was compressed out of the leaf cells, through the conducting system, and then out the end of the stem where it could be collected and weighed. After removing a small and measured quantity of water, the new tension existing in xylem sap was measured as before. With repeated compression of water from the leaf cells and determination each time of the volume removed and the resulting xylem sap tension, the reciprocal of the sap tension could be plotted as a function of the water removed, as illustrated in figure 2. The curve becomes linear after enough water has been removed to drop the turgor pressure to zero. When the turgor pressure was zero, the osmotic pressure of the leaf cell sap was balanced only by the hydrostatic tension of the xylem and extracellular sap. Since the osmotic pressure was proportional to the concentration of active materials in the cell, and since the concentration was inversely proportional to the amount of cell water, the osmotic pressure of the cell and also the hydrostatic tension of the xylem sap was inversely proportional to the water removed from the cell. The line intersects the abscissa at the initial cell volume V_o; *i.e.,* infinite pressure would be required to remove all the cell water from the electrolytes, assuming the semi-permeable mem-

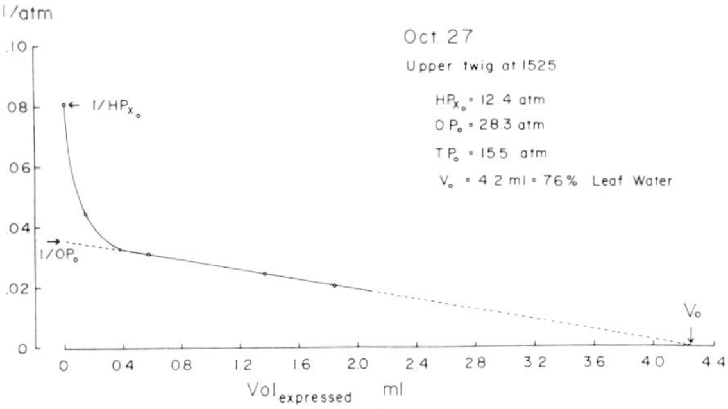

FIG. 2. Reciprocal xylem sap tension *vs* volume expressed from leaf cells from which the initial xylem sap tension (HP_{xo}), average leaf cell osmotic (OP_o) and turgor (TP_o) pressures are determined in red oak.

brane would not be destroyed by the maneuver. Since this line is the zero turgor line, and is always the osmotic pressure for any volume, it intersects the ordinate at the reciprocal of the initial osmotic pressure of the leaf cells. The initial turgor was, as shown in figure 2, the difference between the measured, initial hydrostatic pressure in the xylem sap and the initial osmotic pressure in the leaf cells (1). The melting point depression of the first sample of xylem sap expressed from the leaves (by over-pressure in the chamber) was approximately 0.03°, seldom more. Therefore, the osmotic pressure of the xylem sap (0.3–0.4 atm) was applied as a small correction to the calculation of the turgor pressure.

Results

Turgor pressures were measured directly in the secondary phloem of a mature red oak at 2 levels on the trunk. All penetrations at the upper level were made at 6.3 ± 0.3 meters where the diameter was 13 cm. At 6.6 meters the main trunk divided into 4 limbs of about equal size. All measurements at the lower level were made at 1.5 ± 0.3 meters where the diameter was 34 cm. The difference in hydrostatic pressure due to gravity was about 0.5 atm. The turgor pressures for several punctures at each level at different times and for a variety of weather conditions are shown in table I. The pressures obtained at each level and time are arranged in descending order and not in the order they were obtained. Successive penetrations during 1 measurement period were made by going around the tree in order to reduce the chance that a measurement at 1 site might be affected by the reduction in pressure caused by withdrawing the needle from the preceding penetration. Variations in the pressures did not appear to be attributable to this cause, since the

Table I. *Turgor and Osmotic Pressures in Secondary Phloem of Trunk of Red Oak at an Upper (6.3 ± 0.3 m) and Lower (1.5 ± 0.3 m) Level*

All pressures in atm.

Date, weather, T_{air}	Level	Time	Trunk pressures OP	TP
Oct. 25				
Overcast after 2 cm rain, 17–18°	Upper	1615	—24.3, 24.0, 23.7	19.5, 18.0, 17.6, 16.9, 16.7, 16.1
	Lower	1535	—23.2, 22.9, 22.0	17.4, 16.7, 14 5,
Oct. 26				
Bright sun, breeze 12–13°	Upper	1630	—24.1, 23.8, 23.1	17.2, 14.3, 12.9, 12.0, 11.4, 11.2
	Lower	1530	—22.1, 22.1, 21.5	17.6, 13.3, 12.5, 11.0 8.2, 7.4
clear sky, 7–7.5°	Upper	2250	—25.1, 23 9, 23.7	19.1, 18 1, 15.6, 15.3, 15.3, 12 2
	Lower	2200	—23.2, 22.6, 22.4	19.1, 19.0, 12.2, 11.8, 11.6, 9.5
Oct. 27				
Clear breeze, 13–14°	Upper	1510	—24.5, 24.1	18.6, 18.3, 17.9, 17.9, 14.4
	Lower	1415	—22.9, 22.0	16.7 16.7, 14.7, 12 7, 12.3, 9.5, 8.2
Oct. 28				
Hazy, sunny, breeze 11–14°	Upper	1245	—25.2, 25.2, 24.4	16.9, 16.0, 16.0, 16.0, 15.8, 15.8, 14.9, 14.6, 13.8, 12.0
	Lower	1105	—22.8	16.3, 16.0, 15.2, 15.2, 13.0, 12.0, 11.7, 10.3, 9.6
Oct. 31				
Hazy, sunny, 14–16°	Upper	1350	—24.6, 24.1	17.0, 14.7, 14.1, 14.0, 14.0 13.9, 12.4
	Lower	1255	—22.6, 22.3, 21.9	15.9, 15.8, 15.2, 14.9, 14.0, 13.2, 12 4
Nov. 2				
Cloudy, misty, 14–15°	Upper	1450	—23.6, 23.5, 23.2	19.9, 18.8, 17.1, 16.3, 13.9
	Lower	1400	—22.8, 22.4, 21.6	20.4, 18.4, 15.8, 15.4, 14.0, 13.0, 10.6
Nov. 4				
Cloudy, foggy, rain 15–16°	Upper	1420	—24.2, 23.6, 22.9	18.2, 17.6, 17.5, 17.3, 17.1, 16.7, 16.1, 16.1
	Lower	1245	—22.6, 22.1, 21.7	19.4, 18.8, 18.6, 16.9, 16.6, 16.1, 15.2, 11.9, 10.3
Nov. 10				
Sunny, hazy broken clouds 11–13°	Upper	1245	—23.5, 22.7, 22.4	15.6 15.0, 14.5, 14.1, 13.4, 13.2, 13.2, 12.7
	Lower	1200	—21.4, 21.3, 21.0	14.4, 14.4, 14.2, 13.0, 12.7, 12.0, 11.9, 11.6

highest pressure in a series could occur for a penetration no more than 2 cm lateral to the preceding one. However, penetration 2 cm above or below the preceding one would have an effect.

After a series of penetrations for turgor pressure at each level, penetrations were made to obtain triplicate samples of phloem sap for melting point determinations. These measurements are reported as osmotic pressure in table I, arranged in ascending order for each level and time. The volume of exudate for each penetration usually varied from 5 to 20 mm^3. Occasionally, nothing would exude, but seldom was the yield less than 3 mm^3. The volume did not appear to depend upon the position around the trunk. The yield of phloem exudate appeared to be more copious when the weather conditions favored low transpiration.

Values of the xylem sap pressure by the direct pressure chamber method and values of the average osmotic pressure and turgor pressure of leaf cells obtained by the semi-direct method of the pressure-volumn curve are shown in table II for twigs from 2 levels. The upper twig was taken from a branch on 1 of the 4 limbs at the top of the trunk. The distance from the twig to the trunk was 2 to 3 meters. The lower twig was taken from a branch on the lowest limb of the tree. The twig was 3 to 4 meters from where the limb joined the trunk at 2.3 meters from the ground.

Discussion

The osmotic pressure of the average leaf cell was greatest during conditions favoring transpiration and

Table II. *Average Intracellular Osmotic and Turgor Pressures and Hydrostatic Tension in Xylem in Leaves of Red Oak*

All pressures in atm.

Date	Level	Leaf pressures Time	OP	HP_x	TP
Oct. 25	Upper	1640	—26.1	— 4.0	21.7
	Lower	1640	—26.4	— 3.1	22.9
Oct. 26	Upper	1700	—28.1	— 8.6	19.1
	Lower	1700	—28.4	— 8.3	19.7
	Upper	2315	—25.6	— 1.8	23.4
	Lower	2315	—25.0	— 4.4	21.2
Oct. 27	Upper	1525	—28.3	—12.4	15.5
	Lower	1525	—29.6	—15.4	13.8
Oct. 28	Upper	1330	—29.5	—25.3	3.8
	Lower	1330	—31.2	—22.2	8.6
Oct. 31	Upper	1420	—29.7	—15 5	13.8
	Lower	1420	—30.9	—16.9	13.7
Nov. 2	Upper	1510	—21.5	— 1.0	20.1
	Lower	1510	—24.1	— 1.7	22.0
Nov. 4	Upper	1435	—25.6	— 1.6	23.6
	Lower	1435	—25.3	— 2.0	22.9
Nov. 10	Upper	1310	—25.8	— 9.1	16.3
	Lower	1310	—25.2	—10.2	14.6

photosynthesis. The osmotic pressure was nearly always greater in the leaf than in the upper level of the trunk and decreased further at the lower level of the trunk. However, from these measurements of osmotic pressures alone, no estimate of turgor pressure gradient in the phloem can be determined.

Values of turgor pressure in the secondary phloem of the trunk, as obtained by the direct method, varied over a wide range at each level and time. Is this variation around the tree real, or is it due to a defect in the technique? All errors that could occur in this technique would tend to reduce the true pressure. Among these errors are internal and external leakage of sap from around the needle where the phloem needle seals with the transected phloem sieve tubes. Supposedly, these errors were avoided by reporting only values for which steady state compression was observed. Another possible error would occur if the volume required to achieve compression were more than a few percent of the available volume of fluid at the site of puncture. The diameter and length of the capillary used were selected to eliminate this error. Thus, it would appear that the turgor pressure in the secondary phloem does vary, as measured, over a wide range at each level and time.

If sieve tubes can exist side-by-side with widely different internal pressures, this must reflect their specialization as conduits with low resistance for longitudinal translocation and high resistance for lateral movement of fluid. This feature has been used to explain the sharply delimited differences in the assimilate content of phloem sap between the foliate and the defoliate side of an ash tree (11). Also, tagged compounds introduced into a leaf of a sunflower plant during the period of seed formation supplied a definite sector of the flower head with assimilates (5). Since the turgor pressures in the seive tubes are so variable around the trunk of the oak tree, the existing turgor pressure gradients in this species becomes a will-o'-the wisp when measured as reported here. The most that can be noted in the turgor pressures of the trunk phloem in table I is that the first and subsequent pressures in the upper level tend to be higher than the pressures in the same order in the lower level under conditions expected to favor phloem movement toward the roots.

In a vertical tubular vessel through which fluid is flowing under pressure, the pressure difference between the bottom and top of the fluid column is $P_{bottom} - P_{top} = (L/10 - \rho\ L\ q)$ in atm, where L is the length in meters, ρ the resistance to flow per meter and q is the flowrate (downward when positive and upward when negative). When $q = 0$, the pressure at the bottom is greater by one atmosphere per 10 meters. When $q = 1/10\ \rho$, the pressures are equal at top and bottom and when downward flow exceeds this value the upper pressure must be greater than the lower pressure. A reverse flow, *i.e.* up the vessel, will occur only if the lower pressure exceeds the upper pressure by more than $L/10$ atm. These statements are valid for any pressure in the vessel since they pertain only to pressure differences; and they assume only that the fluid column is continuous, *i.e.* no segmental barriers across which the fluid must be transferred by some active process.

When weather conditions favored moderate to low rates of transpiration and the hydrostatic tension in the xylem was not too high, then the turgor pressure in the average leaf cell was high. If the turgor pressure in the leaf phloem was as high or higher, then a pressure gradient would occur in the phloem system which may favor the movement of phloem sap from the leaves toward the trunk and down the trunk. Since movement of phloem sap was not directly determined in this study, there is no certainty that the favorable gradients actually contributed to phloem sap movement. Also, since the resistance to flow of sap through sections of the phloem system was not determined nor was the magnitude of the flux determined in this study, it is not certain that the turgor pressure gradients in each section of the phloem system were adequate to account for a flux, even under the most favorable pressure gradients.

The turgor pressure gradient from average leaf cell to trunk phloem was most favorable to movement of phloem sap in that direction late in the evening or during a misty or rainy day. Then the turgor pressure difference from leaf to trunk was from 2 to 6 atm for the highest values obtained in the trunk and the difference was greater for the lower values. Nearly always the series of turgor pressures at the upper trunk level were slightly larger than the series at the lower level; that small difference plus approximately 0.5 atm from gravity would favor a downward flux of phloem sap. Only during a rain did a part of the series of turgor pressures at the upper level appear to be less than the lower series, suggesting conditions for low rates of translocation.

On a sunny and breezy day, when the hydrostatic tension in the leaf xylem sap was greatest, the turgor pressure in the average leaf cell was less than in the secondary phloem cells of the trunk. Although the direction of flow of the translocation fluid was not observed, it is unlikely that it was returning from the trunk to the leaves. The more probable interpretation of leaf pressures is that the turgor pressure in the sieve elements of the leaf phloem was greater than the turgor pressure of the average leaf cell. Presumably the companion cells of the sieve elements were secreting sucrose into the sieve tubes under these conditions, increasing their turgor pressure.

The red oak used in this study was in a sheltered site and was the last tree in the vicinity to show color changes in the leaves. The first trace of color change was noted on November 2 and by November 10 nearly one-half of an average leaf was no longer green. On 2 successive nights from November 15 to 17, the ambient temperature fell to $-7°$ and did not rise above 0° during the day. After November 17, the turgor pressure in the trunk phloem was not above 1 atm and no exudate could be obtained. The leaves of a scarlet oak were starting to turn color by October 20 and were fully brown by November 4. The phloem sap in the trunk remained under pressure after the leaves were fully brown but dropped to zero after the freeze of November 15 to 17.

The observations reported here are viewed only as a beginning of yet another approach to a classical problem in vascular physiology. If these methods have merit, it stems from their directness of measurement of turgor pressure, and they should not contradict results obtained by indirect methods. These results on the red oak obtained late in the season indicate that the turgor pressures in the secondary phloem sap and in the leaves are substantially greater than values obtained in the red maple during July and August by indirect methods (3). Also, the turgor pressure in the secondary phloem sap was a larger fraction of its osmotic pressure in the oak than it was in the red maple.

Indications of somewhat larger turgor pressures in the phloem of the trunk at the upper level do not contradict the mass flow theory for translocation of phloem sap first proposed by Munch (4). However, without additional measurements of flow rates and resistance to flow, the results in this report alone cannot indicate that the small turgor pressure gradients observed in the oak are adequate to move the phloem sap toward the trunk or down the trunk. On the other hand, the resistance to flow through the sieve tubes cannot have been great, for repeatedly the sap was observed to exude at a high rate (up to 10 mm^3 in less than a sec) when only a few sieve tubes were transected by the needle point (diam <0.4 mm). The outrush of sap through this needle sized opening in the phloem of oak also indicates that the phloem sap was stored under pressure in vessels with highly elastic walls, *i.e.,* low modulus of elasticity. In such a system, any removal (or addition) of assimilate—sucrose in oaks—from (or to) the sieve tubes by their companion cells, accompanied or followed by a removal (or addition) of water to (or from) the companion cells and nearby xylem sap would effect a much smaller pressure change than if the system were made from rigid tubes. Thus, hydrostatic pressure gradients may be small and difficult to demonstrate in an elastic, low resistance distribution system even with sizeable translocation fluxes from region to region. The difficulty will be compounded when attempting to measure the pressures at 2 levels by random sampling around the trunk, since the organization of the vascular phloem seems to provide a low resistance pathway along the axis of the trunk, or limb or stem, but only high resistance for lateral movement. There is little interchange between 2 parallel translocation streams, and it may be this feature of the distribution system that permits wide variation in the turgor pressure around the trunk at any one level.

Acknowledgments

I thank Mr. Kjell Myhre for technical assistance in this study. The idea to measure directly the phloem sap pressure was partially developed while the author was with the research vessel Alpha Helix on the Rio Negro in Brazil during September 1967. Therefore partial support from NSF Grant GB 4259 is acknowledged.

Literature Cited

1. HAMMEL, H. T. 1967. Freezing of xylem sap without cavitation. Plant Physiol. 42: 55–66.
2. HARTIG, TH. 1860. Beitrag zur physiologischen Forstbotanik. Allgem. Forstz. 36: 257–61.
3. KAUFMANN, M. R. AND P. J. KRAMER. 1967. Phloem water relations and translocation. Plant Physiol. 42: 191–94.
4. MÜNCH, E. 1930. Die Stoffbewegungen in der Pflanze. Gustav Fischer Verlagsbuchhandlung. Jena, Germany.
5. PROKOFYEV, A. A., L. P. ZHDANOVA, AND A. M. SOBOLEV. 1957. Certain regularities in flow of substances from leaves into reproductive organs. Soviet Plant Physiol. 4: 402–08. (English translation)
6. SCHOLANDER, P. F., H. T. HAMMEL, E. D. BRADSTREET, AND E. A. HEMMINGSEN. 1965. Sap pressure in vascular plants. Science 148: 339–46.
7. SCHOLANDER, P. F., L. VAN DAM, J. W. KANWISHER, H. T. HAMMEL, AND M. S. GORDON. 1957. Supercooling and osmoregulation in Arctic fish. J. Cell. Comp. Physiol. 49: 5–24.
8. ZIMMERMANN, M. H. 1957. Translocation of organic substances in trees. I. The nature of the sugars in the sieve tube exudate of trees. Plant Physiol. 32: 288–91.
9. ZIMMERMANN, M. H. 1957. Translocation of organic substances in trees. II. On the translocation mechanism in the phloem of white ash (*Fraxinus americana* L.). Plant Physiol. 32: 399–404.
10. ZIMMERMANN, M. H. 1960. Transport in the phloem. Ann. Rev. Plant Physiol. 11: 167–90.
11. ZIMMERMANN, M. H. 1961. Movement of organic substances in trees. Science 133: 73–79.

Reprinted from PLANT PHYSIOLOGY, Vol. 42, No. 1, January, 1967, pages 67–75.
(PRINTED IN U.S.A.)

Properties of Higher Plant Mitochondria. I. Isolation and Some Characteristics of Tightly-coupled Mitochondria from Dark-grown Mung Bean Hypocotyls[1]

Hiroshi Ikuma[2] and Walter D. Bonner, Jr.
The Johnson Research Foundation, University of Pennsylvania, Philadelphia, Pennsylvania 19104

Received September 6, 1966.

Summary. The mitochondria isolated from dark-grown mung bean hypocotyls oxidize succinate, L-malate, and externally added reduced nicotine adenine dinucleotide (NADH) with good respiratory control. While the pattern of respiration resembles that of animal mitochondria, there are 4 basic differences between the respiratory properties of mung bean and animal mitochondria: A) the ability to oxidize NADH, B) the pattern of succinate and malate oxidation, C) the rate of oxygen uptake, and D) the adenosine-5′-diphosphate to oxygen ratios.

The apparent 'Km' for malate of mung bean mitochondria is about one order higher than that expected from malic dehydrogenase in animal mitochondria, whereas the affinity for phosphate is about 5 times higher with plant mitochondria than rat-liver mitochondria. While the half-maximal stimulation of respiration by adenosine-5′-diphosphate is practically identical to that of animal mitochondria, higher concentrations of adenosine-5′-diphosphate cause some decrease in its stimulating action.

The respiratory control ratio and the ratio of adenosine-5′-diphosphate to oxygen, measured from polarographic traces, have been employed as a biochemical measure to describe the degree of coupling, respiratory control, and intactness of mitochondria [cf. Chance and Williams (10, 14)]. Mitochondrial preparations which show a high degree of both coupling and respiratory control have been isolated from various plant sources by Bonner et al. (4–6, 40), Ohnishi and Hagihara (24–26), Hanson et al. (17), Wiskich et al. (41), Verleur and Uritani (35–37), and by Lance et al. (19). The purpose of this publication is to describe the method of isolation and characterize some respiratory properties of these mitochondria and to compare them with those of animal mitochondria.

In the present study, mung bean hypocotyls (*Phaseolus aureus*, var. Jumbo) were selected as the source of mitochondria (cf. 2, 4, 16, 22, 23), due to ease in obtaining uniform etiolated plant material in a short period of time. Succinate and malate were chosen as primary substrates; the former to represent the succinoxidase system and the latter, the NAD-linked oxidase system. NADH was also included as a substrate since unusual characteristics of plant mitochondria were displayed upon oxidizing this compound with respiratory control.

In the publication to follow, effects of inhibitors and uncouplers on plant mitochondrial respiration will be discussed.

Methods

Mitochondria were isolated from dark-grown mung bean hypocotyls from 4 to 5 days old, in the late exponential to early stationary phase of hypocotyl growth. The seedlings were germinated and grown on perforated aluminum holders suspended in plastic trays containing tap water. The growth chambers were maintained at 28°, and all white light was rigidly excluded.

After separation from roots and leaves, the mung bean hypocotyls were cut into sections 2 to 3 cm long. The cut sections were then ground for 30 seconds in a mortar and pestle with 100 to 200 ml of grinding medium maintaining a ratio of medium to tissue of 2.

During these steps, the pH of the homogenates was maintained at 7.2 by dropwise additions of 6N KOH. The grinding medium contained 0.3 M mannitol, 0.1 % BSA, 0.05 % cysteine, and 1 mM EDTA. All operations were carried out between 0° and +4°.

[1] This work was supported by a grant from the National Science Foundation.

[2] Present Address: Department of Botany, University of Michigan, Ann Arbor, Michigan.

After filtering through muslin, the homogenate was centrifuged for 15 minutes at 1000 × g; the resulting supernatant fraction was then centrifuged for 15 minutes at 10,000 × g. The mitochondrial pellet was washed in a medium consisting of 0.3 M mannitol, 0.1 % BSA, and 0.1 mM EDTA. After a 10-minute centrifuging at 250 × g, the mitochondria were sedimented at 6000 × g for 15 minutes. The supernatant fluid is most conveniently removed from the sedimented mitochondria by aspiration; it is very important to remove all of the fluffy material layered on top of the mitochondria and to gently separate the mitochondrial pellet from the starch which is immediately below it. About 2.5 kg etiolated, 5-day-old hypocotyls will produce approximately 1.5 ml of high-quality mitochondria containing around 150 mg mitochondrial protein.

As for comparison with the data obtained from mung bean hypocotyl mitochondria, rat liver mitochondria were also prepared with a conventional 0.25 M sucrose method (cf. 10, 34).

Oxygen uptake was measured polarographically with a conventional Clark oxygen electrode inserted in a 3 ml lucite chamber. Materials were introduced into the chamber through a movable well with a small opening, and the reaction mixture was stirred by means of a small magnetic stirrer. Respiratory rates were calculated from a recorder trace on the basis of 240 μM O_2 in aerated medium [cf. Chance and Williams (7, 10, 11), Chappell (15)]. Substrates were added to the reaction medium immediately before the measurement and the initial rate was recorded. All measurements were carried out at room temperature (24° ± 1°). All rates of oxygen consumption are expressed as μM O_2/minute/3 ml of reaction medium.

The reaction medium contained 0.3 M mannitol, 0.01 M KCl, 0.01 M potassium phosphate buffer (pH 7.2) and 5 mM $MgCl_2$. In some cases, 0.01 M Tris –HCl buffer (pH 7.2) was added. This buffer, at the concentration used, did not affect the pattern of mitochondrial respiration. A semi-micro Kjeldahl acid digestion method [Umbreit, et al. (34)] was used to calculate the total nitrogen content of mitochondria, and the protein content was estimated on the basis of 16 % nitrogen per protein.

Organic chemicals used in these studies were mannitol, L-malate, sodium succinate, ADP and NADH (Boehringer). All inorganic chemicals were of analytical reagent grade, and solutions were prepared in double-distilled water. The pH of all the substrates and cofactors was adjusted to the vicinity of 7, and the chemicals to be titrated were prepared just before use. The concentrations of NADH and ADP were determined optically at 340 mμ and 260 mμ on the basis of millimolar extinction coefficients of 6.22 and 15.4 respectively. The concentrations of other chemicals were measured gravimetrically.

Results

Critique of the Method of Preparation. The mitochondria are first sedimented at 10,000 × g and, after washing, resedimented at 6000 × g. A relative centrifugal force of 6000 × g was chosen following a series of experiments designed to test mitochondrial quality as a function of sedimentary force. Inspection of table I, which shows the re-

Table I. *Relationship Between Centrifugal Force and Mitochondrial Quality*

The mitochondrial fraction was first collected between 1000 and 10,000 × g, then suspended in the wash medium. Pellets are collected by centrifugation as shown in the first column and the respiration was measured polarographically with 8.3 mM succinate and 150 μM ADP. Second states 3 and 4 rates and second state 3 to 4 transitions are recorded.

Centrifugal force* (× g)	Q_{O_2} (mμmoles O_2/min/mg N) State 3	State 4	ADP : O ratio
0– 250	365	140	1.3
250– 1000	875	235	1.3
250– 3500	935	240	1.4
250– 6000	1270	380	1.3
250– 8000	940	260	1.3
250–10,000	780	280	1.3
1000–10,000	860	360	1.3
3500–10,000	730	240	1.3
6000–10,000	495	155	1.3
8000–10,000	270	95	1.3

* 250 × g spin was 10 minutes and the rest 15 minutes.

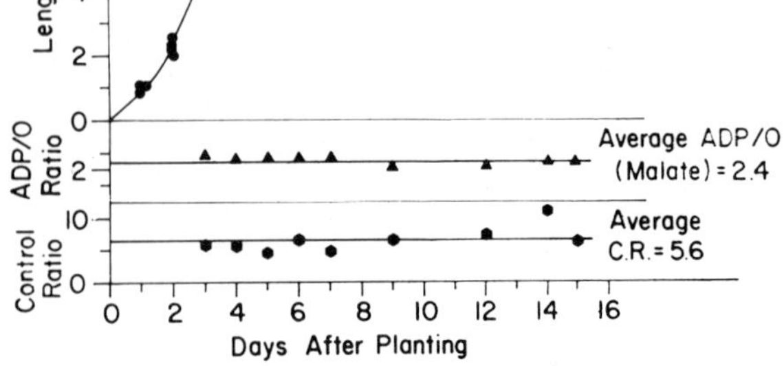

FIG. 1. Activities of isolated mung bean hypocotyl mitochondria plotted as a function of the age of seedlings used for the preparation. The ratio of oxygen uptake rates between state 3 and state 4 is the control ratio (CR).

sults of one such experiment, indicates that in terms of total mitochondria sedimented and of respiratory control, 6000 × *g* is the optimal force. The results further point out the insertion of a low speed spin (250 × *g* for 10 min) is highly important to improve the mitochondrial quality.

In investigations of castor bean mitochondria (1), it was shown that the activity of the isolated mitochondria varied with the age of the plants. For this reason, the activity of mung bean hypocotyl mitochondria was investigated as a function of the age of the seedlings used (fig 1). There are negligible differences between mitochondrial activities, ADP:O ratios, and respiratory control ratios in mitochondria isolated from tissues 3 through 15 days old. No attempts were made to follow activities past 15-day-old seedlings. Since the age of the seedlings seemed unimportant, 5-day-old seedlings were used because of desirable planting schedules and ease of handling.

The properties of tight coupling and respiratory control lasted for some hours after the mitochondria were isolated. Figure 2 shows the ADP:O ratios and respiratory rates exhibited by mung bean hypocotyl mitochondria over a 9-hour period. During this time, the mitochondria were kept in ice and samples were assayed at the times indicated. After

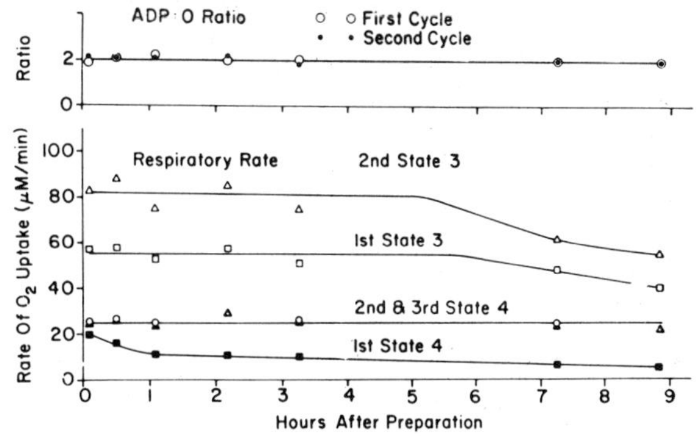

FIG. 2. ADP:O ratios (above) and respiratory rates (below) of isolated mung bean hypocotyl mitochondria as measured over a 9-hour period. Oxygen uptake is expressed as μM/min/3 ml.

5 hours, the state 3 rates and succinate oxidation decline, but the state 4 rates, following the first state 4, remain the same. The ADP:O ratios do not change over the 9-hour period. Similar results were attained with malate as substrate.

Nature of Coupling and Control. Mung bean hypocotyl mitochondria oxidize succinate, malate, and NADH rapidly with good respiratory control. Figure 3 illustrates this in polarographic recorder

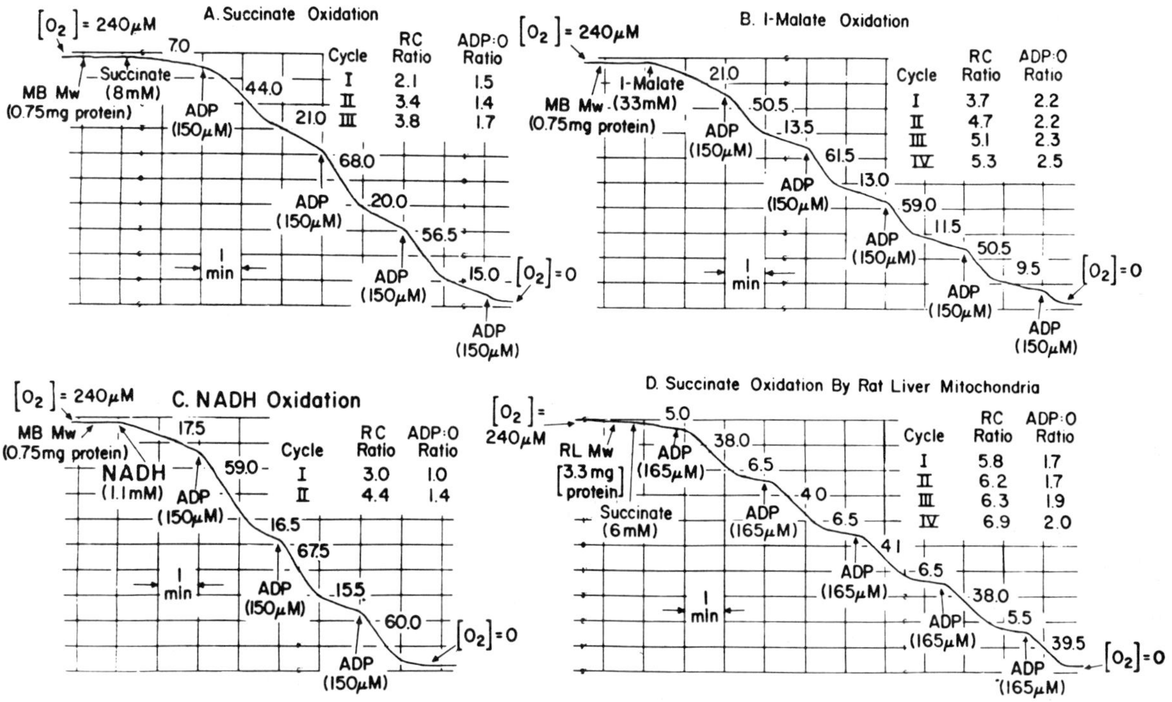

FIG. 3. Pattern of O_2 uptake by mung bean mitochondria (MB Mw) and rat-liver mitochondria (RL Mw). Numbers along the traces are the rates of O_2 uptake (read as downward deflection) in μM/min/3 ml. Respiratory control (RC) ratio and ADP:O ratio calculated as in Chance and Williams (10, 14). A) succinate oxidation by mung bean mitochondria; B) malate oxidation by mung bean mitochondria; C) NADH oxidation by mung bean mitochondria; and D) succinate oxidation by rat-liver mitochondria. Rat-liver mitochondria, isolated with 0.25 M sucrose and tested with the reaction medium for mung bean mitochondria.

traces in which the pattern of succinate oxidation by rat-liver mitochondria is included for comparison. Oxygen uptake by tightly-coupled mung bean mitochondria is clearly stimulated by the addition of ADP and slowed down upon exhaustion of added ADP. Following the terminology of Chance and Williams (10, 14), the ADP-stimulated state is 3, and the ADP-limited state, 4. Immediately after substrate addition, which produces the first state 4, plant mitochondria take up oxygen at a rate different from the rates of subsequent state 4's. With succinate as substrate (fig 3A), the first state 4 is frequently slower than the subsequent state 4's (cf. 40), and upon addition of ADP, the respiration turns slowly into the first state 3. Consequently, the first state 3 rate is always slower than the subsequent state 3 rates. With malate as substrate (fig 3B), however, the first state 4 rate is always faster than subsequent ones. Nevertheless, like succinate oxidation, the first state 3 rate in malate oxidation is always lower than the subsequent state 3 rates. These peculiarities disappear from the second cycle of state 3 to 4 transition. Unlike the oxidation of succinate and malate, NADH oxidation (fig 3C) shows that all state 3 rates are practically identical, as are state 4 rates. It should be pointed out that NADH applied externally is not utilized by tightly-coupled animal mitochondria (12, 14). Under the identical assay procedure, rat-liver mitochondria (fig 3D) do not show the peculiarities discussed above. Some animal mitochondria, however, reveal a characteristic similar to plant mitochondria upon aging (9). These observations suggest that mung bean mitochondria are coupled, but their respiratory properties are somewhat different from those of animal mitochondria (4, 40).

Effect of Mitochondrial Concentration on Respiratory Rates and Control. When the effect of various mitochondrial concentrations on respiratory rates and respiratory control was examined with succinate, malate, or NADH as substrate, the expected linearity was observed in most plots, but the first state 3 rate in succinate oxidation and the first and second state 3 rates of malate oxidation deviated from linearity at higher mitochondrial

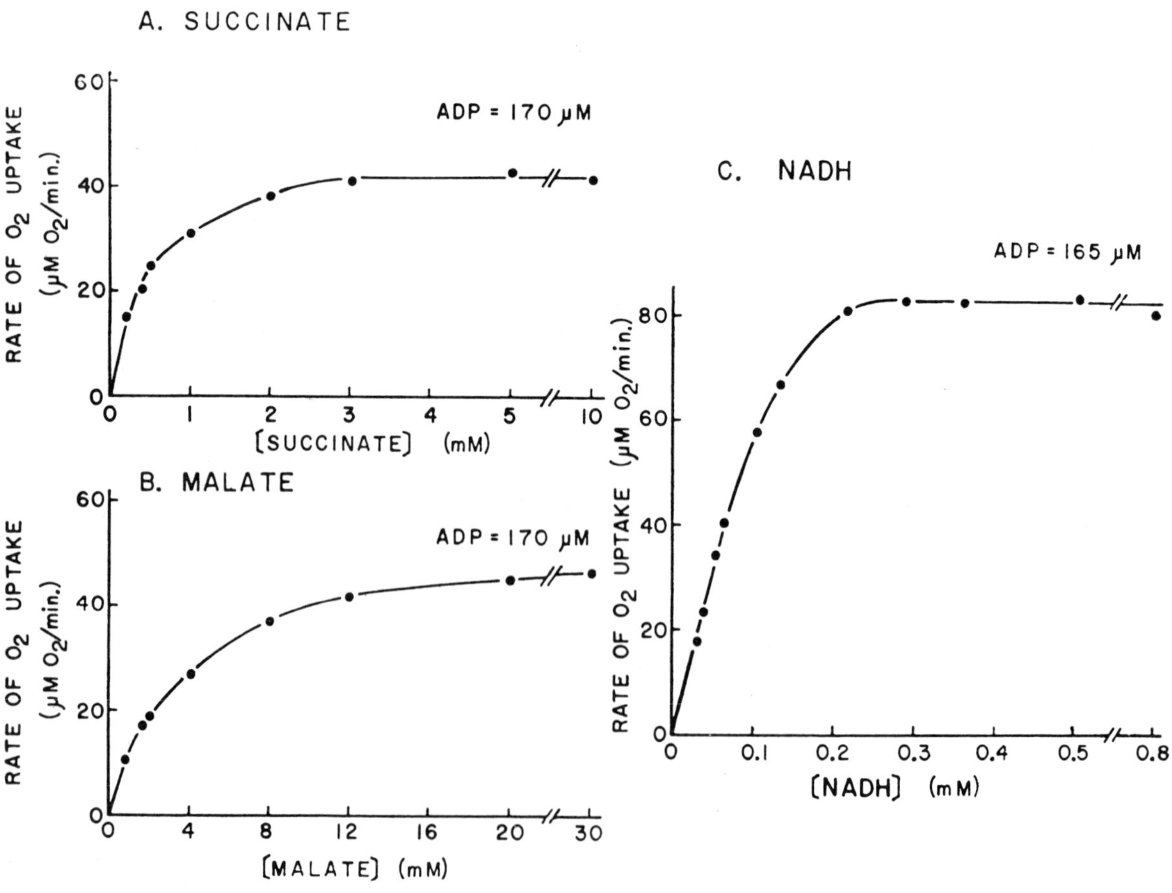

FIG. 4. Affinity for substrate with succinate (A), malate (B), or NADH (C), as substrate. Mitochondrial preparation, different for each measurement. Rate of oxygen uptake is expressed as μM/min/3 ml.

concentrations (> 0.1 mg N). During these studies, it was noticed that average ADP:O ratios were about 75 % of the expected, theoretical values; i.e., 1.6 ± 0.2 with succinate as substrate and 2.3 ± 0.2 with malate. The ADP:O ratio in NADH oxidation increased as the number of state 3 to 4 transitions increases; e.g., the first transition gave 1.2 and the second, 1.4.

Effect of Substrate Concentration on State 3 Rate. The effect of various concentrations of the same 3 substrates on state 3 rate was studied. Mitochondria, ADP, and substrate were successively added to the reaction cuvette, and the initial rate of oxygen uptake after substrate addition was measured. The difference in rate before and after the substrate addition is plotted in figure 4. The apparent Michaelis constants ('Km') for succinate, malate, and NADH are 0.4, 3.0, and 0.07 mM, respectively.

It should be pointed out that the 'Km' for succinate of these mitochondria agrees closely with the data in the literature for Keilin-Hartree heart muscle preparations and for purified succinic dehydrogenase (7, 18, 29, 30, 32). The 'Km' for malate of these mitochondria is in close agreement with NAD-linked malic dehydrogenase from beet root, tobacco leaves, or wheat germ (27, 31, 39). Apparent Km's for NADH, NADH oxidase, or NADH-cytochrome *c* reductase of electron transport particles from animal tissues range from 10 to 30 μM (20, 21, 38).

Effect of Phosphate Concentrations on State 3 Rates. Figure 5 shows the effect of various phosphate concentrations on state 3 rate of succinate, malate, and NADH oxidation. In these experiments, mitochondria, a substrate, and ADP were successively added to the phosphate-free reaction medium (0.01 Tris-HCl buffer substituted for phosphate buffer), and 2 minutes after the ADP addition a given concentration of potassium phosphate buffer (pH 7.2) was added. Some oxygen uptake was noted before phosphate addition; an average of 16 μM O_2/minute with succinate or malate as substrate and 9 μM O_2/minute with NADH. These values were, however, subtracted from the rates measured after phosphate addition in figure 5.

The apparent Michaelis constants are 0.2 mM with all of the substrates used. These values compare fairly well with that obtained by Chance and Hagihara (9) for both succinate and endogenous NADH oxidation by pigeon-heart mitochondria. The 'Km' for phosphate with rat-liver mitochondria was reported as 1 mM (8).

Effect of ADP Concentrations on State 3 Rates. Figure 6 summarizes the results of ADP titration. The same 3 substrates were used in these ex-

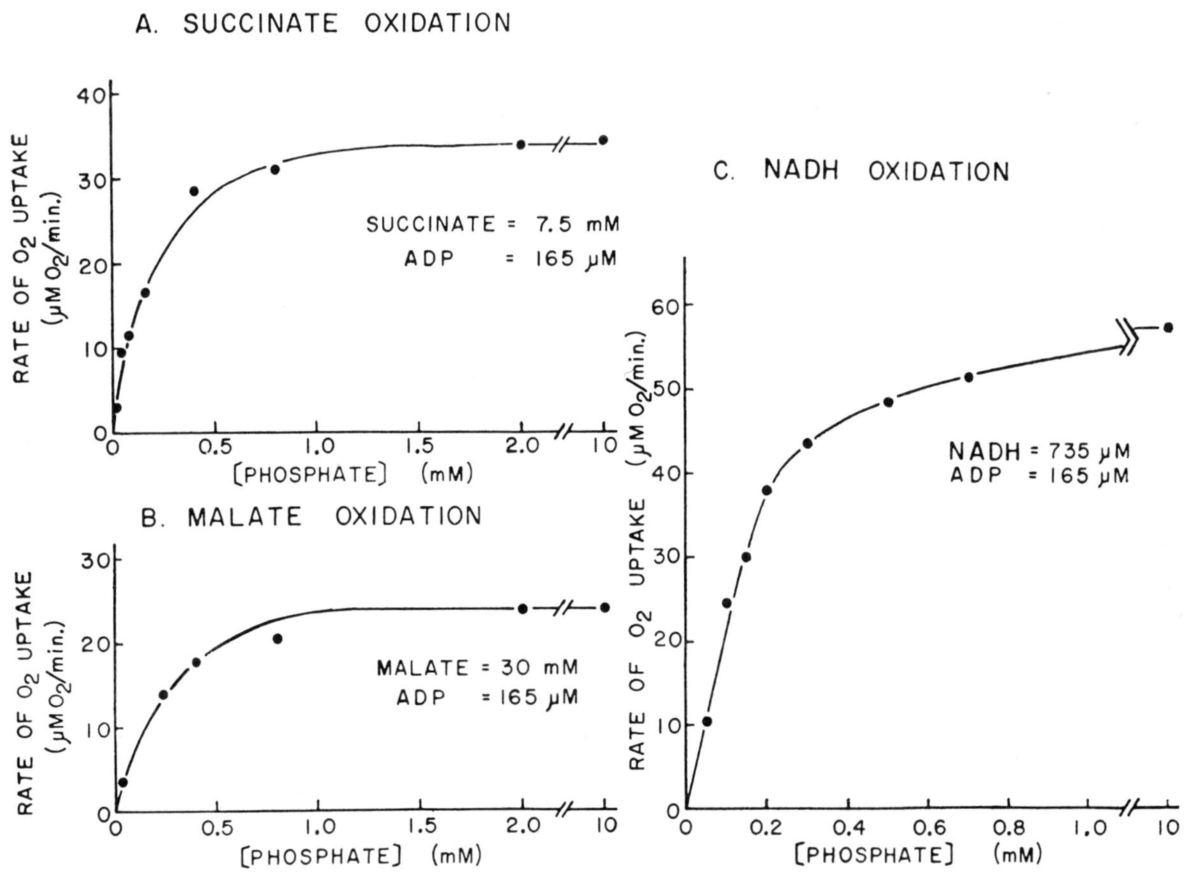

FIG. 5. Affinity for phosphate with succinate (A), malate (B), or NADH (C), as substrate. Rate of oxygen uptake is expressed as μM/min/3 ml.

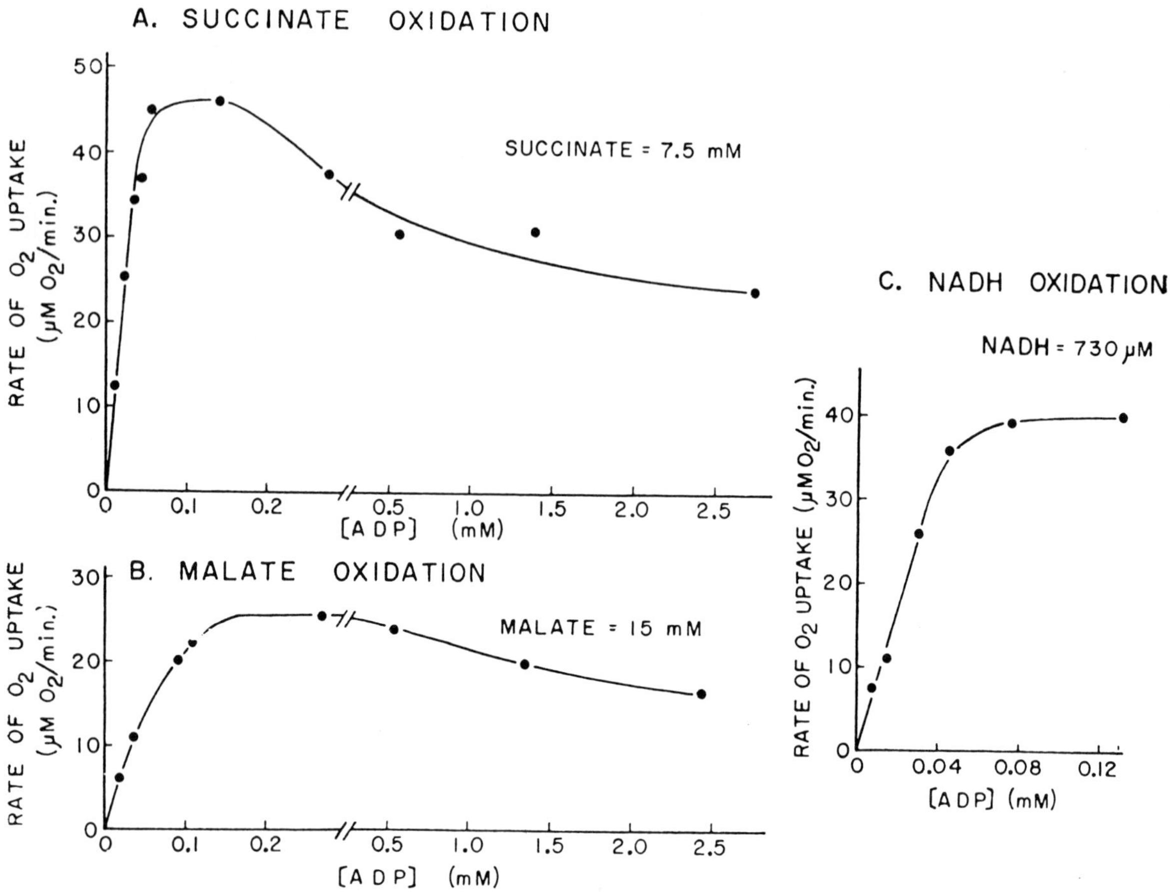

FIG. 6. Affinity for ADP with succinate (A), malate (B), or NADH (C), as substrate. Rate of oxygen uptake is expressed as μM/min/3 ml.

periments, and the sequence and timing of the additions of mitochondria, substrate, and ADP to the reaction cuvette is the same as shown in figure 3. The rates of oxygen uptake detected prior to ADP addition were 24, 16, and 37 μM O_2/minute for the oxidation of succinate, malate, and NADH, respectively. These values were subtracted from the rate after ADP addition and plotted in this figure.

As the ADP concentration increases, the rate of oxygen uptake by mitochondria increases sharply at first, reaches a maximum value at 100 to 200 μM, then decreases gradually at higher concentrations (figs 6A, 6B). The ADP concentrations which cause half-maximal rates are 20 μM with succinate or NADH as substrate and 40 μM with malate. The 20 to 40 μM ADP concentration for half-maximal stimulation of oxygen uptake agree closely with similar data, 20 to 30 μM, obtained with animal mitochondria (8, 9, 11, 13). The pattern of response of plant mitochondria at concentrations higher than 200 μM is unexpected from the similar data from animal mitochondria. Throughout the present study, an ADP concentration of 100 to 200 μM has been selected so that the maximal state 3 rate is secured.

Discussion

Comparison of Oxidation Rates and Phosphorylative Ability Between Plant and Animal Mitochondria. The data presented in figure 3 is tabulated in table II in comparison to similar measurements on rat-liver mitochondria. Within a given state 3 of the respiration by mung bean mitochondria, the rates with different substrates agree closely with one another. The first state 3 rates are slower than the second. The state 4 rates of the plant mitochondria, on the other hand, differ from one another depending on which substrate is used; the first state 4 is, furthermore, markedly different, except with NADH as substrate, from the subsequent state 4's with succinate or malate as substrate.

Rat-liver mitochondria present a different picture. The differences between mung bean and rat-liver mitochondria can be summarized as follows: (1) rat-liver mitochondria are unable to oxidize externally-added NADH, while mung bean mitochondria can with good respiratory control; (2) mung bean mitochondria are capable of oxidizing malate as fast as succinate, while rat-liver mitochondria oxidize malate very poorly in the absence of pyruvate of glutamate; (3) even with

Table II. *Respiratory Rates and ADP:O Ratios by Mung Bean and Rat-liver Mitochondria*

A) Mung bean mitochondria; rates and ratios calculated from the data in figures 4 and 5. B) Rat-liver mitochondria; isolated with 0.25 M sucrose and tested with the same procedure and reaction medium as in figure 4.

Substrate	Rate of O_2 Uptake (mμmoles O_2/min/mg N) in respiratory state* IV_1	III_1	IV_2	III_2	IV_3	ADP:O ratio III_1–IV_2	III_2–IV_3
A. Mung bean mitochondria							
Succinate (8 mM)	213	1000–1250	475	1530	475	1.6	1.6
Malate (30 mM)	565	1140–1380	300	1400–1560	300	2.3	2.3
NADH (0.7 mM)	390	1440	390	1560	390	1.2	1.4
B. Rat-liver mitochondria							
Succinate (3.3 mM)	48	300	56	310	56	1.9	1.9
Malate (3.3 mM)	17	39	21	53	27	2.5	2.7
β-Hydroxybutrate (3.3 mM)	23	114	35	120	35	3.0	2.9

* Roman numerals (III and IV) denote respiratory state 3 and 4; subscripts (1,2, and 3) indicate the first, second, and third. The ratios of III_1 to IV_2 and III_2 to IV_3 give the respiratory control ratios of the first and the second cycles.

succinate as substrate, the Qo_2 (N) rat-liver mitochondria in state 3 is, in general, 4 times slower than that by mung bean mitchondria; (4) there is a distinct difference in rate between the first and second state 4 to 3 transitions in mung bean mitochondria, whereas rat-liver mitochondria show ADP:O ratios very close to theoretical values, while mung bean mitochondria give generally 75 % of the theoretical values.

Tightly-coupled yeast mitochondria, isolated by Ohnishi and Hagihara (24–26), reveal many characteristics similar to mung bean mitochondria. Both types of mitochondria oxidize, with good respiratory control and rates, representative acids of Krebs' cycle as well as externally-added NADH. They differ, however, in that the yeast mitochondria oxidize D- and L-lactate. Another difference is found in the degree of phosphorylation; i.e., the ADP:O ratio in the oxidation of NAD-linked substrates by yeast mitochondria is virtually the same as that in the oxidation of succinate or of externally-added NADH. Spectroscopic characteristics of yeast mitochondria are essentially the same as that of animal mitochondria, while mung bean mitochondria differ markedly from animal mitochondria, especially in the composition of cytochromes *b* and *c* and in the stoichiometric relationship among respiratory carriers (3, 14).

The above comparison among mung bean, yeast, and rat-liver mitochondria points out at least the following 4 possibilities: (1) plant (mung bean and yeast) mitochondria contain the electron transport system in higher concentrations than rat-liver mitochondria, since the Qo_2 values are considerably higher with the former than with the latter; (2) plant mitochondria are less tightly coupled to phosphorylation because of the lower ADP:O ratios; (3) the plant mitochondrial membrane is more permeable to pyridine nucleotides than the animal mitochondrial membrane; and (4) the respiratory properties of yeast mitochondria appear in several respects to be intermediate between animal and plant mitochondria.

Affinities of Substrates and Cofactors of Oxidative Phosphorylation. Table III summarizes the results with mung bean mitochondria (figs 4–6) as well as the data obtained with rat-liver and pigeon-heart mitochondria. The apparent Michaelis con-

Table III. *Affinities of Substrate and Cofactors of Oxidative Phosphorylation by Mung Bean Mitochondria, as Compared with Similar Data Obtained from Animal Mitochondria*

Measurements were made either for oxygen uptake (oxygen electrode) or for NADH oxidation (photometric method).

Substrate	Apparent Michaelis constant for: Substrate (mM)	Phosphate (mM)	ADP (μM)	Reference and remarks
A. Mung bean mitochondria				
Succinate	0.4	0.2	20	figs 6–8
Malate	3.0	0.2	40	figs 6–8
NADH	0.07	0.2	20	figs 6–8
B. Rat-liver mitochondria				
Succinate	0.4	1.2	50	Our own measurements
β-hydroxybutyrate	...	1.0	56	Chance and Hagihara (9)
Glutamate + succinate	...	0.16	...	Chance and Connelly (8)
C. Pigeon heart mitochondria				
Glutamate + succinate	...	0.13	26	Chance and Hagihara (9)

stants for substrates, in general, agree with the values obtained with isolated dehydrogenases, as mentioned earlier. It is, however, of interest to note that malic dehydrogenase, NAD specific, isolated from beet root mitochondria, revealed the 'Km' value about 10 times higher than the malic dehydrogenase in animal mitochondria (39). This characteristic of the plant enzyme is reflected in the 'Km' value for malate with mung bean mitochondria.

The affinity of phosphate to rat-liver mitochondrial respiration is about 1 mM with succinate or β-hydroxybutyrate as substrate, but the affinity increases in the oxidation of succinate in the presence of glutamate (table III). This higher affinity corresponds to the affinity of phosphate to mung bean mitochondrial respiration.

In view of the apparent toxicity of high ADP concentrations in plant mitochondrial respiration (fig 6), we have investigated the effect of ADP on succinate oxidation by rat-liver mitochondria (table III). No toxic effect was detected with these mitochondria and the half-maximum stimulation was 50 μM.

Acknowledgment

The authors are indebted to Mr. Stephen V. Sikes for his technical help which included the isolation of mitochondria.

Literature Cited

1. BEEVERS, H. AND D. A. WALKER. 1956. The oxidative activity of particulate fractions from germinating castor beans. Biochem. J. 62: 114–20.
2. BONNER, J. AND A. MILLERD. 1953. Oxidative phosphorylation by plant mitochondria. Arch. Biochem. Biophys. 42: 135–48.
3. BONNER, W. D., JR. 1964. Plant cytochromes. Sixth International Congress. Abstracts IV, 291–92.
4. BONNER, W. D., JR. A general method for the preparation of plant mitochondria. In: Oxidation and Phosphorylation, R. W. Estabrook and Maynard E. Pullman, eds. Academic Press, Inc., New York, in press.
5. BONNER, W. D., JR. AND T. ITO. 1960. Properties of cauliflower mitochondria. Fed. Proc. 19: 35.
6. BONNER, W. D., JR. AND D. O. VOSS. 1961. Some characteristics of mitochondria extracted from higher plants. Nature 191: 682–84.
7. CHANCE, B. 1952. The kinetics and inhibition of cytochrome components of the succinic dehydrogenase system. I. Activity determinations and purity criteria. J. Biol. Chem. 197: 557–76.
8. CHANCE, B. AND C. M. CONNELLY. 1957. A method for the estimation of the increase in concentration of adenosine diphosphate in muscle saracosomes following a contraction. Nature 179: 1235–37.
9. CHANCE, B. AND B. HAGIHARA. 1963. Direct spectroscopic measurements of interaction of components of the respiratory chain with ATP, ADP, phosphate and uncoupling agents. In: Proceedings of the Fifth International Congress of Biochemistry. Pergamon Press, New York. Vol. 1: p 3.
10. CHANCE, R. AND G. R. WILLIAMS. 1955. A simple and rapid assay on oxidative phosphorylation. Nature 175: 1120.
11. CHANCE, B. AND G. R. WILLIAMS. 1955. Respiratory enzymes in oxidative phosphorylation. I. Kinetics of oxygen utilization. J. Biol. Chem. 217: 383–93.
12. CHANCE, B. AND G. R. WILLIAMS. 1955. Respiratory enzymes in oxidative phosphorylation. II. Difference spectra. J. Biol. Chem. 217: 395–407.
13. CHANCE, B. AND G. R. WILLIAMS. 1955. Respiratory enzymes in oxidative phosphorylation. III. The steady state. J. Biol. Chem. 217: 409–27.
14. CHANCE, B. AND G. R. WILLIAMS. 1956. The respiratory chain and oxidative phosphorylation. Advan. Enzymol. 17: 65–134.
15. CHAPPELL, J. B. 1964. The oxidation of citrate, isocitrate and cis-aconitate by isolated mitochondria. Biochem. J. 90: 225–37.
16. HACKETT, D. P. 1963. Respiratory mechanisms and control in Higher Plant Tissues. Control Mechanisms in Respiration and Fermentation, B. Wright, ed. Ronald Press, New York. p 105–27.
17. HANSON, J. B., S. S. HALHORTA, AND C. D. STONER. 1965. Action of calcium on corn mitochondria. Plant Physiol. 40: 1033–40.
18. HIATT, A. 1961. Preparation and some properties of soluble succinic dehydrogenase from higher plants. Plant Physiol. 36: 552–57.
19. LANCE, C., G. E. HOBSON, R. E. YOUNG, AND J. B. BIALE. 1965. Metabolic Processes in cytoplasmic particles of the avocado fruit. VII. Oxidation and phosphorylation activities throughout the climacteric cycle. Plant Physiol. 40: 1116–23.
20. MACKLER, B. AND D. E. GREEN. 1956. Studies on the electron transport system. III. The properties of two modified forms of ETP. Biochem. Biophys. Acta 21: 6–13.
21. MACKLER, B., R. RAPASKE, P. M. KOHOUT, AND D. E. GREEN. 1954. The properties of DPNH oxidase. Biochem. Biophys. Acta 15: 437–38.
22. MILLERD, ADELE. 1953. Respiratory oxidation of pyruvate by plant mitochondria. Arch. Biochem. Biophys. 42: 149–63.
23. MILLERD, A., J. BONNER, B. AXELROD, AND R. BUNDURSKI. 1951. Oxidative and phosphorylative activity of plant mitochondria. Proc. Natl. Acad. Sci. U. S. 37: 855–62.
24. OHNISHI, T. AND B. HAGIHARA. 1964. Preparation of yeast mitochondria. J. Biochem. Tokyo 55: 584–85.
25. OHNISHI, T. AND B. HAGIHARA. 1964. Preparation of yeast mitochondria by an enzymatic procedure. J. Biochem., Tokyo 56: 484–86.
26. OHNISHI, T., K. KAWAGUCHI, AND B. HAGIHARA. 1966. J. Biol. Chem. In press.
27. PIERPOINT, W. S. 1963. The distribution of succinate dehydrogenase and malate dehydrogenase among components of tobacco leaf extracts. Biochem. J. 88: 120–25.
28. SIEGAL, L. AND S. ENGLARD. 1961. Beef-heart malic dehydrogenases. I. Properties of the enzyme purified from extracts of acetone-dried powders. Biochem. Biophys. Acta 54: 67–76.

29. SINGER, T. P., E. B. KEARNEY, AND V. MASSEY. 1957. Newer Knowledge of Succinic Dehydrogenase. Advan. Enzymol. 18: 65–111.
30. SLATER, E. C. AND W. D. BONNER, JR. 1952. The effect of fluoride on the succinic oxidase system. Biochem. J. 52: 185–95.
31. STAFFORD, H. A. 1956. Dehydrogenase activity of hydroxymalonate and related acids in higher plants. Plant Physiol. 31: 135–41.
32. THORN, M. B. 1953. Inhibition by malonate of succinic dehydrogenase in heart-muscle preparations. Biochem. J. 54: 540–47.
33. THORNE, C. J. R. 1962. Properties of mitochondrial malate dehydrogenases. Biochem. Biophys. Acta 59: 624–33.
34. UMBREIT, W. W., R. H. BURRIS, AND J. F. STAUFFER. 1957. Manometric Technique, Burgess Publishing Company, Minneapolis, Minnesota.
35. VERLEUR, J. D. 1965. Studies on the isolation of mitochondria from potato tuber tissues. Plant Physiol. 40: 1003–07.
36. VERLEUR, J. D. AND I. URITANI. 1965. Mitochondrial activity in healthy white potato tubers and in potato infected with *Ceratocystis fimbriata*. Agri. Biol. Chem. 29: 369–71.
37. VERLEUR, J. D. AND I. URITANI. 1965. Respiratory activity of the mitochondrial fractions isolated from healthy potato tubers and tuber tissue incubated after cutting or infection with *Ceratocystis fimbriata*. Plant Physiol. 40: 1008–12.
38. VERNON, L. P., H. R. MAHLER, AND N. K. SARKAR. 1952. Studies on diphosphopyridine nucleotide cytochrome *c* reduction. III. Kinetic studies. J. Biol. Chem. 199: 599–606.
39. WEDDING, R. T. AND K. M. BLACK. 1963. Kinetics of malic dehydrogenase inhibition by 2,4-Dichlorophenoxyacetic acid. Plant Physiol. 38: 157–64.
40. WISKICH, J. T. AND W. D. BONNER, JR. 1963. Preparation and properties of sweet potato mitochondria. Plant Physiol. 38: 594–604.
41. WISKICH, J. T., R. E. YOUNG, AND J. B. BIALE. 1964. Metabolic processes in cytoplasmic particles of the avocado fruit. VI. Controlled oxidation and coupled phosphorylations. Plant Physiol. 39: 312–22.

ASSOCIATION OF THE GLYOXYLATE CYCLE ENZYMES IN A NOVEL SUBCELLULAR PARTICLE FROM CASTOR BEAN ENDOSPERM

R. W. Breidenbach and Harry Beevers

Department of Biological Sciences
Purdue University
Lafayette, Indiana 47907

Received April 13, 1967

The reactions of the glyoxylate cycle play an essential role in the massive conversion of storage lipid to sucrose during the germination of fatty seeds such as castor bean (Kornberg and Beevers, 1957; Canvin and Beevers, 1961). Previous studies of the intracellular distribution of malate synthetase and isocitrate lyase in plant tissues carrying out this conversion have demonstrated that significant proportions of these enzymes sediment with the mitochondria (Marcus and Velasco, 1960; Tanner and Beevers, 1965). Since, in addition, the enzymes catalyzing reactions common both to the TCA cycle and the glyoxylate cycle are believed to be mitochondrial, the view has grown that *in vivo* the glyoxylate cycle functions in these organelles.

Treatment of the crude mitochondrial fraction on sucrose density gradients has now revealed that enzymes of the glyoxylate cycle are not present in the purified mitochondria; they are specifically associated with a distinct particulate component of higher density. This component will be referred to as glyoxysomes.

MATERIALS AND METHODS

Materials:

Castor bean (*Ricinus communis* L.) variety Baker 296 were a kind gift from Castor Oil Company, Plainview, Texas. The seeds were soaked for approximately twelve hours and then germinated for five days in moist vermiculite at 30°C. The testae, roots, hypocotyl and cotyledons were removed and the endosperm tissue used as starting material.

Methods:

Endosperm tissue (25 g) was homogenized in 100 ml of 0.4 M sucrose, 0.05 M Tris buffer pH 7.5, 0.01 M dithiothreotol, 0.1% bovine serum albumin, 0.01 M KCl, 1×10^{-3} M EDTA, 1×10^{-4} M $MgCl_2$. The tissue was first chopped in a household onion chopper and then vigorously ground for 30 seconds with a mortar and pestle. No abrasive was used. The resulting brei was filtered through cheesecloth and centrifuged at 500 x g for 10 minutes. The supernatant solution was centrifuged at 10,000 g for 10 minutes and the resulting pellet resuspended in 4 to 6 ml of 32% sucrose. Homogenates prepared in this manner gave significantly higher yields of isocitrate lyase in the pellets than had previously been obtained. Portions of this crude particulate suspension containing 15 to 20 mg of protein were then layered on linear sucrose gradients contained in Spinco rotor SW 25-2 cellulose nitrate tubes. The slope of the linear gradients was tripartite, consisting of 16 ml of 60% sucrose, 13 ml grading from 60% to 45% sucrose and 26 ml grading from 45% to 32% sucrose.

The gradients were centrifuged a minimum of four hours at 23,000 r.p.m. in a Spinco L-2-50 ultra-centrifuge and allowed to decelerate without braking. The gradients were fractionated into thirty-five 1 ml fractions numbered consecutively in the order of increasing density. Fractions were maintained at 0-5°C and assayed as quickly as possible.

All enzymes were assayed optically with a dual beam spectrophotometer equipped with a recorder. Isocitrate lyase was assayed by the method of Dixon and Kornberg (1959); fumarase and aconitase by the method of Racker (1950); citrate synthetase at pH 8.0 by the method of Srere et al. (1963); malate synthetase at pH 8.0 by the method of Hoch and Beevers (1966); NADH oxidase by the method of Crane (1957); NAD isocitric dehydrogenase by the method of Kornberg (1955); NADP isocitric dehydrogenase by the method of Ochoa (1955a) and malic dehydrogenase by the method of Ochoa (1955b). All of the above dehydrogenase ssays were run in the presence of 3.3 mM KCN. Succinic dehydrogenase was assayed by the method of Hiatt (1961). The enzyme was incubated with succinate for 15 minutes before initiating the reaction with phenazine methosulphate. Cytochromes were detected by means of difference spectra according to the method described previously (Breidenbach _et al._, 1967). Protein was determined by the method of Lowry _et al._, 1951).

RESULTS

Four bands are discernible after sucrose density gradient centrifugation of crude particulate suspensions (Figures 1 and 2). The uppermost band (A) consists almost exclusively of mitochondria as demonstrated by enzyme composition (Table I) and electron microscopic examination. The next denser band (B) has been tentatively identified as proplastids on the basis of light, U.V. and electron microscopy. The third band (C) is comprised of the particulates referred to as glyoxysomes in this paper. The fourth and densest band (D) is a very minor protein component and has not been characterized.

Figure 2 illustrates the protein distribution profile for a typical gradient separation. The mitochondrial band and the glyoxysome band contain roughly equivalent amounts of protein and together constitute more than half of the total protein layered on the gradient.

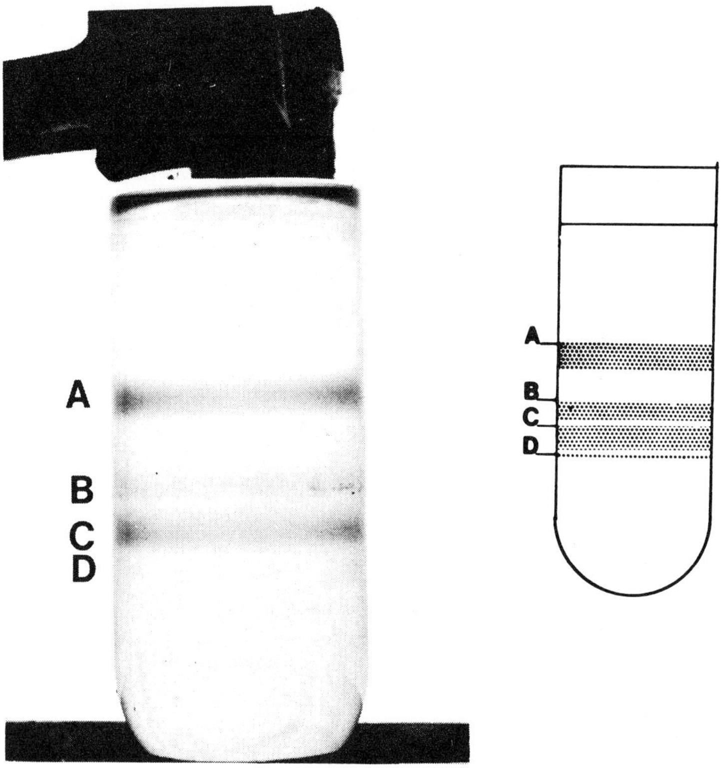

Figure 1. Typical separation of the components of the crude particulate fraction from 5 day old castor bean endosperm on a sucrose density gradient. A. Mitochondria. B. Proplastids. C. Glyoxysomes. D. Unknown component.

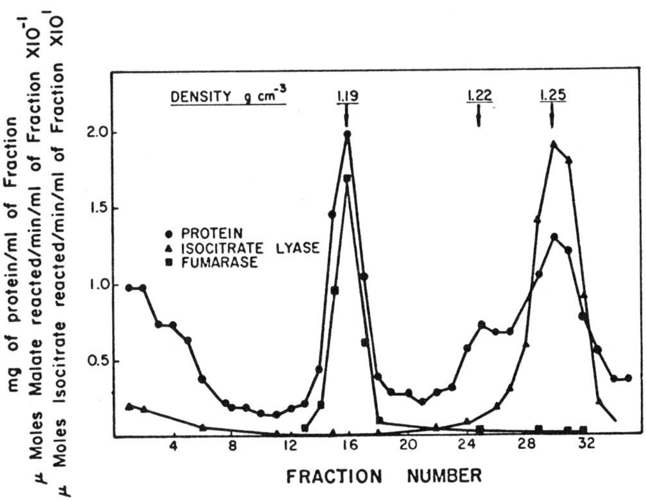

Figure 2. Distribution of protein, fumarase and isocitrate lyase after sucrose density gradient separation of the components of the crude particulate fraction from 5 day old castor bean endosperm.

TABLE I

ENZYME ACTIVITIES IN SUBCELLULAR FRACTIONS

SEPARATED FROM CASTOR BEAN ENDOSPERM

	GLYOXYSOME (Density) 1.25 g cm^{-3}	MITOCHONDRIA (Density) 1.19 g cm^{-3}
	Enzyme Activity (micromoles substrate consumed/min/mg protein)	
Isocitrate lyase	1.16	0
Malate synthetase	1.48	0
Citrate synthetase	1.16	0.72
Malate dehydrogenase (NAD)	22	50
Fumarase	0	1.4
NADH oxidase	0	0.11
Succinic dehydrogenase	0	0.55

The glyoxysome band contains 60-70% of the total isocitrate lyase activity found in the original homogenate. The remainder is about equally distributed between the 500 g pellet and the supernatant solution remaining after sedimenting the crude particulate fraction at 10,000 x g. The distribution of this enzyme on a typical gradient is shown in Figure 2. Clearly the activity is associated with the glyoxysome. Malate synthetase activity shows an identical pattern. Fumarase activity, by contrast, is specifically associated with the mitochondria and is absent from the glyoxysome.

Table I shows the distribution of seven enzyme activities in the separated glyoxysome and mitochondrial fractions. Two of these, isocitrate lyase and malate synthetase are confined to the glyoxysomes and absent from mitochondria. Two, citrate synthetase and malate dehydrogenase are present in both particles. Fumarase, NADH oxidase and succinate dehydrogenase are present in the mitochondria and can not be detected in the glyoxysomes. A further contrast between the two particles is observed in the reduced-oxidized difference spectra. Clear cytochrome peaks were seen in the mitochondrial fraction but none could be detected in the glyoxysomes.

Other enzymes of interest are the isocitrate dehydrogenases and aconitase. Only very low levels of NAD and NADP isocitrate dehydrogenases and aconitase (less than 4% of the isocitrate lyase) were detected in the crude particulate suspension. Considerably greater amounts of NADP isocitrate dehydrogenase and aconitase were present in the original supernatant solution but the aconitase lost activity rapidly even at $0^{\circ}C$.

DISCUSSION

Malate synthetase and isocitrate lyase are specifically associated with a non-mitochondrial particulate component in the castor bean endosperm. Since this particle also contains other enzymes of the glyoxylate cycle it seems probable that this is the site of the reactions <u>in vivo</u>. With the exception of aconitase, which is unstable in these extracts, all of the enzymes needed for the net synthesis of succinate from acetyl CoA units are present in the purified glyoxysome fraction. The source of acetyl CoA in vivo is β-oxidation of the long-chain fatty acids (Yamada and Stumpf, 1965; Rebeiz and Castelfranco, 1964) and the fate of succinate is conversion to sucrose via phosphoenolpyruvate and EMP sequence (Beevers, 1961). The intracellular localization of these reaction sequences and the interplay between them and the glyoxysome requires investigation. The confining of the glyoxylate cycle reactions to a distinct organelle lacking several of the oxidative enzymes of the TCA cycle simplifies the problem of regulating the flow of carbon at the level of isocitrate (Tanner and Beevers, 1965) and also puts the question of development and decline of the enzymes during germination (Beevers, 1961) in a different light.

In previous work with Tetrahymena (Hogg and Kornberg, 1963) it was concluded that the functional enzymes of the glyoxylate cycle were confined to a special mitochondrial fraction. Very recently Müller and Hogg (1967) have shown that malate synthetase and isocitrate lyase are associated with a particle, the peroxisome, which however lacks several other enzymes of the glyoxylate cycle. In Chlorella cells with a functional cycle, isocitrate lyase was associated with a dense particulate fraction (Harrop and Kornberg, 1966). Thus there are precedents from work with other organisms that at least some of the enzymes of the glyoxylate cycle are specially localized.

Although a close comparison with these studies is not possible it is clear that the particles from castor bean endosperm which contain the unique enzymes of the glyoxylate cycle are not mitochondria; they sediment differently and they do not contain fumarase, NADH oxidase, succinate dehydrogenase or cytochromes. Since they do contain the enzymatic equipment for the conversion of acetate to succinate in the glyoxylate cycle we refer to them as glyoxysomes. Similar particles have recently been isolated from watermelon cotyledons in this laboratory. Further characterization of these particles from both castor bean and watermelon storage tissues is in progress.

ACKNOWLEDGMENTS

This investigation was supported in part by a Public Health fellowship 1-F2-GM-23,105-01 from the National Institute of General Medical Sciences, and by NSF Grant GB-2599. We thank Dr. Albert Kahn and Mrs. Dorothy Werderitsh for their generous assistance with the electron and light microscopic phases of this study.

REFERENCES

Beevers, H. Nature 191: 433 (1961).

Breidenbach, R. W., R. S. Criddle and P. Castelfranco. Plant Physiol., in press (1967).

Canvin, D. T. and H. Beevers. J. Biol. Chem. 236: 988 (1961).

Crane, F. L. Plant Physiol. 32: 619 (1957).

Dixon, G. H. and H. L. Kornberg. Biochem. J. 72: 3P (1959).

Harrop, L. C. and H. L. Kornberg. Proc. Royal Soc., Ser. B 166: 11 (1966).

Hiatt, A. J. Plant Physiol. 36: 552 (1961).

Hock, B. and H. Beevers. Z. Pflanzenphysiol. 55: 405 (1966).

Hogg, J. F. and H. L. Kornberg. Biochem. J. 86: 462 (1963).

Kornberg, H. Methods in Enzymol. I: 707, Colowick and Caplan, Eds. (1955).

Kornberg, H. L. and H. Beevers. Biochim. Biophys. Acta 26: 531 (1957).

Lowry, O. H., N. J. Rosebrough, A. L. Farr and R. J. Randall. J. Biol. Chem. 193: 265 (1951).

Marcus, A. and J. Velasco. J. Biol. Chem. 235: 563 (1960).

Müller, M. and J. F. Hogg. Fed. Proc. 26(2): 284 (1967).

Ochoa, S. Methods in Enzymol. I: 735, Colowick and Caplan, Eds. (1955a)

Ochoa, S. Methods in Enzymol. I: 699, Colowick and Caplan, Eds. (1955b)

Racker, E. Biochim. Biophys. Acta 4: 211 (1950).

Rebeiz, T. and P. Castelfranco. Plant Physiol. 39: 932 (1964).

Srere, P. A., H. Brazil and L. Gonen. Acta Chemica Scandinavica 17: S129 (1963).

Tanner, W. and H. Beevers. Z. Pflanzenphysiol. 53: 72 (1965).

Yamada, and P. K. Stumpf. Plant Physiol. 40: 653 (1965).

Reprinted from
Vol 7, No. 3, 1962 BIOCHEMICAL AND BIOPHYSICAL RESEARCH COMMUNICATIONS

CONVERSION OF GLUCOSE TO INOSITOL IN PARSLEY LEAVES

Frank A. Loewus and Stanley Kelly

Western Regional Research Laboratory*, Albany, California

Received March 2, 1962

The idea that inositol+ may be formed by ring closure of glucose was clearly formulated by Fischer in 1945. Since then other investigators have demonstrated an over all conversion of glucose to inositol by using isotopically labeled glucose (Daughaday et al., 1955; Halliday and Anderson, 1955; Charalampous, 1957; Eagle et al., 1960). An internal labeling pattern for inositol formed biosynthetically from specifically labeled glucose by yeast cells has been described, but its complex nature suggested that fragmentation of glucose preceded inositol synthesis (Charalampous, 1957).

Higher plants contain appreciable quantities of inositol. A study of the conversion of glucose to inositol in a tissue of this type would be very desirable. This note describes a set of experiments in which this conversion is demonstrated in the detached parsley (Petroselinum) leaf. A highly specific method of obtaining the internal distribution of radioactivity in inositol is provided by the recent observation that labeled inositol is converted by detached ripening strawberries to galacturonosyl residues (pectin) in appreciable yield (Loewus et al., 1962). Very little redistribution of radioactivity occurs during the conversion. Thus inositol-2-C^{14} is converted to galacturonic acid labeled in carbon 5 (Loewus and Kelly, 1962) as well as to L-arabonosyl residues also labeled in carbon 5 (Loewus and Kelly, unpublished observation).

---

* A laboratory of the Western Utilization Research and Development Division, Agricultural Research Service, U. S. Department of Agriculture.

\+ Inositol refers to myo-inositol, glucose to D-glucose, and galacturonic acid and its derivatives to D-galacturonic acid and its derivatives. Inositol is numbered according to Angyal and Anderson (1959).

In each experiment, three to five detached parsley leaves (total weight, approx. 5 g.) were given 15 μc. of glucose-1-C^{14} (4 μc./mg.) in 0.5 ml. of water through their petioles over a one-half hour period. The leaves were then transferred either to water (Expts. 1 and 2) or to 1% inositol (Expts. 3 and 4) for 42 hours in the light. Unlabeled inositol taken up during this period (Expt. 3, 156 mg. and Expt. 4, 90 mg.) was only partially utilized. The unconsumed portion provided a convenient means of recovering labeled inositol that had been formed from the radioactive glucose.

The labeled leaves were frozen, pulverized, and ground in 40 ml. of 0.1% oxalic acid. This slurry was centrifuged at 12,000 x g. for 10 minutes. The clear supernatant was freed of oxalate with an equivalent of Ca formate and then passed through columns of Dowex 50 (H^+) and Dowex 1 (formate). The neutral effluent was reduced to a light sirup. In Expts. 3 and 4, about 50 to 60% of the inositol taken up by the leaves was recovered from this sirup as a radioactive crystalline residue. This residue was brought to constant specific activity after three recrystallizations from water and ethanol.

These sirups (less the crystalline residues removed from Expts. 3 and 4) were streaked on large sheets of Whatman No. 1 paper and developed 24 hours by descending chromatography in ethyl acetate-pyridine-water (8:2:1) (Williams and Bevenue, 1953) along with appropriate controls. Side strips were removed from the dried sheets and treated with $AgNO_3$ followed by a spray of alcoholic KOH (Trevelyan et al., 1950). Inositol appeared in all experiments as a narrow band midway between the origin and sucrose. Autoradiography revealed a dark streak coincident with inositol in Expts. 1 and 2,but the corresponding region in Expts. 3 and 4 was clear since most of the radioactivity associated with inositol in the latter two experiments was removed by diluent inositol and recovered in the crystalline residue prior to chromatography.

The radioactive inositol was eluted from the chromatograms of Expts. 1 and 2 and rechromatographed in the solvent system described above for 72 hours to obtain a clear separation of this compound from other radioactive constituents. Again the inositol was eluted and assayed chemically (Agranoff et al., 1958). To establish the purity of this material, an aliquot was diluted with authentic non-isotopic inositol, recrystallized, combusted to CO_2, and counted. All of the radioactivity present in the aliquot was accounted for in the recrystallized inositol.

The labeled inositol was administered to detached green strawberries (Fragaria) through their cut stems followed by a period of metabolism in distilled water. The entire berry including stem and hull was ground in cold 70% ethanol. The insoluble residue recovered from this treatment was washed repeatedly with successive portions of 70% ethanol until the radioactivity of the wash was negligible. Pertinent experimental details are given in Table I.

TABLE I

Conversion of Labeled Inositol to Galacturonate by Detached Ripening Strawberries

Experiment Number	1	2	3	4
Amount of inositol administered, mg.	1.9	1.3	65.	30.
C^{14} in inositol administered, dpm.*	34,000	104,000	60,000	68,000
Number of strawberries used	4	4	4	4
Initial fresh weight of berries, g.	9.9	8.7	16.4	14.1
Period of inositol uptake, hours	3	3	45	11
Total period of respiration, hours	51	51	70	70
C^{14} solubilized from washed residues after enzymatic hydrolysis, dpm.*	3,400	7,300	4,300	6,400
C^{14} recovered in galacturonate, dpm.*	1,500	4,000	1,100	3,500
Conversion of inositol to galacturonate (a minimal value), %	4.4	3.9	1.8	5.1

* Disintegrations per minute. Activities of aqueous solutions were measured in a liquid scintillation counter. Specific activities of crystalline materials were obtained with an internal gas phase proportional counter after conversion to CO_2.

Galacturonic acid was recovered by enzymatic hydrolysis of the washed residue followed by ion exchange procedures already described (Loewus, 1961), and its C^{14} distribution among carbon 1, carbons (2+3+4+5), and carbon 6 determined chemically (Seegmiller et al., 1955). Results are listed in Table II.

TABLE II

C^{14} Distribution in Galacturonate

Experiment Number	1	2	3	4
Total combustion to CO_2 (cpm./mmole galacturonate)*	4,800	17,300	2,300	6,800
Degradation of galacturonate (cpm./mmole of fragment)*				
Carbon 1 as HCHO	3,500	15,100	2,100	5,800
Carbons 2+3+4+5 as HCOOH	800	1,000	100	350
Carbon 6 as CO_2	700	2,200	200	750
Sum	5,000	18,300	2,400	7,000
C^{14} in each fragment (%)				
Carbon 1	70	82	88	84
Carbons 2+3+4+5	16	6	4	5
Carbon 6	14	12	8	11

* Values reported are the observed counts per minute/millimole of carbon (80% efficiency) times the number of carbons in the molecule or fragment.

In all experiments carbon 1 of galacturonic acid (corresponding to carbon 6 of inositol) contained most of the C^{14} (70 to 88%). Since the parsley leaves from which the labeled inositol was recovered had been fed glucose-1-C^{14}, carbon 1 of glucose corresponds to carbon 6 of inositol. The appearance of C^{14} in other carbons of the degraded galacturonic acid, especially carbon 6 (corresponding to carbon 1 of inositol), represents a minor redistribution of label, probably resulting from triose equilibration during the conversion of glucose to inositol. Similar labeling patterns have been observed among other carbohydrate products derived from specifically labeled glucose (Shibko and Edelman, 1957; Loewus and Jang, 1958).

The labeling pattern in inositol recovered from parsley leaves given glucose-1-C^{14} supports a cyclization mechanism like the one proposed by Fischer (1945). Such a process might be catalyzed by an aldolase-type reaction in which the intermediate immediately preceeding the formation of a carbon-carbon bond between carbon 1 and carbon 6 could be a phosphorylated 5-keto glucose or a phosphorylated 6-aldo fructose. The ease with which carbon 2 of inositol is oxidized and reduced favors the idea of a 5-keto compound with the configuration of glucose.

In higher plants, oxidative cleavage of the 1,6 bond of inositol results in the formation of galacturonosyl residues as well as other related interconversions (Loewus *et al*., 1962). This means that glucose converted to pectin *via* inositol will have the same carbon chain sequence (carbon 6 of glucose is oxidized) as that observed when uridine diphosphate glucose is oxidized *via* uridine diphosphate dehydrogenase (Strominger and Mapson, 1957). Experiments employing specifically labeled glucose-C^{14} will not distinguish between these pathways.

A detailed study of the C^{14} distribution in galacturonosyl residues, the glucose moiety of sucrose, and the L-ascorbic acid recovered from the same parsley leaves is being completed and will be reported elsewhere.

REFERENCES

Agranoff, B. W., Bradley, R. M., and Brady, R. O., J. Biol. Chem., 233, 1077 (1958).
Angyal, S. J., and Anderson, L., Advances Carbohydrate Chem. 14, 135 (1959).
Charalampous, F. C., J. Biol. Chem., 225, 595 (1957).
Daughaday, W. H., Larner, J., and Hartnett, C., J. Biol. Chem., 212, 869 (1955).
Eagle, H., Agranoll, B. W., and Snell, E. E., J. Biol. Chem., 235, 1891 (1960).
Fischer, H. O. L., Harvey Lectures, 40, 156 (1944-45).
Halliday, J. W., and Anderson, L., J. Biol. Chem., 217, 797 (1955).
Loewus, F. A., Ann. N. Y. Acad. Sci., 92, 57 (1961).
Loewus, F. A., and Jang, R., J. Biol. Chem., 232, 505 (1958).
Loewus, F. A., and Kelly, S., Federation Proceedings, 21, in press (1962).
Loewus, F. A., Kelly, S., and Neufeld, E. F., Proc. Natl. Acad. Sci., 48, in press (1962).
Seegmiller, C. G., Axelrod, B., and McCready, R. M., J. Biol. Chem., 217, 765 (1955).
Shibko, S., and Edelman, J., Biochim. Biophys. Acta, 25, 642 (1957).
Strominger, J. L., and Mapson, L. W., Biochem. J., 66, 567 (1957).
Trevelyan, W. E., Procter, D. P., and Harrison, J. S., Nature, 166, 444 (1950).
Williams, K. T., and Bevenue, A., J. Assn. Official Agr. Chem., 36, 969 (1953).

Changes in Cell Wall Polysaccharides Associated with Growth[1]

Donald J. Nevins[2], Patricia D. English[3], and Peter Albersheim
Department of Chemistry, University of Colorado, Boulder, Colorado 80302

Received January 22, 1968.

Abstract. **Changes in the polysaccharide composition of *Phaseolus vulgaris, P. aureus,* and *Zea mays* cell walls were studied during the first 28 days of seedling development using a gas chromatographic method for the analysis of neutral sugars. Acid hydrolysis of cell wall material from young tissues liberates rhamnose, fucose, arabinose, xylose, mannose, galactose, and glucose which collectively can account for as much as 70 % of the dry weight of the wall. Mature walls in fully expanded tissues of these same plants contain less of these constituents (10 %–20 % of dry wt). Gross differences are observed between developmental patterns of the cell wall in the various parts of a seedling, such as root, stem, and leaf. The general patterns of wall polysaccharide composition change, however, are similar for analogous organs among the varieties of a species. Small but significant differences in the rates of change in sugar composition were detected between varieties of the same species which exhibited different growth patterns. The cell walls of species which are further removed phylogenetically exhibit even more dissimilar developmental patterns. The results demonstrate the dynamic nature of the cell wall during growth as well as the quantitative and qualitative exactness with which the biosynthesis of plant cell walls is regulated.**

It is currently believed that growth and elongation of plant tissues are dependent upon changes in the molecular structure of the cell wall. These alterations permit relaxation of the wall with concomitant tissue expansion and also lead eventually to stabilization of structure in the secondary wall. One of the many functions attributed to the plant cell wall is its role as a barrier between the cell and its external environment, a defense which must be penetrated in pathogen infection. It is known that many plant pathogens attack tissues only at certain stages of development (4, 7). Furthermore, pathogens are known to produce potent polysaccharide degrading enzymes (5, 14, 21, 22). In light of these observations, we were interested in examining the characteristics of wall development during the periods of maximum pathogen susceptibility.

During development, plant cell walls undergo conspicuous changes in structure and form as they elongate, differentiate and mature. In general, these changes in "macro-structure" have been correlated with the development of the cell wall into a static mechanical element at functional maturity. Recently, considerable effort has been devoted to understanding the nature of the developmental changes involved in molecular terms. Since in bacteria many surface-related phenomena, *e.g.* bacteriophage recognition, are properties of the specific molecular organization of the wall and its associated structure (6), we wished to learn whether such molecular specificity was also a general characteristic of the cell walls of higher plants. Our working hypothesis was that a programmed biosynthetic sequence would give rise to a consistent developmental pattern of plant cell wall biogenesis and result in a precise molecular composition of the wall during growth.

It is known that the chemical structure of the cell walls of higher plants is very complex (1, 2), and it has recently been demonstrated that, like the bacterial cell wall, the walls of higher plants exhibit polysaccharide specificity (16). Studies of the changes in polysaccharide composition of cell walls of differentiating xylem in trees have indicated, however, that unlike bacterial cell walls, large compositional changes might be expected to occur in the polysaccharides of plant cell walls during development (20). This observation is supported by the results of another study in which changes in the polysaccharide composition of onion root tips were examined (9). However, not a great deal is known concerning the magnitudes and mechanisms of changes in the wall polysaccharide complement of elongating primary tissues. In this communication the results of an investigation of the changes in the neutral sugar composition of cell wall polysaccharides during seedling development are reported, with particular emphasis being devoted to the development

[1] Supported in part by a grant from the United States Atomic Energy Commission No. AT(11-1)-1426.

[2] Postdoctoral Fellow of the National Institute of General Medical Sciences (National Institutes of Health, No. 5-F2-GM-29. 861-02). Present address: Department of Botany and Plant Pathology, Iowa State University, Ames, Iowa 50010.

[3] Predoctoral Fellow of the United State Department of Health, Education and Welfare under Title IV of the National Defense Education Act.

of walls in varieties known to be variously resistant and susceptible to plant pathogens.

Materials and Methods

Plants utilized in this study include common bean (*Phaseolus vulgaris* L. var. 'Red Kidney', 'Pinto', and 'Small White'), mung bean (*Phaseolus aureus* Roxb.) and corn (*Zea mays* L. var. 'Golden Rocket'). Growth conditions have been described previously and are only briefly reviewed here (16). Seeds were surface sterilized by immersion for 15 minutes in a 0.5 % sodium hypochlorite (10 % "Clorox") solution and planted in a bed of moist vermiculite. The vermiculite was kept moist with a one-half strength Hoagland's nutrient solution (8), and the seedlings were maintained at a temperature of 25 ± 2° and illuminated by "Grow Lux" fluorescent lamps using a 14 hour photoperiod in a 24 hour cycle (16).

Tissue was collected for wall analyses beginning on the fourth day and continuing at irregular intervals until the twenty-eighth day after planting. The sampling frequency was coordinated with the rate of hypocotyl expansion in order to obtain more samples during the period of rapid elongation. The bean plants were subdivided into 4 basic parts: roots, hypocotyls, first internodes, and leaves. Corn plants were divided into roots and individual leaf lamina; leaf sheaths and internodes were not included. (Only the analytical results for the first leaf produced by the corn seedlings are reported here, however.) Other details of the harvesting procedure have been described (16).

Frozen tissue was chopped with a razor blade and then ground 3 times with 3 to 4 volumes of 100 mM potassium phosphate buffer (pH 7.0) in a Waring Blendor. The cell wall material insoluble in this buffer was collected by centrifugation, all of these procedures having been carried out at 2°. The wall material was again ground in the Waring Blendor with 10 volumes of a 1:1 (v/v) mixture of chloroform:methanol, and the wall material collected by filtration through Whatman No. 1 paper in a Buchner funnel. Three final acetone washings were used and the cell walls were air-dried at room temperature. The method used for the preparation of the cell walls for quantitative analysis of the component neutral-sugars by gas-liquid chromatography is described elsewhere (3, 16).

In a separate experiment, cell wall material (approx 20 mg) was assayed for starch contamination by treatment with 10 units of α-amylase (Sigma, Hog Pancreas Type 1-A) for 10 minutes at 25°. One unit of activity is defined as that amount of enzyme which liberates 1 μmole of reducing groups (calculated as maltose) per minute at 25° using soluble starch (Merck) as a substrate. The activity of the enzyme was assayed immediately prior to treatment of the wall samples.

Results

Plant Growth. Direct measurements of tissue expansion as a function of seedling age were used as a basis for growth comparisons in these experiments. Growth curves for the hypocotyls and first

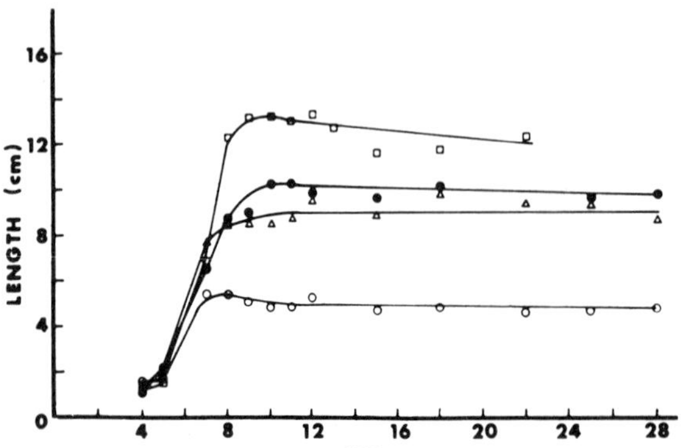

FIG. 1. Lengths of hypocotyls of various beans as a function of time. 'Pinto' □; 'Red Kidney' ●; 'Small White' △; and Mung beans ○.

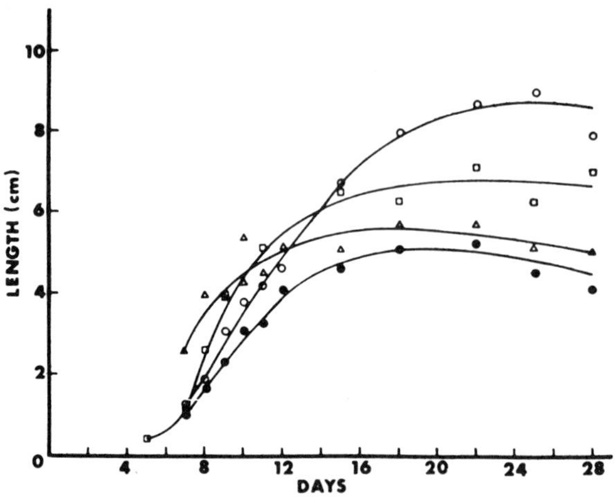

FIG. 2. Lengths of first internodes of various beans as a function of time. Symbols are the same as in figure 1.

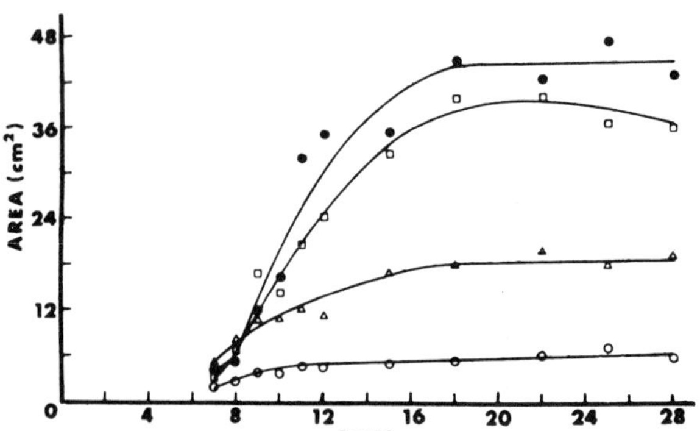

FIG. 3. Area of primary leaflets of various beans as a function of time. Symbols are the same as in figure 1.

RED KIDNEY

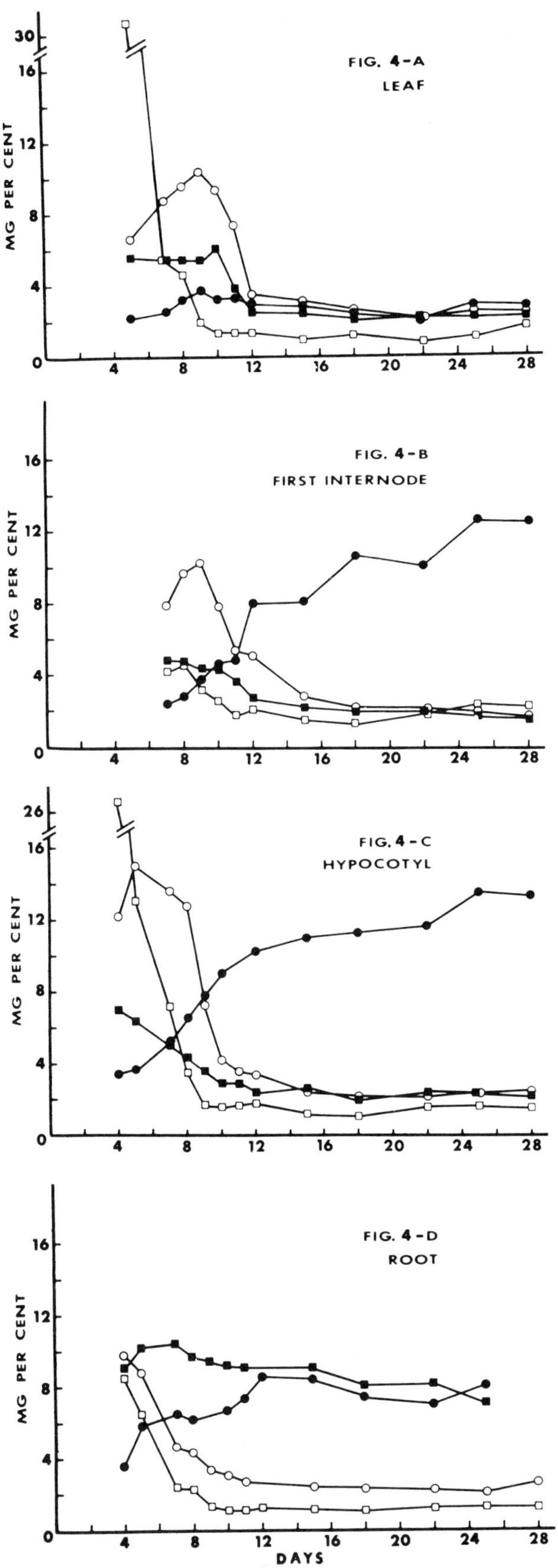

PINTO

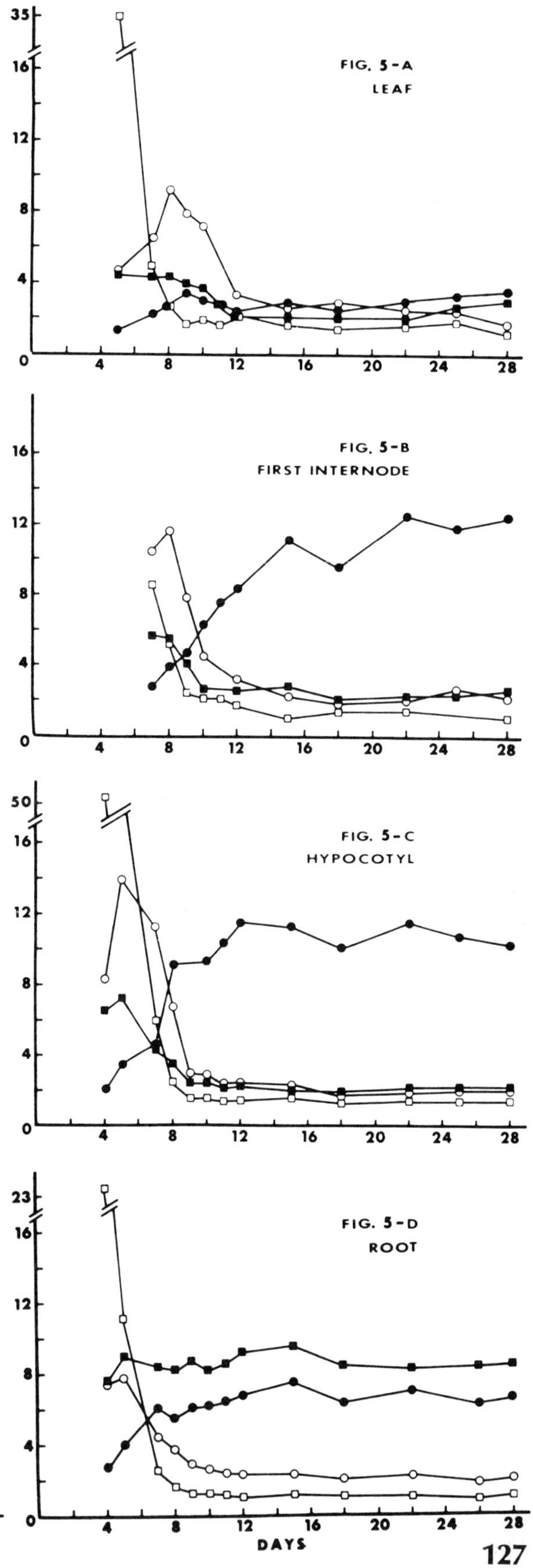

internodes of beans are shown in figures 1 and 2, respectively. Hypocotyl elongation in *P. vulgaris* varieties occurred primarily between the fifth and seventh day in the 'Small White' variety, while in the 'Pinto' and 'Red Kidney' varieties this was observed to take place from the fifth to the eighth days and the fifth to tenth days, respectively. First internode elongation in these 3 varieties followed a developmental pattern similar to that observed with hypocotyl growth, *i.e.*, the 'Small White' variety completed its elongation before that of the 'Pinto' and 'Red Kidney' varieties. Leaf growth in these plants was estimated from the increase in area of individual primary leaves (fig 3). Using these measurements of physical expansion as quantitative estimates of growth, a hierarchy of maturation rates was observed under the environmental conditions employed, with the 'Small White' variety maturing first, the 'Pinto' variety second, and 'Red Kidney' variety being the slowest to mature. It was also clear that in the tissues studied, expansion was essentially complete before the twenty-eighth day.

In contrast to the observations on the *P. vulgaris* varieties, *P. aureus* hypocotyls were observed to elongate during a very limited period (2 days, fig 1), while elongation of the first internodes continued over a longer period (13 days, fig 2). It should be noted that in the case of *P. aureus* one-half of the leaflet expansion was completed by the time of the first sampling (day 7, fig 3).

Cell Wall Polysaccharides. The changes in the neutral sugar composition of bean cell wall polysaccharides during the first 28 days of growth are illustrated in figures 4, 5, 6 and 7. It is obvious from these results that large changes in the cell wall polysaccharides do take place during early growth. Not only are there changes in the relative amounts of the various sugars during this period, but there are significant differences as well in the amounts of the total wall material which are liberated as neutral sugars by acid hydrolysis. Cell walls of 'Pinto' bean hypocotyls (fig 5-C), for example, yield almost 70 % of their total weight as neutral sugars at an age of 4 days, while at an age of 28 days only 18 to 19 % is liberated. Other instances of such yield differences from tissue of varying ages are apparent from the data.

For a species, the changes in the levels of the various sugars of the wall polysaccharides during growth followed similar patterns within a given tissue, although some differences associated with varieties were noted. It should be emphasized that varietal differences were not so much evident in the general pattern of change as they were in the rates of change in sugar composition. This is evident from the data presented in figure 9.

Several consistent patterns of polysaccharide alteration are apparent in cell wall development. One of these is considered in detail. By comparing the changes in hypocotyl cell wall composition of the 3 varieties of *P. vulgaris* and the single variety of *P. aureus* as shown in figures 4C, 5C, 6C and 7C, one observes that galactose, glucose and arabinose are the predominant sugars in the wall before rapid elongation, but that the relative amounts of these sugars decline with age, although at quite different rates. Xylose, on the other hand, constitutes a rather small fraction of the total neutral sugar found in the wall of young tissues, but the amount of this sugar in the wall increases during

Table I. *Amounts of Rhamnose, Fucose, and Mannose in Bean Hypocotyl Cell Walls at Various Ages after Planting*

Plant	Sugar	Day 4	Day 5	Day 10	Day 25
		mg sugar per 100 mg dry wt of wall material			
P. vulgaris 'Red Kidney'	Rhamnose	1.5	1.6	1.2	1.3
	Fucose	0.7	0.5	0.4	0.3
	Mannose	0.8	1.1	1.1	1.0
P. vulgaris 'Pinto'	Rhamnose	1.2	1.7	1.2	1.2
	Fucose	0.3	0.6	0.4	0.3
	Mannose	0.7	0.9	0.9	0.8
P. vulgaris 'Small White'	Rhamnose	1.5	1.8	1.3	1.1
	Fucose	0.5	0.6	0.4	0.2
	Mannose	0.8	1.0	0.9	0.8
P. aureus 'Mung'	Rhamnose	1.5	1.5	1.6	1.4
	Fucose	0.4	0.4	0.4	0.3
	Mannose	0.9	1.1	1.3	1.0

FIGS. 4 and 5. Changes in the neutral sugar composition of cell wall polysaccharides as a function of time. Figures 4 and 5 represent 'Red Kidney' and 'Pinto' varieties respectively, and the letters denote the following: A) leaf, B) first internode, C) hypocotyl, and D) root. The percent yields (mg per 100 mg wall material) are indicated on the ordinate. The various neutral sugars are designated as follows: glucose □; galactose ○; xylose ●; arabinose ■.

SMALL WHITE

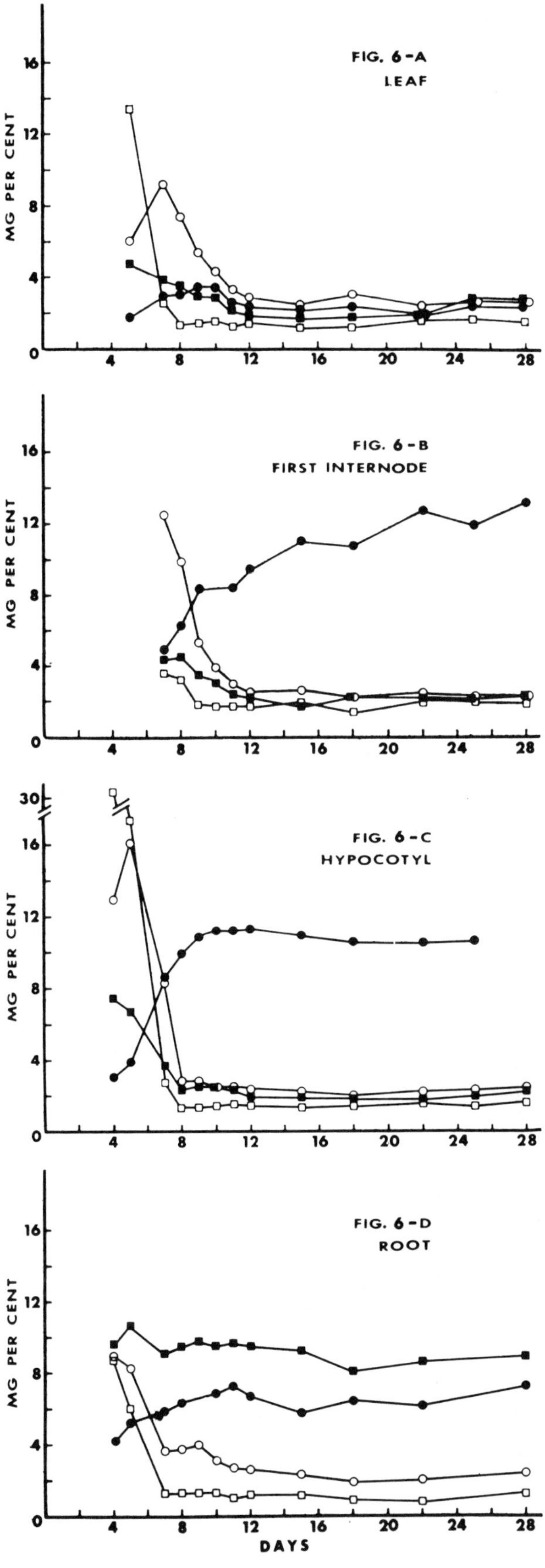

MUNG

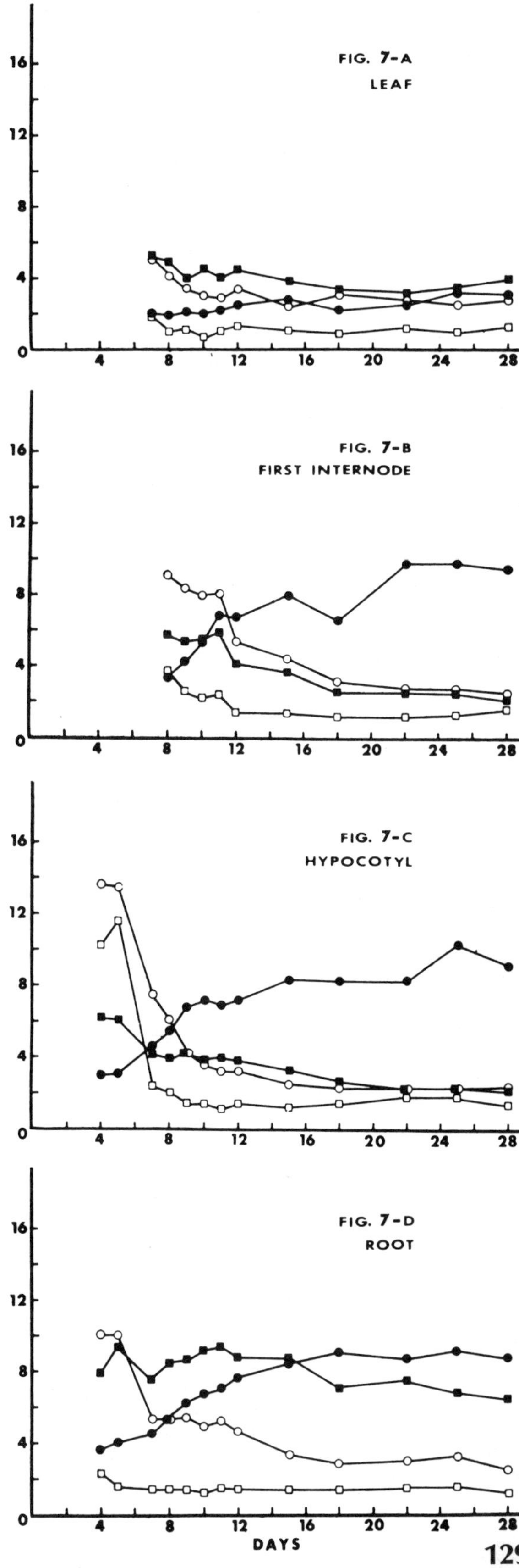

growth. Changes in the other neutral sugars, *viz.* rhamnose, fucose and mannose, which comprise a relatively small portion of the wall polysaccharide, are presented in table I. Because these sugars are present in such small quantities, the analytical reliability of the values obtained is less than that achieved for the more abundant sugars. Nevertheless, relatively large changes in the amounts of these sugars are observed during growth. In general, these sugars are more abundant in the walls of young tissues, decline with time until elongation ceases, and then maintain a stable concentration with respect to the total wall material. The general pattern of change of these 3 sugars in other tissues is similar to that which is observed in hypocotyls.

Although the pattern of change in hypocotyl tissues appears to be similar to that of the first internode, it is by no means a general pattern of cell wall development in all tissues. For example, in the cell walls of leaves (figs 4A, 5A, 6A, 7A) the percent of the dry weight accountable for as xylose is not observed to increase during growth. In roots (figs 4D, 5D, 6D, 7D), unlike the other tissues, arabinose continues as the predominant wall sugar throughout development.

For comparative purposes, the changes in cell wall polysaccharide composition of the first leaf lamina and roots of developing corn seedlings are shown in figures 8A and 8B. In contrast to the observations on the cell walls of bean tissue, it was observed that much of the sugar content of corn root cell walls was unaltered during development. The primary exception to this pattern was observed for glucose, which declines rapidly during the first few days of growth, much as in the case of bean tissue.

Treatment of bean hypocotyl and corn leaf cell wall preparations with α-amylase failed to alter the glucose content of the polysaccharide. While not conclusive, such evidence suggests that the cell wall preparations are not contaminated with starch and that the large amount of glucan in young tissues is indeed a constituent of the cell wall polysaccharide.

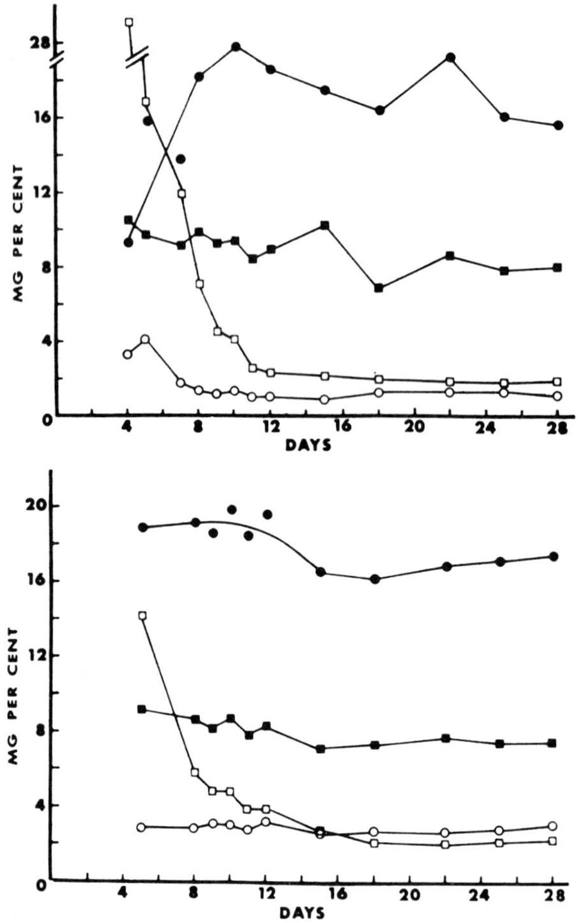

FIG. 8. Changes in the neutral sugar composition of cell wall polysaccharides of corn as a function of time. Figures A and B represent the first leaf lamina and root, respectively. The percent yields (mg per 100 mg wall material) are indicated on the ordinate. The same symbols as in figures 4, 5, 6 and 7 are used to designate the various sugars.

Discussion

The changes in polysaccharide composition observed in this study are consistent with the hypothesis that the development of the higher plant cell wall is regulated and is not a result of the random deposition of wall polysaccharide precursors at the cell surface. The specific molecular structure of the wall may be the result of a programmed sequence of events keyed, at least in part, to the elongation process. In certain instances, kinetic differences in the increase or decrease in wall polysaccharide components in plant varieties are correlated with elongation. From the data of the current report, it is clear that significant differences in the developmental sequence involved in the incorporation of wall polysaccharides do exist between plants which are not closely related genetically, *e.g.*, corn and beans. However, varieties of a single species have very similar developmental sequences.

We have found that walls of sycamore cells

FIGS. 6 and 7. Changes in the neutral sugar composition of cell wall polysaccharides as a function of time. Figures 6 and 7 represent 'Small White' and Mung beans respectively, and the letters denote the following: A) leaf, B) first internode, C) hypocotyl, and D) root. The percent yields (mg per 100 mg wall material) are indicated on the ordinate. The various neutral sugars are designated as follows: glucose □; galactose ○; xylose ●; arabinose ■.

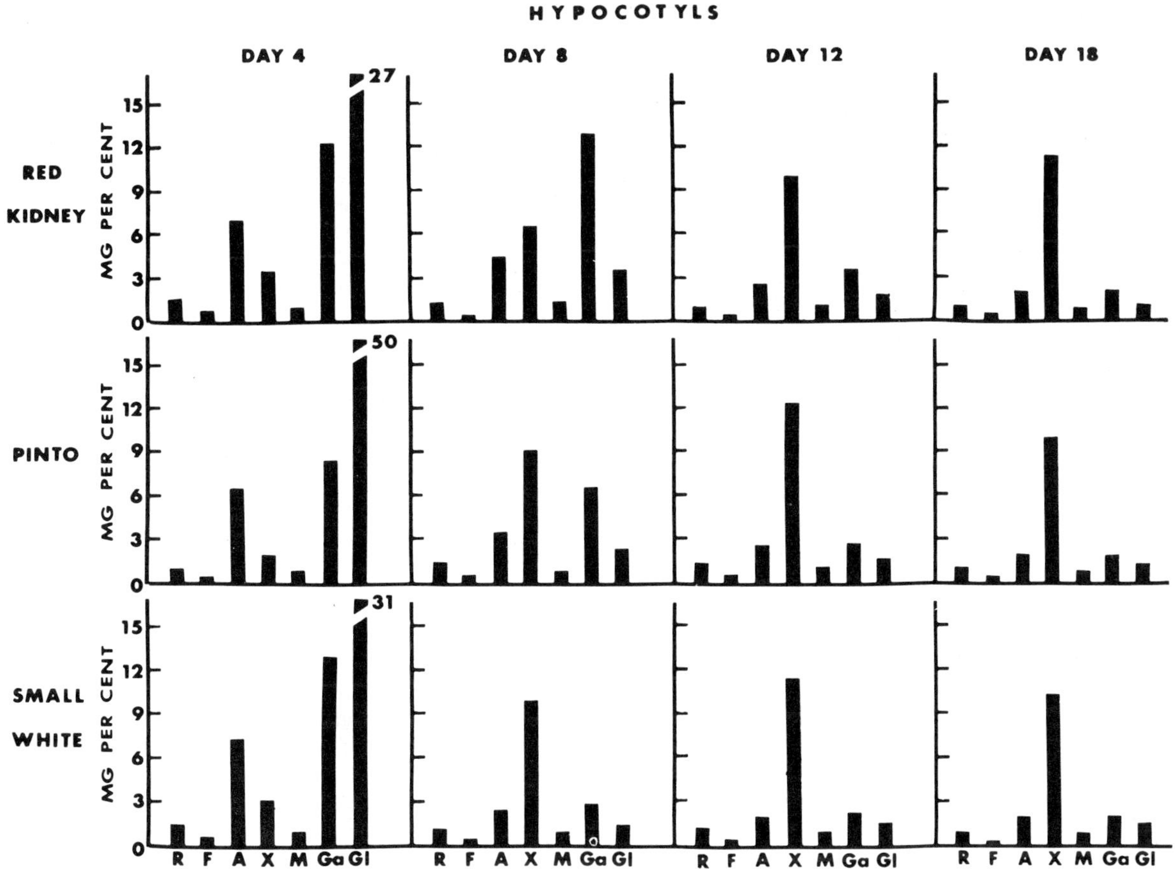

FIG. 9. Neutral sugar composition of hypocotyl cell wall polysaccharides of 3 bean varieties at various times after planting. The percent yields of the various sugars (mg per 100 mg wall material) are given on the ordinate axis. The letter designations on the abscissa represent the component sugars of the wall: R = rhamnose; F = fucose; A = arabinose; X = xylose; M = mannose; Ga = galactose; Gl = glucose.

(*Acer pseudoplatanus* L., cambial origin) contain a considerably different ratio of sugars (16) than do cell walls of bean and corn plants. This was found to be true even at very early stages of plant development where we find that a considerable portion of the differentiated wall is liberated as neutral sugars by acid hydrolysis, just as in the case of undifferentiated cultured cells. The sugar ratio differences may be due, in part, to genetic differences between these distantly related plants, and they may be due to some effect on the biosynthesis of the wall which results from growth in suspension culture (16). It is also conceivable that walls of whole plants would resemble walls of cultured sycamore cells more closely had the plants been sampled at even younger stages than were included in this investigation.

It is probable that at least a part of the observed decline in the overall neutral sugar polysaccharide concentration during growth results from simple dilution of these materials by other wall substances deposited during secondary wall formation and for which we have not analyzed, *i.e.*, cellulose, lignin, *etc.* However, all of the observed changes cannot be attributed to a dilution effect. This is evident from calculations based on the data of figure 4-C ('Red Kidney' hypocotyl cell walls). We find that between days 8 and 10 glucose content decreases at a rate of 0.4 mg $h^{-1}g^{-1}$ dry weight of wall material. The galactose level decreases at a rate of 1.9 mg $h^{-1}g^{-1}$ dry weight of wall material, and arabinose content declines during the same period at a rate of 0.3 mg $h^{-1}g^{-1}$ dry weight of wall material. While the levels of all 3 of these sugar components are decreasing, they do so at differing rates. Obviously, dilution effects have superimposed upon them the effects of different rates of incorporation and/or removal of various polysaccharide components of the wall. The very rapid decline in galactose during the 8 to 10 day period is of interest, as similar effects have been observed during the elongation of pea stem segments (13) and during cell wall growth in the xylem of various trees (20). Although it has been suggested that the decline in galactose content is due to an enzyme of the galactanase type (13), direct evidence for the existence of such an enzyme associated with plant cell walls has not been reported.

In view of several recent reports (10, 11, 12), the rapid decline in the glucose content of young tissues is of particular interest. We have noted that, in most instances, a glucose-containing polymer was the predominant polysaccharide, accounting for from 25 to 50 % of the total wall material, of cell walls in tissues prior to elongation. A large proportion of glucan has also been observed in the non-cellulosic wall polysaccharides of non-elongated oat coleoptiles (17). The only tissues in which we did not observe a large amount of wall glucan were *P. aureus* leaves (fig 7-A) and *P. aureus* roots (fig 7-D). However, at the time of the initial harvest, these leaves were already about one-half expanded (fig 3), and it seems likely that the glucose content of these walls had already undergone the sharp decline by this time. Sycamore cells grown in suspension culture also lack the large amounts of glucan (16) which seem to characterize the cell walls of young plant tissues. Glucose is the only sugar constituent of corn root walls which undergoes any major quantitative change during growth. In the young tissue, this substance, at least in a functional sense, may be a form of "cellulose", although it does not exhibit the acid resistance which characterizes the cellulose of the mature wall. Evidence has recently been presented which indicates that the incorporation of "cellulose" into young cotton fibers is bi-phasic in nature (11) and that quite different degrees of polymerization characterize the 2 products.

Another possible explanation for the high glucose content observed in the tissues of young plants is that of contamination of the wall preparations with starch, an acid labile α-1,4-glucan. However, treatment of the wall preparations with α-amylase had no detectable effect upon the amounts of glucose which were subsequently liberated by acid hydrolysis of the walls. In light of this observation, we conclude that the glucose-rich polysaccharide is not α-1-4-linked, and that it is probably an integral part of the wall structure.

During the early stages of elongation the amounts of glucose present in the wall decline rapidly. Referring again to the data of figure 4-C, one notes that the rate of this decline averages 2.3 mg $h^{-1}g^{-1}$ dry weight of wall material from the fourth through eighth days. An even more striking example of decreasing glucose content is seen in the hypocotyls of 'Pinto' beans (fig 5-C) in which, during the same 4 day period, the decline averaged 5.0 mg $h^{-1}g^{-1}$ dry weight of wall material. A similar decrease in the glucose content of the "hemicellulose" fraction has also been noted following the auxin-induced elongation of oat coleoptiles (17). While it is conceivable that the decline in glucan level, characteristic of young seedlings, is the result of rapid conversion into some other substance, the enzymic degradation and removal from the wall of this material is of equal plausibility. The results of a recent study of wall autolysis in corn indicate that the substrates of the autolysis were exclusively glucans (10), implying that the autolytic cleavage of these glucans represents a native enzymatic process which could contribute to wall relaxation during growth. The results of another recent study, in which non-elongated oat coleoptiles were treated with β-1,3-glucanase is also of interest in this regard (12). Such enzyme treated coleoptiles were observed to elongate to the same extent as those treated with auxin. It would be of interest to establish whether the glucose-rich polymer observed in our studies constitutes a specific substrate for such glucanases.

In addition to those polysaccharide components which decline during growth, the levels of certain sugar components increase during the growth process. In most tissues studied, the galactose levels increase initially, attain a peak, and subsequently decline. The occurrence of temporally controlled incorporation of galactose into a mucopolysaccharide, and of the time-dependent appearance and disappearance of a UDP-galactose:polysaccharide transferase enzyme has been observed during stalk development in the slime mold, *Dictyostelium discoideum* (18, 19). More germane to the observation under discussion is the report of a UDP-galactose:polysaccharide transferase system in the root and hypocotyl tissues of dark-grown *P. aureus* which exhibits maximum activity at a seedling age of 4 to 5 days (15). Our data demonstrate that galactose is most abundant in root and hypocotyl cell walls at the age of 4 to 5 days, suggesting that the reported galactose transfer system is the one responsible for the increased galactan content of these cell walls.

The other neutral sugar which markedly increases in concentration during elongation is xylose, although the rise in this case begins much later than that in galactose content. The rate of increase in xylose level, 0.8 mg $h^{-1}g^{-1}$ dry weight of wall material, during days 5 to 8, in 'Small White' hypocotyls (fig 6-C), is appreciable although probably underestimated because of the increase in other wall materials during this period which are not accounted for in this study (dilution effect). The amounts of xylose largely stabilize once elongation is complete.

Rhamnose, fucose and mannose, which are found in relatively small amounts in the wall, are generally more abundant prior to or during the period of maximum elongation (table I). At least a fraction of the wall mannan is probably bound in linkages which are resistant to the acid hydrolysis treatment employed. Hence, changes in the relative content of this sugar may not be fully detected by our method.

Two interesting deviations from the normal developmental pattern of wall polysaccharides were disclosed in this study. Of particular interest is the apparent absence of an increase in the xylose level of leaf cell walls. The kinetic pattern of wall formation confirms the observation that xylose is present in low concentrations in the mature walls of fully expanded leaves (16). Another situation which might be considered unusual is the apparently continuous incorporation of arabinose into root cell walls throughout growth. It is clear from the data presented here that, within a given plant, the mechanisms for the regulation of wall biosynthesis are uniquely characteristic of the various organs.

Of particular relevance to our own interests is the observation that 3 varieties of bean (*P. vulgaris*) which are known to exhibit differential responses to a plant pathogen (7), undergo cell wall alteration at quite different rates. From the data on hypocotyl composition presented in figure 9, one notes that by day 8 the 'Small White' variety has attained a "mature" wall composition while 'Red Kidney' takes until day 10 to 12 to mature. At days 4 and 18, however, the wall compositions of the 3 varieties are quite similar. Arabinose, galactose and xylose reflect the differing rates of change at the intermediate ages especially well. We have found that a particular bean pathogen, *Colletotrichum lindemuthianum* (Sacc. & Magn.) Brl. & Cav., to which these 3 bean varieties exhibit a differential resistance response, produces extracellular enzymes which liberate primarily arabinose, galactose, and xylose from the bean hypocotyl walls (unpublished data). It may well be that the differential susceptibility responses of these bean varieties are related to their respective abilities to grow rapidly through a cell wall configuration which could serve as an ideal substrate for these polysaccharide degrading enzymes. While such a hypothesis is attractive, it remains to be demonstrated that the enzymes produced by pathogens are indeed specific for these wall substrates, and that they are produced in sufficient quantity to attack the plant effectively during that period when the wall polysaccharides are most susceptible to enzymolysis. This enzyme-substrate interrelationship is presently under investigation.

Acknowledgment

The authors are grateful to Thomas M. Jones for assistance in the preparation of this manuscript.

Literature Cited

1. ALBERSHEIM, P. 1965. The substructure and function of the cell wall. In: Plant Biochemistry. J. Bonner and J. E. Varner, eds. Academic Press, Inc. New York. 151–86.
2. ALBERSHEIM, P. 1965. Biogenesis of the cell wall. In: Plant Biochemistry. J. Bonner and J. E. Varner, eds. Academic Press, Inc. New York. 298–321.
3. ALBERSHEIM, P., D. J. NEVINS, P. D. ENGLISH, AND A. KARR. 1967. A method for the analysis of neutral sugars in plant cell wall polysaccharides by gas chromatography. Carbohydrate Res. 5: 340–45.
4. BATEMAN, D. F. AND R. D. LUMSDEN. 1965. Relation of calcium content and nature of the pectic substances in bean hypocotyls of different ages to susceptibility to an isolate of *Rhizoctonia solani*. Phytopathology 55: 734–38.
5. BATEMAN, D. F. AND R. L. MILLAR. 1966. Pectic enzymes in tissue degradation. Ann. Rev. Phytopathology 4: 119–46.
6. DAVIS, B. D. AND L. WARREN, eds. 1967. The specificity of cell surfaces. Prentice-Hall Inc., Englewood Cliffs, New Jersey. 290 p.
7. GRIFFEY, R. T. AND J. G. LEACH. 1965. The influence of age of tissue on the development of bean anthracnose lesions. Phytopathology 55: 915–18.
8. HOAGLAND, D. R. AND D. I. ARNON (revised by D. I. Arnon). 1950. The water-culture method for growing plants without soil. College of Agriculture, University of California, Berkeley, California, Cir. 347: 32 p.
9. JENSEN, W. A. AND M. ASHTON. 1960. Composition of developing primary wall in onion root tip cells. I. Quantitative analyses. Plant Physiol. 35: 313–23.
10. LEE, S.-H., A. KIVILAAN, AND R. S. BANDURSKI. 1967. *In vitro* autolysis of plant cell walls. Plant Physiol. 42: 968–72.
11. MARX-FIGINI, M. 1966. Comparison of the biosynthesis of cellulose *in vitro* and *in vivo* in cotton balls. Nature 210: 754–55.
12. MASUDA, Y. AND S. WADU. 1967. Effect of β-1,3-glucanase on the elongation growth of oat coleoptile. Botan. Mag. Tokyo 80: 100–02.
13. MATCHETT, W. H. AND J. F. NANCE. 1962. Cell wall breakdown and growth in pea seedling stems. Am. J. Botany 49: 311–19.
14. MCNAB, J. M., D. J. NEVINS, AND P. ALBERSHEIM. 1967. Differential resistance of cell walls of *Acer pseudoplatanus, Triticum vulgare, Hordeum vulgare,* and *Zea mays* to polysaccharide degrading enzymes. Phytopathology 57: 625–31.
15. MCNAB, J. M., C. L. VILLEMEZ, AND P. ALBERSHEIM. 1968. Biosynthesis of galactan by a particulate enzyme preparation from *Phaseolus aureus* seedlings. Biochem. J. 106: 355–60.
16. NEVINS, D. J., P. D. ENGLISH, AND P. ALBERSHEIM. 1967. The specific nature of plant cell wall polysaccharides. Plant Physiol. 42: 900–06.
17. RAY, P. M. 1963. Sugar composition of oat-coleoptile cell walls. Biochem. J. 89: 144–50.
18. SUSSMAN, M. 1967. Evidence for temporal and quantitative control of genetic transcription and translation during slime mold development. Federation Proc. 26: 77–83.
19. SUSSMAN, M. AND M. J. OSBORN. 1964. UDP-galactose polysaccharide transferase in the cellular slime mold, *Dictyostelium discoideum:* appearance and disappearance of activity during cell differentiation. Proc. Natl. Acad. Sci. U.S. 52: 81–87.
20. THORNBER, J. P. AND D. H. NORTHCOTE. 1961. Changes in the chemical composition of a cambial cell during its differentiation into xylem and phloem tissue in trees. 2. Carbohydrate constituents of each main component. Biochem. J. 81: 455–64.
21. VAN SUMERE, C. F., C. VAN-SUMERE-DE PRETER, AND G. A. LEDINGHAM. 1957. Cell-wall-splitting enzymes of *Puccinia graminis* var. *tritici*. Can. J. Microbiol. 3: 761–70.
22. WOOD, R. K. S. 1960. Pectic and cellulolytic enzymes in plant diseases. Ann. Rev. Plant Physiol. 11: 299–322.

Reprinted from PLANT PHYSIOLOGY, Vol. 43, No. 8, August, 1968, pages 1185–1207.
(PRINTED IN U.S.A.)

The Acetylene - Ethylene Assay for N_2 Fixation: Laboratory and Field Evaluation[1]

R. W. F. Hardy, R. D. Holsten, E. K. Jackson, and R. C. Burns
Central Research Department, Experimental Station, E. I. du Pont de Nemours and Company, Wilmington, Delaware 19898

Received April 22, 1968.

Abstract. The methodology, characteristics and application of the sensitive C_2H_2–C_2H_4 assay for N_2 fixation by nitrogenase preparations and bacterial cultures in the laboratory and by legumes and free-living bacteria *in situ* is presented in this comprehensive report. This assay is based on the N_2ase-catalyzed reduction of C_2H_2 to C_2H_4, gas chromatographic isolation of C_2H_2 and C_2H_4, and quantitative measurement with a H_2-flame analyzer. As little as 1 $\mu\mu$mole C_2H_4 can be detected, providing a sensitivity 10^3-fold greater than is possible with ^{15}N analysis.

A simple, rapid and effective procedure utilizing syringe-type assay chambers is described for the analysis of C_2H_2-reducing activity in the field. Applications to field samples included an evaluation of N_2 fixation by commercially grown soybeans based on over 2000 analyses made during the course of the growing season. Assay values reflected the degree of nodulation of soybean plants and indicated a calculated seasonal N_2 fixation rate of 30 to 33 kg N_2 fixed per acre, in good agreement with literature estimates based on Kjeldahl analyses. The assay was successfully applied to measurements of N_2 fixation by other symbionts and by free living soil microorganisms, and was also used to assess the effects of light and temperature on the N_2 fixing activity of soybeans. The validity of measuring N_2 fixation in terms of C_2H_2 reduction was established through extensive comparisons of these activities using defined systems, including purified N_2ase preparations and pure cultures of N_2-fixing bacteria.

With this assay it now becomes possible and practicable to conduct comprehensive surveys of N_2 fixation, to make detailed comparisons among different N_2-fixing symbionts, and to rapidly evaluate the effects of cultural practices and environmental factors on N_2 fixation. The knowledge obtained through extensive application of this assay should provide the basis for efforts leading to the maximum agricultural exploitation of the N_2 fixation reaction.

To meet the imminent crisis in the world food supply (38) it is imperative that the resources of this planet be mobilized as rapidly and effectively as possible. Basic to such mobilization is a knowledge of the magnitude of the dynamic processes in the biosphere which affect the availability of nitrogen, the one element most often limiting in the production of foodstuffs (37). Of paramount importance in this context is the process of biological nitrogen fixation. Just as photosynthesis utilizes the freely available CO_2 of the atmosphere, nitrogen fixation draws on the unlimited supply of atmospheric nitrogen, and its potential role in increasing nitrogen availability has long been recognized. In spite of the importance of N_2 fixation very little accurate information is available to define the quantitative extent to which it occurs in the biosphere (35), and virtually nothing is known concerning the effects of various field practices on N_2 fixation. These gaps in our knowledge are attributable to the absence of effective methods for quantitative measurement of N_2 fixation *in situ*.

In the laboratory, N_2 fixation by living organisms has been measured by Kjeldahl analysis (7), ^{15}N-enrichment assayed by mass spectrometry (7), and ^{13}N-incorporation assayed by radioactive counting (8, 30); N_2 fixation by nitrogenase (N_2ase)[2] in cell-free extracts has been measured by ^{15}N-enrichment (9), ^{13}N-incorporation (8, 30), micro-Conway diffusion technique coupled with titrimetric (26) or colorimetric analysis of NH_3 (14), and N_2-H_2 uptake (27) or H_2 evolution (5) assayed manometrically. These methods are relatively insensitive except for the ^{13}N method, and its application is extremely limited because of its short half-life (10 minutes). Of these procedures only Kjeldahl analysis has been used to an appreciable extent for estimating N_2 fixation in field samples (35), but the method is insensitive and time-consuming. Although isotopic analysis of samples exposed to $^{15}N_2$ in the field has been used to demon-

[1] Contribution No. 1451.

[2] The following abbreviations are used: N_2ase for nitrogenase; DTT for dithiothreitol; TES for N-tris (hydroxymethyl)-methyl-2-amino-ethanesulfonic acid; and MES for 2-(N-morpholino)ethanesulfonic acid; DEAE for diethylaminoethyl.

strate N_2 fixation *in situ,* the complexity and expense of this method have limited widespread application to field studies.

The opportunity for a novel approach to N_2 fixation analysis arose from demonstrations that N_2ase is a most versatile reducing catalyst (15, 20). The following reductions have been reported:

$N_2O \rightarrow N_2 + H_2O$ (17, 21a)
$N_3^- \rightarrow N_2 + NH_3$ (19, 31)
$C_2H_2 \rightarrow C_2H_4$ (13, 19, 23, 32)
$RCCH \rightarrow RCHCH_2$ (16a)
$HCN \rightarrow CH_4 + NH_3$ and tentatively CH_3NH_2 (16a, 19, 22a)
$RCN \rightarrow RCH_3 + NH_3$ (16a)
$RNC \rightarrow CH_4 + C_2H_4 + C_2H_6 + C_3H_8$ etc. (16a, 22a)

Schollhorn and Burris (31) and Dilworth (13) independently observed inhibition of N_2 fixation by C_2H_2 with extracts of *C. pasteurianum*: Schollhorn and Burris (32) established the competitive nature of this inhibition and Dilworth found C_2H_2 to be reduced to C_2H_4 in a reaction analogous to the reduction of N_2 to NH_3. The application of this reaction to a sensitive assay procedure for N_2-fixing activity was proposed by Hardy and Knight (19): "Utilization of the reduction of HCN to CH_4, of $H^{14}CN$ to $^{14}CH_3NH_2$, or of C_2H_2 to C_2H_4, and detection of CH_4 and C_2H_4 by hydrogen flame ionization after gas chromatography or detection of $^{14}CH_3NH_2$ may provide a sensitive new assay for detection of the N_2-fixing system. The gas chromatographic determination makes possible a range of about 10,000 times between minimum and maximum, in contrast to a 20-fold range with the NH_3 assay. C_2H_2 is the preferred assay substrate, since more product is formed because of its requirement for 2 electrons *versus* 6 electrons for HCN". Subsequently, Koch, Evans, and Russell (24, 25), Silver (33), Sloger and Silver (34) and Stewart, Fitzgerald, and Burris (36) have successfully employed C_2H_2 reduction coupled with C_2H_4 detection by H_2 flame ionization as an assay for N_2ase activity.

Since the original proposal by Hardy and Knight (19), the C_2H_2-C_2H_4 assay of N_2-fixing activity has undergone extensive development in this laboratory. This paper reports: 1) methodology of C_2H_2 and C_2H_4 analyses; 2) methodology of the C_2H_2-C_2H_4 assay of N_2-fixing activity *in situ,* and 3) characteristics of C_2H_2 reduction by N_2ase *in vitro* by cultures of N_2-fixing bacteria and by samples of the biosphere *in situ.* The results indicate that the C_2H_2-C_2H_4 assay of N_2-fixing activity is sensitive, universal, specific, rapid, simple, economical, and quantitative. Since this assay has the potential to promote revolutionary fundamental and practical advances, we believe that the C_2H_2-C_2H_4 assay of N_2 fixation represents one of the most important developments in N_2 fixation research. It is emphasized that the adoption of a consistent procedure by the various disciplines, *e.g.,* soil science, agronomy, marine biology, plant biology, microbiology, and biochemistry, which will utilize this method is essential if valid comparisons are to be made among results obtained from various sources.

Methods

Growth of Cells and Preparation of Extracts. Azotobacter vinelandii, ATCC 12518, was grown on nitrogen-free or urea media. Optical densities of bacterial cultures were determined in Will colorimeter tubes with a 650 mμ filter in a Lumetron colorimeter. Cells were broken and N_2ase was purified as previously described for N_2, N_2O, N_3^-, or HCN reduction experiments (5, 17, 18, 19). N_2ase was fractionated into the Mo-Fe protein fraction (also called Enzyme I) and Fe protein fraction (also called Enzyme II) (15) by published procedures (4, 22) modified to produce a discrete fractionation of the 2 components. Fractions were designated as follows: 1) crude extract—supernatant after 35,000 × *g* for 30 minutes; 2) heated extract—crude extract heated for 10 minutes at 60° under 0.5 atm of H_2 and centrifuged at 35,000 × *g* for 1 hour; 3) pre-protamine sulfate precipitate—phosphocellulose resolubilized precipitate between 0.0 to 0.1 mg protamine sulfate per mg protein of heated extract; 4) protamine sulfate precipitate—phosphocellulose resolubilized precipitate between 0.100 to 0.125 mg protamine sulfate per mg protein of heated extract; 5) protamine sulfate supernatant—supernatant after 0.125 mg protamine sulfate per mg protein of heated extract; 6) Mo-Fe protein fraction—fraction of protamine sulfate precipitate eluted from a DEAE-cellulose column in an anaerobic chamber by 0.20 M NaCl + 0.02 M $MgCl_2$ + 5×10^{-4} M DTT in 0.02 M tris•HCl at pH 7.0 following pre-elution with 0.15 M NaCl + 0.02 M $MgCl_2$ + 5×10^{-4} M DTT in 0.02 M tris•HCl at pH 7.0; and 7) Fe protein fraction—fraction eluted from above column by 0.35 M NaCl + 0.02 M $MgCl_2$ + 5×10^{-4} M DTT in 0.02 M tris•HCl at pH 7.0; All fractions were stored anaerobically with fractions 4 and 7 stored at room temperature. Fraction 6 contained Fe and Mo and fraction 7 contained Fe as reported by others (4). Neither fraction was homogeneous by gel electrophoresis.

Clostridium pasteurianum, ATCC 6013, was grown on nitrogen-free or NH_4Cl media (9). Dried cells were broken by autolysis as previously described for N_2, N_2O, N_3^-, or HCN reduction experiments (9, 17, 19). An extremely sharp fractionation of N_2ase into its 2 components was produced by modification of an old procedure (29). Fractions were designated as follows: 1) crude extract—supernatant after 35,000 × *g* for 30 minutes of autolysate; 2) negative phosphate gel preparation—supernatant from crude extract treated with protamine sulfate to remove nucleic acids and with calcium phosphate gel to remove inactive protein (29); 3) Mo-Fe protein fraction—fraction eluted from DEAE-cellulose between 0.1 to 0.3 M NaCl in 0.01 M KPO_4, pH 7.0, after 15 minutes anaerobic batch treatment of a phosphate gel preparation with 8 mg damp DEAE-cellulose per mg protein. This fraction contains Mo and Fe and is stable for months when stored anaerobically at 4°. In some cases it was further purified by anaerobic chromatography on DEAE-cellulose with elution by 0.29 M NaCl in 0.01 M tris•HCl, pH 7.0. This fraction is not homogeneous by gel electrophoresis. 4) Fe-protein fraction—crude extract heated for 10 minutes at 60° under 0.5 atm H_2 and centrifuged at 35,000 × *g* for 30 minutes; this fraction was stored under anaerobic conditions at room temperature. Fraction 3 or 4 could only be obtained from N_2-grown cells.

Clostridium butyricum, Lactobacillus leichmanii, Bacillus subtilis, Serratia marcescens, Escherichia coli, Streptococcus lactis, Saccharomyces cerevisiae, Bacillus cereus var. *mycoides, Pseudomonas aeruginosa, Pseudomonas fluorescens, Aerobacter aerogenes, Staphylococcus epidermidis, Sarcina lutea, Spirillum itersonii, Proteus vulgaris, Alcaligines faecalis,* and *Rhodospirillum rubrum* were the kind gifts of Dr. R. Bailey of the University of Delaware and were grown on appropriate media containing fixed nitrogen. *Rhizobium japonicum*, ATCC 10324, *R. meliloti*, ATCC 10312, *R. leguminosarum*, ATCC 10004, *Rhizobium sp.*, ATCC 10317, and *R. trifolii*, ATCC 10328, were grown on a medium containing in g per liter: K_2HPO_4, 1.0; KH_2PO_4, 1.0; $MgSO_4•7H_2O$, 0.36; $CaSO_4•2H_2O$, 0.17; $FeCl_3•6H_2O$, 0.005; KNO_3, 0.7; yeast extract, 1.0; and mannitol, 3.0 (10).

Growth of Legumes. Field-grown soybeans (*Glycine max* Merr. var. Wayne) were sown on May 15, 1967, in 38-inch rows by a commercial grower in Chester County, Pennsylvania. Standard agricultural practices including seed inoculation with commercial inoculum and recommended additions of phosphorus and potassium but no nitrogen were used. In addition, soybeans and other legumes (*Phaseolus vulgaris, Medicago sativa, Arachis hypogea,* and *Pisum sativum*) were grown in sterilized Perlite using a nitrogen-free nutrient solution (1) in a greenhouse or in controlled environment growth chambers. A normal day-night regime of 16 hours, 24°, and 8 hours, 18° was maintained in Sherer-Gillett Model CEL255-6 chambers operated at maximum light intensity during the light period. Other samples of legumes and soils collected within a radius of 200 miles of Wilmington include samples from the Jordan Fertility Plots through the courtesy of Professors A. Richer and E. S. Lindstrom of Pennsylvania State University, and from the Georgetown Experimental Station through the courtesy of Professor R. Cole of the University of Delaware. Indicated times are Eastern Daylight Saving Time. The date of bud opening was recorded as the date of flowering, and the first indication of leaf yellowing was recorded as the onset of senescence.

Assays. Reductions of N_2 or C_2H_2 by N_2ase preparations or cultures of bacteria were performed in 40 ml incubation vessels sealed with serum caps. For N_2ase preparations dithionite was dissolved in O_2-free water containing a pre-determined quantity of acid or base to produce a final pH of 7. The energy source and reductant were placed in the sidearm, the extract and other components were placed in the main compartment, and the incubation flask was immediately evacuated. After repeated flushing with the indicated gas, the contents of the sidearm were tipped in to initiate the reaction. For cultures of bacteria, the incubation was initiated by the aseptic addition of the bacteria to the sealed incubation vessel containing appropriate medium and gas phase. The reaction mixture or culture was incubated on a rotary shaker at 30° for the indicated time, and the incubation was stopped by the addition of 0.5 ml of 6 N H_2SO_4. Samples of gas phase were analyzed with a mass spectrometer utilizing the initial gas phase as an internal standard or with a H_2-flame ionization detector after gas chromatographic separation (see below). Nitrogen fixation by N_2ase in extracts was measured by titration of NH_3 after micro-diffusion (26), and N_2 fixation by cultures was measured by Kjeldahl analysis of 5 ml aliquots. Deuterated ethylenes were analyzed in a Perkin-Elmer Model 21 infrared spectrophotometer using a 3.3 cm micro gas cell.

Assay of Acetylene and Ethylene. In early work an activated alumina column at 150° and a Perkin-Elmer 880 or 800 gas chromatograph with a dual H_2-flame ionization detector were used (19). Subsequently, a one-eighth inch × 10 foot column containing 20 % ethyl, N′,N′-dimethyl oxalamide on 100 to 120 mesh acid-washed firebrick at 0° with a He flow rate of 30 ml/minute has been found to be most effective for gas chromatographic separation of acetylene and ethylene as well as other saturated and unsaturated hydrocarbons containing up to 4 carbons. Modified Perkin-Elmer F-11 gas chromatographs equipped with H_2-flame ionization detectors are utilized. Representative retention times in minutes are: methane, 0.8; ethane, 1.0; ethylene, 1.1; propane, 1.4; propylene, 1.9; isobutane, 2.1; butane, 2.8; acetylene, 3.8; 1-butene, 4.4; isobutylene, 4.5; allene, 4.8; trans-2-butene, 5.2; cis-2-butene, 6.2; methylacetylene, 10.4. A typical chromatogram of a standard mixture of C_2H_2 (0.1 atmosphere) and C_2H_4 (2.5×10^{-4} atmosphere) is shown in figure 1a, and a chromatogram of C_2H_2 (0.1 atmosphere initial pressure) and C_2H_4 produced by a culture of N_2-grown *Clostridium pasteurianum* in figure 1b. The symmetry of C_2H_2 and C_2H_4 peaks and the absence of other components are indicated in figure 1b. A standard curve of peak height *vs.* C_2H_4 or C_2H_2 content of injected sample (fig 2) demonstrates the linear response and sensitivity of the assay. Less than 10^{-12} moles of C_2H_4 can be detected per injected sample of 200 μl. Ethylene content can be calculated from this standard curve, or

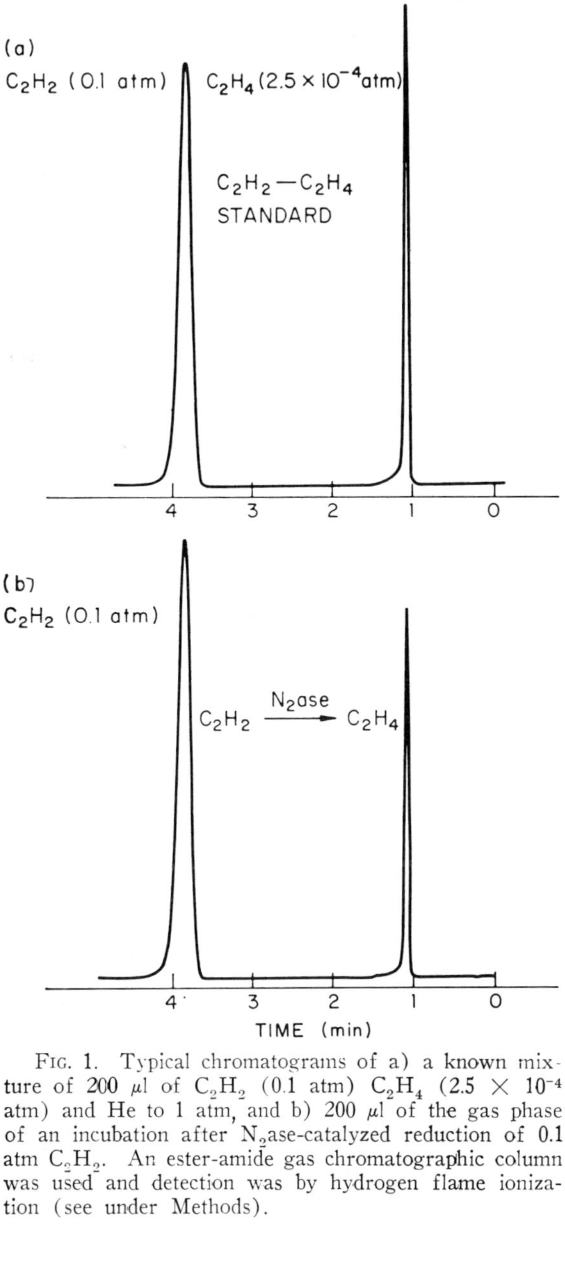

FIG. 1. Typical chromatograms of a) a known mixture of 200 μl of C_2H_2 (0.1 atm) C_2H_4 (2.5×10^{-4} atm) and He to 1 atm, and b) 200 μl of the gas phase of an incubation after N_2ase-catalyzed reduction of 0.1 atm C_2H_2. An ester-amide gas chromatographic column was used and detection was by hydrogen flame ionization (see under Methods).

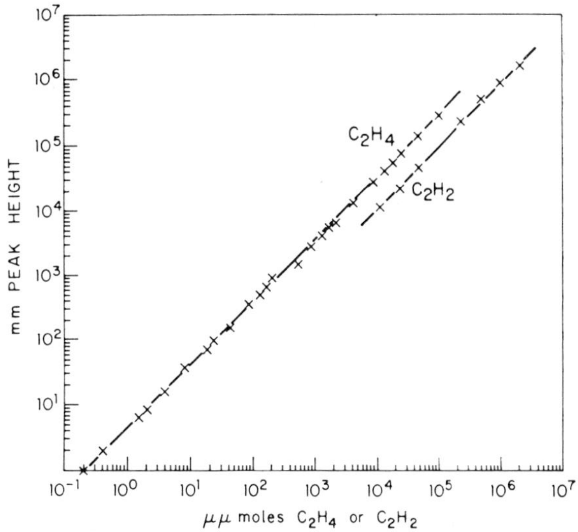

FIG. 2. A standard curve of peak heights of C_2H_2 and C_2H_4 determined with the gas chromatographic system of figure 1.

alternatively the "built-in" internal standard, C_2H_2, can be used since both C_2H_2 and C_2H_4 are determined. Our broad experience with the C_2H_2–C_2H_4 assay (over 2000 samples assayed) indicates that C_2H_2 is a valid and useful internal standard, since with the exception of large nodulated plant roots less than 2 % of the initial C_2H_2 (0.1 atmosphere) is converted to C_2H_4 during a 1 hour incubation.

Reagents. ATP, GTP, CTP, UTP, creatine phosphate, creatine kinase (ATP:creatine phosphotransferase, EC 2.7.1.40), and protamine sulfate were obtained from Sigma Chemical Company; $Na_2S_2O_4$, reagent grade, from Fisher Scientific Company; He, A, N_2, CO, and C_2H_2 as highest purity available from The Matheson Company. Acetone was removed from C_2H_2 by a concentrated H_2SO_4-scrubber and corrections were made for the C_2H_4 content of C_2H_2. The C_2H_4 impurity in C_2H_2 from a given cylinder must be determined daily since it varies inversely with the pressure in the cylinder.

Results

C_2H_2 Reduction by N_2ase in vitro. Reduction of C_2H_2 to C_2H_4 by N_2ase of cell-free extracts of *A. vinelandii* was examined with respect to a wide variety of characteristics, and the striking similarities between N_2 fixation and C_2H_2 reduction are reported in this section.

Requirements and Products of C_2H_2 Reduction. Reduction of C_2H_2 to C_2H_4, like reduction of N_2 to 2 NH_3 (3, 12, 16, 25, 27), requires an enzyme extract containing N_2ase, an energy source, and a reductant (table I). A similar energy and reductant requirement has been reported for reduction of acetylene by extracts of *C. pasteurianum* and soybean bacteroids (13, 25, 29a, 32). Extracts of urea-grown cells do not have N_2ase activity and do not have C_2H_2-reducing activity. No C_2H_4 formation is found in the absence of C_2H_2. Detectable amounts of C_2H_6 or CH_4 are not formed by the complete system capable of reducing C_2H_2 (sensitive analyses indicate that C_2H_6 can be no more than 0.01 % as abundant as C_2H_4); furthermore, ethylene is not reduced to C_2H_6 or CH_4 by *Azotobacter* N_2ase in a complete system. Thus, the N_2ase-catalyzed reaction appears to be quite specific for the reduction of C_2H_2 only to C_2H_4.

Specific Requirement for ATP. The sensitivity of the C_2H_2–C_2H_4 assay of N_2ase permits a deter-

Table I. *Requirements and Product of C_2H_2 Reduction by* Azotobacter *N_2ase*

Complete system contained per ml in μmoles: tris•HCl, 50; creatine phosphate (CrP), 56; ATP, 5; $Na_2S_2O_4$, 20 (all at pH 7.0); and $MgCl_2$, 5; and in mg proteins: heated extract of N_2-grown or NH_3-grown *A. vinelandii*, as indicated, 4; and creatine kinase (CrK), 0.2. Gas volume 36 ml; liquid volume, 4 ml; incubation time, 30 min; temperature, 30°. Aliquots of gas phase assayed by gas chromatography on alumina column.

Incubation system	Gas phase	C_2H_4	C_2H_6	CH_4
			μmoles/incubation	
Requirements				
complete, N_2-grown	0.05 atm C_2H_2 0.95 atm He	24	<0.001	<0.002
Minus enzyme or CRP, CrK, ATP or $Na_2S_2O_4$	0.05 atm C_2H_2 0.95 atm He	<0.001	<0.001	<0.002
Minus C_2H_2	1.0 atm He	<0.001	<0.001	<0.002
Specificity for N_2-grown cells				
complete, N_2-grown	0.1 atm C_2H_2	20.2	<0.001	<0.002
complete, NH_3-grown	0.1 atm C_2H_2	<0.001	<0.001	<0.002
Stability of C_2H_4				
complete, N_2-grown	0.1 atm C_2H_4 0.9 atm He	150	<0.001	<0.002

Table II. *Specificity of Phosphagen Requirement for C_2H_2 Reduction by* Azotobacter *N_2ase*

Complete incubation system, table I, except replacement of creatine phosphate, and creatine kinase by 10 mM ATP, GTP, CTP, or UTP, pH 7.0; 2.2 mg protein of a protamine ppt of N_2ase of *A. vinelandii*. Liquid volume, 2 ml; gas volume, 38 ml; incubation time, 5 min; gas phase, 0.1 atm C_2H_2, 0.9 atm He. Aliquots of gas phase assayed by gas chromatography on ester-amide column.

Phosphagen	mμmoles C_2H_4/incubation
ATP	1440
CTP	0.8
GTP	<0.5
UTP	<0.5

mination of the specificity of its requirement for nucleoside triphosphate. An *Azotobacter* N_2ase preparation was incubated with C_2H_2, reductant and ATP, UTP, CTP, or GTP as the sole energy source (table II). Only ATP supported C_2H_2 reduction indicating that the energy requirement of N_2ase is very specific for ATP. Burns has observed similar phosphagen specificity for the energy-dependent H_2-evolution activity of N_2ase (15).

Enzyme Level. The rate of C_2H_2 reduction is related to N_2ase concentration in a sigmoidal fashion (fig 3a); the plot of the rate of N_2 fixation *vs.* enzyme concentration does not extrapolate linearly to zero enzyme (3, 19) and might show a sigmoidal relationship if the NH_3 assay were sufficiently sensitive to measure NH_3 formation at the lower limits of enzyme concentration. The plot of the rate of C_2H_2 to C_2H_4 reduction *vs.* enzyme is consistent with a 2-component system:

$$\text{Mo-Fe protein} + \text{Fe protein} \rightleftharpoons N_2\text{ase}$$

Time Course. The rate of C_2H_2 reduction *vs.* time is linear for about 45 minutes, and the reaction stops when ATP is exhausted, in complete analogy to the N_2 reduction reaction (fig 3b).

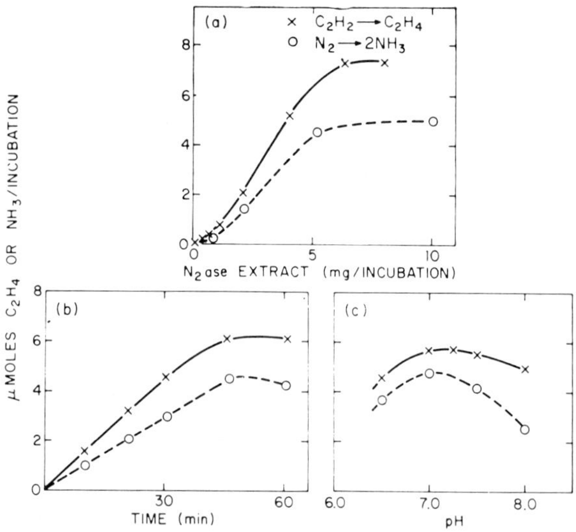

FIG. 3. Reduction by *Azotobacter* N_2ase of $C_2H_2 \rightarrow C_2H_4$ and $N_2 \rightarrow 2NH_3$ *versus* a) enzyme level, b) time, and c) pH. Complete incubation system, table I; liquid volume, 2 ml; gas volume, 38 ml; time, 30 min unless otherwise indicated; gas phase, 0.1 atm C_2H_2 + 0.9 atm He for $C_2H_2 \rightarrow C_2H_4$ and 1 atm of N_2 with 1 atm He as control for $N_2 \rightarrow 2NH_3$. Mixture of TES and MES used to vary pH from 6.5 to 8.0.

pH Maximum. Nitrogenase-catalyzed C_2H_2 reduction was examined from pH 6.5 to pH 8.0 (fig 3c). Maximum activity occurred over a wide range near pH 7.0, which is similar to that observed for N_2 (3) or other reducible substrates of N_2ase (21).

Km of C_2H_2. Nitrogenase is saturated by 0.03 to 0.10 atmosphere C_2H_2 (fig 4a) while 0.5 atmosphere has been observed to inhibit C_2H_2 reduction

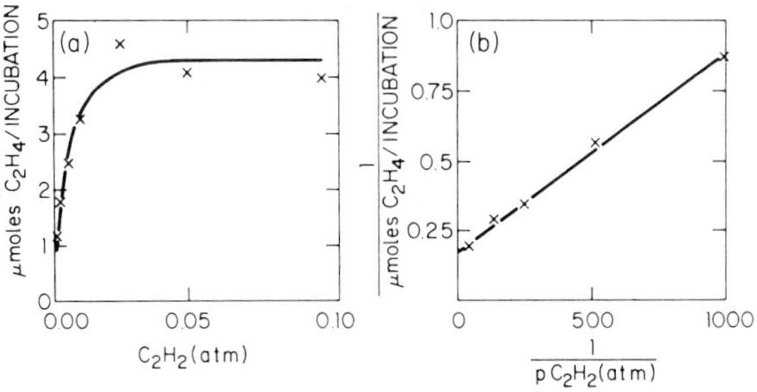

FIG. 4. a) Ethylene formation from C_2H_2 as a function of pC_2H_2 by *Azotobacter* N_2ase, and b) plot of reciprocal velocity *versus* reciprocal pC_2H_2 for determination of K*m* of C_2H_2. Complete incubation system, table I; liquid volume, 2 ml; gas volume, 38 ml; time, 30 min; gas phase, indicated pC_2H_2 plus He to 1 atm.

and reductant-dependent ATPase activity of N_2ase. This inhibition at 0.5 atmosphere C_2H_2 may be due to C_2H_2 or to possible trace impurities in C_2H_2. A typical plot of reciprocal velocity of C_2H_2 reduction *vs.* reciprocal pC_2H_2 is shown in figure 4b. Estimated Michaelis constants of 0.002 to 0.009 atmosphere of C_2H_2, with an average value of 0.004 atmosphere, have been obtained. A tentative K*m* of 0.01 atm has been reported for clostridial

Table III. *Inhibition of ATP-Dependent H_2 Evolution by N_2 and C_2H_2*

Complete system, table I; liquid volume 2 ml; gas volume 38 ml; N_2ase preparation, heated extract of *A. vinelandii,* 7.5 mg; gas phase, as indicated plus He to 1 atm. Hydrogen determined by mass spectrometric analysis of gas phase with He as an internal standard.

Added substrate	H_2 (mμmoles per min per mg protein)	ΔH_2 (mμmoles per min per mg protein)	% Inhibition of H_2 evolution
None	71	0	0
N_2, 0.5 atm	18	53	75
C_2H_2, 0 1 atm	11	61	85

N_2ase (13). Michaelis constants of 0.05 to 0.17 atmosphere of N_2 have been reported for N_2ase (14, 19, 25, 27). Based on partial pressures, the estimated K*m* of C_2H_2 is only about 5 % that of N_2. Based on the calculated concentrations of C_2H_2 and N_2 in an aqueous solution, the estimated K*m* of C_2H_2, 0.1 to 0.3 mM, is similar to that of N_2, 0.03 to 0.1 mM.

Inhibition of H_2 Evolution. The ATP-dependent H_2-evolving activity of N_2ase is decreased by N_2, N_3^-, or N_2O reduction, and the decrease in H_2 evolved is equivalent in electrons to those required for reduction of N_2 to 2 NH_3, N_3^- to N_2 + NH_3, or N_2O to N_2 + H_2O (4, 17, 19, 25). Reduction of C_2H_2 to C_2H_4 also inhibits H_2 evolution by clostridial (13, 32) and *Azotobacter* N_2ase (21). Inhibition by a saturating level of C_2H_2 may be greater than by that of N_2, *e.g.,* 85 % for C_2H_2 and 75 % for N_2 (table III).

Stoichiometry of C_2H_2 Reduction. An excellent balance exists between the concomitant decrease in C_2H_2, increase in C_2H_4, and decrease in H_2 evolution (table IV) during N_2ase-catalyzed reduction of C_2H_2. Thus, reduction of C_2H_2 decreased C_2H_2 by 13.8 μmoles and increased C_2H_4 by 14.5 μmoles, supporting the following relationship:

$$nC_2H_2 \xrightarrow{N_2ase} nC_2H_4$$

The decrease of 12.3 μmoles in H_2 evolution (equivalent to 24.6 μmoles of electrons) produced by C_2H_2 reduction corresponds to the formation of 14.5 μmoles of C_2H_4 (equivalent to 29 μmoles of electrons) and indicates the following electron balance:

$$(H_2 \text{ evolved})_{-C_2H_2} = (H_2 \text{ evolved} + C_2H_4 \text{ formed})_{+C_2H_2}$$

Addition of 0.18 atmosphere CO inhibited C_2H_2 reduction and restored H_2 evolution. Since the loss of C_2H_2 can be accounted for as C_2H_4 and the loss of electrons evolved as H_2 can be accounted for as the electrons required for C_2H_2 reduction, no significant product of C_2H_2 reduction in addition to C_2H_4 is indicated. Furthermore, the equivalence between the decrease in electrons evolved as H_2 and

Table IV. *Stoichiometry of C_2H_2 Reduction by* Azotobacter *N_2ase*

Complete system, table I; liquid volume, 4 ml; gas volume, 36 ml; incubation time, 30 min; aliquots of gas phase assayed by mass spectrometry.

Incubation system	Gas phase	C_2H_2	C_2H_4	C_2H_6	CH_4	H_2
			μmoles/incubation			
Complete	0.02 atm C_2H_2 0 18 atm He	15.5	14.5	0.0	0.0	11.0
Minus enzyme or CrP, CrK, ATP, or $Na_2S_2O_4$	0.02 atm C_2H_2 0.18 atm He	29.3	0.0	0.0	0.0	0.0
Minus C_2H_2	0.2 atm He	0.0	0.0	0.0	0.0	23.3
Complete	0.02 atm C_2H_2 0.18 atm CO	25.0	0.0	0.0	0.0	24.7

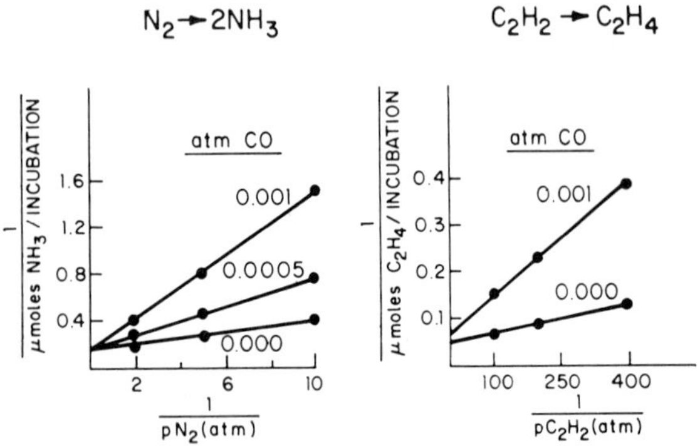

FIG. 5. Competitive inhibition of $C_2H_2 \rightarrow C_2H_4$ and $N_2 \rightarrow 2NH_3$ by CO using *Azotobacter* N_2ase. Complete incubation system, table I; liquid volume, 2 ml; gas volume, 38 ml; time, 30 min; gas phase, indicated pC_2H_2 or pN_2 plus indicated pCO and He or A to 1 atm. N_2 is replaced by A as control for $N_2 \rightarrow 2NH_3$.

the electrons used for ethylene formation indicates that at least the proposed electron-activating site of N_2ase (15, 20) is involved in C_2H_2 reduction, as has been proposed for N_2O and N_3^- reductions (17, 19, 21a).

Inhibition of C_2H_2 Reduction by CO. Carbon monoxide is a competitive inhibitor of N_2 fixation (15, 20, 25a). Figure 5 demonstrates that CO is also a competitive inhibitor of C_2H_2 reduction by *Azotobacter* N_2ase. Furthermore, the similar CO inhibition constants of 2.9×10^{-4} and 3.1×10^{-4} atmosphere for N_2 fixation and C_2H_2 reduction, respectively, provide indirect support that the substrate-complexing site of N_2ase (15, 20) is involved in the reduction of C_2H_2 as well as N_2.

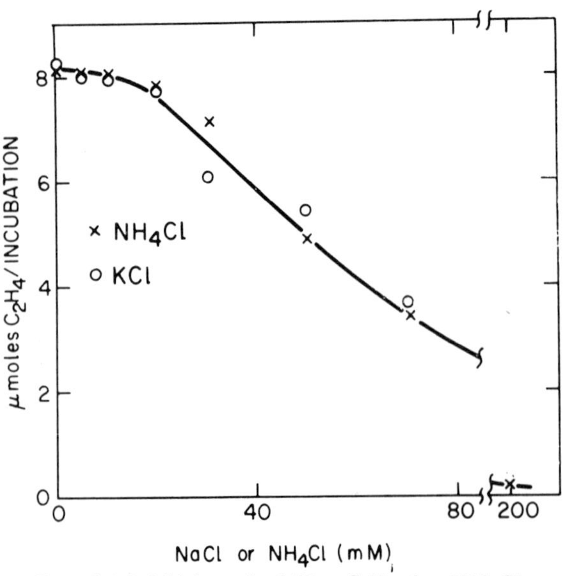

FIG. 6. Inhibition of $C_2H_2 \rightarrow C_2H_4$ by NH_4Cl or NaCl. Complete incubation system with *Azotobacter* N_2ase, table I; liquid volume, 2 ml; gas volume, 38 ml; time, 30 min; gas phase, 0.1 atm C_2H_2, 0.9 atm He; NH_4Cl or KCl as indicated.

Effect of NH_4^+ and Na^+ on $C_2H_2 \rightarrow C_2H_4$ Reduction. Ammonia is the product of N_2 fixation; however, N_2 fixation appears to be relatively insensitive to added NH_4^+ (9). This insensitivity suggests that the product of N_2 fixation does not effectively compete with N_2 for the substrate-complexing site of N_2ase and that NH_4^+ does not control activities of N_2ase associated with electron-activation. Reduction of C_2H_2 to C_2H_4 provides an opportunity to determine if there is a specific effect of NH_4^+ on other N_2ase-catalyzed reductions. Figure 6 indicates no specific inhibition by NH_4^+, since C_2H_2 reduction is equally sensitive to NH_4^+ or Na^+ with 50 % inhibition produced by 50 to 70 mM NH_4Cl or NaCl.

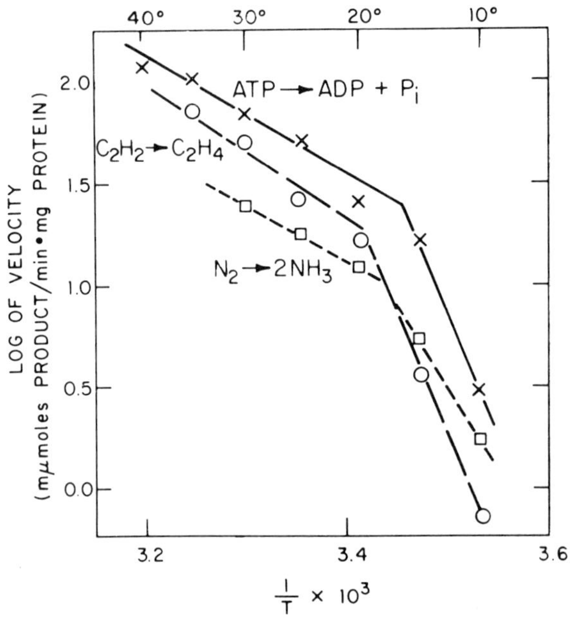

FIG. 7. Arrhenius plots of $C_2H_2 \rightarrow C_2H_4$, $N_2 \rightarrow 2NH_3$ and ATP→ADP + P_i by *Azotobacter* N_2ase in the range 10° to 40°. Complete incubation system, table I; liquid volume, 2 ml; gas volume, 38 ml; time, 30 min for C_2H_2 or N_2 reduction, 15 min for ATP hydrolysis; gas phase, 0.1 atm C_2H_2 plus He to 1 atm for $C_2H_2 \rightarrow C_2H_4$, 1 atm N_2 with 1 atm He as control for $N_2 \rightarrow 2NH_3$, and 1 atm He for ATP→ADP + P_i. Points represent averages of 3 samples.

Activation Energy of C_2H_2 Reduction. The activation energies for reduction of N_2 and other reactions of N_2ase, including ATP-dependent H_2 evolution and reductant-dependent ATPase, have been recently determined (6, 21). A break in the Arrhenius plots for all these activities is observed near 20° with similar but lower activation energies above (13–15 kcal/mole) and similar but higher activation energies below this point (35–50 kcal/mole). Arrhenius plots of C_2H_2 reduction also show a similar break and similar activation energy (fig 7).

Stereochemistry of C_2H_2 Reduction. Ethylene formed from C_2H_2 by reduction by *Azotobacter* N_2ase in a 99.8 % D_2O system was examined by infrared spectrophotometry in order to identify the deuterated species (fig 8). cis-1,2-Dideuteroethylene (843 cm^{-1}) is the major product, as reported for clostridial N_2ase (13). A small amount of mono-

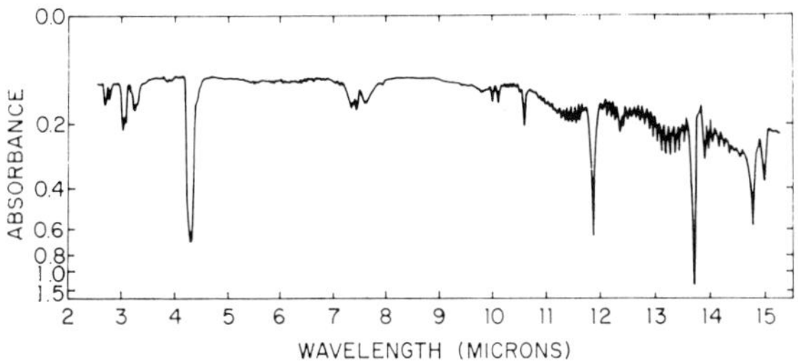

FIG. 8. Infrared spectrum of deuterated ethylenes produced by reduction of C_2H_2 by *Azotobacter* N_2ase. Complete incubation system, table I; liquid volume, 10 ml; gas volume, 30 ml; time, 30 min; gas phase 0.1 atm C_2H_2; all reagents in 99.8 % D_2O and protamine sulfate ppt of N_2ase resuspended in 99.8 % D_2O.

deuteroethylene (1000 cm^{-1}) and a possible trace of trans-1,2-dideuteroethylene (988 cm^{-1}) were found. These results indicate that neither of the original hydrogens of acetylene is replaced during reduction and that acetylene may be complexed to the substrate-complexing site of N_2ase *via* a "side-on" orientation.

Fractionation of N_2ase and N_2-Fixing and C_2H_2-Reducing Activities. Nitrogen-fixing extracts of *A. vinelandii* were fractionated according to the established procedure in this laboratory for N_2ase purification. The N_2- and C_2H_2-reducing activities paralleled each other, and the ratio of C_2H_4 formed to N_2 fixed was found to be in the range of 3 to 4.5 (table V).

Recombination of Mo-Fe Protein and Fe Protein Fractions of N_2ase. Nitrogenase can be separated into 2 protein fractions (4, 22). One contains Fe and Mo and is called the Mo-Fe protein fraction; the other contains Fe and is called the Fe protein fraction. Neither individual protein fraction has biological activity, but N_2ase, the complex formed by the protein fractions, is active for N_2 reduction,

Table V. *C_2H_2 and N_2 Reduction by* Azotobacter *N_2ase Preparations*

Complete incubation system, table I; liquid volume, 2 ml; gas volume, 38 ml; N_2ase preparation, as indicated; gas phase, 0.1 atm C_2H_2, 0.9 atm He for C_2H_2 reduction, 1 atm N_2 with 1 atm He as control for N_2 fixation. C_2H_4 assayed gas chromatographically; NH_3 assayed titrimetrically.

N_2ase preparation	Protein mg/incubation	$C_2H_2 \rightarrow C_2H_4$ µmoles/incubation	$N_2 \rightarrow 2NH_3$ µmoles/incubation	$\frac{C_2H_2 \rightarrow C_2H_4}{N_2 \rightarrow 2NH_3}$
Heated extract	4.3	5.00	1.35	3.7
Pre-protamine precipitate	4.0	<0.002	0.00	
Protamine precipitate	0.91	6.02	1.43	4.2
Protamine supernatant	3.6	<0.002	0.00	

Table VI. *C_2H_2 Reduction by Mo-Fe Protein and Fe Protein Fractions of* Azotobacter *and Clostridial N_2ase*

Complete incubation system, table I; liquid volume, 2 ml; gas volume, 38 ml; incubation time, 30 min; gas phase, 0.1 atm C_2H_2, 0.9 atm He; protein fraction as indicated (see under Methods for preparation and designation of fractions). Data from 2 experiments are shown. C_2H_4 assayed gas chromatographically on ester-amide column.

Fraction	Organism	µmoles C_2H_4/incubation		mg Protein/ incubation
Crude extract	*Clostridium*	1.38		8.7
Mo-Fe protein	"	0.0007	0.0003	1.5
Fe-protein	"	0.0073	0.0003	5.0–8.5
Mo-Fe protein	"	1.72	1.10	1.5
+ Fe protein	"			5.0
Protamine ppt.	*Azotobacter*	7.10		4.2
Mo-Fe protein	"	0.152	0.0008	2.4–4.5
Fe protein	"	0.032	0.0214	2.2–2.9
Mo-Fe protein	"	4.55	2.73	2.4–4.5
+ Fe protein	"			2.2–2.9
Mo-Fe protein	*Azotobacter*	...	0.006	2.4
+ Fe protein	*Clostridium*			8.5
Mo-Fe protein	*Clostridium*	0.102	...	1.5
+ Fe protein	*Azotobacter*			2.9

ATP-dependent H_2 evolution, and reductant-dependent ATPase (4, 22). Nitrogenases of *C. pasteurianum* and *A. vinelandii* were separated into their Mo-Fe protein and Fe protein fractions as described under Methods. Acetylene reduction was determined with the individual fractions, the recombination of the individual fractions from the same species and the cross-combination of the proteins from different species (table VI). No nitrogen-fixing activity remained in the individual fractions, but it was restored by recombination of the fractions. Neither the Mo-Fe protein fraction nor the Fe protein fraction has appreciable C_2H_2-reducing activity, and thus represent the "lowest activity" fractions that have been reported, *e.g.*, our best preparations of Mo-Fe protein (*Clostridium*), Fe protein (*Clostridium*), Mo-Fe protein (*Azotobacter*), and Fe protein (*Azotobacter*) have less than 0.04, 0.02, 0.03, and 0.7 %, respectively, of their recombined activities. Recombination of the 2 protein fractions of *Azotobacter* produces stimulations up to 123-fold, and of *Clostridium* up to 1080-fold. A recent report indicates an enhancement of activity of 5.5-fold by recombination of *Azotobacter* fractions (22). Cross-combination of Mo-Fe protein (*Azotobacter*) + Fe protein (*Clostridium*) or Mo-Fe protein (*Clostridium*) + Fe protein (*Azotobacter*) produces $<\sim 5$ % of the C_2H_2-reducing activity found in the recombination within species experiment. The above cross-combinations have been reported to produce no N_2-fixing activity (11).

These experiments were designed to demonstrate the absence of C_2H_2-reducing activity in each fraction and the presence of this activity in recombined fractions from the same species. Specific activities were not maximized by the addition of an excess of one fraction to a limiting amount of the fraction whose activity is to be maximized.

C_2H_2 Reduction by Bacterial Cells. Characteristics of the reduction of C_2H_2 to C_2H_4 by N_2-fixing cultures of *A. vinelandii* and *C. pasteurianum* are reported in this section. These results indicate the validity of the C_2H_2–C_2H_4 assay of N_2 fixation with living organisms and complement results reported in the previous section with N_2ase preparations from *Azotobacter* and *Clostridium*.

Requirements and Products of C_2H_2 Reduction. Acetylene is reduced to C_2H_4 by N_2-grown cells of *A. vinelandii* and *C. pasteurianum* (table VII). Extracts from these cells contain N_2ase and reduce C_2H_2 to C_2H_4. Control cultures grown on fixed nitrogen sources reduced less than 0.1 % (*Azotobacter*) and less than 5 % (*Clostridium*) the amount of C_2H_2 reduced by the N_2-grown cultures. Extracts from these cells contain little or no N_2ase and do not reduce C_2H_2. Neither ethane nor methane is detected as a product of acetylene reduction; ethylene is not reduced to ethane or methane. The decrease in acetylene during reduction equals the increase in ethylene. The aerobe *Azotobacter* requires aerobic conditions for C_2H_2 reduction, while the anaerobe *Clostridium* reduces acetylene anaerobically. Negligible ethylene is formed in the absence of C_2H_2. Thus, no correction is required for background C_2H_4.

Time Course. Time courses of acetylene reduction by N_2-grown *Azotobacter* and *Clostridium* are shown in figures 9, 10 and 13. The rate of C_2H_2 reduction is constant up to 18 to 20 hours for *Azotobacter* and 6 hours for *Clostridium*. An initial lag is often observed with *Azotobacter*, presumably because of the effect of transfer and dilution. Addition of 40 mM NH_4Cl to N_2-grown cells decreases the rate of C_2H_2 reduction by 95 % after 4 hours (fig 10). This inhibition is in contrast to the effect on N_2ase *in vitro*. The C_2H_2 reduction assay offers a potent method to further define the relationship of N_2 and fixed nitrogen compounds to induction and repression of N_2ase. The results with *Azotobacter* in figure 9 permit calculation of a correlation between C_2H_2 reduction and N_2 fixation. Ethylene formation stops when 0.2 moles of C_2H_4 have been formed for each mole of O_2 initially

Table VII. *Distribution of and Requirements for C_2H_2 Reduction by Cultures of* Azotobacter *and* Clostridium

One ml of culture in early log phase was aseptically added to 4 ml of its nitrogen-free growth media for N_2-grown bacteria and its nitrogen supplemented media for NH_3- or urea-grown bacteria in a sealed incubation vessel of 40 ml total volume containing the indicated gas phase. Incubation time, 16 hr, temperature, 30°. Gas phase assayed chromatographically on ester-amide column.

Organism	Cells ($\times 10^{-6}$)/incubation	Gas phase (atm) A	O_2	C_2H_2	He	µmoles C_2H_4/hr
Azotobacter, N_2-grown	85	0.8	0.2			0.000074
Azotobacter, N_2-grown	85			0.1	0.9	0.030
Azotobacter, N_2-grown	85	0 7	0.2	0.1		1.42
Clostridium, N_2-grown	88				1.0	<0.000001
Clostridium, N_2-grown	88			0.1	0.9	1.05
Azotobacter, N_2-grown	125	0 7	0.2	0.1		1.025
Azotobacter, Urea-grown	240	0 7	0.2	0 1		0.0009
Clostridium, N_2-grown	115			0.1	0 9	2.33
Clostridium, NH_3-grown	210			0.1	0.9	0.12

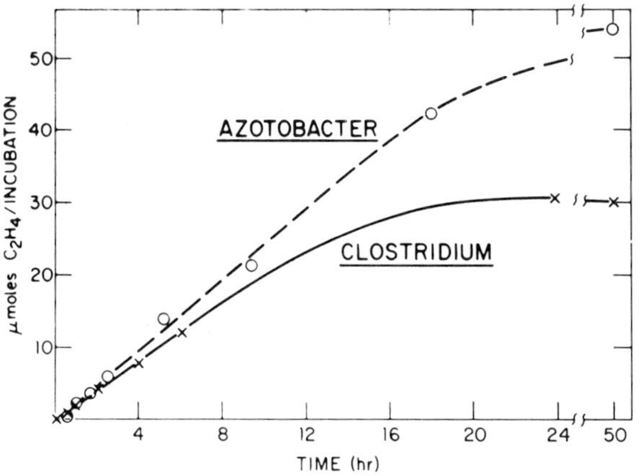

FIG. 9. Time course of $C_2H_2 \rightarrow C_2H_4$ reduction by cultures of N_2-grown *A. vinelandii* and *C. pasteurianum*. Incubation system, table VII; liquid volume, 5 ml; gas volume, 35 ml; temperature, 30°; gas phase, 0.1 atm C_2H_2, 0.2 atm O_2, 0.7 atm A for *Azotobacter* and 0.1 atm C_2H_2, 0.9 atm He for *Clostridium; Azotobacter,* 250 × 10^6 cells; *Clostridium,* 110 × 10^6 cells. Points represent averages of 3 samples.

present. Based on aerobic oxidation of glucose, 1.25 moles of ethylene are formed per mole of glucose oxidized. On the basis of one mole of N_2 fixed per 3 to 4 moles of C_2H_2 reduced, 0.3 to 0.4 mole of N_2 would be fixed per mole of glucose oxidized. This calculated ratio of N_2 fixed per glucose oxidized, based on C_2H_2 reduction, is in reasonable agreement with reported experimental values based on direct measurements of N_2 fixation (35).

Cell Number. A linear relationship exists between cell number and acetylene reduced (table VIII). Ethylene formation measured after 16 hours

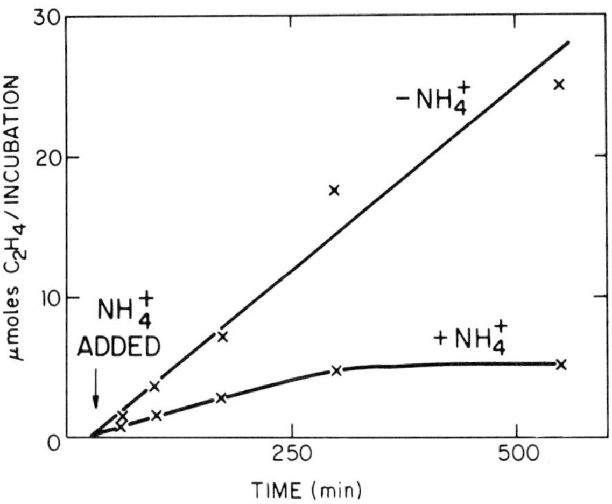

FIG. 10. Effect of NH_4Cl on time course of $C_2H_2 \rightarrow C_2H_4$ reduction by a culture of N_2-grown *A. vinelandii.* Incubation system, figure 9; 300 × 10^6 cells; NH_4Cl added to 3 flasks at 30 min to produce 40 mM NH_4^+ while no addition was made to 3 control flasks.

Table VIII. *Cell Dilution and C_2H_2 Reduction by* Azotobacter *Culture*

Incubation system, table VII, for N_2-grown *Azotobacter*; incubation time, 16 hr; temperature, 30°; gas phase, 0.1 atm C_2H_2, and A:O_2 (0.8:0.2) to 1 atm. Culture serially diluted and cells in original culture counted in a hemocytometer. Ethylene assayed gas chromatographically on ester-amide column.

Cells/incubation	mµmoles C_2H_4/hr incubation	µµmoles C_2H_4/ hr•cell
9,250,000	183	0.020
1,550,000	34	0.022
255,000	6.0	0.024
43,000	0.875	0.021
7,000	0.106	0.015

of incubation at 30° is 0.02 µµmole per hour per *Azotobacter* cell over a 1000-fold range of cell concentration. The extreme sensitivity of the C_2H_2–C_2H_4 assay is indicated; theoretically, as few as 2 to 3 cells produce sufficient C_2H_4 for detection by the H_2-flame ionization system.

Km of C_2H_2. Acetylene saturation of N_2-grown clostridial cells occurs between 0.025 and 0.1 atmosphere (fig 11a), and even 0.5 atmosphere is not inhibitory (fig 11b). The saturation concentration is similar for N_2-grown *Azotobacter,* but these cells

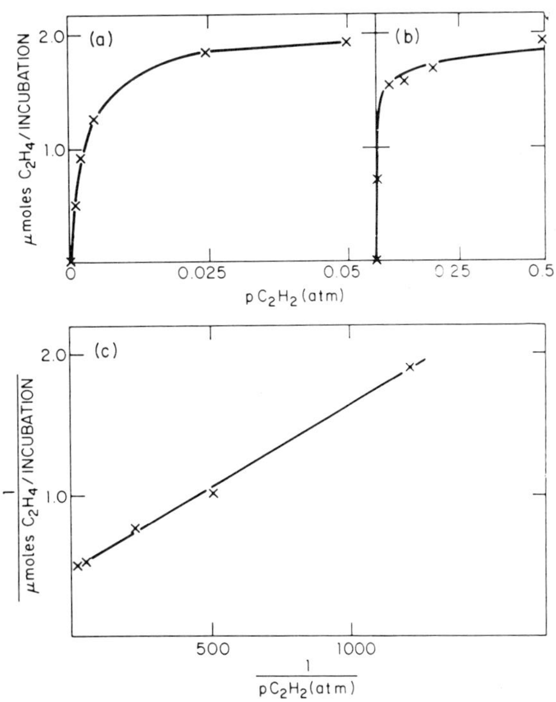

FIG. 11. a) and b). Ethylene formation from C_2H_2 as a function of pC_2H_2 by culture of N_2-grown *C. pasteurianum,* and c) plot of reciprocal velocity *versus* reciprocal pC_2H_2 for determination of *Km* of C_2H_2. Incubation system, table VII; liquid volume, 5 ml; gas phase, indicated pC_2H_2 plus He to 1 atm; points represent averages of 3 incubations.

show an as yet unexplained increase in rate of C_2H_2 reduction at 0.2 and 0.5 atmosphere of C_2H_2. A plot of reciprocal pC_2H_2 *vs.* reciprocal rate of C_2H_2 reduction by clostridial cells is shown in figure 11c. The range of Michaelis constants with *Clostridium* is 0.003 to 0.008 atmosphere C_2H_2 with an average of 0.006, while that for *A. vinelandii* incubated at 0.1 atmosphere or less C_2H_2 is 0.003 to 0.006 atmosphere with an average of 0.005.

Activation Energy of C_2H_2 Reduction. The effect of incubation temperature on C_2H_2 reduction by clostridial cells was determined over the range 10° to 35°. A close analogy with the *in vitro* results on N_2ase is observed which suggests that the limiting factor in growth may be N_2ase activity, and furthermore that this is related specifically to a property of the N_2ase enzyme *per se,* rather than to reactions which furnish energy or reductant to the enzyme. An Arrhenius plot of the cellular activities is shown in figure 12. Results from 20° to 35° form a linear plot with a calculated activation energy of 13 to 15 kcal/mole; results from 10° to 20° are not co-linear with those from 20° to 35°, and a much higher activation energy, approximately 50 kcal/mole, is estimated for the lower temperatures.

C_2H_2 Reduction and N_2 Fixation. Acetylene reduction by N_2-grown *Azotobacter* cultures was compared with N_2 fixation and increase in OD by identical cultures incubated with air under the same conditions. In all cases the 3 parameters measured showed parallel increases with time of incubation. The late-log phase culture (fig 13) did not reduce

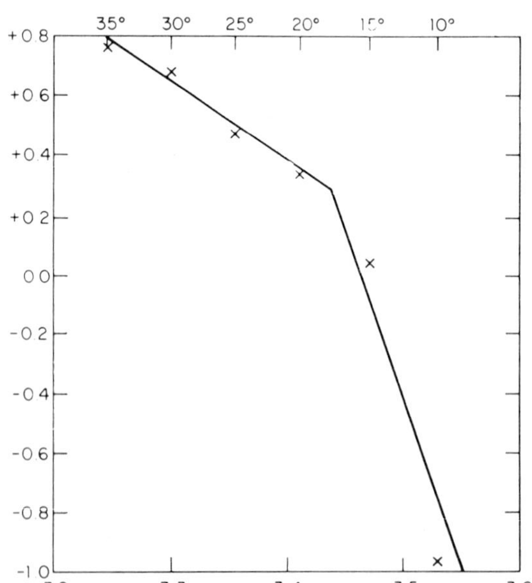

FIG. 12. Arrhenius plot of $C_2H_2 \rightarrow C_2H_4$ by culture of N_2-grown *C. pasteurianum* in the range of 10° to 35°. Incubation system, table VII; liquid volume, 5 ml; gas volume, 35 ml; temperature as indicated; time, 1 hr; gas phase, 0.1 atm C_2H_2, 0.9 atm He.

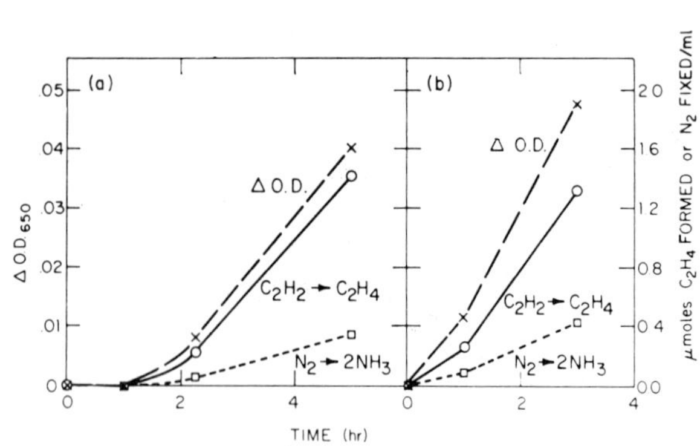

FIG. 13. Acetylene reduction, ΔOD_{650} and N_2 fixation by culture of N_2-grown *A. vinelandii* in a) late log phase, and b) early log phase of growth. Incubation system, table VII; liquid volume, 40 ml; gas volume, 290 ml; temperature, 30°; gas phase, 0.1 atm C_2H_2, 0.2 atm O_2, 0.7 atm A. Initial OD_{650} of late log phase culture after dilution with fresh media was 0.087, while that of early log phase culture was 0.112. Samples of gas phase and culture were analyzed for C_2H_2 and C_2H_4 by gas chromatography and for fixed nitrogen by Kjeldahl analysis at indicated times.

C_2H_2, fix N_2, nor increase in optical density during the initial hour of incubation; in contrast the early log phase culture (fig 13) showed uniformly positive responses during the initial hour. The ratio of moles of N_2 fixed to moles of C_2H_4 formed is 3 to 4.5.

Distribution of C_2H_2-Reducing Activity. The absence of significant C_2H_2-reducing activity in a variety of organisms grown under non-N_2-fixing conditions further establishes the validity of the relationship between C_2H_2-reducing and N_2-fixing ability. Organisms tested included *Clostridium butyricum* under anaerobic conditions on complete

Table IX. *C_2H_2 Reduction by Selected N_2-Fixing and non-N_2-Fixing Bacterial Cultures*

One ml of indicated culture in log phase of growth added to 4 ml of its respective media. Total volume of incubation vessel, 40 ml; gas phase, 0.1 atm C_2H_2, 0.9 atm He for anaerobes, 0.1 atm C_2H_2, 0.2 atm O_2, 0.7 atm A for aerobes; incubation time, 60 min; temperature, 30°. Ethylene assayed gas chromatographically on ester-amide column. Similar results obtained with 0.01 atm C_2H_2.

Organisms	OD_{650}mμ of culture	μmoles C_2H_4/ hr•incubation	N_2-fixing ability
Azotobacter vinelandii	0.30	1.58	+
Clostridium pasteurianum	0.19	1.62	+
Rhizobium japonicum	0.3	0.0002	—
R. melliloti	0.3	<0.00004	—
R. leguminosarum	0.3	<0.00004	—
R. sp. (ATCC 10317)	0.3	<0.0002	—
R. trifolii	0.3	<0.00004	

medium and *Lactobacillus leichmanii, Bacillus subtilis, Bacillus cereus* var. *mycoides, Serratia marcescens, Escherichia coli, Streptococcus lactis, Saccharomyces cerevisiae, Pseudomonas aeruginosa, Pseudomonas fluorescens, Aerobacter aerogenes, Staphylococcus epidermidis, Sarcina lutea, Spirillum itersonii, Proteus vulgaris, Alcaligenes faecalis,* and *Rhodospirillum rubrum* under aerobic conditions on complete medium. All had <0.1 % of the C_2H_2-reducing activity of N_2-grown *A. vinelandii* or *C. pasteurianum.* The absence of C_2H_2-reducing activity (<0.02 % of the *Azotobacter* or clostridial activity) among the various *Rhizobia* grown on the indicated medium (see under Methods) is specifically emphasized, since these data are the most sensitive indication of the inability of cultures of these organisms to fix N_2 (table IX).

C_2H_2 Reduction by Selected Biosphere Samples Assayed in situ. An effective procedure for the *in situ* assay of N_2-fixing activity of the biosphere *via* the C_2H_2–C_2H_4 assay is described in this section. The process is used: 1) to assess requirements, reproducibility, sample variability, and environmental factors that influence C_2H_2-reducing activity of field-grown soybeans, 2) to quantitate C_2H_2-reducing activity of field-grown soybeans through one growing season, and 3) to determine C_2H_2-reducing activities of free-living organisms in the soil and in the hydrosphere. These results represent the first report of the systematic use of C_2H_2 reduction to estimate N_2 fixation during an entire growing season. Since results were obtained during a single season, it is obvious that assays in future seasons will be required to firmly define patterns of N_2 fixation and to establish the effects of variable environmental conditions. Field-grown soybeans were selected for this test because of their agricultural significance. Other biosphere samples were selected to test the universality of the C_2H_2–C_2H_4 assay for determination of N_2 fixation.

Process of in situ *C_2H_2–C_2H_4 Assay*. The 10 steps involved in our *in situ* C_2H_2–C_2H_4 assay process are outlined in figure 14. All the steps are designed to minimize sample alteration and to establish in the assay chamber a micro-ecosystem that is identical in temperature, moisture, porosity, *etc.* to the macro-ecosystem from which the sample is selected. Samples may consist of nodulated plant roots, plant root soil bores, soil bores, or hydrosphere.

Nodulated legumes, after decapitation to eliminate possible background ethylene formation and to decrease the total sample size, were sampled by removal of the complete root system with attached nodules or by a soil bore directly over the tap root. Detached nodules have been used in some preliminary work (23, 33, 34, 36), but they are less active and their collection is time-consuming, particularly in the case of legumes with small nodules, *e.g.*, *Medicago sativa.* We recommend the nodulated root system over the root soil bore system because of: 1) the heterogeneous distribution of nodules per plant and of nodulated plants sown in rows, 2) the injury to nodules by the soil borer, and 3) the lower activity observed with the soil bore technique (fig 18c and 19c). Soil for free-living N_2 fixation is sampled with a soil borer (1-inch × 15-inch graduated in 3-inch steps). Immediately after collection, samples (45 ml) are transferred to the assay chamber, a 50 ml syringe. The diameter of the assay chamber matches the diameter of the soil bore to promote effective gas exchange, addition and removal. Advantages of this assay chamber include: 1) sufficient size to accommodate a representative sample of a nodulated root or 45 ml soil bore plus gas phase, 2) effective and rapid replacement of air with 20 ml additions of desired aerobic [A:O_2(0.8:0.2)], photosynthetic [A:O_2:CO_2(0.8:0.2:0.001)] or anaerobic (A or He) gas phase (step 3), 3) uniform and rapid mixing of 20 ml of C_2H_2 gas mixture (0.2 atm C_2H_2 plus aerobic or anaerobic gas to 1 atm) with sample (steps 5 and 6), and 4) removal and mixing of gas phase at the end of the incubation (step 8), which precludes the requirement for sample inactivation. The determination of C_2H_2 as well as C_2H_4 can provide a useful internal standard for detection and correction of gas leaks, elimination of faulty assays and determination of gas volume of sample in the case of soil bores. Gas volume of soil bores was estimated as 50 % of the volume of the soil bore in the results reported here. It is emphasized that C_2H_2 is a very explosive gas and strict safety measures must be taken to eliminate the possibility of ignition (no smoking) during all steps involving C_2H_2. Observed C_2H_2-reducing activity is converted to a calculated N_2-fixing activity on the basis of the ratio of their electron requirements, *i.e.*, 2 electrons for C_2H_2 reduction and 6 electrons for N_2 reduction. A ratio in the area of 3 to 4.5 is compatible with results of C_2H_2 reduction and N_2 fixation reported in the previous sections. The process described appears to be effective, simple and rapid. More than 450 samples per day have been routinely collected and assayed in the field (steps 1–8) and over 80 samples per hour have been incubated (steps 2–8).

C_2H_2 Reduction by Field-Grown Soybeans—Requirements, Distribution, and Product. Root soil bores or nodulated roots of field-grown soybeans reduced C_2H_2 to C_2H_4 (table X). Neither C_2H_6 nor CH_4 was detected and C_2H_4 was not reduced to C_2H_6 or CH_4. Negligible C_2H_4 (<0.001 % of that with C_2H_2) was formed in the absence of C_2H_2. Root soil bores containing no nodules possessed only a very low level of C_2H_2-reducing activity (0.01–0.02 % of bores containing nodules), equivalent to that of soil bores made 19 inches from the soybean rows or of soil collected around the nodulated roots. Samples containing nodules but not possessing C_2H_2-reducing activity were always found to contain only white nodules. Aerobic conditions are required

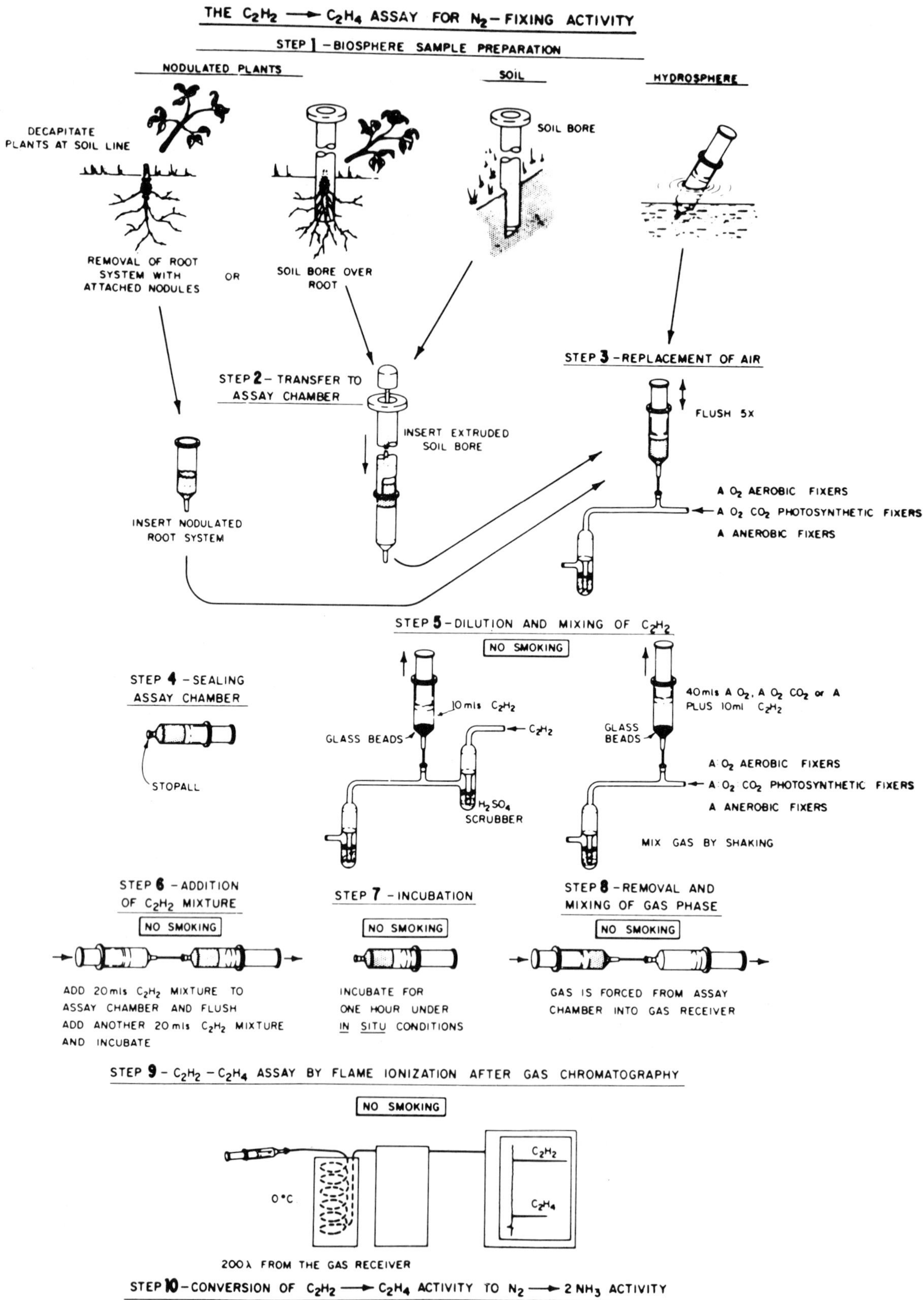

FIG. 14. Steps in $C_2H_2 \rightarrow C_2H_4$ assay for N_2-fixing activity of the biosphere, including samples of nodulated plants, soil, or hydrosphere.

Table X. *Distribution of C_2H_2-Reducing Activity in Soybean Field*

Samples collected between 8 to 9 AM and immediately assayed as outlined in figure 14. Gas phase, 0.1 to 0.2 atm C_2H_2 plus indicated gases $A:O_2$ ($0.8:O_2$), or He to 1 atm; root soil bore volume, 45 ml; total gas phase volume, 40 ml; incubation time, 1 hr.

Sample	Gas phase	μmoles C_2H_4/ day•sample	mμmoles C_2H_4/day •mg fr wt nodule	mg fr wt nodule
Root soil bore	$A:O_2$	0.001	0.001	1030
Root soil bore	$He:C_2H_2$	1.87	3.74	503
Root soil bore	$A:O_2:C_2H_2$	0.026	...	1030
Root soil bore	$A:O_2:C_2H_2$	171	168	0
Nodulated root	$A:O_2:C_2H_2$	195	172	1114
Soil around nodulated root	$A:O_2:C_2H_2$	0.035	...	0
Soil bore between rows 0– 4" deep	$A:O_2:C_2H_2$	0.033	...	0
4– 8" deep	$A:O_2:C_2H_2$	0.030	...	0
8–12" deep	$A:O_2:C_2H_2$	0.028	...	0

for C_2H_2 reduction (table X, ref. 23). Failure to replace air (N_2) with $A:O_2$ results in a 10 to 20 % decrease in C_2H_2 reduction (table XI).

Time Course. The rate of C_2H_2 reduction by nodulated roots or root soil bores of soybeans is constant up to 60 minutes (fig 15) with the standard system described in figure 14. However, it is recommended that heavily nodulated roots be assayed for a shorter time (30 min) since the rate for such

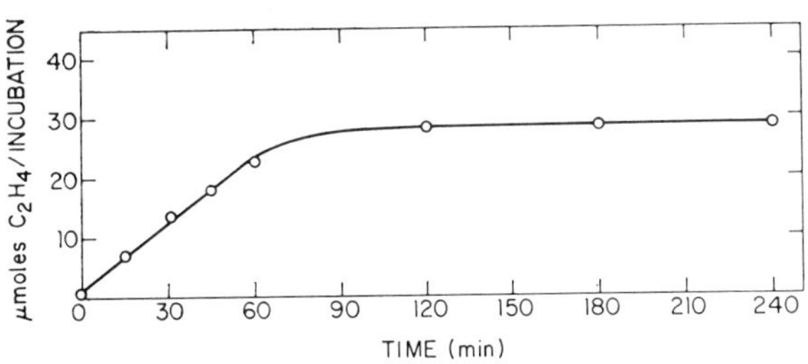

FIG. 15. Time course of $C_2H_2 \rightarrow C_2H_4$ reduction by nodulated soybean roots. Incubation system, figure 14, except that analyses were made from 0 to 240 min. Glass beads were added to assay chambers to facilitate mixing of gas phase.

samples decreases shortly after 60 minutes. This decrease is presumably due to O_2 depletion, since samples which were reflushed with $A:O_2$ and regassed with the C_2H_2 mixture showed activities during a 1 to 2 hour incubation that were comparable to those observed during a 0 to 1 hour incubation (tables XI, XIII).

Interval Between Sampling and Assay. Nodulated roots and root soil bores of soybeans were assayed at 0, 2, 6.5, and 13 hours after sampling (fig 16). Values are expressed on the basis of nodular efficiency, mμmoles C_2H_4 per mg fr wt nodule per day, in order to compensate for the variable nodulation of samples. The results indicate the importance of a minimum interval (0–2 hr) between sampling and assay in order to obtain values which are representative of the *in situ* activity. Consequently, all the results reported in this paper unless otherwise indicated were obtained from assays initiated within 30 minutes after collection.

Sample Variability and Reproducibility. The sample variability with respect to C_2H_2-reducing activity of field-grown soybeans is shown in table

Table XI. *Effect of Air on C_2H_2-Reducing Activity of Soybeans*

Samples were collected between 8 to 9 AM and assayed from 0 to 1 and again from 1 to 2 hr as outlined in figure 14. Air (N_2) replaced $A:O_2$ where indicated.

Sample	Gas phase	μmoles C_2H_4/ day•sample	mμmoles C_2H_2/day •mg fr wt nodule	mg fr wt nodule
Nodulated root				
0–1 hr	$A:O_2:C_2H_2$	659	197	3333
1–2 hr	$A:O_2:C_2H_2$	750	225	3333
0–1 hr	$A:O_2:C_2H_2$	410	91	4511
1–2 hr	$Air:C_2H_2$	328	73	4511
Root soil bore				
0–1 hr	$A:O_2:C_2H_2$	185	79	2340
1–2 hr	$A:O_2:C_2H_2$	190	81	2340
0–1 hr	$A:O_2:C_2H_2$	289	142	2047
1–2 hr	$Air:C_2H_2$	266	130	2047

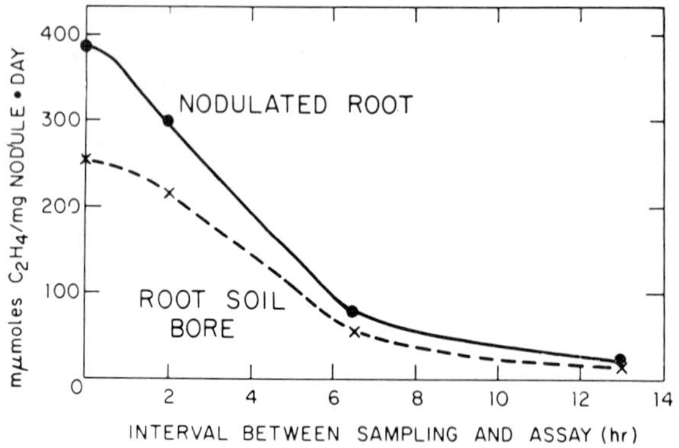

FIG. 16. Reduction of $C_2H_2 \rightarrow C_2H_4$ by nodulated soybean root or soybean root soil bore assayed at various times after collection between 8 to 9 AM. Incubation system, figure 14. Each point represents the average of 5 individual samples.

XII. The standard deviation of C_2H_2-reducing activity of root soil bores or nodulated roots collected at 15, 25, 29, and 44 days post-flowering is 25 to 35 % while that of nodular efficiency is 17 to 22 % for root soil bores and 5 to 10 % for nodulated roots. Thus, expression of activity on the basis of nodular efficiency substantially decreases the variability due to differences in nodule weights.

The excellent reproducibility of C_2H_2-reducing activity of field-grown soybeans is shown in table XIII. Samples collected at various times after flowering show C_2H_2-reducing activities during a 1 to 2 hour assay that are 89 to 107 % of the respective activities determined during a 0 to 1 hour assay.

Km of C_2H_2. Nodulated roots of soybeans are saturated by 0.025 to 0.2 atmosphere C_2H_2 (fig

Table XII. *Sample Variability of Field-Grown Soybeans for C_2H_2 Reduction*

All samples collected from different areas of the same field between 8 to 9 AM on indicated day and assayed immediately as described in figure 14. Each value represents average of 5 samples.

Sampling time (days post-flowering)	μmoles C_2H_4/ day•sample	mμmoles C_2H_4/ mg fr wt nodule•day	mg fr wt nodule/ sample
Root soil bores			
15	33	94	355
	47	72	659
	68	84	812
	43	62	701
Avg	48	79	632
25	105	101	1035
	190	153	1238
	90	180	501
	116	149	776
Avg	125	146	888
Nodulated root			
29	513	176	2908
	514	144	3577
	785	166	4731
Avg	604	162	3739
44	751	226	3319
	412	242	1697
Avg	582	234	2508

Table XIII. *Reproducibility of C_2H_2-Reducing Activity of Field-Grown Soybeans*

Soybean root soil bores or soybean nodulated roots collected between 8 to 9 AM and assayed as in figure 14, from 0 to 1 hr and again from 1 to 2 hr after collection. Each value is average of 5 individual samples.

Sampling time (days post-flowering)	μmoles C_2H_4/sample•day		mμmoles C_2H_4/mg fr wt nodule•day		mg fr wt nodule/ sample
	0–1 hr	1–2 hr	0–1 hr	1–2 hr	
Root soil bore					
0	12	12	74	79	158
0	8	9	71	83	113
2	12	11	65	58	191
44	177	166	56	53	3140
Nodulated root					
44	752	742	226	223	3319

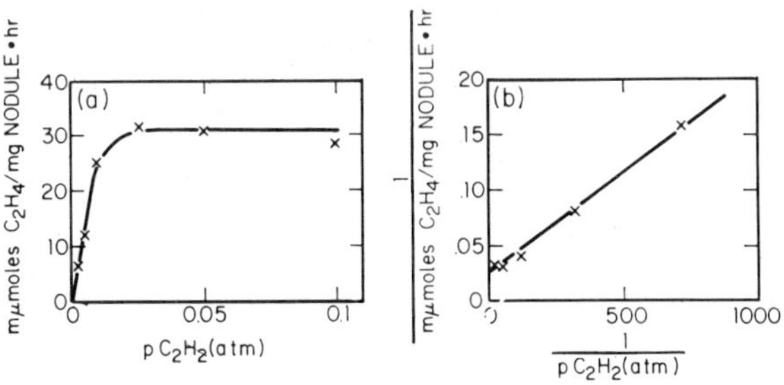

FIG. 17. a) Ethylene formation from C_2H_2 as a function of pC_2H_2 by nodulated roots of soybeans, and b) plot of reciprocal velocity *versus* reciprocal pC_2H_2 for determination of K*m* of C_2H_2. Incubation system, figure 14.

17a); soybean root soil bores showed identical activities when incubated with a C_2H_2 gas mixture of 0.1 to 0.4 atmosphere C_2H_2. Activity was decreased by 0.5 atmosphere C_2H_2. This decrease might be due to direct inhibition by C_2H_2 or to indirect inhibition by the concomitant depletion of O_2. An average K*m* of 0.007 atmosphere C_2H_2 is obtained from a plot of reciprocal pC_2H_2 *vs.* reciprocal rate of C_2H_4 formation. This value is somewhat lower than that reported for excised soybean nodules (23).

Temperature. The effect of temperature of incubation on C_2H_2 reduction by nodulated roots or root soil bores of soybeans was less pronounced than observed with either N_2ase preparations or bacterial cultures. Activity was lower at 10° to 15° and possibly at 35° than at 20° to 30° (table XIV). In contrast, preliminary results suggest that the temperature of growth has a more marked effect on C_2H_2-reducing activity. Nodulated roots from plants maintained in growth cabinets for periods of 1 to 14 days at 30° had only 10 to 20 % of the C_2H_2-reducing activity of those at 20°.

C_2H_2-Reducing Activity and Calculated N_2 Fixation During One Growing Season. Acetylene-reducing activity of field-grown soybeans was determined during a complete growth and maturation cycle. Both the root soil bore (fig 18) and nodulated root (fig 19) techniques were used. Samples were collected on 41 different days for the root soil bore assays and on 27 different days for the nodulated root analyses. The C_2H_2-reducing activity was determined as described in figure 14 utilizing 1 hour incubations. Results are expressed on a 24 hour basis (fig 18a and 19a). The fresh weight of nodules per sample was tabulated (fig 18b and 19b), and the nodular efficiency of C_2H_2 reduction, expressed as mμmoles C_2H_4 formed per mg fresh weight nodule per day, was calculated (fig 18c and 19c). The averages of all analyses on an individual day and during each week are shown.

Nitrogen fixation, as measured by the C_2H_2–C_2H_4 assay, was found to parallel the nitrogen demand of

Table XIV. *Effect of Temperature on C_2H_2 Reduction by Soybeans*

Soybean root soil bores or nodulated roots collected between 8 to 9 AM, flushed with A:O_2 as described in figure 14, equilibrated for 10 min in a water bath at indicated temperature, then incubated for 1 hr after addition of C_2H_2 mixture. Values are the averages of 3 nodulated roots or 5 root soil bores.

Incubation temp	Nodulate root	Root soil bore
deg	*mμmoles C_2H_4/mg fr wt nodule•day*	
10	29	...
15	86	...
20	214	156
25	176	206
30	250	188
35	151	151

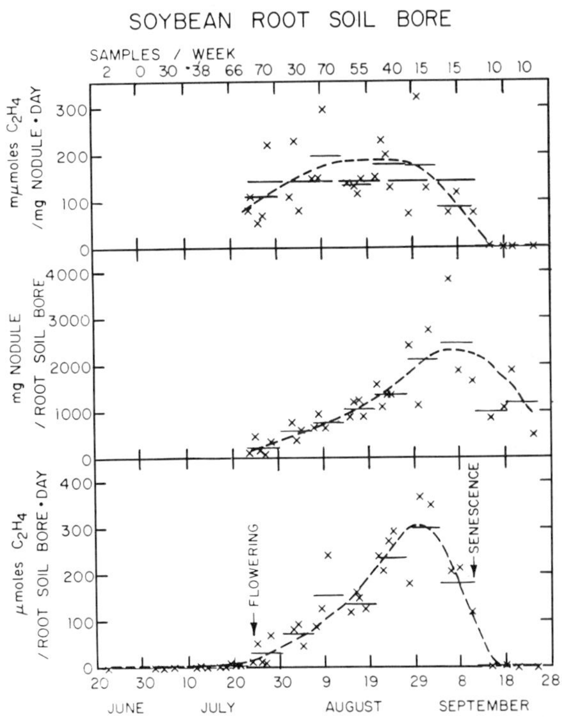

FIG. 18. Summary of a) $C_2H_2 \rightarrow C_2H_4$ reducing activity, b) mg fr wt nodule, and c) mμmoles $C_2H_2 \rightarrow C_2H_4$ per mg fr wt nodule per day by soybean root soil bores at various stages from initiation of activity through flowering and maturation to loss of activity. Incubation system, figure 14; root soil bore volume, 45 ml; gas phase volume, 40 ml. Flowering indicated by initial bud opening and senescence indicated by initial yellowing of leaves. The average of all samples assayed each day (x), during each week — and the average of mμmoles $C_2H_2 \rightarrow C_2H_4$ per mg fr wt nodule per day for the period from flowering to senescence (— — —) are recorded. The number of soil bores of soybean roots assayed each week is recorded. Samples were collected between 8 to 9 AM except for July 27 to 28 and August 9 to 10 when samples were collected at various times as specified in figure 20. Samples were assayed immediately under *in situ* conditions of temperature and moisture. Assay time, 1 hr; results expressed on 24 hr basis.

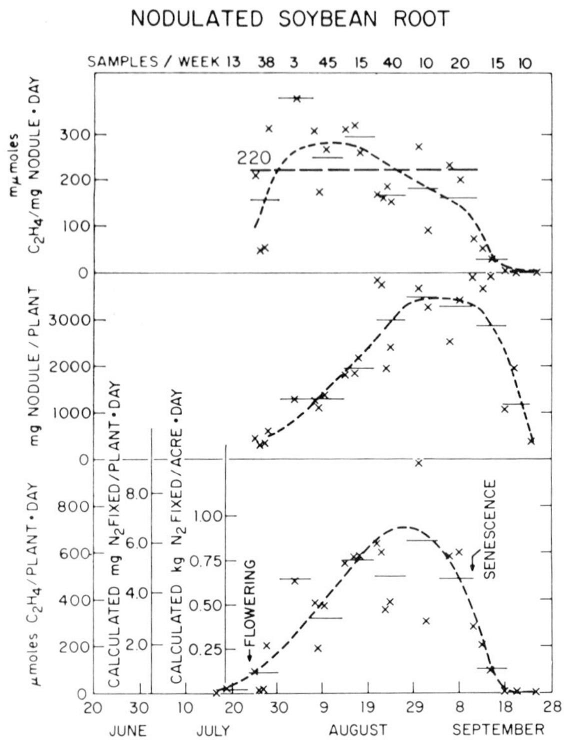

FIG. 19. Summary of a) $C_2H_2 \rightarrow C_2H_4$ reducing activity, b) mg fr wt nodule and c) mμmoles $C_2H_2 \rightarrow C_2H_4$ per mg fr wt nodule per day by nodulated roots of field-grown soybean plants at stages of development similar to those in figure 18. Incubation system, figure 14; gas phase volume, 40 ml. The average of all samples assayed each day (x), each week (——), and the average mμmoles $C_2H_2 \rightarrow C_2H_4$ per mg fr wt nodule per day for the period from flowering to senescence (— — —) are shown. The number of nodulated soybean roots assayed each week is recorded. Sample collection and assay conditions as described in figure 14. Mg N_2 fixed per plant per day and kg N_2 fixed per acre per day are calculated on the theoretical basis of one-third N_2 reduced per C_2H_2 reduced.

the plant. Low C_2H_2-reducing activity occurred until macroscopic flowering was observed, although activity could be detected as early as 32 days before this time; the utilization of residual nitrogen fertilizer during this period may have suppressed N_2-fixing activity to some extent. Following flowering the C_2H_2-reducing activity increased continuously, reflecting the increasing nitrogen requirement for pod formation and filling. Average weekly activity increased from 30 to 299 μmoles C_2H_4 formed per root soil bore per day or 84 to 650 μmoles C_2H_4 formed per plant root per day. The average weekly nodule weight increased from 244 to 2453 mg per root soil bore or 457 to 3478 mg per nodulated plant root. After pod filling was complete the C_2H_2-reducing activity rapidly declined. The decline in nodule weight following senescence lagged behind that of C_2H_2-reducing activity.

The nodule efficiency was relatively constant during the period from flowering to senescence (fig 18c and 19c). The average weekly efficiencies varied from 88 to 196 mμmoles C_2H_4 formed per mg fresh weight nodule per day for root soil bores and from 156 to 378 for nodulated roots. The average nodule efficiency from flowering to senescence was 142 mμmoles C_2H_4 formed per mg fresh weight nodule per day for root soil bores and 220 for nodulated roots.

Summation of the weekly averages of C_2H_2-reducing activity indicates that 22.5 mmoles of C_2H_4 could be formed per plant per season (fig 19a). The calculated N_2 fixation per acre per season is 30 to 33 kg of nitrogen based on: 1) the C_2H_2–C_2H_4 assay, 2) 142,000 plants per acre, and 3) a theoretical conversion factor of one-third N_2 fixed for each C_2H_4 formed.

$$\frac{22.5 \text{ to } 25}{3} \times \frac{28}{10^6} \times 142{,}000 = 30 \text{ to } 33 \text{ kg } N_2 \text{ fixed per acre per season}$$

This calculated value was determined during development of the C_2H_2-C_2H_4 assay and with samples (except for 2 occasions) collected between 8 to 9 AM, a period which may represent less than maximal activity (fig 20). However, this value is in excellent agreement with the average value of 38 kg

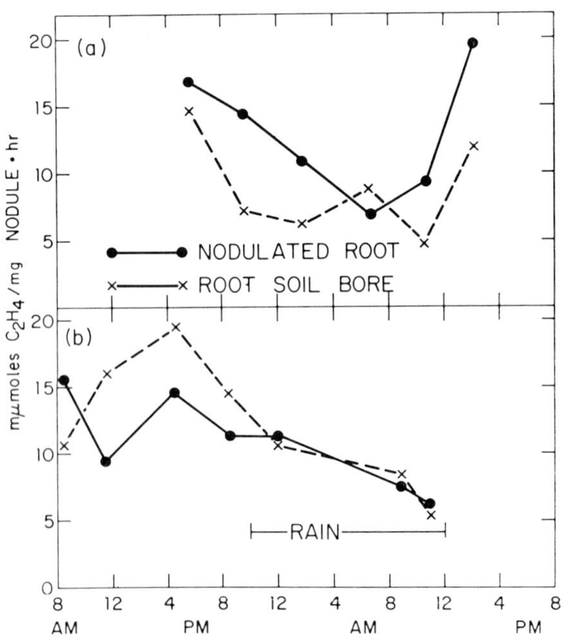

FIG. 20. Diurnal variation of $C_2H_2 \rightarrow C_2H_4$ reduction by nodulated soybean roots and soybean root soil bores collected in the field at indicated times and assayed immediately. a) Samples 3 to 4 days post-flowering, and b) 16 to 17 days post-flowering. Incubation system, figure 14. Each point represents the average of 5 samples. Heavy rainfall occurred during the night and second day of experiment in b).

of nitrogen fixed per acre reported for Kjeldahl and ^{15}N analyses of N_2 fixation by soybeans (35). This correlation provides support for the quantitative reliability of the C_2H_2–C_2H_4 assay performed as outlined in figure 14.

Diurnal Variation. The diurnal variation of C_2H_2-reducing activity of field-grown soybeans was determined at 3 to 4 (fig 20a) and 16 to 17 (fig 20b) days post-flowering. Activities expressed as mμmoles C_2H_4 formed per mg fr wt nodule per hour appear to be maximal for samples collected from noon to 8 PM and minimal for those collected from midnight to 8 AM. Thus, a close relationship between light and N_2-fixing activity is suggested. The effect of light on C_2H_2-reducing activity was further demonstrated with nodulated roots of soybean plants in growth cabinets (fig 21). Control plants maintained on a 16 hour light and 8 hour dark cycle did not show a marked diurnal variation, while experimental plants showed a rapid decline to 30 % of control activity after 17 hours of total darkness, but still had 15 % of control activity after 64 hours of darkness. The initial decline may reflect the depletion of photosynthate, while the residual activity may represent utilization of storage products.

Heavy rainfall eliminated the normal diurnal variation (fig 20b). This effect might be due to increased soil moisture and/or decreased light intensity. Saturation of soybean root soil bores with water decreased their C_2H_2-reducing activities from 170 to 50 mμmoles C_2H_4 per mg fresh weight nodule per day.

Leaf or Pod Removal. Removal of leaves decreased C_2H_2-reducing activity to 12 % of control

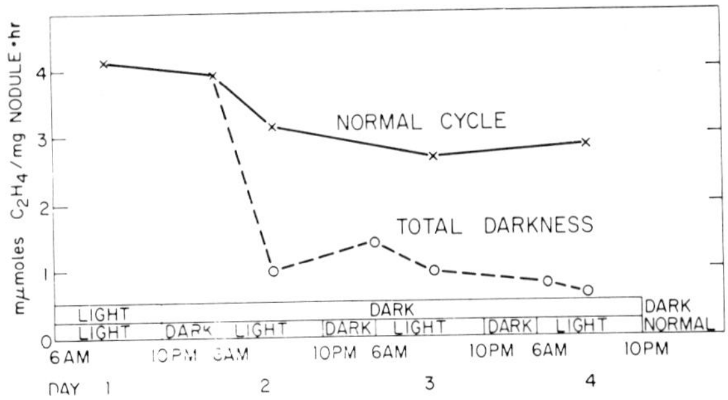

FIG. 21. Effect of light and darkness on $C_2H_2 \rightarrow C_2H_4$ reduction by nodulated soybean roots from plants maintained in plant growth chamber. Incubation system, figure 14. Each point represents the average of 5 individual samples. Control cycle, 6 AM to 10 PM, 24°, with maximum light (see under Methods), and 10 PM to 6 AM, 18°, total darkness. Experimental sample was placed in total darkness with normal temperature cycle at 10 PM of first day of experiment.

after 1 day, and this activity was still only 14 % of control at 10 days after leaf removal.

Removal of pods at 19 days post-flowering did not alter C_2H_2-reducing activity per plant (fig 22a) or mg fresh weight nodule per plant (fig 22b) during the following 10 days. However, during the same period control plants increased both C_2H_2-reducing activity per plant and mg fresh weight nodules per plant. Thus, the magnitude of N_2 fixation reflects the demands of the plant, specifically the pod in this case, for nitrogen.

Varieties. Acetylene reduction provides a technique for the determination of differences in the N_2-fixing activities of different varieties of legumes. A single exploratory experiment was conducted with

Table XV. *C_2H_2 Reduction by Different Varieties of Field-Grown Soybeans*

Nodulated roots of soybean varieties collected between 2 to 3 PM on the same day (114 days after planting) were assayed immediately for C_2H_2 reduction as indicated in figure 14. Soil temperature, 19 to 20°. C1311-Wabash X C1069-Clark X C1069; and UD 61-1806-EC33243 X D49-249L. Each value represents 5 individual roots.

Variety	Stage of development	μmoles C_2H_4/plant•day	mμmoles C_2H_4 /mg fr wt• nodule day
Early-maturing:			
Verde	Yellow, loss of leaves	1.5	1.1
Adelphia	"	1.1	1.5
Clark	"	1.0	1.8
Intermediate-maturing:			
C1311	" + nodule decay	0.8	4.8
C1278	Initial yellowing	7.0	7.9
Kent	"	22.7	17
Delmar	"	30.2	39
Late-maturing:			
Dare	Green, pods still filling	35.5	57
York	"	59.2	98
UD61-1806	"	136	104
Hill	"	162	119

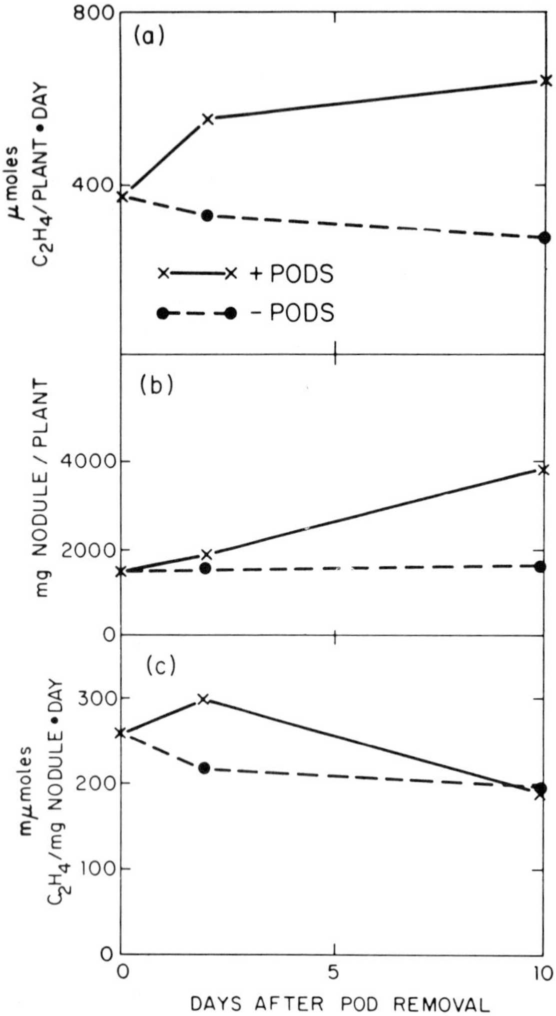

FIG. 22. a to c). Effect of pod removal on $C_2H_2 \rightarrow C_2H_4$ reduction by nodulated soybean root or soybean soil bore. Incubation system, figure 14. Pods were removed 19 days post-flowering.

varieties of soybeans representing different maturation dates. Marked differences found in C_2H_2-reducing activity (table XV) correlated with differences in the stage of maturity at analysis. Varietal differences will be further investigated.

C_2H_2 Reduction by Selected Legumes. Nodulated roots of *Phaseolus vulgaris, Medicago sativa, Arachis hypogea,* and *Pisum sativum* as well as *Glycine max* reduce C_2H_2 to C_2H_4 (table XVI). The nodular efficiencies, mμmoles C_2H_4 formed per mg fresh weight nodule per day, of these legumes were in the range of 106 to 402, similar to those reported by Koch and Evans (23) for excised soybean nodules, but substantially higher than those reported by Sloger and Silver (34) and subsequently by Stewart, Fitzgerald, and Burris (36) for a variety of excised legume and non-legume nodules.

Table XVI. *C_2H_2 Reduction by Selected Legumes*

Nodulated roots of the indicated legumes collected between 8 to 10 AM and immediately assayed as described in figure 14.

Legume	mμmoles C_2H_4/ mg fr wt nodule•day	Calculated kg N_2 fixed/acre•day
Glycine max[1]	220	0.60
Phaseolus vulgaris[2]	106	...
Medicago sativa[2]	161	0.47
Arachis hypogea[2]	402	...
Pisum sativum[2]	304	...

1 Average value from figure 19 for flowering to senescence period.
2 Samples assayed at flowering.

C_2H_2 Reduction by Free-Living Bacteria in Soil. Nitrogen-fixing activities calculated from the C_2H_2-reducing activities of selected soil samples are tabulated in table XVII. The activities varied over 500-fold between locations and suggest that N_2-fixation by free-living soil bacteria varies from negligible to highly significant. Activities under both aerobic and anaerobic conditions were found; the anaerobic activity was higher in the lower soil levels.

Variations in C_2H_2-reducing activity of the Jordan Fertility Plot samples correlated with nutrient applications. The 4 plots receiving recommended additions of N, P, and K showed similar activities, and the average calculated value of N_2 fixation is 0.51 kg of N_2 fixed per acre per day. Plots receiving less or no nitrogen showed, in general, greater activities. The calculated N_2 fixation of the plot receiving only K is 1.52 kg N_2 fixed per acre per day. This is the highest calculated value of N_2 fixation that we have obtained for soil samples analyzed with the C_2H_2–C_2H_4 assay. Samples from all the limed plots had double the activity of those from all the unlimed plots. Initial experiments with soil to which bacterial cultures were added (*Azotobacter* and *Clostridium*) showed negligible increases in C_2H_2-reducing activity.

Miscellaneous Examples of C_2H_2 Reduction. Nitrogen-fixing activity of a variety of miscellaneous samples of the biosphere, including pond water and rumen contents were assayed by C_2H_2 reduction. Samples of pond water reduced C_2H_2 to C_2H_4 when they contained blue-green algae. Negligible C_2H_4 was formed in the absence of C_2H_2. More extensive studies of C_2H_2 reduction by blue-green algae have been communicated recently (36).

Rumen contents from a fistulated steer reduced C_2H_2 to C_2H_4. Ethylene formation was 10-fold greater under anaerobic than aerobic conditions. Methane formation was markedly decreased in the presence of acetylene. The N_2-fixing activity of a rumen calculated on the basis of the anaerobic results was 10 mg N_2 fixed per rumen per day.

Table XVII. *C_2H_2-Reducing Activity and Calculated N_2 Fixation by Soil Samples*

Soil bores were collected from 12 to 4 PM and immediately assayed as indicated in figure 14. Each value represents the average of 6 samples. Soil bores, 2 × 45 ml for 0 to 6" and 1 × 45 ml for 0 to 3" and 6 to 9".

Sample level	Special treatment	mμmoles C_2H_4/sample•day Aerobic	mμmoles C_2H_4/sample•day Anaerobic	Calculated kg N_2 fixed /3" or 6" acre•day Aerobic	Calculated kg N_2 fixed /3" or 6" acre•day Anaerobic
Chester Co., Pa.					
0-3"	...	31	67	0.0020	0.0043
3-6"	...	27	205	0.0018	0.0130
6-9"	...	27	75	0.0018	0.0048
Georgetown, Del.					
0-3"	..	310	155	0.020	0.010
6-9"	...	235	940	0.015	0.106
Jordan fertility plots					
0-6"	+K[1]	11,900	...	1.52	...
0-6"	+N[1]	11,500	...	1.47	...
0-6"	+P[1]	8,700	...	1.12	...
0-6"	No fertilizer	8 600	...	1.10	...
0-6"	+P,K	8,740	...	1.02	...
0-6"	+N,P	7,400	...	0.95	...
0-6"	+N,K	5,850	...	0.75	...
0-6"	+N,P,K,	5,700	...	0.73	...
0-6"	Recommended treatment[2]	4,600	...	0.59	...
0-6"	"	4,200	...	0.54	...
0-6"	"	4,100	...	0.53	...
0-6"	"	2,900	...	0.37	...
0-6"	+3(N), as $NaNO_3$ P,K	2,800	..	0.36	...
0-6"	Excess burned lime	2,260	...	0.29	...
0-6"	+N,P,K	1,400	...	0.27	...
0-6"	Ground bone	1,250	...	0.18	...
0-6"	+3(N) as $(NH_4)_2SO_4$, P,K	1,100	...	0.16	...
0-6"	+Lime	5,850	..	0.75	...
0-6"	—Lime	2,650	...	0.34	...

1 +K = 100 lb K_2O per acre; +N = 24 lb N per acre; +P = 48 lb P_2O_5 per acre.
2 Recommended treatment = 115 lb N, 130 lb P_2O_5,130 lb K_2O per acre.

Discussion

The results reported here support the validity of the C_2H_2–C_2H_4 assay as a sensitive and universal analysis for N_2-fixing activity. The advantages of this assay indicate significant broad applications for measurement of N_2 fixation in both laboratory and field investigations. The essential relationship between C_2H_2-reducing activity and N_2-fixing activity is supported and extended to a most convincing degree by studies within and among cell-free extracts, bacterial cultures, and symbionts.

This relationship is firmly established by detailed experiments with *Azotobacter* preparations which lead to the conclusion that C_2H_2 and N_2 evoke identical responses from N_2ase. Convincingly parallel responses are observed with respect to: requirement for ATP and reductant, linear time course, optimum pH, sigmoidal relationship between rate and enzyme concentration, inhibition of H_2 evolution in an amount equivalent in electrons to those used for C_2H_2 or N_2 reduction, competitive inhibition by CO, relative insensitivity to NH_4^+, activation energy of 13 to 15 kcal/mole above 20° and 35 to 50 kcal/mole below 20°, activity in extracts from N_2- but not urea-grown cells, distribution of activity during fractionation, requirement for both the Mo-Fe and Fe protein fractions of N_2ase, and relative inactivity of the clostridial-*Azotobacter* interspecies recombination of the Mo-Fe and Fe protein fractions. These results provide the most complete correlation between N_2 fixation and C_2H_2 reduction to date. Some of these similarities have been reported for cell-free extracts of bacteroids (24, 25) and *C. pasteurianum* (13, 32). The inhibition of ATP-dependent H_2 evolution by *Azotobacter* N_2ase in an amount equivalent to the formation of ethylene establishes the electron-activating reaction of N_2ase as the source of electrons for C_2H_2 reduction, just as for N_2, N_2O and N_3^- reductions (15, 20). The similar competitive inhibitions of N_2 fixation and C_2H_2 reduction by CO provide strong indirect support for the role of the substrate-complexing site of *Azotobacter* N_2ase for both C_2H_2 and N_2 reduction.

Whole cell experiments are completely consistent with the results obtained *in vitro* and demonstrate the parallel C_2H_2–N_2 relationship in *in vivo* analyses. Thus, cultures of *Azotobacter* or *Clostridium* reduce C_2H_2 to C_2H_4 with characteristics similar to fixation of N_2, which include: anaerobic requirement for *Clostridium* and aerobic requirement for *Azotobacter*, activity in N_2- but not NH_3- or urea-grown cells, time course, and ratio of C_2H_2 reduced to N_2 fixed of 3 to 4.5. In addition, bacterial species without N_2-fixing activity do not possess significant C_2H_2-reducing activity.

Experiments with symbionts establish that the C_2H_2-N_2 correlation is consistently applicable to even these most complex natural N_2-fixing systems. Thus, various legumes also reduce C_2H_2 to C_2H_4 with characteristics similar to fixation of N_2. These include: aerobic requirement, activity only in nodules possessing leghemoglobin, absence of activity in either the root or infecting bacteria, and similar rate of C_2H_4 formation or N_2 fixation per season based on electron requirement.

The characteristics of N_2ase activity, as exemplified by C_2H_2 reduction, are consistent through the entire range of organization studied; results obtained with the most defined *in vitro* system apply with equal validity to even the most complex symbiotic systems. Thus, all systems reduce C_2H_2 to a single significant product, C_2H_4; none reduce substrate C_2H_4; all are saturated by 0.02 to 0.2 atmosphere of C_2H_2; the K*m* values are 0.002 to 0.009 atmosphere, and the calculated activation energies are similar for both N_2ase preparations and bacterial cells with a break in the Arrhenius plots near 20°. A common enzyme, N_2ase, appears to be responsible for C_2H_2 reduction by these diverse systems.

The advantages and disadvantages of the C_2H_2–C_2H_4 assay for N_2 fixation are summarized in table XVIII. The advantages emphasize the superior attributes of this method relative to other assays of N_2 fixation.

Sensitivity of C_2H_4 detection by flame ionization as initially indicated by Hardy and Knight (19) is the critical advantage of the C_2H_2-C_2H_4 assay. The sensitivity of this method is related to other methods of N_2 fixation by the following comparison: the C_2H_2-C_2H_4 method is to the ^{15}N method as the ^{15}N method was to the Kjeldahl method. This sensitivity of the C_2H_2-C_2H_4 assay makes it possible to detect low levels of N_2-fixing activity in biosphere samples, bacterial cultures, or N_2ase preparations, and to investigate variations in N_2-fixing activity with short-term incubations. It is anticipated that the list of N_2-fixing organisms will undergo additions and possibly some deletions because of the C_2H_2-C_2H_4 assay. The only other method with equivalent or greater sensitivity is ^{13}N-incorporation; however, the complex production facilities and short half-life restrict the application of ^{13}N to limited laboratory investigations.

Disadvantages of the C_2H_2-C_2H_4 assay include the indirect nature of the reaction. Although no defined sample has been found with significant C_2H_2-reducing activity and without N_2-fixing activity, the possibility exists for a non-N_2ase-dependent catalysis by biosphere samples of the reduction of C_2H_2 to C_2H_4. The explosive nature of C_2H_2 is emphasized. Laboratories accustomed to utilizing the relatively inert N_2 for ^{15}N or Kjeldahl analysis are cautioned to employ safe practices with C_2H_2.

As recently as 3 years ago Allison (2) found it necessary to state "During the past 50 years much effort has been devoted to the economic evaluation

Table XVIII. *The C_2H_2–C_2H_4 Assay for N_2 Fixation*

Advantages

A) Analytical

Sensitivity. <1 $\mu\mu$mole C_2H_4 per sample is detectable; 10^3 times as sensitive as ^{15}N and 10^6 times as sensitive as Kjeldahl analyses.

Facility. No chemical treatment or manual manipulation of product required, *c.f.* ^{15}N and Kjeldahl analyses which require several time-consuming chemical conversions. Both product and unused reactant are simultaneously recovered in a single, easily secured sample of the gas phase which is directly analyzed.

Specificity. C_2H_4 is separated from CH_4, C_2H_2, C_2H_6.

Internal Standard. C_2H_2 is a natural internal standard and its measurement requires no additional steps.

Storage Stability of Product. C_2H_4 stable indefinitely in simple containers at ambient temperature.

Simplicity. *In situ* assay process and gas chromatography are simple techniques and can be performed by technicians *c.f* ^{15}N mass spectrometric analysis. Few and simple calculations convert raw data to activity values.

Economy. Substrate, assay hardware, and gas chromatographic equipment are inexpensive *c.f.* mass spectrometric equipment and $^{15}N_2$ cost.

Rapidity. Ten complete assays (fig 1) per man hour in the field. Twenty complete gas chromatographic analyses per man hour in the laboratory.

Mobility. All required equipment sufficiently rugged and portable to permit use of a mobile analytical unit.

Sample Conservation. The test material is not sacrificed by atmospheric sampling and sequential samples can be obtained from a single incubation.

B) Biological

Characteristics of $C_2H_2 \rightarrow C_2H_4$ parallel those of $N_2 \rightarrow 2NH_3$.

Universality. N_2ase preparations, and N_2-fixing bacteria, blue-green algae and symbionts reduce $C_2H_2 \rightarrow C_2H_4$, while non-$N_2$-fixing preparations or organisms do not.

Quantitative Relationship of C_2H_2 reduced to N_2 fixed of 3 to 4.

Saturation by a low $pC_2H_2 - Km$ (C_2H_2) of 0.003-0.008 atm *c.f.* $Km(N_2)$ of 0.02 to 0.16 atm.

Specificity of Reaction Product. No significant product other than C_2H_4.

Low Background. Negligible C_2H_4 formed in the absence of C_2H_2 *c.f.* the natural background of ^{15}N.

Metabolic Stability of Product. C_2H_4 is not metabolized *c.f.* the metabolic conversions of inorganic and organic forms of nitrogen.

Disadvantages

Indirect Nature of Reaction. The possibility exists for non-N_2ase catalysis of this reduction by samples of the biosphere.

Explosive Nature of C_2H_2. C_2H_2 is a highly explosive gas *c.f.* N_2 which is relatively inert.

of free-living nitrogen-fixing micro-organisms in soils, but we are nearly as much in the dark now----. There is no sound base upon which to make direct estimate----", while only a year ago the Subcommittee of Production Processes of the United States National Committee for the International Biological Program (37) suggested the following immediate action with respect to nitrogen fixation: "Providing means for the use of available specialized instruments (*e.g.*, mass spectrometers) in the accomplishment of routine analytical determinations." With the advent of the C_2H_2-C_2H_4 assay procedure, these statements are no longer relevant. The C_2H_2-C_2H_4 assay is as crucial for the optimization of biological nitrogen fixation as soil analyses were fundamental to the development of agricultural fertilizer use.

Acknowledgments

The authors are indebted to Mrs. A. Richmond for technical advice on gas chromatographic analysis and to Mrs. B. Musser, Mrs. S. Lidums, Miss C. Madden, Miss M. Parke, Mr. M. Lloyd, and Mr. C. Orth for skilled technical assistance.

Literature Cited

1. Ahmed, S. and H. J. Evans. 1960. Cobalt: a micronutrient element for the growth of soybean plants under symbiotic conditions. Soil Sci. 90: 205–10.
2. Allison, F. E. 1965. Evaluation of incoming and outgoing processes that affect soil nitrogen. In: Soil Nitrogen. W. V. Bartholomew and F. E. Clark, eds. American Society of Agronomy, Incorporated, Madison, Wisconsin. p 573–606.
3. Bulen, W. A., R. C. Burns, and J. R. LeComte. 1965. Nitrogen fixation: hydrosulfite as electron donor with cell-free preparations of *Azotobacter vinelandii* and *Rhodospirillum rubrum*. Proc. Natl. Acad. Sci. U.S. 53: 532–39.
4. Bulen, W. A. and J. R. LeComte. 1966. The nitrogenase system from *Azotobacter*: two-enzyme requirement for N_2 reduction, ATP-dependent H_2 evolution, and ATP hydrolysis. Proc. Natl. Acad. Sci. U.S. 56: 979–86.
5. Bulen, W. A., J. R. LeComte, R. C. Burns, and J. Hinkson. 1965. Nitrogen fixation studies with aerobic and photosynthetic bacteria. In: Non-Heme Iron Proteins: Role in Energy Conversion. A. San Pietro, ed. The Antioch Press, Yellow Springs, Ohio. p 261–74.
6. Burns, R. C. 1968. *Azotobacter* Nitrogenase: activation energy and divalent cation requirement. Biochim. Biophys. Acta: In press.
7. Burris, R. H. and P. W. Wilson. 1957. Methods for measurement of nitrogen fixation. In: Methods in Enzymology, Vol. 4. S. P. Colowick and N. O. Kaplan, eds. Academic Press, New York. p 355–67.
8. Campbell, N. E. R., R. Dular, H. Lees, and K. G. Standing. 1967. The production of $^{13}N_2$ by 50 Mev protons for use in biological nitrogen fixation. Can. J. Microbiol. 13: 587–99.
9. Carnahan, J. E., L. E. Mortenson, H. F. Mower, and J. E. Castle. 1960. Nitrogen fixation in cell-free extracts of *Clostridium pasteurianum*. Biochim. Biophys. Acta 44: 520–35.
10. DeHertogh, A. A., P. A. Mayeux, and H. J. Evans. 1964. The relationship of cobalt requirement to propionate metabolism in *Rhizobium*. J. Biol. Chem. 238: 246–53.
11. Detroy, R. W., D. F. Witz, R. A. Parejko, and P. W. Wilson. 1967. Complementary functioning of two components required for the reduction of N_2 from four nitrogen-fixing bacteria. Science 158: 526–27.
12. D'Eustachio, A. J. and R. W. F. Hardy. 1964. Reductants and electron transport in nitrogen fixation. Biochem. Biophys. Res. Commun. 15: 319–23.
13. Dilworth, M. 1966. Acetylene reduction by nitrogen-fixing preparations from *Clostridium pasteurianum*. Biochim. Biophys. Acta 127: 285–94.
14. Dilworth, M. J., D. Subramanian, T. O. Munson, and R. H. Burris. 1965. The adenosine triphosphate requirement for nitrogen fixation in cell-free extracts of *Clostridium pasteurianum*. Biochim. Biophys. Acta 99: 486–503.
15. Hardy, R. W. F. and R. C. Burns. 1968. Biological Nitrogen Fixation. Ann. Rev. Biochem. 37: 331–58.
16. Hardy, R. W. F. and A. J. D'Eustachio. 1964. The dual role of pyruvate in N_2 fixation. Biochem. Biophys. Res. Commun. 15: 314–19.

16a. Hardy, R. W. F. and E. K. Jackson. 1967. Reduction of model substrates-nitriles and acetylenes—by nitrogenase (N_2ase). Federation Proc. 24: 725.

17. Hardy, R. W. F. and E. Knight, Jr. 1966. Reduction of N_2O by biological N_2-fixing systems. Biochem. Biophys. Res. Commun. 23: 409–14.
18. Hardy, R. W. F. and E. Knight, Jr. 1966. Reductant-dependent adenosine triphosphatase of nitrogen-fixing extracts of *Azotobacter vinelandii*. Biochim. Biophys. Acta 132: 520–31.
19. Hardy, R. W. F. and E. Knight, Jr. 1967. ATP-dependent reduction of azide and HCN by N_2-fixing enzymes of *Azotobacter vinelandii* and *Clostridium pasteurianum*. Biochim. Biophys. Acta 139: 69–90.
20. Hardy, R. W. F. and E. Knight, Jr. 1968. The biochemistry and postulated mechanisms of N_2 fixation. In: Progress in Phytochemistry. L. Reinhold, ed. John Wiley and Sons, Ltd. Sussex, England. p 387–469.
21. Hardy, R. W. F., E. Knight, Jr., and E. K. Jackson. 1967. Substrate versatility of nitrogenase. Bacteriol. Proc. p 112.

21a. Hoch, G. E., K. C. Schneider, and R. H. Burris. 1960. Hydrogen evolution and exchange, and conversion of N_2O to N_2 by soybean root nodules. Biochim. Biophys. Acta 37: 273–79.

22. Kelly, M., R. V. Klucas, and R. H. Burris. 1967. Fractionation and storage of nitrogenase from *Azotobacter vinelandii*. Biochem. J. 105: 3c–5c.

22a. Kelly, M., J. R. Postgate, and R. L. Richards. 1967. Reduction of cyanide and isocyanide by nitrogenase of *Azotobacter chroococcum*. Biochem. J. 102: 1c–2c.

23. KOCH, B. AND H. J. EVANS. 1966. Reduction of acetylene to ethylene by soybean root nodules. Plant Physiol. 41: 1748–49.

24. KOCH, B., H. J. EVANS, AND S. RUSSELL. 1967. Reduction of acetylene and nitrogen gas by breis and cell-free extracts of soybean root nodules. Plant Physiol. 42: 466–68.

25. KOCH, B., H. J. EVANS, AND S. RUSSELL. 1967. Properties of the nitrogenase system in cell-free extracts of bacteroids from soybean root nodules. Proc. Natl. Acad. Sci. U.S. 58: 1343–50.

25a. LIND, C. J. AND P. W. WILSON. 1941. Mechanism of biological nitrogen fixation. VIII. Carbon monoxide as an inhibitor for nitrogen fixation by red clover. J. Am. Chem. Soc. 63: 3511–14.

26. MORTENSON, L. E. 1961. A simple method for measuring nitrogen fixation by cell-free enzyme preparations of *Clostridium pasteurianum*. Anal. Biochem. 2: 216–20.

27. MORTENSON, L. E. 1964. Ferredoxin and ATP, requirements for nitrogen fixation in cell-free extracts of *Clostridium pasteurianum*. Proc. Natl. Acad. Sci. U.S. 52: 272–79.

28. MORTENSON, L. E., J. A. MORRIS, AND D. Y. JENG. 1967. Purification, metal composition and properties of molybdoferredoxin and azoferredoxin, two of the components of the nitrogen-fixing system of *Clostridium pasteurianum*. Biochim. Biophys. Acta 141: 516–22.

29. MORTENSON, L. E., H. F. MOWER, AND J. E. CARNAHAN. 1962. Nitrogen fixation by enzyme preparations. Bacteriol. Rev. 26: 42–50.

29a. MOUSTAFA, E. AND L. E. MORTENSON. 1967. Acetylene reduction by nitrogen fixing extracts of *Clostridium pasteurianum*: ATP requirement and inhibition by ADP. Nature 216: 1241–42.

30. NICHOLAS, D. J. D., D. J. SILVESTER, AND J. F. FOWLER. 1961. Use of radioactive nitrogen in studying nitrogen fixation in bacterial cells and their extracts. Nature 189: 634–36.

31. SCHÖLLHORN, R. AND R. H. BURRIS. 1966. Study of intermediates in nitrogen fixation. Federation Proc. 24: 710.

32. SCHÖLLHORN, R. AND R. H. BURRIS. 1967. Acetylene as a competitive inhibitor of N_2 fixation. Proc. Natl. Acad. Sci. U.S. 58: 213–16.

33. SILVER, W. S. 1967. Biological nitrogen fixation. Science 157: 100–02.

34. SLOGER, C. AND W. S. SILVER. 1967. Biological reductions catalyzed by symbiotic nitrogen-fixing tissues. Bacteriol. Proc. p 112.

35. STEWART, W. D. P. 1966. Nitrogen Fixation in Plants. The Athloe Press, University of London, London.

36. STEWART, W. D. P., G. P. FITZGERALD, AND R. H. BURRIS. 1967. *In situ* studies on N_2 fixation using the acetylene reduction technique. Proc. Natl. Acad. Sci. U.S. 58: 2071–78.

37. United States National Committee for the International Biological Program. 1967. Program Statement of the Subcommittee on Production Processes. National Research Council.

38. United States Government Report of the President's Science Advisory Committee. 1967. The World Food Problem, Vol. 1-3. United States Government Printing Office, Washington, D. C.

Reprinted from PLANT PHYSIOLOGY, Vol. 43, No. 6, June, 1968, pages 865-870.
(PRINTED IN U.S.A.)

Distribution and Metabolism of Protein-Bound Hydroxyproline in an Elongating Tissue, the Avena Coleoptile[1]

Robert Cleland
Department of Botany, University of Washington, Seattle, Washington 98105

Received December 12, 1967.

Abstract. A study has been made of the distribution and metabolism of protein-bound hydroxyproline in an elongating tissue, the excised *Avena* coleoptile. The hydroxyproline-containing proteins of this tissue have been separated into 3 fractions on the basis of their solubilities. The cytoplasmic, trichloroacetic acid-insoluble proteins (S-fraction) contain the bulk of the proline of the cells but only 20 % of the hydroxyproline. The cytoplasm also contains a previously unrecognized trichloroacetic acid-soluble, non-dialyzable fraction (DS-fraction) which is low in proline but contains 20 % of the hydroxyproline. The remaining 60 % of the hydroxyproline is in the wall-bound, cold alkali-soluble fraction (extensin).

Incorporation of free proline into the proline and hydroxyproline of all fractions is linear with time for at least 12 hours. The specific activity of the proline at any time is the same in all 3 fractions while the specific activity of the hydroxyproline is 4-times greater in the S-fraction than in the W-fraction. During a pulse-chase experiment the specific activity of the proline decreases 25 to 40 % in all fractions during the chase. The labeling of hydroxyproline in the wall increases during the chase while that of the DS-fraction remains constant. In the S-fraction, the labeling in hydroxyproline rapidly drops 30 to 35 % during the chase but then remains constant. It is concluded that the majority of the hydroxyproline-proteins in the cytoplasm are not transported to the wall. It is suggested that a sizeable portion of the cytoplasmic hydroxyproline may be located in enzymatic proteins.

The occurrence of protein-bound hydroxyproline in plants (13, 29, 30) and its concentration in the cell wall (7, 16, 18) are now well established. The role of these proteins is still uncertain, although Lamport has suggested (12, 13) that they are structural proteins (extensins) which are involved in the control of cell elongation. If this is so, it will be necessary to understand their properties and metabolism in elongating tissues before we can understand the process of cell elongation.

Little is known about the hydroxyproline-proteins of elongating tissues other than that they are concentrated in the cell wall (5, 13, 19) and that their level is considerably lower than that in callus tissues (5, 10, 13, 19, 29). The metabolism of hydroxyproline-proteins has been extensively studied, to date, only in callus cells (13, 18, 21). The assumption has been made (13) that although callus cells and cells of elongating tissues differ in many ways, they have the same hydroxyproline metabolism.

This investigation into the hydroxyproline metabolism of an elongating tissue, the *Avena* coleoptile, was undertaken for 2 reasons. The primary reason was to provide the information concerning the metabolism and distribution of hydroxyproline-proteins which is necessary for the author's study of the growth-inhibiting effects of free hydroxyproline (3, 4, 6). The second reason was to determine whether the hydroxyproline metabolism of an elongating tissue is necessarily the same as that of callus tissues; *i.e.*, whether conclusions concerning the hydroxyproline-proteins which have been reached with callus tissues can be assumed also to hold for all other tissues. It will be shown that such an assumption is not valid.

Materials and Methods

The experimental material consisted of 14 mm sections cut from 25 to 32 mm long coleoptiles of *Avena sativa*, var. Victory. Seedlings were grown and sections were prepared as detailed earlier (2). Leaves were removed from all sections.

Unless otherwise stated, groups of 100 sections were incubated for 0 to 24 hours in 12 to 20 ml of basal medium that contained K-maleate buffer (2.5 mM, pH 4.7), sucrose (2 % w/v), indoleacetic acid (IAA, 5 μg/ml) and penicillin G (0.1 mM). In some experiments proline-μ-^{14}C (200 μc/μmole) was also present. Incubations were carried out in the dark in beakers which were rotated at 30 rpm on a gyrotory shaker. One of the following 3 methods was then used to prepare the material for analysis.

[1] Supported by Grant GM-12881 from the United States Public Health Service and Grant GB-5385 from the National Science Foundation.

Method A (Whole Tissue). Sections were rinsed with water, extracted 8 times for 5 minutes with 10 ml of boiling 80 % ethanol, washed with 100 % ethanol and dried at 90°.

Method B (Ground Tissue). After rinsing with water, groups of 30 sections were ground in 2 ml of tris buffer (0.05 M, pH 7.5) in an all-glass homogenizer. 4 volumes of absolute ethanol were added and the homogenate was boiled for 5 minutes. The cooled homogenate was then centrifuged and the pellet was extracted 7 times for 5 minutes with 5 ml of boiling 80 % ethanol and then dried. The alternative procedure of precipitating the proteins with 20 % trichloroacetic acid (TCA) at 4°, followed by the TCA wash procedure of Peterson and Greenberg (20) was found to give comparable results.

Method C (Fractionated Tissue). Washed sections were homogenized in 6 ml of tris buffer with 200 μ glass beads in a Virtis "45" homogenizer (11). Walls were separated by filtration through a bed of glass beads and washed by resuspension in water followed by filtration through a new bed of beads. The walls were freed of beads, washed with the TCA series and dried (W-fraction).

The 2 filtrates were combined and TCA-insoluble proteins (S-fraction) were precipitated with 5 % TCA at 4° for 18 hours. After collection by centrifugation, the S-proteins were washed with the TCA series and dried. The supernatant from the centrifugation was dialyzed overnight against 3 changes of distilled water and concentrated. This fraction contained the TCA-soluble proteins (DS-fraction).

Comparable results were obtained when the homogenization was carried out with glycerol instead of tris buffer and when the sections were prehomogenized in an all-glass homogenizer. Omission of the hot TCA wash step, which could be expected to solubilize any collagen-like protein, did not affect the results.

Each of the protein fractions was then hydrolyzed and the proline and hydroxyproline were separated chromatographically and assayed by methods which have already been described in detail (4, 5, 6). Protein was determined by the Kjeldahl technique of Miller and Miller (17).

All experiments were carried out at least 3 times and in most experiments replicates were run.

Results

The presence of protein-bound hydroxyproline in *Avena* coleoptiles has already been reported (4, 10, 19) and is confirmed in this study (table I). Two things should be noted about the data in table I. First, the values for the percent of hydroxyproline in protein and the ratio of hydroxyproline to proline are both considerably lower than the values which have been reported for callus tissues (13, 18). Secondly, although homogenization of the tissue leads to some loss in protein, it does not affect the relative proportion of hydroxyproline, proline, and protein.

The hydroxyproline-proteins of the *Avena* coleoptile have been separated into 3 fractions on the basis of their solubility (table II). The cytoplasmic, TCA-insoluble proteins (S-fraction) contain over 80 % of the bound proline of the cells, but only 20 % of the hydroxyproline. The cytoplasm also contains a previously unrecognized TCA-soluble, non-dialyzable fraction (DS-fraction) which is low in proline but has an additional 20 % of the cells hydroxyproline. The remaining 60 % of the protein-bound hydroxyproline is in the wall fraction (W-fraction).

It should be noted that the wall fraction contains a low but significant amount of proline. Since King and Bayley (10) have reported that they were able to obtain an *Avena* coleoptile wall preparation by essentially identical procedures which was devoid of proline, the possibility was considered that the proline in the W-fraction was due to cytoplasmic contaminants. To test this, aliquots of the W-fraction were subjected to further homogenization and to the NaCl: sodium lauryl sulfate wash procedure of Punnett and Derrenbacker (22). These procedures lowered the protein content of the W-fraction from about 5 % to 1.25 % and lowered the content of both proline and hydroxyproline, but did not significantly alter the H/P ratio. This suggests that this proline is as indigenous to the wall as is the hydroxyproline since procedures which cause a decrease in the proline level also cause a parallel decrease in hydroxyproline.

Table I. *Properties of* Avena *Coleoptile Sections*

Sections were incubated 8 hours in basal medium, then prepared for analysis by method A (whole) or method B (ground). Values are for 1 cm sections.

	Whole	Ground
Dry wt - μg	236	...[1]
Protein - μg	58	46
Proline - μg	2.32	1.90
Proline - % of protein	4.0	4.1
Hydroxyproline - μg	0.21	0.23
H/P × 100	5.2	5.5

1 Value not given due to contamination with glass powder.

Table II. *Distribution of Hydroxyproline and Proline among* Avena *Coleoptile Proteins*

Sections were incubated for 22 hours in basal medium, then fractions prepared by method C. Results are the average of 10 experiments.

Fraction	Proline	Hypro	H/P × 100
	μg/1 cm section		
W-fraction	0.19	0.055	29
S-fraction	1.35	0.019	1
DS-fraction	0.065	0.020	30

Table III. *Removal of Wall-bound Hydroxyproline by NaOH and Formic Acid*

Sections were incubated 20 hours in basal medium + proline-^{14}C (0.5 μc/150 sections). W-fraction then were prepared by method C and aliquots extracted with NaOH (3 times, 1 N, 1 hr, 25°) or formic acid (twice, 1 hr, 100°).

	Hypro/150 sections		Dry wt
	µg	*cpm*	%
Before extraction	7.4	3200	100
After NaOH	0.4	120	55
After formic acid	0	195	51

Olson (18) has shown that the wall fraction of tobacco callus cells contain 2 hydroxyproline-protein fractions; a major fraction (85 %) that is resistant to cold, dilute alkali extraction and a minor component (15 %) which is extractable with cold alkali or hot formic acid. Dougall and Shimbayashi (7) and Lamport (13) have also demonstrated that the bulk of the wall-bound hydroxyproline in callus cells is resistant to cold alkali or formic acid extraction. In contrast, the hydroxyproline-proteins of the *Avena* coleoptile wall are almost totally extracted by both alkali and formic acid (table III). No evidence for a cold alkali-resistant hydroxyproline fraction could be found in *Avena* coleoptiles.

Olson *et al.* (19) have reported that only 40 % of the hydroxyproline of whole *Avena* coleoptile tissues is solubilized by Pronase. This might suggest that the bulk of the hydroxyproline is in a peptide which is resistant to attack by Pronase. Alternatively, the hydroxyproline-proteins of intact tissues may simply be inaccessible to the enzyme. Evidence to support this latter idea is obtained by homogenizing the *Avena* coleoptiles prior to the Pronase treatment. Under these conditions, at least 65 % of the hydroxyproline is solubilized by Pronase (table IV) and in some experiments over 85 % was removed.

We have already shown (4, 6) that free proline rather than free hydroxyproline is the normal precursor of protein-bound hydroxyproline in *Avena* coleoptiles as in other plant (13, 21) and animal

Table IV. *Extraction of Hydroxyproline-Proteins by Pronase*

Sections were incubated 22 hours in basal medium, then prepared by method A (whole) or method B (ground). Tissues then were incubated 16 hours in 0.05 M tris (pH 7.5) ± Pronase (200 μg/ml).

	No Pronase	+ Pronase	% Removed
	µg/1 cm section		
Intact			
Proline	2.31	0.28	88
Hypro	0.12	0.070	42
Ground			
Proline	1.58	0.16	90
Hypro	0.075	0.025	67

systems (26). When *Avena* coleoptiles are incubated with proline-^{14}C, the proline and hydroxyproline of all 3 protein fractions becomes labeled (table V). Time-course studies show that the labeling in each case increases linearly with time for at least 12 hours at which time the proline is exhausted from the medium (fig 1). It should be noted that although the prolines of the 3 fractions differ widely at any time in total labeling (table V), their specific activities are nearly equal (fig 1). In contrast, the specific activities of the hydroxyproline varies markedly between fractions with that of the S-fraction being 4 times higher than that of the W-fraction.

Table V. *Incorporation of Proline-^{14}C into Proline and Hydroxyproline of* Avena *Coleoptile Fractions*

Groups of 100 sections incubated 20 hours in basal medium + 1 μc proline-^{14}C. Tissues prepared by method C.

Fraction	Proline	Hypro	H/P×100
	cpm/100 sections		
W-fraction	24,450	5025	21
S-fraction	180,500	5890	3
DS-fraction	4990	2855	57

The pulse-chase technique has been utilized by Olson (18) to show that most of the hydroxyproline-proteins in the cytoplasm of tobacco callus cells appear to be transferred ultimately to the cell wall. In order to determine whether a similar situation prevails in *Avena* coleoptiles, sections were incubated for 4 hours in a medium that contained proline-^{14}C, and then chased for up to 18 hours in a solution which contained an excess of unlabeled

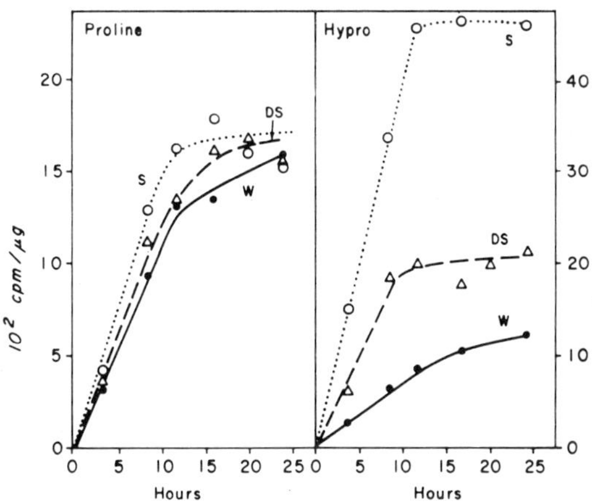

FIG. 1. Time course of proline incorporation into proline and hydroxyproline of *Avena* coleoptile protein fractions. Groups of 100 sections incubated in 13 ml of basal medium + proline-^{14}C (10 μM, 2 μc) for 3 to 24 hours. Tissues prepared by method C. In this experiment the medium was depleted of proline after 12 hours Symbols: W (—●—), S (···○···), DS (---△---).

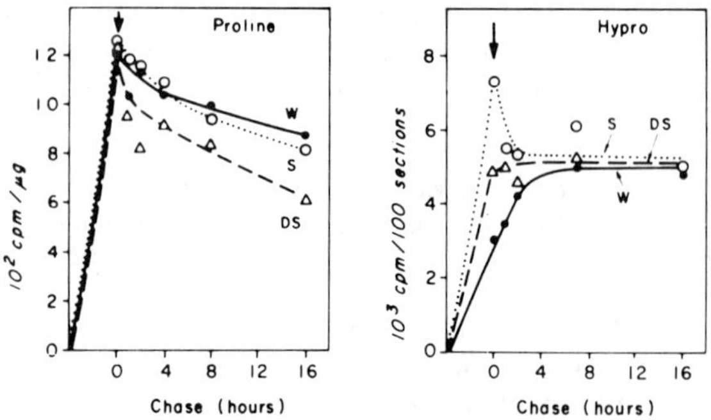

FIG. 2. Changes in labeling of *Avena* coleoptile protein fractions during a pulse-chase experiment. Groups of 150 sections incubated for 4 hours in 12 ml basal medium + proline-^{14}C (1 µM, 4 µc), washed, and placed in groups of 100 in 40 ml basal medium + 100 µM proline. Chase solution replaced after 30 minutes. Tissues prepared by method C. Arrows denote start of chase. Symbols: W (—●—), S (···○···), DS (---△---).

proline (fig 2). During the chase period the specific activity of the proline decreased in all 3 fractions until after 18 hours it was 50 to 75 % of the initial value. Since total proline remains fairly constant in all fractions, this decline would appear to indicate that some turnover of proline-containing proteins does occur in each of the fractions.

The labeling pattern of hydroxyproline presents a different picture (due to the large increase in total hydroxyproline in the W-fraction during the chase the labeling is not expressed in terms of specific activities). During the chase, the labeling of the hydroxyproline of the DS-fraction remains constant while that of the wall fraction continues to increase for 4 to 6 hours. The final labeling of the W-fraction may be 60 to 120 % greater than at the start of the chase. In the S-fraction, the label in hydroxyproline rapidly declines by 30 to 35 %, but then remains constant for the remainder of the chase period. In none of the 4 pulse-chase experiments did the hydroxyproline of the S-fraction lose more than 35 % of its label during the chase period. In each case the total increase in labeled hydroxyproline in the S-fraction was nearly matched by the increase in the W-fraction.

Discussion

It has been shown in this study that the hydroxyproline-proteins of the *Avena* coleoptile can be separated into 3 fractions on the basis of their solubility. The purity of these 3 fractions is difficult to assess and some cross-contamination may occur, but the differences in behavior of their hydroxyproline during the chase period of a pulse-chase experiment suggests that the 3 fractions are distinct. A lack of contamination of the W-fraction with cytoplasmic proteins is also indicated by the fact that while the NaCl: sodium lauryl sulfate extraction removed both proline and hydroxyproline from the W-fraction, the H/P ratio remained unaffected; if cytoplasmic proteins with their much lower H/P ratio had been present, their removal would have caused a marked increase in the H/P ratio. The finding that there was a decrease in the labeling of the proline but not of the hydroxyproline of the W-fraction during a chase period suggests that this fraction contains more than 1 protein and that the proline and hydroxyproline are located in different peptides. This is in agreement with the observation of Lamport (14) that the hydroxyproline-containing peptides of tomato callus walls contain no proline.

The location within the tissue of the S- and DS-fractions is not known, but it is unlikely that they are extracellular (*i.e.*, loosely associated with the wall) since they were always liberated upon disruption of the cells no matter how gentle was the procedure that was used. The DS-fraction (TCA soluble, non-dialyzable), which has not previously been recognized and has apparently been discarded in previous investigations (13, 18, 21), is of some particular interest since it contains 20 % of the hydroxyproline of the tissue and has a H/P ratio that is similar to that of the wall fraction. It is unlikely to consist simply of wall fragments since the labeling of its hydroxyproline remained constant during the chase period while that of the W-fraction continued to increase. Likewise, it is unlikely to be a precursor of wall protein since in that case the labeling of the hydroxyproline would have been expected to decrease during the chase period. It seems more likely that this fraction contains proteins which are chemically similar to the wall proteins but have a different location within the cell.

The metabolism of protein-bound hydroxyproline in the *Avena* coleoptile is similar to that of callus cells (13, 18) in most respects but differs from it in 2 ways. The first concerns the ability of the cytoplasmic hydroxyproline-containing proteins to be transferred to the cell wall. In callus cells the bulk of these proteins undergo such a transfer to the cell wall as judged by the pulse-chase experiments of Olson (18). This is in agreement with the suggestion of Lamport (13) that the hydroxyproline-proteins of plant cells form a single class, the extensins, whose site of action is in the cell wall, and that the hydroxyproline-proteins in the cytoplasm are simply in transit between their site of synthesis and the wall. In contrast, only a minority (30–35 %) of the cytoplasmic hydroxyproline-proteins of the *Avena* coleoptile are transferred to the wall during a pulse-chase experiment, even if it is assumed that all of the decrease in the cytoplasmic fraction and increase in the wall fraction represents transfer of hydroxyproline-proteins. This must mean that there are at least 2 classes of hydroxyproline-proteins; the wall-concentrated extensins and a second group which are located in the cytoplasm and are never transported to the walls. In this connection it is interesting to note that hydroxyproline is reported to be concentrated in the cytoplasmic proteins of brown algae (8) and carrot callus cells (28) and concentrated in the chloroplasts of bean leaves (9). The relative abundance of the 2 classes of hydroxyproline-proteins apparently depends upon the tissue. In callus cells the level of wall-bound extensins and their cytoplasmic precursors is so high that it is difficult to detect the cytoplasmic hydroxyproline-proteins. Because the *Avena* coleoptile has a considerably lower level of wall-bound hydroxyproline it has been possible to demonstrate the existence of the cytoplasmic hydroxyproline-proteins in such a tissue.

Shannon *et al.* (25) have demonstrated that hydroxyproline is present in 3 of 5 isozymes of horse radish peroxidase. This demonstration that hydroxyproline can be present in enzymatic proteins raises the possibility that at least part of the cytoplasmic hydroxyproline in the *Avena* coleoptile is present in enzymatic proteins. It is interesting to note that when Steward and Chang (27) separated the soluble proteins of carrot callus cells by gel electrophoresis they found hydroxyproline in 8 of the 9 protein bands. Although this may simply reflect contamination of the protein bands with a single hydroxyproline-containing protein or fragments of it, it may also mean that hydroxyproline is widely spread among cytoplasmic proteins. However, such a wide distribution of hydroxyproline would not necessarily mean that hydroxyproline exists in a variety of peptides as it has been shown that the same hydroxyproline-arabinose association is present in both horse radish peroxidase (25) and tomato callus cell walls (14, 15). The possibility should be considered that a limited number of hydroxyproline-containing glycopeptides exist which when attached to other proteins or polysaccharides confer some special properties on them such as the ability to be transported across membranes.

The second difference between *Avena* coleoptile and callus cells is the lack of wall-bound, cold alkali-resistant hydroxyproline-proteins in the elongating tissue. This difference may be, in part, the cause of the different growth patterns of these 2 types of cells. Lamport has summarized the evidence (13) which suggests that alkali-resistant hydroxyproline-peptides confer rigidity on the cell wall by cross-linking to arabanogalactans. The slow growth habit of callus cells may be a consequence of the high level of these crosslinks, while the ability of *Avena* coleoptile cells to undergo rapid cell elongation may be a result of the absence of such crosslinks in this tissue.

The fact that the hydroxyproline-proteins of *Avena* coleoptile walls are extracted with cold alkali while those of callus walls are not does not, in itself, show that there is a basic difference between these hydroxyproline-proteins. The ability or lack of ability of a hydroxyproline-protein to be extracted must be due, in large part, to the hemicellulose to which it is attached and the degree of crosslinking. It is already known that there are differences in the hemicellulose components of *Avena* coleoptile (23) and callus cell walls (13, 24). Likewise, the type of polysaccharide to which the hydroxyproline-peptide is linked in tomato callus cells, an arabanogalactan (15), appears to be different from that in corn pericarp, a mucopolysaccharide (1). The chemical relationship between the alkali-extractable hydroxyproline-proteins of the *Avena* coleoptile wall and the alkali-resistant proteins of callus cells cannot be determined until an investigation into the hydroxyproline-sugar linkages in *Avena* coleoptiles is completed.

The difference in hydroxyproline-metabolism between *Avena* coleoptile and callus tissue may indicate a basic difference in hydroxyproline-metabolism between elongating and non-elongating cells, but it is just as likely that these differences are due to the differences in species and in tissues that have been used. More information is needed concerning the hydroxyproline-metabolism of elongating tissues before this can be settled. In any case, it is apparent that the hydroxyproline-metabolism of all tissues is not the same and that conclusions concerning the hydroxyproline-proteins which are reached from studies with callus cells may not be valid for all other tissues.

Acknowledgments

The technical assistance of Mrs. Dorothy Booth is gratefully acknowledged.

Literature Cited

1. Boundy, J. A., J. S. Wall, J. E. Turner, J. H. Woychik, and R. J. Dimler. 1967. A mucopolysaccharide containing hydroxyproline from corn pericarp. J. Biol. Chem. 242: 2410–15.
2. Cleland, R. 1960. Ethionine and auxin-action in *Avena* coleoptile. Plant Physiol. 35: 858–68.
3. Cleland, R. 1967. Inhibition of cell elongation in *Avena* coleoptile by hydroxyproline. Plant Physiol. 42: 271–74.
4. Cleland, R. 1967. Inhibition of formation of protein-bound hydroxyproline by free hydroxyproline in *Avena* coleoptiles. Plant Physiol. 42: 1165–70.
5. Cleland, R. and A. M. Karlsnes. 1967. A possible role of hydroxyproline-containing proteins in the cessation of cell elongation. Plant Physiol. 42: 669–71.
6. Cleland, R. and A. C. Olson. 1967. Metabolism of free hydroxyproline in *Avena* coleoptiles. Biochemistry 6: 32–36.
7. Dougall, D. K. and K. Shimbayashi. 1960. Factors affecting growth of tobacco callus tissue and its incorporation of tyrosine. Plant Physiol. 35: 396-404.
8. Gotelli, I. 1967. The occurrence of hydroxyproline-proteins in algae. M.S. Thesis, University of Washington, Seattle, Washington.

9. KESER, M. 1955. Papierchromatographische Untersuchungen über das auftreten freirer und gebundener Aminosäuren in höheren Pflanzen. Planta 45: 273–88.
10. KING, N. J. AND S. T. BAYLEY. 1965. A preliminary analysis of the proteins of the primary walls of some plants. J. Exptl. Botany 16: 294–303.
11. KIVILAAN, A., T. S. BEAMAN, AND R. S. BANDURSKI. 1959. A partial characterization of maize coleoptile cell wall prepared with the aid of a continually renewable filter. Nature 184: 81–82.
12. LAMPORT, D. T. A. 1963. Oxygen fixation into hydroxyproline of plant cell wall protein. J. Biol. Chem. 238: 1438–40.
13. LAMPORT, D. T. A. 1965. The protein component of primary cell walls. Advan. Botan. Res. 2: 151–218.
14. LAMPORT, D. T. A. 1967. Evidence for a hydroxyproline-*O*-glycosidic crosslink in the plant cell-wall glycoprotein extension. Federation Proc. 26: 608.
15. LAMPORT, D. T. A. 1967. Hydroxyproline-*O*-glycosidic linkage of the plant cell wall glycoprotein extensin. Nature 216: 1322–24.
16. LAMPORT, D. T. A. AND D. H. NORTHCOTE. 1960. Hydroxyproline in primary cell walls of higher plants. Nature 188: 665–66.
17. MILLER, G. L. AND E. MILLER. 1948. Determination of nitrogen in biological materials. Anal. Chem. 20: 481–88.
18. OLSON, A. C. 1964. Proteins and plant cell walls. Proline to hydroxyproline in tobacco suspension cultures. Plant Physiol. 39: 543–50.
19. OLSON, A. C., J. BONNER, AND D. J. MORRE. 1965. Force extension analysis of *Avena* coleoptile cell walls. Planta 66: 126–34.
20. PETERSON, E. A. AND D. M. GREENBERG. 1952. Characteristics of the amino acid incorporating system of liver homogenates. J. Biol. Chem. 194: 359–75.
21. POLLARD, J. K. AND F. C. STEWARD. 1959. The use of ^{14}C-proline by growing cells; its conversion to protein and to hydroxyproline. J. Exptl. Botany 10: 17–32.
22. PUNNETT, T. AND E. C. DERRENBACKER. 1966. The amino acid composition of algal cell walls. J. Gen. Microbiol. 44: 105–14.
23. RAY, P. M. 1963. Sugar composition of oat-coleoptile cell walls. Biochem. J. 89: 114–50.
24. ROBERTS, R. M. AND F. LOEWUS. 1966. Inositol metabolism in plants. III. Conversion of myo-inositol-2-^{3}H to cell wall polysaccharides in sycamore cell culture. Plant Physiol. 41: 1489–98.
25. SHANNON, L. M., E. KAY, AND J. Y. LEW. 1966. Peroxidase isozymes from horseradish roots. I. Isolation and physical properties. J. Biol. Chem. 241: 2166–72.
26. STETTEN, M. R. 1949. Some aspects of the metabolism of hydroxyproline, studied with the aid of isotopic nitrogen. J. Biol. Chem. 181: 31–37.
27. STEWARD, F. C. AND L. O. CHANG. 1963. The incorporation of ^{14}C-proline into the proteins of growing cells. II. J. Exptl. Botany 14: 379–86.
28. STEWARD, F. C., H. W. ISRAEL, AND M. M. SALPETER. 1967. The labeling of carrot cells with ^{3}H-proline: is there a cell-wall protein? Proc. Natl. Acad. Sci. 58: 541–44.
29. STEWARD, F. C., R. H. WETMORE, J. F. THOMPSON, AND J. P. NITSCH. 1954. A quantitative chromatographic study of nitrogenous components of shoot apices. Am. J. Botany 41: 123–34.
30. VANETTEN, C. H., R. W. MILLER, AND I. A. WOLFF. 1963. Amino acid composition of seeds from 200 angiosperm plant species. J. Agr. Food Chem. 11: 399–410.

Reprinted from PLANT PHYSIOLOGY, Vol. 43, No. 4, April, 1968, pages 589–596.
(PRINTED IN U.S.A.)

Transitory Aspects of a Single Protein in Tissues of Solanum tuberosum and Its Coincidence with the Establishment of New Growth[1]

C. A. Ryan[2], O. C. Huisman[3], and R. W. Van Denburgh
Department of Agricultural Chemistry, Washington State University, Pullman, Washington 99163

Received November 6, 1967.

Abstract. **Qualitative and quantitative estimation of a proteinaceous chymotrypsin inhibitor in tissues of potato plants has revealed a transitory existence in all tissues except xylem, flowers, and seeds, where the inhibitor was not found. The distribution pattern in both aerial tissues and in tubers during development and senescence suggests that the concentrations of this protein in all tissues are influenced by meristematic regions of the plants. The transitory existence coinciding with breaking of apical dominance suggests that in the potato plant the protein may have some role in the process of establishing and maintaining meristematic tissue. A search for the protein in other Solanaceae species and in plants from non-related species has proven unsuccessful.**

Studies were recently initiated to acquire information concerning the origin, fate, and function in potato plant tissues of a protein that is a powerful inactivator of proteolytic enzymes (9). This protein was previously isolated and crystallized from potato juice (2, 8) and has been the subject of several reports concerning its chemical properties and inhibitory capacities (2, 3, 6, 7, 8). Recent experiments have revealed that the protein was present not only in potato tubers, where it was originally observed, but could be detected in leaflets of the potato plant (9). The protein could be demonstrated only in carefully chosen leaflets and its existence was found to be transitory (9).

The proteinase inhibitor under study is one of several now known to occur in potatoes and other plant storage organs (4). This inhibitor is specific for chymotrypsin and chymotrypsin-like enzymes, whereas almost all other known proteinase inhibitors from plants are specific for trypsin-like enzymes (4). Data concerning the *in vivo* roles of proteinase inhibitors in plants has been meager. They have been considered as control substances with a capacity to react with endogenous proteolytic enzymes or as protective agents for neutralizing proteolytic enzymes from invading insects and microorganisms. Previous to this report, no attempts have been made to study these inhibitors in tissues other than in the storage organs from which they were isolated.

The present report provides a description of the occurrence and distribution of the specific proteinase inactivator in tissues of entire plants of *Solanum tuberosum* grown in growth chambers, greenhouses, and in the field. The results indicate that the protein is present, but transitory, in all tissues of potato plants except flowers, seeds, and xylem. Its presence and accumulation in tissues consistently coincides with the establishment, or maintainance of, meristematic regions in the plants.

Materials and Methods

Chymotrypsin inhibitor from potatoes [inhibitor I (6)] was 5x crystallized according to the method of Balls and Ryan (2). Rabbit anti-inhibitor I serum was prepared by injecting subcutaneously into rabbits, biweekly, 1 mg 5x crystallized inhibitor I emulsified in complete Freund's adjuvant. Qualitative estimation of the inhibitor protein in standards or in plant extracts was determined by the method of Ouchterloney (5). This method in our hands could detect as little as 8 μg of inhibitor I per ml. Tissues of plants were ground with a mortar and pestle and the whole homogenates, or the juice therefrom, was tested for the presence or-absence of inhibitor I. Although no attempt was made to quantitate exactly the precipitin line, its density was estimated in each test by comparison to known standards.

Juice was obtained by homogenizing tissues with a mortar and pestle and by subsequently squeezing the homogenate using a hand garlic press. Juice that was further clarified by centrifugation will be specified as such in the text.

Quantitative determination of inhibitor I was made using the rapid, sensitive radial diffusion

[1] This work was supported in part by United States Public Health Service Grants 1-K3-GM 17059 and GM 12505. College of Agriculture Scientific Paper #3056, Project 4791.

[2] Research Career Development Awardee of the United States Public Health Service.

[3] Present address: Department of Biochemistry, University of California, Davis, California.

method described by Ryan (7). The method will quantitate 10 μg/ml or more inhibitor I in crude potato extracts.

Inhibitor activities are reported as mg alpha chymotrypsin (3x crystallized from Worthington Biochemical Company) inhibited per mg nitrogen (Kjeldahl). The amount of juice, centrifuged at 2000 $\times$ g for 10 minutes, required to completely inhibit 1 mg chymotrypsin (calculated from the value at 50 % inhibition) was determined by the method of Aldrich and Balls (1). Tyrosine ethyl ester was the substrate at pH 6.3 and 25°.

Russet Burbank potatoes were used for the study except where noted. Potatoes were grown in growth chambers under 1500 ft-c of light in a 14-hour day. The temperature range was 18 to 24°. Plants grown in a greenhouse were grown under natural light during the spring, summer, and fall of 1966. Field potatoes were grown in local soil fertilized with 200 lbs nitrogen per acre.

Results

Anti-inhibitor I serum reacted with purified inhibitor I in the Ouchterloney double diffusion assay to give a single precipitin line (9). When this serum was challenged in this assay system with homogenates or juice from potato plant tissues, a single precipitin line was also found (9). Control serum taken from the rabbit before injections of inhibitor I showed no precipitin line. The high degree of specificity of the serum in detecting the presence of inhibitor I among hundreds of proteins present was the basis for its use in the present study.

Reactivity of Anti-Inhibitor I Serum with Potato Tuber Juice. Nineteen varieties of potatoes were initially tested with the antiserum prepared against inhibitor I purified from Russet Burbank potatoes. Juice from all 19 varieties gave a single precipitin line that was, visually, fully crossreacting with inhibitor I from Russet Burbank tubers. Total inhibitory activity toward alpha chymotrypsin was also tested in the juice from all 19 varieties. The values for all varieties ranged from a high of 1.15 μg chymotrypsin inhibited per mg juice N to a low 0.23 μg/mg N. The mean was 0.51 μg chymotrypsin inhibited per mg juice N. Russet Burbank variety, used for preparation of pure inhibitor, inhibited 0.45 μg chymotrypsin per mg juice N. However, as will be discussed later, the pattern of concentration of inhibitor I within individual potato tubers depended upon the age and condition of the tubers. Thus the total inhibitory activities cannot be used with any certainty as a genetic trait of any given variety. The complete immunological crossreactivity of the proteins however did indicate that the protein among the different varieties is genetically similar.

Determination of Total Chymotrypsin Inhibitory Capacity in Tubers and Roots of the Solanaceae *Family and in Tubers, Roots, and Bulbs of Other Families.* Previous studies have shown that at least 2 proteinaceous inhibitors exist in potato tubers (6). It was considered possible that these inhibitors might be ubiquitous to underground tissues of the Solanaceae family. Juice from nightshade, eggplant, tomato, and pepper roots or stems were tested for inhibitory activity. The presence of chymotrypsin inhibitors was also tested in members of several other plant families that have fleshy underground tissues. Using inhibitor I antiserum, the presence of inhibitor I was determined in centrifuged juice from these tissues by the Ouchterloney (5) method.

Table I gives the plants of the various families and the particular tissue tested. The results in table I show that inhibitor I was found only in the potato although several of the plants contained

Table I. *Inhibition of Chymotrypsin by Juice from Various Tubers, Roots, and Bulbs*

Family	Species	Common name	Tissue	Chymotrypsin inhibited by juice (mg/mg N)	Cross reactivity with inhibitor I antiserum
Solanaceae	*S. tuberosum*	Potato	Tuber	0.45	+
	S. nigrum	Nightshade	Stem	. . .	—
	S. melongena	Eggplant	Stem	. . .	—
	L. esculentum	Tomato	Root	0.11	—
	C. annum	Pepper	Root	0.07	—
Convolvulaceae	*I. batatus*	Sweet Potato	Tuber	0.30	—
Musacea	*M. sapientum*	Banana	Tuber	3.85	—
Arum	*C. antiquorum*	Taro	Tuber	1.85	—
	C. esculenta	Dasheen	Tuber	0	—
Cruciferae	*B. napobrassica*	Rutabaga	Tuber	0.02	—
	B. rapa	Turnip	Tuber	0	—
Ranunculacae	*P. officinalis*	Peony	Tuber	0.02	—
Lilaceae	*A. officinalis*	Asparagus	Tuber	0	—
Iridaceae	*I. florentina*	Iris	Bulb	0	—
Amaryllidaceae	*A. cepa*	Onion	Bulb	0	—

considerable inhibitory activity toward chymotrypsin. Banana tuber juice contained almost 10 times more inhibitory activity than either sweet potatoes or white potatoes. The taro contained a high inhibitory capacity, but its close relative, the dasheen, contained none.

Distribution of Inhibitor I in Potato Tubers. Potato tubers (Russet Burbank) of varying age and development were dissected into small individual sections of tissue and the juice or mascerates of individual samples were tested immunologically for the presence or absence of inhibitor I. In very young attached tubers, 1 to 3 cm in diameter, the inhibitor protein was found evenly distributed throughout all of the tissues. In older attached tubers and in freshly harvested mature tubers, the inhibitor was found to be uniformly located in the external phloem and cortex, *i.e.* the area outside the vascular ring. The concentration decreased significantly just inside of the vascular ring but increased again slightly in the center pith. In a tuber removed from cold storage after 1 year, a somewhat similar pattern was seen, but there was now in the outer phloem and cortex a decreasing gradient from the apical end toward the stem end. The inhibitor in the stem end cortex and center pith had nearly disappeared. In sprouted tubers, the overall concentration of inhibitor I had decreased even more, particularly near the sprouted eye. The sprouts themselves, however, exhibited the presence of inhibitor and the concentration was particularly high at each axil and at the apex of the new sprout. As will be described in a later section, seed pieces planted in the field lost inhibitor I protein completely as sprouts elongated and emerged from the soil. It was concluded that the inhibitor concentration in tubers was definitely changing with storage and its disappearance from the tuber closely coincided with the establishment of the growth of new sprouts.

Using the radial diffusion method, a quantitative estimate of inhibitor I in potato tubers was made. A newly harvested mature potato was selected and a longitudinal slice was removed that was about 1 cm thick and included the area of the stem attachment and the apex. A similar slice was taken from a potato stored 1 year at 4°. Each slice was dissected into the visibly defined areas comprising the tissues outside the vascular ring (cortex and outer phloem), the dense tissue just inside the vascular ring, and the less dense center pith. These tissues were then subdivided into about 30 equivalent sections. The tissue sections were mashed to a pulp with a mortar and pestle and the juice was collected and centrifuged at 2000 $\times$ g for 10 minutes. The juice was analyzed for total protein and total inhibitor I. The results are shown in table II. Whereas the total biuret protein remained relatively constant among all of the tissue sections, the differences in percent inhibitor I between the 2 potatoes were striking, especially at the stem end and center pith.

In order to find if the results from the slices tested typically represented new and stored potatoes, 4 typical tubers were selected from each age group and sections of the stem and apical cortex and of the center pith were removed and the juice recovered and tested as described above. In addition, 4 very young tubers 2 to 3 cm in diameter were detached from parent plants and similarly tested. The cumulative results are shown in table III. Again, with age, the striking change in concentration at the stem end and center pith were seen. The decreasing ratio of percent inhibitor I in the stem versus apical cortical regions is shown in the right-hand column of table III. Coinciding with this ratio was the concentration of inhibitor I in the center pith, which decreased markedly with the age of the tuber. It was also noted that in the very young tuber, the cortical areas exhibited a uniform but lesser concentration than in the new mature tuber. Although inhibitor I in individual tubers can vary considerably, the cumulative results are consistent with the earlier observations that the levels of the inhibitor are undergoing constant change during aging, decreasing near the stem and center pith but remaining in the highest concentra-

Table II. *Summary of the Distribution of Inhibitor I in Tissues of New and Aged Potato Tubers*

Tissue tested	Area tested	New potato: mg Protein per ml juice	New potato: Inhibitor I % of juice protein	Year old potato, 4°: mg Protein per ml juice	Year old potato, 4°: Inhibitor I % of juice protein
	Stem end	12.19	5.17	10.90	0.71
Cortex	Medial area	10.77	2.89	10.65	0.62
	Apical end	9.31	1.56	9.31	3.00
Inner	Stem end	8.71	0.45	12.46	0
phloem	Medial area	7.52	0.60	11.85	0
area	Apical end	8.66	0.51	10.56	0.14
	Stem end	7.22	0.61	10.47	0
Pith	Medial area	7.13	0.52	11.16	0.03
	Apical end	8.21	0.68	8.44	0.07

Table III. *Distribution of Inhibitor I in Cortex and Pith of Potato Tubers in Various Developmental States*

Tubers	Inhibitor I % of juice protein	Ratio % stem/apical
Very young potato tubers 2–3 cm diameter		
Stem cortex	0.74	
Center pith	0.86	1.00
Apical cortex	0.74	
New mature potato tubers		
Stem cortex	1.37	
Center pith	0.75	0.90
Apical cortex	1.53	
Year old mature potato tubers		
Stem cortex	0.42	
Center pith	0.27	0.40
Apical cortex	1.05	

tions near the region of potentially new growth, *i.e.* the apex.

Distribution of Inhibitor I in the Vegetative Potato Plant. The distribution of inhibitor I in vegetative tissue varied depending upon the physiological state of the potato plants. At particular times during development of plants, the protein could be demonstrated qualitatively by the Ouchterloney method in every tissue or organ except in the xylem of the main stem, in flowers and in seeds. The concentration in leaflets, petioles, stem, and roots was generally highest just preceding or during the establishment of new growth, receding or disappearing thereafter.

Inhibitor I was observed quantitatively in leaflets from a young potato plant grown in the growth chamber from true seed and from a young plant also grown in the growth chamber, but from a sprouted potato tuber. When tested, both plants were still developing new rhizomes. Tubers had not yet begun to form. The seedling plant had not visibly begun to break apical dominance whereas the plant from the seed piece was breaking apical dominance at the lower 4 or 5 axils. The small leaflets from the 3 petioles near the apex of both plants were not assayed. These leaflets were small and in all plants tested consistently exhibited the presence of inhibitor I. From the fourth petiole down from the apex to the base of the plant selected, leaflets were assayed for inhibitor I concentration by the radial diffusion method. The crude juice from the terminal and from the leaflet twin nearest the stem on each petiole was tested for inhibitor I. The concentration of inhibitor I in these leaflets at each node are shown in figure 1. A comparison of leaflet inhibitor I between plants shows that the plant grown from the seed piece, just breaking apical dominance at nodes 13 through 16, has much higher levels in most leaflets than the plant grown from true seed. The distribution with respect to leaflet and petiole position is also different. In the seedling plant, not yet breaking apical dominance, inhibitor I is absent or nearly so in the lower leaflets in contrast to the high concentration in the larger plant near the middle lower leaflets. Although the 2 plants described are from different origins, *i.e.* 1 from true seed and 1 from a seed piece, it appears that the stage of development of

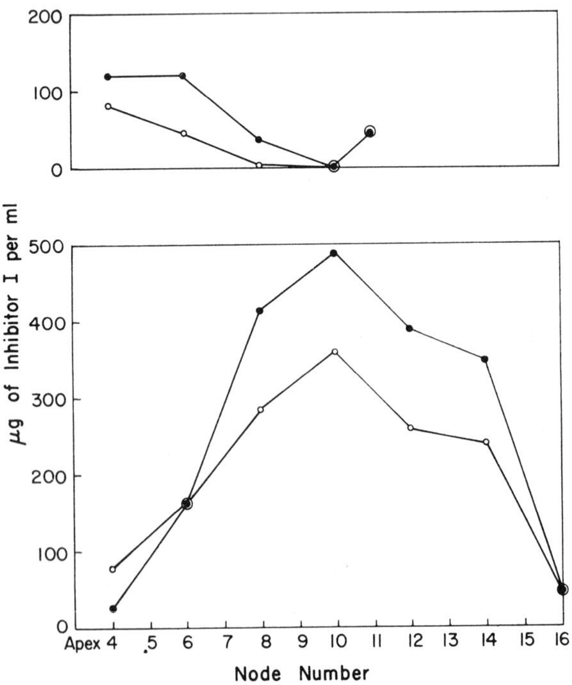

Fig. 1. Quantitative determination of inhibitor I in leaflets of potato plants grown from true seed (top) and from a sprouted tuber piece (bottom). The solid circles (●) represent the μg inhibitor I in juice from leaflets nearest the main stem on petioles of the indicated nodes. Open circles (○) represent μg inhibitor I in juice of the terminal leaflet from each petiole.

the plants, rather than origin, determines the levels of inhibitor I in the tissues.

The youngest leaflets on each petiole, nearest the stem, contain the highest levels of inhibitor as compared to terminal leaflets (fig 1). It has already been mentioned that the youngest leaflets near the apex usually contain inhibitor I. In addition, lateral buds consistently contained inhibitor I no matter what stage of growth or environmental condition the plant was in, except late senescence.

A pattern of transient existence of the inhibitor I in leaflets of very young seedling plants has previously been described (9). In those experiments, inhibitor I appeared in leaflets just preceding new rhizome growth and nearly disappeared after several rhizomes were established (about 50 days after planting). The interesting observation was made that decapitating very young plants before rhizome formation promoted the accumulation of inhibitor I protein in leaflets. In rapidly growing plants where the protein had appeared and subsequently receded, the removal of rhizomes and apex simultaneously caused a striking increase in the numbers of leaflets containing inhibitor I within 48 hours. These results were the first indications that new growth centers were exerting a control upon the accumulation of inhibitor I protein in leaflets. It is now known that seedling plants maintained in the growth chamber beyond about 60 days exhibit an increase in inhibitor I in most leaflets and the plants generally tend to break apical dominance. If the plants are transferred to a greenhouse under natural light in the short-day period of February to April within 60 days of planting they tend to grow with a single stem and apex. Under the short day conditions inhibitor I remains in low concentration throughout most of the tissues of the plants.

Quantitative observations of the accumulation of inhibitor I in seedling plants, induced by excising apices and rhizomes, are reported here using potato plants grown from seed in growth chambers. The plants possessed several actively growing rhizomes but had not yet begun to form tubers (beyond 50 days after planting). Table IV shows quantitatively a typical response of inhibitor I 48 and 96 hours after excision of both the apex and rhizomes. The areas of the plants tested were A) the apical area of the control plant, B) the leaflets from the 3 upper-most petioles nearest the apex, C) the terminal leaflets from the remaining 3 to 4 lower petioles, D) the upper stem corresponding to the upper leaflets, E) the lower stem, F) the roots, and G) the stolons of the control plant. The removal of the apex and stolens caused an increase in the concentration of inhibitor I in the upper leaflets with time and also induced the lower leaflets and lower stem to accumulate inhibitor I (table IV). The progression of the accumulation is apparently from the top of the plant down. The roots did not exhibit the presence of inhibitor in these experiments. The pattern of the accumulation of inhibitor I in these younger plants is somewhat different than the pattern found in older, more mature plants. In young plants, inhibitor I could be observed in stem tissue under the same conditions that it was found in leaflets. In older plants the removal of apices and rhizomes caused an increase in inhibitor I in stem tissue even before an increase could be observed in leaflets. In either case, the energy for the protein synthesis is probably derived from photosynthesis since removal of all leaflets caused an immediate disappearance of inhibitor I in the stems.

In plants just previous to, or during the breaking of apical dominance or establishment of new rhizomes, inhibitor I could sometimes be detected in juice from roots. Under other conditions, it was not possible to demonstrate its presence in root extracts.

Inhibitor I was determined in stem and root

Table IV. *Accumulation of Inhibitor I in* Solanum tuberosum *Seedlings Induced by Excising Apexes, Stolons, and Lateral Buds*

	Inhibitor I, % of soluble protein following excision[1]		
Tissues examined[2]	0 Time control	48 hrs after excision	96 hrs after excision
Apical leaflets	0.21	...	...
Upper leaflets	0.23	0.28	0.48
Lower leaflets	0	0.31	0.26
Upper stem	0.69[3]	0.40	0.31
Lower stem	0	0	0.23
Roots	0	0	0
Stolons	0	...	...

1 90,000 × *g* supernatant of tissue juice. Plants were 6 to 11 cm in height having 9 or 10 petioles and 5 to 9 stolons.

2 Apical leaflets were from 3 small leaflets surrounding the apex. Upper leaflets were the terminal leaflets of the next 3 petioles proceeding down from the apex. Lower leaflets were the terminal leaflets from the lowest 3 or 4 petioles of the plants. The entire stem was equally divided after removing all lateral growth.

3 Included apical stem, initially high in inhibitor I.

Table V. *Accumulation of Inhibitor I in Stem of Tuberizing* Solanum tuberosum *Plants Induced by Removal of Stolons, Tubers, and Apexes*

Plants grown from cuttings had an average height of 46 cm.

Tissues examined	0 Time control	Inhibitor I, % of soluble protein 144 hrs after removal of stolons and apexes 1[1]	2[2]	3[3]
Petioles	0	0	0	0
Upper stem	0	0	0	0.3
Middle stem	0	0	0.8	0.5
Lower stem	0	0.9	1.1	0.7
Underground stem	0	1.8	0.8	0.7
Stolons	0.5	...	...	...
Roots	0.1	0.4	0	0
Tubers	1.4	...	...	...

[1] Plant had broken apical dominance, 6 new stolons.
[2] Plant had broken apical dominance, 3 tubers, no stolons.
[3] All new growth, *i.e.*, lateral buds and stolons removed as they appeared during experiment.

tissues of tuberizing plants grown from cuttings in the greenhouse. In these plants the stolons and attached tubers contained appreciable inhibitor I and it could be demonstrated in the roots (table V, control plant). Removal of the apex and stolons of 3 plants of equivalent age and size induced a considerable accumulation of inhibitor I in the stem after 144 hours. This was especially so near the base of the plants. From 1 of the plants (plant No. 3), the lateral buds and new stolens were continually removed as they grew or reformed. In this plant, the inhibitor could be detected higher up into the stem than in the 2 plants where new lateral growth, stolons or tubers had been allowed to grow. In none of the plants was it possible to detect inhibitor I in the petioles or in leaflets except in close proximity to the apex of the control plant.

Observations of Inhibitor I in Tissues of Field-Grown Potatoes. Because the distribution and concentration of inhibitor I was apparently dependent upon both the age of the plants and the environmental conditions of growth, it was of interest to study inhibitor I in plants in their natural environment under normal growing conditions. A field 50 feet square in the Washington State University Agronomy Farm, Pullman, Washington, was planted with several rows of sprouted Russet Burbank potato seed pieces spaced in the rows about 24 inches apart. The plants were regularly irrigated and were weeded, fertilized, and sprayed with an insecticide at approximately 2-week intervals. The plants grew at what was considered a normal, healthy rate. Seed pieces were planted May 12, and the first plants emerged June 6. Potato plants were analyzed qualitatively for the presence of inhibitor I in tissues every 7 to 10 days from June 8 through senescence. Entire young plants were dissected into individual leaflets, petioles, stem(s), rhizomes, roots, and seed pieces and tested separately. As the plants grew larger, it became im-

Table VI. *The Presence of Inhibitor I in Juice from Tissues of Field Grown Potatoes*

Values were assigned from the densities of precipitin lines formed by the juice with anti-inhibitor I serum in double diffusion assays in agar gels. The precipitin lines were compared to those formed by known inhibitor I standard solutions.

Tissue	Days after planting								
	0	*27*	*35*	*40*	*48*	*61*	*77*	*84*	*90*[2]
Upper leaflets	A[1]	+	+	+	±	±	±	±	—
Lower leaflets	A	A	+	+	—	—	—	—	—
Upper stem	A	A	+	+	—	—	—	—	—
Middle stem	A	A	+	+	±	—	—	±	—
Lower stem	A	A	+	+	±	—	±	—	—
Underground stem	A	+	+	+	+	+	+	+	—
Rhizomes	A	+	±	±	+	+	+	±	—
Tubers	A	A	A	A	+	+	+	+	+
Roots	A	+	±	±	±	±	±	—	—
Seed piece	+	+	±	±	—	—	A	A	A

[1] "A" indicates tissue absent; + represents approximately 50 μg or more inhibitor I per ml juice; ± between 8 and 50 μg per ml; and — indicates not detectable.
[2] Senescent plants.

possible to test the entire plants. Therefore, selected leaflets, stem areas, roots and stolens were tested as representative tissues. The juice from the selected tissues was tested in the Ouchterloney method. Upper leaflets were chosen from petioles near the apices of plants and they contained inhibitor I until late senescence (*i.e.* 90 days after planting). Lower leaflets were sampled randomly from the middle or lower portion of the potato plants. The upper stem was selected near the apex, and the lower stem near the soil level. Table VI shows representative cumulative results of the entire growth cycle of the field-grown potatoes.

The results show a definite pattern of the presence of inhibitor I in the potato plants. The seed piece when planted, and the new growth contained appreciable quantities of the inhibitor. As the plants developed rhizomes, inhibitor I in the seed piece diminished rapidly. The initially high concentration of inhibitor I in aerial tissues receded rapidly and nearly disappeared after 48 days, remaining mainly in the apex and lateral buds. The disappearance coincided with the establishment of new tubers. Underground stems and rhizomes exhibited inhibitor I until late senescence.

In a plant representative of about 70 days growth (when inhibitor I was present in appreciable amounts in underground stems) a transverse section of stem was dissected and the various differentiated tissues were tested for inhibitor I. An interesting pattern was observed that paralleled the distribution in tubers. The outer parenchymal tissue and outer phloem contained relatively high concentrations of inhibitor I. The inner phloem and pith contained a lesser amount. No inhibitor I was detected in the xylem. Its absence in the xylem was confirmed in a separate experiment in which xylem contents, from a stem segment that contained inhibitor I, were extruded through a cut end by exerting a positive hydrostatic pressure on the other end. No inhibitor I could be demonstrated in the extruded xylem juice whereas it could still be demonstrated in the parenchymal tissue of the stem segment.

During growth of the field-grown plants, inhibitor I accumulated in the newly formed tubers. After the tubers matured and the plant leaflets became yellow with senescence, inhibitor I disappeared entirely from the plant except in the mature tubers. Thus a cycle of occurrence of inhibitor I from the tubers to plants to new tubers had taken place.

Discussion

The data establish that inhibitor I is present in high concentrations in potato tubers and can be demonstrated in almost all aerial vegetative tissues. Of interest was the finding that its existence in the tissues was transitory. In the tubers themselves, a striking shift of inhibitor I concentration in individual tissues during development and maturation suggested that this protein was involved in a dynamic aspect of the protein metabolism of the potato plant. The similar but more profound changes in concentration of inhibitor I in aerial vegetative tissues also reinforced the opinion that the protein is somehow involved in the protein metabolism of new growth centers. In all tissues the concentration of inhibitor I was highest just preceding or coincidental with establishment of new growth, be it in the tubers or in vegetative tissues, or induced by removing growth centers. The subsequent recession of its concentration in aerial tissues as new growth centers became established suggests that the new growth centers can influence the levels of inhibitor I in these tissues.

Inhibitor I may be an important factor in the establishment of the new growth. Whether its properties are unique in this respect is not yet known. Other proteins not yet discovered may also exhibit the behavior observed here. The protein itself is apparently not a normal constituent of the other Solanaceae family species tested. However the potato plant shows a lack of inhibitor I during certain stages of development and therefore the other members of the Solanaceae family could possibly possess the gene for producing the protein but its expression may be under strict controls in the plants. Under different conditions of testing, inhibitor I may yet be found in other plant species.

In potato plants containing tubers, the storage organs as a whole always showed the highest levels of any tissues tested. As shown in table II, the stem end of the cortical region of 1 tuber contained over 5% of the total protein as inhibitor I. Individual small localized areas have been assayed that exceed this figure. Yet, with time, the inhibitor disappears from these areas (table III). The continued change in concentration of inhibitor I in the tissues of potato tubers during maturation and storage suggests that the protein may be closely associated with the internal changes in the tuber components that are necessary for the initiation of new growth. These changes may also reflect the presence of substances that can control the expression of genes in cells of these tissues. It is significant that inhibitor I concentration consistently remains high in the apical portion of the tubers, in the area where new sprouts usually appear. In the field-grown potatoes (table VI) inhibitor I entirely disappeared from seed pieces within a few days after the new plants emerged. The seed pieces at this time were still visibly in good condition. The new stem and new aerial growth contained appreciable inhibitor I and continued to exhibit its presence until new stolons and tubers were established. The new tubers then accumulated inhibitor I as the concentration in aerial tissues diminished.

The cumulative data suggest that inhibitor I is present or absent in tissues depending upon the physiological state of the plant. The levels of inhibitor I are influenced by both environment and age as well as the state of development. It is concluded that the biosynthesis and degradation of this protein are regulated within the plant and this regulation to a large extent, and in an unknown way, is influenced directly by the meristematic tissue of the plants. The fact that the protein is found associated with meristematic tissue and that its appearance in other tissues coincides with the breaking of apical dominance strongly suggests a role for the protein in the establishment or maintainance of meristematic tissue.

Experiments to further understand the physiological and biochemical role of this protein, and the nature of the substance or substances involved in the control of its transient existence, are currently in progress.

Acknowledgments

The authors gratefully thank Mr. Charles R. Oldenburg for technical assistance. We also thank Drs. R. Kunkel, W. Hoymen, G. Easton, and S. Locke for samples of potatoes, potato seeds, and potato plants. Our appreciation is extended to Dr. Robert Nilan for the generous use of greenhouse and field space.

Literature Cited

1. Aldrich, R. L., Jr. and A. K. Balls. 1958. The reversibility of inactivation of chymotrypsin. J. Biol. Chem. 233: 1355–58.
2. Balls, A. K. and C. A. Ryan. 1963. Concerning a chymotryptic inhibitor from potatoes and its binding capacity for the enzyme. J. Biol. Chem. 238: 2976–82.
3. Deeds, F., C. A. Ryan, and A. K. Balls. 1964. Pharmacological observations on a chymotryptic inhibitor from potatoes. Proc. Soc. Exptl. Biol. Med. 115: 772–75.
4. Laskowski, M., Sr. and M. Laskowski, Jr. 1954. Naturally occurring trypsin inhibitors. Advan. Protein Chem. 9: 203–42.
5. Ouchterloney, O. 1949. Antigen-antibody reactions in gels. Acta Pathol. Microbiol. Scand. 26: 507.
6. Ryan, C. A. 1966. Chymotrypsin inhibitor I from potatoes: Reactivity with mammalian, plant, bacterial, and fungal proteinases. Biochemistry 5: 1592–96.
7. Ryan, C. A. 1967. The quantitative determination of soluble cellular proteins by radial diffusion in agar gels containing antibodies. Anal. Biochem. 19: 434–40.
8. Ryan, C. A. and A. K. Balls. 1962. An inhibitor of chymotrypsin from *Solanum tuberosum* and its behavior toward trypsin. Proc. Natl. Acad. Sci. U.S. 48: 1839–44.
9. Ryan, C. A. and O. C. Huisman. 1967. Chymotrypsin inhibitor I from potatoes: A transient component in leaves of young potato plants. Nature 214: 1047–49.

Reprinted from PLANT PHYSIOLOGY, Vol. 43, No. 9, September, 1968, pages 1423-1429.
(PRINTED IN U.S.A.)

Species Specificity in the Biosynthesis of Branched Paraffins in Leaves

P. E. Kolattukudy
Department of Biochemistry, The Connecticut Agricultural Experiment Station, New Haven, Connecticut 06504

Received June 3, 1968.

Abstract. Isobutyrate-1-^{14}C and L-isoleucine-U-^{14}C fed through the petiole labeled the surface lipids of broccoli leaves, but the incorporation was much less than from straight chain precursors. Not more than one-third of the ^{14}C incorporated into the surface lipids was found in the C_{29} paraffin and derivatives, whereas more than two-thirds of the ^{14}C from straight chain precursors are usually found in these compounds. The small amount of ^{14}C incorporated into the paraffin fraction was found in the *n*-C_{29} paraffin rather than branched paraffins showing that the ^{14}C in the paraffin must have come from degradation products. Radio gas-liquid chromatography of the saturated fatty acids showed that, in addition to the *n*-C_{16} acid which was formed from both branched precursors, isoleucine-U-^{14}C gave rise to branched C_{15}, C_{17}, and C_{19} fatty acids, and isobutyrate-1-^{14}C gave rise to branched C_{16} and C_{18} acids. Thus the reason for the failure of broccoli leaf to incorporate branched precursors into branched paraffins is not the unavailability of branched fatty acids, but the absolute specificity of the system that synthesizes paraffins, probably the elongation-decarboxylation enzyme complex. Consistent with this view, no labeled branched fatty acids longer than C_{19} could be found in the broccoli leaf. The branched fatty acids were also found in the surface lipids indicating that the epidermal layer of cells did have access to branched chains. Thus the paraffin synthesizing enzyme system is specific for straight chains in broccoli, but the fatty acid synthetase is not.

Hydrocarbons are widely distributed in plants, animals, and microorganisms (3, 17). Although certain organisms have very complicated mixtures of hydrocarbons (2, 13, 15) many contain relatively simple mixtures with one or a few hydrocarbons predominating (3, 10, 20). Normal saturated hydrocarbons, branched saturated hydrocarbons and unsaturated normal hydrocarbons have all been found in living organisms. Some organisms, such as the American cockroach, can synthesize all 3 kinds of hydrocarbons (1) whereas others, such as tobacco leaves, synthesize only saturated normal and branched hydrocarbons (5, 15) while still others such as *Brassica oleracea* leaves synthesize only normal saturated hydrocarbons (6, 20).

Normal paraffins are synthesized from normal saturated fatty acids (7, 8, 10), unsaturated paraffins from unsaturated fatty acids (10), and branched hydrocarbons from branched fatty acids which in turn originate from branched starter pieces derived from branched-chain amino acids (5, 10, 11). What controls the structural characteristic, such as branching and unsaturation, in the paraffins synthesized by a particular organism is not known. Either the availability of the precursors or the innate specificity of the enzyme systems which synthesize the paraffins could determine the nature of the paraffin synthesized.

When *Brassica oleracea,* a plant that normally does not synthesize unsaturated hydrocarbons, was fed radioactive oleic acid, no formation of unsaturated hydrocarbons could be detected (7). This result suggests that the innate specificity rather than the availability of the precursor determines the nature of the hydrocarbons. In the present paper further evidence for such a hypothesis concerning branched paraffins is presented. Preliminary results of these experiments have been reported (12).

Materials and Methods

Plants. Broccoli [*Brassica oleracea,* var. italica (Plenck)] and tobacco plants (*Nicotiana tabacum,* var. Havana seed) were grown in sand on a subirrigated bench in the greenhouse (6).

Chemicals. All of the radiochemicals were purchased from Nuclear-Chicago Corporation. Methyl esters of branched chain fatty acids used as standards were obtained from Applied Science Laboratories, State College, Pennsylvania, straight chain fatty acids and materials for gas chromatography from Analabs, Hamden, Connecticut. Straight chain hydrocarbons used as standards were isolated from *Brassica oleracea* and *Senecio odoris* as described earlier (6, 11) or were purchased from Humphrey Chemical Company, North Haven, Connecticut. Branched paraffins used as standards were isolated from tobacco surface lipids by column and thin-layer chromatography (11).

Incorporation of Labeled Substrates. Labeled compounds were administered to excised leaves as described before (6). Each leaf (about 5 g fr wt) received 50 μc of L-isoleucine-U-^{14}C (290 mc/mmole) or 50 μc of isobutyrate-1-^{14}C (2.0 mc/mmole). At the end of the experimental period (18 hrs for isoleucine and 24 hrs for isobutyrate, both under about 1000 ft-c of light), the surface lipids were isolated by dipping the leaf (excluding the cut surface) in 400 ml chloroform for 30 seconds. The lipids remaining in the leaf (internal lipids) were extracted after cutting up the tissue, with a mixture (2:1 v/v) of chloroform and methanol (11).

Experiments with chopped leaves were done as described before (7). At the end of the incubation period, the material was transferred to a sintered glass funnel and washed with 100 ml of distilled water. Then 100 ml of a mixture (2:1) of chloroform and methanol was added to the funnel, the suspension was stirred for 30 seconds and the solvent mixture was then quickly withdrawn by suction. This solution contained all of the surface lipids (10) which were recovered by the method of Folch *et al.* (4). The lipids remaining in the leaf residue (internal lipids) were extracted with a mixture (2:1) of chloroform and methanol as described before (7).

Isolation of the Products. The surface lipids were fractionated by thin-layer chromatography on silica gel G with benzene as the developing solvent (6). Hydrocarbons were isolated by column chromatography on SilicAR cc-4 100 to 200 mesh (Mallinckrodt Chemical Works, St. Louis, Missouri) with hexane as the eluant (6).

The internal lipids were saponified by refluxing them under nitrogen with 12 % KOH in 95 % ethanol for 2 hours. After acidification, lipids were extracted with chloroform, and the fatty acids were isolated by thin-layer chromatography on 0.5 mm silica gel G plates with hexane:ethyl ether:formic acid (40:10:1) as the solvent system (8). At this stage some radioactive alcohols and hydroxy fatty acids were detected, but these were not further analyzed. Methyl esters of fatty acids were prepared with boron trifluoride-methanol reagent. The saturated esters were separated from the unsaturated by thin-layer chromatography on silver nitrate-silica gel G with 10 % ethyl ether in hexane as the developing solvent (16). The surface lipids were saponified by refluxing them under nitrogen with 12 % KOH in 95 % ethanol for 8 hours. The methyl esters of saturated fatty acids were then prepared and isolated as described above.

Gas-liquid chromatography was carried out on a Perkin-Elmer gas chromatograph equipped with a flame ionization detector and an effluent splitter. Silicone gum rubber (SE-30) was the liquid phase for the analysis of methyl esters and hydrocarbons. Further details of chromatographic conditions are shown in the legends to the figures.

Determination of Radioactivity. Radioactivity in samples of lipid solutions and thin-layer chromatograms was determined as previously described (6, 7). An internal standard of toluene-^{14}C was always used in order to correct for quenching which occurred especially in the colored internal lipid samples. The counting was usually done with 60 % efficiency. The standard deviation was less than 3 %. Radioactivity in the effluent from the gas chromatograph was continuously monitored with a Barber-Colman radioactivity monitor.

Results and Discussion

Branched amino acids are undoubtedly present in all organisms. However, some organisms such as tobacco leaves, synthesize branched chain surface lipids while others (*e.g.* broccoli) make only straight chain surface lipids. In order to understand what factors determine whether or not an organism can synthesize branched surface lipids, the incorporation of ^{14}C labeled branched precursors into the surface lipids of broccoli leaves was studied. Radioactivity from isobutyrate-1-^{14}C and L-isoleucine-U-^{14}C was incorporated into both surface lipids and internal lipids of broccoli leaves (table I). However the incorporation was much less in comparison to that of other substrates such as acetate. Although these results are not strictly comparable with those obtained

Table I. *Incorporation of Isobutyrate-1-^{14}C and L-Isoleucine-U-^{14}C Into the Lipids of Broccoli Leaves*

In experiment 1, 2 g chopped young broccoli leaves were placed in a large Warburg flask (75 ml) containing 6 ml water in which 5.2 μmoles (50 μc) of sodium isobutyrate was dissolved. Incubation temperature 30°, under 2000 ft-c for 6 hr; total uptake was 80 % of administered ^{14}C. Values shown are means from 2 experiments. In experiment 2, each leaf (3-4 g fr wt) received 8 μmoles (16 μc) of sodium isobutyrate through the petiole and metabolized it for 24 hr under 800 ft-c. Values shown are means of 3 experiments. In experiment 3, each leaf (3-4 g fr wt) received 0.17 μmole (50 μc) L-isoleucine through the petiole and metabolized it for 18 hours under 800 ft-c. Values given are means of 2 experiments.

Expt	Substrate	Time	Incorporation, of administered	
			In internal lipids	In surface lipids
		hr	%	%
1	Sodium isobutyrate-1-^{14}C	6	6 5	0.8
2	Sodium isobutyrate-1-^{14}C	24	6.7	0.8
3	L-Isoleucine-U-^{14}C	18	5.0	1.9

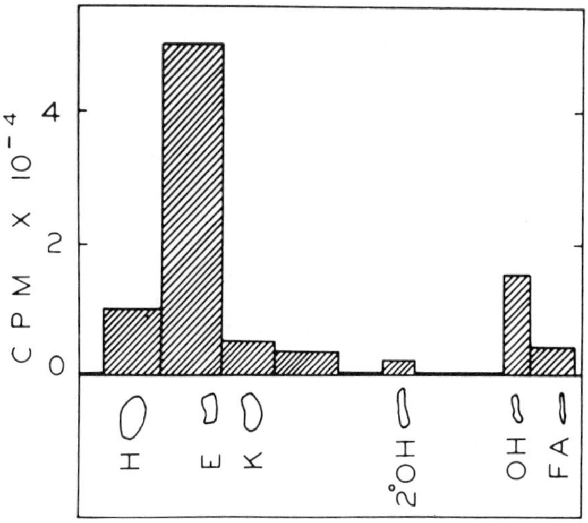

FIG. 1a. Distribution of ^{14}C among the thin-layer chromatographic fractions of surface lipids from broccoli leaves which metabolized L-isoleucine-U-[14]C (expt 3 described in table I). Silica gel G (0.3 mm thickness) plates with benzene as the solvent and 2',7'-dichlorofluorescein as the indicator were used. The areas fluorescing under U.V. light are outlined and labeled as follows: H-hydrocarbon, E-waxy ester, K-ketone, 2 OH-secondary alcohol, OH- primary alcohol, FA-fatty acids and more polar lipids remaining at the origin.

with tobacco leaves (11) the incorporation of branched precursors into surface lipids of broccoli was much smaller than the corresponding incorporation in tobacco.

Surface lipids of *Brassica oleracea* contain many classes of compounds (18, 19, 20). They fall into 2 groups as far as their biosynthesis is concerned—the C_{29} compounds, which include nonacosane, nonacosan-15-one, and nonacosan-15-ol, and the non-C_{29} compounds which include free fatty acids, fatty alcohols and waxy esters (6, 7). Therefore it was necessary to analyze the labeled surface lipids so

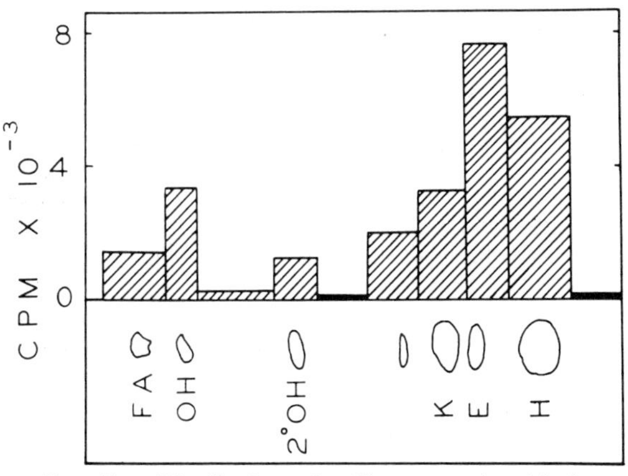

FIG. 1b. Distribution of ^{14}C among the thin-layer chromatographic fractions of surface lipids from excised broccoli leaves which metabolized sodium isobutyrate-1-[14]C (expt 2 described in table I). Chromatography and abbreviations as in figure 1a.

that possible incorporation of branched precursors into the various classes of compounds could be determined. The thin-layer chromatographic analysis of surface lipids isolated from broccoli leaves that metabolized isoleucine-U-^{14}C is shown in figure 1a. The major radioactive components were waxy esters and primary alcohols with much less radioactivity in the hydrocarbons, ketones, and secondary alcohols. Similarly, most of the ^{14}C incorporated into the surface lipids of broccoli from isobutyrate-1-^{14}C was in the non-C_{29} compounds (fig 1b). Comparable results were obtained when leaf slices were used in short-term (few hours) experiments as well as when whole excised leaves were used in relatively long term (24 hr) experiments (figs 1b and c). When leaf slices were fed radioactive substrates the surface lipids were extracted by washing the slices for 30 seconds with a 2:1 mixture of chloroform and methanol which extracted a small quantity of polar lipids in addition to the relatively non-polar surface lipids (fig 1c). In any case about two-thirds of the

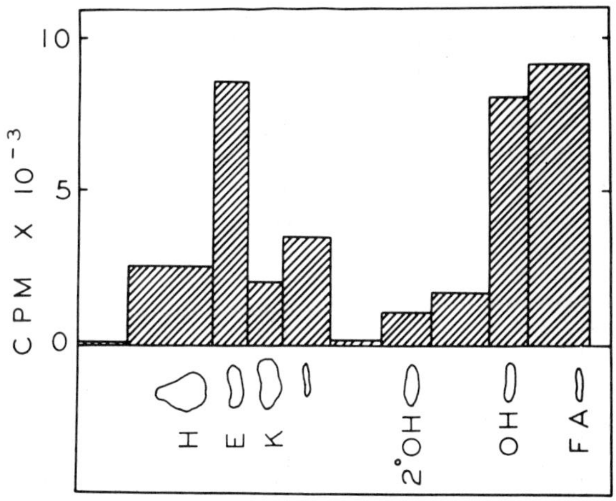

FIG. 1c. Distribution of ^{14}C among the thin-layer chromatographic fractions of surface lipids from chopped broccoli leaves which metabolized sodium isobutyrate-1-[14]C (expt 1 described in table I). Chromatography and abbreviations as in figure 1a.

[14]C in the surface lipids was in non-C_{29} compounds and this distribution of ^{14}C is clearly in contrast to the distribution usually obtained with straight chain precursors (such as C_2–C_{18} fatty acids) because these straight chain substrates were incorporated mostly (at least two-thirds) into the C_{29} compounds (6, 7). The present results could be explained if the branched precursors are essentially excluded from participating in paraffin synthesis, but not from fatty acid synthesis. Furthermore the long branched fatty acids synthesized from the administered branched starter pieces can undergo reduction to the corresponding alcohols and subsequently be esterified.

Tobacco leaf incorporated ^{14}C from L-valine-U-^{14}C and isobutyrate-1-^{14}C into branched C_{29}, C_{31}, and C_{33} paraffins. Similarly L-isoleucine-U-^{14}C labeled branched C_{30} and C_{32} paraffins of tobacco leaves (5, 10, 11). The reason for the absence of branched paraffins in some plants such as broccoli may be that either the amino acid balance is such that the branched amino acids do not provide their catabolic products for lipid synthesis or that the paraffin-synthesizing system specifically excludes branched chains. The considerable incorporation of ^{14}C from branched precursors into the lipids of broccoli (table I) indicates that the branched chain starters are available for lipid synthesis. Hence the small amount of radioactivity found in the hydrocarbons represents either the ability of the hydrocarbon-synthesizing system to utilize limited amounts of available branched chains, or may merely be the result of degradation of a small proportion of the substrate which could provide ^{14}C to the paraffins *via* unbranched catabolic products. If the first possibility is correct then the radioactivity in the hydrocarbons should be found in branched molecules while, according to the second hypothesis, the small amount of ^{14}C found in the paraffin fraction should be present in the normal C_{29} paraffin. Figure 2 shows that the radioactivity in the paraffin fraction of surface lipids derived from L-isoleucine-U-^{14}C was in nC_{29} paraffin. Identical results were obtained with the paraffin fraction derived from isobutyrate-1-^{14}C. If both the possibilities discussed above occurred, radioactivity in the paraffin fraction would

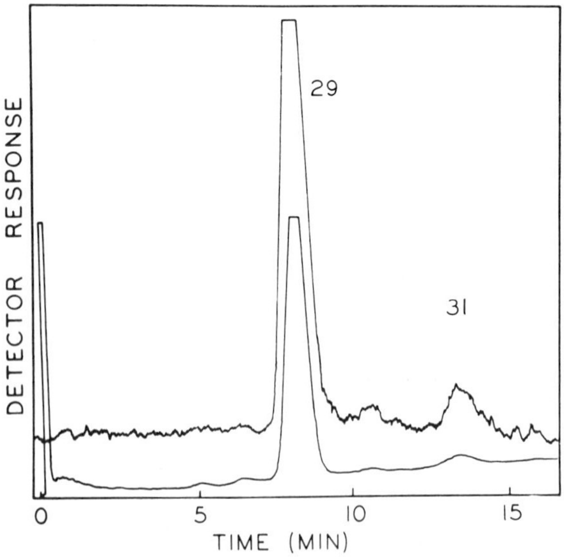

FIG 2. Radio gas-liquid chromatogram of paraffins isolated from surface lipids from broccoli leaves which metabolized L-isoleucine-U-^{14}C (expt 3 of table I). Chain length is indicated by the number on each peak. The top tracing shows the radioactivity monitor response and the bottom tracing the flame ionization detector response. Coiled copper column (4 ft × 0.25 in OD) packed with 5 % silicone gum rubber (SE 30) on 90 to 100 mesh Anakrom SD was used for the gas-liquid chromatography. Temperature of the column and injector were 260° and 360° respectively and the carrier gas argon was supplied at 80 ml/min.

be expected in both normal and branched paraffins. Although a small proportion of radioactive branched C_{29} paraffin would have been difficult to detect with the major C_{29} peak of radioactivity in the case of the isobutyrate experiments, in the case of isoleucine the expected branched paraffins would be C_{28} or C_{30} which would be clearly separated from C_{29} and therefore would have been easily detected if present. Since no branched radioactive paraffin could be found from either precursor, it is clear that the small amount of ^{14}C incorporated into the paraffins of broccoli from the branched precursors represents only the incorporation of degradation products and not intact branched pieces.

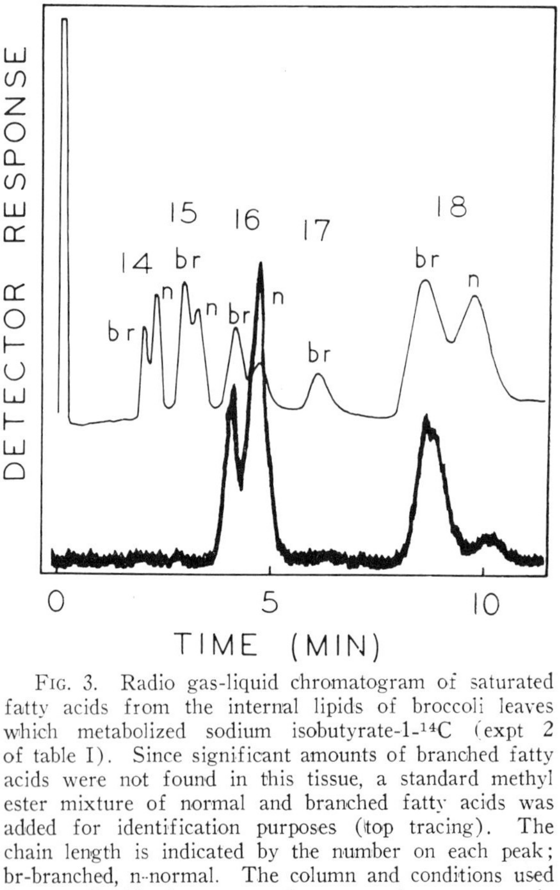

FIG. 3. Radio gas-liquid chromatogram of saturated fatty acids from the internal lipids of broccoli leaves which metabolized sodium isobutyrate-1-^{14}C (expt 2 of table I). Since significant amounts of branched fatty acids were not found in this tissue, a standard methyl ester mixture of normal and branched fatty acids was added for identification purposes (top tracing). The chain length is indicated by the number on each peak; br-branched, n-normal. The column and conditions used for gas-liquid chromatography were the same as in figure 2 but the temperature of the column was 195°.

The inability of broccoli leaves to synthesize branched paraffins from exogenously supplied branched precursors could be explained because the fatty acid synthesizing system of broccoli is so specific for straight chains that the branched starter pieces, such as isobutyrate, are not incorporated into branched fatty acids. If so, then the incorporation of ^{14}C into lipids from the branched precursors should merely represent incorporation of the degradation products; therefore the radioactivity of the fatty acids must be only in the usual straight chain molecules. Alternatively, the absolute specificity for straight chains may reside in the enzyme system which synthesizes paraffins and not in the fatty acid synthetase. In that case a substantial part of the radioactivity in the lipids isolated from the broccoli leaf that metabolized branched precursors would be in the branched fatty acids. In order to distinguish between the 2 possibilities, the saturated fatty acids were isolated from the internal lipids by silver nitrate-silica gel chromatography and then subjected to radio gas-liquid chromatography (fig 3 and 4). Clearly isobutyrate-1-^{14}C labeled branched C_{16} and C_{18} fatty acids (fig 3) and L-isoleucine-U-^{14}C labeled branched C_{15}, C_{17}, and C_{19} fatty acids (fig 4). In fact the incorporation pattern of branched precursors into fatty acids of broccoli was similar to that observed with tobacco (fig 5). These results clearly show that the fatty acid synthesizing system of broccoli is not specific for straight chains, and that it can synthesize branched fatty acids when the branched substrates are available.

Another possibility considered was that the epidermal cells, where paraffin synthesis takes place (11), for some reason do not have access to the branched starter pieces that can be incorporated into the branched fatty acids. However, analysis of the fatty acids from the surface lipids clearly showed

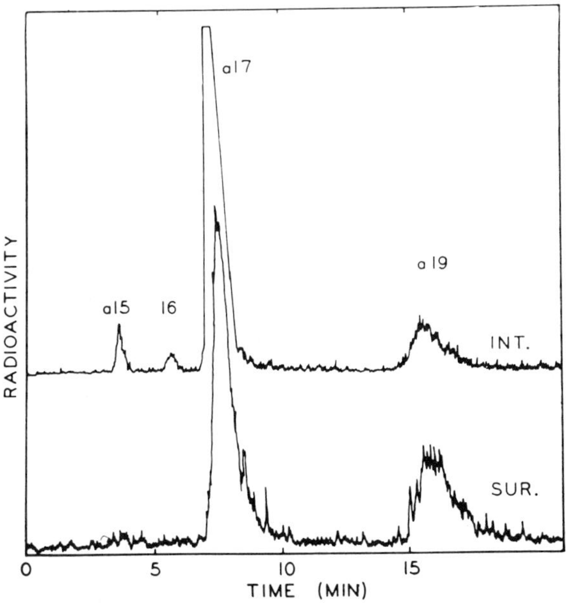

FIG. 4. Radio gas-liquid chromatogram of saturated fatty acids (as methyl esters) from the internal lipids (INT) and surface lipids (SUR) of broccoli leaves which metabolized L-isoleucine-U-^{14}C. The flame ionization detector response did not show significant amounts of branched acids and therefore the sample was mixed with a standard mixture of branched acids for identification purposes. The chain length is indicated by the number on each peak; a-indicates anteiso branch. The gas-liquid chromatography was done as described under figure 3.

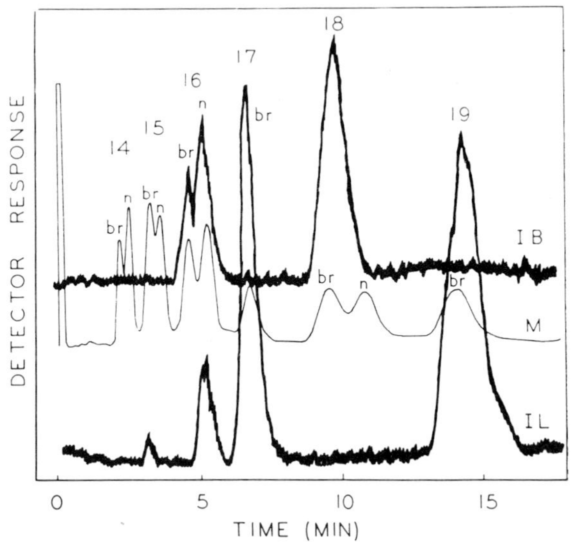

FIG. 5. Radio gas-liquid chromatogram of saturated fatty acids (methyl esters) isolated from tobacco leaves which metabolized isobutyrate-1-^{14}C (1B) and isoleucine-U-^{14}C (IL) (11). The middle tracing (M) shows the flame ionization detector response obtained with an authentic mixture of normal and branched fatty acids (methyl esters). The number on each peak shows the chain length, br-branched, n-normal. The gas-liquid chromatography was done as described under figure 3.

(fig 4) the presence of branched fatty acids on the surface. Hence it is likely that the epidermis did have access to branched chains but that it was unable to synthesize branched paraffins. Thus the paraffin synthesizing system of broccoli, unlike that of tobacco, must be specific for straight chains.

In all cases, some radioactivity was found in the n-C_{16} acid; this ^{14}C must have come from degradation products of the radioactive branched acids fed to the leaves. In tobacco leaves very little radioactivity was incorporated into the unsaturated fatty acids from branched precursors whereas a considerable amount of ^{14}C could be detected in the unsaturated fatty acids of broccoli. Apparently degradation of the branched precursors proceeds to a larger extent in broccoli than in tobacco. However, in spite of the fairly large incorporation of intact branched pieces into fatty acids, no branched paraffins were found in broccoli leaves. So the degradation of the branched acids fed to broccoli leaves undoubtedly is not the reason for the lack of synthesis of branched paraffins.

The incorporation pattern of branched precursors into fatty acids of broccoli differed from that observed with tobacco (fig 5) in the proportion of ^{14}C found in the very long (C_{20}-C_{27}) fatty acids. In tobacco, substantial amounts of the branched precursors were incorporated into the appropriately branched very long (C_{20}-C_{27}) fatty acids that are apparently biosynthetically related to the branched paraffins (10, 11). Since the branched precursors were not incorporated into branched paraffins in broccoli, no incorporation into very long branched acids should be found if the elongation system responsible for the synthesis of both paraffins and very long fatty acids specifically exclude branched chains. In fact no measurable ^{14}C could be detected in the very long fatty acids in broccoli under the present experimental conditions. Thus it appears that the elongation system responsible for the synthesis of very long fatty acids and paraffins excluded branched chains.

Another difference between the ^{14}C distribution in the fatty acids of tobacco and broccoli was that the former showed a higher proportion of ^{14}C in the branched C_{18} acid than the latter. Branched C_{18} acid is apparently synthesized by the chloroplast system and also during the elongation associated with the synthesis of surface lipid. In tobacco the branched C_{18} acid is synthesized by both of these systems whereas in broccoli, where the elongation is specific for straight chains, only the chloroplast system synthesizes this acid. From inhibition studies with trichloroacetic acid on broccoli leaf (8) and studies on fatty acid synthesis in various parts of *Senecio odoris* leaf (11), dual sites of synthesis for straight chain C_{18} acid had been previously postulated. The present work indicates that this may also apply to branched C_{18} acid.

The absolute specificity for straight chains suggests that the hypothetical elongation-decarboxylation enzyme complex can reject molecules when there is a deviation in structure even at the end far removed from the functional end where chain elongation takes place. Although the elongation decarboxylation mechanism can best explain the results of biosynthetic experiments with labeled fatty acids, it is possible that a multienzyme complex, which is also capable of accepting long chain fatty acids, carries out the entire paraffin synthesis starting with acetate or branched starter pieces. If such is the case, the enzyme complex of *B. oleracea* must utilize only acetate and other straight chain starter pieces whereas that of tobacco can use straight chain as well as branched chain starter pieces.

In plants such as tobacco, which is capable of synthesizing branched paraffins, the relative amounts of these substances synthesized may depend on the availability of branched precursors which in turn is controlled by the metabolic balance of the branched amino acids. Although exogenous branched amino acids gave rise to branched fatty acids in broccoli leaves, this tissue normally does not synthesize significant amounts of branched fatty acids. The reason for this is not understood at present; compartmentation could be a possible explanation.

Acknowledgment

I thank Drs. H. B. Vickery, I. Zelitch, and M. Zucker for critical reading of the manuscript, Miss Katherine Clark for preparing the figures, and Mr. G. R. Smith for raising the plants.

Literature Cited

1. Baker, G. L., H. E. Vroman, and J. Padmore. 1963. Hydrocarbons of the American cockroach. Biochem. Biophys. Res. Commun. 13: 360–65.
2. Downing, D. T., Z. H. Kranz, and K. E. Murray. 1960. Studies in waxes. XIV. An investigation of the aliphatic constituents of hydrolyzed wool wax by gas chromatography. Australian J. Chem. 13: 80–94.
3. Eglinton, G. and R. J. Hamilton. 1967. Leaf epicuticular waxes. Science 156: 1322–35.
4. Folch, J., M. Lees, and G. H. Sloane-Stanley. 1957. A simple method for the isolation and purification of total lipids from animal tissues. J. Biol. Chem. 226: 497–509.
5. Kaneda, T. 1967. Biosynthesis of long-chain hydrocarbons. 1. Incorporation of L-valine, L-threonine and L-isoleucine into specific branched-chain hydrocarbons in tobacco. Biochemistry 6: 2023–32.
6. Kolattukudy, P. E. 1965. Biosynthesis of wax in *Brassica oleracea*. Biochemistry 4: 1844–55.
7. Kolattukudy, P. E. 1966. Biosynthesis of wax in *Brassica oleracea*: Relation of fatty acids to wax. Biochemistry 5: 2265–75.
8. Kolattukudy, P. E. 1967. Biosynthesis of paraffins in *Brassica oleracea*: Fatty acid elongation-decarboxylation as a plausible pathway. Phytochemistry 6: 963–75.
9. Kolattukudy, P. E. 1967. Mechanisms of syntheses of waxy esters in broccoli (*Brassica oleracea*). Biochemistry 6: 2705–17.
10. Kolattukudy, P. E. 1968. Biosynthesis of surface lipids. Science 159: 498–505.
11. Kolattukudy, P. E. 1968. Further evidence of an elongation-decarboxylation mechanism in the biosynthesis of paraffins in leaves. Plant Physiol. 43: 375–83.
12. Kolattukudy, P. E. 1968. Control of branching in paraffin biosynthesis. Federation Proc. 27: 524.
13. Kranz, Z. H., J. A. Lamberton, K. E. Murray, and A. H. Redcliffe. 1960. Sugar-cane wax. II. An examination of the constituents of sugarcane cuticle wax by gas chromatography. Australian J. Chem. 13: 498–505.
14. Mold, J. D., R. E. Means, and T. M. Ruth. 1966. The higher fatty acids of flue-cured tobacco. Methyl and cyclohexyl branched acids. Phytochemistry 5: 59–66.
15. Mold, J. D., R. K. Stevens, R. E. Means, and J. M. Ruth. 1963. The paraffin hydrocarbons of tobacco; normal, iso, and anteisohomologs. Biochemistry 2: 605–10.
16. Morris, L. J. 1962. Separation of higher fatty acid, isomers, and vinylogues by thin layer chromatography. Chem. Ind. 1238: 40.
17. Nicolaides, N. 1965. Skin lipids. II. Lipid class composition of samples from various species and anatomical sites. J. Am. Oil Chem. Soc. 42: 691–702.
18. Purdy, S. J. and E. V. Truter. 1963. Constitution of the surface lipid from the leaves of *Brassica oleracea* [var capitata (Winnigstadt)] I. Isolation and quantitative fractionation. Proc. Roy. Soc. London Ser. B. 158: 536–43.

19. Purdy, S. J. and E. V. Truter. 1963. Constitution of the surface lipid from the leaves of *Brassica oleracea* [var capitata (Winnigstadt)]. II. Esters, free primary alcohols and free acids. Proc. Roy. Soc. London Ser. B. 158: 544–52.

20. Purdy, S. J. and E. V. Truter. 1963. Constitution of the surface lipid from the leaves of *Brassica oleracea* [var capitata (Winnigstadt)]. III. Nonacosane and its derivatives. Proc. Roy. Soc. London Ser. B. 158: 553–65.

21. Velick, S. F. and J. English, Jr. 1945. The synthesis and configuration of d-14-methyl palmitic acid and its identity with the natural acid from wool fat. J. Biol. Chem. 160: 473–80.

22. Weitkamp, A. W. 1945. The acidic constituents of degras. A new method of structure elucidation. J. Am. Chem. Soc. 67: 447–54.

23. Wheatley, V. R., D. C. Chow, and F. D. Keenan. 1961. Studies of lipids of dog skin. II. Observations of the lipid metabolism of perfused surviving dog skin. J. Invest. Dermatol. 36: 237–39.

Reprinted from the AMERICAN JOURNAL OF BOTANY, Vol. 53, No. 9, October, 1966
Printed in U.S.A.

Amer. J. Bot. 53(9): 875–882. 1966.

THE DIFFERENTIATION OF PIGMENTATION IN FLOWER PARTS. III. METABOLISM OF SOME EXOGENOUS ANTHOCYANINS BY DETACHED PETALS OF IMPATIENS BALSAMINA[1]

RICHARD L. MANSELL[2] AND CHARLES W. HAGEN, JR.
Botany Department, Indiana University, Bloomington

ABSTRACT

The anthocyanins of mature petals of *Impatiens balsamina* L. are distinct from the pigments found in vegetative tissue. In the red genotype (llHHP^{r}P^r) a sequential elaboration of the characteristic anthocyanins has been previously demonstrated through the examination of buds at successive stages of development. The metabolism of anthocyanins, especially pelargonidin-3-monoglucoside, was examined by infiltration into developing petals of a genetically white strain. This anthocyanin appears to play a central role in the biochemical sequences involved and it has been observed that the genetically white flowers possess the enzymatic potential to metabolize this substrate, producing the same final products which are produced in the red genotype. There is a pattern of change in the relative amounts of each anthocyanin during the incubation period which follows closely the pattern which occurs during normal development of the colored genotypes. This indicates that the enzymes which are normally produced in the colored flowers are also produced in flowers which never produce anthocyanins. The metabolic capabilities of several other genetic strains and the influence of light and puromycin have been examined.

IN A SERIES of studies on the red genotype (llHHP^{r}P^r) of *Impatiens balsamina* L., Hagen (1966 a, b) has demonstrated that there is a sequential elaboration of the characteristic anthocyanins during morphogenesis of the flower buds. This work has provided a partial description of the morphogenesis of a molecule, a developmental sequence which is apparently part of the differentiation of the flower and subject to controls similar to those which regulate other aspects of differentiation.

Past attempts to demonstrate the sequential modification of a particular flavonoid compound in flower petals have been inconclusive (Suomalainen and Keränen, 1961), although Hess (1964) has demonstrated that there is a sequential production of different anthocyanidins during bud maturation in petunia.

The production o pigmentation in petals from flower buds of *Impatiens balsamina* L. may be divided into four periods (Hagen, 1966b). The first period is characterized by the absence of anthocyanins, although other flavonoid compounds are present In the second, pelargonidin-3-monoglucoside appears and is apparently followed by a smaller quantity of the corresponding aglycone. The third period is one of rapid synthesis during which the final products first appear but are accompanied by a confusing wealth of intermediate forms. In the final stage which begins with anthesis, the compounds of intermediate complexity disappear, leaving a mature flower pigmented with acylated-3,5-diglucosides and the other highly substituted forms of pelargonidin.

While this sequence of developmental events suggests a corresponding biosynthetic sequence, there remains confusion concerning the route or routes actually followed during normal flower development. It is impossible from descriptive studies to distinguish those products which are necessary precursors of the final pigments from those which may be accidental by-products or components of secondary pathways. To test the synthetic capabilities of the intact petals, techniques were developed for administering potential substrates to cultured petals. Where the substrates could be prepared in forms compatible with and made available to the intact petals, modification of the substrate provided further information concerning the biosynthetic pathway and the developmental mechanisms which regulate this pathway.

MATERIALS AND METHODS—*Administration of substrate*—Plants of the garden balsam, *Impatiens balsamina* L., were employed in this study. The genotypes used were llHHpp (pink), llhhP^{r}P^r (pink), llHHP^{r}P^r (red) LLhhP^{r}P^r (purple) and llhhpp (white) as originally designated by Davis, Taylor, and Ash (1958). Stage-3 buds (Fig. 1), as defined by Hagen (1966b), were most frequently used as test objects because the descriptive studies had indicated that biosynthetic systems were

[1] Received for publication September 20, 1965.
This investigation is a portion of a dissertation presented to the Graduate School of Indiana University, Bloomington, in partial fulfillment of the requirements for the degree Doctor of Philosophy (1964).
This research was supported in part by National Science Foundation Grants G-6138 and G-21943 awarded to Dr. Charles W. Hagen, Jr.
[2]Present address: United States Air Force School of Aerospace Medicine, Brooks AFB, Texas.

highly active at this stage and because the petals contained sufficient intercellular space to permit substrate infiltration

The petals were dissected out of the buds and as many as 16 were placed in a 40-ml Thunburg tube containing 3–5 ml of the desired infiltration solution. The tissue was submerged with a flattened glass rod and the tube was evacuated to a pressure of 35–50 mm of Hg. After 1 or 2 min, air was permitted to re-enter the tube slowly. This process was repeated to ensure flooding of intercellular spaces with test fluids. The tissue was then removed, blotted, and placed on modified White's solution (Miller, 1961) containing 2% agar in 6-cm petri dishes. Indole-3-acetic acid and kinetin which were included by Miller were omitted from our medium. Sucrose was found to be unnecessary for the metabolic processes under investigation and was omitted. The final *p*H of the medium was 4.5.

The dishes were placed at 18 C under continuous illumination of 800 ft-c provided by 40-w warm-white fluorescent tubes supplemented with 150-w incandescent bulbs. Under these conditions the temperature of the medium in the dishes remained near 24 C. Since the incubation periods were usually 48 hr or less, aseptic procedures were unnecessary.

Chromatography—At various times samples of petals were removed, homogenized in 95% ethanol containing 0.4% HCl (w/v) and filtered through glass fiber filters in Gooch crucibles. Two-dimensional chromatograms were prepared according to the methods described by Hagen (1966a). Absorption spectra were obtained with the Bausch and Lomb Spectronic 505 spectrophotometer.

Preparation of substrate—Pelargonidin-3-monoglucoside was selected for the initial substrate because it seemed to be the first anthocyanin to appear in red flowers of *Impatiens balsamina* and thus apparently occupies a pivotal position in the production of other compounds. It has subsequently proved more readily available and more stable than other compounds which we have tested. Sondheimer and Kertesz (1948) and Robinson and Smith (1955) have reported this compound to be the major anthocyanin of cultivated strawberries and our analyses have confirmed their results. In preparing pigment for infiltration experiments, whole-frozen berries were extracted with *n*-butanol containing 1% HCl (v/v). The pigment was partitioned into an aqueous phase by the addition of about 5 volumes of petroleum ether (bp 60–110 C) and storage at 4 C in the dark (Sondheimer and Karash, 1956). The flask was put in the freezer (-15 C) until the aqueous layer was frozen; then the ether layer was decanted. The frozen layer was allowed to thaw at room temperature and was filtered through glass fiber filters. The filtrate was concentrated to 25% the original volume on a rotary evaporator before being streaked on Whatman No. 3 MM chromatograph paper. The pigment was refined by successive chromatography with two solvent systems (*tert*-butanol, glacial acetic acid, water—3:1:1—and 10% acetic acid). The bands of pigment separated by chromatography were cut from the paper, eluted by first hanging them on glass hangers with the bases of the strips dipping into Harborne's mixture (1958) of methanol, acetic acid, and water (70:5:25). When the pigment had concentrated at the top of the strip, the papers were shortened and the pigment collected by siphoning the same solvent through the paper from an elevated beaker to a lower one.

After collection the pigment was dried, redissolved in a minimal volume of anhydrous absolute ethanol and added to 100 volumes of anhydrous ethyl ether. The pigment came down as a flocculent precipitate and was collected by centrifugation This material was dried in a vacuum desiccator over Drierite for 24 hr, then scraped off as an amorphous powder. If the pigment proved chromatographically pure, it was weighed and dissolved in enough distilled water to make a solution of the desired concentration. In the majority of the experiments an 0.5% (w/v) solution was used. The infiltration solutions were adjusted to a *p*H of 3.0–3.5 because this range showed least tissue alteration and yet maintained this substrate in fairly stable solution.

RESULTS—*Behavior of petals after infiltration*—Because preliminary tests showed little uptake of substrate from liquid media by cultured petals, the vacuum infiltration procedures of McCready and Hassid (1941) and Hutchinson, Roy and Towers (1958) were modified in an effort to secure more intimate contact between the substrate and the petal cells.

Immediately after infiltration the petals are transluscent throughout and appear flaccid. When incubated in the light with the cut ends embedded in an agar medium, the petals usually recover their original turgid condition within 12–24 hr. The petals generally expand rapidly and become opaque as the fluid is absorbed from the intercellular spaces (Fig. 2). The clearing may begin either at the margins or near the center of the tissue, but where the blade of the petal makes contact with the agar surface or the glass of the dish, the area of contact remains saturated. When white petals have been infiltrated with pelargonidin-3-monoglucoside, the color is at first a rather

Fig. 1, 2.—Fig. 1. Photograph depicting the sequential stages of bud development of *Impatiens balsamina* L. In this study, stage-3 buds were most often used.—Fig. 2. Stage-3 petals which have been infiltrated with pelargonidin-3-monoglucoside. The base of the petals are imbedded in nutrient agar. Note the increased expansion after 24 hr.

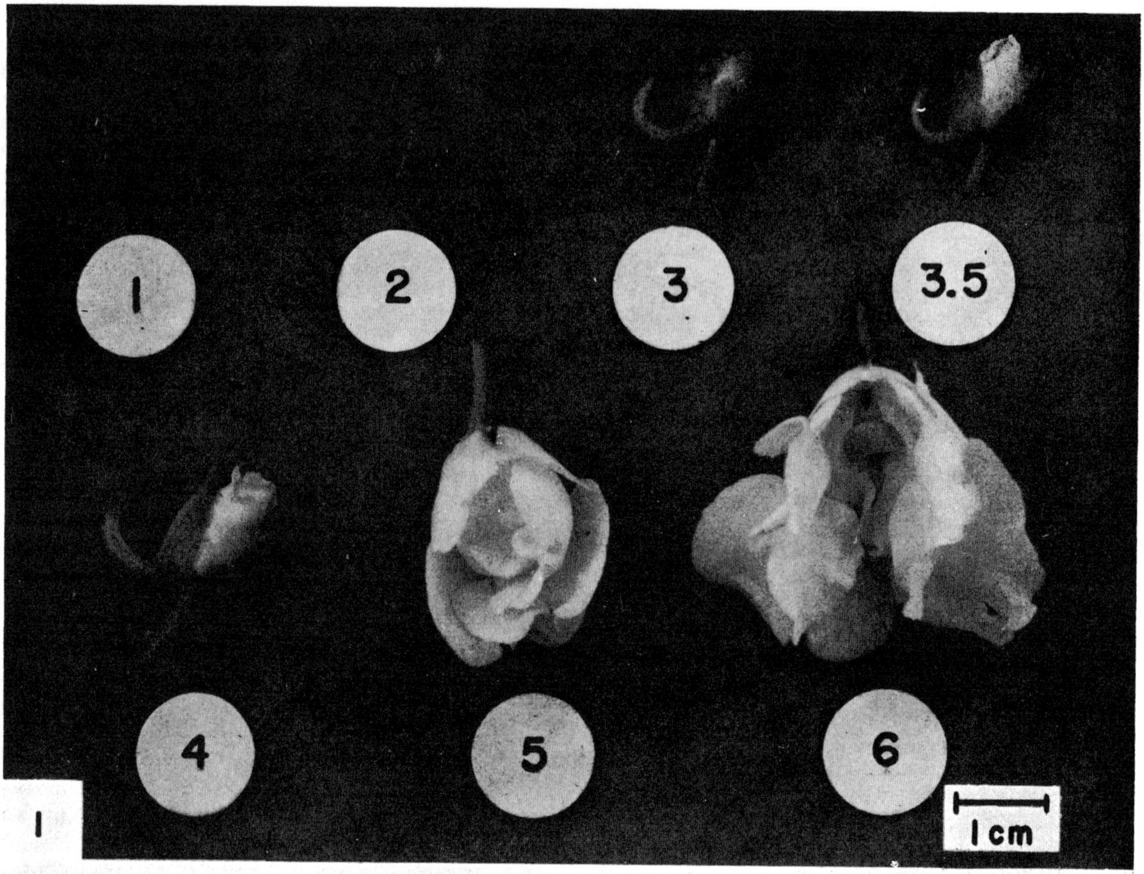

1

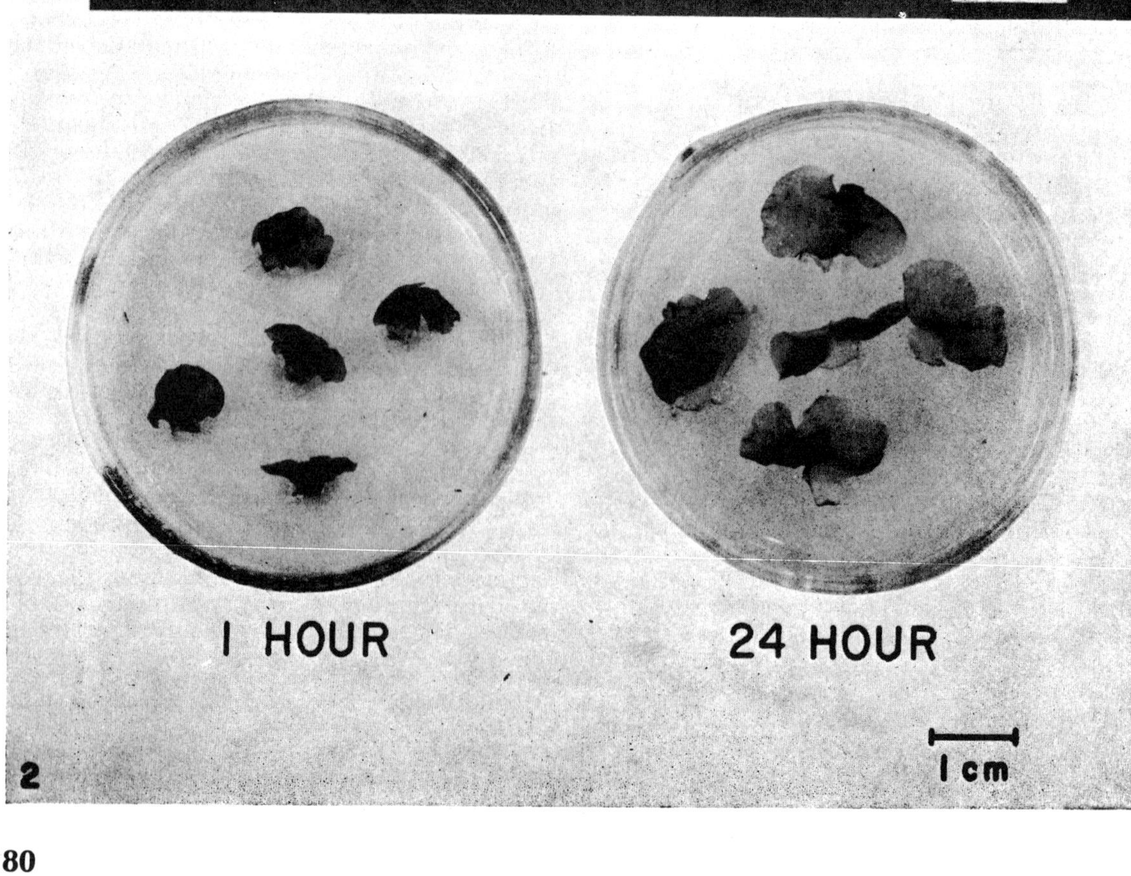

2

dull red, but with the clearing of the intercellular spaces they appear a fresh pink. The color change appears to be associated with the physical distribution of the pigment, for tissues which remain saturated and dull red metabolize the pigment in the same fashion as cleared peta s.

Frequently the clearing is complete within 12 hr, but sometimes 24–48 hr are not sufficient to remove all the intercellular fluid. The rate of clearing appears to depend upon the age and dryness of the agar plates as well as upon the amount of liquid infiltrated. The amount of liquid introduced is in turn affected by uncontrolled variables in the infiltration procedure and upon the degree of expansion which the petal has reached at the time of treatment. We have been unable to devise a quantitatively reproducible infiltration technique.

Metabolism of pelargonidin-3-monoglucoside by white petals—The use of genetically white flowers simplified the identification of products derived from introduced substrates. Successful metabolism by this and other genotypes also permitted conclusions concerning the nature of the pathway of anthocyanin modification during normal development and concerning the regulation of anthocyanin synthesis in flowers.

White petals have produced five different pelargonidin derivatives from pelargonidin-3-monoglucoside. Using the identifying numbers employed by Hagen (1966a, b), these products (Fig. 3–6) are acyl (p-coumaroyl)-3,5-diglucoside (spot 6), 3,5-diglucoside (spot 4), presumptive 3-diglucoside-5-monoglucoside (spot 45), presumptive acyl-3-diglucoside-5-monoglucoside (spot 54), and the aglycone, pelargonidin (spot 41). Spots 45 and 54 have been observed rarely and in small quantities on our chromatograms. The other products of infiltration experiments have been examined spectrophotometrically as well as chromatographically and appear identical with compounds isolated from red petals of the genotype llHHP^{r}P^r. Analyses of hydrolysis products have confirmed the structure of the prominent compounds. Structural formulas of the pigments discussed in the text have been presented in paper II of this series (Hagen, 1966b)

Without quantitative measures, observations were limited to time sequences of appearance of various products and to estimates of relative abundance of introduced substrate and products on chromatograms. At 1 hr after introduction of pelargonidin-3-monoglucoside into white petals (genotype llhhpp), the introduced material is still the major anthocyanin component, but there is considerable aglycone of pelargonidin as well as a trace of pelargonidin-3,5-diglucoside and, sometimes, of the acyl pelargonidin-3,5-diglucoside. As the incubation period lengthens, the amount of residual substrate decreases while there is a marked increase in the amount of acyl-3,5-diglucoside and 3,5-diglucoside. The pelargonidin aglycone decreases gradually below the amount present at 1 hr. After 24–30 hr, both the introduced substrate (spot 5) and the aglycone (spot 41) may be absent or reduced to traces, leaving the acyl-3,5-diglucoside (spot 6) and 3,5-diglucoside (spot 4) as the principal anthocyanins and the most prominent components of the extract. These changes are illustrated in Fig. 3–6. The flavonol pigments (spots 2, 7, 8) are included for comparison because their R_f values and relative concentrations remain constant throughout the incubation period.

The sequence of events in the infiltrated white petals is similar to the sequence described by Hagen (1966b) as the normal developmental pattern for pelargonidin glycosides in the red genotype (llHHP^{r}P^r). Red petals, however produce at intermediate stages two glycosides (spots 15, 52) believed to be acyl-3-glucosides. These have not been detected in white petals infiltrated with pelargonidin-3-monoglucoside.

The influence of light—Light affects the quantity of anthocyanins produced in many systems. Siegelman and Hendricks (1957, 1958) and Downs and Siegelman (1963) have demonstrated at least two light reactions involved in anthocyanin synthesis in turnip seedlings, apple skins and milo seedlings. In red cabbage seedlings, however, there appears to be just one photoreaction (Siegelman and Hendricks, 1957). Klein and Hagen (1961) found light essential for maximum expansion and pigment production in cultured petals of *Impatiens balsamina.*

To test the influence of light on the modification of pelargonidin-3-monoglucoside by white petals, infiltrated petals were cultured in light and in dark at 23 C. No attempt was made to prevent exposure of tissues to light during excision and infiltration.

After 6 hr of culture chromatograms prepared from light- and from dark-grown petals were identical in appearance. Spots 4, 6 were detectable in both extracts, but the introduced material and pelargonidin aglycone predominated. On chromatograms prepared after 24- and 50-hr incubations, spots 4, 6 had increased and there was a concomitant decrease in the residual substrate. At these later periods, however, there was less aglycone in light-grown than in the dark-grown petals. While this difference could indicate a light-dependent conversion of aglycone to glucosylated derivatives, practical laboratory experience suggests that the disappearance of aglycone is associated with the greater sensitivity of anthocyanidins to photodecomposition. Anthocyanins, especially derivatives of pelargonidin, are much more stable in light. Consequently the metabolic changes appear to be the same in light and in darkness. In both situations some aglycone is produced.

We can perhaps exclude pelargonidin aglycone from direct participation in the production of anthocyanins because the light-grown tissues,

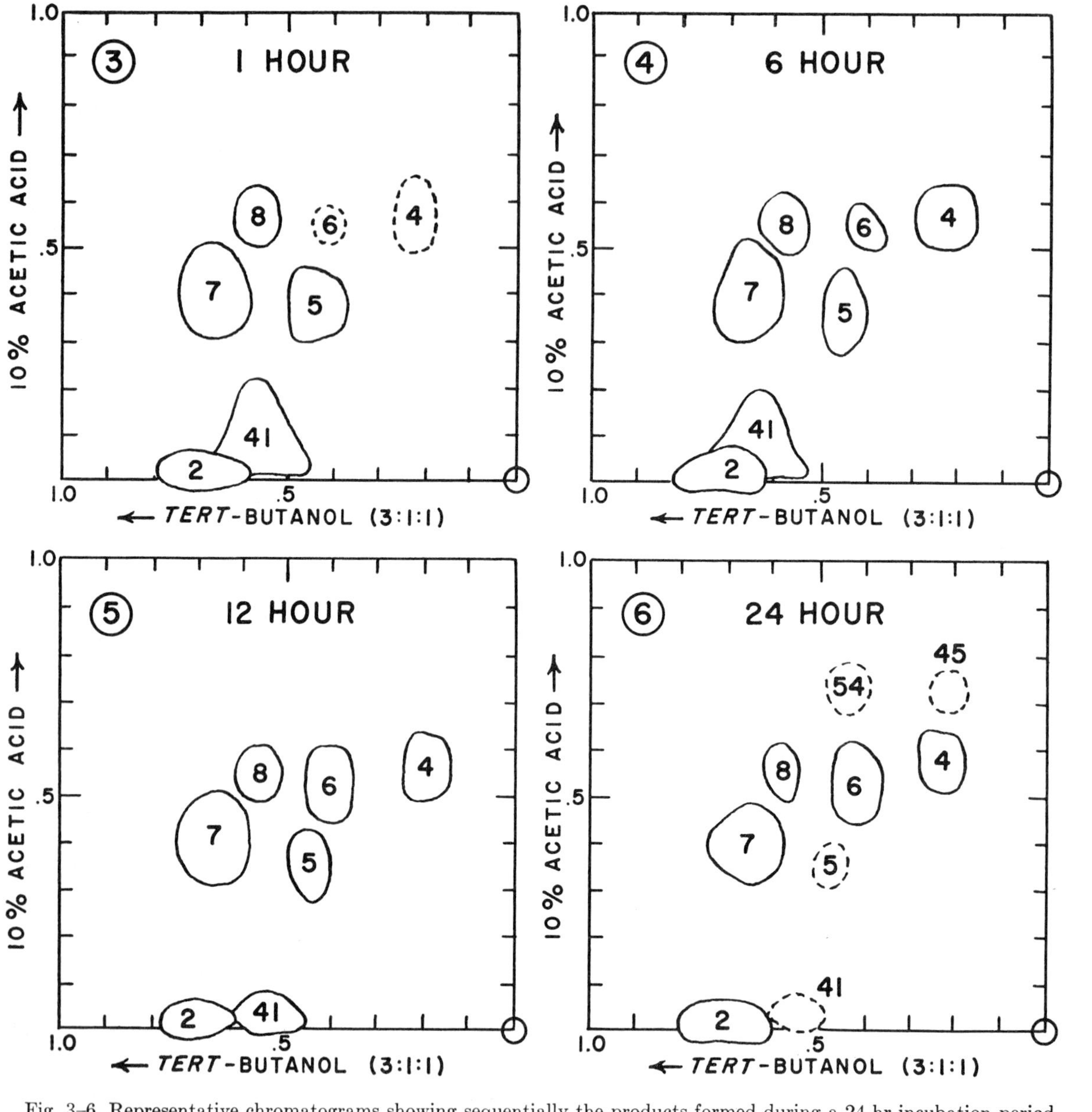

Fig. 3–6. Representative chromatograms showing sequentially the products formed during a 24-hr incubation period by genetically white petals infiltrated with pelargonidin-3-monoglucoside. Spots 2, 7, 8 are flavonols. Spot 41—pelargonidin-3-monoglucoside (substrate); Spot 6—acyl pelargonidin-3, 5-diglucoside; Spot 45—pelargonidin-3-diglucoside-5-monoglucoside; Spot 54—acyl pelargonidin-3-diglucoside-5-monoglucoside. Spots 54 and 45 have been observed only occasionally. Dotted lines represent trace amounts.

in which the aglycone disappears completely, produced no more anthocyanin than dark-grown tissues which retain the aglycone. The appearance of aglycone at early stages, in infiltrated petals as well as in intact red petals, may be attributed to accidental contact with a glucosidase known to be present (Hagen, 1966b). The disappearance at later stages of development could, in both situations, be attributed to photodecomposition.

Metabolism of pelargonidin-3-monoglucoside by other genetic strains—To test aspects of the control of pigment production and modification, the two pink genotypes (llHHpp and llhhP^{r}P^r) and the purple (LLhhP^{r}P^r) were infiltrated with pelargonidin-3-monoglucoside.

Fully open petals of the pink genotypes contain only small quantities of the acyl-3, 5-diglucoside of pelargonidin (Hagen, 1959). Where the dominant genes are present in the same plant (llHHP^{r}P^r), the effect is synergistic, the petals producing more pigment than the sum of the quantities in petals of llHHpp and llhhP^{r}P^r (Alston and Hagen, 1958;

Hagen, 1959). The mature petals of the purple genotype produce derivatives corresponding to the pelargonidin derivatives of the red genotype.

Thus the gene P^r functions to increase anthocyanin production in the presence of genes L and/or H. Also, it seems apparent that the loci L, H, and P influence the quantity and nature of the anthocyanidin nucleus rather than the acylation and glucosylation patterns (Alston and Hagen 1958).

Whether the acylating and glucosylating enzymes are specific as to the anthocyanidin on which they substitute has been answered by the fact that these enzymes are present and active in the white petals which are recessive at all three loci. Gene H, therefore, is not required for the production of the more highly substituted anthocyanins once the 3-monoglucoside has been introduced.

Moreover, when petals of the purple genotype are infiltrated with pelargonidin-3-monoglucoside, there is no noticeable variation from the pattern observed with white flowers. Thus the pelargonidin nucleus is apparently not converted to malvidin in the presence of gene L and the action of L must be at some stage prior to the formation of the 3-monoglucoside derivative and cannot alter the final hydroxylation or methoxylation pattern of the B ring. Here again the substitution pattern of the substrate appears to be unaffected by the presence of genes L, H, or P^r.

Another interpretation of the synergistic effect of genes H and P^r would have these genes controlling separate routes from the 3-monoglucoside level to the final acyl-3,5-diglucoside derivative. However, in none of the experiments involving infiltration of the pink genotypes was there observed any deviation from the pattern produced in the white petals or from the normal developmental pattern of the red genotype.

Metabolism of other anthocyanins—In attempting to further elucidate the enzymatic potential of the genetically white petals, two other anthocyanins, cyanidin-3-monoglucoside and pelargonidin-3,5-diglucoside, were utilized as possible substrates.

Cyanidin-3-monoglucoside was prepared from *I. balsamina* stem tissue (Hagen, 1966a) using the procedure described earlier for pelargonidin-3-monoglucoside. Immediately this compound proved unsatisfactory as a substrate. When the *p*H of the infiltration solution was low enough to render the anthocyanin soluble, the tissue was severely damaged and did not develop normally. If the *p*H was raised above 3.0, the anthocyanin would precipitate. This pigment was tried in several experiments and the only anthocyanin spot detected on the chromatograms was that of the residual substrate.

Pelargonidin-3,5-diglucoside was also tried as a potential metabolite of this system. This anthocyanin was prepared from mature petals of the red genotype. The pigment was infiltrated and the petals expanded and developed slowly. After 48 hr the petals had "cleared" and appeared a light pink. However, chromatographic analysis revealed only the presence of the substrate. There was no production of the 3-monoglucoside, aglycone, or any other anthocyanin during incubation. This experiment has been repeated several times but no other anthocyanin product has been observed. This finding tends to support the suggestion of Hagen (1966b) that the 3,5-diglucoside may not lie on the pathway for the formation of the acyl-3,5-diglucoside, but rather that it may be an artifact produced during extraction and chromatography. The instability of acyl derivatives has been reported by Dodds and Long (1955) and Harborne (1959), and it is possible that the 3,5-diglucoside observed in the red genotype may be derived in a similar manner.

Effect of puromycin—Upon observing that genetically white petals are capable of metabolizing exogenous pelargonidin-3-monoglucoside, the question arose as to whether this enzyme system was constitutive or whether there was a requirement for protein synthesis before metabolism could occur.

The antibiotic, puromycin, has been shown to inhibit protein synthesis in higher plants (Nooden and Th mann, 1963; Click and Hackett, 1963) as well as in microorganisms (Creaser, 1955; Yarmolinsky and de la Haba, 1959).

Stage-3 petals were infiltrated with puromycin solutions up to 10^{-3} M. At intervals from 1–24 hr after the infiltration of the inhibitor, samples were removed, infiltrated with an 0.5% (w/v) solution of pelargonidin-3-monoglucoside and incubated. Samples were extracted and chromatographed at 8 and 24 hr after being infiltrated with anthocyanin.

At present we have been unable to effect inhibition of the anthocyanin metabolism. Whether the puromycin actually inhibits protein synthesis in this particular tissue remains to be substantiated. However, if the enzyme system is constitutive, a protein inhibitor applied to stage-3 buds should have no effect on pre-existing enzymes.

Discussion—Evidence from earlier work (Hagen, 1966a, b) that pigment production in red flowers of *Impatiens balsamina* proceeds through successive extractable intermediate compounds has rested upon descriptive studies. These studies have demonstrated a sequence of events which appears to begin with the 3-monoglucoside and progresses through various intermediates to the p-coumaroyl-3,5-diglucoside as the predominant pigment of mature petals. The experiments described here demonstrate that these observations are not coincidental; that is, the petals are definitely capable of modifying the structure of exogenous pelargonidin-3-monoglucoside. Therefore, this pigment can and probably does serve as a precursor during normal development.

That the variety of intermediates detected during the modification of exogenous substrate is less than during normal development remains to be explained. It is possible that compounds such as spots 16, 52, putative acylated monoglucosides, actually precede the 3-monoglucoside even though they have been detected only in intermediate developmental stages. It is also possible that events proceed too rapidly in cultured petals for the detection of the more transient intermediates. More likely, the acylated monog'ucosides may appear only when reactions leading more directly to final pigments are saturated, thus permitting the 3-monoglucoside to spill over into secondary routes to the same ultimate product or to the more elaborately substituted pelargonidin derivatives which occur in vigorous plants.

The white petals (llhhpp) which were fed pelargonidin-3-monoglucoside differed by two genes (*H* and P^r) from the red petals (llHHP^{r}P^r). The recessive forms of these genes are both leaky, in that neither constitutes an absolute block to pelargonidin pigmentation. The dominant genes are synergistic, producing when together more than the sum of the quantities of the pigment produced by the separate genes in different plants. One interpretation of this situation would have the genes *H* and P^r control different routes to the final pigment, both routes branching from a common precursor with limited pool size. Simultaneous draining into two routes from the limited pool could increase the rate of replenishment of the pool. The existence of two potential intermediates, acyl-3-monoglucoside and 3.5-diglucoside, between the initial 3-monoglucoside and the ultimate acyl-3.5-diglucoside suggested that the 3-monoglucoside might actually be the branching point in the pathway. If this were the case, then infiltration of llHHpp might be expected to lead to different intermediates than infiltration of llhhP^{r}P^r. Because no acyl-3-monoglucosides were formed in either case, we must look to other explanations of the synergism of genes *H* and P^r.

The definition of the route by which acyl-3,5-diglucosides are formed would be eased by clear demonstration that petals could add acid groups to 3,5-diglucosides. Unfortunately this transformation has not yet been achieved in cultivated intact petals. In the case of petals infiltrated with cyanidin-3-monoglucoside, demonstrable incompatibility between our preparations o' substrate and the petal tissues precludes reliance on negative results. Because we have not yet extracted an active synthetic system, the in vitro tests also lack significance.

It is clear from infiltration of purple petals that the malvidin configuration cannot be actively derived from the pelargonidin structure after the monoglucoside stage has been achieved. This was perhaps evident from genetic studies, for acyl diglucosides of pelargonidin and malvidin exist within the same cells in the genotype LLHHP^{r}P^r of *Impatiens balsamina*, while corresponding associations of different pigments occur in other tissues and other species

The ability of genetically white petals to add promptly additional sugar and hydroxycinnamoyl groups to introduced pelargonidin-3-monoglucoside raises questions concerning the initiation of enzymes responsible for these activities. The same reactions do not take place outside of the floral t ssues. Since 3-monoglucosides occur in vegetative tissues, induction by substrate might be excluded a priori unless differentiation in flower parts rendered these cells particularly susceptible to induction. However, white flowers, which could not experience induct on by anthocyanidin-3-monoglucosides, or by immediate precursors to this compound, nevertheless produce p-coumaroyl-3.5-diglucoside within 1 hr after exposure to the introduced 3-monoglucoside. Furthermore, we have been unable to block these activities by the infiltration of puromycin prior to, or simultaneously with, the substrate. The available evidence, therefore, indicates that the enzymes are activated, and genes called into action, by some component of the developmental sequence other than the immediate substrates o those enzymes.

LITERATURE CITED

ALSTON, R. E., AND C. W. HAGEN, JR. 1958. Chemical aspects of the inheritance of flower color in *Impatiens balsamina* L. Genetics 43: 35–47.

CLICK, R. E., AND D. P. HACKETT. 1963. The role of protein and nucleic acid synthesis in the development of respiration in potato tuber slices. Proc. Natl. Acad. Sci. (U.S.) 50: 243–250.

CREASER, E. H. 1955. The induced (adaptive) biosynthesis of β-galactosidase in *Staphylococcus aureus*. J. Gen. Microbiol. 12: 288–297.

DAVIS, D. W., L. A. TAYLOR, AND R. P. ASH. 1958. *Impatiens balsamina* L. The inheritance of floral colors. Genetics 43: 16–34.

DODDS, K. S., AND D. H. LONG. 1955. The inheritance of colour in diploid potatoes I. Types of anthocyanidins and their genetic loci. J. Genet. 53: 136–149.

DOWNS, R. J., AND H. W. SIEGELMAN. 1963. Photocontrol of anthocyanin synthesis in milo seedlings. Plant Physiol. 38: 25–30.

HAGEN, C. W., JR. 1959. Influence of genes controlling flower color on relative quantities of anthocyanins and flavonols of *Impatiens balsamina*. Genetics 44: 787–793.

———. 1966a. The differentiation of pigmentation in flower parts. I. The flavonoid pigments of *Impatiens balsamina*, genotype llHHP^{r}P^r, and their distribution within the plant. Amer. J. Bot. 53: 46–54.

———. 1966b. The differentiation of pigmentation in flower parts. II. Changes in pigments during development of buds in *Impatiens balsamina*, genotype llHHP^{r}P^r. Amer. J. Bot. 53: 54–60.

HARBORNE, J. B. 1958. Spectral methods of characterizing anthocyanins. Biochem. J. 70: 22–28.

———. 1959. Chromatographic identification of anthocyanin pigments. Chrom. Rev. 1: 209–224.

HESS, D. 1964. Die Beteiligung von Ribonucleinsaure an der Gengesteuerten Induktion der Anthocyansynthese bei *Petunia*. Planta 61: 73–89.

Hutchinson, A., C. Roy, and G. H. N. Towers. 1958. Synthesis of phlorin and other phenolic glucosides by plant tissue. Nature 181: 841–842.

Klein, A. O., and C. W. Hagen, Jr. 1961. Anthocyanin production in detached petals of *Impatiens balsamina* L. Plant Physiol. 36: 1–9.

McCready, R. M., and W. Z. Hassid. 1941. Transformation of sugars in barley shoots. Plant Physiol. 16: 599–610.

Miller, C. O. 1961. A kinetin-like compound in maize. Proc. Natl. Acad. Sci. (U.S.) 47: 170–174.

Nooden, L. D., and K. V. Thimann. 1963. Evidence for a requirement for protein synthesis for auxin-induced cell enlargement. Proc. Natl. Acad. Sci. (U.S.) 50: 194–200.

Robinson, R., and H. Smith. 1955. Anthocyanins of the leaf of the copper beech (*Fagus sylvatica*) and the fruit of the cultivated strawberry (*Fragaria virginiana*). Nature 175: 634.

Siegelman, H. W., and S. B. Hendricks. 1957. Photocontrol of anthocyanin formation in turnip and red cabbage seedlings. Plant Physiol. 32: 393–393.

———, and ———. 1958. Photocontrol of anthocyanin synthesis in apple skin. Plant Physiol. 33: 185–190.

Sondheimer, E., and C. B. Karash. 1956. The major anthocyanin pigments of the wild strawberry (*Fragaria vesca*). Nature 178: 648–649.

———, and Z. I. Kertesz. 1948. The anthocyanin of strawberries. J. Amer. Chem. Soc. 70: 3476.

Suomalainen, H., and A. J. A. Keränen. 1961. The first anthocyanins appearing during the ripening of blueberries. Nature 191: 498–499.

Yarmolinsky, M. B., and G. L. de la Haba. 1959. Inhibition by puromycin of amino acid incorporation into protein. Proc. Natl. Acad. Sci. (U.S.) 45: 1721–1729.

3
Photosynthesis

BBA 3783

FUNCTION AND IDENTIFICATION OF TWO PHOTOCHEMICAL SYSTEMS IN PHOTOSYNTHESIS

L. N. M. DUYSENS AND J. AMESZ

Biophysical Laboratory, the State University, Leiden (The Netherlands)*

(Received May 11th, 1962)

SUMMARY

1. Two pigment systems, which we called systems 1 and 2, were identified in the red alga *Porphyridium cruentum*.

2. By means of sensitive differential absorption spectrophotometry, it was observed by simultaneous illumination with two wavelengths of actinic light that the oxidation of a *f*-or *c*-like cytochrome was caused by excitation of system 1 and the reduction of this cytochrome by excitation of system 2. The reduction but not the oxidation was inhibited by not too high concentrations of *N*-ethylurethane, hydroxylamine and by 3-(3,4-dichlorophenyl)-1,1-dimethylurea.

3. On basis of this and other experimental evidence the hypothesis was proposed that upon excitation system 1 reduces CO_2 via phosphopyridine nucleotide and oxidizes cytochrome, and system 2 reduces this cytochrome and produces O_2 by dehydrogenation of water. Presumably one quantum per transported hydrogen equivalent is required for each system. Part of the hydrogen equivalents probably bypass cytochrome *f*.

4. Two methods were proposed for determining action spectra; action spectra of each of the two pigment systems were determined independently by these two methods. Both pigment systems contain chlorophyll *a* and phycoerythrin. The ratios of the activities per incident quantum at the long-wavelength absorption maxima of chlorophyll and phycoerythrin, 680 and 560 mμ, were about 1 for system 1 and about 1/7–1/3 for system 2. The activity of system 1 was greater in light of 680 mμ and smaller in light of 560 mμ than that of system 2.

5. The scheme presented is shown to be consistent with evidence obtained by a variety of experiments by various authors with different species. The scheme is probably valid for photosynthesis in algae and higher plants. In purple bacteria probably only one pigment system, similar to system 1, occurs.

INTRODUCTION

Two pigment systems, which both contain chlorophyll *a*, have been demonstrated in red and blue-green algae[1]. The action spectrum for excitation of chlorophyll *a*

Abbreviations: DCMU, 3-(3,4-dichlorophenyl)-1,1-dimethylurea; PMS, phenazine methosulphate; PNH, reduced phosphopyridine nucleotide.

* Postal address: Nieuwsteeg 18, Leiden (The Netherlands).

fluorescence showed that quanta absorbed by the phycobilins were more efficient in exciting chlorophyll *a* fluorescence than quanta absorbed by chlorophyll *a* itself (*cf.* ref. 2). This proved that part of the chlorophyll *a* was non- or weakly fluorescent and that the energy of quanta absorbed by the phycobilins, phycoerythrin and phycocyanin, was mainly transported to the fluorescent fraction of the chlorophyll. Since the action spectrum of photosynthesis was roughly parallel to the action spectrum of chlorophyll *a* fluorescence, it was concluded that, of the pigments mentioned, the fluorescent fraction of the chlorophyll *a* was directly active in photosynthesis; energy of quanta absorbed by phycobilins was effective in photosynthesis because this energy was transferred to the fluorescent chlorophyll *a*. The above-mentioned non- or weakly fluorescent fraction of chlorophyll *a* was believed to be photosynthetically inactive.

Recently, however, also this non-fluorescent chlorophyll was indicated to be photosynthetically active by experiments[3] on light-induced cytochrome oxidation in the red alga *Porphyridium cruentum*.

Contrary to expectation, the action spectrum for cytochrome oxidation was found to be quite different from that of chlorophyll *a* fluorescence or of photosynthesis: The action spectrum had a pronounced maximum at 680 mμ, due to chlorophyll *a* and much smaller maxima at 560 and 630 mμ, due to the phycobilins. This showed that the cytochrome oxidation was caused by a pigment system different from that containing the fluorescent chlorophyll *a* and thus containing all or part of the non- (or weakly) fluorescent chlorophyll *a*. The relatively high quantum efficiency for cytochrome oxidation (5 quanta absorbed at 680 mμ per cytochrome molecule oxidized) suggests an important role of the weakly fluorescent chlorophyll *a* in photosynthesis. Thus at least two pigment systems occur in Porphyridium with different photochemical functions. For the interpretation of the results to be reported in this paper it is sufficient to postulate only two photochemical activities and two corresponding pigment systems, which we call "system 1" and "system 2", containing amongst others weakly fluorescent chlorophyll a_1 and more strongly fluorescent chlorophyll a_2 respectively.

The kinetics of cytochrome oxidation were found[3] to be different for actinic light of 680 and 560 mμ. This indicated that system 2 interacted in some way with system 1 in the cytochrome oxidation and also that the action spectrum for the cytochrome oxidation does not represent the true action spectrum of system 1 but results from an interaction of both systems.

In this paper experiments will be described concerning this interaction, and other properties of the pigment systems. Action spectra of systems 1 and 2 will be given as determined by two different methods. A hypothesis will be proposed concerning the function of the two systems in the photosynthesis of Porphyridium, and concerning analogous systems in other photosynthesizing organisms.

A condensed report of part of these experiments has been published recently[4].

METHODS AND MATERIALS

Porphyridium cruentum was grown in light of about 2200 lux from white fluorescent lamps (Philips "de Luxe" type 34) for about 6 days in a liquid culture medium[5], bubbled with air enriched with 4% carbon dioxide. Before measurement the algal suspension was either diluted with fresh medium, or concentrated by centrifugation at 500 × *g* during 15 min and resuspension in fresh medium.

By means of an improved difference spectrophotometer, analogous to one described earlier[6], the change in absorbance of the algae was measured at 420 mμ, the wavelength of the maximum of the difference spectrum of the algae. The algae were in a 1-cm Beckman cuvette. The change in absorption was brought about by illumination from the side with actinic beams from a monochromator and from a projector. The monochromator was set to transmit a wavelength band of 4 to 7 mμ half width; in order to obtain illumination homogeneous in colour and intensity, the grating was imaged upon the algae. The projector was provided with a combination of absorption and interference filters, which transmitted a band of about 10 mμ half width. Both actinic beams could be admitted and shut off independently.

The intensity of the beams was measured by means of a calibrated photoelectric cell. Filters in front of the measuring multiplier minimized the actinic, but transmitted the measuring light. The detecting apparatus only responded to modulated light, and only the measuring beam was modulated. The fact that in the presence of 5 volume percent of the inhibitor *N*-ethylurethane the light-induced deflection at 420 mμ was completely suppressed, showed that the deflection with the non-poisoned algae was caused by a change in light transmission, and not by a spurious effect of the apparatus.

The rate of photosynthetic oxygen evolution was measured with an oxygen polarograph with a platinum electrode of about 5 × 5 mm², mounted horizontally. This polarograph was similar in principle to one constructed by HAXO and shortly described by MYERS AND FRENCH[7]; via a direct current amplifier it was connected to a recorder, whose deflections were proportional to changes in electrode current, caused by oxygen evolution upon illumination of a thin layer of algae, deposited on the electrode. The algal layer could be illuminated with two actinic beams, in a way similar to that described for the difference spectrophotometer. The incident intensities were corrected for reflection at the electrode; the reflection coefficients of HAXO AND BLINKS[8] were used. The recorder deflection may be taken to be a linear function of the rate of oxygen evolution.

RESULTS AND INTERPRETATION

Cytochrome oxidation and reduction

As stated in the introduction, the different kinetics of cytochrome oxidation in *Porphyridium cruentum* for actinic light of about 680 and 560 mμ showed that two different photochemical systems interacted in this oxidation. The kind of interaction is clearly demonstrated by the recorder tracing (a) of Fig. 1. Actinic light of 680 mμ causes an oxidation of the cytochrome, as indicated by a decrease in absorption at 420 mμ. When, during the illumination with light of 680 mμ, light of 562 mμ is admitted, the absorption increases, indicating the reduction of cytochrome. If, however, during illumination with light of 680 mμ, extra light of 680 mμ is given, a further decrease in absorption (not shown in the figure) occurs. Similar effects were observed for a piece of thallus of the red alga Porphyra.

The difference spectrum for the increase in absorption occurring upon additional illumination with light of 560 mμ was found to be proportional to the difference spectrum of the decrease occurring upon illumination with light of 680 mμ alone. It was necessary to use a rather weak measuring beam to minimize additional changes in absorption brought about by the measuring light. The proportionality of these difference spectra indicates that both the decrease and the increase of absorption at 420 mμ were caused by the same cytochrome. So far our experiments gave no indications that more than one cytochrome is the cause of the absorption changes.

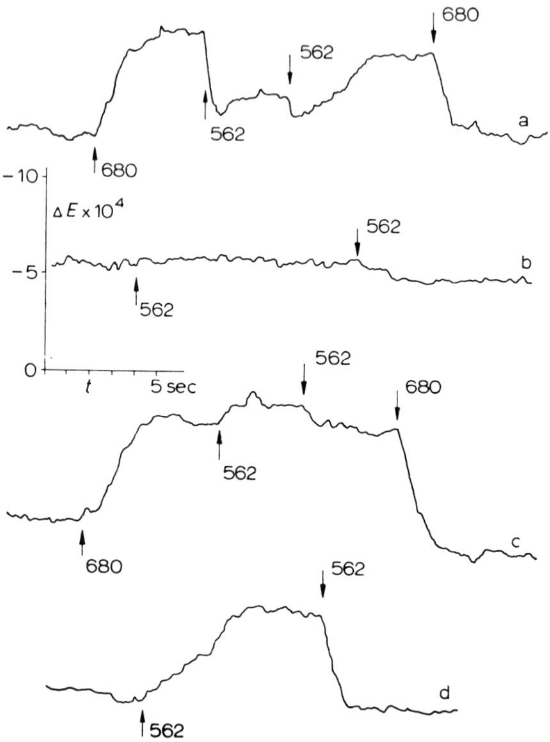

Fig. 1. Kinetics of the changes in absorption of Porphyridium at 420 mμ. Upward- and downward pointing arrows indicate that the actinic light is switched on and off. An upward moving trace indicates a decrease in absorption and the oxidation of a *f*-like cytochrome. Tracings a and b are for non-poisoned algae and tracings c and d for algae poisoned with $1.1 \cdot 10^{-6}$ *M* DCMU. Tracing a shows that light of 680 mμ causes an oxidation of the cytochrome and that subsequent admittance of additional light of 562 mμ causes its reduction. However, as shown by tracing c and d, if light of 562 mμ is admitted to poisoned algae, the cytochrome is oxidized. A 0.9% suspension was used in nutrient medium in a 1-cm cuvette. The intensities of the actinic light were $44 \cdot 10^{-11}$ and $53 \cdot 10^{-11}$ einstein/sec/cm² at 680 and 562 mμ.

Since actinic light of 680 mμ is preponderant in causing cytochrome oxidation and light of 560 mμ in causing reduction, system 2, which has predominant active absorption at 560 mμ, may be taken to be responsible for the reduction of the cytochrome and system 1 for the oxidation. Addition of DCMU completely abolished the reduction of cytochrome by light of 562 mμ (Fig. 1), but did not inhibit the oxdation of cytochrome by light of 680 mμ. Similar effects were observed with 43 m*M* (0.5%, v/v) *N*-ethylurethane, 0.5 m*M* hydroxylamine, 8 m*M* sodium azide or 1.1 m*M* 2,4-dinitrophenol. As mentioned, with a 10 times higher concentration of *N*-ethylurethane, but not of DCMU (the other inhibitors were not tested), also the oxidation of cytochrome was inhibited. In the presence of *N*-ethylurethane, the kinetics of

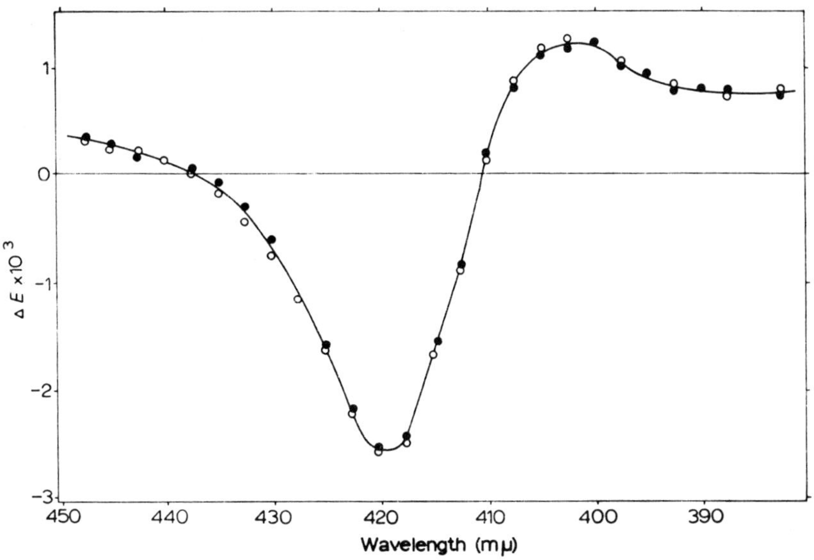

Fig. 2. Difference spectra of the steady state, light minus darkness, for actinic light of 680 (●) and 586 mμ (○) in the presence of $4.3 \cdot 10^{-2}$ *M N*-ethylurethane. The concentration of the suspension was 1 % wet cells. The actinic intensities were $12.5 \cdot 10^{-10}$ and $18.2 \cdot 10^{-10}$ einstein/sec/cm² at 680 and 586 mμ.

cytochrome oxidation and the difference spectrum in the violet region (Fig. 2) were the same, or not noticeably different, upon illumination with light of 586 and of 680 mμ.

These observations indicate that these inhibitors prevented system 2 from interacting with the cytochrome oxidation. Thus the action spectrum for the oxidation of cytochrome in a Porphyridium suspension in the presence of a moderate concen-

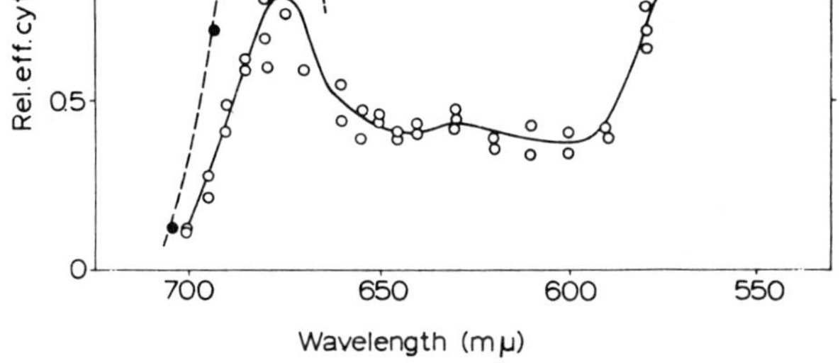

Fig. 3. Action spectra for cytochrome oxidation in the presence of $4.3 \cdot 10^{-2}$ *M N*-ethylurethane (○) and, for other cultures, in the presence of $1.1 \cdot 10^{-6}$ and $1.1 \cdot 10^{-5}$ *M* DCMU (●, △). The absorptances at 680 and 560 mμ of the three suspensions corrected for scattering were 21 and 38 %, 18 and 35 %, and 12.5 and 18.5 %. The three spectra have been normalized to an average value of 1 at 560 mμ. A correction has been applied so that the solid circles and triangles represent a spectrum for an infinitely thin suspension. The same correction would lower the open circles at 680 mμ by about 7 %.

tration of the mentioned inhibitors may be taken to be proportional to the action spectrum of system I. Action spectra in the presence of *N*-ethylurethane and DCMU are shown in Fig. 3. As "action" is taken the difference between the absorption in the dark and in the steady state in the light, at non-saturating light intensities. The reciprocal of the light intensity which causes a certain "action" is plotted as a function of wavelength. Fig. 3 shows that system I contains not only chlorophyll *a*, as was mentioned in the introduction but also phycoerythrin; the active absorbance at the chlorophyll *a* maximum at 680 mμ is approximately equal to the active absorbance at the phycoerythrin maximum of 560 mμ.

System 2, which is the one containing the fluorescent chlorophyll a_2, has a much higher active absorbance at 560 mμ than at 680 or 430 mμ. From points of the action spectrum of chlorophyll *a* fluorescence[1,2] (which spectrum may be assumed to represent approximately the action spectrum of system 2), we calculate the ratio of active absorbances at 560 and 680 mμ to be at least 2.4–6 for different cultures. Thus the ratio of activities of phycoerythrin and chlorophyll *a*, although varying with culture, is much higher in system 2 than in system I.

Photosynthesis measurements

The antagonistic roles of the systems I and 2 in cytochrome oxidation and reduction indicated that excitation of both systems I and 2 is needed for photosynthesis. Since, for moderate light intensities, the rate of photosynthesis is proportional to light intensity, this indicates that at these intensities the rate of back reactions is slow compared to the inter-system rate.

We checked the working hypothesis that photosynthesis proceeds, if a photoproduct of one system reacts with a photoproduct of the other system in stoichiometric proportion. If, for a certain regime of illumination, the rate of one system is lower than that of the other, then the rate of photosynthesis is equal to the rate of the slower system.

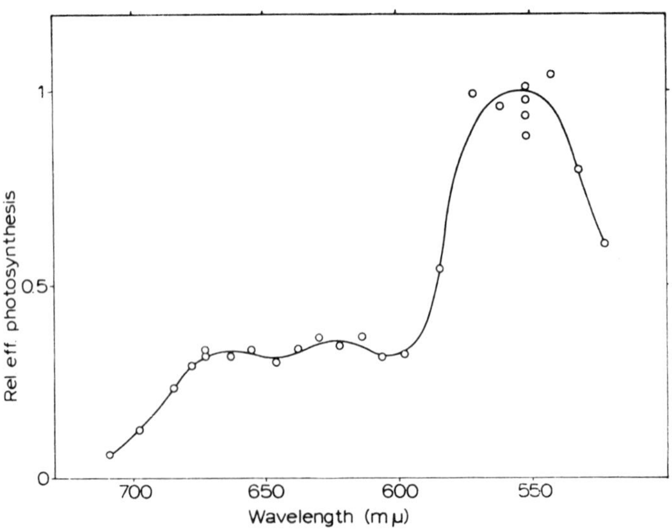

Fig. 4. Action spectrum of photosynthesis. The algae had been exposed to a concentration of $7.5 \cdot 10^{-8}$ *M* DCMU, and were then deposited in a layer of about 10 μ thickness upon the platinum electrode of the oxygen polarograph.

Independent measurements of the action spectra of systems 1 and 2 can then be made as follows. If the action spectrum of photosynthesis is measured with weak light against a constant background of strong but non-saturating light of 560 mμ, in which production by system 2 is in excess, then the action spectrum of photosynthesis will be proportional to the action spectrum of system 1, because the rate of system 1 is at all wavelengths lower than the rate of system 2. If, on the other hand,

TABLE I

RELATIVE EFFICIENCES OR ACTIVITIES FOR EXCITATION OF PHOTOSYNTHESIS, CYTOCHROME OXIDATION AND CHLOROPHYLL *a* FLUORESCENCE

The activities were measured for non-saturating light intensities, and are expressed in relative rate per incident einstein/sec/cm². The intensities are expressed in einstein/sec/cm² × 10^{-10}. Rows 1–7 and rows 10–12 are for the same cultures. The activities for these two cultures are not mutually comparable. S, designates the system which according to our hypothesis is rate-limiting and thus determines the reaction rate; U, is the abbreviation for *N*-ethylurethane. The limits of the ratio's for chlorophyll *a* fluorescence given in row 9, are minimum values calculated from Fig. 7 of ref. 1b. The intensities for the measurements of photosynthesis were roughly $4 \cdot 10^{-10}$ (rows 1 and 2) and $20 \cdot 10^{-10}$ einstein/sec/cm² (rows 3 and 4).

	Type of measurement	*Background* (mμ)	*Background* *Intensity*	*Inhibitor*	"*Activity*" *at* 680 mμ	"*Activity*" *at* 560 mμ	*Ratio* 560/680 mμ	*S*
1	Photosynthesis	680	18		28	196	7.0	2
2	Photosynthesis	562	8	—	88	121	1.37	1
3	Photosynthesis	680	18	DCMU	3.2	24	7.5	2
4	Photosynthesis	562	40	DCMU	2.5	18	7.2	2
5	Cytochrome oxidation	—	—	DCMU			0.93	1
6	Photosynthesis	—	—				4.4	2, 1
7	Absorbance	—	—				1.39	2, 1
8	Photosynthesis	—	—	—			3.0	2, 1
9	Chlorophyll *a* fluorescence	—	—	—			2.4–6	2(1)
					680 mμ	*570 mμ*	*570/680 mμ*	
10	Photosynthesis	—	—	—	20.6	83.5	4.0	2, 1
11	Photosynthesis	—	—	22 m*M* U	9.6	59.5	6.2	2
12	Photosynthesis	—	—	43 m*M* U	4.5	28.8	6.4	2

the action spectrum of photosynthesis is measured against a background of strong light of 680 mμ, then the action spectrum of photosynthesis will be proportional to the action spectrum of system 2.

The above hypothesis predicts also that addition of an amount of DCMU which inhibits appreciably system 2 will cause system 2 to become rate-limiting at all wavelengths. Thus, with DCMU the action spectrum of photosynthesis may be expected to be proportional to that of system 2, independent of the wavelength of background light. Such an action spectrum, measured with the oxygen polarograph is shown in Fig. 4. It shows the high ratio phycoerythrin/chlorophyll *a* characteristic for system 2. This ratio is, as to be expected from our hypothesis, higher for poisoned than for unpoisoned cells as is shown by the last three rows of Table I, which were measured for one culture.

Polarograph recordings of the relative changes in rate of oxygen evolution caused by light of the wavelengths 560 and 680 mμ, added to strong background light of about the same wavelengths, are shown in Fig. 5, and numerical values of steady-state changes are given in Table I. The rate of photosynthesis at 680 mμ of system 1 is about three times higher than the rate of system 2 (Fig. 5 and Table I, rows 1 and 2). The rate at 560 mμ of system 2 is about 1.5 times the rate of system 1 (rows 1 and 2).

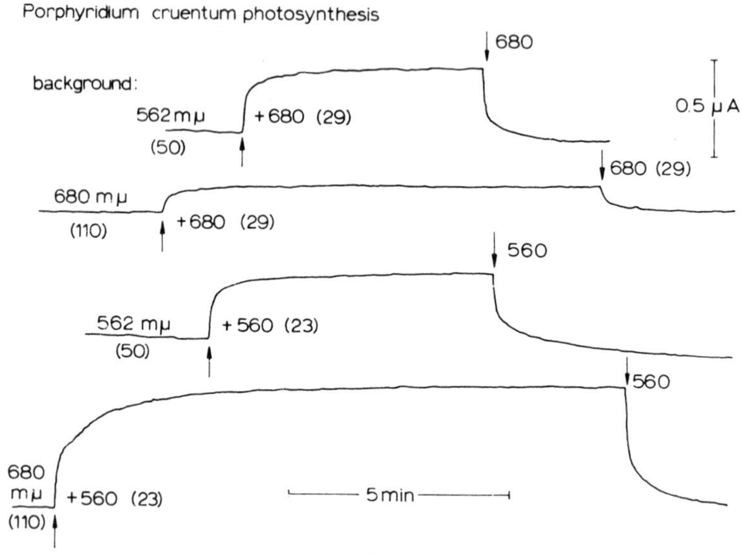

Fig. 5. Time course of photosynthesis as recorded with the oxygen polarograph. The numbers in parentheses written near the numbers indicating the wavelength of the actinic light give the actinic intensity, not corrected for reflection at the platinum surface, in 10^{-11} einstein/sec/cm^2. Upward- and downward-pointing arrows mark the beginning and the end of an illumination period. Each polarograph tracing was recorded in the presence of background light, which was switched on several minutes before the beginning of the tracing, and which was left on during the whole experiment. The two top tracings show that when actinic light of 680 mμ is added to a background of 562 mμ, the increase in the rate of photosynthesis is much higher than the increase when this actinic light is added to a background of 680 mμ (EMERSON effect). The two bottom tracings show a reverse but less pronounced effect for adding actinic light of 560 mμ. The thickness on the algal layer on the electrode was about 10 μ.

This is in accordance with the conclusion that system 2 is rate-limiting at 680 mμ and system 1 at 560 mμ. The ratio of activities of actinic light of 560 and 680 mμ is also given in Table I. The ratios in the presence of DCMU (rows 3 and 4) with a background of light of 562 and 680 mμ appear, as predicted by the hypothesis, to be equal to the ratio in unpoisoned algae with background of 680 mμ. The absolute values are of course lower in the presence of inhibitor. There is, however, a not very large but probably real difference between the ratios for system 1 calculated from photosynthesis measurements, and from measurements of cytochrome oxidation; these ratios are 1.37 (row 2) and 0.93 (row 5). A possible explanation of this discrepancy is given in the discussion.

Summarizing we may say that the ratio of the activities at 680 and 560 mμ is 1/3 to 1/7 for system 2 and about 1 for system 1. There is a strong variation with culture.

DISCUSSION

The experiments reported show that in the red alga *Porphyridium cruentum* two pigment systems are present, which we have called systems 1 and 2. Both systems contain chlorophyll *a* and phycoerythrin but have different photochemical functions. Light of 560 mμ, a wavelength close to the maximum of the absorption spectrum of phycoerythrin, is more active in system 2 per incident quantum, and light of 680 mμ, which is close to the red chlorophyll maximum, more active in system 1. Phycocyanin, which is present in relatively small concentration, and the carotenoids are left out of consideration. Excitation of system 1 causes the oxidation of a *f*- or *c*-like cytochrome; excitation of system 2 causes its reduction.

The simplest hypothesis is that the oxidation of cytochrome by system 1 is accompanied by a reduction of CO_2, probably via phosphopyridine nucleotide, and that the photochemical reduction of cytochrome by system 2 is accompanied by a simultaneous oxidation of water to oxygen. This hypothesis is supported by the observation that the reduction of the cytochrome is inhibited by a very low concentration of DCMU, and by other poisons known to inhibit the "oxygen-evolving reaction" of photosynthesis.

The arrows in the following scheme, which summarizes this hypothesis, indicate the direction of hydrogen or electron transport. The various oxidation reduction couples, which will be discussed below, are, more or less arbitrarily, represented by the oxidized or reduced component. The arrows represent dark reactions except those at both sides of the "systems" which are photochemical reactions, occurring against the potential gradient at the cost of the light energy absorbed by the pigment systems.

$$H_2O \rightarrow \text{system 2} \rightarrow Q \rightarrow \text{cytochrome} \rightarrow P \rightarrow \text{system 1} \rightarrow XH \rightarrow PNH \rightarrow CO_2.$$

Scheme 1.

This scheme is not complete: *e.g.* phosphorylations and the reactions between PNH and CO_2 have been omitted.

Function of the cytochrome and substance Q

At low light intensity the initial rate of cytochrome oxidation upon illumination is higher after a long period of darkness than after a period of illumination[3]. The kinetics of cytochrome changes are in general variable and dependent on the preceding light regime and on culture conditions. This would be understandable if the photosynthetic cytochrome is involved in phosphorylation: variation in the ADP/ATP ratio may influence the oxidation reduction kinetics of the cytochrome, especially if there is, in addition to the cytochrome path, a second path for electrons from system 2 to system 1. The existence of such a second path is indicated by the observation that the quantum efficiency for cytochrome oxidation is presumably at least two times lower than to be expected on basis of our scheme (see DISCUSSION section on *Quantum requirements*).

HILL AND BENDALL[9] recently proposed on admittedly speculative grounds a scheme in which one photochemical system oxidizes cytochrome *f* and a second system reduces cytochrome b_6, which cytochromes are known to be present in chloroplasts. On basis of the E'_0 values of 0.34 and 0.0 V, a dark reaction was postulated between oxidized cytochrome *f* and reduced b_6, coupled with phosphorylation.

As discussed in the section on results, we tried in vain to detect in Porphyridium light-induced changes in the difference spectra in the blue region, which could be attributed to cytochrome b_6. If cytochrome b_6 would have been reduced by system 1 or oxidized by system 2, the changes in b would have been observed unless they were appreciably smaller than the changes in the f-like cytochrome; possibly cytochrome b_6 participates in a non-photosynthetic reaction.

CHANCE AND SAGER's observations[10] of a light-induced oxidation of a b cytochrome in an anaerobic suspension of a pale green mutant of Chlamydomonas might be explained as an indirect oxidation of a respiratory cytochrome by a photosynthetic oxidant. One of us[11] explained in a similar way the light-driven oxidation of phosphopyridine nucleotide, which is observed in this mutant, but not in other photosynthetic cells.

The recently reported[12] oxidation of cytochromes c and f, and reduction of cytochrome b_6 in leaves of *Agrostis stolonifera* occur at too low a rate upon high-intensity illumination to provide convincing evidence that these cytochromes participate in photosynthesis.

Recently WITT *et al.*[13] found by means of an oscillographic technique for absorption measurement that, in a light flash of 10 μsec, the rise time for oxidation of endogenous cytochrome in spinach chloroplasts was about 10^{-3} sec. In another paper[14] it was shown that the shape of the absorption difference spectrum of Chlorella for one actinic intensity of 710 mμ was different from that measured at one intensity of actinic light of 670 mμ. The difference in shape was considered as evidence for the presence of two photochemical systems and this and other evidence was explained by a scheme similar to one given by KOK[15] and ourselves (in ref. 4 and this paper). It was not shown that the difference in shape could not have been caused by a difference in intensity at the two actinic wavelengths.

The substance Q in our scheme is postulated on the basis of studies on the kinetics of the fluorescence of chlorophyll a_2 (see ref. 16): light absorbed by system 1 was found to quench, and light absorbed by system 2 to raise the fluorescence yield. This phenomenon is explained by the hypothesis that oxidized Q, which is reduced by system 2, withdraws the excitation energy from the fluorescence-emitting state of chlorophyll a_2, and that QH, which is (indirectly) reoxidized by system 1, does not withdraw this excitation energy. The kinetics of the fluorescence changes are different from the changes at 420 and 700 mμ, indicating that QH is not identical to reduced cytochrome or P.

Pyridine nucleotide reduction

By means of fluorescence and absorption spectroscopy it was shown that the photoreduction of pyridine nucleotide in intact purple bacteria and algae[17] occurs with good quantum efficiency. If phosphopyridine nucleotide participates in photosynthesis as indicated by our hypothetical scheme, then the action spectrum for phosphopyridine nucleotide reduction should be proportional to the action spectrum of system 1, especially if system 2 is poisoned by DCMU.

Experiments with the blue-green alga *Anacystis nidulans*[18] indicate that DCMU does not appreciably affect the initial reduction rate of phosphopyridine nucleotide. Furthermore, the action spectrum for phosphopyridine nucleotide reduction was found to be approximately proportional to that of system 1. These experiments provide strong support for the postulated function of phosphopyridine nucleotide. The experiments indicated that the quantum requirement for the reduction was about 2.5 per equivalent at 620 mμ, which indicated that the hydrogen transport to CO_2 occurred via reduced phosphopyridine nucleotide.

Photosynthesis and "Emerson effect"

It was known that the quantum yield of photosynthesis markedly decreased for Chlorella[19] beyond 680 mμ and for red algae already beyond 650 mμ (see refs. 1, 8). The long-wavelength absorption maxima of accessory pigments, 650 mμ for chlorophyll *b* in Chlorella and 620 mμ for phycocyanin in red algae suggested to EMERSON[20] that photosynthesis with high quantum yield only occurred, if at least a part of the light was absorbed by the accessory pigments. Thus, light of long wavelengths exclusively absorbed by chlorophyll *a* would have a low quantum yield. If this hypothesis was correct, EMERSON reasoned, then the low quantum yield of long-wavelength light would be enhanced, if light of shorter wavelengths was given which was absorbed by an accessory pigment. It was found indeed, that the quantum yield of photosynthesis in light of two suitably chosen wavelengths simultaneously applied was appreciably higher than the average quantum yield in each wavelength applied separately. This is the so called "EMERSON effect". This discovery clearly shows that under certain conditions more than one photochemical reaction affects the rate of photosynthesis. Although EMERSON's hypothesis was qualitatively consistent with the experimental results, we note here, that qualitative evidence leads us to a different interpretation.

EMERSON's conclusion was partly based on the assumption that the quantum yield of photosynthesis was essentially maximal in the region where the accessory pigments absorbed. From this he concluded that the quantum yield at the longer wavelength beam is enhanced, when illuminating with two beams. KOK AND HOCH[15] recognized that such a maximal quantum yield at shorter wavelengths could not be explained on basis of a theory of two photochemical systems operating in series; they left this difficulty unresolved. Other authors, who proposed schemes in certain respects similar to KOK's and ours, did not mention or try to explain the pronounced discrepancies between action spectra of chlorophyll *a* fluorescence, action spectra of photosynthesis and of chloroplast reactions, photosynthesis enhancement spectra, and action spectra of various photochemical activities, observed by difference spectrophotometry. The failure to explain these discrepancies mainly arose from the incorrect implicite assumption that each action spectrum was equal to the action spectrum of a photochemical system. This led *e.g.* to the incorrect idea that the accessory pigments contributed only to one of the systems and that the action spectrum of the other system had an absorption maximum at about 700 mμ (in Chlorella).

Upon closer examination of the experimental evidence, it appears that the quantum yield of photosynthesis is not necessarily maximal at wavelengths where accessory pigments absorb. For a culture of *Porphyridium cruentum* grown in blue light[21] the quantum yield at 560 mμ, the wavelength of the maximum of the main accessory pigment, phycoerythrin, was about 0.05 as compared to the quantum yield at 650 mμ, which was 0.09. In *Chlorella pyrenoidosa* the quantum yield of photosynthesis may be estimated (from Fig. 5 in ref. 19) to be roughly 10% lower at 650 mμ, where the maximum of the chlorophyll *b* absorption occurs, than at 675 mμ.

As stated in the preceding section, according to our hypothesis the action spectrum of system 1 is proportional to the action spectrum of photosynthesis measured against a strong (but non-saturating) background of light mainly absorbed by system 2, and the action spectrum of system 2 is proportional to the action spectrum of photosynthesis measured against a background of strong light mainly absorbed by system 1. As already mentioned the action spectrum of system 1 thus determined showed somewhat higher relative activity at 560 mμ than the action spectrum for cytochrome oxidation in the presence of poison.

The explanation of this discrepancy might be that in the photosynthesis experiment during the relatively long illumination with strong green background, part of the phycoerythrin moves from pigment system 2 to system 1. Since the efficiency for energy transfer strongly depends upon the distance between the energy-exchanging molecules, a small displacement may cause a phycoerythrin molecule to transfer its excitation energy to chlorophyll a_2 instead of to a_1. The cytochrome experiments were done without preceding green background. The occurrence of an "adaptation" like the one proposed is indicated by a recent observation of BRODY AND BRODY[22], that *Porphyridium cruentum* grown in blue light showed a marked increase in the quantum efficiency of photosynthesis around 560 mμ after a three hours illumination with green light. Other kinds of adaptations are conceivable. A quite different explanation is that a third pigment system occurs, containing mainly chlorophyll *a* which oxidizes cytochrome (and reduces pyridine nucleotide[18]) but does not contribute to photosynthesis. An experimental distinction between these possibilities should be possible. Due to the historical development, the two-wavelengths effects have mostly been plotted as enhancement spectra, *i.e.* action spectra of the EMERSON effect. However, if our hypothesis is correct, enhancement spectra are not suitable for quantitative analysis. Fortunately, there are a few spectra which can be interpreted according to our hypothesis. These spectra were measured by means of an apparatus which automatically plots action spectra of photosynthesis[23]. One of the action spectra for Chlorella was measured with strong light of 700 mμ as a background, the second with a background of strong light of 650 mμ, and the third without any background illumination. Since the first action spectrum shows, just like the action spectrum of chlorophyll *a* fluorescence, a strong participation of chlorophyll *b*, this spectrum may be considered to be proportional to the action spectrum of system 2; the spectrum with 650 mμ background then is that of system 1. The two action spectra with background crossed at 683 mμ. As predicted by our hypothesis, the spectrum without background coincided with the first action spectrum in the region 600–683 mμ and with the second spectrum in the region 683–700 mμ. If we assume that the absorption spectra of systems 2 and 1 are proportional to the first and second action spectrum of photosynthesis, then the quantum yield of photosynthesis, without background should be maximal at 683 mμ, should drop a little below 683 mμ, where the action spectra differ by a relatively small factor, and should drop appreciably beyond 683 mμ, where absorption of system 2 declines much more rapidly towards longer wavelengths than the absorption of system 1. This is consistent with the action spectrum of the quantum efficiency of Chlorella photosynthesis[19].

The action spectrum of system 2 measured with light of 700 mμ as background shows, in contradistinction to the enhancement spectrum, that the absolute contribution of chlorophyll *a* types with maxima at about 670 and 680 mμ (see ref. 23) is much higher than that of chlorophyll *b*. The action spectrum of system 1 predominantly shows participation of a chlorophyll *a* type with a maximum at about 683 mμ and another type with a maximum at about 670 mμ. There may be also a very small contribution of a chlorophyll type with a maximum at about 695 mμ. The interpretation by FRENCH *et al.*[23] themselves is quite different. These authors subtracted the two spectra drawn with different background illumination and stated that the difference spectrum gives the effectiveness spectrum for each of the two pigment systems. RABINOWITCH AND GOVINDJEE[24,25], making the same assumption, concluded that one of the pigment systems in Chlorella mainly contains chlorophyll *b* and a chlorophyll *a* type with a maximum at 670 mμ, and that the other system mainly contained a chlorophyll *a* type with a maximum at about 695 mμ. Also these conclusions are quantitatively completely different from ours.

FRENCH AND FORK[26] interpreted transients in the rate of oxygen evolution in Porphyridium as a light-driven oxygen uptake, superimposed upon photosynthesis. We have observed similar transients, but believe to have minimized their influence upon our experiments (as Fig. 5 indicates) by measuring photosynthesis against a relatively strong background illumination.

It remains to be seen whether the stimulation of oxygen uptake and the wavelength dependence of the saturation rate of photosynthesis[27] are related to the "inhibition" of photosynthesis by light of 730 mμ observed for some algae[28,29]. This inhibition also varies with the wavelength of the photosynthetic illumination.

The EMERSON effect has been observed in species of widely different groups of algae. This suggests that two photochemical systems, with the same photochemical properties as those postulated for *Porphyridium cruentum*, participate in the photosynthesis of all algae and presumably also of higher plants. We do not know of experiments which clearly contradict the postulate that the process of photosynthesis in all algae and higher plants is basically similar to that proposed for Porphyridium. In some species[30] the EMERSON effect was not observed, but this does not necessarily contradict our hypothesis: the EMERSON effect only occurs if system 1 shows excessive absorption at one wavelength and system 2 at the other one. It is quite possible that in some species either system 1 or system 2 is in excess at all wavelengths. In Chlorella, system 2 is in excess at all wavelengths below 683 mμ as is apparent from Fig. 4 in ref. 7.

The decrease in absorption around 700 mμ

KOK *et al.*[15,31] studied, in various species of algae, a reversible light-induced change in absorption in the red region with a maximum between 700 and 708 mμ. Their conclusions may be summarized as follows. A decrease in absorption at about 700 mμ indicates an oxidation of a substance with an E'_0 value of about 0.43 V, which was believed to be a chlorophyll type. An increase in absorption indicates the reduction of the oxidized substance. The oxidation is caused by system 1 (in our terminology) and the reduction by system 2; DCMU inhibits the reduction. The reduced compound was called P 700 by KOK (P in our scheme). The oxidation of P 700 in the presence of DCMU is consistent with the conclusion that P 700 is oxidized by system 1. The observation by WITT *et al.*[13] that in a very short flash the decrease at 700 mμ occurred more rapidly than the cytochrome oxidation is consistent with KOK's conclusion that P(ox) oxidizes cytochrome. The light-induced decrease in absorption in algae at

700 mμ, is analogous to the earlier observed and studied decrease[1,32] in purple bacteria at about 880 mμ. This decrease was attributed to an oxidation of a bacteriochlorophyll-type with an E'_0 value of 0.51 (see ref. 33) or 0.45 (see ref. 34).

These light-induced changes in bacteria may be explained by a scheme similar to scheme 1, in which the reactions at the left side of the cytochrome are replaced by a reaction with a hydrogen donor[35]. The occurrence of only one system in purple bacteria is consistent with the recent observation[36] that in *Rhodopseudomonas spheroides* the action spectrum of phosphopyridine nucleotide reduction at selected wavelengths is proportional to that of bacteriochlorophyll fluorescence.

Chloroplast reactions

If scheme 1 is correct, in principle any oxidized component or "hydrogen donor" of a couple with an oxidation-reduction potential higher than about -0.32 V, the E'_0 value of phosphopyridine nucleotide, can be reduced, and any hydrogen donor of a potential lower that 0.8 V can be oxidized in illuminated chloroplasts. Thus, an oxidation or reduction of an added substance by an illuminated chloroplast preparation does not necessarily mean that this substance participates in photosynthesis, even when this substance is known to occur in the intact cell. Conclusions based on quantitative results obtained with absorption or fluorescence spectrophotometry of intact photosynthesizing cells are more reliable. However, experiments with chloroplast reactions may provide additional evidence and are necessary for study of intermediates which cannot be detected or completely identified by means of spectrophotometry of intact cells.

The following selected chloroplast reactions presumably are related to reactions observed in intact cells with difference spectroscopy, and some provide additional information.

The "Hill reaction" (that is oxygen evolution) with TPN as hydrogen acceptor is accompanied by phosphorylation[37]. Since respiratory cytochromes are concerned in oxidative phosphorylation, the photosynthetic cytochromes may participate in photosynthetic phosphorylation. This phosphorylation may occur between QH and oxidized P. LOSADA *et al.*[38] postulated such a phosphorylation between cytochrome and chlorophyll (a_1). Because of the small difference in oxidation-reduction potentials between cytochrome *f* and P, we prefer to suggest that the phosphorylation occurs between Q and cytochrome *f* (see ref. 9). Chloroplasts to which DCMU had been added are able to photoreduce ascorbic acid in the presence of a catalytic amount of 2,6-dichlorophenolindophenol[39]. This light-driven oxidation–reduction reaction is accompanied by phosphorylation[38].

If ascorbic acid (or 2,6-dichlorophenolindophenol) is oxidized by a substance between Q and P then only system 1 participates in this reaction. Other earlier evidence suggesting that photosynthetic phosphorylation can occur by system 1 without participation of system 2 is the "maximum" at 700 mμ in the "action spectrum" of phosphorylation (see ref. 17, Fig. 11) of a very dense suspension of spinach chloroplasts in the presence of PMS. A comparison of "efficiency spectra" for TPN reduction[40] in the absence and presence of 2,6-dichlorophenolindophenol and ascorbate suggests that the second reaction is only driven by system 1. The "action spectra" mentioned seem rather distorted and more precise data or at least more circumstantial evidence is desirable before definite conclusions could be based on these data alone.

The finding[41] that the EMERSON effect occurs in the Hill reaction of Chlorella with quinone indicates that this hydrogen acceptor is reduced by system 1.

After extraction with petroleum ether the Hill reaction of chloroplasts is inactivated and can be reactivated with the extract[42], the most active component being plastoquinone[43]. It is possible that plastoquinone is an intermediate between XH and the Hill oxidant; in the intact cell plastoquinone might short-circuit the oxidized cytochrome with a reduced component of system 1, *e.g.* XH. The occurrence of a reaction between oxidized cytochrome and a reduced substance is suggested by the observation that when the light-driven reduction of the cytochrome is inhibited by DCMU, nevertheless reduction upon darkening still occurs (traces *c* and *d* of Fig. 1). If we assume that the E'_0 value of plastoquinone is the same (0.1 V) as that of a similar quinone[44], this would be consistent with the postulated role of plastoquinone. This mentioned oxidation–reduction potential and the fact that Q quenches the chlorophyll a_2 fluorescence[16], but QH does not, are consistent with the assumption that Q is a quinone. We concluded that plastoquinone may react at one or at both of the mentioned positions.

Role of chlorophyll a and other pigments

The action spectra for systems 1 and 2 reported in this paper show that in system 1 phycoerythrin and chlorophyll *a* are about equally active per incident quantum, but that in system 2 phycoerythrin shows about 3–7 times higher activity than chlorophyll *a*. The action spectra were found to vary for various cultures. BRODY AND EMERSON[21] found that the relative amounts of phycoerythrin and chlorophyll and the action spectra of photosynthesis varied strongly with the colour and intensity of the light in which the algae were grown, factors which may also have contributed to the variation of our cultures, since these were harvested at varying densities. The action spectrum of chlorophyll *a* fluorescence[1] (see also ref. 2) shows a much higher chlorophyll fluorescence per absorbed quantum for phycoerythrin than for chlorophyll *a*. It follows that mainly the chlorophyll a_2 is responsible for the chlorophyll *a* fluorescence, that the chlorophyll of system 1 is non- or weakly fluorescent, and that phycoerythrin transfers excitation energy mainly to chlorophyll a_2. The approximate proportionality between the action spectra for photosynthesis with the background of strong red light, and the action spectrum for the excitation of chlorophyll *a* fluorescence, indicates that light absorbed by the phycoerythrin of system 2 is passed to photosynthesis only via chlorophyll a_2.

For system 1 we have had as yet no direct evidence for energy transfer from phycoerythrin to chlorophyll a_1, but analogy and simplicity argue in favour of the hypothesis that also in this system light energy absorbed by phycoerythrin is active in photosynthesis only via chlorophyll a_1. The energy transfer between the photosynthetic pigments presumably occurs by the mechanism of induced resonance.

Quantum requirements

Since in algae not much more than 8 quanta are required for the reduction on one CO_2 molecule and the concomitant evolution of one O_2 molecule[45], 4 quanta may be needed by system 1 for transporting these four electrons, and 4 quanta to transport the 4 electrons in system 2. This is consistent with the measured quantum requirement for phosphopyridine nucleotide reduction in Anacystis[18].

It was estimated from the initial rate of cytochrome oxidation that for Porphyridium[3] about 5 and for Anacystis[18] 7–10 quanta of light of 680 mμ were needed for oxidation of one cytochrome molecule. If we assume that one-fourth of the absorptance at 680 mμ is due to system 2, and causes cytochrome reduction, and that the remaining three-fourth are due to system 1, then the quantum requirement for cytochrome oxidation by system 1 is estimated to be 2.5 for Porphyridum and at least 5 for Anacystis (*cf.* also ref. 18). The simplest explanation of this high requirement (the requirement to be expected was 1) is that part of the electrons going from system 2 to 1 bypass the cytochrome.

In most measurements made so far with purple and green bacteria, 8 or more quanta were found necessary to reduce one CO_2 molecule and to oxidize 4 hydrogen equivalents of the hydrogen donor (for references see 46). If, as we concluded, only system 1 is present in bacteria, then one would expect 4 quanta to be sufficient. An explanation of the high quantum requirement so far observed for bacteria would be that more light energy was lost by back reactions or deviated to "cyclic phosphorylation" than in the photosynthesis of algae. Thus, the approximate equality of the quantum requirements of bacterial and algal photosynthesis[46] might be a coincidence. Probably VAN NIEL's hypothesis[47,48], that in bacteria and algae an essentially similar photochemical oxidation–reduction reaction occurs, only applies to system 1.

Photosynthesis as a series of light-driven oxidation–reduction reactions.

If KOK's and our conclusions are correct, the plant has solved the difficult problem of photosynthesis, the reduction of CO_2 and the simultaneous oxidation of water, by means of two light-driven oxidation–reduction reactions in series. The first reaction is the reduction of an intermediate which is able to reduce phosphopyridine nucleotide and the oxidation of another intermediate which is able to oxidize a cytochrome. The second reaction is the oxidation of water to O_2 and the reduction of an intermediate which is able to reduce this cytochrome. The first of these reactions presumably also occurs in photosynthetic bacteria[34,47]. The E'_0 values of the phosphopyridine nucleotide and the P/P(ox) couples are about — 0.3 and 0.5 V. Thus in the first reaction (that of system 1) a potential difference of roughly 0.8 V is generated. This corresponds to a gain of about 18 kcal Gibbs free energy per equivalent, or, if the reaction is a one quantum reaction, to a gain of about 18 kcal per einstein absorbed by system 1. This is lower than the maximum theoretical efficiency of 21 and 29 kcal per absorbed einstein for bacteriochlorophyll and chlorophyll *a*[34]. In the reaction driven by system 2, water with a E'_0 value of 0.8 V is oxidized and another component Q is reduced. Q should have an E'_0 value of less than that of the cytochrome: as stated it might be a quinone with a potential of 0.1 V. The 16 kcal per equivalent needed for overcoming the difference of 0.7 V then can be provided by a one quantum reaction. If the component reduced by system 2 reacts with the component oxidized by system 1, the potential difference is roughly $0.5 - 0.1 = 0.4$ V. This potential difference corresponds to a free energy difference of 9.2 kcal per equivalent. This energy is more than sufficient to generate half a molecule of ATP, the amount found in non-cyclic photophosphorylation[37]. Additional ATP, needed for photosynthesis may be generated by reactions between oxidized and reduced components, formed by one or both light reactions of our scheme, *e.g.* between oxidized cytochrome and XH.

We might speculate that in the course of evolution, perhaps upon exhaustion of suitable hydrogen donors for the dark reduction of CO_2, system 1 was evolved by originally non-photosynthetic organisms to widen the range of utilizable hydrogen donors or to speed up ATP production. Subsequently the photochemical system 2 was evolved, able to oxidize water and to reduce a component which was oxidized by system 1.

Theoretically the free energy available in the lowest excited singlet state of bacteriochlorophyll would be sufficient to drive the reaction of system 2 by a one quantum process. Because, as far as we know, oxygen is not produced by purple bacteria one might speculate that the reactions of system 2, especially the dehydrogenation of water and the evolution of oxygen can only occur at the cost of such high energy losses that the energy of the lowest excited singlet state of bacteriochlorophyll is not sufficient.

ACKNOWLEDGEMENT

This investigation has been supported by a grant from the Netherlands Organisation for the Advancement of Pure Research (Z.W.O.).

REFERENCES

[1a] L. N. M. Duysens, *Nature*, 168 (1951) 548.
[1b] L. N. M. Duysens, *Thesis*, Utrecht, 1952.
[2] C. S. French and V. K. Young, *J. Gen. Physiol.*, 35 (1952) 873.
[3] L. N. M. Duysens, in B. C. Christensen and B. Buchmann, *Progress in Photobiology, Proc. 3rd Intern. Congress on Photobiology 1960*, Elsevier Publ. Co., Amsterdam, 1961, p. 135.
[4] L. N. M. Duysens, J. Amesz and B. M. Kamp, *Nature*, 190 (1961) 510.
[5] M. Brody and R. Emerson, *Am. J. Botany*, 46 (1959) 433.
[6] L. N. M. Duysens, in H. Gaffron, *Research in Photosynthesis*, Interscience, New York, 1957 p. 59.
[7] J. Myers and C. S. French, *J. Gen. Physiol.*, 43 (1960) 723.
[8] F. T. Haxo and L. R. Blinks, *J. Gen. Physiol.*, 33 (1950) 389.
[9] R. Hill and F. Bendall, *Nature*, 186 (1960) 136.
[10] B. Chance and R. Sager, *Plant Physiol.*, 32 (1957) 548.
[11] L. N. M. Duysens, *Proc. 4th Intern. Congress Biochem., Vienna, 1958*, Vol. 13, Pergamon Press, Oxford, 1959, p. 303.
[12] H. Lundegårdh, *Nature*, 192 (1961) 243.
[13] H. T. Witt, A. Müller and B. Rumberg, *Nature*, 192 (1961) 967.
[14] H. T. Witt, B. Rumberg and A. Müller, *Nature*, 191 (1961) 194.
[15] B. Kok and G. Hoch, in W. D. McElroy and B. Glass, *A symposium on Light and Life*, The Johns Hopkins Press, Baltimore, 1961, p. 397.
[16] L. N. M. Duysens and H. E. Sweers, *Plant Cell Physiol.*, in the press.
[17] J. Amesz and L. N. M. Duysens, *Discussions Faraday Soc.*, 27 (1959) 173.
[18] J. Amesz and L. N. M. Duysens, *Biochim. Biophys. Acta*, 64 (1962) 261.
[19] R. Emerson and C. M. Lewis, *Am. J. Botany*, 30 (1943) 165.
[20] R. Emerson, R. Chalmers and C. Cederstrand, *Proc. Nat. Acad. Sci. U.S.*, 43 (1957) 133.
[21] M. Brody and R. Emerson, *J. Gen. Physiol.*, 43 (1959) 251.
[22] M. Brody and S. S. Brody, *Arch. Biochem. Biophys.*, 96 (1962) 354.
[23] C. S. French, J. Myers and G. C. McLeod, in M. B. Allen, *Symposia on Comparative Biology*, Vol. 1, Academic Press, New York, 1960, p. 361.
[24] R. Emerson and E. Rabinowitch, *Plant Physiol.*, 35 (1960) 477.
[25] Govindjee and E. Rabinowitch, *Science*, 132 (1960) 355.
[26] C. S. French and D. C. Fork, *Proc. 5th Intern. Congr. Biochem., Moscow, 1961*, Pergamon Press, Oxford, in the press.
[27] G. C. McLeod, *Plant Physiol.*, 36 (1961) 114.
[28] J. B. Thomas and Govindjee, in W. D. McElroy and B. Glass, *A Symposium on Light and Life*, The Johns Hopkins Press, Baltimore, 1961, p. 475.
[29] Govindjee, E. Rabinowitch and J. B. Thomas, *Biophys. J.*, 1 (1960) 91.

30 R. EMERSON AND R. V. CHALMERS, *Phycol. Soc. Am. News Bull.*, 9 (1958) 51.
31 B. KOK, *Biochim. Biophys. Acta*, 48 (1961) 527.
32 L. N. M. DUYSENS, W. J. HUISKAMP, J. J. VOS AND J. M. VAN DER HART, *Biochim. Biophys. Acta*, 19 (1956) 188.
33 J. C. GOEDHEER, *Brookhaven Symposia in Biol.*, 11 (1958) 325.
34 L. N. M. DUYSENS, *Brookhaven Symposia in Biol.*, 11 (1958) 10.
35 L. N. M. DUYSENS, *Nature*, 173 (1954) 692.
36 J. AMESZ, *Biochim. Biophys. Acta*, submitted for publication.
37 D. I. ARNON, F. R. WHATLEY AND M. B. ALLEN, *Science*, 127 (1958) 1026.
38 M. LOSADA, F. R. WHATLEY AND D. I. ARNON, *Nature*, 190 (1961) 606.
39 L. P. VERNON AND W. S. ZAUGG, *J. Biol. Chem.*, 235 (1960) 2728.
40 D. I. ARNON, M. LOSADA, F. R. WHATLEY, H. Y. TSUJIMOTO, D. O. HALL AND A. A. HORTON, *Proc. Natl. Acad. Sci. U.S.*, 47 (1961) 1314.
41 R. GOVINDJEE, J. B. THOMAS AND E. RABINOWITCH, *Science*, 132 (1960) 421.
42 V. H. LYNCH AND C. S. FRENCH, *Arch. Biochem. Biophys.*, 70 (1957) 382.
43 N. I. BISHOP, *Proc. Natl. Acad. Sci. U.S.*, 45 (1959) 1696.
44 V. MORET, S. PINAMONTI AND E. FORNASARI, *Biochim. Biophys. Acta*, 54 (1961) 381.
45 R. EMERSON, *Ann. Rev. Plant Physiol.*, 9 (1958) 1.
46 H. LARSEN, *On the Microbiology and Biochemistry of the Photosynthetic Green Sulphur Bacteria*, Det. Kgl. Norske Videnskab. Selskab. Forh. Skrifter, NR 1, 1953, p. 199.
47 C. B. VAN NIEL, *Advances in Enzymology*, Vol. 1, Interscience Publishers, Inc., New York, 1941, p. 263.
48 C. B. VAN NIEL, in J. FRANCK AND W. E. LOOMIS, *Photosynthesis in Plants*, The Iowa State College Press, Ames, Iowa, 1949, p. 437.

Biochim. Biophys. Acta, 64 (1962) 243–260

ABSENCE OF DELAYED LIGHT EMISSION IN THE MILLISECOND TIME RANGE FROM A MUTANT OF RHODOPSEUDOMONAS SPHERIOIDES WHICH LACKS FUNCTIONING PHOTOSYNTHETIC REACTION CENTERS

Roderick K. Clayton

C. F. Kettering Research Laboratory, Yellow Springs, Ohio

and

Walter F. Bertsch

Biology Division, Oak Ridge National Laboratory, Oak Ridge, Tennessee 37831

Received December 11, 1964

All photosynthetic organisms emit a long-lived dim glow from the first excited singlet state of their functionally active chlorophyll (Strehler and Arnold, 1951; Bertsch, 1962). Although the physical mechanism of this delayed light emission is unclear, there is ample indirect evidence that the delayed light is closely associated with the mechanism of photosynthesis: (1) both processes have the same action spectrum (Strehler and Arnold, 1951; Arnold and Thompson, 1956; Goedheer, 1962); (2) both processes show a complex interaction between two pigment systems (Bertsch, 1962); (3) both processes have an absorption cross section equal to 300-400 chlorophyll molecules (Arnold, 1955), implying the existence of photosynthetic units; (4) both processes are heat-inactivated at the same temperature; (5) poisons of photosynthetic energy conversion invariably have a drastic effect on the kinetics of delayed light emission (Bertsch et al., 1963).

We now report the first piece of direct evidence that the machinery for photosynthetic energy conversion is responsible for the delayed light emission: in photosynthetic bacteria the emission of delayed light depends on the presence of functional photosynthetic reaction centers.

We have compared the intensity of delayed light emission from wild-type Rhodopseudomonas spherioides to the delayed light emission from a mutant which was isolated by Sistrom et al. (1963) and which is believed to lack photosynthetic reaction centers. The mutant cannot grow photosynthetically even though it has

*Research sponsored jointly by the Charles F. Kettering Research Laboratory and the United States Atomic Energy Commission under contract with the Union Carbide Corporation. This paper is Contribution No. 169 from the Charles F. Kettering Research Laboratory.

a normal complement of photosynthetic pigments. When grown aerobically in the dark, the mutant (PM-8) and the parent (wild-type, strain Ga) have similar absorption spectra, but the bacteriochlorophyll of the mutant serves no discernible photochemical function. The following evidence indicates that the mutant lacks functional photosynthetic reaction centers. The mutant shows none of the reversible light-induced absorbancy changes which occur in the wild-type (Sistrom and Clayton). That is, the mutant shows no photooxidation of photosynthetic cytochromes, no photoreduction of ubiquinone, no shift in carotenoid absorption bands, and no photooxidation of P870 (which is associated with the bacterial photosynthetic reaction center). The mutant does not show the light-induced electron spin resonance signal which is present in the wild type (Sistrom and Clayton). The mutant shows no light-induced transients in fluorescence (Sistrom and Clayton). Chromatophores extracted from the mutant are unable to catalyze either the photooxidation of reduced cytochrome c, or the photooxidation of reduced phenazine methosulfate coupled with reduction of ubiquinone (Clayton and Sistrom, Clayton et al.). Thus the bacteriochlorophyll of the mutant is photosynthetically inert.

The yield of bacteriochlorophyll fluorescence in the mutant is greater than that in the wild-type (Sistrom and Clayton). In contrast, as we shall report here, delayed light is emitted by the wild-type, but not by the mutant.

EXPERIMENTAL

The intensity of delayed light emission, averaged from 1 to 5.5 msec after the center of an exciting flash, was measured with a modification of the Becquerel phosphoroscope. This instrument was constructed by W. F. Bertsch in collaboration with William Arnold and J. B. Davidson at the Oak Ridge National Laboratory. It consisted of a pair of spinning discs mounted on a common shaft, between which the cell suspension was held stationary in a 2-ml cellulose nitrate test tube. In order to provide adequate light-baffling we mounted the discs so that they would spin in narrow slots within an aluminum housing. The cells were illuminated through two oppositely spaced holes in the front disc, so that when the discs were spinning the cells received a flash of exciting light every 6.6 msec. The cells were viewed by a photomultiplier placed behind the rear disc. This rear disc was cut into the shape of a paddle wheel, the blades of which were aligned with the illuminating holes in the front disc. The rear disc thus allowed the cells to be viewed by the photomultiplier for a period (1-5.5 msec) between each exciting flash, but for the duration of each flash the blades of this disc protected the light-measuring equipment from the bright exciting light.

The exciting light consisted of wavelengths between 7200 and 11,000 Å. To obtain this range of wavelengths the emission of a 1500-watt tungsten bulb was filtered through 5 cm of water and a Corning glass filter CS7-69. The filaments of the tungsten lamp were focused on the cells to give an image of about 1 cm^2. With this exciting light, essentially all of the emitted light was in the wavelength region of bacteriochlorophyll fluorescence.

The photomultiplier was an RCA7102, cooled with liquid nitrogen, and operated at either 1260 or 1420 volts. The output of the photomultiplier was smoothed by an electrical filter (10 sec time constant), amplified by a vibrating reed electrometer, and plotted on a Brown recorder.

Cultures of R. spherioides, wild-type strain Ga and the nonphotosynthetic mutant strain PM-8, were grown in the dark under limited aeration as described elsewhere (Sistrom and Clayton).

RESULTS

The figure shows the intensity of delayed light emitted by R. spherioides,

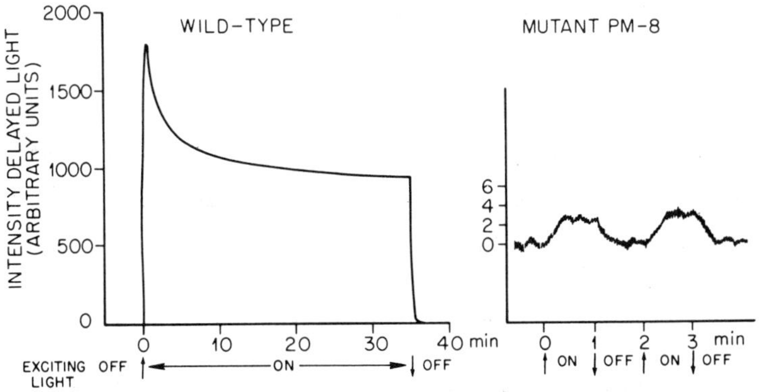

RHODOPSEUDOMONAS SPHERIOIDES

Intensity of delayed light, averaged from 1 to 5.5 msec after excitation, emitted by Rhodopseudomonas spherioides wild-type strain Ga and non-photosynthetic mutant strain PM-8. The same arbitrary units of light emission are used in both graphs. The small signal from strain PM-8 appears to be unrelated to photosynthesis. The two cell suspensions, each containing about 5 X 10^9 cells per ml, were adjusted to the same bacteriochlorophyll concentration, about 2 μg/ml.

wild-type and nonphotosynthetic mutant PM-8. The two cell suspensions were adjusted to the same cell density and had equal concentrations of bacteriochlorophyll. The arbitrary units of intensity of emission in the figure are the same for both cell types; the graph for the mutant represents a much more sensitive measurement than that for the wild-type. It can be seen that the delayed light signal with wild-type cells was 1800 units at the peak of an initial transient and about 900 units in the steady state, whereas the signal with the mutant was only 3 units.

After being heated to 90° C for 4 min, both types of cells gave a delayed light signal of 6 units. With distilled water in place of a cell suspension we observed a signal of about 30 units, attributable to a light leak through the phosphoroscope. With cells in the sample holder the effect of such a leak would be smaller because of absorption and scattering of light by the cells. These results indicate that the 3-unit signal observed with mutant cells (Fig., b) was unrelated to photosynthesis and could be ascribed in part to a light leak.

CONCLUSION

We have measured the delayed light emitted by two strains of photosynthetic bacteria: a nonphotosynthetic mutant of Rhodopseudomonas spherioides and the photosynthetically competent parent (wild-type) strain. The mutant possesses a normal complement of bacteriochlorophyll but lacks functional photosynthetic reaction centers. Delayed light emission, averaged from 1 to 5.5 msec from the center of a flash of exciting light, was found to be more than 300 times as intense in the wild-type (during steady-state photosynthesis) as in the mutant. This result shows that functional reaction centers must be present in order for delayed light to be emitted from these photosynthetic bacteria.

ACKNOWLEDGMENT

We are indebted to Dr. W. R. Sistrom, who isolated the nonphotosynthetic mutant strain of R. spherioides used in these experiments, and to Dr. S. F. Carson, who suggested improvements in the optical system of the phosphoroscope.

REFERENCES

Arnold, W., in F. H. Johnson (Editor), The Luminescence of Biological Systems, American Association for the Advancement of Science, Washington, D. C., 1955, p. 47.
Arnold, W., and Thompson, J., J. Gen. Physiol., 39, 311 (1956).
Bertsch, W. F., Proc. Natl. Acad. Sci. U.S., 48, 2000-2004 (1962)

Bertsch, W. F., Davidson, J. B., and Azzi, J. R., in B. Kok and A. T. Jagendorf (Editors), Photosynthetic Mechanisms of Green Plants, National Academy of Sciences, U.S., Washington, D.C., 1963, p. 701.

Clayton, R. K., Sistrom, W. R., and Zugg, W. S. In preparation.

Clayton, R. K., and Sistrom, W. R., Proc. Natl. Acad. Sci. U.S., in press.

Goedheer, J. C., Biochim. Biophys. Acta, 64, 294 (1962).

Sistrom, W. R., and Clayton, R. K., Biochim. Biophys. Acta, in press.

Sistrom, W. R., Ohlsson, B. M., and Crounse, J., Biochim. Biophys. Acta, 75, 285 (1963).

Strehler, B., and Arnold, W., J. Gen. Physiol., 34, 809 (1951).

THE JOURNAL OF BIOLOGICAL CHEMISTRY
Vol. 240, No. 7, July 1965
Printed in U.S.A.

Light Scattering Changes Associated with the Production of a Possible Intermediate in Photophosphorylation*

G. HIND† AND ANDRE T. JAGENDORF

From the McCollum-Pratt Institute and Biology Department, The Johns Hopkins University, Baltimore, Maryland 21218

(Received for publication, October 19, 1964)

Isolated chloroplasts from spinach (1, 2) and wheat (3, 4) are capable of a dark phosphorylation at the expense of an unknown reactive, labile substance or state formed in an earlier period of illumination. This substance, X_E (1), is optimally stable at low pH, where it may accumulate in the light to the equivalent of 1 mole of adenosine triphosphate for 8 moles of chlorophyll. Such high yields require the presence during illumination of a cofactor catalyzing cyclic electron flow (2).

The criteria of cofactor and low pH dependence have been used in attempts to associate certain other reversible phenomena observed in illuminated chloroplast suspensions with the formation and decay of X_E. Two positive correlations were found (5): (*a*) with a reversible rise in pH of the suspending medium and (*b*) with a turbidity change thought to reflect an altered conformation of the particles.

Packer (6, 7) and Itoh, Izawa, and Shibata (8) have shown that chloroplasts scatter more light after treatment with strong red light, and that the original low scatter state is restored during a subsequent dark period. The relation of these effects to electron flow and photophosphorylation at pH values between 7.5 and 8.0 was noted by Packer. Dilley and Vernon (9) extended these observations to the acid pH range and proposed that such changes are an indication of the presence of a high energy intermediate, probably X_E. Our comparative data on the kinetics of light scattering and of X_E to be reported here will show that the absorbance changes are secondary effects of the high energy condition.

EXPERIMENTAL PROCEDURE

Fresh turgid spinach was obtained from the market. Whole chloroplasts were prepared by maceration of the laminae in sucrose medium, as described previously (1). Resuspension for 15 min in Tris, 20 mM, pH 8.0, containing NaCl, 10 mM, effected hypotonic swelling of the particles, which were then collected by centrifugation at 1000 × *g* for 10 min, washed in this same medium, and finally resuspended in Tris, 5 mM; 3,3-dimethylglutarate, 5 mM; and NaCl, 40 mM (Tris-dimethylglutarate-NaCl buffer). This buffer mixture was present in all illuminated samples. Light scattering changes were observed at 90° from the measuring beam in a Brice-Phoenix light scattering photometer, model 1000-D. The cuvette, 1 × 1 × 4 cm, was illuminated from above by the beam of a 100-watt microscope lamp filtered through a Kodak Wratten No. 70 filter (cutting off wave lengths below 640 mμ) and a Corning CS1-50 heat filter. The photocell was protected from actinic light with a Corning CS4-96 filter, which transmitted only the 546 mμ mercury line of the measuring beam. The sample was maintained at 5° in a water-cooled holder. Supplementary measurements of turbidity (narrow angle scatter) were made with similar optics in the sample compartment of a Cary model 14 recording spectrophotometer. In this instance the reference cuvette contained either the same suspension as the sample or, for calibration and observation of dark scattering levels, an acidified pyocyanine solution. The samples were again maintained at 5°. Both instruments were operated at high gain and minimal measuring beam intensity. Results from turbidity changes were essentially identical with those obtained from 90° scattering measurements provided the pigment concentration was low enough (not more than 50 μg of chlorophyll per ml) to exclude interference from specific light-induced absorption changes (10). For measurements at 546 mμ the upper limit of chloroplast concentration is determined by the desirability of using saturating light intensities; the lower limit may be set by the need to obtain comparative data for X_E with the same chlorophyll concentration, since this parameter influences X_E kinetics (1). Two-stage phosphorylation experiments were conducted as previously described (1), but with the phosphate content of the dark reaction mixture raised from 0.5 to 2.0 μmoles per ml (5).

Pyocyanine was prepared from *N*-methylphenazonium methosulfate and recrystallized (11).

RESULTS

Wave Length Dependency

As noted by Dilley and Vernon (9), the increase in absorbance of chloroplasts at pH 6 when illuminated with red light can be observed at all wave lengths from 245 to 740 mμ. In our experience this increase, obviously due to light scattering increments, could amount to 25% of the dark absorbance at 546 mμ, the wave length used routinely for assay of turbidity. We found the largest absorbance changes to occur on the long wave length shoulders of the major chloroplast absorption bands, as is expected for scattering spectra (10, 12). Our observations differ from those of Dilley and Vernon in that we never observed a

* Contribution No. 440 from the McCollum-Pratt Institute. This research was supported by Grant GM 03923 from the National Institutes of Health, and by a grant from the C. F. Kettering Foundation. Many exploratory experiments were conducted at the Johnson Foundation, University of Pennsylvania, in cooperation with Drs. W. D. Bonner, B. Chance, N. Bryant, and H. Schleyer, to whom we are indebted. We thank Dr. Chance for provision of facilities. Marie F. Smith provided invaluable technical assistance.

† Present address, Department of Biology, Brookhaven National Laboratory, Upton, New York.

light-induced decrease in absorbance at the specific wave length of 435 mμ.

Extent of Scattering Changes

The earlier investigations of Packer (6, 7) were based on measurements of light scattered at 90° from the measuring beam, as a function of illumination at pH 8.0 in the presence of activated ATPase or of phosphorylating reagents. It was of interest to compare the relative magnitudes of these responses to the low pH-dependent changes reported here. In the absence of phosphorylating reagents there is (Fig. 1) a 3% rise in scattering level due to illumination at pH 8.0. Lowering the pH to 6.0 increases this response to about 30%, while the presence of cysteine-activated ATPase can also increase it to about 25%. Although the kinetics of the ATPase-driven response is slower, the maximum scattering levels appear to have a common limitation.

Influence of Cofactor

As noted by Dilley and Vernon (9), the rate at which absorbance changes occur responds to added oxidation-reduction dyes in a manner similar to the rates of photophosphorylation. Fig. 2 shows, in addition, that the amplitude of the absorbance change is the same whether pyocyanine or FMN is added, or no cofactor is present. This result stands in contrast to those with X_E in which the illuminated steady state level is 60 to 100% greater with pyocyanine than with FMN, or without added cofactors. This suggests that the conformational change cannot be directly quantitated with X_E.

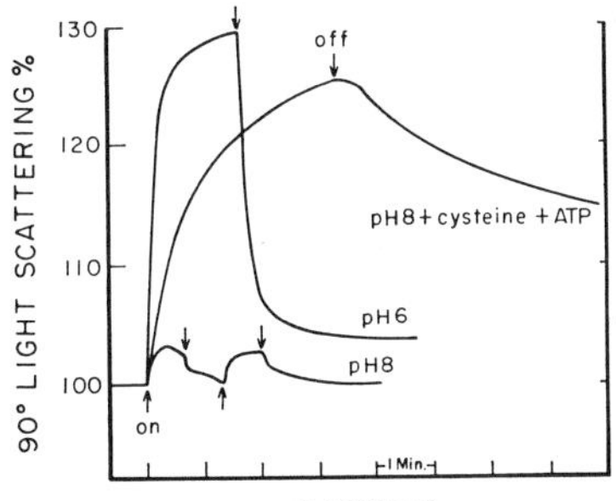

FIG. 1. High scattering levels achieved by illumination at low pH, or at high pH in the presence of cysteine-activated ATPase. Samples contained Tris-dimethylglutarate-NaCl buffer at the pH indicated, 3 mM $MgCl_2$, 15 μM pyocyanine, and 25 μg of chlorophyll per ml. The ATPase system had, in addition, 1.7 mM ATP and 0.08 M cysteine.

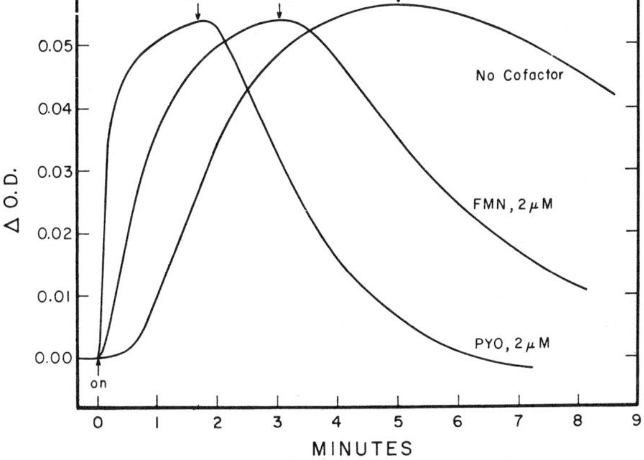

FIG. 2. Effect of FMN and pyocyanine (*PYO*) on the low pH absorbance changes. Data were obtained in a Cary spectrophotometer with 20 μg of chlorophyll per ml in Tris-dimethylglutarate-NaCl buffer at pH 6.0.

TABLE I

Influence of cofactor concentration on rise in optical density

Chloroplasts giving 55 μM total chlorophyll concentration were illuminated at pH 6.0 with rate-saturating red light, in the various cofactor environments. The maximum change in optical density in continuous light was 0.14. The lag was estimated from the projection of the line of maximum rise rate back on to the time axis. Higher cofactor levels did not cause any further changes.

Cofactor	Cofactor concentration	Maximal rise in rate of ΔA per min	Initial lag
	μM		sec
Pyocyanine	0.00	0.04	36
	0.05	0.09	22
	0.50	0.54	4.0
	5.00	0.72	1.7
FMN	0.00	0.04	36
	0.05	0.06	33
	0.50	0.11	8.5
	5.C0	0.11	6.0

The "on" and "off" responses in Fig. 2 are doubly inflected. When the length of the lag period was estimated by extrapolation of the linear rate back to the base-line, the extent of the lag was found to be diminished by the presence of oxidation-reduction cofactors, proportional to their concentration, and hence presumably to the electron flow rate. In these quantitative experiments we found an inverse relation between the maximum rate of change of absorbance and the length of the initial lag at various cofactor concentration (Table I). It was of interest that as little as 0.05 μM pyocyanine could double the reaction rate and significantly reduce the lag. In earlier studies with photophosphorylation no effect of pyocyanine was seen at or below 1.0 μM (13). Concentrations of pyocyanine and FMN needed to saturate the rate of the light scattering change are similar to those needed for photophosphorylation (13, 14). We would like to suggest that light scattering changes are more sensitive to small amounts of pyocyanine because they may be more intimately linked to electron transfer than is the esterification of inorganic phosphate.

Light Intensity

Dilley and Vernon (9) previously reported the effect of light intensity on the initial rate of absorbance increase. We find in addition that decreased illumination extends the "on" lag. Fig. 3 shows this effect with optimal concentration of cofactor at pH 7.4, where the initial lag is greater than that at pH 6.0, and higher light intensities are needed to achieve the same rate and degree of absorbance change. The effect of 40% light intensity at pH 7.4 is duplicated by 10% light intensity at pH 6; 1% at pH 7.4 is equivalent to 0.1% at pH 6, etc. These differences can be ascribed to a faster "decay" rate (postillumination decrease in absorbance) at pH 7.4 than at pH 6. Otherwise the results are similar at both pH levels, and we would like to point out that lowering the light intensity not only increases the length of the lag and slows down the rate of formation, but also diminishes the total extent of the absorbance change. Such diminution of the total change is *not* accomplished by removal of the cofactor. This is partly because the absence (or lowered concentration) of the cofactor slows both the light-driven reaction and the competing dark decay processes (in so far as these are measured by the "off" response rate), whereas a lower light intensity will reduce only the rate of the light-associated reactions.

Lag in Absorbance Changes, and Buffer-induced Lag in X_E Formation

The lag at low light or cofactor levels might be explained if the observed absorbance increase required a threshold level of a light-induced high energy intermediate (*i.e.* X_E) for its inception. X_E would then need to be formed at a critical minimal rate, in view of competing dark decay processes, to support a measurable rate of light scattering changes. An explanation similar to this was offered by Shen and Shen (3) to account for the very much reduced efficiency of phosphorylation at low light intensities; it might also be invoked to account for the lag in the time course of phosphorylation at low light intensities found by Kahn (15).

Conditions were therefore arranged for measuring X_E formation during the lag period in the scattering responses either in the absence of cofactor or with limiting (0.15 μM) pyocyanine. Surprisingly, under these conditions the time courses of X_E formation (Fig. 4) also showed a lag. Both with and without the low level of pyocyanine, the lag amounted to about 10 sec, although in the cofactorless reaction mixture there was an initial fast rise of X_E to a very low plateau level.

The lag in X_E turned out to be caused by the buffer, which in this case was 2.5 mM dimethylglutarate. In the absence of a buffer, no such lag appeared (Fig. 4). The lag could also be induced by 2.5 mM maleate, both these acids having a pK of 6.2 The inhibition was not caused by 2.5 mM benzoate (pK 4.2)

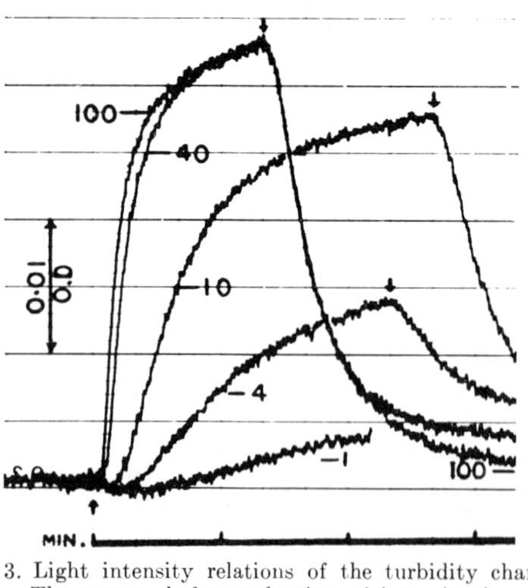

FIG. 3. Light intensity relations of the turbidity change at pH 7.4. The source, before reduction of intensity by neutral filters with percentage of transmission as indicated, was 22,000 foot-candles white (tungsten) light. Illumination was provided from above, in a reaction volume of 2.7 ml in a cuvette 1 cm square, containing 50 μg of chlorophyll. The cofactor was pyocyanine at 15 μM.

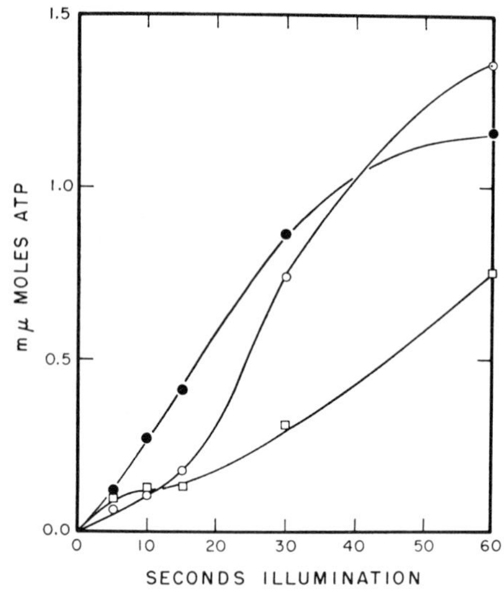

FIG. 4. Time course of cofactor-limited X_E formation. Chloroplasts giving 55 μM total chlorophyll concentration were illuminated at pH 6.0 with rate-saturating red light. The *lower curve* (□——□) contained Tris-dimethylglutarate-NaCl buffer as in "Experimental Procedure," but no cofactor; the *middle curve* (O——O) contained both Tris-dimethylglutarate NaCl buffer and 0.15 μM pyocyanine; and the *upper curve* (●——●) contained pyocyanine but no buffer.

acetate (pK 4.7), fumarate (pK 4.4), phthalate (pK 5.4), glycine (pK 2.5, 9.6), or citrate (pK 5.4) buffers. Although histidine (pK 6.1) inhibited 40%, this is probably from a different cause since its effect was seen even when added after illumination. Dimethylglutarate induction of a lag is not observed at rate-saturating cofactor levels.

On going to 10 mM, glycine and citrate inhibited the yield of X_E about 25% without inducing a lag; benzoate inhibited more than 50% (light stage only), and histidine and dimethylglutarate inhibited 100%, whether added in the light or in the dark.

The light absorbance changes were found to behave very differently from X_E with respect to maleate and to dimethylglutarate. First, neither maleate nor dimethylglutarate at 10 mM was inhibitory, although they completely eliminated X_E recovery as ATP. Second, complete omission of these buffers did *not* eliminate the lag in light scattering changes. (Omission of buffer did cut the amplitude of the absorbance changes in half, a phenomenon probably similar to the smaller absorbance increments in phosphate, compared to Tris or succinate buffers, reported by Dilley and Vernon (9).) Thus the existence of the long lag period is unique to light scattering, even when the X_E formation starts almost immediately. Examination of Fig. 4 suggests that light scattering changes start only after X_E has risen to the level of about 0.5 mμmole of recoverable ATP, under present conditions.

Postirradiation Overshoot; Effect of pH Variation

If the actinic beam is removed from a suspension after only a 5-sec illumination, when the rate of absorbance increase with 25 μM pyocyanine is close to maximal, the optical density will continue to rise in the dark for up to 30 sec before turning downward.

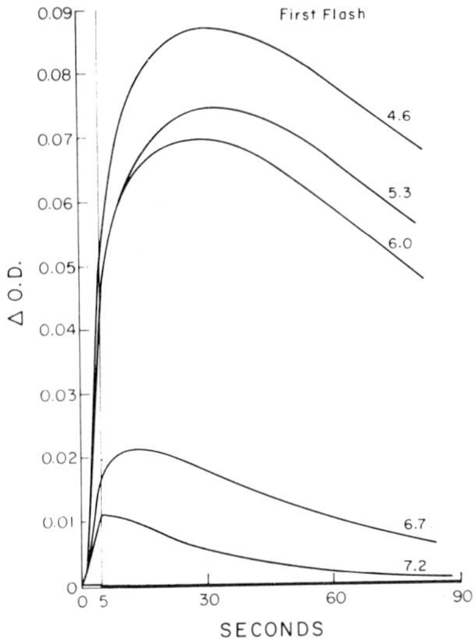

FIG. 5. Change in absorbance produced by a 5-sec flash at saturating light intensity. Reaction mixture contained 50 μg of chlorophyll in 3.0 ml; pyocyanine concentration was 25 μM. The maximum absorbance change under continued illumination amounted to 0.16 optical density units.

TABLE II

Effect of pH on 90° light scattering

Chloroplasts giving 40 μg of chlorophyll per ml were illuminated in Tris-dimethylglutarate-NaCl buffer containing 25 μM pyocyanine. Increase of pyocyanine to 100 μM at pH 4.6 did not affect the values quoted, and the apparent optimum at pH 5.3 in Lines A, B, C, and D is observed also with FMN cofactor.

Line	Measurement	pH					
		4.6	5.3	6.0	6.7	7.2	7.7
	Maximal rate of rise (% excess scatter per 5 sec)	24	30	26	9.5	4.7	
	Half rise time (sec)	7.5	7.0	9.0	16	32	
A	Yield from 5-sec flash (% excess scatter due to light)	16	19	12	5.4	4.0	1.3
B	Maximal amplitude (% excess scatter due to light)	51	62	62	57	49	7
	A/B × 100	31	31	19	10	8	
	Maximal return rate (% loss of excess scatter per 5 sec)	2.0	2.0	2.8	5.0	5.4	
C	Half-return time (sec)	60	80	63	34	23	
D	"Off" response lag (sec)	10	40	15	5	2	0
E	Initial dark scattering level (% excess over pH 7.7 value)	60	34	16	11	6.7	0
B + E	Maximal scatter (% excess over pH 7.7)	111	96	78	68	56	7

TABLE III

Effect of pH on X_E

Chloroplasts giving 40 μg of chlorophyll per ml were illuminated as for Table II. Adjustment of pH was made by dilution of a concentrated particle suspension in the appropriate buffer, just before use. Samples for determining the dark X_E level were kept in total darkness at pH 8.0 and 5° for 1 hour before use, and were manipulated in total darkness. Tables II and III were derived from simultaneous study of the same preparations.

Line	Measurement	pH					
		4.6	5.3	6.0	6.7	7.2	7.7
A′	Yield from 5-sec flash (mμmoles of ATP per mg of chlorophyll)	15	17	18	7.1	4.3	1.9
B′	Maximal yield (mμmoles of ATP per mg of chlorophyll)	52	54	37	7.8	4.3	1.9
	A′/B′ × 100	29	32	49	91	100	100
C′	Half-decay time (sec)	42	63	25	8.0	2.5	<2
D′	Decay lag period (sec)	0	0	0	0	0	0
E′	Initial dark X_E level (mμmoles of ATP per mg of chlorophyll)	5.5	1.4	1.6	1.6	0.6	0
B′ + E′	Maximal X_E (mμmoles of ATP per mg of chlorophyll)	58	55	39	8.0	4.9	1.9

The extent of the upward dark excursion is greatest at pH 4.6 and is zero at or above pH 7.2 (Fig. 5).

The scattering response was found to decrease between successive flashes at pH 4.6, and even somewhat at pH 6.0. Below pH 6.0 the optical density change is only partially reversible in the dark, and so the second flashes, having a higher base-line, in fact reach the same maximum optical density as the first flash.

Table II lists further effects of pH variation on the form of the scattering response, and Table III gives like data for X_E. Owing to the complexity of scattering kinetics, both maximal rates and half-times are presented. The half-time is taken after excluding any initial lag period.

It is clear that the yield of both X_E and the light scattering change, induced by a 5-sec flash of light, drop off sharply above pH 6.0 (Line A, Table II; Line A′, Table III). In the case of X_E the formation curve terminates sooner at the higher pH, and 5 sec are sufficient to give the maximal yield at pH 6.7 (see Line A′/B′ × 100, Table III). On the other hand, the formation curve of the light scattering change has been greatly slowed down by the higher pH values, but the maximum yield (Line B, Table II) is practically unchanged until pH 7.7 is achieved. Thus for light scattering, the 5-sec flash produces only 10% of the total yield at pH 6.7 or above (see Line A/B × 100, Table III). Looking at this comparison in another way, we note that the half-formation times at pH 7.2, for instance, are 32 sec for light scattering, and less than 5 sec for X_E.

An interesting analogy may be drawn between the effect of

high pH (Table II) and of removal of cofactor (Fig. 2) on light scattering. Both of these conditions have major effects on the rate of formation, with little or no diminution of the total extent of the change.

When the light is turned off, there is no perceptible lag in X_E decay, and logarithmic decay plots are linear; however, the scattering change may take 15 sec to begin reversion (Line D, Table II). The extreme case is the "overshoot" in light scattering after a 5-sec flash of light (Fig. 5). Again, when simply the "off" response half-times are considered, the scattering change response is 10 times slower than that for X_E at pH 7.2 (Line C, Table II; Line C′, Table III) and marked disparities also occur at the lower pH values. Thus both "on" and "off" responses of X_E are faster than the corresponding changes in light scattering.

Induction of Absorbance Increase and of X_E in Darkness

During investigations on the pH dependence of X_E kinetics it was observed that dark controls always had significantly higher values as the pH was decreased down to the point of chloroplast precipitation. It was possible to show that this dark X_E was not formed during manipulation of the samples and could be generated in total darkness by injection of chloroplasts into the precalculated amount of acid. These chloroplasts then form some ATP when returned to pH 8.0 in the presence of ADP, phosphate, and magnesium (Line E′, Table III). Pyocyanine is not required for the demonstration of dark X_E or scattering levels, and formation of ATP can be inhibited either by making the chloroplasts alkaline again without ADP and phosphate, or by adding ammonia or other uncouplers.

Line E of Table II shows the dark scattering levels obtained by substituting a solution of dye in the reference cell of the spectrophotometer and using the pH 7.7 value as an arbitrary zero. A comparison of Tables II and III shows a parallel between the dark X_E and scattering levels, the sharp rise below pH 5.3 being particularly characteristic. Below pH 4.6 the X_E yield diminishes, and irreversible scattering changes set in. However, the scattering changes found at pH 4.6 are completely reversed by making the medium alkaline again. Although the chloroplasts appear to be agglutinated by acid treatment and are removed by centrifugation at a lower gravitational force, they can be redispersed after being returned to the pH 6 to 8 range.

Inhibition by Sucrose

Lehninger (16) observed that the ATP-induced contraction of thyroxine-swollen mitochondria is strongly inhibited by a variety of polyols, including sucrose. Sucrose also inhibits the light-driven absorbance increase at low pH in isolated chloroplasts (Table IV). The inhibition cannot be ascribed to a decrease in the difference between the refractive index of the medium and that of the particles, since (*a*) sucrose increases the extent of light scattering (or absorbance) in the dark and (*b*) the kinetics is altered by sucrose. The "on" response becomes biphasic (a sharp initial rise followed by a gradual increase lasting many minutes), and the "off" response becomes slower, with a characteristic incomplete reversion toward the initial level. Table IV shows that X_E accumulation is only slightly sensitive to concentrations of sucrose which strongly inhibit absorbance changes.

Temperature Dependence

We have described the effect of temperature on the kinetics of X_E formation and decay (1). Values of Q_{10} for the conformational changes were obtained by illumination of chloroplast suspensions in a water-cooled sample holder inside the Brice-Phoenix photometer. Sample temperature was monitored with a thermistor inside the cuvette, and was varied between illuminations by changing the coolant. Stirring was manual and intermittent.

Several light scattering curves were obtained, passing either from 5° to 15° or from 15° to 25° and vice versa, with each sample of chloroplasts. The average of five Q_{10} determinations with each temperature pair showed a Q_{10} of 2.0 to 2.7 for the maximum decay rate and 2.2 to 3.5 for half-decay times, depending on the cofactor (FMN or pyocyanine) and temperature range. Corresponding Q_{10} values for the forward reactions were between 0.9 and 1.1, for either the maximum forward rate or the half-rise time.

TABLE IV

Effect of sucrose on absorbance changes at pH 6

X_E was assayed at pH 6.2, with the indicated sucrose concentrations present in the light. Sucrose was added in the dark reaction to give a concentration of 0.29 M after injection, for all samples. Illumination for 30 sec gave maximal X_E yields, the control value being 14 mμmoles of ATP/200 μg of chlorophyll.

	Sucrose concentration		
	0.0 M	0.32 M	0.56 M
Absorbance (light)	0.799	0.898	0.904
Absorbance (dark)	0.765	0.886	0.897
Difference	0.034	0.012	0.007
Relative absorbance	100	35	21
X_E yield, %	100	93	89

DISCUSSION

Observations on the reversible increase in light scattering by illuminated chloroplasts at pH 6 (5, 9) have been confirmed and extended in the present work. In mitochondria incubated with ATP, magnesium, and bovine serum albumin (17), the physical basis for a turbidity increase is a contraction of the particles and extrusion of water. Contraction of whole chloroplasts when illuminated has been reported from a number of laboratories (8, 9, 18–20).

In our experience, the hypotonically shocked chloroplasts of the present studies, when measured in a Coulter counter, show light-induced shrinkage that is somewhat too slow to correspond directly to the rapid initial phase of absorbance increments. A change in chloroplast refractivity could also result from internal rearrangement of lamellar components. In either case, the basic physical cause of enhanced light scattering would be a conformational change.

As noted previously (9), the general outlines of pH and cofactor effects suggest that the conformational change is related to X_E. Changes in the level of X_E might therefore be expected to find a direct reflection in the light scattering increments. Elsewhere we have shown that the X_E level may be depressed by adding phosphate and ADP (21). Our observations on the light scattering changes agree with those of Dilley and Vernon (9), who found inhibition on adding phosphate and ADP. On the other hand, the hydrolysis of ATP by the light- and sulfhydryl-dependent, induced ATPase reaction of chloroplasts (22, 23) increases the yield of both X_E[1] and of light scattering incre-

[1] Unpublished data.

ments (Fig. 1 and Reference 6). Data of this sort support a close connection between light scattering changes and X_E.

However, a careful examination of a large part of our present data shows, contrary to the suggestion embodied in the scheme of Dilley and Vernon (9), that the absorbance change cannot be simply a physical measurement of the high energy state itself. In the first place, the use of pyocyanine doubles the total yield of X_E, but increases only the rate of the absorbance increase, not the total amount (Fig. 2). In the second place, there is a very distinct lag of up to 30 or 60 sec in both the "on" and "off" responses of the absorbance change, especially prominent at low cofactor or light levels (Fig. 3); and these lags do not occur in X_E formation or decay if the use of dimethylglutarate buffer is avoided. In the third place, the absorbance change is hardly affected by pH values from 6.7 to 7.2, where X_E yield is inhibited approximately 85 to 90%. This is a reflection of the generally faster kinetics for X_E than that for the light scattering changes. The difference is emphasized by higher pH values so that by pH 7.2 X_E formation is 6 times faster, and its decay 10 times faster, than those of absorbance. The fourth datum is that of the overshoot in which light scattering continues to increase during the 30 or 60 sec following a 5-sec flash of light. This never happens in X_E kinetics. All of these facts and some additional effects of uncouplers (24) prove that the absorbance change is secondary to, but probably dependent on, the high energy state X_E.

The fact that sucrose inhibits light scattering changes, but not X_E accumulation, further rules out the alternative possibility that X_E could represent energy stored in some strained conformation of the chloroplasts, showing up directly in the light scattering relations of the plastids.

The actual relationship between absorbance changes and X_E depends in part on the nature of X_E itself. We have suggested that X_E might be a *trans*-membrane proton gradient with light inducing an internally acid pH (5, 25). Alternatively, it could be a distinct chemical entity containing a high energy bond. In the latter case, the pH dependency of X_E and some of the slower kinetics of the pH shift or of light scattering might be explained by the following formulation.

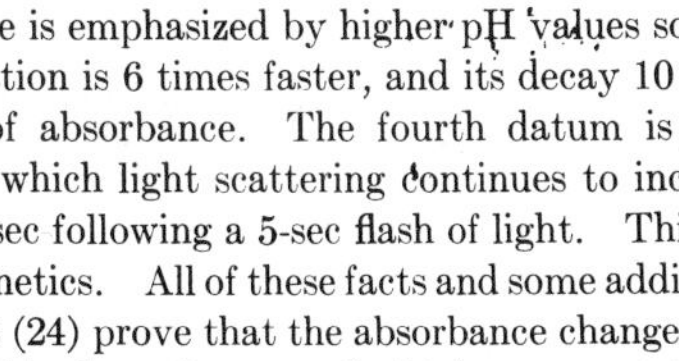

The scheme suggests that some of the decline in photophosphorylation rate at low pH might be due to competition by hydrogen ions for the substrate X_E. In favoring the protonated form HX_E^+, a lower pH would also lead to a decrease in the rate of breakdown (k_d) by lowering the X_E concentration. If it is assumed that the light scattering changes are due to a conformational change in the macromolecule of which X_E is a functional group, the slower scattering kinetics might be partially ascribed to a slow formation of HX_E^+ from X_E. Even more likely is a lag in alteration of the structure following formation of HX_E^+.

If the mechanism corresponds to the above, then the conformational change might occur without any necessary major loss in the energy of X_E; it would simply be a passive and secondary consequence of the presence of X_E. This comment applies as well if X_E is an internal H^+ pool: the hydrogen ion accumulation could conceivably lead to changes in structure without a major loss of energy. Present data do not rule out, however, the alternative in which the conformational change occurs at the expense of the energy content of X_E.

The concept of X_E as a *trans*-membrane pH gradient is favored by two new observations reported here. One is the temporary inhibition of X_E formation by 2.5 mM dimethylglutarate, at low rates of electron flow. If some of the anion penetrates inside the pertinent double membrane enclosures before the light is turned on, it might neutralize the first group of protons entering because of illumination. This would temporarily delay the establishment of a pH gradient across the membrane until the anion buffering capacity was expended. Other experiments showed that the dimethylglutarate effect requires several seconds of preincubation with chloroplasts in the dark, suggesting a slow penetration phase. Induction of the lag under these conditions is restricted to dimethylglutarate and maleate and hence may be a function of their pK of 6.2.

The second new observation is the induction of some X_E and of greater absorbance in darkness by making chloroplasts acid. If the membranes are at least a bit leaky, sufficient protons might enter the internal space in darkness at pH 4.6 so that, when they are injected into pH 8.0 buffer, the artificially created pH gradient would be able to cause phosphorylation of ADP. None of these events would then depend on light-induced electron flow. The internal hydrogen ions at pH 4.6 could easily be responsible for the dark absorbance increments found here. We must note, of course, that this is only one out of several possible interpretations of the discovery of dark X_E.

The turbidity of mitochondrial suspensions is also highly pH-dependent. Novelli, Nanni, and Baldini (26) demonstrated reversible shrinkage on acidification. The time course was rather longer than any in this study and so perhaps corresponds with the slow responses of Itoh *et al.* (8), which involve water redistribution.

The Q_{10} for the light-induced absorbance increase is anomalously low, especially since some values fall just below 1.0, implying negative activation energy. The explanation is most likely that these initial rates are composite functions of forward and back reactions, the latter having a slightly higher Q_{10}. The rate of the "on" response is therefore not truly representative of the forward process, and reservations must be attached to conclusions derived from inhibitor or activator studies of the forward reaction. These reservations do not apply to the dark decay reactions. Calculations based on the treatment used in Reference 1 for finding true formation rates have been unsuccessful.

Neubert, Foster, and Lehninger (27) have described a passive swelling and contraction cycle of isolated mitochondria due to variation in osmolarity of the medium. The rate-limiting step was ascribed to physical diffusion on account of its low Q_{10} (1.2 to 1.8) and insensitivity to inhibitors of electron flow. We consider it unlikely that a diffusion process accounts for the low Q_{10} noted here, because (*a*) the scattering change is dependent on electron flow, (*b*) the dark reversion has a high Q_{10}, and (*c*) the temperature coefficient for X_E formation is also fairly low (1.5). However, if the rate-limiting step in absorbance increases is subsequent to electron flow (as suggested by saturation at low levels of cofactors), then a diffusion process might still be in control.

A question of considerable interest is the occurrence of conformational changes *in vivo*, and their significance for control mechanisms in photosynthesis. One estimate of the pH of the cytoplasm of nongreen plant cells is 6.8 to 7.0 (28), and estimates for animal cytoplasm run from 6.0 to 7.3 (29). If this is the pH range of leaf cells, we would expect chloroplasts to be almost fully shrunken in the light and just slightly so in darkness (Table II); and the electron microscope observations of Kushida *et al.* (30) show that this indeed may be the case. There are many interesting but as yet unproven possibilities for the significance of shrinkage in these membrane-containing particles; it seems reasonable to expect functional relationships between electron transport and other enzymatic components to alter as they are all brought closer together. In this case it follows that photophosphorylation at pH 8.0 *in vitro*, unaccompanied by an initial conformational change, may be of little significance relative to energy transduction *in vivo*.

SUMMARY

The induction of light scattering and absorbance increments in a chloroplast suspension by strong light is maximal near pH 6.0. The extent of these changes is almost as great as the extent of those caused by light at pH 8 in the presence of cysteine-activated adenosine triphosphatase. At pH 6 the light scattering changes are positively correlated with formation of a high energy state of the chloroplasts (X_E), and the two respond in a similar way to variations in light intensity and to the addition of cofactors. However, discrepancies appear in kinetics since X_E formation and decay are generally faster and more immediate than those of absorbance. Other differences are noted in response to pH in the region from pH 6.7 to 7.2; to sucrose; and to a 5-sec flash of light. It is concluded that the scattering change may be formed at the expense of X_E, but cannot be simply an alternative indication of the existence of the high energy condition.

Measurement of the temperature coefficients for these absorbance changes suggests that the formation kinetics is complex and does not show true initial rates.

At pH 4.6 both an increase in absorbance and some X_E are formed in complete darkness. In the light the initial formation of X_E can be delayed by two organic acids, possibly because of their buffering capacity at pH 6.2. These results are considered to be consistent with the concept that X_E is a *trans*-membrane pH gradient.

REFERENCES

1. Hind, G., and Jagendorf, A. T., *Proc. Natl. Acad. Sci. U. S.*, **49**, 715 (1963).
2. Hind, G., and Jagendorf, A. T., *Z. Naturforsch., Pt. b*, **18**, 689 (1963).
3. Shen, Y. K., and Shen, G. M., *Sci. Sinica (Peking)*, **11**, 1097 (1962).
4. Shen, Y. K., Li, Y. T., Shen, G. M., and Yin, H. C., *Acta Biochem. et Biophys. Sinica*, **3**, 278 (1963).
5. Jagendorf, A. T., and Hind, G., *Natl. Acad. Sci.-Natl. Research Council Publ. No. 1145*, 599 (1963).
6. Packer, L., *Natl. Acad. Sci.-Natl. Research Council Publ. No. 1145*, 587 (1963).
7. Packer, L., *Biochim. et Biophys. Acta*, **75**, 12 (1963).
8. Itoh, M., Izawa, S., and Shibata, K., *Biochim. et Biophys. Acta*, **66**, 319 (1963).
9. Dilley, R. A., and Vernon, L. P., *Biochemistry*, **3**, 817 (1964).
10. Latimer, P., in *Studies of microalgae and photosynthetic bacteria, Japanese Society of Plant Physiologists*, University of Tokyo Press, Tokyo, 1963, p. 213.
11. McIlwain, H., *J. Chem. Soc.*, 1704 (1937).
12. Prickett, W. F., Bryant, F. D., and Latimer, P., *Natl. Acad. Sci.-Natl. Research Council Publ. No. 1145*, 711 (1963).
13. Hill, R., and Walker, D. A., *Plant Physiol.*, **34**, 240 (1959).
14. Jagendorf, A. T., and Avron, M., *J. Biol. Chem.*, **231**, 277 (1958).
15. Kahn, J. S., *Arch. Biochem. Biophys.*, **98**, 100 (1962).
16. Lehninger, A. L., *J. Biochem. (Tokyo)*, **49**, 553 (1961).
17. Lehninger, A. L., *J. Biol. Chem.*, **234**, 2465 (1959).
18. Izawa, S., Itoh, M., and Shibata, K., *Biochim. et Biophys. Acta*, **75**, 349 (1963).
19. Ohnishi, T., *J. Biochem. (Tokyo)*, **55**, 494 (1964).
20. Dilley, R. A., *Biochem. and Biophys. Research Communs.*, **17**, 716 (1964).
21. Jagendorf, A. T., and Hind, G., *Biochem. and Biophys. Research Communs.*, **18**, 702 (1965).
22. Petrack, B., and Lipmann, F., in W. D. McElroy and H. B. Glass, (Editors), *Light and life*, Johns Hopkins Press, Baltimore, 1961, pp. 621–630.
23. Hoch, G., and Martin, I., *Biochem. and Biophys. Research Communs.*, **12**, 223 (1963).
24. Hind, G., and Jagendorf, A. T., *J. Biol. Chem.*, **240**, 3202 (1965).
25. Neumann, J., and Jagendorf, A. T., *Arch. Biochem. Biophys.*, **107**, 109 (1964).
26. Novelli, A., Nanni, G., and Baldini, I., *Ital. J. Biochem.*, **13**, 247 (1963).
27. Neubert, D., Foster, G. V., and Lehninger, A. L., *Biochim. et Biophys. Acta*, **60**, 492 (1962).
28. Chambers, R., and Kerr, T., *J. Cellular Comp. Physiol.*, **2**, 105 (1932).
29. Novelli, A., Nanni, G., and Baldini, I., *Experientia*, **18**, 558 (1962).
30. Kushida, H., Itoh, M., Izawa, S., and Shibata, K., *Biochim. et Biophys. Acta*, **79**, 201 (1964).

Reprinted from
Biochimica et Biophysica Acta
Elsevier Publishing Company
Amsterdam
Printed in The Netherlands

BBA 45720

THE STOICHIOMETRIC RELATION OF PHOSPHORYLATION TO ELECTRON TRANSPORT IN ISOLATED CHLOROPLASTS

S. IZAWA AND NORMAN E. GOOD

Department of Biology, Queens University, Kingston, Ontario (Canada) and Department of Botany and Plant Pathology, Michigan State University, East Lansing, Mich. (U.S.A.)

(Received April 22nd, 1968)

SUMMARY

Electron transport in chloroplasts can proceed by two pathways. One pathway is dependent on the presence of ADP and orthophosphate and is responsible for photophosphorylation. This pathway is sensitive to the photophosphorylation inhibitor phlorizin, to antimycin A and to *p*-chloromercuribenzoate (PCMB). The other pathway is independent of phosphorylation, insensitive to phlorizin, relatively insensitive to antimycin A and insensitive to PCMB. Apparently amines increase the electron transport rate by uncoupling the antimycin-sensitive phosphorylating pathway while carbonylcyanide phenylhydrazones uncouple by increasing the activity of the antimycin-resistant non-phosphorylating pathway.

During photophosphorylation the electron transport rate normally exceeds the corresponding rate measured in the absence of phosphate and, under many conditions, the amount of ATP formed is proportional to the additional electron transport which occurs in the presence of phosphate. Indeed there is often a rather precise mole for mole correspondence between ATP formation and the accompanying increase in ferricyanide reduction. This is true when phosphorylation conditions are optimal and it is also true when the phosphorylation rate is varied over a wide range by limiting concentrations of phosphate, by the addition of antimycin A, by the addition of PCMB, or by the addition of carbonylcyanide phenylhydrazones. It is not true unless the rate of electron transport is controlled at the level of the phosphorylation reaction. Thus the correspondence between phosphorylation and extra electron transport is no longer observed when electron transport is limited by low light intensities or by the addition of phenylureas. Under these conditions phosphorylation may greatly exceed the increase in electron transport associated with phosphorylation. Moreover the correspondence is not always found when the chloroplasts have been partially uncoupled by amines such as Tris or methylamine.

Our observations suggest that, at high light intensities and in the absence of amine uncouplers, the phosphorylating electron transport is superimposed on a constant non-phosphorylating electron transport. If this is so the phosphorylating process must yield two molecules of ATP for every pair of electron it transfers.

* Abbreviations: PCMB, *p*-chloromercuribenzoate; CCCP, carbonylcyanide 3-chlorophenylhydrazone, *i.e.*, ketomalononitrile 3-chlorophenylhydrazone; DCMU, 3(3,4-dichlorophenyl)-1,1-dimethylurea; P/e_2, the ratio of ATP molecules formed to pairs of electrons transported; TES, *N*-Tris(hydroxymethyl)methyltaurine; tricine, *N*-Tris(hydroxymethyl)methylglycine.

INTRODUCTION

It has been widely accepted that the theoretical maximum efficiency of "non-cyclic" photophosphorylation is 1 ATP molecule formed for each pair of electrons transported[1,2]. This implies that "non-cyclic" phosphorylation is coupled to a single oxidation–reduction reaction involving 2 electrons. Moreover the ratio of 1 ATP to 1 electron pair means that only 1 ATP molecule can be formed in the reduction of a molecule of $NADP^+$, an amount of ATP which is insufficient for reduction of CO_2 by the Calvin–Benson cycle[3] or by any plausible modification thereof. Consequently much effort has been expended in a search for a second kind of phosphorylation which must perforce utilize a "cyclic" (*i.e.* an as yet unmeasured) electron flow.

Recent observations have shown that the value of $P/e_2 = 1.0$ is probably wrong and therefore most of the reasons for postulating a "cyclic" phosphorylation mechanism *in vivo* have disappeared. In the first place, overall values of P/e_2 much higher than 1.0 are routinely observed in our laboratories[4] and in other laboratories*. This in itself suggests that the theoretical maximum value may be 2.0. In the second place, the overall values do not take into account the fact that two kinds of electron transport, phosphorylating and non-phosphorylating, may occur simultaneously in chloroplasts. We know that isolated chloroplasts are capable of very considerable rates of electron transport in the absence of phosphorylation and there is no reason for supposing that this process must completely cease during phosphorylation. Therefore the overall stoichiometry may represent the average efficiency of two processes, only one of which phosphorylates ADP. In the absence of any knowledge of the relative magnitudes of these two concurrent processes, the efficiency of the phosphorylation process itself cannot be deduced. Nor does the average efficiency in isolated chloroplasts tell us anything about the *in vivo* efficiency of phosphorylation since we cannot assume that the non-phosphorylating and phosphorylating pathways of electron transport have the same relative activities *in vivo* and *in vitro*. Indeed it is possible that the whole of the non-phosphorylating electron transport *in vitro* results from a partial uncoupling of the chloroplasts which occurs during their isolation. It follows that the overall *in vivo* efficiency of photophosphorylation might be as high as the *in vitro* efficiency of the phosphorylating part of the electron transport of isolated chloroplasts.

This paper describes experimental conditions under which the phosphorylating and non-phosphorylating electron transport are probably additive. In other words the phosphorylating electron transport seems to be superimposed on a independent and constant non-phosphorylating process. By taking advantage of these conditions we have attempted to determine the real efficiency of the phosphorylating component of the electron transport.

MATERIALS AND METHODS

Chloroplasts

Leaves of spinach (*Spinacea oleracea* L.), obtained from a local market, were ground in a Waring Blendor for 5 sec in a medium containing 0.3 M NaCl, 2 mM

* Personal communications from M. AVRON and H. BALTSCHEFFSKY.

$MgCl_2$ and 0.04 M tricine adjusted to pH 7.8 with NaOH. The homogenate was squeezed through cheesecloth and centrifuged at 3000 × *g* for 4 min. The pellet was resuspended in a medium containing 0.2 M sucrose, 3 mM $MgCl_2$ and 0.03 M *N*-Tris-(hydroxymethyl)methylglycine (tricine) adjusted to pH 7.3 with NaOH. The suspension was centrifuged at about 1200 × *g* for 1 min to remove cell debris, then centrifuged again at 3000 × *g* for 4 min. The washed chloroplasts were finally taken up in a small amount of the same sucrose–tricine–$MgCl_2$ medium. (Chloroplasts used in *p*-chloromercuribenzoate (PCMB) experiments were prepared and tested in *N*-Tris-(hydroxymethyl)methyltaurine (TES) instead of tricine since tricine complexes mercuric ions very strongly.) All operations were conducted in a cold room at or as near to 0° as possible. The stock suspension of chloroplasts thus prepared was very stable and could be stored at 0° for more than 12 h without serious loss of activity.

In one experiment (see Table I), chloroplasts of white mustard (*Brassica hirta* Moench) and tobacco (*Nicotiana tobacum* L.) were employed. These were prepared in the same manner except that the leaves were ground for 20 sec.

Chemicals

Phlorizin was recrystallized from aqueous ethanol after charcoal (Norit) treatment in order to remove impurities which reduced a significant amount of ferricyanide and had weak uncoupling effects. The purified phlorizin (0.01 M) was dissolved in tricine buffer (0.02 M, pH 8.0) immediately before use. Ethanolic solutions of phlorizin were avoided since the combination of phlorizin and traces of ethanol often produced a slight uncoupling or an unspecific inhibition of electron transport. 3(3,4-Dichlorophenyl)-1,1-dimethylurea (DCMU) and carbonylcyanide 3-chlorophenylhydrazone (CCCP) were dissolved at appropriate concentrations in 50 % aqueous ethylene glycol.

TABLE I

RATES OF FERRICYANIDE REDUCTION AND ATP FORMATION AS AFFECTED BY LIGHT INTENSITY, AMINES AND CHLOROPLAST SOURCE

The rates are given in μmoles/h per mg chlorophyll. Observed values of P/e_2 represent the average efficiencies of the phosphorylating and non-phosphorylating transport systems. The computed value of P/e_2 is the efficiency of the phosphorylating part of the transport, calculated by assuming the rate of non-phosphorylating transport to be the same in the presence and absence of phosphorylation ($+P_i$ and $-P_i$). As explained in the text this assumption is not valid at low light intensities or in the presence of amines. Control conditions as described in MATERIALS AND METHODS.

Chloroplast source	*Reaction conditions*	*Ferricyanide reduction*			*ATP formation*	P/e_2	
		$+P_i$	$-P_i$	$(+P_i/-P_i)$		*Observed*	*Calculated*
Spinach	Control	403	175	(2.30)	221	1.10	1.95
Spinach	+0.5 mM methylamine	504	418	(1.29)	200	0.74	3.27
Spinach	Low light*	172	140	(1.34)	81	0.94	5.05
Mustard	Control	564	242	(2.33)	310	1.10	1.92
Mustard	Assayed in 0.05 M Tris–HCl (pH 8.2)	545	340	(1.60)	264	0.97	2.58
Mustard	Aged and assayed in 0.05 M Tris–HCl**	215	168	(1.28)	75	0.70	3.20
Tobacco	Control	468	210	(2.23)	264	1.13	2.05

* Light intensity approx. 14 kergs/cm² per sec.

** Chloroplasts aged in Tris buffer (pH 7.4) for 5 h at 0°. Assayed at pH 8.2.

Antimycin A was dissolved in a methanol–ethylene glycol mixture (2:3, v/v). In all cases the final concentrations of the organic solvents were kept as low as possible.

Measurements

The standard reaction mixture (2.0 ml) contained, in µmoles: sucrose, 300; tricine (pH 8.4) (or TES, pH 8.2, for PCMB tests), 80; $Na_2H^{32}PO_4$, 20 (if used); ADP, 2; $MgCl_2$, 8; potassium ferricyanide, 0.8; and chloroplasts containing 40 µg chlorophyll. Reactions were carried out at 19° in a standard rectangular 1.0-cm cuvette illuminated with red light (approx. 600–700 mµ). Preincubation for temperature equilibration was for 3 min unless otherwise noted. The actinic light was obtained by passing the beam of a 500 W (incandescent) projector lamp through a dilute $CuSO_4$ solution and a pair of Corning glass filters (Nos. 2424 and 4600). The intensity of the actinic beam was about 570 kergs/cm² per sec. This intensity was enough to saturate the ferricyanide reducing system. Lower light intensities were obtained by placing calibrated screens in the light beam. Ferricyanide reduction was continuously monitored by recording changes in the $A_{420\,m\mu}$ of the reaction mixture. Scattered red actinic light was kept from entering the spectrophotometric device by inserting appropriate blue filters before the photomultiplier tube. In all instances the total change in $A_{420\,m\mu}$ agreed very closely with the amount of ferricyanide present at the beginning of the period of illumination. Therefore it seems safe to equate absorbance change to ferricyanide reduction. ATP was measured as residual radioactivity after extraction of the remaining orthophosphate as phosphomolybdate by the method of AVRON[5], either as soon as all of the ferricyanide had been reduced (3–5 min) or when the reaction was very slow, after a known portion of the ferricyanide had been reduced*. There were negligible amounts of radioactivity in the dark controls and in DCMU-treated, illuminated controls.

RESULTS

Conditions under which the number of moles of ATP formed is equal to the number of additional moles of ferricyanide reduced during phosphorylation (Figs. 1–6)

A consistent correspondence between the rate of ATP formation and additional electron transport associated with phosphorylation strongly implies that the phosphorylating electron transport is superimposed on a constant non-phosphorylating electron transport. This, in turn, implies that the two processes are parallel but do not compete. Such absence of competition is to be expected if the overall rate of transport is determined by the combined capacities of the two processes to carry electrons. The initial carrier enzymes of both systems should then be saturated and therefore the two processes should be additive. Figs. 1–6 illustrate a series of experiments which seem to satisfy these conditions. The potential activity of the chloroplasts was such

* This technique does not actually measure the amount of ATP produced but rather the amount of orthophosphate converted into other forms during illumination. It therefore measures electron transport-dependent phosphorylation but not necessarily the phosphorylation of ADP. However, the phosphorylation reactions described herein were completely dependent on the addition of ADP and, on the basis of chromatography of the reaction mixture, yielded ATP as the preponderant radioactive product. We cannot yet preclude the possibility that a very small proportion of the radioactivity of the orthophosphate was incorporated into ADP instead of ATP as claimed by DEL CAMPO, RAMIREZ AND ARNON[6].

that the rate of electron transport increased several-fold when uncouplers were added. Therefore the rate of electron transport must indeed have been controlled at the level of the phosphorylation reaction.

Phlorizin inhibits both ATP synthesis and that part of the electron transport which is induced by the addition of ADP and phosphate, without in any way affecting the rate of electron transport in the absence of phosphate[7,8,12]. Regardless of the concentration of phlorizin and the rate of phosphorylation, 1 molecule of ATP is formed for every extra ferricyanide ion reduced in response to the addition of phosphate (see Fig. 1). We may also state this relationship in another way which begs the question of the activity of the non-phosphorylating pathway of electron transport during phosphorylation: for every ferricyanide ion which is not reduced because of phlorizin, a molecule of ATP is not formed.

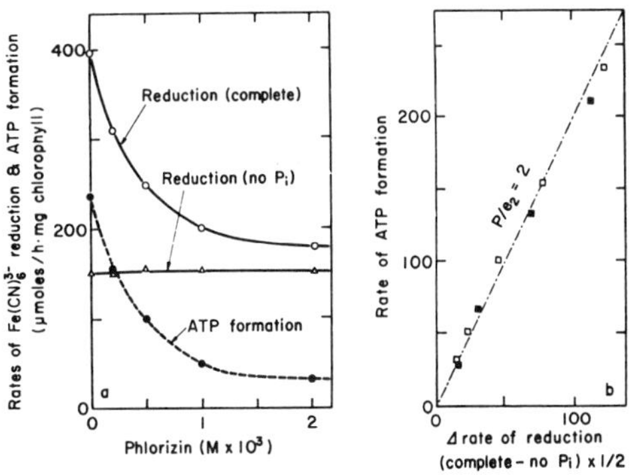

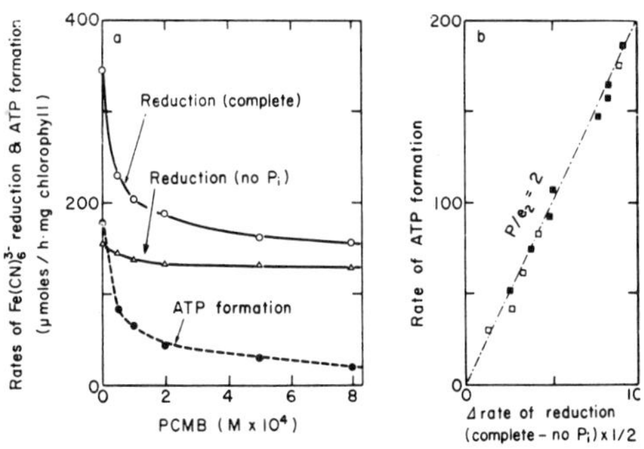

Fig. 1. Effects of phlorizin on electron transport and phosphorylation in spinach chloroplasts. For details of reaction conditions see MATERIALS AND METHODS. Phosphorylating electron transport (complete) was measured in the presence of ADP, phosphate (P_i) and Mg^{2+}. Non-phosphorylating electron transport (no P_i) was measured in the same system lacking only phosphate. b. □, values computed from the data of (a); ■, values computed from a duplicate experiment. The dotted line is not drawn through the points but rather represents the location of the points predicted by assuming that each additional pair of electrons carried during phosphorylation results in the formation of 2 molecules of ATP.

Fig. 2. Effects of PCMB on electron transport and phosphorylation. Data are presented as in Fig. 1.

PCMB is known to inhibit photophosphorylation[9] and it has been suggested that the phosphorylation reaction is more sensitive to this sulphydryl reagent than is (uncoupled) electron transport[10]. We find that PCMB at low concentration behaves much as does phlorizin: it inhibits photophosphorylation and the phosphorylation-associated part of the electron transport (Fig. 2). And, as with phlorizin, 1 ATP molecule is always found for every additional electron transferred to ferricyanide when the rate of phosphorylation is varied with PCMB.

As was pointed out in a earlier publication[11], phosphorylation may be reduced to any desired extent by the use of sub-optimal concentrations of phosphate, and the varying amounts of ATP formed are then equal to the varying amounts of extra ferricyanide reduced as the result of the addition of the phosphate. Fig. 3 confirms the earlier observation and extends it by showing that the same relationship is observed when phosphorylation rates are controlled by combinations of phlorizin and limited phosphate.

Antimycin A can also inhibit photophosphorylation and the phosphorylation-associated increase in electron transport without inhibiting the non-phosphorylating transport (see Fig. 4). (Antimycin A also has some uncoupling action since it causes a slight increase in the rate of electron transport in the absence of phosphate.) However, antimycin differs from phlorizin in that it also inhibits amine-uncoupled electron transport[12]. Again at all levels of inhibition the amount of ATP formed is equal to the extra ferricyanide reduced during phosphorylation.

Carbonylcyanide phenylhydrazones prevent ATP synthesis in a different way, neither inhibiting phosphorylation directly nor inhibiting phosphorylation-coupled electron transport. Rather they seem to open an alternative, non-phosphorylating

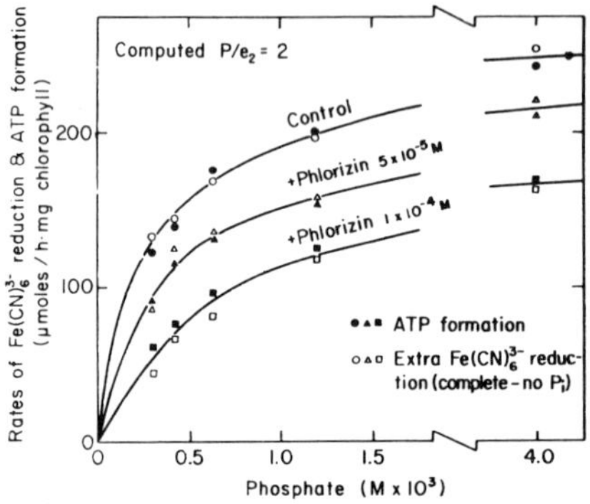

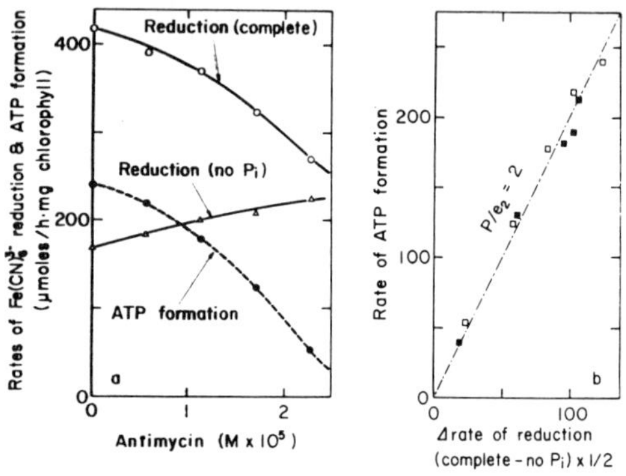

Fig. 3. Effects of phosphate concentration on electron transport and phosphorylation in the presence and absence of phlorizin. Methods as in Fig. 1. Under all of these conditions and at all levels of activity of the phosphorylation system, the number of molecules of ATP formed closely approximated the number of additional ferricyanide ions reduced during phosphorylation ($P/e_2 = 2.0$).

Fig. 4. Effects of antimycin A on electron transport and phosphorylation. Conditions were as in Fig. 1 except that the chloroplasts were preincubated with the inhibitor for 5 min. Data are presented as in Fig. 1.

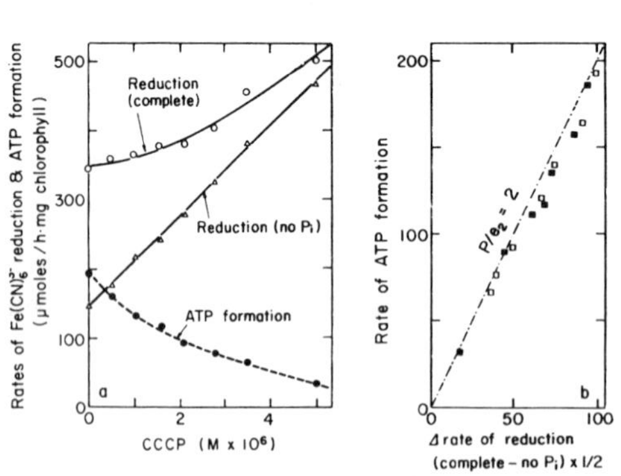

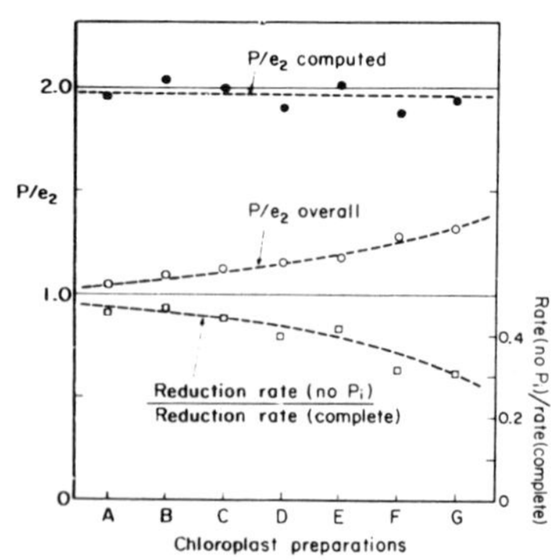

Fig. 5. Effects of CCCP on electron transport and phosphorylation. Data are presented as in Fig. 1.

Fig. 6. Variations among chloroplast preparations in overall efficiency of phosphorylation (P/e_2 overall), the ratio of non-phosphorylating (no P_i) electron transport to total transport during phosphorylation (complete), and the computed efficiency of the phosphorylating part of the electron transport (P/e_2 computed). Computation is based on the assumption that the non-phosphorylating electron transport observed in the absence of phosphate continues at the same rate during phosphorylation. Values are plotted in order of increasing overall efficiencies. Despite considerable differences in overall P/e_2, the computed efficiency of the phosphorylating component of the electron transport is almost constant at a value of P/e_2 close to 2.0.

pathway for electron transport. The induced by-pass of the phosphorylation site resembles the endogenous "basal" electron transport process in being relatively insensitive to antimycin A (ref. 12). Once again the amount of ATP formed is equal on a molar basis to the extra ferricyanide reduced during phosphorylation, no matter how severely the phosphorylation is curtailed by CCCP (Fig. 5). Details of the differential effects of antimycin A on amine- and CCCP-uncoupled electron transport will be described elsewhere.

The amount of inadvertent uncoupling probably depends on the method of preparation of the chloroplasts and on the condition of the leaves from which the chloroplasts have been prepared. For historical reasons the resulting non-phosphorylating electron transport is referred to as the "basal" transport—perhaps a happy choice of terms since it does seem that this process forms a base on which phosphorylating electron transport is superimposed. In any event, the degree of uncoupling and hence the "basal" non-phosphorylating electron transport varies considerably from one preparation to another (see Fig. 6). As a result the average efficiency of phosphorylation (the observed or overall P/e_2) also varies considerably. However, in every instance the amount of phosphorylation corresponds to the extra electron transport which occurs during phosphorylation. The ratio of ATP molecules to extra electrons transported is close to 1.0 (or $P/e_2 = 2.0$) in each preparation.

Conditions under which the number of moles of ATP formed is greater than the additional number of moles of ferricyanide reduced during phosphorylation (Figs. 7–9)

Ratios of P/e_2 "corrected" in the manner described above for non-phosphorylating electron transport are not always consistent and are sometimes unacceptably high. Indeed, as STILLER AND VENNESLAND[13] and PUNNETT[14] have pointed out, values thus computed can be infinite! Consequently it is incumbent on us to explain why the computation is sometimes invalid if we are to assert that it is sometimes valid. As we have already suggested, non-cyclic electron transport in active chloroplasts at high light intensities must be controlled at the level of the phosphorylation reaction since uncouplers of phosphorylation can produce dramatic increases in electron transport rates. It is probable that under such conditions the carriers supplying the phosphorylating and non-phosphorylating pathways are both saturated by a more than adequate supply of photochemically produced electrons, and therefore one would not predict any interaction between the two processes. However, the situation is completely changed if the supply of electrons becomes for any reason a rate-determining factor. Low light intensities, DCMU-poisoning, and the ageing of chloroplasts under unfavorable conditions all inhibit electron transport at or near the level of the photochemical reactions. The phosphorylating and non-phosphorylating processes are then forced to compete for the limited supply of electrons and, under these conditions, the introduction of the phosphorylating process does decrease the activity of the non-phosphorylating process. This is particularly obvious in the limiting case where light intensities are so low or other conditions are so unfavorable that the rate of electron transport is not at all enhanced by situations which permit phosphorylation. Since phosphorylation at such low light intensities may still proceed with reasonable overall efficiency it is clear that the phosphorylation-coupled process must occur at the expense of the non-phosphorylating process which is correspondingly diminished. Computations based on a supposedly constant rate of non-phosphorylating electron transport are patently invalid if the rate-determining process is not associated with phosphorylation.

Fig. 7 shows the effects of various light intensities on rates of phosphorylation, and on rates of ferricyanide reduction in the presence and absence of phosphorylation. (Our results differ from those of other workers[15–17] in that we find quite high overall phosphorylation efficiencies even at the lowest intensities. Possibly the mild uncoupling effect of the Tris buffer employed by others becomes more significant as electron transport rates are drastically reduced.) Clearly the non-phosphorylating component of the electron transport is not constant under these conditions and therefore it is not possible to estimate the activity of the phosphorylating part of the electron transport except in terms of ATP formed.

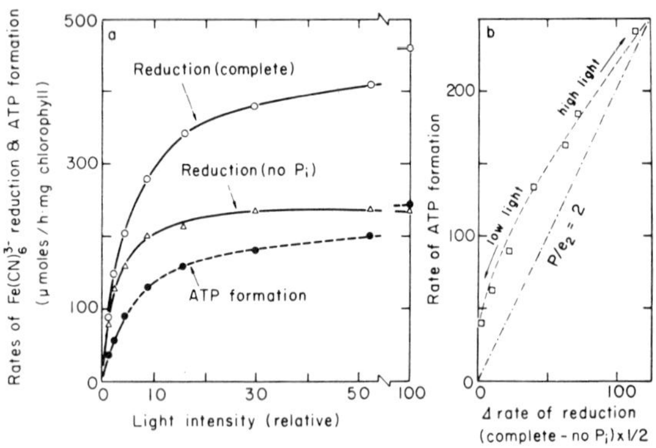

Fig. 7. Effects of different light intensities on electron transport and phosphorylation. The intensity at relative value 100 was approx. 570 kergs/cm^2 per sec (600–700 mμ). Other conditions as in Fig. 1. At low light intensities the overall efficiency of phosphorylation is almost as high as at high light intensities, yet there is little or no increase in the rate of electron transport on the addition of phosphorylation substrates. Consequently phosphorylating electron transport must occur at the expense of the non-phosphorylating process which is correspondingly decreased. Thus "correcting" by subtracting the basal electron transport as in (b) is not a valid procedure and yields meaningless high values for the computed P/e_2.

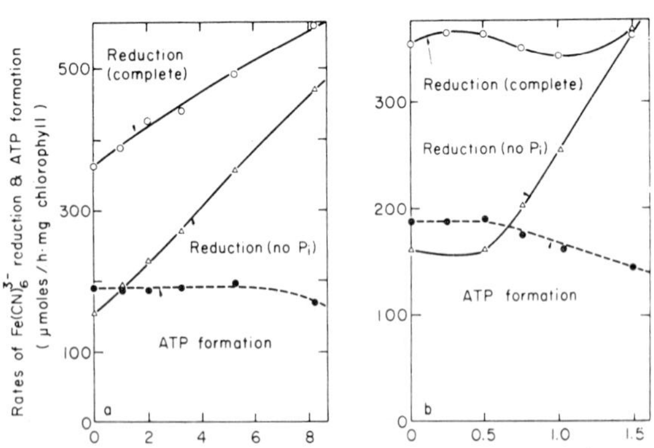

Fig. 8. Effects of methylamine hydrochloride (a) and brucine (b) on electron transport and phosphorylation. Conditions as in Fig. 1. At the relatively low concentrations employed in the experiment, these amine uncouplers do not seriously inhibit phosphorylation. Nevertheless they do greatly enhance the non-phosphorylating (no P_i) electron transport. Thus the amine-uncoupled component of the non-phosphorylating electron transport seems to compete very unfavorably with the phosphorylating electron transport. It is therefore not surprising that the number of extra electrons transported during phosphorylation falls far short of the number of molecules of ATP formed.

Since adding DCMU to chloroplast suspensions is tantamount in terms of electron transport to reducing the light intensity, it is not surprising that computations based on the erroneous assumption of a constant rate of non-phosphorylating transport are also meaningless when phosphorylation is limited by this inhibitor[6].

Uncoupling by amines also renders invalid our procedure of subtracting a supposedly constant non-phosphorylating electron transport. Fig. 8 shows the effects of methylamine and the new uncoupler brucine, a complex alicylic amine, on electron transport and phosphorylation. In this case the interaction between the phosphorylating and non-phosphorylating processes is more complicated. Since low concentrations of amines enhance the rate of non-phosphorylating electron transport (rate without phosphate) but do not greatly decrease phosphorylation, it may be deduced that amine-uncoupled electron transport competes very poorly with phosphorylating electron transport. (In contrast see CCCP-uncoupling, Fig. 5.) When we consider, further, that the site of amine-uncoupling lies on the same antimycin-sensitive pathway as phosphorylation[12], we can readily imagine that amine-uncoupled and phosphorylating electron transport share a common rate-limiting step which is not involved in the "basal" or CCCP-uncoupled transport. For some such reason—a shared bottleneck and competition which favors the phosphorylation process—it seems that electron transport by the amine-uncoupled pathway is much slower when phosphorylation occurs. This partial suppression of amine-induced electron transport during phosphorylation is shown very clearly by a corresponding suppression of the large chloroplast volume changes which are always associated with amine-uncoupled electron transport (S. IZAWA, unpublished observations). Computations of the kind which yielded consistent values for the efficiency of phosphorylating electron transport in Figs. 1–6 therefore cannot be used when amines are present since the non-phosphorylating component of the electron transport is conspicuously not constant. This applies even to Tris buffer, which is a highly polar amine with a weak uncoupling action[18]. Chloroplasts aged and assayed in Tris buffer probably cannot be used for computation of the "corrected" P/e_2 for two reasons: Tris-induced uncoupling and Tris-induced inhibition of electron transport[19] may both invalidate the procedure (see Table I).

DISCUSSION

Fig. 9 summarizes our hypotheses regarding the various pathways for electron transport. Chloroplasts as prepared by current methods always exhibit a "basal" electron transport. This antimycin-resistant, non-phosphorylating transport represents a variable proportion of the total transport capacity. Under optimal conditions (active chloroplasts, high light intensities and no uncoupling amines) ATP formation is closely proportional to the difference between the capacity of the basal transport system, as determined in the absence of phosphate, and the total transport capacity as determined in the presence of phosphate. This relationship is most readily interpreted in terms of a continuing non-phosphorylating electron transport on which the phosphorylating transport is superimposed. In other words the two processes do not compete under the optimal conditions enumerated. Such absence of competition could arise in two ways: saturation of potentially competing enzymes, or a physical separation of the sites of the two processes which precluded competition. However, examination of the behavior of chloroplasts at low light intensities or in the presence of DCMU shows that competition for the same electrons is indeed possible. Therefore the concept of separate and effectively isolated phosphorylating and non-phosphorylating sites must be abandoned.

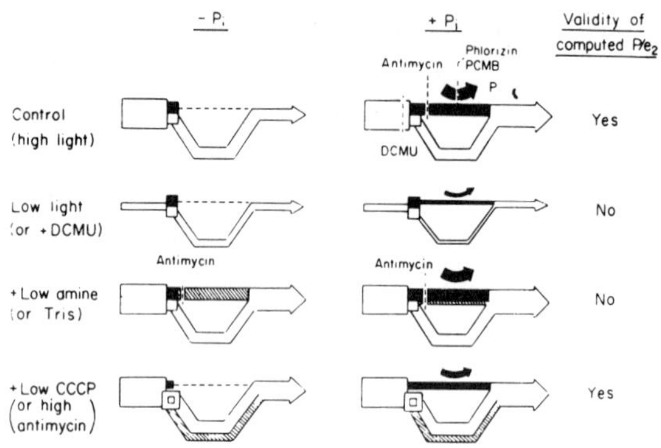

Fig. 9. Diagrammatic representation of the various electron transport processes discussed. Width of arrows is intended to represent electron transport rates. Solid squares represent electron-accepting sites associated with phosphorylating transport while open squares represent similar (damaged?) sites associated with non-phosphorylating transport. At high light intensities both types of sites are saturated with the result that the phosphorylating and non-phosphorylating processes are additive. At low light intensities the supply of electrons becomes limiting, the two types of sites are forced to compete, and the processes are no longer additive. Amines uncouple the antimycin A sensitive phosphorylating pathway but the amine-uncoupled transport competes unfavorably with the phosphorylating process and may be largely suppressed during phosphorylation. Hence the non-phosphorylating and phosphorylating electron transport are not additive when amines are present. In contrast, the uncoupler CCCP seems to convert phosphorylating sites into non-phosphorylating sites, but both types of sites remain saturated at high light intensities; hence the phosphorylating and non-phosphorylating processes are once more additive.

Naturally we cannot expect to estimate the activity of the phosphorylating pathways on the basis of a constant non-phosphorylating transport when the two processes are quite evidently competing; we should not expect to and do not find ATP formation proportional to the additional electron transport induced by phosphorylation conditions when the electron transport rate is controlled at some site other than the phosphorylation site. This situation typically arises when sub-optimal light intensities are used and when the chloroplasts have been badly treated in various ways. Another typical but often overlooked instance sometimes occurs when $NADP^+$ is employed as electron acceptor instead of ferricyanide or flavins or viologens; the much lower rate of electron transport and the poor stimulation of electron transport by uncouplers or ADP and phosphate point to the fact that the terminal steps in $NADP^+$ reduction often limit the overall process.

Our tentative conclusion regarding the maximum efficiency of photophosphorylation is based on two independent considerations: (1) Most models of phosphorylating mechanisms predict a stoichiometry which is represented by a whole number and it is unlikely that the overall value of P/e_2 in photophosphorylation is as low as 1.0. And (2) over a wide range of conditions 2 ATP molecules are formed for each additional pair of electrons transported during phosphorylation. It would be well to consider each of these points in some detail.

(1) Our first point has been challenged by DEL CAMPO, RAMIREZ AND ARNON[6]

on the ground that 5 % or more of the radioactivity assumed to be in ATP may be associated with ADP. However, we would like to point out that almost 30 % of the radioactivity would have to be in ADP to explain our data[4] if the true value of P/e_2 is 1.0. In view of the virtual absence of ADP–P_i exchange reactions in chloroplasts reported by others[20] this seems exceedingly unlikely. Moreover, if ADP does become labeled to a significant extent, the incorporation of orthophosphate therein is totally dependent on electron transport and thus represents but another (and very minor) electron transport-coupled phosphorylation. However, there is a much more serious objection to all computations of the stoichiometry of photophosphorylation. This resides in the difficulty of being sure that all of the electron transport has been measured, that no hidden "cyclic" or "pseudocyclic" electron transport has occurred. Indeed, it may well be that the existence of a considerable amount of unrecognized electron transport was responsible for the high P/e_2 ratios observed by LYNN AND BROWN[21]. Nevertheless, it should be stressed at this point that the phosphorylation reactions we have discussed in this paper and in ref. 4 were ferricyanide-dependent. That is to say, one can subtract the endogenous cyclic or pseudocyclic phosphorylation which occurs in the absence of ferricyanide (5–7 % of the ferricyanide rate) and still measure ratios of P/e_2 which are consistently in excess of 1.0. Therefore, if one is to invoke cyclic phosphorylation to explain the excess over 1.0 one must postulate a ferricyanide-dependent cyclic phosphorylation. In view of the fact that all known types of cyclic and pseudocyclic phosphorylation seem to be inhibited by substrate level concentrations of ferricyanide[1,22,23], this seems improbable.

(2) It is difficult to explain the linear relation between phosphorylation and the phosphorylation-associated increase in electron transport except in terms of additive phosphorylating and non-phosphorylating electron transport processes. If the correspondence between phosphorylation and increased electron transport had been observed over a narrow range of phosphorylation rates, or if it had only been encountered when the phosphorylation was varied in a single way, one might reasonably have attributed the correspondence to a fortuitous balance of several factors. In fact the correspondence is observed over a wide range of phosphorylation rates when the various rates have been obtained in a wide variety of demonstrably different ways. Consequently it is difficult to plead happenstance. On the other hand, if we accept the fact that the non-phosphorylating and phosphorylating processes are additive, it follows that the non-phosphorylating electron transport observed in the absence of phosphate may be legitimately subtracted from the total electron transport in order to compute the activity of the phosphorylation process. And if one accepts the validity of such operations it follows that the phosphorylating efficiency of the phosphorylating part of the electron transport must be 2 ATP molecules formed for every electron pair transported.

ACKNOWLEDGEMENTS

This work was supported in part by Grant (GB 4568) from the National Science Foundation of the U.S. and in part by a grant (A-3743) from the National Research Council of Canada and is published as Journal Article 4373 of the Michigan Agricultural Experiment Station.

REFERENCES

1 D. I. Arnon, F. B. Whatley and M. B. Allen, *Science*, 127 (1958) 1026.
2 D. W. Krogmann, A. T. Jagendorf and M. Avron, *Plant Physiol.*, 34 (1959) 272.
3 M. Calvin and J. A. Bassham, *The Path of Carbon in Photosynthesis*, Prentice-Hall, New York, 1957.
4 G. D. Winget, S. Izawa and N. E. Good, *Biochem. Biophys. Res. Commun.*, 21 (1965) 438.
5 M. Avron, *Biochim. Biophys. Acta*, 40 (1960) 257.
6 F. F. Del Campo, J. M. Ramirez and D. I. Arnon, *J. Biol. Chem.*, 243 (1968) 2805.
7 S. Izawa, G. D. Winget and N. E. Good, *Biochem. Biophys. Res. Commun.*, 22 (1966) 223.
8 N. Good, S. Izawa and G. Hind, in R. Sanadi, *Current Topics in Bioenergetics*, Vol. 1, Academic Press, New York, 1966, p. 75.
9 D. I. Arnon, M. B. Allen and F. R. Whatley, *Biochim. Biophys. Acta*, 20 (1956) 449.
10 A. T. Jagendorf and M. Avron, *Arch. Biochem. Biophys.*, 80 (1959) 246.
11 N. E. Good, *Nature*, 188 (1960) 661.
12 S. Izawa, T. N. Connolly, G. D. Winget and N. E. Good, *Brookhaven Symp. Biol.*, 19 (1966) 169.
13 M. Stiller and B. Vennesland, *Nature*, 191 (1961) 677.
14 T. Punnett, *Photosynthetic Mechanisms in Green Plants*, Publication 1145, National Academy of Sciences-National Research Council, Washington D.C., 1963, p. 619.
15 H. C. Yin, Y. K. Shen, G. M. Shen, S. Y. Yang and K. S. Chui, *Sci. Sinica Peking*, 10 (1961) 976.
16 H. Turner, C. C. Black and M. Gibbs, *J. Biol. Chem.*, 237 (1962) 577.
17 H. Sakurai, M. Nishimura and A. Takamiya, *Plant Cell Physiol.*, 6 (1965) 309.
18 N. E. Good, G. D. Winget, W. Winter, T. N. Connolly, S. Izawa and R. M. M. Singh, *Biochemistry*, 5 (1966) 467.
19 L. P. Vernon and W. S. Zaugg, *J. Biol. Chem.*, 235 (1960) 2728.
20 C. Camerli and M. Avron, *European J. Biochem.*, 2 (1967) 318.
21 W. S. Lynn and R. S. Brown, *J. Biol. Chem.*, 242 (1967) 412.
22 M. Avron and A. T. Jagendorf, *J. Biol. Chem.*, 234 (1959) 1315.
23 G. Hind and A. T. Jagendorf, *Z. Naturforsch.*, 18b (1963) 689.

Biochim. Biophys. Acta, 162 (1968) 380–391

Observations on the Reducing Side of the O_2-Evolving Photoact*

Bessel Kok, Shmuel Malkin, Olga Owens, and Biff Forbush**
Research Institute for Advanced Studies, Baltimore, Maryland 21227

Figure 1 shows the familiar scheme of photosynthesis. The left side shows a number of known chloroplast intermediates arranged according to their normal potential, indicative of their possible site of participation in electron transport. The intermediates marked with an asterisk occur in a concentration of one per 500 chlorophyll molecules, which correspond to the "photosynthetic unit," as originally revealed by the flashing light experiments of Emerson and Arnold.[1] The right side shows the two photoacts operating in series and interconnected by dark steps presumably coupled to phosphorylation. This presentation provides further support for the depicted interplay of two equivalent, series-connected photoreactions but strengthens our earlier doubts[2] concerning the locus of ATP formation in chloroplasts shown in Figure 1: between P, the photooxidant of system I, and Q, the photoreductant of system II.

In our present concept of photosystem I,[3] light quanta absorbed in a unit of a few hundred sensitized chlorophyll molecules flow to a small number (~20) of long-wavelength chlorophylls (C 700) and from these to a trapping center (P 700). Upon excitation an electron is transferred from P 700 (+0.43 V) to its reaction partner X which is a very strong reductant ($E_0' \leqslant -0.6$ V). Electron donors to photooxidized P 700 are plastocyanin (Cu) and cytochrome *f* (Fe).

During the last few years we have concentrated on photosystem II, which evolves oxygen and simultaneously generates reducing power of a medium potential (Q in Figure 1). So far, our observations have mainly been concerned with the "reducing end" of this photosystem, of which the fluorescence yield is a useful indicator. Of great value to us was *Scenedesmus* mutant No. 8,† which was kindly furnished by Dr. N. Bishop.[4]

In Figure 2 the fluorescence excitation spectra observed with suspensions of normal *Scenedesmus* cells and mutant cells of equal chlorophyll concentration are compared. Whereas the excitation spectrum measured with the normal cells shows a severe long-wavelength drop, especially in the presence of DCMU, such a drop is lacking in the mutant. Mutant No. 8 thus allows the observation of many system II characteristics without interference from system I.

---

*This work was supported in part by the National Institutes of Health, the Air Force Office of Scientific Research and School of Aerospace Medicine, the National Aeronautics and Space Administration, and the Charles F. Kettering Foundation.

**Present address: The Weizmann Institute, Rehovoth, Israel.

†Properties of mutant No. 8 are described by various authors in ref. 5. It lacks a long-wave absorption band C 700, photocatalyst P 700, and the narrow EPR signal. Whole cells are unable to reduce CO_2, and cell-free preparations are unable to photoreduce low potential oxidants.

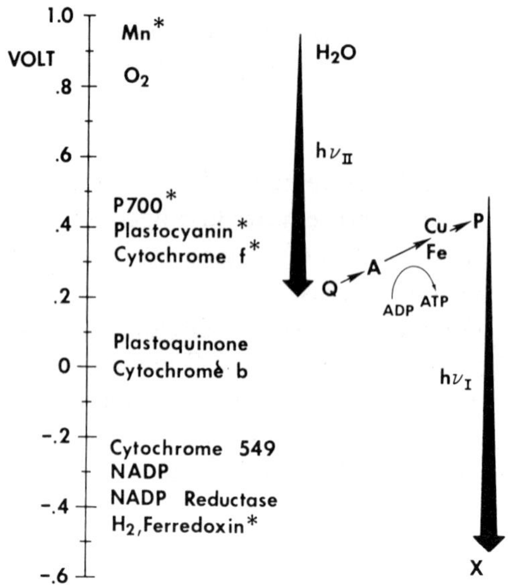

Figure 1. Chloroplast intermediates arranged according to their normal potentials and the series scheme of photosynthesis.

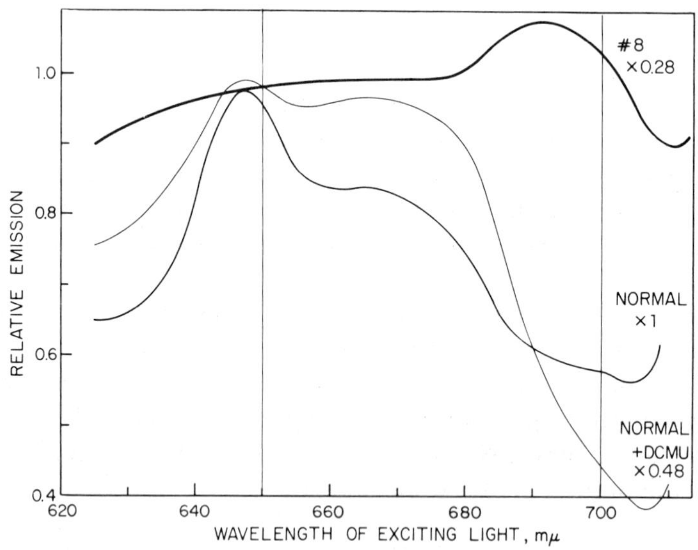

Figure 2. Fluoroscence (λ>730 mμ) excitation spectra of *Scenedesmus* mutant No. 8 and of normal *Scenedesmus* in presence and absence of DCMU.

THE FLUORESCENCE RISE CURVE

In the experiment of Figure 3 (full line) a suspension of dark-adapted or far-red illuminated spinach chloroplasts was placed in a fluorometer and cooled with liquid nitrogen. Blue light was switched on and the fluorescence emission (680 to 710 mμ) recorded as a function of time. Initially the emission is low (F_0), and it rises exponentially towards a high value (F_{max}). This increase of the fluorescence yield during il-

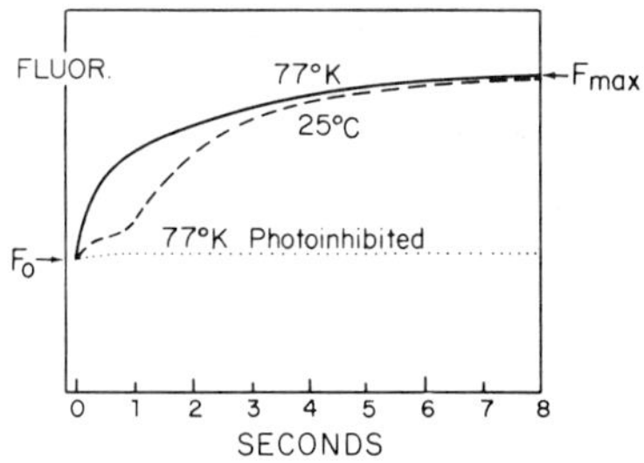

Figure 3. Fluorescence rise curves observed upon illumination with the same intensity of green light (10 nanoeinstein cm^{-2} sec^{-1}). The chloroplast suspensions had been exposed at 25°C to 740-mμ light for 3 min.

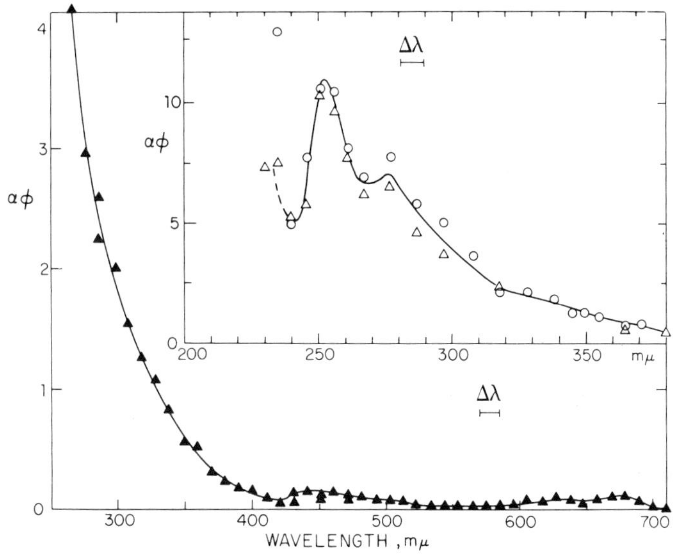

Figure 4. Action (αφ) spectrum of photoinhibition.[7]

lumination reflects the photoconversion of quenching centers (denoted Q after Duysens[6]). These centers collect light energy of photosystem II and are at the same time converted into a nonquenching state, unable to trap excitation. Once the traps are converted, the sensitizing pigment no longer finds an outlet for absorbed photons, and an increase of its fluorescence yield results. The fluorescence yield at any moment thus reflects the redox state of Q.

The dotted curve in Figure 3 shows that in chloroplasts which have been exposed to UV light, so that photosynthesis is inhibited, the fluorescence yield does not increase above the initial value F_0. Figure 4 shows measurements recently made by Jones.[7] Plotted is the action spectrum of this photoinhibition, which was shown to involve two distinct photoprocesses, one sensitized by visible, the other by ultraviolet light. According to this action spectrum, the UV effect might well be the photodestruction of a quinone, and the traps revealed in the experiment of Figure 3 could be tentatively identified with a quinonelike substance.

The dashed line in Figure 3 is a kinetic curve of fluorescence observed at room temperature with the same material and exciting light used for the low temperature

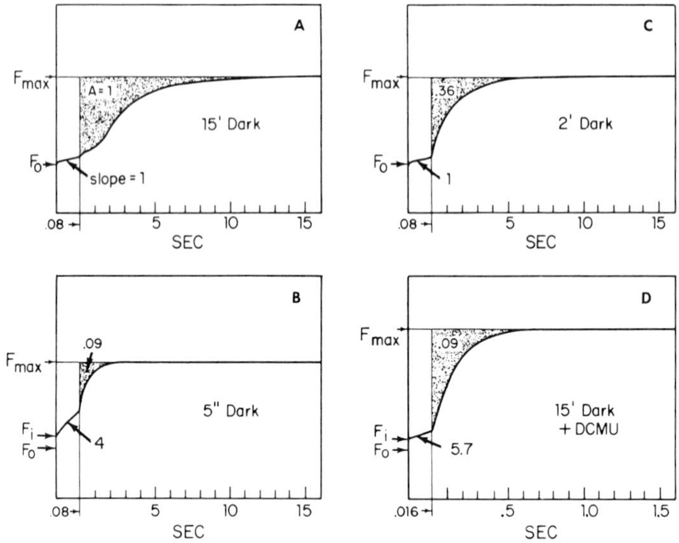

Figure 5. Rise curves of fluorescence observed at room temperature after indicated dark periods. Chloroplasts without added oxidant; illumination: green light 510 to 640 mμ. During the initial phase of the rise curve the oscilloscope time base was accelerated. Note faster time base used for experiment *D* with slope and area values normalized to match the slower time base.

experiment. At 25°C the rise from F_0 to F_{max} is biphasic and reflects a larger pool of oxidant than the exponential low temperature curve.

The results of a similar experiment are plotted in Figure 5*A*. The oscilloscope time base was accelerated during the initial 80 msec after the shutter was opened to allow a better view of the initial value of the fluorescence (F_i). The time it took to bring the fluorescence from F_0 to F_{max} proved inversely proportional to the exciting intensity over a considerable range. This validity of the $I \times t$ law indicates a process with a constant quantum yield in which a definite amount of photoproduct is formed. The shaded area in Figure 5*A* is a measure of this amount of photoproduct, which is equivalent to the pool of internal chloroplast oxidants. If a chemical reductant (e.g., hydrosulfite) is added to the chloroplasts before the exposure, the fluorescence is high ($F_i = F_{max}$) from the start of illumination and no additional rise occurs. This is one of several indications that a photoreduction is involved:

$$Q \xrightarrow{h\nu} Q^- .$$

If, after F_{max} is attained, the light is shut off for a few seconds and the experiment repeated, rise curve 5*B* is observed; a longer dark time between exposures yields curve 5*C*. Only after 10 to 20 min of darkness is the restoration complete and curve 5*A* again observed. Note in Figure 5*C* that the initial fluorescence yield F_i, which reflects the redox state of Q at the end of the dark period, is back to its minimum value F_0 long before the area is restored (the curve rises fast to F_{max}). To explain this, we assume a secondary pool of internal chloroplast oxidant(s) which we denote A. Pool A reoxidizes Q^- in a dark step so that in the light F_{max} is attained only when both Q and A are completely reduced. The area of rise curve 5*A* thus reflects both pools Q and A. In the dark, Q^- is reoxidized ($F_i \rightarrow F_0$) before a signifi-

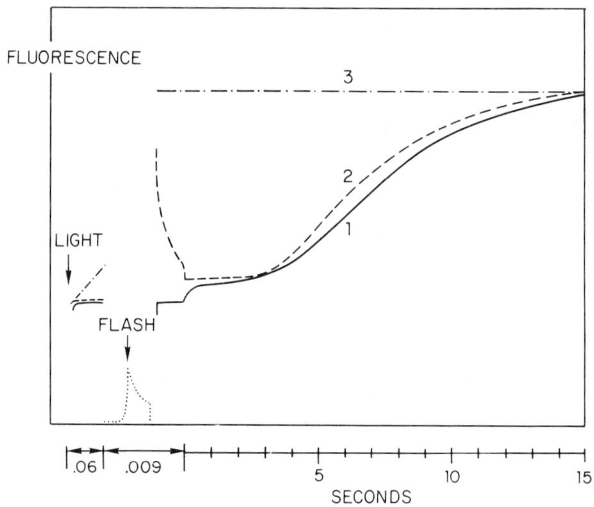

Figure 6. Effect of a brief flash upon the fluorescence induction. (1) Control experiment as in Figure 5*A*; (2) flash given at indicated moment; (3) repeated in the presence of DCMU. Curves recorded with three-phase time base as indicated.

cant amount of A is reoxidized. In the presence of excess external oxidants such as ferricyanide, the fluorescence never attains F_{max}. These agents keep Q in its oxidized form, either directly (e.g., in mutant No. 8) or via A and photosystem I (P 700^+ oxidizes A^-, see Figure 1). This explains why in far-red light, which sensitizes system I but not system II, restoration of the complete rise curve proceeds many times faster than in darkness. This oxidation of Q by far-red light does not occur in mutant No. 8. In normal cells or chloroplasts the reoxidation is inhibited by DCMU (see Duysens[6]).

It proved possible[3,8] to determine the following: (a) The total amount of internal system II oxidant in the chloroplast. The concentration of Q+A in different materials was between 1/20 and 1/60 the total chlorophyll concentration. The slow rise ("long induction") of the fluorescence curve thus is a mirror image of the O_2 gush (1 O_2/150 Chl $\simeq$1 eq/40 Chl)[9,10] observed at the onset of illumination, pool Q^-+A^- being the reduced end products of this reaction. (b) The quantum yield of the photoreduction of Q and A: ~2 quanta of green, yellow, or red light were needed to ph oreduce one equivalent of oxidant. (c) The quantum yield of the photooxidation of Q^- and A^- by far-red light: a value close to unity was found.

These quantum efficiencies support the series scheme of photosynthesis if it is assumed that in long-wavelength light all absorbed quanta drive photosystem I and act with a quantum yield of one, whereas in short-wavelength light only half the absorbed quanta reach system II and they also act with a quantum yield of unity.

The rise curve of Figure 5*D* was observed with dark-adapted chloroplasts in the presence of DCMU. The fluorescence rises very fast, indicating an area ~1/10 the area of the normal curve: a pool of only 1 eq per 500 Chl.[cf.11] According to Morin[12] such a small pool is also seen if the exciting light is extremely strong, regardless of the presence of DCMU.

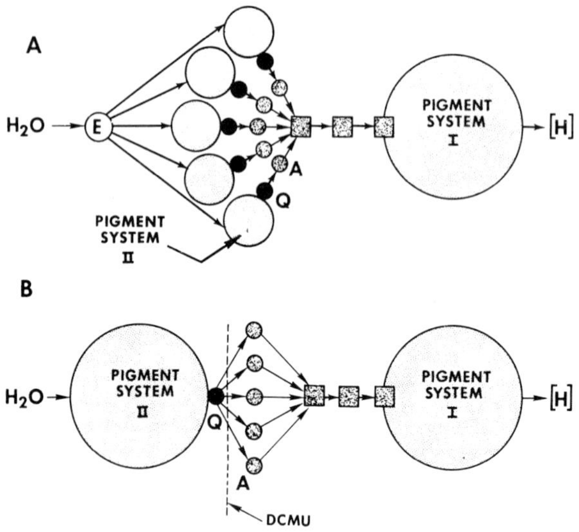

Figure 7. Two schemes to explain fluorescence rise curves (see text for details). Dotted spheres indicate pigment units and squares, oxidized products of system I. Arrows indicate direction of electron flow.

Figure 6 shows further experiments pertaining to this small pool. Kinetic fluorescence curves similar to the one shown in Figure 5*A* were recorded on an oscilloscope screen. A three-phase time base was used as indicated. In experiment 2, after the initial low fluorescence F_0 was seen, a strong, $\leqslant$1-msec light flash was given. Rotating discs, one in front of the flash tube and one in front of the photomultiplier, were used to cut off the tail of the flash and to avoid its being "seen" by the photocathode. Immediately after the disc shutter is opened ($\sim$1 msec after the flash) the fluorescence is high, but it decays within a few msec to a value slightly above F_0. The right part of the trace, recorded with a slow time base, reveals the slow filling of pools Q and A by the weak exciting light beam. Comparison with the curve 1 (Figure 6), from a parallel experiment made without a flash, shows that the slow part of the rise curve was not changed much by the flash. Obviously a single short flash cannot fill more than 10% of the total Q+A pool. Curve 3 is from a parallel experiment made in the presence of DCMU. A brief flash again raises the fluorescence to its maximum value, but no subsequent return to the low yield occurs.

The presence of such a small pool in photosystem II (i.e., between O_2 and intermediate A) is well known: in a single brief flash no more than 1 O_2 is liberated per 2000 chlorophylls (1 eq/$\sim$500 Chl).

At present it is not easy to devise a scheme which encompasses all the discussed observations. Two attempts are shown in Figure 7. Scheme *A* (Malkin[13]) assumes the small pool (E) to be at the oxidizing side of photoact II and the number of traps Q to be large (1 per $\sim$80 Chl), in fact equal to the number of A molecules. The detailed model further assumes separate and linear chains in which cross-transfer of photons between pigment units and of electrons between the Q's and A's of different units is forbidden. Consequently the reaction

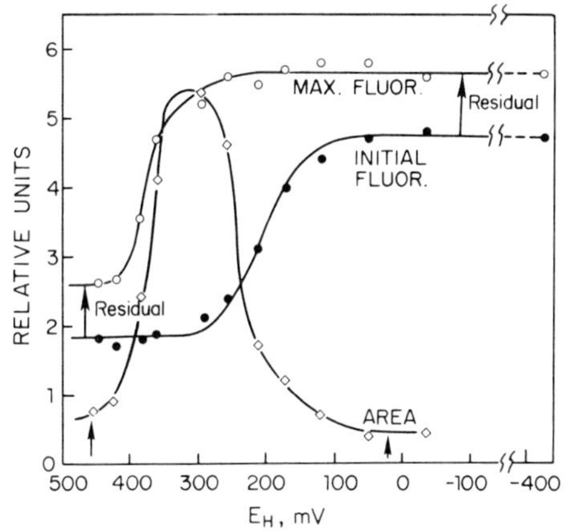

Figure 8. Effect of redox potential on initial fluorescence yield (●), maximum fluorescence yield (○), and area bounded by the rise curve (◊). Cell-free particles of *Scenedesmus* mutant No. 8 suspended in 0.05 *M* phosphate buffer, *p*H 7.5, were equilibrated in dark at indicated potential for 15 min before a rise curve in green light was recorded.

$$Q\,A \xrightarrow{h\nu_{II}} Q^{-}A \xrightarrow{k} Q\,A^{-} \xrightarrow{h\nu_{II}} Q^{-}A^{-}$$

is first order. This model describes the "long induction" (Figures 3 and 5*A*) in detail, the inflection of the rise curve reflecting dark reaction k (at 20°C, $k=40\ sec^{-1}$; at 0°C, $k \cong 2\ sec^{-1}$; at -196°C, $k=0$). The scheme needs additional assumptions to explain the "short induction" (Figures 5*D* and 6).

Scheme *B* assumes that the pigment complex of system II has the same size as that of system I: the concentration of quencher Q being the small pool of 1/500 Chl while A accounts for nearly all of the large pool (Q+A).[see 9] It accounts for the "short induction," i.e., the quick dark return of F_i in Figures 5 and 6, if a reaction time of ~1 msec is assumed for k in the reactions

$$Q \xrightarrow{h\nu_{II}} Q^{-} \quad \text{and} \quad Q^{-}+A \xrightarrow{k} Q+A^{-}.$$

Rate limitation by dark reaction k explains the fast rise of fluorescence in very strong light and in the presence of DCMU, if it is assumed that this poison blocks the reoxidation of Q^- by A. This scheme needs additional assumptions to explain the "long induction" and the biphasic course of the rise curve at room temperature (Figures 3 and 5*A*).

A detailed discussion of these hypotheses and additional pertinent data is beyond the scope of this paper. Although Malkin strongly prefers scheme *A*, the observations described in the following sections will be discussed within the framework of the simpler scheme *B*.

NORMAL POTENTIALS OF Q AND A COMPUTED FROM THE FLUORESCENCE YIELD

Isolated chloroplasts without added oxidants were equilibrated in darkness at defined redox potentials. Suddenly light was given and the fluorescence rise curve

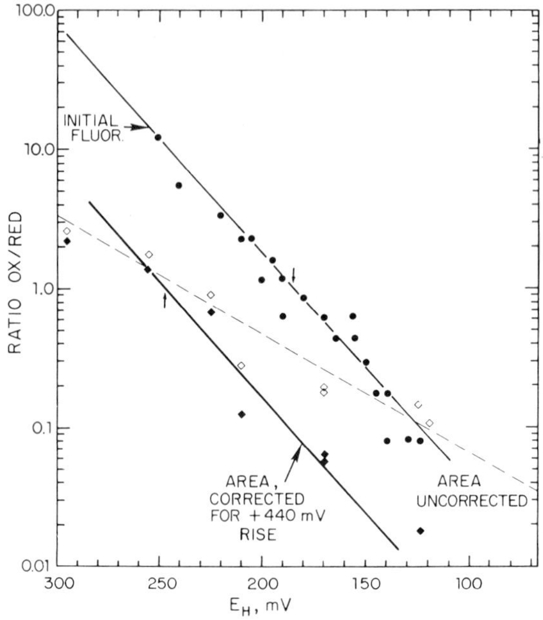

Figure 9. Data similar to those of Figure 8 plotted as ratio oxidized/reduced vs potential, midpoint potentials (ox/red = 1) indicated by arrows. ●, Initial fluorescence (Q); ◊ ---, area (A); ♦, area values corrected for "residual" rise curve observed at high and low potentials (cf. arrows in Figure 8).

recorded. The initial fluorescence yield (F_i, Figure 5) should reflect the dark equilibrated redox state of Q. In the simplest case, when F_i is halfway between F_0 and F_{max}, Q is half oxidized and half reduced. Such a titration of F_i, shown in Figure 8, reveals a midpoint potential of about 180 mV. This value is in good agreement with the one obtained earlier by an entirely different approach, the photoreduction of various oxidants by particles prepared from *Scenedesmus* mutant No. 8.[14] By the same reasoning, one can measure the area of the rise curve (large pool Q+A in Figure 7) as a function of potential and so titrate a midpoint potential of this pool. The results of such a titration, also shown in Figure 8, indicate a higher midpoint potential than that of the F_i titration, about 230 mV.

In Figure 9, similar data are plotted according to the Nernst equation. The F_i plot clearly shows a single electron transfer step ($Q \rightarrow Q^-$) with a midpoint potential of ~180 mV. This midpoint potential proved to be unaffected by *p*H (6 to 9), which further agrees with the expectation that the primary event of system II involves the photochemical transfer of an electron only. We should stress that our titration data reveal minimum values: if cross-transfer of photons between units occurred, fluorescence would not be a linear function of the redox state of Q; therefore, the observed midpoint potential for Q would be too low. Application of Joliot's formula and cross-transfer probability factor yields a parallel displacement of the linear plots in Figure 9 by +20 mV and thus would lead to an E_0' value for Q of ~+200 mV.

Plots of area vs potential were more difficult to obtain and evaluate, especially in fresh whole chloroplasts. In several cases the application of a correction, which will be discussed below, yielded a quite reasonable fit with the Nernst equation. The values in Figure 9, obtained in this way, indicate a single electron transfer and a midpoint potential for intermediate A of $\sim +240$ mV. The available data show no significant variation of the maximum area (size of pool A) and of the midpoint potential between *p*H 5.5 and 8.5. The ~40-mV difference between their normal potential values signifies an equilibrium constant of ~5 for the reaction

$$Q^- + A \rightarrow Q + A^-$$

RESIDUAL PHOTOPROCESS(ES)

The arrows in Figure 8 indicate the correction applied in Figure 9 to the titration of area (component A) for the conversion of the dashed line into the solid line: at potentials on both the oxidized and the reduced side of the titration curve there still occurs a rise of fluorescence when the light is switched on. For our correction these "high" and "low" potential rises were assumed to be one and the same phenomenon: another trapping process, unaffected by redox potential in the range between $+500$ and <0 mV. This assumption would imply an alternative interpretation for the first phase of the biphasic (25°C) rise curve shown in Figures 3 and 5*A*. At this time we can offer only a preliminary evaluation of this "residual" fluorescence, which in some materials accounts for a considerable fraction (up to 50%) of the total emission.

The "high" and "low" potential rises have several features in common, such as their respective sizes and the dark restoration kinetics [a fast partial restoration followed by a slow (~40 sec) phase]. To account for some other aspects – such as the response to intensity – the assumption of a single process may be insufficient. Because of its occurrence in *Scenedesmus* mutant No. 8, although to a relatively smaller extent, this residual photoprocess appears distinct from system II (as described in Figure 7) as well as from system I. Possibly it is related to the "photoactivation" of system II.[9,14,17]

KINETICS OF DPIP REDUCTION AND OXIDATION

Figure 10 shows a different approach to the analysis of the "reducing end" of photosystem II and its interrelation with the "oxidizing end" of system I. The reduction of dichlorophenolindophenol by chloroplasts is viewed during and after a brief, intense light flash in the manner of Witt and Müller.[18] The upper figure shows that the flash very rapidly reduces a certain amount of dye to level D_1. However, most of this reduced dye is reoxidized within a few milliseconds; thereafter a second, slow, wave of dye reduction is observed which reaches a final level D_2, about half the level D_1. The lower figure shows that, if the same experiment is performed with spinach chloroplasts in the presence of 10 μM DCMU, only the first fast reduction is seen, followed by the return reaction. With cell-free particles of *Scenedesmus* mutant No. 8, the fast spike is completely absent and only the slow, secondary rise is observed. If DCMU is added to mutant particles, no DPIP reduction is seen at all. The interpretation of this experiment is straightforward: the photoreductants of both photoacts I and II reduce DPIP. Strong reductant X (see Figure 1) reduces the dye with ex-

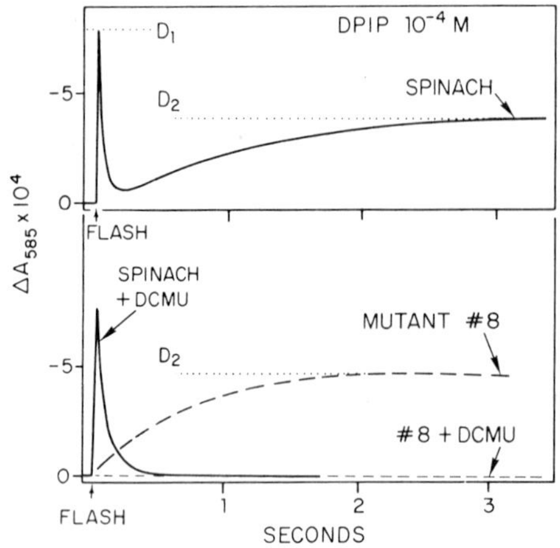

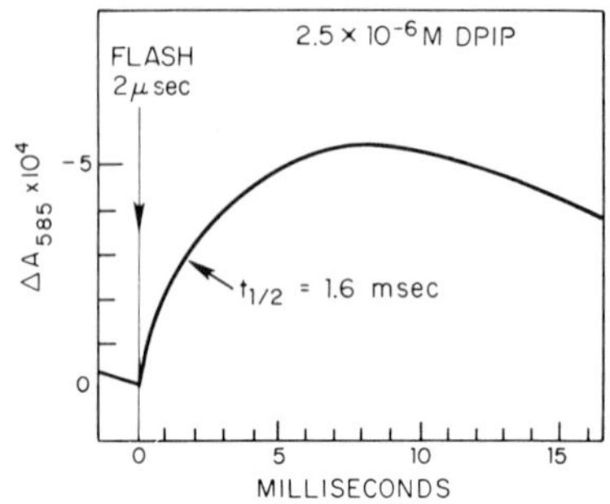

Figure 10 (Left). Flash-induced reduction of DPIP by spinach chloroplasts and particles of mutant No. 8 in presence and absence of DCMU. Data obtained by repetitive observation and signal averaging.

Figure 11 (Right). Detailed observation of the fast reduction of DPIP by the photoreductant of system I. Repetitive observation with spinach chloroplasts.

treme rapidity, but the reduced indophenol reacts back with photooxidized P 700 or cytochrome *f*. System II also reduces dye, but this process is much slower with a half-time of several seconds. Also, since $D_2 \cong \frac{1}{2} D_1$, it appears that the yield per flash in system II is only about half that in system I. However, this difference may well be an artifact found only in isolated chloroplasts (see also Witt et al.[19, Fig. 25]).

Figure 11 shows the results of an experiment on the fast photoreduction by photosystem I. With a DPIP concentration of 2.5 μM, the reaction was half complete 1.6 msec after the 2-μsec flash. Similar experiments in which the DPIP concentration was varied showed the reaction

$$X^- + DPIP \xrightarrow{k_5} DPIP^- + X$$

(see Figure 13) to depend linearly on DPIP concentration. From this a half-time of 40 μsec was extrapolated for a concentration of 0.1 mM DPIP. Similarly the rate of reaction k_6 (the quick return reaction from level D_1 in Figure 10) depended linearly upon the concentration of DPIPH. For instance, with 0.8 mM dye and an excess of ascorbate, the half-time was 2 msec.

Further kinetic analyses were carried out on the reduction and reoxidation of DPIP by experiments such as that of Figure 12, which shows the time course of DPIP reduction in continuous strong light with particles of mutant No. 8. At the onset of illumination there is an immediate reduction of DPIP, but upon darkening the rate suddenly becomes much lower and decreases further until a final level of reduced DPIP is attained. The break of the slope upon darkening indicates a sizable pool of reducing power which survives after illumination and reacts slowly with DPIP. This pool amounted to ~1 eq/70 Chl, and we identify it with A in scheme *B* of Figure 7. This indirect reduction of DPIP by A is also seen in Figure 10 (D_2). During the light, DPIP reacts more rapidly with another intermediate, most likely

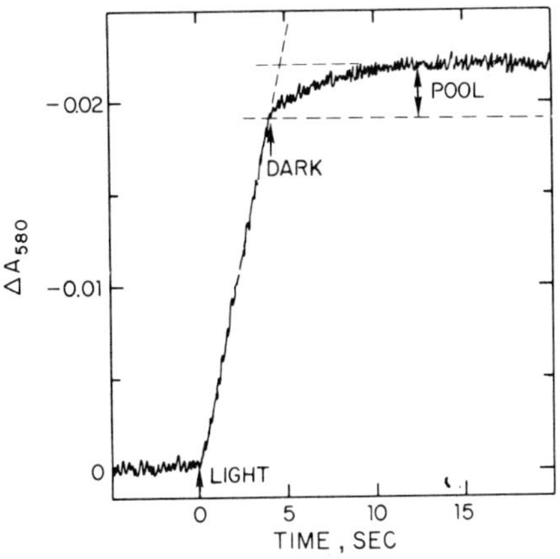

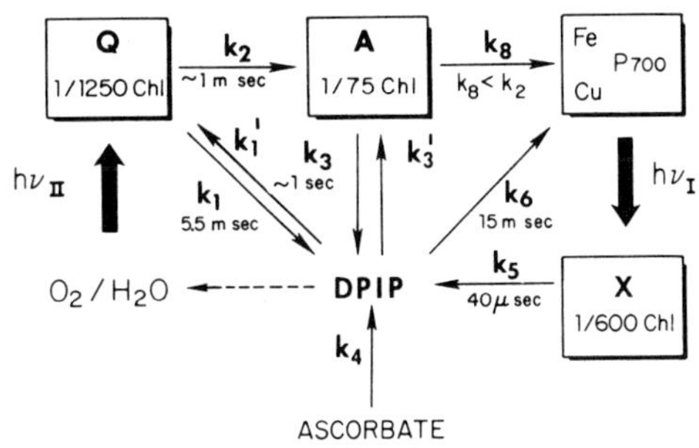

Figure 12 (Left). Time course of the reduction of DPIP by particles of mutant No. 8 during and after a brief period of illumination with a saturating intensity.

Figure 13 (Right). Reaction rates of oxidized or reduced DPIP with various endogenous chloroplast oxidants and reductants, normalized as first-order half-times for 0.1 m*M* DPIP or DPIPH. The DCMU-sensitive reaction between Q and A (k_2) is viewed in the experiment of Figure 6 as the quick return of the flash-induced high fluorescence yield.

Q in scheme *B*. Without further experimental details, our present picture of the various reactions of DPIP in the chloroplast system is summarized in Figure 13. All time constants are normalized as first-order half-times for 0.1 m*M* DPIP or $DPIPH_2$. The data proved consistent with the assumption of a small primary pool of photoreductant in system II (Q) which feeds into secondary large pool A, which in turn feeds into the oxidants of photosystem I (scheme *B*).

Note that DPIP can be reduced at three places: extremely rapidly by the reductant of photosystem I (k_5; 40 μsec); much more slowly by the primary reductant of system II (k_1; a few msec); and very slowly by the large secondary pool between the two photoacts (k_3; about 1 sec).

Conversely, $DPIPH_2$ can also be oxidized at several places: by system I (k_6; 15 msec), and by Q and A. We have no direct observations yet concerning rates k_1' and k_3', but the normal potential values of Q and A reported in the preceding section predict that k_1/k_1' and k_3/k_3' are not very far from unity. In any case, entry at A would be extremely slow, whereas entry at Q could be of the order of 10 msec. No evidence has been found for entry of DPIPH at the oxidizing end of system II.

A REMARK ON THE SITE OF PHOTOPHOSPHORYLATION

The data presented in Figures 9 and 10 argue against the possibility that in chloroplasts ATP formation is coupled to the dark step(s) between the two photoacts. (a) Based on the currently accepted values for the free energy of ATP, the potential gaps between intermediate A and plastocyanin or cytochrome *f* (~140 mV) and between the latter compounds and P 700 (~60 mV) are insufficient. (b) The evidence is that in cyclic phosphorylation with DPIP only, or the noncyclic process in the presence of NADP, DCMU, or $DPIPH_2$, the latter agent donates electrons

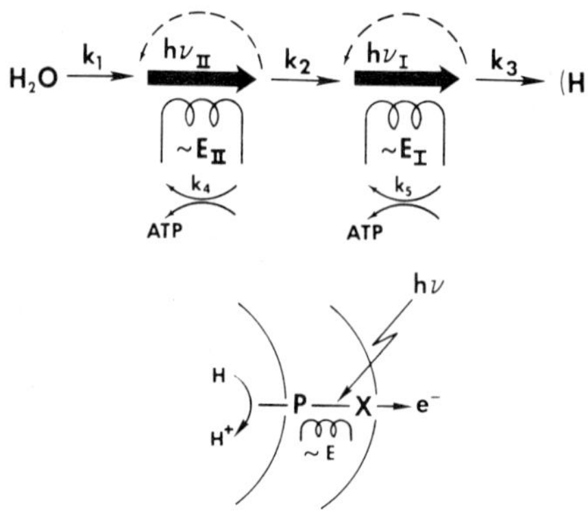

Figure 14. Electron flow path in two "series"-connected photoacts incorporating "parallel"-connected sites of photophosphorylation. Coils indicate high energy state of chloroplast matrix. Lower figure also incorporates proton accumulation within membrane; P and X symbolize reaction partners of a photoact.

directly to cytochrome *f* or plastocyanin: whereas at 0.5 m*M* DPIPH (optimal in the donor system[20,21]) reaction k_6 (Figure 13) takes ~3 msec, k_8 is reported to take ~10 msec even under conditions such that pool A might be assumed to be fully reduced.[17] Furthermore, the fluorescence data (Figures 5 and 6) lead to the conclusion that DCMU blocks reactions k_1 and k_2. This would imply that entry of DPIP into system I is via reaction k_6 rather than via the sequence k_1, k_2, k_8, which might have included a phosphorylation site.

The rigid coupling between electron transport and phosphorylation and the constant P/2*e* ratios, observed especially in weak light, suggest a stoichiometric association between the two processes. An oxidative site, analogous to the mitochondrial ones, in series with the photoacts, as originally suggested by Hill and Bendall,[22] would fill this requirement. However, none of the three possible loci (k_1, k_2, k_3) shown in the scheme of Figure 14, dark steps on either end or in the middle of the electron flow path, appears to fit the evidence. The difficulties would be compounded if one were to accommodate two sites – or a P/2*e* ratio >1 (Izawa et al.[23]). Several presentations in this meeting, such as those of Jagendorf and Mayne, appear to allow an alternative view: that photophosphorylation occurs in *parallel* rather than in *series* with one or with both photoacts, indicated as $\sim E_I$ and $\sim E_{II}$ in Figure 14.

It is assumed that a photoact, besides transferring an electron against the thermodynamic gradient, simultaneously generates a high energy state. A subsequent photoact cannot occur unless this high energy state is released, which imposes a rate limitation. The slow spontaneous release is accelerated by ADP + P_i or uncouplers (k_4, k_5). Such a parallel mechanism thus can account for the stoichiometry and coupling observed in open phosphorylations. At the same time it can explain ATP generation in case a photoact is completely short-circuited in terms of electron flow (as can be the case with DPIP or PMS, cf. dashed arrows in Figure 14). The crude scheme needs to be filled out in more detail to account for the reported interrelations between photochemical electron transfer, matrix changes, proton exchange, luminescence, and phosphorylation.

In the context of the Mitchell[24] theory, this high energy state would be a proton gradient, formed when electrons are moved across the chloroplast membrane. An alternative and perhaps more attractive possibility is the direct photochemical induction of a high energy state (~E), such as a conformation change of the chloroplast matrix.

REFERENCES

1. Emerson, R. and Arnold, W., *J. Gen. Physiol.* **15**, 391 (1932).
2. Kok, B., Cooper, B., and Yang, L., in *Studies on Microalgae and Photosynthetic Bacteria*, Special Issue of *Plant Cell Physiol. Tokyo*, p. 373, Univ. of Tokyo Press, 1963.
3. Kok, B., in *Currents In Photosynthesis*, p. 383, J.B. Thomas and J.C. Goedheer, Editors, Ad. Donker, Rotterdam, 1966.
4. Bishop, N., *Record Chem. Progr. Kresge-Hooker Sci. Lib.* **25**, 181 (1964).
5. Kok, B., in *Photosynthetic Mechanisms of Green Plants, Natl. Acad. Sci.–Natl. Res. Council Publ.* **1145**, 45 (1963).
6. Duysens, L.N.M., *Progr. Biophys. Mol. Biol.* **14**, 1 (1964).
7. Jones, L.W. and Kok, B., *Plant Physiol.* **41**, 1037 (1966).
8. Malkin, S. and Kok, B., *Biochim. Biophys. Acta* **126**, 413 (1966).
9. Joliot, P., *Biochim. Biophys. Acta* **102**, 116 (1965).
10. de Kouchkovsky, Y., Thesis, University of Paris, 1963.
11. Kautsky, H., Appel, W., and Amann, H., *Biochem. Z.* **332**, 277 (1960).
12. Morin, P., *J. Chim. Phys.* **61**, 674 (1964).
13. Malkin, S., *Biochim. Biophys. Acta* **126**, 433 (1966).
14. Kok, B. and Datko, E.A., *Plant Physiol.* **40**, 1171 (1965).
15. Joliot, A. and Joliot, P., *Compt. Rend.* **258**, 4622 (1964).
16. Joliot, P., *Biochim. Biophys. Acta* **102**, 135 (1965).
17. Kok, B. and Cheniae, G.M., in *Current Topics in Bioenergetics*, p. 1, R. Sanadi, Editor, Academic Press, New York, 1966.
18. Witt, H.T. and Müller, A., *Z. Physik. Chem. Frankfurt* **21**, 1 (1959).
19. Witt, H.T., Rumberg, B., Schmidt-Mende, P., Siggel, U., Skerra, B., Vater, J., and Weikard, J., *Angew. Chem. Intern. Ed. Engl.* **4**, 799 (1965).
20. Keister, D.L., *J. Biol. Chem.* **240**, 2673 (1965).
21. Avron, M., *Biochim. Biophys. Acta* **17**, 430 (1964).
22. Hill, R. and Bendall, F., *Nature* **186**, 136 (1960).
23. Izawa, S., Winget, G.D., and Good, N.E., *Biochem. Biophys. Res. Commun.* **22**, 223 (1966).
24. Mitchell, P., *Chemiosmotic Coupling in Oxidative and Photosynthetic Phosphorylation*, Glynn Research Ltd., Bodmin, Cornwall, 1966.

DISCUSSION

Avron: Is it really necessary to put the "real Q," the one present at a ratio of one per 500 to 2000 chlorophylls, on the system I side of system II? It seems to me that the same results would be observed if one assumed that this Q is located on the oxygen site of system II. One should observe a high fluorescence yield independently of whether one blocks the transfer of an electron to the trap or one blocks the incoming electron into the chlorophyll.

Joliot: If one measures the steady state concentrations of oxidized Q and A in the light as a function of wavelength, one finds that these concentrations are maximum in system I light and minimum in system II light. This means that Q and A are located between systems I and II.

Avron: Izawa and Good clearly showed that only one DCMU per 2000 chlorophyll molecules was perfectly sufficient to block completely electron transfer in chloroplasts. This just about forces one to find a component present at such a low concentration which is essential for all electron flow in chloroplasts.

Kok: We did a "pool size" measurement with chloroplasts and found a ratio of 1/1250 for Q/chlorophyll. In chloroplasts we find too small a "unit." Witt cannot, with system II, reduce more than half the P 700, which shows the same phenomenon: somehow, in making chloroplasts, one may lose half of system II. On whole cells, we have no data.

JOLIOT: The pool Q or E is certainly larger in intact algae than in isolated chloroplasts.

KOK: I think the pool size is the same in both systems. If one adds DCMU to chloroplasts, one can still see a fluorescence rise curve. It is a "small" rise curve. It is pool Q, not A. This agrees with Duysens, that DCMU blocks between Q and A.

JOLIOT: But in the presence of DCMU, you observe only the activation reaction and not oxygen evolution.

Studies on the Function of Manganese in Photosynthesis*

G.M. Cheniae and I.F. Martin
Research Institute for Advanced Studies, Baltimore, Maryland 21227

INTRODUCTION

For some time, since the investigations of Pirson,[1] Kessler et al.,[2,3,5] and Eyster et al.,[4] the implication has been in the literature that manganese is in some way involved in photosynthesis or the Hill reaction. In most instances, past observations on the effects of manganese deficiency on these reactions have been made with tissues relatively aged or in many cases of uncertain age. Thus, the effects of Mn deficiency observed were possibly secondary rather than primary. To date there are no completely convincing data to indicate that manganese *directly* participates in a primary reaction of either photosynthesis or the Hill reaction. Nevertheless many schemes of photosynthesis dogmatically include manganese as a catalyst on the oxidant side of photosystem II. Only within the past few years has there been renewed interest in the direct participation, if any, of manganese in photosynthesis. The data presented briefly here represent some of our attempts to evaluate critically the function of manganese in photosynthesis, and to examine the manner in which manganese is bound within the chloroplast.

METHODS

The alga *Scenedesmus* D_3 has been used throughout, both for whole-cell studies and as a source of chloroplasts. Manganese deficiencies resulting in ~90% loss of growth or photosynthesis can be obtained within 48 hr when "Specpure" chemicals are used as the source of microelements. Unless otherwise stated, quantum yield measurements (relative and absolute) were made with an O_2 electrode. Absorption measurements were made with an Ulbricht integrating sphere. Radioactivity measurements of ^{54}Mn were made with a NaI scintillation well counter calibrated to isolate the ^{54}Mn γ emission.

RESULTS

Effect of Mn Deficiency on Photosynthesis, Whole-Cell and Chloroplast Hill Reaction, and Photosystem I Reactions

Deficiency of manganese results in effects dependent upon the level of deficiency. At low levels of deficiency the effect observed upon either photosynthesis or the

*This research was supported by the National Aeronautics and Space Administration (NASw-747).

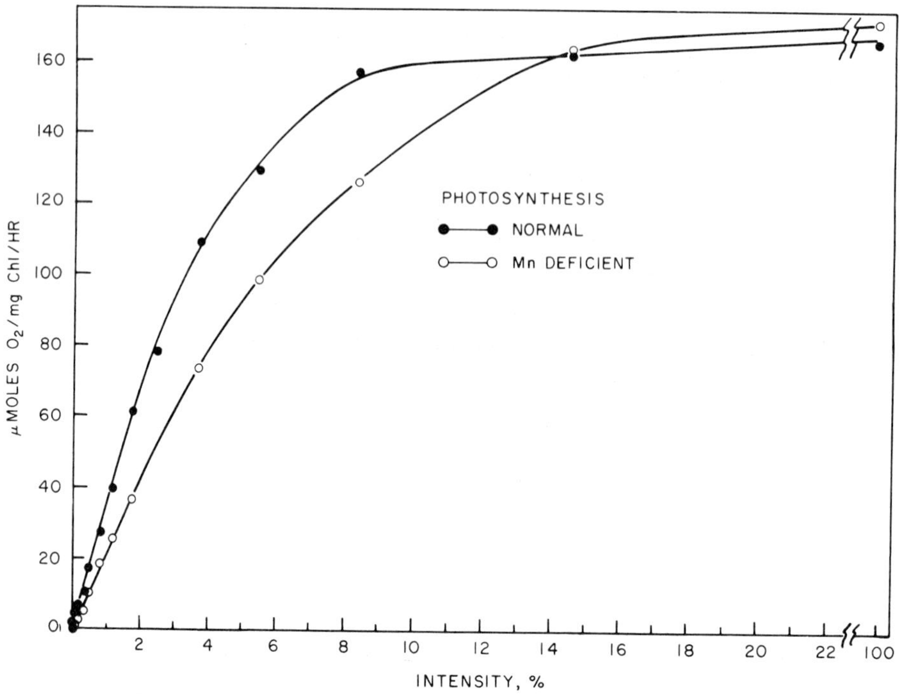

Figure 1. Effect of moderate Mn deficiency on photosynthesis. Photosynthesis measurements were made with a Clark O_2 electrode with a 0.5-mil Teflon type A membrane. *Scenedesmus* D_3 cells (48 hr culture) containing 20 μg total chlorophyll were illuminated at 25° in 0.1 *M* Na_2CO_3–0.1 *M* $NaHCO_3$, *p*H 9.4, saturated with 5% CO_2–air. Total volume of vessel was 1.13 ml. Irradiation was of wavelength >500 mμ. Full intensity corresponded to 4000 foot-candles. Values are corrected for dark respiration.

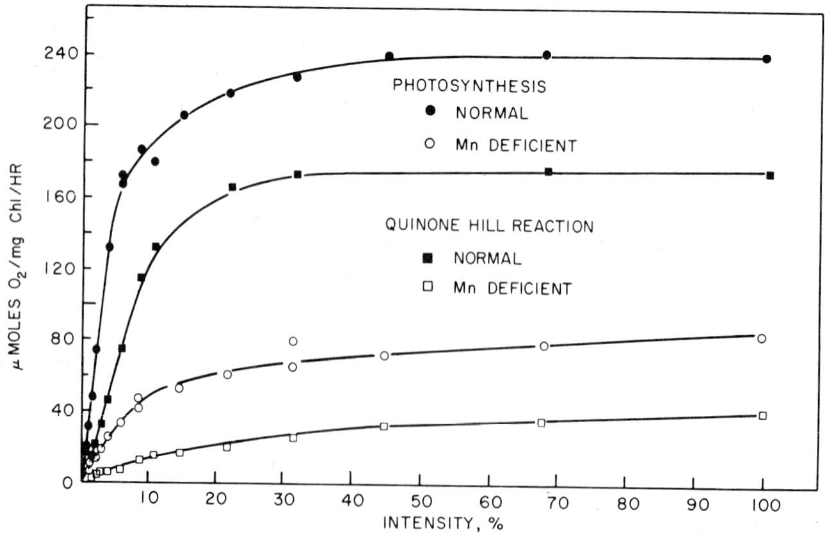

Figure 2. Effect of extreme Mn deficiency on photosynthesis and quinone Hill reaction. Conditions as for Figure 1 except that extremely Mn-deficient cells (48 hr culture) were used where noted. Quinone Hill reaction was run at *p*H 6.5 with freshly sublimed benzoquinone.

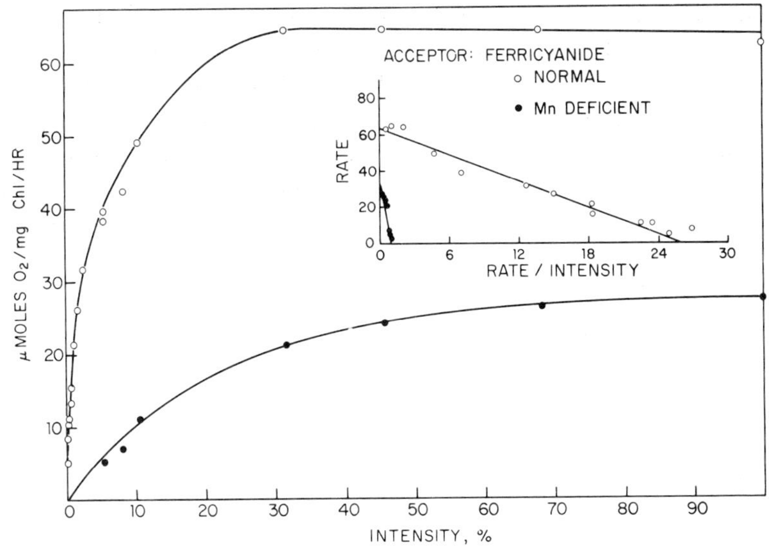

Figure 3. Hill reaction activity of chloroplast particles from cells used in experiments of Figure 2. Reaction vessel contained the following in μmoles: Tris-HCl, *p*H 7.2, 50; NaCl, 10; and FeCN, 1; in total volume of 1.13 ml containing 20 μg total chlorophyll. For other details see Figure 1.

quinone Hill reaction is upon K_1, the rate constant for the limiting light reaction[8] (Figure 1). With a high level of deficiency, still with 48-hr cells, two effects are observed both in whole cells (Figure 2) and in particles derived from the same culture (Figure 3). Manifestation of the effects of deficiency is now seen on both K_1 and K_d for the Hill reaction, with the same order of magnitude in particles as in whole cells. The primary effect of extreme deficiency observed by this type of kinetic analysis is, however, on the K_1 rate constant. Flashing-light kinetic analyses are needed to resolve completely the dilemma of whether or not the deficiency directly effects a light reaction. The fact that the absolute quantum yield per O_2 is 16 to 20 quanta (650 to 680 mμ) for the Hill reaction in normal particles gives some assurance that particle preparation is not creating artifacts that seriously jeopardize a kinetic analysis. In the Hill reaction, the same results are obtained whether one uses high potential acceptors such as FeCN, cytochrome *c*, or quinone or low potential acceptors such as viologen or NADP. Such an observation lends support to the concept that the effect of Mn deficiency is primarily upon photosystem II.[7,9]

This concept is further supported by data from relative quantum yield measurements with Mn-deficient or normal particles on reactions sensitized by photosystem I. With particles showing essentially no Hill activity, the photooxidation of reduced cytochrome *c* or the photoreduction of NADP mediated by ascorbate-DPIP shows quantum yields very similar to those with particles from nondeficient *Scenedesmus* (Figure 4). However, such experiments do not preclude the possibility that within photosystem I a site of manganese function might exist with an affinity constant manyfold greater than that within photosystem II. Since it is impossible to culture cells autotrophically in complete absence of manganese, we sought another means of eliminating any possibility of a function of manganese within photosystem I.

Chloroplast particles obtained from uniformly ^{54}Mn-labeled cells bind the incorporated manganese tenaciously. By several procedures, however, the simplest of which is a 50° heat treatment, it is possible to remove all manganese from the particle without loss of photosystem I activity. There seems to be no direct correlation between loss of Mn and loss of Hill activity as judged by the dissimilarity in their kinetics (Figure 5). It is clear, however, that photosystem I activity (photoox-

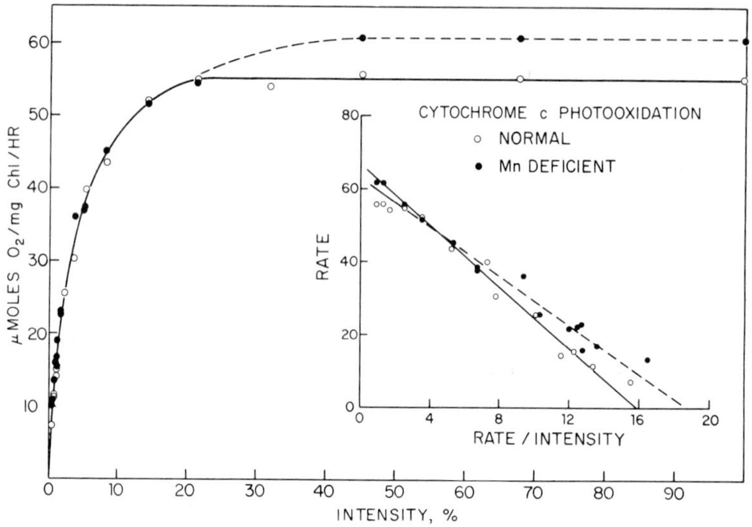

Figure 4. Reduced cytochrome *c* photooxidation with digitonin-treated particles from normal and extremely Mn-deficient algae used in experiments of Figure 2. The reaction mixture contained in μmoles: K phosphate buffer, *p*H 6.8, 10; $MgCl_2$, 7.5; NaCl, 17.5; methylviologen, 0.1; KCN, 0.2; plastocyanin, 0.005; and 1.0 mg cytochrome *c* (92% reduced) and 20 μg total chlorophyll in 1.13 ml. For other details see Figure 1.

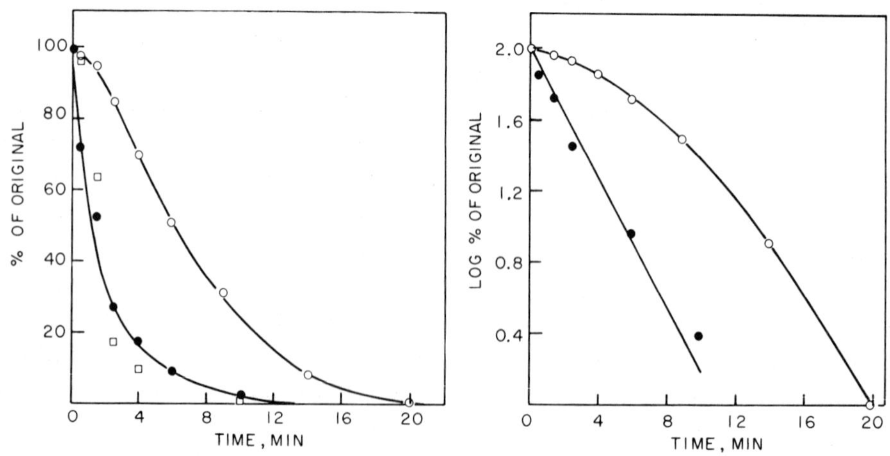

Figure 5. Effect of treatment at 50° on loss of Hill activity and ^{54}Mn-labeled particles from uniformly ^{54}Mn-labeled *Scenedesmus*. See Figure 3 for details. At indicated time, aliquots of particles were added to 0.02 *M* phosphate, *p*H 6.7, containing 1 m*M* versene at 4°, and particles were removed by centrifugation. ^{54}Mn was determined as described under Methods. ○, Particle-bound manganese; ●, Hill activity with ferricyanide; □, Hill activity with NADP.

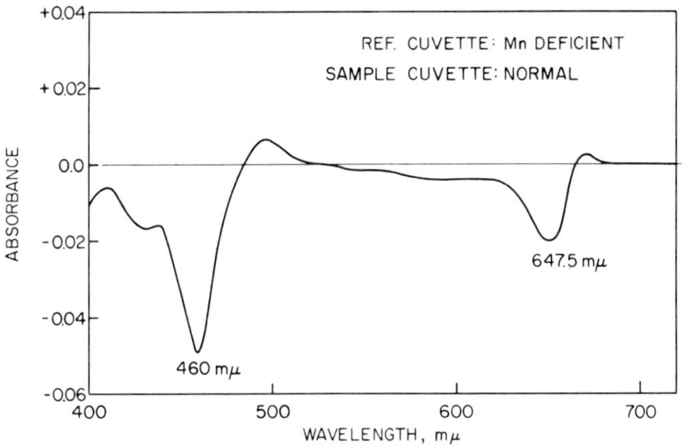

Figure 6. Difference spectrum of 80% acetone extract of particles from normal and extremely Mn-deficient *Scenedesmus*. Absorbance at 663 mμ in both sample and reference cuvettes was 0.615. Trace recorded on Cary Model 15.

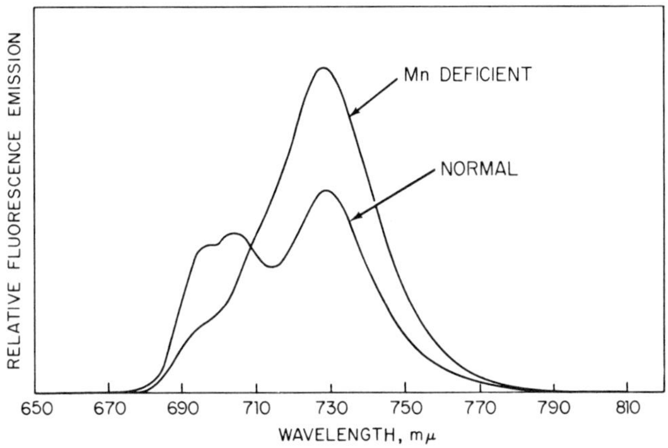

Figure 7. Relative fluorescence emission at −196° of normal and extremely Mn-deficient *Scenedesmus* D_3. Samples had equal amounts of total chlorophyll. Particles from such cells gave essentially the same results.

idation of reduced cytochrome *c*[10] or $DPIPH_2$ photoreduction of NADP[11]) is not influenced in either quantum yield or saturation rate by complete removal of chloroplast-bound manganese (Figure 4).

The effects of manganese deficiencies on photosynthesis or the Hill reaction are observed without a decrease in total chlorophyll *b* or carotenoids. In fact, in agreement with reports of Spencer and Possingham,[6] we consistently find in extremely Mn-deficient particles a small but significantly *greater* amount of chlorophyll *b* and possibly carotenoids than in normal particles (Figure 6). Conditions of deficiency affecting only the rate-limiting light reaction result in no change in the Chl *a*/Chl *b* ratio. These results indicate that the effect of deficiency most probably is not a gross alteration of pigment composition.

Low temperature (−196°) fluorescence emission[12,13] analyses do, however, reveal striking differences between the deficient and normal particles (Figure 7).

Both the F_{685} and F_{698} bands are diminished, whereas the F_{730} band is increased in comparison with these emissions in normal particles. Upon restoration of Hill activity in deficient algae by addition of manganese, the relative emissions of F_{685}, F_{698}, and F_{730} are restored to their "normal" ratios. More data are needed, however, to establish clearly that these effects on fluorescence emission are specifically and directly related to manganese and not simply a result of differences in particle size.

Restoration of Hill Activity in Deficient Cultures by Light and Mn^{2+} in the Absence of Protein and Net Chlorophyll Synthesis

Many workers in the past have observed restoration of photosynthesis or Hill reaction activity in Mn-deficient algae by addition of manganese salts to whole cells. Such experiments do not indicate whether the restorative effect is direct by binding of manganese to an "apoenzyme" or indirect by an increase in general cellular material syntheses dependent upon manganese.

To delineate some of these possibilities, restoration experiments were made under conditions such that protein synthesis was severely limited (~90%) by cycloheximide (Actidione).[14,16] The results obtained were independent of the time (15 min to ~4 hr) needed for complete restoration. In addition to inhibiting the incorporation of ^{14}C-phenylalanine into TCA-precipitable protein, this inhibitor also effectively blocks net chlorophyll synthesis[17] (Figure 8). Despite these limitations upon the cell,

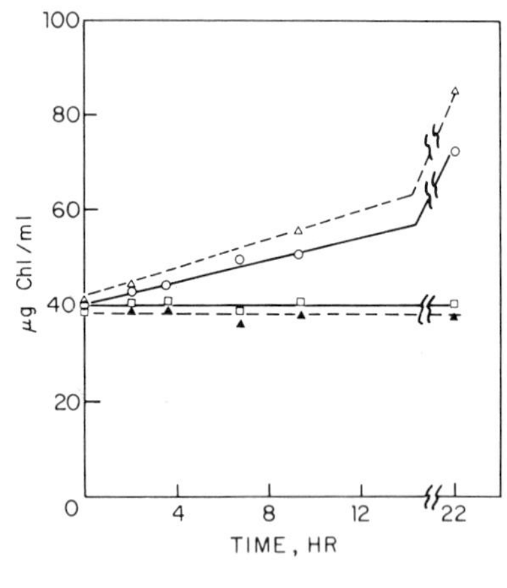

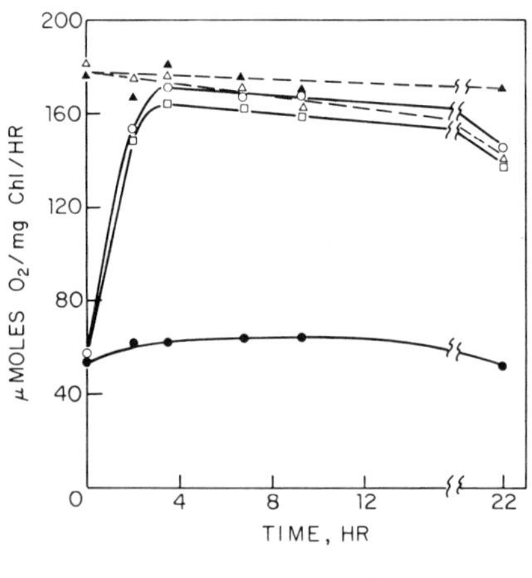

Figure 8 (Left). Course of chlorophyll synthesis during reactivation of Mn-deficient *Scenedesmus*. △, Illuminated, nondeficient; ▲, illuminated nondeficient cells with cycloheximide (5 μg/ml) added; □, illuminated deficient cells with cycloheximide (5μg/ml) and Mn^{2+} added, or deficient cells in darkness with Mn^{2+}; ○, illuminated deficient cells with Mn^{2+} added.

Figure 9 (Right). Reactivation of quinone Hill reaction by light and Mn^{2+} in absence of net chlorophyll and protein synthesis. ●, Deficient cells illuminated in absence of Mn^{2+}, or with Mn^{2+} and without illumination; ○ deficient cells illuminated, Mn^{2+} added; □, deficient cells illuminated, Mn^{2+} and cycloheximide (5 μg/ml) added; ▲, nondeficient cells; △, nondeficient cells with cycloheximide. Cells were suspended in growth medium containing all trace elements including Co^{2+} and VO_3^-. At zero time Mn^{2+} ("Specpure," 5 μM) was added. At indicated times, aliquots were removed and assayed for quinone Hill reaction, total chlorophyll, and incorporation of ^{14}C-phenylalanine into TCA-precipitable protein. Protein synthesis was inhibited ~90% by cycloheximide throughout the course of the experiment.

no inhibitory effect on restoration of the Hill reaction is observed (Figure 9). These observations lend support to the hypothesis of a direct function of manganese in photosynthesis. However, until *in vitro* restoration of Hill activity can be shown, the question remains unresolved.

The requirement of light for restoration (Figure 9), which under the experimental conditions employed is always observed, is in conflict with the report of Arnon.[18] This light requirement has not been investigated in detail, but exposure time seems to be more important than intensity. The light requirement cannot be explained simply as causing a large increase in the flux of Mn^{2+} ions in the cell. It could possibly be related to (a) oxidation of Mn^{2+} to an "active" oxidation state or (b) a configurational change of a manganese apoenzyme enabling it to bind its manganese coenzyme. At this time no data are available to exclude these or other speculative hypotheses.

Nature of Chloroplast-Bound Manganese

The literature contains speculations concerning the manner in which manganese is bound within the chloroplasts. Suggestions have been made that manganese is bound within a "special chlorophyll" or a "special cytochrome." To date, however, there is little information to assess these hypotheses, nor is there agreement concerning the amount of manganese per unit of chlorophyll. If manganese indeed is a functional component within photosystem II, additional data are needed on (1) nature of binding, (2) amount, and (3) valency state(s) of the metal ion. With use of the ^{54}Mn-chloroplasts from uniformly labeled *Scenedesmus*, attempts have been made to answer some of these questions.

The data of Table 1 represent the distribution of ^{54}Mn within fractions of particles that had been dialyzed overnight against deionized water. Extractions at −15° (to minimize degradation of a possible Mn-chlorophyll) with solvents to remove lipid revealed only a very small percentage of the total radioactivity associated with the lipid-chlorophyll fraction.[19] Partitioning of the lipid phase with water resulted in complete loss of all ^{54}Mn into the aqueous phase, which indicated no association of Mn within a chlorophyll molecule. The residue from lipid extraction contained virtually all the manganese. This fraction undoubtedly is not homogeneous, but because of the high percentage of nitrogen in the dry weight we have labeled it protein.

Table 1

Distribution of ^{54}Mn in Fractions From Chloroplast Particles From Uniformly ^{54}Mn-Labeled *Scenedesmus*

		% of dry weight	
	Total Mn, cpm	Chlorophyll	Nitrogen
Chloroplast suspension	372,400	6.3	10.7
Protein fraction	352,000	0.0	16.5
Lipid fraction	890	22.0	1.1

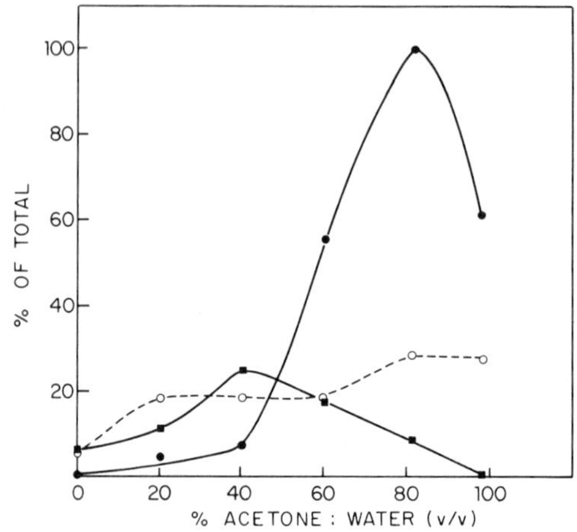

Figure 10. Effect of acetone on extraction of chlorophyll and ^{54}Mn from particles obtained from uniformly ^{54}Mn-labeled *Scenedesmus.* ●, Chlorophyll; ■, ^{54}Mn in acetone extracts; ○, ^{54}Mn extractable by buffer following acetone extraction.

Additional evidence against the hypothesis of a manganese-chlorophyll is shown in Figure 10. Here it is shown that with increasing acetone ($-15°$) concentration it is possible to extract all chlorophyll without extracting any appreciable amount of manganese. Thus, although it seems clear that there is no mangano-chlorophyll, the possibility of a protein-Mn-chlorophyll complex is not excluded.

The amounts of manganese per chlorophyll given by our data center around the number 46 moles chlorophyll per gram-atom manganese, a value near the mean of values previously reported in the literature. Curiously, this ratio of Chl/Mn approximates the pool size responsible for the O_2 gush[20] observed by many workers, i.e., $\sim 1\ O_2/150$ Chl[21,22] or ~ 1 eq/40 Chl. This ratio of Chl/Mn is found regardless of the chemical composition of the buffer used in the isolation of particles, and, more importantly, independent of the total phosphate concentration in the particle. This latter observation is considered pertinent since mitochondria, chromatophores, and chloroplasts are known to precipitate certain cations as their phosphates. This, of course, would lead to erroneously high values for the Chl/Mn ratio. Furthermore, if insoluble manganese phosphates constituted a sizable portion of the total chloroplast manganese pool, then breakage of the chloroplast by sonication and subsequent centrifugations would lead to separation of the more dense insoluble phosphates from the less dense lipid-protein material. Our data do not indicate this, but do indicate that sonication alone is insufficient to break the manganese bonding.

Manganese binding in chloroplasts of *Scenedesmus* has been investigated with respect to parameters such as *p*H, ionic strength, and effects of detergents, protein denaturants, and chelating agents. Such information was considered necessary for work towards isolation of the manganese carrier.

At *p*H values of 6.3 to 6.5 (Figure 11) the manganese associated with the particles is not easily removed by washing or dialysis. On either side of this narrow *p*H range, however, the binding of manganese is decreased considerably. The manganese released by adverse *p*H is not precipitable by $(NH_4)_2SO_4$ and is dialyzable,

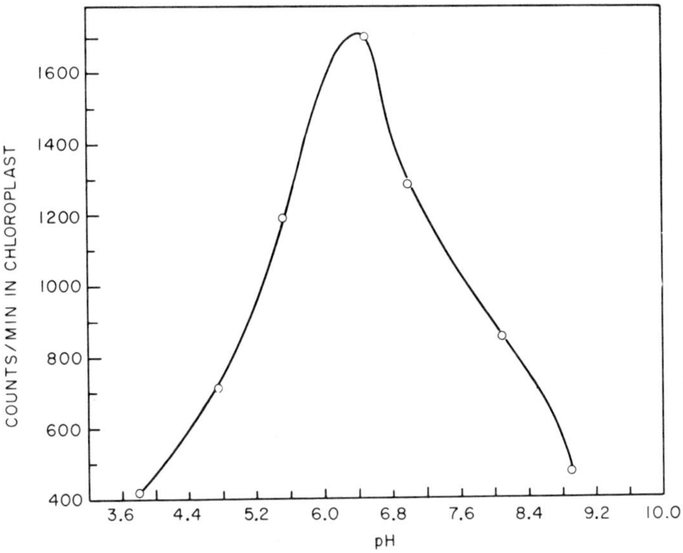

Figure 11. Effect of *p*H on loss of chloroplast-bound manganese. Particles incubated at 5° for 21 hr in 0.02 *M* buffer and then removed by centrifugation. A total of 2200 cpm was used in each.

Table 2

Effect of Washing of Chloroplast Particles on Removal of Manganese

Suspending medium	Total Mn in particle, cpm: Original	Total Mn in particle, cpm: Final	% of total Mn removed
0.02 *M* K phosphate, *p*H 7.0	8984	7209	19.8
0.02 *M* K phosphate, *p*H 8.1	9127	6412	29.6
0.02 *M* K phosphate, *p*H 6.5	9230	8660	6.5
" " " " " " + 1 m*M* EDTA	9301	8596	8.0
" " " " " " + 0.1 m*M* hydroxyquinoline	9285	8913	4.0
" " " " " " + 0.1 m*M* *o*-phenanthroline	9369	8669	7.5
" " " " " " + 1 m*M* KCN	9185	7451	18.8
" " " " " " + 0.1 m*M* diethyldithiocarbamate	9232	8639	6.5
" " " " " " + 0.1 m*M* $MnCl_2$	9334	8111	13.0
" " " " " " + 0.01 m*M* $KMnO_4$	9286	8656	6.6

which indicates that the manganese released in this manner is not protein bound. It is curious that the *p*H range for stability of manganese binding is similar to the *p*H range for stability of Hill activity. However, we have not made any measurements indicating a correlation between the half-lives of the two.

At *p*H 6.5 chelating agents such as those shown in Table 2, with the possible exception of KCN, do not effectively release the particle-bound manganese. In addition, little exchange is found between the ^{54}Mn-particle-bound manganese and the

Mn^{2+} and MnO_4^-, an observation consistent with the fact that we have not been able to exchange $^{54}Mn^{2+}$ into unlabeled chloroplasts under a wide variety of different conditions. These observations are, in general, consistent with those of Spencer and Possingham[6] and point to an extremely low dissociation value for chloroplast manganese. However, upon prolonged storage at 4° or −15° but not at −196° or under conditions common for protein denaturation, the chelating agents become increasingly more effective in removing the manganese.

In fact, simple incubation at neutral *p*H of particles with denaturants such as urea, thiourea, or guanidine results in a time, concentration, and temperature dependent release of manganese. The data of Figure 12 show an ~60% loss of bound manganese within 10 min by 6 *M* urea treatment at 4°. At room temperature the release is more rapid and essentially complete. High salt concentrations, such as saturated ammonium sulfate, do not result in release of manganese. Such results suggest a binding of manganese by protein, which is released upon denaturation.

Although it is possible to solubilize the manganese protein in micellar form with Triton (Figure 13), digitonin and phosphatides are ineffective. Micelles prepared with Triton can be broken with $(NH_4)_2SO_4$ to form a lipoprotein layer and a clear hypophasic layer. The manganese is found associated with the lipoprotein layer unless (again) the micelles have been treated with urea, guanidine, 5 min of 50° heat, or other means generally used for protein denaturation. At −196° the micelles may be stored, but overnight storage at −15°, 4°, or 23° results in extensive (70 to 100%) loss of the manganese to the hypophase upon ammonium sulfate treatment. Such behavior points to extreme lability of the manganese-protein in the micelle.

By other procedures it has been possible to obtain a protein containing manganese in solution and demonstrate adsorption onto DEAE with subsequent elution at

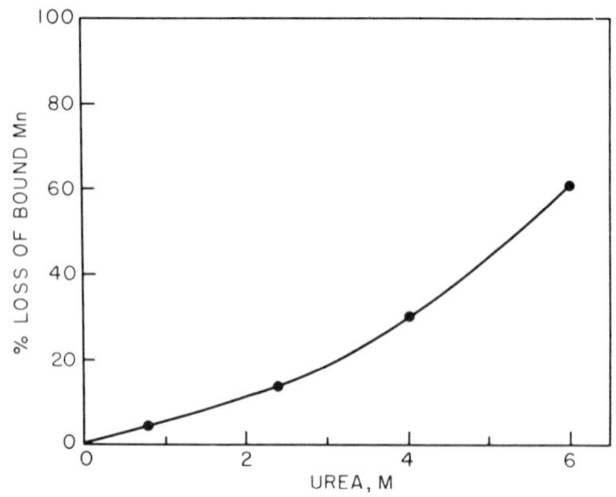

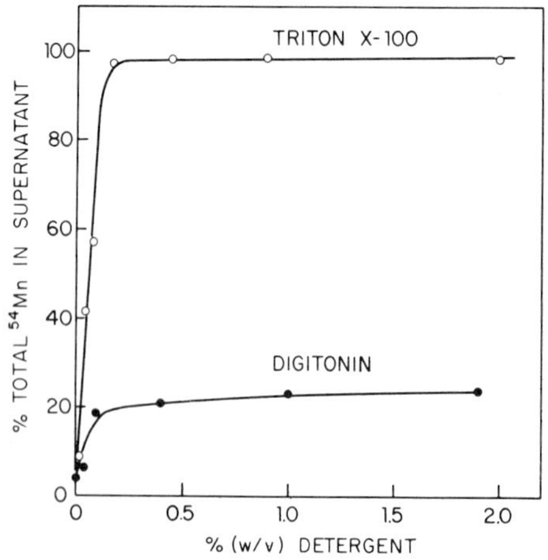

Figure 12 (Left). Effect of urea concentration on loss of particle-bound manganese. Labeled particles were incubated for 10 min at 4° in 0.02 *M* K phosphate, *p*H 6.7, containing urea at specified concentrations and then removed by centrifugation.

Figure 13 (Right). Detergent treatment of ^{54}Mn-labeled particles. Particles (100 μg Chl) were incubated at 4° for 1 hr with the indicated concentration of detergent and then centrifuged at 144,000×g for 1 hr.

high ionic strengths. The great instability of the soluble protein and the lack of a specific "enzymatic" assay has hampered purification of such a protein(s) and critical assessment of its function within photosystem II.

REFERENCES

1. Pirson, A., *Z. Botan.* **31**, 193 (1937).
2. Kessler, E., *Arch. Biochem. Biophys.* **59**; 527 (1955).
3. Kessler, E., Arthur, W., and Brugger, J.E., *Arch. Biochem. Biophys.* **71**, 326 (1957).
4. Eyster, C., Brown, T.E., Tanner, H.A., and Hood, S.L., *Plant Physiol.* **33**, 235 (1958).
5. Kessler, E., Moraw, R., Rumberg, B., and Witt, H.T., *Biochim. Biophys. Acta* **43**, 134 (1960).
6. Spencer, D. and Possingham, J.V., *Australian J. Biol. Sci.* **13**, 441 (1960).
7. Spencer, D. and Possingham, J.V., *Biochim. Biophys. Acta* **52**, 379 (1961).
8. Lumry, R. and Spikes, J.D., in *Research in Photosynthesis*, p. 373, H. Gaffron, Editor, Interscience, New York, 1957.
9. Bishop, N.I. and Gaffron, H., *Biochem. Biophys. Res. Commun.* **8**, 471 (1962).
10. Kok, B., Rurainski, H.J., and Harmon, E.A., *Plant Physiol.* **39**, 513 (1964).
11. Hoch, G. and Martin, I., *Arch. Biochem. Biophys.* **102**, 430 (1963).
12. Brody, S.S. and Brody, M., *Arch. Biochem. Biophys.* **95**, 521 (1961).
13. Kok, B., in *Photosynthetic Mechanisms of Green Plants, Natl. Acad. Sci.–Natl. Res. Council Publ.* **1145**, 45 (1963).
14. Kerridge, D., *J. Gen. Microbiol.* **19**, 497 (1958).
15. Siegler, M.R. and Sisler, H.D., *Biochim. Biophys. Acta* **87**, 70 (1964).
16. Siegler, M.R. and Sisler, H.D., *Biochim. Biophys. Acta* **87**, 83 (1964).
17. Kirk, J.T.O. and Allen, R.L., *Biochem. Biophys. Res. Commun.* **21**, 523 (1965).
18. Arnon, D.I., in *Trace Elements*, p. 1, C.A. Lamb et al., Editors, Academic Press, New York, 1957.
19. Park, R.B. and Pon, N.G., *J. Mol. Biol.* **6**, 105 (1963).
20. Blinks, L.R. and Skow, R.K., *Proc. Natl. Acad. Sci. U.S.* **24**, 420 (1938).
21. de Kouchkovsky, Y. and Briantais, J.M., in *Photosynthetic Mechanisms of Green Plants, Natl. Acad. Sci.–Natl. Res. Council Publ.* **1145**, 362 (1963).
22. Fork, D.C., *Plant Physiol.* **38**, 323 (1963).

DISCUSSION

Homann: Our experiments, in general, agree very well with yours. We found the *p*H dependence of the leakage of Mn from chloroplasts to vary with the suspension medium. For example, Hill activity of chloroplast fragments is preserved in dilute saline even at *p*H 4; manganese does not leak out. In other suspension media, however, manganese leaks out at this *p*H, and, correspondingly, some Hill activity is lost. Also, we found the incorporation of manganese into algae to be 50% inhibited by $\sim 2\ \mu M$ Cl-CCP and about 30% inhibited by anaerobiosis.

Cheniae: Our results with Cl-CCP differ from yours.

Homann: All right, but the incorporation of manganese is also inhibited by anaerobiosis, and dinitrophenol acts like Cl-CCP. In some experiments we found a slight stimulatory effect of light on the incorporation, but in most cases there was no effect. With regard to the problem of extracting manganese from chloroplasts of higher plants, these chloroplasts are usually loaded with excess manganese, which is gradually lost into the suspension medium during fragmentation of the chloroplasts.

Cheniae: We do not find that.

Homann: One loses very little of this manganese while preparing "broken chloroplasts." The main losses occur when the grana are fragmented. The situation is different if the chloroplasts contain manganese only in a concentration of about the value you mentioned, i.e., one manganese atom per 100 molecules of chlorophyll. This manganese will not leave the particles during fragmentation. Thus the manganese content per chlorophyll of the very small fragments is independent of the original manganese content of the chloroplasts. There appear to be two types of manganese in the chloroplasts, one bound to the lamellae and one located between the lamellae and the grana. The transport of manganese across the cell membrane is the first step of manganese uptake. Chloroplasts certainly must use a mechanism for the transport of manganese across their membrane. Then there must be a mechanism to accumulate manganese in the grana, and finally that type of manganese we cannot extract has to be bound to some site on the lamella. One of these steps seems to require light; which do you think it is?

CHENIAE: I can only speculate. With particles from deficient algae we have not as yet achieved full restoration of Hill activity; they exhibit a pronounced bend-off in rate. We interpret this to be due to an irreversible photooxidation of traps. Such behavior complicates restoration experiments if indeed there is a light requirement for restoration of Hill activity in isolated particles. If the light requirement can be shown with isolated chloroplasts, then one can speculate on a light reaction resulting in oxidation of the inactive manganous ion state to a higher valency state with subsequent binding to its active site. Alternatively, the light may cause a configurational change of a protein opening the site for manganese to bind. But this is only speculation; we have no evidence.

Reprinted from
Biochimica et Biophysica Acta
Elsevier Publishing Company
Amsterdam
Printed in The Netherlands

BBA 45379

TEMPERATURE AND PHOTOSYNTHESIS

I. SOME EFFECTS OF TEMPERATURE ON CARBON DIOXIDE FIXATION BY ISOLATED CHLOROPLASTS

C. W. BALDRY, C. BUCKE AND D. A. WALKER
Department of Botany, Imperial College, London (Great Britain)
(Received April 18th, 1966)

SUMMARY

1. Isolated pea chloroplasts were used to determine effects of temperature on photosynthetic CO_2 fixation in saturating light and high CO_2.

2. Increases in temperature between 5° and 30° increased the maximum rate and shortened an initial induction period. The maximum rate was not an exponential function of temperature.

3. Above 20°, values for the Q_{10} were less than 2. Below 15° they were greater than 2 and increased progressively, with decreasing temperature, to values as high as 9.

4. The results, obtained with a purely photosynthetic system, support earlier work with intact organisms which led some investigators to conclude that photosynthesis did not obey the Arrhenius law. It is suggested that this may be a consequence of the cyclic nature of the reactions involved.

INTRODUCTION

Following the work of BLACKMAN[1] and others it became apparent that photosynthesis involved both light and dark reactions. The primary photochemical events may be assumed to be independent of temperature ($Q_{10} = 1$), and in general this is true of photosynthesis in weak light where the rate of the overall process is limited by the photochemical component[1,2]. At higher light intensities the rate is increased by increases in temperature but this only becomes fully evident in the presence of high CO_2 (ref. 2). Experiments carried out under these conditions have given results which seem to fall into distinct groups. Some workers (*e.g.* NODDACK AND KOPP[3]) found that there was a linear relationship between the logarithm of the maximum velocity of photosynthesis and the reciprocal of temperature on the absolute scale, so that $\log V_{max} = f(1/T°K)$ in accordance with the Arrhenius law. Others (*e.g.* EMERSON[4]) found this relationship to be non-linear and that the Arrhenius law was not valid over any substantial part of the temperature range between 0° and 30°. WARBURG[5] and YABUSOE[6], whose results also fall into this category, held that the rate of photosynthesis was more or less directly proportional to temperature, in contrast to respiration, where in common with other dark enzymic reactions, the rate tends to increase in an exponential fashion with a Q_{10} of about 2.

Much of the work on temperature in relation to photosynthesis has been based on the measurement of O_2 evolution by intact cells and the assumption that there is fixed a stoichiometry of 1 molecule of O_2 evolved for each molecule of CO_2 fixed. It has always been necessary therefore to make estimates of the extent to which respiration contributes to the observed gaseous exchange. This has frequently involved the further assumption that respiration continues essentially unchanged in passing from dark to light and that the respective volumetric relationships between O_2 and CO_2 are independent of temperature. These assumptions are not unreasonable, particularly since at high light intensities it is clear that photosynthesis contributes much more to gaseous exchange than respiration. Nevertheless, there would be a distinct advantage in making measurements with an isolated photosynthetic unit which did not respire. This would be especially true with higher plants where use of the intact leaf brings the added complication of stomata which themselves respond to changes in light intensity, CO_2 concentration and temperature. Isolated chloroplasts would have none of these drawbacks, though there would certainly be doubts about the extent to which the isolated chloroplasts adequately represents whole-cell photosynthesis. Recent improvements in the rate of CO_2 fixation by isolated chloroplasts[7,8] have encouraged us to believe that their use might help to clarify the rather inadequately defined relationships between temperature and rates of whole-cell photosynthesis which exist at present.

METHODS

Plant material

Peas (var. Laxton Superb) were grown for 14 days in moist vermiculite in a box which was illuminated for 11 h out of each 24 by five 40-W 240-V fluorescent tubes (Ecko daylight) at a distance of 30 inch. This gave a light intensity of approx. 150 ft. candles. The temperature was maintained at 18° ± 1° throughout.

CO_2 fixation

This was estimated by using ^{14}C as before[7,8,10].

Illumination and temperature

Reaction mixtures were illuminated in saturating light (about 8000 ft. candles) from 1.5-W 240-V quartz–iodine lamps fitted with parabolic reflectors. A red perspex filter which transmitted 90 % of the incident light above 595 mμ was inserted between the light source and the reaction vessels. In each experiment four different temperatures were employed; this was achieved by using four separate, perspex baths, each with an independent heating unit, immersed in a single large tank with a pyrex glass floor. Cold water at 3° circulating through this tank served both to cool the reaction baths and to act as a heat filter to absorb the heat from the four quartz–iodine lamps. Temperatures were maintained to ± 0.25°.

Chloroplasts and reaction mixtures

Chloroplasts were isolated from 50 g of pea tissue in media containing sucrose and illuminated in 0.3-ml mixtures (with added sugar phosphates) as before[7,8,10] in 15-ml stoppered centrifuge tubes. Mixtures were completed, immediately prior to illumination, by the addition of sodium carbonate followed by chloroplasts and illuminated at 1-min intervals.

RESULTS

Figs. 1–3 are progress curves for CO_2 fixation by isolated chloroplasts in saturating red light. Reactions were carried out in Tricine–NaOH (see ref. 9) at pH 7.5 and sufficient sucrose (see legends) to maintain the integrity of the chloroplast's envelope[10]. Each 0.3-ml reaction mixture (see METHODS) initially contained a total of 2.8 μmoles of added carbonate which, under our conditions, was at least 5 times that required for half saturation.

Reactions were carried out in an illuminated bath (see METHODS) which provides four temperatures simultaneously. Each figure (Figs. 1–3) shows values obtained with a single chloroplast preparation and each point represents a value obtained with a separate reaction mixture. Simultaneous illumination allowed the determinations to be completed within 1 h of the disruption of the leaf and for any given set of four points (*e.g.* Fig. 3, time 10 min, temp. 10°, 15°, 20° and 25°) the chloroplasts had exactly the same history to within 3 min (the time taken to start the four reactions at 1-min intervals). Once isolated from the leaf, chloroplasts commence to lose their capacity to fix CO_2 (ref. 11). Errors attributable to this source are thus minimised by the above procedure. That there was not a larger decline in fixation after relatively

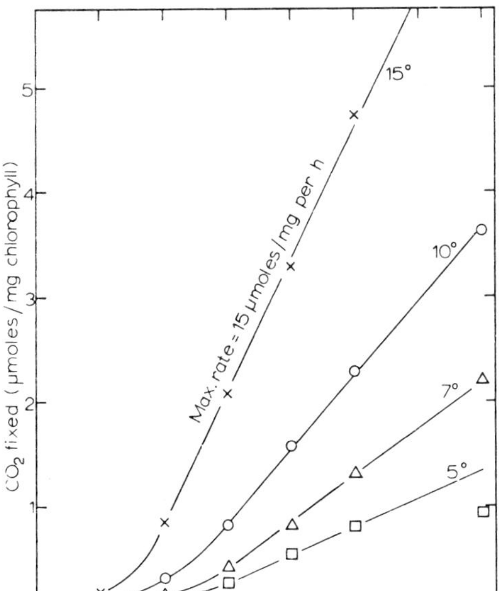

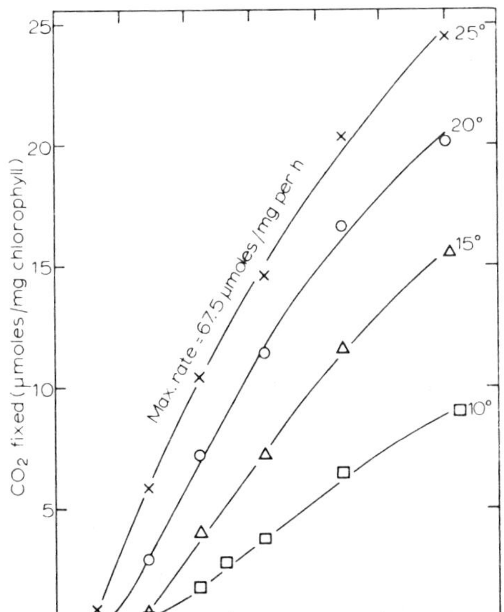

Fig. 1. Progress curves for CO_2 fixation by illuminated chloroplasts at different temperatures. Each point represents the CO_2 fixed in samples taken from separate 0.3-ml reaction mixtures (see METHODS) to which were initially added ribose 5-phosphate, 2 μmoles; $MnCl_2$, 0.5 μmole; $MgCl_2$, 0.25 μmole; EDTA, 0.5 μmole; inorganic phosphate, 0.5 μmole; Tricine buffer, 7.5 μmoles (pH 7.5); carbonate–bicarbonate, 2.95 μmoles (containing 20 μC ^{14}C); reduced glutathione, 1.0 μmole; sucrose, 30 μmoles, and 0.1 ml of chloroplasts in resuspending medium containing 27 μg chlorophyll.

Fig. 2. See Fig. 1. 0.1 ml of chloroplasts in resuspending medium containing 33 μg chlorophyll.

prolonged incubation (Figs. 1–3) implies that the pea chloroplasts were reasonably stable in the reaction mixtures used, even at temperatures as high as 30°. The initial lag or induction period is a characteristic of the system (see DISCUSSION and ref. 10). It should be noted that the reaction tubes were not preincubated at different temperatures before illumination but that the volume of reactants was so small (0.3 ml) that temperature equilibration would be extremely rapid. Similar lag periods have been demonstrated in mixtures which have been allowed to equilibrate before illumination[10].

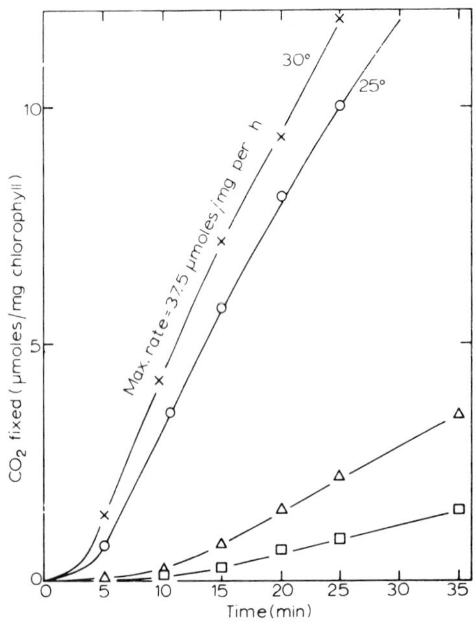

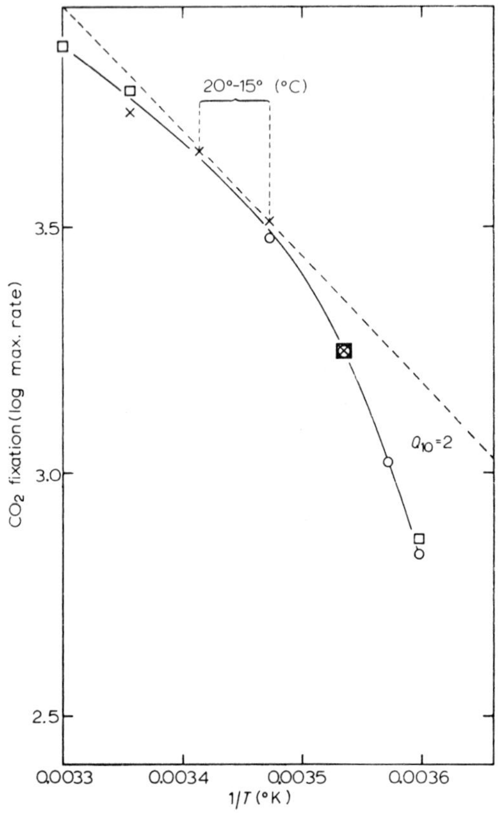

Fig. 3. See Fig. 1. 0.1 ml of chloroplasts in resuspending medium containing 25 μg chlorophyll. □---□, 5°; △—△, 10°.

Fig. 4. Rate of photosynthesis as a function of temperature in the range 5–30°. Data from Figs. 1–3. For purposes of comparison the 10° values from all three figures were made equal and the remaining values adjusted proportionately. Between 15° and 25° the value for the temperature coefficient was approx. 2 and the dotted line shows the relationship which would have been obtained had the Arrhenius law been obeyed in this way over the whole range.

Figs. 1–3 show a common pattern of response to temperature which was also observed in other experiments. Increases in temperature shortened the initial lag period and increased the maximal rate. The greatest effect on the rate was produced at the lower end of the temperature range (see *e.g.* Fig. 3 where the difference in rates between 5° and 10° is clearly larger than that between 25° and 30°). This progressive decrease in the response to temperature in going from 5° to 30° can also be seen in Figs. 1 and 2 and becomes even more apparent (Fig. 4) when the log of the maximum rate is plotted against $1/T$(°K). In plotting Fig. 4 the maximum rates at 10° (Figs.

1–3) were made equal and the other values multiplied by the same factor then fell on the same curve. The dotted line is an approximation of the result which would have been obtained if the rate had doubled for each 10° rise in temperature ($Q_{10} = 2$) and it can be seen that this coincides with the experimental values only between 15° and 20°. The Q_{10} decreased above 20° and increased progressively below 15°. The "Q_{10}" for temperature differences of less than 10° may also be calculated from the relationship

$$\ln Q_{10} = \frac{10 \ln (v_1/v_2)}{T_1 - T_2}$$

Values obtained in this way are given in Table I where they are compared with those obtained for Chlorella by other workers.

TABLE I

"Q_{10}" VALUES FOR PHOTOSYNTHETIC CO_2 FIXATION BY ISOLATED CHLOROPLASTS COMPARED WITH SIMILAR VALUES FOR PHOTOSYNTHESIS BY INTACT ORGANISMS

The values for chloroplasts were calculated from data given in Figs. 1–3.

Temperature (°C)	*Emerson*[4] *Chlorella*	*Emerson and Green*[15] *Gigartina*	*Warburg*[5] *Chlorella*	*Noddack and Kopp*[3] *Chlorella*	*This paper Pea chloroplasts*
30					
26			1.6		1.28
25	1.54				
20					1.5
16					1.91
15	2.5	1.7	2.1		
12				2.0	2.9
10	4.8	4.2			
8					5.3
7			4.3		
6	7.6	13.5			9.0
5					
4					
0					

TABLE II

RATES OF PHOTOSYNTHETIC CO_2 FIXATION BY ISOLATED CHLOROPLASTS COMPARED WITH THOSE BY WHOLE PLANTS

Reference	*Material*	*Temperature*	CO_2	CO_2 *fixation* (*µmoles/mg chlorophyll per h*)
WILSTATTER AND STOLL (see ref. 2)	*Helianthus annus*	20°	5%	318
This paper	Pea chloroplasts			
	Best rate (Fig. 2)	25°	(see	67.5
	Usual rate	20°	text)	20–40
RABINOWITCH[2]	Land plants	Natural habitat, moderate climate, favourable conditions.		80–120

In three experiments using very low light (approx. 10 ft. candles before the insertion of the red filter) the maximum rate was not increased by raising the temperature from 10° to 25°.

In Table II the maximum rates of CO_2 fixation by illuminated pea chloroplasts observed in this work are compared with some for whole tissues. The value for *Helianthus annus* is an absolute maximum[2]. The value for "land plants" is an attempt to derive an average figure from the results of a great many workers for a large number of species[2].

DISCUSSION

Any consideration of the effects of temperature on CO_2 fixation by isolated chloroplasts must take into account the partial reactions involved and the extent to which isolated plastids constitute an intact photosynthetic system. The photosynthetic cycle may be divided into three parts; one concerned with carboxylation, a second with reduction and a third with the regeneration of the CO_2 acceptor. Since we have supplied substrate levels of ribose 5-phosphate to ensure maximum rates of fixation it is pertinent to ask whether our chloroplasts can successfully catalyse all three parts of the cycle or whether we are simply measuring the conversion of ribose 5-phosphate to ribulose 1,5-diphosphate and its subsequent carboxylation. The evidence for the complete cycle is as follows.

(1) Isolated pea chloroplasts, with intact envelopes, will commence to fix CO_2 in the complete absence of added sugar phosphates, added coenzymes such as ATP or NADP, or added protein fractions such as ferredoxin. The rate is slow at first and reaches a maximum, which may be ten times that of the initial, only after a prolonged lag period (as much as 15 min)[10]. This implies that they are able to build up the constituents of the Calvin cycle in an autocatalytic manner until saturation is reached. Working with spinach chloroplasts, which behave similarly[12], HAVIR AND GIBBS[13] have established that ^{14}C initially introduced into C-1 of 3-phosphoglycerate is rapidly distributed between the carbon atoms of this and other products in a manner which is consistent with the operation of the cycle.

(2) The initial lag may be shortened by the addition of "catalytic" quantities of cycle intermediates[10]. These are "catalytic" in the sense that the increment in fixation which they cause is greater than that which would result from their stoichiometric conversion to ribulose 1,5-diphosphate and its subsequent carboxylation. Although the absolute maximum rates (two to three times the unstimulated rates) are obtained only in the presence of substrate levels of ribose 5-phosphate[10] the rates with "catalytic" levels are often only slightly slower[10].

(3) Chromatography[10,14] has shown that ^{14}C and ^{32}P are incorporated into sugar phosphates by isolated pea chloroplasts. Dihydroxyacetone phosphate is a major product, implying that the chloroplasts retain a capacity for reduction, and labelled pentose and hexose monophosphates and diphosphates are indicative of the regenerative phases of the cycle.

The failure of the isolated plastids to match the rates of fixation by the intact leaf (Table II) remains a matter for concern though we feel that it is most unlikely that we have yet succeeded in devising the most favourable reaction mixture. In the present context it is reassuring that preparations with somewhat different levels of activity showed a remarkably constant response to temperature (*cf.* Figs. 1–3 with Fig. 4).

Our results are based entirely on the rates of CO_2 fixation by isolated chloroplasts. They therefore represent what is at present the best practical realisation of a photosynthetic system, uncomplicated by respiration and other physiological processes of whole leaves. They confirm the findings of WARBURG[5], YABUSOE[6], EMERSON[4] and EMERSON AND GREEN[15] (Table I) which were obtained with whole organisms and rule out the possibility that the high values for Q_{10} at low temperatures were a consequence of unwarranted assumptions about the contribution of other cellular processes. They provide no support for the views of NODDACK AND KOPP[3] and others (see INTRODUCTION). The view of EMERSON[4] that a sigmoid curve is obtained when rate is plotted against temperature is correct, but this curve is so attenuated that it would scarcely conflict with WARBURG's suggestion that the rate of photosynthesis is a linear rather than an exponential function of temperature. Certainly, a general statement that, in high light and high CO_2, the rate of photosynthesis is directly proportional to the temperature would be in accord with the present results.

The decline in Q_{10} above 20° is almost certainly attributable to heat inactivation. The theoretical implications of high Q_{10}'s in photosynthesis have been widely discussed, mostly in terms of a straight series of consecutive reactions (for a summary see RABINOWITCH[2]). An alternative, and much more plausible explanation (to which we subscribe), takes into account the cyclic nature[13,16] of CO_2 fixation and the possibility of autocatalysis or positive feed-back which then arises. This concept is analysed in the following paper by Dr. M. J. SELWYN.

ACKNOWLEDGEMENTS

This work was supported by the Scientific Research Council. Special apparatus was provided by the Royal Society and the University of London Central Research Fund. We are most grateful to Dr. R. HILL, F. R. S. and Professor C. P. WHITTINGHAM for their advice and criticism.

REFERENCES

1 F. F. BLACKMAN, *Ann. Botany*, 19 (1905) 281.
2 E. I. RABINOWITCH, *Photosynthesis and Related Processes*, Vol. 2, Interscience, New York, 1956, p. 1211.
3 W. NODDACK AND C. KOPP, *Z. Physik. Chem. Leipzig*, A187 (1940) 79.
4 R. EMERSON, *J. Gen. Physiol.*, 12 (1929) 623.
5 O. WARBURG, *Biochem. Z.*, 100 (1919) 258.
6 M. YABUSOE, *Biochem. Z.*, 152 (1924) 498.
7 D. A. WALKER, *Biochem. J.*, 92 (1964) 22c.
8 D. A. WALKER, *Plant Physiol.*, 40 (1965) 1157.
9 N. E. GOOD, *Arch. Biochem. Biophys.*, 96 (1962) 653.
10 D. A. WALKER, *Carbon Dioxide Fixation in Isolated Chloroplasts, Proc. NATO Ad. Study Inst. "Biochem. of the Chloroplast"*, Aberystwyth, 1965, Vol. 2, Academic Press, New York, 1966, p. 53.
11 M. GIBBS, C. C. BLACK AND B. KOK, *Biochim. Biophys. Acta*, 52 (1961) 474.
12 E. S. BAMBERGER AND M. GIBBS, *Plant Physiol.*, 40 (1965) 919.
13 E. A. HAVIR AND M. GIBBS, *J. Biol. Chem.*, 238 (1963) 3183.
14 C. W. BALDRY, C. BUCKE AND D. A. WALKER, *Nature*, 210 (1966) 793.
15 R. EMERSON AND L. GREEN, *J. Gen. Physiol.*, 17 (1934) 817.
16 J. A. BASSHAM AND M. CALVIN, *The Path of Carbon in Photosynthesis*, Prentice-Hall, Englewood Cliffs, N.J., 1957, p. 56.

Biochem. J. (1966) **101**, 103
Printed in Great Britain

Photosynthesis by Sugar-cane Leaves

A NEW CARBOXYLATION REACTION AND THE PATHWAY OF SUGAR FORMATION

By M. D. HATCH and C. R. SLACK
David North Plant Research Centre, The Colonial Sugar Refining Co. Ltd., Indooroopilly, Queensland, Australia

(*Received 7 March* 1966)

1. Radioactive products in detached leaf segments were examined after periods of steady-state photosynthesis in $^{14}CO_2$. 2. After exposure to $^{14}CO_2$ for approx. 1sec. more than 93% of the fixed radioactivity was located in malate, aspartate and oxaloacetate. After longer periods large proportions of the radioactivity appeared in 3-phosphoglycerate, hexose monophosphates and sucrose. Similar results were obtained with leaves still attached to the plant. 3. Radioactivity appeared first in C-4 of the dicarboxylic acids and C-1 of 3-phosphoglycerate. The labelling pattern in hexoses was consistent with their formation from 3-phosphoglycerate. 4. The reaction giving rise to C_4 dicarboxylic acid appears to be the only quantitatively significant carboxylation reaction. 5. Evidence is provided that the radioactivity incorporated into the C_4 dicarboxylic acid pool is transferred to sugars via 3-phosphoglycerate. A scheme is proposed to account for these observations.

Several observations which appear to be inconsistent with the scheme for photosynthesis proposed by Calvin and co-workers (Calvin & Bassham, 1962) have been considered in recent reviews by Stiller (1962) and Bassham (1964). The low activity of ribulose 1,5-diphosphate carboxylase in some tissues relative to the rate of photosynthesis, and the high concentrations of bicarbonate required by the enzyme for maximum activity, raise doubts about its quantitative importance in photosynthetic fixation of carbon dioxide (Racker, 1957; Peterkofsky & Racker, 1961; Stiller, 1962). Recently Kortschak, Hartt, & Burr (1965) reported that malate and aspartate were the major labelled products formed in sugar-cane leaves during short periods of photosynthesis in $^{14}CO_2$. The present studies confirm this observation with another sugar-cane variety and provide information about the nature of the primary carboxylation reaction and the pathway of sugar formation.

MATERIALS

3-Phosphoglyceric acid (tricyclohexylammonium salt) and wheat-germ acid phosphatase were purchased from Sigma Chemical Co. (St Louis, Mo., U.S.A.). Yeast invertase was obtained from Difco Laboratories (Detroit, Mich., U.S.A.). $Ba^{14}CO_3$ (26·2mc/m-mole), L-[U-^{14}C]malic acid (15·1mc/m-mole) and L-[U-^{14}C]aspartic acid (6·1mc/m-mole) were obtained from The Radiochemical Centre (Amersham, Bucks.).

Leaf tissue. Field-grown sugar cane (hybrid variety, Pindar) was used. Sections (approx. 18cm. long, 1·5cm. wide) containing no midrib tissue were cut from the youngest fully expanded leaf. All cuts were made under water to prevent the entry of air into the vascular tissue.

METHODS

Photosynthesis in $^{14}CO_2$. A rectangular Perspex box (19cm. × 16cm. × 9cm.), sufficiently large to accommodate six leaf segments, served as a photosynthesis chamber. The top of the chamber was detachable and split in half lengthwise. Leaf segments were clamped vertically between the halves with about 1cm. protruding from the chamber and the bases of the segments dipping into water. Soft rubber on the adjoining edges of the top provided a seal around the leaf segments but allowed them to be withdrawn from the chamber. Air, saturated with water vapour, was pumped through the chamber and the air in the chamber was circulated by a high-speed fan. Before the addition of $^{14}CO_2$ the segments were illuminated for at least 45min. by a 400w Phillips HPL lamp. The light-intensity at the leaf surface, measured with a selenium cell, was 8200ft.-candles. From a calibration curve of the selenium cell against a Kipp thermopile this intensity was equivalent to 0·39cal./cm.^{2}/min.

Immediately before the introduction of $^{14}CO_2$ the air supply to the chamber was removed and exit holes were sealed with tape; $^{14}CO_2$ (0·86mc) was injected from a syringe bringing the CO_2 concentration to 0·055%. At intervals individual leaf segments were transferred to 50ml. of boiling 80% (v/v) ethanol and kept at this temperature for 3min. In certain experiments the segments were transferred to methanol–chloroform–2·0M-formic acid (12:5:3, by vol.) at −80°, then held at −15° for 24hr. (Bieleski & Young, 1963).

A second procedure allowed both shorter treatments with

$^{14}CO_2$ and the killing of leaves while still exposed to $^{14}CO_2$ and light. After pre-illumination in the chamber leaf segments were placed in a large test tube (vol. 60 ml.). $^{14}CO_2$ (45 μc) was injected and then boiling 80% (v/v) ethanol was poured into the tube either simultaneously or after 1–2 sec.

Extraction of tissue. Segments killed in the methanol–chloroform–formic acid mixture were extracted as described by Bieleski & Young (1963). Segments killed in boiling ethanol were homogenized and the insoluble material was re-extracted in sequence with 20 ml. of boiling 80% (v/v) ethanol, and twice with 15 ml. of boiling water. Water and ethanol extracts were combined. The remaining insoluble material was washed with a large volume of water followed by ethanol and then dried at 70° to a constant weight. Hereafter this material will be referred to as 'residue'. Radioactivity incorporated into segments is expressed as counts/min./100 mg. of the residue.

Counting procedures. Samples of 0·1 ml. or less of the combined extracts containing at least 1000 counts/min. were counted in a liquid-scintillating counter (model N664A; Ekco Electronics). A toluene–phosphor–ethanol solution (Ziegler, Chleck & Brinkerhoff, 1957) was used for these aqueous samples. The counting efficiency was 55% and quenching was not significant.

Weighed samples (approx. 10 mg.) of the dried residue were counted in steel planchets with a Geiger-Müller tube. In the range used counts were proportional to the weight of the sample. The counting efficiency of this method was determined by liquid-scintillation counting of similar residue samples suspended in toluene containing phosphor and 2% (w/v) thixin.

Identification and estimation of radioactivity in individual compounds

Chromatography solvents and counting procedure. For identifying individual components and determining the amount of the radioactivity in each compound samples of extracts were chromatographed on paper with one or more of the following solvent systems: A, butan-1-ol–propionic acid–water (10:5:7, by vol.) (Benson *et al.* 1950); B, pentan-1-ol saturated with 5 M-formic acid (Aronoff, 1956, p. 122); C, propan-1-ol–aq. NH_3 (sp.gr. 0·90)–water (6:3:1, by vol.) (Bieleski & Young, 1963); D, ethyl acetate–pyridine–water (8:2:1, by vol.) (White & Secor, 1953); E, phenol saturated with water (Benson *et al.* 1950). For subsequent references to these solvents the letter prefixing each will be used. Whatman no. 1 paper was used and for solvents A, C and E the paper was previously washed with oxalic acid.

The location and the amount of radioactivity in individual compounds or groups of compounds was determined with a chromatogram strip-counter and the area of individual peaks measured with a planimeter. When quantities of individual compounds were required samples of extracts were chromatographed as a band and eluted from developed chromatograms.

Marker compounds on chromatograms were detected as follows: organic acids and amino acids with ninhydrin (Aronoff, 1956, p. 120), phosphorylated compounds with ammonium molybdate (Bandurski & Axelrod, 1951) and sugars with *p*-anisidine.

Malic acid. When extracts were chromatographed in solvents A and B a peak was obtained which co-chromatographed with L-malate. With solvent B this was the only radioactive compound which moved a significant distance from the origin. The radioactivity eluted from chromatograms developed in solvent B co-chromatographed with L-malate in solvents A, C and E. When samples of this compound were treated with a fumarate hydratase preparation from wheat-germ 18–20% of the radioactivity was located in fumaric acid at equilibrium. Similar values were obtained with L-[U-^{14}C]malate.

Aspartic acid. A compound which ran as a separate peak corresponding to L-aspartate was detected with solvent A. The eluted radioactivity co-chromatographed with L-aspartate in solvents B and E, and the compound was completely degraded by treatment with ninhydrin (Greenberg & Rothstein, 1957).

Phosphorylated compounds. With solvent A a separate area of radioactivity was obtained which corresponded in mobility to 3-phosphoglycerate and hexose monophosphates. Occasionally a very small separate peak was observed with a mobility corresponding to fructose 1,6-diphosphate and ribulose 1,5-diphosphate.

When the major peak was eluted and co-chromatographed with marker compounds in solvent C the radioactivity corresponded with 3-phosphoglycerate, glucose 6-phosphate and fructose 6-phosphate. After treatment with acid phosphatase all the radioactivity was recovered in compounds which co-chromatographed with glyceric acid, glucose and fructose in solvents B, C and D. In some experiments traces of glyceric acid were detected when the original extracts were chromatographed in solvents A and B.

Sucrose. After longer periods of exposure to $^{14}CO_2$ a labelled compound appeared which chromatographed with sucrose in solvent A. This compound co-chromatographed with sucrose in solvent D and after treatment with invertase the radioactivity was located equally between glucose and fructose.

Proportions of radioactivity in individual compounds. The proportion of the total radioactivity in malate was calculated from chromatograms developed in solvent B and checked with solvent A and the proportions of aspartate and sucrose were obtained with solvent A. After determining the proportion for the total phosphorylated compounds with solvent A the amount of radioactivity in 3-phosphoglycerate and glucose 6-phosphate plus fructose 6-phosphate was calculated from chromatograms run in solvents B or D after treatment of this fraction with phosphatase.

Oxaloacetic acid. With the standard procedure employed radioactive oxaloacetate was not detected. Hence the following procedure was used. Leaves were exposed to $^{14}CO_2$ in a test tube as previously described and after approx. 1 sec. photosynthesis was stopped by pouring one of the following solutions into the tube: (1) 80% (v/v) ethanol containing 10 mg. of 2,4-dinitrophenylhydrazine and 0·2 N-HCl at −80°, (2) the same solution at 82°, (3) the same solution at 82° but without 2,4-dinitrophenylhydrazine. The subsequent procedure was the same as already described except that the concentrated extracts were extracted with chloroform to obtain phenylhydrazones (Aronoff, 1956, p. 133). The radioactivity in the chloroform extract was determined and samples were chromatographed with authentic oxaloacetate 2,4-dinitrophenylhydrazone in the following solvents: butan-1-ol–water–ethanol (5:4:1, by vol.), butan-1-ol saturated with M-$NaHCO_3$, 0·1 M-potassium glycinate, pH 8·4, and butan-1-ol–ethanol–0·5N-NH_3 (7:1:2, by vol.) described by Block, Durram & Zweig (1958). With the chloroform extract from the extract of leaf killed at −80° at least 90% of the radioactivity co-chromatographed with the marker in all solvents. When a portion of the chloroform extract was mixed with authentic 2,4-dinitrophenylhydrazone of oxaloacetate and recrystallized twice the specific activity of the derivative remained constant.

Insoluble compounds. The insoluble residue was treated with 72% (v/v) H_2SO_4 to hydrolyse cellulose (Jermyn, 1955). Essentially all the radioactivity was recovered in a compound which co-chromatographed with glucose in solvents C and D. When the material was hydrolysed with 6N-HCl (Pirie, 1955) no radioactivity was detected in amino acids.

Degradation of labelled products

Malic acid. Radioactivity in the C-1 plus C-4 and the C-2 plus C-3 of malate was determined by MnO_2 oxidation (Friedemann & Kendall, 1928). Samples of the CO_2 and acetaldehyde, trapped in sodium hydroxide and sodium bisulphite respectively, were counted in the liquid-scintillation counter by using the procedure described for aqueous samples. The C-1 of malate was determined by a modification of the Von Peckmann reaction described by Racusen & Aronoff (1953). Trials with L-[U-^{14}C]malate established the exact conditions for the complete release of the C-1 as CO. Samples of malate containing at least 35,000 disintegrations/min. were dried in a combustion flask with L-malate carrier and heated at 80° for 45 min. with 5 ml. of 100% H_2SO_4 in the combustion apparatus for a Nuclear–Chicago Dynacon electrometer. The CO released was then flushed into a 250 ml. ion chamber and counted. Similar samples were subjected to combustion with Van Slyke reagent to determine the total radioactivity.

Aspartic acid. Radioactivity in the C-1 plus C-4 and the C-2 plus C-3 of aspartate was determined by treatment with ninhydrin (Greenberg & Rothstein, 1957).

Glyceric acid. Glyceric acid was isolated from 3-phosphoglycerate by chromatography in solvent B after treatment with acid phosphatase. The glycerate was decarboxylated by oxidation with ceric sulphate (Zelitch, 1965). Radioactivity in the CO_2, trapped in N-NaOH, and in the residual acetic acid was determined with a liquid-scintillation counter.

Hexoses. After treatment of sucrose with invertase and hexose monophosphates with acid phosphatase the resulting glucose and fructose were isolated from chromatograms developed in solvent D and converted into their common osazone. This derivative was degraded to give the C-1, C-2 and C-3 as the mesoxaldehyde osazone, the C-4 and C-5 as formic acid and the C-6 as formaldehyde (Aronoff, 1956, p. 110).

RESULTS

Incorporation of radioactivity from $^{14}CO_2$ as a function of time. Experimental conditions were chosen to produce as nearly as possible a steady state of photosynthesis under physiological conditions of concentration of carbon dioxide and light-intensity. Studies by K. T. Glasziou & J. C. Waldron in our Laboratory have shown that the rate of photosynthesis of leaf segments, prepared as described in the Methods section, is low when first exposed to light. The rate increases to a steady value after about 30 min. and is maintained for several hours. In our experiments leaves were exposed to light in a humidified stream of air for at least 45 min. before the addition of $^{14}CO_2$. The intensity of visible light at the leaf surface was approx. 40% of that in direct sunlight.

The total fixation of $^{14}CO_2$ (Fig. 1) and the radioactivity in individual compounds (Figs. 2 and 3) were determined for leaves exposed to $^{14}CO_2$ for periods of up to 150 sec. Over this period the rate of $^{14}CO_2$ fixation remained linear. The changing proportions of radioactivity in individual compounds is shown in Fig. 2. Three trends are apparent, the continued fall of the first-labelled products, malate and aspartate, the rise then fall of 3-phosphoglycerate and hexose monophosphates, and the steady increase of the percentage of the total radioactivity in the end products sucrose and a glucan. The plot of total radioactivity in individual compounds (Fig. 3) shows the asymptotic trend of the early products and intermediates and the exponential rise in the radioactivity associated with end products. When a similar experiment was carried out in the dark the rate of $^{14}CO_2$ fixation was only 0·4% of that in the light. Even after 5 min. in $^{14}CO_2$ radioactivity was located almost entirely in aspartate and malate whereas no radioactivity was detected in 3-phosphoglycerate or hexose monophosphates. Our conditions of leaf pretreatment and light-intensity differed from those used by Kortschak *et al.* (1965). However, our

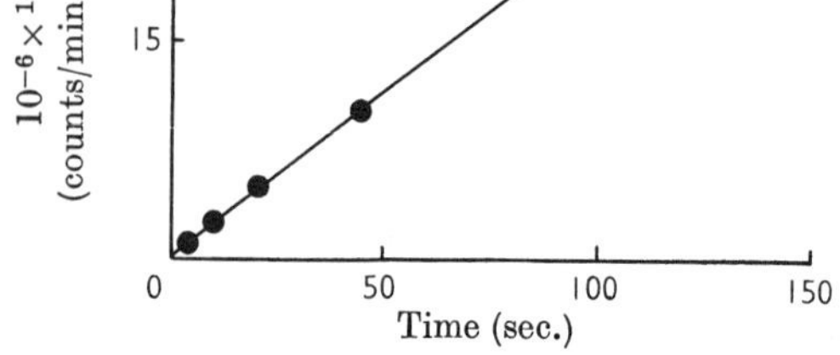

Fig. 1. Total ^{14}C incorporated into leaf segments as a function of time. The radioactivity in the residue is included and the counts are for a 55% counting efficiency. Other details are described in the Methods section.

results with the variety Pindar were qualitatively similar to those obtained by these workers with the variety H37–1933.

The effect of different killing procedures on the distribution of radioactivity in individual compounds was examined (Table 1). Results were very similar when leaves were killed either in boiling 80% (v/v) ethanol or in the methanol–chloroform–formic acid mixture at −80°. In these experiments leaf segments were transferred from the chamber to the killing mixture. To determine if any modification of the labelling pattern occurred during transfer, leaf segments were exposed to $^{14}CO_2$ in test tubes and

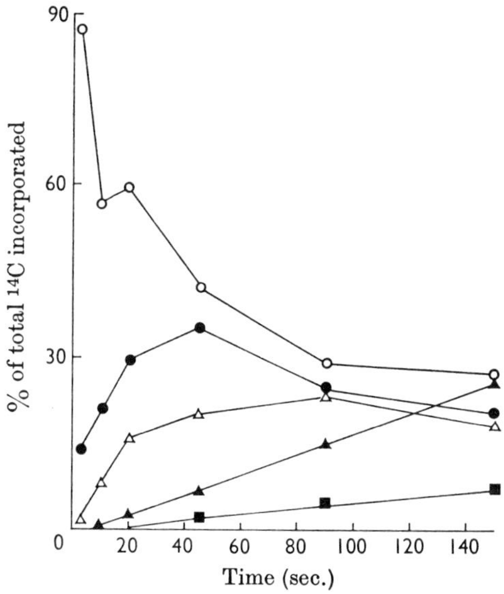

Fig. 2. Proportion of the total radioactivity in individual compounds for periods in $^{14}CO_2$ up to 150 sec. Data are from the same experiment as described in Fig. 1. Details of procedures for indentifying and counting individual compounds are described in the Methods section. ○, Malate + aspartate; ●, 3-phosphoglycerate; △, hexose monophosphates; ▲, sucrose; ■, glucan.

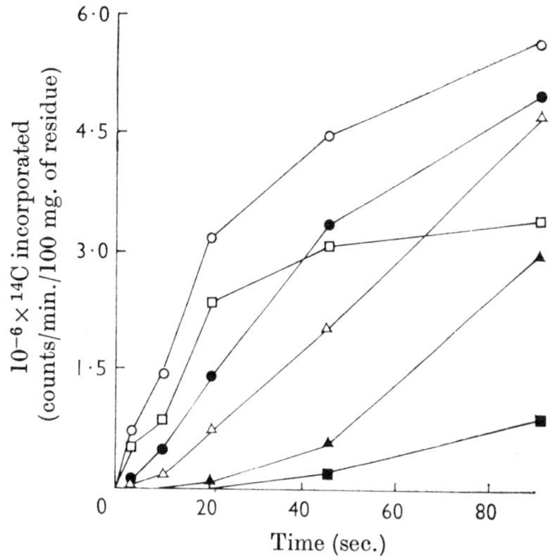

Fig. 3. Time-course for the total radioactivity incorporated into individual compounds. Data are from the same experiment as described in Figs. 1 and 2. ○, Malate + aspartate; □, malate; ●, 3-phosphoglycerate; △, hexose monophosphates; ▲, sucrose; ■, glucan.

Table 1. *Effect of different killing procedures on the distribution of radioactivity in individual compounds*

In Expt. 1 leaf segments were supplied with $^{14}CO_2$ in the chamber, and at the times shown segments were removed and placed in boiling 80% (v/v) ethanol or methanol–chloroform–formic acid (MCF) at −80°. In Expt. 2 leaves were supplied with $^{14}CO_2$ in large test-tubes and after 4 sec. one was killed by adding boiling 80 % (v/v) ethanol and the other was transferred to another tube containing the same mixture. Other details are described in the Methods section.

Expt. no.	Time in $^{14}CO_2$ (sec.)	$10^{-6}\times$ Total ^{14}C incorporated (counts/min./100 mg. of residue)	Killing method	Percentage of total ^{14}C in individual compounds					
				Malate	Aspartate	3-Phosphoglycerate	Hexose monophosphates	Sucrose	Glucan
1	30	8·9	Ethanol	30	15	36	14·5	2·5	1·0
	30	10·1	MCF	32	16	34	15	2·0	1·0
	60	19·9	Ethanol	26	8·5	27	25	8·5	4·0
	60	19·3	MCF	27	10	30	23	6·0	4·5
2	4	2·6	Ethanol added to tube	53	25	15	7	0	0
	4	2·3	Leaf removed to ethanol	56	25	14	5	0	0

then killed either by the direct addition of boiling 80% (v/v) ethanol or by transferring them to the same mixture. The two procedures gave almost identical results (Table 1).

To confirm that the results obtained were not due to an alteration of the normal photosynthetic process caused by detaching and cutting the leaves experiments were conducted with leaves on the plant. The apical 25 cm. of the youngest fully-expanded leaf of plants was exposed to $^{14}CO_2$ in a sealed chamber or, for short-term experiments, an open test tube (see the Methods section). A time-course study over a period from 4 sec. to 90 sec. in $^{14}CO_2$ gave results which were almost identical with those obtained with leaf sections. When an attached leaf was exposed to $^{14}CO_2$ for 2 sec. and killed by the direct addition of boiling 80% (v/v) ethanol the proportions of the incorporated radioactivity in malate, aspartate, 3-phosphoglycerate and hexose monophosphates were 54%, 37%, 7% and 2% respectively.

We established that oxaloacetate would be completely destroyed during our isolation and chromatography procedures. Hence attempts were made to identify radioactive oxaloacetate in leaf extracts as its 2,4-dinitrophenylhydrazone. Only killing at −80° in the presence of the reagent gave reproducible results (Table 2). A much smaller proportion of the total radioactivity was recovered in the derivative when leaves were killed in boiling 80% (v/v) ethanol containing 2,4-dinitrophenylhydrazine.

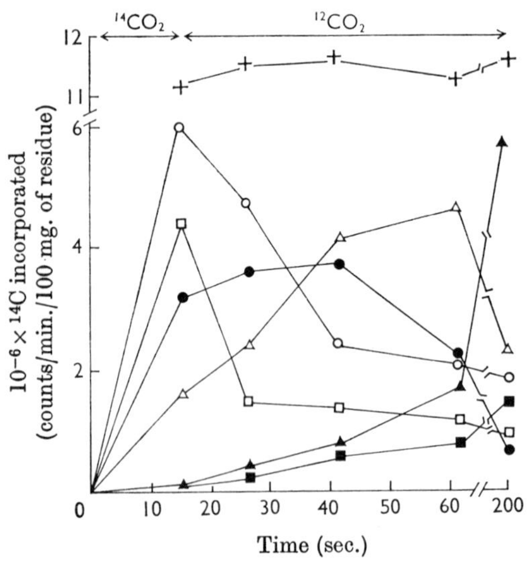

Fig. 4. Changes in the distribution of radioactivity after transfer of leaf segments from $^{14}CO_2$ to carbon dioxide. After 15 sec. in $^{14}CO_2$ leaf segments were removed simultaneously from the chamber to a stream of air without changing the distance from the light source. At intervals individual segments were killed and analysed as described in the Methods section. ○, Malate+aspartate; □, malate; ●, 3-phosphoglycerate; △, hexose monophosphates; ▲, sucrose; ■, glucan; +, total radioactivity.

Table 2. *Radioactivity in oxaloacetate after different killing procedures*

Leaves were exposed to $^{14}CO_2$ for approx. 1 sec. and then killed either at 82° or −80° in the presence of 2,4-dinitrophenylhydrazine or without the reagent (see the Methods section). Oxaloacetate was isolated and counted as its 2,4-dinitrophenylhydrazone.

	Percentage of total ^{14}C		
Compound	With reagent at −80°	With reagent at 82°	Without reagent at 82°
Malate	54	57	47
Aspartate	36	32	42
Oxaloacetate	3·4	0·6	0
3-Phosphoglycerate	5·2	8·0	6·0
Other compounds	3·0	2·4	5·0

Table 3. *Changes in the distribution of radioactivity in leaf segments after transfer from* $^{14}CO_2$ *in the light to carbon dioxide in the dark*

Illuminated segments were exposed to $^{14}CO_2$ for 9 sec. At 9 sec. the chamber was darkened with a shutter and humidified air was flushed through at a rate of 1 l./sec. After 3 sec. the first leaf segment was killed. Radioactive compounds were extracted and analysed as described in the Methods section.

	Percentage of total ^{14}C in individual compounds						
Time in the dark and carbon dioxide (sec.)	Malate	Aspartate	3-Phosphoglycerate	Hexose monophosphates	Sucrose	Alanine	Glucan
3	35	26	29	9	0	0	0·6
20	37	24	24	12	1	2	0·7
37	35	38	16	8	1	2·5	0·5
83	34	42	10	8	0	6	0·6

Changes in the distribution of radioactivity after transfer from $^{14}CO_2$ *to air.* When leaf segments were transferred from $^{14}CO_2$ to air but maintained at the same light-intensity there was no loss of radioactivity during the following 200sec. (Fig. 4). During the first 40sec. most of the radioactivity in malate and aspartate was transferred to the hexose monophosphate pool whereas the label in 3-phosphoglycerate increased only slightly. Subsequently radioactivity appeared in sucrose and a glucan at the expense of hexose monophosphates and 3-phosphoglycerate. The relatively constant amount of radioactivity associated with malate and aspartate during the period from 40sec. to 200sec. was probably due to label in the C-1, C-2 and C-3 (see Table 4). To further test the possibility that $^{14}CO_2$ may be released and then re-fixed during these transformations a period in $^{14}CO_2$ was followed by a period in 5% (v/v) unlabelled carbon dioxide in air. Transformations similar to those shown in Fig. 4 were observed and there was no significant loss of radioactivity.

When leaves were transferred from $^{14}CO_2$ in the light to unlabelled carbon dioxide in the dark there was little change in distribution of radioactivity in individual compounds during the first 20sec. (Table 3). Between 20sec. and 83sec. the proportion of radioactivity in 3-phosphoglycerate declined from 24% to 10%. Accompanying this fall was an increase in the radioactivity in aspartate and alanine. The total radioactivity in the leaf segments remained constant during the period in unlabelled carbon dioxide.

Radioactivity in individual carbon atoms of photosynthetic products. The primary labelling of malic acid occurred in the C-4 position (Table 4). After 20sec. the total radioactivity in this carbon became relatively constant. At this time C-4 still contained 80% of the total radioactivity, but label was not detected in the C-2 and C-3 positions until 45sec. The distribution of radioactivity in the C-1 plus C-4 and the C-2 plus C-3 of aspartate was almost identical with that of malate. At 3sec. nearly all the radioactivity in 3-phosphoglycerate was located in the carboxyl carbon atom whereas the compound was almost uniformly labelled at 150sec. It is noteworthy that, with respect to the C-1, the C-2 plus C-3 of 3-phosphoglycerate became labelled much more rapidly than the C-2 plus C-3 of malate.

Glucose and fructose from hexose monophosphates and sucrose were converted into their common osazone and then partially degraded (Table 5). Almost half of the total radioactivity was located in the C-1, C-2 plus C-3 positions after 20sec. and 45sec. The C-6 contained only a small part of the radioactivity in C-4, C-5 plus C-6 at 20sec. but this proportion doubled by 45sec.

Table 4. *Radioactivity in individual carbon atoms of aspartate, malate and 3-phosphoglycerate*

Compounds used for these analyses were from the same experiment as described in Figs. 1, 2 and 3. Procedures for isolating and degrading the compounds are described in the Methods section. Counts/min. are expressed per 100mg. of residue.

Period in $^{14}CO_2$ (sec.)	Aspartic acid: C-1 plus C-4: % of total ^{14}C	Aspartic acid: C-1 plus C-4: $10^{-4}\times$ counts/min.	Malic acid: C-1: % of total ^{14}C	Malic acid: C-1: $10^{-4}\times$ counts/min.	Malic acid: C-2 plus C-3: % of total ^{14}C	Malic acid: C-2 plus C-3: % of C-1, C-2 plus C-3	Malic acid: C-2 plus C-3: $10^{-4}\times$ counts/min.	Malic acid: C-4: % of total ^{14}C	Malic acid: C-4: $10^{-4}\times$ counts/min.	Glyceric acid: C-1: % of total ^{14}C	Glyceric acid: C-1: $10^{-4}\times$ counts/min.	Glyceric acid: C-2 plus C-3: % of total ^{14}C	Glyceric acid: C-2 plus C-3: $10^{-4}\times$ counts/min.
3	98	21·8	10	4·7	0	0	0	90	42·5	98	11·5	2	0·24
10	98	49·5	16	15·1	0	0	0	84	76·0	96	49·4	4	2·0
20	97	83·0	20	46·0	0	0	0	80	185	83	120	17	24·6
45	96	132	29	91·0	4	14	14·1	67	208	61	212	39	135
90	89	198	27	97·0	15	35	51·0	58	200	52	258	48	240
150	77	276	33	200	25	43	152	42	253	45	333	55	415

Table 5. *Distribution of radioactivity in the carbon atoms of hexoses derived from sucrose and hexose monophosphates*

Compounds were obtained from the same extracts used for the studies described in Figs. 1, 2 and 3 and Table 4. Procedures used to isolate and degrade the hexoses are described in the Methods section.

Period in $^{14}CO_2$ (sec.)	Origin of hexose	C-1, C-2, C-3 (as % of total ^{14}C)	C-6 (as % of ^{14}C in C-4, C-5, C-6)	C-4 plus C-5 (as % of ^{14}C in C-4, C-5, C-6)
20	Sucrose	52	11	89
20	Hexose monophosphate	46	10	90
45	Sucrose	45	18	82
45	Hexose monophosphate	45	20	80

DISCUSSION

Our aim was to examine the radioactive products formed from $^{14}CO_2$ under steady-state conditions for photosynthesis and at a light-intensity and concentration of carbon dioxide as close to physiological conditions as possible. Under these conditions we assumed that only physiologically significant carboxylation reactions were operative. For determining the sequence of chemical events a number of other assumptions are necessary. The major assumptions are that pool sizes remain constant and that the absolute pool size of intermediates is sufficiently large to allow the radioactivity to be detected. Other factors which must be considered are labelling of compounds by side reactions from the main path, secondary labelling of intermediates by cycling, and labelling of independent second pools.

With these points in mind we interpret the time-studies to indicate that the first product of the major carboxylation reaction is either malate or oxaloacetate and that aspartate is probably formed by a side reaction from these products. The ^{14}C then moves from the C_4 dicarboxylic acid pool to 3-phosphoglycerate and subsequently to sucrose and a glucan via the hexose phosphates. Since the major products of $^{14}CO_2$ fixation in the dark are malate and aspartate the same carboxylation reaction may be operating for both processes. In the dark the rate of fixation of carbon dioxide is only 0·4% the rate in light and no radioactive phosphorylated compounds were detected. Hence light not only produces a 250-fold increase in the rate of fixation of carbon dioxide but is also necessary for the transfer of radioactivity from the C_4 dicarboxylic acid pool into 3-phosphoglycerate.

Support for this interpretation was obtained from experiments in which a period in $^{14}CO_2$ in the light was followed by periods in unlabelled carbon dioxide in either light or dark. In both cases the total radioactivity in leaf segments remained constant during the period in unlabelled carbon dioxide. In the light the radioactivity in malate and aspartate moved into other pools in the sequence already suggested. In contrast radioactivity in the C_4 dicarboxylic acid pool increased significantly in the dark and there was no movement of ^{14}C into hexose monophosphates or sucrose. The increase of radioactivity in the dicarboxylic acid pool and the appearance of label in alanine occurred at the expense of 3-phosphoglycerate. It appears unlikely that the transfer of carbon from the dicarboxylic acids to 3-phosphoglycerate in the light proceeds by a process involving decarboxylation and re-fixation of carbon dioxide.

The feature of photosynthesis is a net gain of carbon. This would be achieved only if the transfer of carbon from the C_4 dicarboxylic acids to 3-phosphoglycerate proceeded without the release and loss of carbon dioxide. Known pathways for this transformation involve the decarboxylation of oxaloacetate or malate with the appearance of their C-1, C-2 and C-3, as the C-1, C-2 and C-3 respectively, of 3-phosphoglycerate. Besides the problem of loss of carbon dioxide, the operation of this pathway is inconsistent with the fact that the labelling of the C-2 and C-3, relative to the C-1, is much more rapid for 3-phosphoglycerate than for malate. Rather, our results indicate that the C-4 of the C_4 dicarboxylic acids is the source of the C-1 of 3-phosphoglycerate. The earliest labelling of malate occurs in the C-4 and the total radioactivity in this carbon atom approaches an optimum value more rapidly than the radioactivity in the C-1 of 3-phosphoglycerate. The simplest interpretation is that the C-4 carboxyl group of a C_4 dicarboxylic acid is transferred by a transcarboxylation reaction to an acceptor substance, the latter providing the C-2 and C-3 of 3-phosphoglycerate. The carbon dioxide-acceptor for the initial carboxylation reaction could be regenerated from the remaining C_3 compound.

A definitive conclusion about the nature of the primary carboxylation reaction cannot be derived from the available data. Enzymes which carboxylate pyruvate to give malate, and phosphopyruvate to give oxaloacetate, are known to occur in leaves. They are malate dehydrogenase (decarboxylating) (EC 1.1.1.40) ('malic enzyme') and phosphopyruvate carboxylase (EC 4.1.1.31). The operation of another enzyme, ATP-oxaloacetate carboxy-lyase (Mazelis & Vennesland, 1957) would not be favoured by the high ATP/ADP ratio which should prevail in the light. However, our data indicate that the C_4 dicarboxylation acids are in rapid equilibrium so that the primary carbon dioxide-fixation product need not be the acid involved in the subsequent transcarboxylation reaction.

Oxaloacetate is known to participate in several trans-carboxylation reactions (Wood & Stjernholm, 1961; Hülsmann, 1962) and therefore appears the more likely carboxyl donor. The carboxyl acceptor may be either a C_2 compound such as glycolaldehyde phosphate or the C_5 compound of the Calvin cycle, ribulose 1,5-diphosphate. The latter possibility is appealing since the subsequent steps may then be identical with those of the Calvin cycle (Calvin & Bassham, 1962). The rates of labelling of 3-phosphoglycerate and hexose monophosphates, and the distribution of radioactivity in their individual carbon atoms, are consistent with this possibility.

The above interpretations are summarized in Scheme 1, which shows two cyclic processes linked by a transcarboxylation reaction. The role of one cycle is to fix carbon dioxide and to provide carboxyl groups to react with a carboxyl acceptor in the other cycle. The latter cycle operates to regenerate the carboxyl acceptor for the transcarboxylation reaction. The relative rates of labelling of the C-2 and C-3, compared with the C-1, for malate and 3-phosphoglycerate is consistent with some of the latter compound being converted into the C_3 carbon dioxide-acceptor. While this movement of carbon is not a basically essential part of the scheme it may serve to replace carbon lost from the first cycle by side-reactions.

The possibility that the pathway outlined in Scheme 1 is peculiar to sugar cane appears remote. Hence the question why similar labelling patterns have not been observed during studies on other species arises. If malate and aspartate are not essential intermediates they may be labelled only if sufficient malate dehydrogenase and transaminase are present in the chloroplasts. Detection of radioactive oxaloacetate would depend upon an adequate pool size and preventing its breakdown during extraction. The only procedure we found satisfactory for measuring the radioactivity in oxaloacetate involved killing off leaves at −80° in the presence of 2,4-dinitrophenylhydrazine. Hence, if standard killing procedures are used, the operation of the pathway described in Scheme 1 could give

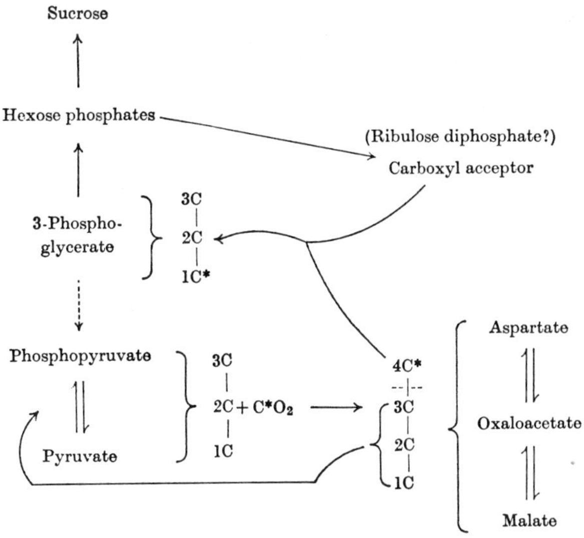

Scheme 1. Proposed pathway for photosynthetic fixation of carbon dioxide in sugar-cane leaves. The broken arrow indicates a minor pathway.

3-phosphoglycerate as the first detectable radioactive product.

REFERENCES

Aronoff, S. (1956). *Techniques of Radiochemistry*. Ames: Iowa State College Press.

Bandurski, R. S. & Axelrod, B. (1951). *J. biol. Chem.* **193**, 405.

Bassham, J. A. (1964). *Annu. Rev. Plant Physiol.* **15**, 101.

Benson, A. A., Bassham, J. A., Calvin, M., Goodale, T. C., Haas, V. A. & Stepka, W. (1950). *J. Amer. chem. Soc.* **72**, 1710.

Bieleski, R. L. & Young, R. E. (1963). *Analyt. Biochem.* **6**, 54.

Block, R. J., Durram, E. L. & Zweig, G. (1958). *Paper Chromatography and Paper Electrophoresis*, p. 239. New York: Academic Press Inc.

Calvin, M. & Bassham, J. A. (1962). *The Photosynthesis of Carbon Compounds*, p. 10. New York: W. A. Benjamin Inc.

Friedemann, T. E. & Kendall, A. I. (1928). *J. biol. Chem.* **82**, 23.

Greenberg, D. M. & Rothstein, M. (1957). In *Methods in Enzymology*, vol. 4, p. 713. Ed. by Colowick, S. P. & Kaplan, N. O. New York: Academic Press Inc.

Hülsmann, W. C. (1962). *Biochim. biophys. Acta*, **62**, 620.

Jermyn, M. A. (1955). In *Modern Methods of Plant Analysis*, vol. 2, p. 205. Ed. by Peach, K. & Tracey, M. V. Berlin: Springer-Verlag.

Kortschak, H. P., Hartt, C. E. & Burr, G. O. (1965). *Plant Physiol.* **40**, 209.

Mazelis, M. & Vennesland, B. (1957). *Plant Physiol.* **32**, 591.

Peterkofsky, A. & Racker, E. (1961). *Plant Physiol.* **36**, 409.
Pirie, N. W. (1955). In *Modern Methods of Plant Analysis*, vol. 4, p. 30. Ed. by Peach, K. & Tracey, M. V. Berlin: Springer-Verlag.
Racker, E. (1957). *Arch. Biochem. Biophys.* **69**, 300.
Racusen, D. W. & Aronoff, S. (1953). *Arch. Biochem. Biophys.* **42**, 25.
Stiller, M. (1962). *Annu. Rev. Plant Physiol.* **13**, 151.
White, L. M. & Secor, G. E. (1953). *Arch. Biochem. Biophys.* **43**, 60.
Wood, H. G. & Stjernholm, R. (1961). *Proc. nat. Acad. Sci., Wash.*, **47**, 289.
Zelitch, I. (1965). *J. biol. Chem.* **240**, 1869.
Ziegler, C. A., Chleck, D. J. & Brinkerhoff, J. (1957). In *Liquid Scintillation Counting*, p. 158. Ed. by Bell, G. C. & Hayes, F. N. Oxford: Pergamon Press Ltd.

Amer. J. Bot. 55(8): 875–883. 1968.

CHLOROPLAST SPECIALIZATION IN DICOTYLEDONS POSSESSING THE C_4-DICARBOXYLIC ACID PATHWAY OF PHOTOSYNTHETIC CO_2 FIXATION[1]

W. M. Laetsch

Department of Botany, University of California, Berkeley

ABSTRACT

The C_4-dicarboxylic-acid pathway of photosynthetic CO_2 fixation found in tropical grasses has recently been demonstrated in members of the Amaranthaceae and Chenopodiaceae. In the tropical grasses this CO_2-fixation pathway is correlated with specialized leaf anatomy and chloroplast structure. This investigation was undertaken to determine if leaf cells of some representatives of these other families had structural features similar to those of tropical grasses. The leaf anatomy of *Amaranthus edulis* and a variety of *Atriplex* species is very similar and it resembles that of grasses such as sugar cane. Prominent bundle sheaths are surrounded by a layer of palisade cells. Bundle-sheath cells of *Am. edulis* have large chloroplasts containing much starch, but the chloroplasts have grana. The palisade cells have much smaller chloroplasts containing very little starch. The bundle-sheath cell chloroplasts of *At. lentiformis* have grana, their profiles tend to be ovoid, and they contain abundant starch grains. The palisade cell chloroplasts contain little starch and their profiles are discoid. The bundle-sheath cells of both species contain mitochondria which are much larger than those in the palisade cells. The chloroplasts in both types of cells in both species have a highly developed peripheral reticulum. This reticulum is composed of anastomosing tubules which are contiguous with the inner plastid membrane. The leaf anatomy and cell ultrastructure of these dicots are similar to those of the tropical grasses possessing this new photosynthetic carbon-fixation pathway. These morphological features are interpreted as adaptations for the rapid transport of precursors and end products of photosynthesis. A hypothesis is presented stating that the unique morphological and biochemical characters of these plants represent adaptations for efficient and rapid carbon fixation in environments where water stress frequently limits photosynthesis.

Hatch and Slack (1966) demonstrated that the photosynthetic carbon fixation pathway in sugar cane differs from the conventional Calvin cycle. After 1 sec in $C^{14}O_2$, 93% of the fixed radioactivity was found in oxaloacetate, malate, and aspartate. They have investigated carbon fixation pathways in a variety of monocotyledon and dicotyledon species, and the "cane-type" or C_4-dicarboxylic acid pathway of photosynthetic carbon fixation has been found in species of the following genera of grasses: *Zea*, *Sorghum*, *Paspalum*, *Axonopus*, *Digitaria*, *Chloris*, *Eragrostis* (Hatch, Slack, and Johnson, 1967). A species of *Cyperus* also possessed this pathway. Phosphopyruvate carboxylase is considered to be the major photosynthetic CO_2-fixing enzyme in these taxa (Slack and Hatch, 1967). These grasses possess other features in common. They are characterized by their high rates of photosynthesis at high light intensities (Burr et al., 1957; Hesketh and Moss, 1963; El-Sharkawy and Hesketh, 1965), by their failure to "leak" respiratory CO_2 during the light period (El-Sharkawy, Loomis, and Williams, 1967), and by low CO_2 compensation values (Moss, 1962). In addition, Dr. Bruce Smith (Division of Geology, California Institute of Technology) has recently shown that these tropical grasses have much higher $^{13}C/^{12}C$ ratios than temperate grasses (personal communication). It has also been known for many years that the chloroplasts in the bundle-sheath cells of tropical grasses differed from the chloroplasts in the surrounding mesophyll cells. Rhoades and Carvalho (1944) demonstrated this in maize and sorghum on the basis of light microscopy, but the details of these dimorphic chloroplasts were only revealed by studies with the electron microscope. These were made on maize by Hodge, McLean, and Mercer (1955), on a variety of tropical grasses by Johnson (1964), and on sugar cane by Laetsch, Stetler, and Vlitos (1965). In these plants the mature chloroplasts in the promi-

[1] Received 24 October 1967.
This work was supported by a grant from Tate and Lyle, Ltd., and by NSF Grant GB-6066.
The technical assistance of Janet Burr and Ian Price is gratefully acknowledged.

[The Journal for August (55: 745–874) was issued 1 August 1968]
American Journal of Botany, Vol. 55: No. 8, 1968

nent bundle-sheath cells around the vascular tissue are large, lack grana, and have many starch grains. The chloroplasts in the mesophyll cells surrounding the bundle sheath are smaller, possess grana, and have few starch grains.

The "cane-type" photosynthesis has recently been found in *Amaranthus Palmeri* Wats., *Am. viridis* L., *Atriplex semibacatta* Brown. Prodr. (Hatch, personal communication) and in *At. spongiosa* F.v.M. (Osmond, 1967). *Amaranthus* also has photosynthetic rates paralleling those of certain tropical grasses (El-Sharkawy and Hesketh, 1965). Tregunna and Downton (1967) have reported CO_2 compensation values in several species of *Amaranthus* and one species of *Atriplex* which are similar to those of tropical grasses, and Hesketh (1967) has shown that photosynthetic CO_2 assimilation in *Am. palmeri* and several tropical grasses is not enhanced by the absence of oxygen. *Amaranthus edulis* does not "leak" respiratory CO_2 during the light period (El-Sharkawy et al., 1967), and Smith (personal communication) has demonstrated that the $^{13}C/^{12}C$ ratios of *Am. edulis* and *At. canescens* are similar to the tropical grasses.

The investigation reported here was undertaken to determine if the leaf anatomy and leaf-cell ultrastructure of representatives of these dicot genera illustrated any parallelism with structural features of tropical grass leaves.

Materials and methods—*Amaranthus edulis* Speg. was grown from seed kindly supplied by Prof. Robert Loomis of the University of California, Davis. Leaves were used from young plants grown in a constant temperature room (25 C) under fluorescent light of 800 ft-c. Leaves of *Atriplex lentiformis* (Torr.) Wats. were obtained from mature plants growing in the University Botanical Garden. Leaves of both species were prepared for light microscopy by fixing in Craf III, embedding in paraffin, and sectioning at 10 μ. The sections were stained with safranin and fast green. Leaf tissue was prepared for electron microscopy by fixing overnight in cold 3.0% glutaraldehyde and cacodylate buffer (*p*H 6.8) after which it was washed in cold buffer, and placed overnight in cold OsO_4 (2.0% aq sol). The tissue was taken through an acetone-water series and stored overnight in cold 70% acetone and 1.0% uranyl nitrate. Following dehydration in acetone, the tissue was embedded in epon (Luft, 1961), sectioned with a diamond knife, and mounted on uncoated grids. The sections were stained in lead citrate (Reynolds, 1963) for 3 min and viewed with a Zeiss Em-9.

Results—Figure 1 illustrates the leaf anatomy of *Am. edulis*. The numerous vascular bundles are surrounded by a prominent bundle sheath, and the bundle sheath is encircled by a concentric layer of palisade cells. The bundle-sheath cells contain large centripetally arranged chloroplasts. The palisade cells have smaller chloroplasts, and they are randomly distributed in the cells. The anatomy of the leaf is strikingly like that of a grass such as sugar cane. The anatomy of *At. lentiformis* (Fig. 2) is like that of *Am. edulis* in its essential features. The leaf anatomy of *At. semibaccata, At. polycarpa* (Torr.) Wats., *At. canescens* (Pursh) Hutt., *At. hymenelytra* (Torr.) Wats. and *At. argentea* Nutt. was also examined. In leaf anatomy and chloroplast size these species were very similar to *At. lentiformis*. *Atriplex argentea* is the only annual in this group, and they are all halophytes found primarily in California, the Great Basin, and the Southwest (Hall and Clements, 1923). *Atriplex semibaccata* is native to Australia, but it is now common in arid regions of California. Leaves of *At. rosea* L. and *At. hortensis* L., on the other hand, did not possess the prominent chloroplast-containing bundle sheath or the concentrically arranged palisade. Both of these species are probably of Eurasian origin and presently enjoy a wide distribution as annual weeds in Central and Southern Europe as well as in North America (Hegi, 1906).

A bundle-sheath cell of *Am. edulis* is seen in Fig. 3. The chloroplasts are large (the curved one in the center is 24 μ long), contain grana, and have many starch grains. They are clustered in the centripetal part of the cell with the large mitochondria. Single membrane-bound bodies with an electron-dense interior matrix (identified by arrow and question mark) are frequently observed.

Cells from the concentrically arranged palisade tissue of *Am. edulis* are seen in Fig. 4. The chloroplasts are 6–8 μ in length; they contain grana but have very little starch. These chloroplasts, as well as those in the bundle-sheath cells, possess a peripheral reticulum of anastomosing tubules. An enlarged view of this peripheral reticulum is seen in the insert of Fig. 4.

Details of the large mitochondria of the bundle-sheath cells of *Am. edulis* are seen in Fig. 5. The cristae are vesicular and numerous. An unidenti-

Fig. 1–3.—Fig. 1. Cross section of leaf of *Amaranthus edulis*. The cells of the bundle sheath (BS) contain large chloroplasts. The bundle sheath is surrounded by a layer of palisade cells, ×252.—Fig. 2. Cross section of leaf of *Atriplex lentiformis*. The bundle-sheath (BS) and palisade cells are arranged much like those of *Amaranthus*, ×252.—Fig. 3. A bundle-sheath cell of *Amaranthus edulis*. The large chloroplasts are filled with starch (S). The chloroplasts and mitochondria occupy a typically centripetal position. A number of unidentified single membrane-bound bodies (?) are present, ×5,280.

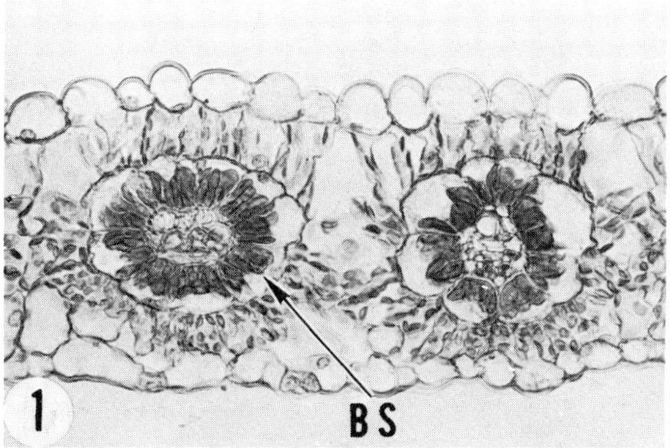
BS
1

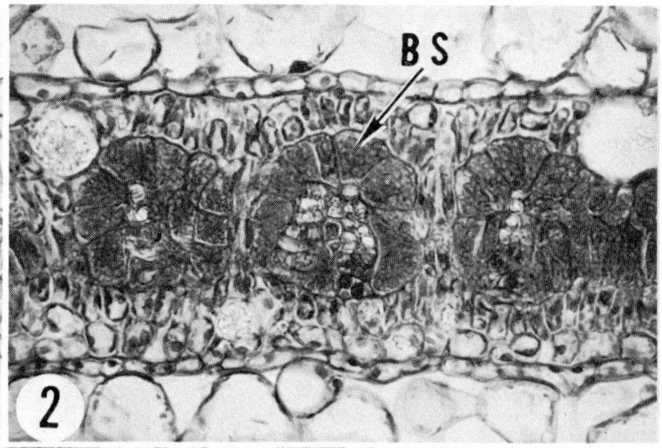
BS
2

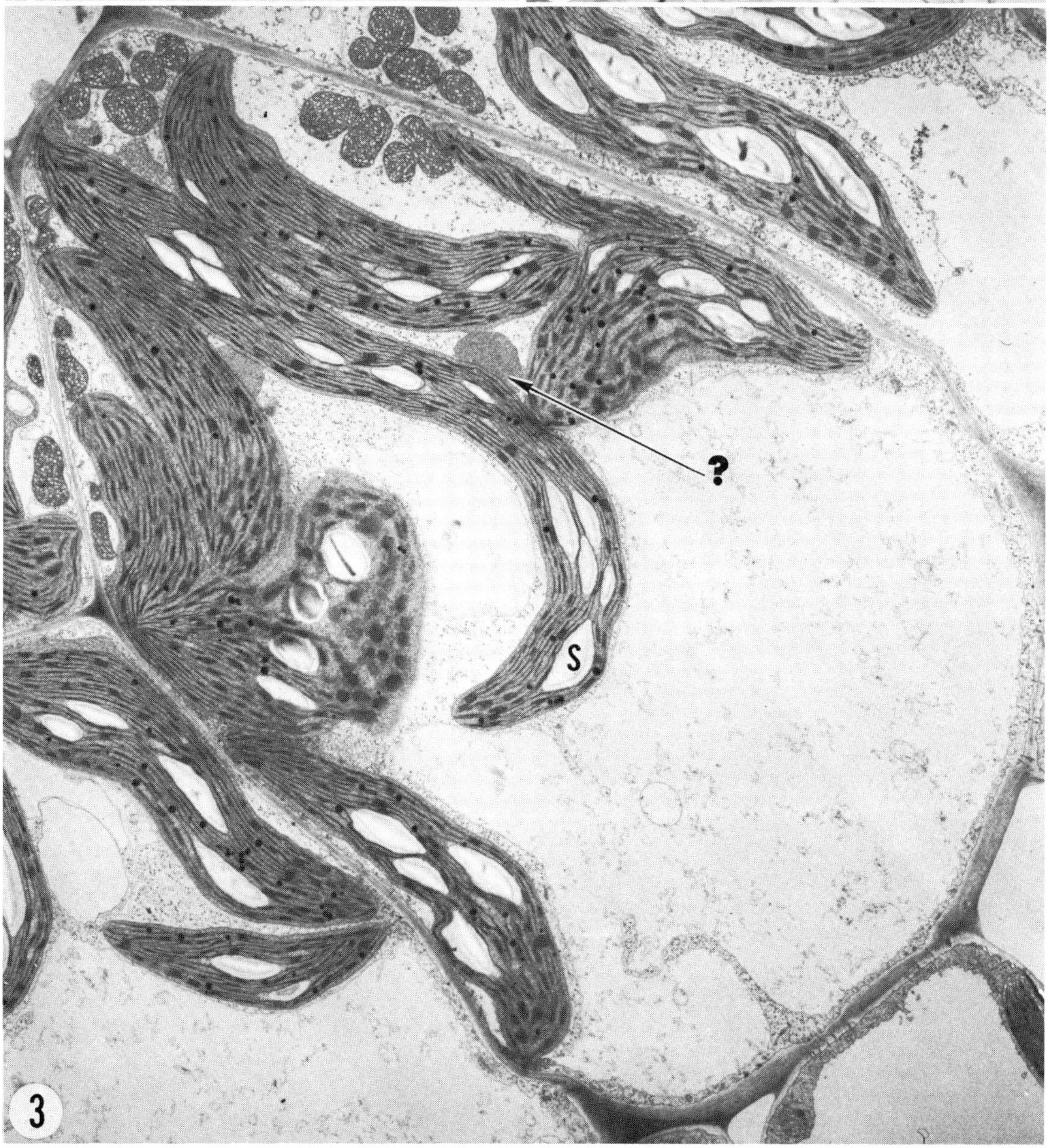
?
S
3

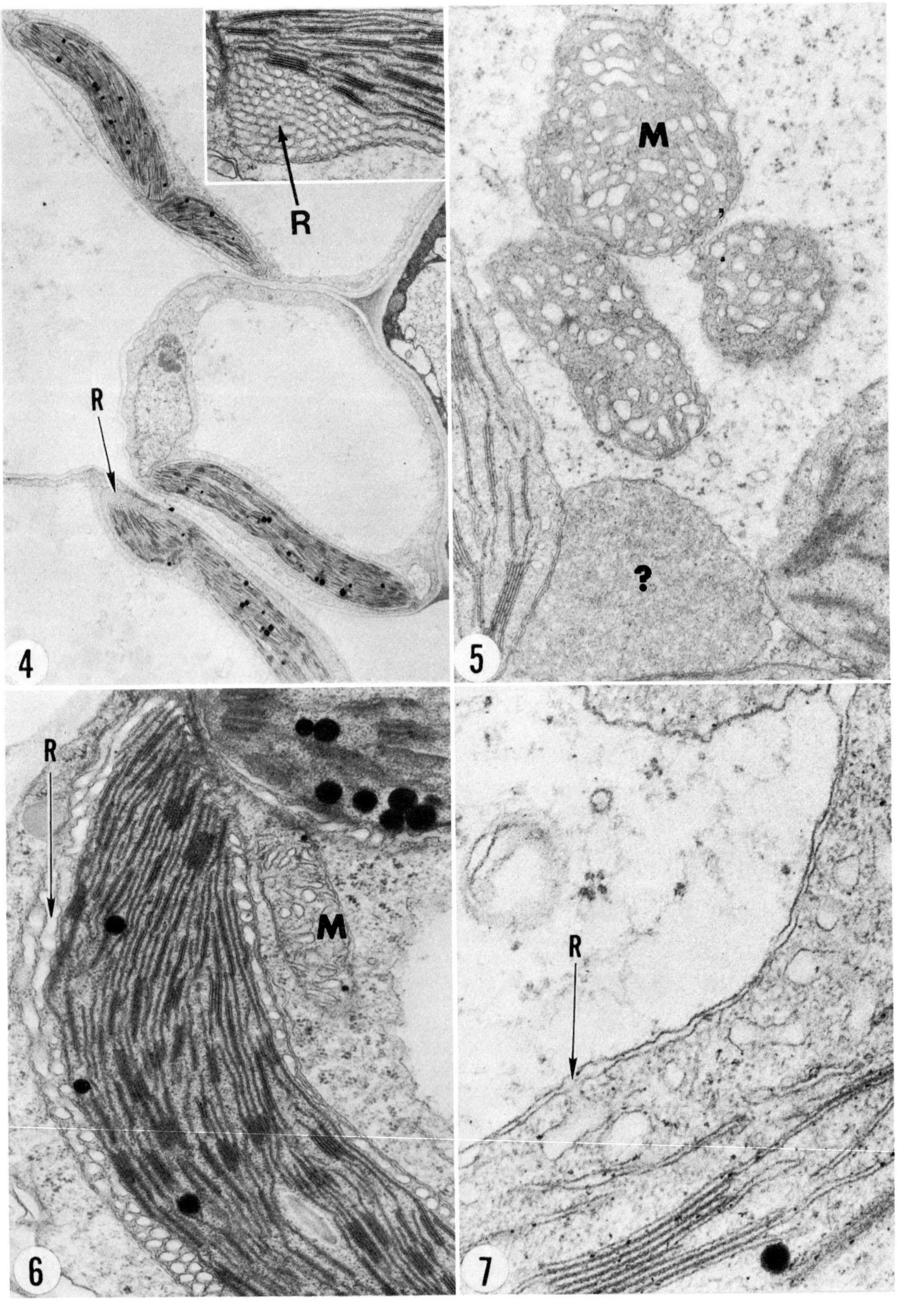
R
R
4
M
?
5
R
M
6
R
7

fied, single membrane-bound body is also seen in this figure. The contrast between the mitochondria in the bundle-sheath and those in the palisade cells is seen by comparing Fig. 5, 6. The palisade-cell chloroplast in Fig. 6 also illustrates the peripheral reticulum. The channels and vesicles appearing in the same section lend strong support to the interpretation of these structures as a network of tubules. The connection between the tubules of this reticulum and the inner plastid membrane is clearly seen in the chloroplast shown in Fig. 7.

Portions of adjoining bundle-sheath cells of *At. lentiformis* are illustrated in Fig. 8. The chloroplasts possess grana and store starch. They are usually much "rounder" than the mesophyll-cell chloroplasts. The mitochondria are large and possess numerous cristae. Unidentified, single membrane-bound bodies are also observed in these cells. The chloroplasts in the palisade cells (Fig. 9) are somewhat flatter than those in the bundle-sheath cells, and they contain little starch. Grana are rudimentary and seldom contain more than two or three thylakoids in a grana stack. The peripheral reticulum is present in these chloroplasts, but it is not as well developed as in *Am. edulis* chloroplasts.

The size difference between the mitochondria in the bundle-sheath and palisade cells of *Am. edulis* is also found in *At. lentiformis* and is illustrated in Fig. 10. The cell at the top of this electron micrograph is from the bundle sheath and the cell at the bottom is from the palisade tissue. The difference in size and internal complexity of these mitochondria is readily apparent. The chloroplast in the palisade cell in Fig. 10 has a sizable starch grain, but the occurrence of starch grains of this size in such chloroplasts is much less frequent than in bundle-sheath-cell chloroplasts.

Discussion—Two species of dicotyledons from closely related families possess similar features of leaf anatomy and cell ultrastructure. In addition, the leaf anatomy of five species of *Atriplex* was very similar to the species of *Atriplex* whose leaf-cell ultrastructure was examined. These structural features have also been found in certain members of the grass tribes Andropogoneae and Maydeae. The bundle-sheath cells in the two species discussed in this paper have chloroplasts which store large amounts of starch. In this respect, they resemble the chloroplasts in the bundle-sheath cells of grasses such as sugar cane and corn. Unlike the bundle-sheath chloroplasts in the latter plants, they possess grana. The partially developed chloroplasts of corn possess grana, but they are almost completely lost in the mature chloroplast (Johnson, 1964). Recent observations indicate that chloroplasts in bundle-sheath cells of immature tissue of sugar cane leaves also possess grana. As the chloroplasts mature, the grana are lost; thus it would appear that the structurally less complex bundle-sheath chloroplasts of the tropical grasses are a result of reduction and might represent a greater degree of specialization than is found in similar chloroplasts in the dicots.

The mitochondria in the bundle-sheath cells of the dicot species are considerably larger than the mitochondria in the palisade cells and this feature has not been observed in the published ultrastructure studies of the grasses. The palisade cells of the two dicot species are arranged in a concentric layer around the bundle sheaths and this feature is reminiscent of the mesophyll in sugar cane and other related grasses. The sizes of the chloroplasts and mitochondria in the palisade cells are similar to those in many other plant species. The palisade-cell chloroplasts contain very little starch and this feature is also found in sugar cane and corn mesophyll cells. The chloroplasts in both bundle-sheath and palisade cells in *Am. edulis* and *At. lentiformis* possess a peripheral reticulum. A similar reticulum has also been observed in corn (Shumway and Weier, 1967; Rosado-Alberio, Weier, and Stocking, 1967) and in sugar cane (Laetsch, unpublished). This highly developed reticulum has not to my knowledge been observed in chloroplasts from other species of plants.

The chloroplasts in the bundle-sheath cells of these two dicot species can be distinguished from the palisade chloroplasts on the basis of size or shape and starch-storing capacity. They possess grana; hence they differ in this respect from the bundle-sheath chloroplasts of certain grasses. The situation in *At. lentiformis* is almost the reverse of that found in certain grasses because its palisade-cell chloroplasts have poorly developed grana. Aspects of mitochondrial size, and perhaps internal structure, in the two cell layers of those dicots are other features which apparently differ from the situation in certain grasses.

The leaves of the plants which have been found to possess the C_4-dicarboxylic-acid pathway of photosynthetic CO_2 fixation possess a

Fig. 4–7.—Fig. 4. Palisade cells of *Amaranthus edulis*. The chloroplasts have a highly developed peripheral reticulum (R), ×5,280. Insert: enlarged view of peripheral reticulum (R), ×22,750.—Fig. 5. Large mitochondria (M) of bundle-sheath cell of *Am. edulis*. Note single membrane-bound body adjacent to chloroplasts, ×28,000.—Fig. 6. Chloroplast from palisade cell of *Am. edulis*. It is apparent that the peripheral reticulum (R) is composed of tubules. Compare the size and structure of the mitochondrion (M) with the mitochondria illustrated in Fig. 5, ×26,450.—Fig. 7. The tubules of the peripheral reticulum (R) are connected with the inner chloroplast membrane. This illustration is an enlarged view of a portion of a chloroplast from a bundle-sheath cell of *Am. edulis*, ×29 750.

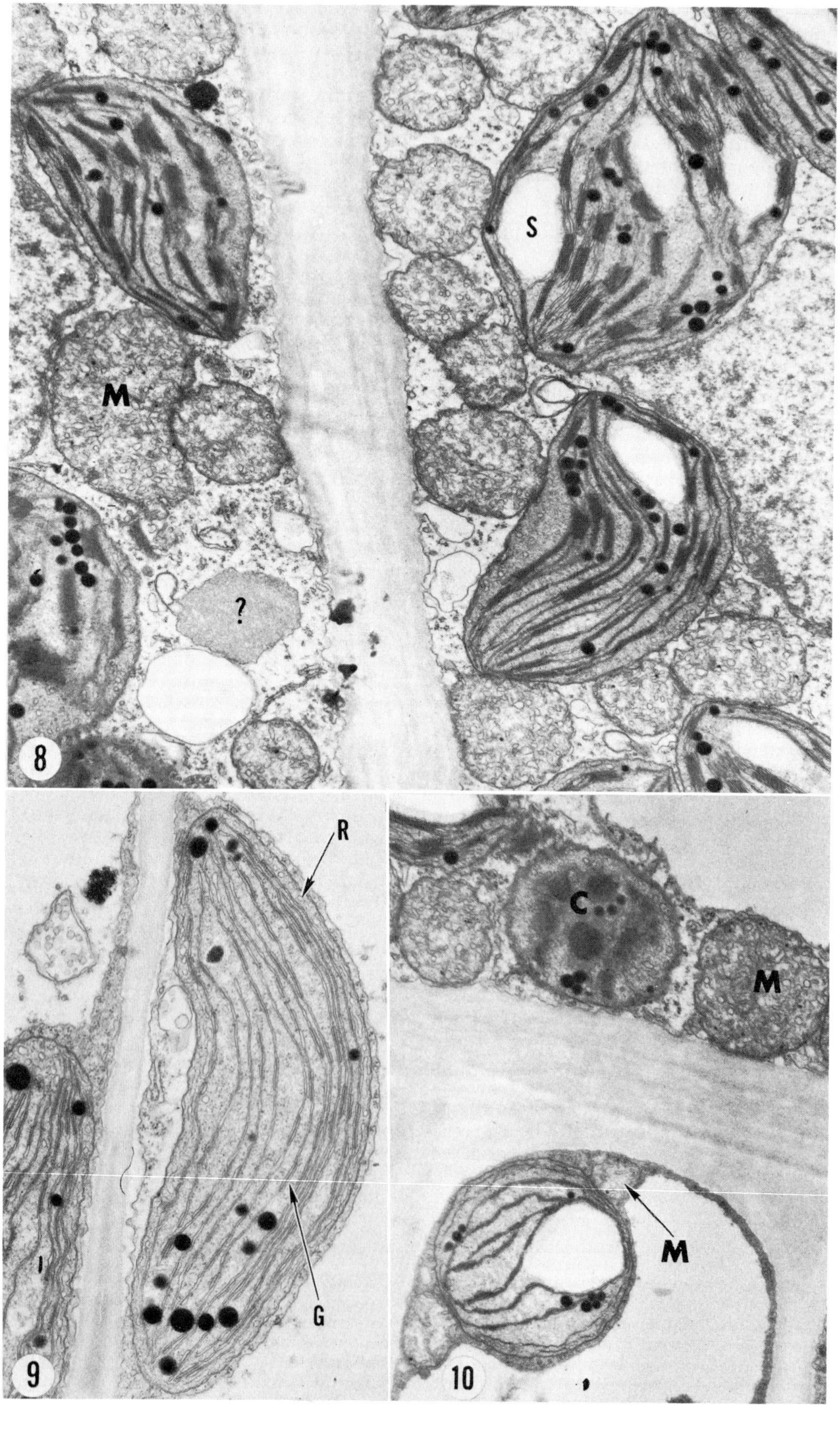
M
S
?
8
R
G
9
C
M
M
10

complex of similar anatomical and cytological characters. A rather remarkable example of biochemical and morphological convergence between widely separated taxa is before us, and it will be interesting to find if this set of distinguishing characters is widespread in the plant kingdom. Experience to date suggests that in this search an examination of gross leaf anatomy should provide valid markers concerning the specific carbon-fixation pathway followed by particular species.

These dicots share a pattern of characters at both the biochemical and structural level with particular groups in the Gramineae. It should be asked how these characters are related and what their biological significance is. Even though the data are incomplete, it is profitable to propose a hypothesis unifying the scattered observations. Such a hypothesis can at least provide a testing ground for further work in this field.

The grasses possessing the C_4-dicarboxylic-acid photosynthetic carbon-fixation pathway, resistance to photosynthetic photosaturation, and dimorphic chloroplasts belong to taxa which are presumed to be of tropical origin (Stebbins, 1956). The Amaranthaceae are also thought to be of tropical origin (Lawrence, 1951). *Atriplex lentiformis* and the other species of this genus with similar leaf anatomy are native to hot, arid regions in the United States and Australia. All of these taxa originated in areas of high light intensity and at least seasonal dry periods. The hypothesis proposed here considers that the biochemical, physiological, and structural features shared by the plants discussed are adaptations to high light intensity in an environment in which water stress could frequently be the limiting factor for photosynthesis. Photosaturation could be a critical factor in photosynthesis if the time available for photosynthesis were severely limited because of water stress. The ability of the plants in question to maintain a maximum rate of photosynthesis under high light intensity could be related to their leaf anatomy and chloroplast specialization. The large size of the bundle-sheath chloroplasts and their ability to store large amounts of starch suggest that they act as "sinks" for carbohydrate end products of photosynthesis. Their proximity to the vascular tissue is an efficient arrangement for the rapid transport of reactants and end products of photosynthesis. The concentric arrangement of the palisade and mesophyll cells around the bundle sheaths provides a minimal distance for the transport of photosynthetic materials from one cell layer to the next. Most cells in this tissue are no more than one cell layer removed from vascular tissue. The small amount of starch found in these palisade- and mesophyll-cell chloroplasts relative to bundle-sheath-cell chloroplasts suggests a rapid transport of photosynthetic products. The large mitochondria in the dicot bundle-sheath cells could provide the energy for such a transport system. The peripheral reticulum of the chloroplasts in both kinds of cells consists of a network of tubules leading from the interior of the plastid to the space between the limiting membranes. It is tempting to speculate that this reticulum is another adaptation for rapid transport.

According to the arguments above, the "normal" chloroplasts in the palisade cells could maintain a maximum rate of photosynthesis under high light intensity because the rapid transport of photosynthetic intermediates to the bundle-sheath chloroplasts would prevent an accumulation of end products. This interpretation would be consistent with the observations of Hesketh (1963) that the differences in photosynthetic rates between species are influenced by the kinetics of the dark reactions. The production of oxaloacetate rather than phosphoglyceric acid as the first product of photosynthetic carbon fixation could be another adaptation for efficiency in photosynthesis because the phosphoenol-pyruvate-carboxylase-enzyme system has a greater affinity for CO_2 than the carboxy dismutase system (Walker, 1962). In an arid environment water stress could frequently limit photosynthesis during the light period, and a rapid method of carbon fixation would have a definite selective advantage. Brown (1958) suggested that leaf anatomy of the grasses was correlated with their environment. He made the particular point that the leaf anatomy of chloridoid grasses, some members of which show the characteristics outlined above, might be correlated with high temperature and high light intensity and might also be related to drought resistance.

It is known that sugar cane and maize can remove far more CO_2 from a closed system than many plants (Forrester, Krotkov, and Nelson, 1966). This has been attributed to a lack of enzymes involved in photorespiration (Moss, 1967), a fact that might help to explain the high

Fig. 8–10.—Fig. 8. Adjacent bundle-sheath cells of *Atriplex lentiformis*. The chloroplasts contain starch (S). The mitochondria (M) are large and single membrane-bound bodies (?) are present, ×15,600.—Fig. 9. Palisade-cell chloroplast from *At. lentiformis*. A peripheral reticulum (R) is present, but is not as highly developed as in the chloroplasts illustrated in Fig. 4, 6. The grana (G) generally consist of two thylakoids. Starch is frequently absent in these chloroplasts, ×22,750.—Fig. 10. Portions of an adjacent bundle-sheath cell (top) and a palisade cell (bottom) are illustrated. Compare the size of the mitochondria (M) in the two cells. The chloroplast section (C) in the bundle-sheath cell was made through the edge of a chloroplast, ×15,600.

photosynthetic efficiency of the tropical grasses. The relationship between the structural features discussed above and this apparant lack of a photorespiratory system is not at all clear. It is expected that further work on this latter system, as well as on other elements in the set of correlates characteristic of these plants, will modify the proposed hypothesis.

It should be pointed out that the carbon-fixation pathway in question is very similar to the crassulacean acid metabolism (CAM) followed by many succulent plants (Ranson and Thomas, 1960; Walker, 1962). These plants live in arid environments and their ability to fix CO_2 in the dark permits the closure of stomata during the day. Malate is the end-product of the dark-fixation of CO_2 in CAM, and it apparently serves as a reservoir of fixed carbon which can be utilized in Calvin cycle reactions during the light period. It has been suggested by Osmond (1967) that the dicarboxylic acids produced during "cane-type" photosynthesis could serve as temporary carbon sinks. A parallel situation with CAM could then prevail during periods of water stress when the stomata were closed. It is tempting to speculate that the C_4-dicarboxylic acid pathway of photosynthesis in *Atriplex* might possess more than a surface relationship to CAM. The halophytic species of *Atriplex* have a considerable osmotic problem to solve, and the fixation of carbon into organic acids in this genus may have more to do with regulating osmotic pressures in the cell vacuole than with efficient carbon assimilation. The convergence of morphological and biochemical features in these various taxa could reflect the evolution of similar means for different ends.

The hypothesis that the photosynthetic pathways and structural features of the groups of plants discussed above are adaptations for efficient and rapid carbon fixation in arid environments can be readily tested. If proven to be valid, this hypothesis could help to interpret the origin and evolution of the plants in question.

LITERATURE CITED

Brown, W. V. 1958. Leaf anatomy in grass systematics. Bot. Gaz. 5: 517–521.

Burr, G. O., C. E. Hartt, H. W. Brodie, T. Tanimoto, H. P. Kortschak, D. Takahashi, F. M. Ashton, and R. E. Coleman. 1957. The sugar cane plant. Annu. Rev. Plant Physiol. 8: 275–308.

El-Sharkawy, M., and J. Hesketh. 1965. Photosynthesis among species in relation to characteristics of leaf anatomy and CO_2 diffusion resistances. Crop Sci. 5: 517–521.

———, R. S. Loomis, and W. A. Williams. 1967. Apparent reassimilation of respiratory carbon dioxide by different plant species. Physiol. Plantarum 20: 171–186.

Forrester, M. L., G. Krotkov, and C. D. Nelson. 1966. Effect of oxygen on photosynthesis, photorespiration, and respiration in detached leaves. I. Soybean.
II. Corn and other monocotyledons. Plant Physiol. 41: 422–431.

Hall, H. M., and F. E. Clements. 1923. The phylogenetic method in taxonomy. The North American species of *Artemisia, Chrysothamnus*, and *Atriplex*. Carnegie Institution, Washington, D.C. Publication no. 326.

Hatch, M. D., and C. R. Slack. 1966. Photosynthesis by sugar cane leaves. A new carboxylation reaction and the pathway of sugar formation. Biochem. J. 101: 103–111.

———, ———, and H. S. Johnson. 1967. Further studies on a new pathway of photosynthetic carbon dioxide fixation in sugar cane and its occurrence in other plant species. Biochem. J. 102: 417–422.

Hegi, G. 1906. Illustrierte Flora von Mittel-Europa. J. F. Lehmann, Munchen.

Hesketh, J. D. 1963. Limitations to photosynthesis responsible for differences among species. Crop Sci. 3: 493–496.

———. 1967. Enhancement of photosynthetic CO_2 assimilation in the absence of oxygen, as dependent upon species and temperature. Planta 76: 371–374.

———, and D. N. Moss. 1963. Variation in the response of photosynthesis to light. Crop Sci. 3: 107–110.

Hodge, A. J., J. D. McLean, and F. V. Mercer. 1955. Ultrastructure of the lamellae and grana in the chloroplast of *Zea mays* L. Biophys. Biochem. Cytol. 1: 605–614.

Johnson, Sr. M. C. 1964. An electron microscope study of the photosynthetic apparatus in plants with special reference to the Gramineae. Ph.D. thesis, Univ. Texas, Austin.

Laetsch, W. M., D. A. Stetler, and A. J. Vlitos. 1965. The ultrastructure of sugar cane chloroplasts. Z. Pflanzenphysiol. 54: 472–474.

Lawrence, G. H. 1951. Taxonomy of vascular plants. Macmillan Co., New York.

Luft, J. H. 1961. Improvements in epoxy resin embedding methods. Biophys. Biochem. Cytol. 9: 409–414.

Moss, D. N. 1962. The limiting carbon dioxide concentration for photosynthesis. Nature 193: 587.

———. 1967. High activity of the glycolic acid oxidase system in tobacco leaves. Plant Physiol. 42: 1463–1464.

Osmond, B. 1967. β-carboxylation during photosynthesis in *Atriplex*. Biochem. Biophys. Acta 141: 197–199.

Ranson, S. L., and M. Thomas. 1960. Crassulacean acid metabolism. Annu. Rev. Plant Physiol. 11: 81–110.

Reynolds, E. S. 1963. The use of lead citrate at high *p*H as an electron-opaque stain in electron microscopy. J. Cell Biol. 17: 208–212.

Rhoades, M. M., and A. Carvalho. 1944. The function and structure of the parenchyma sheath plastids of the maize leaf. Bull. Torrey Bot. Club 71: 335–346.

Rosado-Alberio, J., T. E. Weier, and C. R. Stocking. 1967. Chloroplast ultrastructure in some chloro-

phyll mutants of corn. Plant Physiol. 42 (Suppl): 27.
SHUMWAY, L. K., AND T. E. WEIER. 1967. The chloroplast structure of iojap maize. Amer. J. Bot. 54: 773–780.
SLACK, C. R., AND M. D. HATCH. 1967. Comparative studies on the activity of carboxylases and other enzymes in relation to the new pathway of photosynthetic carbon dioxide fixation in tropical grasses. Biochem. J. 103: 660–665.
STEBBINS, G. L. 1956. Cytogenetics and evolution of the grass family. Amer. J. Bot. 43: 890–905.
TREGUNNA, E. B., AND J. DOWNTON. 1967. Carbon dioxide compensation in members of the Amaranthaceae and some related families. Can. J. Bot. 45: 2385–2387.
WALKER, D. A. 1962. Pyruvate carboxylation and plant metabolism. Biol. Rev. 37: 215–256.

Reprinted from PLANT PHYSIOLOGY, Vol. 41, No. 5, May, 1966, pages 885–890.
(PRINTED IN U.S.A.)

Regulation of Photosynthetic Capacity in Chlamydomonas mundana[1]

George K. Russell[2, 3] and Martin Gibbs
Department of Biology, Brandeis University, Waltham, Massachusetts

Received December 28, 1965.

Summary. A regulatory system has been described in the obligately phototrophic green alga *Chlamydomonas mundana.* Cells grown in acetate media are unable to fix carbon dioxide in the light but carry out a photoassimilation of acetate to carbohydrate; cells cultured with carbon dioxide as the sole source of cellular carbon carry out typical green plant photosynthesis. The control appears to take place at the level of the reductive pentose phosphate cycle. The presence of sodium acetate in the medium strongly inhibits formation of ribulose-1,5-diphosphate carboxylase, ribulose-5-phosphate kinase, and one of the 2 fructose-1,6-diphosphate aldolase activities of the cell. Ribose-5-phosphate isomerase is present in higher activity in autotrophic cells. Changes in the levels of triose phosphate dehydrogenase were also noted. The total pigment content of the cell and the photosynthetic electron transport reactions are not altered under different conditions of growth.

Little information is available in the literature concerning the presence or absence of metabolic control systems affecting photosynthetic capacity in photosynthetic green algae. A variety of sugars, organic acids and alcohols can serve as carbon sources for growth (4), but the extent to which the organic carbon source may bring about adaptive changes in the photosynthetic capacity of the cell through inhibition and/or repression mechanisms is unknown.

In at least 2 cases it has been shown that the CO_2 fixing and O_2 evolving capacity of the cell is not altered by the nature of the growth medium and the environmental conditions under which the cells are grown. *Chlorella vulgaris* cells form normal amounts of chlorophyll when cultured in either the light or dark in media containing glucose, and are photosynthetically competent under all conditions of growth (16). Similar observations have been made for the wild-type strain of *Chlamydomonas reinhardi* (11). The chloroplast seems to be a constituitive organelle in these 2 algal strains.

Photosynthetic O_2 evolution and CO_2 fixation by whole cells of the obligately phototrophic green alga *Chlamydomonas mundana* grown in a medium containing sodium acetate have been reported to be very sluggish (5, 6). However, since the organism is capable of autotrophic growth with CO_2 as the sole source of cellular carbon (with a doubling time of approximately 8 hrs), it seemed apparent that the autotrophic cells must be able to fix CO_2 at a high rate, and that acetate was probably effecting some change in the photosynthetic capacity of the cell.

The purpose of the present communication is to demonstrate that autotrophically grown cells of *C. mundana* fix CO_2 and evolve O_2 in the light at high rates, and that the presence of sodium acetate in the medium appears to inhibit the synthesis of several enzymes of the reductive pentose-P cycle. Acetate apparently has no effect on the photosynthetic electron transport system.

Methods and Materials

Organism and Culture Methods. A slant of the unicellular green alga *Chlamydomonas mundana* Gerloff var. *astigmata* nov. var. (Mojave strain) was obtained from Dr. R. W. Eppley. Cells were cultured at 32° in liquid media described previously (25) to which 1.0 μg vitamin B_{12} and 100 μg thiamine hydrochloride were added per liter. Sodium acetate (0.3 %) was added where required. Cultures were grown in continuous light at 2000 ft-c. Best growth was obtained when a mixture of 5 % CO_2:95 % N_2 was bubbled through the medium.

Carbon Dioxide Fixation. CO_2 fixation was measured as the incorporation of $NaH^{14}CO_3$. Reactions were carried out in standard 17 ml Warburg flasks in 1.0 ml reaction mixtures. After exposure to

[1] This investigation was aided by grants from the National Science Foundation and the United States Air Force through the Air Force Office of Scientific Research of the Air Research and Development Command, under Contract AF-49(638)789. A preliminary report of these findings has been published (18).
[2] National Science Foundation postdoctoral fellow.
[3] Present address: Department of Biology, Princeton University, Princeton, New Jersey.

2000 ft-c for varying periods of time, the reactions were terminated by the addition of 0.3 ml 0.5 N HCl. Aliquots were plated in duplicate at infinite thinness and counted in a Nuclear-Chicago gas flow counter.

Photosynthetic Electron Transport Reactions. The photoreduction of TPN and 2,6-dichlorophenol indophenol by isolated chloroplast fragments of *C. mundana* was measured in the model 14 Cary recording spectrophotometer as described by Levine and Smillie (14), using the near infrared lamp as the source of actinic and measuring light. Photosynthetic phosphorylation was measured according to the technique of Russell and Levine (19). Photosynthetic pyridine nucleotide reductase (PPNR) was prepared from spinach according to the method of San Pietro (21). Chloroplast fragments were prepared by sonication of whole cells of *C. mundana* for 60 seconds in the MSE (Measuring and Scientific Equipment Company) sonic oscillator (10 kc). Chlorophyll concentrations were determined after Arnon (1).

Preparation of Cell-Free Extracts. Cell-free extracts of the organism were prepared by sonication of whole cells for 2 minutes with the MSE sonicator. The extracts were clarified by centrifugation for 10 minutes at 10,000 × *g* and the clear supernatant fluids were used for the enzyme assays. Soluble protein was measured by the biuret method (10).

Enzyme Assays. Ribulose-1,5-diP carboxylase was measured according to Fuller et al. (8). The reaction mixture contained *Chlamydomonas* extract, $NaH^{14}CO_3$ (10 μmoles, 5–10 μc), $MgCl_2$ (20 μmoles), EDTA (0.1 μmole), GSH (12.5 μmoles), and 0.25 μmole ribulose-1,5-diP in a total volume of 2.0 ml. The reactions were run for 10 minutes, terminated with 0.5 N HCl and assayed for counts fixed as described above.

Fructose-1, 6-diP aldolase was measured by coupling with triose-P isomerase and α-glycerol-P dehydrogenase (28). The colorimetric assay of Sibley and Lehninger (22) was also used for the determination of aldolase activity.

DPN and TPN dependent triose-P dehydrogenase activities were measured in the reverse direction according to the method of Wu and Racker (28).

Ribose-5-P isomerase was assayed after Axelrod (2). A standard curve was prepared using purified spinach phosphoriboisomerase (2).

Ribulose-5-P kinase was measured according to Wu and Racker (28). Commercial pyruvate kinase and lactic dehydrogenase (Sigma) were used in the assay. Corrections were made for the endogenous reaction in the absence of substrate due to the presence of adenosine triphosphatase in the crude extracts.

Results

The data in table I clearly demonstrate that autotrophic cells of *Chlamydomonas mundana* are able to carry out green plant photosynthesis. High rates of CO_2 fixation and O_2 evolution have consistently been measured. The rate of CO_2 fixation by auto-

Table I. *Light Dependent CO_2 Fixation and O_2 Evolution by Autotrophic and Photoheterotrophic Cells of* C. mundana

CO_2 fixation was measured as the incorporation of $NaH^{14}CO_3$. Reaction mixtures containing whole cells of *C. mundana* equivalent to 100 to 150 μg chlorophyll were incubated in 0.02 M $NaH^{14}CO_3$ in the light (2000 ft-c) and dark for 10 minutes at 30°. Unlabeled sodium acetate was added to the appropriate flasks as indicated below. The data are corrected for dark fixation.

Photosynthetic O_2 evolution was measured manometrically in $NaHCO_3/KHCO_3$ buffer. The conditions were identical to those given above for the measurement of CO_2 fixation.

Carbon source for growth	Additions to reaction mixture	CO_2 Fixation (μmoles/hr per mg chl)	O_2 Evolution (μmoles/hr per mg chl
Acetate	None	1.5	3.0
Acetate	0.1 % Acetate	1.7	...
Acetate	0.2 % Acetate	1.3	...
CO_2	None	133	80
CO_2	0.1 % Acetate	101	...
CO_2	0.2 % Acetate	96	...

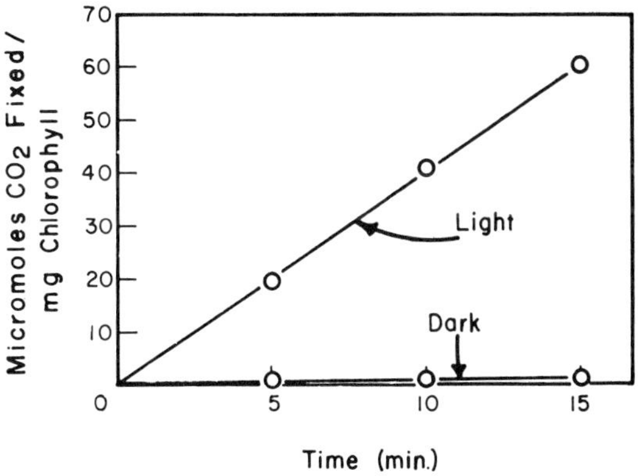

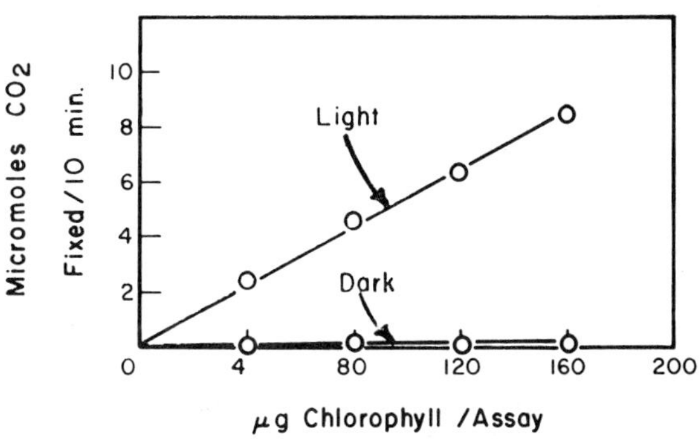

FIG. 1. CO_2 fixation by autotrophic cells of *C. mundana* as a function of time. Each flask contained whole cells equivalent to 87 μg chlorophyll.

FIG. 2. CO_2 fixation by autotrophic cells of *C. mundana* as a function of chlorophyll concentration. The reactions were run for 10 minutes.

trophic cells is linear with respect to time for at least 15 minutes (fig 1) and is linear with respect to chlorophyll concentration up to 160 μg chlorophyll per reaction mixture under the conditions of the assay (fig 2). The highest rate of CO_2 fixation measured with whole cells of *C. mundana* is 300 μmoles per hour per mg chlorophyll (fig 2) and is more than sufficient to account for the observed growth rate in inorganic media.

The results recorded in table I also demonstrate that cells grown with acetate are essentially unable to fix CO_2 and evolve O_2 in the light, in confirmation of Eppley et al. (5, 6). The addition of sodium acetate to the reaction mixture did not restore CO_2 fixing capacity to cells grown on acetate, and had only a small inhibitory effect on the rate of CO_2 fixation by autotrophic cells.

The ability to fix CO_2 is rapidly restored to acetate grown cells upon transfer of the cells to inorganic media (fig 3). Within 6 hours high rates of CO_2 incorporation were observed. In control experiments in which acetate grown cells were resuspended in acetate media, there was no increase in CO_2 fixing ability. It should be noted that little if any growth (measured as increase of dry wt) took place during the first 5 hours of the adaptation from acetate to CO_2. Only after 5 and one-half or 6 hours was an increase in dry weight observed.

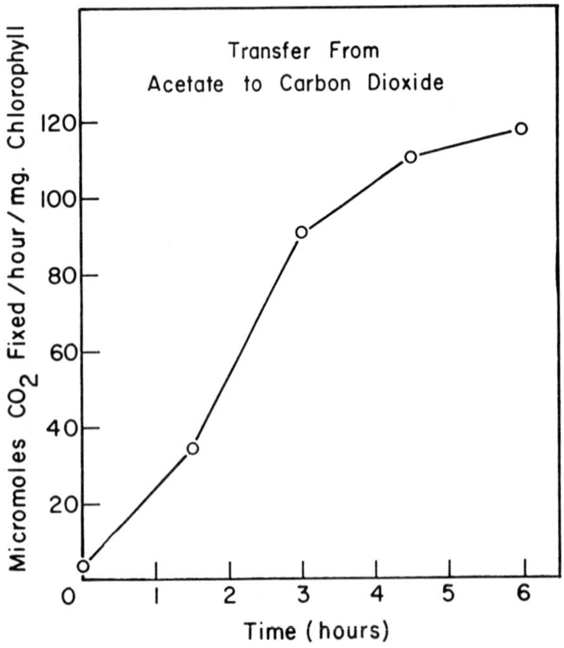

FIG. 3. CO_2 fixation by whole cells of *C. mundana* after transfer from acetate medium to inorganic medium. Acetate cells were harvested, washed once with inorganic medium, and resuspended in inorganic medium under the usual growth conditions. At various times samples were removed and assayed for their ability to fix CO_2. Each reaction mixture contained whole cells equivalent to 50 to 100 μg chlorophyll. The reactions were carried out for 10 minutes. The values given in the figure are corrected for dark fixation.

It seemed that acetate might exert its effect either at the level of the photosynthetic electron transport system or at the level of the reductive pentose-P cycle, the pathway of CO_2 fixation in photosynthesis. Chloroplast fragments of both acetate grown cells and autotrophic cells were found to carry out photosynthetic electron transport reactions at roughly similar rates (table II). The Hill reaction (measured as the photoreduction of 2, 6-dichlorophenol indophenol), the photoreduction of TPN in the presence of spinach PPNR, and cyclic photosynthetic phosphorylation with phenazine methosulfate took place at approximately the same rates with preparations of both kinds of cells. The presence of acetate in the growth medium does not appear to exert a major effect on the photosynthetic electron transport system of the cell. In addition, there is no major difference in the chlorophyll content of cells grown under both sets of conditions; acetate cells and autotrophic cells both contain about 2.5 to 3.0 % of the dry weight of the cell as chlorophyll.

Table II. *Photosynthetic Reactions by Chloroplast Fragments of Autotrophic and Photoheterotrophic Cells*

To assay for TPN production, the reaction mixture (0.8 ml) contained chloroplast fragments (10–15 μg chlorophyll), an excess of purified spinach PPNR, and the following in μmoles: TPN, 0.2; $MgCl_2$, 0.015; and Tris buffer (pH 7.5), 30. The reaction was measured as the increase of OD at 340 mμ in the Cary recording spectrophometer.

The Hill reaction was assayed as follows: the reaction mixture (0.8 ml) contained chloroplast fragments (3–4 μg chlorophyll) and the following in μmoles: phosphate buffer (pH 6.8), 10; KCl, 5; and 2,6-dichlorophenol indophenol, 0.05. The reaction was measured as the decrease of absorption at 600 mμ in the Cary recording spectrophotometer.

Photophosphorylation was assayed as follows: the reaction mixture (3.0 ml) contained chloroplast fragments (100 μg chlorophyll) and the following in μmoles: Tris buffer (pH 7.5), 100; $MgCl_2$, 10; K_2HPO_4, 10; ADP, 10; and phenazine methosulfate, 0.15. The reaction was measured as the disappearance of Pi (19).

These data represent the average of 2 assays.

Reaction	Carbon source for growth Acetate	 CO_2
	(μmoles/hr per mg chl)	
Photoreduction of TPN	20	30
Hill reaction	46	68
Photosynthetic phosphorylation	100	110

A survey of some of the enzymes of the reductive pentose-P cycle revealed major differences between acetate grown and autotrophic cells. In contrast to autotrophically grown cells, only low levels of ribulose-5-P kinase and ribulose-1, 5-diP carboxylase were detected in extracts of cells grown with acetate (table III). Ribulose-5-P kinase was difficult to measure because of the presence of adenosine triphos-

Table III. *Enzyme Activities in Extracts of Autotrophic and Photoheterotrophic Cells*

The values given in the table represent the average of at least 2 experiments with different batches of cells.

Enzyme	Carbon source for growth Acetate	CO_2
	(μmoles of substrate consumed/hr per mg protein	
Ribulose-1,5-diP carboxylase	0.02	0.6
Ribulose-5-P kinase	0.1	4
Ribose-5-P isomerase	39	167
Triose-P dehydrogenase (TPN)	3	7
Triose-P dehydrogenase (DPN)	11	17
Fe^{+2} dependent aldolase	9	<1
Nonmetal aldolase	<0.5	15

phatase activity in the crude extracts; the data for this activity are less reliable than the data for ribulose-1, 5-diP carboxylase. Mixing experiments give no evidence for an inhibitor of ribulose-1, 5-diP carboxylase in extracts of acetate cells, or for an activator of the enzyme in autotrophic cells (table IV). Ribulose-1, 5-diP carboxylase is either not present in acetate cells or is present in a totally inactive form.

Fructose-1, 6-diP aldolase activity was detected in extracts of both types of cells, but it appears that the cell contains 2 distinct enzymes, one of which is formed under each set of growth conditions. Autotrophic cells contain an aldolase which has a high affinity for fructose-1, 6-diP, is not inhibited by chelating agents, and is not stimulated by Fe^{+2} (table V). Acetate cells contain an enzyme which is inhibited by EDTA and *o*-phenanthroline, is markedly stimulated by Fe^{+2}, and has a much lower affinity for fructose-1, 6-diP in crude extracts (table V).

Discussion

The results reported here demonstrate very clearly that the inclusion of an organic carbon source in the medium of *C. mundana* brings about adaptive changes in the photosynthetic capacity of the cell. In the presence of sodium acetate, several of the enzymes involved in the reduction of CO_2 to carbohydrate are formed in greatly reduced amounts. Under these conditions, the cells carry out a light dependent assimilation of acetate (6), similar to that reported for *Rhodospirillum rubrum* (23), other photosynthetic bacteria (9), and the acetate requiring obligately phototrophic green alga *Chlamydobotrys*

Table IV. *Comparison of Ribulose-1,5-diP Carboxylase Activities*

Each reaction mixture contained extract as indicated above, 80 μmoles Tris buffer (pH 8.0), 20 μmoles $MgCl_2$, 0.1 μmole EDTA, 12.5 μmoles GSH, 0.25 μmole ribulose-1,5-diP, and 10 μmoles $NaHCO_3$ in a total volume of 2.0 ml. The values given below are corrected for fixation in the absence of added substrate.

Source of extract	Mg protein in extract	Enzyme activity mμmoles fixed/10 min
Acetate cells	0.57	0.5
Acetate cells	1.15	1.3
CO_2 cells	0.82	58
CO_2 cells	1.64	125
Acetate cells	0.57	
plus		62
CO_2 cells	0.82	

Table V. *Comparison of 2 Fructose-1,6-diP Aldolases from Chlamydomonas*

The effects of EDTA, *o*-phenanthroline, *p*-CMB and Fe^{+2} were measured using the colorimetric assay of Sibley and Lehninger (22). Each reaction mixture contained *Chlamydomonas* extract, Tris buffer (100 μmoles), hydrazine (140 μmoles), cysteine (10 μmoles), fructose-1,6-diP (10 μmoles), and water to a total volume of 2.5 ml. Fe^{+2} was added as $Fe(NH_4)_2(SO_4)_2$. The reactions were run for 10 minutes and killed with 2.0 ml 10 % trichloroacetic acid. Appropriate aliquots were assayed for hydrazone formation as described in reference 22. Michaelis constants were determined using the spectrophotometric assay of Wu and Racker (28). Each reaction mixture contained extract, DPNH (0.2 μmole), excess triose phosphate isomerase and α-glycerol phosphate dehydrogenase (Sigma), cysteine (5 μmoles), and varying amounts of substrate. 10^{-4} M $Fe(NH_4)_2(SO_4)_2$ was added to the assay of extracts of acetate grown cells. Higher concentrations of Fe^{+2} interfered with the assay.

	Carbon source for growth Acetate	CO_2
K_m (Fructose-1,6-diP)	2×10^{-3} M	5×10^{-5} M
Effect of 10^{-3} M EDTA	95–100 % Inhibition	None
Effect of 10^{-3} M *o*-phenanthroline	95–100 % Inhibition	None
Effect of 10^{-3} M Fe^{+2}	3–5-fold stimulation	None
Effect of 10^{5-} M *p*-CMB	15 % Inhibition	97 % Inhibition

(17). Eppley et al. (6) have reported the adaptive formation of isocitrate lyase, an enzyme of the glyoxylate cycle, in cells of *C. mundana* cultured with acetate. Extracts of autotrophic cells show very little isocitrate lyase activity. Upon transfer to inorganic media, the cells quickly adapt to typical green plant photosynthesis and are capable of growth with CO_2 as the sole source of carbon. The evidence presented here suggests that acetate or some metabolite derived from acetate inhibits enzyme formation, but definitive evidence for *de novo* synthesis during adaptation has not been given. Acetate or an acetate derivative may inhibit enzyme action.

Ribulose-1, 5-diP carboxylase has been implicated as the enzyme directly responsible for the photosynthetic fixation of CO_2 both through kinetic analysis of the soluble products of CO_2 fixation (3) and, more recently, through investigation of mutant strain of *Chlamydomonas reinhardi* devoid of carboxylase activity (15). The results reported in this paper indicate that a regulatory system exists in *Chlamydomonas mundana* controlling the formation of ribulose-1, 5-diP carboxylase and probably ribulose-5-P kinase as well as one of the 2 fructose-1, 6-diP aldolase activities of the cell. The absence of these activities during growth on acetate probably spares the cell metabolic energy needed for the synthesis of these enzymes (if the control is at the level of enzyme synthesis), or permits more efficient utilization of energy for carbon assimilation (if the control is at the level of enzyme activity).

Changes in the levels of ribulose-1, 5-diP carboxylase brought about by organic carbon sources have been reported in several photosynthetic bacteria (8, 9, 13) and the chemosynthetic bacterium, *Micrococcus denitrificans* (12). Fuller and Gibbs (7) reported that acetate grown cells of *Chlorella variegata* contain very low levels of the enzyme. In this latter case, however, acetate brings about repression of the entire chloroplast, since acetate grown cells of *Chlorella variegata* are essentially devoid of chlorophyll. Dark grown *Euglena gracilis* cells contain low levels of the enzyme and are pigmentless (7). The unique feature of the *C. mundana* system reported in this paper, at least with respect to photosynthetic algae, is that acetate specifically affects the formation of several of the enzymes of CO_2 fixation without affecting the total chlorophyll content of the cell or the ability of the cell to carry out photosynthetic electron transport reactions. Since *C. mundana* has an absolute requirement for light for growth, the photosynthetic electron transport system may be the major source of ATP and/or TPNH for the cell.

Ribose-5-P isomerase was detected in extracts of cells cultured under both growth conditions, but was present in 5-fold higher activity in autotrophic cells. The enzyme may function in both the oxidative and reductive pentose-P cycle in cells cultured with CO_2.

Both DPN and TPN dependent triose-P dehydrogenase activities are present in *C. mundana* under both conditions of growth and may be essential for photosynthetic as well as photoheterotrophic growth, although TPN-dependent triose-P dehydrogenase is usually considered to be a component of the photosynthetic CO_2 fixation cycle (3). The activity of triose-P dehydrogenase is reduced in acetate grown cells, but not to as great an extent as ribulose-1, 5-diP carboxylase and ribulose-5-P kinase. Hudock (private communication) has recently demonstrated that TPN-linked triose-P dehydrogenase is formed in the y-2 mutant of *Chlamydomonas reinhardi* (yellow in the dark; green in the light) under conditions where ribulose-diP carboxylase and PPNR are greatly reduced in activity (11). The enzyme may not function in photosynthesis or may be involved in other metabolic reactions as well as reactions of photosynthesis.

The demonstration of 2 aldolase activities in *Chlamydomonas mundana* extends the findings of Rutter (20), who has found 2 aldolases in *Euglena gracilis* and *Chlamydomonas reinhardi*. In *C. mundana* the Fe^{+2} activated enzyme appears to be involved in the conversion of acetate to carbohydrate through the glyoxylate cycle (6) and reversal of glycolysis. The enzyme is very similar to the aldolase of the blue-green alga *Anacystis nidulans* and the photosynthetic bacterium *Rhodopseudomonas spheroides* (26, 27). Under conditions of autotrophic growth, only small amounts of this enzyme are formed as judged by elution of aldolase activity from ion exchange columns (Russell and Gibbs, unpublished observations). Cells capable of carrying out green plant photosynthesis contain an enzyme not dependent on a metal for activity with a relatively high affinity for the substrate. Only trace amounts of the enzyme are found in acetate grown cells (table III). Stumpf (24) has described a nonmetal aldolase in pea seeds. This enzyme is not sensitive to *p*-CMB in contrast to the enzyme from autotrophic *Chlamydomonas* (table V).

Further investigation is needed to establish both the nature and specificity of the control system reported here. A study of the adaptation process in the presence of inhibitors of protein synthesis should indicate the nature of control; a comprehensive survey of the effect of a wide variety of organic carbon sources should provide information for an understanding of the specificity of control.

Literature Cited

1. ARNON, D. I. 1949. Copper enzymes in isolated chloroplast. Plant Physiol. 24: 1–15.
2. AXELROD, B. 1955. In: Methods in Enzymology, Vol. I. S. P. Colowick and N. O. Kaplan, eds. Academic Press, New York. p 363–66.
3. BASSHAM, J. A. AND M. CALVIN. 1957. The Path of Carbon in Photosynthesis. Prentice-Hall, Englewood Cliffs, New Jersey.
4. DANFORTH, W. F. 1962. Substrate assimilation and heterotrophy. In: Physiology and Biochemistry of Algae. R. Lewin, ed. Academic Press, New York. p 99–123.

5. Eppley, R. W. and F. M. Maciasr. 1962. Rapid growth of sewage lagoon *Chlamydomonas* with acetate. Physiol. Plantarum 15: 72–79.
6. Eppley, R. W., R. Gee, and P. Saltman. 1963. Photometabolism of acetate by *Chlamydomonas mundana*. Physiol. Plantarum 16: 777–92.
7. Fuller, R. C. and M. Gibbs. 1959. Intracellular and phylogenetic distribution of ribulose-1,5-diphosphate carboxylase and D-glyceraldehyde-3-phosphate dehydrogenases. Plant Physiol. 34: 324–29.
8. Fuller, R. C., R. M. Smillie, E. C. Sisler, and H. L. Kornberg. 1961. Carbon metabolism in *Chromatium*. J. Biol. Chem. 236: 2140–49.
9. Gest, H. 1963. Metabolic Aspects of Bacterial Photosynthesis. H. Gest, A. San Pietro, and L. P. Vernon, eds. Antioch Press, Yellow Springs. p 129–50.
10. Gornall, A. G., C. J. Bardawill, and M. M. David. 1949. Determination of serum proteins by means of the biuret reaction. J. Biol. Chem. 177: 751–66.
11. Hudock, G. A. and R. P. Levine. 1964. Regulation of photosynthesis in *Chlamydomonas reinhardi*. Plant Physiol. 39: 889–97.
12. Kornberg, H. L., J. F. Collins, and D. Bigley. 1960. The influence of growth substrates on metabolic pathways in *Micrococcus denitrificans*. Biochim. Biophys. Acta 39: 9–24.
13. Lascelles, J. 1960. The formation of ribulose-1,5-diphosphate carboxylase by growing cultures of *Athiorhodaceae*. J. Gen. Microbiol. 23: 499–510.
14. Levine, R. P. and R. M. Smillie. 1962. The pathway of triphosphopyridine nucleotide photoreduction in *Chlamydomonas reinhardi*. Proc. Natl Acad. Sci. U. S. 48: 417–21.
15. Levine, R. P. and R. K. Togasaki. 1965. A mutant strain of *Chlamydomonas reinhardi* lacking ribulose diphosphate carboxylase activity. Proc. Natl Acad. Sci. U. S. 53: 987–90.
16. Myers, J. 1940. A study of the pigments produced in darkness by certain green algae. Plant Physiol. 15: 575–88.
17. Pringsheim, E. G. and W. Wiessner. 1961. Ernährung and Stoffwechsel von *Chlamydobotrys* (Volvocales) Arch. Mikrobiol. 40: 231–46.
18. Russell, G. K. and M. Gibbs. 1964. Regulation by acetate of photosynthetic capacity in *Chlamydomonas mundana*. Plant Physiol. 39: lxv.
19. Russell, G. K. and R. P. Levine. 1963. Photosynthetic reactions by chloroplast fragments of *Chlamydomonas reinhardi*. Plant Physiol. 38: 542–45.
20. Rutter, W. J. 1964. Evolution of aldolase. Federation Proc. 23: 1248–57.
21. San Pietro, A. and H. M. Lang. 1958. Photosynthetic pyridine nucleotide reductase. I. Partial purification and properties of the enzyme from spinach. J. Biol. Chem. 231: 211–29.
22. Sibley, J. A. and A. L. Lehninger. 1949. Determination of aldolase in animal tissues. J. Biol. Chem. 177: 859–72.
23. Stanier, R. Y., M. Doudoroff, R. Kunisawa, and R. Contopoulou. 1959. The role of organic substrates in bacterial photosynthesis. Proc. Natl Acad. Sci. U. S. 45: 1246–60.
24. Stumpf, P. K. 1948. Carbohydrate metabolism in higher plants. I. Pea aldolase. J. Biol. Chem. 176: 233–41.
25. Sueoka, N. 1960. Mitotic replication of deoxyribonucleic acid in *Chlamydomonas reinhardi*. Proc. Natl Acad. Sci. U. S. 46: 83–91.
26. van Baalen, C. 1965. Aldolase in blue-green algae. Nature 206: 193–94.
27. Willard, J. M., M. Schulman, and M. Gibbs. 1965. Aldolase in *Anacystis nidulans* and *Rhodopseudomonas spheroides*. Nature 206: 195.
28. Wu, R. and E. Racker. 1959. Regulatory mechanisms in carbohydrate metabolism. III. Limiting factors in glycolysis of ascites tumor cells. J. Biol. Chem. 234: 1029–35.

Reprinted from PLANT PHYSIOLOGY, Vol. 41, No. 10, December, 1966, pages 1623-1631.
(PRINTED IN U.S.A.)

Increased Rate of Net Photosynthetic Carbon Dioxide Uptake Caused by the Inhibition of Glycolate Oxidase

Israel Zelitch

Department of Biochemistry, The Connecticut Agricultural Experiment Station, New Haven, Connecticut 06504

Received July 22, 1966.

Summary. There is considerable variation among species in their rate of photorespiration, and photorespiration increases greatly at higher temperatures. The addition of an inhibitor of glycolate oxidase, α-hydroxy-2-pyridinemethanesulfonic acid, to tobacco leaf disks at 35° stimulated photosynthetic $^{14}CO_2$ uptake at least 3-fold, but $^{14}CO_2$ uptake was not changed by the inhibitor at 25°. The inhibitor did not increase photosynthesis in maize leaf disks at either temperature.

The evolution of CO_2 from glycolate was greatly enhanced in tobacco at 35° compared with 25°. Labeling of the glycolate of tobacco with glycolate-1-^{14}C and -2-^{14}C showed that the increased CO_2 evolved in the light (photorespiration) arose specifically from the carboxyl-carbon atom of glycolate. Maize, a species known to have a negligible photorespiration, produced $^{14}CO_2$ poorly from glycolate-1-^{14}C in comparison to tobacco.

Acetate-1-^{14}C, a substrate metabolized by dark respiration, produced similar amounts of $^{14}CO_2$ in the light in both tobacco and maize. This respiration was changed little relative to photosynthesis by increasing temperature.

Most plants, such as tobacco, have a high photorespiration. The loss of fixed carbon causes an increase in the internal concentration of CO_2 especially at higher temperatures, and results in a lower CO_2 concentration gradient and therefore a lower net photosynthetic CO_2 uptake. Some species, like maize, have a negligible photorespiration and are thus more efficient photosynthetically. The use of an inhibitor of the oxidation of glycolate, the substrate for photorespiration, changed tobacco so that it behaved photosynthetically like maize. Thus high rates of photorespiration may limit the net CO_2 uptake in many plant species.

Ample evidence exists that photosynthetic tissues have a respiration in the light that results in CO_2 evolution (photorespiration) which occurs by reactions different from those which occur in darkness. Under normal conditions of photosynthesis, this photorespiration is often greater than the dark respiration, whether measured by O_2 uptake or CO_2 evolution. This paper is concerned with the nature of the substrate of this photorespiration and with the role of photorespiration in net CO_2 assimilation.

Decker (1) observed that when leaves of several species including tobacco were transferred from the light to darkness, there was a burst of CO_2 evolution before CO_2 output resumed a lower steady rate. The CO_2 burst undoubtedly represented an overshoot of the greater CO_2 evolution that occurred in the light. By measuring O_2 uptake with the aid of $^{18}O_2$ labeling, Hoch, Owens, and Kok (6), have shown that several species of algae have a photorespiration which increases with increasing light intensity and that this is superimposed upon and presumably different in mechanism from the dark respiration. Ozbun, Volk, and Jackson (13) also found a greater O_2 uptake by bean leaves in the light using mass spectrometric techniques.

Forrester, Krotkov, and Nelson (3, 4) recently found that photorespiration in soybean leaves was greatly dependent on the O_2 concentration in the atmosphere, whereas dark respiration was essentially unaffected at O_2 concentrations as low as 1 %. Thus photorespiration must occur by reactions that are different from those concerned with dark respiration. Maize had an insignificant photorespiration at all concentrations of O_2 tested.

Several lines of evidence indicate that photorespiration differs in intensity among various species. Moss (11) measured CO_2 evolution in the light in a CO_2-free atmosphere and observed that, in a number of species including tobacco, CO_2 evolution was at least 50 % greater than dark respiration. In maize, significant CO_2 evolution could not be detected under the same conditions, again emphasizing the difference in photorespiration between these species.

When a plant is placed in turbulent air in a chamber with strong light, the CO_2 concentration in the atmosphere surrounding the plant reaches a steady state (the CO_2 compensation point). Certain plants known to be efficient in their photosynthesis, including maize and sugar cane, were found by Moss (9) to have a CO_2 compensation point close to zero. Most other species have a CO_2 compensation point in excess of 50 ppm. Therefore the value of the CO_2 compensation point correlates well with the previously indicated knowledge about photorespiration in various species. In such species where the internal CO_2 concentration is considerable, it is reasonable to assume that photorespiration can cause a diminished net CO_2 uptake from the atmosphere through the loss of fixed carbon which decreases the CO_2 concentration gradient on the pathway from the air to the chloroplast.

Other workers (5, 15) have described the effect of temperature on the CO_2 compensation point in a number of species and found the value to be at least twice as great at 35° as at 25°. Meidner (7) also showed there was an effect of temperature on the CO_2 compensation point of outdoor grown maize although the quantities were small, i.e. 2.7 ppm at 25° and 5.5 ppm at 35°.

The net CO_2 uptake by maize is twice as great at 30° as compared to 20° when water is not limiting (10). In tobacco, however, CO_2 uptake is less at 30° to 35° than at 25° (personal communication from D. N. Moss). Egle and Schenk (2) have shown that the CO_2 compensation point in *Pelargonium* leaves varied greatly with temperature. At 25° they found about 80 ppm and at 35° 185 ppm. Thus over this 10° range, photorespiration must have initially increased at a rate faster than the rate of CO_2 uptake by the chloroplasts. These workers also observed a 20 % inhibition in net CO_2 uptake at 35° compared to 25°.

Although direct evidence on the nature of the specific substrate for photorespiration was then lacking, I previously suggested (21) on the basis of a number of correlations found in the literature that the CO_2 produced could come from glycolate by the glycolate oxidase reaction as follows:

$$CH_2OH\text{-}C^*OOH + O_2 \rightarrow CHO\text{-}C^*OOH + H_2O_2 \quad (1)$$

$$H_2O_2 \rightarrow HCOOH + C^*O_2 + H_2O \quad (2)$$

If this were the source of the CO_2 produced by leaf tissues with a high photorespiration, inhibition at either reaction 1 or 2 would be expected to lower the internal concentration of CO_2. Thus plants with a high photorespiration could be anticipated to show increased rates of CO_2 uptake with inhibitor because losses of fixed carbon would be diminished and the internal CO_2 concentration would be decreased. This effect would be magnified at warmer temperatures if photorespiration increases greatly. Thus one might be able to convert a plant which behaves like tobacco into one like maize and to obtain increased net CO_2 uptake experimentally at higher temperatures. Tregunna (16) recently accomplished the reverse effect when he supplied maize leaves with riboflavin phosphate to increase the CO_2 compensation point and caused a decreased CO_2 uptake.

By adding an effective inhibitor of glycolate oxidase, α-hydroxy-2-pyridinemethanesulfonic acid (19), to tobacco leaf disks, I have now found that there is a 3-fold increase in CO_2 uptake at 35° while at 25° CO_2 uptake is the same with or without inhibitor. Further experiments confirm that the carboxyl-carbon of glycolate is the primary source of the CO_2 produced in photorespiration in a species with a high rate of photorespiration. These experiments also permit a quantitative estimation of the role of photorespiration in the process of net CO_2 uptake.

Materials and Methods

Plants. Tobacco (*Nicotiana tabacum,* variety Havana seed) and maize (*Zea mays,* hybrid Penn 602A) were grown in sand in a subirrigated bench in the greenhouse. Large leaves were excised and placed with their bases in water in the dark for at least several hours before disks were cut from symmetrical positions with a sharp punch. Six disks, 1.6 cm in diameter, were threaded on to a length of nylon thread to facilitate rapid handling and placed in large (75 ml) Warburg vessels containing 1 to 2 ml water. The vessels were shaken at 120 oscillations per minute for about 1 hour in the light (2000 ft-c supplied by tungsten lamps above and a mirror directly below the vessels) in a constant temperature bath with the vessels open to the air.

Photosynthetic CO_2 Uptake. At the end of the preliminary light period, the water in control vessels was sometimes removed and replaced with fresh water. In vessels with inhibitor, the water was removed and quickly replaced with a solution of α-hydroxy-2-pyridinemethanesulfonic acid and $NaH^{14}CO_3$ was added to the sidearm. The manometer taps were closed after 30 seconds and after a period of 2 or 3 minutes in light, $^{14}CO_2$ of known concentration and radioactivity was liberated into the gas phase by the addition of 10 N sulfuric acid to the solution of $NaH^{14}CO_3$ in the double sidearm of the Warburg vessel. At the end of the experiment, the leaf disks on their threads were rapidly taken from the vessels and plunged into 20 ml of boiling 20 % (v/v) ethyl alcohol, and the mixture was boiled for 3 minutes. The killed disks were homogenized, the suspension made to 25 ml, and the radioactivity was determined by scintillation counting on a 0.05 ml sample in 10 ml of a solution containing 6 g of 2,5-diphenyloxazole and 100 mg of *p-bis* [2-(5-phenyloxazolyl)] benzene in 1 liter of toluene and 500 ml of 95% ethyl alcohol. A small quenching correction was made.

Assimilated-^{14}C and $^{14}CO_2$ Production. When a radioactive substrate was supplied to the leaf disks, this was always done after the disks had been kept for a preliminary period in the light in water. Then the water was replaced with 1.0 ml of radioactive solution. When glycolate-^{14}C was added, the $^{14}CO_2$ produced was collected by passing air saturated with water vapor at the same temperature through the Warburg vessels continually at a rate of 200 ml per minute and trapping the $^{14}CO_2$ in the air stream with aeration tubes in 25 ml of 1 M ethylenediamine. The radioactivity in the ethylenediamine solution was determined by scintillation counting. Control experiments demonstrated that released $^{14}CO_2$ was trapped quantitatively under the conditions used.

At the end of the experiment, the disks were removed and were briefly immersed in a large volume of water to remove adhering ^{14}C-substrate before being killed in boiling 20% ethyl alcohol. After grinding the disks with a glass homogenizer, radioactivity in the homogenate was determined, and the suspension was centrifuged. The combined supernatant fluid and washings were placed on a column of Dowex-1 acetate, and the glycolic acid fraction was collected by elution with 4 N acetic acid (19). In some experiments the glycolic acid concentration was determined colorimetrically on samples taken from the glycolic acid fraction (19). The amount of assimilated-^{14}C was calculated from the total ^{14}C in the homogenate less the ^{14}C in glycolic acid.

When acetate-1-^{14}C was the substrate, the experiments were carried out exactly as with glycolate-^{14}C. The acetate-^{14}C remaining in the leaf disks at the end of the experiment was also determined by placing the leaf extract on a column of Dowex-1 acetate. The acetic acid-^{14}C was eluted with 4 N acetic acid and appears in the eluate just prior to the glycolic acid fraction.

Radioactive Substrates. These were obtained from Volk Radiochemical Company. Before use, the glycolic acid-1-^{14}C and -2-^{14}C were each placed on Dowex-1 acetate columns, and the glycolic acid fraction was collected as previously described. The fraction was dried in a stream of air at 45°. Carrier glycolic acid was added, and the acid was neutralized to about pH 5 with solid $KHCO_3$ and made to a known volume. Potassium acetate-1-^{14}C adjusted to pH 5.2 was used.

Results

Effect of Temperature on the CO_2 Compensation Point for Tobacco. In those species that have a significant photorespiration, the CO_2 compensation point is considerably greater at 35° than at 25°. Figure 1 demonstrates that this generalization also holds true for tobacco. The CO_2 compensation point at 25° was 48 ppm and at 35° was 80 ppm. As others have found, there was a sharp rise above 30°. This increase in the compensation point could be caused by an increase in the internal CO_2 concentration from dark as well as photorespiration, or by a decrease in the rate of CO_2 uptake by the chloroplasts as the temperature is raised. The results with maize (10) as well as the increase in net photosynthesis in tobacco that is shown later in this paper indicate that the latter explanation is most unlikely. Thus one can conclude that increased CO_2 compensation points at higher temperatures are caused by a relatively faster production of CO_2 by respiration, mostly photorespiration, rather than by decreased rates of photosynthesis.

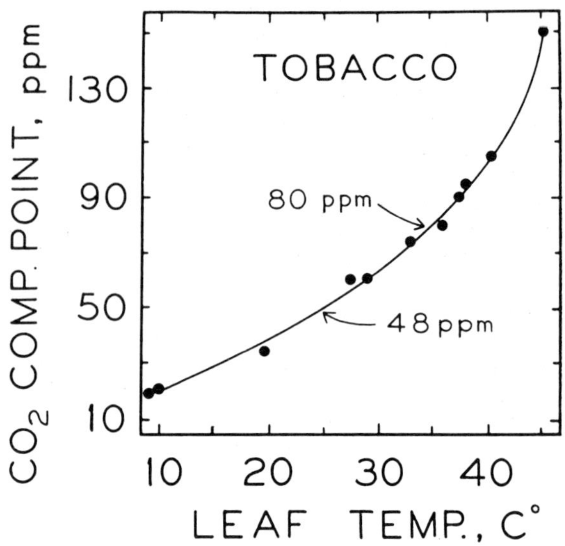

FIG. 1. Effect of temperature on the CO_2 compensation point of excised tobacco leaves. The leaves were placed in light (6000 ft-c from an incandescent source) in a chamber in which the air was circulated and the CO_2 concentration in the atmosphere was determined with an infrared CO_2 analyzer (10). The relative humidity was about 90%. Leaf temperatures were determined by thermocouples attached to the leaf. The steady state CO_2 concentrations were established in about 30 min.

Effect of Glycolate Oxidase Inhibitor on $^{14}CO_2$ Uptake. In previous experiments with tobacco leaf disks in light (22), I have shown that the addition of glycolate oxidase inhibitor at 30° had no significant effect on the $^{14}CO_2$ uptake. From the CO_2 compensation points, one would anticipate that photorespiration in this tissue would greatly increase above 30°. Accordingly, the rate of $^{14}CO_2$ uptake by tobacco leaf disks was measured in the presence and absence of an effective concentration of inhibitor at 25° and at 35°. Although this inhibitor is known to inhibit opening of leaf stomata (20), control experiments showed that there was no closing effect on stomatal widths as measured by silicone rubber impressions for at least 20 minutes at 25° and for as long as 40 minutes at 35° under the conditions of the photosynthesis experiments.

Table I shows that at 35° the rate of $^{14}CO_2$ uptake in the presence of inhibitor was stimulated in the 6 experiments from 2.1- to 6.0-fold, with a mean increase of 3.8-fold. In the 3 experiments carried out at 25°, the inhibitor changed the $^{14}CO_2$ uptake from 0.5- to 1.2-fold, with a mean of 0.8-fold, results similar to those previously observed at 30° (22). The $^{14}CO_2$ uptake took place in a closed system and varying amounts of the available $^{14}CO_2$ were assimilated in the different experiments. The photosynthetic rates given in table I therefore represent minimal values. Thus under these experimental conditions there was always greater net photosynthesis at 35° but not at 25°. As shown in Experiments No. 4, 5, and 6, disks in water did not have a greater photosynthetic rate at 35° than at 25°. Since the rate of synthesis of glycolic acid did not increase sufficiently at the higher temperature in the presence of inhibitor, the effect of the inhibitor in bringing about higher rates of $^{14}CO_2$ uptake at warmer temperatures must be related to a slowing of the normal production of CO_2 in the glycolate oxidase reaction.

Similar experiments on the effect of temperature on the rate of $^{14}CO_2$ uptake by maize leaves were also carried out for comparative purposes (table II). In maize, the α-hydroxysulfonate did not cause an increase in $^{14}CO_2$ uptake, and was in fact somewhat inhibitory and more variable at both 25° and 35° in contrast to the results with tobacco. The rate of glycolate synthesis in maize was considerably lower than in tobacco. Nevertheless, significant quantities of glycolate were produced in maize, so that the differences in the photorespiration of these 2 tissues, if glycolate is involved, is more likely determined by the metabolic fate of this substrate rather than its synthesis.

To account for the results in table I a balance sheet (table III) has been prepared which provides a basis for further experimental tests of the role of glycolate metabolism in photorespiration. The table was constructed as follows: lines 4 and 5 give values of $^{14}CO_2$ uptake that are typical of

Table I. *Effect of α-Hydroxysulfonate on $^{14}CO_2$ Uptake by Tobacco Leaf Disks in Light at 25° and 35°*

After a preliminary period in the light at the temperature indicated, the fluid in the Warburg vessels was replaced with 0.01 M α-hydroxy-2-pyridinemethanesulfonic acid (the inhibitor), and after 2 min 3 μmoles of $^{14}CO_2$ (total of 4 μmoles in the gas phase to make the initial concentration about 0.12 %) containing 502,000 cpm were liberated for the time shown. In Expt. No. 4, 72 % of the available $^{14}CO_2$ was taken up by the disks with inhibitor at 35°, the highest percentage of any experiment.

The glycolic acid formed in the presence of inhibitor, in μmoles hr^{-1} g fr wt^{-1}, at 25° and 35° respectively was: Expt. No. 4, 15.7 and 29.9; Expt. No. 5, 26.9 and 25.8; Expt. No. 6, 5.3 and 9.0.

		μmoles $^{14}CO_2$ uptake per hr per g fr wt			
		25°		35°	
Expt No	Min in $^{14}CO_2$	Disks in water	Disks in inhibitor	Disks in water	Disks in inhibitor
1	5			31	66
2	2.5			45	268
3	3			47	203
4	5*	56	67	61	181
5	2.5	136	62	58	128
6**	2.5	39	34	37	188

* The exposure to $^{14}CO_2$ was 5 min at 25° and 4 min at 35°.
** The concentration of inhibitor in this experiment was 0.005 M.

Table II. *Effect of α-Hydroxysulfonate on $^{14}CO_2$ Uptake by Maize Leaf Disks at 25° and 35° in Light*

The experiments were carried out in a manner similar to those in table I, except that 2.5 μmoles of $^{14}CO_2$ (total of 3.5 μmoles in the gas phase to make the initial concentration about 0.10 %) were liberated for the time shown. The concentration of α-hydroxy-2-pyridinemethanesulfonic acid in Experiment No. 2, 3 and 4 was 0.005 M.

The glycolic acid formed in the presence of inhibitor, in μmoles hr^{-1} g fr wt^{-1}, was: Expt. No. 1, 3.6; Expt. No. 4, 3.5 at 25° and 4.9 at 35°.

		μmoles $^{14}CO_2$ uptake per hr per g fr wt			
		25°		35°	
Expt No	Min in $^{14}CO_2$	Disks in water	Disks in inhibitor	Disks in water	Disks in inhibitor
1	5			51	9.8
2	4			78	77
3	3	78	75		
4	5	40	11	57	28

Table III. *Model of CO_2 Budget for Tobacco Leaf in Light at 25° and 35°*

	μmoles CO_2 per hr per g fr wt 25°	35°
(1) Gross CO_2 uptake	—100	—200
(2) CO_2 output, dark respiration	+10	+20
(3) CO_2 output, photorespiration	+15	+115
(4) Net CO_2 uptake observed:	—75	—65
(5) When photorespiration inhibited, net CO_2 uptake observed:	—90	—180

observed rates. The rate of dark respiration at 25° (line 2) is similar to a published value (19) on the assumption that dark respiration continues at the same rate in the light. Moss has shown (11) that photorespiration in tobacco at 25° exceeds dark respiration and this is shown in line 3. It is reasonable to assume that the dark respiration is approximately doubled by an increase of 10°, a typical value for enzymatic reactions. I assume that the rates of enzymatic reactions concerned with CO_2 uptake also double at 35° compared with 25°, since such an increase occurs normally in maize between 20° and 30° (10). To account for the typical results in lines 4 and 5, it follows that the photorespiration (line 3) must increase about 8-fold over this range of 10°.

Table III shows that photorespiration accounts for about 60 % of the gross CO_2 uptake at 35°. If the CO_2 output resulting from photorespiration were inhibited, there would be little change in the apparent photosynthesis at 25° (lines 4, 5). If the CO_2 output caused by photorespiration were inhibited at 35°, however, the model shows the expected 3-fold increase in the measured CO_2 uptake (table I).

This model was tested further by supplying leaf disks with tracer quantities of glycolate, the suspected substrate of photorespiration, as well as with acetate, a substrate known to be metabolized by dark respiration (through the citric acid cycle), in order to see if the results predicted from table III would be realized.

Effect of Temperature on Metabolism of Glycolate-1-^{14}C and Glycolate-2-^{14}C. Table III and figure 2 predict that if the carboxyl-carbon of glycolate is the primary source of the enhanced CO_2 produced at higher temperatures by photorespiration, the $^{14}CO_2$ released from glycolate-1-^{14}C should be about 8-fold greater at 35° than at 25°. However, the rate of gross CO_2 uptake would be increased 2-fold at the higher temperature. Accordingly, the observed increase in $^{14}CO_2$ output from glycolate-1-^{14}C might only be expected to be approximately 4 times as great at 35° as at 25°. The results were calculated on the basis of the amount of assimilated-^{14}C, because this permits a more meaningful comparison between experiments when the amount of ^{14}C-substrate taken up by the leaf disks may vary. The assimilated-^{14}C is the total radioactivity in substances other than glycolate. The results were therefore expressed as the ratio of the $^{14}CO_2$ produced to the assimilated-^{14}C.

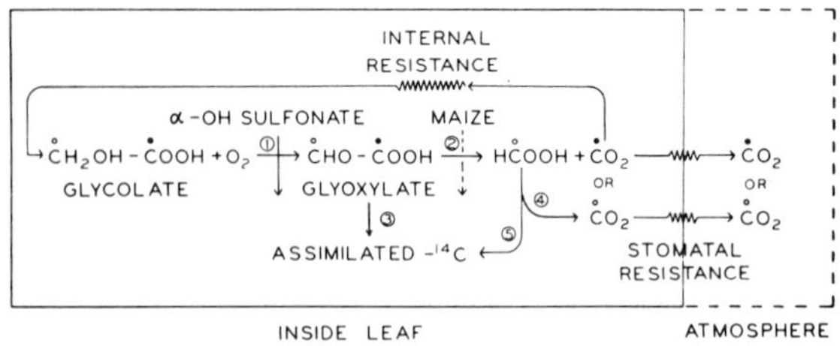

FIG. 2. Representation of the biochemical reactions of glycolic acid in relation to the CO_2 produced by photorespiration. Glycolate-1-^{14}C (●) or 2-^{14}C (O) was added to leaf disks in the light and the $^{14}CO_2$ in the atmosphere was swept out by a rapid stream of air and trapped in alkali. Reaction 1 is the glycolate oxidase reaction that is inhibited by α-hydroxysulfonate. Reaction 2 is non-enzymatic, requiring H_2O_2, and is presumably slow in maize relative to Reaction 3. Reaction 3 represents all of the biochemical reactions whereby glyoxylate is converted into other products. Assimilated-^{14}C was determined from the total ^{14}C in the tissue less that of the glycolate-^{14}C. Reaction 4 represents the formate dehydrogenase reaction, known to be present in leaves. Reaction 5 indicates the conversion of formate into products other than $^{14}CO_2$. Stomata must be open in these experiments to minimize the stomatal diffusive resistance and thus aid the release of $^{14}CO_2$.

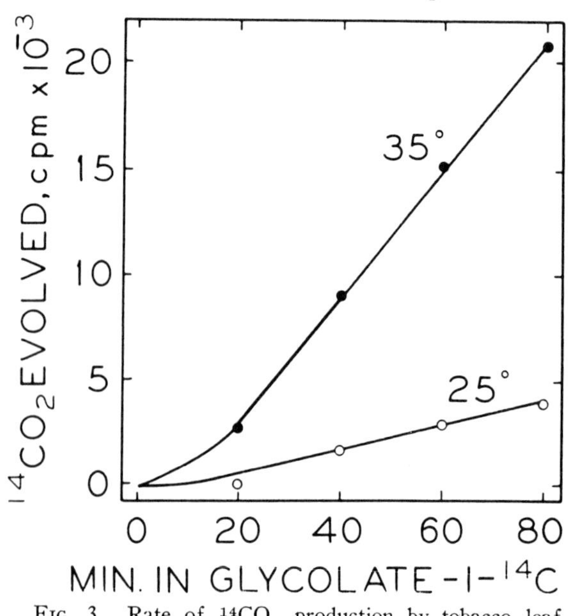

FIG. 3. Rate of $^{14}CO_2$ production by tobacco leaf disks in light from glycolate-1-^{14}C at 25° and 35°. Six leaf disks were shaken in the light in water for about 1 hr, then the fluid was replaced by 1.0 ml of K glycolate-1-^{14}C (0.04 M containing 1.85 × 10^6 cpm). Moistened air at the same temperature was passed through the vessels at a rate of 200 ml per min, and the $^{14}CO_2$ evolved was trapped in 25 ml of 1 M ethylenediamine.

Table IV. *Effect of Temperature on Metabolism of Glycolate-1-¹⁴C and Glycolate-2-¹⁴C by Tobacco Leaf Disks in Light*

Experiment 4 was carried out in darkness. In an experiment carried out like those in the table, it was found that after 60 min in the light in H_2O followed by 60 min in 0.04 M K glycolate at pH 5.0 the mean stomatal width was 7.3 μ at 25° and 7.2 at 35°.

Expt No	Position of glycolate-^{14}C	Temp, °	Total ^{14}C supplied, 10^6 cpm	Time, min	$^{14}CO_2$ released, cpm	Assimilated-^{14}C, cpm	Ratio $^{14}CO_2$ released: ^{14}C-assimilated
1	-1-	25	1.04	80	1380	18 200	0.076
	-1-	35	1.04	80	9250	29 600	0 31
	-2-	25	1.00	80	750	12 400	0.060
	-2-	35	1.00	80	3630	28 200	0.13
2	-1-	25	1 04	60	1000	16 200	0.062
	-1-	35	1 04	60	4700	25 300	0 19
	-2-	25	1.00	60	900	13 400	0 067
	-2-	35	1.00	60	1900	25 500	0.075
3	-1-	25	2 56	60	3250	32,600	0 10
	-1-	35	2 56	60	17,300	55 800	0 31
	-2-	25	2.89	60	1500	23,300	0 064
	-2-	35	2.89	60	6130	66,100	0 093
4	-1-	25	2.51	60	18,500	20 500	0 90
	-1-	35	2.51	60	65 700	39 600	1.7
	-2-	25	2.79	60	1450	55 700	0 026
	-2-	35	2.79	60	6290	122 000	0 052

Glycolate-2-¹⁴C may serve as a control, since the ratio with this substrate should not change significantly (fig 2) if the methylene-carbon of glycolate does not serve as the primary source of the increased CO_2 in photorespiration at higher temperatures. Figure 2 also indicates that if reaction 2 were blocked in maize, in the light little $^{14}CO_2$ would be produced compared with tobacco.

Experiments to test this hypothesis with glycolate-¹⁴C are shown in figure 3 and table IV. The rate of $^{14}CO_2$ release from glycolate-1-¹⁴C was constant after a short lag period and was greatly dependent on the temperature. No correction for assimilated-¹⁴C was made in figure 3. In Experiment 1 (table IV), the ratio of the $^{14}CO_2$ produced to the assimilated-¹⁴C at 35° compared to 25° with glycolate-1-¹⁴C increased 4.1-fold, and in Experiments 2 and 3 the ratio was 3.1 times as great. These figures agree reasonably well with the anticipated increase of about 4-fold expected from the model in table III. The yield of $^{14}CO_2$ from glycolate-2-¹⁴C was similar at 25° to that produced from glycolate-1-¹⁴C but there was no great increase in $^{14}CO_2$ from methylene-labeled glycolate at 35°. In the dark (Experiment 4), at least 6 times as much $^{14}CO_2$ was evolved from the carboxyl-carbon as in light, and the ratio increased less in the dark than in light when the temperature was raised although there could not have been significant $^{14}CO_2$ uptake in darkness. Thus the greatly enhanced $^{14}CO_2$ output observed from glycolate-1-¹⁴C during photorespiration at higher temperatures probably does not represent all of the $^{14}CO_2$ produced within the tissue. The small change in ratio with glycolate-2-¹⁴C as substrate in the light serves as a control for comparison.

If the carboxyl-carbon of glycolate is the primary source of CO_2 in photorespiration, one would anticipate that little $^{14}CO_2$ would be liberated from glycolate-1-¹⁴C by a leaf such as maize which normally has only a slight photorespiration (fig 2). The rate of $^{14}CO_2$ should be small from either carbon atom of glycolate, and should be relatively independent of temperature. This prediction was confirmed by the data in table V. Maize leaves assimilated as much ¹⁴C from glycolate-¹⁴C as did tobacco, but the quantity of $^{14}CO_2$ produced from glycolate-1-¹⁴C at 35°, for example, was only 2 % as much in maize as in tobacco. There was also a relatively small effect of the position of the label or of the temperature on the quantity of $^{14}CO_2$ produced in maize in comparison with tobacco.

Acetate-1-¹⁴C Metabolism in Light. The model in table III shows that if a substrate known to be oxidized to CO_2 by dark respiration mechanisms were labeled with ¹⁴C, the rate of $^{14}CO_2$ released in the light would not be greatly affected by a change in temperature from 25° to 35°. Thus at 35°, one would expect an increase in the rate of $^{14}CO_2$ production of 2-fold compared with 25°, but the gross CO_2 uptake would also be twice as rapid. Therefore table III predicts that the ratio of the $^{14}CO_2$ produced would be the same at 35° as at 25°. Acetate-1-¹⁴C was therefore added to leaf disks of tobacco and maize in the same manner as in the

Table V. *Effect of Temperature on Metabolism of Glycolate-1-^{14}C and Glycolate-2-^{14}C by Maize Leaf Disks in Light*

Expt No	Position of glycolate-^{14}C	Temp, °	Total ^{14}C supplied, 10^6 cpm	Time, min	$^{14}CO_2$ released, cpm	Assimilated-^{14}C, cpm	Ratio $^{14}CO_2$ released: ^{14}C-assimilated
1	-1-	25	1.04	60	400	27,400	0.02
	-1-	35	1.04	60	200	54,100	0.004
	-2-	25	1.00	60	200	18,700	0.01
	-2-	35	1.00	60	100	46,300	0.002
2	-1-	25	9.78	60	2500	194,000	0.01
	-1-	35	9.78	60	1250	475,000	0.003
	-2-	25	10.4	60	1000	228 000	0.004
	-2-	35	10.4	60	700	348,000	0.002

Table VI. *Effect of Temperature on Metabolism of Acetate-1-^{14}C by Tobacco and Maize Leaf Disks in Light*

Expt No	Species	Temp, °	Total ^{14}C, 10^6 cpm	Time, min	$^{14}CO_2$ released, cpm	Assimilated-^{14}C cpm	Ratio $^{14}CO_2$ released: ^{14}C-assimilated
1	Tobacco	25	3.95	80	4280	226,000	0.019
	Tobacco	35	3.95	80	7020	284,000	0.025
2	Tobacco	25	0.790	60	500	16,800	0.03
	Tobacco	35	0.790	60	1000	27.500	0.04
	Maize	25	0.790	60	200	8100	0.02
	Maize	35	0.790	60	400	9400	0.04
3	Tobacco	25	13.0	60	6900	143,000	0.048
	Tobacco	35	13.0	60	14 000	308 000	0.045
	Maize	25	13.0	60	5600	...	...
	Maize	35	13 0	60	2900	61 000	0.048

previous experiments with glycolate-^{14}C, and the results were expressed in the same manner. Table VI (Experiments 1, 2, 3) shows that the ratio of $^{14}CO_2$ produced to the ^{14}C-assimilated was about the same in tobacco at 25° and 35°. Experiments 2 and 3 demonstrate that the relative proportion of $^{14}CO_2$ produced in tobacco and maize was also the same at both temperatures. This is in sharp contrast to the results obtained when the glycolic acid in the leaf was labeled with glycolate-1-^{14}C (tables IV and V).

Discussion

Glycolate is an important early product of photosynthesis produced from CO_2 only in the light (22). There are a number of suggestions that glycolic acid serves as the primary substrate for photorespiration, and is hence the main source of CO_2 inside the leaf in most species including tobacco. Species known to be efficient in photosynthetic CO_2 uptake appear to lack the reactions whereby CO_2 is produced from glycolate. Moss (12) has recently found that addition of a glycolate oxidase inhibitor, α-hydroxy-2-pyridinemethanesulfonic acid, to tobacco leaves slowed the photorespiration but did not affect the dark respiration. The quantity of glycolate formed in the light agreed well with the amount that would normally have been needed to produce CO_2 in the leaf without inhibitor. Photorespiration increases relative to photosynthesis especially at temperatures above 30°. When the oxidation of glycolate was inhibited in tobacco leaf disks at 35°, there was at least a 3-fold increase in the rate of photosynthetic CO_2 uptake. The inhibitor did not stimulate photosynthesis in tobacco at 25° or in maize at either temperature. This increase in photosynthesis in tobacco was undoubtedly brought about by inhibiting the CO_2 output from glycolate, so that fixed carbon was conserved and the CO_2 concentration gradient between the atmosphere and the chloroplast was thereby increased. Since maize has only a negligible photorespiration, large increases in CO_2 uptake with higher temperatures occur normally (10). Increases in gross photosynthesis also occur at higher temperatures in tobacco, but this is masked by the high photorespiration.

The CO_2 budget that was constructed to account for the results with tobacco (table III) indicates that at 35° the rate of CO_2 produced by photorespiration would represent 60 % as much as the rate of the gross CO_2 uptake by the chloroplasts. Thus in most plants, especially at higher temperature, the loss of assimilated carbon by photorespiration is probably the limiting factor in net photosynthetic CO_2 uptake. The anticipated increase in $^{14}CO_2$

formation at 35° compared to 25° was found with glycolate-1-^{14}C as substrate but not with glycolate-2-^{14}C. In the dark, $^{14}CO_2$ ratios increased equally at the higher temperature from both carbon atoms of glycolate.

The experiments in which glycolate-^{14}C was used and the $^{14}CO_2$ released was measured were carried out under conditions that were optimal for stomatal opening. Control experiments demonstrated that the addition of glycolate solution had a tendency to cause even somewhat larger stomatal widths compared with leaf disks floated on water alone. These conditions were essential to permit the demonstration that the carboxyl-carbon of glycolate was the specific source of the increased CO_2 output in photorespiration at higher temperatures in tobacco but not maize (tables IV, V). Acetate oxidation did not fulfill these requirements (table VI). Another attempt was made to test the role of glycolate in photorespiration by use of an inhibitor that might block CO_2 uptake in the light but not photorespiration. For example 3-(4-chlorophenyl)-1, 1-dimethylurea (CMU), at 5 $\times$ 10^{-5} M, completely inhibited $^{14}CO_2$ uptake by tobacco leaf disks. This concentration also closed stomata significantly, hence this inhibitor was not suitable to permit a valid test of the role of glycolate in photorespiration, since CO_2 produced would likely be refixed when stomatal diffusive resistance is high (fig 2).

It is known that if leaf tissues are adequately hydrated, stomata open wider in the light at higher temperatures (14, 18). Also in tobacco leaf, the stomata open wider at such higher temperatures although it is clear that in a closed system the internal concentration of CO_2 must increase greatly as the leaf is warmed. Accordingly, the hypothesis (8) that closing of stomata is the result of increased concentrations of CO_2 within the tissues leads to certain difficulties which are presently not fully resolved.

Maize assimilated glycolate effectively but did not produce CO_2 from this substrate in the light (fig 2). In maize, the further oxidation of glyoxylate is presumably slow relative to the rate at which it is converted into other products. Tregunna (16) suggested that the prosthetic group of glycolate oxidase (riboflavin phosphate) is dissociated in maize so that glycolate oxidase activity is limiting, hence only small amounts of CO_2 are produced. Data in table V, however, indicate that glycolate is oxidized effectively in vivo. Hence the failure to produce CO_2 by this reaction must result from a block as shown in figure 2, since reactions for the direct utilization of glycolate are unknown.

The high rate of CO_2 evolution by photorespiration, especially at elevated temperatures, serves to emphasize that the internal turnover of carbon must be considerable in many species. These differences in the photorespiration among species may well account for the high photosynthetic efficiency of maize, as compared with plants like tobacco, that has often been observed (17). The glycolate oxidase reaction is presumably essential for plants growing at lower temperatures but appears to be harmful at warmer temperatures when the reaction products are altered. The demonstration in the laboratory that the efficiency in absorbing CO_2 from the air by a plant with a high photorespiration (tobacco) can be made similar to one with a low photorespiration (maize), provides another approach to the possibility of obtaining large increases in photosynthetic yields.

Acknowledgments

I thank my colleague Dale N. Moss for carrying out the experiment on the CO_2 compensation point, Bonnie Gambs for skillful technical assistance, and my associates for their helpful discussion.

Literature Cited

1. DECKER, J. P. 1955. A rapid, postillumination deceleration of respiration in green leaves. Plant Physiol. 30: 82–84.
2. EGLE, K. AND W. SCHENK. 1953. Der einfluss der temperatur auf die lage des CO_2-kompensationspunktes. Planta 43: 83–97.
3. FORRESTER, M. L., G. KROTKOV, AND C. D. NELSON. 1966. Effect of oxygen on photosynthesis, photorespiration and respiration in detached leaves. I. Soybean. Plant Physiol. 41: 422–27.
4. FORRESTER, M. L., G. KROTKOV, AND C. D. NELSON. 1966. Effect of oxygen on photosynthesis, photorespiration and respiration in detached leaves. II. Corn and other monocotyledons. Plant Physiol. 41: 428–31.
5. HEATH, O. V. S. AND B. ORCHARD. 1957. Temperature effects on the minimum intercellular space CO_2 concentration "Γ". Nature 180: 180–82.
6. HOCH, G., O. v. H. OWENS, AND B. KOK. 1963. Photosynthesis and respiration. Arch. Biochem. Biophys. 101: 171–80.
7. MEIDNER, H. 1962. The minimum intercellular-space CO_2-concentration of maize leaves and its influence on stomatal movements. J. Exptl. Botany 13: 284–93.
8. MEIDNER, H. AND T. A. MANSFIELD. 1965. Stomatal responses to illumination. Biol. Rev. 40: 483–509.
9. MOSS, D. N. 1962. The limiting carbon dioxide concentration for photosynthesis. Nature 193: 587.
10. MOSS, D. N. 1963. The effect of environment on gas exchange of leaves. In: Stomata and Water Relations in Plants. Conn. Agr. Expt. Sta. Bull. 664.
11. MOSS, D. N. 1966. Respiration of leaves in light and darkness. Crop Sci. 6: 351–54.
12. MOSS, D. N. 1966. Glycolate as a substrate for photorespiration. Plant Physiol. 41: xxxviii.
13. OZBUN, J. L., R. J. VOLK, AND W. A. JACKSON. 1964. Effects of light and darkness on gaseous exchange of bean leaves. Plant Physiol. 39: 523–27.

14. STALFELT, M. G. 1962. The effect of temperature on opening of the stomatal cells. Physiol. Plantarum 15: 772–79.
15. THOMAS, M. D., R. H. HENDRICKS, AND E. R. HILL. 1944. Apparent equilibrium between photosynthesis and respiration in an unrenewed atmosphere. Plant Physiol. 19: 370–76.
16. TREGUNNA, B. 1966. Flavin mononucleotide control of glycolic acid oxidase and photorespiration in corn leaves. Science 151: 1239–41.
17. WAGGONER, P. E., D. N. MOSS, AND J. D. HESKETH. 1963. Radiation in the plant environment and photosynthesis. Agron. J. 55: 36–39.
18. WALKER, D. A. AND I. ZELITCH. 1963. Some effects of metabolic inhibitors, temperature, and anaerobic conditions on stomatal movement. Plant Physiol. 38: 390–96.
19. ZELITCH, I. 1959. The relationship of glycolic acid to respiration and photosynthesis in tobacco leaves. J. Biol. Chem. 234: 3077–81.
20. ZELITCH, I. 1961. Biochemical control of stomatal opening in leaves. Proc. Natl. Acad. Sci. U. S. 47: 1423–33.
21. ZELITCH, I. 1964. Organic acids and respiration in photosynthetic tissues. Ann. Rev. Plant Physiol. 15: 121–42.
22. ZELITCH, I. 1965. The relation of glycolic acid synthesis to the primary photosynthetic carboxylation reaction in leaves. J. Biol. Chem. 240: 1869–76.

4
Regulation by Light; Rhythms

Reprinted for private circulation from
BOTAN. GAZ. 125(4):227–231. 1964.

ROLES OF ACTIVE PHYTOCHROME IN CONTROL OF FLOWERING OF XANTHIUM PENSYLVANICUM

H. A. BORTHWICK AND R. J. DOWNS
ARS, USDA, Plant Industry Station, Beltsville, Maryland

ABSTRACT

Effects of active phytochrome (P_{fr}) on flowering in cocklebur, *Xanthium pensylvanicum* Wallr., subjected to unusually short photoperiods were examined. For interpretation of the results, P_{fr} is considered to be an enzyme that catalyzes a reaction that controls many phenomena including flowering. In cocklebur plants phytochrome is in the P_{fr} form at the end of the photoperiod. After photoperiods of 1½ or 2 hr the P_{fr}, although present, does not act during the first 6 or 8 hr of darkness. This failure is attributed to delay in arrival of substrate for P_{fr} action. Between the eighth and tenth hours P_{fr} acts to promote flowering. After the tenth hour it is no longer present, and, if it is then reintroduced by a red irradiation, it inhibits flowering. Dark reversion of P_{fr} is completed simultaneously with completion of its flower-promoting action and is regarded as a consequence of its action. The amount of radiant energy received by cockleburs during a 3-hr photoperiod determines whether P_{fr} action is necessary for promotion of flowering, the flower-promoting effect being exhibited only after photoperiods of low irradiance.

The P_{fr} left in cockleburs at the end of 2-hr photoperiods thus plays no role essential to flowering during the first 8 hr of darkness. During the next 2 hr it promotes flowering, and during several hours after that, P_{fr} is absent. If reintroduced during that time, it would be inhibitory.

Introduction

Reversible control of flowering of cocklebur, *Xanthium pensylvanicum* Wallr., by red and far-red radiations was first reported by BORTHWICK, HENDRICKS, and PARKER (1952) and was described in further detail by DOWNS (1956). Those reversibility studies were performed with plants subjected to 8-hr or longer photoperiods and involved brief irradiations with red or far red near the middle of the accompanying long dark periods. In such experiments red radiation inhibited and far red repromoted flowering. Flowering of cocklebur plants subjected to photoperiods much shorter than 8 hr is also photoreversibly controlled by red and far red, but photoreversibility is displayed when the irradiations occur in the early hours of the dark period and the plant responses to the two kinds of radiation are opposite to those previously reported (BORTHWICK *et al.*, 1952; DOWNS, 1956). These reversible responses of cockleburs on very short photoperiods are very similar to those of seedlings of *Pharbitis nil* Chois., as described by NAKAYAMA (1958) and TAKIMOTO and IKEDA (1959), and to those of *Kalanchoe blossfeldiana* v. Poelln., as described by FREDERICQ (1963). In fact, FREDERICQ's results with *Kalanchoe* and his later ones with *Pharbitis* (1964) stimulated our interest in looking for similar responses in cocklebur.

The purpose of this paper is to describe experimental conditions that lead to these inhibitory effects of brief far-red irradiations on flowering and their reversal by red. The results are interpreted in terms of action by P_{fr}, the far-red-absorbing form of phytochrome.

Material and methods

The plants for most of the experiments were grown in the greenhouse until they had six or eight leaves. During growth they were subjected to the natural light at the particular time of year supplemented by 3 hr of incandescent-filament light from 11 PM to 2 AM. A supplemental illuminance of approximately 40 ft-c applied in this way insured vegetative development of the plants regardless of the season.

Plants of suitable size were brought to plant growth rooms for the period of experimental treatment, after which they were again returned to their original growing conditions in the greenhouse. The terminal buds were dissected 10 days after the beginning of the experiment, and the stage of flowering was indicated by numbers 0 to 7 as used by DOWNS (1956). Individual lots in some experiments were three but in most six plants each.

TABLE 1

EFFECTS OF FAR-RED IRRADIATIONS AFTER $1\frac{1}{2}$- OR 2-HR PHOTOPERIODS ON FLOWERING OF COCKLEBUR[a]

PHOTO-PERIOD (HR)	PLANTS TESTED PER TREATMENT (NUMBER)	MEAN STAGE OF FLOWERING[b]	
		No far red	4 or more min far red
$1\frac{1}{2}$	41	4.5	0.3
2	27	4.6	0.5

[a] Flowering stages for each photoperiod are averages from seven experiments.

[b] Flowering stages as given by DOWNS (1956).

TABLE 2

EFFECTS ON FLOWERING OF COCKLEBUR AND ITS FAR-RED INHIBITION CAUSED BY DIFFERENT NUMBERS OF DAILY $1\frac{1}{2}$-HR PHOTOPERIODS[a]

DAILY $1\frac{1}{2}$-HR PHOTOPERIODS (NO.)	MEAN STAGE FLOWERING[b]	
	No far red	8 min far red
1	3	0
2	6	0
3	5	<1
4	6	0
5	5	<1

[a] Three plants per lot.

[b] Flowering stages as given by DOWNS (1956).

The illuminance during the experimental photoperiods, unless indicated otherwise, was about 2,500 ft-c from standard cool-white fluorescent lamps. In one experiment sunlight was used during the short photoperiod. For red- and far-red-irradiation treatments filtered sources described by DOWNS, BORTHWICK, and PIRINGER (1958) were used. Except as indicated, the red and far-red irradiations followed immediately after the end of the photoperiods.

Numbers and lengths of photoperiods differed in different experiments and are indicated for each. In some experiments that involved a single short photoperiodic cycle a conditioning period of about 24 hr was inserted between the last non-inductive photoperiod and the inductive one. During this period the plants were held in darkness, interrupted at frequent intervals with a few minutes of incandescent-filament light of low intensity. The purpose of this procedure was to separate the last natural light period from the very short experimental photoperiods and thus to minimize possible effects resulting from residual photosynthetic reserves carried over from pretreatment.

Experimental results

Cocklebur plants subjected to $1\frac{1}{2}$- or 2-hr photoperiods in a growth room flower well but not quite so vigorously as they do on longer photoperiods. When such plants are irradiated briefly with far red immediately after these short photoperiods, however, their flowering is completely prevented, or their vigor is sharply reduced. This kind of response has been observed in numerous experiments, seven of which are summarized in table 1. The short photoperiods in some of those experiments were given for only 1 day but in others for 2 or more days.

In one experiment $1\frac{1}{2}$-hr photoperiods were given for various periods from 1 to 5 days (table 2). Far red immediately after the photoperiod was almost completely inhibitory to flowering in all lots. Flowering of plants that received no far red was about the same in lots that received two to five successive cycles. Plants of the one lot that received only one short photoperiod flowered less vigorously than the others but sufficiently well to permit detection of an inhibitory effect of far red.

The inhibitory effects of far red on flowering of cockleburs became less evident when photoperiods of fluorescent light longer than 2 hr were tested (table 3). Essentially the same inhibitory effects of far red after short photoperiods of sunlight were also exhibited. Inhibition was strongest after 1- and 2-hr photoperiods of sunshine but was still detected after 3-hr ones.

The reason for the sudden change of effect with a slight increase in length of photoperiod was sought in an experiment in which both light intensity and photoperiod were varied. Plants were subjected to $1\frac{1}{2}$- and 3-hr photoperiods with illuminance of 2,000, 1,000, and 500 ft-c of unfiltered cool-white fluorescent light on 2 successive days (table 4). Far red applied after the photoperiod completely inhibited flowering of nearly all plants on $1\frac{1}{2}$-hr photoperiods. Its inhibitory effect on flowering of the 3-hr lots increased as light intensity decreased.

The inhibitory effects on flowering, potentiated by far red applied after 1½- or 2-hr photoperiods, were almost completely nullified by subsequently applied red treatments (table 5). When the far-red and red irradiations occurred as much as 2 hr after the end of the photoperiods, no change in effectiveness of the treatments was noted. In another experiment (data not presented) 4-min irradiations with red or far red were given at times varying from 0 to 5½ hr after a single 1½-hr photoperiod. The flowering stages of the untreated dark-control plants and of plants irradiated 4 min with red at various times were alike and averaged stage 3 or slightly higher.

TABLE 3

EFFECTS ON FLOWERING OF COCKLEBUR CAUSED BY FAR-RED IRRADIATIONS AFTER PHOTOPERIODS OF VARIOUS DURATIONS[a]

Fluorescent light (hr)	Mean stage of flowering[b] after indicated far-red irradiations			
	0 min	½ min	2 min	8 min
½	0	0	0	0
1	2	0	0	0
1½	5	2	0	0
2	5	4	<1	0
2½	7	6	5	5
3	6	6	6	6

[a] Repeated on three successive days; four plants per lot.
[b] Flowering stages as given by DOWNS (1956).

TABLE 4

EFFECTS ON FLOWERING OF COCKLEBUR CAUSED BY FAR-RED OR FAR-RED PLUS RED IRRADIATIONS GIVEN IMMEDIATELY AFTER PHOTOPERIODS OF DIFFERENT DURATION AND ILLUMINANCE[a]

Duration and illuminance of photoperiod	Mean stage of flowering[b]		
	No far red	8 min far red	8 min far red+4 min red
1½ hr:			
2,000 ft-c	5	<1	5
1,000 ft-c	4	<1	4
500 ft-c	3	0	3
3 hr:			
2,000 ft-c	6	4	6
1,000 ft-c	6	3	6
500 ft-c	5	<1	5

[a] Repeated two successive days; three plants per lot.
[b] Flowering stages as given by DOWNS (1956).

Plants treated with far red for 4 min averaged stage 2 or less in all cases.

Still greater intervals between the end of the photoperiod and the red or far-red treatments were tested in another experiment (table 6). The far-red treatments given during the first 8 hr of darkness were almost completely inhibitory to flowering, but during the second 8 hr they were only slightly inhibitory. Red treatments during the first 8 hr had relatively little inhibitory effect on flowering, but they had a sharply inhibitory effect when given during the second 8 hr.

TABLE 5

EFFECTS ON FLOWERING OF COCKLEBUR CAUSED BY IRRADIATIONS WITH FAR RED, OR FAR RED FOLLOWED BY RED, AFTER EACH 1½- OR 2-HR PHOTOPERIOD[a]

Photoperiod (hr)	Time of far-red irradiations (hr after photoperiod)	Mean stage of flowering[b] after indicated irradiations	
		8 min far red	8 min far red+8 min red
1½	½	1	3
1½	1	<1	4
1½	2	<1	5
2	0	<1	4
2	½	<1	5
2	1	0	4
2	2	<1	4

[a] Repeated two successive days; three plants per lot.
[b] Flowering stages as given by DOWNS (1956); flowering in dark controls was stage 5 for both photoperiods.

TABLE 6

EFFECTS ON FLOWERING OF COCKLEBUR CAUSED BY RED OR FAR-RED IRRADIATIONS OF 4 MIN AT VARIOUS TIMES AFTER A SINGLE 1½-HR PHOTOPERIOD[a]

Dark interval between photoperiod and red or far-red irradiance (hr)	Mean stage of flowering[b]	
	Red	Far red
0	....	0
2	3	1
4	3	<1
6	3	<1
8	2	0
10	0	3
12	0	2
16	1	2

[a] Five plants per lot.
[b] Flowering stages as given by DOWNS (1956); flowering in dark control was stage 4.

The red reversal of the inhibitory effect of far red on flowering was examined as a function of time between the far-red and red treatments (table 7). Red light effectively reinduced flowering during the first 5 hr, became progressively less effective during the sixth to eighth hours, and was without detectable effect after 8 hr.

In a multiple reversal experiment the plants were irradiated with red and far red either at the beginning or near the middle of the dark period. The photoperiod treatments consisted of 2 hr of light and 22 hr of darkness and were continued for 2 successive days. The red and far-red irradiations were 4 min each and were applied either at the beginning or the middle of each 22-hr dark period in sequences indicated in table 8. Complete reversibility resulted from treatments at the beginning of the dark period and complete irreversibility from those made in the middle of the dark period.

TABLE 7

EFFECTS ON FLOWERING OF COCKLEBUR CAUSED BY TREATMENTS IN WHICH A 4-MIN FAR-RED IRRADIATION GIVEN IMMEDIATELY AFTER EACH 2-HR PHOTOPERIOD IS SEPARATED FROM A SUBSEQUENT 4-MIN RED IRRADIATION BY VARIOUS PERIODS OF DARKNESS[a]

Dark interval between far-red and red irradiations (hr)	Mean stage of flowering[b]
0	4
½	5
1	4
2	5
3	4
4	3
5	4
6	3
7	2
8	2
11½	<1
15	0

[a] Repeated two successive days; six plants per lot.

[b] Flowering stages as given by DOWNS (1956); flowering in far-red control was stage 0.

Discussion

The experiments reported here were done with daily inductive photoperiods quite different in length from those ordinarily used. Under these conditions the responses of cockleburs to red and far-red treatments differed in many details from those previously encountered, and these differences shed new light on the ways in which phytochrome effects control of flowering.

In the discussion that follows, phytochrome in the P_{fr} form is regarded as an enzyme that effects biological control of many phenomena, including flowering (HENDRICKS and BORTHWICK, 1963, 1964). The enzymatic reaction is not known, but some of the details of how and when it operates are shown by the results presented.

If P_{fr} is an enzyme, physiological response of a plant to red light results from enzymatic action of the P_{fr} thereby generated. Similarly, response to far-red irradiation results from far-red conversion of P_{fr} to P_r, thereby preventing the enzymatic action of P_{fr} from occurring.

In many of the experiments reported herein (table 1) flowering was inhibited by far red at the close of a 1½-hr photoperiod. Conclusions are that under these experimental conditions (*a*) P_{fr} was present at the close of the photoperiod, and (*b*) P_{fr}, if not removed, acted during the dark period to promote flowering. Essentially these same conclusions were reached by CUMMING (1963), who used a somewhat different experimental approach with *Chenopodium rubrum* L.

TABLE 8

EFFECTS ON FLOWERING OF COCKLEBUR CAUSED BY SUCCESSIVE ALTERNATING 4-MIN RED AND FAR-RED IRRADIATIONS AT THE BEGINNING OR MIDDLE OF EACH 22-HR DARK PERIOD[a]

Time of dark period and kind of irradiation	Mean stage of flowering[b]
Beginning of 22-hr dark period:	
FR	0
FR+R	3
FR+R+FR	0
FR+R+FR+R	4
Middle of 22-hr dark period:	
R	0
R+FR	0
R+FR+R	0
R+FR+R+FR	0

[a] Repeated two successive days; six plants per lot.

[b] Flowering stages as given by DOWNS (1956); flowering in dark control was stage 4.

In other experiments (table 6) flowering of plants on 1½-hr photoperiods was inhibited by red light given late in the dark period. The conclusions in this case are that (*a*) P_{fr} was not present late in the dark period, and (*b*) it acted to inhibit flowering if it was introduced into the plant by red light at that time. Successful flowering of a cocklebur plant on a 1½-hour photoperiod thus depends on the presence and flower-promoting action of P_{fr} sometime during the first part of the dark period and on its absence and, therefore, lack of action later in the same dark period. These conclusions raise questions as to when the flower-promoting action of P_{fr} begins and when it ends, and whether the transition from its promotive to inhibitory effects is abrupt or gradual.

Results in table 6 show that the flower-promoting function of P_{fr} was completed between the eighth and tenth hours. Since red light had essentially no inhibitory effect at the eighth hour and was completely inhibitory 2 hr later, the transition from a promotive to an inhibitory effect of P_{fr} on flowering was fairly abrupt.

The question of when the promotive effect of P_{fr} began under these particular conditions is answered by the results in table 7. The removal of P_{fr} at the beginning of darkness by far red did not inhibit flowering, provided the P_{fr} was returned by a subsequent red irradiance at or before the eighth hour of darkness. The action of P_{fr}, as judged by promotion of flowering, thus appears to be restricted to a period of 2 hr or less, starting at about the eighth hour of darkness. Its failure to act during the first 6 hr is thought to result from a lack of substrate for the enzymatic action of P_{fr} during that period. This delayed action of P_{fr} indicates that the substrate, although undoubtedly derived from photosynthesis, is not an immediate product.

Since the effects of P_{fr} action on flowering of the plants in these experiments shifted abruptly from promotive to inhibitory, it follows that completion of the promotive action of P_{fr} and the dark reversion of P_{fr} to P_r were simultaneous events. This simultaneity indicates that dark reversion and action of P_{fr}, whatever its nature, are linked; that is, P_{fr} as a consequence of its action, is converted to P_r.

In cocklebur the need for P_{fr} to promote flowering is displayed only after photoperiods of low or intermediate radiant energy. For example, this need increased abruptly in plants on 3-hr photoperiods when the illumination was decreased from 2,000 to 500 ft-c (table 4). The need for P_{fr} to promote flowering also appeared abruptly when the daylength at a fixed light intensity was decreased from 3 to 2 hr per day. Although a difference in radiant energy during the photoperiod resulted in a difference in response to P_{fr} in the early part of the night, such a difference in energy appeared to have no effect on response to P_{fr} in the middle of the night. P_{fr} at that time was very inhibitory regardless of energy under the photoperiod.

In these cocklebur experiments flowering response was not reversed when alternating red and far-red irradiations were given in the middle of the night following a 2-hr photoperiod (table 8). Similar results were obtained under somewhat different photoperiodic conditions by NAKAYAMA (1958), NAKAYAMA, BORTHWICK, and HENDRICKS (1960), and PURVES (1961). NAKAYAMA *et al.* (1960) ascribed this result to failure of P_{fr} to undergo conversion when irradiated by far red.

FREDERICQ (1964) found, however, that reversal of the flowering response in *Pharbitis* in the middle of the dark period could be accomplished provided the successive red and far-red irradiations were completed in 1 min. He was able to inhibit flowering by 30 sec of red light. When he followed this immediately with 30 sec of far red, flowering capability was restored, but when 2 min of darkness intervened between the red and far-red irradiations, flowering failed. In FREDERICQ's experiment accomplishment of the flower-inhibiting action of P_{fr} was thus completed in more than 1 and less than 3 min. This result of FREDERICQ strongly suggests that reversal of flowering of cocklebur in the middle of a 22-hr dark period might also be possible, provided the irradiation energies were adequately high and the over-all period of red–far-red irradiation very brief. This possibility was not appreciated at the time these experiments were performed but will be tested.

LITERATURE CITED

BORTHWICK, H. A., S. B. HENDRICKS, and M. W. PARKER. 1952. The reaction controlling floral initiation. Proc. Natl. Acad. Sci. **38**:929–934.

CUMMING, B. G. 1963. Evidence of a requirement for phytochrome-P_{fr} in the floral initiation of *Chenopodium rubrum*. Canad. J. Botany **41**:901–926.

DOWNS, R. J. 1956. Photoreversibility of flower initiation. Plant Physiol. **31**:279–284.

DOWNS, R. J., H. A. BORTHWICK, and A. A. PIRINGER. 1958. Comparison of incandescent and fluorescent lamps for lengthening photoperiods. Proc. Am. Soc. Hort. Sci. **71**: 568–578.

FREDERICQ, H. 1963. Flower formation in *Kalanchoe blossfeldiana* by very short photoperiods under light of different quality. Nature **198**:101–102.

———. 1964. Conditions determining the effects of far-red and red irradiations on the flowering response of *Pharbitis nil*. Plant Physiol. (In press.)

HENDRICKS, S. B., and H. A. BORTHWICK. 1963. Control of plant growth by light. Pp. 233–261 *in* Environmental control of plant growth. Ed., L. T. EVANS. Academic Press, New York.

———, and ———. 1964. The physiological functions of phytochrome. *In* Biochemistry of plant pigments. Ed., T. W. GOODWIN. Academic Press, London. (In press.)

NAKAYAMA, S. 1958. Photoreversible control of flowering at the start of the inductive dark period in *Pharbitis nil*. Ecol. Rev. **14**:325–326.

NAKAYAMA, S., H. A. BORTHWICK, and S. B. HENDRICKS. 1960. Failure of photoreversible control of flowering in *Pharbitis nil*. BOTAN. GAZ. **121**:237–243.

PURVES, W. K. 1961. Dark reactions in the flowering of *Lemna perpusilla* 6746. Planta **56**:684–690.

TAKIMOTO, A., and K. IKEDA. 1959. Studies on the light controlling flower initiation of *Pharbitis nil*. I. Intensity and quality of the light preceding the inductive dark period. Botan. Mag. Tokyo **72**:137–145.

Reprinted for private circulation from
BOT. GAZ. 127(1):1–5. 1966.

Printed in U.S.A.

PHYTOCHROME ACTION IN TOMATO-SEED GERMINATION[1]

A. L. MANCINELLI,[2] H. A. BORTHWICK, AND S. B. HENDRICKS
Crops Research Division and Soil and Water Conservation Research Division,
Agricultural Research Service, Beltsville, Maryland

ABSTRACT

Dark-germinating seeds of Ace and Porte tomato varieties can be inhibited by far-red radiation. Germination of inhibited seed is promoted by red radiation and in turn is reversed by far-red radiation, indicative of control by phytochrome. In the Ace variety initial inhibition requires only a single short far-red exposure. Prolonged far-red irradiation extending over many hours is required for inhibition of Porte seeds. The prolonged exposure can be replaced by cyclical far-red irradiation for 3.3% of the total time. This effect is considered as arising in part from conversion of P_{fr} to P_r as phytochrome appears in the seed by rehydration. Effects are also evident of exhaustion of endogenous factors other than P_{fr} needed for germination when P_{fr} is maintained at a low level.

Introduction

Germination of several kinds of seeds and spores is clearly dependent on the action of phytochrome, but such action is absent or not obvious in others. Some seeds, including those of many cultivated plants, germinate equally well in light or darkness. In a few, however, a light requirement has been induced either by holding the seed at temperatures unfavorable for germination (BORTHWICK *et al.*, 1954) or by subjecting them to several hours of continuous or cyclic far-red irradiation at a favorable temperature (MANCINELLI and BORTHWICK, 1964).

Several extremes of response might be expected among seeds. In some the point of P_{fr} control in the germination process might have been passed prior to storage. In others the seed upon drying might contain a mixture of P_r and P_{fr} adequate for germination upon rehydration. Such seed could possibly be inhibited by far-red irradiation adequate to convert P_{fr} to P_r. If the rehydration is slow, the irradiation would need to be prolonged and thus could be confused with a high radiant-energy requirement (MOHR, 1962). Only lettuce seeds have been adequately examined with respect to these possibilities (MANCINELLI and BORTHWICK, 1964).

Observations on tomato seed are reported here. Germination of these seeds is inhibited by low temperatures (10–15 C) and promoted by higher temperatures (25–30 C) in contrast to lettuce seed. Although sensitivity to light has been observed (TOOLE, 1961), most tomato seed will germinate in darkness.

Material and methods

Seeds of Porte and Ace tomato varieties from the 1963 harvest were planted in Petri dishes on filter paper wetted with distilled water and, unless otherwise indicated, were held at 20 C and in darkness except for the period of exposure to light. Counts of germination were usually made 3 days after the last irradiation, and dark controls (DC) were included in all experiments.

Far-red light was obtained by filtering the radiation from incandescent-filament lamps through two

[1] Under a research contract from the Campbell Soup Co., Camden, N.J.

[2] Present address: Department of Botany, Columbia University, New York, N.Y.

layers each of dark-blue (Du Pont 300 PC dark blue)[3] and red (Du Pont 300 MSC red) cellophane, or through a FRF-700 Plexiglas filter (Westlake Plastics Co.). Red light was obtained by filtering the radiation from cool-white fluorescent lamps through two layers of the red cellophane. The distribution of the spectral energy of these sources was measured with a spectroradiometer (DOWNS *et al.*, 1964).

TABLE 1

INHIBITION OF GERMINATION OF ACE TOMATO SEED BY BRIEF FAR-RED (FR) IRRADIANCE AND ITS REVERSAL BY BRIEF RED (R) IRRADIANCE[a]

Treatment	Germination (%)
Dark control	87
FR	28
FR, 0 hr D, R	87
FR, 12 hr D, R	58
FR, 36 hr D, R	45
FR, 72 hr D, R	37
FR, 192 hr D, R	36

[a] FR, 16 min after 9 hr imbibition, 14.0 μw cm^{-2} mμ^{-1} at 740 mμ; R, 1 min 0.3 μw cm^{-2} mμ^{-1} at 660 mμ at indicated time; D, darkness in hours.

TABLE 2

INHIBITION OF GERMINATION OF PORTE TOMATO SEEDS BY FAR-RED (FR) IRRADIANCE AND ITS IMMEDIATE REVERSAL BY RED (R) IRRADIANCE

FAR-RED IRRADIATION (DAYS)[a]	PER CENT GERMINATION AFTER RED IRRADIATION[b]			
	0 min	0.25 min	1 min	16 min
2	20	85	93	89
3	3	90	92	93
4	2	61	71	71
5	0	1	0	0

[a] Far red, 1.6 μw cm^{-2} mμ^{-1} at 740 mμ for indicated days.

[b] Red, 2.4 μw cm^{-2} mμ^{-1} at 660 mμ for indicated minutes immediately after far-red irradiation.

Energies are indicated at 660 mμ for the red source and at 740 mμ for the far-red source. In the spectrograph irradiations, imbibed seeds were placed in the image plane of the spectrograph for the time required and then returned to darkness. The distribution of the spectral energy of the radiation from the spectrograph, which varied by about 15% over the region of 600–780 mμ, was measured with a thermopile.

Results

GERMINATION IN DARKNESS.—Both Ace and Porte seeds germinate in high percentage in darkness (tables 1, 3). They germinate equally well under natural light, artificial light from cool-white fluorescent lamps, and red light.

PROPERTIES OF LIGHT REQUIREMENT INDUCED BY FAR-RED IRRADIATIONS.—Exposure of the tomato seeds to far-red radiation results in the inhibition of germination and in the induction of a light requirement (tables 1, 2). Germination is reduced markedly during a dark-incubation period of 12 hr or more following far-red irradiation before exposure to red radiation (table 1). If the red irradiation is followed by a further brief exposure to far red, the seeds again fail to germinate (table 3). Thus, the induced light requirement shows the typical red, far-red reversibility of a phytochrome-controlled response. The induction of the light requirement requires different periods of exposure to far red for the two varieties.

RESPONSE TO BRIEF FAR-RED IRRADIATIONS.—Ace seeds are effectively inhibited by brief exposures to far-red radiation (table 1) as well as by prolonged exposures. A single exposure for 1 min at 14.3 μw cm^{-2} mμ^{-1} (740 mμ) reduced the germination to 24% (DC 92%). The response to far red changes with time and reaches a maximum 9–12 hr after the start of imbibition (fig. 1). At the time of maximum response no appreciable difference is found between the responses induced by a brief irradiation and those induced by 6 hr of continuous or intermittent irradiations. During the first 6 hr of imbibition some differences were observed: Cyclic far-red irradiation extended for 6 hr, with a total energy less than one-half that applied in a single 10-min exposure, is more effective than the latter and is as effective as a continuous 6-hr irradiation with a total energy about eight times higher. The response induced by a brief far-red irradiation is completely reversed by red applied immediately (table 1) or shortly after far red. If, after the far-red treatment, the seeds are

TABLE 3

INHIBITION OF GERMINATION OF PORTE TOMATO SEEDS BY 2 DAYS OF FAR-RED (FR),[a] ITS REVERSAL BY RED (R),[b] AND ITS REINDUCTION BY A BRIEF FAR-RED[c] IRRADIANCE AFTER VARIOUS HOURS OF DARKNESS (D)

Treatment	Germination (%)
Dark control	92
2 days FR[a]	1
2 days FR, R[b]	89
2 days FR, R, 0 hr D, FR[c]	8
2 days FR, R, 2 hr D, FR	11
2 days FR, R, 9 hr D, FR	20
2 days FR, R, 15 hr D, FR	55
2 days FR, R, 21 hr D, FR	71

[a] 1.6 μw cm^{-2} mμ^{-1} at 740 mμ.

[b] 3 min 2.4 μw cm^{-2} mμ^{-1} at 660 mμ.

[c] 5 min 4.3 μw cm^{-2} mμ^{-1} at 740 mμ.

[3] Trade names are given as part of the exact experimental conditions and not as an endorsement of products over those of other manufacturers.

incubated for several days in darkness before application of the red irradiation, the repromotion of germination is lower than when red is applied immediately after far red (table 1). An approximate action spectrum indicated that only the far-red region was highly effective in inducing inhibition, with a very low efficiency in the blue region and none elsewhere between 400 and 800 mμ. Irradiations at several wavelengths in the red and far-red regions showed that the peak for inhibition was between 730 and 750 mμ, and the peak for repromotion was around 660 mμ (figs. 2, 3).

Response to prolonged far-red irradiations.—Germination of Porte seeds is effectively inhibited by prolonged continuous or intermittent far red irradiation (table 2, fig. 5). Exposures of 10, 30, 60, and 180 min (results not presented) reduced germination only to 70%–78% (DC 91%). Seeds exposed to prolonged periods of far-red radiation did not germinate during the exposure nor during a dark-incubation period following the close of the far-red irradiation. Exposure to red light at the close of the far-red treatment repromoted germination during the subsequent period of incubation in darkness (table 2). The amount of red energy required to promote germination of the far-red-treated seeds increases with increasing duration and intensity of the far-red irradiation (fig. 4). The germination was not repromoted after 5 days of irradiation (table 2). Response to far red changes with time, and the maximum response to a 12-hr irradiation was obtained for irradiation starting 12–18 hr after start of inhibition (table 4).

Germination is almost completely inhibited when the seeds are exposed to prolonged intermittent far-red irradiation (fig. 5). When 1 min of red radiation is included, the results depend on its position relative

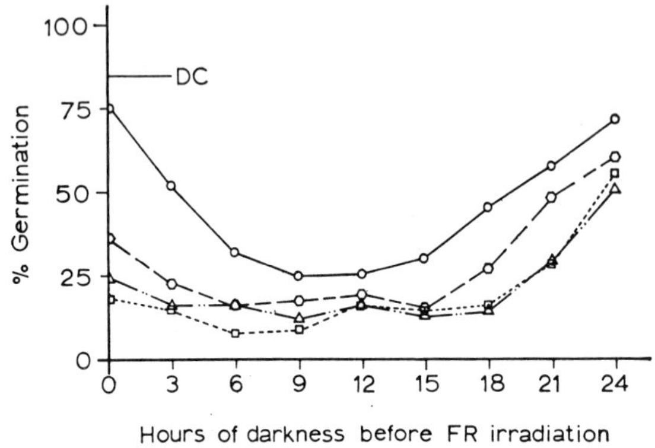

Fig. 1.—Action of far-red (FR) irradiation on germination of tomato (Ace) seeds. *Circles*, 10 min FR, 14.3 μw cm^{-2} mμ^{-1} at 740; total energy 8.5 × 10^{-3} joules. *Triangles*, 6 hr continuous FR, 1.6 μw cm^{-2} mμ^{-1} at 740; total energy 34.6 × 10^{-3} joules. *Hexagons*, 6 hr cyclic FR, 1.9 μw cm^{-2} mμ^{-1} at 740; total energy 4.1 × 10^{-3} joules. *Squares*, 6 hr cyclic FR, 14.3 μw cm^{-2} mμ^{-1} at 740; total energy 31.0 × 10^{-3} joules. Cyclic FR means, 1 min FR followed by 9 min dark every 10 min.

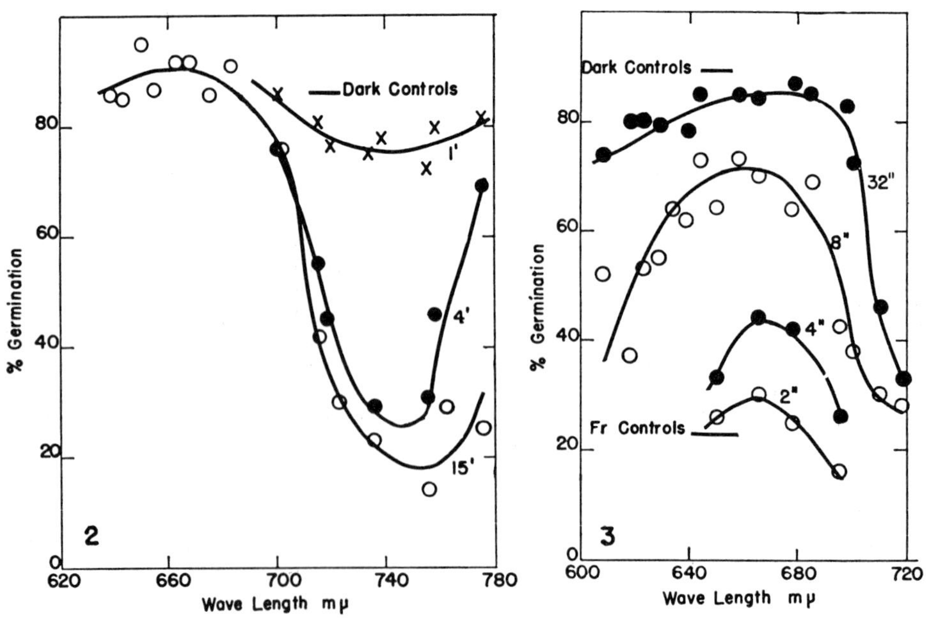

Figs. 2, 3.—Fig. 2, inhibition of germination in tomato seeds (Ace) in light of 620–780 mμ. Exposures of 1, 4, and 15 min were made after 9-hr imbibition. Intensities of radiation were 132, 148, and 152 μw cm^{-2} mμ^{-1} at 650, 700, and 740 mμ, respectively. Fig. 3, repromotion of germination in tomato seeds (Ace) in light of 600–720 mμ following 9-hr imbibition and a saturating far-red (FR) irradiation. Intensities of radiation were 132, 148, and 152 μw cm^{-2} mμ^{-1} at 650, 700, and 740 mμ, respectively.

to far red in each cycle (fig. 5). Germination is inhibited when red is given immediately before far red and repromoted when red immediately follows far red. Intermediate germination percentages are obtained with cycles such as 1 min red, 14 min dark, 1 min far red, 14 min dark. When the seeds, after several hours of such intermittent radiation, are moved to darkness, the germination percentage at the end of the dark-incubation period depends not only on the relative positions of red and far red in each cycle but also on the type of irradiation they received last (fig. 6). High temperatures appear to counteract the action of far-red irradiations (table 5).

Discussion

The need for the presence of phytochrome in the P_{fr} form for many seeds to germinate contrasts with its apparent lack of control over seeds of other species, varieties, or even individuals of the same lot in dark controls. Two possibilities arise in this regard with respect to P_{fr} action. There may be alternative reactions leading to germination in seeds, or P_{fr}

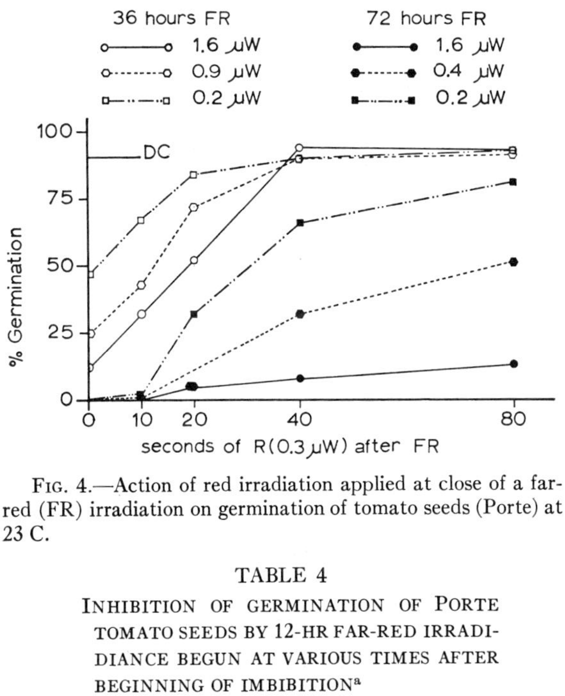

FIG. 4.—Action of red irradiation applied at close of a far-red (FR) irradiation on germination of tomato seeds (Porte) at 23 C.

TABLE 4

INHIBITION OF GERMINATION OF PORTE TOMATO SEEDS BY 12-HR FAR-RED IRRADIANCE BEGUN AT VARIOUS TIMES AFTER BEGINNING OF IMBIBITION[a]

PERIOD OF DARKNESS BEFORE FAR-RED IRRADIATION (HR)	PER CENT GERMINATION		
	Continuous	Cycle A	Cycle B
0	56	61	59
3	37	43	47
6	34	35	38
12	20	19	23
18	17	21	25
24	46	55	60

[a] Continuous, 1.6 μw cm^{-2} mμ^{-1} at 740 mμ supplied continuouly; Cycle A, 14.3 μw cm^{-2} mμ^{-1} at 740 mμ for 1 min of every 15 min supplied cyclically; Cycle B, 3.4 μw cm^{-2} mμ^{-1} at 740 mμ for 1 min of every 15 min supplied cyclically.

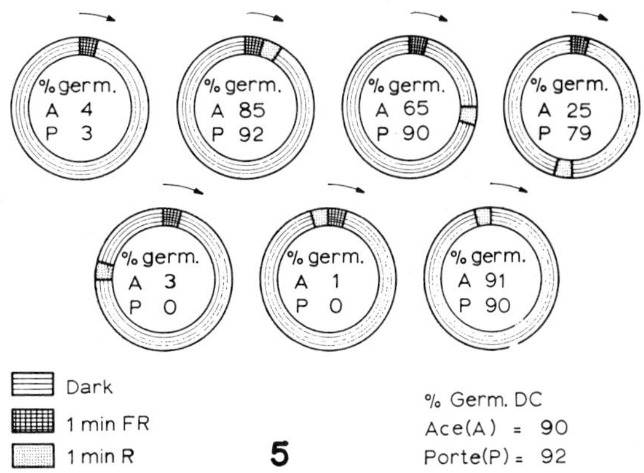

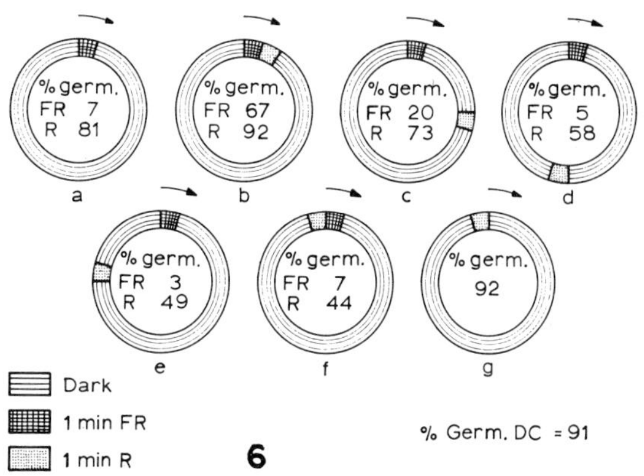

FIGS. 5, 6.—Fig. 5, action of cyclic irradiation (30-min cycle) applied for 4 days on germination of Ace and Porte tomato seeds at 20 C. Germinated seeds were counted at close of fourth day of treatment. Fig. 6, action of cyclic irradiation (10-min cycle) on germination of tomato seeds (Ace). Seeds were exposed to cyclic irradiation for 40 hr and then moved to darkness after receiving a last irradiation of either red (R) or far red (FR). Germinated seeds were counted 3 days after close of irradiations.

TABLE 5

INFLUENCE OF VARIOUS TEMPERATURES ON GERMINATION OF PORTE TOMATO SEED FOLLOWING DIFFERENT 2-DAY TREATMENTS AT 20 C

INITIAL TREATMENT 2 DAYS AT 20 C[a]	PER CENT GERMINATION AFTER 3 DAYS AT VARIOUS TEMPERATURES		
	17.5 C	22.5 C	27.5 C
Darkness	87	96	92
Far red	0	4	54
Far red, 1 min red	42	96	92

[a] Far red, 1.6 μw cm^{-2} mμ^{-1} at 740 mμ continuously for 2 days; red, 2.4 μw cm^{-2} mμ^{-1} at 660 mμ at end of FR irradiation.

alone might be effective. In the latter case, evidence of P_{fr} action might be obscured because of limited possibilities afforded by methods of physiological study. The findings on tomato-seed germination bear on the latter possibility.

The original search in 1955 for a light-sensitive lot of tomato seed (E. H. TOOLE, personal communication) was extensive before one was found, suggesting that the specific lot was an exception rather than the rule. Accordingly, in this work with tomato, two dark-germinating varieties were selected without particular discrimination. The behavior of the Ace variety is quite straightforward in interpretation. A single short exposure to far-red radiation of seeds imbibed in darkness induces sensitivity to red light. The spectral dependence, the energies required, and the reversibility of the response show it to be controlled by phytochrome (figs. 2, 3). Dry Ace seeds contain a mixture of P_r and P_{fr} with the latter sufficient in amount to induce germination upon rehydration. The energy requirements for response in the regions of 650 and 740 mμ show that less than 20% of the phytochrome must be in the P_{fr} form for 50% of the seed to germinate (note CUMMING, HENDRICKS, and BORTHWICK [1965] for table showing phytochrome conversion as a function of radiant energy). The action of far-red radiation indicates that after 9 hr of imbibition in darkness the average P_{fr} content of the seed is more than 60%. Moreover, at least 75% of the total phytochrome is rehydrated by the ninth hour.

Porte variety seeds display a more complex response. Exposures to as much as 3 hr of far-red radiation after 15 hr imbibition reduced germination only from 91% to 76%. Exposure to 12 hr of far-red radiation starting at the eighteenth hour reduced germination to 17%. Far-red radiation must act over a far longer time to reduce germination than is required in the Ace variety. After 2 days irradiation, the low germination can be fully reversed by brief red radiation, which in turn is reversed by brief far-red radiation. Eventual control by P_{fr} is thus demonstrated.

The cyclic lighting experiments (fig. 5, table 4) in which low-intensity radiation is used for only 3.3% of the time show that the controlling photoreaction needs only a low energy and thus is compatible with phytochrome conversion. If each far-red exposure of 1 min in a 30-min cycle is followed immediately by a 1-min exposure to red radiation (fig. 5), germination is unaffected, while in the reverse order it is completely suppressed. Control through phytochrome action instead of a possible high-energy reaction is thereby established.

Considerable question remains, however, as to how prolonged far-red radiation reduces germination by action on phytochrome in seeds that would otherwise germinate in darkness. If rates of phytochrome rehydration alone are involved, the results indicate that they differ by more than tenfold for the two varieties. A short far-red irradiation of Ace variety seeds at the ninth hour of darkness, which reduces germination to 28%, is fully reversed by an immediate short exposure to red radiation (table 1). Delay of 36 hr between the two irradiations, on the other hand, leads to only 45% germination; that is, less than half the seeds are reversed (table 1) over a total treatment period of less than 2 days. A cyclic exposure to far-red radiation for 40 hr (fig. 6), which reduces germination to 7%, is fully reversed by red radiation (81%). If an interval is used between far-red and red irradiation in a cycle (fig. 6), germination is reduced to 49% following final red irradiation, provided that P_{fr} is present for one-fourth of each cycle. The conclusion from these several observations is that P_{fr} at a high level causes germination, but at a much lower P_{fr} level, attained by continued rehydration, endogenous factors (other than phytochrome) required for germination are exhausted.

With the Porte seeds the action of far-red radiation, which gives about 1% P_{fr}, is adequate after 4 days to produce a non-germinating condition. In each cycle, with P_{fr} present, the Porte variety requires a shorter period than the Ace variety to induce a light response (fig. 5). The several findings are in accord with the Porte variety having a high level of the factors for response (substrate) which must be reduced to below a critical value before phytochrome becomes controlling.

LITERATURE CITED

BORTHWICK, H. A., S. B. HENDRICKS, E. H. TOOLE, and VIVIAN K. TOOLE. 1954. Action of light on lettuce seed germination. BOT. GAZ. **115**:205–225.

CUMMING, B. G., S. B. HENDRICKS, and H. A. BORTHWICK. 1965. Rhythmic flowering responses and phytochrome changes in a selection of *Chenopodium rubrum*. Can. J. Bot. **43**.825–853.

DOWNS, R. J., K. H. NORRIS, W. A. BAILEY, and H. H. KLUETER. 1964. Measurement of irradiance for plant growth and development. Amer. Soc. Hort. Sci., Proc. **85**:663–671.

MANCINELLI, A. L., and H. A. BORTHWICK. 1964. Photocontrol of germination and phytochrome reaction in dark-germinating seeds of *Lactuca sativa* L. Ann. Bot. Rome. **28**:9–24.

MOHR, H. 1962. Primary effects of light on growth. Ann. Rev. Plant Physiol. **13**:465–488.

TOOLE, E. H. 1961. The effect of light and other variables on the control of seed germination. Int. Seed Test. Ass., Proc. **26**:659–673.

Reprinted from the PROCEEDINGS OF THE NATIONAL ACADEMY OF SCIENCES
Vol. 59, No. 2, pp. 376–380. February, 1968.

*A RAPID PHOTOREVERSIBLE RESPONSE OF BARLEY ROOT TIPS IN THE PRESENCE OF 3-INDOLEACETIC ACID**

BY TAKUMA TANADA

AGRICULTURAL RESEARCH SERVICE, U.S. DEPARTMENT OF AGRICULTURE,
BELTSVILLE, MARYLAND

Communicated by S. B. Hendricks, December 20, 1967

For several years I have observed that root tips of some species of plants adhere to glass surfaces in water. Examination of this phenomenon shows that under suitable conditions the root tips of barley adhere to glass within 30 seconds when irradiated with red light and detach from the glass within 30 seconds when irradiated with far-red light. This rapid reversibility has been repeated for five or more cycles in numerous trials with different lots of seedlings. 3-Indoleacetic acid at very low concentrations and several other compounds are required for the reversible process.

Methods and Materials.—About 14 gm of barley seeds (*Hordeum vulgare* L., var *Compana*) are washed several times with deionized water. They are then spread over cheesecloth supported on a stainless steel wire net hung inside a porcelain crock. Six liters of deionized water from which chlorine has been removed by passing through charcoal are poured into the crock until the net just touches the water surface. The seeds are held in darkness for 2 days at 23°C with the water aerated continuously.

After 2 days, the seedlings are exposed to dim white light (0.3 ft-c) from a tungsten filament lamp for at least 1 hr before use in a constant-temperature room at 23°C. The dim light is used in subsequent handling of seedlings. Roots, after gentle blotting on filter paper, are placed on a ruled plastic block and 1-mm sections are quickly cut with a stainless steel razor blade from tips of the first two seminal roots of a seedling. Ten tips are washed into a Pyrex glass beaker (tall form) of 200-ml capacity containing about 25 ml of deionized water. The water is poured off after a few seconds and 10 ml of solution of the following composition is added immediately to the beaker: sodium adenosine-5′-triphosphate (ATP), 2.5×10^{-6} *M;* 3-indoleacetic acid (IAA), 10^{-8} *M;* L-ascorbic acid (AA), 10^{-6} *M;* $MnCl_2$, 2×10^{-6} *M;* $MgCl_2$, 2×10^{-4} *M;* KCl, 10^{-4} *M;* and HCl, 5×10^{-5} *M*. Solutions of the organic chemicals are freshly prepared each day. The pH of the final solution is 5.0.

The beaker is immediately placed in red light. After mixing with a gentle rotary motion for about 10 sec, the tips are allowed to settle and remain undisturbed in the central portion of the bottom for 20 sec. The beaker is then raised a few millimeters on one edge and very gently swirled 3 or 4 times. Sudden or jerky movements are avoided. The number of tips adhering, or showing signs of adhering, to the glass surface is then counted. Tips that do not adhere move smoothly away from the spot on which they have settled. After the gentle swirling motion, the tips are left undisturbed for about 5 sec. This process of dislodging, counting, and settling is repeated three times. A good count of the average number of tips adhering to the glass can be obtained by the end of the fourth dislodging. After 1 min in red light, the tips are exposed to far-red light for 1 min during which time the procedure done in red light is repeated. The red and far-red irradiation cycle is carried out for 5 cycles.

The source of red light is a 15-watt daylight fluorescent tube filtered with 2 layers of red cellophane. Both ends of the tube are covered with 15 cm of black plastic. The beaker is placed 55 cm from the tube where the radiant flux is 7.9 $\mu w/cm^2$ in the region of 600–700 nm.[1] The far-red light source is a 60-watt tungsten filament lamp covered with 6 sheets of red cellophane and 2 sheets of blue cellophane. The lamp is inverted so that the far-red light passes through 5.5 cm of water contained in a glass dish with a watch-glass cover. The beaker is placed on the cover at a distance of 10 cm from the lamp where the radiant flux is 260 μw/cm^2 in the region of 700–750 nm. The tips can be observed best in the far-red light by suitable positioning of black paper. These types of light sources are in routine use for phytochrome studies at Beltsville.

The same beaker is used throughout an experiment. It is washed in a biodegradable detergent and rinsed several times with tap water followed by deionized water. The charge on the glass surface responsible for root tip adherence appears to be associated with phosphate. Beakers are treated once with a dilute solution of Na_2HPO and washed several times with deionized water before use. An acid wash removes the adsorbed phosphate ion, but the detergent appears to have little or no effect on it.

Results.—Results from one experiment showing the essentiality of the various components of the solution for reversibility in red and far-red light are listed in Table 1. Reversibility of attachment is manifest in the complete solution. Without ATP the root tips do not detach in far-red light. The ATP cannot be replaced by ADP. IAA is necessary for both attachment in red light and detachment in far-red light. AA also seems to be necessary for both processes. Mn^{++} is essential only for adherence in red light. Both processes in red and far-red light require Mg^{++} which cannot be replaced by Ca^{++}. K^+ is required for detachment in far-red light, and Na^+ cannot substitute for it.

The effect of increasing ATP concentration on detachment of root tips in red and far-red light is shown by the results described in Table 2. Again, detachment does not occur in far-red light unless ATP is present. However, when the ATP concentration exceeds an optimum value (2.5×10^{-6} *M* in this experiment), the tips tend to adhere less and less even in red light.

The effect of increasing IAA concentration on adherence of root tips in red and far-red light can be seen in Table 3. When IAA is absent, attachment and detachment become progressively poorer as the irradiation cycles are run. A very low concentration of 2×10^{-10} *M* enables a large proportion of tips to adhere in red light, but detachment in far-red is not seen until a concentration of about 5 ×

TABLE 1. *Essentiality of various compounds for the reversible adhesion of barley root tips to glass during one-minute red (R) or far-red (FR) irradiation.*

Media composition	Number of Root Tips Adhering to Glass				
	Cycle 1 R–FR	Cycle 2 R–FR	Cycle 3 R–FR	Cycle 4 R–FR	Cycle 5 R–FR
Complete	10–4	10–3	9–3	8–1	8–1
−ATP	9–10	10–10	10–10	10–10	10–10
−ATP, + ADP	9–9	10–9	9–8	8–7	8–8
−IAA	5–5	5–4	3–3	3–2	2–1
−AA	8–6	8–7	6–5	4–4	3–3
$-MnCl_2$	6–3	5–1	4–0	5–2	4–0
$-MgCl_2$	8–6	6–5	5–5	3–4	3–2
$-MgCl_2$, + $CaCl_2$	7–7	7–5	5–5	4–4	4–3
−KCl	10–10	10–9	9–9	9–8	9–9
−KCl, + NaCl	10–10	9–8	9–9	9–8	8–8

Length of root tip: 1 mm. Ten tips per run.

Composition of complete medium: ATP, adenosine-5′-triphosphate (2.5×10^{-6} *M*); IAA, 3-indoleacetic acid (10^{-8} *M*); AA, L-ascorbic acid (10^{-6} *M*); $MnCl_2$ (2×10^{-6} *M*); $MgCl_2$ (2×10^{-4} *M*), KCl (10^{-4} *M*), HCl (5×10^{-5} *M*).

Substitutions: ADP, adenosine-5′-diphosphate (2.5×10^{-6} *M*); $CaCl_2$ (2×10^{-4} *M*), NaCl (10^{-4} *M*).

R, red light; FR, far-red light.

The first number in each column indicates number of tips adhering in red light, second for number in far-red light.

TABLE 2. *Effect of increasing ATP concentration on the reversible adhesion of barley root tips to glass during one-minute red (R) or far-red (FR) irradiation.*

ATP concentration (M)	Number of Root Tips Adhering to Glass				
	Cycle 1 R–FR	Cycle 2 R–FR	Cycle 3 R–FR	Cycle 4 R–FR	Cycle 5 R–FR
0	10–9	10–9	9–10	9–9	9–9
1×10^{-6}	10–8	9–7	9–7	9–8	9–7
2×10^{-6}	10–7	10–8	10–7	9–6	10–6
2.5×10^{-6}	9–3	9–2	8–2	7–0	7–0
3×10^{-6}	9–3	8–1	8–0	7–0	6–1
4×10^{-6}	6–2	4–0	4–0	4–0	2–0

Except for the concentrations of ATP, tips were irradiated in the medium described as "complete" in Table 1.

TABLE 3. *Effect of increasing 3-indoleacetic acid concentration on the adhesion of barley root tips to glass during one-minute red (R) or far-red (FR) irradiation.*

IAA concentration (M)	Number of Root Tips Adhering to Glass				
	Cycle 1 R–FR	Cycle 2 R–FR	Cycle 3 R–FR	Cycle 4 R–FR	Cycle 5 R–FR
0	8–7	5–3	5–3	3–2	2–1
2×10^{-10}	8–7	6–6	6–6	7–6	7–5
5×10^{-10}	9–9	9–10	10–8	10–8	8–6
1×10^{-9}	10–8	10–9	9–7	9–7	8–6
2×10^{-9}	10–9	10–8	9–7	9–5	8–3
5×10^{-9}	10–6	9–5	9–2	8–1	8–0
1×10^{-8}	9–3	10–3	10–1	8–1	8–0

Except for the concentrations of IAA, tips were irradiated in the medium described as "complete" in Table 1.

10^{-9} M is reached. Good reversibility is achieved at about 10^{-8} M concentration. Results from some unreported experiments indicate that at higher concentrations of IAA the tips adhere to glass even in far-red light.

After the experiments reported in Tables 1–3 were completed, it was found that when the deionized water used to make up the solution is placed in sunlight for about one week (in a greenhouse), the tips respond to a much lower concentration of IAA. Some results are reported in Table 4. The tips without IAA do not show a pronounced lack of it until after several cycles. These results suggest that something (probably chlorine) in the water destroyed some of the endogenous and exogenous IAA in the previous experiments. The response of the tips

TABLE 4. *Effect of water quality on the influence of indoleacetic acid on attachment of barley root tips to glass during one-minute red (R) or far-red (FR) irradiation.*

IAA concentration (M)	Number of Tips Adhering to Glass						
	Cycle 1 R–FR	Cycle 2 R–FR	Cycle 3 R–FR	Cycle 4 R–FR	Cycle 5 R–FR	Cycle 6 R–FR	Cycle 7 R–FR
0	10–8	10–6	8–5	7–5	7–4	7–5	5–4
10^{-10}	10–6	10–3	10–3	10–3	9–4	8–6	8–6
2×10^{-10}	10–4	10–3	10–2	9–2	9–3	—	—
3×10^{-10}	10–4	10–2	10–2	10–2	10–1	9–1	8–1
5×10^{-10}	10–4	10–4	10–3	10–4	10–4	9–4	9–3
10^{-9}	10–7	10–7	10–5	10–6	9–5	10–4	—
5×10^{-9}	10–10	10–9	10–9	10–8	10–8	—	—

Deionized water, exposed to sunlight for 1 week, was used to make up solution. ATP concentration, 4×10^{-6} M; concentrations of other compounds as shown in Table 1.

in far-red light to IAA can now be seen when the concentration is as low as 2×10^{-10} M. A higher concentration than 10^{-9} M causes the tips to adhere in both red and far-red light.

Discussion.—The photoreversible attachment and detachment of barley root tips on a glass surface in light corresponding to the spectral regions of phytochrome action indicate that the phenomenon is mediated by phytochrome. The specificities of the several compounds, which correspond with known physiological actions, and their requirements in physiologically effective concentrations make it unlikely that the system is an artificial one. In addition, it looks as if phytochrome expression is controlled not only by light, but is modified by endogenous levels of IAA, ATP, and AA.

Several reports of fast phytochrome-mediated responses in plants have been recently reviewed by Hillman.[2] Up to now the earliest measurable result from phytochrome activity is that reported by Fondeville *et al.*[3] who have shown that the effect of red light on the closing movement of *Mimosa* leaflet can be seen five minutes after irradiation. Hendricks and Borthwick[4] in a detailed treatment of rapid responses suggest that phytochrome controls permeability of plant cell membranes. Because the phenomenon reported here can be detected within 15 seconds from the beginning of exposure time and is largely completed within 30 seconds, it is quite likely an initial expression of phytochrome action.

A molecular explanation of the reversible attachment is possible in terms of known biochemical and biophysical principles. Because phosphate causes the glass surface to be negatively charged, adherence of the tips to the surface suggests that they are positively charged. Detachment takes place when the tips become either neutral or negatively charged. The changes in charge could arise from molecular conformation or configuration changes in the protein moiety of phytochrome, particularly about its isoelectric point. Jones[5] and Jones[6] suggest a structural change in a contractile protein to explain the action of ATP on detachment of animal cells from glass and other cells. They postulate that this protein is at the cell surface, and that in the relaxed state it has positive charges which cause the cells to adhere to glass and other cells. Upon the addition of ATP, the protein contracts and the positive charges are decreased, resulting in less adhesion to negatively charged surfaces. ADP and other relaxing factors reestablish adhesiveness.

An IAA requirement in this phenomenon was unexpected. An external source of IAA was found to be necessary only after the seedlings were exposed for some time to dim white light. A review of the literature gives only a slight inkling of a possible connection between phytochrome and IAA. Liverman and Bonner[7] reported an interaction between IAA and red light on enhancing growth of *Avena* coleoptile. They suggest that several processes effected by red light and IAA may share in a common mechanism. Recently, Furuya and Torrey[8] have reported that IAA-induced lateral root initiation in pea root segments is inhibited by red light. The red-light inhibition is reversed by far-red light. They suggest a possible relationship between IAA action and phytochrome action. The results presented in this report clearly implicate IAA as an essential component of the phytochrome system.

A requirement for AA also became apparent only after the seedlings were exposed for some time to dim white light. While AA is a reducing agent, its low effective concentrations (about 10^{-6} M) suggest that it probably serves in some other capacity. Fawcett[9] has reviewed several papers showing a relationship between the actions of IAA and AA.

A promotion of a charge by weak light seems implicit in some investigations of Haupt.[10] He found that weak light induced the attachment of chloroplast to the cell wall of an alga, *Vaucheria*. In addition, Haupt's observations suggest that phytochrome is located in the periphery of the cell. The system reported in this paper also indicates that phytochrome is located near or in the cell membrane. Its reaction to external compounds and its quick response (within 30 sec) upon their addition point to such a location. If phytochrome is in the membrane, it could very well control permeability as suggested by Hendricks and Borthwick.[4]

When repeating these experiments, some practice will possibly be required in rotating the beaker with the right force. A concentration study of ATP is helpful to acquire the necessary skill. Also, because of differences in experimental setup, the quantities of the necessary compounds might vary considerably from those reported here. This is likely to be true especially for IAA and possibly for AA and ATP.

Preliminary results show that a similar system is present in mung bean root tips (*Phaseolus aureus*) with one notable difference. In addition to the seven factors listed in Table 1, mung bean root tips require Ca^{++}.

Summary.—A very rapid photoreversible attachment of barley root tips to wet glass surfaces has been found to be mediated by phytochrome. Adhesion of the tips to glass can be seen within 30 seconds in red light. They are released within 30 seconds in far-red light. Both processes are reversible for several light cycles. 3-Indoleacetic acid in very low concentrations is necessary for reversibility. Adenosine-5′-triphosphate, ascorbic acid, Mn^{++}, Mg^{++}, and K^{+} are also required in the system. This phenomenon is probably a result of the first physiological response to the change in form of phytochrome induced by radiation. 3-Indoleacetic acid is essential for this phytochrome action.

* This investigation was supported in part by the U.S. Atomic Energy Commission.

[1] Measurement of radiant energy was made by K. H. Norris (MQRD; ARS, USDA, Beltsville, Maryland).

[2] Hillman, W. S., *Ann. Rev. Plant Physiol.*, **18**, 301 (1967).
[3] Fondeville, J. C., H. A. Borthwick, and S. B. Hendricks, *Planta*, **69**, 357 (1966).
[4] Hendricks, S. B., and H. A. Borthwick, these PROCEEDINGS, **58**, 2125 (1967).
[5] Jones, B. M., *Nature*, **212**, 362 (1966).
[6] Jones, P. C. T., *Nature*, **212**, 365 (1966).
[7] Liverman, J. L., and J. Bonner, these PROCEEDINGS, **39**, 905 (1953).
[8] Furuya, M., and J. G. Torrey, *Plant Physiol.*, **39**, 987 (1964).
[9] Fawcett, C. H., *Ann. Rev. Plant Physiol.*, **12**, 345 (1961).
[10] Haupt, W., *Ann. Rev. Plant Physiol.*, **16**, 267 (1965).

PHYSIOLOGIA PLANTARUM, VOL. 20. 1967, 128–138

Action Spectra for Light-induced Elongation in Fern Protonemata

By

JOHN H. MILLER and PAULINE M. MILLER

Department of Bacteriology and Botany
Syracuse University, Syracuse, New York

(Received September 15, 1966)

Abstract

The action spectrum for promotion of elongation of protonemata of *Onoclea sensibilis* has peaks at 400–420, 580–600 and 640–660 nm. The largest growth increments at saturating light doses are produced by yellow and far-red light. Elongation induced by yellow and far-red irradiation persists in old as well as young filaments, while red-light promotion is found only in young filaments. The growth promotion caused by yellow light is partially reversed by red light down to the level of growth produced by red irradiation alone. Elongation of rhizoids is under reversible red, far-red control, while yellow light is inactive. A model is proposed and discussed in which the light-sensitive elongation of filaments is accounted for by the presence of three distinct photoreceptors: phytochrome; a pigment absorbing yellow light, P_{580}; and a pigment absorbing blue light, P_{420}.

Introduction

If spores of the fern *Onoclea sensibilis* are given a minimum exposure to light to permit germination and then are placed in darkness, they grow as filamentous protonemata. Miller and Miller (1963) showed that both red light (660 nm) and far-red light (720 nm) accelerated filament elongation, and at saturating light doses far-red induced a higher growth rate than did red. Photo-reversibility was found in the sense that the rate of filament elongation depended on whether red or far-red light was given last in a sequential series of irradiations. Thus red irradiation following far-red reduced elongation to the level produced by red irradiation alone. Any sequence of irradiation however, promoted elongation above that which took place in darkness.

The effectiveness of red and far-red light and the occurrence of mutual photoreversibility suggested the involvement of phytochrome in filament elongation, although the type of control differed so significantly from the usual phytochrome systems that serious difficulties faced any attempt to account for the growth data in terms of the known properties of phytochrome. All major data (Butler, Hendricks and Siegelman 1965, Hendricks and Borthwick 1965) suggest that in dark-grown plants phytochrome exists solely in the inactive, red-absorbing form (P_R); absorption of red light converts P_R to the physiologically active, far-red-absorbing form (P_{FR}), and within limits the physiological response is proportional to the concentration of P_{FR}. Far-red irradiation converts P_{FR} almost totally to the inactive P_R. To account for the response of fern filaments to red and far-red light assumptions must be made which are in complete disaccord with the experimentally determined characteristics of phytochrome. The situation in ferns is further complicated, since the elongation of rhizoids in dark-grown protonemata and cell division in 2-dimensional prothallia are controlled by a typical phytochrome system (Miller and Miller 1961, 1963), which corresponds to that found in higher plants.

The present paper grew out of attempts to clarify the unusual red, far-red effects on filament elongation. The action spectra and related growth data obtained during this study indicate the presence in O. sensibilis gametophytes of another pigment, distinct from phytochrome, which also is active in regulating filament elongation. The joint presence of this new system and phytochrome makes possible a coherent explanation of the growth responses which previously we reported.

Materials and Methods

Spore material and growth conditions. — Spores were collected in the vicinity of Syracuse, New York in the fall of 1965. The methods of spore sterilization and liberation from the sporangia are described in Miller and Miller (1961). The liquid growth medium for all experiments was Knop's solution plus 1 % sucrose (Miller and Miller 1964). The medium was distributed in 25 × 100 mm aluminum capped culture tubes, about 10 ml per tube, and was autoclave sterilized for 10 min at 1.05 kg pressure.

Sterile spores were sown on the surface of the sterilized medium and were exposed for 24 h to ca. 2000 lux white fluorescent light at 25–26°C. The spores then were transferred to darkness at 25–26°C and allowed to grow for 72 h. Experimental light treatments were given at the end of this time, and after irradiation the cultures were returned to darkness for an additional 24 h of growth. Measurement of filament length was then made. A sample of gametophytes was mounted in 70 % ethyl alcohol on a microscope slide and covered with a cover slip. The cover slips were ringed with heavy mineral oil so that the slides could be kept for several days. The image of the gametophytes was projected at a magnification of 200× onto a sheet of paper, using a Bausch and Lomb microprojector, and the lengths of a random sample of 50 plants per light treatment were measured with a flexible plastic ruler.

Since spores from different plants showed some variability in dark-growth rate,

Table 1. *Monochrometer settings, auxilliary glass color filters, and typical energy output at each wavelength setting.*

λ	Corning glass color filters	μW · cm⁻²
400	5–59	0.113
420	5–59	0.195
440	5–59	0.285
460	5–59	0.317
480	5–59	0.270
500	4–97 + 3–71	0.443
520	4–97 + 3–70	0.449
540	4–97 + 3–69	0.430
560	3–67	0.369
580	3–67	0.851
600	2–62	0.088
620	2–61	0.142
640	2–61	0.198
660	2–59	0.194
680	2–64	0.188
700	2–64	1.035
720	2–64	1.110
740	2–64	1.269

all spores were pretested, and only those were used which reached an average length of at least 180 μ by the end of the 72 h dark-growth period.

Determination of action spectra. — A Bausch and Lomb model 33-86-02 diffraction grating monochrometer with a tungsten source was used to provide monochromatic irradiation. The exit slit of the monochrometer was set at 1.5 mm, and for each wave length setting appropriate Corning glass color filters were used at the exit slit to eliminate unwanted wave lengths. Between 600 and 680 nm the intensity of irradiation was reduced by placing a screen of two layers of cheesecloth at the exit slit. Light intensities at the position of the plant material were measured with an Isco model SR spectroradiometer. During the experiments the intensity of irradiation was always determined immediately following irradiation of the plant material. Table 1 gives the wave length settings, along with the glass filters used, and a typical set of light intensities (as $\mu W \cdot cm^{-2}$) measured during one action spectrum determination.

For each wavelength plants were irradiated for different times ranging generally from 10 s to 10 min but up to 20 min for some wave lengths. After 24 h the increments in length above non-irradiated controls (Δl) were determined. The quantum intensity of irradiation was calculated from the relationship: $\text{quanta} \cdot cm^{-2} \cdot s^{-1} = \frac{\lambda}{1987} (\mu W \cdot cm^{-2}) \times 10^{12}$, λ expressed in Ångström units. The incident $\text{quanta} \cdot cm^{-2}$ were calculated for each irradiation by multiplying $\text{quanta} \cdot cm^{-2} \cdot s^{-1}$ by the time in s of irradiation. Δl's were then plotted against log $\text{quanta} \cdot cm^{-2}$ for each wavelength, and straight lines were fitted to the data by the method of least squares. The regression equations were solved for log $\text{quanta} \cdot cm^{-2}$ when $\Delta l = 0.01$ mm and figures converted to $\text{quanta} \cdot cm^{-2}$. Action spectra were drawn by plotting vs λ the reciprocal of $\text{quanta} \cdot cm^{-2}$ necessary to give a growth increment of 0.01 mm.

Results

The practice of irradiating plants for different times at a constant intensity in order to construct dose-response curves was justified by the maintenance of reciprocity within the range of incident energies used. Reciprocity was tested at 4 wave lengths, 400, 560, 660 and 720 nm, at which there was either an action maximum or a maximum of growth promotion at saturating light dose. For each of these wave lengths plants were irradiated at different incident energies for various times. The results are plotted in Figure 1 a and b as Δl vs log quanta · cm^{-2}. The results obtained at different incident energies are drawn with distinct symbols. In each case the response (Δl) depended upon the total dose and not the intensity or time of irradiation.

Complete sets of dose-response curves in 20 nm steps from 400–740 nm were run 4 times, and additional curves were determined for some wavelengths in the region of action maxima. The action spectrum derived from these data is given in Figure 2. Each point is the average of 4–7 independent determinations, and the vertical lines indicate the magnitude of plus or minus one standard error of the mean, $S.E._{\bar{x}} = \frac{S}{\sqrt{N}}$; $S^2 = \sum \frac{1}{N-1} (x-\bar{x})^2$. Light of any wave length from 400–740 nm promoted filament elongation, but 3 clear action maxima were found at 400–420, 580–600 and 640–660 nm.

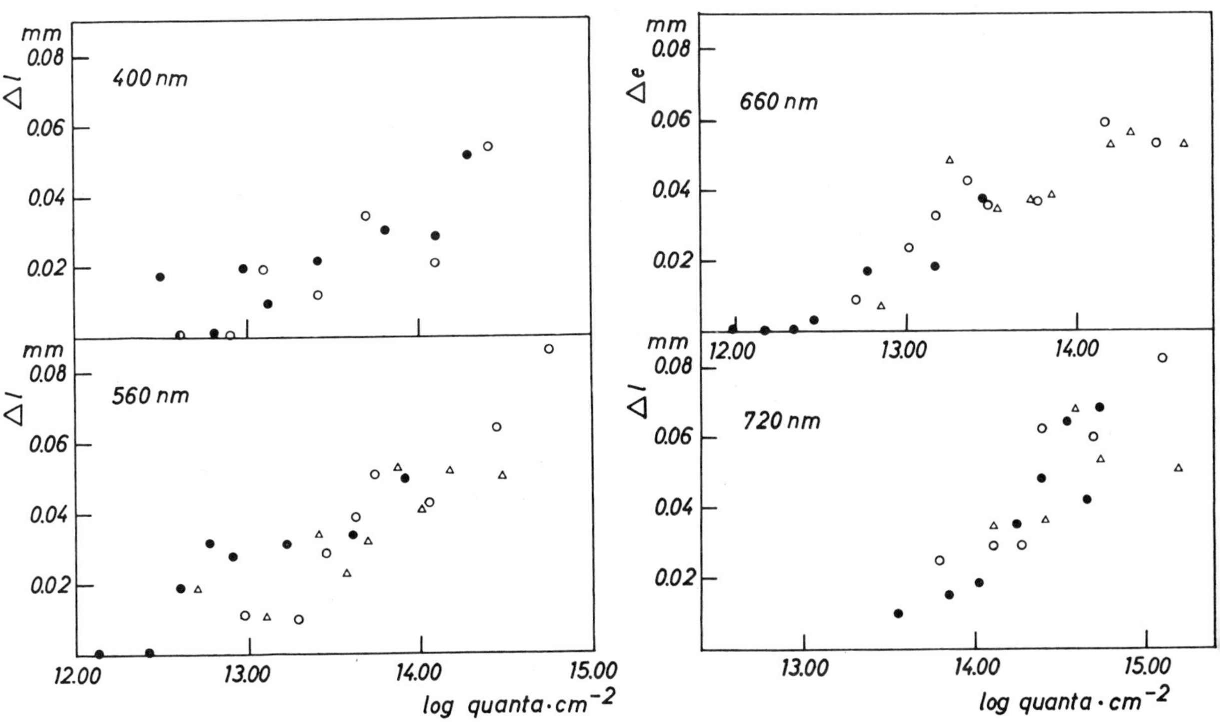

Figure 1. *Test for reciprocity at 400, 560, 660 and 720 nm.* Growth increment (Δl) plotted against log quanta · cm^{-2}. The different symbols represent irradiation at different incident intensities as follows (in μW · cm^{-2}): For 400 nm; ● = 0.108, ○ = 0.210. For 560 nm; ● = 0.049, ○ = 0.35, △ = 0.915. For 660 nm; ● = 0.015, ○ = 0.154, △ = 1.090. For 720 nm; ● = 0.162, ○ = 1.150, △ = 3.510.

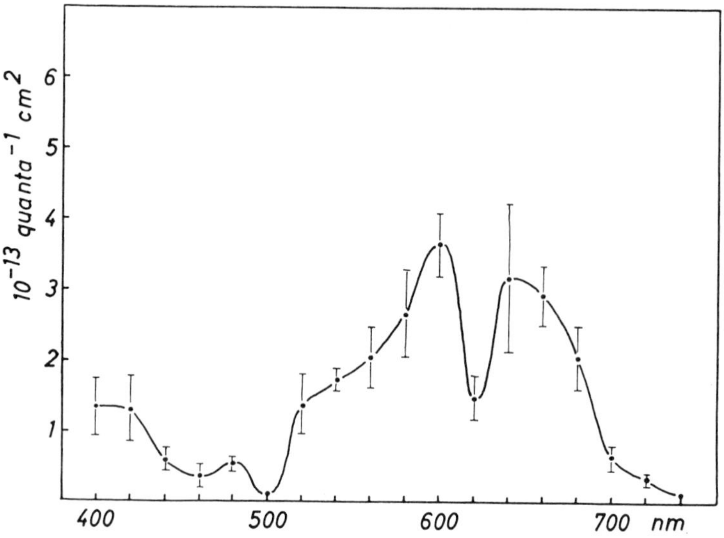

Figure 2. *Action spectrum for promotion of elongation of protonemata plotted as 1/quanta · cm^{-2} to give a growth increment* (Δl) *of 0.01 mm vs* · λ. 4–7 determinations averaged for each λ; ± one standard error of the mean indicated about each point.

Minimum effectiveness was at 500 nm and 620 nm, and the quantum effectiveness declined rapidly on the long-wave-length side of 660 nm. The effectiveness of yellow light was unexpected, but the magnitude of the action maximum in the region 580–600 nm demonstrated that yellow light was equal to red light in effectively promoting filament elongation. Substantial activity was found throughout the region 520–600 nm.

A somewhat different graph was obtained by plotting against λ the maximum elongation increment (Δl_{max}) produced by saturating irradiation. This sort of plot, a similar form of which was used by Mohr (1956), is not a true action spectrum, but it does yield information which is essential to the interpretation which will be presented of the action spectra. In Figure 3 5–8 determinations of Δl_{max} were averaged for each point and plus or minus

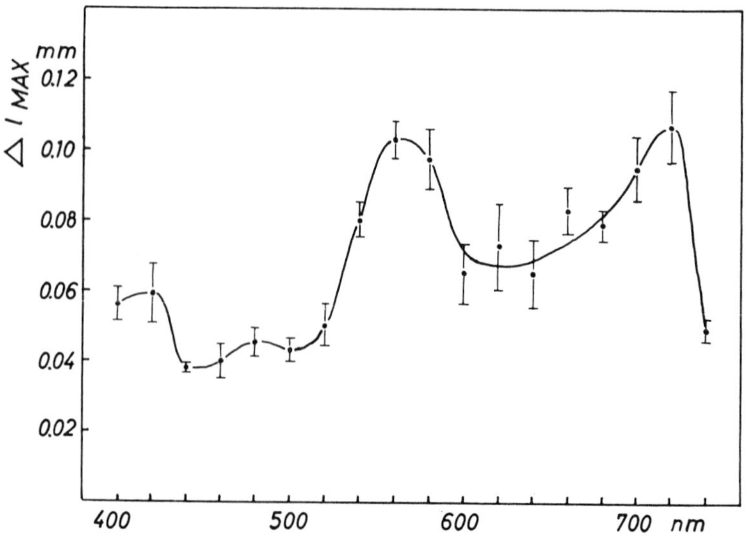

Figure 3. *Maximum growth promotion following a single saturating irradiation* (Δl_{max}) *vs* λ. 5–8 determinations averaged for each λ; ± one standard error of the mean indicated about each point.

Table 2. *Filament growth following a single saturating irradiation with red, yellow or yellow followed by red.* Each figure is the average of a sample of 50 plants. Duplicate treatments.

Treatment	Δl (mm)
R	0.065, 0.062
Y	0.094, 0.106
Y + R	0.053, 0.068

one S.E.$_{\bar{x}}$ is shown about each point. Δl_{max} at saturating irradiation varied with wave length, a fact which is meaningful in itself. Furthermore, the highest Δl_{max} was found at 560–580 and at 720 nm, and was equal for the 2 wavelengths. Special note should be made of the low quantum effectiveness of 720 nm (Figure 2) compared to the high Δl_{max} at this wave length. The lower Δl_{max} at 660 nm than at 720 nm confirmed our previous report of the different growth increments caused by red and far-red light (Miller and Miller 1963). In the region 400–420 nm there occurred a third maximum of Δl_{max} which was lower than those at 560 and 720 nm.

The action maximum in the violet (Figure 2) is matched by a peak in the graph of Δl_{max} vs $\cdot \lambda$ (Figure 3). The action peak at 580–600 nm corresponds to a Δl_{max} peak at slightly shorter wave lengths of 560–580 nm. Comment on a possible reason for this wave length difference is given in the discussion. In contrast the Δl_{max} peak at 720 nm falls at a wavelength of low quantum effectiveness, while the 640–660 nm action maximum is produced by wavelengths which give a lower Δl_{max} than either 720 or 560 nm.

In the following experiments saturating irradiation at various wavelengths means a 15 min exposure at incident energies comparable to those listed in Table 1. 560 nm was used as yellow light, since the highest Δl_{max} was consistently produced at this wave length.

The growth promotion resulting from irradiation with yellow light was reversible by subsequent irradiation with red light. Table 2 gives the results of an experiment in which filaments were irradiated with saturating doses of 560 or 660 nm and the combination of 560 followed immediately by 660. Each treatment was run in duplicate. It is apparent from the Δl figures that red light reversed the yellow light promotion down to the level produced by red light alone.

Table 3. *Rhizoid growth following various irradiations.* Results of 3 separate experiments. Each entry is the average of a sample of 50 plants.

Treatment	Δl (mm)			
	I	II	III	Average
R	0.107	0.092	0.077	0.092
FR	−0.026	0.018	0.037	0.010
Y	0.030	0.021	0.008	0.020
R + FR	0.020	−0.009	0.033	0.014
R + Y	0.084		0.082	0.083

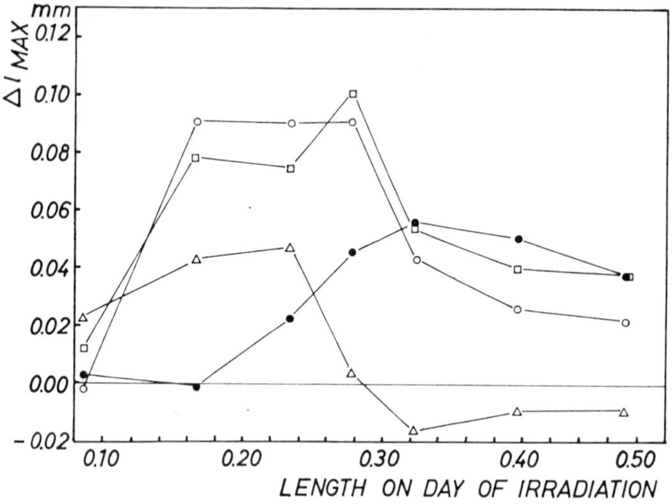

Figure 4. *Maximum growth promotion following single saturating irradiation at 420, 560, 660 and 720 nm plotted against filament length at the time of irradiation.* Each point is the average of a sample of 50 plants. ● = = 420 nm; ○ = 560 nm; △ = = 660 nm; □ = 720 nm.

Earlier we reported (Miller and Miller 1963) that the elongation of rhizoids seemed to be under a simple control by phytochrome. Irradiation with red light promoted rhizoid elongation. Far-red irradiation alone was inactive, but the promotive effect of red light was reversed by subsequent far-red irradiation. Since rhizoid elongation was not promoted by far-red light as was filament elongation, the question arose whether filaments and rhizoids might show a similar difference in response to yellow light. Cultures were irradiated with saturating doses of 560, 660 and 720 nm and with various sequences of one irradiation immediately following another. For experiments with rhizoids irradiation was performed after 48 rather than 72 h of dark growth. The results of 3 experiments are given in Table 3 listing the light treatment and the increment in rhizoid length caused by irradiation and the average Δl for the 3 experiments. While red light produced elongation which was reversed by subsequent far-red irradiation, yellow light was ineffective. It promoted rhizoid elongation only slightly, if at all, and did not significantly reverse the promotion caused by red light.

Another puzzling aspect of the red and far-red effects on filament elongation was the change from promotion to inhibition by red light as the filaments aged, while far-red light always promoted elongation regardless of filament age (Miller and Miller 1963).

Figure 4 shows essentially the changing patterns of response to blue, yellow, red and far-red light with increasing filament age. Cultures were allowed to grow in darkness for different times (from 2–8 days) before they were irradiated with a single, saturating exposure to 420, 560, 660 or 720 nm. The average size at time of irradiation was measured and the Δl_{max} after 24 h additional growth. In Figure 4 Δl_{max} is plotted against the filament length at the time of irradiation. Three different response patterns were found. Blue light did not promote filaments less than about 0.180 mm; its effect increased with increasing length (age) at the time of irradiation to a maximum effectiveness at about 0.30 mm. Red light gave significant promotion the earliest of any of the different wavelengths, but its effect rapidly decreased and actually became inhibitory in filaments greater than about 0.240 mm. This is the age-dependent change in response to red light described by Miller and Wright (1961). Yellow and far-red light exhibited patterns similar to each other. Little effect was produced in filaments less than 0.100 mm, but promotion rapidly increased and remained high in filaments up to about 0.280 mm, after which the promotive effects of these wavelengths declined gradually.

Discussion

The hypothesis which we propose to account for the effects of light on protonemal elongation will first be presented, followed by a justification of its separate points. The model attempts only to define the number and characteristics of the photoreceptor systems which mediate the growth responses to different wave lengths of light.

At least 3 photoreceptors control the elongation of protonemata. One of these is phytochrome, the other 2 are pigments which absorb maximally at 400–420 nm and 580–600 nm respectively. For convenience they will be designated at P_{420} and P_{580}. The absorption of light by any of these pigments leads to an increase in filament elongation compared to filaments growing in total darkness.

The effects of red and far-red light on filament elongation which earlier we suggested might result from an unusual type of phytochrome (Miller and Miller 1963) are now seen to result from the presence of 2 distinct systems, phytochrome and P_{580}. P_{580} is assumed to have a long-wave-length absorption tail which extends at least to 720 nm. The action of various wave lengths may best be understood by considering dark-grown filaments which are irradiated with different colors of monochromatic light. Red irradiation (660 nm) alone is absorbed and acts through phytochrome. Yellow light alone acts through P_{580}. Far-red (720 nm) produces its growth effect through absorption by the long-wave-length tail of P_{580}; in effect the actions of yellow and far-red should be regarded as identical. When red irradiation is followed by far-red, the action of far-red is two-fold; the phytochrome-mediated red light effect is reversed, but in addition P_{580} is activated to produce the growth characteristic of far-red irradiation. When far-red irradiation is followed by red, the growth mediated by P_{580} absorption of far-red is reversed, and phytochrome is simultaneously activated to produce growth characteristic of red irradiation alone.

In the foregoing model it is proposed that phytochrome accounts for the red-light-induced filament elongation. However, since far-red light also promotes elongation (through absorption by P_{580}) the characteristic of photoreversibility cannot be observed, and admittedly the evidence for phytochrome participation in filament elongation is circumstantial. The action peak at 640–660 nm is consonant with phytochrome action. The photoreversible red, far-red control of rhizoid elongation (Miller and Miller 1963), and cell division in 2-dimensional prothallia (Miller and Miller 1961) demonstrate that phytochrome is in fact present in the plants. The simplest interpretation of filament growth is that photoreversible phytochrome is one controlling system, but reversal by far-red of the physiological response to red light is masked by a simultaneous far-red activation of P_{580}.

Some justification is required for proposing a distinct photomorphogenetic pigment absorbing yellow light. It could be objected that yellow light is simply being absorbed by and acting through phytochrome. Phytochrome does absorb light throughout the visible range, and Pratt and Briggs (1966) showed in corn that irradiation with all wavelengths from about 520–680 nm established a photostationary phytochrome state of $P_{FR} : P_R = 80 : 20$. Thus all light within this range should at saturating doses cause the same physiological response as red light, although the quantum effectiveness of other wavelengths is lower than that of red light. The evidence against the possibility that yellow light affects filament growth through phytochrome seems conclusive. The action spectrum (Figure 2) shows a definite maximum at 580–600 nm, and the quantum effectiveness of yellow light is as high as 660 nm. Neither the absorption spectrum of purified phytochrome nor the action spectrum for the conversion of P_R to P_{FR} shows any such feature (Siegelman and Firer 1964, Pratt and Briggs 1966). Secondly (Figure 3) at saturating doses yellow light produces a higher growth increment in filaments than does red light, which could not happen if phytochrome alone were involved. Finally, if phytochrome were the sole pigment, any characteristic effect of red light should be mimicked by other phytochrome-absorbed wavelengths in the range which establishes the same photostationary state (520–680). This is not true of yellow light which does not promote rhizoid elongation, although a photoreversible red, far-red control is found (Table 3). The persistance of yellow light promotion of elongation in older filaments in contrast to the change to ineffectiveness and inhibition by red light (Figure 4) also shows that the two wavelengths must have distinct actions.

The identity of the effects of yellow and far-red light is supported by several pieces of evidence. Comparable growth increments follow saturating irradiation with both yellow and far-red light, each higher than the growth following red-light irradiation (Figure 3). The quantum effectiveness of far-red, however, is only about $^1/_{10}$ the quantum effectiveness of yellow light (Figure 2). This is consistent with the absorption of far-red by a long-wavelength absorption tail of P_{580}. The growth promotion caused by yellow light is reversed by red light down to the level caused by red light alone (Table 2), displaying the same relationship as does far-red to red irradiation (Miller and Miller 1963). Where in rhizoids far-red light does not promote elongation, yellow-light also is inactive (Table 3). The implication of this result is that P_{580} either is not present or is inactive in rhizoids. Finally both far-red and yellow light give the same pattern of promotion of filament elongation with respect to filament age (Figure 4).

While the red-light reversal of yellow-light-induced filament elongation is consistent with the proposed identity of the effects of yellow and far-red, the nature of the interaction is unknown. In fact the term reversal, in the sense it is used for phytochrome, may be incorrectly applied here. The data could as well result from an inhibition by red light of the yellow-light-induced growth. The interaction does offer a possible explanation for the fact that the action maximum in yellow is at 600 nm while the Δl_{max} peak is at somewhat shorter wavelengths (560–580 nm). Perhaps irradiation at 600 nm is nearest the absorption maximum of P_{580}, but in addition is more strongly absorbed by the red-absorbing pigment (phytochrome?) which limits the maximum response to yellow light. Irradiation at 560 allows fuller response to P_{580} stimulation, since it is further away from the reversing red region.

The separation of P_{420} as a separate system is based on studies of the differing interactions of blue light and red or far-red light with growth substances (Miller and Miller 1965) and with temperature (Miller and Miller 1966), and the pattern of blue-light-promotion with filament age, which is different from red light and yellow or far-red (Figure 4).

Reports of growth effects of green to yellow light on some lower plants have been made several times in the past. Since generally long periods of irradiation were used, and true action spectra were not published, the inference of a distinct photoreceptor cannot be made from these observations. However, some of the data strongly suggest the presence in other material of our proposed P_{580}. Using fairly broad wave lengths bands Stephan (1928) and Orth (1937) found maxima in length growth in yellow light for some ferns and mosses. Mohr (1956) employed interference filters for more precisely defined monochromatic light. Under 6 days of continuous monochromatic irradiation maximal filamentous growth of *Dryopteris filix-mas* gametophytes was produced by wave lengths in the green-yellow region and by far-red (Mohr 1956; Figure 12). Somewhat lower growth was found under red light, but all wave lengths from about 520 to 740 nm promoted filamentous growth above the dark-grown controls. Within this range our graph of Δl_{max} vs λ (Figure 3) is similar to Mohr's data. The resemblance suggests that P_{580} is present in the species on which Mohr worked. (The difference in our data and Mohr's below 500 nm results from the conversion to 2-dimensional growth under conditions of continuous irradiation as used by Mohr and the persistence of filamentous growth under periodic, low intensity irradiation, cf. Mohr 1956, Miller and Miller 1964). Recently Valanne (1966) showed that the growth of *Funaria hygometrica* was most strongly promoted by yellow-green and far-red light, although relatively long periods of irradiation were required to obtain the maximum effects. Steiner (1964) found promotion of germination in spores of *Spaerocarpus donellii* extending into the yellow-green. However, he attributed this to absorption by phytochrome. An action spectrum published by Halldal (1960) for induced negative photoaxis in the green flagellate *Platymonas subcordiformis* may indicate a photoreceptor similar to P_{580}. A narrow band in the yellow peaking at 590 to 600 nm was found to be active.

The work reported in this paper was supported by grants GB-371 and GB-4285 from the National Science Foundation.

References

Butler, W. L., Hendricks, S. B. & Siegelman, H. W.: Purification and properties of phytochrome. — *In* Chemistry and Biochemistry of Plant Pigments (ed., T. W. Goodwin), pp. 197–210. Academic Press, London and New York. 1965.

Halldal, P.: Action spectra of induced phototactic response changes in Platymonas. — Physiol. Plant. 13: 726–735. 1960.

Hendricks, S. B. & Borthwick, H. A.: The physiological functions of phytochrome. — *In* Chemistry and Biochemistry of Plant Pigments (ed., T. W. Goodwin), p. 405–436. 1965.

Miller, J. H. & Miller, P. M.: The effect of different light conditions and sucrose on the growth and development of the gametophyte of the fern, Onoclea sensibilis. — Amer. J. Bot. 48: 154–159. 1961.

— — Effects of red and far-red illumination on the elongation of fern protonemata and rhizoids. — Plant and Cell Physiol. 4: 65–72. 1963.

— — Blue light in the development of fern gametophytes and its interaction with far-red and red light. — Amer. J. Bot. 51: 329–334. 1964.

— — The relationship between the promotion of elongation of fern protonemata by light and growth substances. — *Ibid.* 52: 871–876. 1965.

— & Wright, D. R.: An age-dependent change in the response of fern gametophytes to red light. — Science 134: 1629. 1961.

Miller, P. M. & Miller, J. H.: Temperature dependence of the effects of light and auxin on the elongation of fern protonemata. — Plant and Cell Physiol. 7: 485–488. 1966.

Mohr, H.: Die Abhängigkeit des Protonemawachstums und der Protonemapolarität bei Farnen vom Licht. — Planta 47: 127–158. 1956.

Orth, R.: Zur Keimungsphysiologie der Farnsporen in verschiedenen Spektralbereichen. — Jb. Bot. 84: 358–426. 1937.

Pratt, L. H. & Briggs, W. R.: Photochemical and non-photochemical reactions of phytochrome in vivo. — Plant Physiol. 41: 467–474. 1966.

Siegelman, H. W. & Firer, E. M.: Purification of phytochrome from oat seedlings. — Biochem. 3: 418–423. 1964.

Steiner, A. M.: Der Einfluss von Licht und Temperatur auf die Sporenkeimung bei Sphaerocarpus donnellii Aust. (Hepaticae). — Ztschr. Bot. 52: 245–282. 1964.

Stephan, J.: Untersuchungen über die Lichtwirkung bestimmter Spektralbezirke und bekannter Strahlungsintensität auf Farne und Moose. — Planta 5: 381–443. 1928.

Valanne, N.: The germination phases of moss spores and their control by light. — Ann. Bot. Fennici 3: 1–60. 1966.

Amer. J. Bot. 54(10): 1241–1249. 1967.

EFFECTS OF RADIATION ON FLOWERING OF HYOSCYAMUS NIGER

M. J. Schneider, H. A. Borthwick, and S. B. Hendricks

Agricultural Research Service, U. S. Department of Agriculture, Beltsville, Maryland

ABSTRACT

Flowering and stem-lengthening responses of the long-day plant *Hyoscyamus niger* L. were followed under various photo- and dark-period combinations. Phytochrome action was identified by its photoreversibility, spectral region of response, and requisite levels of irradiance. Flowering control was also affected by another photoreaction having an action maximum at 710–720 nm. This reaction, while observable over appropriate short periods, is best displayed by protracted irradiation. The effectiveness of this reaction has a half-life of the order of 1–4 min. Responses to it can be observed over an intensity range of more than 100-fold. The two photoreactions interplay in flowering control depending on light quality during photoperiods. The interplay is a prominent factor affecting the adequacy of fluorescent and incandescent-filament radiation in plant growth experiments.

The roles of light in growth and flowering are inadequately resolved. This is particularly true for flowering of long-day plants, which depends markedly on light quality during the photoperiod. Brief interruptions of long nights with red radiation adequate for conversion of phytochrome are effective in flowering control of many short-day plants, but these treatments are less effective with long-day ones.

Varied elongation of normal green plants growing in light of different colors was noted by Senebier (1799). It was studied in detail by Pfeffer (1904), who recognized distinctly different effects of blue and red radiation. Trumpf (1924) studied the radiation effects in more detail. Effectiveness of blue light in suppressing elongation of stems was reported by Popp (1926), Shirley (1929), and Roodenburg (1940). When plant growth rooms came into use, radiation from an incandescent-filament lamp added to that from a carbon arc was found to be required for optimum plant growth (Parker and Borthwick, 1949). A need for blue and far-red radiation was also recognized when radiations of restricted spectral quality from fluorescent lamps were used throughout the photoperiods (van der Veen, 1950; Wassink, Sluijsmans and Stolwijk, 1950; Stolwijk and Zeevaart, 1955; Vince, 1956).

Effectiveness of blue radiation in controlling flowering of plants grown on natural days extended with light of various colors was noted by Funke (1931, 1948). Later studies with both blue and far-red radiations were made by Meijer (1959) and van der Veen and Meijer (1959). They established that both types of radiation, when adequately intense, are effective in producing long-day responses of both short- and long-day plants. Phytochrome control of flowering attained by short irradiations was also noted by van der Veen and Meijer. Effectiveness of red-light control of flowering of *Hyoscyamus niger* L. was reported by Parker, Hendricks, and Borthwick (1950).

Observations on flowering and other morphological changes of the long-day plant *H. niger* are reported here. Clear-cut effects of phytochrome action and a photoreaction requiring high energy (HER) are shown. The latter is an important

[1] Received for publication 23 February 1967.

reaction in control of flowering in this long-day plant.

PLANT MATERIALS AND METHODS—*H. niger* plants were grown in a greenhouse from seed increased from an inbred annual strain developed by G. Melchers and A. Lang. Plants were subjected to 8-hr photoperiods of natural light from the time of germination and were used when 2 or 3 months old and still in 3-inch pots. Response levels varied somewhat with age and size of plants, which were similar among lots in individual experiments.

Experiments were performed in the greenhouse and associated chambers or after plants were transferred to growth chambers. In the greenhouse experiments, plants received daily 8-hr photoperiods of natural light from 8 AM to 4 PM. Irradiations of various types, durations, and placements were used in the subsequent 16-hr daily periods between 4 PM and 8 AM.

Plants for spectrographic and several other experiments were transferred from the greenhouse to growth chambers about a day prior to use. Fluorescent lamps were usually used in the chambers (ca. 2,000 ft-c illumination at plant height), but in some instances they were supplemented with incandescent-filament lamps (ca. 200 ft-c at plant height). The plants in flowering experiments were dissected after 19 to 21 days' treatment, unless otherwise noted. Numbers of flowers and stem lengths were measured in some experiments. Comparisons of specific numerical values are valid only within, rather than between, experiments because of differences in ages and development of plants.

Dark periods were variously interrupted with unfiltered radiation from incandescent (about 50 ft-c = 0.20 mw/cm^2 at 660–680 nm), or BCJ (60-w incandescent-filament lamp used at 1.5 m, providing radiation >690 nm [0.10 mw/cm^2 at 700–720]), or filtered far-red (referred to as FR, 0.6 mw/cm^2 between 710–800 nm) sources. Details of interruptions are given as needed for the various experiments.

ACTION SPECTRUM METHODS—Experiments were performed with plants that had received 8 hr each day of fluorescent plus incandescent radiation in growth chambers. The plants were transferred at the ends of the photoperiods to an interference-filter arrangement, where they were irradiated during the middle 8 hr of daily 16-hr dark periods.

Intensities of radiation required for the high-energy responses are high and t mes of exposure are long. The energy source must operate unattended for many hours each day over a period of several weeks. These conditions were adequately met by the equipment shown in Fig. 1.

The interference filters were either single- or double-band pass types with transmission-band widths of about 20 nm at 0.1 maximum transmission. The radiation sources were 35-w incandescent-filament lamps with internal parabolic mirrors of about 1.5 π steradians aperture. Band-

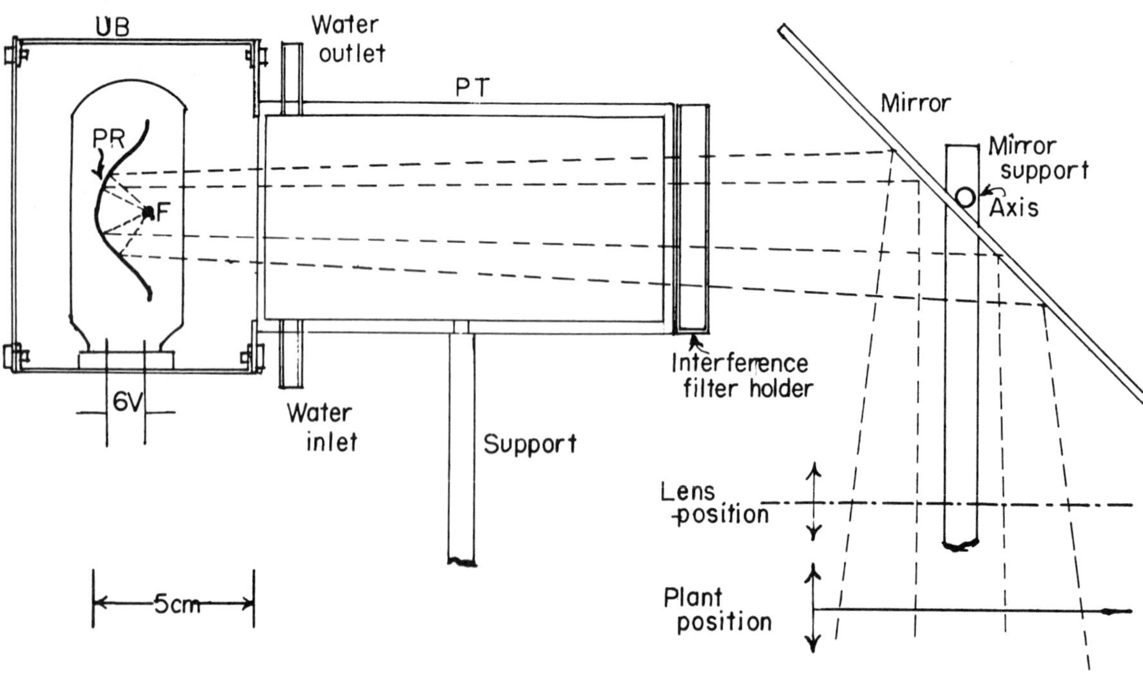

Fig. 1. Arrangement of interference filters for use with an incandescent filament lamp with an internal parabolic reflector (PR). A utility box (UB) and a plastic tube (PT) are shown, while F indicates the filament.

pass intensities in the region of 570–800 nm over areas adequate for one plant (50 cm²) were 1.0 mw/cm². This could be reduced to any desired level by increasing the distance from the source in the somewhat divergent beam (about 6°) or by the use of neutral screens. The beam was somewhat inhomogeneous in intensity because of the filament shadow which was out of focus. This can be avoided when desired by placing a double-convex lens as shown in Fig. 1. Intensities of radiation were measured with a thermopile calibrated against a National Bureau of Standards standard lamp.

TABLE 1. *Effects of kind of light source during daily 16-hr photoperiods on initiation of flowers and lengthening of stems of Hyoscyamus niger*

Light source during 16-hr photoperiod	Illuminance (ft-c)	Mean flowers per plant[a]		Mean length of stem[a]	
		E15 (no.)	E16 (no.)	E15 (mm)	E16 (mm)
Fluorescent[b]	1,000	5.0	3.5	16	16
Fluorescent	2,500	6.1	3.1	28	17
Fluorescent + incandescent	1,200	8.4	7.6	78	43

[a] Experiment E15, five plants per lot; Experiment E16, six plants per lot.
[b] Cool-white fluorescent.

TABLE 2. *Effects of intensity of illumination from incandescent-filament lamps during the middle 6 hr of daily 16-hr dark periods[a] on production of flowers and lengthening of stems of Hyoscyamus niger*

Illuminance (ft-c)	Mean flowers per plant		Mean stem length per plant	
	E11 (no.)	E14 (no.)	E11 (mm)	E14 (mm)
160	10.1	—	69	—
40	8.0	9.8	59	117
20	8.8	—	66	—
10	7.5	9.0	59	82
5	7.0	8.5	46	54
2.5	—	6.1	—	29
1.2	—	1.8	—	11
0.0	0.0	0.0	0	0

[a] Eight-hour photoperiods of natural light in greenhouse from 8 AM to 4 PM.

Blue radiation in the region of 400–480 nm was obtained by use of dark-blue cellophane as filters of fluorescent radiation in growth chambers. The maximum irradiance was 0.44 mw/cm² with two layers of blue cellophane. Fluorescent radiation filtered through four layers of red cellophane served as a broad source, 590–680 nm, of radiation for comparison with the blue. Irradiances were about 0.53 mw/cm². The level of far-red radiation in the region of 700–760 nm transmitted by these filters with the fluorescent source was 0.06 mw/cm².

EXPERIMENTS: SOME DETAILS, AND RESULTS—*Flowering and stem-lengthening responses (Table 1, 2)*—Flowering and stem lengthening are simultaneously expressed within the degree of precision used in these various experiments. This is evident in flowering of most long-day plants, which remain as rosettes under short-day conditions.

Evidence for a high-energy reaction (Table 1, 2)—A high-energy reaction in the far-red is indicated by responses to changes in radiation quality during 16-hr photoperiods (Table 1). A 1.5-fold increase in intensity of irradiation only slightly affected flowering. A small amount of incandescent radiation, increasing the relative energy in the far-red, added to the lower level of fluorescent light, however, had a marked promotive effect. We speak of the high-energy reaction as *an* HER because more than one may occur and the results reported here do not bear on that question.

Enhanced responses were also shown with increases in intensities of incandescent radiation used during the middle 6 hr of what otherwise

TABLE 3. *Effects of various irradiations for 8 min at the beginning of darkness on flowering and stem lengthening of Hyoscyamus niger*

Photoperiod Dark (hr)	Photoperiod Light (hr)	Mean flowers per plant[a]				Mean length of stem			
		Control (no.)	Incand. (no.)	BCJ (no.)	FR (no.)	Control (mm)	Incand. (mm)	BCJ (mm)	FR (mm)
12	12	0.0	0.0	0.0	0.0	0	0	0	0
11	13	0.0	0.0	0.0	0.0	0	0	0	3
10	14	0.8	1.4	2.2	2.4	4	8	8	9
9	15	3.6	4.4	4.4	4.2	13	23	25	27
8	16	3.6	5.2	5.2	5.4	26	47	40	40
7	17	3.0	5.6	5.2	6.0	19	43	45	50

[a] Five plants per lot. Fluorescent radiation only during the photoperiod. Treatments were ended after 14 days and plants were returned to 8-hr photoperiods in natural light for 6 days prior to dissection.

TABLE 4. *Effects of sequence of natural day and fluorescent light (60 ft-c) during 16-hr photoperiods and of brief illuminations from BCJ lamps at various times during 8-hr dark periods on production of flowers and lengthening of stems of Hyoscyamus niger*

Light source		Location of BCJ[a] in 8-hr dark period (hr)	Mean flowers per plant[b] (no.)	Mean stem length[b] (mm)
First 8 hr	Second 8 hr			
Fluorescent	Natural	Control (no BCJ)	6.0	37
"	"	1	7.2	49
"	"	3	7.2	53
"	"	5	7.2	58
"	"	7	6.3	34
Natural	Fluorescent	Control (no BCJ)	3.0	9
"	"	1	6.3	39
"	"	3	6.2	41
"	"	5	4.2	15
"	"	7	3.7	13

[a] Fifteen-min duration.
[b] Six plants per lot.

were 16-hr dark periods (Table 2). The various intensities were obtained by placing groups of plants at appropriate distances from a single lamp. Illuminations of 5 ft-c or less led to significant responses and it is not evident that saturation was reached at 160 ft-c, the highest level tested.

Evidence for phytochrome-mediated responses (Table 3, 4, 5)—Lowering the percentage of the P_{fr} form is promotive under many conditions but is without effect or inhibitory in others. Flowering and stem lengthening are promoted over controls by reducing the amount of P_{fr} after 14–17-hr light periods (Table 3). The incandescent, BCJ, and far-red sources set the P_{fr} levels progressively lower. However, response differences induced by irradiations with the three sources were not significant.

Flowering response depended on sequence of daily 8-hr periods of natural light (n), fluorescent illumination (f) and darkness (d), when combined in the two possible combinations (*nfd* or *fnd*) and with or without 15-min dark-period interruptions with radiation from BCJ lamps at various times (Table 4). The schedules gave different sequences of P_{fr} levels. The sequence *fnd*, or high P_{fr}, moderate P_{fr}, darkness, led to higher flowering (6.0) than did the reverse sequence (3.0). Interruptions of darkness at various times by BCJ enhanced flowering in both series to varying degrees. The enhancement was less in the *fnd* than in the *nfd* sequence and in both was less when BCJ was applied toward the end of the dark period.

Photoreversibility by 4 min of red and far-red radiation, in that sequence, at various hours in the dark period was observed in plants on 16-hr fluorescent-light periods (1,000 ft-c), which establishes predominant P_{fr} (Table 5). Flowering and stem lengthening were increased if P_{fr} was lowered by far-red irradiation at any time during the dark period, and the responses were fully reversed by immediate replacement of P_{fr}.

Combined high-energy and phytochrome responses (Table 6, 7)—A combined high-energy and phytochrome action on flowering is evident. Plants on 12-hr photoperiods with 2,500-ft-c fluorescent and 200-ft-c incandescent-filament radiation were given 15-min interruptions with unfiltered fluorescent radiation near the middle of 12-hr dark periods. The interruptions enhanced P_{fr} relative to P_r. A significant increase of flowering occurred with 160 ft-c, which is adequate for phytochrome equilibration (Table 6). About 1,800 ft-c, however, was required for a maximum effect. If a far-red irradiation followed the fluorescent light, flowering was reduced (Table 7). Choices of photoperiod durations as well as times, durations, and intensities of the irradiations in the interruptions were crucial factors in these experiments.

Cyclic lighting (Table 8)—Plants receiving 8 hr of natural light daily in the greenhouse were subjected to a regular sequence of interruptions with incandescent-filament illumination of 40 ft-c during the middle 6 hr of the dark period. The light was on for 10% of a cycle, so that total irradiance was at constant energy for each cycle length. Controls were irradiated continuously for the 6 hr. The values for flowering under 1,800-sec cycles (3 min light, 27 min darkness, repeated for 6 hr) were significantly low, as also were the ones on 2- and 4-sec cycles. Maximum flowering was attained with cycles of 1 min (6 sec light, 54 sec darkness) to 4 min. Results are best compared within a given experiment. They are most striking in Experiment 22 (Table 8) for unknown reasons but possibly connected with conditions of plant production.

TABLE 5. *Effects of irradiations[a] with far-red, or far-red followed by red at various times during daily 8-hr dark periods on production of flowers and lengthening of stems of Hyoscyamus niger*

Response measured and hr irradiated	Control		Far-red		Far-red + red	
	E3	E10	E3	E10	E3	E10
Mean[b] flowers (no.)						
0	1.0	3.7	3.5	5.0	2.5	4.2
1	2.2	3.7	4.7	4.5	3.0	4.5
2	2.5	4.7	5.3	5.2	2.5	3.2
3	1.3	4.5	3.3	5.5	1.8	3.2
4	1.0	3.5	4.3	4.4	1.3	4.1
5	1.2	3.2	3.3	4.0	0.3	3.0
6	1.8	4.2	1.5	4.2	2.0	3.2
7	1.7	2.7	4.0	3.5	1.8	2.7
Average	1.6	3.8	3.7	4.5	1.9	3.5
Mean stem length (mm)						
0	6	19	20	30	12	19
1	11	13	26	21	16	18
2	14	19	30	31	12	12
3	10	17	15	32	9	18
4	7	15	21	21	8	14
5	8	13	16	18	6	13
6	8	16	8	22	5	14
7	9	10	16	12	9	11
Average	9	15	19	24	10	15

[a] Photoperiods 16-hr fluorescent with far-red and red irradiations of 4 min each.
[b] Six plants per lot.

High-energy action spectra (Fig. 2, Table 9)—Plants were given daily 8-hr photoperiods with fluorescent radiation followed by 4 hr of darkness. They were then exposed for 8 hr under one of the interference filters at an intensity of 0.66 mw/cm² and returned to darkness for 4 hr. Only single plants could be used at each station, but two lots differing by 12 hr in the start of the photoperiod were included in each experiment. The initial rosettes of plants used in this experiment were about 3 inches in diam. Plants for each lot were independently selected for uniformity. Flowering and stem-lengthening results after 20 days' treatment in one experiment are shown in Fig. 2. In repeated experiments at the same (0.66 mw/cm²) or slightly lower intensities, maxima for flowering responses were always found at 710- and 719-nm-band centers, with some observable response at 675, 645, and 730 nm.

TABLE 6. *Effect of 15-min interruptions of 12-hr dark periods with various intensities of fluorescent illumination[a] on production of flowers and lengthening of stems of Hyoscyamus niger*

Intensity of fluorescent interruption (ft-c)	Mean flower buds[b] (no.)	Mean stem length (mm)
2 700	4.3	33
2,400	4.2	35
2,000	4.5	36
1,800	4.3	39
1,100	3.8	28
350	3.8	23
160	2.8	17
Dark control	0.0	>1

[a] Illumination during daily 12-hr photoperiods was from a mixture of fluorescent and incandescent filament lamps.
[b] Six plants per lot.

TABLE 7. *Effects of interruptions of daily 12-hr dark periods[a] with different intensities and durations of fluorescent illumination followed in some cases by far-red radiation on production of flowers and lengthening of stems of Hyoscyamus niger*

Duration and intensity of fluorescent illumination	Mean flowers per plant		Mean stem length	
	No FR (no.)	5 min FR (no.)	No FR (mm)	5 min FR (mm)
5 min				
300 ft-c	0.8	0.2	8	3
2,700 ft-c	6.5	1.0	49	10
15 min				
300 ft-c	0.7	0.3	8	5
2,700 ft-c	7.5	3.7	58	28

[a] Daily 12-hr photoperiods with fluorescent (2,000 ft-c) and incandescent (200 ft-c) illumination.

TABLE 8. *Effects of incandescent illumination applied 10% of light-dark cycles of various durations during the middle 6 hr of daily 16-hr dark periods on production of flowers and lengthening of stems of Hyoscyamus niger*

Length of light-dark cycle	Values for indicated experiments						Means for 6 experiments
	E11	E14	E17	E22	E25	E41	
Mean[a] flowers (no.)							
1,800 sec	3.3	8.2	1.5	1.3	6.3	3.2	4.0
1,200 "	—	9.5	6.2	3.2	5.7	6.2	6.5
600 "	4.5	9.2	7.5	4.7	8.0	4.8	6.5
240 "	7.2	9.2	6.8	6.8	7.5	6.7	7.4
60 "	7.5	9.7	7.8	7.2	7.5	5.2	7.5
30 "	5.7	9.0	8.7	5.0	8.3	5.3	7.0
8 "	5.5	9.0	8.3	4.0	8.7	5.8	6.9
4 "	4.3	8.3	5.3	3.2	6.8	4.2	5.4
2 "	5.3	8.8	7.7	1.7	8.0	6.5	6.3
Control	8.0	9.8	10.0	9.5	9.2	8.0	9.1
Mean stem length (mm)							
1,800 sec	15	27	6	8	9	14	13
1,200 "	—	59	30	12	19	35	31
600 "	23	54	52	17	35	30	35
240 "	39	68	40	32	31	38	41
60 "	36	64	48	35	39	34	43
30 "	34	51	52	26	53	27	40
8 "	27	53	48	15	34	38	36
4 "	21	45	28	12	23	26	26
2 "	23	44	48	7	40	46	35
Control	59	117	117	73	72	77	86

[a] Six plants per lot.

The action in response to irradiation with a broad band in the blue region of the spectrum (400–480 nm) is the order of magnitude (Table 9) of the action in the red (590–680 nm).

Leaf-blade and petiole responses (*Table 10*)—The hyponastic attitudes of leaves were seen to vary after various regimens of light and darkness. Changes were detectable in periods as short as 1 day in contrast to flowering responses, which required 14 or more days. Etiolation responses, measured as ratios of leaf blade to petiole lengths, were also detectable in 1 day but were generally measured later. Ratios determined at 20 days are listed in Table 10. The ratio is seen to be between 1.0–1.2 in the region 710–750 nm and about 1.5–2.0 in the region 576–675 nm. The ratio is independent of the flowering response, since the wavelength band centered at 750 nm does not cause flowering and that at 719 nm does, yet both bands give the same ratio, 1.2. Although the P_{fr} absorption for phytochrome is near 725 nm, the incident energies in the region 710–750 nm all lead to low P_{fr} levels, probably $<1\%$ of total P, and all elicit a similar blade-petiole ratio. Thus the etiolation phenomenon responds to irradiations effective in the phototransformation of phytochrome. Moreover, the hyponastic attitudes of the leaves in darkness were found to be photoreversible and mediated by phytochrome. Results of many experiments support and extend these findings, which are not further considered here.

TABLE 9. *Comparative effects of a broad blue spectral band during the middle 8 hr of daily 16-hr dark periods on flowering and stem lengthening of Hyoscyamus niger*

Spectral energy (nm)	Energy (mw/cm²)	Flowers (no.)	Stem length (mm)
400–480	0.44	4	73
	0.33	3	33
	0.19	0	6
	0.08	0	2
590–680	0.44	3	28
700–760	0.06	0	7
Darkness	0.00	0	0

DISCUSSION—An action spectrum (Parker et al., 1950) was published for promotion of flowering of *H. niger*. This action spectrum is compatible with phytochrome being a pigment involved and, under the particular experimental conditions, with the far-red form (P_{fr}) being promotive of flowering. The conditions in the previous work were approximately those used on plants for which results are shown in Table 6, except that carbon-arc with supplementary incandescent-filament radiation was used in the photoperiods. Under the conditions for which results in Table 5

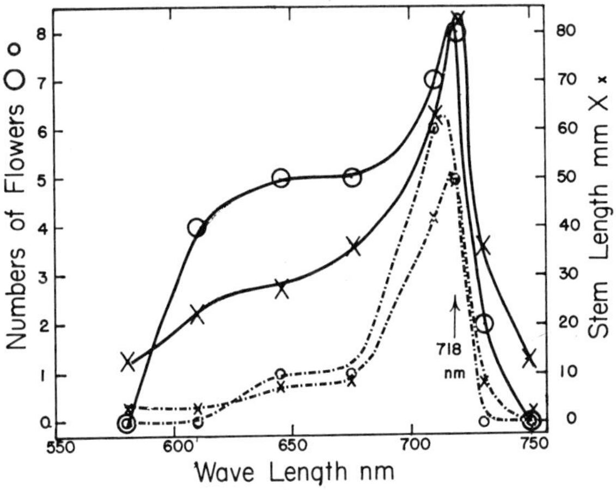

Fig. 2. Flowering and stem lengthening of *Hyoscyamus niger* in response to indicated wavelengths at 0.66 mw/cm² during the middle 8 hr of daily 16-hr dark periods. Solid lines and broken lines represent separate lots of plants in an experiment.

were obtained, flowering is suppressed by the presence of P_{fr}. The action of P_{fr} is photoreversible. The requirement for the presence of P_{fr} for flowering when one set of experimental conditions is imposed and for the absence of P_{fr} when another set exists is not unprecedented. It is also displayed in short-day plants such as *Pharbitis* (Fredericq, 1964) and *Xanthium* (Borthwick and Downs, 1964).

A high-energy reaction (HER) with a quite narrow action maximum at 710–720 nm and with an action of similar magnitude integrated over the region of 400–480 nm is prominent as a factor in flowering of *H. niger*. The existence of such an HER also controlling flowering does not confound the prior work because of differences in the energy levels used. Evidence for an HER was previously obtained by Stolwijk and Zeevaart (1955) and by van der Veen and Meijer (1959). They used radiation of limited spectral region as the main photoperiod for *H. niger* and found that flowering was enhanced by blue or far-red (near infrared). Although normal aspects of photosynthesis are confounded in the several experiments, the results are qualitative evidence of a flowering requirement. These findings were extended in control of flowering of the long-day plant *Lolium temulentum* L. by Evans, Borthwick, and Hendricks (1965) by use of photoperiod extensions with radiation in broadly limited spectral regions. Their paper discusses the pertinent literature. A degree of flowering control of long-day plants by phytochrome superimposed on an HER is evident in these several works.

With the purposes of the various experiments indicated under Results, only the essence of the findings need be considered here. The main result is the action spectrum (Fig. 2) for an HER promotion of flowering of *H. niger*. The maximum is at 710–720 nm, with a shoulder extending through the red to about 610 nm. Any phytochrome action is confounded in this result because in wavelength regions <700 nm, P_{fr} is greatly predominant, and in regions >700 nm it is reduced to <1% total P. Under conditions shown in Table 7, high levels of P_{fr} are promotive of flowering and low levels are inhibitory. Consequently, it is thought that the response in the red region shown in Fig. 2 is enhanced by the high level of P_{fr} inherently established by the incident radiation. The response at the maximum in the far-red region is diminished because of the low level of P_{fr} inherently established. However, the red response is not caused by phytochrome conversion alone, because it has a high-energy requirement. Our work with blue radiation (Table 9), while not seriously confounded by photosynthesis, does not particularly add to previous findings about the effectiveness of blue radiation in promoting flowering of *H. niger*.

TABLE 10. *An action spectrum for leaf-blade and petiole lengthening of Hyoscyamus niger. Plants were those for which flowering and stem-lengthening results are shown in Fig. 2*

Wavelength (nm)	Flowers (no.)	Length ratio blade/petiole	Amount of P_{fr} (%)
750	0	1.21	<1
730	1	1.06	<1
719	7	1.25	<1
710	7	1.07	<1
675	3	1.45	>80
645	3	1.84	>80
610	2	2.07	>80
576	0	1.91	>80

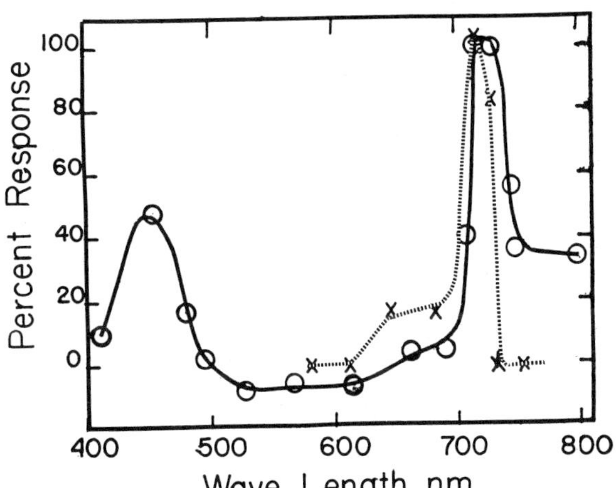

Fig. 3. High-energy action spectra for flowering (x) *H. niger*, and the plumular hook opening (o) of *L. sativa* (after Mohr and Noble, 1960).

The HER for flowering of *H. niger* is compared in Fig. 3 with action curves for plumular hook opening of *Lactuca sativa* L. 'Grand Rapids' (Mohr and Noble, 1960). It is also similar to the action spectra for cotyledon enlargement in *Sinapis alba* L. (Mohr, 1959), and anthocyanin production in *Brassica rapa* L. 'Purple Top White Globe' (Siegelman and Hendricks, 1957). Curve shape and positions of action maxima are the important factors in the comparisons because relative degrees of action vary with effectiveness for action as can be seen from the flowering results for two lots of plants shown in Fig. 2.

Phytochrome control of flowering is evident from the reversibility of potentiated effects of red and far-red radiation under the treatment schedule of Table 5. Removal of P_{fr}, in general, enhances flowering, particularly if the dark period is short (8 hr) and follows an extended period of fluorescent radiation, which lacks an appreciable far-red component (Table 4).

In fact, an HER is best displayed when a photoperiod lacking in far-red radiation is augmented by radiation containing prominent far-red (or possibly blue radiation, if we had worked with it alone instead of the fluorescent source). This need for an HER is quite likely the underlying reason for maintaining a far-red component in radiation sources to insure optimum growth of many plants under growth-room conditions.

The nature of the pigment effective in an HER has been discussed without a compelling answer being found. Possibilities are: (1) the photosynthetic system, or (2) an action of phytochrome. The first of these suggests that an HER might be a result of some aspect of photosynthesis. Bertsch and Mohr's (1965) argument that an HER is shown by etiolated plants with poorly organized photosynthetic apparatus is against a photosynthetic involvement. Attempts to show ways in which phytochrome under continuous excitation could lead to an HER (Hartmann, 1966), while somewhat encouraging, involve questionable assumptions, as also does an earlier paper from this laboratory (Hendricks and Borthwick, 1959).

Plants requiring long days for flowering seem to be ones in particular need of the products of an HER over a considerable part of each day. The cyclic responses (Table 8) indicate that the half-life of these products is of the order of 1–4 min. P_{fr} action in darkness can either inhibit or enhance flowering depending on the kind of radiation and duration of the photoperiod. In general, lowering P_{fr} enhances flowering when radiation during the photoperiod is low in far-red.

LITERATURE CITED

Bertsch, W., and H. Mohr. 1965. Die Unabhängigkeit der Lichtinduzierten Anthocyansynthese von der Photosynthese. Planta 65: 17–26.

Borthwick, H. A., and R. J. Downs. 1964. Roles of active phytochrome in control of flowering of *Xanthium pensylvanicum*. Bot. Gaz. 125: 227–231.

Evans, L. T., H. A. Borthwick, and S. B. Hendricks. 1965. Inflorescence initiation in *Lolium temulentum* L. VII. The spectral dependence of induction. Austral. J. Biol. Sci. 18: 745–762.

Fredericq, H. 1964. Conditions determining effects of far-red and red irradiations on flowering response of *Pharbitis nil*. Plant Physiol. 39: 812–816.

Funke, G. L. 1931. On the influence of light of different wave-lengths on the growth of plants. Rec. Trav. Bot. Neerl. 28: 431–485.

———. 1948. The photoperiodicity of flowering under short day with supplemental light of different wave lengths. Lotsya 1: 79–82.

Hartmann, K. M. 1966. A general hypothesis to interpret "high energy phenomena" of photomorphogenesis on the basis of phytochrome. Photochem. Photobiol. 5: 349–366.

Hendricks, S. B., and H. A. Borthwick. 1959. Photocontrol of plant development by the simultaneous excitations of two interconvertible pigments. Proc. Nat. Acad. Sci. (U.S.) 45: 344–349.

Meijer, G. 1959. The spectral dependence of flowering and elongation. Acta Bot. Neerl. 8: 189–246.

Mohr, H. 1959. Der Lichteinfluss auf das Wachstum der Keimblätter bei *Sinapis alba* L. Planta 53: 219–245.

———, and A. Noble. 1960. Die Steuerung der Schliessung und Öffnung des Plumula-Hakens bei Keimlingen von *Lactuca sativa* durch Sichtbare Strahlung. Planta 55: 327–342.

Parker, M. W., and H. A. Borthwick. 1949. Growth and composition of Biloxi soybean grown in a controlled environment with radiation from different carbon-arc sources. Plant Physiol. 24: 345–358.

———, S. B. Hendricks, and H. A. Borthwick. 1950. Action spectrum for the photoperiodic control of floral initiation of the long-day plant *Hyoscyamus niger*. Bot. Gaz. 111: 242–252.

Pfeffer, W. 1904. Pflanzenphysiologie 2. Aufl. Bd II, S 117. Wilhelm Engelmann, Leipzig.

Popp, H. W. 1926. A physiological study of the effect of light of various ranges of wave length on the growth of plants. Amer. J. Bot. 13: 706–735.

Roodenburg, J. W. M. 1940. Das Verhalten von Pflanzen in verschiedenfarbigem Licht. Rec. Trav. Bot. Neerl. 37: 301–376.

Senebier, J. 1799. Physiologie Végétale, Geneva. Vol. 3.

Shirley, H. L. 1929. The influence of light intensity and light quality upon the growth of plants. Amer. J. Bot. 16: 354–389.

Siegelman, H. W., and S. B. Hendricks. 1957. Photocontrol of anthocyanin formation in turnip and red cabbage seedlings. Plant Physiol. 32: 393–398.

Stolwijk, J. A. J., and J. A. D. Zeevaart. 1955. Wave length dependence of different light reactions governing flowering in *Hyoscyamus niger*. Proc. Ned. Akad. Wetensch. 58: 386–396.

Trumpf, C. 1924. Ueber den Einfluss intermittierender Belichtung auf das Etiolement der Pflanzen. Bot. Arch. 5: 381–410.

Veen, R. van der. 1950. A small greenhouse with artificial lighting for studying plant growth under reproducible conditions. Philips Tech. Rev. 12: 1–7.

———, and G. Meijer. 1959. Light and plant growth. Macmillan, New York.

VINCE, D. 1956. Studies on the effects of light quality on the growth and development of plants. II. Formative effects in *Lycopersicon esculentum* and *Pisum sativum*. J. Hort. Sci. 31: 16–24.

WASSINK, E. C., C. M. J. SLUIJSMANS, AND J. A. J. STOLWIJK. 1950. On some photoperiodic and formative effects of coloured light in *Brassica rapa, F oleifera*, subf. *annua*. Proc. K. Ned. Akad. Wetensch. 53: 1466–1475.

Photochemistry and Photobiology. 1966 Vol. 5, pp. 397–406. Pergamon Press Ltd. Printed in Great Britain.

KINETIC STUDIES TO INTERPRET 'HIGH-ENERGY PHENOMENA' OF PHOTOMORPHOGENESIS ON THE BASIS OF PHYTOCHROME

E. WAGNER and H. MOHR

Department of Botany, University of Freiburg/Br., Germany

Abstract—Kinetic studies with the mustard seedling (*Sinapis alba* L.) support the hypothesis that the so-called 'high energy reaction' of photomorphogenesis can be understood solely on the basis of phytochrome. Light-induced anthocyanin synthesis (a typical 'positive' photoresponse[1] and light dependent inhibition of hypocotyl lengthening (a typical 'negative' photoresponse[1]) have been investigated. In order to explain the experimental data we have to assume that there are two different types of phytochrome 730 which differ greatly as far as their resistance to irreversible destruction is concerned. The existence of these two different types of phytochrome 730 has already been postulated on the basis of spectrophotometric measurements *in vivo*.[2]

OUR AIM is to interpret the high-energy phenomena of photomorphogenesis on the basis of phytochrome at least as far as the far-red peak of action is concerned. Our experimental system is the mustard seedling, *Sinapis alba* L.; this seedling has been used since 1957[3] for a great number of experiments on photomorphogenesis.[4]

The dashed curve on Fig. 1 will indicate the action spectrum for the induction of a photoresponse in the case of a 'short-time irradiation'. We call it therefore an 'induction spectrum'. The solid curve, on the other hand, represents the type of action spectrum which we observe in the case of 'long-time irradiation', e.g. 6 hr or 12 hr continuous irradiation. The problem is to explain this type of action spectrum exclusively on the basis of phytochrome. The other possible hypothesis—that short-time irradiation and long-time

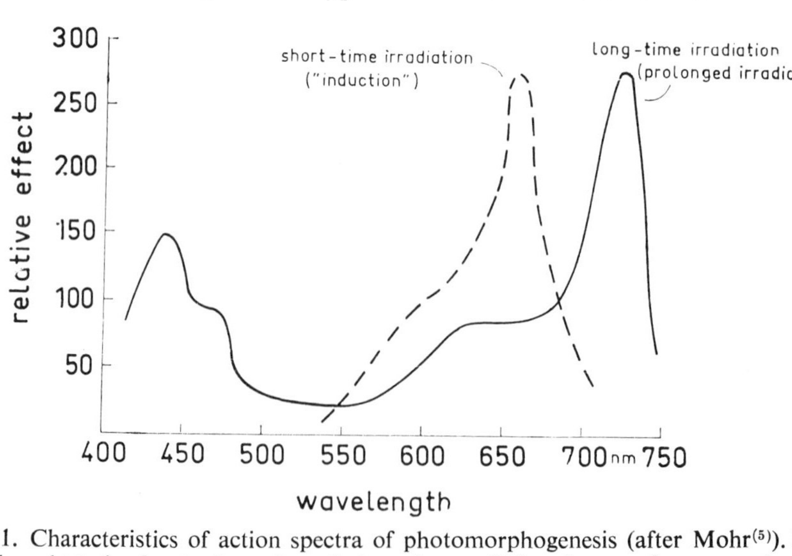

FIG. 1. Characteristics of action spectra of photomorphogenesis (after Mohr[5]). Both types are thought to be due to the action of phytochrome 730, at least as far as wavelengths above 500 nm are concerned.

irradiation act through different photoreceptors—was followed for some time.[5] It has turned out, however, not to be very satisfactory in the light of more recent experimental data.[6] Therefore we have been trying to approach the problem of high-energy phenomena in a different way.

Butler and co-workers[2] have suggested that in a seedling there are probably two types of phytochrome 730 (P_{730}). The larger part is very unstable; it is readily destroyed. A small part, however, seems to be resistant to destruction. Left in darkness, it reverts slowly to P_{660} in a thermal reaction. If we further assume that both types of P_{730} can exert the same physiological function, we are able to explain—at least in principle—the high-energy phenomena on the basis of phytochrome.

Our hypothesis concerning the high-energy phenomena is expressed qualitatively on the next two figures. First, we present a rough sketch which is drawn to illustrate the changes in the pool size of P_{730} during and after a relatively brief irradiation with red or far-red, e.g. 30 min (Fig. 2). Because the biological response is a function of pool size

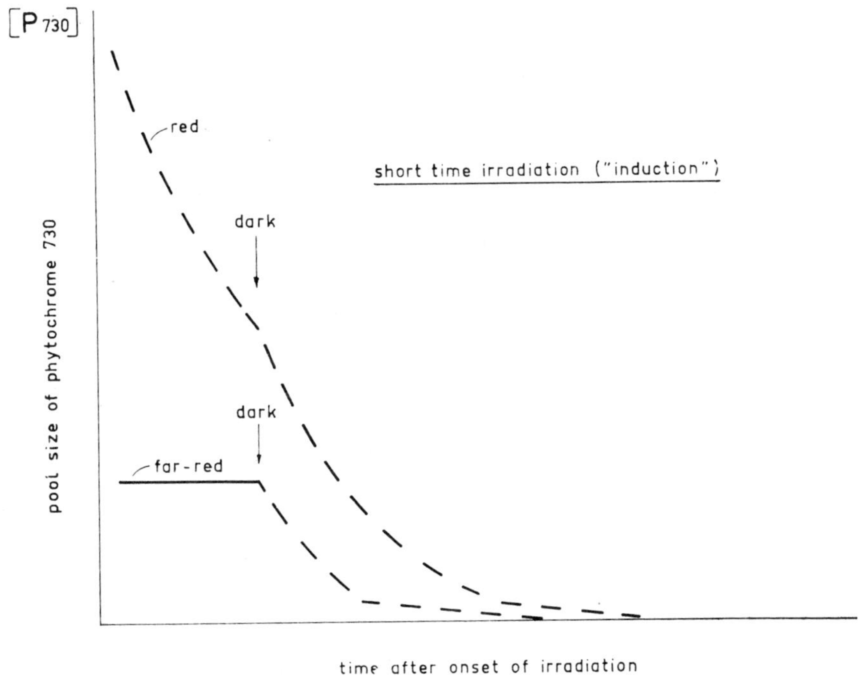

FIG. 2. Hypothetical changes of pool size of phytochrome 730 during and after a relatively brief period of irradiation with red or far-red. The co-ordinates must not be regarded as linear (experimental background[2, 7, 8]).

and time of maintenance of P_{730}, red is far more effective than far-red under these conditions. Figure 3 illustrates the situation in the case of continued irradiation. In red light, the pool size of P_{730} will decrease from a high initial value to a low steady-state value within a short period of time, for the following reason: in the red about 80 per cent of the total phytochrome will be present as P_{730}. Most of this P_{730} is destructable. It will rapidly disappear through irreversible destruction. The pool of P_{660} will soon be nearly emptied because in the photostationary state in red light, 80 per cent of the total phytochrome will always be present as P_{730}. The low steady-state level of P_{730} which is finally reached in the red is thought to be exclusively due to this type of P_{730} which is resistant to destruction. In continuous far-red, on the other hand, only a small percentage (less than 5) of the total

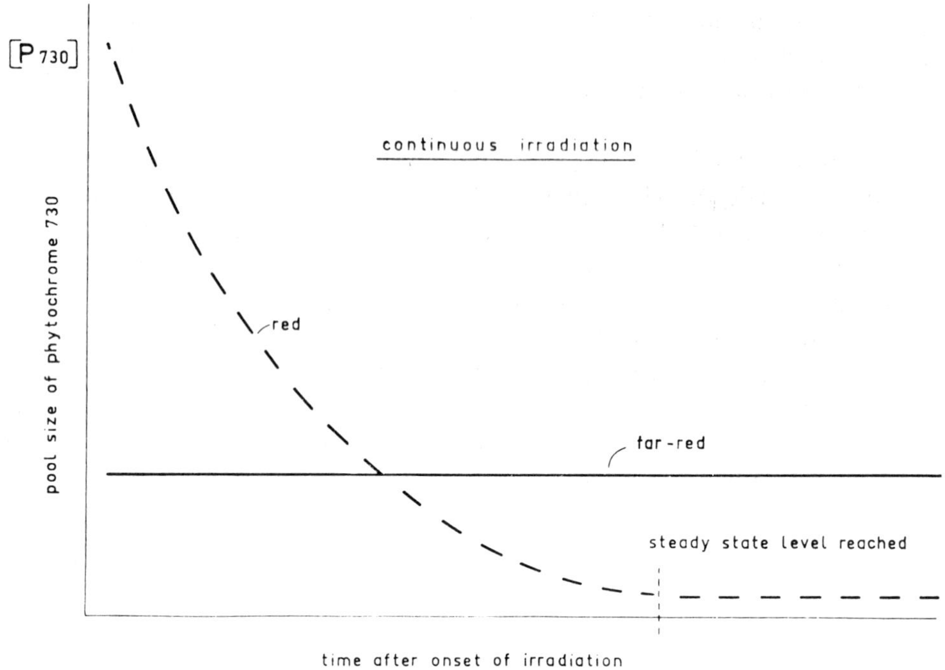

FIG. 3. Hypothetical changes of pool size of phytochrome 730 during continuous irradiation with red or far-red. The co-ordinates must not be regarded as linear (experimental background[2, 7, 8]).

phytochrome will be present as P_{730}. This part is subject to destruction but, due to the large pool size of P_{660}, it can be replaced continuously. Also in the case where there is no rapid *de novo* synthesis of P_{660}, the pool size of P_{730} will be virtually constant for some time. This level of P_{730} which is maintained in the far-red is higher than the steady-state level in the red. The result is that with long-time irradiation, far-red is more effective than red.

This hypothesis has been tested using light-induced anthocyanin synthesis of the mustard seedling. The experimental data verify our hypothesis. A few of them will be presented now.

(a) If we induce anthocyanin synthesis with different periods of red and far-red, we find that the red and far-red curves cross as would be expected from the hypothesis (Fig. 4). In this case anthocyanin was extracted 48 hr after the onset of irradiation. To check the hypothesis further we used continuous irradiation and determined the amount of anthocyanin which was formed during the course of irradiation. We call the resulting curves 'time courses' of anthocyanin synthesis.

(b) Under continuous light the accumulation of anthocyanin is linear with time over at least 18 hr, after the lag phase of 3–4 hr (Fig. 5). This is true probably for all wavelengths. If the P_{730} which is present in continuous red is different from most of the P_{730} which is present in continuous far-red could it be that the intensity dependence of the light effect is different in red and far-red? This is indeed the case (Fig. 6). The data on this figure virtually represent the slopes of the linear part of the time course of anthocyanin accumulation as a function of quantum flux density. It seems that in red as well as in far-red a logarithmic function results at least in the middle part of the curve. The dependence on intensity, however, is very different in red and in far-red. On the other hand, it is

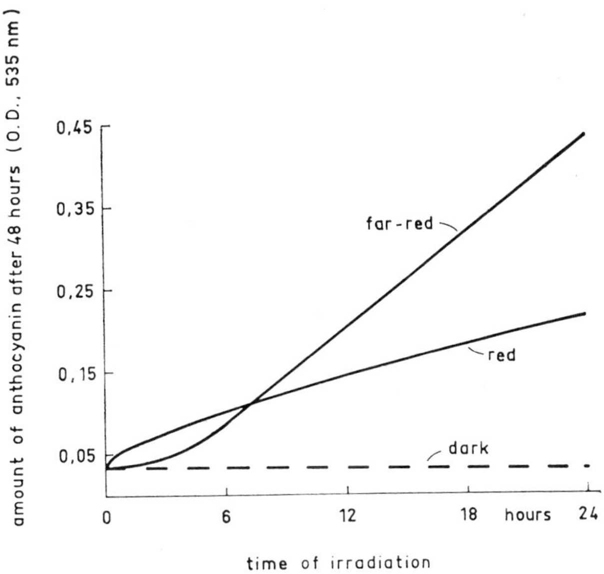

FIG. 4. Induction of anthocyanin synthesis with different periods of red (700 ergs/cm²·sec) and far-red (6000 ergs/cm²·sec). Onset of irradiation 24 hr after sowing: extraction of anthocyanin 72 hr after sowing (after Bertsch and Mohr[9]).

the same in far-red and in blue. These remarkable results may indicate that the P_{730} which is present in far-red and in blue will be more or less of the same type, whereas in continuous red only the other stable type of P_{730} will be present in the seedling. The two types of P_{730} do apparently differ as far as the intensity dependence of the physiological function of P_{730} is concerned. It must be mentioned that the coincidence of the intensity curves in blue and far-red is hard to understand if in the photoequilibrium in the blue, about 35–50 per cent of the total phytochrome, is supposed to be present as P_{730} as is indicated by Butler's data.[10] This percentage might be lower.

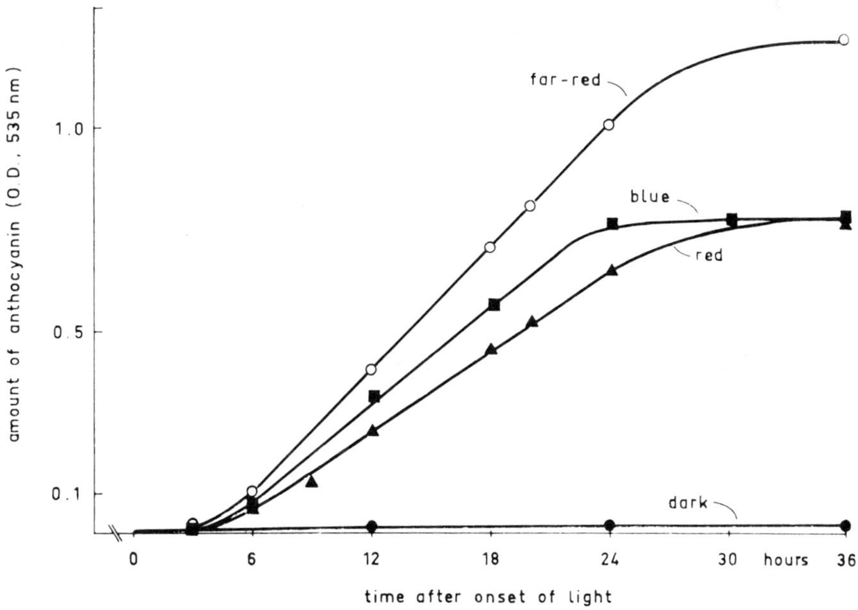

FIG. 5. Time courses of anthocyanin synthesis under red, far-red and blue irradiation. Onset of light 36 hr after sowing. Irradiances: red 650 ergs/cm²·sec; blue 1000 ergs/cm²·sec; far-red 3300 ergs/cm²·sec.

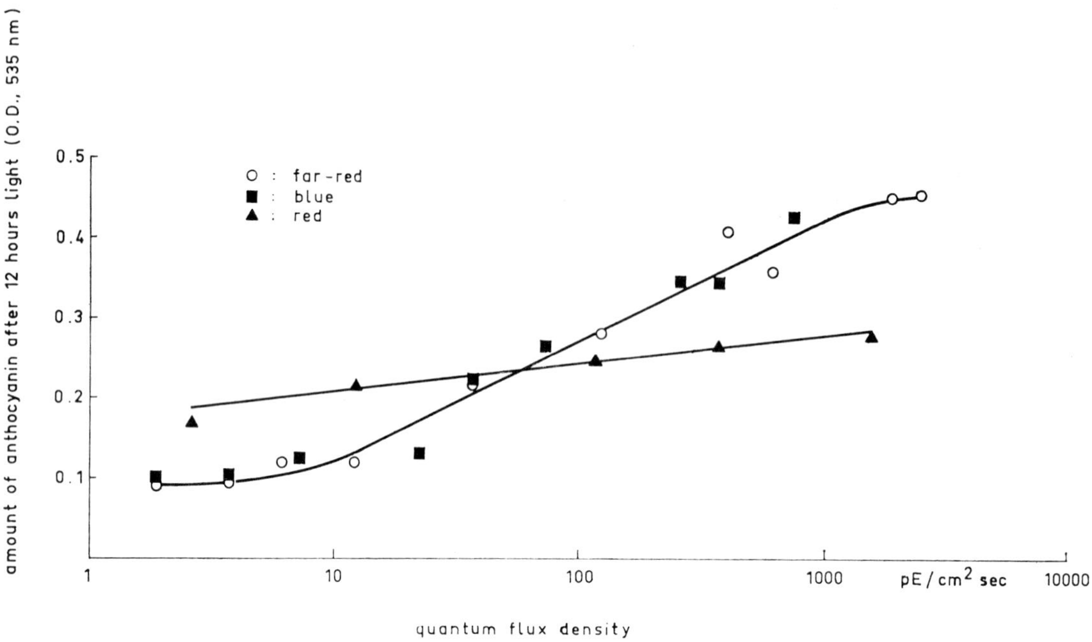

FIG. 6. The dependence of anthocyanin synthesis on quantum flux density in red, far-red and blue. Onset of light 36 hr after sowing. Extraction of anthocyanin 12 hr after the onset of light. The amount of anthocyanin virtually represents the slope of the linear part of the time course of anthocyanin synthesis because the lag-phases are virtually the same under all conditions of irradiation. Six hours after the onset of light the linear part of the time course seems to be reached under all light treatments.

(c) Figure 7 will give us more critical information. We keep the seedlings in red or far-red light for 12 hr. Then the lights are turned off and the time-course of anthocyanin production is followed further. If our hypothesis is correct, the time-course in continuous

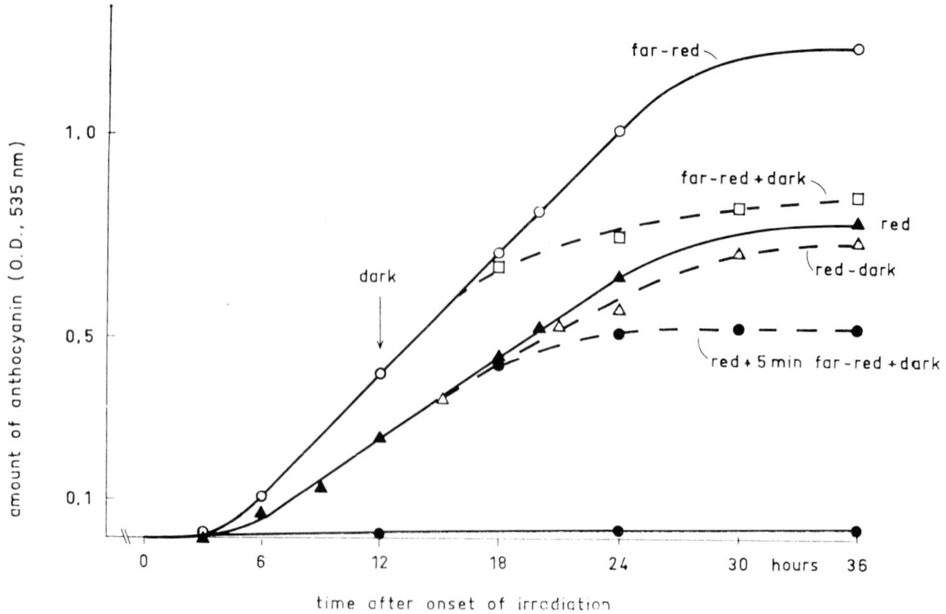

FIG. 7 Time courses of anthocyanin synthesis in continuous red, continuous far-red and in darkness following an irradiation period of 12 hr. Onset of light 36 hr after sowing (irradiances, Fig. 5).

red should only be slightly above the time-course under the regimen red + dark, because the P_{730}, which is resistant to irreversible destruction, will only slowly revert to P_{660} in a thermal reaction. This prediction is clearly verified. The time-course of anthocyanin accumulation in those seedlings, however, which had been irradiated for 12 hr with far-red should be far below the time-course which we observe in continuous far-red because most of the P_{730} which is present at the end of the 12 hr of far-red is not stable against irreversible destruction and will thus disappear very rapidly when the seedlings are kept in the dark. Only a very small amount will be of the stable type. The data show that this prediction too is verified. If at the end of 12 hr of continuous red we irradiate with a dose of far-red, which is saturating as far as the photoequilibrium is concerned, we will transform more than 95 per cent of the stable P_{730} back to P_{660}. A simple calculation allows the prediction that the time courses of anthocyanin accumulation under the programme far-red + dark and the programme red + 5 min far-red + dark must be very similar if the basic assumptions of our hypothesis are correct. The satisfactory verification of our prediction is evident.

(d) Another test of the hypothesis can be made in the following way (Fig. 8). After an irradiation of 6 hr, more anthocyanin is present under far-red than under red irradiation.

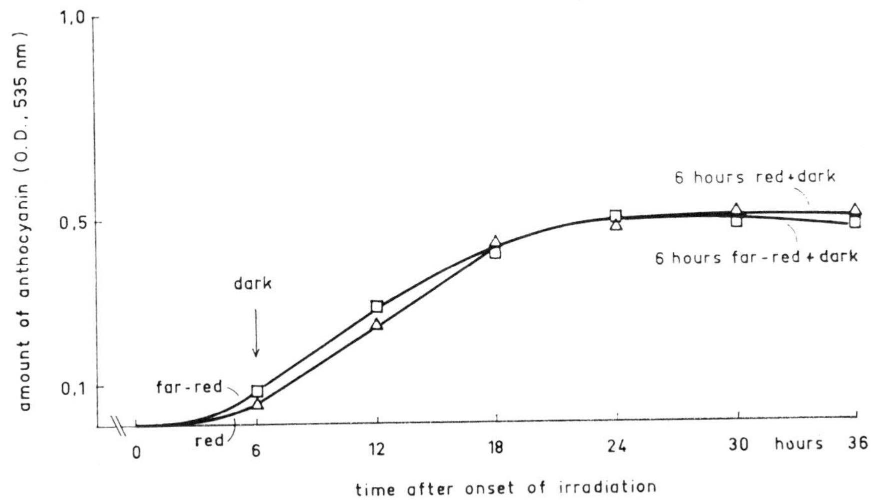

FIG. 8. Time courses of anthocyanin synthesis in darkness after a 6 hr irradiation with red or far-red. Onset of irradiation 36 hr after sowing (irradiances, Fig. 5).

The time courses of anthocyanin formation in the following darkness should be different; the time courses should cross. They do, indeed, 36 hr after the onset of light those seedlings which had been preirradiated with red almost certainly contain more anthocyanin than those seedlings which had been preirradiated with 6 hr far-red. The standard errors of the data upon which these curves are based are so small that these conclusions are reasonable.

Just one remark: These curves may be a warning for all those who derive conclusions from induction experiments without knowing the details of the kinetics. If we had determined anthocyanin contents only 12 hr after the onset of light, we would have found that far-red is more effective than red. On the other hand, if we had determined the anthocyanin contents only 36 hr after the onset of light we would have found that red is more effective than far-red.

(e) We now pursue a little further the question whether the action of far-red and the action of blue in connection with 'high-energy phenomena' can be explained in the same way. This seems indeed to be possible as the following data will indicate. Our data on intensity dependence (Fig. 6) offer the opportunity to determine a certain quantum flux density where the contents of anthocyanin after 12 hr of continuous red, blue or far-red are the same. We use this quantum flux density for the following experiment (Fig. 9).

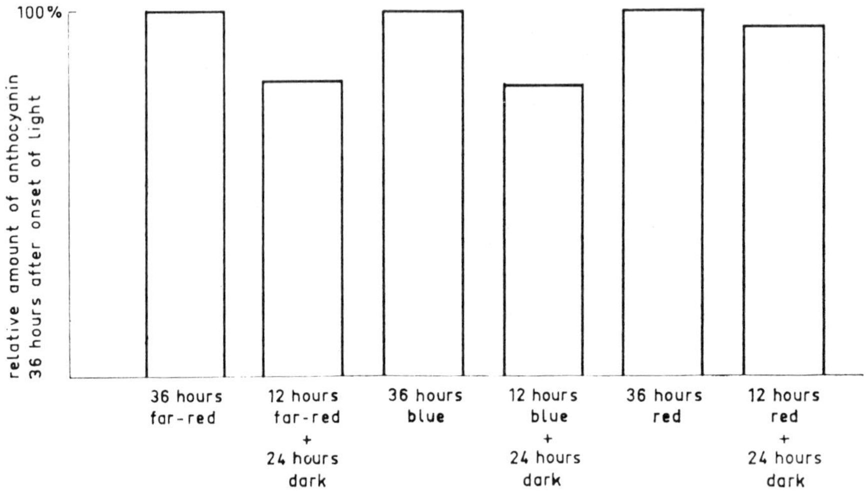

FIG. 9. Seedlings are irradiated with red, far-red or blue for 12 hr with a quantum flux density of 55 $PE/\text{cm}^2 \cdot \text{sec}$ (they will then have the same content of anthocyanin, Fig. 6). After a further 24 hr in darkness the anthocyanin is extracted. This amount of anthocyanin is compared with that amount (100 per cent) which has been formed in those control seedlings which received irradiation for 36 hr.

After an irradiation of 12 hr in either red, far-red or blue, all seedlings contain the same amount of anthocyanin. We now determine the percentage of anthocyanin which they contain 24 hr later in darkness compared with the amount they would accumulate if kept continuously in the light. The amount of anthocyanin which is present in those seedlings which had been kept for 36 hr in continuous light is indicated as 100 per cent. One can recognize on Fig. 9 that these data confirm our earlier statement that the actions of blue and far-red show a remarkably similar pattern. The data on Table 1 further indicate that far-red and blue will establish a similar photostationary state of the phytochrome system. Blue and far-red impose the same degree of reversal if after 12 hr of red we irradiate with

TABLE 1. THE REVERTING ƎFFECTS OF FAR-RED AND BLUE, FOLLOWING RED RADIATION, ARE COMPARED. QUANTUM FLUX DENSITY OF RED, FAR-RED AND OF BLUE: 55 $PE/\text{cm}^2 \cdot \text{SEC}$(FIG. 6). ONSET OF RED 36 HR AFTER SOWING. EXTRACTION OF ANTHOCYANIN 36 HR AFTER THE ONSET OF RED

Treatment	Amount of anthocyanin (O.D., 535 nm)
12 hr red + 24 hr dark	0,480
12 hr red + 5 min far-red + 24 hr dark	0,411
12 hr red + 5 min blue + 24 hr dark	0,411

5 min of far-red or of blue. These data and the foregoing ones probably mean that the behaviour of phytochrome under continuous blue is not very different from that under continuous far-red. The peaks of action in the blue and in the far-red in the case of long-time irradiation are thus understandable as are also the identical intensity curves for blue and far-red. The P_{730} which is present in the photostationary state in blue and far-red is to a large extent of the unstable sort. The relative concentration of P_{730} in the photostationary state seems to be similar in blue and far-red. It cannot be excluded, however, that besides the formation and maintenance of P_{730} some other photoreaction may occur in the blue as is indicated by data from other systems and laboratories.[11]

(f) The general hypothesis is that we are dealing with two sorts of P_{730} in the mustard seedling which differ, among other properties, in the degree of resistance against irreversible destruction. This general hypothesis can be further tested using other photoresponses of the mustard seedling. We only mention one example: hypocotyl lengthening. In this case, P_{730} causes an inhibition, not a promotion as in the case of anthocyanin synthesis. Figure 10 will illustrate the basic phenomena: after the onset of far-red or red light, the

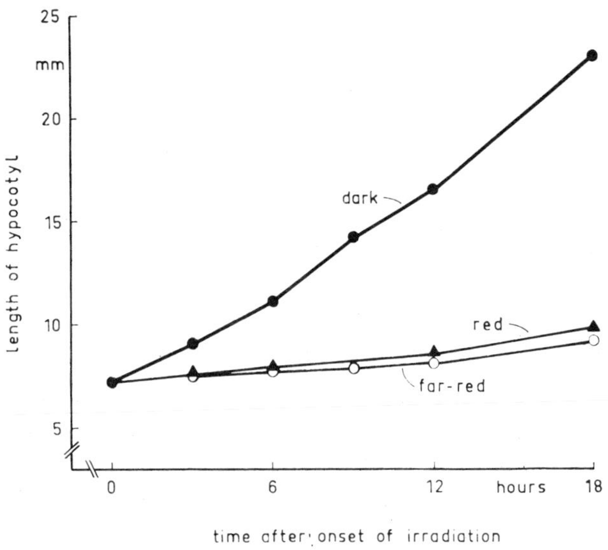

FIG. 10. Time courses of hypocotyl lengthening in continuous red, in continuous far-red and in darkness. Onset of irradiation 36 hr after sowing (irradiances, Fig. 5) (after Lüthy[12]).

rate of lengthening is rapidly reduced. A low but relatively constant rate is again reached 20–40 min after the onset of light. If we use suitable quantum flux densities in red and far-red, we can obtain similar time courses of hypocotyl elongation in far-red and red. In the present case red is slightly less effective than far-red at least some time after the onset of light. If the effect of far-red is mainly exerted through the destructible sort of P_{730} and the effect of red mainly through the resistant sort of P_{730}, one would expect that —if we turn the light off—the increase of the rate of elongation would be faster in those seedlings which had received far-red than in those which had received red. This is indeed the case. Figure 11 shows some pertinent data which can readily be understood on the basis of our hypothesis. The readaptation of the rate of elongation to dark conditions occurs later if the preirradiation was done with red, also under our conditions where red was less effective than far-red.

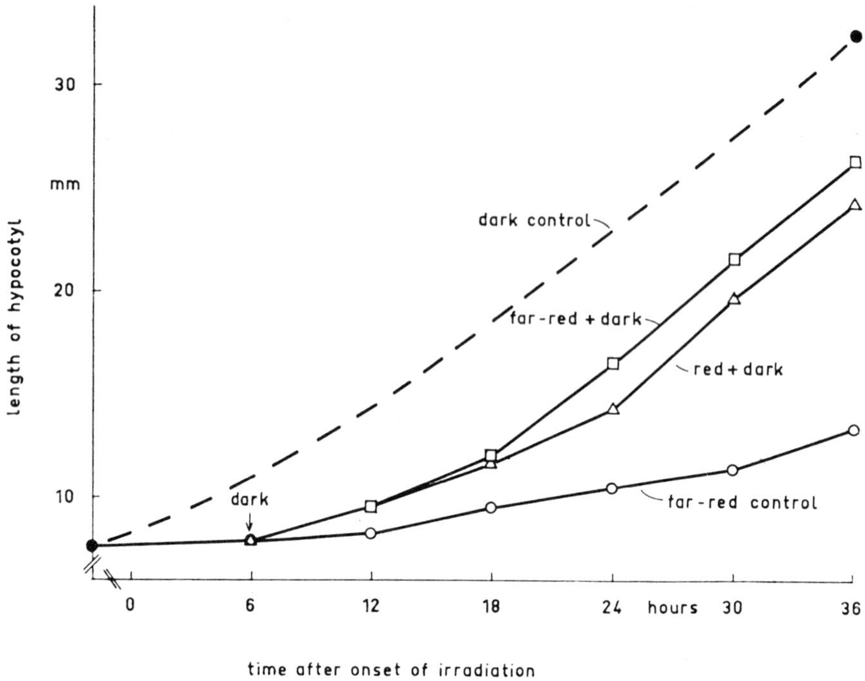

FIG. 11. Time courses of hypocotyl lengthening in darkness following an irradiation period of 6 hr with red or far-red. Onset of irradiation 36 hr after sowing (after Schultz[13]).

CONCLUSIONS

We conclude that kinetical studies with the mustard seedling support the hypothesis that the so-called 'high-energy phenomena' of photomorphogenesis can be understood solely on the basis of phytochrome. We have to assume that there are two different types of P_{730} which differ greatly as far as their resistance to irreversible destruction is concerned. The existence of these two different types of P_{730} has already been postulated on the basis of differential spectrophotometry *in vivo*. We have to assume that both types do exert the same physiological function, but differ markedly as far as the influence of quantum flux density on the physiological function is concerned. There is only a slight influence of changes in quantum flux density on the action of the stable P_{730} whereas the physiological function of the destructible P_{730} does show a strong dependence on quantum flux density. We have only checked hitherto two photoresponses of the mustard seedling; it seems possible, however, that all 'high-energy phenomena' of the mustard seedling can eventually be understood in this way. In our opinion the only difference between the lettuce seedling —which has been throughly investigated by Dr. Hartmann in connection with 'high-energy phenomena'—and the mustard seedling is the following: The fraction of P_{730} which is stable against irreversible destruction is much larger in the mustard than in the lettuce seedling.

Acknowledgements—We are grateful to G. Lüthy and to W. Schultz for allowing us to use their unpublished data. Our work has been supported by the 'Deutsche Forschungsgemeinschaft' and the 'Stiftung Volkswagenwerk'.

REFERENCES

1. H. MOHR, *Z. Pflanzenphysiol.* **54,** 63 (1966).
2. W. L. BUTLER and H. C. LANE, *Plant Physiol.* **40,** 13 (1965).
3. H. MOHR, *Planta* **49,** 389 (1957).
4. H. MOHR, *Naturwiss. Rundschau.* **8,** 101 (1965).

5. H. MOHR, *Biol. Rev.* **39,** 87 (1964).
6. H. MOHR, E. WAGNER and K. M. HARTMANN, *Naturwiss.* **52,** 209 (1965).
7. W. L. BUTLER, *Quart. Rev. Biol.* **39,** 6 (1964).
8. M. FURUYA and W. S. HILLMAN, *Planta* **63,** 31 (1964).
9. W. BERTSCH and H. MOHR, *Planta* **65,** 245 (1965).
10. W. L. BUTLER, S. B. HENDRICKS and H. W. SIEGELMAN, *Photochem. Photobiol.* **3,** 521 (1964).
11. R. GRILL and D. VINCE, *Planta* **67,** 122 (1965).
12. G. LÜTHY, unpublished data.
13. W. SCHULTZ, unpublished data.

DISCUSSION

DR. VINCE: If you assume that in blue the photoequilibrium is less far in the direction of P_{fr} than Butler's data indicate, how do you account for the high efficiency of short pretreatment with blue light in the phytochrome destruction effect in turnip which is in accordance with Butler's data.

DR. MOHR: There are several aspects of this problem. First, we should like to remind you that in the best action spectra of phytochrome induced effects (e.g. Withrow *et al.*, 1957; opening of the bean hook) the quantum responsivity in the blue is very much lower than in the red as long as one works within thc range where reciprocity holds. In simpler terms: blue radiation is generally much less effective in the biological test than red light as far as the induction of a response with short-time irradiation is concerned. These conclusions have been confirmed by the action spectrum of the photochemical transformation of P_r (Butler *et al.*, 1964).

Secondly, if one irradiates until the photostationary state is reached one would indeed find a different action of blue and far-red, if the data on photoequilibria are correct. Our remark that a percentage of 35–50 per cent P_{fr} in the photostationary state in the blue might be too high for the mustard seedling should be taken as a suggestion to measure these ratios also in this plant.

DR. HARTMANN: On the basis of my phytochrome model, I see the possibility of identifying the 'unstable' P_{fr} from Wagner with the 'free P_{fr}', and the 'stable P_{fr}' with the 'bound P_{fr}'$=P_{fr}\,X$. Doing this, we are confronted with the difficulty of explaining the intensity dependence, low in the red, high in the far-red and the blue. After the final equation (7) of my proposed model one can calculate for example the slope ratio a/a_{max} of dose-response curves for the data $K_p=0{\cdot}035$ and $t/\tau'=20$ and the assumption that $\Delta\mu_\lambda=$ constant.

λ (nm)	a/a_{max}
740	0
726	0,8
720	0,9
710	1,0
700	0,2
690	0,02
660	0
430	0,04

The results give a good approach for the situation in hypocotyl lengthening for the long wave part (see Figs. 9 and 10). The intensity dependence in blue however is too low. The assumption, that $(P_{fr})/(P)$ is lower in blue as is calculated from the extract data (BUTLER, SIEGELMAN and MILLER, *Biochem. J.* **3,** 851 (1964)) and the assumption that $\Delta\mu_\lambda \sim \nu$ can account for this deviation. For the case of anthocyanin synthesis in *Sinapis alba* we only have to do the calculation with a greater K_p and a lower t/τ'. Arguing on this basis, the biological response in blue results from a low ϵ_{fr} and a high $\Delta\mu_\lambda$, whereas in the far-red it results from a high ϵ_{fr} and a low $\Delta\mu_\lambda$.

This means that the same biological response is produced through different physical situations.

DR. ROLLIN: La vitesse de photodestruction de P_{fr} en lumière rouge explique que l'action; de la lumière rouge est inférieure à celle de la lumière rouge sombre. Ceci ne semble pas expliquer l'action de la lumière chez *Phacelia tanacetifolia* (Anthocyanin synthesis!).

8 min red light + 3 hr darkness + 8 min red light + . . . (during 12 hr)			0,150
12 hr red light			0,150
6 hr far-red light			0,250
12 hr red light + 6 hr far-red light	} additive		0,390
6 hr far-red light + 12 hr red light	} action		0,390
8 min red light : about 10^{-2} J/cm²	=	200 ergs/cm²·sec	
far-red light :	=	8000 ergs/cm²·sec	

DR. MOHR: It could be that both combinations (red + far-red and far-red + red) lead to the same amount of anthocyanin because the far-red irradiance is so high. It might be easier to interpret experiments with a lower irradiance in the far-red.

Reprinted from PLANT PHYSIOLOGY, Vol. 41, No. 7, September, 1966, pages 1159–1166.
(PRINTED IN U.S.A.)

The Physiological Versus the Spectrophotometric Status of Phytochrome in Corn Coleoptiles[1]

Winslow R. Briggs and Hyangju P. Chon
Department of Biological Sciences, Stanford University, Stanford, California

Received April 18, 1966.

Summary. The influence of red light in altering the phototropic sensitivity of corn coleoptiles (*Zea mays* L., cultivar Burpee Barbecue Hybrid) is compared with the spectrophotometric status of the phytochrome they contain. The distribution of measurable phytochrome corresponds roughly with the distribution of sensitivity to red light for physiological change. Both phytochrome concentration and red light sensitivity are maximal in the coleoptile tips. Red light pretreatments which reduce total phytochrome by about 50 %, however, do not alter subsequent red light sensitivity of the phototropic system. Dosages of red light sufficient to saturate the physiological system are two orders of magnitude too small to induce measurable phytochrome transformation. The log-dosage-response curves for physiological change and for phytochrome transformation do not have the same slopes. The time course for appearance, main-concentration of the far-red-absorbing form of phytochrome over a broad range of tenance, and decay of the physiological response is independent of the measurable concentrations. The following hypothesis is proposed: the phytochrome mediating the alteration in phototropic sensitivity is only a small proportion of the total present. This small active fraction is physically and kinetically independent of the bulk measurable, and is packaged in some manner which facilitates its transformation in both directions.

There have been several attempts to correlate the physiological responses of etiolated plants to red and far-red light with the in vivo spectrophotometric status of the pigment phytochrome. Treatment of etiolated seedlings with small amounts of red light in most cases induces measurable physiological responses. Likewise, far-red light normally reverses the effect of red if it is administered sufficiently soon after the red. The physiological consequences of red treatment are usually attributed to a red-induced transformation of the red-absorbing form of phytochrome, P_R to the physiologically active far red-absorbing form, P_{FR}. Thus, one might expect the magnitude of a physiological response to be related in some simple way to the amount of P_{FR} spectrophotometrically measurable within the plant material in question. One might also expect some simple correlation between the amount of reversibility of a red-induced response by far-red light and the amount of P_{FR} present.

In most cases thus far examined one or both of the expectations have not been realized. For instance, freshly excised roots of *Pisum* seedlings contain detectable amounts of phytochrome, but failed to respond in a measurable way to red or far-red light. After 2 weeks of sterile culture in the dark, however, red light strongly suppressed lateral root initiation whereas phytochrome was no longer detectable (12, 13). The red-induced inhibition of growth of stem sections from etiolated peas can be reversed by far-red light long after all measurable P_{FR} has disappeared (16). Likewise, the red-induced opening of hypocotyl hooks of *Phaseolus* can be reversed with far-red light in the absence of detectable P_{FR} (11). A superficially less paradoxical example is that of the red-induced promotion of elongation of apical *Avena* coleoptile segments (18). In this last case, far-red reversibility is lost while P_{FR} remains detectable. The authors account for this last discrepancy with the reasonable hypothesis that following red light treatment, the elongation process eventually progresses beyond photochemical control, despite the continuing presence of P_{FR}. The hypothesis unfortunately does not bear on the other cases mentioned above. Hopkins and Hillman (18) cite several other cases of discrepancy between the physiological response of plants to light and the known properties of phytochrome in these plants.

The present paper considers another case in which the response of an etiolated plant to red light can not readily be related to the spectrophotometric status of its phytochrome and presents an hypothesis consistent with the cases thus far examined. Several investigators have recently studied the sub-

[1] This investigation was supported in part by grants G-21530 and GB-2846 from the National Science Foundation and a grant from Research Corporation to the senior author.

stantial influence of red light on the subsequent phototropic sensitivity of *Avena* and *Zea* coleoptiles (1, 2, 3, 4, 9, 23). Several lines of evidence strongly implicate phytochrome in this complex phenomenon. The action spectrum for red-induced alteration of phototropic sensitivity of *Zea* (9) is very much like the action spectra for many other phytochrome-mediated processes (14) including the action spectrum for phytochrome transformation in vivo in *Zea* coleoptiles (20). All of these action spectra are quite similar to the absorption spectrum of purified oat phytochrome (21). In both *Zea* and *Avena* the red light dosage required for alteration of phototropic sensitivity is extremely low and far-red reversibility has been demonstrated (3, 9).

Phytochrome is readily measured in the isolated coleoptiles of etiolated *Zea* seedlings (5, 20). Furthermore, many quantitative data are available concerning red light-induced changes in the phototropic sensitivity of these and of *Avena* coleoptiles (3, 9, 23). Thus a comparison of the physiological status of this particular phototropic system under various conditions of red light treatment with the status of the photometrically detectable phytochrome is particularly appropriate. A brief report of some of the work has appeared elsewhere (19).

Materials and Methods

Corn (*Zea mays* L.), Burpee cultivar Barbecue Hybrid, lot 29234 was grown as described elsewhere (9). The procedures for red-light treatment of rows of seedlings, phototropic induction, recording and measuring phototropic curvatures, and for statistical treatment of the data have also been described elsewhere (9).

Fifteen w green fluorescent bulbs wrapped in 2 layers of amber and 1 layer of green cellulose acetate provided the sole illumination for experimental manipulation. A 100 w high pressure dc mercury arc used with a 436 mμ interference filter and a 5 cm water filter provided blue light for phototropic induction. Low intensity red light was obtained from a Bausch and Lomb grating monochromator with a tungsten source and a supplementary filter to eliminate second order wavelengths. Higher intensity red light was obtained from a 500 w slide projector with a 5 cm water filter, an appropriate interference filter, and appropriate supplementary filters. In several experiments red light was administered to seedlings for 3 hours starting 46 hours after the start of germination, to suppress first internode growth. The red light for this purpose was obtained from 2 concentric circular fluorescent lamps, 32 and 40 w, and 3 layers of red cellulose acetate. The intensity of this source at the level of the plants was approximately 1.9×10^3 ergs cm^{-2} sec^{-1}. In certain experiments, red light for alteration of phototropic sensitivity was obtained from 40 w ruby-red incandescent bulbs, used with a 5 cm water filter. The techniques used for measuring light energy are described elsewhere (9, 23).

In another paper (9), optimal conditions for determining the influence of a given red-light treatment on subsequent phototropic sensitivity are described. These conditions were met for the experiments described below, and are briefly summarized. Red light was given to plants approximately 80 hours of age (growth at 25° ± 1°). The plants were placed in total darkness until 1 hour had elapsed from the beginning of the red treatment, at which time they were phototropically induced. The hour's wait was sufficient for completion of the consequences of the red treatment for phototropism (9). The dose of blue light used for phototropic induction was 1.5×10^{-9} Einsteins cm^{-2}. Phototropic curvatures were allowed to develop for 100 minutes following the start of phototropic induction, and were then shadowgraphed. All operations beginning with the red light treatment were carried out in a dark room maintained at 24° ± 1°. Under these conditions, phototropic curvature in the absence of red light pretreatment was normally about 3°. With a red-light dosage sufficient to induce a maximum change in phototropic sensitivity, the curvatures normally exceeded 20°.

The spectral reversibility of phytochrome was measured in vivo using a dual wavelength photometer (Ratiospect, Model R-2, Agricultural Specialty Company). This instrument is functionally identical with that described by Butler et al. (7). One can determine not only relative amounts of phytochrome in a given sample but also the percentages of the pigment present as P_R and P_{FR}, following various light treatments. Samples were measured in an aluminum cell with a cross-sectional area of 1 cm^2, and were maintained at 0° during measurement to prevent thermo-chemical loss of P_{FR} (7, 20). Preparation and handling of samples is described in another paper (5). All samples to be compared with each other were of similar weight, allowing direct comparison of relative phytochrome content between them (5).

Results

Localization of region of maximum red-light sensitivity. Under the growing conditions used in this study, the coleoptiles of the corn cultivar Barbecue Hybrid elongate more rapidly than the primary leaves within them. In seedlings approximately 80 hours old, the tip of the primary leaf is at least 6 mm below the coleoptile apex. Thus one can irradiate the coleoptile tip alone without directly illuminating primary leaf tissue or being concerned about light screening by the primary leaf. One can also speedily obtain large numbers of excised coleoptile apices without laboriously removing primary leaf fragments.

The region of maximum sensitivity to red light was determined by irradiating seedlings unilaterally with red light (source: Bausch & Lomb monochromator, 660 mμ) after first shielding various regions with black cardboard. Thus one could irradiate only the top 3 mm of the coleoptile, or, using a 5 mm horizontal slit in the cardboard, the center 5 mm of the coleoptiles (1.3–1.8 cm long) including, of course, the underlying primary leaf tissue, or the center 5 mm of the first internode (suppressed, 0.5–0.8 cm long). Since the plants were 50 cm from the monochromator exit slit, and the black card a maximum of 5 mm from the plants, fairly sharp delimitation of the irradiated area was possible. The red light dosage in all cases was 1.8×10^{-11} Einsteins cm^{-2} (intensity : 3.0×10^{-12} Ein cm^{-2} sec^{-1}; time: 60 sec). This dosage was selected since previous work (9) showed that it was just sufficient to saturate the physiological system mediating red-induced alteration in phototropic sensitivity.

Table I presents the combined results of 3 such experiments, each done on a different day. Rows of plants were in each case given the appropriate red-light treatment and then stored, phototropically induced, and measured as described above. Each number in the second column of table I represents the mean curvature of a single row of 8 or 9 plants. These mean curvatures are averaged in the third column, and percent response to red light (tip-irradiated = 100 %, dark control = 0 %) is shown in the fourth column.

Table I. *Distribution of Red-Light Sensitivity in Various Parts of Corn Seedlings*

Red-light dosage: 1.8×10^{-11} Einsteins cm^{-2}, 660 mμ.
Phototropic dosage: 1.5×10^{-9} Einsteins cm^{-2}, 436 mμ.

Region exposed	Mean curvature degrees	Avg	% response
None	1.0		
	4.1	3.2	0
	3.9		
	3.8		
Whole plant	18.9		
	20.8		
	22.7	20.7	95
	20.1		
	19.8		
	22.0		
Coleoptile, tip, 3 mm	24.3		
	21.7		
	24.3	21.7	100
	20.5		
	20.8		
	18.5		
Coleoptile, center, 5 mm	15.3		
	13.0		
	11.6	12.8	52
	12.0		
	12.6		
	12.4		
First internode, center, 5 mm	7.5		
	6.6	6.8	19
	6.0		
	7.0		

Table II. *Distribution of Phytochrome in Corn Seedlings Grown Without First Internode Suppression*

Region of seedling	Δ(ΔOD)/g fr wt	Avg	% of coleoptile tip
Coleoptile, tip, 3 mm	0.17		
	0.16	0.160	100
	0.14		
Coleoptile, center, 5 mm (with primary leaf)	0.048		
	0.073	0.061	38
	0.063		
Coleoptile, center, 5 mm (without primary leaf)	0.055		
	0.006	0.0045	3
	0.070		
First internode, center, 5 mm	0.003		
	0.006	0.0045	3
Node, 5 mm	0.093		
	0.070		
	0.090	0.089	56
	0.064		

Exposure of the apical 3 mm of the coleoptiles to the red-light dosage used was just as effective as exposing the entire seedlings. The Student *t* test did not reveal a significant difference between the phototropic responses of these 2 samples in any of the 3 experiments done. Irradiation of the central 5 mm of the coleoptiles (including the underlying primary leaf) with this dosage of red light was considerably less effective. The curvatures thus obtained differed significantly (99 % confidence level) from those obtained either without prior red-light treatment or with irradiation of whole seedlings or 3 mm tips. Irradiation of the central 5 mm of the first internodes caused only a very small shift in phototropic sensitivity. The curvatures differed from those of unirradiated controls at the 90 % confidence level. It is clear that in this *Zea* cultivar, the coleoptile tip is the most sensitive region for red-induced alteration in phototropic sensitivity, and not the primary leaf tip, as suggested by Blaauw-Jansen (2) for *Avena*.

Relationship between the distribution of red-light sensitivity and the distribution of phytochrome. Although Briggs and Siegelman (5) had already explored the distribution of phytochrome in etiolated seedlings of one corn cultivar (USDA 13), it was necessary to do such a study on Barbecue Hybrid at the precise age and under the precise growing conditions used for the phototropic experiments. All samples assayed for phytochrome weighed 0.3 g.

Table II shows the results of 2 experiments in which completely etiolated seedlings (no suppression of first internodes) 83 hours of age were examined. Phytochrome reversibility is expressed as Δ(Δ OD) /g fresh weight as in earlier studies (5) The extreme tip of the coleoptile contains almost twice as much phytochrome per g fresh weight as any other region examined. While it is difficult to assess the importance of the difference in the amount of phytochrome per g found in the central 5 mm sections of the coleoptile with or without primary leaf tissue present, it is noteworthy that this region, approximately half as sensitive to red light as the coleoptile apex, also contains approximately half as much phytochrome. The central 5 mm of the first internodes contained only a very small amount of phytochrome. The coleoptilar node contained somewhat more than the central region of the coleoptile. Thus the distribution of phytochrome in Barbecue Hybrid corn is qualitatively like that described earlier (5) for the corn cultivar USDA 13.

Column 4 of table II shows phytochrome reversibility for various regions of the seedlings as a percentage of that found in the 3 mm coleoptile tips. A comparison of these figures with the percentage figures in table I shows a fairly good correspondence between the distribution of phytochrome and the distribution of sensitivity to red light. Some of the discrepancies may be attributed to inherent difficulties in the rather cumbersome phototropic assay. It is also possible that the amount of reversible phytochrome in the coleoptile tips is in excess of that needed for induction of the maximum phototropic sensitivity change. Thus, red-light sensitivity in the lower portions of the seedlings might appear proportionally higher than the actual amount of phytochrome present might suggest. However, the correspondence of phytochrome distribution with red-light sensitivity may be fortuitous. Decreased sensitivity to red light from the apex to the base of the seedling may reflect increasing distance between the phytochrome transformed and the phototropic system affected by this transformation.

The reduction of total measurable phytochrome in etiolated corn seedlings by red light is well documented (6, 7, 10, 20). If phytochrome concentration is of prime importance in determining relative sensitivity to red light, one would expect a treatment which reduced the phytochrome concentration to reduce or at least in some way alter red-light sensitivity. Two previous studies (3, 9) had shown that the suppression of first internode elongation in corn with red light at about the end of the second day of growth did not measurably affect the phototropic sensitivity of the plants 2 or 3 days later. It was therefore important to determine whether the phytochrome concentration and distribution in plants with suppression of the internodes (as was the case with the phototropic experiments above) differed in any way from that in non-suppressed plants. Conceivably, phytochrome lost during and after the red-light treatment for first internode suppression might be fully replaced by the usual time of the phototropic experiments.

Table III shows the distribution of phytochrome

Table III. *Distribution of Phytochrome in Corn Seedlings Grown With First Internode Suppression*

Region of seedling	$\Delta(\Delta OD)$/g fr wt	Avg	% of coleoptile tip
Coleoptile, tip, 3 mm	0.082 0.071	0.077	100
Coleoptile, center, 5 mm (without primary leaf)	0.020 0.031 0.030	0.027	35
First internode, center, 5 mm	0.0066 0.0034	0.005	7
Node, 5 mm	0.023 0.033 0.039	0.032	42

in seedlings that had had their first internodes suppressed. Comparison with table II shows that for any given region, suppression of the first internodes as described generally reduces measurable phytochrome to approximately half that found in non-suppressed plants. However, the distribution pattern is almost identical with that in non-suppressed plants (with the exception of the first internodes). Thus corn, like *Avena* (5), shows a change in phytochrome concentration following first internode suppression, but little change in distribution pattern.

Since the quantitative aspects of sensitivity to red light, in terms of subsequent phototropic behavior, are unchanged by reducing the coleoptile phytochrome concentration by 50 %, it seemed reasonable that only a portion of the coleoptilar phytochrome might be involved in the observed physiological response. Thus more detailed comparison of phytochrome status versus physiological response seemed justified.

Dosage-response relationships for induction of physiological response versus phytochrome transformation. Detailed dosage-response relationships for the effect of a number of wavelengths of red light in altering phototropic sensitivity were already available. The left-hand curve in figure 1, from previous work (9), shows such relationships for light of 660 mμ. These results were obtained from plants that had not had their first internodes suppressed. The threshold red-light dosage is approximately 10^{-12} Einsteins cm^{-2}, and saturation is achieved with approximately 10^{-11} Einsteins cm^{-2}. A similar curve obtained using red light of wavelength 670 mμ (not shown) is almost the same as that shown for 660 mμ (9). Thus it was of great interest to determine how much phytochrome transformation was required to achieve both threshold and saturation for the physiological response.

Rows of plants (without first internode suppression) were therefore arranged for red-light treatment in exactly the same manner as for the phototropic experiments. However, the various red-light treatments were administered at 5°, instead

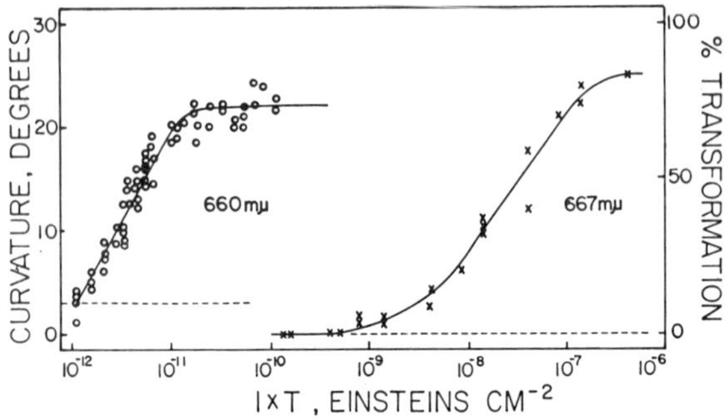

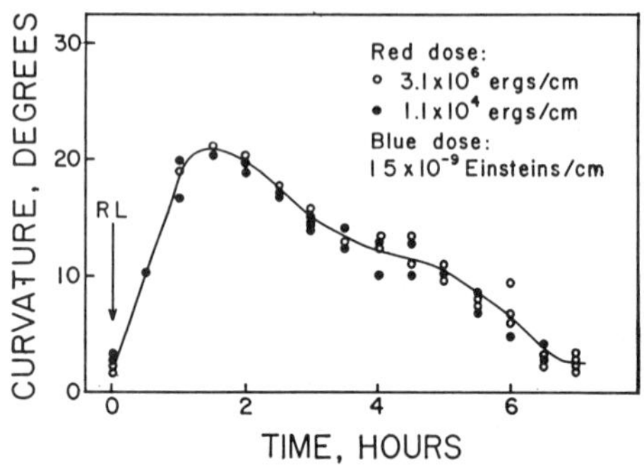

FIG. 1. (top) Dosage-response curves for alteration of phototropic sensitivity (circles) and for phytochrome transformation (crosses). Dotted lines, dark controls.

FIG. 2. (bottom) Appearance, persistence, and eventual decay of red-light-induced alteration in phototropic sensitivity following 2 different dosages of red light. Closed circles, red-light dosage 7×10^{3} ergs cm^{-2}. Open circles, red-light dosage 1.87×10^{6} ergs cm^{-2}. RL, red-light treatment. Blue-light dosage for phototropic induction, 1.5×10^{-9} Einsteins cm^{-2}.

of 25° as in the phototropic experiments, to minimize dark destruction of P_{FR} (7, 20), and instead of receiving phototropic induction, the coleoptile tips (apical 7 mm) were immediately harvested for phytochrome measurement. The right-hand curve in figure 1 presents the dosage-response relationships for phytochrome transformation obtained under these conditions with light of wavelength 667 mμ. Surprisingly, a red-light dosage just sufficient to saturate the physiological system is 2 orders of magnitude too small to induce measurable phytochrome transformation. The 2 curves shown in figure 1 clearly indicate that the threshold values are at least 3 orders of magnitude apart, with saturation values at least 4 orders of magnitude apart [because of overlapping absorption of P_R and P_{FR} at 667 mμ, 80 % represents the maximum possible transformation (see 8)].

Persistence of physiological response as a function of red-light dosage. The results shown in figure 1 are open to 2 interpretations. First, induction of the physiological response might merely require transformation of a minute fraction of a percent of the total phytochrome. In this event, one might expect red-light treatments transforming relatively large amounts of phytochrome to P_{FR} to induce physiological changes which would persist longer than those induced by red-light treatments transforming only very small fractions of the total pigment. In the former case, P_{FR}, whatever its eventual fate, destruction or reversion to P_R, would remain present in the tissue in substantial quantity longer than in the latter case. Earlier work with another corn cultivar, Burpee Golden Cross Bantam (3), had already shown that the phototropic sensitivity changes induced by relatively massive dosages of red light (3.3×10^{3} ergs cm^{-2} sec^{-1}, source: 40 w ruby red incandescent bulbs, exposure time: 2 hr) did not persist indefinitely but began to disappear approximately 1 hour after the end of the red treatment. Second, the phytochrome associated with the physiological response might be in some way sequestered from the bulk of the phytochrome measurable, and physically and kinetically independent from it.

Thus the appearance, persistence, and eventual decline of the phototropic sensitivity change were measured following 2 different red-light treatments: first, one which was sufficient to saturate the physiological response but insufficient to transform a measurable amount of phytochrome; and second, sufficient both to saturate the physiological response and to transform a significant fraction of the measurable phytochrome. For these experiments, the light source was a ruby-red incandescent bulb with a 5 cm water filter. The 2 dosages were 7×10^{3} ergs cm^{-2} (exposure time: 5 sec) and 1.87×10^{6} ergs cm^{-2} (exposure time: 6 min). The coleoptile tips (7 mm) of certain rows of seedlings were harvested immediately following red treatment and assayed for relative phytochrome transformation. Although the red treatment occurred at 25°, the harvested tips were quickly placed in a small beaker surrounded with ice to minimize any dark reactions involving the P_{FR} produced (20). The absence of any detectable change in total phytochrome concentration following either red-light treatment (table IV) indicates that this technique was successful. An initial Δ OD was obtained and then the change in Δ OD induced by a saturating dosage of far-red light measured. Inspection of the dark controls in table IV shows that the far-red source used for this purpose, an old Baird-Atomic 725 mμ interference filter, was sufficiently imperfect that it induced the formation of over 11 % P_{FR}. (A negative Δ OD indicates formation of P_{FR}, and a positive Δ OD indicates formation of P_R.) Total phytochrome was calculated for each sample by assuming that the far-red source always yielded 11 % P_{FR} at saturation and that the red source yielded 80 % (8). Thus the Δ OD first induced by saturating far-red light could be used directly to calculate the amount of P_{FR} initially present in the sample. Table IV shows that the lower dosage of red light administered to

Table IV. *Relative Phytochrome Transformation in Corn Coleoptiles by Two Dosages of Red Light*

The dosages were: 7 × 10³ ergs cm⁻² (low dosage) and 1.87 × 10⁶ ergs cm⁻² (high dosage).

Treatment		Total Δ(Δ OD)	ΔOD induced by initial saturating far-red light	% P_{FR} (avg)
Darkness		0.0266	−0.003	
		0.0200	−0.001	
		0.0313	−0.005	
	avg	0.0260	−0.003	0
Low red-light dosage		0.0264	−0.003	
		0.0235	−0.003	
		0.0249	−0.000	
	avg	0.0249	−0.002	3*
High red-light dosage		0.0271	+0.0085	
		0.0286	+0.0075	
		0.0274	+0.0085	
	avg	0.0277	+0.0082	47

* Well within range of instrument noise.

the intact seedlings was insufficient to transform an amount of phytochrome measurable with our instrument. The higher dosage, however, transformed approximately 47 %.

Other rows of seedlings were either phototropically induced immediately after red-light treatment or were allowed to remain in the dark for up to 7.5 hours preceding phototropic induction. The curvatures obtained from phototropic induction at various times following the 2 red-light treatments are shown in figure 2. With either dosage, the phototropic sensitivity change started immediately, reaching a maximum after about 1 hour, and remaining there for an additional hour. Furthermore, the disappearance of the effect followed the same time course whether the dosage of red light had been large or small. The *t* test showed no significant difference between the 2 red-light treatments at any time following the irradiation.

Discussion

In interpreting the above results, several points need to be considered. First, the dosage of red light needed for saturation of the physiological response is 100-fold too small to induce measurable phytochrome transformation. Second, the measurable physiological consequence of red-light treatment is far-red-reversible (3, 9) though even the best far-red filters leave at least a small fraction of phytochrome present as P_{FR} (7, 16). For example, the actinic far-red light source in the Ratio pect, used for the phytochrome measurements above, was sufficiently imperfect to induce the formation of over 10 % P_{FR} in dark-grown coleoptiles (table IV). Furthermore, reversibility of the red light-induced phototropic sensitivity change in *Avena* (3) was achieved using an incandescent lamp and Corning filter 7-69, similar to filter 4B of Butler et al. (7). This filter must have induced the steady-state formation of at least 1 % P_{FR} since the rate of destruction of phytochrome in dark-grown corn at these low concentrations is limited by P_{FR} concentration, and filter 4B allowed phytochrome destruction to proceed between one-half and one-third as rapidly as filter 2B which set up a measurable steady-state of 3 % P_{FR}. Third, the slope of the log-linear portion of the physiological dosage-response curve is over 2-fold steeper than that for phytochrome. Fourth, the time course for appearance, maintenance, and subsequent decay of the physiological change is independent of measurable P_{FR} concentration ranging from an amount undetectable with current instrumentation to 47 %. Previous work bearing on the fourth point showed that the physiological change in another corn cultivar (Burpee Golden Cross Bantam) also persisted at its maximum for only an hour, as in the present experiments, even following continuous red-light treatment for 2 hours (total dosage: 2.4 × 10⁷ ergs cm⁻²) (3).

If the large discrepancy between the physiological dosage-response curve and that for phytochrome transformation were to be explained by the hypothesis that transformation of only an extremely small fraction of a homogeneous population of phytochrome molecules is sufficient to saturate the physiological response, 3 conditions must be met. First, far-red light, which at best leaves behind a finite amount of P_{FR} at saturation should not reverse the effect of red light, and in fact should be as effective as red in inducing the physiological change. Second, the slopes of the 2 dosage-response curves should be identical despite the difference in threshold values, since transformation of measurable phytochrome in corn is known to obey the kinetics of a first-order photochemical reaction (20). Third, there should be some relationship between the amount of P_{FR} formed by a given red-light dosage and the magnitude and persistence of the subsequent physiological change. Though failure to meet the third condition alone might be open to several alternative explanations, failure to meet all 3 conditions leads to the following hypotheses. The phytochrome which mediates the physiological response must be functionally separate from the bulk which one can measure, and must be present in a concentration too small to detect with currently available instrumentation. Furthermore, it seems likely that this active phytochrome is sequestered in some manner that facilitates its transformation in either direction.

Several other observations have bearing on these hypotheses. Wilkins (22) has shown red light to have profound effects on the subsequent geotropic sensitivity of *Avena* coleoptiles. The dosage required for a threshold response (3.3 × 10⁻¹³ Einsteins cm⁻²) is even lower than that required for the phototropic system described above. Furthermore, the geotropic effects are fully far-red reversible. Like the phototropic system, saturation of the response is reached over a single order of magnitude of dosages. Although a detailed dosage-response curve for phytochrome transformation in oat coleoptiles, obtained in the same manner as the one shown above for corn coleoptiles, is not available, our own experience with corn and oats (e.g. 5) indicates that measurable transformation of phytochrome requires roughly comparable light dosages in each. Thus the same dosage-response discrepancy found for the phototropic system in corn undoubtedly exists for the geotropic system in oats. Though precise data are not available for oats, the red light-induced alteration in their phototropic sensitivity resembles that found in corn in being at least partially reversible by far-red light (see above) despite requiring a very small red-light dosage (3). Finally the action spectrum for phytochrome transformation in corn (20) is similar to but not identical with that for the induction of a shift in phototropic sensitivity (9). The ratio of relative quantum efficiencies between 660 and 670 mμ is 0.81 for phytochrome transformation, but is 1.1 for the physiological response. A more detailed comparison of these 2 action spectra is needed to determine the magnitude and significance of this difference. All of these observations support the hypothesis that there is something different about a very small portion of the total phytochrome involved in alteration of the various tropic sensitivities mentioned.

The discrepancies between physiologically assayable and photometrically measurable phytochrome described here for phototropic sensitivity of corn (probably also found in the red-sensitive systems altering geotropic and phototropic sensitivity in *Avena*) differ from the systems described in the introduction in one important respect. These systems all show a requirement for a substantial amount of measurable P_{FR} before physiological responses can be detected. Hillman (16) has proposed a segregation of active from bulk phytochrome in *Pisum* to account for his observation of significant far-red reversibility of red-induced stem-segment growth inhibition long after P_{FR} is photometrically undetectable. More recently, he has shown that varying exposure time with pure red light to produce different percentages of P_{FR} (17) causes physiological consequences for *Pisum* stem sections and *Avena* coleoptile apical segments just like those obtained using filter combinations selected to give these percentages of P_{FR} at photostationary equilibrium, as was done in the earlier experiments (16, 18). Once again, at least the initial physiological response was correlated with conversion of a substantial portion of measurable phytochrome. Thus none of these results requires the hypothesis of a small sequestered pool of phytochrome packaged in some way which facilitates transformation.

Butler et al. (6, 7) showed several years ago that the phytochrome found in vivo in etiolated corn differed from that found in cauliflower heads grown and stored in the light. P_{FR} disappeared without measurable reversion to P_R in corn, while it appeared only to revert to P_R in cauliflower (see also 15). Thus measurable phytochromes which behave quite differently from each other in vivo are known. Unfortunately, however, no obvious phytochrome-mediated response has been described for cauliflower heads. Therefore comparison of the physiological versus the spectrophotometric status of this phytochrome is not feasible.

Of the discrepancies noted in various systems between measurable phytochrome and physiological status, only the phototropic and geotropic systems show the large apparent difference between the amount of light required for physiological response and the amount needed for phytochrome transformation. These systems all respond to minute amounts of red light and all respond extremely rapidly (3, 9, 22). Thus there is perhaps something unique about the relationships between phytochrome and these tropic responses, involving both facilitation of transformation and expediting of the consequences of forming P_{FR} that is not found in the slower growth-promotion or growth-inhibition systems.

The above results suggest that the fairly good correspondence between the distribution of phytochrome and the distribution of sensitivity to red light is not particularly significant. It may simply mean that the distribution of the packaged phytochrome suggested above corresponds roughly to the distribution of bulk phytochrome, and both may be directly correlated with the amount of cytoplasm in the various regions studied. If these 2 hypothesized species of phytochrome are interconvertible, such interconversion must be relatively slow, since physiological response is at such complete variance with measurable pigment transformation.

Acknowledgment

The authors are extremely grateful to Dr. William S. Hillman for several long and fruitful discussions of the problems considered in this paper. They are also grateful to Miss Barbara Platz for valuable technical assistance.

Literature Cited

1. Blaauw, O. H. and G. Blaauw-Jansen. 1964. The influence of red light on the phototropism of *Avena* coleoptiles. Acta Botan. Neerl. 13: 541–52.
2. Blaauw-Jansen, G. 1959. The influence of red and far-red light on growth and phototropism of the *Avena* seedling. Acta Botan. Neerl. **8**: 1–39.
3. Briggs, W. R. 1963. Red light, auxin relationships and the phototropic responses of corn and oat coleoptiles. Am. J. Botany **50**: 196–207.
4. Briggs, W. R. 1963. The phototropic responses of higher plants. Ann. Rev. Plant Physiol. **14**: 311–52.

5. Briggs, W. R. and H. W. Siegelman. 1965. Distribution of phytochrome in etiolated seedlings. Plant Physiol. 40: 934–41.
6. Butler, W. L. and H. C. Lane. 1965. Dark transformations of phytochrome in vivo II. Plant Physiol. 40: 13–17.
7. Butler, W. L., H. C. Lane, and H. W. Siegelman. 1963. Nonphotochemical transformations of phytochrome in vivo. Plant Physiol. 38: 514–19.
8. Butler, W. L., H. W. Siegelman, and C. O. Miller. 1964. Denaturation of phytochrome. Biochem. 3: 851–57.
9. Chon, H. P. and W. R. Briggs. Phytochrome-mediated alteration of phototropic sensitivity of corn coleoptiles. Plant Physiol. In press.
10. DeLint, P. J. A. L. and C. J. P. Spruit. 1963. Phytochrome destruction following illumination of mesocotyls of *Zea mays* L. Mededel. Landbouwhogeschool, Wageningen 63: 1–7.
11. Edwards, J. L. and W. H. Klein. 1964. Relationship of phytochrome concentration of physiological responses. Plant Physiol. 39: I.
12. Furuya, M. and W. S. Hillman. 1964. Observations on spectrophotometrically assayable phytochrome in vivo in etiolated *Pisum* seedlings. Planta 63: 31–42.
13. Furuya, M. and J. G. Torrey. 1964. The reversible inhibition by red and far-red light of auxin-induced lateral root initiation in isolated pea roots. Plant Physiol. 39: 987–91.
14. Hendricks, S. B. and H. A. Borthwick. 1965. The physiological functions of phytochrome. In: Biochemistry of Plant Pigments, T. A. Goodwin, ed. Academic Press, London, p 519–49.
15. Hillman, W. S. 1964. Phytochrome levels detectable by in vivo spectrophotometry in plants parts grown or stored in the light. Am. J. Botany 51: 1102–07.
16. Hillman, W. S. 1965. Phytochrome conversion by brief illumination and the subsequent elongation of etiolated *Pisum* stem segments. Physiol. Plantarum 18: 346–58.
17. Hillman, W. S. 1966. Responses of *Avena* and *Pisum* tissues to phytochrome. Plant Physiol. 41: 907–08.
18. Hopkins, W. G. and W. S. Hillman. 1966. Relationships between phytochrome state and photosensitive growth of *Avena* coleoptile segments. Plant Physiol. 41: 593–98.
19. Hughes, A. P. 1965. Phytochrome discussion. In: Recent Progress in Photobiology, E. J. Bowen, ed. Blackwell, p 219–22.
20. Pratt, L. H. and W. R. Briggs. 1965. Photochemical and nonphotochemical reactions of phytochrome in vivo. Plant Physiol. 41: 467–74.
21. Siegelman, H. W. and E. M. Firer. 1964. Purification of phytochrome from oat seedlings. Biochem. 3: 418–23.
22. Wilkins, M. B. 1965. Red light and the geotropic response of the *Avena* coleoptile. Plant Physiol. 40: 24–34.
23. Zimmerman, B. K. and W. R. Briggs. 1963. Phototropic dosage-response curves for oat coleoptiles. Plant Physiol. 38: 248–53.

Reprinted from PLANT PHYSIOLOGY, Vol. 43, No. 5, May, 1968, pages 823–826.
(PRINTED IN U.S.A.)

Response of Tissue with Different Phytochrome Contents to Various Initial Photostationary States[1]

L. Raymond Fox and William S. Hillman
Biology Department, Brookhaven National Laboratory, Upton, New York 11973

Received January 19, 1968.

Abstract. **Pretreatment of etiolated pea plants with red light and with red combined with far-red light produced morphologically similar plants having 4-fold differences in spectrophotometrically detectable phytochrome. Stem segments from the variously pretreated plants respond in the same way to different percentage conversions of phytochrome to P_{FR}. These results suggest that the P_{FR}/P_R ratio, rather than the concentration of P_{FR}, governs pea stem segment elongation. However, the ratio hypothesis does not explain contradictions between spectrophotometric and physiological assays previously obtained with this tissue, nor does it explain similar contradictions obtained in other systems. The only hypothesis consistent with the data to date is that of the existence of bulk and active phytochrome fractions, with the latter present in insufficient quantities to be spectrophotometrically detectable.**

Attempts to correlate phytochrome content and state, assayed spectrophotometrically, with physiological response have led to conflicting observations. In some cases there is agreement between spectrophotometric and physiological assays (1, 2, 11). On the other hand, some tissues, such as young, intact pea roots (7), and cauliflower heads, parsnip roots, and artichoke receptacles (8) contain large amounts of spectrophotometrically detachable phytochrome yet exhibit no known light responses. A number of contradictions between spectrophotometric and physiological assays have also been reported (1, 4, 9, 12, see also 10). Such contradictions have prompted us to see whether or not differences in the amount of spectrophotometrically detectable phytochrome might be reflected in differences in phytochrome responses. The phytochrome content of etiolated seedlings is easily reduced by brief exposures to red light (3, 7, 10). However, such exposures also bring about photomorphogensis. Since differences in phytochrome responses might be expected on grounds other than phytochrome content when etiolated and de-etiolated tissues are compared, it seemed desirable to obtain morphologically similar plants which contained different amounts of phytochrome. This was made feasible by the observation that in dicot seedlings phytochrome content remains relatively constant under continuous far-red illumination, even though such illumination is highly effective in de-etiolation (5).

[1] Research carried out at Brookhaven National Laboratory under the auspices of the United States Atomic Energy Commission.

Materials and Methods

Seeds of *Pisum sativum* L. cultivar "Alaska" were obtained from Asgrow Seed Company, New Haven, Connecticut. They were soaked for 4 hours and then sown in either polyethylene basins (19 × 29 × 10 cm) for routine growth assays, or in cardboard ice-cream cartons (9 × 9 cm) for determining the effect of various light pretreatments on de-etiolation. Basins or cartons were nearly filled with saturated vermiculite (Zonolite No. 1), which also covered the seeds to a depth of 3 to 6 cm. A 1/10 strength solution of Hutner's medium was used for both soaking the seeds and saturating the vermiculite. Hutner's medium was prepared as previously reported (7). After planting, the seeds were allowed to develop at 26 to 27° in darkness for 6 or 7 days.

For growth assays, 10-mm segments were cut just below the apical hooks of plants with developing third internodes 12 to 48 mm in length. The segments were randomized in large petri dishes containing Na_2HPO_4-KH_2PO_4 buffer (0.02 M with respect to PO_4), pH 6.2 to 6.4. After all had been cut, lots of 15 or 20 were placed in 50 ml beakers containing 3.0 ml of a medium consisting of buffer, 3×10^{-5} M $Co(NO_3)_2$, and 2% (w/v) sucrose. Following light treatments the beakers were covered with small petri dishes, and the segments were allowed to elongate in darkness at 26 to 27° for 20 hours, after which they were measured to the nearest 0.5 mm.

Standard red light sources consisted of 2 cool white fluorescent tubes (Sylvania F15T8/CW) behind a 0.3 cm thickness of red Plexiglas (Rohm and Haas 2444). Far-red light was provided by four 15-watt and one 7 and one-half watt incandescent light bulbs over 8 to 10 cm of water and a 0.3 cm thickness of Rohm and Haas V-58015 "black" Plexiglas. The light sources and filters used to achieve various photostationary state levels of the far-red absorbing form of phytochrome have been described elsewhere (9); 10 minute exposures were used. All manipulations were performed under dim green safelights (9).

Spectrophotometric assays of phytochrome content were performed as previously described (7, 9) on samples of segments cut for the growth assay. Each sample of 10 segments was cooled to the temperature of ice and irradiated for approximately 10 minutes with cool white fluorescent light to saturate protochlorophyll conversion in sections from plants having received light pretreatments. The low temperature prevented significant changes in phytochrome content during the illumination with fluorescent light. Segments were then cut in half and packed to a sample length of 1 cm in 6-mm diameter aluminum cuvettes. The packed cuvettes were kept on crushed ice until and during the measurement.

Results

Light pretreatments involving a number of schedules of red light, far-red light, and combinations of them were tested with the objective of optimizing differences in phytochrome content while maintaining the same state of morphological de-etiolation. For studies of this type, the seedlings in ice-cream cartons were allowed to develop in darkness at 26 to 27° for 6 days. At that time 10 seedlings per carton were selected so that the leaf scale marking initiation of the third internode was still in the apical hook region, and the remaining seedlings were cut from the carton. Replicate lots were then given various light pretreatments for different lengths of time, following which the apical buds were excised and weighed individually, and the lengths of the third internodes measured. Using stem growth inhibition and apical bud weight increase as measures of de-etiolation, it was found that a schedule (R) consisting of 3 minutes of red light every 2 hours alone, and a second schedule (RF) consisting of (R) plus continuous far-red light, caused the same degree of de-etiolation when given for 14 or 18 hours (fig 1) while maximizing the differences in phytochrome content. Pretreatment with continuous far-red light alone also caused a similar degree of de-etiolation; however, R and RF pretreated plants were used in subsequent experiments because of the similar action of these light regimes on chlorophyll synthesis, which conceivably might have modified the results of the phytochrome assays.

For a comparison of the phytochrome response of tissues subjected to R and RF light schedules, the pretreatments were administered to etiolated seedlings in basins. The R and RF pretreatments were begun when the seedlings were 6 days of age, and a total pretreatment time of 16 hours was used. The

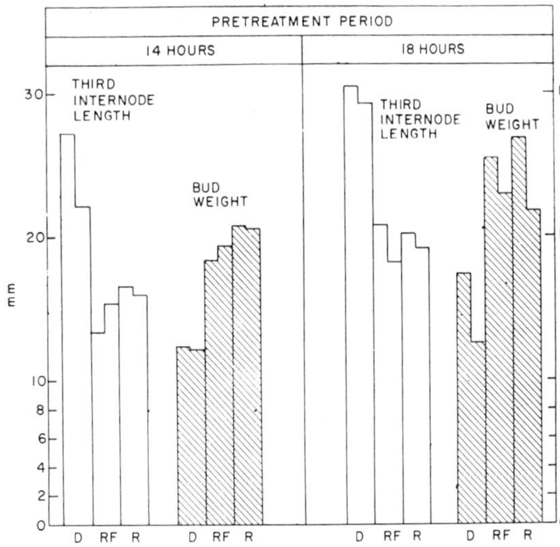

FIG. 1. De-etiolation by R and RF light schedules for 14 and 18 hours. Each half bar represents the mean of 10 plants. D) dark controls; R) 3 minutes red light every 2 hours; RF) R plus continuous far-red light.

last red exposure in both R and RF pretreatments was given approximately 2 hours before the end of the pretreatment period. Those plants pretreated with the R regime received 15 minutes of far-red light at the end of the pretreatment period, to ensure that the phytochrome in both R and RF plants was in the same state. A third "pretreatment," darkness (D), was also administered to seedlings in basins. These plants were held in darkness for the duration of the pretreatment period. Sections were cut from the pretreated plants immediately following the end of the pretreatment period. The R and RF pretreatments reduced the spectrophotometrically detectable phytochrome, compared to that observed in etiolated plants of the same age, to about 20 and 80 %, respectively (table I). Thus, even though plants pretreated with R and RF light regimes were morphologically similar (fig 1), they showed a 4-fold difference in spectrophotometrically detectable phytochrome. The various P_{FR} contents brought about by photostationary state illumination of segments from D, R, and RF pretreated plants are indicated by the "relative P_{FR} units" of table I. These values are obtained by setting the phytochrome of the dark controls equal to 100, so that an assumed full conversion of this to P_{FR} would give 100 units.

Initial phytochrome conversions to P_{FR} of from 5 to 88 % were established by exposing segments under photostationary state filters for 10 minutes. The effects of such exposures on the subsequent growth of tissues from the 3 groups of pretreated plants are shown in figure 2. Each point represents the mean of 30 or 40 segments. From 5 % P_{FR} upward, elongation decreased in a linear fashion with the log of the percent P_{FR} established. These results are summarized in figure 3A, where each point represents the mean value of the points shown in figure 2.

Table I. *The Effect of Dark, R, and RF Pretreatments on the Phytochrome Content of* Pisum *Hypocotyl Tissue*

	Phytochrome content[1] Dark pretreatment	RF pretreatment	R pretreatment
		$\Delta(\Delta$ OD$) \times 10^3$	
	59.0	45.0	9.5
	66.0	48.5	10.3
	55.7	43.5	12.0
	53.0	45.5	11.0
	64.0		
	58.0		
Means ± SE	59.3 ± 2.0	45.6 ± 1.0	10.7 ± 0.5
Percent	100	77	18
Relative P_{FR} units[2] for photostationary states of:			
5 %	5	4	1
20 %	20	15	4
44 %	44	34	8
88 %	88	68	16

1 Each value for phytochrome content represents the mean of 3 determinations on the same sample. Samples of 10 segments, 10 mm in length packed to 1 cm sample thickness in 6-mm diameter aluminum cuvettes.
2 Relative P_{FR} units obtained by setting the phytochrome content of dark controls to 100, so that an assumed full conversion of this to P_{FR} would give 100 units.

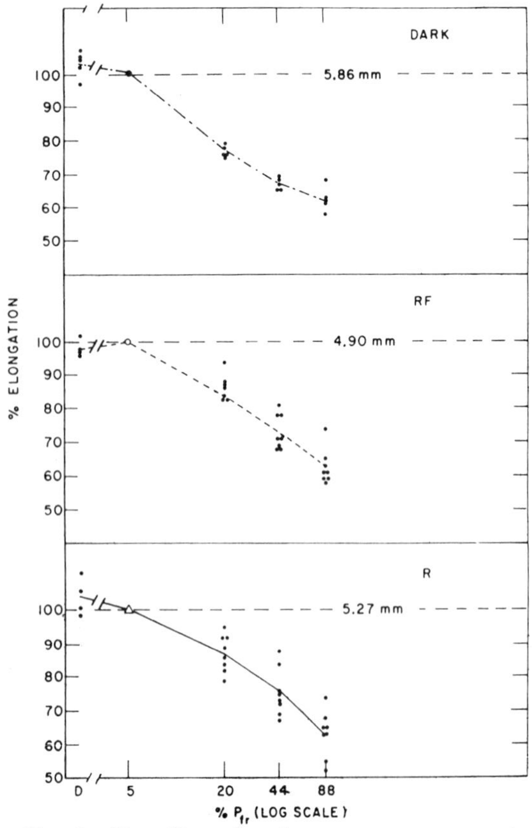

FIG. 2. The effect of various percentage conversions of phytochrome to P_{FR} on the elongation of segments from Dark, R, and RF pretreated plants. Each point represents the mean of 30 or 40 segments, initially 10 mm in length. The various % P_{FR} levels were established by 10 minutes exposure through appropriate filters (9). Elongation measured after 20 hours of growth.

In figure 3A it is clear that the segments from R and RF pretreated plants responded to various percentages of phytochrome conversion to P_{FR} in the same way, even though the RF pretreated plants contained approximately 4 times as much phyto-

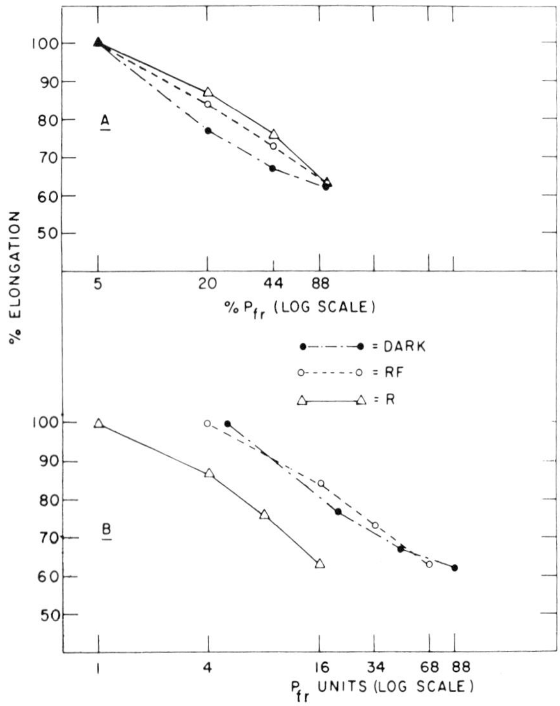

FIG. 3. The effect of various % P_{FR} levels (A) and relative P_{FR} units (B) on the elongation of segments from Dark, R, and RF pretreated plants. Each point represents the mean of the appropriate points in figure 2. For method of determining relative P_{FR} units, see text and table I.

chrome as did the R pretreated plants. However, in figure 3B the data are plotted in terms of relative P_{FR} units, as described in table I. In this plot the results for segments from R pretreated plants are now quite separate from those for the D and RF pretreated material.

Discussion

Since, for sections from any one group of pretreated plants, there appears to be a linear relationship between elongation and the log of the percentage P_{FR} established (fig 2), one might conclude that the amount of P_{FR} initially present governed the degree of growth inhibition. Such correlations have been reported previously (*e.g.,* 11, 12). However, when the growth inhibiting effects of increasing percentages of P_{FR} are compared among groups, it is clear that sections from morphologically similar (R and RF pretreated) plants responded in the same way (fig 3A), even though there was a 4-fold difference in the amount of spectrophotometrically detectable phytochrome (table I). In figure 3B the data are plotted in terms of relative P_{FR} units, so as to account for the difference in phytochrome content between R and RF pretreated plants. Here again, it appears that elongation is not regulated by the absolute amount of P_{FR} produced, but rather by the proportion of P_{FR} produced relative to the total phytochrome present. This contention is supported by the fact that the maximum inhibition observed, with an initial P_{FR} level of 88 % was the same (*ca.* 35 %) regardless of phytochrome content, and by the similarity of the slopes of the plots (fig 3A). Because of the variability in growth of pea stem segments, a large number of experiments would be required to ascertain whether or not the slight differences between segments from etiolated and de-etiolated plants are significant (fig 3A).

Apparently there is no discernible relationship between the amount of phytochrome detected spectrophotometrically and that which is physiologically active, at least in this system. These results call to mind explanations proposed for several phytochrome "paradoxes" (10). One such paradox, the *Pisum* paradox, involves pea stem segments which respond to far-red light in the absence of spectrophotometrically detectable P_{FR} (9). Another paradox, in *Zea,* involves a response to red light which is saturated by an undetectably small percentage of P_{FR}, yet is reversible by far-red filters establishing detectable amounts of P_{FR} (1, 4). The first paradox has also been observed for *Phaseolus* hypocotyl hook opening (6, 12). The concept of "bulk" and "active" phytochrome fractions has been invoked to explain these paradoxes. Bulk phytochrome represents that phytochrome which is spectrophotometrically detectable but physiologically inert, while the active fraction is present in such a small proportion of the total that no information concerning it can be obtained spectrophotometrically. To explain the present results, the only further assumption required is that the photostationary state illuminations affect both bulk and active fractions the same way.

In conclusion, an attempt to determine rigorously whether or not physiological effects are due to P_{FR} concentration, in some sense, or to its proportion of the total phytochrome present, seems to lead to the conclusion that the latter and not the former is the controlling factor. Considering these experiments alone, one would reject the widely-accepted idea that P_{FR} is the only active form of phytochrome, and propose instead a mechanism in which the P_{FR}/P_R ratio was significant because both forms are active and perhaps compete. Some such ratio mechanism, however, would still fail to explain the paradoxes discussed earlier. The bulk-and-active hypothesis remains the only one fully consistent, if uninformatively so, with all the data so far accumulated.

Acknowledgments

The authors are indebted to Rosemarie Dearing and Helen Kelly for technical assistance.

Literature Cited

1. Briggs, W. R. and H. P. Chon. 1966. The physiological versus the spectrophotometric status of phytochrome in corn coleoptiles. Plant Physiol. 41: 1159–66.
2. Briggs, W. R. and H. W. Siegelman. 1965. Distribution of phytochrome in etiolated seedlings. Plant Physiol. 40: 934–41.
3. Butler, W. L., H. C. Lane, and H. W. Siegelman. 1963. Nonphotochemical transformation in phytochrome *in vivo.* Plant Physiol. 38: 514–19.
4. Chon, H. P. and W. R. Briggs. 1966. Effect of red light on the phototropic sensitivity of corn coleoptiles. Plant Physiol. 41: 1715–24.
5. Clarkson, D. T. and W. S. Hillman. 1967. Stability of phytochrome concentration in dicotyledenous tissues under continuous far-red light. Planta 75: 286–90.
6. Edwards, J. L. and W. H. Klein. 1964. Relationship of phytochrome concentration of physiological responses. Plant Physiol. 39 (Suppl.): 1.
7. Furuya, M. and W. S. Hillman. 1964. Observations on spectrophotometrically assayable phytochrome *in vivo* in etiolated *Pisum* seedlings. Planta 63: 31–42.
8. Hillman, W. S. 1964. Phytochrome levels detectable by *in vivo* spectrophotometry in plant parts grown or stored in the light. Am. J. Botany 51: 1102–07.
9. Hillman, W. S. 1965. Phytochrome conversion by brief illumination and the subsequent elongation of etiolated *Pisum* stem segments. Physiol. Plantarum 18: 346–58.
10. Hillman, W. S. 1967. The physiology of phytochrome. Ann. Rev. Plant Physiol. 18: 301–24.
11. Hopkins, W. G. and W. S. Hillman. 1966. Relationships between phytochrome state and photosensitive growth of *Avena* coleoptile segments. Plant Physiol. 41: 593–98.
12. Klein, W. H., J. L. Edwards, and W. Shropshire, Jr. 1967. Spectrophotometric measurements of phytochrome *in vivo* and their correlation with photomorphogenic responses of *Phaseolus.* Plant Physiol. 42: 264–70.

Denaturation of Phytochrome

W. L. Butler, H. W. Siegelman, and C. O. Miller*

From the Market Quality Research Division, Agricultural Marketing Service; Crops Research Division, Agricultural Research Service, U. S. Department of Agriculture, Beltsville, Maryland; and the Botany Department, University of Indiana

Received February 3, 1964

The absorption spectra of phytochrome were found to be sensitive to the association between the chromophoric group and the protein moiety. The effects of protein denaturation were studied by measuring the absorption spectra of both forms of the photochromic pigment system. One form of phytochrome, P_{FR}, was much more susceptible to denaturation by urea, to attack by pronase and trypsin, and to sulfhydryl-reacting reagents than the other form, P_R. The results suggest that light-induced interconversions between P_R and P_{FR} involve changes of protein conformation as well as changes of the chromophoric group.

Action spectra of the effects of light on vàrious aspects of plant growth and development (Borthwick and Hendricks, 1960) revealed the presence and action of a red-far-red photochromic pigment. The pigment, called phytochrome, was detected spectrophotometrically in intact plant tissue and was extracted as a soluble protein (Butler *et al.*, 1959). Procedures for purifying the chromoprotein have been described (Siegelman and Firer, 1964). The reversible photoconversion between the two forms of phytochrome[1] is retained in the purified solutions.

The best source of phytochrome has been dark-grown seedling tissue. The phytochrome extracted from different species of seedlings, however, showed differences in ease of purification and in resistance to denaturation. In particular, the phytochrome extracted from barley seedlings was less stable than that from maize or oats, and differences in the absorption spectra of barley phytochrome suggested that the chromoprotein might be altered by the purification procedures. These observations led to the studies of denaturation reported below. The term "denaturation" is used in a very general sense to indicate an unknown alteration of the protein.

The absorption spectra of both forms of phytochrome were found to depend upon the state of the protein. The absorption bands in the long-wavelength region of the spectrum, where other proteins do not absorb, provide a specific tag for phytochrome molecules and permit the phytochrome protein to be studied in the presence of large amounts of other proteins. In the present paper the absorption spectra have been used to indicate the progress and degree of denaturation. Complete denaturation was assumed when the red and far-red absorption bands of the chromophores disappeared. This operational assay for denaturation is thus limited to those parts of the protein which influence the chromophoric group.

* Supported by a research grant (G-24052) from the National Science Foundation while on sabbatical leave.

[1] Abbreviations used in this work: P_R, red-absorbing form of phytochrome; P_{FR}, far-red-absorbing form of phytochrome; EDTA, ethylenediaminetetraacetic acid.

$$P_R \underset{\text{far-red light}}{\overset{\text{red light}}{\rightleftarrows}} P_{FR}$$

Materials and Methods

Phytochrome was purified through the Sephadex G-200 step according to the procedures described (Siegelman and Firer, 1964). Unless otherwise noted, the phytochrome was purified from dark-grown oat

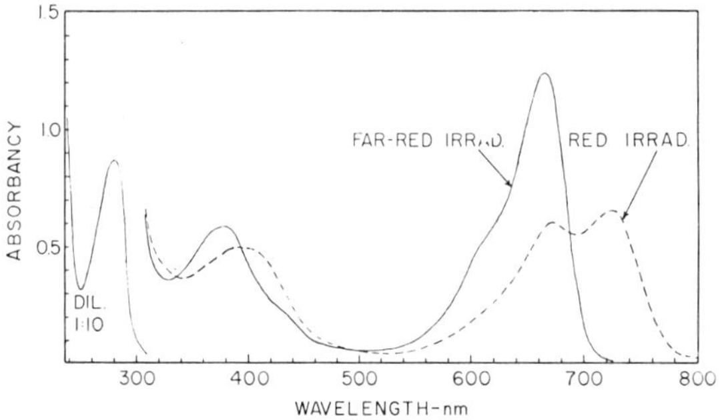

FIG. 1.—Absorption spectra of a solution of oat phytochrome, following irradiation with red and far-red light. The spectrum at wavelengths shorter than 300 mμ (nm) was measured on a solution diluted 10-fold.

seedlings and was dissolved in 0.1 M potassium phosphate buffer, pH 7.8, which contained 0.05 M 2-mercaptoethanol and 0.001 M EDTA. Most of the solutions used in these studies had a protein content of about 20 mg/ml of which 10% or less was phytochrome protein. The phytochrome solution used for the spectra in Figure 1 was taken through an additional stage of purification which consisted of absorption onto and gradient elution from a brushite column.

Effects of urea were determined by adding a 10 M solution, adjusted to pH 7.8 with potassium phosphate buffer, to an equal volume of the phytochrome solution. p-Mercuribenzoate and N-ethyl maleimide were generally made up to 5×10^{-2} M and were diluted 10-fold by addition to the phytochrome solution. Pronase (California Corp. for Biochemical Research) and trypsin (Worthington Biochemical Corp.) were made up to 1.0 mg/ml and were diluted 10-fold in the phytochrome sample. The experiments with pronase and trypsin and the experiment on dark conversion were carried out at room temperature. In all other experiments the phytochrome solution was kept near 0°.

Allophycocyanin and phycocyanin were purified from *Smithora naidum* according to the method of Haxo *et al.* (1955).

With the exception of Figures 1 and 4, spectrophotometric measurements were made with a single-beam recording spectrophotometer similar to one previously described (Norris and Butler, 1961). (The spectra in Figs. 1 and 4 were measured with a Cary recording spectrophotometer.) The sample was contained in a vertical cylindrical cell which could be surrounded by ice. The vertical optical path permitted the addition of reagents and dilution of the sample without changing the number of pigment molecules in the measuring beam.

Actinic sources of red and far-red light were available for irradiating the sample in the spectrophotometer. These sources consisted of a tungsten lamp with either a 650-mμ interference filter or a far-red cut-on filter which transmitted wavelengths longer than 730 mμ. The sample was irradiated with these sources for about 1 minute, which was more than sufficient to drive the $P_R \rightleftarrows P_{FR}$ photoconversion to completion in either direction. The sequence of the spectral measurements is indicated in the figures if it is pertinent to the design of the experiment.

RESULTS

Figure 1 shows the absorption spectra of a solution of phytochrome following irradiation with the red and far-red actinic sources. The absolute absorption spectra indicate that the absorption maxima of the two forms are at 665 and 725 mμ (nm) rather than at the previously reported values of 660 and 730 mμ (Hendricks *et al.*, 1962) which were obtained from difference spectra. The far-red actinic light converts practically all of the phytochrome to P_R, but the red actinic light does not appear to convert all of the P_R to P_{FR}. It will be shown that a certain amount of P_R remains at the photostationary state in red light due to the absorption of P_{FR} in the red region. Many of the absorption spectra (*vide infra*) also have shoulders at about 630 mμ which vary in magnitude among different preparations. These absorption bands probably are due to inactive protochlorophyll which is present in etiolated tissue.

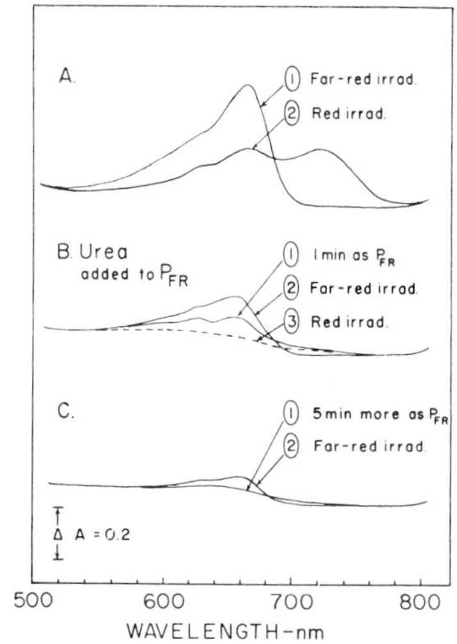

FIG. 2.—Effect of 5 M urea. (A) Absorption spectra of original solution; (B) 1 minute after adding urea to P_{FR}; (C) phytochrome kept in urea 5 minutes more as P_{FR}.

*P and P**.—The ratio of the maximal reversible absorbancy change in the far-red region to that in the red region, $\Delta A_{FR}/\Delta A_R$, in the phytochrome solution purified from oats (Fig. 1) is 1.0. In earlier work, with phytochrome purified from dark-grown barley seedlings, we noted that this ratio was 1.0 *in vivo* and at early stages of purification but tended to decrease at higher stages of purification to about 0.7–0.5. In some cases in which the phytochrome appeared to be mildly denatured, as indicated by a decrease of solubility, the ratio for both the soluble and insoluble phytochrome was less than 0.2 (Hendricks *et al.*, 1962). Difference spectra of the preparations with a low ratio showed that the difference band in the far-red region shifted to shorter wavelength as the ratio decreased. The variation of the $\Delta A_{FR}/\Delta A_R$ ratio at various stages of purification of barley phytochrome suggested that an alteration of the phytochrome protein resulted in an altered phytochrome, P*, which was still photoreversible. The absorption spectrum of P_R* was similar to P_R but P_{FR}* had a shorter wavelength-absorption maximum and a lower extinction coefficient than P_{FR}. Phytochrome may be denatured to various degrees. P* is not meant as definite molecular species with a fixed $\Delta A_{FR}/\Delta A_R$ ratio, but rather the sum of molecules in various degrees of denaturation. As denaturation proceeds, the absorbancy of the far-red absorption band decreases, finally going to zero, in which case the red-absorbing form cannot be regenerated by far-red light.

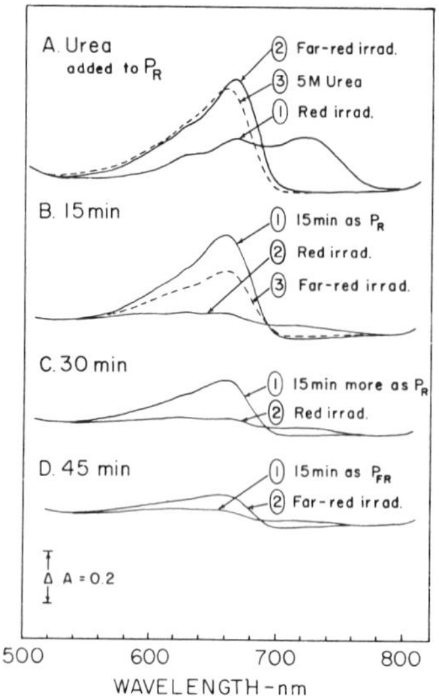

FIG. 3.—Effect of 5 M urea. (A) Curves 1 and 2, original solution; curve 3, 1 minute after adding urea to P_R; (B) phytochrome kept in dark 15 minutes more as P_R; (C) phytochrome of (3-B) kept in dark 15 minutes more as P_R; (D) phytochrome kept in dark an additional 15 minutes as P_{FR}.

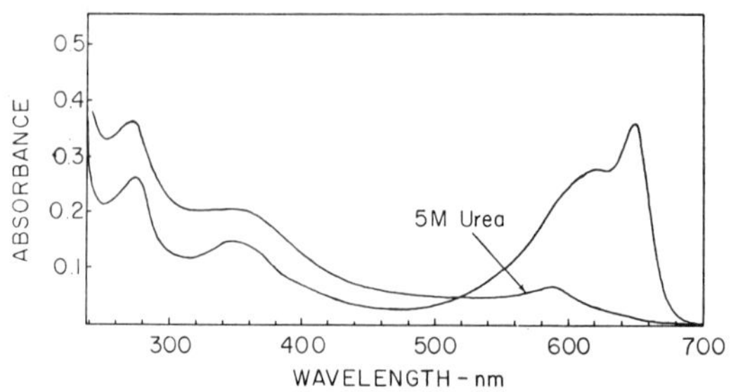

FIG. 4.—Absorption spectra of allophycocyanin before and after addition of urea.

Transformation of P to P by Urea*.—The native phytochrome can be converted to the P* system by denaturing the protein with urea. Figure 2 shows the effect of 5 M urea on a phytochrome solution that had a $\Delta A_{FR}/\Delta A_R$ ratio between 0.9 and 1.0. The addition of urea to P_{FR} caused an immediate (within 1 minute) decrease in the P_{FR}-absorption band (Fig. 2B, curve 1). The P_R, which was present at the photostationary state in red light before the addition of urea, remained. The phytochrome was still photoreversible. Irradiating the sample with far-red and red light produced absorbancy changes with a $\Delta A_{FR}/\Delta A_R$ ratio between 0.1 and 0.2, which is typical of conversions between P_R^* and P_{FR}^*. If the sample remained as P_{FR}^* in the presence of urea for 5 minutes (Fig. 2,C), the absorption bands and the reversible absorbancy changes decreased further.

Addition of urea to P_R does not cause the rapid deterioration of the absorption bands that occurred in Figure 2 where the urea was added to P_{FR}. The addition of urea to P_R (Fig. 3) caused an immediate shift of the absorption maximum from 665 mμ to 660 mμ with only a slight loss of absorption. The absorption spectrum did not undergo further change during the next 15 minutes (Fig. 3 B, curve 1). In other experiments the 660-mμ band was stable in urea for over an hour.

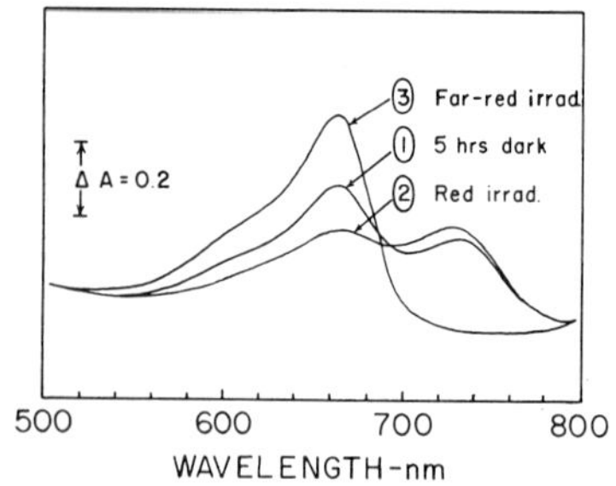

FIG. 5.—Absorption spectra of barley phytochrome. Curve 1, absorption spectrum after standing at room temperature in the dark for 5 hours after an irradiation with red light. Curve 2, absorption spectrum after subsequent irradiation with red light. Curve 3, absorption spectrum after irradiation with far-red light.

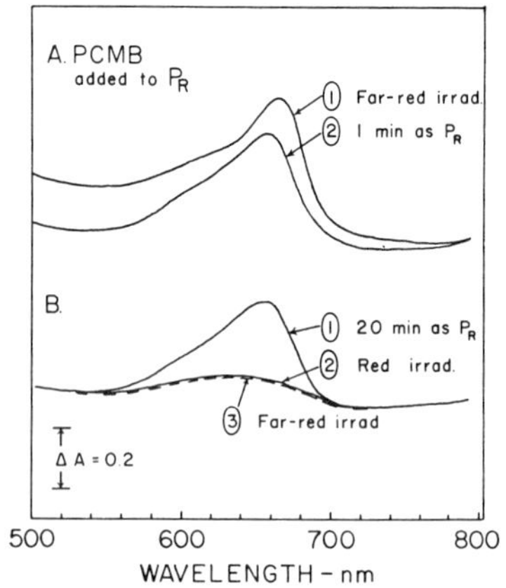

FIG. 6.—Effect of 5 × 10^{-3} M *p*-mercuribenzoate (PCMB). (A) *p*-Mercuribenzoate added to P_R; (B) phytochrome kept in dark for 20 minutes as P_R. The phytochrome solution did not contain mercaptoethanol.

Irradiating the sample with red light, however, gave a typical P_R^*-to-P_{FR}^* conversion. The urea denatured P_R to P_R^*, as indicated by the 665- to 660-mμ shift of the absorption maximum and the transformation to the low-extinction P_{FR}^* on irradiation with red light, but P_R^* was stable in urea. Irradiation of P_{FR}^* with far-red light did not produce as much P_R^* as was originally present, but the P_R^* that was formed was still stable in 5 M urea. (Compare curve 3, part B, with curve 1, part C, of Fig. 3, measured 15 minutes later.) The absorption bands were less, however, (Fig. 3D) after the solution stood for 15 minutes as P_{FR}^*.

The action of 5 M urea on P_R was reversed by removing the urea. Urea was added to a solution of phytochrome ($\Delta A_{FR}/\Delta A_R$ = 0.9) in the P_R form. The shift of the absorption maximum from 665 to 660 mμ indicated that P_R was denatured to P_R.* After 10 minutes in the P_R^* form the urea was rapidly removed by passing the solution through Sephadex G-25, all operations being carried out in darkness or with a dim green safe-light. The absorption spectra of the collected fractions showed that P_R was quantitatively regained from the column. The phytochrome was in the 665-mμ-absorbing P_R form which, on irradiation with red light, converted to P_{FR} with a $\Delta A_{FR}/\Delta A_R$ ratio of 0.9. The results from a similar experiment in which urea was added to and removed from P_{FR} showed that approximately 30% of the phytochrome,

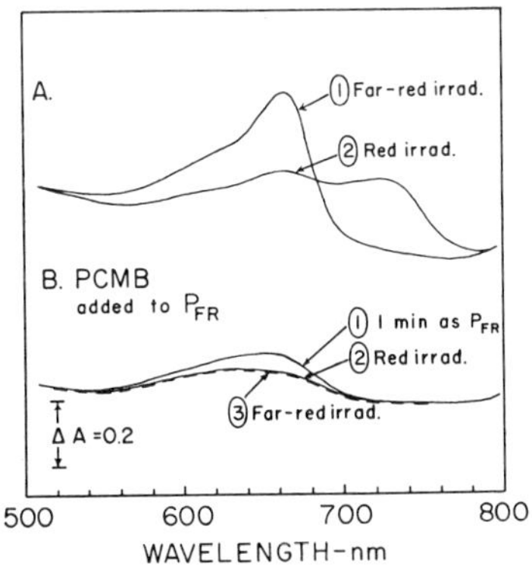

FIG. 7.—Effect of 5×10^{-3} M *p*-mercuribenzoate (PCMB). (A) Original solution; (B) *p*-Mercuribenzoate added to P_{FR}. The phytochrome solution did not contain mercaptoethanol.

with a $\Delta A_{FR}/\Delta A_R$ ratio close to unity, was regained from the column.

Denaturation of Allophycocyanin by Urea.—The action of urea on the absorption spectrum of allophycocyanin was examined because of the similarity between the absorption spectra of allophycocyanin and P_R. Ó hEocha (1963) and Jones and Fujimori (1961) showed that the absorption bands of C-phycocyanin, allophycocyanin, and phycoerythrin were decreased by denaturation. Our results with allophycocyanin, which are similar to those reported by Ó hEocha, are shown in Figure 4. Addition of urea to a concentration of 5 M completely abolished the main 650-mμ absorption band within a few minutes. C-Phycocyanin behaved in a similar manner. We were not able to restore the chromophore absorption by rapidly diluting the urea-treated sample 10-fold or by removing the urea by gel filtration. It is somewhat surprising that the absorption spectra of these chromophores, which are noted for their stability (the chromophoric groups are split from the protein by hydrolysis in 12 N HCl), are readily destroyed by urea, presumably through denaturation of the protein.

Dark Transformation of P_{FR} to P_R*.*—Phytochrome solutions were examined for dark conversions. Bonner (1962), working with solutions of phytochrome purified from dark-grown pea seedlings, reported that the far-red-absorbing form reverted to the red-absorbing form in the dark. A similar transformation was shown with barley preparations which appeared to be partially denatured (Hendricks *et al.*, 1962). The highly purified solutions of phytochrome from oats, such as that of Figure 1, exhibited no transformation in the dark when freshly prepared. However, after aging or after mild denaturation, the oat phytochrome often showed some degree of dark conversion. The ratio of the reversible absorbancy changes, $\Delta A_{FR}/\Delta A_R$, for the phytochrome which transforms in the dark indicated that the transformation was from P_{FR}* to P_R*.

The dark conversion *in vitro*, obtained at room temperature with a phytochrome sample purified from maize, is shown in Figure 5. Curve 1 is the absorption spectrum of the solution 5 hours after an irradiation with red light. The dark period was extended to 5 hours to ensure that the dark reaction had gone to completion. Most of the transformation occurred in the first hour. Irradiating the sample with red light after the dark period (curve 2, Fig. 5) restored the absorption spectrum to that of the initial red-irradiated sample and showed that the phytochrome which trans-

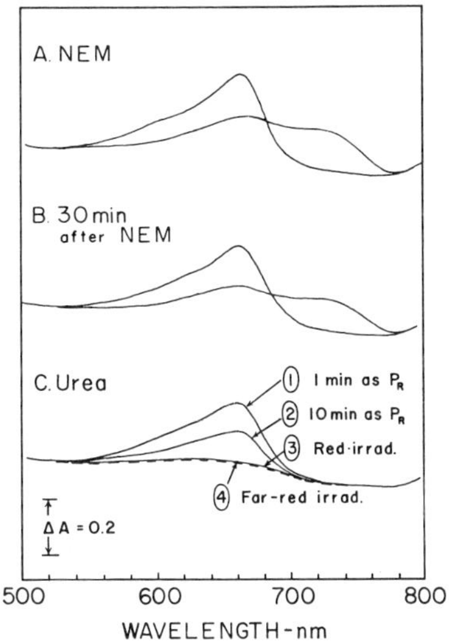

FIG. 8.—Effect of 5×10^{-3} M *N*-ethylmaleimide (NEM). (A) Original solution; (B) 30 minutes after addition of *N*-ethylmaleimide; phytochrome kept 20 minutes as P_R and 10 minutes as P_{FR}; (C) curve 1, urea added to P_R (5 M); curve 2, kept 10 minutes in dark as P_R*; curve 3, irradiated with red light; and curve 4, irradiated with far-red light. The phytochrome solution did not contain mercaptoethanol.

formed in the dark had a $\Delta A_{FR}/\Delta A_R$ ratio of about 0.2. Apparently the solution, which had a $\Delta A_{FR}/\Delta A_R$ ratio of 0.8–0.9 for the photoreversible absorbancy changes, was a mixture of P and P*. There could also have been some denaturation of P to P* during the prolonged dark period at room temperature. P_{FR}* reverted to P_R* during the dark period, but P_{FR} remained.

Effect of Sulfhydryl-reacting Reagents.—The influence of phytochrome SH groups on the absorption bands was examined by adding *p*-mercuribenzoate, a reagent which reacts with SH groups, to phytochrome solutions which were free of mercaptoethanol. The addition of *p*-mercuribenzoate to a concentration of 5×10^{-3} M (Fig. 6) had little effect on the red-absorption band other than to shift the absorption maximum to a slightly shorter wavelength. This change is similar to the addition of urea to P_R and may represent a denaturation of P_R to P_R*. The nonspecific change in the baseline slope of the spectra in Figure 6A is probably caused by a change in the scattering characteristics of the particular protein solution. A similar change of slope is also often observed on the addition of urea to P_R (though not in Fig. 3). After the initial spectral shift, the red-absorption band was not altered by standing for 20 minutes in 5×10^{-3} M *p*-mercuribenzoate (curve 1, Fig. 6 B). On irradiation with red light, however, the red-absorption band was lost without the concomitant formation of any absorption band in the far-red. An immediate irradiation with far-red light did not re-form any of the red-absorption band. The *p*-mercuribenzoate attacked the P_{FR} chromophore as soon as it was formed. Thus the integrity of a specific SH group (or groups) is required for the characteristic absorption band of the P_{FR} chromophore. In higher concentrations *p*-mercuribenzoate can attack the chromophore of P_R or P_R*, although relatively slowly. In one experiment in which *p*-mercuribenzoate was added to P_R to a concentration of 10^{-2} M, the red-absorption band gradually decreased to zero over a period of 30 minutes.

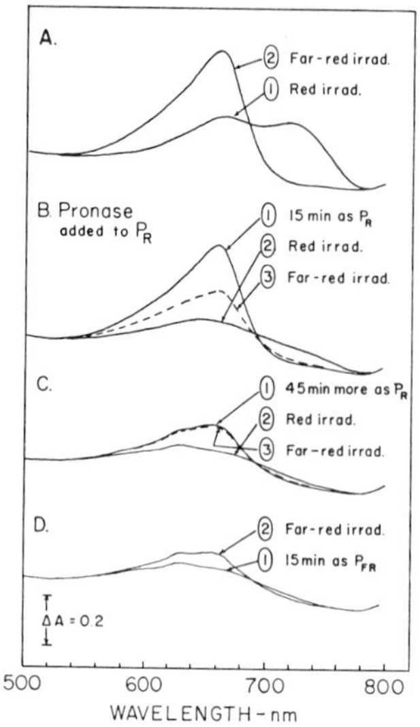

FIG. 9.—Effect of 0.1 mg/ml pronase at room temperature. (A) Original solution; (B) 15 minutes after adding pronase to P_R; (C) phytochrome of (3-B) kept 45 minutes as P_R*; (D) phytochrome kept additional 15 minutes as P_{FR}.

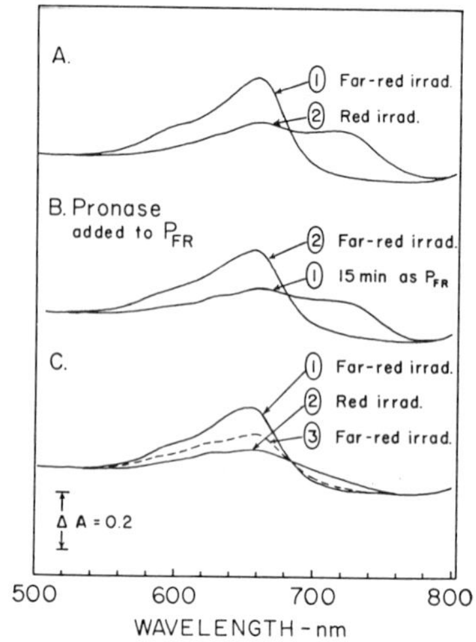

FIG. 10.—Effect of 0.1 mg/ml pronase at room temperature. (A) Original solution; (B) 15 minutes after pronase was added to P_{FR}; (C) immediately following (B).

When *p*-mercuribenzoate was added to solution which had been irradiated with red light (Fig. 7), the absorption band of P_{FR} was immediately lost. However, the P_R which was present at the photostationary state in curve 2 of Figure 7A remained. Irradiating the sample with red light (Fig. 7B) converted the remaining P_R* and completed the destruction of the photoreversible pigment. The spectra indicate that 10–20% of the phytochrome at the photostationary state in red light is P_R.

The amount of *p*-mercuribenzoate required to effect these changes varied to some extent with different phytochrome solutions, in part because of varying amounts of nonphytochrome protein which also binds the *p*-mercuribenzoate. In most of the solutions used in these studies, the phytochrome protein was in the order of 10% of the total protein. The susceptibility of the phytochrome protein to the denaturation of P to P* also appears to play a role. We have had a phytochrome solution in which 10^{-2} M *p*-mercuribenzoate had no effect on the absorption bands of either P_R or P_{FR}. With this solution the *p*-mercuribenzoate was not able to denature the phytochrome to the P* system.

Another reagent which reacts with SH groups, *N*-ethyl maleimide, had little effect on phytochrome when present at a concentration of 5×10^{-3} M. The photoreversible absorbancy changes decreased only about 25% after standing in 5×10^{-3} M *N*-ethyl maleimide for 30 minutes (Fig. 8B), during which time the solution was irradiated with both red and far-red light. After addition of urea, however, the *N*-ethyl maleimide acted like *p*-mercuribenzoate. In Figure 8C the absorption band of P_R*, in the presence of 5×10^{-3} M *N*-ethyl maleimide and 5 M urea, decreased about 50% in 10 minutes, which was similar to the action of 10^{-2} M *p*-mercuribenzoate on P_R. On irradiating the solution with red light, all absorption was lost immediately and no red-absorption band could be re-formed by a subsequent irradiation with far-red light. In another experiment, addition of *N*-ethyl maleimide (5×10^{-3} M) to P_{FR}*, after the addition of urea (5 M), resulted in the immediate loss of the P_{FR}* absorption. Thus, the SH group or groups required for the absorption bands of phytochrome are susceptible to attack when phytochrome is in the partially denatured P_{FR}* form.

Urea attacks the chromophore of P_{FR}* more rapidly in the absence of mercaptoethanol, although not so rapidly as *N*-ethyl maleimide plus urea. We noted previously with partially denatured phytochrome solutions from barley that P_{FR}* was much more stable than P_R* and that mercaptoethanol exerted a protective effect on P_{FR}*. Solutions of native phytochrome are stable in the absence of mercaptoethanol. These observations also suggest that specific SH groups are much more susceptible in the P_{FR}* form than in the P_R*, P_R, or P_{FR} forms.

Effects of Proteolytic Enzymes.—The effects of pronase and trypsin on phytochrome were studied at room temperature at a concentration of 0.1 mg/ml. This concentration gave a measurable effect in a reasonably short period of time. Fifteen minutes after adding pronase to P_R, the red-absorption band showed little change except for a slight shift to shorter wavelength. Irradiation with red light, however, produced the low-extinction P_{FR}* (curve 2, Fig. 9 B). Thus, as in the case of urea, pronase altered the protein of P_R to give P_R*, which immediately transformed to P_{FR}* on irradiation. Also, the chromophore was relatively stable in the P_R* form but deteriorated with time in the P_{FR}* form.

When pronase was added to P_{FR} the expected conversion to P_{FR}* was not observed. Figure 10, part B, curve 1, shows that the absorption spectrum of the red-irradiated solution changed very little in 15 minutes after the addition of pronase to P_{FR}. Irradiation of the solution with far-red light after 15 minutes produced the typical P_{FR}-to-P_R conversion. An immediate reirradiation with red light (Fig. 10C), however, produced the low-extinction P_{FR}*. Thus, the formation of P_{FR}* required not only the proteolytic enzyme, but also the photoconversion of P_R or P_R*.

The action of trypsin was essentially the same as that of pronase. The similarity between the effects of the proteolytic enzymes and of urea on the chromophore will be examined in the discussion.

Denaturation by Acid and Base.—The chromophores of P_R and P_{FR} are stable in the *p*H range between 5 and 9. The protein in the solution generally precipitates below pH 6, but the phytochrome is fully photoreversible. At pH values below 5 and above 9 both forms begin to lose absorbancy so that the $\Delta A_{FR}/\Delta A_R$ ratio remains about the same. In one experiment 80% of the photoreversible absorbancy changes were lost at pH 4.4 but, on bringing the pH back to 7.1, the reversible absorbancy changes were restored to 85% of their original value. If the acid treatment is more drastic, the loss of absorbancy is irreversible. In another experiment the photoreversible absorbancy changes were irreversibly lost at pH 10.8.

Precipitation of phytochrome by 5% trichloroacetic acid gives a slightly blue precipitate which has a broad flat absorption band through the 580- to 680-mμ region similar to that shown by the denatured solutions. The absorption spectrum of this precipitate is not affected by light.

Discussion

The purpose of the denaturation studies was to examine some molecular properties of phytochrome. The results with urea show that the chromophores of P_R and P_{FR} have different sensitivities to denaturation. P_R is denatured to P_R^*, but no further loss of chromophore occurs so long as the phytochrome remains as P_R^*. If P_R^* is converted to P_{FR}^* by light, or if urea is added to P_{FR} to form P_{FR}^*, the chromophore deteriorates with time. The deterioration can be prevented by returning P_{FR}^* to P_R^*. The different susceptibilities of P_R and P_{FR} to denaturation suggest differences of protein conformation.

The effect of proteolytic enzymes on the chromophore of P_R is similar to the effect of urea on P_R. The absorption maximum shifts to a slightly lower wavelength, indicating a change of P_R to P_R^*, but no further change of the chromophore is observed as long as it remains P_R^*. On irradiation of the chromophore with red light, P_{FR}^* is formed. P_{FR}^* deteriorates with time unless it is returned to the stable P_R^*. The greater resistance of P_R^* to the effects of the proteolytic enzymes also suggests a difference of protein conformation. The proteolytic enzymes and urea appear to have similar effects on the absorption spectra of chromophores. The chromophore change probably reflects a general loosening of the protein structure.

The effect of the proteolytic enzymes on P_{FR}, however, is somewhat different from that of urea. The addition of urea to P_{FR} rapidly transformed it to P_{FR}^*, whereas the action of pronase had very little effect on the chromophore of P_{FR}. Pronase had acted on the protein of P_{FR}, however, because the red-far-red irradiation sequence immediately produced the low-extinction P_{FR}^*. Apparently the peptide bonds broken by pronase had no direct influence on the chromophore of P_{FR}, but the protein conformational changes involved in the photoconversion were sufficient to loosen the protein structure in the region of the chromophore after the bonds had been broken.

Other cases are known where the state of a prosthetic group affects the conformation of the attached protein. Fridovich (1962) presented evidence that the protein conformation of ferricytochrome c was different from that of ferrocytochrome c. Nozaki *et al.* (1958) have also reported that ferrocytochrome c was more resistant to proteolytic digestion than was ferricytochrome c.

The effect of *p*-mercuribenzoate indicates that a specific SH group (or groups) is intimately involved with the chromophore in the P_{FR} form. The observation that urea attacks the chromophore of P_{FR}^* more rapidly in the absence of mercaptoethanol also suggests the involvement of SH groups. Urea probably causes an unfolding of the protein. The experiments with *N*-ethyl maleimide showed that this SH reagent could not readily attack the specific SH groups in the native P_{FR} but could if P_{FR} were partially denatured to P_{FR}^* by urea, presumably because denaturation opened the protein. The opening of the protein by urea is accompanied by a decrease in the far-red absorption band. As denaturation proceeds the absorption band of P_{FR}^* progressively decreases, presumably as the protein progressively opens or unfolds. When the absorbancy of P_{FR}^* goes to zero the ability to regenerate P_R^* is lost.

The molecular properties of phytochrome, which we observed in the denaturation studies, may relate to the physiological action of phytochrome. It has been postulated on physiological grounds that P_{FR} is the active form of an enzyme (Borthwick and Hendricks, 1960), while P_R is inactive. The apparent change of protein conformation which accompanies the interconversions between P_R and P_{FR} is attractive for such a hypothesis. Furthermore, our attention is drawn to the specific SH group or groups which are involved with the chromophore of P_{FR}. The SH groups in some way determine the absorbancy of the chromophore and the chromophore somehow controls the accessibility of the SH groups. The SH groups are much more susceptible to attack in the presumably active P_{FR} form than in the inactive P_R form. Perhaps the substrate for phytochrome must complex with these SH groups, and the act of enzyme activation involves making the SH groups available. If a substrate-SH complex did occur we would expect a change in the absorption spectrum because of the sensitivity of the chromophore to the SH groups. This could afford an experimental means of looking for the substrate. The absorption band of the complex might vanish as it does when the SH groups complex with *p*-mercuribenzoate. This would have the physiological implication that phytochrome would not be sensitive to light while acting on the substrate.

The phytochrome system has some similarities to rhodopsin. Light-induced changes of rhodopsin are not normally reversible, but at temperatures between $-196°$ and $-140°$ the pigment is reversibly photochromic (Yoshizawa and Wald, 1963): rhodopsin (λ_{max} = 498 mμ) ⇄ pre-lumirhodopsin (λ_{max} = 518 mμ). Light of 440 mμ favors the forward reaction while light of 600 mμ drives the reaction to the left. These changes apparently involve the isomerization of 11-*cis*-retenene to the all-*trans* form with little change of the protein. Transformations of phytochrome do not occur at these temperatures, perhaps because a protein conformational change which is obligatory for the transformation is prevented by the rigidity of the medium. At higher temperatures, pre-lumirhodopsin goes over to lumirhodopsin and further to metarhodopsin, presumably as the protein opens. In the transformation of rhodopsin to metarhodopsin, two additional SH groups become exposed (Wald and Brown, 1953). Wald *et al.* (1963) have suggested that rhodopsin may be a proenzyme which is transformed by light to the active enzyme, metarhodopsin.

Changes of protein conformation resulting from changes of a prosthetic group appear to be the rule rather than the exception. Phytochrome, once it is available in sufficient purity, should be an ideal protein for such studies because of its property of photoreversibility.

References

Bonner, B. A. (1962), *Plant Physiol.* Suppl. *37*, xxvii.
Borthwick, H. A., and Hendricks, S. B. (1960), *Science 132*, 1223.

Butler, W. L., Norris, K. H., Siegelman, H. W., and Hendricks, S. B. (1959), *Proc. Natl. Acad. Sci. U. S. 45*, 1703.

Fridovich, I. (1962), *J. Biol. Chem. 237*, 584.

Haxo, F. T., Ó hEocha, C., and Norris, P. (1955), *Arch. Biochem. Biophys. 54*, 162.

Hendricks, S. B., Butler, W. L., and Siegelman, H. W. (1962), *J. Phys. Chem. 66*, 2550.

Jones, R. F., and Fujimori, E. (1961), *Physiol. Plantarum 14*, 253.

Norris, K. H., and Butler, W. L. (1961), *IRE Trans. Bio-Med. Electron. 8*, 153.

Nozaki, M., Yomanaka, T., Horro, T., and Okunuki, K. (1958), *J. Biochem. 45*, 815.

Ó hEocha, C. (1963), *Biochemistry 2*, 375.

Siegelman, H. W., and Firer, E. (1964), *Biochemistry 3*, 418.

Wald, W., and Brown, P. K. (1953), *J. Gen. Physiol. 35*, 797.

Wald, W., Brown, P. K., and Gibbons, I. R. (1963), *J. Opt. Soc. Am. 53*, 20.

Yoshizawa, T., and Wald, W. (1963), *Nature 197*, 1279.

Second Positive Phototropism in the Avena Coleoptile

Marylee Everett[1, 2] and Kenneth V. Thimann
Division of Natural Sciences, University of California, Santa Cruz, California

Received June 7, 1968.

Abstract. **A method has been developed whereby the second positive phototropism can be observed separately from the first positive and negative phototropic responses which also occur in oat coleoptiles. Although the second positive phototropic response has often been referred to as the base response, photoreception for it is shown to occur mainly in the apical 3 mm of the coleoptile. The Bunsen-Roscoe reciprocity law, so typical of first positive phototropism, does not apply to the second positive responses, and the amount of curvature increases linearly with the duration of the stimulus. However, although this linear proportionality between stimulus duration and response is the major factor determining response at all intensities tested, the intensity of the stimulus does influence the response somewhat. The action spectrum for the response shows no activity above 510 nm and has peaks at 375 and 450 nm. In all but one particular it closely resembles the action spectrum for the first positive phototropism, and it is concluded that the same, or similar, pigments may well be the photoreceptors for both types of response. The identity of this blue light absorbing pigment is not known.**

Despite the availability of action spectra for the first positive phototropism in oats (15, 17), the photoreceptor for this response has not been definitely characterized (9, 17). Furthermore, the early experiments on phototropism demonstrated that in addition to first positive phototropic curvatures, oat coleoptiles also exhibit negative and second positive types of response (1, 4). From a study of the dose-response curves for phototropism in *Avena* and of theoretical models for these data, Zimmerman and Briggs (19, 20) suggested that *Avena* coleoptiles possess 3 different photoreceptor systems for the 3 types of phototropic response—first positive, negative, and second positive. If this be the case, then the photoreceptive pigment for the second positive type of curvature would be different from that for the first positive phototropism and might be identified from an action spectrum for the second positive response. For this reason, and in order to characterize the little-known second positive response more clearly, the present experiments on second positive phototropism in coleoptiles of *Avena* were undertaken.

Although the second positive phototropic response, which occurs with large stimulus energies or long stimulus durations, has not been extensively studied, there are strong indications that the Bunsen-Roscoe reciprocity law, which holds for first positive phototropism, is not applicable to it (2, 5, 17). This law states that the response depends only on the total energy of the light dose and is not influenced by the intensity or duration of the stimulus. Thimann and Curry (17) suggested that curvatures of the second positive phototropic type are determined mainly by the duration of the stimulus, and Briggs (2) provided strong evidence for this hypothesis by showing that second positive responses of oat coleoptiles to stimuli of equal energy, delivered with various intensities of light, increased with increasing stimulus duration. Furthermore, when Zimmerman and Briggs (20) subtracted the responses predicted by their theoretical models for first positive and negative phototropism from their phototropic dose-response curves, the remaining responses, assumed to be of the second positive type, increased linearly with stimulus duration and were not affected by a 10-fold variation in intensity. Unfortunately the overlap of first positive responses by second positive ones when low intensity light is used (1, 17, 19) has prevented a direct study of the characteristics of the second positive phototropic response. For this reason it seemed essential to develop a method of isolating second positive curvatures, and thus to study them alone. In this paper such a method is described and used to elucidate both some general properties of the system and its action spectrum.

Materials and Methods

Seeds of *Avena sativa* var. Victory, obtained from the U.S.D.A. Branch Experiment Station at Aberdeen, Idaho, were husked and soaked for 2 hr in distilled water. The soaked seeds were planted individually on 1.5 % agar slanted in 1.5 ml vials, which were placed in deep petri dishes, covered to maintain a high humidity. The dishes were placed

---

[1] Supported by a pre-doctoral fellowship from the NSF.
[2] Present address: Biological Laboratories, Harvard University, Cambridge, Massachusetts.

in a growth room maintained at 25° ± 1° and exposed to 22 to 24 hr of red light at a distance of about 50 cm from a 25 watt ruby red bulb, where the intensity was about 1.5 × 10^4 ergs cm^{-2} sec^{-1}. This treatment completely inhibited the growth of the mesocotyl. The plants were then placed in a light-tight cabinet in the growth room. At 70 to 72 hr from the time of soaking, when the coleoptiles were 2.5 to 3.0 cm tall, they were exposed to 1 to 2 hr of the same red light to bring the sensitivity changes caused by red light to a maximum (5, 19). Exposure to the stimulus followed this treatment.

The seedlings were selected for straightness and positioned (in their vials) in racks so that the narrow side of the coleoptile would be exposed to the stimulus beam. Ten plants at a time were exposed to each stimulus, the entire coleoptile being illuminated unless otherwise stated. Monochromatic light, of band width 5 nm, was obtained from a 150 watt xenon lamp and a Bausch and Lomb high-intensity grating monochromator. The monochromator, located in a part of the growth room partitioned from the rest by a black curtain, was encased in a light-tight box provided with a shutter. Intensity was usually varied by changing the distance of the plants from the exit slit of the monochromator, but in a few cases neutral density filters were employed. For stimulus times of less than 1 sec the shutter mechanism was used, while for longer stimuli the shutter was operated by hand and timed with a stop watch. If only the bases of the plants were to be exposed, aluminum foil caps of the appropriate size were placed on the coleoptiles during stimulation.

Immediately after exposure to the stimulus, the vials containing the plants were rotated 90° around a vertical axis and placed in position before a film holder containing Kodabromide paper. A shadowgraph was taken with phototropically inactive green light; the plants were left in position, and a second shadowgraph was taken 100 min after the time of the start of the stimulus. The double shadowgraphs were measured twice with a goniometer; the 2 sets of values for each plant were compared and averaged, and if a discrepancy of more than 3° occurred in the 2 measurements of each plant, it was corrected by a third measurement. Standard errors for the means of these measurements were no more than 2°, and rarely more than 1.5°. Each value in the plotted data rests on measurements of not less than 50 plants.

A Reeder vacuum thermocouple connected to a Keithley microvoltammeter was used to measure light intensities. Absolute values of intensity were determined with an 8-junction bismuth-silver Eppley thermopile which had been calibrated against a standard lamp by the manufacturer. Variation of intensity with distance from the monochromator was determined with an Evans electroselenium microphotometer used with a neutral density filter. Since the readings on the photometer indicated that the intensity of the light decreased as the square of the distance from the monochromator (except much closer to the monochromator than any plants were placed), intensities in the range where this relation was valid were calculated using this assumption.

Results

Isolation of the Response. In preliminary experiments the dose-response curves for 5 intensities of 475 nm light were determined. The results are plotted in figure 1 and are in agreement with the data of Zimmerman and Briggs (19) on the shape of the dose-response curves at a series of intensities of stimulus light. Negative responses were not observed because the stimulus intensities reached with the narrow slit width were not sufficiently high. Since the first positive responses depend only on the total stimulus energy, the 5 curves coincide in the region of the lower energy doses. In the higher energy region where the second positive responses occur, the curves do not overlap, and the second positive curvature is initiated at increasingly lower energies as the stimulus intensity decreases. The explanation of the onset of second positive curvature at progressively decreasing energy levels is given by the theory that the second positive responses depend primarily on the stimulus *duration,* irrespective of the total energy contained in the stimulus. For instance, the time required to deliver the dose of 1000 ergs/cm^2 decreases from over 1600 sec at the lowest intensity to 21 sec at the highest intensity. With long stimulus durations the second positive responses overlap those of the first positive, and the difficulty of studying responses observed in this way is clearly illustrated in figure 1.

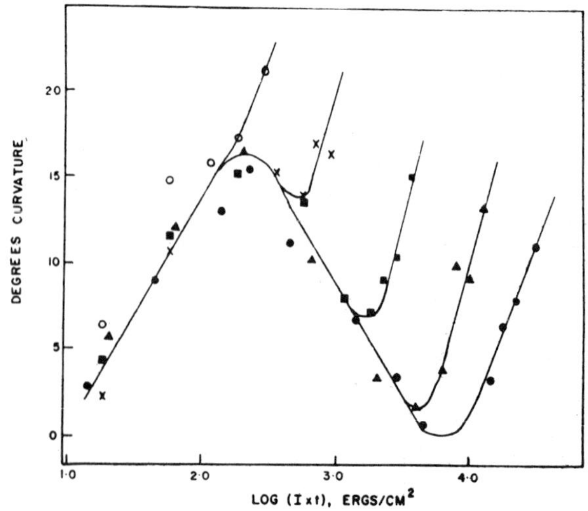

FIG. 1. Dose-response curves for phototropism in oat coleoptiles at 5 different intensities of 475 nm light. ●) 48 ergs cm^{-2} sec^{-1} (80 I in fig 3), ▲) 24 ergs cm^{-2} sec^{-1} (40 I in fig 3), ■) 6 ergs cm^{-2} sec^{-1} (10 I in fig 3), ×) 1.5 ergs cm^{-2} sec^{-1} (2.5 I in fig 3), ○) 0.6 ergs cm^{-2} sec^{-1} (I in fig 3).

There is 1 situation in which the second positive responses can be isolated, namely when high intensity light is used for the stimulus. Such a situation is illustrated by the responses of the coleoptiles to the highest intensity of light used in figure 1. Here there is an "indifferent" responsive region between first and second positive curvatures, the plants showing no response to a stimulus of this energy. First positive responses occur only with stimulus energies lower than those in this indifferent zone; and because the intensity being used is high, the stimulus durations are not long enough to cause significant second positive curvatures. It is possible, however, to use this system to obtain isolated second positive responses by following a high intensity dose with one of a lower intensity. Thus, if a high intensity light dose, causing an indifferent response, is given unilaterally to the experimental plants and is immediately followed by a second unilateral stimulus of any intensity, lasting several hundred sec, the plants now respond with a positive curvature. Such curvatures take place when the sum of the energies of the 2 stimuli is greater than the energies that cause first positive or negative curvatures. Evidence will be presented below that these curvatures have the characteristics of second positive responses as described above. The initial stimulus used in most of the experiments was a dose of 4800 ergs/cm² of 475 nm light, delivered in 100 sec. The second stimulus was usually given for 200 sec or more. In the action spectrum experiments, the initial dose was changed to 5700 ergs/cm², delivered in 100 sec, at 470 nm.

Characteristics of the Response. Authors have occasionally used the terminology "tip response" and "base response" in referring to first and second positive curvatures, respectively, to suggest that photoreception and response for the 2 types of phototropism occur in different regions of the coleoptile (18). However, in the experiments of Arisz (1) and of Zimmerman and Briggs (19) only the apical 3 mm tips of the coleoptiles were stimulated, but nevertheless second positive responses were observed. Thus, as pointed out previously (17), the term "base response" is inappropriate for second positive phototropism. In addition, the "base responses" observed with 260 to 330 nm light (7) differ in dose-response characteristics from second positive phototropism so that these responses may well represent a type of phototropic curvature distinct from those occurring with wavelengths of light above 350 nm.

With the experimental method used here, a comparison of the responses of plants bearing 3 mm apical caps with those of plants stimulated along their entire length showed that light-reception in the tip of the coleoptile accounts for 75 to 80 % of the responses observed in these experiments. For instance, in 1 experiment capped plants responded to a stimulus consisting of the initial 100 sec high intensity dose followed by 400 sec of 6 ergs cm^{-2} sec^{-1} at 475 nm with a curvature of 3.7° ± 0.9° while plants without caps curved 14.5° ± 1.5° to the same stimulus combination. Only with illumination lasting 1000 sec or more did capped plants give curvatures of over 10°. Thus, the responses described in these experiments are primarily tip responses, and the photoreception and resulting response that undoubtedly can occur in the base of the coleoptile during some second positive curvatures (1, 17) play a minor role here.

The characteristics of the response in plants stimulated with the 2-dose method described above were examined at 1 wavelength, 475 nm. Figure 2

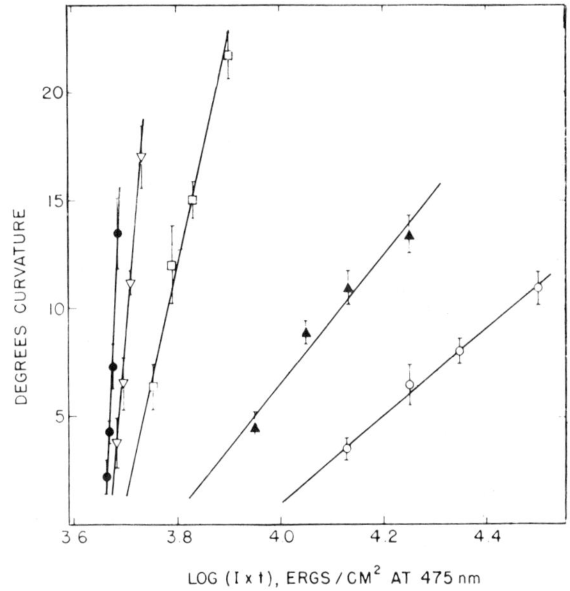

FIG. 2. Phototropic responses of oat coleoptiles stimulated with an initial dose of 4800 ergs/cm² at 475 nm [log A$I \times t$) = 3.68], and a second dose delivered in 5 intensities of 475 nm light: ○) 48 ergs cm^{-2} sec^{-1} ▲) 24 ergs cm^{-2} sec^{-1}, □) 6 ergs cm^{-2} sec^{-1}, △) 1.5 ergs cm^{-2} sec^{-1}, ●) 0.6 ergs cm^{-2} sec^{-1}. Stimuli of 4 durations were given at each intensity. The responses are plotted against the logarithm of the total energy of the stimulus. Standard errors are indicated.

shows the responses to the second doses at 5 intensities, as a function of the logarithm of the total stimulus energy. All the plants received the initial dose of 4800 ergs/cm² [log ($I \times t$) = 3.68], and were then exposed to a second stimulus, lasting 200 to 600 sec, with 475 nm light. When the curvature is thus plotted against dose, it is evident that the reciprocity law does not apply to these responses; equal stimulus energies do not produce equal curvatures. At any 1 intensity the response increases with increasing dose and therefore with increasing stimulus duration.

The indication that second positive responses depend on the stimulus duration is directly tested in figure 3, where the data shown in figure 2 are

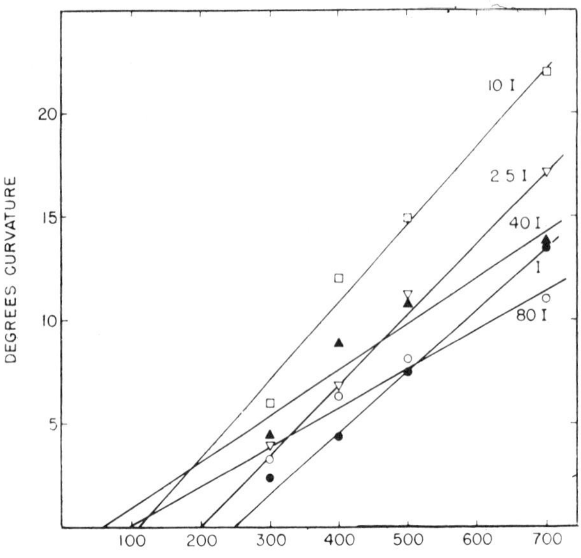

FIG. 3. Responses of oat coleoptiles plotted against the stimulus duration. The initial 100 sec of all the stimuli gave the same dose of 4800 ergs/cm² at 475 nm. The second parts of the stimuli were delivered with 5 intensities of 475 nm light: ○) 48 ergs cm^{-2} sec^{-1}, ▲) 24 ergs cm^{-2} sec^{-1}, □) 6 ergs cm^{-2} sec^{-1}, △) 1.5 ergs cm^{-2} sec^{-1}, ●) 0.6 ergs cm^{-2} sec^{-1}. Lines drawn through the data for each intensity were calculated by the least squares method. Standard errors for these data are indicated in figure 2.

re-plotted against the duration of the stimulus. The first 100 sec represent the duration of the initial stimulus received by all the plants, bringing them to the zone of indifference. For any 1 intensity of the second dose the response is seen to increase linearly with stimulus duration. However, the intensity does modify the slope or X-intercept of the line relating response to stimulus duration, so that all the data points do not fall on 1 line. To illustrate the influence of stimulus intensity on second positive phototropism, the responses of plants to stimuli of constant duration have been plotted against the logarithm of the intensity of the second stimulus. The result is shown in figure 4. Response is maximal at an intensity of 6 ergs cm^{-2} sec^{-1}, higher and lower intensities causing decreased responses. The decrease in response at high intensities is not caused by heat radiating from the monochromator, since water filters placed in front of the monochromator made no difference in the response. It must be emphasized that at any 1 intensity of the second dose, stimulus duration is the factor determining the amount of curvature, and response increases linearly with duration. The modifying effect (on this linear relationship) of the intensity of the stimulus has not been observed previously, although it is possible that the decreased phototropic responsiveness reported with very large light doses (1) is a manifestation of this phenomenon. On the other hand the existence of a second indifferent and a third positive region of the phototropic dose-response curve (4) has not been substantiated in these experiments.

The Action Spectrum. In the highest intensity region, where curvature decreases with increasing intensity, the spectral sensitivity of the response could not be investigated thoroughly since sufficiently high intensity light was not available throughout the spectrum. However, measurements of the responses of coleoptiles to a quantum flux for the second stimulus of 3.5×10^{12} quanta/cm² sec^{-1}, lasting 400 sec, indicated that light of wavelength between 440 and 480 nm was most effective in bringing about the decrease in response, and that at this intensity 495 nm and 355 nm light did not appear to cause any depression in response.

The action spectrum for second positive phototropism was determined in the low intensity stimulus region, where response increases with increasing intensity. Since the duration of the stimulus is the major factor determining the response, the durations of the stimuli in all these experiments were held constant by using an initial high intensity stimulus of 100 sec, followed by a second stimulus of 400 sec. To determine the effect of the intensity of the second stimulus on the second positive phototropic response at all the wavelengths investigated, 2 quantum fluxes, 0.53×10^{12} and 1.2×10^{12} quanta cm^{-2} sec^{-1}, were employed at each wavelength. The wavelengths were spaced every 20 nm from 355 to 435 nm and every 10 nm from 445 to 495 nm. The experiments at each wavelength and quantum flux were repeated 4 or 5 times, and the data are presented in table I. As shown, the change in response for a change in quantum flux of 1 log unit is found to be reasonably constant with wavelength in this intensity region. The values for the slope calculated at 455 and 475 nm are much smaller than the others,

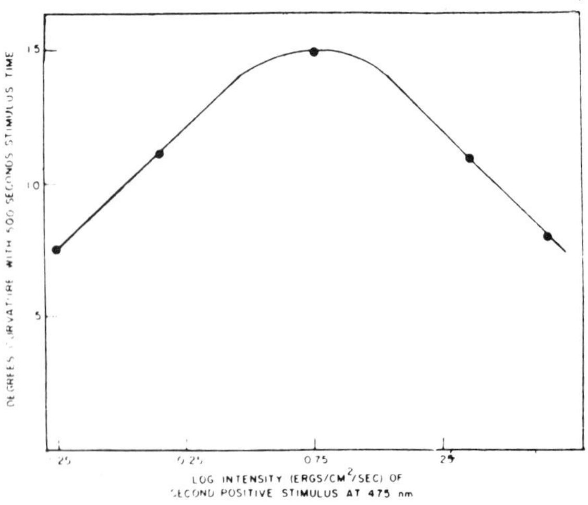

FIG. 4. Variation of responses to stimuli of fixed (500 sec) duration at 5 intensities of 475 nm light. Initial 100 sec stimulus of 48 ergs cm^{-2} sec^{-1} received by all the plants. The response is plotted against the logarithm of the intensity of the second stimulus (400 sec duration). Standard errors are indicated in figure 2.

Table I. *Variation of Response With Intensity and Wavelength of the Second Stimulus*

Response is measured in degrees curvature ± standard error. The slopes are determined in the intensity region where response is increasing with intensity. An initial 100 second stimulus of 57 ergs cm^{-2} sec^{-1} at 470 nm was given to all the plants.

Wavelength	Response to 400 sec of 0.53×10^{12} quanta/cm^2 sec^{-1}	Response to 400 sec of 1.2×10^{12} quanta cm^{-2} sec^{-1}	Slope: Response/log (Intensity)
nm	*deg of curvature*		
355	7.9 ± 0.8	...	...
375	12.6 ± 0.9	19.3 ± 1.5	19.1
395	11.5 ± 1.3	17.3 ± 1.4	16.6
415	12.9 ± 0.6	18.2 ± 1.7	15.1
435	14.9 ± 1.2	20.2 ± 0.9	15.1
445	17.7 ± 1.0	22.0 ± 0.9	12.3
455	17.8 ± 1.1	20.7 ± 1.7	8.3[1]
465	17.4 ± 0.6	22.4 ± 0.6	14.3
475	16.0 ± 0.8	17.4 ± 2.1	4.0[1]
485	14.1 ± 0.9	...	...
495	6.8 ± 0.8	...	...
Avg slope			15.4

[1] Values not included in calculation of average slope (see text).

and in these cases it seems reasonable to assume that the higher quantum flux initiated the processes causing the decreased curvatures observed at high intensities.

For the action spectrum, the data on the responses of the coleoptiles to the 2 quantum fluxes were used to extrapolate or interpolate to the quantum flux required for a standard response. The standard response chosen was 15°, and the quanta contained in a second stimulus, lasting 400 sec, required to produce this standard response were determined graphically from the data on the responses to the 2 quantum fluxes investigated. The reciprocal of this value is plotted against wavelength in the action spectrum, figure 5. At the 2 wavelengths, 455 and 475 nm, where the slopes of the lines relating response to logarithm of stimulus intensity deviated greatly from the other values, the quantum flux required for the standard response was determined by assuming that the slope of the line relating response to quantum flux through the point at 0.53×10^{12} quanta cm^{-2} sec^{-1} was the average slope of 15.4°/log unit.

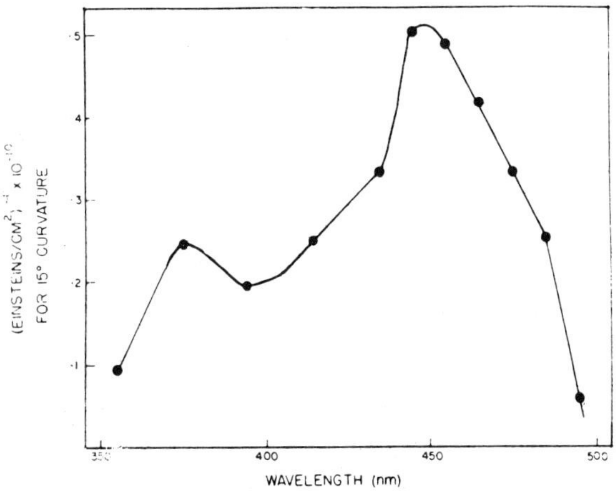

FIG. 5. Action spectrum for the second positive phototropic response in *Avena* coleoptiles. The reciprocal of the energy, in Einsteins/cm^2, of the second stimulus, when delivered in 400 sec, for a 15° curvature is plotted against the wavelength of the second stimulus.

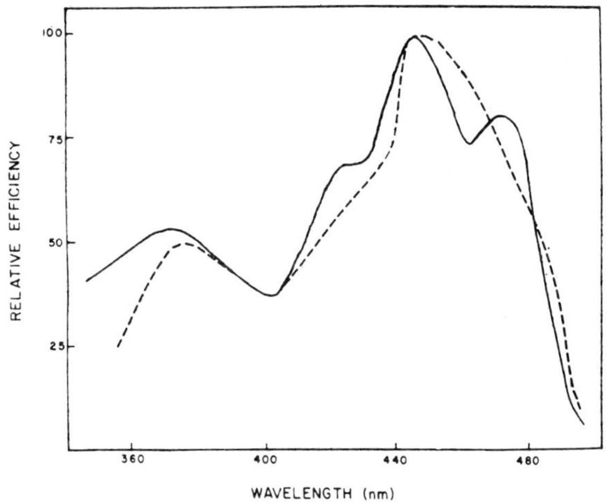

FIG. 6. A comparison of the action spectrum for first positive phototropism in oat coleoptiles, solid line from (16), with the action spectrum for second positive phototropism (dashed line).

In the action spectrum shown, the peaks fall at about 375 and 450 nm, and no activity is observed above 510 nm. The resemblance of this action spectrum to the 1 for first positive phototropism in *Avena* (17) is striking, and the 2 are graphed together in figure 6. In both action spectra the peaks in the near ultraviolet are based on observations at only a few wavelengths, so the details of the shape in this spectral region are not well defined. Both action spectra have a valley at 400 nm, but above this point certain differences are discernible.

Because of the lengthy experiments involved, the points for the second positive action spectrum are further apart than those for the first positive response. For this reason a shoulder at 425 nm in the second positive action spectrum may well have been missed. The major peak in the action spectrum for first positive phototropism is at 445 nm and for the second positive phototropism it is at 445 to 455 nm. The ratio of the height of the near ultraviolet peak to this one is in both cases very close to 1:2. The absence of a second peak around 475 nm in the second positive action spectrum is the major discrepancy between the 2 action spectra. Unfortunately there is a large change in the lamp output at about this wavelength. Additional experiments on the second positive response in this spectral region were carried out using a second stimulus of 0.53×10^{12} quanta cm^{-2} sec^{-1} lasting 400 sec at the wavelengths 465, 470, and 475 nm. The experiment was repeated 6 times, and the results are shown in table II. These experiments indicate that the second positive phototropic action spectrum may well have a second peak around 475 nm. Since the presence of such a peak has not yet been satisfactorily demonstrated in detailed experiments with varying stimulus intensities, however, it has not been included in the action spectrum for the response. Nevertheless, the similarities between the 2 action spectra are strong, and we conclude that the same, or very similar, pigments appear to be involved in photoreception for these 2 phototropic responses.

Table II. *Additional Action Spectrum Data*

Responses to an initial sec stimulus of 57 ergs cm^{-2} sec^{-1} at 470 nm plus 400 sec of 0.53×10^{12} quanta cm^{-2} sec^{-1} at the test wavelength.

Wavelength	Curvature ± standard error
nm	*deg*
465	16.8° ± 1.3 a[1]
470	13.6° ± 0.4 b[1]
475	17.8° ± 1.4 a[1]

[1] a, b Values with different letters are significantly different at the 95 % confidence level.

Discussion

The method of isolating second positive phototropism used in these experiments has permitted a characterization of the effects of stimulus duration, intensity, and wavelength on the response. Haupt (10) and Meyer zu Bentrup (12), working on the photo-induction of polarity of germination in *Fucus* zygotes and *Equisetum* spores, used a method similar to the one developed in these experiments. The dose-response characteristics of the low and the high energy types of polarity induction are similar to those for first and second positive phototropism in coleoptiles, and the wavelengths of light effective for the responses are also in the blue and near ultraviolet. However, studies on the spectral sensitivity of the polarity induction processes indicated that the low and high energy responses might have different action spectra (12), while it appears that the 2 kinds of positive phototropism have similar action spectra.

The experiments reported here agree with earlier evidence (2, 5, 20) that second positive phototropic responses are determined by the duration of the stimulus rather than by the total energy of the stimulus. Zimmerman and Briggs (20) reported that a 10-fold change in intensity had no influence, detectable with their method, on the linear relationship between response and stimulus duration. However, the present experiments cover a larger intensity range with a more direct method and bring out the fact that the stimulus intensity does indeed have some modifying effect on the second positive response. The means whereby stimulus intensity and duration affect the response are not known. Perhaps the decreased responsiveness at high intensity is caused by bleaching or photo-oxidation of the photoreceptor. Zimmerman and Briggs (20) suggested that the second positive response is determined by the length of time during which a photoequilibrium is maintained between an inactive and an active form of the photoreceptor. A modification of this scheme in which the equilibrium concentration of the active form of the photoreceptor is intensity-dependent would provide 1 possible mechanism to explain the secondary effect of intensity.

In these experiments the minimum stimulus duration required for initiation of second positive phototropic bending is 100 sec. Curry (5) estimated that a duration of about 240 sec was required, while Zimmerman and Briggs' (20) data indicated that the second positive type of response was initiated with the start of the stimulus. These differences are probably related to the different methods used to observe the response. With the 2 dose method used in these experiments the intensity and duration of the initial dose might influence the point of onset of the second positive responses, but it was not found possible to test this factor thoroughly.

The similarity between the action spectra for first and second positive phototropism accords strikingly with the demonstration of Pickard and Thimann (13) that lateral transport of auxin occurs during both types of response in coleoptiles. There are 2 possible interpretations of these facts. One is that the 2 photoreceptors are indeed the same, the apparent discrepancy at 475 nm being due to experimental imperfections. The oat coleoptile would then present the remarkable phenomenon of having 2 different types of response (each with its characteristic kinetics) both mediated by the same photoreceptor and brought about by the same effector, namely auxin asymmetry. However, the great differences in dose-response characteristics between the 2 types of positive phototropism, as well as the opposite effects of red light pretreatment (in decreasing the sensitivity of the first positive and increasing that of the second positive) (19, 20) have not yet permitted the formulation of a model in which first and second positive phototropism are related to the same photoreceptive mechanism. The alternative interpretation of the data is that the photoreceptors in fact differ slightly; perhaps the same pigment is conjugated with 2 different proteins. Zimmerman and Briggs (20) indeed suggested separate photoreceptor systems for first and second positive responses. Each system would have its own dose-response characteristics, but the end result, lateral transport of auxin, would be the same. One would be tempted to draw an analogy with mammalian vision where 2 different types of photoreceptors, the rods and the cones, operate in low or high light intensities respectively, but both bring about the sensation of light, and both contain very closely related pigments.

The phototropic photoreceptor has not been identified from the action spectrum for first positive phototropism. The same must at present be the situation for the second positive photoreceptor. The problem has been discussed in detail in references (3), (9), and (17). Suffice it to say here that neither type of pigment, flavin, or carotenoid, suggested as the photoreceptor, has an absorption spectrum precisely matching the action spectra. The idea, discussed by Briggs (3) and Thimann (16), that flavins and carotenoids participate together in light absorption for the phototropic response is, indeed, reasonable, but proof for this or any other theory is lacking.

The identity of the photoreceptor for blue-light-reactions such as phototropism is an important unsolved question in plant physiology. Reactions with action spectra similar to those for phototropism in oats are found throughout the plant kingdom; a few examples of such responses are phototropism in *Phycomyces* (6, 8) whose action spectrum is virtually identical with that for first positive phototropism in *Avena,* photo-stimulation of respiration in *Chlorella* (11) with peaks at about 375 and 460 nm, light-stimulated carotenoid synthesis in *Fusarium* (14) and chloroplast phototaxis in *Lemna* (21) with peaks at 382, 452, and 485 nm. In none of these has the photoreceptor been unequivocally identified. The experiments reported here add yet another response to the list of these blue-light-reactions and provide an additional reason for continued efforts to identify this photoreceptor.

Acknowledgments

The authors acknowledge the continued interest of Dr. W. R. Briggs and his assistance in valuable and critical discussions.

Literature Cited

1. ARIZ, W. H. 1915. Untersuchungen über den Phototropismus. Rec. Trav. Botan. Néerl. 12: 44–216.
2. BRIGGS, W. R. 1960. Light dosage and phototropic responses of corn and oat coleoptiles. Plant Physiol. 35: 951–61.
3. BRIGGS, W. R. 1963. The phototropic responses of higher plants. Ann. Rev. Plant Physiol. 14: 311–62.
4. BUY, H. G. DU AND E. NUERNBERGK. 1934. Phototropismus und Wachstum der Pflanzen. II. Ergeb. Biol. 10: 207–322.
5. CURRY, G. M. 1957. Studies on the spectral sensitivity of phototropism. Thesis, Harvard University, Cambridge, Massachusetts.
6. CURRY, G. M. AND H. GRUEN. 1959. Action spectra for the positive and negative phototropism of *Phycomyces* sporangiophores. Proc. Natl. Acad. Sci. U. S. 45: 797–804.
7. CURRY, G. M., K. V. THIMANN, AND P. M. RAY. 1956. The base response of *Avena* seedlings to the ultra violet. Physiol. Plantarum 9: 429–40.
8. DELBRÜCK, M. AND W. SHROPSHIRE. 1960. Action and transmission spectra of *Phycomyces.* Plant Physiol. 35: 194–204.
9. GALSTON, A. 1959. Phototropism of stems, roots, and coleoptiles. In: Handbuch der Pflanzenphysiologie, Vol. 17/1. W. Ruhland, ed. Springer, Berlin. p 492–529.
10. HAUPT, W. 1958. Über den Primärvorgang bei der polarisierenden Wirkung des Lichtes auf keimende *Equisetum*-sporen. Planta 51: 74–83.
11. KOWALLIK, W. 1967. Action spectrum for an enhancement of endogenous respiration by light in *Chlorella.* Plant Physiol. 42: 672–76.
12. MEYER Z BENTRUP, F. 1963. Vergleichende Untersuchungen zur polaritätsinduktion durch das Licht an der *Equisetum* spore und der *Fucus*-zygote. Planta 59: 472–91.
13. PICKARD, B. G. AND K. V. THIMANN. 1964. Transport and distribution of auxin during tropistic response. II. The lateral migration of auxin in phototropism of coleoptiles. Plant Physiol. 39: 341–50.
14. RAU, W. 1967. Untersuchungen über die Lichtabhängige Carotenoid-synthese. I. Das Wirkungsspektrum von *Fusarium aquaeductuum.* Planta 72: 14–28.
15. SHROPSHIRE, W. AND R. WITHROW. 1958. Action spectrum of phototropic tip curvature of *Avena.* Plant Physiol. 33: 360–65.
16. THIMANN, K. V. 1967. Phototropism. In: Comprehensive Biochemistry, Vol. 27. M. Florkin and E. Stotz, eds. Elsevier Publishing Company, Amsterdam. p 1–29.
17. THIMANN, K. V. AND G. M. CURRY. 1960. Phototropism and phototaxis. In: Comparative Biochemistry, Vol. 1. M. Florkin and H. Mason, eds. Academic Press, New York. p 243–306.
18. WENT, F. 1956. Phototropism. In: Radiation Biology, Vol. III. A. Hollaender, ed. McGraw Hill Book Company, Inc., New York. p 463–78.
19. ZIMMERMAN, B. AND W. R. BRIGGS. 1963a. Phototropic dosage-response curves for oat coleoptiles. Plant Physiol. 38: 248–53.
20. ZIMMERMAN, B. AND W. R. BRIGGS. 1963b. A kinetic model for phototropic responses of oat coleoptiles. Plant Physiol. 38: 253–61.
21. ZURZYCKI, J. 1962. The action spectrum for the light dependent movements of chloroplasts in *Lemna trisulca* L. Acta Soc. Botan. Polon. 31: 489–538.

Sonderabdruck aus der ZEITSCHRIFT FÜR NATURFORSCHUNG Bd. 7b, Heft 9/10, 1952
Verlag der Zeitschrift für Naturforschung, Tübingen, Mathildenstr. 29

A critical experiment on Bünning's theory of Photoperiodism

By D. J. Carr

Max-Planck-Institut für Biologie, Abt. Melchers, Tübingen

(Z. Naturforschg. **7 b**, 570—571 [1952]; eingeg. am 2. Okt. 1952)

The theory of photoperiodism propounded by Bünning, which assumes the existence in plants of an endogenous component of diurnal rhythms of activity is the only one of the many theories which have so far been put forward which explains the effects of daylength on both long-day and short-day plants in a uniform manner. Although this theory has had notable successes in predicting and accounting for the various and complex effects so far discovered in photoperiodic experiments, it has always been possible to explain these effects without assuming the existence of an endogenous rhythm in plants. Claes and Lang[1], however, performed experiments on the long-day plant *Hyoscyamus niger* in which they used 48-hour cycles consisting of a 7-hour main light period and 41 hours of darkness. At intervals of 1 hour successive experimental groups of plants received another period of light lasting two hours, interrupting the long dark period. It was found that the flower-inducing effect of this light break reached optimal values at two points, one near the beginning and one near the end of the dark period. No flower-inducing effect was found when the light-break occurred near the middle of the dark period, in striking contrast to the light-break effect with 24-hour cycles. This result can either be explained in terms of Bünnings theory or on an alternative basis suggested by the authors. They assumed that a light-break may interact with a main light period when the interval between them is not too great, and that it is indifferent whether the light break precedes or follows the main light period. Similar results were obtained by Carr[2] with the short-day plants *Xanthium saccharatum* and *Perilla ocymoides*, but here the light effect at the two optima is inhibitory of flowering and light at the middle of the long dark period stimulates, or at least is not inhibitory to flowering. Both Claes and Lang, and Carr concluded that a critical experiment discriminating between the two possible explanations of this phenomenon could be made using a 72-hour cycle with a short main light period as before, but extending the dark period by another 24 hours. On Bünnings theory light breaks during such a very long dark period should show 3 optima in their promoting effect on flowering in long-day plants and inhibiting effect in short-day plants. On the alternative explanation of Claes and Lang only 2 optima, situated near the beginning and near the end of the dark period would be expected.

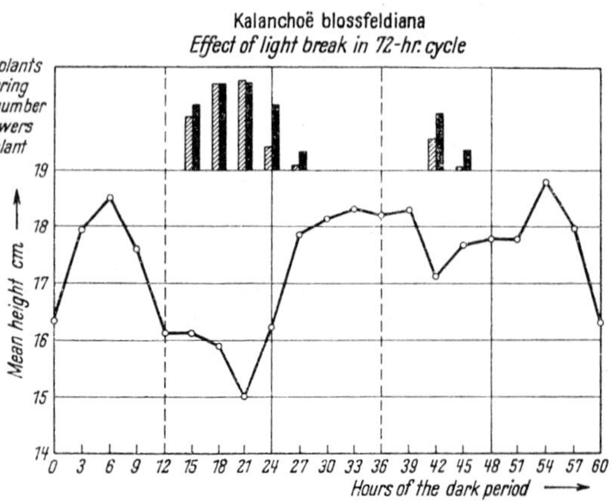

Hatched columns - Mean No. of flowers per plant
Shaded " - No. of plants flowering per group

Such an experiment was carried out this summer at the Max-Planck-Institut für Biologie, Abt. Melchers, in Tübingen[3] using the short-day plant *Kalanchoe blossfeldiana*. Seeds kindly supplied by Prof. Harder were sown on 17/4/1952 under long-day (c. 18 hours of light per day) conditions and the plants were kept continuously under these conditions up to the time of the experiment, when they had formed 7 pairs of expanded leaves. The plants were distributed into 21 groups of 9 plants each. One group (Gr. 21) was maintained under long-day conditions. Beginning on 5/8 the other groups were exposed to day-light in the greenhouse from 7.30 a.m. to 7.0 p.m.

[1] H. Claes u. A. Lang, Z. Naturforschg. **2b**, 56 [1947].

[2] D. J. Carr, Physiol. Plantarum **5**, 78—84 [1952].

[3] The author wishes to express his gratitude to Prof. Melchers for placing the facilities of his Institute at his disposal and to the Max-Planck-Gesellschaft for providing financially for his stay in Tübingen. He also wishes to thank the staff and other workers at the Institute for their invaluable co-operation in carrying out the experimental treatments.

and then removed to dark cabinets kept at 21° C and a relative humidity of 90%. Every 3 hours thereafter a group of plants was removed from the dark cabinet and exposed for 1 hour to light of 1000 Lux intensity (at plant height) from warm-type fluorescent lamps (Osram HNG)[4], at a controlled temperature of 21° C and atmospheric humidity of 90%. The plants were then returned to the dark cabinet for the rest of the 60,5-hour-long dark period. It was thus possible to test the effectiveness of the light-break in inhibiting or promoting flowering at successive intervals of 3 hours during the prolonged dark period. After 9 cycles of this treatment the plants were returned on 1/9 to longday conditions in the greenhouse. About two weeks after the end of the experimental treatment certain groups of plants were seen to be developing inflorescences. The plants are still under observation but the results are such that in view of the importance of this experiment it has been decided to publish an interim report.

[4] The author and the Max-Planck-Institut f. Biologie thank Osram G. m. b. H., K. G., for supplying the fluorescence lamps.

On 27/9 the macroscopic flower buds on each plant were counted and the heights of the plants measured. The data for the number of plants flowering, the mean number of flowers per plant and the mean plant height in each group are presented in the diagram.

The results show clearly that flowering was induced only in groups 5—8 and in groups 14 and 15; that the degree of flowering as measured by the mean number of flowers per plant reaches an optimal value in groups 6 and 7 and in group 14; and that the mean heights of the plants show 3 clear optima, in groups 2, groups 11—13 and group 18. Thus whether we regard the effect of the light break on flowering or on vegetative growth the results agree closely with expectations from Bünnings theory and not with those from Claes and Lang's suggestions. This experiment constitutes a critical test of the theory of Bünning which must therefore be regarded as finally and decisively proved. It is desirable however that the experiment should be repeated with other short-day plants and with long-day plants and it is hoped that this will be done in the near future.

Reprinted from PLANT PHYSIOLOGY, Vol. 43, No. 12, December, 1968, pages 1883–1886.
(PRINTED IN U.S.A.)

The Circadian Rhythm of Leaf Movement of Coleus blumei x C. frederici, a Short Day Plant. I. Under Constant Light Conditions[1, 2]

Ruth Halaban
Department of Biology, Princeton University, Princeton, New Jersey 08540

Received March 15, 1968.

Abstract. **A new instrument for the recording of leaf movement rhythm is described. *Coleus blumei* x *C. frederici*, a short day plant, exhibits a circadian rhythm of leaf movement. The period length of the free running rhythm is shortest in continuous darkness and is increased with an increase in the light intensity. The amplitude of the rhythm tends to damp in continuous bright light.**

Circadian rhythms are those biological rhythms with a period length of approximately 24 hr in constant conditions of light and temperature; the period length is temperature compensated. Most of the current knowledge of circadian rhythms in higher plants comes from studies of the rhythm of leaf movement in day-neutral species of *Phaseolus* (1, 2, 6, 7, 11) and the petal movement of the short-day-plant *Kalanchoe blossfeldiana* (3, 4, 13). It is the purpose of this paper to report the circadian rhythm of leaf movement in a short-day-plant, *Coleus blumei* x *C. frederici*, under different light intensities.

Several techniques for recording leaf movement were tested by this author. These were the kymograph and leverage system (1, 2), time lapse photography (5, 7), and the strain gauge method (8). These methods were found unsatisfactory for the present study and therefore a new one was developed, which is presented below.

Materials and Methods

Method for Leaf Movement Recording. The height of the leaf is measured hourly by periodically raising a motor driven contact until it completes a circuit by touching an electrical contact mounted on the leaf. Synchronously, a second motor drives a pen which records the height at which contact was made, on pressure-sensitive paper mounted on a rotating kymograph.

The height sensing unit (fig 1a), located inside the plant growth chamber, consists of 2 electrical contacts and a reversible motor. The motor rotates a 12-inch long vertical lead screw which raises a gold-foil electrical contact until it touches the second electrical contact, a 15 mg 0.002 inch diameter gold wire suspended at the tip of the leaf. The contact reverses the synchronous motor and causes the driven gold foil to descend to its original reference point. No electrical current passes to the leaf itself.

The recording unit located outside the growth chamber consists of a similar lead screw to which a pen is attached, a reversible motor, and a drum covered with pressure sensitive recording paper (fig 1b). The pen travels vertically along the lead screw and registers the distance on the recording paper. By changing the relative speed of the height-sensing and recording motors one can either amplify or reduce the height of the recording. The drum rotates once per week. In test records made at 1 min intervals a maximum difference of about 0.5 % of the average height of the leaf movement was observed. Figure 2 is a sample of leaf movement recordings; the shortest vertical line represents the minimum position of the leaf, whereas the longest one represents the maximum height of the leaf. The time of the minimum leaf position is used as the phase reference point in the circadian cycle; it can be determined accurately to ± 1.0 hr. Tracings of the original leaf movement recordings are used in further graphs.

Plant Material and Light Conditions. A clone of *Coleus blumei* x *C. frederici* was used (a gift from Professor Laibach). Plants propagated by cuttings were grown in the greenhouse under con-

---

[1] The material presented in this paper is a part of a dissertation submitted to the Department of Biology, Princeton University in partial fulfillment of the requirements for the degree of Doctor of Philosophy.

[2] Supported in part by a NSF grant to Dr. W. P. Jacobs.

tinuous illumination; during the night, light was supplied by 500 watt incandescent bulbs. Plants were transferred to the growth chamber (Model E0025, National Appliance Company) when they reached the 4 to 6 internode stage (not counting unfolded leaves). Only one 12 to 14 cm-long leaf on each plant was used for recording; the leaf below was cut off to prevent interference with the recording (fig 1a).

A light-dark cycle of 12 hr light, 12 hr dark was maintained in the growth chamber for 3 to 4 days before subjecting the plants to the experimental conditions.

In addition to light-dark cycles, the following conditions were studied: 1) Constant darkness; 2) Constant darkness + 4 % w/v sucrose and a trace of Phaltan, a commercial fungicide. The solution of sucrose + Phaltan (200 ml) was added to each pot (7 inches in diameter) on the first day in constant dark; 3) Constant dim light of 10 ft-c; 4) Constant dim light of 30 ft-c; 5) Constant bright light of 1300 ft-c. The light intensity, as measured with a foot-candle meter (Weston illumination meter Model 756, Daystrom, Incorporated), was approximately 1300 ft-c at the plant level during the main light period. The lamps used were 32 cool-white fluorescent bulbs (Sylvania F9612/CW/VHO/L) and 12 60-watt incandescent bulbs, all separated from the plants by heat-reducing transparent plastic sheet. Spectral energy distribution curve for the cool-white fluorescent bulbs is given by Thomas and Dunn (12). The lights were turned on in the light-dark regime at 0800 eastern standard time (EST) and went off at 2000 EST. The transition from dark to light and *vice versa* occurred in stages and took about 30 min.

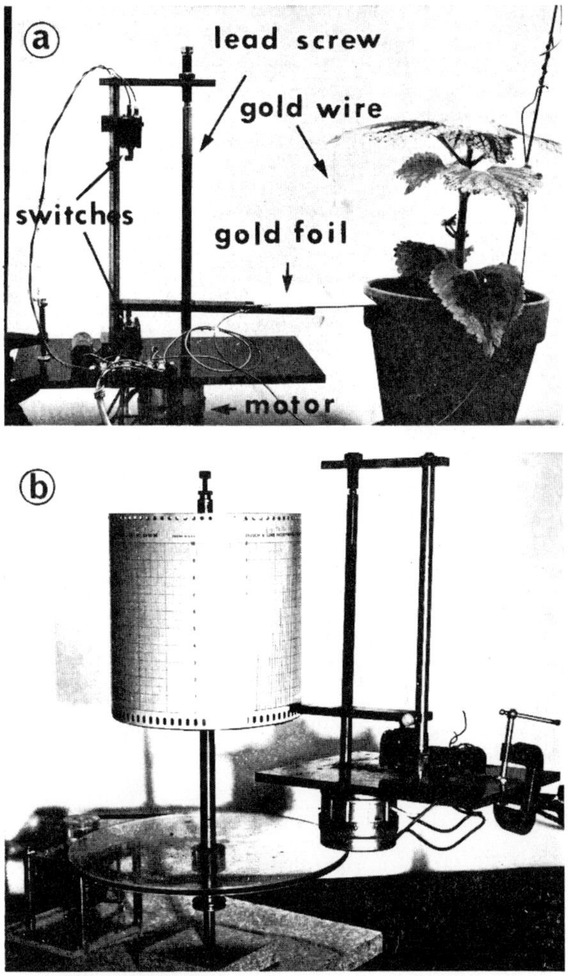

FIG. 1. The leaf movement recording apparatus: a) A photograph of *Coleus blumei* x *C. frederici* and its height sensing unit; b) The recording unit. For further details see text.

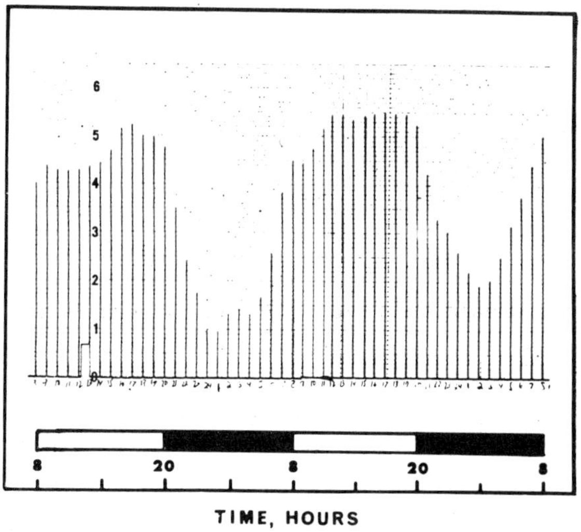

FIG. 2. A sample of actual leaf movement recording of *Coleus* under light-dark conditions. Distance between vertical numbers is 13 mm. Each black bar on the abscissa represents a 12 hr dark period ending at 0800 eastern standard time.

Light intensity of 30 ft-c was obtained by reducing the number of operating bulbs in the growth chamber and by the addition of a darkened cheesecloth below the bulbs. Light intensity of 10 ft-c was obtained by using a 20 watt fluorescent bulb (Arc Ray).

Temperature in the growth chamber was maintained constantly at 21 ± 1°.

Table I. *Values of the Free Running Period* (τ) *of Leaf Movement Under Different Light Conditions*

Light conditions	τ ± Standard deviation of the slope
Constant darkness (DD)	22.67 ± 0.06
Constant darkness + sucrose	22.33 ± 0.06
Constant dim light (10 ft-c)	23.06 ± 0.02
Constant dim light (30 ft-c)	23.86 ± 0.08
Constant light (1300 ft-c)	24.79 ± 0.01

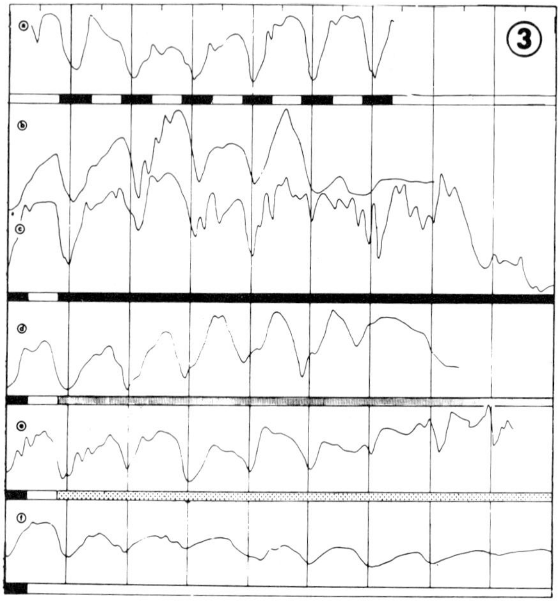

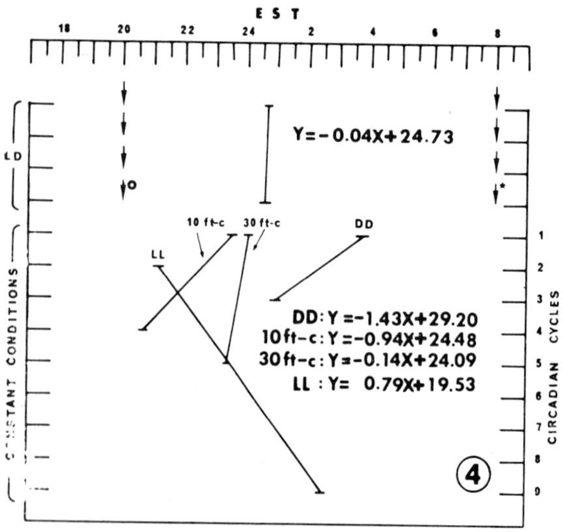

FIG. 3, a to f. Nyctinastic leaf movement tracings of *Coleus* under different light conditions. Vertical lines are 24 hr apart and indicate hr 2400 eastern standard time. Each recording is a typical representative from a group of 5 plants. a) Under light-dark cycles of 12 hr light (empty bars) and 12 hr dark (black bars). b) Plant entrained under 12 hr light and 12 hr dark then released to constant darkness. c) Plant released to constant darkness; sucrose was added to the pot. d) Plant released to constant dim light of 10 ft-c. e) Plant released to constant dim light of 30 ft-c. f) Plant released to constant bright light of 1300 ft-c.

FIG. 4. Calculated regression lines for different light conditions fitted to the minimum leaf positions in successive days (indicated by the vertical divisions). Numbers on top scale indicate eastern standard time (EST) in hr. LD = light dark cycles of 12 hr light and 12 hr dark, with light turned off at 2000 EST (arrows on the left), and turned on at 0800 EST (arrows on right). LL: Plants were transferred from the light-dark regime to constant bright light of 1300 ft-c which started at 0800 EST (arrow at right that is marked by an asterisk). DD: Plants were transferred from the light-dark regime to constant dark which started at 2000 EST (arrow at left marked by an open circle). Regression line for DD + sucrose was not significantly different from DD, and therefore was omitted. 10 ft-c and 30 ft-c: Plants were transferred from the light-dark regime to constant dim light of either 10 ft-c or 30 ft-c, starting at 2000 EST (arrow marked by an open circle).

Results

Tracings of the pattern of the movement of an individual leaf of 1 plant in each of the above listed conditions are represented in figure 3a to f. Regression lines were fitted to the data (fig 4) to estimate the period of the circadian rhythm under the different conditions and table I tabulates these values.

A statistical analysis of the slopes of the regression lines (table II) shows that the slope of the regression lines in constant dark compared to constant dark + sucrose and of 2 different runs in constant light of 1300 ft-c are not significantly different. The purpose of supplying sucrose to plants kept in constant darkness was to prolong the leaf movement rhythm for more than 2 days by providing a source of metabolic energy. However, as the sample in figure 3c indicates, the pattern of the rhythm is highly erratic, and, furthermore, the period length for individual plants is inconsistent, varying between 21 and 26 hr. The addition of sucrose to the pot changes the pattern of the leaf movement (perhaps by raising the osmotic pressure in the pot) and, therefore, is not similar to constant darkness alone. The slopes of regression lines for constant darkness, constant dim light of 10 and 30 ft-c and bright light are significantly different from each other.

Table II. *Statistical Test for the Significance of Period Differences in Different Light Conditions*

The coefficients of regression lines fitted to time of minimum leaf position on successive days (fig 4) are the measure of the period length. Student's t-values are calculated for the conditions indicated in the table.

Treatment	t-values
DD + sucrose *vs.* DD	0.930
DD *vs.* dim LL 10 ft-c	2.477[1]
DD *vs.* LL 30 ft-c	3.070[2]
DD *vs.* LL 1300 ft-c	5.700[2]
dim LL 10 ft-c *vs.* dim LL 30 ft-c	4.850[2]
dim LL 10 ft-c *vs.* LL 1300 ft-c	−11.220[2]
dim LL 30 ft-c *vs.* LL 1300 ft-c	4.480[2]
LL 1300 ft-c *vs.* repeat LL 1300 ft-c	1.028

1 Periods significantly different at the 95 % confidence level.

2 Periods significantly different at the 99 % confidence level.

Discussion

Table I indicates that the endogenous rhythm of leaf movement in *Coleus blumei* x *C. frederici* persists in constant light as well as in constant dark with a free running period (τ) of about 24 hr. It also shows that the value of the free running period (τ), as calculated from the regression lines, depends on the light intensity; it increases with increasing light intensity. The effect of light intensity on leaf movement is species specific rather than general. The rhythm in *Phaseolus* persists under continuous light of different intensities without any damping effect or significant changes in period length (7, 11). *Kalanchoe* petal movement gradually damps under increasing light intensities whereas the free running period shortens (3). The rhythm of leaf movement in *Canavalia ensiformis* damps under continuous illumination of high intensity with no significant difference in the period from that in constant darkness (10). Persistence of leaf movement rhythm under light of high intensity was also observed in *Portulaca grandiflora* (9).

The tracing of the leaf movement in *Coleus* shows another feature (fig 3, b to f): the amplitude of the rhythm is greatest in constant darkness and damps with increasing light intensities.

From both the tracings of leaf movement and the regression lines for constant darkness and constant dark + sucrose (fig 3, b to c, and 4), it is evident that there is a phase delay of 3.6 and 2.5 hr, respectively, in the position of the minimum point on Day 1 compared to Day 0 (these values are achieved by subtracting the average time of minimum leaf position on Day 0 from that on Day 1). The same calculation, applied to conditions of constant light of 1300 ft-c, shows that there is a phase advance of 3.1 hr, which is achieved by the second cycle in constant light conditions (fig 3f and 4). One can suppose, therefore, that in a rhythm entrained by a light-dark cycle, the "light-off" signal initiates a phase delay of 3 hr (the average value of the 2 constant darkness regimes) which is followed, after the "light-on" signal, by a phase advance of 3 hr.

Acknowledgments

I am indebted to Drs. V. G. Bruce and W. P. Jacobs for their criticism during the course of the experimentation and for their valuable suggestions while writing the manuscript. I gratefully acknowledge the advice and valuable discussions of Dr. C. S. Pittendrigh. Mr. R. Charydczak invented and built the leaf movement apparatus.

Literature Cited

1. Bünning, E. and I. Moser. 1966. Response-Kurven bei der circadianen Rhythmik von *Phaseolus*. Planta 69: 101–10.
2. Bünning, E. and R. Zimmer. 1962. Zur Deutung der Phasenverscshiebungen und "Transients" nach exogener Störung endogener Rhythmen. Planta 59: 1–14.
3. Bünsow, R. 1953. Endogene Tagesrhythmik und Photoperiodismus bei *Kalanchoë blossfeldiana*. Planta 42: 220–52.
4. Engelmann, W. and H. W. Honegger. 1967. Versuche zur Phasenverschiebung endogener Rhythmen: Blutenbewegung von *Kalanchoë blossfeldiana*. Z. Naturforsch. 22B: 200–04.
5. Holdsworth, M. 1964. The leaf movement of soybean, a short day plant. J. Exptl. Botany 15: 391–98.
6. Hoshizaki, T. and H. C. Hamner. 1966. Computer analysis of the leaf movements of *Phaseolus vulgaris*, variety 'Pinto', grown under constant conditions. Plant Physiol. 41: xlii-xliii.
7. Hoshizaki, T. and K. C. Hamner. 1964. Circadian leaf movements: Persistence in bean plants grown in continuous high-intensity light. Science 144: 1240–41.
8. Hoshizaki, T. and K. Yokayama. 1965. Recording leaf movements with a strain gauge. Nature 207: 880–81.
9. Karvé, A. D. and S. G. Jigajinni. 1965. Circadian rhythm in *Portulaca*. Z. Pflanzenphysiologie 53: 169–72.
10. Kleinhoonte, A. 1932. Untersuchungen über die autonomen Bewegungen der Primärblätter von *Canavalia ensiformis* DC. Jahrb. Wiss. Botan. 75: 679–725.
11. Moser, I. 1962. Phasenverschiebungen der endogenen Tagesrhythmik bei *Phaseolus* durch Temperatur- und Lichtintensitätsänderungen. Planta 58: 199–219.
12. Thomas, A. S., Jr. and S. Dunn. 1967. Plant growth with new fluorescent lamps. I. Fresh and dry weight yield of tomato seedlings. Planta 72: 198–207.
13. Zimmer, R. 1962. Phasenverschiebung und andere Störlichtwirkungen auf die endogenen tagesperiodischen Blütenblattbewegungen von *Kalanchoë blossfeldiana*. Planta 58: 283–300.

The Flowering Response of Coleus in Relation to Photoperiod and the Circadian Rhythm of Leaf Movement[1, 2]

Ruth Halaban
Department of Biology, Princeton University, Princeton, New Jersey 08540

Received March 15, 1968.

Abstract. **The flowering response of *Coleus frederici* and *Coleus blumei* x *C. frederici* is dependent on the photoperiod; both plants have a critical day length of about 12 hr. The inductive phase, defined as the period when light signals inhibit floral development, started 10 hr after the onset of darkness under 4 and 8-hr photoperiods, and 8 hr after the onset of darkness under a 12-hr photoperiod. However, a fixed temporal relationship between the inductive phase and the minimum leaf position was observed for *Coleus frederici.* The inductive phase always started 5 hr after the minimum leaf position. This evidence supports the theory that a circadian clock participates in the time measurement process of photoperiodic floral induction.**

It is generally agreed that photoperiodism in both plants and animals is a phenomenon which involves a process of time measurement (see, for examle, 16, 23). However, the mechanism of the time measurement process is still disputed. To explain photoperiodic induction of flowering in higher plants, the 2 principal theories that have been advanced are the hour glass and the coincidence theory. The first one advocates a process that is initiated at the beginning of each dark period, mainly considered to be a pigment conversion process (*i.e.*, Phytochrome FR to Phytochrome R), which measures the length of time required to attain the critical day length (1, 16).

The coincidence theory hypothesizes that a circadian rhythm is used by the organism as a time scale (2, 3, 6, 7, 10, 13, 19, 20, 21). In nature light entrains the rhythm and controls its phase. The coincidence of an inductive phase of the rhythm with the main light period controls the future development of the organism.

One difficulty in providing evidence for the theory that the time measurement process in photoperiodism involves circadian rhythms stems from the fact that in very few instances have both phenomena been studied with the same organism. In higher plants, a circadian pattern in floral induction has been demonstrated in *Xanthium* (22), *Pharbitis* (14), Soybean (10), *Chenopodium* (6, 7), and *Lemna* (12, 13); However, there is no report in the literature of an overt circadian rhythm in these plants. On the other hand, most of the literature dealing with the nature of the circadian rhythm in plants is based on the day neutral plant, *Phaseolus multiflorus* (4). The only plant in which both phenomena have been studied is a short day plant *Kalanchoe blossfeldiana,* that exhibits an overt circadian rhythm of petal movement (5, 8). The experiments with *Kalanchoe* have neither proven nor disproven that the time measurement process of the photoperiodic response involves the clock which controls the circadian rhythm.

It is the purpose of this paper to present experimental evidence to support the model suggested by Pittendrigh and Minis (21) in which the inductive phase (the time of the cycle when light has a photoperiodically inductive effect) is correlated with the phase of the circadian rhythm (leaf movement). The results cannot be explained by a simple hour glass mechanism. Furthermore, they agree with Pittendrigh and Minis's model (21) in that the inductive phase lies at the end of the main dark period rather than at its beginning as was suggested by Bunning (2, 3).

Materials and Methods

C. frederici and *C. blumei* x *C. frederici* (later designated as *C. hyb.*) plants were propagated by cuttings and grown in the greenhouse under continuous light as described before (9).

Most of the experiments were conducted in growth chambers at 21 ± 1° (9). The light signals, of 200 ft-c intensity, were provided by a 40 watt cool-white fluorescent tube and a 150 watt incandescent bulb. For the experiments in which *C. frederici* was grown under 4- and 8-hr daylengths with night interruption by light signals, special cabinets were used. For these experiments the temperature was 28 ± 2° and a 40 watt cool-white fluorescent bulb (F36 T12 CW GE) provided about 200 ft-c of illumination.

At the end of the experiments, the apical meristems of the plants were examined under a dissecting

---

[1] This work was partially supported by a NSF grant to Dr. W. P. Jacobs.

[2] The material presented here is a part of a dissertation submitted to the Department of Biology of Princeton University in partial fulfillment of the requirements for the degree of Doctor of Philosophy.

microscope for the determination of floral induction. Vegetative apices of *Coleus* exhibit the typical appearance of a flat meristem. Development of the reproductive phase was first indicated by a broadening of the apex, which later develops successive pairs of bracts. When 2 bracts are fully formed and another 2 are just starting to emerge, the length of the transformed apex is about 0.2 mm. When 6 bracts can be detected, the length is about 0.3 mm. In a 0.4 mm long inflorescence there are 8 to 10 bracts; later on the bracts are all tightly folded one above the other and it is difficult to count them precisely any more. Most of the experiments were duplicated and provided similar results.

Recording of leaf movement was done as described before (9).

Results

Flowering and Leaf Movement Rhythm Under Various Photoperiods. The purpose of the following experiments was to determine the critical day-length for floral induction of both *C. frederici* and *C. blumei* x *C. frederici* and in parallel to follow the pattern of their leaf movement rhythm under different daylengths. Kribben (15) reported that both species are short-day-plants with a critical day length of 12 to 13 hr. Since his experiments were done in the greenhouse under natural light conditions, I repeated them under controlled environmental conditions.

Figure 1 indicates that the critical day length for *C. frederici* was above 12 hr and just below 13 hr; *C. hyb.* reacted similarly. *C. frederici* plants grown under a 13-hr photoperiod showed a very early stage in the formation of a reproductive apex, *i.e.*, broad-

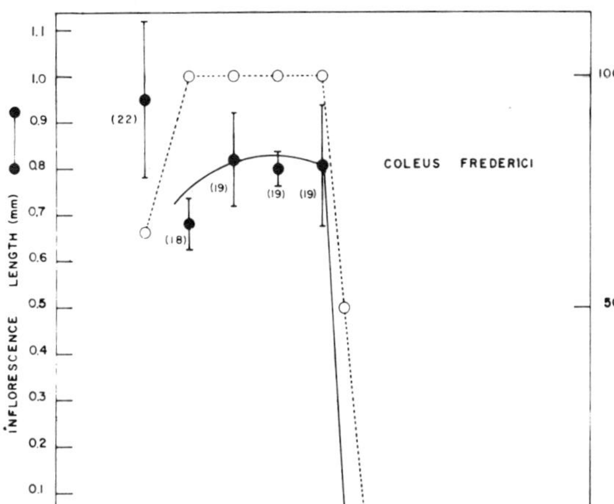

FIG. 1. Flowering response of *Coleus frederici* under 24-hr light-dark cycles of different daylengths. The number in parentheses near each closed circle indicates the number of cycles the plants experienced under that photoperiod. Each point is the average of about 15 plants.

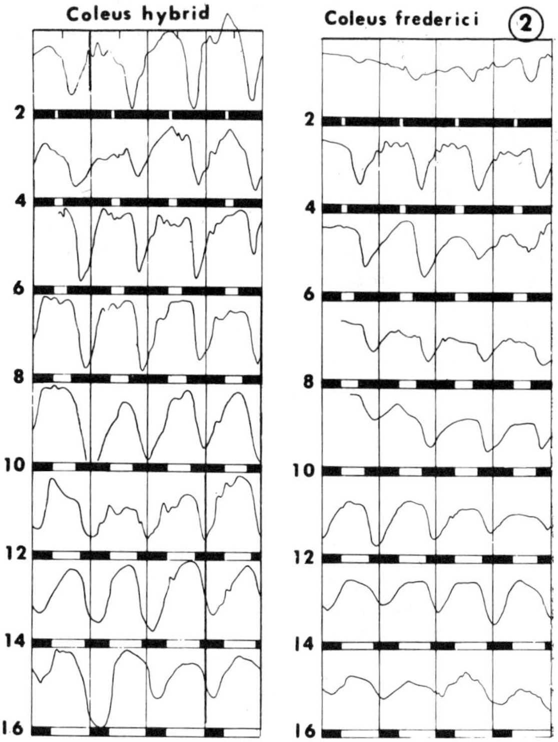

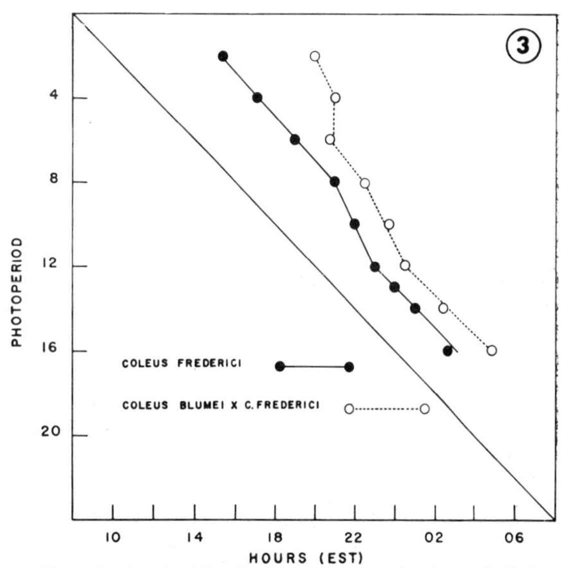

FIG. 2. (top) The leaf movement rhythm of *Coleus blumei* x *C. frederici* and *C. frederici* under 24-hr light-dark cycles of different photoperiods. The onset of light occurs at hr 0800 eastern standard time. Vertical guide lines indicate hr 2400. Numbers on the left side indicate the daylengths in hr.

FIG. 3. (bottom) Plot of the time of mean minimum leaf position of *C. frederici* and *C. hyb.* under different daylengths (ordinate). Diagonal line designates the onset of darkness. Each point is an average of 3 to 5 plants for 4 cycles, starting on the second day in the light-dark conditions. The standard error of each point is within the range of ± 0.00 to ± 0.35.

ening of the apex, but did not progress beyond this stage even after 29 cycles. When plants at that stage were transferred to the greenhouse, they stayed vegetative for a period of 2 months.

The phase of the circadian rhythm as measured by the time of minimum leaf position changed slightly relative to the onset of the dark period. The time between the onset of dark and the minimum leaf position was about 5 hr under 2 to 8-hr photoperiods and about 3 hr under 12 to 16-hr photoperiods in the case of *C. frederici.*

A similar trend was observed for *C. hyb.* as shown in figure 3.

The phase relationship between the minimum leaf position and the light cycle persisted for 4 cycles in constant dim light. This supports the assumption that the phase of the overt rhythm indicates the phase of the light sensitive oscillator.

Flowering and Leaf Movement Rhythm Under Various Daylengths With Night Interruptions by Light Signals. The following experiments were designed to locate the inductive phase, *i.e.,* the clock time when illumination inhibits flowering, under 4, 8, and 12-hr daylengths. It was expected that the inductive phase would be determined by the circadian rhythm of leaf movement. Since *C. frederici* is faster to flower than *C. hyb.* (19 cycles compared to 40), experiments were restricted to *C. frederici.*

Figures 4 to 6 show the shift of the inductive phase in different photoperiods. While inhibition of floral induction by light signals started 10 hr after the onset of dark under 4 and 8-hr daylengths, it started 8 hr after the onset of dark for a 12-hr daylength. However, the onset of floral inhibition always occurs 5 hr after the time of the minimum leaf position under each photoperiod.

In addition, there was a lengthening of the inductive phase under increasing photoperiods. One can estimate that the duration of the inductive phase under a 4-hr photoperiod is no more than 1 hr, while under a 8-hr photoperiod it is probably over 1.5 hr, and for a 12-hr photoperiod it is clearly over 2.5 hr. Estimates are based on those portions of the cycles covered by the light signals which inhibited flowering maximally.

The interruption of the dark period with a light signal (table I) had a very slight effect on the phase of the circadian rhythm when the light signal was given after the time of the minimum leaf position.

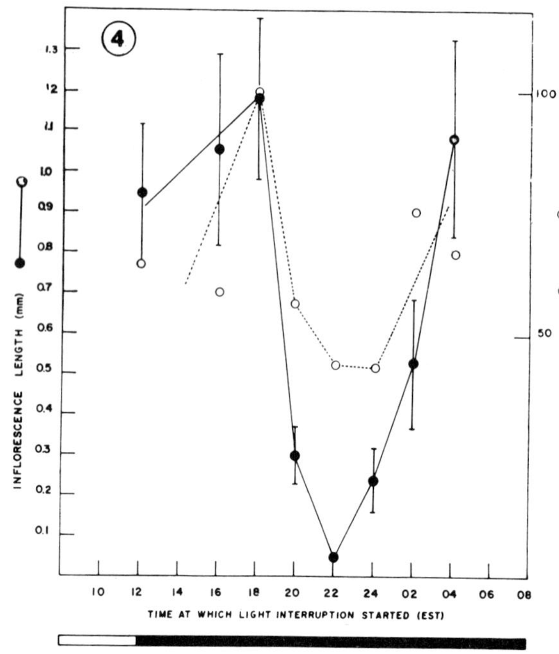

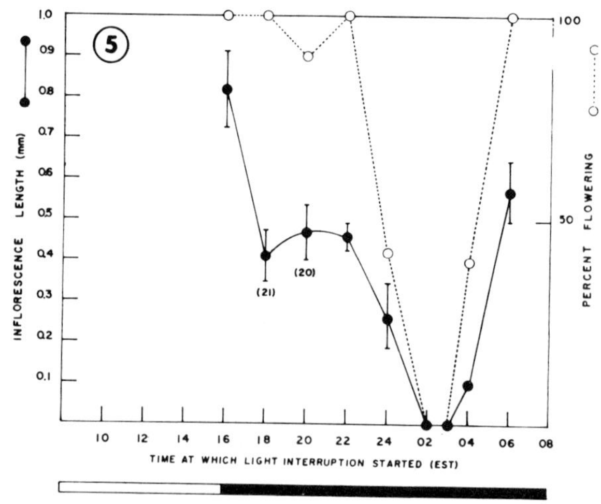

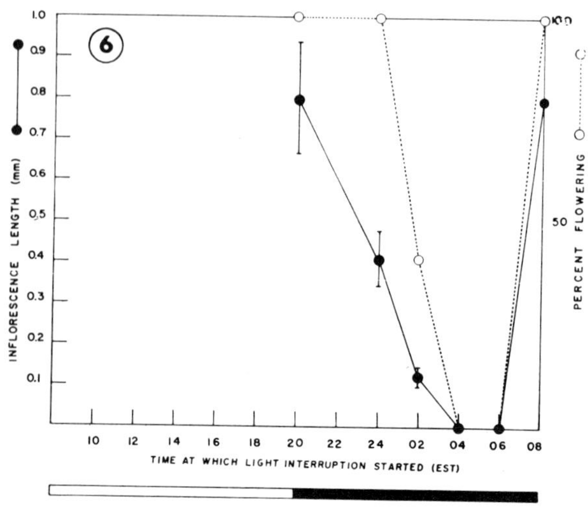

Fig. 4. (top) Flowering response of *C. frederici* under a 4-hr photoperiod with 60 min light signals interrupting the night at different clock time (indicated by the bottom scale as EST). Empty and dark bars indicate the light and dark periods of the cycle, respectively.

Fig. 5. (middle) Flowering response of *C. frederici* under 8-hr photoperiod with 30 min light signals interrupting the night at different clock times.

Fig. 6. (bottom) Flowering response of *C. frederici* under 12-hr photoperiod with 30 min light signals interrupting the night at different clock times.

On the other hand, when the light signal was given before the minimum position of the leaves, the phase of the rhythm was delayed in such a way that the termination of the light signal simulated the comparable full photoperiod. For example, *C. frederici* plants under a 4-hr daylength with a light signal at hr 1500 to 1600 eastern standard time showed a mean minimum leaf position at hr 20•31 ± 0.19 simulating a full photoperiod of 8:16.

An advance of the phase occurred in only 1 case, when the light signal started about the time of minimum leaf position. In this case an advance of about 4 hr was observed.

All the light signals that inhibit flowering fall after the minimum leaf position, so they do not affect the phase of the circadian rhythm; when light signals do shift the rhythm, they do not effect a photoperiodic response since the direction and the amount of phase shift is such that the inductive phase remains in the dark.

Conclusions and Discussion

The literature on plant photoperiodism has shown that flowering may follow a circadian pattern, and gives an indication that the control of flowering might depend on the circadian rhythm, but it does not show direct evidence for a correlation between the 2 phenomena. It is demonstrated here, for the first time, that the inductive phase of the short day plant *C. frederici* has a fixed phase relationship to the phase of the circadian rhythm of the plant. Under different photoperiods maximum inhibition of floral induction by light signals always started 5 hr after the minimum leaf position (but 10 hr from onset of darkness under 4 and 8-hr photoperiods and 8 hr under a 12-hr photoperiod). If an hour glass mechanism were operating here as the time measurement process, the inductive phase would be timed from the onset of darkness and would thus

Table I. *Time of Minimum Leaf Position in a 24-hr Light-Dark Cycle When the Night is Interrupted With a Light Signal*

Each value is the mean of 2 to 4 plants for 3 to 5 cycles.

Photoperiod	Time at which light signal is commenced EST	Mean time of minima EST ± S.E.
a. *Coleus blumei* x *C. frederici*		
8-hr	. . .	22.26 ± 0.38
8-hr	2000	23.67 ± 0.18
8-hr	2400	22.50 ± 0.37
4-hr	. .	21.00 ± 0.36
4-hr	2200	18.38 ± 0.19
b. *Coleus frederici*		
8-hr	. . .	21.00 ± 0.00
8-hr	1830	22.00 ± 0.04
4-hr	. . .	17.12 ± 0.33
4-hr	1500	20.31 ± 0.19
4-hr	2200	16.25 ± 0.29

show a fixed relation to it under different photoperiods [as was suggested, for example, by Bonner (1)].

However, it should be noted that such a correlation has been shown in insect photoperiodism, mainly with *Pectinophora gossypiella* (18, 21) and bird photoperiodism (11, 17). These results follow the main idea expressed before, namely that the position of the inductive phase relates to the phase of the overt circadian rhythm.

The results shown in figures 4 to 6 also suggest that the duration of the inductive phase lengthens under increasing photoperiods. Considering these 2 effects of prolonged photoperiods on the inductive phase, *i.e.* delaying the phase and increasing its duration, one could predict that under a 13-hr photoperiod, part of the inductive phase will be already illuminated by the beginning of the main photoperiod, and therefore inhibition of floral development takes place.

The Precision of the Clock. Although photoperiodic induction of flowering is considered to be an all-or-none phenomenon, the developmental rate of the reproductive stage appears to be a quantitative one (1). Quantitative reduction in floral development was observed under photoperiods only slightly longer than the critical daylength as for example a 13-hr photoperiod in the case of *Coleus* plants (fig 1), or as a result of experiments with light signals falling near the inductive phase (fig 4). The idea generally proposed to explain this phenomenon was that, under such circumstances, less "florigen" is produced during each cycle (1).

Another explanation is suggested here, in which the precision of the biological clock is considered.

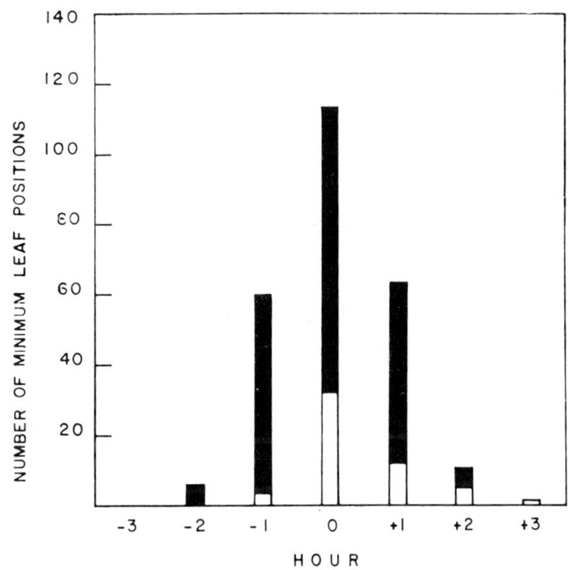

FIG. 7. Frequency distribution illustrating variation in time of minimum leaf position in light-dark cycles relative to the time of most frequent occurrences of the minima (Hr zero). Empty bars: *Coleus frederici;* Dark bars: *Coleus blumei* x *C. frederici.*

Examination of the circadian rhythm of leaf movement of individual plants for successive cycles under full photoperiod shows that the time of the minimum leaf position is not precisely the same in each cycle. Analysis of variance has shown that variability in time of minimum leaf position for individual plants on successive days is the same as the variability between plants on the same day (over 99 % level of confidence). This justifies pooling the data for minimum leaf position recorded under different photoperiods, designating as time 0 the time where the peak in number of minima occurred. The plot of these data (fig 7) shows that they have a bell shaped distribution around the mean. Assuming that the inductive phase is determined by the phase of the circadian rhythm, the existence of small errors or fluctuations in the timing of the inductive phase may play a role under photoperiods which are close to the critical daylength. Induction of flowering under such circumstances will occur only when the circadian rhythm is slightly advanced; in all the slightly delayed cycles the terminal part of the inductive phase will be illuminated by the succeeding main light period. For example, with a 13-hr photoperiod the observed developmental stage after 29 cycles is a broad apex for *Coleus frederici* plants. On the other hand, this developmental stage may occur with as few as 15 or 16 cycles of maximally inductive photoperiods. Consequently, it is suggested that only 15 to 16 cycles of the 29 given with a 13-hr photoperiod were fully inductive and the rest were entirely non-inductive. This reasoning may also explain why complete inhibition of floral induction did not occur in the night-interruption experiments with a 4-hr photoperiod. In these cases, the 1 hr duration of the inductive phase might be so short that the light signal "missed" the inductive phase occasionally, thus resulting in partially inductive cycles.

Acknowledgments

I am grateful to Dr. V. G. Bruce for helpful suggestions and criticism during the course of this work and in the preparation of the manuscript and to W. P. Jacobs for his support and critical reading of the manuscript. I gratefully acknowledge also the advice and valuable discussions of Dr. C. S. Pittendrigh.

Literature Cited

1. BONNER, J. 1959. The photoperiodic process. In: Photoperiodism and Related Phenomena in Plants and Animals. R. B. Withrow, ed. A.A.A.S., Washington. 245–54.
2. BÜNNING, E. 1936. Die endogene Tagesrhythmik als Grundlage der photoperiodischen Reaktion. Ber. Deut. Botan. Ges. 54: 590–607.
3. BÜNNING, E. 1960. Circadian rhythms and the time measurement in photoperiodism. Cold Spring Harbor Symp. Quant. Biol. 25: 249–56.
4. BÜNNING, E. 1967. The Physiological Clock. Springer-Verlag, Incorporated, New York. 167 p.
5. BÜNSOW, R. C. 1960. The circadian rhythm of photoperiodic responsiveness in *Kalanchoë*. Cold Spring Harbor Symp. Quant. Biol. 25: 257–60.
6. CUMMING, B. G., S. B. HENDRICKS, AND H. A. BORTHWICK. 1965. Rhythmic flowering response and phytochrome in a selection of *Chenopodium rubrum*. Can. J. Botany 43: 825–53.
7. CUMMING, B. G. 1967. Circadian rhythmic flowering responses in *Chenopodium rubrum*. Effects of glucose and sucrose. Can. J. Botany 45: 2173–93.
8. ENGLEMANN, W. AND H. W. HONEGGER. 1967. Versuche zur Phasenverschiebung endogener Rhythmen: Blütenbewegung van *Kalanchoë blossfeldiana*. Z. Naturforsch. 22B: 200–04.
9. HALABAN, R. 1968. The circadian rhythm of leaf movement of *Coleus blumei* x *C. frederici*, a short day plant. I. Under constant light conditions. Plant Physiol. 43: 1883–86.
10. HAMNER, K. C. 1960. Photoperiodism and circadian rhythms. Cold Spring Harbor Symp. Quant. Biol. 25: 269–77.
11. HAMNER, W. H. AND J. T. ENRIGHT. 1967. Relationships between photoperiodism and circadian rhythms of activity in the house finch. J. Exptl. Biol. 46: 43–61.
12. HILLMAN, W. S. 1964. The Physiology of Flowering. Holt, Rinehart, and Winston. New York, Chicago, San Francisco, Toronto, and London. 164 p.
13. HILLMAN, W. S. 1964. Endogenous circadian rhythms and the response of *Lemna perpusilla* to skeleton photoperiods. Am. Naturalist 98: 323–28.
14. IMAMURA, S. I. 1967. Physiology of Flowering in *Pharbitis nil*. Japanese Soc. Plant Physiol., Tokyo, Japan. 150 p.
15. KRIBBEN, F. J. 1955. Zu den Theorien des Photoperiodismus. Beitr. Biol. Pflanzen 31: 296–311.
16. LANG, A. 1965. Progressiveness and contagiousness in plant differentiation and development. Encycl. Plant Physiol. W. Ruhland, ed. 409–23.
17. MENAKER, M. AND A. ESKIN. 1967. Circadian clock in photoperiodic time measurement: a test of the Bünning hypothesis. Science 157: 1182–85.
18. MINIS, D. H. 1965. Parallel peculiarities in the entrainment of a circadian rhythm and photoperiodic induction in the pink boll worm (*Pectinophora gossypiella*). In: Circadian Clocks. J. Aschoff, ed. North Holland Publishing Company, Amsterdam. 333–43.
19. PITTENDRIGH, C. S. 1966. The circadian oscillation in *Drosophila pseudoobscura* pupae: A model for the photoperiodic clock. Z. Pflanzenphysiol. 54: 275–307.
20. PITTENDRIGH, C. S. AND V. G. BRUCE. 1959. Daily rhythms as coupled oscillator system and their relation to thermoperiodism and photoperiodism. In: Photoperiodism and Related Phenomena in Plants and Animals. R. B. Withrow, ed. A.A.A.S., Washington. 475–505.
21. PITTENDRIGH, C. S. AND D. H. MINIS. 1964. The entrainment of circadian oscillations by light and their role as photoperiodic clocks. Am. Naturalist 98: 261–94.
22. SALISBURY, F. B. 1963. The Flowering Process. The Macmillan Company, New York. 234 p.
23. WITHROW, R. B. 1959. Photoperiodism and Related Phenomena in Plants and Animals. A.A.A.S., Washington. 903 p.

The Circadian Rhythm in Photosynthesis in *Acetabularia* in the Presence of Actinomycin D, Puromycin, and Chloramphenicol

BEATRICE M. SWEENEY, CHARLES F. TUFFLI, JR., and RICHARD H. RUBIN

From the Department of Biology, Yale University, New Haven, Connecticut

ABSTRACT Anucleate *Acetabularia crenulata* shows a circadian rhythm in photosynthesis. In this study, an oxygen electrode was employed to measure this photosynthetic rhythm in the presence and absence of the inhibitors, actinomycin D, chloramphenicol, and puromycin. High concentrations of the inhibitors were used: actinomycin D, 20–40 $\mu g\ ml^{-1}$; puromycin, 30 and 100 $\mu g\ ml^{-1}$; and chloramphenicol, 250 $\mu g\ ml^{-1}$. The effectiveness of these inhibitors on protein synthesis was also measured under the same conditions used for the determination of rhythmicity. In spite of large effects of all three inhibitors on the incorporation of ^{14}C leucine, no effect on the period or the phase of the photosynthetic rhythm was observed. The higher concentration of puromycin and chloramphenicol produced toxic effects which were expressed as a reduction in the amount of photosynthesis, but rhythmicity was still apparent. After 3 or 4 days' exposure to actinomycin, *Acetabularia* became resistant to its effect. Recovery was also observed in the ability to incorporate leucine. The implications of these results for theories of the basic oscillator responsible for circadian rhythmicity are discussed.

INTRODUCTION

In both plant and animal cells, regular fluctuations in the rate of one, or even of several physiological processes may occur, although the environmental conditions are not changing. If such patterns repeat at intervals of about 24 hr in constant light and temperature, these fluctuations are called circadian rhythms. At the present time it is not known what controls these oscillations, but it is quite unlikely that they are governed by any environmental cyclic cues arising from the rotation of the earth, since all such cues would of necessity show exactly a 24 hr period, while the period of many circadian rhythms can be appreciably different from 24 hr. Furthermore it is not uncommon for the period to change by some small though discernible amount when the temperature is changed. Therefore it seems likely that the

Reprinted from THE JOURNAL OF GENERAL PHYSIOLOGY, 1967, Vol. 50, No. 3, pp. 647–659, *Printed in U.S.A.*

mechanism which determines rhythmicity is endogenous rather than exogenous. However, in spite of extensive effort, this mechanism has not been identified. Many inhibitors which are specific for certain metabolic pathways do not appear to affect rhythmicity. The processes by which we measure the rhythm may often be partially inhibited without any change in the period or phase of the rhythm itself. This lack of feedback from the processes which show rhythmicity to the underlying rhythmic mechanism is quite characteristic of many circadian rhythms. In *Gonyaulax* for example, rhythmicity is expressed in luminescence, photosynthesis, and cell division, and all these processes show identical behavior of period and phase, as if they were all cued from a single basic oscillator. However, specific inhibitors of photosynthesis and cell division have no effect on either period or phase of the rhythm in luminescence. With most inhibitors, fluctuations can still be seen as long as any measurable metabolism remains. One exception to this generalization comes from the experiments of Karakashian and Hastings (12, 13). They found that actinomycin D inhibited the rhythm in luminescent glow in *Gonyaulax* but did not entirely suppress the glow. From this evidence they suggested that DNA-dependent RNA synthesis may be involved in the mechanism underlying this rhythmicity. Whether the site of the effect was within the primary oscillator or in the coupling sequence which must intervene between the oscillator and the actual production of the glow was not determined. Rhythmicity in photosynthesis in *Gonyaulax* was not inhibited by the same concentration of actinomycin D nor in fact by a concentration ten times higher, and the rhythms in glow and photosynthesis were neither one inhibited by chloramphenicol. Thus the meaning of the effect of actinomycin on the glow rhythm was difficult to interpret.

The nucleus is certainly not required for the continuation of rhythmicity in *Acetabularia*, since Sweeney and Haxo (21) showed that removing the nucleus in this giant cell, the cytoplasm of which is so remarkably persistent, had not the slightest effect on the rhythm in photosynthesis which this alga shows very clearly (18, 19, 21). Of course, DNA-dependent RNA synthesis was not eliminated from consideration by these experiments since the chloroplasts of *Acetabularia* have been shown to contain DNA (10), and this DNA has been shown to function in the synthesis of RNA and protein (16). It is probable that mitochondria also contain functional DNA and RNA.

The experiments reported here were designed to examine the effect of inhibitors of RNA and protein synthesis on the rhythm of photosynthesis in enucleated *Acetabularia*, in order to determine whether or not steps in this process in the cytoplasm could be implicated in the causation of rhythmicity.

MATERIALS AND METHODS

Acetabularia crenulata used in this study was collected either from the Florida Keys or from Jamaica, W. I. Plants with caps that had not yet formed reproductive cysts were selected and were enucleated immediately by cutting off the basal portion. This procedure was adopted since plants showed a strong tendency to undergo nuclear division and form cysts within the cap on being brought into the laboratory, and, of course, such plants cannot be enucleated. Epiphytes were removed under the dissecting microscope. The plants, with caps about 1 cm in diameter and stems 1.5 cm long, were maintained in covered dishes containing autoclaved sea water in an artificial light-dark cycle (12 hr light at 250 ft-c and 12 hr darkness) at a temperature of 16–17°C. Plants were not used for experiments until at least 2 wk after enucleation and any with cysts at this time were discarded. In later experiments, plants grown from cysts in the laboratory were used. These plants were not contaminated with other algae, but were not bacteria-free.

The oxygen evolved in photosynthesis was measured by the method of Schweiger, Wallraff, and Schweiger (18), using either a Jarrell-Ash or a Beckman oxygen electrode. Four or eight *Acetabularia* cells were enclosed in a glass-stoppered tube to the base of which a stopcock and capillary had been attached. Sterile sea water, equilibrated for 15 min with 5% oxygen in nitrogen, was added and the tube was stoppered tightly. The plants were then exposed to light from a cool white fluorescent source (500 ft-c for 1 hr at 17°C). At the end of the light exposure, the oxygen in the tube was measured by introducing the water into a small closed chamber with a tiny magnetic stirrer and the electrode. The volume of this chamber was 3.5 ml and that of the water sample was 12 ml so that it was possible to rinse the electrode chamber thoroughly between measurements. This whole apparatus was immersed in a water bath at 16°C. Two oxygen measurements were made from each tube and two to three replicate tubes for each treatment were used. Three tubes without plant material but filled with sea water at the same time and measured at the same time as tubes containing plants served as blanks. The difference between the oxygen content of the tubes with *Acetabularia* and these blanks was used as the measure of the rate of photosynthesis. Measurements were usually made at 1100–1200 and at 2300–2400 each day during the course of an experiment. The method measures net photosynthesis. However, no correction was made for respiration since preliminary experiments had shown oxygen uptake in darkness to be small relative to photosynthesis and constant with time, not rhythmic.

That the measurements did not themselves affect rhythmicity was determined in experiments in continuous light when a different group of plants was used for each oxygen determination (Fig. 4).

When inhibitors were used, they were usually first added at the beginning of a dark period. This procedure was adopted to minimize inactivation by light in the case of actinomycin D. The oxygen measurement during the following night was omitted. Inhibitors were never present during the hour long measurement of photosynthesis. Thus colored inhibitors such as actinomycin D did not interfere with light absorption for photosynthesis. Fresh inhibitor solution was added at the end of each period of oxygen measurement.

Between measurements of photosynthesis, the plants, still in the stoppered tubes but immersed in fresh sea water or in the inhibitor solution in sea water, were returned to the maintenance conditions, either a light-dark cycle or continuous light (50 ft-c) at 16–17°C.

The incorporation of L-leucine uniformly labeled with ^{14}C into protein was measured by an adaptation of the method of Mans and Novelli (14). Young enucleated plants, with a total length of about 300 mm per aliquot, were pretreated with inhibitors for various times, either in alternating light and darkness, or in continuous light of 45 ft-c intensity at 16°C. They were then exposed to radioactive leucine, 0.5 mc in 10 ml sea water containing 5×10^{-4} M penicillin G, with or without inhibitors for 5 hr at 28°C in light (800 ft-c intensity). Plants were rinsed three times with sea water and homogenized in sea water. The volume of each homogenate was made up to 1 ml, and 0.1 or 0.2 ml aliquots were pipetted to discs of No. 5 Whatman paper impaled on pins. After drying briefly in air, the discs were placed in 10% cold TCA containing 0.1 M nonradioactive leucine and allowed to stand overnight at 4°C. The filters were then treated with 5% TCA at 90°C for 30 min to remove nucleic acids and subsequently dried in ethanol and ether. The radioactivity was counted in an Ansitron liquid scintillation spectrometer in 10 ml Bray's solution.

Actinomycin D was the gift of Merck, Sharp and Dohme Research Laboratory, chloramphenicol was obtained through the courtesy of Parke, Davis and Company, and puromycin was purchased from Sigma Chemical Company. L-Leucine uniformly labeled with ^{14}C, specific activity 94.7 mc/mM, was obtained from Calbiochem.

RESULTS

When grown in a light-dark cycle, *Acetabularia* showed a marked rhythm in photosynthesis as previously described (21). Light-dark cycles are not necessary for the manifestation of this rhythm (18, 21), but they serve to entrain the rhythm so that the phase can be predicted. Previous experiments have shown that on a light-dark cycle (12 hr light; 12 hr darkness), the maximum photosynthesis occurs in the middle of the light period. The first experiments on the effect of inhibitors on the photosynthesis of *Acetabularia* were thus carried out using plants on this light-dark cycle (Fig. 1). The rate of photosynthesis was reduced by the time the first measurement was taken after the addition of actinomycin D (20 μg ml^{-1}), an indication that the inhibitor was able to penetrate into the chloroplast. The amount of inhibition appeared to be greater during the day and this effect was progressive, so that little rhythm was observed on the 3rd day after addition of the inhibitor. On the 4th day, however, although fresh actinomycin D was added after each measurement as usual, the effect of the inhibitor began to diminish, and by the 7th day an appreciable rhythm was restored. This recovery could not be attributed to destruction of actinomycin in the sea water solution since freshly prepared solutions were no more inhibitory than those stored at 4°C in the dark before use. Doubling the actinomycin D concentration produced neither greater nor more persistent inhibition. Lower concentrations of actinomycin D (0.1, 0.5, and 1.0 μg ml^{-1}) inhibited neither photosynthesis nor rhythmicity.

Puromycin was used as an inhibitor in one experiment in which the plants were kept on light-dark cycles. At the concentration used (30 μg ml^{-1}), this inhibitor showed no effect on either photosynthesis or rhythmicity. Chloramphenicol (250 μg ml^{-1}) did not affect rhythmicity in *Acetabularia* kept on a light-dark schedule, but like actinomycin, brought about a reduction in photosynthesis.

Parallel experiments in which the incorporation of labeled leucine into protein was measured after 3 or 7 days' exposure to actinomycin D (20 μg ml^{-1}), puromycin (30 μg ml^{-1}), and chloramphenicol (250 μg ml^{-1}) in alter-

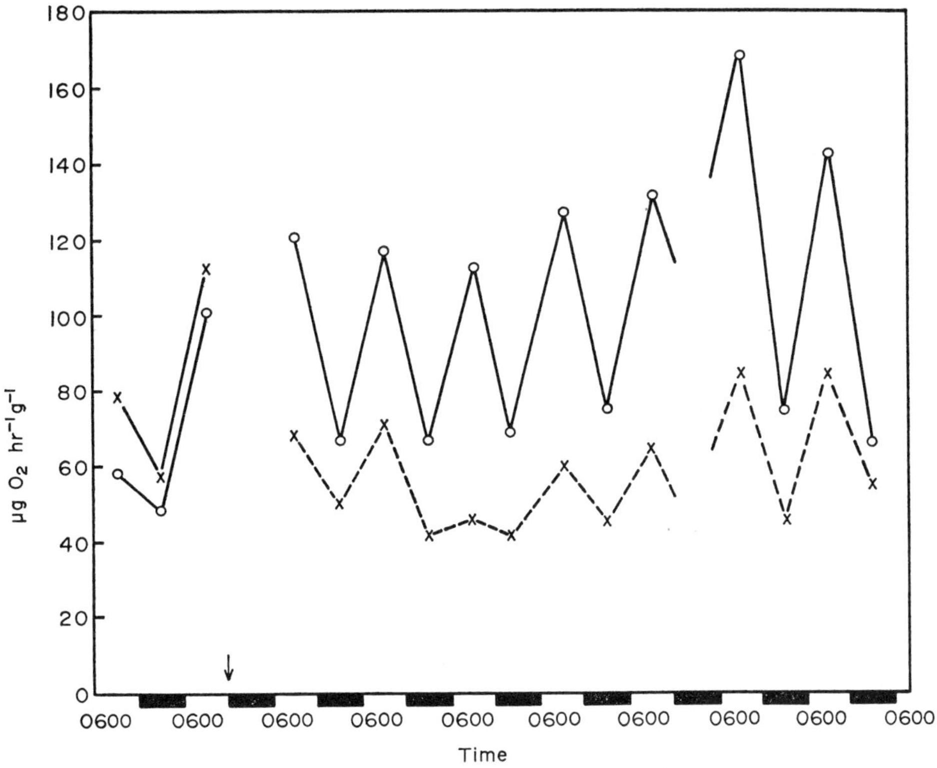

FIGURE 1. The effect of actinomycin D on the rhythm in photosynthesis in *Acetabularia crenulata* in alternating light and darkness. Plants with actinomycin D, 20 μg ml^{-1} (×), control plants without actinomycin D (open circles). The actinomycin was first added at the arrow and again after every subsequent oxygen determination. The black bars on the abscissa each represent a 12 hr dark period ending at 0600 eastern standard time. Ordinate is net oxygen evolution at 500 ft-c and 17°C. The points represent averages of three groups of four plants each. All plants had caps and were anucleate.

nating light and darkness (Table I, Experiment I), showed that protein synthesis had been inhibited by these substances 46, 37, and 44% respectively. The smaller effect of puromycin is consistent with the lack of inhibition of photosynthesis by this inhibitor at this concentration. After about 7 days' treatment in actinomycin, no inhibition of incorporation was found, in keeping with the recovery also observed in photosynthesis.

That the inhibitors at the concentrations employed did not prevent the light-dark cycle from entraining the rhythm in photosynthesis was shown in experiments in which plants received one double light period in the presence and absence of inhibitors. This procedure brings about an inversion of the light-dark cycles. Also none of the inhibitors prevented the rhythm from becoming reversed. Thus in experiments in which the plants were exposed to a light-dark cycle, even in the presence of inhibitors the phase will be held fixed by this cycle. Complete inhibition of rhythmicity would be observable. In order to be able to observe changes in period or phase, however, an experimental regime without entraining signals is necessary. Hence, experiments were carried out in which *Acetabularia* was maintained in continuous light and this light was of relatively low intensity (45–50 ft-c) to prevent the dis-

TABLE I

THE INCORPORATION OF L-LEUCINE u^{14}C INTO PROTEIN BY ENUCLEATED *Acetabularia crenulata* IN THE PRESENCE OF ACTINOMYCIN D, PUROMYCIN, AND CHLORAMPHENICOL

The figures are for the incorporation by 300 mm of fragments at 28°C. in light (800 ft-c) for 5 hr in the presence of 5×10^{-4} M penicillin G.

Experiment:	IA-LD		IB-LD		II-LL	
Pretreatment duration, *hr*	63		159		87.5	
Inhibitor	*cmp*	−%	*cmp*	−%	*cmp*	−%
None	39,685		55,460		35,800	
Actinomycin D						
20 μg ml^{-1}	21,655	46	66,925	0	16,900	53
40 μg ml^{-1}	—		—		20,300	44
Chloramphenicol						
250 μg ml^{-1}	22,140	44	—		19,200	46
Puromycin						
30 μg ml^{-1}	25,155	37	—		—	
100 μg ml^{-1}			—		6,750	81

appearance of rhythmicity (Figs. 2 and 3). Actinomycin D (20 μg per ml), puromycin (100 μg/ml), or chloramphenicol (250 μg per ml) was introduced and replaced as in the experiments described above. In the presence of actinomycin D, photosynthesis was at first inhibited (Fig. 2), more during the day phase than during the night phase, so that rhythmicity was almost lost during the second cycle. Recovery from the effect of actinomycin D also again occurred. Chloramphenicol inhibited photosynthesis but rhythmicity continued in the presence of this inhibitor.

In the course of the experiment shown in Fig. 2, measurements were made every 12 hr. Since the same plants were used repeatedly it was possible that exposure to the bright light might alter the subsequent course of the rhythm in photosynthesis. To test this possibility, an experiment was carried out in which new plants were used for the determination of each point and the resulting curve was compared with one obtained by repeated measurement of the same plants (Fig. 4). No major phase discrepancy was observed. In both this experiment and that of Fig. 2 the period of the rhythm in photosynthesis was unusually long, almost 40 hr. Perhaps this is because the temperature is abnormally low for such a tropical species.

To follow more closely the course of the rhythm in continuous light in the presence of inhibitors, plants were exposed to actinomycin D (20 μg ml^{-1}), puromycin (100 μg ml^{-1}), and chloramphenicol (250 μg ml^{-1}) for 85 hr at 45 ft-c light intensity before oxygen measurements were made. Frequent

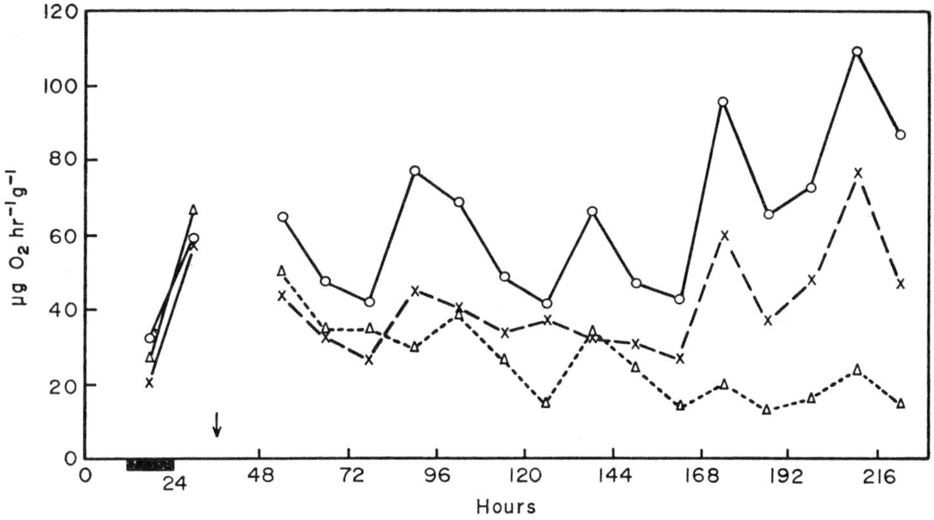

FIGURE 2. The effect of actinomycin and chloramphenicol on the rhythm in photosynthesis in anucleate *Acetabularia crenulata* in continuous light, 50 ft-c. Actinomycin D, 20 μg ml^{-1} (×), chloramphenicol, 250 μg ml^{-1} (open triangles), control (open circles). Inhibitors were first added at arrow and again after every subsequent oxygen determination. The arrow also represents the beginning of the exposure to continuous light of 50 ft-c intensity, ordinate is net oxygen evolution at 17°C and 500 ft-c. The points represent averages of three groups of four plants each.

determinations were then made over the course of several days with special attention to determining the phase in each case. The results are shown in Fig. 3. It is clear that the phase of plants pretreated with actinomycin D and chloramphenicol is not different from that of untreated plants. In an experiment carried out in conjunction with that of Fig. 3, the incorporation of leucine after 87½ hr exposure to the same inhibitors was appreciably inhibited (Table I, Experiment II). Inhibition was in no case complete, however. Doubling the concentration of actinomycin D did not increase the inhibition, as was to be expected from the observation reported above that the inhibition of photosynthesis was no greater at 40 than at 20 μg ml^{-1} actinomycin.

To summarize, then, protein synthesis in enucleated *Acetabularia crenulata*, as in other kinds of cells which have been investigated, appears to be inhibited by actinomycin D, chloramphenicol, and puromycin. The effect of actinomycin is assumed to be indirect via an inhibition of the synthesis of mRNA.

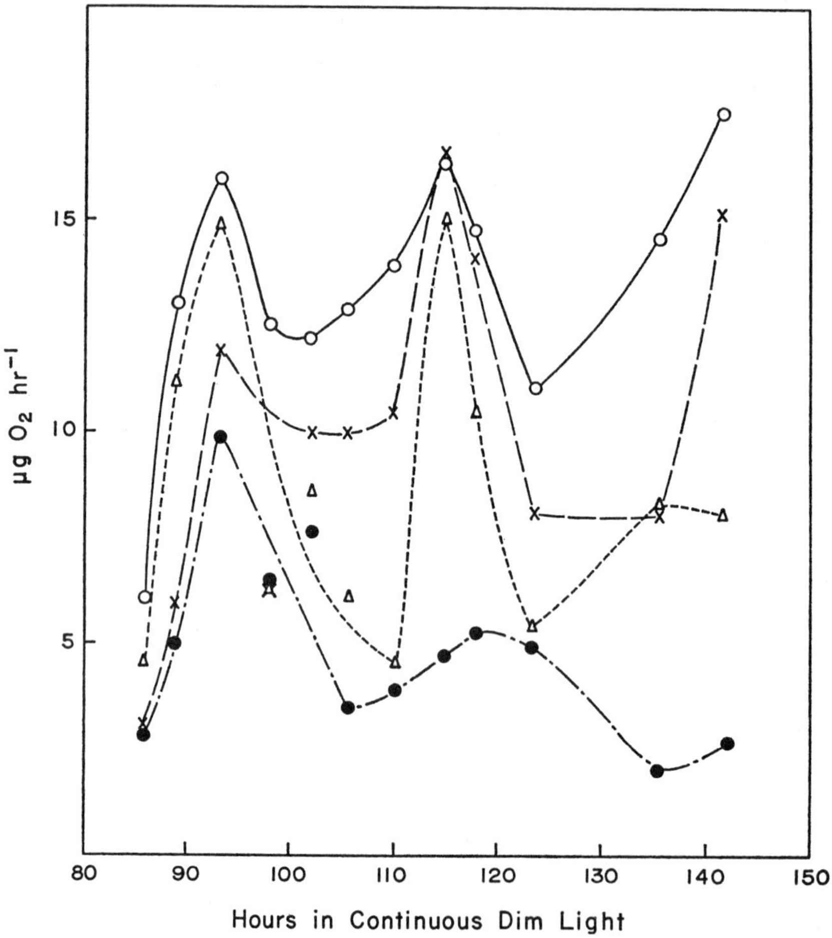

FIGURE 3. The rhythm in photosynthesis in anucleate *Acetabularia crenulata* in continuous light of 45 ft-c intensity during the 4th, 5th, and 6th days in the presence of actinomycin D, 20 μg ml^{-1}(×); puromycin, 100 μg ml^{-1} (filled circles); or chloramphenicol, 250 μg ml^{-1} (open triangles). Control without inhibitors (open circles). Abscissa, time in hours since the plants were placed in continuous light and the inhibitor added. Each point is the average oxygen evolution of two groups of young plants without caps. Each group originally contained a total of 300 mm of plants.

The presence of these inhibitors, however, neither brings about the loss of rhythmicity, nor is the period or the phase of the rhythm in photosynthesis appreciably altered.

DISCUSSION

The most interesting problem in the study of circadian rhythms at the present time is the identification of an oscillator at the molecular level which all evidence indicates must exist within the individual cell and by which the rhythms which we observe in various physiological functions are controlled. There is almost no positive evidence concerning the nature of this oscillator, but it is insensitive to most specific inhibitors as well as to temperature level. However, the evidence of Karakashian and Hastings (12, 13) that actinomycin D inhibits rhythmicity of the luminescent glow in *Gonyaulax* without completely

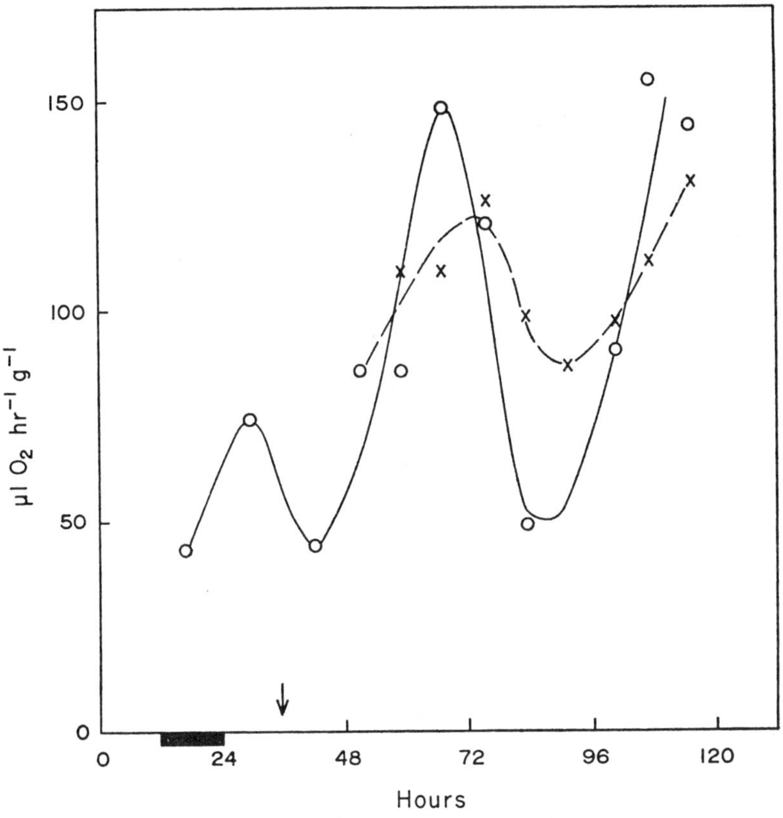

FIGURE 4. A comparison of the rhythm in photosynthesis in anucleate *Acetabularia crenulata* plants with caps measured repeatedly with one group of plants as in experiments shown in Figs. 1–3 (×); or using a different group of plants for each point (open circles). Plants were placed in continuous light, 50 ft-c, at the arrow on the abscissa. Each group contained eight plants with caps.

extinguishing this glow provided a suggestion that the synthesis of messenger RNA on a DNA template might have some connection with the postulated oscillator, although the lack of inhibition by chloramphenicol and other inhibitors of protein synthesis which act at sites removed from the synthesis of RNA seemed to rule out the intervention of protein synthesis as a whole.

Two other studies have yielded evidence of the involvement of an actinomycin-sensitive step in the primary oscillator. Strumwasser (20) found that an isolated neuron from the parietovisceral ganglion of the sea hare *Aplysia* in constant conditions produces a burst of spikes of intracellular electrical potential about the time that the light would have gone on in the light-dark cycle to which the animal had been accustomed, and another smaller spike about 24 hr later. Hence this cell shows circadian behavior. When actinomycin is injected into the neuron before the first spike has occurred, this spike is induced at once, and the next appears 24 hr after the injection of actinomycin. Adding the inhibitor after the first spike has no effect. Strumwasser interprets his results to mean that actinomycin displaces prematurely from its DNA template an RNA which carries the information for the synthesis of a depolarizing substance. Since the rhythm in spike output started afresh after the actinomycin injection, the cyclic production of this mRNA could be the basic oscillator. This possibility has not yet been examined further, however.

One additional study concerns the effect of actinomycin on rhythmicity. This is an unpublished investigation of Vanden Driessche on the photosynthetic rhythm in *Acetabularia mediterranea*, which Brachet mentions very briefly (3, 4). He states that when rhythmicity is lost after long exposures to bright continuous light, it cannot be reinstated if actinomycin is present. Such a finding could mean only that new mRNA is necessary for rebuilding some component of the basic oscillator, but is not required for its day-to-day function. Incidentally, Vanden Driessche also reports that she is able to correlate changes in chloroplast shape with the photosynthetic rhythm in *Acetabularia mediterranea* (22). However, we have not been able to find rhythmic morphological changes in *Acetabularia crenulata* by light microscopy and photomicrography.

The results of the experiments reported here show that in *Acetabularia crenulata* without a nucleus, rhythmicity can continue essentially unchanged in the presence of high concentrations of the inhibitors, actinomycin D, puromycin, and chloramphenicol, concentrations which do affect protein synthesis in this alga. In the face of evidence to the contrary, is it justified to cite this as evidence that steps in the synthesis of mRNA and protein are hence not a part of the basic oscillator responsible for this rhythm?

The first problem concerns the type and degree of inhibition brought about by these inhibitors in this tissue. Eisenstadt and Brawerman (8) have shown that the incorporation of labeled amino acids into protein in isolated chloroplasts from *Euglena* is sensitive to puromycin, less so to actinomycin, and insensitive to chloramphenicol. In *Acetabularia*, actinomycin C and D (10–40 μg ml^{-1}) and puromycin (3–30 μg ml^{-1}) cause marked reduction in growth and cap formation (2, 5, 6, 17, 23). The morphology of the nucleus and chloroplasts appears to be modified by the presence of actinomycin D (10–20 μg ml^{-1}) and puromycin (10 μg ml^{-1}) (1, 3, 6). Therefore there seems to be no doubt that these inhibitors can penetrate into the cytoplasm of this alga. The chloroplasts of *Acetabularia* have been shown to contain DNA (10), and this DNA can function as an RNA template in isolated chloroplasts, since RNA can be synthesized in such chloroplasts (15). That this RNA synthesis is inhibited by actinomycin D is suggested by the inhibition of uracil incorporation in the presence of this inhibitor (16). Brachet and Goffeau (7, 11) have demonstrated that chloroplasts isolated from anucleate fragments of *Acetabularia mediterranea* do incorporate labeled amino acids, and that this incorporation is inhibited by actinomycin D, chloramphenicol, and puromycin. However, high concentrations were required and the inhibition was incomplete. Our experiments on the incorporation of label from leucine ^{14}C by intact anucleate cells also show that all these inhibitors at the concentrations employed cause marked although not complete inhibition.

The finding that actinomycin, puromycin, and chloramphenicol do not inhibit protein synthesis 100% is almost universal (11, 16). In most cases, the maximum inhibition is about 50%, and increasing the inhibitor concentration does not increase the inhibition. An exception is the inhibition of the incorporation of label from amino acids by very high concentrations of puromycin (167 $\mu g\ ml^{-1}$) which Goffeau and Brachet (11) report may reach 100% under special conditions. Several explanations for a maximum inhibition of 50% have been suggested. Schweiger and Berger (16) find that a part of the label from uracil is not rendered soluble by RNAase and hence may be in substances other than RNA. Long lived mRNA's have been postulated in *Acetabularia* to account for the long survival of this cell in the anucleate condition. Such mRNA's may be protected from attack in stable complexes.

The presence of bacteria in all the plant material except that of Gibor (9, 10) renders the interpretation more difficult, although infection was controlled by penicillin and the use of short exposure times. The fact that Gibor (9) obtained inhibitions of similar magnitude with actinomycin D in bacteria-free *Acetabularia* suggests that the intervention by bacteria is not quantitatively important.

The second question concerns what one would expect to observe were the basic oscillator inhibited by inhibitors of protein synthesis. Effects at the level of the oscillator must be distinguished from those on processes which are only controlled by this oscillator, photosynthesis in this case. Thus only changes in period and phase are usually considered significant in this respect. In certain cases, however, the disappearance of rhythmicity may be detected, provided the process by which rhythmicity is measured is not itself completely inhibited, as in the experiments of Karakashian and Hastings with actinomycin D (12, 13). Our experiments on plants in a light-dark cycle (Fig. 1) were designed with the hope of detecting a similar effect of actinomycin. However, in *Acetabularia* the complete inhibition of rhythmicity was not observed, and partial inhibition under these conditions cannot be considered of any significance for the basic oscillator so long as any rhythmicity can be detected. The effects indicate rather an inhibition of photosynthesis.

The rather unprecedented recovery from the effects of actinomycin shown by both photosynthesis and leucine incorporation warrants brief comment. Possibly an enzyme or microorganism capable of breaking down actinomycin is induced after 4–5 days in the presence of this substance. As a result of this phenomenon, it is not possible to observe the inhibitory effects of actinomycin for an unlimited time.

The failure to observe 100% inhibition of rhythmicity is not surprising, in view of the observation that protein synthesis shows only partial inhibition. But what sort of effect might one expect to observe were the basic oscillator, like protein synthesis, 50% inhibited? Such an inhibition in a step basic to the oscillator might be expected to slow its operation so that its period would be lengthened. The experiments using plants kept in continuous light were designed to detect changes in period in the rhythm of photosynthesis. A period 50% longer than usual would certainly have been detected, but no such change was observed. If the basic oscillator is compensated with respect to chemically active substances, as it appears to be for temperature, then a much smaller increase or even a decrease in period might result from 50% inhibition. Small changes in period accumulate over the course of a number of cycles, however, and hence would be detected in an experiment running a relatively long time, like those reported here. Such changes were not found.

What then can be concluded from these experiments concerning the nature of the basic oscillator? Since the synthesis of protein is not completely inhibited by any of the inhibitors used, the participation of a special inhibitor-resistant path of protein synthesis in the clock cannot be eliminated. However, it seems unlikely that this resistant path could involve protected mRNA complexes, since if their activity is cyclic, they would be expected to become unprotected at some time in this cycle. Participation of an inhibitor-sensitive step in protein synthesis in rhythmicity should be manifest in changes in period, and none was observed. Therefore it seems likely that we should look elsewhere than in the steps leading to RNA and protein synthesis for the primary oscillator which determines the circadian rhythm in photosynthesis in *Acetabularia crenulata*, and perhaps other circadian rhythms as well.

A portion of this research was supported by a grant from the National Science Foundation which the authors gratefully acknowledge.

Received for publication 28 March 1966.

REFERENCES

1. Bolouкнére Presburg, M. 1965. Effet de l'actinomycine D sur l'ultrastructure des chloroplastes et du noyau d'*Acetabularia mediterranea*. *J. Microscopie*. **4**:363.
2. Brachet, J. 1963. Effect of puromycin on morphogenesis in amphibian eggs and *Acetabularia mediterranea*. *Nature*. **199**:714.
3. Brachet, J. 1965. Acetabularia. *Endeavour*. **14**:155.
4. Brachet, J. 1965. Le contrôle de la synthèse des protéines en l'absence du noyau cellulaire. Faits et hypothèses. *Bull. Acad. Roy. Belg., Ser. 5*. **51**:257.
5. Brachet, J., and H. Denis. 1963. Effects of actinomycin D on morphogenesis. *Nature*. **198**:205.

6. BRACHET, J., H. DENIS, and F. DE VITRY. 1964. The effects of actinomycin D and puromycin on morphogenesis in amphibians and *Acetabularia mediterranea. Develop. Biol.* **9**:398.
7. BRACHET, J., and A. GOFFEAU. 1964. Le rôle des acides déoxyribonucleiques (DNA) dans la synthese des protéines chloroplastiques chez *Acetabularia. Compt. Rend.* **259**:2899.
8. EISENSTADT, J., and G. BRAWERMAN. 1963. The incorporation of amino acids into the protein of chloroplasts and chloroplast ribosomes of *Euglena gracilis. Biochim. Biophys. Acta.* **76**:319.
9. GIBOR, A., and S. GRANICK. 1964. Plastids and mitochondria: inheritable systems. *Science.* **145**:890.
10. GIBOR, A., and M. IZAWA. 1963. The DNA content of the chloroplasts of Acetabularia. *Proc. Nat. Acad. Sci. U.S.* **50**:1164.
11. GOFFEAU, A., and J. BRACHET. 1965. Desoxyribonucleic acid-dependent incorporation of amino acids into the proteins of chloroplasts isolated from anucleate *Acetabularia* fragments. *Biochim. Biophys. Acta.* **95**:302.
12. KARAKASHIAN, M. W., and J. W. HASTINGS. 1962. The inhibition of a biological clock by actinomycin D. *Proc. Nat. Acad. Sci. U.S.* **48**:2130.
13. KARAKASHIAN, M. W., and J. W. HASTINGS. 1963. The effects of inhibitors of macromolecular biosynthesis upon the persistent rhythm of luminescence in *Gonyaulax. J. Gen. Physiol.* **47**:1.
14. MANS, R. J., and G. D. NOVELLI. 1961. Measurement of the incorporation of radioactive amino acids in protein by a filter-paper disc method. *Arch. Biochem. Biophys.* **94**:48.
15. NAORA, H., H. NAORA, and J. BRACHET. 1960. Studies on independent synthesis of cytoplasmic ribonucleic acids in *Acetabularia mediterranea. J. Gen. Physiol.* **43**:1083.
16. SCHWEIGER, H. G., and S. BERGER. 1964. DNA-dependent RNA synthesis in chloroplasts of Acetabularia. *Biochim. Biophys. Acta.* **87**:533.
17. SCHWEIGER, H. G., and E. SCHWEIGER. 1963. Zur Wirkung von Actinomycin C auf Acetabularia. *Naturwissenschaften.* **50**:620.
18. SCHWEIGER, E., H. G. WALLRAFF, and H. G. SCHWEIGER. 1964. Uber tagesperiodische Schwankungen der Sauerstoffebilanz kernhaltiger und kernloser *Acetabularia mediterranea. Z. Naturforsch.* **19b**:499.
19. SCHWEIGER, E., H. G. WALLRAFF, and H. G. SCHWEIGER. 1964. Endogenous circadian rhythms in cytoplasm of Acetabularia: influence of the nucleus. *Science.* **146**:658.
20. STRUMWASSER, F. 1965. The demonstration and manipulation of a circadian rhythm in a single neuron. *In* Circadian Clocks. J. Aschoff, editor. North-Holland Publishing Co., Amsterdam.442.
21. SWEENEY, B. M., and F. T. HAXO. 1961. Persistence of a photosynthetic rhythm in enucleated Acetabularia. *Science.* **134**:1361.
22. VANDEN DRIESSCHE, T. 1966. Circadian rhythms in *Acetabularia*: photosynthetic capacity and chloroplast shape. *Exptl. Cell Res.* **42**:18.
23. ZETSCHE, K. 1964. Der Einfluss von Actinomycin D auf die Abgabe morphogenetischer Substanzen aus dem Zellkern von *Acetabularia mediterranea. Naturwissenschaften.* **51**:18.

5
Chemical Growth Regulation and Hormones

Reprinted from PLANT PHYSIOLOGY, Vol. 39 No. 3, May, 1964, pages 360–364.

(PRINTED IN U.S.A.)

Enhancement by Auxin of Ribonucleic Acid Synthesis in Excised Soybean Hypocotyl Tissue [1, 2, 3]

Joe L. Key and Jack C. Shannon

Department of Botany and Plant Pathology, Purdue University, Lafayette, Indiana

Many changes in RNA metabolism occur during normal (3, 4, 9, 16, 27, 28) and auxin-modified growth (1, 2, 7, 8, 10, 11, 12, 13, 14, 25) of plant cells. Skoog (18) suggested in 1954 that the action of auxin in regulating growth is concerned with nucleic acid metabolism. Yet, there is no evidence that the influence of auxin on RNA metabolism is more primary than other metabolic responses of auxin.

The experiments reported in this and the following paper (11) were done to gain additional information on the regulation by auxin of growth and development of plant cells as related to RNA metabolism. In general, growth promotive concentrations of auxin stimulated C^{14}-nucleotide incorporation into RNA of elongating cells whereas inhibitory concentrations decreased incorporation. In fully elongated cells, auxin induced a net synthesis of RNA, the increase occurring primarily in the ribosomal fraction. The data reported provide evidence for the DNA dependence of ribosomal RNA synthesis.

Materials and Methods

Soybean seeds (*Glycine max*, var. Hawkeye) were planted between layers of Krum moistened with distilled water in 22 × 33 cm Pyrex baking dishes covered with Saran wrap (perforated for aeration). After 48 hours at 27 to 29° in the dark, the Saran cover was removed, and 150 ml of solution containing 1×10^{-3} M $CaCl_2$, 3×10^{-4} M $MgCl_2$, and 3×10^{-3} M KCl were added to each tray. Experimental tissue was harvested after an additional 24 hours of growth.

Tissue sections were incubated in Erlenmeyer flasks in a 30° water-bath shaker. After incubation, the sections were blotted to remove excess moisture, weighed, and homogenized in ice-cold deionized water containing one drop of antifoam, unless otherwise stated. The homogenates were filtered through glass wool, and aliquots removed for RNA determinations. In experiments where C^{14}-nucleotide was added to the incubation medium, aliquots of the homogenates were removed, dried, and counted for total nucleotide uptake. RNA analyses were made by a modified Smillie and Krotkov procedure (19). The aliquots were made to 0.2 N with respect to $HClO_4$, thoroughly mixed, and centrifuged at 1100 × *g* for 10 minutes. The pellets were then suspended and twice washed in 0.2 N $HClO_4$ and once in methanol containing 0.02 M formic acid. All steps including centrifugation were carried out at 2 to 4°. The washed pellets were twice extracted at 37° for 30 minutes in a 2: 2: 1 mixture of ethanol-ether-chloroform to remove lipids. RNA was hydrolyzed in 0.3 N KOH for 18 hours at 37°.

---

[1] Aug. 5, 1963.

[2] This work was supported by Public Health Service Research Grant GM 10157–03 from the National Institutes of Health.

[3] Purdue Agriculture Experiment Station Journal Paper 2186.

After chilling, $HClO_4$ was added to a final concentration of about 0.3 N followed by centrifugation to remove the $KClO_4$ precipitate, protein, and DNA. Absorbance of samples of RNA was measured at 260 and 290 mμ with a Zeiss PMQII Spectrophotometer. The absorption difference was referred to a standard conversion factor obtained in the following manner. Samples of RNA treated as above were added to a column of Dōwex-1-formate, and the component nucleotides were eluted by a formic acid gradient and collected in a Rinco collector. Each sample was read at 260 mμ, and from the known A_{260} values the quantity of each of the 4 nucleotides in the sample of RNA was determined. These quantities (based on 6 separate determinations) were referred to the initial 260 to 290 mμ absorption difference and used in calculating the conversion factor of 48: $(A_{260}–A_{290})$ (48) = μg RNA/ml of sample read.

After determining the RNA, each sample was neutralized with KOH to pH 4.5 to 5.5. After chilling, the $KClO_4$ precipitate was removed by centrifugation, and aliquots were plated, dried, and counted for determination of C^{14}-nucleotide incorporation into RNA. Counting was done with an automatic gas-flow counter equipped with a micromil window.

Table I

Effect of Auxin on Growth and RNA Metabolism in Excised Soybean Hypocotyl Tissue

0.75, 1.25, and 1.5 g of tissue section 1, 2, and 3, respectively, were incubated for 12 hours at 30° in a water bath shaker in a volume of 5 ml of solution containing 5 × 10^{-3} M KH_2PO_4 (neutralized to pH 6.0 with NH_4OH), 1 % sucrose, 20 μg/ml streptomycin, K salt of 2,4-D (pH 6.0), and 0.125 μC ADP-8-C^{14} (9.2 μc/mg).

μg/ml 2,4-D	% increase fr wt	mg RNA/ g fr wt*	cpm/g fr wt in RNA	cpm/mg RNA
Tissue Section 1**				
0	45	2.84	5,330	1,875
5	71	2.98	7,155	2,405
25	67	2.95	7,020	2,380
100	57	2.87	5,200	1,810
500	24	2.87	2,660	925
Tissue Section 2**				
0	32	1.07	2,300	2,145
5	102	1.21	3,950	3,260
25	95	1.22	3,680	3,020
100	73	1.14	3,175	2,780
500	20	1.01	1,310	1,095
Tissue Section 3**				
0	9	0.58	2,270	3,915
5	26	0.76	4,925	6,480
25	25	0.75	5,600	6,800
100	20	0.72	4,815	6,700
500	7	0.63	1,355	2,150

* The RNA values compare to an initial RNA content of 3.02, 1.27, and 0.59 mg per g initial fresh weight of tissue sections 1, 2, and 3, respectively. All data are based on initial fresh weight. Data are averages of 3 closely duplicating experiments.

** Tissue section 1, 2, and 3 refer to distance from cotyledon where hypocotyl was sectioned and represent tissue of increasing mean cell age. Section 1, 0.0 to 0.5 cm below cotyledon; Section 2, 0.5 to 1.5 cm below cotyledon; Section 3, 1.5 to 3.5 cm below cotyledon.

ADP-8-C^{14} and ATP-8-C^{14} were purchased from Schwarz Bio-Research with a specific activity of 9.2 and 3.9 μc/mg, respectively. These compounds were used interchangeably based on their availability. The chemical form in which the C^{14} entered the tissue is not known.

Results

The incorporation of C^{14}-nucleotide into RNA of excised soybean hypocotyl tissue was enhanced by concentrations of auxin, 2,4-dichlorophenoxyacetic acid (2,4-D), which promoted expansive growth of these cells (table I). Inhibition of C^{14}-nucleotide incorporation occurred at growth inhibitory levels of auxin. [During this investigation we discovered that previous results (8, table III) were complicated by bacterial contamination and are corrected by data in table I]. The magnitude of the auxin-induced changes in RNA synthesis depended upon the physiological age of the tissue as well as the auxin concentration (table I). No net increase in RNA occurred in meristematic and rapidly expanding cells (sections 1, 2) following auxin treatment even though incorporation of C^{14}-nucleotide was enhanced. In fully elongated cells (section 3), auxin induced a net increase in RNA (25 to 30% in most 12-hour experiments). Although higher concentrations are re-

Table II

Comparison of Effects of 2,4-D and IAA on RNA Synthesis in Mature Soybean Hypocotyl Tissue

1.5 g of tissue (section 3) were incubated in 5 ml solution containing 0.25 μc ATP-8-C^{14} (3.9 μc/mg) as described in table I.

Treatment	Mg RNA/g fr wt	cpm/g fr wt in RNA	cpm/mg RNA
None	0.49	3,620	7,400
25 μg/ml 2,4-D	0.66	8,770	13,280
25 μg/ml IAA	0.56	5,870	10,490
100 μg/ml IAA	0.63	8,900	14,100
Initial	0.49	...	...

Table III

Time Study on Auxin-induced Synthesis of RNA in Mature Soybean Hypocotyl Tissue

Incubation conditions same as in table I except that 0.25 μc of ADP-8-C^{14} (9.2 μc/mg) was used in a volume of 5 ml with 1.5 g of tissue. Initial RNA content was 0.61 mg/g initial fr wt. Tissue section 3 was used.

Incubation Interval (hr)	Mg RNA/ g fr wt	ADP-8-C^{14} in RNA cpm/g fr wt	cpm in RNA/ 3 hr interval	cpm/mg total RNA
0 2,4-D				
0–3	0.60	1,105	1,105	1,840
0–6	0.61	2,473	1,368	4,050
0–9	0.60	3,612	1,139	6,020
0–12	0.59	4,332	920	7,650
25 μg/ml 2,4-D				
0–3	0.61	1,290	1,290	2,115
0–6	0.66	4,406	3,116	6,680
0–9	0.70	7,535	3,129	10,750
0–12	0.75	10,698	3,163	14,250

Table IV

Cytoplasmic Distribution of RNA in Homogenates of Excised Soybean Hypocotyl Tissue

4 g of tissue (section 3) were incubated in 10 ml of solution containing 2 μc ADP-8-C^{14} (9.2 μc/mg) as described in table I.

Sub-cellular* Fraction	Medium 1**		Medium 2**	
	mg RNA/ g fr wt	cpm/mg RNA	mg RNA/ g fr wt	cpm/mg RNA
		0 2,4-D		
Nuclei rich	0.065	15,200	0.027	20,600
Mitochondria rich	0.060	8,390	0.024	6,350
Ribosome rich	0.284	6,650	0.302	6,350
Soluble	0.058	7,160	0.097	7,670
		25 μg/ml 2,4-D		
Nuclei rich	0.070	21,570	0.028	24,250
Mitochondria rich	0.077	15,940	0.026	15,400
Ribosome rich	0.389	13,320	0.419	13,020
Soluble	0.078	14,720	0.128	14,870

* Sub-cellular fraction: 1, 0–6,000 × *g* for 15 min (nuclei rich); 2, 6–20,000 × *g* for 20 min (mitochondria rich); 3, 20–100,000 × *g* for 120 min (ribosome rich); 4, 100,000 × *g* supernatant (soluble).

** Medium 1,-tissue homogenized in 0.5 M sucrose containing 0.01 Tris Cl buffer (pH 7.5). Medium 2,-tissue homogenized in 0.5 M sucrose containing 0.01 M Tris Cl buffer (pH 7.5), 0.001 M $MgCl_2$, and 0.5 % deoxycholate.

quired, IAA elicited the same general enhancement of RNA synthesis as 2,4-D (table II). The increase in fresh weight was associated primarily with cell elongation in sections 1 and 2 and with radial enlargement in section 3.

After an initial lag, the synthesis of RNA as measured by either C^{14}-nucleotide incorporation or by RNA determination proceeded at a linear rate for at least 9 hours in auxin-treated tissue (table III). However, the rate of RNA synthesis in untreated tissue as measured by C^{14}-nucleotide incorporation, sharply declined after 6 hours of incubation. A similar decline in the rate of C^{14}-nucleotide incorporation into RNA of expanding cells occurred concommitant with a decreased growth rate. At optimum concentration, auxin maintained higher rates of both growth and RNA synthesis.

The increase in RNA following auxin treatment occurred primarily in the microsome fraction of the cell (table IV). The same result was obtained when deoxycholate was included in the homogenization medium. Since the increase in RNA was primarily in fraction 3 in the presence of the detergent, it is assumed that there was a net synthesis of ribosomes.

Time-course studies during the linear phase of RNA synthesis showed that at early times a major portion of the C^{14}-RNA was associated with the nuclei-rich fraction (table V). The relative amount of C^{14}-RNA in this fraction progressively declined, although the total amount of C^{14}-RNA increased. The relative decrease in C^{14}-RNA in the nuclear fraction was similar to the increase in the proportion of C^{14}-RNA associated with the ribosomal fraction. The relative amount of C^{14}-RNA in fractions 2 and 4 thus remained essentially constant.

Actinomycin D, which effectively inhibits DNA-dependent RNA synthesis in bacteria (6), inhibited

Table V

Time Study on Cytoplasmic Distribution of C^{14}-RNA in Auxin-treated Soybean Hypocotyl Tissue

4 g of tissue (section 3) were incubated in 10 ml of solution containing 2 μc ADP-8-C^{14} (9.2 μc/mg) and 25 μg/ml 2,4-D. ADP-8-C^{14} was added at appropriate time after 3 hours of preincubation in 2,4-D with all tissue being harvested 12 hours after start of incubation. Homogenates were made in medium 1 and differentially centrifuged as described in table IV.

Exposure time to ADP-8-C^{14} (hr)	Subcellular Fraction				Total cpm in RNA
	Nuclei rich	Mito-chondria rich	Ribo-some rich	Solu-ble	
1.5	44	16	25	15	882
3	27	16	40	18	1,880
6	13	16	51	20	5,403
9	9	17	56	19	8,135

Table VI

Inhibition by Actinomycin D of RNA Synthesis in Excised Soybean Hypocotyl Tissue

1.5 g of tissue (section 3) were incubated in 5 ml of solution containing 0.25 μc ADP-8-C^{14} (9.2 μc/mg) and 25 μg/ml 2,4-D as described in table II. Initial RNA content was 0.56 mg/g fresh weight.

Actinomycin (μg/ml)	mg RNA/ g fr wt	ADP-8-C^{14} in tissue cpm/g fr wt	ADP-8-C^{14} in RNA cpm/g fr wt	% Inhibition by actinomycin of ADP-8-C^{14} incorporation into RNA
0.0	0.70	37,300	13,870	...
0.5	0.60	32,500	7,410	46.6
1.0	0.57	31,300	4,160	70.0
10.0	0.54	30,300	1,070	92.4

Table VII

Cellular Distribution of RNA and C^{14}-RNA in Soybean Hypocotyl Tissue as Affected by Actinomycin D

4.0 g of soybean hypocotyl tissue (section 3) incubated for 12 hours in 10 ml of solution containing 1 μc ATP-8-C^{14} (3.9 μc/mg), 25 μg/ml 2,4-D, and 10 μg/ml actinomycin D as described in table I. Tissue was homogenized in medium 2 and differentially centrifuged as described in table IV.

Cellular fraction	mg RNA/g fr wt		cpm/g fr wt in RNA		cpm/mg RNA		Ratio total C^{14} in RNA
	− Actinomycin	+	− Actinomycin	+	− Actinomycin	+	− Act. D / + Act. D
Nuclei rich	0.029	0.030	694	93	23,900	3,100	7.5
Mitochondria rich	0.026	0.020	414	36	15,900	1,800	11.5
Ribosome rich	0.414	0.309	5,540	742	13,400	2,400	7.5
Soluble	0.137	0.123	1,970	531	14,400	4,310	3.7
Total	0.606	0.482	8,618	1,402	14,200	2,910	6.1

C^{14}-nucleotide incorporation into RNA by 85 to 90 % at a concentration of 10 μg/ml (table VI). At a concentration of 0.5 μg/ml, actinomycin D inhibited RNA synthesis by 50 %. In the presence of actinomycin D, at a concentration which inhibited C^{14}-nucleotide incorporation by about 85 %, ribosomal RNA was the only cellular fraction of RNA which was significantly affected by the chemical (table VII). Also, the incorporation of C^{14}-nucleotide into supernatant RNA, although inhibited by actinomycin D, was relatively less affected than incorporation into RNA of other cellular fractions.

Discussion

Auxin (2,4-D) at growth-promoting concentrations enhanced C^{14}-nucleotide incorporation into RNA of the elongating zones of excised soybean hypocotyl tissue, whereas inhibitory concentrations decreased incorporation. The increase in incorporation was associated with a slightly higher RNA content than in comparable untreated tissue although somewhat lower than the initial RNA content.

In regions of the hypocotyl of fully elongated cells, auxin, at concentrations of 5 to 100 μg/ml, induced a 25 to 30 % net increase in RNA during 12 hours of excised incubation. IAA at higher concentrations caused a similar enhancement of RNA synthesis. After an initial lag, the synthesis of RNA in auxin-treated tissue proceeded at a linear rate for at least 9 hours. Differential centrifugation experiments showed that the major increase in RNA occurred in the ribosomal fraction. However, all RNA fractions showed about a twofold increase in specific activity relative to RNA from untreated tissue.

Many lines of evidence support the view that most cellular RNA is synthesized within the nucleus (15, 30), but there are reports that anucleated *Acetabularia crenulata* synthesize ribosomal RNA (20, 24). Ts'O and Sato (22) reported data from time course experiments suggestive of a transfer of labeled (P^{32}) RNA from the nucleus to the ribosomal fraction of the cell (however, no net increase in ribosomal RNA was realized). Results from time course experiments with our system (under conditions where there was a 30 to 40 % increase in ribosomal RNA) indicate a net transfer of C^{14}-RNA from the nuclear fraction into the ribosomal fraction. This net transfer occurred under such conditions that the other fractions of RNA maintained a constant proportion of C^{14}-RNA over the period. Thus, the proportion of C^{14}-RNA associated with the nuclear and ribosomal fractions showed a reciprocal relationship.

Actinomycin D, at concentrations as low as 10 μg/ml, inhibited the incorporation of C^{14}-nucleotide into RNA by 85 to 90 %. The actinomycin treatment also completely eliminated the net increase in ribosomal RNA following auxin treatment. There appears to be preferential labeling of supernatant RNA in the presence of actinomycin, presumably because of the end group addition of C^{14}-adenylate to s-RNA (21). Inasmuch as actinomycin D specifically blocks DNA-dependent RNA synthesis (6), most, if not all of the RNA synthesis in excised soybean hypocotyl tissue must occur on a DNA template, perhaps within the nucleus. A chromatin system as described for pea embryo nuclei (5) probably carries out the synthesis of RNA. Yankofsky and Spiegelman (29) recently presented evidence, based on specific hybridizations, for participation of DNA in ribosomal RNA formation.

Although it is clear that auxin has some pronounced effects on RNA metabolism, the significance of these responses are presently not fully understood. It is apparent, however, from data presented in the following paper (11), that functional RNA (and in turn protein) synthesis is essential for expansive growth of cells to proceed. The possibility that auxin enhances the growth rate by stimulating RNA synthesis seems reasonable, although more direct evidence is needed. The actual response to auxin varies somewhat with different tissues. In rapidly elongating sections of soybean hypocotyl there is only a slight decrease in RNA content during excised growth (table I) whereas in corn mesocotyl there is considerable RNA breakdown (10). Growth-promoting concentrations of auxin enhance RNA breakdown in corn mesocotyl (10) while causing the maintenance of a slightly higher level of RNA in soybean hypocotyl (table I). The net effect of auxin on C^{14}-nucleotide incorporation into RNA of these tissues is also quantitatively different. Again these differences are not fully understood, but they may well be related to differences in ribonuclease activity. Shannon and Hanson (17) have shown the ribonuclease content to increase about twofold during excised growth of corn mesocotyl tissue; with auxin the increase is about threefold. There is little or no change in ribonuclease activity in soybean hypocotyl tissue following excised growth either in the presence or absence of auxin (unpublished). Also under the assay conditions used the specific activity of ribonuclease is initially much lower in soybean than in corn tissue (26). Thus these differences in the effects of auxin on RNA metabolism in excised corn and soybean tissues may be related to the differential responses obtained with ribonuclease, presumably the enzyme responsible for the degradation of RNA in these tissues.

Summary

Indoleacetic acid and 2,4-dichlorophenoxyacetic acid, at concentrations which promoted cell elongation, enhanced C^{14}-nucleotide incorporation into RNA of excised soybean hypocotyl tissue, whereas inhibitory levels decreased incorporation. In fully elongated cells, auxin induced a 25 to 30 % net increase in RNA, primarily ribosomal. Although different concentrations are required, 2,4-D and IAA produced similar effects on RNA metabolism. Results from time course studies on C^{14}-nucleotide incorporation into RNA, in conjunction with differential centrifugation of tissue homogenates, indicate that a net transfer of C^{14}-RNA from the nucleus to the ribosomes occurred. Moreover, the synthesis of RNA was inhibited by 85 to 90 % by actinomycin D. No increase in ribosomal RNA occurred in auxin treated tissue in the presence of the antibiotic. From the known action of actinomycin D, most of the RNA in soybean hypocotyl cells must be synthesized on a DNA template.

Acknowledgments

The authors are grateful to Mrs. Linda Rainford for technical assistance. The actinomycin D was kindly supplied by Merck, Sharpe, and Dohme, Rahway, New Jersey.

Literature Cited

1. Basler, E. and K. Nakazawa. 1961. Some effects of 2,4-D on nucleic acids of cotton cotyledon tissue. Botan. Gaz. 122: 228–32.
2. Biswas, B. B. and S. P. Sen. 1959. Relationships between auxins and nucleic acid synthesis in coleoptile tissue. Nature 183: 1824–25.
3. Cherry, J. H. 1962. Ribonucleic acid metabolism during growth of excised root tips from normal and X-irradiated corn seeds. Biochem. Biophys. Acta 55: 487–94.
4. Heyes, J. K. 1960. Nucleic acid changes during cell expansion in the root. Proc. Roy. Soc. London 152: 218–30.
5. Huang, Ru-chih C., N. Maheshrwari, and J. Bonner. 1960. Enzymatic synthesis of RNA. Biochem. Biophys. Res. Commun. 3: 689–94.
6. Hurwitz, J., J. J. Furth, M. Malamy, and M. Alexander. 1962. The inhibition of the enzymatic synthesis of RNA and DNA by actinomycin D and proflavin. Proc. Natl. Acad. Sci. 48: 1222–30.
7. Izawa, M. 1961. Effect of IAA on protein and ribonucleic synthesis in cultured bean germ-axes. Botan. Mag. Tokyo 74: 98–103.
8. Key, J. L. and J. B. Hanson. 1961. Some effects of 2,4-D on soluble nucleotides and nucleic acid of soybean seedlings. Plant Physiol. 36: 145–52.
9. Key, J. L., J. B. Hanson, H. A. Lund, and A. E. Vatter. 1961. Changes in cytoplasmic particulates accompanying growth in the mesocotyl of *Zea mays*. Crop Sci. 1: 5–8.
10. Key, J. L. 1963. Studies on 2,4-D-induced changes in RNA metabolism in excised corn mesocotyl tissue. Weeds 11: 177–81.
11. Key, J. L. 1964. RNA and protein synthesis as essential processes for cell elongation. Plant Physiol. 39: 365–70.
12. Masuda, Y. 1959. Role of cellular RNA in the growth response of Avena coleoptile to auxin. Physiol. Plantarum 12: 324–35.
13. Masuda, Y. 1960. Physiological significance of RNA in the growth promoting action of auxin. J. of Inst. Polytech. (Osaka City University) Series D, 11: 1–23.
14. Ormrod, D. P. and W. A. Williams. 1960. Phosphorus metabolism of *Trifolium hirtum* All. as affected by 2,4-D. Plant Physiol. 35: 81-87.
15. Prescott, D. M. 1962. Synthetic processes in the cell nucleus. II. Nucleic acid and protein metabolism in the macronucleus of 2 ciliated protozoa. J. Histochem. Cytochem. 10: 145–53.
16. Setterfield, G. 1961. Structure and composition of plant-cell organelles in relation to growth and development. Can. J. Botany 39: 469–89.
17. Shannon, J. C. and J. B. Hanson. 1962. Ribonuclease activity as a function of growth in the corn mesocotyl. Plant Physiol. suppl. 37: xx.
18. Skoog, F. 1954. Substances involved in normal growth and differentiation of plants. Brookhaven Symp. Biol. 6(BNL 258): 1–21.
19. Smillie, R. M. and G. Krotkov. 1960. The estimation of nucleic acids in some algae and higher plants. Can. J. Botany 38: 31–49.
20. Sutter, R. P., Sandra Whitman, and G. Webster. 1961. Cytoplasmic formation of the ribonucleic acid of ribosomes. Biochem. Biophys. Acta. 49: 233–34.
21. Tamaoki, L. and G. C. Mueller. 1962. Synthesis of nuclear and cytoplasmic RNA of Hela cells and the effects of actinomycin D. Biochem. Biophys. Res. Commun. 9: 451–54.
22. Ts'O, P. O. P. and C. S. Sato. 1959. Synthesis of RNA in plants. II. Phosphate-^{32} P incorporation into the RNA of the pea epicotyls. Exptl. Cell Res. 17: 237–45.
23. Ts'O, P. O. P. 1962. The ribosomes-ribonucleoprotein particles. Ann. Rev. Plant Physiol. 13: 45–80.
24. Webster, G., Sandra Whitman, and R. Heintz. 1962. Cytoplasmic formation of ribosomes. Plant Physiol. suppl. 37: xx.
25. West, S. H., J. B. Hanson, and J. L. Key. 1960. Effect of 2,4-D on the nucleic acid and protein content of seedling tissue. Weeds 8: 333–40.
26. Wilson, C. and J. C. Shannon. 1963. The distribution of ribonucleases in corn, cucumber, and soybean seedlings. Effects of isolation media. Biochem. Biophys. Acta. 68: 311–13.
27. Woodstock, L. W. and F. Skoog. 1960. Relationships between growth rates and nucleic acid contents in the roots of inbred lines of corn. Am. J. Botany 47: 713–16.
28. Woodstock, L. W. and F. Skoog. 1962. Distributions of growth, nucleic acids, and nucleic acid synthesis in seedling roots of *Zea mays*. Am. J. Botany 49: 623–33.
29. Yankofsky, S. A. and S. Spielgelman. 1963. Distinct cistrons for the 2 ribosomal RNA components. Proc. Natl. Acad. Sci. 49: 538–44.
30. Zalokar, M. 1960. Sites of protein and RNA synthesis in the cell. Exptl. Cell Res. 19: 559–76.

Reprinted from Science, November 10, 1967, Vol. 158, No. 3802, pages 791-793

Induction of Coiling in Tendrils by Auxin and Carbon Dioxide

Abstract. *Symmetric application of indole-3-acetic acid, CO_2, or, to a lesser extent, ethylene can substitute for the contact stimulus in inducing coiling in the tendrils of* Marah fabaceus. *In the case of auxin, treatment of the apical few millimeters results in strong, permanent coiling throughout the length of the tendril. The speed of the response to CO_2 is comparable to that to tactile stimuli. A possible mechanism for thigmotropism is outlined.*

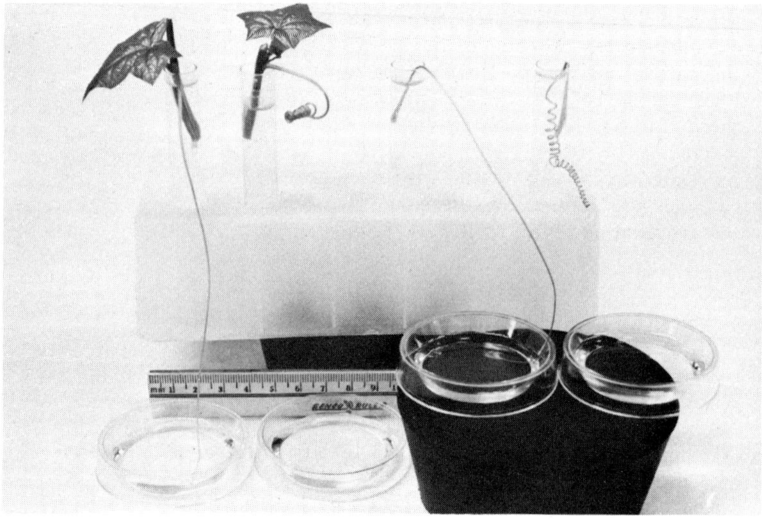

Fig. 2. Tendrils of *Marah fabaceus* 15 hours after placement with tips immersed a few millimeters in water (first and third from left) or in IAA at 150 mg/liter (second and fourth from left). The coiling of the IAA-treated tendrils has caused them to lift high out of the solution.

Accumulated experimental evidence (*1*) suggests that, when a tendril receives a contact stimulus, the growth rate of the stimulated side falls while that of the opposite side rises sharply. While this differential growth would of itself result in coiling, it has recently been proposed (*2*) that in addition a contractile protein may play a part in the initial phase. The mechanism effecting the differential growth response is unknown. Though it does not appear to have been established that auxin participates, Boresch (*3*) advanced a theory according to which contact induces a transverse transport of hormone from the stimulated to the opposite side, analogous to the transverse displacement of auxin in phototropism and geotropism.

In the experiments I report, strong coiling of tendrils was induced by symmetric application of indole-3-acetic acid (IAA); auxin can thus replace the contact stimulus. The experimental material was the wild Californian cucumber, *Marah fabaceus* (Naud) Greene (*4*), which was observed to be more sensitive than peas. In one type of experiment, tendrils excised at their extreme bases were gently floated on water in shallow pans. After an interval of at least 30 minutes to check whether or not manipulation had itself caused a response, IAA was added to the pans to a final concentration of from 2 to 150 mg/liter. Responses were detectable in all IAA concentrations after 20 to 30 minutes. With IAA at 20 mg/liter or above, overnight immersion produced tightly coiled spirals, the control tendrils remaining straight (Fig. 1). Even when the control tendrils received a brief contact stimulus before placement in water and consequently curled at first, they were largely uncurled overnight by the straightening reaction that follows the initial response if the contact is not sustained (*1*). It was immaterial which side of the tendril floated uppermost on the auxin solution. Prolonged immersion in these high concentrations of IAA produced no discernible toxic effects, the tendrils remaining highly turgid; I suggest that their permeability to the auxin is extremely low.

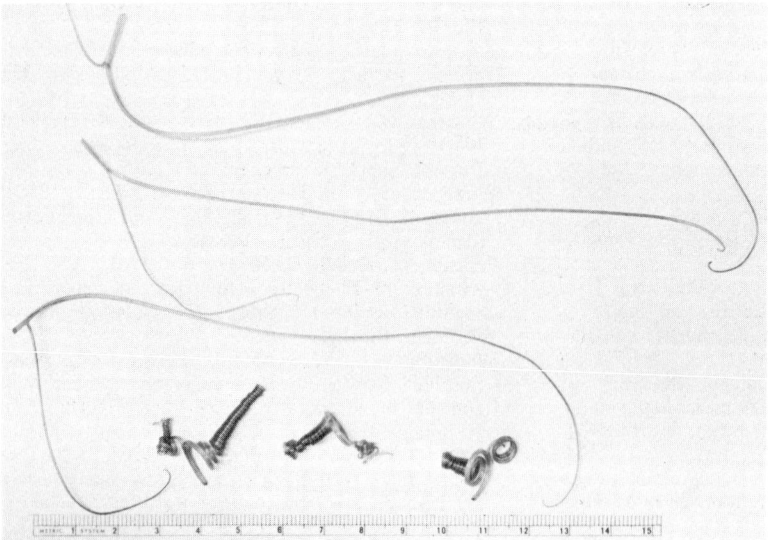

Fig. 1. Tendrils of *Marah fabaceus* after floating for 15 hours on water (top) or on IAA at 100 mg/liter (bottom).

The probable reason why Jaffe and Galston (*5*) observed only slight effect of symmetrically applied IAA is that they sought the enhancement of a tactile stimulus, whereas in our experiments IAA substituted for the latter.

In a second type of experiment, tendrils, either excised or still attached to their branches, were placed overnight with their tips submerged a few millimeters in either IAA solutions (100 to 200 mg/liter) or water. Extremely strong coiling throughout the length of the IAA-treated tendrils (Fig. 2) indicated basipetal transport of the auxin.

That the effect of IAA did not depend on the low pH of the solutions was shown when the latter were buffered at pH 4.0 or 6.5 with citrate.

The pH-4.0 buffer, however, produced slight coiling on its own, the response becoming evident after about 25 minutes in the case of floating tendrils. Gibberellin proved completely ineffective.

Ethylene was slightly effective; CO_2, extremely so. For experiments with these gases, either excised tendrils or sections of stem with tendrils attached were placed with cut ends in water for several hours, preferably overnight, to achieve a high state of turgor. (It has been observed that turgor plays a very influential role in the thigmotropic response.) In the case of ethylene they were then transferred to closed chambers into which the gas was injected. Ethylene at 10 to 20 parts per million produced slight but definite curling; the weakness of the response, compared with the coiling induced by IAA (Figs. 1 and 2) makes it unlikely that the latter was in fact caused by ethylene produced in vivo under the influence of high concentrations of IAA.

Table 1. Time course of coiling of fully turgid tendrils of wild cucumber when gassed with O_2 or with CO_2.

Gas	Increase in curvature (deg) after (min):			
	2.5	5	7.5	10
O_2	0	0	20 ± 31	20 ± 31
CO_2	0	75 ± 78	405 ± 150	555 ± 196

In the experiments with CO_2 the fully turgid tendrils were placed in chambers through which streams of gas (bubbled first through water) were passed. Table 1 shows that, whereas O_2 was almost totally ineffective, CO_2 elicited vigorous curling that started within 5 minutes—close to the average time for response to tactile stimuli. It is likely that this effect of CO_2 is intimately related to its action as a remarkably efficient and quick-acting promoter of extension growth (*6*).

The strong coiling in response to symmetrically applied IAA and CO_2 reported here argues against Boresch's theory that an asymmetric distribution of auxin, effected by contact, causes the thigmotropic response. [The possibility that CO_2 brings about an asymmetric release of auxin within the tissues may be discounted, since growth induced by CO_2 and by IAA differs widely in susceptibility to inhibitors (*6*).] I therefore propose that the thigmotropic response is due, not to an asymmetric distribution of auxin, but to an asymmetric response to auxin. The situation may closely parallel that proposed for the various tissues of the pea epicotyl (*7*). Under the influence of the very low concentrations of auxin present in an unstimulated tendril, the latter grows more or less straight. Higher concentrations, however, produce curvature, either because of differences in the shape of the growth-concentration curves for the tissues on the two sides of the tendril, or because the maximum capacity for growth on the two sides differs.

I suggest the following mechanism for thigmotropism: The contact stimulus gives rise to an action potential in the tendril [observed by Umrath (*8*)]. According to modern theory, this action potential is associated with changed permeability of cell membranes to ions, and a consequent redistribution of ions between, for instance, the vacuole and the cytoplasm. The changed ionic environment brings about an increase in free auxin (by either releasing bound auxin or promoting synthesis) to which the two sides of the tendril respond asymmetrically.

Two of my findings possibly weigh against this proposal: the relatively slow response of the tendrils to external auxin (though this is probably due to slow penetration, as I have noted), and the fact that in its early stages the response to auxin in the floating experiments often did not resemble in form contact coiling. An alternative suggestion is that the auxin-induced coiling represents not the primary swift response to touch—which is soon reversed by the straightening reaction (see *1*) and which may involve a contractile protein (*2*)—but the continued curvature under a sustained contact stimulus which leads to a permanent grasp of the support. On the basis of the experiment shown in Fig. 2, I suggest that this permanent grasp (as well as the coiling that subsequently occurs along the entire length of the tendril, bringing the stem closer to the support) involves a supply of auxin translocated basipetally from the point of contact.

It is noteworthy that the effect of CO_2 resembles the tactile response both in form and in speed. Since CO_2 action is comparable to that of IAA in that it produces a decrease in wall pressure (*9*), this interesting finding appears to favor the first of my two proposals.

LEONORA REINHOLD*
Division of Natural Sciences, University of California, Santa Cruz

References and Notes

1. E. Bünning, in *Encyclopedia of Plant Physiology*, W. Ruhland, Ed. (1959), vol. 17/1, p. 254.
2. M. J. Jaffe and A. W. Galston, *Plant Physiol.* **41**, 1152 (1966); **42**, 845 (1967).
3. G. Borgström, *The Transverse Reactions of Plants* (Williams & Norgate, London, 1939).
4. A specimen has been deposited in the herbarium of the University of California, Berkeley.
5. M. J. Jaffe and A. W. Galston, *Plant Physiol.* **41**, 1014 (1966).
6. M. L. Evans, P. M. Ray, L. Reinhold, in preparation.
7. K. V. Thimann and C. L. Schneider, *Amer. J. Botan.* **25**, 627 (1938).
8. K. Umrath, *Planta Berlin* **23**, 47 (1934).
9. L. Reinhold and Z. Glinka, *Plant Physiol.* **41**, 39 (1966).
10. Supported by NSF grant GB 4337 to K. V. Thimann. I thank K. V. Thimann for his advice and hospitality; also P. M. Ray for discussions and for continuously supplying the plant material.
* Permanent address: Botany Department, Hebrew University, Jerusalem, Israel.

5 September 1967

Timing of the Auxin Response in Coleoptiles and Its Implications Regarding Auxin Action

MICHAEL L. EVANS and PETER M. RAY

From the Division of Natural Sciences, University of California, Santa Cruz, California 95060. Dr. Evans's present address is Department of Biology, Kalamazoo College, Kalamazoo, Michigan 49001. Dr. Ray's present address is Department of Biological Sciences, Stanford University, Stanford, California 94305

ABSTRACT The timing of the auxin response was followed in oat and corn coleoptile tissue by a sensitive optical method in which the elongation of about a dozen coleoptile segments was recorded automatically. The response possesses a latent period of about 10 min at 23°C, which is extended by low concentrations of KCN or by reducing the temperature, but is not extended by pretreatments with actinomycin D, puromycin, or cycloheximide at concentrations that partially inhibit the elongation response. Analysis of the data indicates that auxin probably does not act on the elongation of these tissues by promoting the synthesis of informational RNA or of enzymatic protein. Not excluded is the possibility that auxin acts at the translational level to induce synthesis of a structural protein, such as cell wall protein or membrane protein. While the data do not provide direct support for this hypothesis, the speed with which cycloheximide inhibits elongation suggests that continual protein synthesis may be important in the mechanism of cell wall expansion.

INTRODUCTION

A considerable amount of recent research effort has been directed toward showing that the developmental effects of hormones such as auxin can be ascribed to an activation of gene transcription and specific protein synthesis (5, 41, 42, 47, 62–64, and references there cited). The principal lines of evidence for this are (*a*) inhibition of hormonal responses by antibiotics that inhibit RNA and protein synthesis, (*b*) promotion by hormones of incorporation of labeled precursors into RNA and protein, and (*c*) promotion of synthesis of particular enzymes by hormones. It is widely recognized that these kinds of evidence are somewhat inconclusive as regards the actual means by which hormones influence development. The antibiotics will inhibit the developmental process if it depends in any way upon RNA or protein synthesis,

Reprinted from THE JOURNAL OF GENERAL PHYSIOLOGY, 1969, Vol. 53, No. 1, pp. 1–20 *Printed in U.S.A.*

whether or not the hormone works by gene activation. It is to be expected that promoted growth will in general be accompanied by increased synthesis of RNA and protein even if the hormone does not act directly upon these processes. And although many inductive effects of hormones on enzyme synthesis are known in animals and several in plants (23, 33), it is a separate question whether such enzyme increases are the causal basis of particular physiological responses such as the cell elongation response to auxin.

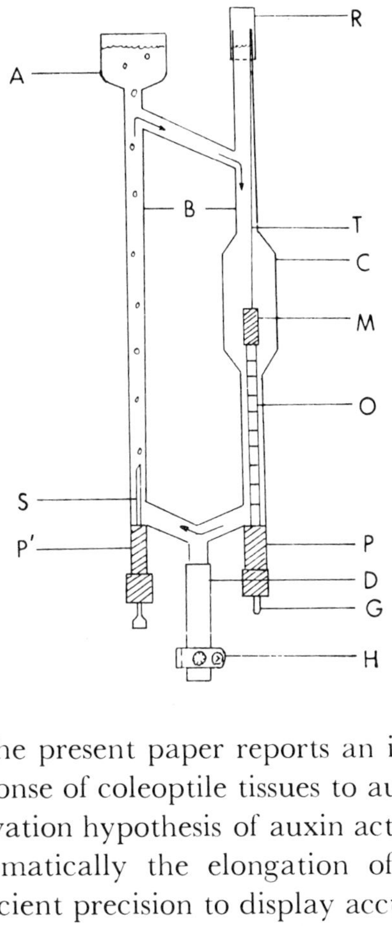

FIGURE 1. Growth measurement chamber, made from Pyrex glass. *A*, overflow reservoir, gas exit vent, and funnel for filling chamber. *B*, 7 mm O.D. glass tubing. *C*, spectrophotometer cuvette (10 mm square). *D*, drain outlet, Tygon tubing. *H*, screw clamp. *G*, glass plug. *M*, weight. *O*, row of coleoptile segments. *P*, *P'*, Teflon plugs. *S*, No. 25 syringe needle. *T*, thread.

The present paper reports an investigation of the timing of the elongation response of coleoptile tissues to auxin, conducted in order to explore the gene activation hypothesis of auxin action. A method was developed for recording automatically the elongation of about a dozen coleoptile segments with sufficient precision to display accurately the course of their elongation during the latent period of about 10 min that was known from previous work (52) to be involved in auxin action, and to permit determination of the effects of various treatments upon the time course of the auxin response.

MATERIALS AND METHODS

Plant Material Oats (*Avena sativa* var. Victory) were grown in an arrangement similar to that of Wiegand and Schrank (66). Immediately after planting, the dish containing the seeds was placed in a refrigerator at about 4°C overnight; the next morning, the dish was removed to a dark room kept at 26°C and placed under a 25 w ruby red bulb about 18 inches above the dish for the next 24 hr. The seedlings were subsequently allowed to grow in complete darkness for 48 hr. Under dim red light segments 8 mm long were cut from coleoptiles 2.5–3.0 cm long, beginning 3 mm from the tip, using a double-bladed cutting device. The leaf was removed from each segment.

When corn coleoptiles were used, corn seeds (*Zea mays* var. Golden Bantam) were soaked for 2 hr in distilled water and grown in vermiculite in a dark room at 26°C. They were given red light as described above during the first 24 hr and then allowed

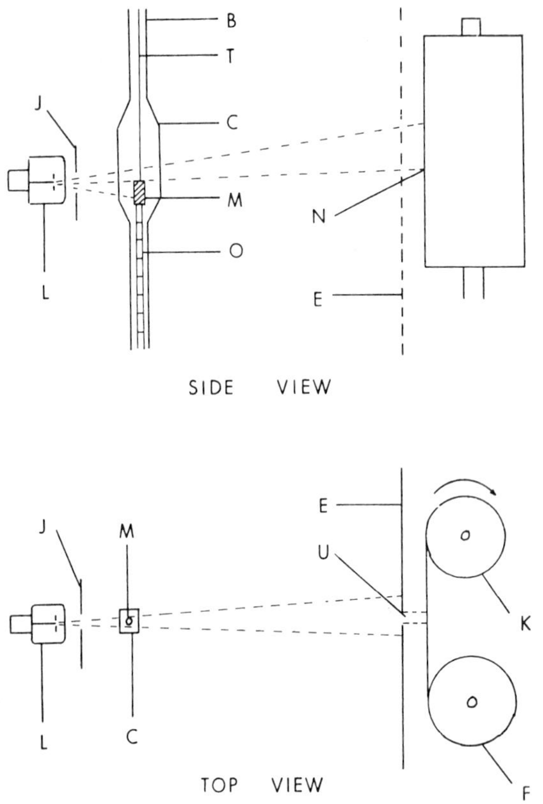

FIGURE 2. Schematic of shadowgraphic growth-recording apparatus (not to scale). *B*, *C*, *M*, *O*, *T*, parts of the growth measurement chamber as indicated for Fig. 1. *E*, baffle, with slit (*U*). *F*, roll of photographic paper. *J*, diaphragm. *K*, kymograph drum. *L*, zirconium arc lamp. *N*, upper edge of shadow of *M*.

to grow in complete darkness to a height of about 3 cm. Coleoptile segments were cut in the same manner as described above for oats.

Method of Measurement A piece of cotton thread about 40 cm in length was passed through a small hole in the Teflon plug *P* shown in Fig. 1. A knot was tied in the end of the thread to prevent it from slipping through the hole. The hollow coleoptile segments were then strung on the thread one at a time until a column of (usually) 13 segments was obtained. A weight consisting of a piece of dark red Pyrex glass capillary tubing weighing 92 mg, about 2 mm in diameter and 10 mm long, was then threaded so that it rested on the uppermost coleoptile segment. The Teflon plug (*P*) with thread (*T*) bearing the coleoptile segments (*O*) was inserted into the growth measurement chamber as shown in Fig. 1. The glass weight (*M*) fell within a spectrophotometer cuvette (*C*). The upper end of the thread was dropped over the upper edge of the glass tube above the cuvette and was secured by slipping a tight-fitting rubber sleeve (R) over the glass tube and thread. The hole in Teflon plug P was stoppered with a small glass rod (G). Opening D was fitted with a small piece of Tygon tubing and a pinch clamp (H); this tube served as a drain. A syringe needle (S) fitted tightly into Teflon plug P' and served to gas the solution within the chamber with prepurified grade oxygen. The gassing caused the solution to circulate within the chamber as indicated by the arrows in Fig. 1. The total volume of the chamber was 14 ml. The chamber could be drained (via opening D) and refilled (via funnel A) with a different growth medium in a few seconds.

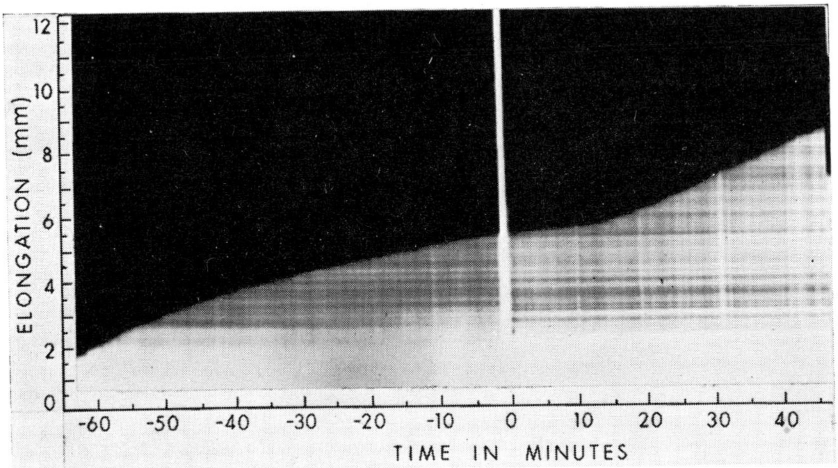

FIGURE 3. Record produced by the shadowgraphic apparatus (reproduced 0.25 times original size). Medium was changed from water to 3 μg/ml. IAA at the time corresponding to the vertical white line. The downward displacement of the shadow at the end of the record, made for calibration of optical magnification, represents 1.5 mm of actual vertical displacement. The rapid elongation that was occurring at the beginning of the record is the result of tactile stimulation of the oat coleoptile segments as explained in Methods.

The elongation of the entire column of coleoptile segments was recorded shadowgraphically as diagrammed in Fig. 2. A 2 w zirconium arc lamp (L) (Sylvania Type GZ2U Model C2) was positioned within the body tube of a horizontal microscope in place of the objective lens. At a distance of about 1 m from the stand supporting the growth chamber, a baffle (E) with a vertical slit (U) 2 mm wide was placed. Behind the baffle was located a spool (F) bearing a roll of Kodak polycontrast photographic paper (203 mm wide) and a multispeed kymograph drum (K), positioned as illustrated in the lower part of Fig. 2. The strip of photographic paper was pulled slowly by the kymograph drum past the slit. The speed of travel was usually 3.08 mm/min.

Light from the zirconium arc, which constitutes a nearly point source (0.127 mm in diameter), was limited by the diaphragm (J) so that a narrow cone of light passed through the cuvette (C) and fell upon the weight (M) but not upon the coleoptile segments. A sharp shadow (N) of the weight, within the circle of light, was thereby cast upon the slit as shown in Fig. 2. The light exposed that portion of the photographic paper (behind the slit) that lay above the edge of the shadow. As the segments grew, this boundary moved upward and traced a record of the elongation which, after development of the photographic paper, appeared as in the example shown in Fig. 3. The vertical white line shown in Fig. 3 is an event marker and was produced by covering the slit during a solution change. All other growth curves presented in this paper are tracings of original records of the same kind.

In this recording method the elongation of the coleoptile segments was multiplied by the optical lever arm *LN/LM*. This magnification was calibrated at the end of each experiment, by moving the lamp up or down a known distance using a vertical

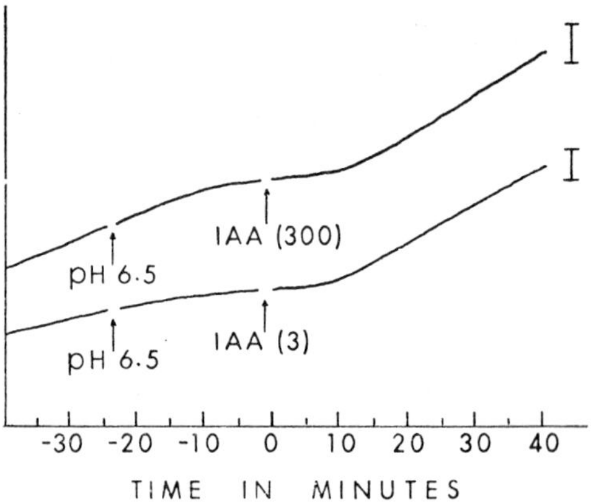

FIGURE 4. Effect of widely different IAA concentrations on timing of elongation response of oat coleoptile segments. At the first arrow water in chamber was replaced with Na citrate buffer, 10^{-3} M, pH 6.5; at the second arrow this was replaced with the same medium containing either 3 or 300 μg/ml IAA as indicated. The vertical bar by each curve in this and subsequent figures represents 1.0 mm of elongation, for that particular record, for the entire row of coleoptile segments.

micrometer screw in the lamp mounting, thus causing a displacement of the shadow boundary which appeared on the photographic record (cf. Fig. 3). In most experiments the optical magnification was between 15 and 20. This calibration allowed the absolute elongation to be calculated from the shadowgraph record. With each growth curve a vertical bar is given which indicates 1 mm of elongation (by the entire row of coleoptiles) for that growth record.

Chemicals Actinomycin D was generously supplied by Merck, Sharp and Dohme, West Point, Pa. Puromycin and cycloheximide were obtained from Nutritional Biochemicals Corporation, Cleveland, Ohio. Tissue was treated with antibiotics in neutral, unbuffered solution.

Preconditioning of Segments Oat coleoptile segments were normally floated on distilled water (and/or in some pretreatment medium) for at least 30 min after having been strung on the thread, prior to insertion into the growth measurement chamber and commencement of recording. This preconditioning was necessary because immediately after being strung on the thread the segments showed a burst of rapid elongation which lasted about half an hour. This is illustrated in the left-hand part of Figs. 3 and 7 (C and D), which are records that were begun shortly after the segments had been strung on the thread. During this period of rapid elongation a promotion by auxin could not be obtained. Investigation indicated that this initial growth burst was not due to disappearance of endogenous auxin nor to manipulations accompanying insertion into the growth chamber but was the result of mechanical stimulation of the coleoptile segments by the thread when they were moved along it. Analogous effects of tactile stimuli on coleoptile elongation have been described previously (6, 13). We found no such tactile effect on the elongation of corn coleoptile segments,

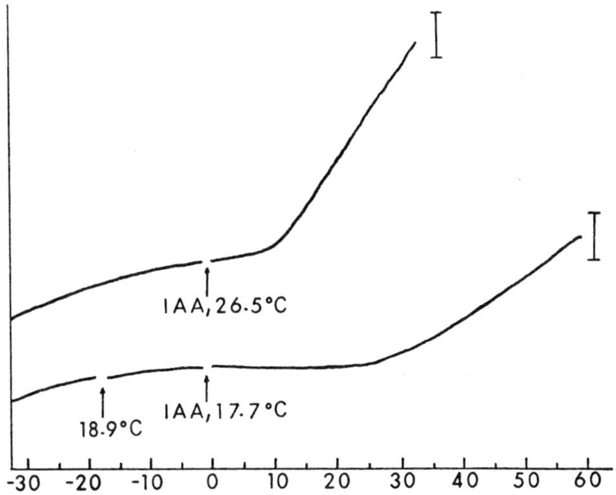

FIGURE 5. Timing of the auxin response at 17.7°C as compared with 26.5°C. Temperatures were maintained by continuously circulating through the growth measurement chamber water or IAA solution (3 μg/ml) held in large reservoirs in a constant temperature bath. Temperatures monitored using a copper-constantan thermocouple placed adjacent to the row of coleoptile segments. In upper curve, medium changed from water at 26.5°C to IAA at 26.5°C at the arrow. In lower curve, medium changed from water at 26.5°C to water at 18.9°C (first arrow) and then to IAA at 17.7°C (second arrow). In this experiment each curve was obtained using 10 coleoptile segments.

which may be because their much larger diameter results in little contact between the thread and the surface of the coleoptile's central cavity.

RESULTS

Timing of the Auxin Response Fig. 3 illustrates the response of oat coleoptile segments to 3 μg/ml IAA. After the addition of IAA, the segments continue to grow at a low rate for about 10 min, at which time a rather sudden increase in elongation rate occurs. Within about 3 additional min a rate of elongation about six times the previous rate is attained, the rate thereupon becoming steady. This type of result has been obtained many times in the course of the present work.

The auxin response of corn coleoptiles, illustrated in Fig. 7 A, was similar in possessing an absolute latent period (period of no increase in rate) of about

10 min. At the end of this period the elongation rate rose rapidly, but did not attain a steady maximum value until about 25 min after the beginning of exposure to IAA. This difference could be due to the much greater tissue thickness of corn coleoptiles with a resultant greater time dependence of penetration of IAA into the interior.

Fig. 4 shows that in buffered media there is no difference in the timing of the response of oat coleoptile segments to IAA at concentrations of 3 and 300 μg/ml. Similar results were obtained at intermediate concentrations of IAA. This finding agrees with earlier observations using single coleoptile segments (52).

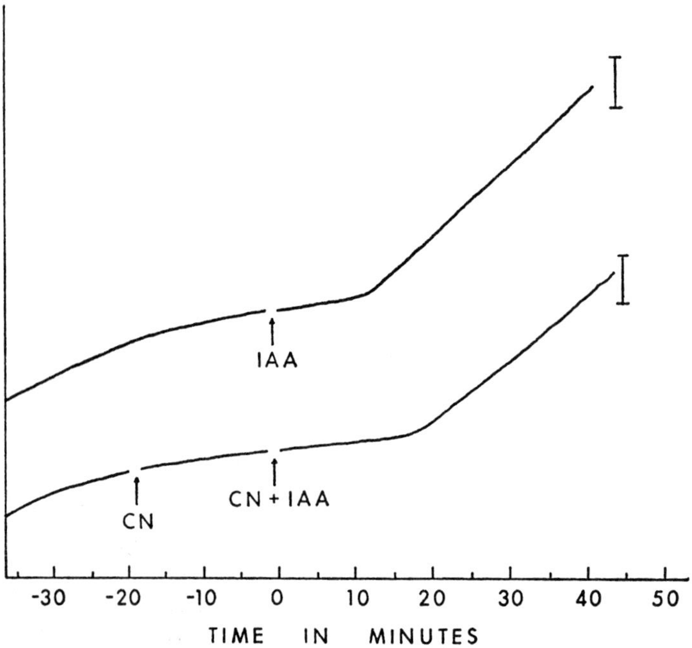

FIGURE 6. Effect of 5×10^{-5} M KCN on timing of auxin response. Upper curve shows control. Medium changed from water to 3 μg/ml IAA at arrow. Lower curve, medium changed from water to cyanide at first arrow and then, after 20 min, to cyanide plus 3 μg/ml IAA (second arrow).

In unbuffered media an elongation response to 300 μg/ml IAA was found to occur within 1 min. This rapid effect proved to be caused by the low pH of this medium. This effect of pH on elongation has been investigated and will be considered in a later paper.

Figs. 5 and 6 show the effect of cyanide and low temperature on the length of the latent period. The absolute latent period in response to auxin at 17.7°C was almost three times as long (23 min vs. 8 min) as the absolute latent period at 26.5°C. A 20 min pretreatment with 5×10^{-5} M KCN nearly doubled the latent period while reducing the magnitude of the response to auxin by 25%.

Effect of Actinomycin D Fig. 7 (A-D) shows the effect of a 1 hr pretreatment with 20 μg/ml actinomycin D on the response to IAA in corn and oat coleoptile segments. Upon transfer to IAA, the maximum rates of elongation attained by pretreated segments were 55 and 32% of the control

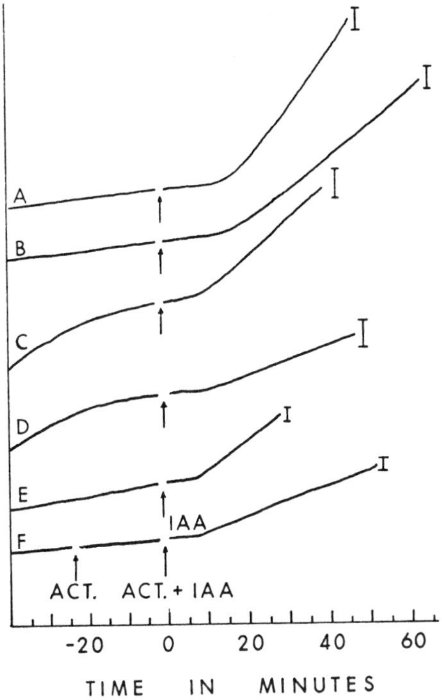

FIGURE 7. Effect of actinomycin D on response of coleoptile segments to IAA. A and B, response of corn coleoptile segments to 3 μg/ml IAA (added at arrow) after pretreatment for 1 hr in either water (curve A) or 20 μg/ml actinomycin D (curve B); segments were pretreated outside the growth measurement chamber, then introduced into the chamber, which contained water only, just before the record was begun. C and D, response of oat coleoptile segments to 3 μg/ml IAA with or without pretreatment similar to that in A and B, respectively; the part of the record prior to the arrow shows response to the tactile stimulation of placing segments onto thread, as explained in Methods. E and F, comparison of oat coleoptile segments transferred from water to IAA (curve E), and from water to 20 μg/ml actinomycin D and then, after about 20 min, to the same medium containing 10 μg/ml IAA (curve F).

rates, respectively. There was no difference in the length of the latent period between control and actinomycin-pretreated segments, with either corn or oats. Similar results were obtained if segments were pretreated with actinomycin in the growth chamber and then transferred to actinomycin plus IAA, as illustrated in Fig. 7 (E and F).

Table I summarizes data from several experiments with actinomycin. A 20 min pretreatment with 20 μg/ml was almost as effective as a 60 min pretreatment with the same concentration. Inhibition of elongation by actinomycin D was in no case accompanied by a significant change in the length of the latent period.

Inhibition of oat coleoptile elongation by actinomycin was not reversible within a 50 min period, after transfer from actinomycin plus IAA to IAA alone. This explains why actinomycin pretreatment of the segments only prior to

TABLE I

EFFECT OF ANTIBIOTICS ON LATENT PERIOD AND MAGNITUDE OF AUXIN RESPONSE IN COLEOPTILE SEGMENTS

Experiment No. and antibiotic	Pretreatment	Latent period		Auxin-promoted elongation rate		Auxin-promoted elongation rate of treated segments, as per cent of control
		Control	Treated	Control	Treated	
	min	*min**		*mm/hr × segment*		
Actinomycin D (20 $\mu g/ml$)						
1	60	15.9	17.8	1.30	0.72	55
2	60	11.3	11.7	0.62	0.20	32
3	20	8.8	10.0	0.89	0.53	60
4	20	11.7	11.0	0.70	0.28	40
5	20	11.4	11.0	0.88	0.43	49
Cycloheximide (5 $\mu g/ml$)						
6	10	6.8	5.9	0.68	0.29	43
7	15	14.6	7.8	0.55	0.24	44
8	15	12.3	7.2	0.41	0.21	51
9	20	10.4	3.3	0.62	0.21	34
10	20	13.3	5.5	0.35	0.20	57
Puromycin (8×10^{-4} M)						
11	180	12.4	12.6	1.00	0.36	36

* Estimated as the time at which elongation rate had become one-half of its steady final value.

Coleoptile segments (oats, except experiment 1 in which corn was used) were pretreated for the indicated times in antibiotics, while controls were pretreated with water; they were then treated with 10 $\mu g/ml$ IAA containing, in the case of the antibiotic-pretreated segments, the same concentration of antibiotic as in the pretreatment; except in experiments 1, 2, and 11 in which segments were treated as described for Fig. 7 (curves A-D). Elongation was followed for at least 40 min after addition of IAA.

their introduction into the growth chamber was just as inhibitory as transfer from actinomycin pretreatment to actinomycin plus IAA (Fig. 7).

In several experiments actinomycin was added after segments had responded to IAA. Actinomycin caused no inhibition of elongation within at least 60 min after its addition to IAA-promoted segments.

These findings agree in general with some observations on single coleoptile segments given by Masuda and Wada (43), although they found a much slower effect of actinomycin pretreatment. The difference may possibly be due to the fact that our 20 min pretreatments were given with the coleoptiles totally submerged in the medium, rather than floating on it. Unpublished experiments using our oat coleoptile material show that inhibition of RNA synthesis by actinomycin D becomes fully established within 30 min after its addition.

Cycloheximide and Puromycin Table I summarizes the results of five experiments in which oat coleoptile segments were pretreated with 5 μg/ml cyclo-

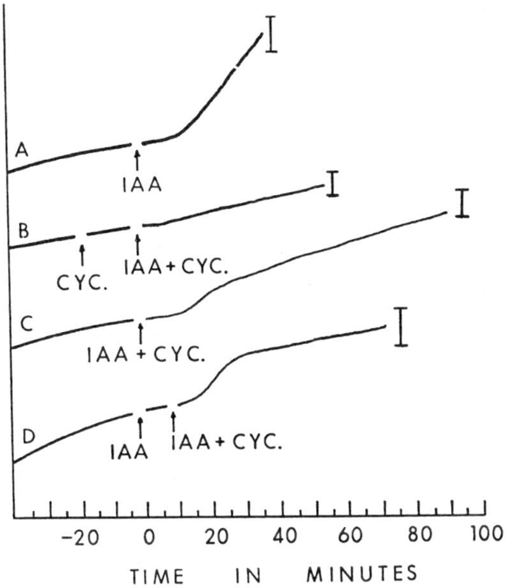

FIGURE 8. Effect of cycloheximide on auxin-induced elongation of oat coleoptile segments. Cycloheximide was added before (curve B), at the same time (curve C), and after (curve D) the IAA. Curve A shows the control (IAA only). Chamber contained water prior to the additions indicated. Cycloheximide concentration was 5 μg/ml in B and 10 μg/ml in C and D; IAA concentration was 10 μg/ml in B and 3 μg/ml in A, C, and D.

heximide and then transferred to cycloheximide plus IAA. The inhibitory action of cycloheximide on elongation was accompanied by no lengthening of the latent period of auxin action; to the contrary, the latent periods found for the cycloheximide-treated segments were in all cases less than those of the controls.

In the experiment shown in Fig. 8, cycloheximide was added before, at the same time as, or after the addition of IAA. Inhibition began within 15 min after addition of cycloheximide and became complete within 20 min. Similar experiments demonstrated that cycloheximide, unlike actinomycin, is capable of inhibiting within 15 min even when given after the establishment of auxin-promoted elongation. Also in contrast to actinomycin, inhibition by cycloheximide was completely reversible within 30 min after removal of the segments from the antibiotic.

Inhibition by puromycin had no effect on the latent period, after a pretreatment that reduced by 64% the elongation response of oat coleoptile segments to auxin. These data are given in Table I.

DISCUSSION

Since the latent period for auxin action on elongation is independent of the externally supplied concentration of IAA over a wide range, whereas the rate of IAA absorption is strongly concentration-dependent (7, 26, 49), the latent period appears not to be mainly time required for uptake of IAA by the cells. However, this does not exclude the possibility that a concentration-independent transport of auxin within the cells might contribute to the timing of the response, nor is the conclusion valid if the auxin response were to depend upon the concentration ratio between different cell or tissue compartments rather than simply upon concentration at a site of auxin action.

Since cyanide and reduced temperature extend the latent period, metabolic processes seem to be involved. The latent period (minus any time required for auxin uptake) may be regarded as the time that is required for that process or property which is directly affected by auxin, to give rise to an effect on the cell wall and thus on the rate of cell enlargment via whatever metabolic and transport processes couple the latter to the former (the primary action of auxin).

The possibility has been suggested (45, 65) that IAA may be physiologically active only after biochemical conversion to a different compound; in this event the time required for such conversion would also contribute to the latent period.

In order to better examine the significance of the observations regarding the latent period, we first consider the kind of timing that is to be expected for changes in output of a generalized reaction sequence

$$q_o \xrightarrow{k_o} q_1 \xrightarrow{k_1} q_2 \xrightarrow{k_2} \cdots \xrightarrow{k_{n-1}} q_n \xrightarrow{k_n} \text{output} \qquad (1)$$

in response to a step-up change at the start of the sequence.

In this model it is assumed that the reaction steps are irreversible and first order, the differential equation for concentration of each intermediate $[q_i]$ being

$$\frac{d[q_i]}{dt} = k_{(i-1)}[q_{(i-1)}] - k_i[q_i]. \qquad (2)$$

$[q_o]$ is assumed to be constant; the step-up change is initiated by an increase in $[q_o]$ or k_o (or both). The *n* steps include whatever kinetically significant transport processes are involved in the operation of the reaction chain. A particular q_i may be either a substrate *or* a catalyst (enzyme, coenzyme, template, etc.) for the succeeding step, but in the latter case k_i in equation (2) is the rate coefficient for inactivation of q_i whereas the rate at the $(i + 1)$th step [which produces $q_{(i+1)}$] is to be written $K_i[q_i]$ where K_i is the specific catalytic activity (turnover number) of q_i at the prevailing concentrations of substrates and other cofactors of the $(i + 1)$th step.

As an illustrative case, all the k_i's (and K_i's) are now assumed to be equal, so that n represents the number of "kinetically comparable" steps. This assumption is justifiable because steps whose k_i's are much larger than the rest will not contribute significantly to the timing of the response; the assumption of k_i's all equal greatly simplifies equation (3). The assumption of irreversibility is likewise made in the interest of mathematical simplicity, and is justifiable because reversible steps can exert a significant controlling effect on the timing only if they lie relatively far from

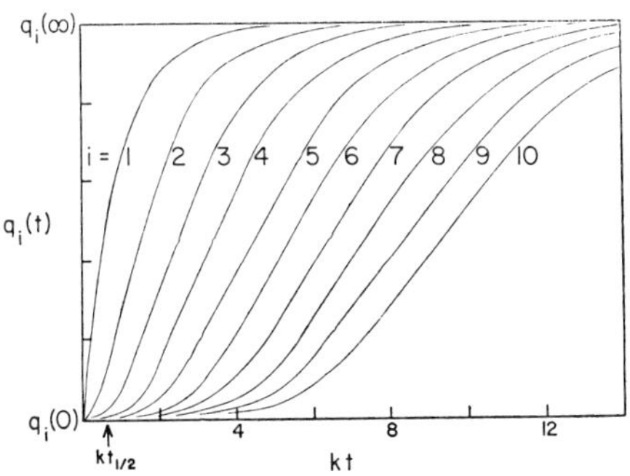

FIGURE 9. Time courses of concentration change predicted by equation (3) for intermediates in a linear pathway as represented in equation (1), following a step-up change at the beginning of the pathway, under simplifying assumptions explained in the text. $q_i(0)$ is initial concentration, $q_i(\infty)$ is final steady-state concentration, and the number i given by the curves indicates the number of kinetically comparable steps preceding the compound in question.

equilibrium (cf. references 27, 28). The assumption of first-order dependence of the reactions upon their substrates (equation [2]), also adopted for mathematical convenience, is not critical to the use that we shall make of the results. We are interested mainly in the general form of the response as a function of the number of steps, not in the detailed shape of the individual curves for concentration change.

The set of equations (2) for the q_i's can then be integrated to give the time course for increase in q_n following the step-up change:

$$[q_n](t) = [q_o]\left(1 - e^{-kt}\left[1 + kt + \frac{k^2t^2}{2!} + \frac{k^3t^3}{3!} + \cdots + \frac{k^{n-1}t^{n-1}}{(n-1)!}\right]\right) \quad (3)$$

These time courses for different numbers of steps n are drawn in Fig. 9.

The curves in Fig. 9 show that an appreciable absolute latent period (time interval between the step-up and the beginning of increase in output) is to be expected if four or more kinetically comparable steps are involved in the chain of response.

Another source of absolute latency in the response will be the time required to assemble from their monomers complete chains of polymers such as RNA or protein, if the formation of these is involved in the response. These different sources of latency should have additive effects.

In Fig. 9 one can also visualize the relationship between the time that is required to attain a steady-state response [i.e., to attain q_n (∞)] and the metabolic "half-life" $t_{\frac{1}{2}}$ of the components q_i as would be measured in turnover or decay experiments. This half-life, which is $(\ln 2)/k_i$, is equal (if all k_i's are equal) to the time for the first intermediate q_1 to attain half of its steady-state concentration change in the step-up response, and is indicated as $t_{\frac{1}{2}}$ in Fig. 9. It is clear that the time required to attain a steady state is at least five times $t_{\frac{1}{2}}$ even for q_1, and is still several times longer if the sequence involves several steps. More generally, it can be concluded that to attain a steady-state response requires at least five times the half-life of the component q_i that has the *longest* half-life (cf. reference 19).

The Gene Activation Hypothesis If auxin causes cell enlargment by inducing the synthesis of certain species of RNA which code(s) for protein that promotes the cell enlargement process, it follows that within the latent period is subsumed the time required for the induced synthesis of this informational RNA and its homologous protein(s), as well as for other obligatory steps such as the transport of mRNA (messenger RNA) from DNA template to ribosomes, transport of the growth-specific protein(s) or of metabolites derived from the action thereof to the cell wall, and the interaction of these with cell wall polymers in some manner.

On the principle that the epigenetic system should have a relaxation time orders of magnitude greater than that of the metabolic system (19), we expect that, under the gene activation hypothesis of auxin action, the latent period will be determined mainly by the timing of induced RNA and protein synthesis.

Induction and derepression of the synthesis of various enzymes in *E. coli* and other bacteria involve latent periods of 2–10 min. In certain cases this latency is known to be determined mainly by the time required to complete the transcription of the relevant cistrons (2, 3, 29, 36, 67); it is lengthened by conditions that slow down RNA synthesis. Assembly of a messenger RNA molecule probably requires at least as long a time in higher organisms.

In animal systems the time required for assembly of complete protein chains has been estimated as 1.5–4 min in particular instances (14, 39). Goodwin (19) concluded that a typical time for assembly of a complete protein molecule by cells of higher organisms would be 5 min. It is probable that in a plant cell, as in bacteria, the time required for mRNA assembly is even longer than that for assembly of the homologous polypeptide.

These values for the latency of completion of transcription and translation thus fall in the same range as the absolute latent period of about 10 min that is seen in the elongation response to IAA. If the latter were in fact determined by the former, we would expect that actinomycin and cycloheximide, which slow down the rate of transcription (53, and personal communication from E. Reich, 1968) and translation (37, 38), respectively, would extend the latent period of the growth response. However, these agents, at concentrations giving partial inhibition of the final growth response, did not extend the latent period significantly. This implies either that IAA does not act by gene activation or that the transcription and translation responsible for any growth-specific protein that is induced by IAA are completed in a small fraction of 10 min. This at least weakens the gene activation hypothesis because it conflicts with available indications of the time required for transcription and translation in higher organisms.

We then want to know how the gene activation hypothesis would account for the fact that at the conclusion of the latent period of about 10 min the elongation rate of oat coleoptiles rises within 2–3 min to a maximum and does not increase further. Two alternative explanations can be considered: During the latent period, (*a*) steady-state levels of the growth-specific mRNA and protein(s) are attained (*steady-state hypothesis*) or (*b*) the level of growth-specific mRNA and/or protein rises, as a result of its auxin-induced synthesis, to the point that some step in the chain of processes connecting transcription with cell enlargement becomes rate-limiting (*saturation hypothesis*).

Steady-State Hypothesis If a steady state in respect to mRNA or protein is attainable because its rate of degradation or inactivation increases with its concentration while its rate of formation does not, as expressed in equation (2), then the time required to attain the steady state will be determined only by its metabolic half-life (i.e. by k_i in equation 2) and not by its rate of synthesis (56, 57). Thus we would expect that the latent period would not be extended by antibiotic treatments giving partial inhibiton of RNA or protein synthesis and of the growth response; this conforms with the observations presented.

Attainment of such a steady state requires, as previously mentioned, at least five times the half-life of the component having the longest half-life. Therefore, by the steady-state hypothesis the mRNA and specific protein that are induced by IAA must both have half-lives of not greater than about 3 min.

While the half-life of certain kinds of mRNA in bacteria has been estimated as about 2 min (e.g., reference 36), such a short half-life is in serious conflict with available evidence about the stability of mRNA in higher organisms, which indicates half-lives of hours to days (8, 12, 16, 20, 22, 25, 30, 32, 44, 53, 58, 59). In the present case, treatment with actinomycin D *after* IAA did not cause inhibition of elongation within the following hour, although the inhibitory effect of actinomycin D on RNA synthesis becomes fully established within 0.5 hr. Similar results with other auxin-sensitive growing tissues have been interpreted as indicating that the growth-specific mRNA has a lifetime of hours (10, 47).

Likewise the requirement for a half-life of the order of 3 min for the growth-specific protein supposedly induced by IAA is incompatible with the extensive evidence regarding protein turnover in higher organisms, which indicates half-lives of hours to days (1, 18, 23, 24, 54, 55).

Saturation Hypothesis If the IAA-induced mRNA and protein were stable enough to build up, by the end of the latent period, to the point at which some step beyond transcription and translation became saturated and rate-limiting, then a steady growth rate would ensue, but not a true steady state (except at steps beyond the "saturated" step), because IAA-induced mRNA and protein would continue to accumulate. If synthesis of RNA or protein were partially inhibited, the growth-saturating levels would still be attainable, but merely in a longer period of time, provided the inhibitions were not too severe; thus the latent period would be extended. It follows a fortiori that the latent period should be extended if RNA or protein synthesis were inhibited strongly enough so that the final growth response would be reduced; i.e., when the rate-saturating level of mRNA or protein cannot be attained. This is clearly inconsistent with the observed effects of antibiotics on the latent period.

One might suggest that the length of the latent period is actually determined mainly by the dynamics of steps that follow transcription and translation, the saturation of the later steps by these earlier ones taking place in such a small fraction of the normal latent period that the extension of the latent period by inhibitors is not detectable. But this seems unsatisfactory since it demands a complete response of the epigenetic system in a time that is short compared even with what appears to be required to complete the assembly of single RNA and protein molecules.

The observations conflict, therefore, with the gene activation hypothesis of auxin action, and we conclude that this hypothesis is probably not a correct explanation of the promotion of cell elongation in coleoptile tissue by auxin. As a number of authors have pointed out, the inhibition of a hormonal response by inhibitors of RNA and protein synthesis may equally well be ascribed to a requirement for the availability of particular gene products either in order for the hormone to act, or in the mechanism of the hormone-sensitive physiological process. In order to maintain, against the kinetic evidence, the gene activation hypothesis for the growth response of coleoptiles to auxin, it would be essential first to demonstrate a substantial effect of IAA on RNA synthesis in this material *within the latent period* of 10 min, and to show that this promoted synthesis involves RNA that has a half-life of only a few minutes.

Translational Control Auxin could act at the translational level to promote synthesis of growth-specific protein(s), as has been proposed regarding certain animal hormones (17, 33, 60). If we suppose that the induced proteins are effective upon elongation by virtue of catalytic (enzymatic) activity, the alternatives of steady-state and saturation hypotheses again have to be considered. As discussed previously, the saturation hypothesis must be rejected, from the fact that puromycin and cycloheximide did not extend the latent period, and the steady-state hypothesis seems to be virtually ruled out because it requires that the half-life of growth-active protein(s) be of the order of 3 min.

However, the observation that cycloheximide, when added after IAA, causes a strong inhibition of elongation in 15 min might be interpreted as indicating a short half-life for a protein essential to the cell enlargment process, as has been concluded in an analogous instance involving an animal hormone effect (17). But there is an alternative explanation that does *not* conflict with plausible conceptions about protein stability, namely, that cycloheximide inhibits rapidly because elongation depends upon *continual synthesis* of new protein (rather than upon catalytic activity of some unstable protein previously synthesized). This type of hypothesis has been suggested to explain the rapid effect of cycloheximide in certain other systems (31, 40, 46). It may then be proposed that auxin promotes the rate of formation of this protein.

By this hypothesis the steady rate of elongation, reached after auxin application, requires not the attainment of a new steady-state level of growth-active protein, but merely the attainment of a steady rate of formation of such protein. Since stability of the protein is irrelevant to the timing of the response, this obviates the contingency that defeats the steady-state hypothesis in its previously discussed form. That translational level promotion of protein synthesis by auxin could take place within 10 min is made plausible by such phenomena as the induction of protein synthesis by fertilization in sea urchin eggs (15) and the induction of nitrate reductase formation by nitrate in *Aspergillus* (11), both of which have latent periods of about 10 min.

The protein synthesis envisaged under the continual formation hypothesis cannot represent simple accumulation of protoplasmic protein, as might naturally be assumed in a bacterial or animal growth system, because such accumulation cannot cause expansion of the cell wall which is critical to the plant cell enlargment phenomenon being measured, and in addition it is known that total protein does not increase during auxin-induced elongation of coleoptile segments (4). The growth-active protein must be assumed to be a minor component of total protein. Moreover, it is inherent in the continual formation hypothesis, and essential to its superiority over the steady-state hypothesis, that the growth-active protein function only at the time of its formation; i.e., not possess catalytic activity. This implies a structural protein, and excludes a wide range of suggested mechanisms of auxin action that envisage induction of enzymes that degrade or otherwise modify the chemical structure and mechanical properties of the cell wall.

One possibly structural protein that comes immediately to mind under the continual formation hypothesis is cell wall protein (35). Promotion by auxin of incorporation of precursors into hydroxyproline-rich cytoplasmic and cell wall protein has been reported (9, 34). This and other results have suggested that this protein may be involved in the process of cell wall expansion; however, the available evidence indicates that the immediate synthesis of this protein is not required for cell enlargment in coleoptiles (9).

Another class of structural protein that may be considered is membrane protein. Formation and secretion of cell wall polysaccharides may be involved in the mechanism of cell wall expansion (51). Polysaccharide synthesis and transport are believed to occur via the Golgi system (48). Formation of Golgi vesicles or analogous transport structures might well require continued synthesis of structural components of membranes and thus might account for a dependence of cell enlargment upon continued protein synthesis.

However, lest we uncritically accept as a fact the hypothesis that continual protein synthesis is required in the mechanism of cell enlargement, it would be well to bear in mind the possiblility that the cell possesses secondary regulatory mechanisms that brake the process of cell enlargment if and when the cell is unable to synthesize protein. Since such a feedback loop would probably be advantageous it might well have been selected for; if it existed, experimental results with inhibitors of protein synthesis could lead us erroneously to think that protein synthesis participates directly in the mechanism of cell wall expansion.

If elongation does depend upon continual synthesis of structural protein this does *not* necessarily mean that auxin promotes by activating protein synthesis. The latter hypothesis should be seriously considered only if it can be shown that auxins promote synthesis of certain proteins maximally within 10 min; published evidence is a long way from meeting this requirement (e.g., references 62, 63). The observation that cycloheximide shortened the latent period for the auxin response of coleoptiles suggests the contrary possibility that auxin action actually competes with protein synthesis; for example, for energy or for metabolites.

From the comparative point of view it is striking that a wide variety of hormonal effects on enzyme or protein synthesis that are known in animal and plant systems exhibit latent periods of about 1–2 hr, with even longer periods required to attain a steady rate of increase or a steady state (23, 33, and references there cited). The elongation response of coleoptiles to auxin, attaining completion as it does in little over 10 min, would thus be quite exceptional if it were in fact a hormonal activation of protein synthesis (whether at the transcriptional or translational level). This is even less probable for the promotion by auxin of protoplasmic streaming in coleoptile cells, which can occur in less than a minute (61). While there is no proof that the protoplasmic streaming effect leads to the effect on elongation, the idea that influence on a transport process may be the way in which auxin affects elongation is suggested by the fact that various nonspecific stimuli can cause an elongation effect similar to that of auxin; for example, the tactile effect noted above under Methods, and the effect of cyanide (50) or CO_2 (21).

We thank K. V. Thimann and Leonora Reinhold for their interest in and discussion of this work, A. Kelley and F. H. Clauser for help with the mathematical aspect, and M. Van Ghent for expert glasswork in constructing the growth-measuring chamber.
This work was supported by a grant to Dr. Ray from the National Science Foundation.

Received for publication 29 July 1968.

REFERENCES

1. Afridi, M. M. R K., and E. J. Hewitt. 1964. The inducible formation and stability of nitrate reductase in higher plants. *J. Exp. Bot.* **15**:251.
2. Alpers, D. H., and G. M. Tomkins. 1966. Sequential transcription of the genes of the lactose operon and its regulation by protein synthesis. *J. Biol. Chem.* **241**:4434.
3. Baker, R. F., and C. Yanofsky. 1968. The periodicity of RNA polymerase initiations: a new regulatory feature of transcription. *Proc. Nat. Acad. Sci. U. S. A.* **60**:313.
4. Boroughs, H., and J. Bonner. 1953. Effects of indoleacetic acid on metabolic pathways. *Arch. Biochem. Biophys.* **46**:279.
5. Briquet, M. V., J. R. Decallonne, R. R. Lambert, and A. L. Wiaux. 1967. DMPA and Avena coleoptiles RNA turn-over. *Physiol. Plant.* **20**:337.
6. Bünning, E. 1959. Die thigmonastischen und thigmotropischen Reaktionen. *Handb. Pflanzenphysiol.* **17**(1):254.
7. Christie, A. E., and A. C. Leopold. 1965. Entry and exit of indoleacetic acid in corn coleoptiles. *Plant Cell Physiol.* **6**:453.
8. Chroboczek, H., and J. H. Cherry. 1966. Characterization of nucleic acids in peanut cotyledons. *J. Mol. Biol.* **19**:28.
9. Cleland, R. 1967. Inhibition of formation of protein-bound hydroxyproline by free hydroxyproline in Avena coleoptiles. *Plant Physiol.* **42**:1165.
10. Courtney, J. S., D. J. Morre, and J. L. Key. 1967. Inhibition of RNA synthesis and auxin-induced cell wall extensibility and growth by actinomycin-D. *Plant Physiol.* **42**:434.
11. Cove, D. J. 1967. Kinetic studies of the induction of nitrate reductase and cytochrome *c* reductase in the fungus Aspergillus nidulans. *Biochem. J.* **104**:1033.
12. Cozzone, A., and G. Marchis-Mouren. 1967. Messenger ribonucleic acid stability in rat pancreas and liver. *Biochemistry.* **6**:3911.
13. Darwin, C. 1881. The Power of Movement in Plants. DaCapo Press Edition, New York. 1966.
14. Dintzis, H. M. 1961. Assembly of the peptide chains of hemoglobin. *Proc. Nat. Acad. Sci. U. S. A.* **47**:247.
15. Epel, D. 1967. Protein synthesis in sea urchin eggs: a "late" response to fertilization. *Proc. Nat. Acad. Sci. U. S. A.* **57**:899.
16. Fristom, J. W., L. Brothers, V. Mancebo, and D. Stewart. 1968. Aspects of RNA and protein synthesis in imaginal discs of *Drosophila melanogaster. Mol. Gen. Genet.* **102**:1.
17. Garren, L. D., R. L. Ney, and W. W. Davis. 1965. Studies on the role of protein synthesis in the regulation of corticosterone production by adrenocorticotropic hormone. *Proc. Nat. Acad. Sci. U. S. A.* **53**:1443.
18. Glasziou, K. T., J. C. Waldron, and T. A. Bull. 1966. Control of invertase synthesis in sugar cane. *Plant Physiol.* **41**:282.
19. Goodwin, B. C. 1963. Temporal Organization in Cells. Academic Press, Inc., New York.

20. HARRIS, H. 1963. Nuclear ribonucleic acid. *Progr. Nucl. Acid Res.* **2**:20.
21. HARRISON, A. 1965. Auxanometer experiments on extension growth of *Avena* coleoptiles in different carbon dioxide concentrations. *Physiol. Plant.* **18**:321.
22. HEMLEBEN-VIELHABEN, V. 1966. Characterization of rapidly labelled nucleic acids in tissues of plant seedlings. *Z. Naturforsch.* **21b**:983.
23. HOLLAENDER, A. Editor. 1965. Hormonal control of protein biosynthesis. *J. Cell. Comp. Physiol.* **66**(Suppl. 1):1.
24. INGLE, J., K. W. JOY, and R. H. HAGEMAN. 1966. The regulation of activity of the enzymes involved in the assimilation of nitrate by higher plants. *Biochem. J.* **100**:577.
25. INGLE, J., J. L. KEY, and R. E. HOLM. 1965. The demonstration and characterization of a DNA-like RNA in excised plant tissue. *J. Mol. Biol.* **11**:730.
26. JOHNSON, M. P., and J. BONNER. 1956. The uptake of auxin by plant tissue. *Physiol. Plant.* **9**:102.
27. KACSER, H. 1957. Some physico-chemical aspects of biological organization. *In* The Strategy of the Genes. C. H. Waddington, editor. Allen and Unwin, Ltd., London.
28. KACSER, H. 1963. The kinetic structure of organisms. *In* Biological Organization at the Cellular and Supercellular Level. R. J. C. Harris, editor. Academic Press, Inc., New York.
29. KAEMPFER, R. O. R., and B. MAGASANIK. 1967. Mechanism of β-galactosidase induction in *Escherichia coli. J. Mol. Biol.* **27**:475.
30. KEY, J. L., and J. INGLE. 1964. Requirement for the synthesis of DNA-like RNA for growth of excised plant tissue. *Proc. Nat. Acad. Sci. U. S. A.* **52**:1382.
31. KIRK, J. T. O. 1968. Studies on the dependence of chlorophyll synthesis on protein synthesis in *Euglena gracilis*, together with a nomogram for determination of chlorophyll concentration. *Planta.* **78**:200.
32. KOCH, G., G. TRANS, and H. KUBINSKI. 1966. Synthese und Umsatz von Ribonucleinsäuren in suspendierten Säugetierzellen. II. Einfluss von Actinomycin. *Z. Naturforsch.* **21b**:967.
33. KORNER, A. 1967. Ribonucleic acid and hormonal control of protein synthesis. *Progr. Biophys. Mol. Biol.* **17**:61.
34. KURAISHI, S., S. UEMATSU, and T. YAMAKI. 1967. Auxin-induced incorporation of proline in mung bean hypocotyls. *Plant Cell Physiol.* **8**:527.
35. LAMPORT, D. T. A. 1965. The protein component of primary cell walls. *Advan. Bot. Res.* **2**:151.
36. LIEVE, L., and V. KOLLIN. 1967. Synthesis, utilization and degradation of lactose operon mRNA in *Escherichia coli. J. Mol. Biol.* **24**:247.
37. LIN, C. Y., and J. L. KEY. 1967. Dissociation and reassembly of polyribosomes in relation to protein synthesis in the soybean root. *J. Mol. Biol.* **26**:237.
38. LIN, S. Y., R. D. MOSTELLER, and B. HARDESTY. 1966. The mechanism of sodium fluoride and cycloheximide inhibition of hemoglobin biosynthesis in the cell-free reticulocyte system. *J. Mol. Biol.* **21**:51.
39. LOFTFIELD, R. B., and E. A. EIGNER. 1958. The time required for the synthesis of a ferritin molecule in rat liver. *J. Biol. Chem.* **231**:925
40. MANGAN, F. R., G. E. NEAL, and D. C. WILLIAMS. 1968. The effect of cycloheximide on ribonucleic acid polymerase activity in rat liver nuclei. *Biochem. J.* **108**:44 P.
41. MASUDA, Y., and E. TANIMOTO. 1967. Effect of auxin and antiauxin on the growth and RNA synthesis of etiolated pea internode. *Plant Cell Physiol.* **8**:459.
42. MASUDA, Y., E. TANIMOTO, and S. WADA. 1967. Auxin-stimulated RNA synthesis in oat coleoptile cells. *Physiol. Plant.* **20**:713.
43. MASUDA, Y., and S. WADA. 1966. Requirement of RNA for the auxin-induced elongation of oat coleoptile. *Physiol. Plant.* **19**:1055.
44. MATSUMOTO, H., M. KOBAYASHI, and T. TAKAHASHI. 1967. Some characteristics of rapidly labelled RNA in jack bean leaves. *Physiol. Plant.* **20**:927.

45. Moyed, H. S., and V. Tuli. 1968. The oxindole pathway of IAA metabolism and the action of auxins. *In* Biochemistry and Physiology of Plant Growth Substances. F. Wightman and G. Setterfield, editor. Runge Press, Ottawa. In press.
46. Nicolette, J. A., and G. C. Mueller. 1966. *In vitro* regulation of RNA polymerase in estrogen-treated uteri. *Biochem. Biophys. Res. Commun.* **24**:851.
47. Nooden, L. D. 1968. Studies on the role of RNA synthesis in auxin induction of cell enlargement. *Plant. Physiol.* **43**:140.
48. Pickett-Heaps, J. D. 1967. Further observations on the Golgi apparatus and its function in cells of the wheat seedling. *J. Ultrastruct. Res.* **18**:287.
49. Poole, R. J., and K. V. Thimann. 1964. Uptake of indole-3-acetic acid and indole-3-acetonitrile by *Avena* coleoptile sections. *Plant Physiol.* **39**:98.
50. Ray, P. M. 1961. Hormonal regulation of plant cell growth. *In* Control Mechanisms in Cellular Processes. D. M. Bonner, editor. Ronald Press, New York.
51. Ray, P. M. 1967. Radioautographic study of cell wall deposition in growing plant cells. *J. Cell Biol.* **35**:659.
52. Ray, P. M., and A. W. Ruesink. 1962. Kinetic experiments on the nature of the growth mechanism in oat coleoptile cells. *Develop. Biol.* **4**:377.
53. Reich, E., and I. H. Goldberg. 1964. Actinomycin and nucleic acid function. *Progr. Nucl. Acid Res. Mol. Biol.* **3**:184.
54. Schimke, R. T. 1967. Protein turnover and the regulation of enzyme levels in rat liver. *Nat. Cancer Inst. Monogr.* **27**:301.
55. Schimke, R. T. 1969. Regulation of protein degradation in mammalian tissue. *In* Mammalian Protein Metabolism. H. N. Munro, editor. Academic Press, Inc., New York. **4**: In press.
56. Schimke, R. T., E. W. Sweeney, and C. M. Berlin. 1964. An analysis of the kinetics of rat liver tryptophan pyrrolase induction: the significance of both enzyme synthesis and degradation. *Biochem. Biophys. Res. Commun.* **15**:214.
57. Segal, H. L., and Y. S. Kim. 1963. Glucocorticoid stimulation of the biosynthesis of glutamic-alanine transaminase. *Proc. Nat. Acad. Sci. U. S. A.* **50**:912.
58. Shearer, R. W., and B. J. McCarthy. 1967. Evidence for ribonucleic acid molecules restricted to the cell nucleus. *Biochemistry.* **6**:283.
59. Singer, M., and P. Leder. 1966. Messenger RNA: an evaluation. *Annu. Rev. Biochem.* **35**(1): 195.
60. Sokoloff, L., C. M. Francis, and P. L. Campbell. 1964. Thyroxine stimulation of amino acid incorporation into protein independent of any action on messenger RNA synthesis. *Proc. Nat. Acad. Sci. U. S. A.* **52**:728.
61. Thimann, K. V., and B. M. Sweeney. 1937. The effect of auxins upon protoplasmic streaming. *J. Gen. Physiol.* **21**:123.
62. Trewavas, A. J. 1968. Effects of 3-indolylacetic acid on the metabolism of ribonucleic acid and protein in etiolated subapical sections of *Pisum sativum. Arch. Biochem. Biophys.* **123**: 324.
63. Trewavas, A. 1968. The effect of 3-indolylacetic acid on the levels of polysomes in etiolated pea tissue. *Phytochemistry.* **7**:673.
64. Truelsen, T. A. 1967. Indoleacetic acid–induced decrease of the ribonuclease activity *in vivo. Physiol. Plant.* **20**:1112.
65. Tuli, V., and H. S. Moyed. 1967. Inhibitory oxidation products of indole-3-acetic acid. *Plant Physiol.* **42**:425.
66. Wiegand, O. F., and A. R. Schrank. 1959. Regimen for growing uniform *Avena* coleoptiles. *Bot. Gaz.* **121**:106.
67. Zimmerman, R. A., and C. Levinthal. 1967. Messenger RNA and RNA transcription time. *J. Mol. Biol.* **30**:349.

Reprinted from the PROCEEDINGS OF THE NATIONAL ACADEMY OF SCIENCES
Vol. 55, No. 2, pp. 262–269. February, 1966.

*THE INTERACTION BETWEEN AUXIN AND ETHYLENE AND ITS ROLE IN PLANT GROWTH**

BY STANLEY P. BURG AND ELLEN A. BURG

UNIVERSITY OF MIAMI SCHOOL OF MEDICINE

Communicated by Kenneth V. Thimann, December 6, 1965

Auxin induces ethylene formation in several plants[1, 2] and in segments cut from roots, stems, and fruits.[3] If ethylene alters the metabolism of auxin, as has been suggested,[2, 4, 5] then this auxin-induced ethylene formation might interfere with studies of the uptake, transport, and destruction of the applied growth hormone. Moreover, the newly synthesized ethylene would mask or modify a response to applied gas and thus complicate any investigation of the effect of ethylene on auxin metabolism. It will be shown that all of these problems arise under the conditions commonly used to study auxin metabolism, and the significance of this finding will be discussed with special reference to a postulated mechanism of ethylene action.

Materials.—Pea seeds (*Pisum sativum var. Alaska*) were soaked for 5 hr and planted in vermiculite if the seedlings were to be etiolated, or in potting soil if light-grown. The plants were kept at 23 ± 3°C either in darkness with occasional dim red light for 7 days, or on a 16-hr light cycle under Gro-lights for 14 days. Sections were then cut from the apical part of the uppermost internode. Other seedlings were planted in the same manner as etiolated pea plants, except that corn (*Zea mays var. Barbecue*) was sterilized in 10% Clorox for 10 min and rinsed for 3 hr before it was sown. Oat (*Avena sativa L. var. Victory*) and corn coleoptiles were harvested at 80 hr, sunflower (*Helianthus annus L.*) after 5 days, and sections were cut 2–3 mm below the cotyledons or tip.

Methods.—Straight growth test: Groups of ten 10-mm sections were placed in 10 ml of solution contained in 125-ml flasks. The medium typically consisted of 10^{-6} *M* indole-3-acetate (IAA), 2% sucrose, 5×10^{-2} *M* potassium phosphate buffer (pH 6.8), and 5×10^{-6} *M* $CoCl_2$ in glass-distilled water. To add ethylene, the flasks were sealed with vaccine caps and a known quantity of a calibrated ethylene-air mixture was injected. The flasks were gently shaken in the dark at 23°C, and at various times the fresh weight and length of the sections were determined. In some studies the tissue was shadowgraphed with green light, and stem curvature measured from the film with a protractor. Ethylene production was estimated by removing samples of air from the flasks and gas-chromatographing them on an alumina column.[6]

Transport and destruction of C^{14}-IAA: Polar transport through 7-mm stem sections was determined by applying apical donor blocks containing carboxyl-labeled C^{14}-IAA (8mC/mM) in 3% agar, and 1.5% agar receiver blocks. After 2 hr the receiver blocks were remelted, dried on a grooved planchette, and counted with a 30% efficient automatic gas-flow counter; the radioactivity remaining in the stem segments was extracted by crushing them on a planchette with 1 ml of 50% ethanol. Plating a homogeneous suspension of duplicate tissue at less than 2 mg/cm^2 resulted in the same count, so it was concluded that self-absorption was not a complication. All transport experiments were carried out in sealed desiccators at 100% relative humidity, and enough ethylene was added to bring the concentration to 100 ppm. Polar transport was also studied indirectly by incubating 15-mm pea stem sections in media containing 10^{-6} *M* C^{14}-IAA (8mC/mM). At various time intervals the sections were rinsed, cut into 5 equal lengths, and the radioactivity per mg dry weight was determined for each. The destruction of auxin which occurred during the incubation was appraised by counting the $C^{14}O_2$ released. Air samples removed from the flasks were passed through a dry-ice freezing trap into an evacuated 250-ml Cary-Tolbert ionization chamber, and radioactivity was determined by the rate of charge method using a Cary electrometer. When tissue was omitted from the flasks, 98% less volatile radioactivity was produced in 24 hr, and when samples of air from around the tissue were pretreated with 5% KOH, all counts were removed, indicating that they were due solely to $C^{14}O_2$ evolved by the stem sections. For lateral transport studies ten 15-mm corn coleoptile sections were mounted horizontally with their vascular bundles oriented in a vertical plane, a donor block containing C^{14}-IAA was applied apically, and a receiver block at the base.[7] After 135 min each section was marked on its upper side, the apical 5 mm discarded, and the next 5 mm split into upper and lower halves whose radioactivity per mg dry weight was determined. The receiver block was also counted. The method for pea tissue was similar, except that twenty 7-mm sections were used, the apical 2 mm discarded, and usually the segments were split into upper, middle, and lower zones (representing 25.1 ± 2.6, 49.3 ± 6.3, and 25.5 ± 2.4% of the dry weight, respectively) instead of upper and lower halves. Curvature was determined with a protractor after shadowgraphing 15-mm sections at the end of the same transport period.

Results and Discussion.—Auxin-induced ethylene formation in pea stem sections: Pea stem sections elongate best when they are supplied with 10^{-6} *M* IAA, whereas their fresh weight continues to increase until ten times that level of growth hormone is applied (*8*, Fig. 1). Those concentrations of IAA which inhibit elongation also

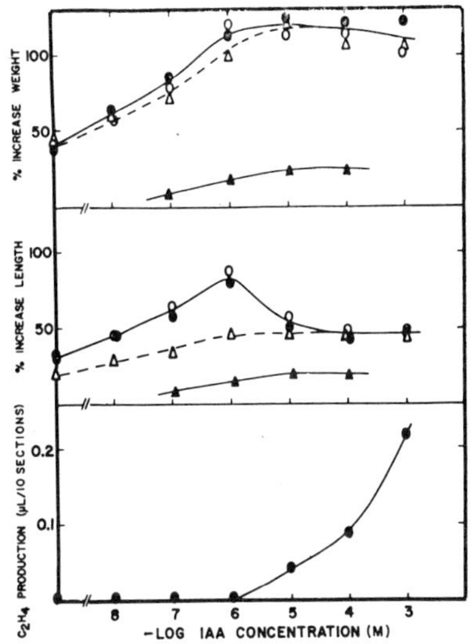

FIG. 1.—*Upper:* Effects of IAA and C_2H_4 on the increase in fresh weight of etiolated pea stem sections. After 3 hr, control in sealed flasks (▲—▲). After 18 hr, control in Petri dishes (○—○); control in sealed flasks (●—●); and treated with 10 ppm C_2H_4 (△---△). *Middle:* Effect of IAA and C_2H_4 on elongation. Symbols same as above. *Lower:* C_2H_4 production during 18 hr. All flasks contained 2% sucrose, 5×10^{-2} *M* potassium phosphate buffer (pH 6.8), and 5×10^{-2} *M* $CoCl_2$.

stimulate the tissue to produce ethylene (Figs. 1 and 2), and since there is a close correlation between the intensity of ethylene production and the extent of the inhibition, it is natural to wonder whether ethylene might be contributing to the auxin response. To answer this question, sections were incubated in various concentrations of auxin in the presence and absence of applied ethylene, so that the effects of auxin and ethylene could be compared. Using 10^{-6} *M* IAA (the highest concentration which does not stimulate ethylene formation), it was found that elongation is markedly inhibited by application of ethylene, whereas the increase in fresh weight is only slightly retarded (Fig. 3). These changes are caused by an isodiametric expansion of cells in the cortical region of gassed tissue in contrast to their normal longitudinal mode of expansion.[9] High concentrations of auxin cause the same effect, however (Fig. 1), and the resultant tissue is morphologically and cytologically indistinguishable from that gassed with ethylene.[5] This suggests that the ethylene might be the cause of the swelling and inhibition of extension growth heretofore attributed to high levels of auxin, and in order to ex-

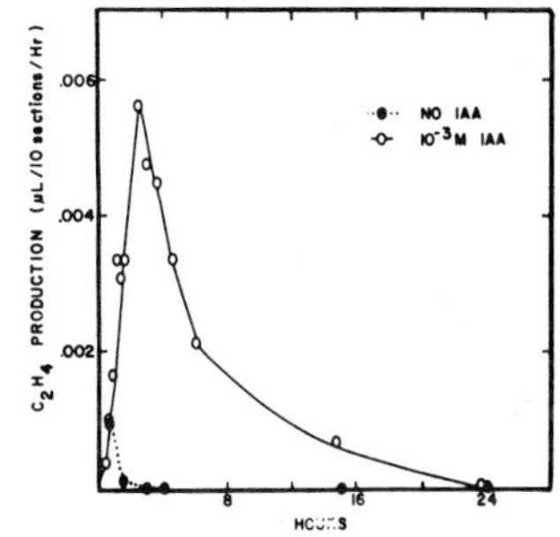

FIG. 2.—IAA-induced C_2H_4 formation in etiolated pea stem sections. Medium is described in legend to Fig. 1. A detectable acceleration in ethylene formation also occurs within a few hours after 10^{-4} or 10^{-5} *M* IAA is applied.

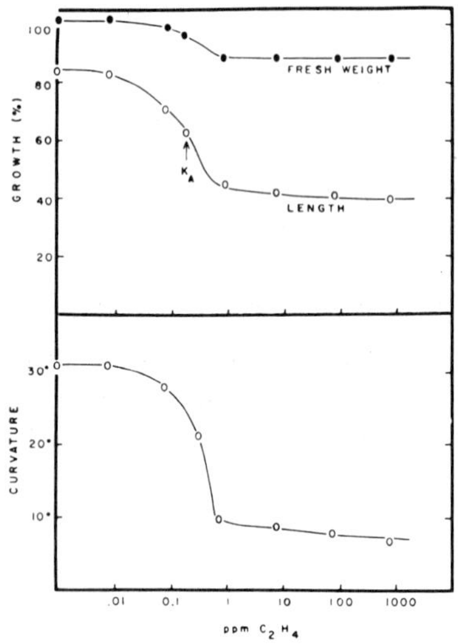

FIG. 3.—*Upper:* Relationship between % increase in fresh weight, % elongation, and the concentration of C_2H_4 present during an 18-hr experiment with etiolated pea stem sections. Flasks contained 10^{-6} M IAA and the medium described in the legend to Fig. 1. K_A is the concentration of C_2H_4 causing a half-maximal effect. *Lower:* Curvature in the same stem segments after 2.5 hr have elapsed. The initial random curvature of controls was 8°.

amine this possibility three types of experiments were devised in which the effect of endogenous ethylene was minimized or eliminated. Each showed that auxin alone does not inhibit the growth of pea tissue: (1) Because ethylene has very little effect on the final fresh weight of growing pea stem sections (Fig. 3), if the fresh weight were to decrease at a high IAA concentration, this would have to be due to auxin action *per se.* No such decrease occurs (*8*, Fig. 1); instead, the fresh weight reaches its maximum value at 10^{-5} M IAA and is little changed at higher auxin levels. (2) After auxin application, endogenous ethylene cannot affect the growth of pea stem sections for at least 4 hr because ethylene formation is not stimulated for 1 hr (Fig. 2) and the tissue does not respond to applied gas for 3 hr (Fig. 4). Measurements taken at 3 hr (Fig. 1) show that the rate of growth, evaluated either as increase in fresh weight or length, approaches a maximum at 10^{-5} M IAA with no decline at higher levels of auxin. (3) When pea stem segments are treated with 10 ppm ethylene, their rate of elongation is reduced by 50 per cent, and additional ethylene has no further effect (Fig. 3). Under these conditions, ethylene produced in response to auxin would also be without effect; hence, an inhibition due to auxin could be readily discerned. No such inhibition occurs when the IAA concentration is raised; instead, extension growth saturates at 10^{-5} M IAA (Fig. 1) just as in the two previous examples. Thus, three different approaches yield the same conclusion; only when auxin induces ethylene formation and the gas has had sufficient time to act, does IAA lead indirectly to an inhibition of elongation. It is important to note that ethylene does not have to accumulate in the ambient atmosphere for tissue to be inhibited by the gas which it produces. This is proved by the fact that the dependence of growth upon IAA concentration is the same regardless of whether pea stem sections are incubated in sealed flasks or in Petri dishes from which ethylene readily escapes (cf. open vs. closed circles, Fig. 1).

Does ethylene mediate the entire response to a high concentration of auxin? If so, the ethylene present within the tissue at each concentration of auxin should quantitatively account for the inhibition of elongation which results. Stem segments containing more than 1 ppm ethylene cannot be affected by added ethylene because they must already have responded maximally to their endogenous gas (Fig. 3). Therefore, 1 ppm ethylene must be present within sections growing in just over 10^{-5} M IAA, for this is the lowest concentration of auxin which renders tissue completely unresponsive to applied gas (Fig. 1). But a concentration of 1 ppm ethylene ought to bring about a 50 per cent reduction in the rate of elongation, and if this is exactly the change caused by application of 10^{-5} M or more auxin, it may be concluded that ethylene is responsible for the entire inhibition heretofore ascribed to auxin. Application of $>10^{-5}$ *M* IAA does cause a 50 per cent decline in the rate of elongation (Fig. 1): when the auxin concentration is increased to 2×10^{-6} *M* or higher, elongation is progressively retarded until a maximum inhibition of 50 per cent results at an IAA concentration just in excess of 10^{-5} *M* (Fig. 1). Since the final length of pea stem sections would have been almost the same at 10^{-6} *M* IAA and all higher concentrations if there were no effect from endogenous ethylene (see results at 3 hr, Fig. 1), it is clear that *ethylene alone* is responsible for the inhibition of elongation which occurs at "supraoptimal" concentrations of auxin. By an analogous argument it can be shown that the swelling which accompanies this inhibition is also due solely to ethylene. At low IAA concentrations the ratio between per cent increase in fresh weight and per cent increase in length is constant at an average value of 1.3, whereas at auxin concentrations $>10^{-5}$ *M* the tissue swells and the ratio changes to 2.6. Regardless of the IAA concentration, when 1 ppm ethylene is applied, the ratio is always 2.6, and therefore it is clear that ethylene, not auxin, causes the swelling.

Auxin-induced ethylene formation in other tissues: The entire growth inhibition in etiolated *sunflower* stem sections at a high concentration of IAA is due to auxin-induced ethylene formation. The argument which proves this is similar to that advanced for peas, except that 10 ppm ethylene is required to inhibit the growth of sunflower maximally (about 30%), and as a result this tissue does not become insensitive to applied gas until 10^{-3} *M* IAA is present and ethylene evolved at a relatively high rate. Added ethylene inhibits elongation and the increase in fresh weight to almost the same extent, and this explains why high concentrations of auxin inhibit both weight and length in sunflower. The growth of *light-grown pea plants* does not saturate at any applicable concentration of IAA.[8] In part, this is because they are considerably less responsive to ethylene than etiolated seedlings and produce the gas at a slower rate in the presence of 10^{-5} or 10^{-4} *M* IAA. As a result, even though green tissue swells when it responds to its own ethylene, the inhibition of elongation due to auxin-induced ethylene formation is never sufficient to offset the auxin-induced increase in length. Auxin stimulates *corn* and *oat* coleoptiles to evolve ethylene, but the production does not continually increase as the IAA concentration is raised. Instead, ethylene formation and growth reach their highest rates at 10^{-5} *M* IAA, and then both decline as the auxin concentration is increased further. Only 1 ppm ethylene is needed to inhibit the growth of corn and oat coleoptiles maximally, and in both cases length and weight decline in unison by about 40 per cent. Since corn produces large quantities of ethylene and becomes insensitive to applied gas when the concentration of IAA is raised to 10^{-6} *M*, the tissue must respond completely to its endogenous ethylene at that and higher auxin levels. However, the growth inhibition at high auxin levels is not due to ethylene action, for ethylene production *decreases* as this inhibition sets in, and the response to applied ethylene is not large enough to account for the precipitous decline in growth which occurs at $>10^{-4}$ *M* IAA. The situation is similar in *Avena*. In fact it is unlikely that *Avena* coleoptiles ever respond substantially to the ethylene they produce, for even when 10^{-5} *M* IAA is present and ethylene formed most rapidly, the rate of evolution is so low that the tissue still reacts to applied gas. It must be concluded that auxin-induced growth inhibitions occur for basically different reasons in coleoptiles and seedlings such as sunflower and peas.

Effect of ethylene on auxin synthesis: Within a few hours after ethylene is applied to *Vicia faba*,[4] pea,[5] and *Avena* seedlings,[4, 5, 10] the amount of diffusible auxin declines. Three explanations for this phenomenon may be envisioned: ethylene might enhance the destruction of auxin, inhibit its polar transport, and/or reduce its rate of synthesis. Ethylene did not stimulate pea sections to release more $C^{14}O_2$ from carboxyl-labeled IAA at any time during a 24-hr period. These studies were carried out using a medium containing 10^{-6} *M* IAA since greater amounts of auxin stimulate ethylene formation and make it impossible to determine what the rate of destruction would have been in the absence of the gas. When 15-mm pea stem segments were incubated in this medium, addition of ethylene did not alter the total uptake of auxin or its distribution along the segments. Since radioactivity entered through both ends and distributed as though it were being pumped from apex to base, it would appear that ethylene has no effect on polar auxin transport. Studies using donor and receiver blocks substantiate this conclusion by showing that application of ethylene does not change auxin uptake or the velocity and capacity of the polar transport system in etiolated pea, corn, or sunflower stem sections during a 3-hr period. Similar results have been reported for *Avena* and pea tissue.[4, 5, 10] Since ethylene has no effect on either the destruction or polar transport of IAA in pea stem sections, it must be concluded that less diffusible auxin is present in gassed tissue because the rate of auxin synthesis is lowered. This result, coupled with the studies on auxin-induced ethylene formation, suggests the possibility of a feedback mechanism in which each hormone controls the other's rate of synthesis. (See *Note added in proof.*)

Effect of ethylene on tropisms and lateral auxin transport: When pea seedlings are placed on their sides, the stem starts to curve upward within 2 hr and the roots downward within 3–4 hr, whereas if ethylene is present, both continue to grow in a horizontal direction. Many plants react similarly,[11, 12] but ethylene does not prevent geotropic curving of *Avena* or corn coleoptiles. Pea stem segments afford a convenient system in which to study the influence of ethylene on geotropic sensitivity. Not only does the gas prevent them from curving in response to gravity (Fig. 5), but in addition it inhibits a spontaneous curvature from developing when the sections are incubated in liquid medium (Figs. 3 and 6). If 10^{-6} *M* IAA is present, pea tissue begins to curve immediately, even though the flasks are con-

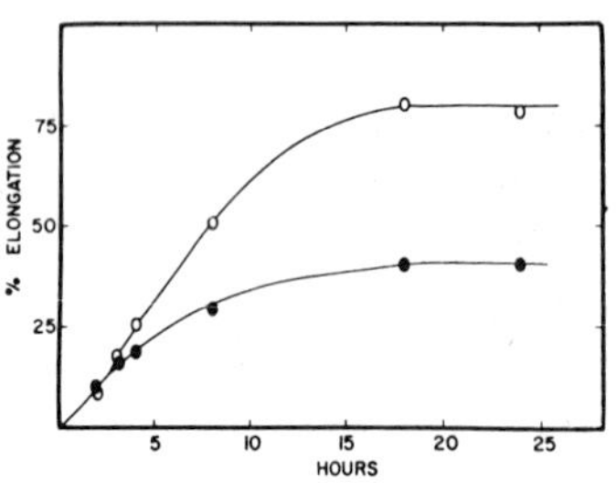

FIG. 4.—Effect of C_2H_4 on the elongation of etiolated pea stem segments. Control (○—○) and 5 ppm C_2H_4 (●—●). Flasks contained 10^{-6} *M* IAA and the medium described in the legend to Fig. 1.

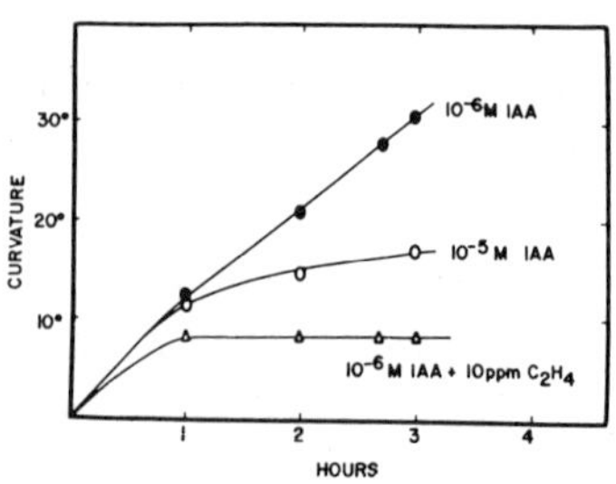

FIG. 5.—Effect of 10 ppm C_2H_4 on the geotropic curvature of 15-mm etiolated pea stem sections. Tissue was mounted horizontally with an apical donor block containing IAA, and a basal receiver block. Similar results were obtained with sunflower.

tinuously shaken; the curvature reaches a maximum within 3 hr, and thereafter straightening gradually occurs (Fig. 6). If IAA is omitted (or growth inhibited by azide, iodoacetate, etc.), the curvature develops more slowly, and when ethylene is present, the bending never takes place. This apparently spontaneous tropism may be analogous to a stem circumnutation which pea and other seedlings exhibit under certain conditions,[12, 13] and this also ceases shortly after ethylene is applied.[12] Curvature development and elongation are inhibited in perfect unison as the ethylene concentration is raised (Fig. 3), suggesting that inhibition of both spontaneous and geotropic curvatures might be due to the effect of ethylene on elongation, but this cannot be the case because curvature is arrested by ethylene several hours before the gas begins to slow the rate of elongation (cf. Figs. 4, 5, and 6). For this reason the effect of ethylene on lateral auxin transport was investigated. Using 10^{-6} M IAA in the donor block (Table 1), it was found that after 135 min the specific activity of the upper quarter of the sections was 28.4 per cent that of the lower quarter. (This method is preferable to the usual procedure of comparing upper and lower halves because, unlike coleoptiles, solid sections contain most of their tissue mass in the center of the stem where the auxin gradient must be very slight. As a result, after 135 min the specific activity of the upper half of pea sections was 37.5 ± 2.7 per cent that of the lower half, a much smaller differential than that obtained in experiments with quarters.) When ethylene is present, the uptake and polar transport of auxin are not affected, but lateral transport is almost completely abolished (Table 1), indicating that lateral and polar auxin transport are mediated by completely separate systems. This result agrees with that of van der Laan[4] who found that auxin did not redistribute laterally in the stems of *Vicia faba* seedlings which had been grown in the presence of ethylene until they experienced a horizontal nutation. It may be concluded that ethylene causes ageotropism and horizontal nutation by inhibiting lateral transport of auxin and, in addition, we should like to suggest (as did Borgström[14]) that all tropistic responses to ethylene are mediated by the same mechanism. Many of these responses to ethylene are

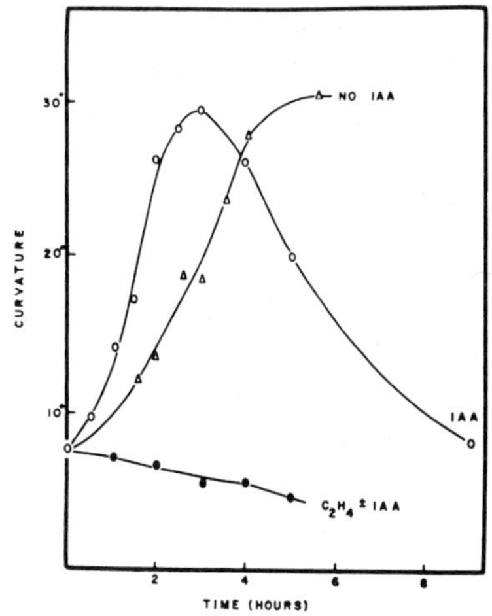

FIG. 6.—Effect of C_2H_4 on the development of a spontaneous curvature in 10-mm etiolated pea stem sections. Flasks contained the medium described in the legend to Fig. 1, supplemented as indicated with 10 ppm C_2H_4 and/or 10^{-6} M IAA. Tissue was continuously shaken. Similar results were obtained with sunflower stem sections.

TABLE 1

EFFECT OF ETHYLENE ON THE LATERAL TRANSPORT OF C^{14}-IAA IN PEA STEM SECTIONS

	10^{-6} M IAA	10^{-6} M IAA + 100 ppm C_2H_4	10^{-}[illegible] M IAA
Uptake (cpm/mg in intact segment)	24.6 ± 4.6	22.8 ± 6.4	191 ± 17
Relative specific activity*			
Upper	0.58 ± 0.07	0.91 ± 0.08	0.79 ± 0.13
Middle	0.99 ± 0.07	1.0 ± 0.07	1.05 ± 0.02
Lower	1.46 ± 0.10	1.11 ± 0.13	1.10 ± 0.02
Polar transport (cpm/10 sections)	6.1 ± 1.4	5.3 ± 1.0	60.0 ± 4.9

* Specific activity of each zone relative to that of the intact segment (average of 8 expts.).

duplicated by applying high concentrations of growth hormone. Auxin must accomplish this by stimulating ethylene formation, and if this is true, pea sections should not carry out lateral transport or curve in response to gravity if too much auxin is present in the donor block. When the concentration of IAA is raised from 10^{-6} to 10^{-5} *M*, uptake and polar transport are increased tenfold but, as expected, lateral transport and curvature are markedly inhibited (Table 1 and Fig. 5). An attempt to show that ethylene inhibits lateral auxin transport in corn coleoptiles gave negative results. Ethylene did not affect the uptake of auxin, using 10^{-6} or 10^{-5} *M* in the donor block, and it did not change the distribution of IAA in the tissue. It may be argued that these experiments are not conclusive because corn coleoptiles produce so much ethylene in the presence of 10^{-6} *M* IAA that they cannot respond to applied gas. This is a valid objection, but the studies with corn do show that ethylene could cause very little inhibition of lateral transport since a large auxin gradient arises in spite of the presence of the gas. Similarly, van der Laan[4] found that ethylene does not alter lateral auxin transport in the *Avena* coleoptile, and his experiment cannot be complicated by auxin-induced ethylene formation since *Avena* never produces so much ethylene that it fails to react to applied gas. These studies add further support to the proposal that all tropistic responses to ethylene are due to an inhibition of lateral auxin transport, for in corn and *Avena* coleoptiles where ethylene does not inhibit lateral transport, it does not produce any known tropistic effect.

Summary.—IAA induces ethylene formation in a variety of stem sections. The ethylene in turn causes the swelling and inhibition of extension growth in pea tissue, and the inhibition of growth in sunflower sections, which heretofore has been attributed directly to a high concentration of auxin. In pea stem sections ethylene leads to an immediate cessation of lateral auxin transport which is the basis for all tropistic responses to the gas.

Note added in proof: A paper in press [Morgan, P. W., and H. W. Gausman, *Plant Physiol.* (1966)] reports that polar auxin transport is markedly inhibited in stem sections cut from cotton or cowpea plants which have been fumigated with ethylene. We find that etiolated pea seedlings behave similarly; both the velocity and capacity of the polar transport system are reduced by 25% between the 4th and 5th hour after ethylene is applied to pea plants. When pregassing is continued for 18 hr, the inhibition of transport becomes almost complete, whereas auxin uptake is not affected. Since experiments cited in the present communication and elsewhere[4, 5, 10] have proved that ethylene is not a direct inhibitor of polar transport, the new findings indicate that the gas either prevents the development of the polar transport system in apical cells or else it causes a slow deterioration of the existing system. This might account for the loss of diffusible auxin which occurs within a few hours after seedlings are treated with ethylene,[4, 5, 10] and if so, it would not be necessary to postulate an effect of ethylene on auxin synthesis.

The authors would like to express their appreciation to Dr. K. V. Thimann for many helpful discussions in the course of this work, to Dr. A. C. Leopold for kindly providing the C^{14}-IAA and collaborating in the studies on polar transport, and to Mr. Robert Cuevas for his technical assistance.

* This investigation was supported by research grant EF-00214 from the U.S. Public Health Service, Division of Environmental Engineering and Food Protection, and was carried out while S. P. Burg was the recipient of Career Research Development Award 1-K3-GM-6871 from the USPHS.

[1] Zimmerman, P. W., and F. Wilcoxon, *Contrib. Boyce Thompson Inst.*, **7**, 209 (1935).

[2] Hall, W. C., and P. W. Morgan, *Régulateurs Naturels de la Croissance Végétale* (Paris: C.N.R.S., 1964), p. 727.

[3] Abeles, F. B., and B. Rubenstein, *Plant Physiol.*, **39**, 963 (1964).

[4] van der Laan, P. A., *Rec. Trav. Bot. Neerl.*, **31**, 691 (1934).

[5] Michener, D. H., *Am. J. Botany*, **25**, 711 (1938).

[6] Burg, S. P. and E. A. Burg, *Plant Physiol.*, **37**, 179 (1962).

[7] Gillespie, B., and K. V. Thimann, *Plant Physiol.*, **38**, 214 (1963).

[8] Galston, A. W., and R. Kaur, in *Light and Life*, ed. W. D. McElroy and B. Glass (Baltimore: Johns Hopkins Press, 1961), p. 687.

[9] Burg, S. P., and E. A. Burg, *Science*, **148**, 1190 (1965).

[10] von Guttenberg, H., and E. Steinmetz, *Pharmazie*, **2**, 17 (1947).

[11] Neljubow, D., *Ber. Deut. Botan. Ges.*, **29**, 97 (1911).

[12] Crocker, W., P. W. Zimmerman, and A. E. Hitchcock, *Contrib. Boyce Thompson Inst.*, **7**, 231 (1935).

[13] Galston, A. W., A. A. Tuttle, and P. J. Penny, *Am. J. Botany*, **51**, 853 (1964).

[14] Borgström, G., *Kgl. Fysiograf. Sällskap. Lund Förh.*, **9**, 135 (1939).

Reprinted from SCIENCE, August 10, 1951, Vol. 114, No. 2954, pages 159–161.

Plant Growth-stimulating Hormones in Immature Bean Seeds

John W. Mitchell, Dorothy P. Skaggs, and W. Powell Anderson

Division of Fruit and Vegetable Crops and Diseases, Bureau of Plant Industry, Soils, and Agricultural Engineering, USDA, Beltsville, Maryland

Immature seeds of plants such as apple, corn, rye, and wheat contain extractable phytohormones (*1–6*). Presence of these substances in seed extracts has been detected through their accelerating effect on the elongation of cells in oat coleoptiles and through their stimulating effect on the growth of tomato and pepper fruits. In studying these seed hormones, investigators have most often applied them to plants of different species from those from which they were taken.

In the present investigation, which was carried out under controlled greenhouse conditions, ether-soluble hormones obtained from the immature seeds of Black Valentine bean plants were applied to seedling plants of the same variety. Vegetative growth was accelerated in this way, and the plants were made to develop flowers 3–5 days earlier than untreated ones. The magnitude of the response reflected in vegetative growth varied, depending on the stage of development of the seeds from which the extracts were made.

Under the conditions of the experiment, the amount of ether-soluble hormones increased rapidly in the seeds after the fourth day following pollination, reaching a maximum on about the seventh or eighth day, and then decreased rapidly to an immeasurable amount on about the fifteenth day. The most potent extracts were obtained from seeds 3–4 mm long taken from pods that varied from 12 to 13 cm in length. Maximum rate of pod growth (increase in pod length) occurred during the first 8–10 days following pollination. Growth of the seeds (fresh weight) was most rapid subsequent to this period. The peak of extractable hormones was more closely associated with pod growth than with seed growth, but this does not mean that the hormones necessarily influenced growth of the pods (Fig. 1). The lack of extractable hormones in the case of relatively mature seeds was not due to the thickened seed coats, since these seeds, when chopped into small pieces and soaked in ether, failed to yield a measurable amount of hormones.

The extracts were prepared by soaking a freshly collected sample of whole seeds in moist ether at room temperature for 18–24 hr. After the ether was separated and evaporated, lanolin was mixed with the residue to afford the desired concentration of extract in the fatty carrier. In testing this method, the residue obtained from evaporating ether alone was mixed with lanolin and the mixture was applied to test plants, with no apparent effect.

In measuring the potency (cell-elongating activity) of an extract, lanolin paste containing a known amount of the extract was applied quantitatively and

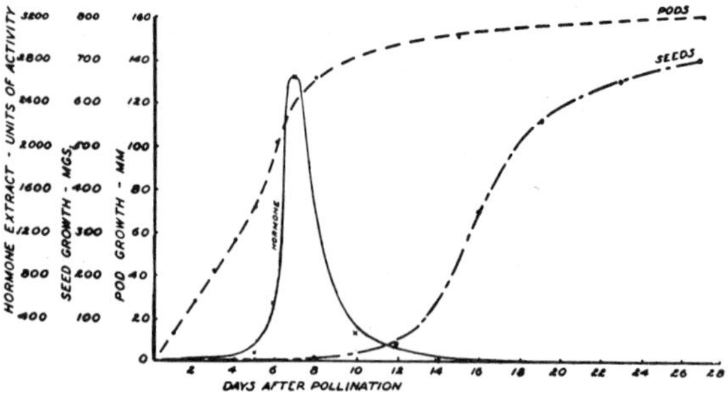

FIG. 1. Growth of bean pods and seeds in terms of length and fresh weight, respectively, compared with amount of extractable cell-elongating hormones obtained from the seeds at intervals following pollination.

unilaterally to the first internode of each seedling. The concentration of extract in the lanolin was varied during repeated trials, and the weight of extract required to cause a slight curvature of the stems was determined. Total activity of an extract was calculated by multiplying this value by the weight of extract obtained per seed. This method was used to measure the amount of extractable hormones in seeds at different stages of development (Fig. 1).

In some experiments, approximately 125 μg of the crude extract was applied in lanolin as a band around the first internodes of 10 seedlings. Ten comparable untreated plants were used for comparison. The extract stimulated elongation of the treated internodes and of all internodes above the treated ones, thus causing the stems to elongate much more rapidly than did those of the untreated plants (Table 1 and Fig. 2). The time of flowering was advanced 3–5 days, and the expansion of trifoliate leaves proceeded at a much more rapid rate on the treated than on the untreated plants. The extract affected stem growth much less when applied to relatively mature bean plants having several expanded trifoliate leaves than when applied to seedlings.

The extract stimulated growth of the seedlings when applied to the young succulent portions of their

TABLE 1

AVERAGE LENGTH (MM) OF INTERNODES OF BEAN PLANTS TREATED WITH 125 μG CRUDE SEED EXTRACT, COMPARED AT INTERVALS AFTER TREATMENT WITH AVERAGE LENGTH OF COMPARABLE INTERNODES OF UNTREATED PLANTS

Comparison	No. days					Internode
	2	5	6	7	8	
Check	39	52	53	54	54	First
Treated	89	115	116	117	117	
Check	0	10	19	28	32	Second
Treated	0	99	111	120	120	
Check	0	0	0	0	10	Third
Treated	0	5	32	74	95	
Check	0	0	0	0	0	Fourth
Treated	0	0	0	23	57	
Check	0	0	0	0	0	Fifth
Treated	0	0	0	0	10	

FIG. 2. Growth of bean plant (Black Valentine variety) following application of 125 μg hormone extract from immature seeds of other bean plants of the same variety, right. Comparable untreated plant, left. Photographed 3 weeks after treatment.

stems, but had no apparent effect when applied in equal amounts to the more mature part of their stems (hypocotyls) near the soil level. It failed to accelerate stem elongation of the seedlings when applied to either the upper or the lower surface of the primary leaves. Apparently hormones in the extract were not translocated from these leaves in sufficient amounts to bring about an observable effect in the stems.

Expansion of the primary leaves of seedlings was accelerated following application of the extract to a limited area on their upper or lower surfaces. Approximately 13 μg of the crude extract was applied in lanolin along the midribs of leaves, so as to cover an area of approximately 10 mm^2 on each leaf. The treated area was near the petiole attachment, in some instances on the ventral surface and in others on the dorsal surface. Elongation of the treated segment of midrib was accelerated. In addition, expansion of the treated leaves was increased so that their final area exceeded that of untreated ones by an average of 29.8% within a period of 6 days following treatment.

Untreated bean plants of the Black Valentine variety developed relatively short stems when grown under light supplied solely by means of daylight fluorescent tubes. In spite of the tendency to be short under these conditions, stems of seedlings treated with approximately 125 μg of the extract grew to be 3 times the height of untreated ones during the first 5 days of exposure to light from this source. Ten treated and 10 untreated plants were used in making the comparison, and the intensity of light used was 700–800 ft-c.

Since indolacetic acid accelerates cell elongation, the response of the plants to this compound was compared with that induced by means of the extract. Response to the extract was different in several respects from that which resulted from application of the acid. In comparing the responses, two series of treatments, comprised of 9 dosage levels of the acid and 9 of the extract, were prepared, with lanolin as a carrier. The dosages applied ranged from 125 to .013 μg/plant in each series. In treating, the paste was applied as a band around the first internode of each plant. Twenty-four hr after treatment, first internodes of plants treated with the optimum cell-elongating dose of the acid increased in length an average of 318% more than did those of untreated plants. During the same period, first internodes of seedlings treated with the most effective dose of seed extract increased in length an average of 870% more than did the untreated ones. Indolacetic acid brought about root initiation at the higher dosage levels, but the extract failed to cause this type of response at the highest dosage level used. The stimulating effect of the acid was confined to the treated first internode, whereas the extract not only accelerated elongation of this internode but also stimulated elongation of all internodes that developed above this segment of stem.

The extract had a systemic growth-stimulating effect when applied in minute amounts to the cotyledons of bean seeds that had just broken their coats in the process of germinating. Before the treatment was applied, the coats were removed from seeds that had developed radicles about 2.5 cm long. Approximately 13 μg of the extract was applied in lanolin as a thin layer covering about one fourth of the dorsal surface of each cotyledon. The treated seeds were replanted, along with an equal number of comparable but untreated ones from which the coats had also been removed.

Within 2½ days, growth of plants from treated seeds was noticeably more vigorous than that of the controls. This advantage persisted, and after 10 days the average lengths of the hypocotyls and first, second, and third internodes of plants that developed from the treated seeds were 10.9, 48.5, 65.8, and 100% greater, respectively, than the lengths of comparable parts of plants from untreated seeds. The area of each leaf that developed during this period was increased by the seed treatment. Ten days after treatment, the plants from treated seeds had a total leaf area that was 27.6% greater than that of plants from untreated seeds.

Additional experiments are necessary to determine the effect of the seed extract on seed production of bean plants and to learn whether the growth of other crop plants, such as cereals, can be stimulated by phytohormones obtained from their seeds.

References

1. AVERY, G. S., JR., BERGER, J., and SHALUCHA, B. *Am. J. Botany*, **29**, 612, 765 (1942).
2. GUSTAFSON, F. G. *Ibid.*, **26**, 189 (1939).
3. HAAGEN-SMIT, A. J., *et al. Ibid.*, **33**, 118 (1946).
4. HATCHER, E. S. J. *Ann. Botany*, **9**, 235 (1945).
5. LUCKWILL, L. C. *J. Hort. Sci.*, **24**, 32 (1948).
6. WHITTWER, S. *Missouri Research Bull. No. 371* (1943).

Hormonal Control of Enzyme Synthesis: On the Mode of Action of Gibberellic Acid and Abscisin in Aleurone Layers of Barley[1]

Maarten J. Chrispeels[2] and J. E. Varner
MSU/AEC Plant Research Laboratory, Michigan State University
East Lansing, Michigan 48823

Received March 31, 1967.

Summary. Gibberellic acid (GA) enhances the synthesis of α-amylase and ribonuclease in isolated aleurone layers and this process is inhibited by abscisin. Removal of gibberellic acid in mid-course of α-amylase production results in a slowing down of α-amylase synthesis, suggesting a continued requirement of GA for enzyme synthesis. This is paralleled by a continuous requirement for RNA synthesis. Addition of 6-methylpurine or 8-azaguanine in mid-course results in an inhibition of α-amylase synthesis within 3 to 4 hours. However, actinomycin D added in mid-course is almost without effect. This is not due to its failure to enter the cells, because it does inhibit ^{14}C-uridine incorporation at this stage. Addition of abscisin to aleurone layers which are synthesizing α-amylase results in an inhibition of this synthesis within 2 to 3 hours. Cycloheximide on the other hand inhibits enzyme synthesis immediately upon its addition. These data are consistent with the hypothesis that the expression of the GA effect requires the synthesis of enzyme-specific RNA molecules. The similarity in the kinetics of inhibition between abscisin on the one hand and 8-azaguanine or 6-methylpurine on the other suggests that abscisin may exert its action by inhibiting the synthesis of these enzyme-specific RNA molecules or by preventing their incorporation into an active enzyme-synthesising unit.

Gibberellic acid greatly enhances the synthesis of amylolytic enzymes (and several other hydrolytic enzymes) in aleurone cells of barley (20, 23, 27), wild oats (16), and rice (19). We have studied in detail the GA-enhanced synthesis and release of α-amylase and ribonuclease (4) and of protease (Jacobsen and Varner, in preparation) by isolated aleurone layers. The manifestation of this action of gibberellic acid is inhibited by inhibitors of protein synthesis (*p*-fluorophenylalanine and cycloheximide) and of ribonucleic acid synthesis (actinomycin D and 8-azaguanine). Evidence obtained by a new method indicates that the appearance of α-amylase (6) and protease (Jacobsen and Varner, in preparation) is due to the de novo synthesis of all the enzyme which is produced. This de novo synthesis is dependent on the synthesis of one or more species of ribonucleic acid.

Recently we reported that abscisin (abscisin II or dormin) inhibits the GA-enhanced synthesis of α-amylase and the release of free amino acids by aleurone layers (3). The inhibition of enzyme synthesis by abscisin can be partially overcome by the addition of a larger amount of GA. The kinetics of this interaction are neither competitive, nor non-competitive. The complexity of the system makes an interpretation of the kinetics data impossible. The inhibition of enzyme synthesis by abscisin is different from, and much more specific than, that obtained by metabolic inhibitors. Indeed, abscisin does not affect respiration or phosphorylation; neither does it inhibit the incorporation of radioactive precursors into protein (other than the hydrolases) and RNA (3).

An interaction between GA and abscisin has been reported in several systems. Abscisin inhibits the expansive growth of leaf sections of corn and this inhibition is reversed by GA (21). It also inhibits the GA-enhanced elongation of normal and dwarf peas (21). The elongation (normal and GA-enhanced) of lentil epicotyls is accompanied by DNA synthesis (16); both elongation and DNA synthesis are inhibited by abscisin and this inhibition can be partially reversed by the addition of GA (Chrispeels, unpublished).

We have studied in greater detail the effects of GA and abscisin on barley aleurone layers, as well as the effect of RNA-synthesis-inhibitors on α-amylase synthesis in order to learn more about the mode of action of the 2 hormones.

---

[1] This work was supported by the United States Atomic Energy Commission under contract No. AT (11-1) 1338.

[2] Present address: John Muir College, Department of Biology, UCSD, La Jolla, California 92307.

Materials and Methods

Most of the materials and methods used in these experiments have been described in detail in our previous paper (4). Synthetic d,l-abscisin, a gift from Dr. J. van Overbeek (Shell), was used in all the experiments reported here. A stock solution of 0.2 mM abscisin in water was kept frozen or in the refrigerator.

The incorporation of ^{14}C-leucine and ^{14}C-uridine was measured in the following way: Aleurone layers were incubated as usual and 1 μc of ^{14}C-leucine (200 mc/mmole) or of ^{14}C-uridine (25 mc/mmole) (both purchased from Schwarz Bio-research, Inc.) was added, and incorporation was allowed to proceed for the appropriate period of time. The aleurone layers were then rinsed 3 times with 1 mM cold, carrier leucine or uridine. When ^{14}C-leucine was used the aleurone layers were homogenized in the cold in 5 ml of 0.2 N NaCl in the usual way, and the extracts were centrifuged for 10 minutes at 2000 × *g*. An aliquot of 0.2 ml of the supernatant was plated on a glass planchet to determine the total uptake. Another aliquot of 0.25 ml of the salt soluble proteins was mixed with an equal volume of 10 mM leucine and precipitated with 2 volumes of 15 % trichloroacetic acid. The precipitated proteins were collected by filtration on Millipore filters and washed with 5 % trichloroacetic acid. The filters were dried and counted in a gas flow counter. When ^{14}C-uridine was used the aleurone layers were homogenized in the cold in 5 ml of 1 M NaCl with 5 mg of carrier RNA. Incorporation of the precursor into the salt-soluble nucleoproteins was determined as described above, except that carrier uridine was used. This method gives results similar to those obtained with more elaborate extraction procedures for nucleic acids.

Results

Removal of Gibberellic Acid in Midcourse of α-Amylase Production. The addition of GA to isolated aleurone layers of barley results in a linear synthesis of α-amylase after a lag-period of 7 to 9 hours. When aleurone layers are incubated in 0.05 μM GA for 11 hours, frequent rinsing over a period of 1 hour to remove as much GA as possible results in a substantial reduction of the α-amylase synthesized in the next 12 hour period (4). Gibberellic acid can therefore not be considered as a trigger because its presence is required during the lag-period as well as during the period of α-amylase synthesis, if maximal enzyme synthesis is to be obtained. In the present experiments the removal of GA was improved by rinsing the aleurone layers 4 times at half-hour intervals with a medium containing 20 mM $CaCl_2$ and 1 mM sodium acetate buffer pH 4.8. At the end of the rinsing period 1 set of aleurone layers was further incubated with GA and 1 without GA. The results of such an experiment are shown in figure 1. It is apparent that α-amylase synthesis is severely inhibited within a few hours, if GA is not added again after it has been removed.

When aleurone layers are incubated without GA for 10 hours, and 0.1 μM GA is added, no α-amylase is produced in the first 5 to 7 hours indicating that GA is required to overcome the lag-period. The lag-period can be shortened, but not abolished by first incubating the aleurone layers in a low concentration of GA (2.5 mμM) and adding a higher amount of GA (0.5 μM) after 10 hours (fig 2). After such an increase in GA concentration there is a lag-period of 3 to 4 hours before the aleurone layers synthesize α-amylase at a rate equivalent to the one obtained with the higher GA concentration.

Once the lag-period has been overcome it cannot be reintroduced by a prolonged incubation of the aleurone layers in the absence of GA. Aleurone layers were incubated for 8 hours in 0.05 μM GA, and GA was removed as described above. The aleurone layers were further incubated without GA for 6 hours. The rate of enzyme synthesis had slowed down considerably at this time. Subsequent addition of GA resulted in an immediate resumption of α-amylase synthesis at the control rate (fig 3).

Inhibition of GA Enhanced Synthesis of α-Amylase and Ribonuclease by Abscisin. The GA-enhanced production of α-amylase by aleurone layers is inhibited by abscisin (table I). When aleurone layers are incubated with 0.1 μM GA and 1 μM abscisin, very little enzyme is synthesized. Abscisin also inhibits the small amount of α-amylase synthesis which occurs in the absence of GA. Addition of a larger amount of GA reverses this inhibition by abscisin, as shown earlier (3) and as demonstrated in a different way in table II. Aleurone layers were incubated with increasing amount of GA (from 0.1 mμM to 0.1 mM) and either without abscisin or with 0.05 μM or 5 μM abscisin. The inhibitory effect of 0.05 μM abscisin is almost completely reversed by 1 μM GA. When 5 μM abscisin are used reversal by GA never exceeds 30 % regardless of the amount of GA used. This demonstrates that reversal is not merely a matter of the molar ratio of GA and abscisin.

Abscisin inhibits the GA enhanced production of ribonuclease in a similar fashion (table III). The

Table I. *Inhibition of α-Amylase Synthesis by Abscisin*

Ten aleurone layers were incubated with buffer, 20 mM of $CaCl_2$ and with or without 0.1 μM GA and the concentrations of abscisin indicated.

Treatment	α-Amylase per 10 aleurone layers
	μg
Control	40
Control + 0.5 μM abscisin	8
GA	475
GA + 1 μM abscisin	67

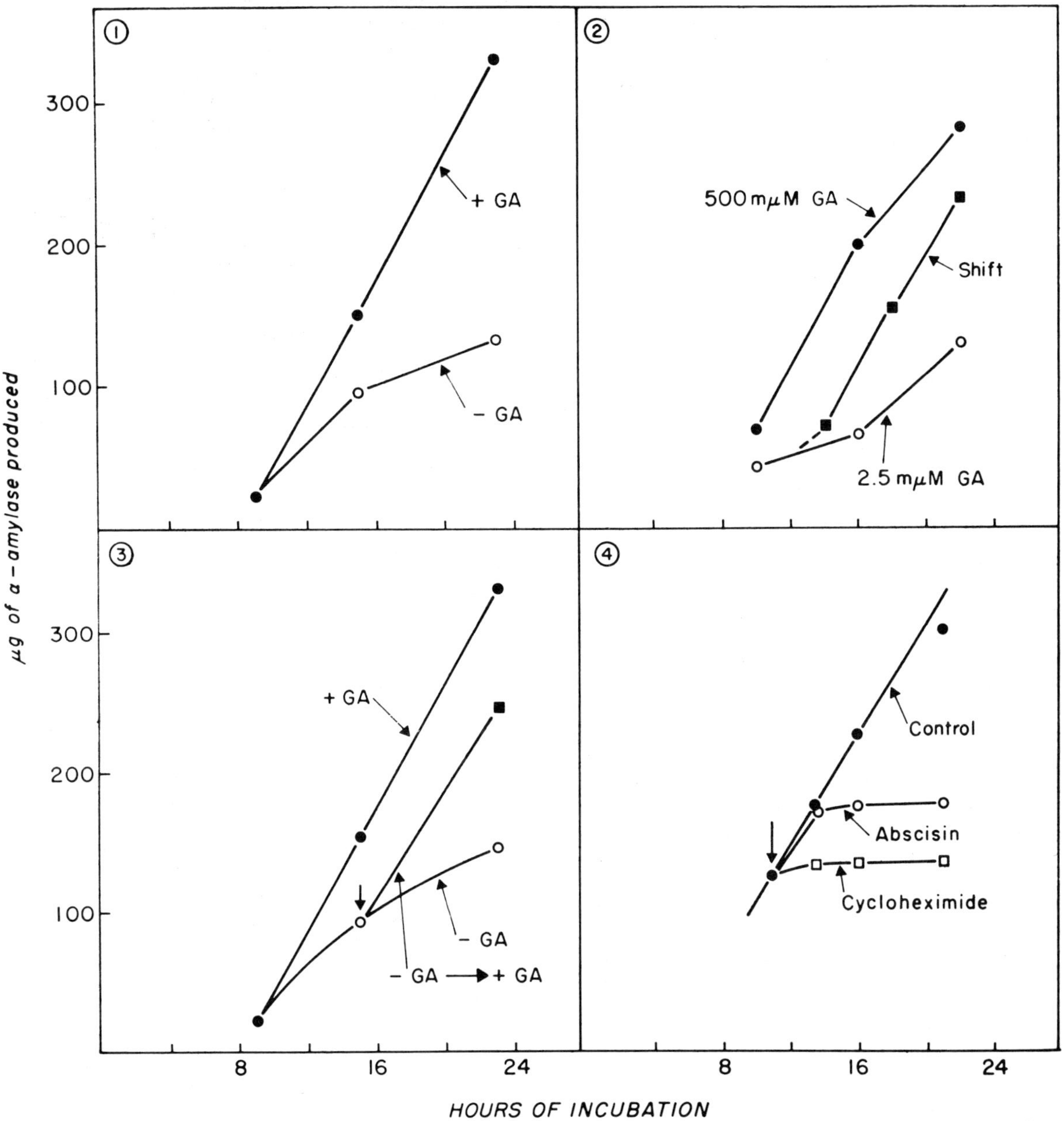

FIG. 1. Effect of removing GA at the end of the lag-period. Aleurone layers were incubated for 7 hours in 0.5 μM GA. GA was then removed by 4 consecutive one-half-hour rinses. The aleurone layers were further incubated either with GA, or without GA, and α-amylase synthesis was measured 6 and 14 hours later.

FIG. 2. Shift from a low to a high concentration of GA. Aleurone layers were incubated in 2.5 mμM GA or in 500 mμM GA and α-amylase synthesis was measured after 10, 16 and 22 hours. To 1 set of flasks the GA concentration was increased to 500 mμM after the aleurone layers had been in 2.5 mμM GA for 10 hours (Shift).

FIG. 3. Effect of adding GA after it has been withheld for 6 hours. All conditions as in figure 1, except that GA was added again at 15 hours to aleurone layers from which it had been withheld at 9 hours. Total α-amylase synthesis was measured 15 and 23 hours after the start of the incubation.

FIG. 4. Mid-course inhibition of α-amylase synthesis by abscisin and cycloheximide. Aleurone layers were incubated in 0.1 μM for 11 hours. At this time abscisin (5μM) or cycloheximide (10 μg/ml) was added and α-amylase synthesis was measured 2 and one-half, 5 and 10 hours later.

Table II. *Reversal by Gibberellic Acid of the Abscisin Inhibition of α-Amylase Synthesis*

Ten half aleurone layers were incubated in buffer with 20 mM $CaCl_2$ and the concentrations of GA and abscisin indicated. Activity of α-amylase was measured in the medium and a tissue-extract after 24 hours.

GA concentration	α-Amylase per 10 aleurone layers, Abscisin concentration 0	0.05 μM	5 μM
	μg	μg	μg
0.1 mμM	73	10	...
1 mμM	248	71	...
10 mμM	522	280	22
0.1 μM	638	525	101
1 μM	662	584	189
10 μM	692	...	185
0.1 mM	580	...	113

increase in ribonuclease which takes place in isolated aleurone layers in the absence of added GA, or in half seeds preincubating on moist sand (4) is also inhibited by abscisin.

In the experiments discussed above the 2 hormones, GA and abscisin, were added simultaneously. However, abscisin is also effective as an inhibitor of α-amylase synthesis if it is added in mid-course, while enzyme synthesis is in progress (fig 4). Aleurone layers were incubated for 11 hours in 0.1 μM GA and 5 μM abscisin was added at that time. Further synthesis of α-amylase was inhibited completely after a delay of 2 to 3 hours. If both hormones had been added at these concentrations, at the same time, at the beginning of the incubation, α-amylase synthesis would have been inhibited completely. Addition of cycloheximide (10 μg/ml) at this time results in an immediate inhibition of enzyme synthesis. It was also found that both the production and the release of ribonuclease can be inhibited by the addition of abscisin in mid-course (20–24 hrs after GA).

Requirement for RNA Synthesis. Actinomycin D and various base analogues which inhibit RNA synthesis prevent α-amylase formation if they are added at the same time as the hormone (4, 23, 24). The synthesis of α-amylase is inhibited to a much lesser extent if actinomycin D is added 4 hours after GA, and is almost unaffected if the antibiotic is added 8 hours after the hormone (table IV). A different effect is observed with 6-methyl-purine, a potent inhibitor of all RNA synthesis (11). When aleurone layers are incubated with 1 μM GA and 0.1 mM 6-methylpurine α-amylase synthesis is inhibited by 90 %. However, if the analogue is added 4 or 8 hours after the hormone enzyme synthesis is inhibited by only 70 % and 45 % respectively (table V). Larger inhibitions are obtained with 1 mM 6-methylpurine, and α-amylase synthesis is still inhibited by 66 % if this concentration of the analogue is added 8 hours after GA.

The failure of actinomycin D to inhibit α-amylase synthesis if the antibiotic is added 8 hours after the hormone could be due to its inability to inhibit RNA synthesis in general at this stage or to its inability to inhibit the RNA fraction which

Table III. *Inhibition of Ribonuclease Production by Abscisin*

Ten aleurone layers were incubated in buffer, 20 mM of $CaCl_2$ and with or without 2 mμM GA and the concentrations of abscisin indicated. Ribonuclease activity was measured after 48 hours in the medium and an extract of the tissue. Initial refers to the amount of ribonuclease present in the aleurone layers at the start of incubation.

Treatment	Units of ribonuclease per 10 aleurone layers
Initial	29
Control	108
Control + 0.5 μM abscisin	61
GA	256
GA + 5 μM abscisin	53

Table IV. *Inhibition of α-Amylase Synthesis and ^{14}C-Uridine Incorporation by Actinomycin D Added 4 and 8 Hours After GA*

Ten aleurone layers were incubated in 0.1 μM GA and after 4 or 8 hours actinomycin D (100 μg/ml) was added. Enzyme synthesis was measured at the end of the 24 hour incubation period. ^{14}C-Uridine (1 μc/flask) was added 4 hours after actinomycin D and incorporation was allowed to proceed over a 4 hour period.

Treatment	α-Amylase per 10 aleurone layers	^{14}C-Uridine incorporated
	μg	cpm
GA	359	1380
GA + Act. D after 4 hrs	212	
GA + Act. D after 8 hrs	325	466

Table V. *Inhibition of α-Amylase Synthesis by 6-Methylpurine*

Ten aleurone layers were incubated in buffer, 20 mM, $CaCl_2$ and 0.5 μM GA. 6-Methylpurine was added at the same time as GA or 4 or 8 hours later and total α-amylase production was measured after 24 hours incubation.

Treatment: Time of addition of 6-methylpurine	α-Amylase per 10 aleurone layers: 0.1 mM Of 6-methylpurine	1.0 mM Of 6-methylpurine
	μg	μg
0 hrs	38	9
4 hrs	115	55
8 hrs	208	140
Control	384	426

is required for α-amylase synthesis. To test the first possibility aleurone layers were incubated in 0.1 μM GA, and 100 μg/ml of actinomycin D were added after 8 hours. After 4 more hours 1 μc of ^{14}C-uridine was added to each flask containing 10 aleurone layers and incorporation was allowed to proceed for 4 hours. During this period actinomycin D inhibited the incorporation of ^{14}C-uridine by 66 % (table IV). We can assume that the inhibition of RNA synthesis was even greater because some of the incorporation represents turnover of the terminal ends of the transfer RNA molecules. (It was shown by Chandra and Varner (2) that there is a rapid conversion of uridine to cytidine in this tissue). It appears then that actinomycin D is a good inhibitor of ^{14}C-uridine incorporation whether it is added at the same time as GA (4) or 8 hours after the hormone. This leads to the conclusion that the inability of actinomycin D to inhibit α-amylase synthesis must be due to its inability to prevent the synthesis of the specific RNA fraction which is required for continued α-amylase synthesis.

Table VI. *Sensitivity of α-Amylase Synthesis to Actinomycin D and 6-Methylpurine After Removal of GA*

Ten aleurone layers were incubated in buffer, 20 mM $CaCl_2$ and 0.5 μM GA for 9 hours. GA was removed by 4 subsequent one-half-hour rinsings in a medium without GA. The aleurone layers were further incubated for 11 hours without GA or with 0.5 μM GA and actinomycin D (100 μg/ml) or 6-methylpurine (0.2 mM). Initial refers to the amount of α-amylase present in the aleurone layers after the removal of GA (at the end of the washing out procedure).

Treatment	α-Amylase per 10 aleurone layers
	μg
Initial	61
—GA	128
+GA	285
+GA + actinomycin D	221
+GA + 6-methylpurine	153

We have shown above that GA is not a trigger, but is required continuously for maximal α-amylase synthesis. The following experiment was done to test whether the α-amylase synthesis which is dependent on the second addition of GA also requires continued RNA synthesis. Aleurone layers were incubated for 10 hours in 0.05 μM GA and the GA was removed by 4 successive one-half-hour rinses. The aleurone layers were then further incubated without GA, or with GA and with actinomycin D (100 μg/ml) or 6-methylpurine (0.2 mM). The results show (table VI) that the α-amylase synthesis which is dependent on the second addition of GA is not very sensitive to actinomycin D but is very strongly inhibited by 6-methylpurine. Synthesis of α-amylase under these circumstances proceeds without delay upon the addition of GA, even if GA has been withheld for many hours (fig 3). This indicates that the aleurone layers are capable of a rapid synthesis of the RNA fraction necessary for α-amylase synthesis. When aleurone layers are incubated with GA there is normally a 7 to 9 hour lag before α-amylase synthesis begins.

The observations discussed above suggest that this lag is not necessitated by the requirement for the synthesis of the metabolically unstable RNA fraction, the synthesis of which must accompany α-amylase synthesis and can be inhibited by 6-methylpurine. It seems likely that other biochemical processes are associated with the lag-period and must occur before α-amylase synthesis can start. The possibility that a more stable RNA needs to be synthesized during the lag-period can, of course, not be ruled out.

To test whether the requirement for RNA synthesis can be satisfied by incubating for a long time with a small amount of GA, half seeds were preincubated for 3 days on sterile sand moistened with GA (0.01 or 0.1 μM) instead of water. The aleurone layers were removed from the starchy endosperm and further incubated with GA (1 μM) or with GA and actinomycin D (50 μg/ml). The data show (table VII) that actinomycin D inhibits enzyme synthesis to approximately the same extent, whether the half seeds were preincubated in water

Table VII. *Inhibition of α-Amylase Synthesis by Actinomycin D After Preincubation of Half Seeds in Gibberellic Acid*

Half seeds were preincubated in water or in GA (0.01 μM or 0.1 μM) for 3 days. The aleurone layers were removed and further incubated in buffer, 20 mM, $CaCl_2$ and 1 μM GA, and with or without actinomycin D (50 μg/ml). 0 hours refers to the amount of α-amylase present in the aleurone layers at the start of the incubation.

Treatment and time	α-Amylase per 10 aleurone layers Preincubation in water	Preincubation in 0.01 μM GA	Preincubation in 0.1 μM GA
	μg	μg	μg
0 hrs GA	11	22	64
8 hrs GA	35	64	136
24 hrs GA	408	445	505
24 hrs GA + actinomycin D	240	270	278

or in GA. As a result of preincubation in 0.1 μM GA α-amylase synthesis is initiated at a slow rate (compare α-amylase synthesis at 0 and 8 hrs after preincubation in 0.1 μM GA with preincubation in water), but this does not obviate the necessity for the initiation of more RNA synthesis if a higher concentration of GA is added.

Comparison of the Effect of Abscisin and Inhibitors of RNA Synthesis. We have shown that the addition of cycloheximide, a potent inhibitor of protein synthesis at the level of translation, to aleurone layers results in an immediate cessation of α-amylase synthesis and also of the incorporation of ^{14}C-leucine into the cellular proteins of aleurone layers (Varner, unpublished). Abscisin, on the other hand, inhibits enzyme synthesis after a lag of 2 and one-half to 3 hours. Experiments with 6-methylpurine show a continuous requirement for RNA synthesis during the period of α-amylase synthesis. It seemed, therefore, of interest to study the mid-course inhibition of α-amylase synthesis by abscisin and base analogues and compare their kinetics.

Aleurone layers were incubated in 0.5 μM GA for 11 hours to induce the GA enhanced synthesis of α-amylase. At this stage the medium was withdrawn, the aleurone layers were rinsed and further incubated for 12 hours in a medium containing GA (0.5 μM) or GA and various inhibitors (10 μM abscisin, 0.5 or 5 mM 6-methylpurine, 5 mM 8-azaguanine and 5 mM 5-azacytidine). The results in table VIII indicate that abscisin and 5 mM 6-methylpurine give the same amount of inhibition (74 %) while 0.5 mM 6-methylpurine and 5 mM 8-azaguanine or 5-azacytidine are less inhibitory (58–39 %). The kinetics of inhibition of α-amylase synthesis by abscisin and the base analogues are very similar (fig 6). Abscisin inhibits α-amylase synthesis completely within 2 and one-half to 3 hours, as shown earlier (fig 4). The base analogues also

Table VIII. *Mid-course Inhibition of α-Amylase by Abscisin and Base Analogues*

Aleurone layers were incubated in buffer, 20 mM $CaCl_2$ and 0.5 μM GA for 11 hours. The medium was drawn off and the aleurone layers were rinsed. The aleurone layers were further incubated for 12 hours in 0.5 μM GA together with the inhibitors, as indicated. The figures represent the amount of α-amylase produced in the 12 hour incubation.

Treatment	α-Amylase per 10 aleurone layers
	μg
Control	317
Abscisin (10 μM)	82
6-Methylpurine (0.5 mM)	156
6-Methylpurine (5 mM)	84
5-Azacytidine (5 mM)	194
8-Azaguanine (5 mM)	132

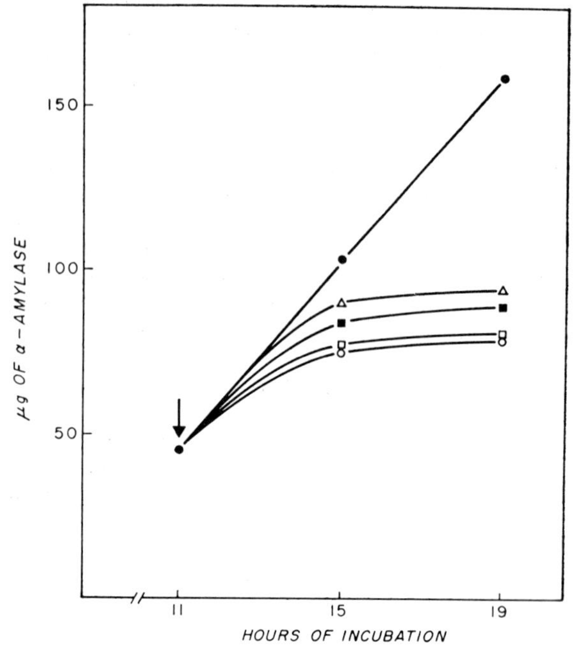

FIG. 5. Mid-course inhibition of α-amylase synthesis by abscisin, 6-methylpurine and 8-azaguanine. Aleurone layers were incubated in 0.05 μM GA for 11 hours. At this time the medium was removed, the aleurone layers were rinsed, and they were further incubated with 0.05 μM GA (●—●), or with GA and 10 μM abscisin (○—○), 5 mM 6-methylpurine (□—□), 0.5 μM 6-methylpurine (■—■), or 5 mM 8-azaguanine (△—△).

inhibit α-amylase synthesis completely within 2 and one-half to 4 hours depending on the concentration used. The incorporation of ^{14}C-uridine by the aleurone layers is inhibited by 1 mM 6-methylpurine in the following way: 30 to 35 % within 2 and one-half hours of its addition and 65 to 70 % after 4 hours of its addition. However, if the 6-methylpurine concentration is increased to 5 mM the inhibition of ^{14}C-uridine incorporation is 65 to 70 % within 2 and one-half hours. This may account for the shorter lag time in the inhibition of α-amylase synthesis observed with the higher concentration of 6-methylpurine. The incorporation of ^{14}C-leucine by the aleurone layers into the cellular proteins is inhibited by only 30 %, 4 hours after the addition of 1 mM 6-methylpurine. This is probably an indirect effect mediated through its inhibition of RNA synthesis. We therefore conclude that the rapid inhibition of α-amylase synthesis caused by 6-methylpurine is not due to a direct inhibition of protein synthesis.

Discussion

It is not yet possible to decide whether these hormones work at the level of transcription or at the level of translation. The data presented here, as well as the earlier results obtained by Varner and coworkers (4, 22, 23, 24), are consistent with the hypothesis that GA exerts its control at the level of the gene, to bring about the synthesis of an RNA fraction specific for the proteins being synthesized. The similarity in kinetics for the inhibition of α-amylase synthesis by abscisin and 6-methylpurine or 8-azaguanine suggests that abscisin exerts its control by inhibiting the synthesis of such a specific RNA fraction. However, much caution must be used in basing a model on the mode of action of the hormones on a study of the effect of inhibitors alone. The data are equally consistent with a control mechanism at the level of translation with a requirement for continued RNA synthesis.

Many hormonal effects are inhibited by actinomycin D (1, 9, 13, 17, 20, 25), and this has often been interpreted as evidence that the hormones affect the synthesis of messenger RNA in a direct way while all that can be concluded with certainty is, that some kind of DNA directed RNA synthesis must be allowed to go on before the hormonal effect can express itself at the level of protein synthesis or of morphogenesis. Hybridization experiments with bacterial nucleic acids have shown regions of complementarity between DNA and messenger RNA, ribosomal RNA, and transfer RNA suggesting that the synthesis of all 3 major classes of cytoplasmic RNA is DNA directed. The synthesis of any or all of these 3 may be required for the hormonal effect to become evident.

A second reason for caution is that the hormone-enhanced enzyme synthesis may result in a more rapid turnover of one of the RNA's involved in the synthesis of these enzymes (e.g. messenger RNA or a particular transfer RNA). This would result in a requirement for RNA synthesis, and such a process would be inhibited by actinomycin D or by base analogues, even though the initial effect of the hormone is not at the level of transcription.

The principal, easily observed effect of GA on the aleurone layers is the enhancement of the synthesis of several hydrolytic enzymes. RNA synthesis must be allowed to occur in order to obtain this effect. A control mechanism at either transcription or translation would probably involve the synthesis of a new RNA. There does not appear to exist a great reserve of this critical RNA fraction. Indeed, base analogues inhibit α-amylase synthesis within 2 and one-half to 4 hours and continued RNA synthesis must accompany enzyme synthesis. We can also conclude that this RNA fraction is synthesized very rapidly since addition of GA to aleurone layers from which this hormone has been withheld for several hours results in an immediate resumption of α-amylase synthesis which is completely dependent on continued RNA synthesis.

The principal, easily observed effect of abscisin is the inhibition of hydrolase synthesis (both normal and GA-enhanced). The kinetics of this inhibition resemble those obtained with the base analogues 6-methylpurine and 8-azaguanine. In the presence of 10 μM abscisin or 5 mM 6-methylpurine, at this concentration the base analogue inhibits ^{14}C-uridine incorporation by 70 % within 2 and one-half hours, α-amylase synthesis is completely inhibited within 2 to 3 hours. With 0.5 mM 6-methylpurine or 5 mM 8-azaguanine it takes somewhat longer before inhibition is complete. The incorporation of ^{14}C-leucine is inhibited to a much smaller extent (30 %) than the incorporation of ^{14}C-uridine (70 %) when aleurone layers are incubated for 4 hours in 1 mM 6-methylpurine. This suggests that the inhibition of protein synthesis is a result of the inhibition of RNA synthesis, brought about by the decay of short-lived messengers. This was confirmed by A. Morris (private communication) who showed that 10 mM 6-methylpurine has no effect on the incorporation of ^{14}C-valine into the proteins of rabbit reticulocytes, indicating that the analogue has no direct effect on protein synthesis in this system. (Reticulocytes have no nucleus and protein synthesis is independent of continued RNA synthesis). The 2 and one-half to 3 hour delay in the inhibition of α-amylase synthesis which is observed with abscisin as well as with the base analogues could be due to the presence of a sufficient amount of the critical RNA fraction necessary to sustain enzyme synthesis for this period of time.

However, similar kinetics might be expected if abscisin exerted its action at the level of translation by inhibiting the incorporation of this RNA into an active enzyme-synthesizing unit. In this case the delay would reflect the half-life of this active complex.

The inhibition of α-amylase synthesis by abscisin can be overcome, at least partially, by the addition of a larger amount of GA. However, if too high a concentration of abscisin is used, no reversal is obtained. Similar observations were made by J. van Overbeek (personal communication) who studied the inhibition of growth of *Lemna minor* by abscisin and the reversal of this inhibition by benzyladenine. Reversal by benzyladenine only takes place if growth has not been completely inhibited by abscisin.

A large number of studies indicate that plant hormones increase the rate of protein and RNA synthesis in the tissues to which they are applied. In some systems this can be measured by an increase in the total amount of protein and RNA, while in others there is only an increase in the rate of incorporation of radioactive precursors into these macromolecules. Hormone application always results in a general derepression of RNA synthesis and in no instance has a system been found in which the hormone preferentially increases the synthesis of 1 particular RNA fraction.

Gibberellic acid does not enhance the incorporation of ^{14}C-uridine into the RNA[3] or of ^{14}C-leucine into the cellular proteins of isolated aleurone layers, dissected from the half seeds after the customary 3-day preincubation period (Chandra, Chrispeels and Varner, unpublished); neither does abscisin inhibit these processes (3). Although the hormones appear to have no effect on protein synthesis as measured by precursor incorporation, they do affect the synthesis of specific enzymes. This observation has not been matched at the RNA level, and we have been unable to find a specific RNA fraction (after extraction of the RNA and chromatography on methylated albumen kieselguhr columns), the synthesis of which is evoked by GA and inhibited by abscisin. Recent improvements in the methods used to extract nucleic acids from animal tissues (12) and to separate the various RNA's (5, 14) will be applied in our search for such an RNA fraction. The possibility that GA and abscisin exert their control at the level of translation can be checked by studying the effect which these hormones have on the in vitro synthesis of α-amylase by a subcellular fraction isolated from aleurone layers.

Acknowledgments

We acknowledge with pleasure the skillful technical assistance of Miss Nancy Joseph. We thank R. A. Nilan for supplying the Himalaya variety of barley, and A. Morris for determining the effect of 6-methylpurine on amino acid incorporation in rabbit reticulocytes.

The actinomycin D was a gift from Merck, Inc.

Literature Cited

1. Abeles, F. B. and R. E. Holm. 1966. Enhancement of RNA synthesis, protein synthesis, and abscision by ethylene. Plant Physiol. 41: 1337–42.
2. Chandra, G. R. and J. E. Varner. 1965. Gibberellic acid controlled metabolism of RNA in aleurone cells of barley. Biochim. Biophys. Acta 180: 583–92.
3. Chrispeels, M. J. and J. E. Varner. 1966. Inhibition of gibberellic acid induced formation of α-amylase by abscision II. Nature 212: 1066–67.
4. Chrispeels, M. J. and J. E. Varner. 1967. Gibberellic acid-enhanced synthesis and release of amylase and ribonuclease by isolated barley aleurone layers. Plant Physiol. 42: In press.
5. Ellem, K. A. O. 1966. Some properties of mammalian DNA-like RNA isolated by chromatography on methylated bovine serum albumin-kieselguhr columns. J. Mol. Biol. 20: 283–305.
6. Filner, P. and J. E. Varner. A simple and unequivocal test for de novo synthesis of enzymes: density labeling of barley α-amylase with H_2O^{18}. Proc. Natl. Acad. In press.
7. Fletcher, R. A. and D. J. Osborne. 1966. Gibberellin, as a regulator of protein and ribonucleic acid synthesis during senescence in leaf cells of *Taraxacum officinale*. Can. J. Botany 44: 739–45.
8. Hamilton, T. H., R. J. Moore, A. F. Rumsey, A. R. Means, and A. R. Schrank. 1965. Stimulation of synthesis of ribonucleic acid in subapical sections of *Avena* coleoptile by indolyl-3-acetic acid. Nature 208: 1180–83.
9. Key, J. L. 1964. Ribonucleic acid and protein synthesis as essential processes for cell elongation. Plant Physiol. 39: 365–70.
10. Key, J. L. 1966. Effect of purine and pyrimidine analogues on growth and RNA metabolism in the soybean hypocotyl, the selective action of 5-fluorouracil. Plant Physiol. 41: 1257–64.
11. Key, J. L. and J. Shannon. 1964. Enhancement by auxin of ribonucleic acid synthesis in excised soybean hypocotyl tissue. Plant Physiol. 39: 360–64.
12. Kirby, K. S. 1965. Isolation and characterization of ribosomal ribonucleic acid. Biochem. J. 96: 266–69.
13. Leaver, C. J. and J. Edelman. 1965. Nucleic acid synthesis in carrot tissue slices. Biochem. J. 95: 27 p.
14. Loening, V. E. 1967. The fractionation of high-molecular-weight ribonucleic acid by polyacrylamide-gel electrophoresis. Biochem. J. 102: 251–57.
15. Naylor, J. M. 1966. Dormancy studies in seed of *Avena fatua*. 5. On the response of aleurone cells to gibberellic acid. Can. J. Botany 44: 19–32.
16. Nitsan, J. and A. Lang. 1966. DNA synthesis in the elongating nondividing cells of the lentil epicotyl and its promotion by gibberellin. Plant Physiol. 41: 965–70.
17. Noodén, L. D. and K. V. Thimann. 1963. Evidence for a requirement for protein synthesis for auxin-induced cell enlargement. Proc. Natl. Acad. Sci. 50: 194–200.
18. Ogawa, Y. and S. Imamura. 1965. Effect of plant extracts and gibberellin A_3 on α-amylase production in embryoless rice endosperm in relation to growth promoting activity. Proc. Japan Acad. 41: 842–45.
19. Paleg, L. 1960. Physiological effects of gibberellic acid. I. On carbohydrate metabolism and amylase activity of barley endosperm. Plant Physiol. 35: 293–99.
20. Roychoudhury, R. and S. P. Sen. 1964. Studies on the mechanism of auxin action: auxin regulation of nucleic acid metabolism in pea internodes and coconut milk nuclei. Physiol. Plantarum 17: 352–62.

[3] Chandra and Varner (2) and Varner et al. (24) reported a GA-enhanced incorporation of radioactive precursors into the RNA of aleurone layers 8 to 24 hours after the start of imbibition of half seeds. Naylor (15) observed a GA-enhanced incorporation of ^{3}H-cytidine into the nuclei of aleurone cells from wild oats, 24 hours after the start of imbibition. It seems likely that in these cases GA is accelerating processes which are associated with inhibition or with biochemical changes which follow it. It is exactly for this reason, to avoid all biochemical changes associated with imbibition, that half seeds are preincubated for 3 days before we study the effects of GA and abscisin on aleurone layers.

21. THOMAS, T. H., P. F. WAREING, AND P. M. ROBINSON. 1965. Action of the sycamore 'Dormin' as a gibberellin antagonist. Nature 205: 1270–72.

22. VARNER, J. E. 1964. Gibberellic acid controlled synthesis of α-amylase in barley endosperm. Plant Physiol. 39: 413–15.

23. VARNER, J. E. AND G. R. CHANDRA. 1964. Hormonal control of enzyme synthesis in barley endosperm. Proc. Natl. Acad. Sci. U. S. 52: 100–06.

24. VARNER, J. E., G. R. CHANDRA, AND M. J. CHRISPEELS. 1965. Gibberellic acid-controlled synthesis of α-amylase in barley endosperm. J. Cellular Comp. Physiol. 66, suppl. 1: 55–68.

25. VENIS, M. A. 1964. Induction of enzymatic activity of indolyl-3-acetic acid and its dependence on synthesis of ribonucleic acid. Nature 202: 900–01.

26. YOMO, H. 1960. Studies on the α-amylase activating substances. IV. On the amylase activating action of gibberellin. Hakko Kyokaishi, 18: 600–02 (C.A. 55: 26145, 1961).

Reduction of the Gibberellin Content of Pharbitis Seeds by CCC and After-Effects in the Progeny[1]

Jan A. D. Zeevaart

Department of Biology, McMaster University, Hamilton, Ontario, Canada
and
MSU/AEC Plant Research Laboratory, Michigan State University, East Lansing, Michigan

Received December 20, 1965.

Summary. Plants of *Pharbitis nil* were treated with the growth retardant (2-chloroethyl) trimethylammonium chloride (CCC) shortly before and after anthesis. Fresh and dry weight of immature seeds were not affected by the CCC treatment.

The level of gibberellin-like activity in *Pharbitis* seeds as compared to control seeds was strongly reduced by CCC application. The progenies of the treated plants also had a much reduced GA content in the seedling stage. These results are interpreted to indicate that CCC blocks gibberellin biosynthesis in higher plants, as it does in the fungus *Fusarium.*

CCC applied via the roots accumulated in the immature seeds and was carried over to the following generation. Consequently, growth of CCC progenies was dwarfed and flower formation inhibited. Both phenomena were overcome by application of gibberellin A_3.

Three gibberellin-like substances (called fractions I, II, and III) were present in *Pharbitis* seeds and could be separated by thin-layer chromatography. All 3 fractions were also present in seeds treated with CCC. Fractions II and III were present in much higher quantities than fraction I. Both fractions II and III promoted growth of d5 corn but only fraction II was active in dwarf peas grown under red light.

Plant growth-retarding chemicals form a group of chemically unrelated compounds which have the common characteristic of inhibiting stem elongation (4). This effect is usually overcome by the application of gibberellin (GA).

A breakthrough concerning the mode of action of growth retardants was made by the discovery that Amo-1618 [2′-isopropyl-4′-(trimethylammonium chloride)-5′-methylphenyl piperidine-1-carboxylate] and CCC [(2-chloroethyl) trimethylammonium chloride] inhibit GA biosynthesis in the fungus *Fusarium moniliforme* (10, 15). This led to the hypothesis that growth retardants also block GA production in higher plants, thus causing a GA-deficiency and subsequent dwarfed growth. In support of this idea Harada and Lang (7) found good agreement between the effect of CCC analogs on growth of higher plants and their capacity to block GA production in *Fusarium.* Furthermore, Baldev et al. (1) showed that the GA content in developing pea seeds was significantly reduced by Amo-1618.

Immature seeds of *Pharbitis nil* are known to have a high content of GA-like substances (16). Moreover, *Pharbitis* plants respond readily with dwarfed growth to applications of CCC (21). In order to test the hypothesis that CCC exerts its effects on plant growth by blocking GA production, a study of the effect of CCC on the GA content of *Pharbitis* was undertaken. The evidence to be presented in this paper shows that CCC inhibits the accumulation of GA-like substances in *Pharbitis* seeds without affecting seed development. A preliminary report of this work has been presented previously (22).

Material and Methods

Plant Growth. Seedlings of the short-day plant *Pharbitis nil,* strain Violet, were grown in 450-ml plastic beakers filled with a mixture of equal volumes vermiculite and coarse sand. They were watered daily with half-strength Hoagland solution containing 10 mg/liter Fe as sequestrene NAFe 13 % Fe (Geigy Industrial Chemicals).

On the fourth day after germination the seedlings were transferred to a growth room kept at a constant temperature of 25°. In order to induce flower formation the photoperiod was 10 hours and consisted of 8 hours fluorescent light (Sylvania cool white VHO fluorescent tubes, approximately 900 ft-c at plant level), followed by 2 hours incandescent light (150 ft-c).

[1] This work was supported by the National Research Council of Canada and by the United States Atomic Energy Commission, Contract No. AT (11-1) 1338.

Progenies of plants treated with chemicals were grown in 240-ml plastic beakers under a light regime of 16 hours fluorescent light, supplemented with 4 hours incandescent light at 23°.

Application of CCC. Treatments with CCC were started shortly before opening of the first flowers, usually around 30 days after germination. The first treatment consisted of an application of 100 mg CCC (20 ml of a 5000 mg/liter solution, i.e. 3.2×10^{-2} M) via the roots. Treatments were repeated every fifth day with doses of 50 mg of CCC until a total of either 250 or 350 mg had been applied per plant in different experiments. Control plants were treated with distilled water.

Seed Harvest. Flowers were labeled with paper tags on the morning of opening. The fruits were collected at designated times after anthesis, and the fresh weight per seed was determined. The seeds were next frozen in liquid nitrogen, lyophilized, and the dry weight determined. They were stored in stoppered vials at −20°.

Seed Extraction. Samples of 5 seeds were extracted several times with methanol at 2° and subsequently twice at room temperature until the residue was colorless. The combined extract was evaporated at room temperature to dryness under reduced pressure.

Thin-layer Chromatography. The residue of seed extracts was applied in a small volume of methanol as a 1-cm wide band to the origin of a 20 × 20 cm silica gel H thin-layer plate, which was approximately 400 μ thick. The plate was developed to 15 cm in a mixture of chloroform-ethyl acetate-acetic acid (60:40:5) as the solvent (19). The chromatogram was divided into 10 equal zones, the silica gel scraped off and eluted twice with wet ethyl acetate and once with 95 % (v/v) methanol. The eluate was evaporated and each sample taken up in 1 ml water containing 0.05 % Tween 20 for bioassay.

Extraction of Seedlings. Seedlings were cut off at ground level, frozen in liquid nitrogen and lyophilized. The dry material was homogenized with methanol in a Waring blendor at 2° and extracted 3 times. Lipid material was removed by mixing equal volumes of petroleum ether (B.R. 38–58°) and methanol extract at 2° and adding water to make 80 % methanol. Partitioning was repeated until the petroleum ether was colorless. The aqueous methanol was concentrated under reduced pressure until a water residue was left and 1 M potassium phosphate buffer (pH=8.2) added to give a 0.1 M concentration. The buffer was partitioned twice against ethyl acetate and then adjusted with 6 N HCl to pH 2.5. The acidic fraction was prepared by partitioning 3x against ethyl acetate, drying over sodium sulfate and evaporating under reduced pressure. The acidic fraction was further purified by TLC as described for seed extracts.

Bioassays. The dwarf corn d5-mutant served as the main bioassay plant. Segregating material supplied by Dr. Phinney was used in some assays. In others homozygous d5 (sibbed seed from GA_3-treated parents) was used as assay material.

Additional testing was done on dl-corn, and on dwarf peas, Progress No. 9, dwarfed by either red light, or by 200 mg/liter Amo-1618 in darkness. The procedures and growing conditions for all assays were similar to those described by Kende and Lang (9). The results are expressed as total GA_3-equivalents per seed.

CCC Determinations in Seeds. Methanol extracts of seeds were concentrated to a known volume (usually 2 or 3 ml) and spotted with a microsyringe on a cellulose powder MN 300 thin-layer plate (2) along with standard series of CCC and choline chloride. The plate was developed to 10 cm in a mixture of butanol-acetic acid-water (4:1:1.8). After spraying with modified Dragendorff reagent (20) the CCC in the seed extract was determined semi-quantitatively by visual comparison with the standard series.

For preparative work the methanol extract from 10 seeds was purified by descending chromatography on Whatman 3 MM paper with butanol-acetic acid-water (4:1:1.8) as the solvent. A zone at the R_F value for CCC (0.57) was eluted with 80 % (v/v) ethanol. The eluate was re-chromatographed on Whatman 3 MM using chloroform-methanol-water (75:22:3) as the solvent (2). The zone at the R_F value for CCC (0.39) was now eluted with water. The eluate was made up to a volume of 100 ml, applied to *Pharbitis* seedlings via the roots (21), and compared with the growth retardation caused by a series of CCC concentrations ranging from 0, 4, 8, ...to 40 mg/liter.

Results

Gibberellin Content of Seeds. The results for fresh and dry weight of *Pharbitis* seeds harvested from plants treated with CCC and from control plants are presented in figure 1. It is clear that CCC treat-

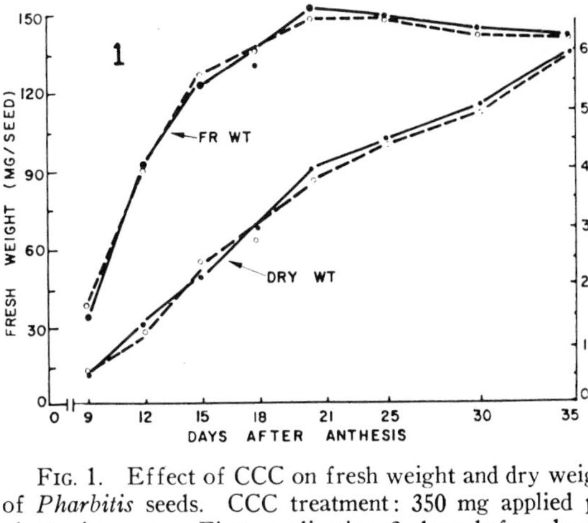

FIG. 1. Effect of CCC on fresh weight and dry weight of *Pharbitis* seeds. CCC treatment: 350 mg applied per plant via roots. First application 2 days before beginning of flowering. ○- - - -○, Control; ●———●, 350 mg CCC applied per plant.

ment had no effect on seed weight. Fresh weight reached a maximum around 21 days after anthesis, then started to decrease. Dry weight increased almost linearly with time for at least 35 days. Harvesting was discontinued when the color of the seed coat turned black at 35 days after anthesis.

Results of thin-layer chromatograms of seed extracts assayed on d5 corn are given in figure 2. Two major peaks of GA-like activity were detected at R_F 0.2 to 0.3 (fraction II) and R_F 0.4 to 0.6 (fraction III). Considerable activity was also found in the zone which included the origin (fraction I).

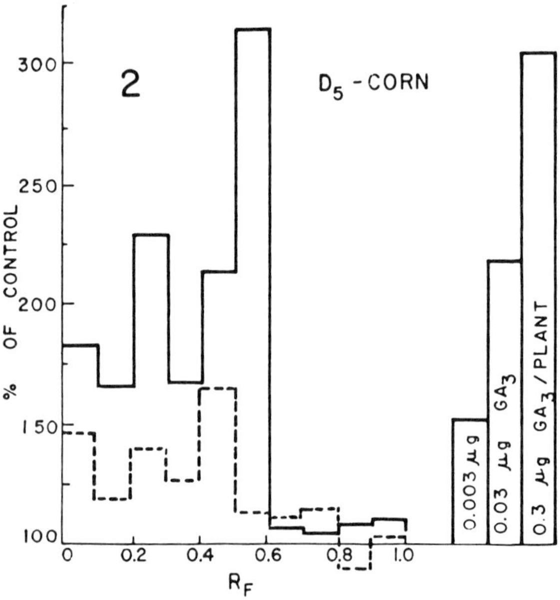

FIG. 2. Thin-layer chromatograms of *Pharbitis* seed extracts. Solvent: Chloroform-ethyl acetate-acetic acid (60:40:5). Seeds harvested 25 days after anthesis. Ten fractions were tested on d5 corn. Length of first + second leaf sheath in control: 41 mm. ———, Control seeds; - - - - - -, 350 mg CCC applied per plant.

All 3 fractions of GA-like substances were also present in seeds from CCC treated plants, but in much reduced quantities (fig 2). The total amount of GA-like material in the control was 0.94 µg GA_3 equivalents per seed and in + CCC only 0.02 µg per seed, whereas the dry weights per seed were 44.0 and 45.2 mg, respectively. It is clear from these results that the content of GA-like substances present in *Pharbitis* seeds was greatly reduced by CCC treatment, although the seed weight was not affected at all.

The extracts which yielded the results presented in figure 2 were further tested on dwarf peas grown under red light (fig 3). Extra stem elongation was obtained only after application of fraction II whereas fraction III was almost completely inactive. The total amount of GA-like material in fraction II (R_F 0.2–0.3) of the control as calculated in figures 2 and 3 was 0.44 µg and 0.20 µg GA_3 equivalents, respectively. Thus, this fraction was about equally active in both assays. On the other hand, fraction III was highly active in the d5 corn assay (fig 2), but practically inactive when tested on dwarf peas (fig 3). Expressed in GA_3 equivalents, fraction III of the control contained 3.8 and 0.015 µg, respectively, as determined in the 2 bioassays.

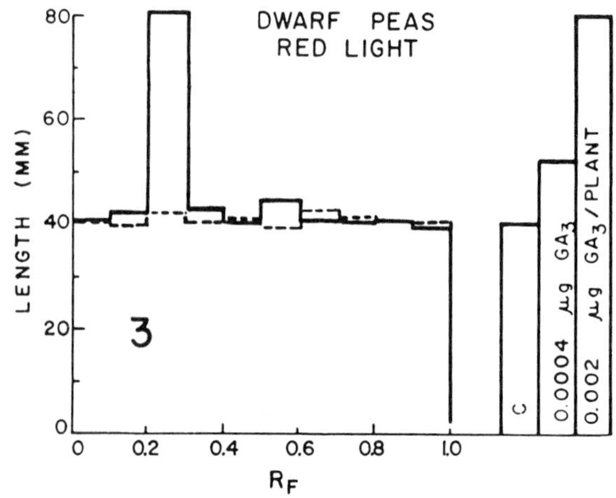

FIG. 3. Same thin-layer chromatograms as in figure 2. Fractions tested on dwarf peas grown in red light. ———, Control seeds; - - - - -, 350 mg CCC applied per plant.

Since highest GA activity was obtained on d5 corn, further quantitative determinations were carried out using this plant for bioassay. Average values of GA determinations made on duplicate samples of seeds harvested from 9 to 35 days after anthesis are plotted in figure 4. GA-like activity was quite high in 9-day-old seeds, then declined until the fifteenth day, and increased to a second maximum at 21 days which coincided with maximal fresh weight (fig 1). The peak at 21 days was due to the high activity in fraction III. GA content of CCC-treated seeds decreased steadily after 9 days and was always much lower than in control seeds.

Gibberellin Content of Seedlings. It was reported previously (21) that the content of GA-like substances in *Pharbitis* seedlings was not affected by CCC when harvested 40 hours after the beginning of treatment. As will be shown below the growth retardant CCC is carried over in considerable quantities by the seeds to the following generation. So, in these seedlings CCC is present from the onset of germination and thus GA production might be blocked more effectively than in case the chemical is applied via the roots at a later stage.

Control and CCC progenies were grown under continuous light and harvested on the fifth day after planting. The results in table I show that the weight of the CCC progeny was slightly less than that of the control. This was due to the shorter hypocotyls. The average hypocotyl length was 50 and 38 mm in the control and CCC progeny, respectively. Acidic fractions purified by means of thin-layer chroma-

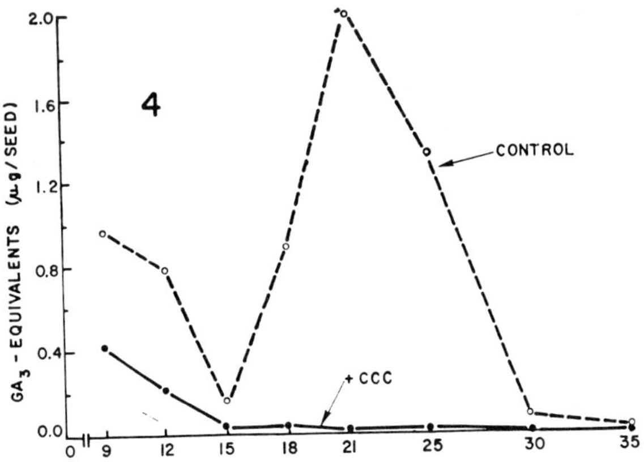

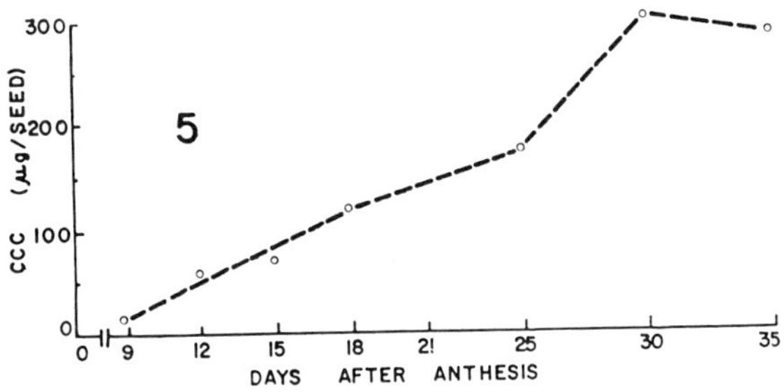

FIG. 4. Effect of CCC on GA-like activity in immature *Pharbitis* seeds as detected in d5 corn bioassay. Compare with figure 1. Points are averages of determinations on 2 samples, each containing 5 seeds.

FIG 5. Accumulation of CCC in *Pharbitis* seeds after treatment of parent plants with 350 mg CCC per plant. Compare with figure 1.

tography were assayed on d5 corn. Results in table I show that the total GA-like activity in the acidic fraction of the CCC progeny was reduced by 79 %. In all seedling extracts tested the activity present in fraction II was higher than in fraction III.

Growth and Flower Formation of Progenies. Comparison of stem growth of control and CCC progenies in long days revealed (fig 6) that the latter exhibited a dwarfed growth habit and reached a linear growth rate much later than the control. Such a growth curve is typical for CCC-treated seedlings [compare fig 7 in (21)].

Flower formation of CCC progenies was strongly inhibited. Typical results are shown in table II. No terminal flower buds were formed in the CCC progeny after exposure to one long night. A dose of 0.01 μg GA_3 per plant was sufficient to restore the percentage of plants with terminal buds to that of the control. However, at least 10X more GA_3 was required to overcome suppression of internode length. Again these responses are similar to those exhibited by *Pharbitis* seedlings to which CCC has been applied via the roots (21). Therefore, these results supply indirect evidence for the presence of CCC in the progeny of treated plants.

CCC Content of Seeds. Some seedlings of 1 CCC progeny exhibited paling between the veins of cotyledons, a symptom typical for high CCC doses (21). In view of this and other observations described above seeds were tested for the presence of CCC by thin-layer chromatography of methanol extracts. Spraying thin-layer chromatograms with Dragendorff's reagent revealed 1 purple spot on a yellow background in extracts of control seeds. The R_F of 0.44 corresponded to that of choline chloride.

Table I. *Effect of CCC Treatment on the Gibberellin-like Activity in the Progeny*

Plants from which the CCC-progeny was harvested received 350 mg CCC each. Seedlings grown under continuous light from fluorescent tubes at an intensity of 900 ft-c at 23°. Seedlings harvested 5 days after planting, 72 seedlings per treatment. Thin-layer chromatograms of acidic fractions tested on d5 corn.

	Fr wt	Dry wt	GA_3-equivalents/72 plants	
Treatment	g	g	μg	%
Control	59.0	5.46	0.435	100
CCC progeny	55.5	5.21	0.091	21

Table II. *Effect of CCC Treatment on Flower Formation and Internode Length of the Progeny*

Plants from which CCC progeny was harvested received 250 mg CCC each. Short day (SD) was a dark period at 27°. GA_3 was applied to plumule before long night. Nine plants were used per treatment.

Treatment	% Plants with terminal flower buds	No. of flower buds per plant	Length of 1st + 2nd internode, mm
Control	78	6.9	39.3
Control + 250 mg/liter CCC*)	0	4.6	16.3
CCC progeny, 1 SD	0	3.2	16.9
CCC progeny, 2 SD	22	5.1	18.2
CCC progeny, 1 SD, 0.01 μg GA_3	78	5.6	23.7
CCC progeny, 1 SD, 0.1 μg GA3	100	6.2	35.8

* Seedlings on a solution of 250 mg/liter CCC for 24 hours prior to long night.

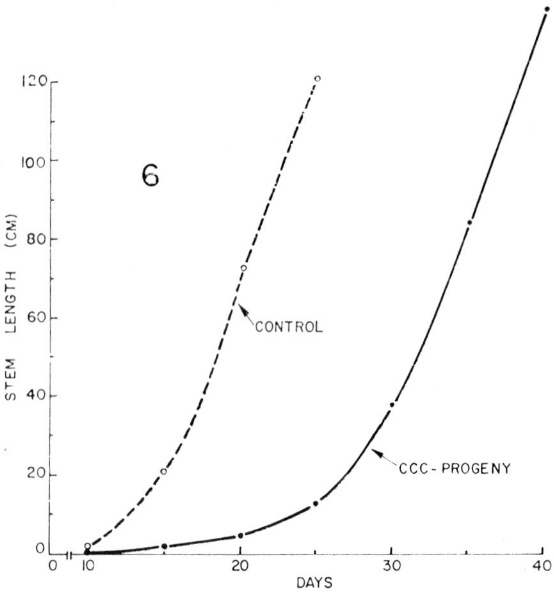

FIG. 6. Stem growth of *Pharbitis* progenies in long days as affected by CCC treatment of parent plants. Each parent plant received 350 mg CCC.

A second spot with a carmine-red color and an R_F of 0.55 was observed only in extracts of seeds harvested from treated plants. Co-chromatographed CCC had the same R_F value and exhibited the same color after spraying with Dragendorff's reagent. It was therefore assumed that this compound represented CCC which had accumulated in seeds of treated plants.

Semi-quantitative determinations of the total amount of CCC present per seed at different ages are plotted in figure 5. In 9-day-old seeds 18 μg CCC was present per seed. This increased gradually with time to approximately 300 μg CCC per seed at 30 days after anthesis.

The biological activity as a growth retardant of the CCC-like material extracted from seeds was further tested as described under methods. The eluates supposedly containing CCC caused inhibition of *Pharbitis* stem growth as compared with the corresponding eluates of control chromatograms. The CCC concentrations in the eluates were determined by interpolation on the CCC standard series. By this method 15- and 30-day-old seeds had a content of 105 and 240 μg CCC per seed, respectively. These figures are in good agreement with those obtained by means of thin-layer chromatography (fig 5). It can be concluded from these data that CCC applications to *Pharbitis* plants during seed development result in accumulation of the growth retardant in the seeds and transmission to the following generation.

Discussion

The results described here clearly demonstrate that CCC applied to *Pharbitis* plants shortly before and after anthesis did not affect seed growth, but greatly reduced the amount of GA-like substances occurring in the seeds. Michniewicz (13), too, has observed a reduction in the level of endogenous GA by CCC in embryos of winter wheat. As argued by Baldev et al. (1) the possibility of GA-destruction due to the presence of CCC remains, but this is unlikely in view of results with *Fusarium* (7, 10, 15). The present data can therefore be best interpreted to indicate that CCC inhibits the production of GA in higher plants. This agrees with the finding that the effects of CCC on growth and flower formation in *Pharbitis* can be completely overcome by applied GA_3 as shown earlier (21) and in this paper (table II). However, as discussed by Harada and Lang (7) other effects of growth retardants on plant growth cannot be ruled out.

Dennis et al. (5) have pinpointed the cyclization of *trans*-geranylgeranyl pyrophosphate to (-)-kaurene as the enzymic site of inhibition of GA biosynthesis by Amo-1618. However, CCC failed to inhibit (-)-kaurene formation. This suggests that CCC inhibition occurs beyond the cyclization step. Thus, in the presence of CCC one might expect the accumulation of this or related intermediates of the GA biosynthetic pathway. Since (-)-kaurene and some of its derivatives are active in d5-corn (18), thin-layer chromatography of methanol extracts of seeds +CCC might yield a new zone of activity, but this was not observed (fig 2). It is conceivable, however, that these GA precursors would escape detection as they are relatively inactive in promoting growth of the d5-mutant as compared to GA_3 (18).

Since the GA content of seeds was strongly reduced by CCC whereas growth was unaffected, it seems logical to assume that only a minor fraction of GA plays a physiological role in the growth of *Pharbitis* seeds. Admittedly, 9-day-old seeds from plants treated with CCC had still a relatively high GA content (fig 4). In order to reduce this further, in 1 experiment, treatment with CCC was started 7 days before first open flowers. As a result all fruits aborted shortly after flowering. This indicates that GA does play a role in seed and/or fruit development, most likely during the early stages.

Although GA-like substances accumulate in *Pharbitis* seeds to a high level (2 μg GA_3 equivalents per seed), they could be synthesized at other sites such as roots (3, 17), shoot apices (8, 11), or young leaves (11) and be transported to the seeds. However, Baldev et al. (1) found that pea seeds in excised pods also accumulated GA-like substances. And Graebe et al. (6) showed that *Echinocystis* endosperm contains all the enzymes for the biosynthesis of (-)-kauren-19-ol, a GA precursor. This evidence strongly suggests that seeds and fruits, usually rich sources of GA, are sites of GA biosynthesis.

CCC applied via the roots accumulated in the seeds and resulted in dwarfed growth of the progeny. With Amo-1618 after-effects have been observed in the following 2 generations of bean plants (12). In

Pharbitis no dwarfed growth was observed in the second progeny.

When seed extracts were chromatographed on silica gel H plates, CCC remained at the origin. Thus, CCC could have interfered in the bioassay only with the activity in fraction I (R_F 0.0–0.1). Since the highest possible concentration of CCC in fraction I (1500 μg/ml in 30-day-old seeds) reduced the response to GA_3 in the d5 corn test by only 16 %, this cannot be the cause for the decreased content of GA-like material in CCC treated seeds.

The reduced GA content in *Pharbitis* seedlings with CCC might be due to inhibition of de novo GA biosynthesis after germination. But it is also possible that at least part of the GA-like substances present in seedlings had been produced in the immature seeds. This latter alternative is supported by Murakami's observation (14) that the GA-content of etiolated *Pharbitis* seedlings decreased after germination.

Reduced GA content of seedlings was accompanied by dwarfed growth (fig 6). It is interesting to mention in this connection that the acidic material obtained from seedlings of the dwarf strain Kidachi contained the same amount of GA-like material as normal Violet seedlings (unpublished results). This suggests that inhibitors may play a role in the dwarfed growth of Kidachi seedlings.

Murakami (14) and Ogawa (16) have analyzed the relationship between growth of *Pharbitis* seeds and the content of GA-like substances. Our data agree with those of Ogawa and show that the GA content is maximal when fresh weight reaches its maximum.

Ogawa (16) separated GA-like material present in *Pharbitis* seeds by means of paper chromatography and distinguished 3 active factors according to R_F values in the solvent system isopropanol-7 N ammonium hydroxide-water (8:1:1). His factor I (highest R_F value) appears to be identical with our fraction II. This substance is highly active in promoting growth of d5 and dl corn, and of peas dwarfed by either red light, or by Amo-1618 in darkness. Preliminary results obtained with thin-layer chromatography (unpublished) indicate that its R_F value is slightly lower than those of GA_1 and GA_3.

The properties of Ogawa's (16) factor II and our fraction III indicate that they are probably identical. This material was present in highest amounts around 20 days after anthesis. It is highly active in promoting growth of d5 corn and moderately active in dl corn. However, fraction III is inactive in promoting growth of dwarf peas grown in red light (fig 3), but, like GA_5 (9), it is quite active when applied to etiolated and chemically dwarfed peas. Fraction III also co-chromatographed with GA_5 on silica gel thin layers in 2 different solvent systems. Thus, biological effects (9) and chromatographic properties suggest that fraction III of *Pharbitis* may be identical with GA_5.

Fraction I (Ogawa's factor III) was always present in relatively small amounts as compared to fractions II and III. It is the most polar one of the 3 GA-like substances present in *Pharbitis* seeds. This material was only partially recovered from silica gel H after 2 washings with wet ethyl acetate. Additional activity was released after subsequent elution with 95 % methanol.

Biological activity of fractions II and III was also tested in *Pharbitis* seedlings treated with CCC. In such seedlings formation of terminal flower buds is the most sensitive response to GA (21). Using this as a criterion for GA activity, both fractions II and III were approximately equally active on the basis of GA_3 equivalents applied per plant.

Acknowledgment

I thank Dr. B. O. Phinney, University of California, Los Angeles, for supplying d5 corn.

Literature Cited

1. Baldev, B., A. Lang, and A. O. Agatep. 1965. Gibberellin production in pea seeds developing in excised pods: Effect of growth retardant AMO-1618. Science 147: 155–57.
2. Bayzer, H. 1964. Dünnschichtchromatographische Trennung quarternärer Ammoniumverbindungen auf Celluloseschichten. Experientia 20: 233.
3. Carr, D. J., D. M. Reid, and K. G. M. Skene. 1964. The supply of gibberellins from the root to the shoot. Planta 63: 382–92.
4. Cathey, H. M. 1964. Physiology of growth retarding chemicals. Ann. Rev. Plant Physiol. 15: 271–302.
5. Dennis, D. T., C. D. Upper, and C. A. West. 1965. An enzymic site of inhibition of gibberellin biosyntheisis by Amo-1618 and other plant growth retardants. Plant Physiol. 40: 948–52.
6. Graebe, J., D. T. Dennis, C. D. Upper, and C. A. West. 1965. Biosynthesis of gibberellins. I. The biosynthesis of (-)-kaurene, (-)-kauren-19-1 and *trans*-geranylgeraniol in endosperm-nucellus of *Echinocystis macrocarpa* Greene. J. Biol. Chem. 240: 1847–54.
7. Harada, H. and A. Lang. 1965. Effect of some (2-chloroethyl) trimethylammonium chloride analogs and other growth retardants on gibberellin biosynthesis in *Fusarium moniliforme*. Plant Physiol. 40: 176–83.
8. Jones, R. L. and I. D. J. Phillips. 1964. Agar-diffusion technique for estimating gibberellin production by plant organs. Nature 204: 497–99.
9. Kende, H. and A. Lang 1964. Gibberellins and light inhibition of stem growth in peas. Plant Physiol. 39: 435–40.
10. Kende, H., H. Ninnemann, and A. Lang. 1963. Inhibition of gibberellic acid biosynthesis in *Fusarium moniliforme* by Amo-1618 and CCC. Naturwissenschaften 50: 599–600.
11. Lockhart, J. A. 1964. Physiological studies on light sensitive stem growth. Planta 62: 97–115.

12. MARTH, P. C., W. H. PRESTON, JR., AND J. W. MITCHELL. 1953. Growth-controlling effects of some quaternary ammonium compounds on various species of plants. Botan. Gaz. 115: 200–04.
13. MICHNIEWICZ, M. 1965. Inhibitory effect of 2-Chloroethyl trimethylammonium chloride (CCC) on vernalization in winter wheat. Naturwissenschaften 52: 88.
14. MURAKAMI, Y. 1961. Paper-chromatographic studies on change in gibberellins during seed development and germination in *Pharbitis nil*. Botan. Mag. 74: 241–47.
15. NINNEMANN, H., J. A. D. ZEEVAART, H. KENDE, AND A. LANG. 1964. The plant growth retardant CCC as inhibitor of gibberellin biosynthesis in *Fusarium moniliforme*. Planta 61: 229–35.
16. OGAWA, Y. 1963. Gibberellin-like substances occurring in the seed of *Pharbitis nil* Chois. and their change in contents during the seed development. Plant Cell Physiol. 4: 217–25.
17. PHILLIPS, I. D. J. AND R. L. JONES. 1964. Gibberellin-like activity in bleeding sap of root systems of *Helianthus annuus* detected by a new dwarf epicotyl assay and other methods. Planta 63: 269–78.
18. PHINNEY, B. O., P. R. JEFFERIES, M. KATSUMI, AND C. A. HENRICK. 1964. The biological activity of Kaurene and related compounds. Plant Physiol. 39: xxvii.
19. SEMBDNER, G., R. GROSS, AND K. SCHREIBER. 1962. Die Dünnschichtchromatographie von Gibberellinen. Experientia 18: 584.
20. THIES, H. AND F. W. REUTHER. 1954. Ein Reagens zum Nachweis von Alkaloiden auf Papierchromatogrammen. Naturwissenschaften 41: 230–31.
21. ZEEVAART, J. A. D. 1964. Effects of the growth retardant CCC on floral initiation and growth in *Pharbitis nil*. Plant Physiol. 39: 402–08.
22. ZEEVAART, J. A. D. 1965. The growth retardant CCC as an inhibitor of gibberellin production in immature *Pharbitis* seeds. Plant Physiol. 40: lxxv.

Reprinted from PLANT PHYSIOLOGY, Vol. 43, No. 4, April, 1968, pages 629–634.
(PRINTED IN U.S.A.)

Extractable and Diffusible Gibberellins from Light- and Dark-grown Pea Seedlings[1]

R. L. Jones[2] and A. Lang
MSU/AEC Plant Research Laboratory, Michigan State University, East Lansing, Michigan 48823

Received November 9, 1967.

Abstract. **Gibberellins were obtained from light- and dark-grown peas by solvent extraction and agar diffusion. Both A_5- and A_1-like gibberellins were obtained by extraction; however, by diffusion only the A_1-like gibberellin was found. There was no significant quantitative difference in the levels of diffusible or extractable gibberellin obtained from light- and dark-grown tall and dwarf peas. Several possible explanations for the discrepancy between diffusible and extractable gibberellin were investigated. Of these, only 1 was supported by experimental evidence, namely, that GA_5 can be converted to GA_1.**

Determinations of the gibberellin content of plants have been generally made by extraction of the active substances with solvents, followed by partial purification and separation by means of solvent partition and chromatography. More recently, a method has been developed to obtain gibberellins from plant material by diffusion into agar (9). Diffusion can be followed by partitioning and/or chromatography, using the same methods as after extraction.

In the case of auxin, where the diffusion method has been used for a long time, it is well known that extraction and diffusion may yield quite divergent results and measure different aspects of the auxin metabolism of the plant. In general, diffusion permits the measurement of auxin production while extraction measures the total hormone which is present in the tissue at a given moment, but tells little about its origin and function. In the case of gibberellin, the diffusion technique, coupled with the use of inhibitors of gibberellin biosynthesis, has permitted an unequivocal demonstration that gibberellins are synthesized in shoot buds (specifically, the young, not yet unfolded leaves) and root tips of sunflower plants (10, 11).

It appeared to us interesting to extend the examination of endogenous gibberellins by the diffusion technique to more plants, and to compare results obtained by diffusion and extraction, as such comparison may add to our understanding of the function and metabolism of endogenous gibberellins. The plant material chosen was pea seedlings, this choice being based on 2 considerations. Firstly, the gibberellin content of pea seedlings has already been measured by several authors, using extraction techniques (13, 15, 16, 17, 22, 25), although the results have been divergent. Secondly, there are strong suggestions that gibberellins play an important part in endogenous regulation of growth in the pea shoot, and that they perform this function in some kind of interaction with light. Briefly and summarily, the essential facts are as follows (see 6, 13, 18, 19): A) Peas occur in normal (tall) and dwarf varieties, this difference being mainly determined by the alleles of 1 gene. B) Both normals and dwarfs grow at approximately the same rates in the dark. C) In either type, growth is reduced by light, but the effect is considerably greater in the dwarfs than in the normals. D) Applied gibberellin has little if any effect on growth in the dark but causes marked growth promotion in the light. E) The response of dwarfs to applied gibberellin (when grown in the light) is much greater than that of the normals, resulting in obliteration or reduction of the growth differences between the 2 types. Thus, it appears that the inhibitory effect of light on stem growth in pea plants is based on some reduction of the effective gibberellin level in the plants, and that the difference between normals and dwarfs arises because this reduction is greater in the latter than in the former. The effect of light on growth in peas is mediated by phytochrome, red light causing the inhibition of growth, far-red light negating the effect of red (18, 19).

Materials and Methods

Plant Material. Normal (Alaska) and dwarf (Progress No. 9) peas (*Pisum sativum* L.) obtained from Asgrow, New Haven, Connecticut, were allowed to imbibe at room temperature for 18 hours, sown in vermiculite and grown either in darkness

[1] Research carried out under United States Atomic Energy Commission Contract No. AT (11-1)-1338.
[2] Present address: Department of Botany, University of California, Berkeley, California 94720.

or in continuous white light (900 ft-c) at 23° ± 1°. They were watered with half-strength Hoagland solution. The plants were harvested for both diffusion and extraction when 8 to 10 days old. Seedlings treated with growth retardants were grown as described above except that they were transferred to plastic boxes containing half-strength Hoagland nutrient solution on the fourth day after sowing.

Application of Gibberellins and Growth Retardants. Both GA_5 and GA_1 were applied to the apical buds of seedlings in an aqueous solution containing 0.05 % Tween 20. The growth retardant AMO1618 [2′-isopropyl-4′-(trimethylammonium chloride) 5′-methylphenyl piperidine-1-carboxylate] was applied as a 200 mg/liter solution in half strength Hoagland solution via the roots.

Gibberellin Diffusion. Agar diffusates were obtained from pea apices using the techniques described previously (9, 10). Agar blocks were prepared from a 1.5 % aqueous solution of agar which was poured into 10 mm × 2 m glass tubes. After solidifying, the agar plug was removed from the tubes and cut into blocks 10 mm diameter × 5 mm height.

Following the diffusion period, the agar blocks were extracted by freezing. The frozen agar blocks were flooded with methanol and the methanolic extract obtained from the agar was evaporated on a flash evaporator until the aqueous phase remained. This was partitioned against ethyl acetate at pH 2.5 and dried over sodium sulfate prior to chromatography.

Gibberellin Extraction. Seedlings and seedling tips were frozen with liquid nitrogen, lyophilized, and homogenized with methanol in a Waring Blendor. The methanol was allowed to extract for 24 hours at 2°. Lipid material was removed by mixing petroleum ether (boiling range 30–60°) with the methanol extract which had been adjusted to 80 % by the addition of water. The aqueous methanol was evaporated on a flash evaporator, the remaining aqueous phase adjusted to pH 9.5 with 1 N NaOH and partitioned twice against ethyl acetate. This ethyl acetate was discarded and the remaining aqueous phase was acidified to pH 2.5 with 1 N NCL and partitioned against ethyl acetate 3 times. The acidic ethyl acetate fraction was dried over sodium sulfate prior to further purification by thin layer chromatography (TLC).

Chromatography. Both agar block and tissue extracts were purified by TLC prior to bioassay. Acidic fractions were reduced to dryness, redissolved in a small volume of ethyl acetate and applied as a band to the origin of a 20 × 20 cm silica gel (H) thin-layer plate which was approximately 250 μ thick. The plates were developed in the following solvent systems: Chloroform/ethyl acetate/acetic acid (60:40:5 v/v); benzene/acetic acid/water (8:3:5 v/v), lower phase plus 20 % ethyl acetate; isopropyl ether/acetic acid (95:5 v/v).

Following development, the plates were divided into 10 equal zones between origin and solvent front. Each zone was scraped off and eluted 3 times with wet ethyl acetate. The eluates were reduced to dryness and redissolved in water containing 0.05 % Tween 20 (polyoxyethylenesorbitan monolaurate).

Bioassays. Eluates were assayed with the dwarf corn (d5 mutant) seedling test, the lettuce hypocotyl test (Frankland and Wareing, 4) and the barley half-seed test; in the latter case, the activity of alpha amylase was determined directly as described by Jones and Varner (12).

Results

When apical buds from light and dark-grown, normal and dwarf peas were excised and allowed to diffuse on agar blocks for 20 hours, bioassay with d5 corn of extracts following chromatography in a solvent mixture of benzene/acetic acid/water (8:3:5) with 20 % ethyl acetate added to the lower phase indicated 1 zone of growth promotion (fig 1), corresponding to the position of GA_1 and/or GA_5.

Extraction of identical apical buds, however, indicated both the presence of GA_1 and GA_5 (table I), confirming the observations made by Kende and Lang (13). There was no quantitative difference in the levels of diffusible or extractable gibberellin in the apices of either normal or dwarf peas grown in light or dark (fig 1, table I).

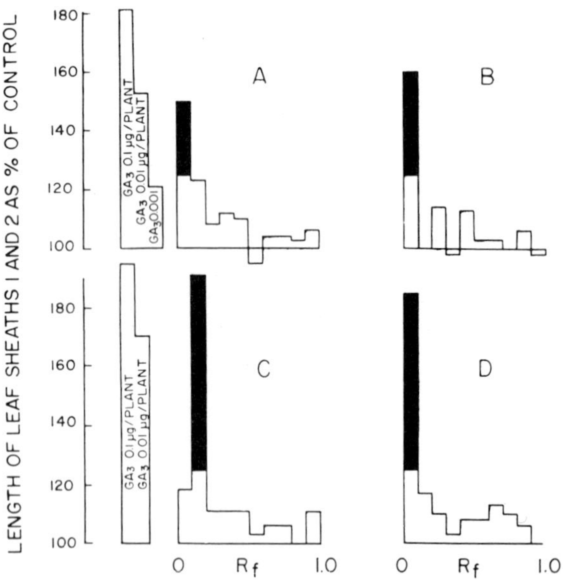

FIG. 1. Diffusates from 100 apices each of light- and dark-grown normal and dwarf pea seedlings. Extracts bioassayed with d_5 corn. TLC solvent; di-isopropyl ether: acetic acid (95.5 v/v). A) Normal, dark-grown; fresh weights = 6.1 g. B) Dwarf dark-grown; fresh weights = 6.3 g. C) Normal, light-grown; fresh weights = 10.6 g. D) Dwarf, light-grown; fresh weights = 11.5 g. Base of darkened area indicates significant biological activity at the 5 % level of risk.

Table I. *Amount of GA Obtained from* Pisum sativum *Apical Buds by Diffusion and Extraction*

		Light		Dark	
		Normal	Dwarf	Normal	Dwarf
	GA_1	1.70	2.30	1.90	2.30
Diffusion	GA_5	0	0	0	0
Total		1.70	2.30	1.90	2.30
	GA_1	0.90	0.60	1.00	0.80
Extraction	GA_5	0.40	0.45	0.52	0.36
Total		1.30	1.05	1.52	1.16

Because of the apparent absence of the GA_5-like substance from diffusates of apical buds of peas, intensive examinations were made of agar diffusates using various chromatographic and bioassay techniques. The results are summarized in table II. It can be seen that no significant growth response was obtained by those zones of the chromatograms corresponding to the known position of GA_5.

Several reasons can be visualized for the failure to find a GA_5-like substance in diffusates, as opposed to extracts. The substance may be unable to move through the pea internode. This possibility was tested using conventional donor/receiver agar block experiments. Gibberellins A_1 and A_5 were incorporated into agar blocks and placed at the apical end of 15-mm-long sections from the subapical internode obtained from light-grown normal and dwarf peas. The sections were then allowed to stand on receiver blocks of plain agar and following a 20-hour diffusion period the receiver blocks were removed and extracted. Bioassays of extracts indicated that both GA_1 and GA_5 diffused readily through such isolated internode sections (table III). In fact, GA_5 would appear to move more rapidly than GA_1 through pea internode sections.

An alternative explanation for the discrepancy between diffusible and extractable gibberellins may lie in the incomplete removal of the gibberellin from the extracts of the aqueous agar blocks.

Table II. *Bioassay of Diffusates from Agar Blocks Following Chromatography in 3 Different Solvent Systems*

Solvent systems: 1) Benzene/acetic acid/water (8:3:5), lower phase 20 % ethyl acetate (v/v). 2) Di-isopropyl ether/acetic acid (95.5 v/v). 3) Chloroform/ethyl acetate/acetic acid (60:40:5 v/v).

Solvent system	Assay	$R_F GA_3$	$R_F GA_5$	Zones of growth promotion
	d_5 Corn	0.00	0.35	0.0–0.1
1	Barley	0.00	0.34	0.0–0.1
	Lettuce	0.00	0.35	0.0–0.1
	d_5 Corn	0.07	0.3	0.0–0.2
2	Barley	0.05	0.31	0.0–0.2
	Lettuce	0.07	0.32	0.0–0.3
	d_5 Corn	0.18	0.54	0.2–0.3
3	Barley	0.16	0.55	0.2–0.3
	Lettuce	0.16	0.54	0.2–0.3

Table III. *Transport of GA_5 and GA_1 Through 15 mm-Long Light-Grown Pea Internode Sections*

Section source	Donor block[1]	GA_3 equivalent[1] in receiver blocks
Normal	GA_1 0.50 μg	0.045 μg
	GA_5 0.50 μg	0.110 μg
Dwarf	GA_1 0.50 μg	0.015 μg
	GA_5 0.50 μg	0.065 μg

[1] 50 Donor blocks treatment.

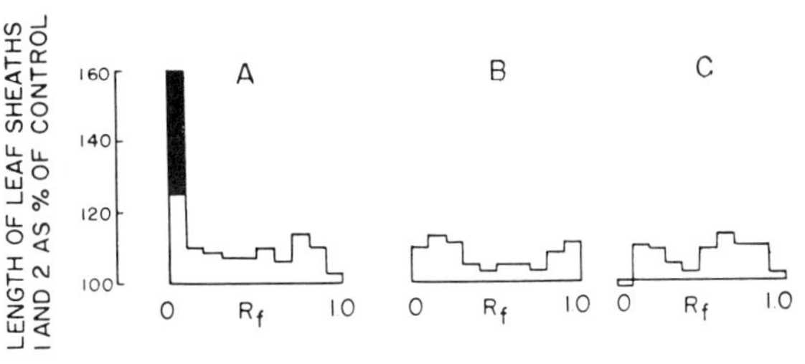

FIG. 2. Hydrolysis of aqueous extracts of the agar blocks. Bioassay, d_5 corn. TLC solvent, di-isopropyl ether; acetic acid (95.5 v/v). A) Acidic ethyl acetate phase from 130 light grown normal pea apices. B) Aqueous phase from 125 light-grown normal pea apices following hydrolysis with ficin. C) Aqueous phase from 130 light-grown normal pea apices following hydrolysis wih 0.5 N HCl. Base of darkened area indicates significant biological activity at the 5 % level of risk.

Several workers have shown that gibberellins can exist as bound complexes (7, 21), the free gibberellins being obtained only after hydrolysis of the extract with acid or with certain proteases (21). It has also been shown that these 'bound' gibberellins remain in the aqueous phase during partition of extracts with conventional solvents. However, bioassay of the aqueous phase of agar block extracts following hydrolysis with dilute acid and the protease ficin (21) indicated no biological activity (fig 2). This result was not surprising as it has already been shown that GA_5 can be readily obtained from light and dark-grown peas by conventional extraction and partition techniques (13, table I).

It may be argued that GA_5 is inactivated during diffusion, either at the cut surface of the excised bud or in the agar block. This phenomenon has been encountered during the diffusion of auxin from excised organs (5), but there is no evidence for GA destruction during diffusion. Indeed, the results obtained in the transport experiments indicate that preferential destruction of GA_5 does not occur during diffusion. It must be pointed out, however, that much more GA_5 was used in the donor/receiver block experiments than would normally be found in internode sections; consequently, any inactivating mechanism present could be overwhelmed.

In an examination of the discrepancy between "free" (diffusable) and extractable auxin, van Overbeek (23) suggested that the extractable form of auxin was a precursor of diffusible auxin. Following this argument, the discrepancy between diffusible and extractable gibberellin in peas could be explained by postulating that GA_5 is a non-diffusable but extractable precursor of GA_1. Phinney and West (24) have suggested that GA_5 is a normal intermediate in the biosynthesis of GA_1 in corn. This hypothesis has received support by circumstantial evidence obtained by Jones (8) and Brian *et al.* (2). Similarly, Macmillan *et al.* (20) suggested that GA_5 was a possible intermediate in the biosynthesis of gibberellins A_1, A_3, A_8, and A_6.

In order to test the hypothesis that GA_5 is an intermediate in the biosynthesis of GA_1 in peas, the following experiments were performed. Four-day-old, light-grown normal pea seedlings were transferred to plastic boxes containing a mixture of half-strength Hoagland and AMO1618 (200 mg/l). Six days following treatment with AMO1618, 0.05 μg GA_5 was applied to the apical bud of each seedling. Plants were harvested 24 and 48 hours following the application of GA_5, frozen with liquid nitrogen, lyophilized and extracted as described in the Methods section. Bioassay of the extracts with d5 dwarf corn seedlings following chromatography in chloroform/ethyl acetate/acetic acid (60:40:5, v/v) indicated 2 distinct regions of growth promotion, 1 corresponding to the position of applied GA_5, the other to that of GA_1 (fig 3). Seedlings treated with AMO1618 alone possessed no significant levels of extractable gibberellin (fig 3). Also, chromatography of the GA_5 sample applied to the seedlings followed by bioassay indicated only 1 zone of growth promotion, at the R_F of GA_5.

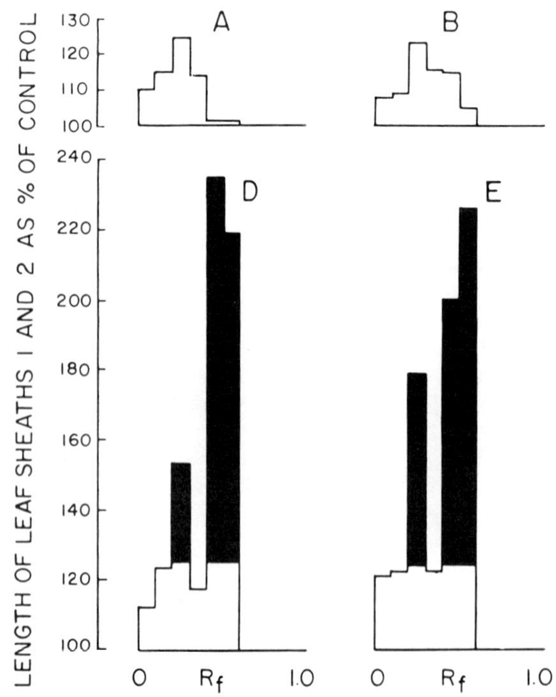

FIG. 3. Extracts from normal, light-grown pea seedlings treated with AMO1618 (A,B) and with both AMO1618 and GA_5 (D,E). Samples taken 24 hours (A,D) and 48 hours (B,E) following GA_5 application. Bioassay, d_5 corn. TLC solvent, chloroform: ethyl acetate: acetic acid (60:40:5 v/v). Base of darkened area indicates significant biological activity at the 5 % level of risk.

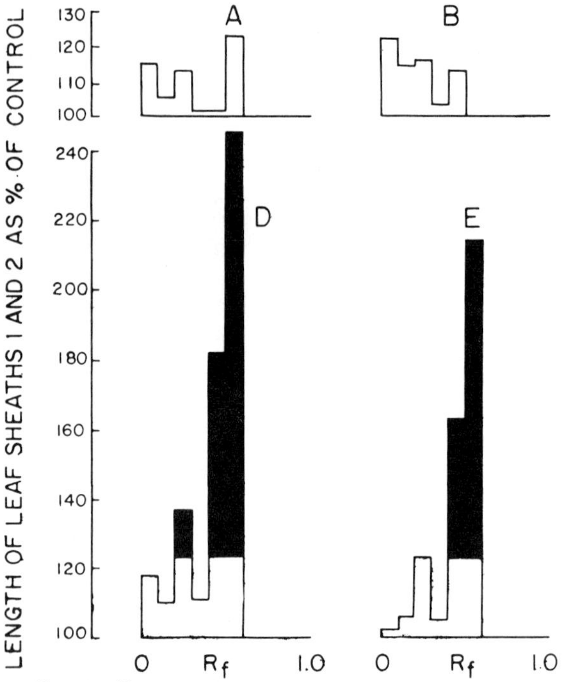

FIG. 4. Extracts from normal, dark-grown pea seedlings treated with AMO1618 (A,B) and both AMO1618 and GA_5 (D,E). Samples taken 24 hours (A,D) and 48 hours (B,E) following GA_5 application. Bioassay, d_5 corn. TLC solvent, chloroform: ethyl acetate: acetic acid (60:40:5 v/v). Base of darkened area indicates significant biological activity at the 5 % level of risk.

These experiments were repeated using dark-grown seedlings and the results are shown in figure 4. Although there is no qualitative difference in the results obtained from the experiments with light or dark-grown seedlings, there appears to be a quantitative difference. This feature was consistent when the experiments were repeated, and can be best expressed in terms of the amount of GA_5 converted to the GA_1-like component. In 3 experiments using light-grown seedlings the range of conversion of GA_5 to GA_1 was between 10 to 19 %, whilst in a similar number of experiments using dark-grown seedlings the range was 4 to 8 %.

Discussion

Our experiments have yielded 2 principal results: 1) Pea seedlings contain—as has been shown before—2 kinds of extractable, acidic gibberellins, 1 similar to GA_1 and the other to GA_5, but only 1 of these, namely, the GA_1-like factor, can be obtained by diffusion; 2) there is no quantitative difference in the levels of either extractable or diffusible gibberellins obtained from seedlings of a normal (tall) and a dwarf cultivar, and in neither case was there a difference whether the plants had been grown in the dark or in the light.

The former finding is undoubtedly surprising but it underlines the experience, long known in work on auxin, that extraction procedures alone may give a 1-sided and, hence, insufficient picture of the hormone status of a plant.

Several explanations can be suggested for the absence of GA_5 in the diffusates from pea shoot tips. GA_5 may be unable to diffuse through pea stem tissue; it may be present in a bound, non-mobile form; it may be inactivated at the cut surface of the diffusing organ or in the receiver agar block; it may be a non-mobile precursor of GA_1.

All of these possibilities were tested but positive evidence was found only for the latter; after application of GA_5 to pea seedlings a certain amount of activity was found at an R_F characteristic for GA_1. Thus, it is possible that GA_5 is a natural precursor of GA_1 in the pea plant. Such a conclusion would be in agreement with existing, although circumstantial evidence in literature that GA_5 may be a precursor of GA_1 (and other gibberellins) (2, 8, 20, 24). It is also supported by the fact that the observed appearance of GA_1 after application of GA_5 took place in pea seedlings treated with AMO1618. This growth retardant has been shown to inhibit gibberellin biosynthesis in the fungus *Fusarium moniliforme* (14), in developing pea seeds (1), in the endosperm of *Echinocystis macrocarpa* (3), and in seeds of *Pharbitis nil* (26). Dennis *et al.* (3) have shown that the inhibition occurs at the point of cyclization of transgeranylgeraniol pyrophosphate to (−)-kaurene. It therefore seems unlikely that GA_1 extracted from pea seedlings treated with both GA_5 and AMO is a product of the biosynthetic machinery of the plant for gibberellins, operating on residual gibberellin precursors present in the plant. Seedlings treated with AMO alone did not yield GA_1 or any other gibberellin-like material whatsoever.

However, it must be emphasized that the quantitative aspects of this interpretation require further investigation. The amounts of GA_5 applied to the plants in the conversion experiments were much higher (about 100×) than the amounts that can be extracted from the same tissue. On the other hand, the conversion factor was not very high, maximally 19 % after 48 hours. It is possible that the enzymatic system which would be responsible for the conversion was overloaded and that the conversion factor is much higher when only endogenous GA_5 is available as substrate. But it also cannot be ruled out that even if endogenous conversion of GA_5 to GA_1 is occurring in pea plants it is not the sole or main pathway of synthesis of the latter gibberellin. It may be possible to obtain a more precise and quantitative answer to this question by the use of labelled GA_5; it is hoped that such experiments can be carried out in near future.

Our results on the quantities of extractable and diffusible gibberellins in pea seedlings are in agreement with those of the extraction experiments of Kende and Lang (13). Lockhart (18, 19) has suggested that the effect of light on growth in peas, and its reversal by applied gibberellin, may be explained in 3 ways: reduction of gibberellin synthesis in the plant; enhancement of gibberellin destruction in the plant; reduction of the responsiveness of the tissue to gibberellin. The results of Kende and Lang (13) argue against the former 2 possibilities and in favor of the third, and our results support this general conclusion.

Kende and Lang (13) had specifically shown that it is the response of the pea seedlings to GA_5 or the endogenous GA_5-like factor that is affected by light, the response to GA_1 and the GA_1-like endogenous gibberellin being the same in light and dark. If GA_5 were the sole or principal precursor of GA_1 in the plant and if the conversion were reduced by light this could offer an explanation for the findings of Kende and Lang. However, as already emphasized, it is premature to attribute to GA_5 a role as the predominant GA_1 precursor in pea seedlings. In this context it must be noted that the observed conversion of applied GA_5 to GA_1 was, if anything, greater in the light than in the dark (10–19 versus 4–8 %). Moreover, if conversion of physiologically inactive GA_5 to active GA_1 were the critical event one would expect that the GA_1 content of the seedlings would be lower in the light than in the dark; this, however, was clearly not the case. Thus, our results, while supporting the general explanation proposed by Kende and Lang (13), do not provide a more specific explanation, and further studies are necessary.

Our data are in disagreement with those of Kohler (15, 16, 17) who reports that light- and dark-grown dwarf pea seedlings, dark-grown normal seedlings, and light-grown but retardant-treated normal seedlings have identical endogenous gibberellin levels but that the level in light-grown normals is more than 10 times as high. He explains the ability of the normals to make relatively good growth in light with an enhancement of gibberellin synthesis by light. We have no explanation for this discrepancy, and a comparison is difficult as Kohler expresses his results on a *per* seedling basis and does not give data on a weight basis. It should also be borne in mind that all data of Kohler refer to the GA_5-like material; he was unable to obtain significant activities in the fractions which should contain the GA_1-like factor. With our methods we have never experienced such a problem. However, it may be pointed out that Kohler (15) cut his seedlings above the substrate and apparently extracted them *in toto,* that is, he determined the gibberellin content of normal seedlings that (according to our experience) were about 20 to 25 cm tall, and of dwarf seedlings only about 4 to 5 cm tall. It seems possible, assuming extractable gibberellins are rather uniformly distributed throughout the shoot of pea seedlings (see Radley, 25) that if the gibberellin content were expressed on a weight basis the difference between normals and dwarf seedlings would be greatly reduced.

Literature Cited

1. BALDEV, B., A. LANG, AND A. O. AGATEP. 1965. Gibberellin production in pea seeds developing in excised pods; effect of growth retardant AMO-1618. Science 147: 155–57.
2. BRIAN, P. W., H. G. HEMMING, AND D. LOWE. 1964. Comparative potency of nine gibberellins. Ann. Botany London 28: 369–89.
3. DENNIS, D. T., C. D. UPPER, AND C. A. WEST. 1965. An enzymatic site of inhibition of gibberellin biosynthesis by AMO-1618 and other plant growth retardants. Plant Physiol. 40: 948–52.
4. FRANKLAND, B. AND P. F. WAREING. 1960. Effect of gibberellic acid on hypocotyl growth of lettuce seedlings. Nature 185: 255–56.
5. GALSTON, A. W. 1959. Studies on indoleacetic acid oxidase inhibitor and its relation to photomorphogenesis. In: Photoperiodism in Plants and Animals. R. B. Withrow, ed. Am. Assoc. Advan. Sci., Washington, D. C. p 137–57.
6. GORTER, C. J. 1961. Dwarfism of peas and the action of gibberellic acid. Physiol. Plantarum 14: 332–43.
7. HAYASHI, F. AND L. RAPPAPORT. 1962. Gibberellin-like activity of neutral and acidic substances in the potato tuber. Nature 195: 617–18.
8. JONES, D. F. 1964. Examination of the gibberellins of *Zea mays* and *Phaseolus multiflorus* using thin layer chromatography. Nature 202: 1309–10.
9. JONES, R. L. AND I. D. J. PHILLIPS. 1964. Agar diffusion technique for estimating gibberellin production by plant organs. Nature 204: 497–99.
10. JONES, R. L. AND I. D. J. PHILLIPS. 1966a. Organs of gibberellin synthesis in light-grown sunflower plants. Plant Physiol. 41: 1381–86.
11. JONES, R. L. AND I. D. J. PHILLIPS. 1966b. Effect of CCC on the gibberellin content of excised sunflower organs. Planta 72: 53–59.
12. JONES, R. L. AND J. E. VARNER. 1966. The bioassay of gibberellin. Planta 72: 155–61.
13. KENDE, H. AND A. LANG. 1964. Gibberellins and light inhibition of stem growth in peas. Plant Physiol. 39: 435–40.
14. KENDE, H., H. NINNEMANN, AND A. LANG. 1963. Inhibition of gibberellic acid biosynthesis in *Fusarium moniliforme* by AMO-1618 and CCC. Naturwissenschaften 50: 599–600.
15. KOHLER, D. 1965a. Uber den Gibberellingehalt von Zwerg- und Normalerbsen im Rotlicht und die Wirkung von Chlorchlinchlorid auf das Wachstum der Erbsen. Planta 65: 218–24.
16. KOHLER, D. 1965b. Die Wirkung von Schwachem Rotlicht und Chlorcholinchlorid auf den Gibberellingehalt Normaler Erbsensamlinge und die Ursache der Unterschiedlichen Empfindlichkeit von Zwerg- und Normalerbsensamlingen gegen ihr Eigenes Gibberellin. Planta 66: 44–54.
17. KOHLER, D. 1966. Die Abhangigkeit der Gibberellin Produktion von Normalerbsen vom Phytochromsystem. Planta 69: 27–33.
18. LOCKHART, J. A. 1956. Reversal of the light inhibition of pea stem growth by the gibberellins. Proc. Natl. Acad. Sci. U. S. 42: 841–48.
19. LOCKHART, J. A. 1959. Studies on the mechanism of stem growth inhibition by visible radiation. Plant Physiol. 34: 457–60.
20. MACMILLAN, J., J. C. SEATON, AND P. J. SUTER. Plant hormones II: Isolation and structures of gibberellin A_6 and A_8. Tetrahedron 18: 349–55.
21. MCCOMB, A. J. 1961. 'Bound' gibberellin in mature runner bean seeds. Nature 192: 575–76.
22. MCCOMB, A. J. AND D. J. CARR. 1958. Evidence from a dwarf pea bioassay for naturally recurring gibberellins in the growing plant. Nature 181: 1548–49.
23. VAN OVERBEEK, J. 1941. A quantitative study of auxin and its precursor in coleoptiles. Am. J. Botany 28:1–10.
24. PHINNEY, B. O. AND C. A. WEST. 1960. Gibberellins as native plant growth regulators. Ann. Rev. Plant Physiol. 11: 411–36.
25. RADLEY, M. 1958. The distribution of substances similar to gibberellic acid in higher plants. Ann. Botany London 22: 297–307.
26. ZEEVAART, J. A. D. 1966. Reduction of the gibberellin content of *Pharbitis* seeds by CCC and after effects of progeny. Plant Physiol. 41: 856–62.

On the Significance of Cytokinin Incorporation into RNA

Hans Kende and James E. Tavares

MSU/AEC Plant Research Laboratory, Michigan State University, East Lansing, Michigan 48823

Received April 18, 1968.

Abstract. **The clarification of the following 2 questions was attempted: (a) are cytokinins precursors in the formation of sRNA, (b) is the observed incorporation of cytokinins into sRNA related to the action of the hormone? Although *Escherichia coli* contains cytokinins in its sRNA, no cytokinin auxotroph mutants of *E. coli* could be found and the statistical probability for the existence of such mutants is extremely low. This suggests that cytokinins are not precursors in the synthesis of sRNA. A radioactive cytokinin, 6-benzylamino-9-methylpurine was synthesized and it was tested whether or not it is incorporated into sRNA of soybean callus tissue. Masking the 9-position of the purine inhibited the incorporation of this cytokinin into RNA while not affecting its biological activity. This is taken as an indication that the observed incorporation of cytokinins such as benzyladenine into sRNA is not related to the action of this hormone.**

Substances with cytokinin activity have been isolated from the hydrolysates of the soluble RNA (sRNA) of microorganisms, animals and plants (1, 3, 12, 13, 20, 24, 26). The minor base 6-(γ,γ-dimethylallylamino)purine (= *N*-6-isopentenyladenine or IPA) was found in seryl transfer RNA (tRNA) of yeast at the position following the presumed anticodon (3, 31). This compound has high cytokinin activity (14) and is, chemically, closely related to the naturally occurring cytokinin zeatin, 6-(4-hydroxy-3-methylbut-*trans*-2-enyl)aminopurine (21). IPA was also isolated from sRNA of spinach and peas and the *cis*-isomer of zeatin from the sRNA of the same 2 plants and of corn (12). The presence of cytokinins seems to be restricted to sRNA since no such compounds were found in hydrolysates of ribosomal RNA (26).

Recently, evidence was presented that soybean and tobacco callus cultures, which require exogenous cytokinins for growth, incorporate one such growth regulator, 6-benzylaminopurine (benzyladenine), into sRNA (7). Chromatography of the labeled sRNA on *O*-benzoyldiethylaminoethyl cellulose showed that only 1 of the 6 fractions obtained was radioactive, and the bulk of this radioactivity appeared to be associated with the nucleotide of benzyladenine (8).

The incorporation of a cytokinin into sRNA and the natural occurrence of cytokinins in sRNA led to the question whether this hormone is exerting its biological effect through its presence in certain RNA species. As one possibility, it was suggested that the incorporation of benzyladenine into sRNA may confer amino acid transfer activity to certain tRNAs (7). Regulation at the level of amino acid transfer is well known in microorganisms (27) and similar control mechanisms can be visualized in plants.

There are, however, serious arguments against the hypothesis that cytokinins act by being incorporated into RNA, either at very specific positions or, more generally, in place of methylated bases as suggested by Fox (7). From microbial systems it is known that tRNA is first synthesized by polymerization of purine and pyrimidine bases and that methylation occurs on the finished RNA chain (4). There is also evidence that pseudouridine is formed by intramolecular rearrangement within the polymerized sRNA (30). If similar biosynthetic mechanisms are functional in plants, cytokinins are not precursors in the formation of tRNA. In this respect, it is interesting to note that ethanolic extracts of corn kernels contain the *trans*-isomer of zeatin (21) while the sRNA hydrolysates from the same source yield the *cis*-isomer (12), suggesting that zeatin is not a precursor in the synthesis of sRNA. There is, however, one possibility which could reconcile the contradictions discussed above; cytokinins may not be incorporated as a whole into RNA but the side-chain may be transferred to an adenine in the RNA polymer (7).

The following 2 questions need to be answered at this point: A) are cytokinins precursors in the formation of RNA, and B) is the incorporation of benzyladenine into sRNA related to the action of the hormone? An attempt was made to answer the first question by screening for cytokinin-requiring mutants of *Escherichia coli* which contains cytokinins in its sRNA (26). Such a mutant would be a useful tool for studying the formation of cytokinin-containing sRNA species. In order to answer the second question, experiments were designed by which one might separate the hormonal action of cytokinins from their incorporation into RNA. A radioactive analogue of benzyladenine, 6-benzylamino-9-methylpurine, was synthesized and its biological activity was established. It was hoped that, by masking the 9-position of the purine ring with a methyl group, formation of a nucleotide and hence incorporation into RNA could be prevented.

Materials and Methods

Bacterial Mutants. According to the method of Benzer and Champe (2), rapidly growing cultures of *Escherichia coli* [C-medium (25) with 1 % glucose; approximately 5 × 10^8 cells per ml] were treated with 0.1 ml of diethylsulfate per 20 ml of culture. The culture was not agitated or aerated, but rather the diethylsulfate was allowed to diffuse slowly into the medium. One ml of this bacterial suspension was then transferred to 20 ml of fresh medium and allowed to reach 5 × 10^8 cells per ml before a second exposure to diethylsulfate was made. The mutagenic procedure was repeated for 3 consecutive treatments. Mutants were concentrated by using the penicillin method (10). The bacteria were centrifuged, resuspended in minimal medium without test substances, and were allowed to grow for 90 to 120 minutes to eliminate stored metabolites in the mutants. After this, the culture was treated with penicillin at 2000 units per ml for 90 minutes. The bacteria were then centrifuged, resuspended in 0.5 % sodium chloride solution with 5 % glycerine, and frozen. This procedure aided in breaking up the partially lysed cells. Individual mutants were selected and tested by replica plating (19). Auxotrophs were sought for the following test substances: L-isoleucine and L-arginine supplied at 0.5 mM, and adenine, kinetin (6-furfurylaminopurine) and IPA supplied at 50 μM. Cytokinins at this concentration appear to have no effect on the growth rate of the bacteria.

Synthesis of 6-Benzylamino-9-methylpurine. 6-Benzylamino-9-methylpurine-benzyl-7-^{14}C (fig 1A) was obtained by refluxing 14 μmoles of benzylamine-7-^{14}C HCl (specific activity 12 mc/mmole, International Chemical and Nuclear Corp.), 14 μmoles of 6-chloro-9-methylpurine (J. T. Baker Chemical Co.) and 0.2 ml triethylamine in 5 ml *n*-butanol for 19 hours. Addition of triethylamine facilitates the condensation reaction. The reaction mixture was taken to dryness and resuspended in water. The aqueous solution was partitioned 10 times against diethyl ether, the combined ether phases were dried with sodium sulfate and evaporated. The residue was chromatographed with water on washed Whatman No. 3 paper, and the U.V.-absorbing band was eluted with 90 % (v/v) ethanol. The purity of the product was checked by thin layer chromatography.

Growth of Callus. Callus originally derived from soybean cotyledons [*Glycine max* (L.) Merr. cv. Acme] was grown in 125 ml Erlenmeyer flasks on 20 ml of Miller's medium (23). Cytokinins were added prior to autoclaving. The tissue cultures were kept in darkness at 27° and were harvested 2 to 3 weeks after start of the experiment.

Extraction of Nucleic Acids. Nucleic acids were extracted according to Kirby (18) as modified by Johri and Varner (personal communication): 6 g of tissue was ground in a mortar in 5 ml sodium aminosalicylate (SAS), 5 ml phenol-cresol mixture and some sand. The homogenate was mixed vigorously and centrifuged at 10,000 rpm (Sorvall, head No. HB-4) for 10 minutes. The aqueous phase was removed and the extraction of the phenol phase was repeated twice using 5 ml SAS and an additional 2 ml phenol-cresol each time. The homogenates were centrifuged and the aqueous phases combined. Sodium chloride (30 mg/ml) was added to the pooled aqueous fraction which was extracted then with one-half of its volume of phenol-cresol mixture. The extract was again centrifuged and the aqueous phase was carefully separated from the phenol. The nucleic acids were precipitated with 2 volumes of ethanol-cresol mixture (−20°, 4 hr) and collected by centrifugation at 10,000 rpm for 10 minutes. The nucleic acids were resuspended and dialyzed for 20 hours against 3 changes of 1000 ml tris-HCl buffer (0.01 M, pH 7.6) containing 0.001 M $MgCl_2$ and 0.01 M KCl. Results obtained with the SAS extraction method do not differ from those obtained with the sodium laurylsulfate extraction method (16).

Chromatography of Nucleic Acids. Methylated Albumin-Kieselguhr chromatography (MAK) was performed according to Mandell and Hershey (22). The nucleic acids were eluted with a linear gradient of 0.3 to 1.6 M NaCl in phosphate buffer (200 ml each). Fractions of 3 ml were collected and the optical density at 260 nm was measured with a Beckman DB spectrophotometer.

Counting of Radioactivity. Three consecutive chromatographic fractions were combined into one: to this, 150 μg carrier DNA and 2 ml trichloroacetic acid (50 %, w/v) were added. The nucleic acids were precipitated at 2° for 4 to 5 hours. The precipitates were collected on Millipore filters (HA 0.45 μ) and washed with an excess of ice-cold trichloroacetic acid (5 %, w/v). The filters were dried and each sample was counted twice for 20 minutes in a Packard 3375 liquid scintillation spectrometer using Bray's solution as scintillator (6). Identical results were obtained when every chromatographic fraction was dried and counted directly in Bray's solution.

Results

The Probability of Cytokinin Mutants in E. coli. The efficiency of the mutagenic treatment was tested by selecting for arginine, isoleucine, and adenine mutants, besides cytokinin auxotrophs. Table I summarizes the result of one such experiment. Assuming that the mutability of each gene is the same when treated with a "random" mutagen such as diethylsulfate, any measure of the frequency of a known

Table I. *Frequency of Arginine, Adenine, Isoleucine, Threonine, and Cytokinin Mutants in* E. coli *Treated with Diethylsulfate*

Total number of colonies tested	10,831
No. of arginine mutants	119
No. of adenine mutants	16
No. of isoleucine mutants	18
No. of isoleucine mutants that are also threonine mutants	16
No. of IPA mutants	0
No. of kinetin mutants	0

one-gene mutation will allow a statistical estimate of the probability of finding another one-gene mutation.

To estimate the probability of cytokinin mutations, a further screening for isoleucine and IPA auxotrophs was made. Any isoleucine mutant of *E. coli* which does not grow on threonine must be a single-gene mutation as threonine is the immediate precursor of isoleucine. While 240 isoleucine-requiring mutants were found which did not grow on a medium containing threonine, not 1 single IPA mutant could be isolated. Using the Poisson distribution test, the probability of finding a IPA auxotroph mutated in 1 gene would be

$$P = \frac{e^{-m}\ (m)^x}{x!}$$

where P = probability of IPA mutation, x = number of IPA mutations, and m = number of isoleucine mutants. Thus $P = e^{-m} = e^{-240}$, or $P = 1.6 \times 10^{-105}$. Therefore, the probability is very small indeed that a cytokinin auxotroph exists in *E. coli.*

The Fate of 6-Benzylamino-9-methylpurine in Soybean Calluses. 6-Benzylamino-9-methylpurine stimulated cell division in soybean tissue cultures to the same extent as or even more than benzyladenine (table II). Activation of growth of axillary buds by 6-benzylamino-9-methylpurine has been reported by Guern (11).

Table II. *Biological Activity of 6-Benzylamino-9-methylpurine*

Each figure represents the average fresh weight of 12 soybean calluses grown for 21 days.

Treatment	Fr wt per callus
	mg
Benzyladenine (1 μM)[1]	786
Benzyladenine (3 μM)	1053
Benzyladenine (10 μM)	950
6-Benzylamino-9-methylpurine (1 μM)[2]	1125
6-Benzylamino-9-methylpurine (3 μM)	1158
6-Benzylamino-9-methylpurine (10 μM)	850
Control	123

1 Nutritional Biochemicals Corporation.

2 Non-radioactive sample synthesized and crystallized by the authors; the purity was verified by thin-layer and gas-liquid chromatography.

In initial experiments, soybean calluses were grown on media containing benzyladenine-^{14}C or 6-benzylamino-9-methylpurine-^{14}C of the same specific activity (12 mc/mmole), and the incorporation of radioactivity into RNA was compared. Benzyladenine-benzyl-7-^{14}C was synthesized as described earlier (5) but it was further purified by chromatography on Whatman No. 3 paper using 0.01 N HCl as solvent. The total trichloroacetic acid-precipitable nucleic acids from calluses (6.4 g) grown for 2 weeks on benzyladenine-^{14}C (3 μM) showed an incorporation of 68 cpm. In the same experiment and under identical conditions, the total nucleic acid extract of calluses (6.56 g) grown for 2 weeks on 6-benzylamino-9-methylpurine-^{14}C (3 μM) contained 4.5 cpm. Similar results were obtained in other experiments.

If 6-benzylamino-9-methylpurine-^{14}C was given at the optimal concentration (1 μM), no incorporation of radioactivity into RNA was observed (fig 1B). Identical results were obtained in 5 experiments with calluses derived from different explants, with different times of incubation (14–21 days), and with different concentrations of the hormone (up to 3 μM).

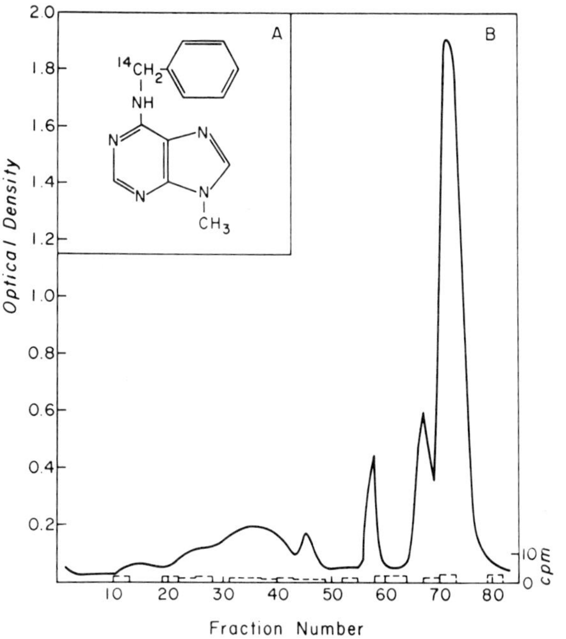

FIG. 1. A) Structure of 6-benzylamino-9-methylpurine-benzyl-7-^{14}C. B) MAK chromatography of nucleic acids extracted from 10 soybean calluses (6.336 g) grown for 17 days on a medium containing 6-benzylamino-9-methylpurine-7-^{14}C (1 μM). Weight of 10 benzyladenine-treated (1 μM) calluses grown under identical conditions 5.895 g, 10 control calluses 0.814 g. Solid line: optical density at 260 nm. Broken line: radioactivity. For the determination of the radioactivity the contents of 3 neighboring tubes were combined (fractions 11-82). All samples were counted twice and the counts were never significantly above background.

Discussion

The fact that no cytokinin auxotroph was found in *E. coli* is an indication that IPA is not a precursor in the synthesis of sRNA. This is in agreement with the hypothesis that the isopentyl group is added onto the macromolecule by substitution in a manner similar to that demonstrated for the methylation of nucleic acids (4).[1] A transfer of the side chain from IPA to an adenine in the sRNA seems also unlikely since there appears to be no need for the synthesis of IPA in *E. coli.*

The argument can be made that only kinetin and IPA were tested and that neither of these 2 compounds can substitute for the yet unidentified cytokinin in *E. coli* sRNA. However, there is evidence that IPA may be a widely occurring minor base since it has been isolated from the sRNA of mammals (13), plants (12), yeast (3, 13), and bacteria (24).

As with most negative results, the lack of a cytokinin mutant can only serve as supporting but not as conclusive evidence that cytokinins are not precursors in the synthesis of sRNA.

Methylation at the 9-position of the purine ring appears to inhibit the incorporation of this cytokinin into RNA while not affecting its biological activity. There was also no indication of the transfer of the labeled benzyl group to sRNA. The specific activity of the 6-benzylamino-9-methylpurine used was sufficient to detect the incorporation of 1 cytokinin molecule per molecule of 1 tRNA species. This was estimated from the amount of radioactivity associated with 1 tRNA species charged with a ^{14}C-amino acid, taking into account the specific activity of the amino acid and the optical density of the whole sRNA peak chromatographed on a MAK column (29).

Our experiments and the considerations discussed earlier in this paper make it seem unlikely that cytokinins exert their regulatory activities through incorporation into RNA. Results obtained in experiments on cytokinin-controlled bud formation in moss protonemata suggest that cytokinins interact with their site of action by loose, probably non-covalent bonds (5). A similar hypothesis was advanced for the binding of gibberellins (9, 17) and auxins (15) to their receptor sites. Evidence has been presented that estrogens bind to a hormone-specific protein in rat uteri most likely by non-covalent linkage (28). The finding and cellular localization of hormone-specific binding sites for any of the plant hormones would be an important step towards elucidating the mechanism of action of plant hormones.

[1] Note added in proof: Fittler, Kline, and Hall [Biochem. Biophys. Res. Commun. 31: 571–76 (1968)] reported recently that a cell-free extract from rat liver or yeast catalyzes the incorporation of mevalonic acid or isopentenyl pyrophosphate into pre-formed tRNA to form *N*-6-isopentenyladenosine.

Acknowledgments

The generous gift of 6-(γ,γ-dimethylallylamino) purine by Dr. N. J. Leonard (University of Illinois, Urbana) is gratefully acknowledged. We thank Dr. Renate deZacks and Mr. Eric Berke for their competent technical assistance, and express our appreciation to Dr. Rainer Hertel for helpful suggestions concerning bacterial mutagenesis. This work was supported by the United States Atomic Energy Commission under Contract AT(11-1)-1338.

Literature Cited

1. BELLAMY, A. R. 1966. Cytokinins in ribonucleic acids. Nature 211: 1093–95.
2. BENZER, S. AND S. P. CHAMPE. 1961. Ambivalent rII mutants of phage T4. Proc. Natl. Acad. Sci. 47: 1025–38.
3. BIEMANN, K., S. TSUNAKAWA, J. SONNENBICHLER, H. FELDMANN, D. DÜTTING, AND H. G. ZACHAU. 1966. Struktur eines ungewöhnlichen Nucleotids aus serine-spezifischer Transfer-Ribonucleinsäure. Angew. Chem. 78: 600–01.
4. BOREK, E. AND P. R. SRINIVASAN. 1966. The methylation of nucleic acids. Ann. Rev. Biochem. 35: 275–98.
5. BRANDES, H. AND H. KENDE. 1968. Studies on cytokinin-controlled bud formation in moss protonemata. Plant Physiol. 43: 827–37.
6. BRAY, G. A. 1960. A simple efficient liquid scintillator for counting aqueous solutions in a liquid scintillation counter. Anal. Biochem. 1: 279–85.
7. FOX, J. E. 1966. Incorporation of a kinin, N,6-benzyladenine into soluble RNA. Plant Physiol. 41: 75–82.
8. FOX, J. E. AND C. CHEN. 1967. Characterization of labeled ribonucleic acid from tissue grown on ^{14}C-containing cytokinins. J. Biol. Chem. 242: 4490–94.
9. GINZBURG, C. AND H. KENDE. 1968. Studies on intracellular localization of radioactive gibberellin. In: Biochemistry and Physiology of Plant Growth Substances. Runge Press, Ottawa. In press.
10. GORINI, L. AND H. KAUFMAN. 1960. Selecting bacterial mutants by the penicillin method. Science 131: 604–05.
11. GUERN, J. 1968. Metabolism, transport and biological activity of some cytokinins. In: Biochemistry and Physiology of Plant Growth Substances. Runge Press, Ottawa. In press.
12. HALL, R. H., L. CSONKA, H. DAVID, AND B. McLENNAN. 1967. Cytokinins in the soluble RNA of plant tissues. Science 156: 69–71.
13. HALL, R. H., M. J. ROBINS, L. STASIUK, AND R. THEDFORD. 1966. Isolation of ^{6}N-(γ,γ dimethylallyl)adenosine from soluble ribonucleic acid. J. Am. Chem. Soc. 88: 2614–15.
14. HAMZI, H. Q. AND F. SKOOG. 1964. Kinetin-like growth-promoting activity of 1- substituted adenines [1-benzyl-6-aminopurine and 1-(γ,γ-dimethylallyl)-6-aminopurine]. Proc. Natl. Acad. Sci. 51: 76–83.

15. HERTEL, R. AND R. FLORY. 1968. Auxin movement in corn coleoptiles. Planta. In press.
16. INGLE, J., J. L. KEY, AND R. E. HOLM. 1965. Demonstration and characterization of a DNA-like RNA in excised plant tissue. J. Mol. Biol. 11: 730–46.
17. KENDE, H. 1967. Preparation of radioactive gibberellin A_1 and its metabolism in dwarf peas. Plant Physiol. 42: 1612–18.
18. KIRBY, K. S. 1965. Isolation and characterization of ribosomal ribonucleic acid. Biochem. J. 96: 266–69.
19. LEDERBERG, J. AND E. M. LEDERBERG. 1952. Replica plating and selection of bacterial mutants. J. Bacteriol. 63: 399–406.
20. LETHAM, D. S. AND R. K. RALPH. 1967. A cytokinin in soluble RNA from a higher plant. Life Sciences 6: 387–94.
21. LETHAM, D. S., J. S. SHANNON, AND I. R. C. McDONALD. 1967. Regulators of cell division in plant tissues. III. The identity of zeatin. Tetrahedron. 23: 479–86.
22. MANDELL, J. D. AND A. D. HERSHEY. 1960. A fractionating column for analysis of nucleic acids. Anal. Biochem. 1: 66–77.
23. MILLER, C. O. 1965. Evidence for the natural occurrence of zeatin and derivatives: compounds from maize which promote cell division. Proc. Natl. Acad. Sci. 54: 1052–58.
24. PETERKOFSKY, A. 1968. The incorporation of mevalonic acid into the N6-(Δ^2-isopentenyl)adenosine of transfer ribonucleic acid in *Lactobacillus acidophilus*. Biochemistry 6: 472–81.
25. ROBERTS, R. B., D. B. COWIE, P. H. ABELSON, E. T. BOLTON, AND R. J. BRITTEN. 1955. Studies of biosynthesis in *Escherichia coli.* Carnegie Inst. Wash. Publ. 607. p 5.
26. SKOOG, F., D. J. ARMSTRONG, J. D. CHERAYIL, A. E. HAMPEL, AND R. M. BOCK. 1966. Cytokinin activity: Localization in transfer RNA preparations. Science 154: 1354–56.
27. STENT, G. S. 1964. The operon: on its third anniversary. Science 144: 816–20.
28. TOFT, D., G. SHYAMALA, AND J. GORSKI. 1967. A receptor molecule for estrogens; studies using a cell-free system. Proc. Natl. Acad. Sci. 57: 1740–43.
29. VOLD, B. S. AND P. S. SYPHERD. 1968. Modification in transfer RNA during the differentiation of wheat seedlings. Proc. Natl. Acad. Sci. 59: 453–58.
30. WEISS, S. B. AND J. LEGAULT-DEMARE. 1965. Pseudouridine formation: evidence for RNA as an intermediate. Science 149: 429–31.
31. ZACHAU, H. G., D. DÜTTING, AND H. FELDMANN. 1966. Nucleotidsequenzen zweier serinspezifischer Transfer-Ribonucleinsäuren. Angew. Chem. 78: 392–93.

Abscisic Acid: A New Name for Abscisin II (Dormin)

29 MARCH 1968

Abscisin II was the name given to the second of two abscission-accelerating substances isolated (*1*) from cotton fruit. The same substance was subsequently isolated (*2*) from sycamore leaves as the result of a search for a "dormin" [an endogenous substance inducing dormancy (*3*)]. Since then, the substance has been identified in a large number of higher plants. The structure of abscisin II has been determined and confirmed by synthesis (*4*); structure and correct absolute configuration (*5*) are shown by the insert.

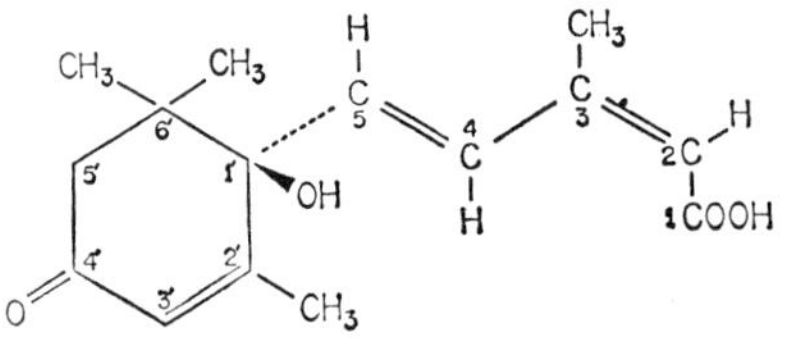

(S)-ABSCISIC ACID

Some confusion has developed from the use of the two names, abscisin II and dormin, derived from the abscission and dormancy effects of the compound. The more recent discovery of several other physiological effects has emphasized the need for an agreed terminology.

We now propose the term abscisic acid as a reasonable and useful compromise which gives an indication of the compound's chemical nature, facilitates naming derivatives, and is close enough to the original name to avoid confusion arising from the change.

We suggest that authors specify racemic abscisic acid when they use it in experiments by calling it (*RS*)-abscisic acid, and that the naturally occurring enantiomorph be called (*S*)-abscisic acid when it is necessary to draw attention to the stereochemistry. As an abbreviation for abscisic acid, we propose ABA. For specifying positions within the molecule, the numbering system shown in the figure (which conforms to the systematic name for the substance) is recommended. We propose that the question of renaming abscisin I be deferred until its structure is known.

The foregoing proposals were agreed at the Sixth International Conference on Plant Growth Substances (Ottawa, 1967) and are published in the proceedings of that conference.

F. T. ADDICOTT, J. L. LYON
K. OHKUMA, W. E. THIESSEN
University of California, Davis
H. R. CARNS
U.S. Agricultural Research Service, Beltsville, Maryland
O. E. SMITH
University of California, Riverside
J. W. CORNFORTH
B. V. MILBORROW, G. RYBACK
Shell Research, Limited, Milstead Laboratory, Sittingbourne, United Kingdom
P. F. WAREING
University College of Wales, Aberystwyth, United Kingdom

References

1. W.-C. Liu and H. R. Carns, *Science* **134**, 384 (1961); K. Ohkuma, J. L. Lyon, F. T. Addicott, O. E. Smith, *ibid.* **142**, 1592 (1963); F. T. Addicott, H. R. Carns, J. L. Lyon, O. E. Smith, J. L. McMeans, *Colloq. Int. Centre Nat. Rech. Sci. Paris* **123**, 687 (1964).
2. J. W. Cornforth, B. V. Milborrow, G. Ryback, P. F. Wareing, *Nature* **205**, 1269 (1965).
3. C. F. Eagles and P. F. Wareing, *ibid.* **199**, 874 (1963); P. F. Wareing, C. F. Eagles, P. M. Robinson, *Colloq. Int. Centre Nat. Rech. Sci. Paris* **123**, 377 (1964).
4. K. Ohkuma, F. T. Addicott, O. E. Smith, W. E. Thiessen, *Tetrahedron Letters* **1965**, 2529 (1965); J. W. Cornforth, B. V. Milborrow, G. Ryback, *Nature* **206**, 715 (1965).
5. J. W. Cornforth, W. Draber, B. V. Milborrow, G. Ryback, *Chem. Commun.* **1967**, 114 (1967).

26 December 1968

Reprinted from the
AUSTRALIAN JOURNAL OF BIOLOGICAL SCIENCES

ABSCISIN II AND SOME HORMONE-REGULATED PLANT RESPONSES

By D. Aspinall,* L. G. Paleg,* and F. T. Addicott*†

[*Manuscript received May* 8, 1967]

Summary

The activity of (±)-abscisin II [(±)AbII] and its interaction with a number of plant growth regulators in the control of lettuce germination, lettuce hypocotyl and radicle elongation, cucumber seedling growth, radish leaf senescence, and barley vernalization were examined. The interactions between (±)AbII and the promoting hormones were of two general types. In gibberellic acid (GA_3)-promoted lettuce germination and kinetin-controlled leaf senescence, the effects of low concentrations of (±)AbII were completely overcome by high concentrations of the other substance. In the second type of interaction, (±)AbII was inhibitory only in the presence of high concentrations of the promoter (kinetin in lettuce germination and allogibberic acid in lettuce radicle elongation). Cucumber radicle elongation, on the other hand, was promoted by (±)AbII in the presence of a mixture of GA_4 and GA_7.

These various responses suggest that linking the mechanism of AbII action with that of one growth hormone alone may be too restrictive.

I. Introduction

The involvement of naturally occurring inhibitory substances in the regulation of plant growth and development has been often suggested, particularly in relation to dormancy, and several substances have been implicated (Evanari 1949; see also Bentley 1958). However, it was only with the isolation of abscisin II (AbII) from immature cotton (*Gossypium hirsutum*) fruits (Ohkuma *et al.* 1963) that an inhibitory substance active at concentrations comparable with the known plant hormones was obtained.

It has since been shown that AbII is the active inhibitor in sycamore (*Acer pseudoplatanus*) leaves (Cornforth *et al.* 1965) and that it is present in a number of other plants (Cornforth, Milborrow, and Ryback 1966). It has been suggested that AbII has a hormonal function in cotton fruit abscission (Addicott *et al.* 1964) and regulates the photoperiodic control of dormancy in birch (Eagles and Wareing 1963, 1964). In addition, AbII inhibits flowering in some long-day plants (Evans 1966; El-Antably, Wareing, and Hillman 1967) and promotes it in some short-day plants, accelerates the senescence of leaf disks from many species, and inhibits bud

* Department of Plant Physiology, Waite Agricultural Research Institute, University of Adelaide.

† Present address: Department of Agronomy, University of California, Davis, California.

growth on potato tubers though promoting tuber formation on potato plants (El-Antably, Wareing, and Hillman 1967).

The possible importance of the abscisins in the regulation of plant growth has led to experiments, and some speculation, on its mode of action. Addicott *et al.* (1964) demonstrated inhibition of auxin-mediated *Avena* coleoptile growth by AbII, and Thomas, Wareing, and Robinson (1965) found that such inhibition could be overcome by gibberellic acid (GA_3) but not auxin, although the coleoptile is responsive to auxin in the absence of AbII. Thomas, Wareing, and Robinson (1965) also found that their preparations of AbII reduced the elongation of tall (but not dwarf) maize (*Zea mays*) leaf sections and that this inhibition could be overcome by GA_3. In peas (*Pisum sativum*), however, it was the response to GA_3, not normal growth, which was inhibited by these extracts. In another GA_3-controlled response, the release of α-amylase by the barley aleurone layer, AbII was also inhibitory although the inhibition was not clearly competitive (Chrispeels and Varner 1966). Some of these responses could conceivably be due to inhibition of GA_3 synthesis, as has been suggested for the growth retardants (Paleg *et al.* 1965). A similar interpretation was suggested by the inverse relationship between AbII activity and gibberellin activity in sycamore and birch seedlings, although Thomas *et al.* (1965) found no effect of their extracts on GA_3 production by *Gibberella fujikuroi*.

The spectrum of AbII activities led Thomas *et al.* (1965) to suggest that it acts as a gibberellin antagonist *in vivo*, although they reported inhibition with sycamore cambial cell culture which was not reversed by GA_3 and only partially reversed by indoleacetic acid or kinetin. The question of whether AbII is specifically inhibitory to GA_3-controlled responses or whether it inhibits responses controlled by other gibberellins or by other classes of hormones is of importance in evaluating its function in the plant. In the present investigation the activity of AbII has been assessed in responses promoted by GA_3, by other gibberellins and related compounds, by GA_3 or kinetin, or by kinetin alone.

II. Methods

Synthetic ($\pm$)-abscisin II [($\pm$)AbII] was used in all experiments. The sample was obtained through the courtesy of J. van Overbeek and Shell Research Ltd., Sittingbourne, Kent; it is considered a mixture of two enantiomorphs (Cornforth, Milborrow, and Ryback 1965). Allogibberic acid, GA_3, and a mixture of GA_4 and GA_7 were obtained through the courtesy of Imperial Chemical Industries Ltd.

(*a*) *Lettuce Germination*

In all the lettuce germination experiments, 50 seeds (*Lactuca sativa* cv. Great Lakes) were placed on one filter paper in a 5-cm Petri dish with 1 ml of test solution per dish, at least two dishes being used per treatment. Germination, in the dark or light, was at 25°C and was recorded after 48 hr unless otherwise stated. Far-red light was provided by filtering incandescent light through a Westlake FRF 700 plastic filter (250 W incandescent source at 2 ft from seeds); red light was provided by a Philips 40 W red fluorescent tube (2 ft from seeds). A seed was recorded as having germinated when the tip of the radicle was visible. In one experiment in which seeds were exposed to ($\pm$)AbII for limited periods, ($\pm$)AbII was removed from these seeds at the end of the exposure period by rinsing three times in distilled water.

(*b*) *Lettuce Hypocotyl and Radicle Extension*

Lettuce seeds (cv. Great Lakes) were pregerminated for 48 hr in the dark at 20°C. At the end of this period, seedlings were selected for uniformity and placed, 10 per 10-cm Petri dish, on two filter papers and 4 ml of test solution. The dishes were placed in clear plastic sleeves to prevent evaporation and incubated for a further 72 hr at 25°C in light (Philips 40 W fluorescent tubes, 900 f.c.). At the end of this period, hypocotyls and radicles were measured to the nearest millimetre. Replication of dishes was threefold and mean values were used in all statistical calculations.

(*c*) *Cucumber Hypocotyl and Radicle Extension*

Cucumber seed (*Cucumis sativus* cv. Long Green) were pregerminated in distilled water at 30°C in light (as for lettuce experiments) for 72 hr. The seedlings were then selected for uniformity and transferred to filter paper in plastic dishes, 20 per dish, with 20 ml of test solution. The dishes were covered with a sheet of glass and incubated in light for a further 48 hr, when the hypocotyls and radicles of the seedlings were measured to the nearest millimetre. Replication was fourfold.

(*d*) *Senescence of Leaf Disks*

Mature, but not senescent, leaves of radish (*Raphanus sativus* cv. Icicle) plants grown in a glasshouse were selected for uniformity. Disks (1 cm diam.) were cut from these leaves, avoiding main veins, and five disks were placed on filter paper in 5-cm Petri dishes with 1 ml of test solution. The disks were incubated in the dark at 20°C for 48 or 96 hr. Chlorophyll was extracted by briefly boiling the disks with two 4-ml portions of 80% ethanol. The extracts were adjusted to 5 ml with 80% ethanol and their absorbances determined at 660 mμ. Each treatment was replicated at least twice.

(*e*) *Vernalization of Barley*

Barley grains (*Hordeum vulgare* cv. Pioneer), which requires 6 weeks vernalization for maximum promotion of flowering (Aspinall, unpublished data), were germinated at 20°C for 12 hr in the presence of GA_3 and ($\pm$)AbII [all combinations of 0, 1, 10, and 100 μg/ml GA_3 with 0, 0·25, 2·5, and 25 μg/ml ($\pm$)AbII]. At the end of this period the grains were surface-dried, weighed, and placed in tared plastic Petri dishes. Aliquots of the GA_3 or ($\pm$)AbII or both solutions were then added to the appropriate dishes to bring the grains to 55% moisture content. The dishes were placed at 3°C for 5 weeks, distilled water being added weekly to maintain the original water content. Following vernalization, the grains were transplanted into pots of soil in a glasshouse. During plant growth the temperature of the glasshouse was maintained above 15°C, and the plants were illuminated continuously with incandescent light (*c*. 100 f.c.). Plants were harvested after 3 and 7 weeks growth and the apices were dissected to observe the extent of floral development.

III. Results

(a) Lettuce Germination

Although the nature of the response is still uncertain, lettuce seed germination has been shown to be influenced by a variety of factors including red light, gibberellins, and cytokinins. Thus the seed provides a valuable system for evaluating abscisin as an inhibitor of the action of several promotive agents. The cultivar used, Great Lakes, normally germinates well in darkness and is unresponsive to red light unless the seeds are previously inhibited by exposure to far-red light.

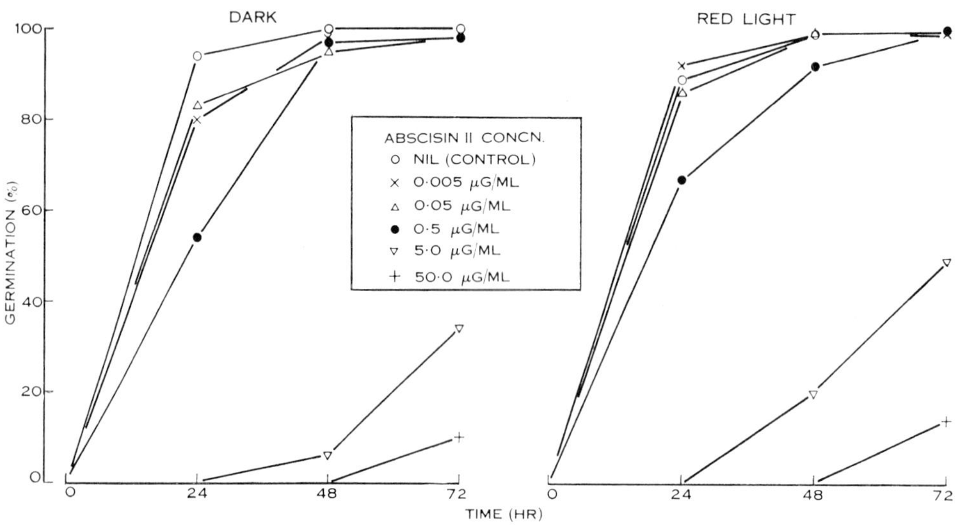

Fig. 1.—(±)-Abscisin II and the germination of lettuce (cv. Great Lakes) seed in the dark or in continuous red light.

Seeds were placed in red light or darkness at various concentrations of (±)AbII and germination was recorded daily under green light for 3 days. Under both conditions the lower concentrations slightly retarded germination (Fig. 1). At the higher concentrations germination was strongly inhibited but there was some germination (12%) by 72 hr, suggesting that the inhibition was not permanent. There was no evidence of an interaction between the effects of (±)AbII and red light.

An attempt was made to associate the inhibitory action of (±)AbII with definite periods of the germination process. Imbibed seeds were placed under far-red light for 48 hr to inhibit germination and some treatments were then illuminated for 5 min with red light. All treatments were then placed in the dark for a further 24 hr. During the experiment the seeds were exposed either for long periods (48–72 hr) to 0·5 μg/ml (±)AbII or for short periods (2 hr) to a higher concentration (5 μg/ml). Table 1 gives a description of the treatments and the resulting germination. Far-red light drastically reduced germination. (±)AbII at both concentrations and at all times significantly inhibited germination unless applied after 50 hr. This indicates that the abscisin-sensitive process(es) are probably completed by 2–4 hr after red light exposure. The inhibition caused by (±)AbII applied prior to the red light may indicate that the substance was not completely removed by the washing procedure or that the inhibited seed only partially recovered the ability to germinate.

The interaction of the effects of AbII with GA_3 or kinetin was explored by exposing seed to far-red light for 24 hr in the presence of (±)AbII and GA_3 or kinetin and then placing them in the dark for a further 24 hr. High (±)AbII levels completely

Table 1

(±)-ABSCISIN II AND SOME HORMONE-REGULATED PLANT RESPONSES

Germination of lettuce (cv. Great Lakes) seeds following different periods of exposure to (±)-abscisin II. All seeds were held for 48 hr in far-red light followed in some cases by 5 min illumination by red light and then a further 24 hr in darkness

Red Light Treatment	Period of Exposure to (±)-Abscisin II (hr from start of expt.)	(±)-Abscisin II Concentration (μg/ml)	Germination at 72 hr (%)
−	0	0	8·6
−	0–72	0·5	5·4
−	0–48	0·5	9·2
+	0	0	76·1
+	0–72	0·5	60·2 ‡
+	0–48	0·5	61·7 ‡
+	0–48*	0·5	63·5 ‡
+	46–48	5	64·3 ‡
+	47–49	5	65·3 ‡
+	48–50†	5	62·0 ‡
+	50–52	5	70·9 §
+	52–54	5	77·9 §

* Including red light illumination.

† Excluding red light illumination.

‡ These values significantly different ($P < 0 \cdot 05$) from value obtained in red light but in absence of (±)AbII.

§ These values not significantly different from value obtained in red light but in absence of (±)AbII.

suppressed the response to all GA_3 concentrations (Fig. 2). Lower levels were less effective. Kinetin was more efficient than GA_3 in reversing the far-red inhibition of germination (Fig. 3). The two lower (±)AbII concentrations showed no significant counteraction of the kinetin promotion although an intermediate concentration was slightly effective, and the highest concentration suppressed the response to kinetin.

(*b*) *Lettuce Hypocotyl and Radicle Extension*

The inhibition of lettuce hypocotyl elongation by light can be overcome by GA_3. On the other hand, radicle elongation is relatively insensitive to GA_3 but is promoted by an allied compound, allogibberic acid (Paleg *et al.* 1964). The interaction of (±)AbII with GA_3 and allogibberic acid in the control of these responses provided information on the comparative effectiveness of (±)AbII and led to a valuable comparison with its effects on germination.

Hypocotyl elongation produced by high levels of GA_3 (Fig. 4) was inhibited only 20% by the highest level of (±)AbII. There was no response to (±)AbII in the absence of GA_3.

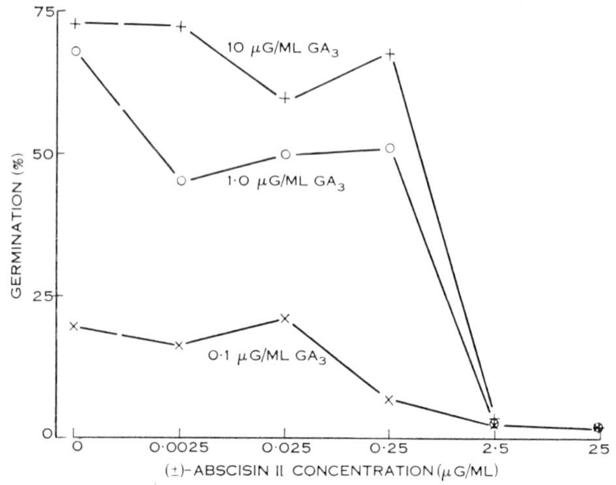

Fig. 2.—Interaction between gibberellic acid (GA_3) at the concentrations indicated and (±)-abscisin II in the control of far-red inhibited lettuce seed germination.

The intermediate, but not the high, GA_3 concentration stimulated radicle elongation when compared with water controls [19·7 and 16·8 mm for GA_3 concentrations of 1 μg/ml and 100 μg/ml respectively; water control, 15·4 mm

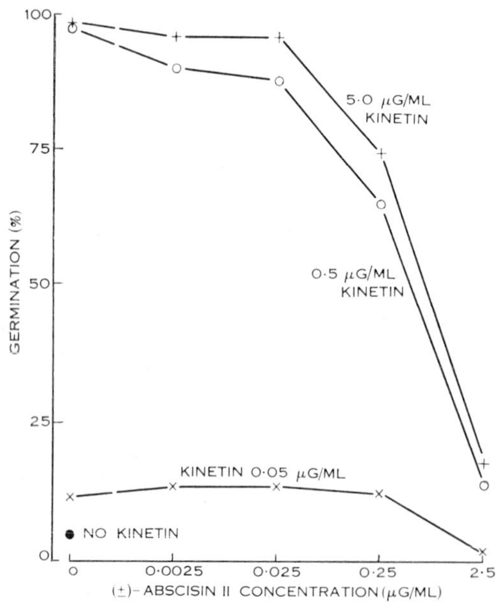

Fig. 3.—Interaction between kinetin and (±)-abscisin II in the control of far-red inhibited lettuce seed germination.

significant difference 1·5 mm ($P = 0{\cdot}05$)—see Fig. 5]. With both the water controls and the GA_3-treated lettuce seedlings, (±)AbII stimulated radicle elongation at the lower and retarded it at the highest concentration, and there was no evidence of a statistically significant interaction between (±)AbII and GA_3. Radicle elongation was inhibited rather more by the (±)AbII than hypocotyl elongation in the same seedlings, but this may be merely a manifestation of the greater growth potential of the radicle.

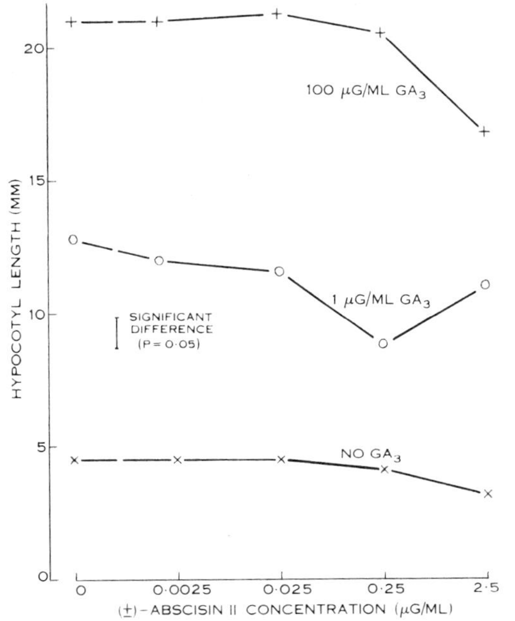

Fig. 4.—Interaction between gibberellic acid (GA_3) at the concentrations indicated and (±)-abscisin II in the control of hypocotyl elongation of lettuce grown in light.

In an initial experiment with allogibberic acid and (±)AbII, lettuce seeds were germinated in the dark for 48 hr before being transferred to the test solutions and incubated in the light for a further 3 days. With both radicles and hypocotyls,

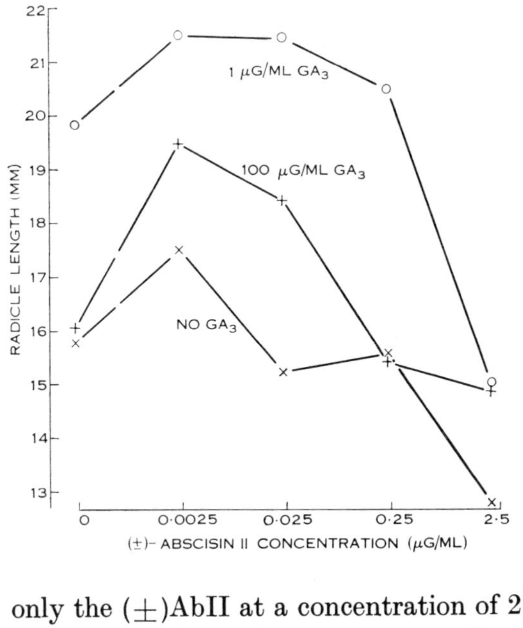

Fig. 5.—Interaction between gibberellic acid (GA_3) at the concentrations indicated and (±)-abscisin II in the control of radicle elongation of lettuce grown in light.

only the (±)AbII at a concentration of 2·5 μg/ml inhibited elongation. In a further experiment, the seeds were grown in the light throughout in an attempt to maximize the effects of allogibberic acid. Allogibberic acid at higher concentrations stimulated radicle elongation (Fig. 6), but the higher concentrations of (±)AbII were inhibitory. In this case, low concentrations of both allogibberic and (±)AbII produced a slight synergistic promotion of elongation.

(c) *Cucumber Hypocotyl and Radicle Elongation*

The interactions between (±)AbII and GA_3 and allogibberic acid in the lettuce seedling suggested that exploration of a similar system sensitive to another gibberellin also might be worth while. Accordingly the effects of (±)AbII on the growth of

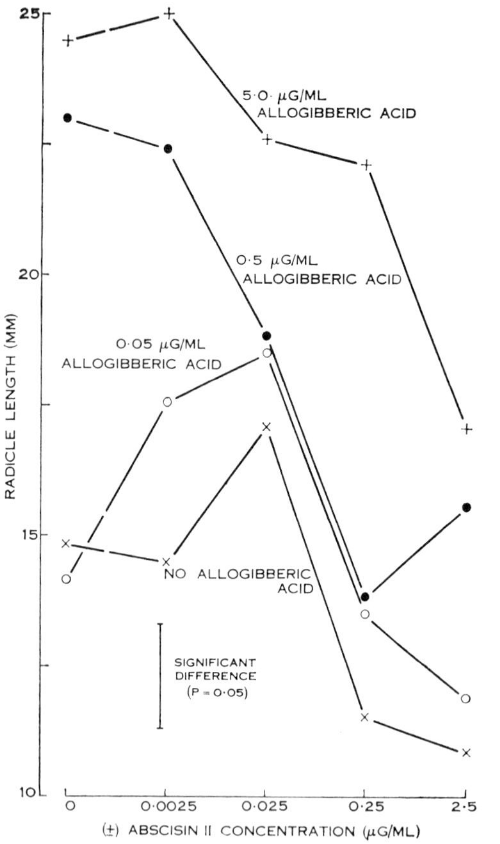

Fig. 6.—Interaction between allogibberic acid and (±)-abscisin II in the control of radicle elongation of lettuce grown in light.

cucumber seedlings in the presence of a mixture of GA_4 and GA_7 was examined. This mixture produced significant hypocotyl elongation (Fig. 7) and, in this system also, the gibberellins interacted with (±)AbII. In the absence of, or at low concentrations of, the gibberellins, the abscisin slightly inhibited elongation. However, at higher gibberellin concentrations, it markedly promoted elongation. This response is in contrast with the effect of GA_3 and (±)AbII on lettuce hypocotyls (Fig. 4).

Cucumber radicle elongation (Fig. 8) was even more markedly stimulated by low levels of (±)AbII than was cucumber hypocotyl elongation. The interaction between the abscisin and the gibberellins did not attain statistical significance although, as with the hypocotyls, the promotive effects of the abscisin were most marked in the presence of gibberellins; indeed, in its absence, elongation was inhibited by the gibberellins. In this tissue, then, we have the interesting situation of abscisin overcoming the growth *inhibiting* effects of gibberellins.

(*d*) *Senescence of Leaf Disks*

The experiments with lettuce germination showed a kinetin-promoted response inhibited by abscisin. Therefore, it was of interest to confirm this result in a very different system, sensitive to cytokinin, but in which growth and cell division are

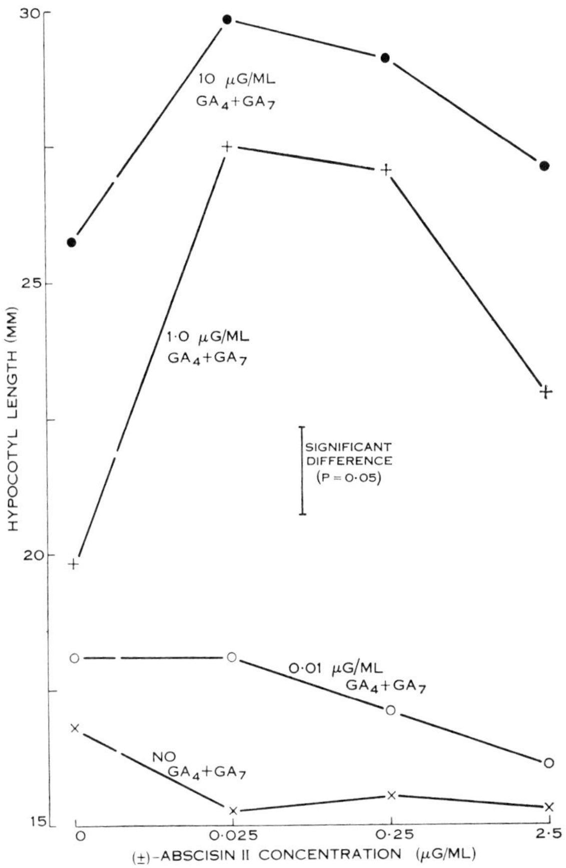

Fig. 7.—Control of cucumber hypocotyl extension by GA_4+GA_7 mixture and (±)-abscisin II.

apparently not involved. Accordingly the interaction of the effects of kinetin and of (±)AbII on chlorophyll retention in radish leaf disks was examined (Loeffler and van Overbeek 1964).

In the initial experiment leaf disks were incubated for 96 hr in the dark at 20°C. There was considerable senescence in the disks incubated in water alone (Fig. 9) and this was not greatly increased by (±)AbII, possibly because chlorophyll loss was virtually complete even in the control series. Kinetin led to the retention of chlorophyll; this response was counteracted by (±)AbII at low kinetin concentrations but not at the higher kinetin concentrations. Thus it appears that the acceleration of senescence by (±)AbII can be overcome by kinetin. In a further experiment over a shorter time interval (48 hr), kinetin alone had only a small effect on chlorophyll retention (Fig. 10) as little senescence occurred in the control disks. However, (±)AbII reduced chlorophyll retention at low kinetin levels, confirming the results of the first experiment.

In view of the evidence that abscisin can produce GA_3 sensitivity in tissue usually unresponsive to gibberellin (Thomas *et al.* 1965), the response of leaf disks incubated for 48 hr in (±)AbII and GA_3 was explored. (Fig. 11). GA_3 clearly over-

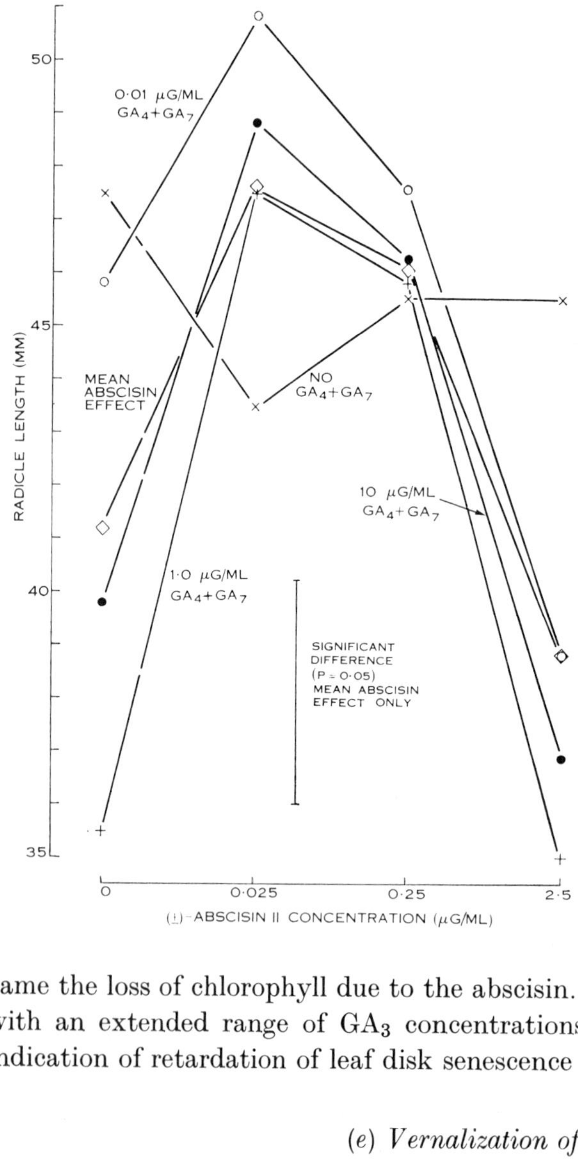

Fig. 8.—Control of cucumber radicle elongation by GA_4+GA_7 mixture and (±)-abscisin II.

came the loss of chlorophyll due to the abscisin. In this and in a further experiment with an extended range of GA_3 concentrations incubated for 72 hr, there was no indication of retardation of leaf disk senescence by GA_3 in the absence of (±)AbII.

(*e*) *Vernalization of Barley*

The growth of barley seedlings shortly after emergence was considerably affected by pretreatment with GA_3 (e.g. first leaf length: 0 GA_3, 11·7 cm; 1 μg GA_3, 15·0 cm; 10 μg GA_3, 15·2 cm; 100 μg GA_3, 17·0 cm) but there was no measurable effect of (±)AbII. This response pattern continued throughout the growth of the plants. At the final harvest (7 weeks after sowing), (±)AbII had had no consistent effect on any of the parameters measured: stage of development, number of leaves on main axis, apex length, and stem length. Although GA_3 had very little effect on stage of inflorescence development or the number of leaves on the main axis, apex length was reduced (0 GA_3, 35 mm; 100μ g/ml GA_3, 30 mm) as was stem length

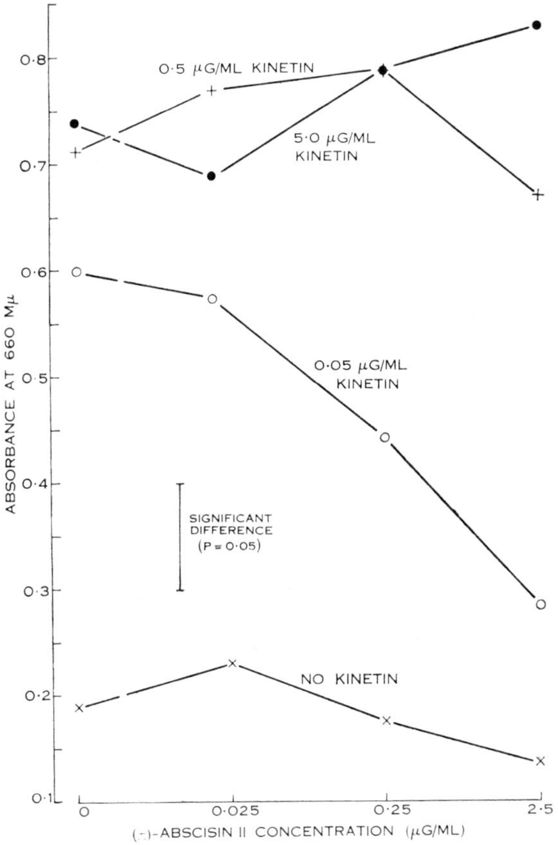

Fig. 9.—Interaction between kinetin and (±)-abscisin II in the retardation of chlorophyll degradation in excised radish leaf disks. Disks incubated for 96 hr in darkness.

(0 GA_3, 34·2 cm; 100 μg/ml GA_3, 29·7 cm). These small, but statistically significant, effects of high GA_3 concentration resemble the inhibitory effects of GA_3 on flowering

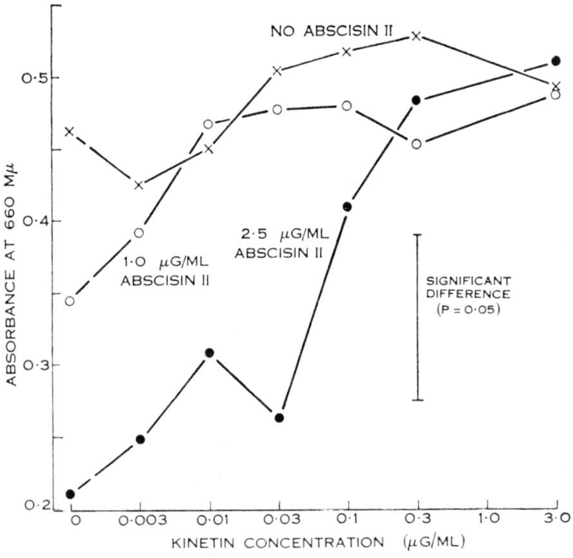

Fig. 10.—Interaction between kinetin and (±)-abscisin II in the retardation of chlorophyll degradation in excised radish leaf disks. Disks incubated for 48 hr in darkness.

in barley previously described (Paleg and Aspinall 1958). There was no evidence that these treatments with (±)AbII or GA_3 had any effect on the vernalization process.

IV. Discussion

The present results add considerably to the spectrum of responses to (±)AbII, particularly the inhibition responses. In our experiments (±)AbII inhibited processes promoted by GA_3 (lettuce hypocotyl growth), other gibberellins (cucumber hypocotyl growth), gibbanes (lettuce radicle growth), gibberellin or cytokinin (lettuce germination), or cytokinin alone (radish leaf senescence). This wide range of inhibitory activity may indicate that (±)AbII does not specifically inhibit one particular hormone-controlled process. The earlier suggestion that the substance acts as a gibberellin antagonist *in vivo* (Thomas *et al.* 1965), although attractive, would seem to be inadequate in view of this range of activity. Although it is possible that a

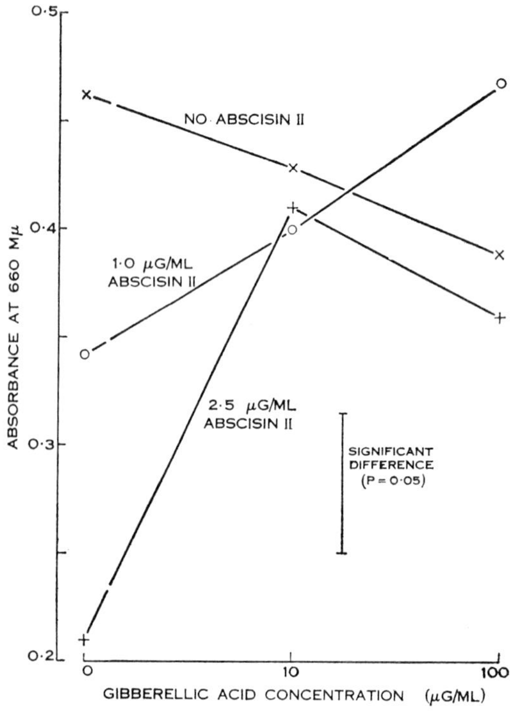

Fig. 11.—Interaction between gibberellic acid and (±)-abscisin II in the retardation of chlorophyll degradation in excised radish leaf disks. Disks incubated for 48 hr in darkness.

gibberellin-controlled step is involved in the responses mediated by the other hormones, such an explanation would require that the effects of abscisin be reversed by gibberellin only. This is not so in radish leaf senescence where both kinetin and GA_3 are effective.

The interactions between (±)AbII and the promoting hormones were of two types in the processes we investigated. In (1) GA_3-controlled lettuce germination and (2) kinetin-controlled leaf senescence, the effects of low concentrations of (±)AbII were completely overcome by high concentrations of the promoting substance. In the second type of interaction, (±)AbII was inhibitory only in the presence of high concentrations of the promoter in (1) lettuce germination controlled by kinetin, and (2) lettuce-radicle elongation controlled by allogibberic acid. Similar interactions have been reported by Thomas *et al.* (1965); the first type in the elongation of tall maize leaf sections, and the release of reducing sugars from the barley endosperm; the second in pea seedling elongation. It is apparent that neither the process nor the growth promoter involved is specific to the type of abscisin interaction.

Abscisin II is not unique as a naturally occurring growth-inhibiting substance; indeed many such substances have been reported (Evanari 1949). Of these inhibitors coumarin and naringenin are perhaps best known, and it is of interest to compare their activities with those of (±)AbII. Coumarin, which has been isolated from a number of plants (Evanari 1949), will inhibit germination in lettuce seeds, and this inhibition can be overcome by light (Nutile 1945), GA_3 (Mayer 1959), a combination of red light and kinetin (Khan and Tolbert 1965), kinetin alone (Khan and Tolbert 1965), or cycocel (Khan and Tolbert 1966), depending upon the variety of lettuce and the conditions of the experiment. On the other hand, coumarin inhibition of lettuce hypocotyl extension does not seem to be reversed by GA_3 (Mayer 1959; Phillips 1962). Coumarin will also inhibit the release of reducing sugars from cereal endosperm under the influence of GA_3, this inhibition not being reversed by higher GA_3 concentrations (Paleg 1963). Coumarin is not as potent an inhibitor of lettuce germination as (±)AbII; 10^{-4}M coumarin reduced germination in the dark to 13·5% (Mayer 1959) whereas 2×10^{-6}M (±)AbII reduced germination to 6% in the present experiments.

Naringenin, another flavonoid inhibitor, has been found in dormant peach flower buds (Hendershott and Walker 1959). The interactions between naringenin and auxin and gibberellin have been investigated by Phillips (1962) using several growth assays.

In discussing the role of inhibitors in plant processes, stress is frequently laid on a specific competitive action of the inhibitor under discussion. In this connexion (±)AbII, coumarin, and naringenin each seem to approach a competitive relationship with GA_3 in the control of lettuce germination, at least insofar as the effects of a low abscisin concentration can be overcome by higher GA_3 concentrations. On the other hand, lettuce hypocotyl extension was inhibited in a non-competitive manner by each compound. It would appear that competitive or non-competitive inhibition is as much a characteristic of the biological system under consideration as of the precise inhibitor-hormone combination. This is particularly evident since the effects of (±)AbII in leaf disk senescence can be overcome by a completely different substance: kinetin.

Though there are obvious similarities in the actions of the three inhibitors there are also evident differences. (±)AbII seems to be effective at considerably lower concentrations than the others, and its structure suggests that its biosynthetic pathway involves the linking together of isoprenoid units. These features emphasize the possibility that, in fact, AbII may be the forerunner of a new group of biologically important controlling substances.

Further, in complex systems, such as hormone-promoted growth responses, where the parameter measured may be a multiplicity of steps removed from the mechanism of hormone action, the steps at which an inhibitor may produce an effect are numerous, and there is no *a priori* reason for believing that an inhibitor will always produce its effect at the same step in all responses elicited by that hormone. Thus, in considering the role of an endogenous growth inhibitor such as abscisin, it may even be too restrictive to link its action solely with that of one growth promoter.

V. References

ADDICOTT, F. T., CARNS, H. R., LYON, J. L., SMITH, O. E., and MCMEANS, J. L. (1964).—On the physiology of abscisins. In "Régulateurs Naturels de la Croissance Végétale." pp. 687–703. (Cent. natn. Rech. scient.: Paris.)

BENTLEY, J. A. (1958).—The naturally-occurring auxins and inhibitors. *A. Rev. Pl. Physiol.* **9**, 47–80.

CHRISPEELS, M. J., and VARNER, J. E. (1966).—Inhibition of gibberellic acid induced formation of α-amylase by abscisin II. *Nature, Lond.* **212**, 1066–7.

CORNFORTH, J. W., MILBORROW, B. V., and RYBACK, G. (1965).—Synthesis of (±)-abscisin II. *Nature, Lond.* **206**, 715.

CORNFORTH, J. W., MILBORROW, B. V., and RYBACK, G. (1966).—Identification and estimation of (+)-abscisin II (dormin) in plant extracts by spectropolarimetry. *Nature, Lond.* **210**, 627–8.

CORNFORTH, J. W., MILBORROW, B. V., RYBACK, G., and WAREING, P. F. (1965).—Identity of sycamore 'dormin' with abscisin II. *Nature, Lond.* **205**, 1269–70.

EAGLES, C. F., and WAREING, P. F. (1963).—Dormancy regulators in woody plants. *Nature, Lond.* **199**, 874.

EAGLES, C. F., and WAREING, P. F. (1964).—The role of growth substances in the regulation of bud dormancy. *Physiologia Pl.* **17**, 697–709.

EL-ANTABLY, H. M. M., WAREING, P. F., and HILLMAN, J. (1967).—Some physiological responses to D,L,-abscisin (dormin). *Planta* **73**, 74–90.

EVANS, L. T. (1966).—Abscisin II: inhibitory effect on flower induction in a long-day plant. *Science, N.Y.* **151**, 107–8.

EVANARI, M. (1949).—Germination inhibitors. *Bot. Rev.* **15**, 153–86.

HENDERSHOTT, C. H., and WALKER, D. R. (1959).—Identification of a growth inhibitor from extracts of dormant peach flower buds. *Science, N.Y.* **130**, 798–800.

KHAN, A. A., and TOLBERT, N. E. (1965).—Reversal of inhibitors of seed germination by red light plus kinetin. *Physiologia Pl.* **18**, 41–3.

KHAN, A. A., and TOLBERT, N. E. (1966).—Light-controlled cycocel reversal of coumarin inhibition of lettuce seed germination and root growth. *Physiologia Pl.* **19**, 76–80.

LOEFFLER, J. E., and OVERBEEK, J. VAN (1964).—Kinin activity in coconut milk. In "Régulateurs Naturels de la Croissance Végétale." pp. 77–82. (Cent. natn. Rech. scient.: Paris.)

MAYER, A. M. (1959).—Joint action of gibberellic acid and coumarin in germination. *Nature, Lond.* **184**, 826–7.

NUTILE, G. E. (1945).—Inducing dormancy in lettuce seed with coumarin. *Pl. Physiol.* **20**, 433–42.

OHKUMA, K., LYON, J. L., ADDICOTT, F. T., and SMITH, O. E. (1963).—Abscisin II, an abscission-accelerating substance from young cotton fruit. *Science, N.Y.* **142**, 1592–3.

PALEG, L. (1963).—The influence of coumarin and gibberellic acid on cereal endosperm. Proc. Int. Symp. Physiology, Ecology, and Biochemistry of Germination, Greifswald.

PALEG, L. G., and ASPINALL, D. (1958).—Inhibition of the development of the barley spike by gibberellic acid. *Nature, Lond.* **181**, 1743–4.

PALEG, L., ASPINALL, D., COOMBE, B., and NICHOLLS, P. (1964).—Physiological effects of gibberellic acid. VI. Other gibberellins in three test systems. *Pl. Physiol.* **39**, 286–90.

PALEG, L., KENDE, H., NINNEMANN, H., and LANG, A. (1965).—Physiological effects of gibberellic acid. VIII. Growth retardants on barley endosperm. *Pl. Physiol.* **40**, 165–9.

PHILLIPS, I. D. J. (1962).—Some interactions of gibberellic acid with naringenin (5,7,4′-trihydroxy flavanone) in the control of dormancy and growth in plants. *J. exp. Bot.* **13**, 213–26.

THOMAS, T. H. WAREING, P. F., and ROBINSON, P. M. (1965).—Action of the sycamore 'dormin' as a gibberellin antagonist. *Nature, Lond.* **205**, 1270–2.

Growth and Cellular Differentiation in the Wheat Coleoptile (*Triticum vulgare*)

II. FACTORS INFLUENCING THE GROWTH RESPONSE TO GIBBERELLIC ACID, KINETIN AND INDOLYL-3-ACETIC ACID

S. T. C. WRIGHT

Agricultural Research Council, Plant Growth Substance and Systemic Fungicide Unit, Wye College, University of London, Nr. Ashford, Kent

WITH THREE FIGURES IN THE TEXT

Received 5 May 1965

ABSTRACT

The sequential growth response of excised wheat coleoptiles during growth and cellular differentiation to solutions of gibberellic acid, kinetin, and indolyl-3-acetic acid is further analysed in this paper.

Excised young coleoptiles which possess a low potential growth-rate (controls) respond to gibberellic acid and kinetin, whereas excised older coleoptiles which have a high potential growth-rate respond to indolyl-3-acetic acid. The difference between these results and those reported for pea-stem sections is discussed.

To obtain a well-defined separation of the sequential growth response to gibberellic acid and kinetin it is necessary to grow the wheat coleoptiles at 22·5° C. At a higher temperature (25° C) the two optimal responses tend to overlap. It is suggested that this difference is due to a delayed onset of cell division at the lower of the two temperatures.

Substances diffusing from the primary leaves contained within the excised coleoptile do not interfere with the sequential growth response of the coleoptile. Differential uptake of the growth regulators into the older coleoptiles does not explain the lower growth response stimulated in these coleoptiles by gibberellic acid and kinetin as compared to the high activity of indolyl-3-acetic acid.

The possible importance of gibberellic acid in primary cell-wall growth is discussed.

INTRODUCTION

EARLIER publications (Wright, 1961 *a*, *b*) showed that the growth of the intact wheat coleoptile can be divided into three phases: (i) an early period of cell enlargement (approx. 0–30 hours after sowing), (ii) a period of cell enlargement accompanied by cell division (approx. 30–60 hours after sowing), and (iii) a final period of cell enlargement (approx. 60–120 hours after sowing). Details were also given of how the wheat coleoptile responds sequentially to gibberellic acid, kinetin, and indolyl-3-acetic acid (IAA)

Journal of Experimental Botany, Vol. 17 No. 50, pp. 165–76, February 1966.

during its growth and cellular differentiation (Wright, 1961*a*). The greatest sensitivity to gibberellic acid was found to be correlated with the period of cell enlargement which precedes cell division, whereas kinetin elicited the greatest response from coleoptiles in which many of the cells were about to divide or were dividing. IAA seemed to be important in the final cell-enlargement phase. In this paper it is proposed to enlarge upon the practical details of these experiments and to give the results of experiments which were carried out to test the validity of the sequential growth response.

Since the earlier work was published, van Overbeek (1962) has postulated that these three classes of growth regulators may be concerned in a sequential manner in the growth of fruit.

MATERIALS AND METHODS

Coleoptile culture method. Wheat grains (var. Eclipse) were soaked for two hours in distilled water and then sown, with their embryos oriented uppermost, on moist filter paper in 9-cm Petri-dishes. The Petri-dishes (without their lids) were transferred to a box ($25 \times 21 \times 9$ in.) with a humid atmosphere (90 per cent relative humidity) and maintained at 22·5° C in darkness until required.

Method of testing response to growth-regulating substances. Whole coleoptiles were excised using the epiblast as a guide (Wright, 1961*b*) and the coleoptiles placed on moist filter paper until sufficient numbers had been cut for the experiment. Coleoptiles were measured under a binocular microscope with a micrometer eyepiece or with a millimetre rule to the nearest 0·5 mm, depending on age. Five coleoptiles were placed on a 2×1 in. strip of Whatman No. 1 filter paper and inserted in a $3 \times \frac{1}{2}$ in. test-tube containing 0·6 ml of the test solution. (The purpose of the filter-paper strip was to prevent the small coleoptiles from drying out on the sides of the tube.) Five, occasionally four, replicates of five coleoptiles were used per treatment. The tubes were sealed with 'Parafilm' in which a small hole was pierced near the centre to allow ventilation. The tubes were rotated at 1 r.p.m. on a klinostat in dim red light (emission > 645 mμ) at 26° C for 24 hours. From the mean length of the five coleoptiles before and after treatment the percentage increase in mean length of the coleoptiles was calculated for each replicate (i.e. percentage increase over original length).

Method of preparing growth-substance solutions. An aqueous solution of gibberellic acid (10^{-4} M $= 3{\cdot}46$ mg/100 ml) was stored at 2° C for up to four weeks. Kinetin crystals do not dissolve at room temperature and it was necessary to heat the solution in a water bath at 95° C for five minutes. Fresh solutions were prepared for each test since the solutions crystallized out in the refrigerator ($3{\cdot}3 \times 10^{-4}$ M $= 7{\cdot}16$ mg/100 ml). IAA was kept as a 10^{-3} M aqueous solution of the sodium salt at 2° C for up to four weeks (17·5 mg IAA crystals dissolved in distilled water with 1 ml M/10 NaOH and made up to 100 ml).

RESULTS

(i) *Analysis of the sequential growth response data.* In earlier experiments (Wright, 1961*a*) dosage-response curves for gibberellic acid, kinetin, and IAA were determined for a series of coleoptile ages for each growth regulator from 18 to 78 hours after sowing. Fig. 1*a* and 1*b* show dosage-response

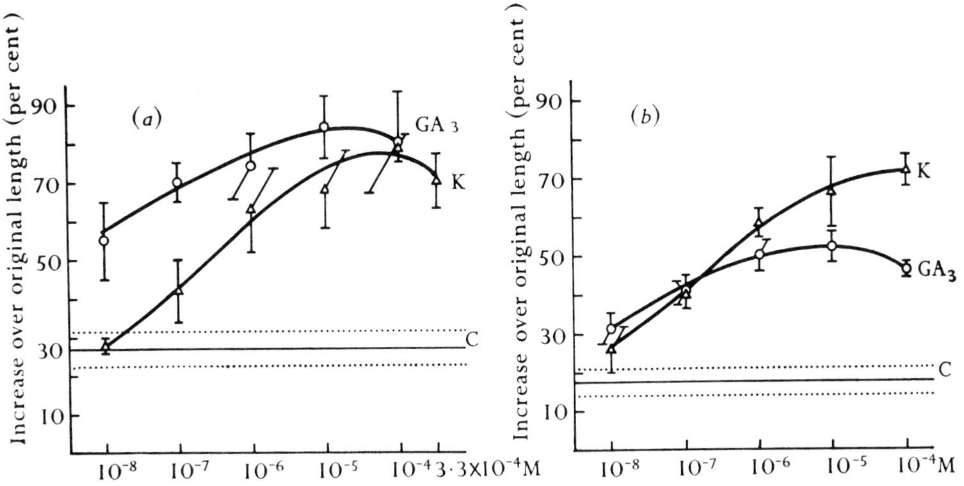

FIG. 1. Dosage-response curves for gibberellic acid and kinetin of coleoptiles excised (*a*) 18 hours after germination, (*b*) 36 hours after germination. The vertical lines indicate the 2 S.E. limits.

○ = gibberellic acid; Δ = kinetin)

curves for gibberellic acid and kinetin of coleoptiles excised 18 hours and 36 hours after sowing (dosage-response curves for IAA are not shown) and by referring to such curves it was possible to determine the concentrations giving the optimal growth response over a range of coleoptile ages for each growth regulator. It was found that there was little change in the optimal concentration of each of the three growth regulators with increasing age of the coleoptile. The optimal concentrations were 10^{-5} M for gibberellic acid, 10^{-4} M for kinetin and 10^{-4} M for IAA. By using only the optimal concentration for each growth regulator it was possible to determine the growth response of the coleoptile at 18, 30, 42, 54, 66, and 78 hours after sowing in one large-scale experiment and thereby eliminate any inter-experimental variation in coleoptile sensitivity (Wright, 1961*a*). The results which show the sequential growth response of the excised wheat coleoptile to gibberellic acid, kinetin, and finally IAA are reproduced again in Fig. 2*b*, with the growth response expressed as percentage increment over control.[1] When the results are plotted as percentage increase over original length they appear as in Fig. 2*a*. This figure shows clearly that the large growth responses to gibberellic acid and kinetin which occur in young coleoptiles (Fig. 2*b*) are partly

[1] Percentage increment over control = percentage increase over original length of treated coleoptiles − percentage increase over original length of control coleoptiles.

a reflection of the low potential growth-rate of the excised coleoptile at this stage as illustrated by the growth of the water controls (Fig. 2*a*).

(ii) *The reproducibility of the sequential growth response to gibberellic acid, kinetin, and IAA.* The sequential growth response illustrated in Fig. 2*b* has been repeated ten times in this laboratory, using wheat (var. Eclipse) from the harvests of several different years. The results have always shown that

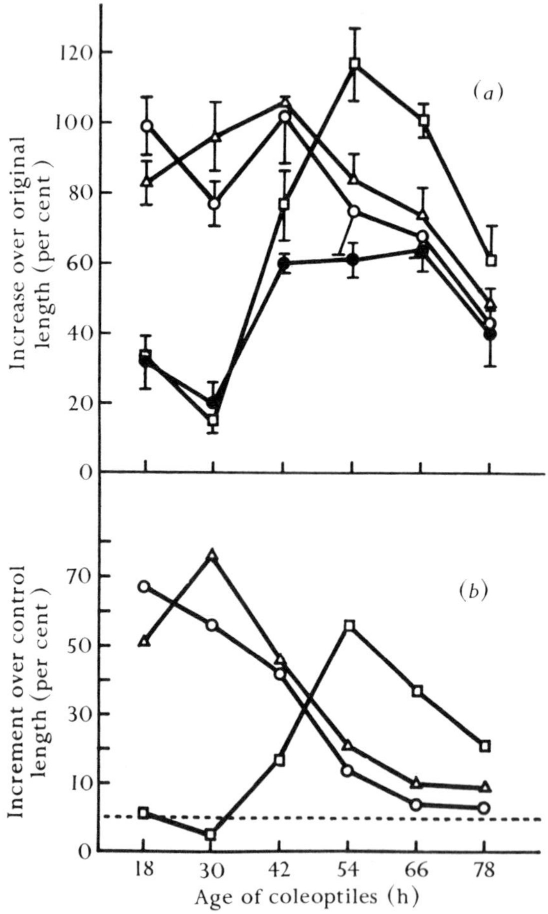

FIG. 2. The response of the wheat coleoptile to exogenous solutions of gibberellic acid, kinetin, and indolyl-3-acetic acid: (*a*) with growth responses expressed as percentage increase over original length; and (*b*) growth responses expressed as percentage increment over control length, to illustrate the sequential growth response. The vertical lines indicate the 2 S.E. limits.

● = water controls; ○ = gibberellic acid, 10^{-5} M; △ = kinetin, 10^{-4} M; □ = indolyl-3-acetic acid, 10^{-4} M)

a response to gibberellic acid and kinetin precedes a response to IAA. However, the time elapsing between the gibberellic acid and kinetin peaks has been found to vary and on some occasions the two peaks have almost coincided. The reason for this variation in the time of these peak responses has been partially elucidated.

Two treatments which had no effect on the position in time of the gibberellic acid and kinetin peaks were (i) exposure to red light as compared with darkness during the period of treatment in the klinostat, (ii) ventilation of the test-tubes containing the coleoptiles by piercing a hole in the centre of the 'Parafilm' cover, compared with no ventilation in the klinostat. It must be noted that there was a need for ventilation in tubes containing the larger coleoptiles (i.e. 60 hours and above) to obtain a maximum response to IAA.

The third treatment was a comparison between two growing temperatures. Normally, all coleoptiles were grown at 22·5° C (see Methods) and these

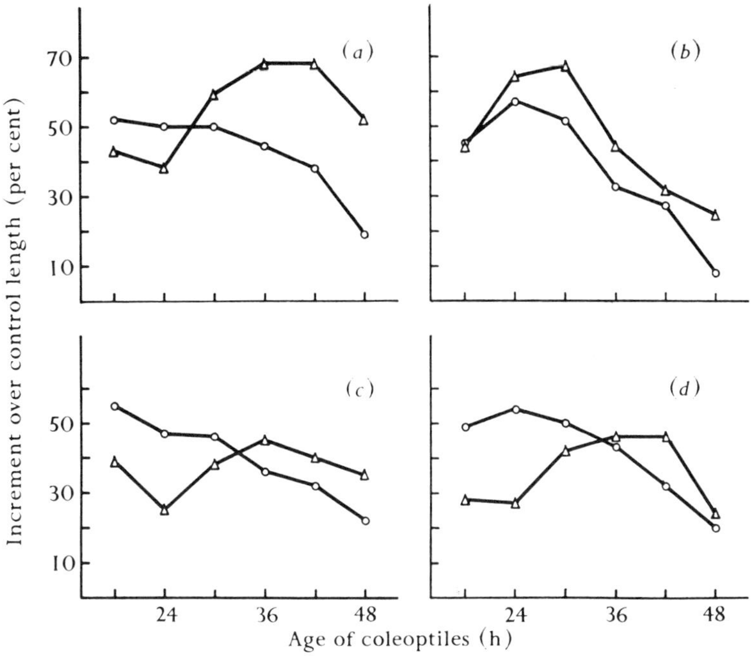

FIG. 3. The reproducibility of the gibberellic acid and kinetin sequence (*a*), (*c*), and (*d*) with coleoptiles grown at 22·5° C; (*b*) with the coleoptiles grown at 25° C.
○ = gibberellic acid, 10^{-5} M; Δ = kinetin, 10^{-4} M)

when excised gave a peak response to kinetin somewhere between 36 and 42 hours after sowing (Fig. 3*a*). When the coleoptiles were grown at 25° C, however, the kinetin peak response was found to occur between 24 and 30 hours after sowing and almost overlapped the gibberellic acid peak (Fig. 3*b*). (In the experiment represented in Fig. 2*b* the coleoptiles were grown at 23° C and the peak response to kinetin was at 30 hours.) Thus the growing temperature was found to be very critical and a difference of 2·5° C produced distinctly different results. Fig. 3*c* and 3*d* represent repeat experiments of the experiment in Fig. 3*a*, and as can be seen there was generally a falling off in the growth response stimulated by gibberellic acid with increasing coleoptile age (except in Fig. 3*d* where there was a peak response at twenty-four hours after sowing). Kinetin gave a peak response at approximately 36 hours after sowing in all three experiments.

(iii) *Effect of substances diffusing from the primary leaves enclosed within the coleoptile on the growth response of the coleoptile to gibberellic acid, kinetin, and IAA.* A possible criticism of the experiments which demonstrated the sequential growth response concerns the fact that the primary leaves were retained within the coleoptiles during the period of growth-substance treatment. For technical reasons the leaves were retained since it was found to be impossible, except with very young coleoptiles (i.e. 18 hours after sowing), to remove the leaves without damaging the coleoptiles. Now it is conceivable that during the early growth of the primary leaves within the coleoptile a substance is produced at a certain growth stage which could diffuse into the medium when excised and react synergistically or antagonistically with a growth regulator in coleoptile growth. For instance if a synergist to kinetin was produced by the leaves at 36 hours after sowing it would lead to the observed optimal growth response of the excised coleoptile to kinetin at this stage. The following experiments were carried out to determine whether such substances diffuse from the primary leaves and influence the response of the excised coleoptile to the growth regulators. These determinations were made at the three critical stages of coleoptile growth (i.e. 18 hours after sowing) which is the stage responding most actively to gibberellic acid; 36 hours after sowing when the coleoptile is most sensitive to kinetin (Fig. 3*a*, *c*, and *d*); and 66 hours after germination when the coleoptile responds to IAA.

With the youngest coleoptiles (18 hours after sowing) four treatments were given, (i) the primary leaves were left undisturbed, (ii) the primary leaves were teased out of the coleoptile but retained in the test-tube with the treatment solution, (iii) the primary leaves were teased out and discarded, and (iv) as (iii) except that leaves from five 36-hour coleoptiles were placed in the test-tube. The last treatment was included to see if any substance diffusing from 36-hour leaves might cause the 18-hour coleoptile to respond preferentially to kinetin. The results shown in Table 1 indicate that the response of the coleoptiles to gibberellic acid, kinetin, IAA, and distilled water in treatments ii, iii, and iv was not significantly different from control coleoptiles (i.e. treatment i). When the leaves were teased from the coleoptiles as in treatments ii, iii, and iv the kinetin response was slightly lower than that given by gibberellic acid (i.e. 56, 59, and 57 per cent as compared with 65, 64, and 60 per cent); these sets of figures were significantly different at the 5 per cent level whereas in treatment i although the kinetin response was lower than the gibberellic acid response there was no significant difference.

With coleoptiles excised 36 hours after sowing it was found to be impracticable to tease out the primary leaves since the coleoptiles were too long (i.e. 3 mm). Three treatments were chosen, (i) control coleoptiles with their enclosed primary leaves left undisturbed, (ii) coleoptiles with their enclosed primary leaves plus an extra set of primary leaves dissected from five 36-hour coleoptiles, (iii) coleoptiles with their enclosed primary leaves plus leaves from ten 18-hour coleoptiles. The growth response to gibberellic acid, kinetin, IAA, and distilled water in treatments ii and iii was not significantly different from the controls (i) (Table 1). The growth response to kinetin was always greater than that given by gibberellic acid whatever the treatment and

Table I

Age of coleoptiles when excised	Treatment	Gibberellic acid 10^{-5} M: % increase over original length ± 2 S.E.	Gibberellic acid 10^{-5} M: Significance from treatment (i)	Kinetin 10^{-4} M: % increase over original length ± 2 S.E.	Kinetin 10^{-4} M: Significance from treatment (i)	IAA 10^{-4} M: % increase over original length ± 2 S.E.	IAA 10^{-4} M: Significance from treatment (i)	Water controls: % increase over original length ± 2 S.E.	Water controls: Significance from treatment (i)
18 hr	(i) With enclosed primary leaves (*c*)	67 ± 6·72	. .	62 ± 10·00	. .	23 ± 5·61	. .	21 ± 3·04	. .
	(ii) With primary leaves teased out and retained	65 ± 4·30	N.S.	56 ± 3·28	N.S.	23 ± 4·44	N.S.	24 ± 3·20	N.S.
	(iii) With primary leaves teased out and discarded	64 ± 8·68	N.S.	59 ± 8·36	N.S.	21 ± 5·34	N.S.	29 ± 8·10	N.S.
	(iv) With primary leaves teased out and discarded but leaves from five 36-hr coleoptiles inserted in the test-tube	60 ± 5·14	N.S.	57 ± 7·96	N.S.	25 ± 3·10	N.S.	24 ± 8·62	N.S.
36 hr	(i) With enclosed primary leaves (*c*)	67 ± 8·34	. .	83 ± 6·44	. .	30 ± 10·80	. .	35 ± 6·70	. .
	(ii) With enclosed primary leaves + extra leaves from five 36-hr coleoptiles	73 ± 4·62	N.S.	85 ± 6·16	N.S.	32 ± 8·94	N.S.	30 ± 6·00	N.S.
	(iii) With enclosed primary leaves + extra leaves from ten 18-hr coleoptiles	72 ± 3·96	N.S.	89 ± 7·72	N.S.	24 ± 4·72	N.S.	39 ± 4·50	N.S.
66 hr	(i) Sections (1 cm) with enclosed primary leaves (*c*)	62 ± 4·96	. .	62 ± 2·36	. .	112 ± 5·12	. .	51 ± 2·52	. .
	(ii) Sections with enclosed primary leaves pushed out and discarded	50 ± 0·88	**	56 ± 2·52	*	106 ± 4·48	N.S.	40 ± 2·96	**
	(iii) As (ii), but the coleoptile sections threaded on glass rods	50 ± 3·48	*	40 ± 4·48	***	96 ± 4·64	**	31 ± 3·20	***

* $P = 0·05$; ** $P = 0·01$; *** $P = 0·001$

this again suggested that the retention of the primary leaves was not the cause nor in any way influenced the sequential growth response.

In the larger coleoptile (i.e. 66 hours after sowing) the effect of the primary leaves was investigated by using coleoptile sections. Three treatments were employed using 1-cm sections cut 3 mm from the tip of the coleoptiles, (i) the leaves were left within the sections (controls), (ii) the enclosed leaves were pushed out with the aid of fine glass rods and the leaf sections discarded, (iii) the enclosed leaves were pushed out with glass rods and the coleoptile sections left threaded on them during treatment. The removal of the primary leaves (treatments ii and iii) resulted in a significant reduction in the growth response to all three growth regulators and also in the water controls. When the sections were left threaded on glass rods (treatment iii) the growth response was greatly impaired, particularly in the water- and kinetin-treated sections. Nevertheless, whether the primary leaves were present or absent, IAA was always the most active growth regulator at this stage.

TABLE 2

Treatment of 60-hr coleoptiles		% increase over original length (±2 S.E.)	% increment over control length
Water controls	Whole coleoptiles	59±4·92	0
	Cut in two	87±6·10	0
Gibberellic acid 10^{-5} M	Whole coleoptiles	77±5·80	18
	Cut in two	102±11·20	15
Kinetin 10^{-4} M	Whole coleoptiles	86±6·10	27
	Cut in two	86±10·00	−1
IAA 10^{-4} M	Whole coleoptiles	116±15·06	57
	Cut in two	131±9·00	44

(iv) *Influence of the rate of uptake of the growth regulators by the older coleoptiles on their response to gibberellic acid, kinetin, and IAA.* To test this effect coleoptiles excised 60 hours after sowing were cut in half and the total response of the two halves compared with the response of whole coleoptiles. The results are given in Table 2. The most pronounced difference between the two treatments was in the water controls. Here the cut coleoptiles grew significantly more ($P = 0{\cdot}001$) than whole coleoptiles. Gibberellic acid and IAA gave a greater final length when the coleoptile was cut in two, whereas the kinetin response was the same whether cut or not cut. When the results are expressed as percentage increment over control length the gibberellic acid, kinetin, and IAA responses are higher in the whole coleoptiles.

DISCUSSION

Purves and Hillman (1958) cut sections from Alaska pea stems at various distances from the apical hook and measured their growth response after treatment in solutions of gibberellic acid and IAA. They found sections near the apex responded particularly to gibberellic acid (i.e. young cells), whereas sections more distant (i.e. older cells) were more sensitive to IAA. Furthermore, the sections which were very responsive to gibberellic acid also possessed a high potential ('endogenous') growth-rate whereas those sensitive to IAA had a low potential growth-rate. In the excised wheat coleoptile the pattern of response to gibberellic acid and IAA with increasing cell age was similar to that of pea sections (Fig. 2*b*), but there was a complete reversal as regards the correlation between potential growth-rate and growth-substance activity. Excised wheat coleoptiles with low potential growth-rates responded most actively to gibberellic acid and those with high potential growth-rates responded most actively to IAA (Fig. 2*a*). Thus it would seem that the potential growth-rate of a tissue is not a guide to how it will respond to gibberellic acid or IAA.

The low potential growth-rate of the young excised coleoptile would suggest that a gibberellin and/or a kinin may be severely limiting growth in these early stages. Later when the potential growth-rate increases (i.e. at 42 hours after sowing the excised coleoptile grows 60 per cent above the original length) it begins to respond to IAA. This increase in the potential growth-rate could indicate a build up in the levels of naturally occurring gibberellins and kinins.

When the wheat grain germinates the embryo imbibes water and the coleoptile cells begin to expand. This expansion process can be stimulated in excised young coleoptiles by gibberellic acid. Thus in the wheat coleoptiles and also in the young pea stems studied by Purves and Hillman (1958) gibberellins are probably required before the early growth of cell walls can begin. Other studies which suggest the probable importance of gibberellic acid in early cell-wall growth include those of Wareing (1958), who investigated the effects of gibberellic acid and IAA upon cambial activity of shoots of sycamore, poplar, and ash. Gibberellic acid stimulated cell division in the cambium and gave rise to a zone of new tissue consisting of very thin-walled fusiform cells with no lignification (primary cell walls?). Where IAA alone had been applied, new xylem elements could clearly be discerned on the periphery of the old wood. However, the new elements did not form a continuous ring around the stem but isolated groups of vessels occurred at intervals. When gibberellic acid and IAA were applied together the cambium produced a zone of fully differentiated xylem cells. Thus in these experiments IAA appears to be essential for xylem differentiation whereas gibberellic acid and possibly IAA are required for primary cell-wall growth. Then there is the observation of Pecket (1960) that gibberellic acid leads to an increase in root-hair production on excised pea roots. Root hairs resemble rapidly growing cell walls (i.e. primary cell walls) in respect of their high proportion of pectic substances and these results, like those above, suggest that gibberellic acid may be important in early cell-wall growth.

The role of kinetin in the sequential growth response is rather an enigma since although its optimal activity coincides with the period of cell multiplication, the response that it elicits in the excised coleoptile is one of cell elongation and not cell division (Wright, 1961*a*). If one of the results of kinetin action is to cause a release of energy by the breakdown of food reserves (Wright, 1963), then it is conceivable that in the excised coleoptile this energy is channelled into the cell-elongation process, whereas in the intact coleoptile it might be used in cell division in the presence of some other factor or factors (these we should need to assume diffuse from the coleoptile into the medium when it is excised). The age at which coleoptiles give the maximum kinetin response was found to be altered quite markedly by the germination temperature (Fig. 3*a* and *b*). Haber and Luippold (1960) who have investigated the processes of cell elongation and cell division in the root tips of germinating lettuce seed have shown that a low temperature of germination (10° C) causes cell elongation to precede cell division, whereas, at a higher temperature (26° C) cell division and cell elongation take place together. These findings may be applicable to the wheat coleoptile and explain the shift in the optimal kinetin response when the growing temperature is changed. At the lower temperature (22·5° C) cell elongation may precede cell division, whereas at the higher temperature (25° C) the two processes may overlap.

The experiments detailed in Table 1 suggested that substances diffusing from the primary leaves within the excised whole coleoptiles did not interact with gibberellic acid, kinetin, or IAA during the period of treatment in the klinostat. When the primary leaves were absent from the 66-hour coleoptile sections (Table 1, treatments ii and iii) there was a general lowering of the growth response. This was probably due to unavoidable damage to the coleoptile sections when the leaves were ejected. The effect of increasing the number of cut surfaces by cutting the older coleoptiles in two (60-hour, Table 2) decreased the growth response to all three growth regulators but in particular the response to kinetin (per cent increment over control length). The results suggested that the low growth response produced by gibberellic acid and kinetin in 60-hour coleoptiles was probably not attributable to any difficulty in the rate of uptake of these substances into the older coleoptiles.

When the original paper describing the sequential growth response of the wheat coleoptile to gibberellic acid, kinetin and IAA was published in 1961 there were several reports in the literature of the isolation of naturally occurring kinins from rapidly dividing plant tissues suggesting kinins could be concerned in controlling normal plant growth (some recent references include: Zwar and Skoog, 1963; Letham, 1964; Maïa, 1964; Powell and Pratt, 1964; Zwar *et al.*, 1964). This view has been strengthened recently by the identification of the kinin of sweet-corn kernels as 6-(4:hydroxy-3-methylbut-2-enyl)-aminopurine (Letham *et al.*, 1964; Shaw and Wilson, 1964) and by the studies in which kinin-like substances, capable of delaying leaf senescence, have been shown to be present in the bleeding sap of vines (Loeffler and van Overbeek, 1964) and in root exudates of sunflower seedlings (Kende, 1964). Furthermore, Osborne and Black (1964) have reported the basipetally polar transport of benzyladenine through sections of young petioles of bean, the rate of transport being increased twofold in the presence of IAA. Thus it seems probable that naturally occurring kinins may be transported in plants like true hormones and synthesized in the root rather like the caulocalines postulated by Went (1938), except that kinins would appear to have wider functions than those ascribed to caulocalines.

As can be seen from Fig. 2*b* there is a considerable spread and overlap of the growth response to the three growth regulators with coleoptile age which suggest that the orderly growth and cellular differentiation of the wheat coleoptile is determined by a sequence of growth substance mixtures ever shifting in composition (Wright, 1961*a*). It is intended to examine the effect of mixtures of gibberellic acid, kinetin, and IAA on coleoptile growth.

ACKNOWLEDGEMENTS

The author wishes to thank Professor R. L. Wain, F.R.S. for his support and encouragement of this work and Mr. P. Cozens and Miss C. Johns for their painstaking technical assistance.

LITERATURE CITED

HABER, A. H., and LUIPPOLD, H. J., 1960. Separation of Mechanisms Initiating Cell Division and Cell Expansion in Lettuce Seed Germination. *Plant Physiol.* **35,** 168–73.

KENDE, H., 1964. Preservation of Chlorophyll in Leaf Sections by Substances Obtained from Root Exudates. *Science,* **145,** 1066–7.

LETHAM, D. S., 1964. Isolation of a Kinin from Plum Fruitlets and Other Tissues. Régulateurs Naturels de la Croissance Végétale. *Centre National de la Recherche Scientifique,* 109–19.

——, SHANNON, J. S., and McDONALD, I. R., 1964. The Structure of Zeatin, a Factor Inducing Cell Division. *Proc. Chem. Soc.* 230–1.

LOEFFLER, J. E., and VAN OVERBEEK, J., 1964. Kinin Activity in Coconut Milk. Régulateurs naturels de la croissance végétale. *Centre National de la Recherche Scientifique,* 77–83.

MAÏA, E., 1964. Isolement d'un facteur de division cellulaire à partir du jus de tomate. Ibid. 103–9.

OSBORNE, D., and BLACK, M. K., 1964. Polar Transport of a Kinin, Benzyladenine. *Nature,* London, **201,** 97.

PECKET, R. C., 1960. Effects of Gibberellic Acid on Excised Pea Roots. Ibid. **185,** 114–15.

POWELL, L. E., and PRATT, C., 1964. Kinins in the Embryo and Endosperm of *Prunus persica.* Ibid. **204,** 602.

PURVES, W. K., and HILLMAN, W. S., 1958. Response of Pea Sections to Indoleacetic Acid, Gibberellic Acid, and Sucrose as Affected by Length and Distance from Apex. *Physiol. Plantarum,* **11,** 29–35.

SHAW, G., and WILSON, D. V., 1964. A Synthesis of Zeatin. *Proc. Chem. Soc.* 231.

VAN OVERBEEK, J., 1962. Endogenous Regulators of Fruit Growth. *Proc. of Campbell Soup Company Plant Science Symposium,* 37–58.

WAREING, P. F., 1958. Interaction between Indole-acetic Acid and Gibberellic Acid in Cambial Activity. *Nature,* London, **181,** 1744–5.

WENT, F. W., 1938. Specific Factors other than Auxin Affecting Growth and Root Formation. *Plant Physiol.* **13,** 55–80.

WRIGHT, S. T. C., 1961*a*. A Sequential Growth Response to Gibberellic Acid, Kinetin and Indolyl-3-acetic Acid in the Wheat Coleoptile (*Triticum vulgare*). *Nature,* London, **190,** 699–700.

—— 1961*b*. Growth and Cellular Differentiation in the Wheat Coleoptile (*Triticum vulgare*). I. Estimation of Cell Number, Cell Volume and certain Nitrogenous Constituents. *J. Exp. Bot.* **12,** 303–18.

Wright, S. T. C., 1963. Cellular Differentiation at the Molecular Level with Special Reference to Proteins. *Symp. Soc. exp. Biol.* **17,** 18–39.

Zwar, J. A., and Skoog, F., 1963. Promotion of Cell Division by Extracts from Pea Seedlings. *Aust. J. Biol. Sci.* **16,** 129–39.

——, Bruce, M. I., Bottomley, W., and Kefford, N. P., 1964. A Comparison of the Kinins of Apple and Coconut Milk and of the Modifications Produced by Purification Procedures. Régulateurs Naturels de la Croissance Végétale. *Centre National de la Recherche Scientifique,* 123–30.

Anemia phyllitidis: Inducibility of Physiological State Antagonistic to Antheridium Formation

Abstract. *It has been reported that a preparation obtained from prothalli of the fern species* Anemia phyllitidis *controls antheridium formation in this species. If this same preparation is applied at the spore stage it induces a physiological state antagonistic to antheridium formation. The inhibitory state is inducible only at a very early stage of development, but, once induced, it remains manifest at much later stages. Quite possibly, the inhibitory state arises in response to the factor that induces formation of antheridia. The inhibitory state may have an important function in fern development.*

Prothalli of *Pteridium aquilinum* elaborate a hormone that controls antheridium formation in many if not all species of the fern family Polypodiaceae (*1, 2*). This hormone has been isolated and termed an antheridiogen (*3*). Native antheridiogens have also been demonstrated in the fern species *Blechnum gibbum, Lygodium japonicum, Anemia phyllitidis* (*4*), and *Anemia hirsuta.*

The antheridiogen of *P. aquilinum* does not affect *A. phyllitidis*, a species belonging to the fern family Schizaeaceae. The prothalli of this species elaborate a different antheridiogen (A_{An}, *4*). The phytohormone gibberellic acid (A_3) also induces formation of antheridia in *A. phyllitidis* (*5, 6*). However, like A_{An} (*4*), A_3 has no effect on a species used to assay for the antheridiogen of *P. aquilinum* (*2*). Other gibberellins also cause antheridium formation in *A. phyllitidis*, some of them at an exceedingly low concentration (*7, 8*). The antheridiogen A_{An} does not seem to be identical with any of the several tested gibberellins (*9*).

The sensitivity of fern prothalli to antheridiogen varies with the developmental stage (*1–6*). If *Anemia* prothalli are exposed to A_{An} at the 0-day stage (before, or immediately after, spore inoculation), then the interval between the appearance of antheridia at the highest concentration of the antheridiogen and that at the lowest concentration is 14 days. In contrast, the corresponding interval is approximately 1 day if A_{An} is supplied at the 11-day stage (*10*). If we take this interval as a basis for comparison, the sensitivity of 0-day-old prothalli to differences in antheridiogen concentration exceeds that of 11-day-old prothalli by a factor of about 14. The delayed appearance of antheridia in prothalli exposed at the 0-day stage might be interpreted to mean that the juvenile prothalli are refractory to the action of A_{An}. If this were true, prothalli exposed at the 0-day stage would be expected to form antheridia no later than the time at which they are formed by prothalli exposed at the 11-day stage. Instead, prothalli exposed to the lower concentrations of A_{An} at the 0-day stage formed antheridia much later than prothalli exposed at the 11-day stage did, and they also formed a much larger number of vegetative cells before they attained the antheridial phase (*10*) (Table 2). Clearly, an inhibitory state becomes operative in the juvenile prothallus, greatly delaying the appearance of antheridia but permitting the continuance of vegetative cell division. This inhibitory state may not be an obligatory characteristic of development but may be, instead, induced by the preparation containing A_{An}. On this assumption, inducibility of the inhibitory state decays before the prothallus attains the 11-day stage, but, once it is induced in the juvenile prothallus, the inhibitory state remains manifest beyond the stage at which it ceases to be inducible.

For all experiments, protnalli were cultured in 50-ml erlenmeyer flasks on 10 ml of an agar-solidified, inorganic medium. Conditions of culture and the method of spore sterilization have been described (*11*). The antheridiogen A_{An} was applied as the liquid medium harvested from 7-week-old cultures of *A. phyllitidis*. It induced formation of antheridia to a dilution of one part in a thousand in most spore samples; it is termed the active preparation below.

Prothalli were first grown for 11 days, some on nonsupplemented medium, others on medium supplemented with dilutions of the active preparation at full strength ranging from 1:100 to 1:3000. Eleven days after spore inoculation, prothalli grown on the nonsupplemented medium were transferred to the 11-day-old cultures of prothalli grown on supplemented medium. Prothalli on agar slivers about 60 mm² and 2 mm thick were transferred from the nonsupplemented cultures (two slivers per flask) to areas of the agar surface free of prothalli in the supplemented cultures (Table 1, experiment 1).

Table 1. Differential behavior of prothalli exposed to A_{An} at 0-day and 11-day stages upon transfer to same medium. Recorded numbers are averages of numbers of individuals bearing antheridia in each of four samples of 30 observed prothalli. A, grown first on supplemented medium; B, grown first on nonsupplemented medium.

Dilution, *Anemia* medium	Days after transfer: A, 2	A, 6	A, 12	B, 2	B, 12
	Experiment 1				
1:100	0.25	2.75	26.00	25.25	30.00
1:300	.00	0.00	16.00	25.25	28.00
1:1000	.00	.00	1.25	9.25	14.00
1:3000	.00	.00	0.00	0.00	0.00
	Experiment 2				
1:100	0.00	1.75	16.75	19.75	29.50
1:300	.00	0.00	9.25	16.50	22.00
1:1000	.00	.00	2.00	3.25	8.50
1:3000	.00	.00	0.00	0.00	0.00

In another experiment, prothalli at the 11-day stage were transferred to culture flasks with freshly prepared medium supplemented with various dilutions of the active preparation. Each flask received two agar slivers with prothalli first grown for 11 days on medium supplemented with the active preparation at the same dilution and two agar slivers with prothalli grown for 11 days on nonsupplemented medium (Table 1, experiment 2).

Prothalli exposed to A_{An} at the 0-day stage form antheridia later than do prothalli exposed at the 11-day stage, even if they are cultured on the same medium (Table 1). Clearly, the results are consistent with the hypothesis that exposure at the 0-day stage causes the induction of a physiological state inhibitory to antheridium formation. However, the possibility that an inhibitory state of very brief duration becomes operative even in the unexposed prothallus must be considered. If this were true, exposure of the 0-day-old prothallus to the preparation containing A_{An} would be effective by greatly lengthening the duration of the inhibitory state.

I studied at what stage of development the inhibitory state ceases to be inducible. *Anemia* prothalli were exposed to A_{An} at the 0-day (immediately after spore inoculation), 4-day, 6-day, and 11-day stages. Three-tenths milliliter of active preparation was added at strengths calculated to give dilutions in the medium of 1:30, 1:100, 1:300, 1:1000, and 1:3000. In prothalli exposed at the 0-day stage, the interval between the appearance of antheridia at the highest concentration of A_{An} and the appearance of antheridia at the lowest concentration is more than 14 days (Table 2). In contrast, in prothalli exposed at the 4-day stage, the corresponding interval is about 1 day. Clearly, the inducibility of the inhibitory state has already largely disappeared in most of the 4-day-old prothalli. However, with development from the 4-day stage to the 11-day stage, the interval between exposure to A_{An} and the emergence of antheridium initials is shortened by a roughly equal amount at all concentrations of A_{An} from about 3 days to less than 2 days (Table 2). It cannot yet be decided whether the observed decay of the high sensitivity to differences in concentrations of A_{An} between the 0-day stage and the 4-day stage has the same physiological basis as the subsequent, largely concentration-insensitive reduction of the lag period.

The terminal (apical) cells of juvenile individuals give rise to an antheridium only in response to the highest concentration of antheridiogen if the prothalli are exposed to A_3 or A_{An} at the 0-day stage (*5, 11*). At lower concentrations, the sex organs arise from more basal cells. In contrast, many of the prothalli exposed at the 4-day stage form terminal antheridia even at the lowest effective concentration of A_{An}, that is, at a concentration 10 to 30 times lower than that causing formation of terminal antheridia by prothalli that are exposed at the 0-day stage. Clearly, the inhibitory state induced in the juvenile prothallus has a strong effect on the apical cell.

Chromatography was used to study the nature of the substance that induces the inhibitory state. Descending chromatography on Whatman filter paper No. 1 with solvent mixtures 1 (isopropanol, water and ammonia, 80:10:10) and 2 (water and isopropanol, 4:1) (*3*) was chosen. Fifty microliters of an ethanol extract with an activity ten times that of the active preparation at full strength were spotted. Chromatography was stopped when the solvent front had reached a distance of 24 or 48 cm from the origin. The chromatograms were cut into 10 or 20 sections, respectively, each section being eluted with 6 ml of water. Eluate (5.1 ml) from each section of the chromatogram was used for assay, 1.7 ml being added to each of three culture flasks. Two sets of assay flasks were inoculated with spores of *A. phyllitidis*, one set being observed at 12 days for antheridium formation, the other set being observed at 20 days for antheridium formation. To the third set of assay flasks were added two agar slivers with 9-day-old prothalli grown on nonsupplemented medium. These prothalli were observed for antheridium formation 3 days after transfer (12 days after inoculation).

Table 3. Attempt to separate a factor inducing the physiological state antagonistic to antheridium formation from A_{An}. Solvent mixture 1: isopropanol, water, and ammonia, 8 : 1 : 1; solvent mixture 2: water and isopropanol, 8:2. Conditions of assay: (1) exposed to A_{An} at 9-day stage; observed for antheridia at 12-day stage. (2) Exposed at 0-day stage; observed at 12-day stage. (3) Exposed at 0-day stage; observed at 20-day stage. Recorded numbers are averages of individuals bearing antheridia in each of two samples of 30 prothalli.

Section of chromatogram	Condition of assay 1	2	3
	Solvent mixture 1		
1–7, 9–10	0.0	0.0	0.0
8	18.5	.0	6.5
	Solvent mixture 2		
1–17, 20	0.0	0.0	0.0
18	22.5	.0	8.5
19	2.0	.0	0.0

The relative flow rates for the antheridium-inducing activity (Table 3) corresponded to those reported earlier for the two solvent mixtures used (*11*). Antheridia arose on prothalli exposed to A_{An} at the 9-day stage and assayed at the 12-day stage, and on prothalli exposed at 0-day stage and assayed at the 20-day stage. In contrast, with prothalli exposed at the 0-day stage and assayed at 12-day stage none of the eluates caused antheridium formation within the specified period of time. The results are interpreted to mean that the sections of the chromatogram which caused antheridium induction also caused the induction of the physiological state inhibitory to antheridium formation. Clearly, a substance inducing the inhibitory state was not separated from A_{An}. The two solvent mixtures used led to the chromatographic separation of the three native fern antheridiogens thus far demonstrated (*12*). The possibility cannot be excluded that the inhibitory state is induced by A_{An}.

Such a hypothesis seems paradoxical. The described phenomenon might be functional in preventing the precocious attainment of the sexual phase. However, the inducibility of the inhibitory state has already largely decayed at the 4-day stage. The prothalli developed from the germinated spores then comprised an average of 1.5 green vegetative cells (only cells seen outside the spore coat were counted). The first spores germinated 3 days after inoculation. Quite likely, the peak in the inducibility of the inhibitory state begins to decline even before the spore germinates. It may be stressed that A_3 (*5*) and medium harvested from *Anemia* cultures (*11*) not only cause antheridium formation but also can cancel the light requirement for spore germination. In all likelihood, the native factor that induces germination in the dark is identical with A_{An} (*11*). In nature, recently shed spores are probably exposed to A_{An} secreted by prothalli developed from spores shed earlier. Accordingly, I propose that the inducibility of a state antagonistic to antheridium formation may be functional in the prevention of "hybrid metabolism," that is, the simultaneous occurrence of physiological states leading to germination and to formation of antheridia. A consideration of the above situation is valid regardless of whether or not light promotes germination by causing the synthesis of A_{An}. At a genetic level such a phenomenon would have its parallel in the finding that activation at one gene site may lead to suppression at another.

Table 2. Relation of A_{An} concentration to onset of antheridium formation in prothalli exposed to A_{An} at the 0-day, 4-day, 6-day, and the 11-day stage. Recorded numbers are averages of numbers of individuals bearing antheridia in each of four samples of 30 observed prothalli.

Dilution, *Anemia* medium	Days after exposure at stage: 0-day						4-day		6-day		11-day	
	4	8	12	16	20	22	2	3	1	2	1	2
1:30	0.0	8.75					0.0	13.25	0.0	7.50	1.25	25.25
1:100	.0	0.25	0.25	2.75			.5	11.75	.0	7.75	0.75	24.75
1:300	.0	.00	.00	0.00	4.50	9.75	.0	7.50	.0	7.25	.00	23.50
1:1000	.0	.00	.00	.00	0.00	2.50	.0	1.75	.0	2.25	.25	16.25
1:3000	.0	.00	.00	.00	0.00	0.00	.0	0.00	.0	0.00	.00	0.50
Control	.0	.00	.00	.00	.00	.00	.0	.00	.0	.00	.00	.00

ULRICH NÄF

Laboratory of Plant Morphogenesis, Manhattan College, Bronx, New York

References and Notes

1. W. Döpp, *Ber. Deut. Bot. Ges.* **63**, 139 (1950); *ibid.* **72**, 11 (1959).
2. ———, *Planta* **58**, 483 (1962).
3. R. B. Pringle, U. Näf, A. C. Braun, *Nature* **186**, 1006 (1960); R. B. Pringle, *Science* **133**, 284 (1961).
4. U. Näf, *Nature* **184**, 798 (1959); *J. Linn. Soc. London (Bot.)* **58**, 321 (1962); *Ann. Rev. Plant Physiol.* **13**, 507 (1962).
5. H. Schraudolf, *Biol. Zentralbl.* **81**, 731 (1962).
6. ———, *Planta* **68**, 335 (1966).
7. ———, *Nature* **201**, 98 (1964).
8. B. Voeller, *Science* **143**, 373 (1964).
9. ———, in *Régulateurs naturels de la croissance végétale*, J. P. Nitsch, Ed. (Centre national de la recherche scientifique, Gif sur Yvette, 1964), p. 665.
10. U. Näf, *Z. Pflanzenphysiol.*, in press.
11. ———, *Physiol. Plant.* **19**, 1079 (1966).
12. ———, in preparation.

13. Aided by NSF grant GB-381. The laboratory is supported by PHS grant RC-1193, Damon Runyon Memorial Grant 710, and the Alfred and Christine Sonntag Foundation. I thank my colleagues for criticism and Mr. Tsun-San Chang for expert technical assistance.

Reprinted from Science, March 17, 1961, Vol. 133, No. 3455, page 756

Preparation of a Floral Initiating Extract from Xanthium

Abstract. An extract from lyophilized tissue of flowering *Xanthium* (cocklebur) plants initiates floral development when applied to *Xanthium* test plants maintained on long-day photoperiodic conditions. Details of the preparation of the extract are described.

Hamner and Bonner (*1*) have submitted the first evidence of a hormone formed in *Xanthium* at the site of photoperiodic perception, the leaf, and later utilized by the bud in the transition from the vegetative to the flowering condition. They also record attempts to find a floral initiating preparation in extracts from flowering plant material. At the same time they tested numerous chemicals for activity. Although they did occasionally find evidence of floral-initiating activity in their plant extracts, they were unable to demonstrate any consistent pattern or to repeat their results with any confidence.

Table 1. Flowering response of *Xanthium* after application of extract.

Flowering response	Numerical av. of flowering response (*6*)
Untreated control plants	
All ten plants vegetative	0.0
Extract-treated plants	
Five plants vegetative (one plant, stage one; two plants, stage two; two plants, stage three)	1.1

Since that time there have been occasional published reports of preparations exhibiting floral-initiating properties. Such reports have not met the test of consistent repetition, with the exception of the now well-known effect of gibberellic acid in stimulating floral development in many long-day photoperiodically sensitive plants. Lang (*2*) suggests that gibberellic acid is probably not the primary floral-initiating stimulus, but that it acts indirectly in stimulating flowering. The action of gibberellic acid in promoting more rapid flowering in *Xanthium* (*3*) also appears to be an indirect effect.

The experiment described in the present report is the last of four successful demonstrations of floral initiation in *Xanthium* plants resulting from applications of as yet unknown substance(s) extracted from the branch parts of flowering plants. Our preparation has been independently checked by B. H. Carpenter and K. C. Hamner (*4*), and the results are in accord with our own.

Flowering branch tips were harvested from indigenous *Xanthium strumarium* L. var. *canadense* (Mill.) T. and G. (*5*) of the Long Beach area. The plants were picked when the staminate terminal inflorescence was from ½ to 1 cm in diameter. Each branch carried three to five mature leaves. The fresh material was frozen in liquid nitrogen and broken into fine fragments while in the frozen state. Care was taken that the material remained frozen until it had been thoroughly dried in a laboratory vacuum lyophilizer. After lyophilization, the material was placed in sealed containers and stored in a deep freeze at −20°C.

The lyophilized material was extracted with absolute methanol under a partial vacuum sufficient to maintain the temperature of boiling methanol between −5° and −10°C. A modified Soxhlet apparatus was used with a condenser containing a dry ice-acetone slurry as a coolant. The methanol solvent was removed from the extract by evaporation at room temperature or below in a Rinco apparatus attached to a water aspirator system. The product was a dark green, tarry residue which was mixed to homogeneity with anhydrous (USP) lanolin.

Two grams of the residue mixed with 17 g of lanolin was applied to the underside of the leaf surface of ten test plants. Ten control plants were treated in like fashion with 17 g of pure lanolin. During the 14 days subsequent to the application of the lanolin preparations, all test plants and control plants were maintained on a precise 8-hour dark and 16-hour (500 ft-ca intensity) light regime. On the 14th day after application, the terminal bud of each plant was dissected to ascertain the flowering response. The flowering stages were numerically evaluated (*6*), and the results are presented in Table 1. Three earlier experiments had yielded essentially similar results.

It is our belief that these results constitute the first reproducible demonstrations of floral initiation in a short-day plant as the direct result of an extract prepared from the tissues of flowering plants. It is suggested that this is a crude extract of the flowering stimulus. Currently, efforts are being directed toward further concentration and characterization of the active entity in this extract (*7*).

Richard G. Lincoln
Darwin L. Mayfield
Alan Cunningham
Departments of Botany and *Chemistry, Long Beach State College, Long Beach, California*

References and Notes

1. K. C. Hamner and J. Bonner, *Botan. Gaz.* **100**, 388 (1938).
2. A. Lang, *Proc. Natl. Acad. Sci. U.S.* **43**, 709 (1957).
3. R. G. Lincoln and K. C. Hamner, *Plant Physiol.* **33**, 101 (1958).
4. A preparation of 2 g of extract mixed in 17 g of lanolin was given to B. H. Carpenter and K. C. Hamner, Botany Department, University of California at Los Angeles. The mixture was applied to ten test plants of *Xanthium;* ten comparable plants served as controls. All plants were maintained on long-day conditions for 17 days, after which they were dissected. The treated plants were flowering at the 50-percent level. All controls were vegetative.
5. C. L. Hitchcock, A. Cronquist, M. Ownbey, J. W. Thompson, *Vascular Plants of the Pacific Northwest* (Univ. of Washington Press, Seattle, 1955), part 5.
6. Results were evaluated in accordance with a numerical scale based on the diameter and morphological stage of development of the terminal staminate inflorescence. Vegetative plants were rated as zero on the scale. The first morphological change in the stem apex that could be clearly recognized as flowering was assigned a value of 1.0. A flowering apex measuring 0.25 mm in diameter was evaluated as 2.0. An additional increment of 1.0 was allowed for each 0.25 mm increase in the diameter of the developing inflorescence.
7. This investigation was supported by National Science Foundation grants (Nos. 7252 and 11482). We gratefully acknowledge the technical aid and assistance of Marda L. West and Donald Siems.

10 November 1960

6
Cell, Organism, Environment

Reprinted from PLANT PHYSIOLOGY, Vol. 43, No. 8, August, 1968, pages 1169–1184.
(PRINTED IN U.S.A.)

Growth Physics in Nitella: a Method for Continuous in Vivo Analysis of Extensibility Based on a Micro-manometer Technique for Turgor Pressure[1]

Paul B. Green
Department of Biology, University of Pennsylvania, Philadelphia, Pennsylvania 19104

Received February 12, 1968.

Abstract. The view that the plant cell grows by the yielding of the cell wall to turgor pressure can be expressed in the equation: rate = cell extensibility × turgor. All growth rate responses can in principle be resolved into changes in the 2 latter variables. Extensibility will relate primarily to the yielding properties of the cell wall, turgor primarily to solute uptake or production. Use of this simple relationship *in vivo* requires that at least 2 of the 3 variables be measured in a growing cell. Extensibility is not amenable to direct measurement. Data on rate and turgor for single *Nitella* cells can, however, be continuously gathered to permit calculation of extensibility (rate/turgor). Rate is accurately obtained from measurements on time-lapse film. Turgor is estimated in the same cell, to within 0.1 atm or less, by measurement of the ability of the cell to compress gas trapped in the closed end of a capillary the open end of which is in the cell vacuole. The method is independent of osmotic equilibrium. It operates continuously for several days, over a several fold increase in cell length, and has response time of less than one minute. Rapid changes in turgor brought on by changes in tonicity of the medium, show that extensibility, as defined above, is not constant but has a value of zero unless the cell has about 80 % of normal turgor. Because elastic changes are small, extensibility relates to growth. Over long periods of treatment in a variety of osmotica the threshold value for extensibility and growth is seen to fall to lower values to permit resumption of growth at reduced turgor. A brief period of rapid growth (5× normal) follows the return to normal turgor. All variables then become normal and the cycle can be repeated. The cell remains essentially at osmotic equilibrium, even while growing at 5× the normal rate. The method has potential for detailed *in vivo* analyses of "wall softening."

The growth of the plant cell may be regarded, instantaneously at least, as the yielding of the cell wall to the stresses present in it. In the *Nitella* internode and other cells not subject to tissue tensions, the stresses can be assumed to be proportional to turgor pressure. This allows the rate of growth (R) to be viewed simply as the product of the yielding tendency of the cell (Ex_g or gross extensibility) and turgor pressure (T).

$$R\ (\text{fraction/hr}) = Ex_g\ [\text{fraction/hr})/\text{atm}] \cdot T\ (\text{atm})$$

All changes in growth rate then become resolvable into changes in one or both terms. Such changes in individual terms, once clearly identified, can then be studied with regard to their origin in structure or metabolism.

Equation (I), despite its brevity, is remarkably difficult to solve in growing cells since only the rate term is readily measured while enlargement is occurring (21, 22). Extensibility is not amenable to direct measurement because unilateral stretching of the growing cell would alter the usual stress pattern and disturb the growth process. Extensibility can be calculated using equation (I), provided information on turgor can be combined with data on rate in the same material.

Turgor estimates for the growing cell are often equivocal because equilibrium methods such as plasmolysis and plasmometry give only an upper limit for the value during growth. In some tissues the value during enlargement is estimated to be as much as 2 atmospheres below the equilibrium figure (32). Even if near-equilibrium conditions do obtain during growth, as might be expected in single cells growing under water, the growth response of interest may be rapid compared to the rate at which diffusion brings

[1] Supported by National Science Foundation Grants GB 2575 and 6055X.

concentrations to the new equilibrium required for a measurement (8, 32). Use of equation (I) to analyze growth in tissues is made particularly difficult by the possibility that the various cell types in the tissue are not equivalent with regard to generating pressure and yielding to it. For these reasons a method for turgor in single cells which is rapid, continuous, independent of osmotic equilibrium, and compatible with growth is of some utility. This report describes such a method based on measurement of a cell's capacity to compress gas trapped in a capillary (16). Data on turgor, when combined with information on the growth rate of the same cell, permit the continuous solving of equation (I) and the calculation of gross extensibility (Ex_g).

Note that while turgor clearly can be affected by changes in extensibility (*e.g.* increased extensibility under conditions of low permeability to water can reduce turgor (21, 22), this does not mean that the 2 variables cannot be measured independently. The cell's ability to compress gas gives its internal pressure regardless of the many factors governing this pressure. Extensibility can thus be defined simply as the quotient of the existing rate and the existing turgor pressure.

Ex_g is given a subscript to distinguish it from related expressions. Generally, extensibility relates the rate of yielding of a material to the applied stress. For many materials the relationship can be written so that extensibility is a constant. This often involves incorporating the complexity of yielding behavior into the stress term. For example, in quasi-plastic flow the stress above a certain threshold is raised to a power and then multiplied by a constant extensibility term to give the rate of yield (28). Because there is no *a priori* reason to expect the extensibility of a growing cell to fall into one of the rheological relationships established for inert substances, it seems reasonable here to keep the stress term a simple pressure and to express complex yielding behavior, when it appears, in some elaboration of the extensibility term.

Plant cells typically show constant growth rate at constant turgor, so Ex_g will of necessity have the characteristics of a reciprocal viscosity. The similarity to viscous flow may well be only formal because, if metabolism is allowed to play a role, one can envision a variety of complex yielding systems which show this kind of behavior (29).

Ex_g, when measured over short periods of time and over small deformations, will reflect the instantaneous yielding properties of the cell wall, properties which may be maintained by metabolism occurring in, or at least acting on, the wall (30). The behavior of Ex_g *in vivo* therefore can be compared with the yielding properties of killed cells or isolated wall strips subjected to known artificial stresses. Provided the physical data on isolated material derive from small deformations at physiological rates, and provided correction can be made for the fact that a unilateral force instead of a multi-lateral pressure is applied in isolated material, differences in yielding properties from those seen *in vivo* must reflect the absence of chemical or metabolic factors normally present in the cell.

By the present method Ex_g can be evaluated after rapid shifts in turgor to give information on the stress:strain-rate relationship in the growing cell. Continuation of the analysis over longer periods reveals sizable shifts in these properties.

Materials and Methods

Principle. When a fine glass capillary is fused at one end and submerged, air is trapped inside it. If the open end is then inserted into a cell and the wound sealed, the trapped gas is compressed in accord with pressure inside the cell. Because of the constancy of pressure times volume, cell pressure is readily calculated. Turgor pressure is 1 atmosphere less than the pressure in the bubble. The small tendency of capillarity to compress the gas can be measured beforehand and subtracted from final turgor values. Adaptation of this principle for use on the growing cell, under conditions where the growth rate of the same cell could be simultaneously followed, involved many problems. The major physical one was that the trapped gas dissolved into the capillary fluid, at varying rates depending on turgor, so that volume changes over long periods of time were not a reliable basis for pressure calculation. A short-term volume change, brought on by a vacuum pump, was used for calibration.

Tazawa's cell compression method for turgor (34) and Virgin's resonance frequency method (38) also do not require osmotic equilibrium. Arens (1) has described 2 methods to estimate turgor in *Nitella.* In one, the original compressibility of a cell was restored, after loss of turgor, by inflating the cell through a canula. The gas pressure required was equal to the original pressure of the cell interior. In the other, the mid-portion of the cell was gripped and the position of one protruding cell-end (a) carefully noted. The cell was allowed to wilt and gas pressure was then applied to the space around the other cell-end (b) so as to restore the original configuration at (a). Values of 3 to 10 atmospheres were obtained. Recently Villegas (37) has used a capillary inserted in a *Valonia* cell to balance cell pressure with gas pressure. Pressure changes associated with small volume changes could be estimated. Gutknecht (17) has altered turgor in *Valonia* by connecting a source of hydrostatic pressure to the cell interior by a capillary. Apparently none of these methods has been adapted for continuous use on growing material.

Detailed Method. The following considerations governed the adaptation of the capillary method to the growing cell. All data necessary for determination of both growth rate and cell pressure would be photographed (time-lapse). This meant that the cell and capillary had to be in the same plane, preferably side by side. Calibration was to be done periodically and solution changing, over a period of 48 hours, was to be automatic. Because calibration involved a large drop below atmospheric pressure, the system, including automated valves, had to be vacuum tight.

The major biological problem involved the cell's reaction to the presence of the capillary which, at first, either resulted in failure of the cell to grow or in failure of the capillary to function (due to plugging). Success was obtained only after extreme measures were taken to insure mechanical rigidity of the cell and capillary both during insertion and the protracted growth period.

a) *Calibration of Cell Pressure.* The pressure in a bubble can be determined by its volume change in response to transitory changes in its pressure. Normally 1 atm of the pressure on the bubble is from the atmosphere, the rest being due to turgor. The former component can be altered by known amounts by placing the entire system, plant and capillary, in a strong vessel and applying a pump. From the change in volume one obtains the pressure (P_B) in the bubble.

$$P_B = \frac{\Delta P \, V_2}{V_1 - V_2} \qquad \text{(II)}$$

We gave a decrement in pressure, ΔP, of minus two-third atm for 3 minutes every half hour. V_1 was taken as the mean of 2 volumes: 5 minutes before and 5 minutes after the pulse. V_2 was the maximum volume during the pulse.

Turgor pressure, T, is:

$$T = [P_B - C] - 1 \qquad \text{(III)}$$

where C is capillarity, roughly 0.05 atm, measured by the extent of water entry (gas compression) when the capillary was submerged. Turgor was computed from measurements punched on computer cards. For pressure values between calibration pulses, the volume of the bubble could be used because the shrinkage was steady for a given turgor. No variation in streaming rate or growth rate was associated with pump action.

The response of bubble length to pump action and to the addition of osmoticum to the medium is seen in figure 3 where length is measured at 2 minute intervals. Upon a fall in turgor the bubble increases both its length and its response to the pump. The bubble can accurately respond to a change in pressure of two-third atm within 1 minute. Accuracy of pressure drop was maintained by a Cartesian diver type manostat (Greiner Science Company, New York).

The Growth Vessel and Automated Solution Changing. The plant grew in a continuously illuminated pressure vessel and was photographed through a flat window. The vessel was made from a large "0-ring" joint by glass-blowing. It was clamped to a heavy lathe tool-holding device to permit precise and stable adjustment of the vessel's position. An outlet above the fluid level in the vessel went to a vacuum pump, an inlet was provided below the 0-ring, and a drain was at the base (fig 1). Solutions were changed by draining and refilling, all movements by gravity. Valves were operated by air-driven linear actuators controlled by solenoid valves. Automation of both calibration and solution change involved combined use of a 1 rev/hr multiple cam ("gang") timer and a stepping switch. One cam activated a vacuum pump for 3 minutes every half-hour for calibration (see above). During every hour a second cam closed 1 of 2 gaps in a circuit which could open the drain valve, and shortly thereafter a third cam closed a similar gap in an inlet valve circuit for a few minutes. The valves would actually open only during those hours when the stepper closed the second gap. The stepper was advanced 1 step per hour by a fourth cam on the gang timer. Thus the major functions, being all timed by the same rotating shaft, could not get out of phase. Typically, normal growth medium was alternated with growth media of increased tonicity, using various osmotica such as sugar or polyethylene glycols (Carbowax). The Carbowax (Union Carbide Company) had been passed through an ion-exchange column. Concentration of growth medium was kept constant, independent of osmoticium concentration. A large flask of normal medium was repeatedly used to fill the growth vessel during 1 run. Smaller flasks of altered medium (used once each) were drained completely to fill the growth vessel. This arrangement permitted 7 or 8 solution changes with only 4 flasks and valves. Timing had to be adjusted in advance to allow for variation in viscosity of the various solutions, particularly those containing Carbowax.

Toxicity was avoided by using polyethylene for tubing, valves, and fittings and by making all glue junctions with DeKhotinsky cement (Central Scientific Company, St. Louis). The 0-ring itself was toxic so the solutions were kept from it.

The Capillary. A fish-hook shaped capillary was used in order to bring the compressed bubble alongside the growing cell surface for photography. At full turgor only about one-sixth of the capillary was filled with gas. To have the length of the bubble roughly correspond to the length of the young cell in order to maximize accuracy, the capillary was made much longer than the cell, then bent. An internal diameter of about 40 μ was found to make the bubble readily visible on the film and to transfer only a modest volume of cell sap into the capillary. Cell diameter was about 200 μ so the capillary cross-section was about 4 % that of the young cell. Initial sap dilution ran about 3 to 6 %. One mm diameter thin-wall glass capillary (Corning Glass Works, Corning, New York) was hand-drawn over a broad flame to give several cm of thin capillary of essentially constant bore. The glass was then treated with Siliclad (Clay, Adams, Inc., New York, New York) to prevent later segmentation of the gas

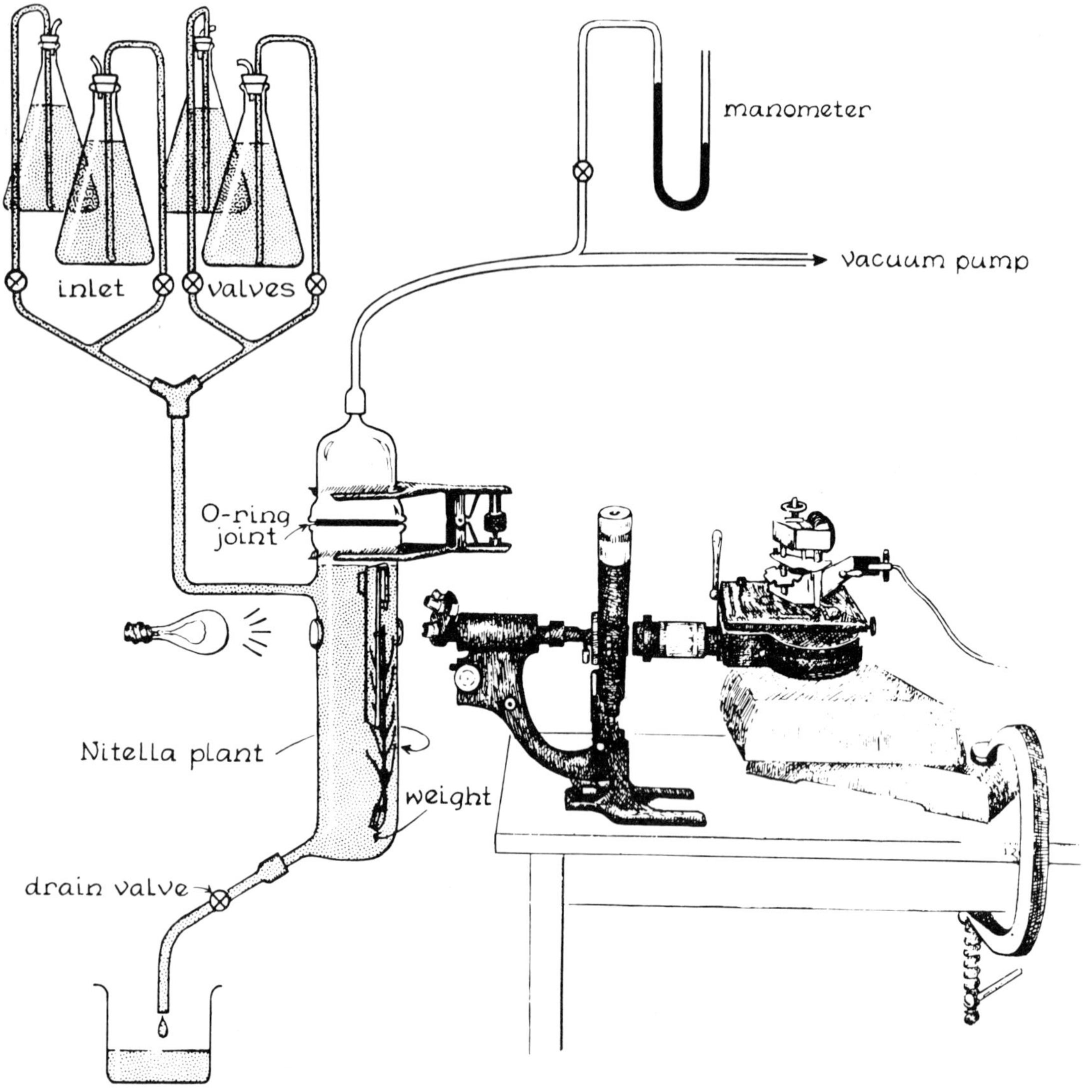

FIG. 1. General experimental set-up to permit continuous time-lapse photography (right) of the plant while it is growing in the pressure vessel. The vessel is held by a clamp (not shown) fixed to a machinist's tool holder with screw adjustments in 3 dimensions. The vacuum pump acts at one-half hour intervals for calibration of cell pressure (see text). Solution changing was on a drain-and-refill basis, automated by circulating electric valves through a gang timer and stepping switch. The mercury manometer was occasionally used to check the level of vacuum delivered by the pump.

bubble. The 40 μ bore was ideal for the reading part, but was too large for insertion. A special tip was fashioned with a DeFonbrune microforge (Aloe Science Company, St. Louis). The platinum filament was coated with glass to prevent toxic vapors from being deposited on the capillary. The capillary was fused at 2 points 12 mm apart and a region just above the lower fusion was heated to form an extremely thin-walled (3 μ) swelling. A weight was added below and the swelling stretched, by heating, into a delicate, and thin-walled extension. After bending into a hook shape, the capillary was glued to a cover-glass chip. The tip was broken with tweezers to give a roughly oblique cut tip of diameter 20 μ. The mounted capillary was then returned to the forge and the broken tip touched to the filament and stretched and smoothed to a form like that of a syringe needle. Just before insertion it was dipped in water, then 20 % "Bio-nox" detergent (Triton X-100, Rohm and Haas Chemical Corporation, Philadelphia, Pa.) to minimize the cell's reaction to insertion.

The Holder. A complex holder was used for both the insertion, carried out in a large Petri dish under a dissection microscope, and for prolonged observation in the pressure vessel. See figure 2.

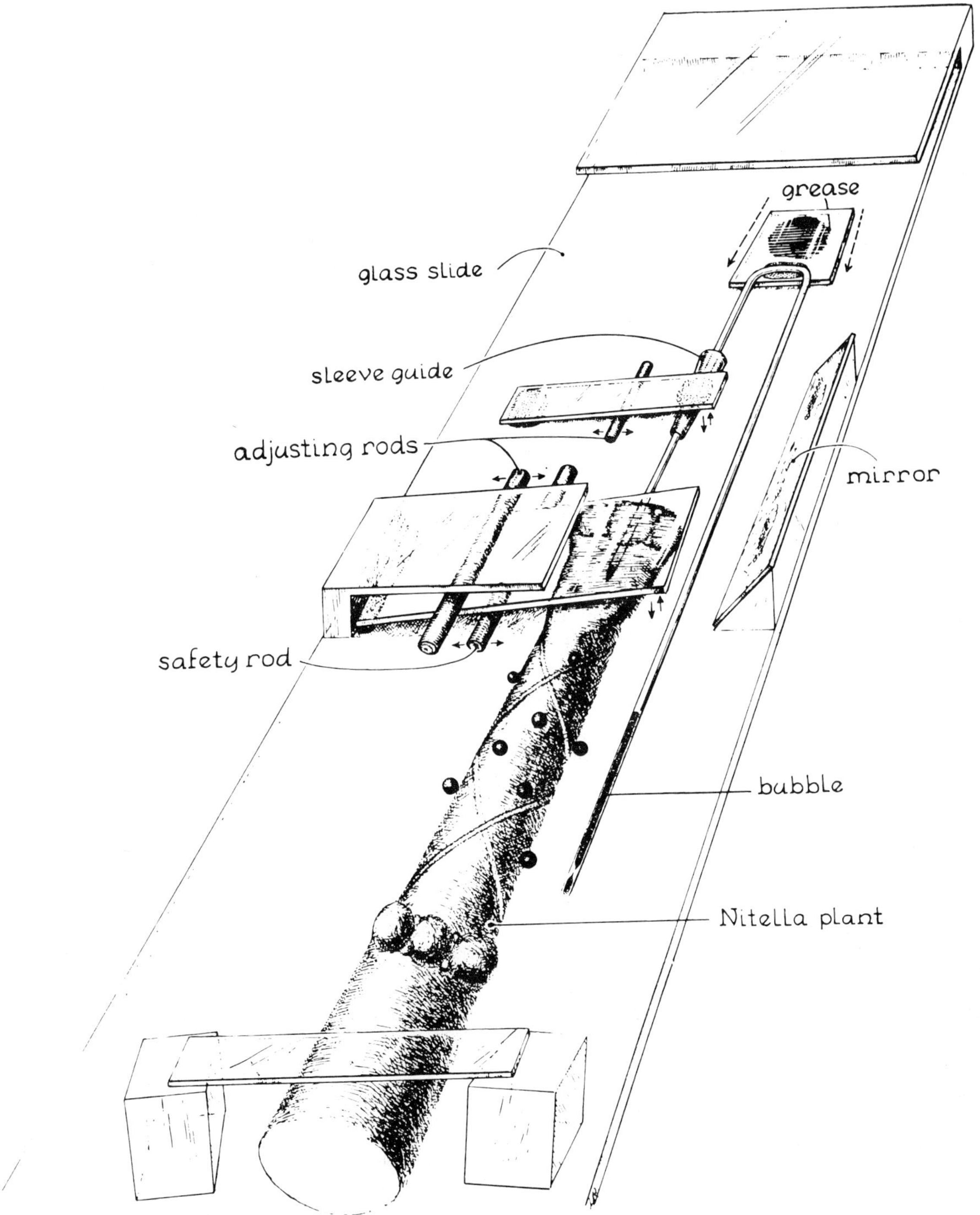

FIG. 2. A glass holder, assembled from a slide and cover glass pieces, used to hold the cell during insertion of the capillary and during prolonged growth in the pressure vessel. Adjustment of the position of the cell and the sleeve guide permitted advance alignment of the capillary tip with the central axis of the cell. The cell was firmly compressed at full turgor as shown; further compression upon reduction of turgor was prevented by the safety rod. When the holder and plant were in the vessel (fig 1), the field of the camera included the region of the cell showing the marking beads (on cell, left of bubble) and the adjacent part of the capillary containing the bubble. The structure at the base aids in the straightness of growth of the cell.

It hooked onto a horizontal bar glued inside the pressure vessel. The holder gripped the plant by visibly compressing the upper end of the cell of interest. It also served to hold an adjustable sleeve-guide by which the capillary tip was aligned precisely with the axis of rotation of the cell. In the horizontal plane the alignment was made by placing the cell in the proper position before compressing. The vertical alignment was made by adjusting the elevation of the sleeve guide with a small rod (fig 2) while observing through a 45° mirror which was part of the holder. The mirror was made by evaporating chromium onto a cover-glass chip.

The Cell. A plant with an upper internode about 5 mm long was chosen. The shoot tip and laterals at both ends of the chosen cell were removed with a glass needle. This bare upper end (node) was compressed and accepted the capillary. A small weight was attached to keep the cell vertical during growth. Curiously, this eliminated the "twisting" component of growth and facilitated analysis.

Insertion. The capillary-bearing chip was positioned so it could slide along the holder on a layer of stopcock grease. The capillary tip was directed into the sleeve guide by pushing the chip with forceps (fig 2). A protection bar was put under the cell-compressing lever to prevent any further compression upon reduced turgor and the capillary tip was pushed several hundred microns into the cell. In about one-fourth of the cases cytoplasmic streaming did not stop during insertion. In the others it resumed within 30 seconds. Good alignment, complete rigidity throughout insertion and growth, and a well-fashioned capillary tip were essential. Otherwise the cytoplasm built up at the wound region and gradually covered the capillary tip. Often it would lay down a cell wall (found to be birefringent) over all the glass inside the cell, rendering the capillary topologically outside (and useless). Cell growth rate was not optimal during such cytoplasmic reaction.

Cell Marking and the Growth Rate. The extension of the cell surface can be measured by following the separation of any 2 points on the cell surface because growth is uniformly distributed (12). The fractional increase in length, per unit time, is the relative rate. It is measured as the slope on a curve relating the natural logarithm of separation to time. For marks, small anion-exchange resin beads in the hydroxyl form (15) are ideal optically. Marks were applied by hand with a hair, the cell being in ion-free water. The use of a defined salt medium (in which chloride competed with the wall for the beads' binding groups) along with periodic draining and filling of vessel, lead to frequent loss of marks. An excess had to be applied initially. Those beads that remained on throughout were seen in the movie to remain firmly fixed to the wall during growth. The camera took 1 frame per minute, recording behavior of the cell surface and the bubble. Key dimensions on the developed film (Kodachrome II, 16 mm) were measured on a Vanguard Motion Analyzer (Vanguard Instrument Corporation, Melville, New York). *L* was the distance between the centers of 2 widely separated resin beads. A computer averaged measurements for 1 frame and provided ln *L*.

Elasticity and the Measurement of Growth. Total elongation, as measured on the film, is the sum of elastic and permanent (irreversible) extension. There is considerable evidence that the elastic changes in *Nitella* are small enough to be ignored in all but the most severe shifts in turgor. Kamiya *et al.* (18) have shown that the total shrinkage in length of a live *Nitella* cell is about 0.8 % and the shrinkage upon a fall in turgor of 2.4 atm, nearly

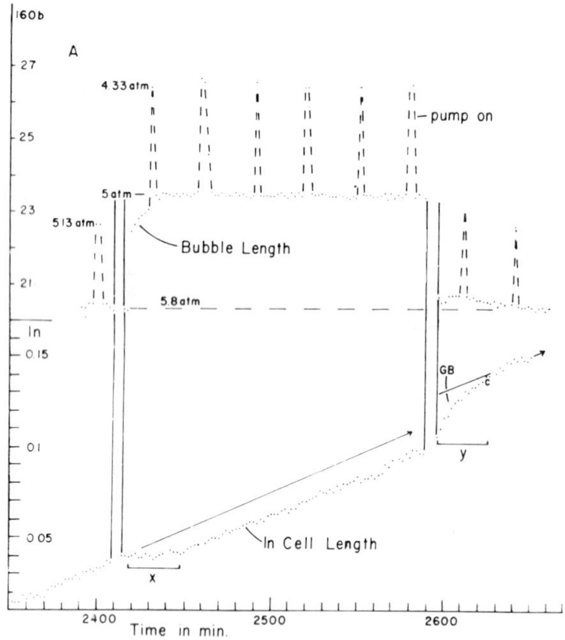

FIG. 3. Analysis at 2 minute intervals of capillary-bubble length (above) and in cell length (below). The intermittent expansion of the bubble, due to the action of a vacuum pump, is shown by vertical dashed lines. This permits calculation of the pressure inside the bubble (see text). (These calibration values, converted to turgor pressure, are plotted in fig 5, middle treatment). After a change to a medium of higher tonicity (2410 min, vertical solid lines), there is an initial delay of at least 2 minutes before the response of the bubble is seen. There is a further lag of at least 10 minutes before a stable bubble length is reached. That these delays are due to slow establishment of the new trans-membrane concentration gradient is indicated by the fact that they are long compared to the response time of the bubble to hydrostatic pressure change. A similar delay is seen upon return to higher turgor (2590 min). Growth halts for a brief period (x) at the start of inhibition, but then normal rate is rapidly restored. Upon return to essentially full turgor, a growth burst (GB) ensues, leading to a net increase in cell length over that anticipated in the absence of inhibition (long arrow). Duration of the burst (y) is about equal to the initial lag (x).

equal to the largest drop imposed by us, is only 0.1 %. Elastic changes took about 2 to 3 minutes to be three-fourths complete. In our material shrinkage could not be detected upon a fall in turgor of 0.8 atm (fig 3). A fall in turgor of about 3 atm did lead to about 0.4 % shrinkage over about 10 minutes on some trials, but to no shrinkage on others (fig 6). These shrinkages are small compared to length changes used to determine rates and are ignored unless otherwise specified.

Culturing. Stock cultures were maintained by alternate growth (2 wk periods) in autoclaved soil extract (100 ml garden soil in 6 l distilled water autoclaved 20 min) and in a modified Forsberg (10) medium. It contained, per liter, $Ca(NO_2)_2$, 80 mg; $MgSO_4 \cdot 7H_2O$, 100 mg; $NaHCO_3$, 60 mg; KNO_3, 100 mg; NH_4Cl, 20 mg; Na_2SiO_3, 12.5 mg; KCl, 100 mg; K_2HPO_4, 1 mg. The tris buffer used by Forsberg was omitted because it supported bacterial growth. Because the plants grew in large volumes which were replaced every 6 hours or less, buffering was not important. pH was adjusted to 6.5 with HCl. Trace elements, individually chaelated with sodium nitrilotriacetate were added, at 1 ml/l from stock solutions: $FeSO_4$, 2.5 g/l; $ZnCl_2$, 526 mg/l; $MnCl_2$, 1.15 mg/l; $CuCl_2$, 21.3 mg/l; H_3BO_3, 295 mg/l; Na_2MoO_4, 60 mg/l; $Co(NO_3)_2$, 30 mg/l. The modified Forsberg medium supported growth for 2 weeks and was used for experiments. It did not sustain good growth indefinitely as did soil. The Forsberg solution did not support growth of any of the visible contaminants, including *Oedogonium,* so by innoculating soil medium with the terminal 10 cm tips of Forsberg medium plants, the cultures were kept free of visible contamination.

Osmotica. To alter turgor it is desired to add a solute whose sole action is to lower the water potential of the medium. There is always the possibility that any soluble organic molecule will enter the metabolism of the cell to either stimulate or inhibit growth. In non-sterile systems such as the present one, effects mediated by bacterial activity are hard to exclude. The traditional osmotica, mannitol and polyethylene glycols, were used in the present study and it was found that when incorporated into growth medium they could support bacterial growth after days of standing.

Certain inorganic molecules cannot serve as energy sources but may have other disadvantages. KCl would be expected to effect the resting potential of the membrane. Further, potassium and calcium ions are known to have direct "loosening" and "hardening" effects on *Nitella* cell walls (27). None-the-less, the cell response to KCl does not differ markedly from that to organic osmotica. (KCl does not enter unusually rapidly to alter turgor).

Limitations. The method is capable of resolving any growth response into turgor and extensibility components. The major limitation is that the growth response caused by the presence of an osmoticum may be due to changes other than the simple change in the colligative properties of the medium. The most striking example of this was found with Carbowax 400 (A low mol wt polyethylene glycol, Union Carbide Company, New York). Presence of this compound reduced turgor and initially stopped growth. The cell then not only recovered its original rate but exceeded it by 80 %, still at reduced turgor! The data to be presented will therefore primarily illustrate the method and only tentatively characterize the response of the cell to turgor change. Some confidence is justified in the immediate aspects of a response and in those slower aspects that are common to osmotica of diverse chemical nature.

Results

Possible Departure From Osmotic Equilibrium — Delays in Mixing. A rough test for whether the growing cell is in osmotic equilibrium is whether a decrease in tonicity of the medium surrounding a non-growing cell is fully matched by an increase in turgor inside the cell as it resumes growth. A discrepancy would show a departure from equilibrium. In figure 6, below, the length of the vertical arrows gives the change in tonicity of medium, the vertical shift in the large dots, the shift in cell turgor. The differences are small fractions of an atmosphere and, after considerable growth has occurred, could be due to rapid sap dilution. It was thought possible that detailed study of bubble length might detect finite departures from osmotic equilibrium during the bursts of rapid growth after the end of inhibition. The detailed record in figure 3 does show slightly reduced pressure (by 0.08 atm) during the growth burst compared to values for steady elongation. This discrepancy apparently has its origin however in the delay required for the establishment of a new concentration gradient across the cell membrane. This general problem is discussed by Dainty (8). There is a roughly 20 minute delay in fully establishing a new turgor pressure when osmoticum is added and growth stops (fig 3, 2412–2432 min). This delay exists despite the presence of considerable differences in density of the solutions being changed; these differences would promote mixing by convection. Here the pressure is "too high" to roughly the same extent for about the same amount of time as it was "too low" above. Part of the slightly reduced pressure during the growth burst could be the result of rapid sap dilution during the growth burst (see fig 6, third treatment). It is concluded that departure from osmotic equilibrium is certainly less than 0.08 atm and probably infinitesimal. This result is not unexpected in light of the high water permeability (7–12 μ/min/atm) of mature *Nitella* cells (35).

Growth Response. In all experiments a growing cell was alternately placed in normal medium and in normal medium supplemented with osmoticum. The periods in normal medium were 4 to 6 hours.

The data are presented with the 2 directly measured variables, ln L and turgor pressure, plotted against time. This gives an over-all view of the course of the experiment, the magnitude of the growth response, *etc.* The growth rate is the slope on the ln L curve and is given at representative points as r. Cell extensibility, Ex_g, is the quotient: rate/turgor and is given at the same points as e. In figure 6 the variable ln diameter is also given.

Initially a detailed study using one osmoticum was planned. Concern with the potential side effects of all osmotica has lead, however, to a broader and less intensive effort. The response to various osmotica will be briefly described and the common features summarized.

In figure 4 a cell is periodically subjected to increasing concentrations of Carbowax 4000 in Forsberg medium, reducing its turgor by 0.3, 1.0, and 1.8 atm respectively. In all cases the cell's initial response is to stop elongating. Then, after periods of time which increase with larger loss of tugor, growth resumes. This recovery is primarily due to

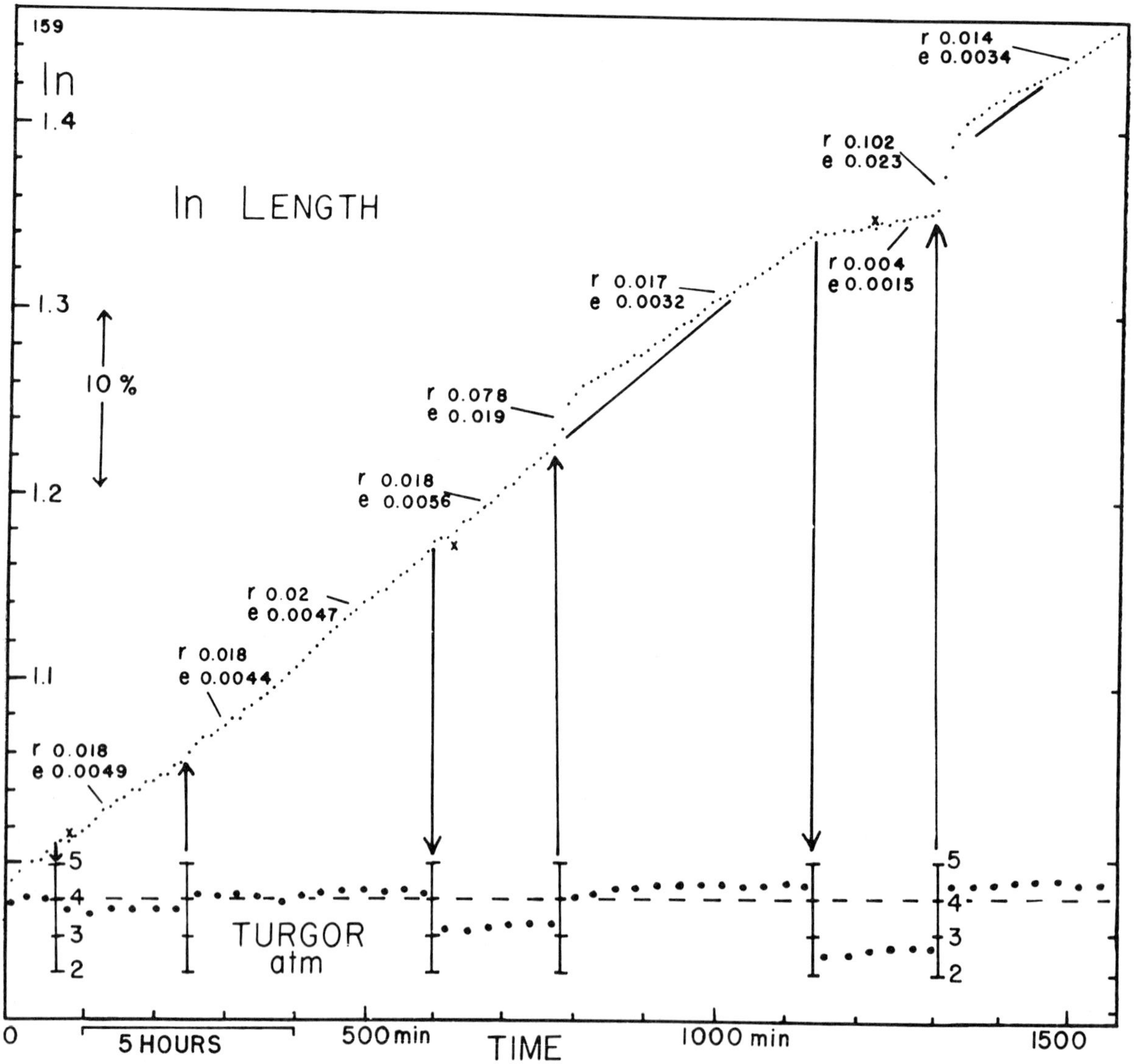

FIG. 4. Time course of an experiment with periods of osmotic inhibition of constant duration but increasing severity. The natural logarithm of length (between 2 marks) is given at 10 minute intervals, by the small dots. Use the upper vertical scale. The large dots and the small inserted scales give the turgor of the same cell at the same time. The relative rate of growth is the slope of the ln length curve and is given as r at various points. The gross extensibility (e) of the cell is also given. Descending arrows show start of osmotic inhibition, ascending arrows, its release. Note slight sensitivity to a small drop in turgor, resumption of normal growth rate upon moderate reduction of turgor (2nd inhibition) which is followed by a growth burst. A greater drop in turgor (3rd inhibition) almost totally inhibits growth, but is followed by a growth burst. Osmoticum: Carbowax 4000. Small x's show resumption of growth.

a change in Ex_0, not turgor. After the 2 smaller losses in turgor the normal rate is regained but after a loss of 1.8 atm only partial recovery is seen during the time available. In all cases a rapid elongation follows the return to osmoticum-free medium. The wall is seen to be highly extensible. The maximum rate attained and the amount of added length achieved during these "growth bursts" appear to increase as a function of the time the cell has spent at reduced (or zero) growth rate. A similar record is seen in figure 5 where the osmoticum was the same but the medium was soil extract. Here a pronounced net increase in length, due to inhibition plus recovery, is seen after the second inhibition. The over-shoot was very striking in runs with Carbowax 400, where a bacterial bloom eventually appeared, so its origin may be in extraneous metabolic factors combined with the change in turgor.

An increase in size of the growth burst with increasing time of growth inhibition (but no overshoot) is seen in figure 6 where the osmoticum was mannitol and where the duration, not the severity, of osmotic inhibition was varied. Recovery of rate appears only after about 4 hours into the longest inhibition. During this long inhibition an increase in turgor pressure is clearly seen. This increase is about that expected if solute increase had continued undiminished during the inhibition of volume increase. Extremely high rate and extensibility are noted upon return to normal medium. A fourth period of inhibition (at 2200 min) shows that the growth burst does not increase as a function of the number of inhibitory periods (a possibility not otherwise excluded) and that the rapid increase in length of the large growth burst is inelastic, at least after several hours.

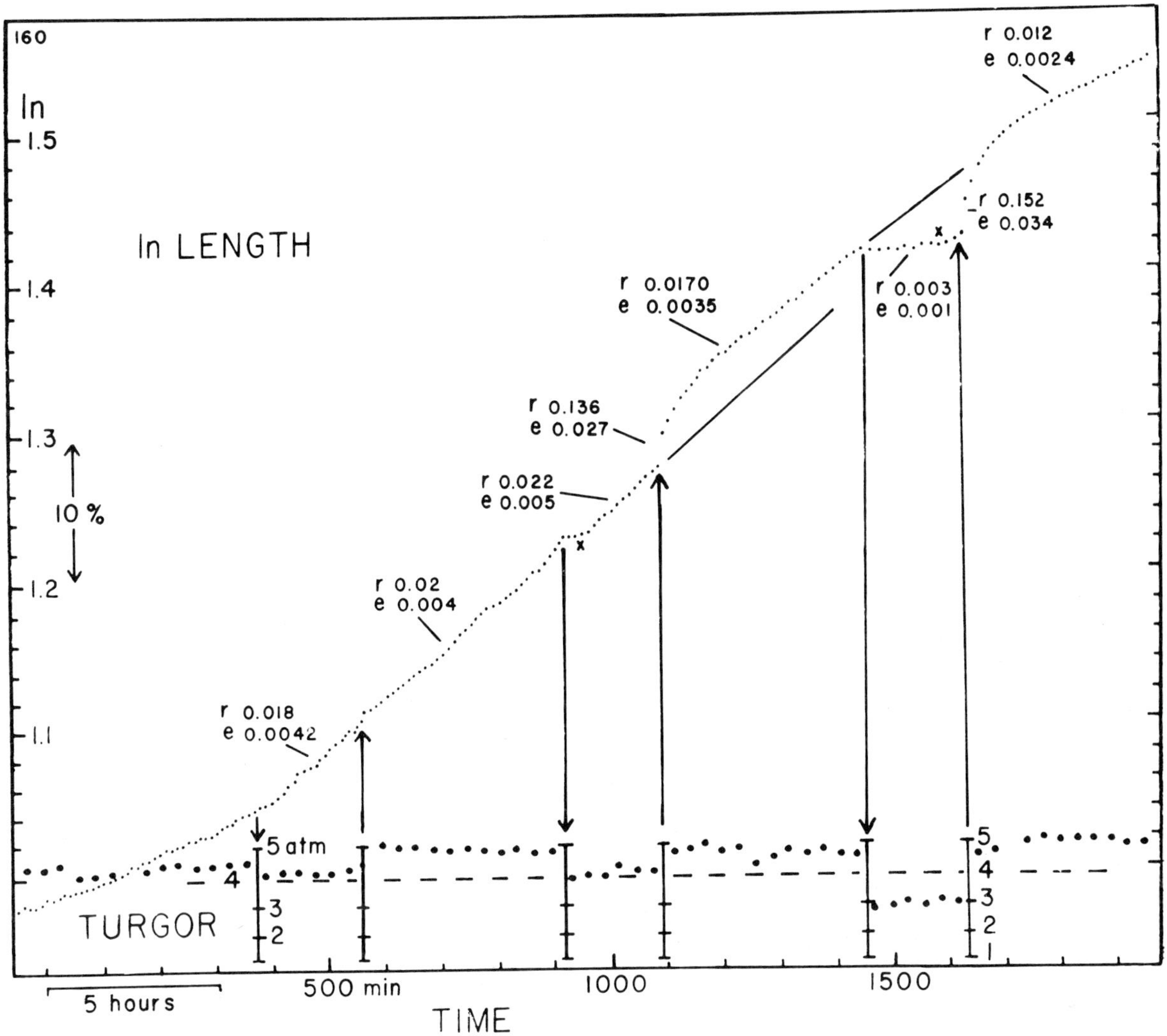

FIG. 5. Legend and experimental procedure the same as in figure 4. The second inhibition led to a pronounced net increase in cell length over that which would have been reached if the cell had been undisturbed. Note that the growth burst of the third inhibition gives a cell extensibility about 10 times that typical for the steady growth periods before and after the inhibition. Osmoticum: Carbowax 4000.

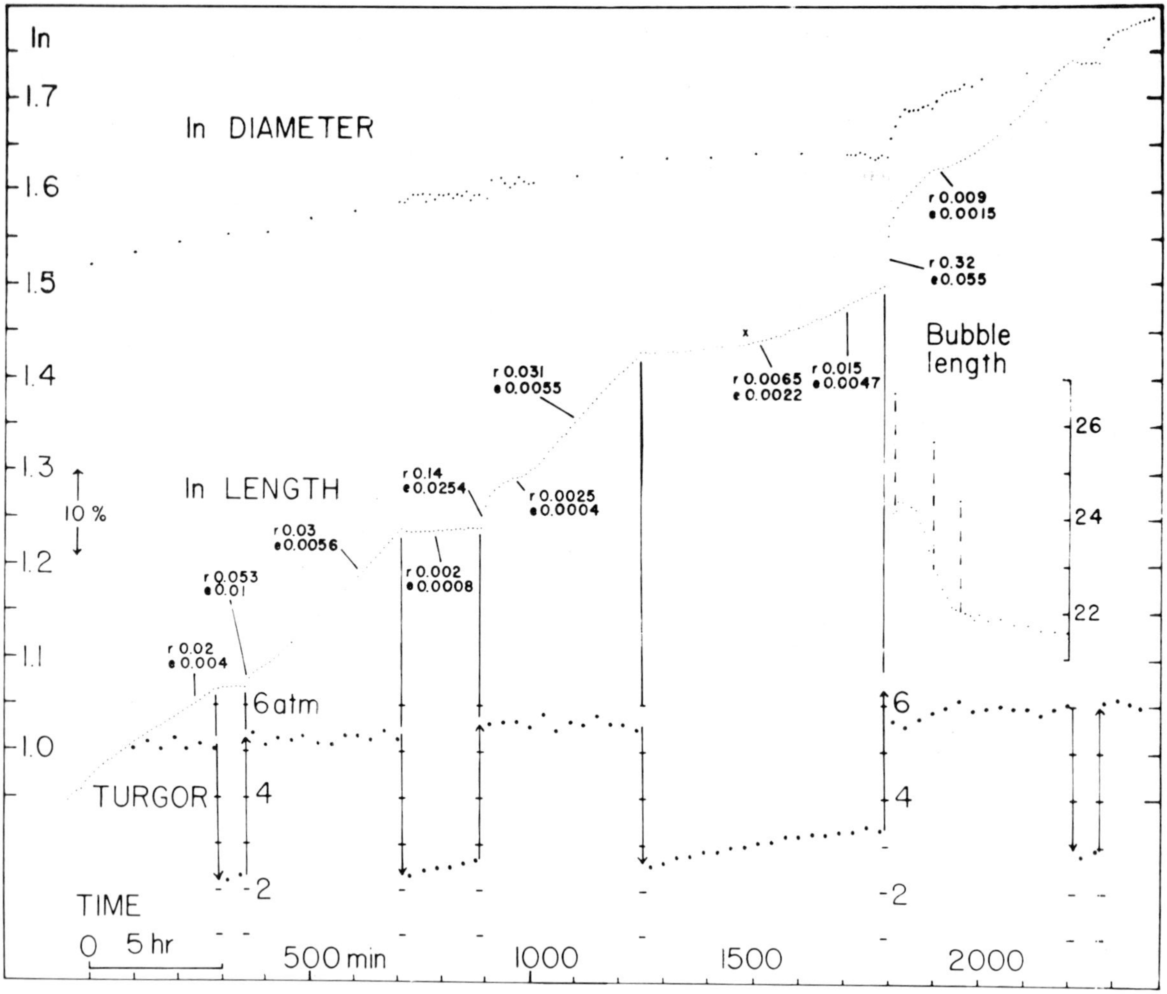

FIG. 6. Symbols and scales as in figure 4 and 5. An experiment where duration, not severity, of osmotic inhibition is varied. Here 1, 3, 9, and 1 hour of the same osmotic inhibition is given, with 6 hours of growth in normal medium between treatments. Note the progressive increase in size of the growth burst with increasing time of inhibition. The maximum slope at the start of each burst also increases. The turgor slowly increases throughout, but much more rapidly during inhibition. The turgor increases as if solute accumulation were continuing at the normal rate while volume increase was inhibited. Note partial recovery of growth rate after 4 hours of the third inhibition. ln Diameter is plotted above. Note the sharp increase during the largest growth burst (1800 min). Normally the increase in ln diameter is about 0.2 that of ln length. Over the first 40 minutes of the growth burst, it is about 0.5. The length of the vertical arrows corresponds to the change in turgor predicted by change in the tonicity of the medium (mannitol). That the change in cell turgor nearly corresponds to this, after a period at osmotic equilibrium during inhibition, indicates that a near-equilibrium condition persists even during the growth bursts. Bubble length measurements show lag in attainment of the steady trans-membrane concentration gradient after a change of solution.

Figure 7 shows the response to turgor changes where the osmoticum was KCl. The typical initial halt in growth, the eventual recovery of much of the previous growth rate, and the rapid elongation following an increase in turgor are evident.

The similarity of response among ionic and uncharged osmotica is somewhat unexpected, particularly if the resting potential of the cell membrane is significant in the growth process. This is not improbable since membrane fusions, particularly of Golgi vesicles, are implicated in the growth response. A fairly direct demonstration of this is found in geotropic response of the rhizoids of *Chara* (33). The concentration of KCl used would virtually abolish the resting potential if artificial pond water were the normal medium (19, 20). Because our normal medium contained KCl in sizeable amount, the resting potential would be reduced by roughly half, according to a simplified equation for resting potential given by Giese (11). Relatively non-polar

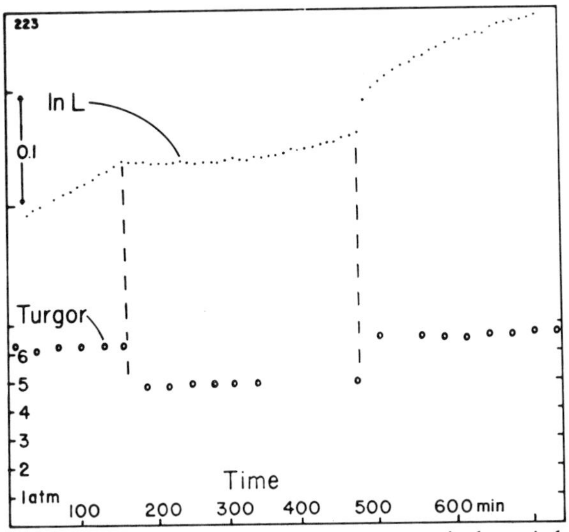

FIG. 7. Course of growth response to a single period of osmotic inhibition where the osmoticum is KCl. The time required for resumption of growth is somewhat longer than typical for other osmotica and the growth burst is somewhat reduced. Otherwise the response is similar to uncharged osmotica.

solutes, even ethanol, can also lower the resting potential. Because KCl and Polyethylene glycol, (a fat solvent) give a response similar to that elicited by mannitol, it appears that the value of the resting potential may not be a highly significant factor in the growth process. This could be confirmed by adapting the micro-manometer to serve as a micro-electrode to monitor resting potential concurrently with pressure, but this modification has not yet been tried.

The well-known "wall loosening" properties of potassium are not evident in the comparison of responses, perhaps because our normal medium contained sufficient KCl to saturate those wall components responsible for the effect.

In light of the similarity of response to various added solutes, there is some justification in concluding that the typical growth reactions reflect the cell's adjustment to a change in turgor *per se* rather than to various chemical and electrical properties of the osmotica. The adjustments, which tend to restore normal growth rate after shifts in turgor in either direction, are made primarily in extensibility, not turgor.

The Directed Character of Expansion. In *Nitella* the ratio of relative rate of increase in length to that in diameter is normally about 4.5 (13). During the largest growth burst observed (fig 6) changes in diameter were also measured. The ratio temporarily fell to about 2.0, so it appears that high Ex_o correlates with reduced physical anisotropy of yield in the wall. That extension occurring far above the minimum yield stress should show reduced anisotropy was predicted in the model for *Nitella* wall growth developed by Probine and Barber (28).

Inhibition of Helical Growth. The extension of the *Nitella* cell is normally accompanied by a pronounced rotation of the top of the cell relative to its base (12). In the present experiments a small weight (0.07 g when submerged) was tied to the base of the plant to keep it vertical. Because the cell was gripped at its top, this tension spanned the observed cell. This vertical tension, small compared to the 3.5 g exerted by turgor on the end walls of the growing cell, virtually eliminated the rotational component. This facilitated the following of individual marks. That such a small weight is effective is remarkable because it acts at a small angle (about 15°) to the direction of helical growth. One of the osmotica tested, 1,2 di-methoxy ethane, when present at 0.1 M, virtually eliminated twisting in unclamped cells.

Discussion

The results permit preliminary examination of the working relationship: relative growth rate = extensibility × turgor, as set forth in equation (I). The equation implies that shifts in turgor should lead to proportional shifts in growth rate. This clearly does not happen because the cell responds to a small drop in turgor by cessation of growth then a gradual resumption of the normal rate. This and other behavior shows that the variables in equation (I) are interdependent. The observed behavior could be the result of complexity in either extensibility or turgor (or both). Fortunately, the variation in turgor, beyond that brought on by changes in tonicity of the external medium, is relatively small. A minor increase in turgor (due to continued solute uptake or production) can be seen during the resumption of growth after a turgor drop (fig 6). A small decrease in turgor due to rapid sap dilution during a growth burst probably contributes to the decline to normal growth rate. These changes, however, are small compared to the concurrent ones in extensibility. It will develop that the value of the extensibility term is a function of turgor, growth rate, and time.

Cell Extensibility and the Cell Wall. Extensibility of a cell has been shown by Lockhart (21, 22) to be a function of several variables. It is inversely proportional to wall thickness, and proportional to cell radius and cell wall extensibility. Wall thickness in *Nitella* is known to increase comparatively slowly (40 %/day) during mechanical inhibition of growth (14). Cell extensibility may change several fold in an hour. Since cell extensibility increases during inhibition in the present experiments, wall thickness change is actually in the wrong direction to explain the extensibility changes. It is therefore not considered a primary factor. Cell radius can be measured and its total variation over 2 days (increase of about 15 %) is very small compared to changes in cell extensibility. It is concluded that cell extensibility is primarily a function of extensibility of the cell wall.

Two Stages of Growth Response to Turgor Shift. The first step in an effort to modify equation (I) to permit prediction of growth behavior is to look for broad characteristics common to all or almost all the responses observed. Study of the reaction of the growing *Nitella* cell to sudden shifts in turgor (fig 3, 4, 5, 6, 7) reveals that its growth rate can be characterized by 2 stages of response. The immediate change in rate following a shift in turgor in *either* direction appears as an exaggeration of the effect that would be predicted by equation (I). That is, a moderate (25 %) drop in turgor leads to much larger (100 %) drop in growth rate; a moderate increase in turgor (33 %) in a growing cell (fig 4, 5, middle) temporarily leads to a much greater (up to several hundred percent) increase in growth rate. This rapid response is followed by a slower compensatory adjustment which tends to restore the normal rate: normal growth may be resumed after being stopped by reduced turgor; the high rates of growth seen immediately after an upward shift in turgor are gradually restored to normal (fig 8A).

Somewhat similar rate changes, but without rate recovery at reduced turgor, are found in non auxin-treated *Avena* coleoptile sections temporarily exposed to osmoticum (29). A transient reduction in rate, a return to normal, and a brief recovery period at a rapid rate are associated with the temporary exposure of growing *Phycomyces* sporangiophores to longitudinal stretch (9). The same general kinetics are found upon exposure of a light-grown sporangiophore to a period of darkness (3). This sort of rate pattern, seen here in the response of *Nitella* to transitory osmotic inhibition, can be analyzed in terms of turgor and extensibility.

A Preliminary Model. The immediate rate response (either decrease or increase) seen upon change in turgor could be based on a single relationship between Ex_g (or rate) and turgor which has a threshold value and, above this, has a positive slope. See figure 8B. The slow response could involve a transposition of this relation. During the typical experimental cycle the rapid and slow responses alternate. A normally growing cell would be at position 1 in figure 8A, B, with a turgor of 5 atm. Upon a rapid loss of turgor to 3 atm (position 2) growth would stop because this turgor is below the threshold. One assumption sufficient to give restored growth at reduced turgor involves a lateral displacement of the original curve, to the left, until the normal rate is resumed (position 3). This is one aspect of the slow response. The sudden return to 5 atm of turgor would give, temporarily, a very high rate of growth (position 4) and then the slow response would shift the curve to the right until normal rate and extensibility were restored (position 5).

The major aspects of this scheme that seem established are the disproportionate shift in rate upon change in turgor (in both directions) and the resumption of growth at low turgor. Data are insufficient to show that relationship between Ex_g (or rate) and turgor is of any specific form (such as the linear one in the diagram) and that this form is unchanged as the threshold value is displaced

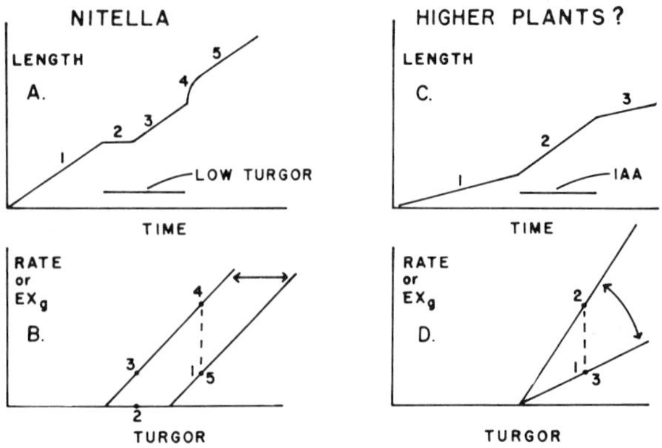

FIG. 8. A diagram to illustrate 2 types of growth response and their possible origin in different kinds of "wall softening." As a simplification, rate is considered roughly proportional to extensibility, Ex_g. *A* shows the typical sequence of length change in *Nitella* through a temporary reduction in turgor. This is accounted for in the model *B* where ordinary growth finds Ex_g near the base of an arbitrarily linear curve relating Ex_g to turgor (postion 1). A drop in turgor, position 2, reduces Ex_g and rate to zero. A shifting of the original curve to the left supports normal growth at position 3. Restoration of the original turgor causes very high extensibility and rate (position 4). A shift of the curve to its original position ultimately restores the normal growth rate (position 5). In C a transient stimulation of growth in higher plants by auxin is diagrammed. Its possible explanation, in terms of Ex_g and turgor, as inferred from the literature, is shown in *D*. The mechanism here is a change in slope of the curve without shifting the threshold. Note that the variation of rate and Ex_g is identical along the dashed lines in *B* and *D* but the underlying mechanism of rate change would be different The present method has the potential of describing the relation between Ex_g and turgor pressure through various growth responses.

Note added in proof: The work of Evans (Ph.D. Thesis, University of California, Santa Cruz, 1967) shows that auxin-grown *Avena* sections have a solute content of 8.2 atm compared to 10.1 atm for non-treated material. The threshold turgor for growth, however, was 3.4 atm and 7.8 atm respectively. Because the treatments were prolonged it is not known whether this reduction in minimum yield stress (and an apparent lateral shifting of the extensibility-*vs*-turgor relation) is the immediate action of IAA on wall properties or is the result of considerable elongation in auxin. Only in the latter case could diagrams C and D apply to *Avena* coleoptiles. The 2 model growth responses contrasted here never-the-less illustrate the potential utility of *in vivo* analysis of the extensibility-turgor relation in the study of wall softening.

along the turgor axis. This can be tested by breaking up the large turgor shifts into small steps. The various components of this tentative model can be examined for their generality and, where possible, for their physical or metabolic basis.

a) *A Threshold Value of Turgor for Growth and Extensibility.* All osmotica tested show that in *Nitella* a small drop in turgor leads to a cessation of growth and thus a reduction of Ex_0 to zero. As seen at the left in figure 4 a drop in turgor of about 0.4 atm is sufficient. Since turgor was 4 atm the threshold value is roughly 90 % of full turgor. A threshold concentration (above the plasmolytic value) of osmoticum sufficient to stop growth, has often been reported for other material. Cleland (4) found that 0.25 M mannitol inhibited almost all growth in *Avena* coleoptile sections while 0.12 M inhibited almost all of the auxin-stimulated component of growth. Ray and Ruesink (32) showed that mannitol of 0.2 to 0.3 M stopped growth in tissues whose cell contents were about 0.4 osmolal. Bennet-Clark (2) found that KCl at a concentration of about 5.4 atm was sufficient to halt growth in material with an internal solute concentration of about 13 atm. While more dilute solutions may momentarily halt growth, this is not pertinent because the effect is probably a rapid elastic response which briefly compensates the extension accomplished by growth.

In a study of the gradual "creep" of excised *Nitella* cell wall strips Probine and Preston (27) found a threshold for extension at about 35 % of the stress normally experienced by the wall *in vivo*. Our data indicate a considerably higher threshold. It would be interesting to know if the *in vivo* shift in minimum yield stress would be reflected in the properties of wall strips from cells killed before and after adjustment to osmotic inhibition.

b) *Extensibility Above the Threshold.* The *in vivo* quantitative relation between increase in turgor and increase in Ex_0, above the threshold, cannot be described by the data presently available. Upward shifts in turgor of varying amount, all starting from a given low turgor at which growth was occurring, could be used to discover the relation between increased turgor and increased extensibility, from that given turgor. Perhaps the curves would be similar for various turgors. At present no *in vivo* data is available. Probine and Preston, however, (27) have studied the mechanical properties of excised longitudinal strips of *Nitella* cell wall and found that, after a variable amount of "set" following application of a weight, the strips slowly yielded to the weight. Their data correlated well qualitatively with the normal growth process in that strips from rapidly growing cells crept relatively rapidly. In absolute terms the strips yielded more rapidly to the applied stress than they had been yielding to the equivalent longitudinal component of turgor stress. [*E.g.*, a 9 % yield at normal stress over 99 min was observed in a strip from cell that had grown 9.5 % over 1 day (1,440 min)]. The difference may reflect the fact that the strip could decrease its lateral dimension while the intact growing wall continuously increases its lateral dimension. A role for creep in the normal growth process was also supported by their finding that transverse strips showed negligible creep, correlating with the small increase in this direction *in vivo*.

Their data were scattered above the threshold but a plot of percent extension (ordinate) *vs.* load gave curves which rose roughly linearly. As already noted the threshold value was considerably lower than that seen *in vivo*. Such a relation, if reversibly shifted as in figure 8B, could give the typical response (fig 8A). Their data were later used in a theoretical model for *Nitella* wall growth that gave a linear increase in rate of movement of individual microfibrils, with stress, above a threshold stress (28).

The rapid "exaggerating" growth response to a shift in turgor in *Nitella* appears to have a counterpart in *Avena* coleoptile tissue. In auxin grown material, above a certain value for turgor, a shift in turgor will yield a relatively larger shift in growth rate. *E.g.*, a 14 % increase in apparent turgor (8.4–9.6 atm) gives a 39 % increase in rate over 4 hours as calculated from data of Cleland (4). Ex_0 would also increase over this range. Unlike the data from *Nitella* there is growth below the break in the curve. This may reflect osmoregulation during the 4 hours. In osmotic inhibition of coleoptile section growth there appears to be little growth recovery by lowering of the yield threshold of the wall. In auxin-free medium, however, some action on the wall during inhibition is indicated by the rapid elongation that occurs upon restoration of full turgor (29). Because the yield point does not seem to be lowered, this softening may take the form seen in figure 8D. Here the increased extensibility would be due to a change in slope of the Ex_0-turgor relation rather than to a shift in the curve with constant slope as in figure 8B. In *Avena* the elastic component of elongation is greater than in *Nitella*; it would be included in Ex_0

c) *Movement of the Yield Threshold.* Movement of minimum yield stress to lower values of turgor is a widespread process. It undoubtedly occurs in many cases of initiation of expansion such as the branching of hyphae, rhizoid formation in the *Fucus* zygote, *etc.* Presumably the lowering involves a chemical breaking of cross-links in the wall so as to concentrate stress on the remaining links, causing them to give. If this process is reversible, a reduction in the concentration of the lytic agent would raise the yield value. An alternate mechanism is available, however, for the raising of the yield value. Higher plant tissues (and many inert materials) show "strain hardening", an increase in yield value coupled to the amount of stretch the material has undergone (5, 24). The metabolic nature of the lowering and the physical (or metabolic) nature of the raising of the yield point is not known for *Nitella*. A rough similarity in the duration of the reduction and raising of the yield point in figure 3 suggests both might be metabolic.

Few higher plant tissue display yield point lowering, as evidenced in resumed growth at reduced turgor. In *Avena*, full restoration of growth rate at reduced turgor appears not to occur. After complete cessation, the amount of subsequent elongation over several hours is not much more than a recovery of elastic shrinkage (2, 26). The small amount of rate recovery after growth cessation, weakly enhanced by auxin at 5 ppm but not at 1 ppm (2), may be due to osmoregulation rather than to a change in wall properties because deplasmolysis is observed in strongly inhibited sections (26). Sucrose promotes increase in internal solute content and recovery.

Recovery of the full normal growth rate in a solution of osmoticum (0.19 M) strong enough to initially halt growth has been reported for potato by Thimann *et al.* (their table I) (36). This recovery only took place in the presence of auxin and appeared to require 2 days. Data on internal solute concentration were not given so it is not known if recovery was due to recovery of turgor and IAA-increased extensibility or a lowering of a threshold for extensibility (or a combination).

Two Modes of Wall Softening. One can recognize, even at the present abstract level of analysis, 2 extreme kinds of "softening." In terms of growth rate and turgor alone a shift from position 1 to position 4 in figure 8B would be indistinguishable from the shift from position 1 to 2 in figure 8D. The basis of the rate change in terms of the whole Ex_o-turgor relation could be quite different. Combinations of these 2 extremes can readily be imagined.

Growth Physics in Higher Plant Tissues. In comparison to the situation in *Nitella* the physical properties of higher plant tissues, manifested in their resistance to imposed stretch, appear less adequate to explain, in themselves, the *in vivo* responses to shifts in turgor. There are major mechanical differences between growth and imposed stretch, especially in killed tissue. A *decrease* in volume and surface area may accompany stretch (24). At any event, the high threshold of turgor needed for rapid growth *in vivo* is not matched by a high minimum yield stress in the artificially stretched tissue (25). The minimum yield force for increase in area in nearly plasmolysed mung bean is only about 5 grams (with IAA) to about 30 grams (without previous incubation in IAA). The hypocotyl's longitudinal component of normal turgor was calculated to be 235 g (24). Further, the *in vivo* relation between rate and turgor, which might be expected to correspond to a strain-rate *vs.* stress relation found in imposed stretch, does not obtain. Strain-rate and stress are almost independent. Starting at a rate 15 × the normal one for *Avena*, an increase in rate of imposed extension of 100-fold is accompanied by roughly a 15 % decrease in the yielding tendency of coleoptile tissue not previously grown in auxin (6, fig 8). With previous growth in auxin the decrease is roughly 25 %. Lockhart (23) found only small viscous properties in mung bean. A 5× increase in rate of stretch gave only a 50 % increase in the work of deformation. Temperature effects were minimal. The failure to find prominent viscous properties in stretch tissue has lead Cleland (personal commun.) to the view that growth occurs in a series of pulses statistically spread throughout the wall.

A pulse is started by the metabolic lowering of the yield threshold of the wall. When this falls below the existing stresses generated by turgor, the wall yields. The process of yielding however is coupled to an increase in the yield threshold, the phenomenon of strain-hardening. This ultimately raises the threshold above existing stresses to end the pulse. This process is called strain-hardening plastic deformation (SHPD) (5, 6). The action of auxin, which could potentially increase rate in many ways (*e.g.* increasing the frequency of pulses, rate of lowering of yield threshold, *etc.*), may be to change the coupling between extension (not rate of extension) and increase in the yield threshold of the wall. In stretched tissues it is found that this coupling is changed by auxin treatment so as to give more extension per unit increase in the yield threshold. Even below its threshold value for growth, increased turgor promotes the increase in compliance brought on by incubation in auxin (7). The present method cannot distinguish between viscous flow and either SHPD or a chemorheological mode of wall yielding (29, 31). The major component of the pulse growth mechanism that is obviously present in *Nitella* is a capacity to lower the yield threshold. This is seen in the resumption of growth at reduced turgor, a phenomenon not prominent in *Avena*. In the coleoptile it is therefore assumed that yield point lowering can occur only above a very high threshold value of turgor (Cleland, personal commun.).

In summary, there appears to be a high turgor threshold for growth and Ex_o in many plant systems, including *Nitella*. The physical nature of yielding above this threshold is not clear but stretching experiments on *Nitella* cell walls indicate it could be a viscous process while stretch studies on higher plant tissues find little viscosity and inappropriately small auxin effects on it. The prominent auxin effect in stretch experiments is a reduction in strain-hardening, a process essentially independent of rate. To incorporate strain-hardening as a rate limiting step *in vivo* a pulsed metabolic lowering of the wall's yield threshold during pulses of growth must be assumed. This lowering is, curiously, a striking feature of *Nitella's* slow *in vivo* response to reduced turgor but it is not noteworthy in the response of growing higher plant tissues to reduced turgor. It is suggested that further *in vivo* characterization of the extensibility-turgor relationship, in the manner diagrammed in figure 8, can resolve "wall softening" into sub-processes and thereby promote understanding of the growth process in terms of structure and metabolism.

Acknowledgment

Conceptual and technical advice from Dr. Peter M. Ray (University of California, Santa Cruz), Dr. Albert List (Drexel Institute of Technology, Philadelphia), and Dr. Gordon Ellis (University of Pennsylvania), was incorporated into the present project. The technical assistance of Mrs. Shirley Serota, Mr. James Buggy, and Mr. Edward Jahn is gratefully acknowledged.

Literature Cited

1. Arens, K. 1939. Bestimmung des Turgordruckes an einer Einzelzelle mit dem Manometer. Planta 30: 113–17.
2. Bennet-Clark, T. A. 1956. Salt accumulation and mode of action of auxin. A preliminary hypothesis. In: Chemistry and Mode of Action of Plant Growth Substances. R. L. Wain and F. Wightman, eds. Butterworth and Company, London.
3. Castle, E. S. 1966. Light responses of *Phycomyces*. Science 154: 1416–20.
4. Cleland, R. 1959. Effect of osmotic concentration on auxin-action and on irreversible and reversible expansion of the *Avena* coleoptile. Physiol. Plantarum 12: 809–25.
5. Cleland, R. 1967. Extensibility of isolated cell walls: measurement and changes during cell elongation. Planta 74: 197–209.
6. Cleland, R. 1967. Auxin and the mechanical properties of the cell wall. Ann. N. Y. Acad. Sci. 144: 3–18.
7. Cleland, R. 1967. A dual role of turgor pressure in auxin-induced cell elongation in *Avena* coleoptiles. Planta 77: 182–91.
8. Dainty, J. 1963. The polar permeability of plant cell membranes to water. Protoplasma 57: 220–28.
9. Dennison, D. S. and C. Roth. 1967. *Phycomyces* sporangiophores: fungal stretch receptors. Science 156: 1386–88.
10. Forsberg, C. 1965. Axenic culture of *Chara globularis* Thuill. and *Chara zeylanica* Wild. Life Sciences 4: 225–26.
11. Giese, A. C. 1962. Cell Physiology. W. B. Saunders Company, Philadelphia, Pennsylvania.
12. Green, P. B. 1954. The spiral growth pattern of the cell wall in *Nitella axillaris*. Am. J. Botany 41: 403–09.
13. Green, P. B. 1963. Cell walls and the geometry of plant growth. In: Meristems and Differentiation. Brookhaven Symp. Biol. 16: 203–17.
14. Green, P. B. 1963. On mechanisms of elongation. In: Cytodifferentiation and Macromolecular Synthesis. M. Locke, ed. Academic Press, New York.
15. Green, P. B. 1965. Anion-exchange resin spheres as marking material for wet cell surfaces. Exptl. Cell Res. 40: 195–96.
16. Green, P. B. and F. W. Stanton. 1967. Turgor pressure: direct manometric measurement in single cells of *Nitella*. Science 155: 1675–76.
17. Gutknecht, J. 1968. Salt transport in *Valonia*: inhibition of potassium uptake by small hydrostatic pressure. Science 160: 68–70.
18. Kamiya, N., M. Tazawa, and T. Takata. 1963. The relation of turgor pressure to cell volume in *Nitella* with special reference to mechanical properties of the cell wall. Protoplasma 57: 501–21.
19. Kishimoto, U. 1959. Electrical characteristics of *Chara corallina*. Ann. Rep. Sci. Works Fac. Sci. Osaka Univ. 7: 115–46.
20. Kishimoto, U., R. Nagai, and M. Tazawa. 1965. Plasmalemma potential in *Nitella*. Plant Cell Physiol. 6: 519–28.
21. Lockhart, J. A. 1965. An analysis of irreversible plant cell elongation. J. Theoret. Biol. 8: 264–75.
22. Lockhart, J. A. 1965. Cell extension. In: Plant Biochemistry. J. Bonner and J. Varner, eds. Academic Press, New York.
23. Lockhart, J. A. 1967. Physical nature of irreversible deformation of plant cells. Plant Physiol. 42: 1545–52.
24. Lockhart, J. A., C. Bretz, and R. Kenner. 1967. An analysis of cell-wall extension. Ann. N. Y. Acad. Sci. 144: 19–33.
25. Olson, A., J. Bonner, and D. J. Morre. 1965. Force extension analysis of *Avena* coleoptile cell walls. Planta 66: 126–34.
26. Ordin, L., T. H. Applewhite, and J. Bonner. 1956. Auxin induced water uptake by *Avena* coleoptile sections. Plant Physiol. 31: 44–53.
27. Probine, M. C. and R. D. Preston. 1962. Cell growth and the structure and mechanical properties of the wall in internodal cells of *Nitella opaca*. II. Mechanical properties of the walls. J. Exptl. Botany 13: 111–27.
28. Probine, M. C. and N. F. Barber. 1966. The structure and plastic properties of the cell wall of *Nitella* in relation to extension growth. Australian J. Biol. Sci. 19: 439–57.
29. Ray, P. M. 1961. Hormonal regulation of plant cell growth. In: Control Mechanisms in Cellular Processes. D. M. Bonner, ed. Ronald Press, New York.
30. Ray, P. M. 1967. Radioautographic study of cell wall deposition in growing plant cells. J. Cell Biol. 35: 659–74.
31. Ray, P. M. and A. W. Ruesink. 1962. Kinetic experiments on the nature of the growth mechanism in oat coleoptile cells. Devel. Biol. 4: 377–79.
32. Ray, P. M. and A. W. Ruesink. 1963. Osmotic behavior of oat coleoptile tissue in relation to growth. J. Gen. Physiol. 47: 83–101.
33. Sievers, A. 1967. Elektronenmikroskopische Untersuchungen zur geotropischen Reaction. III. Die transversale Polarisierung der Rhizoidspitze von *Chara foetida* nach 5 bis 10 Minuten Horizontallage. Z. Pflanzenphysiol. 57: 462–73.
34. Tazawa, M. 1957. Neue Methode zur Messung des Osmotischen Wertes einer Zelle. Protoplasma 48: 342–59.

35. TAZAWA, M. AND N. KAMIYA. 1965. Water relations of Characean internodal cell. Ann. Rep. Biol. Works Fac. Sci. Osaka Univ. 13: 123–57.
36. THIMANN, K. V., G. M. LOOS, AND E. SAMUEL. 1960. Penetration of mannitol into potato disks. Plant Physiol. 35: 848–53.
37. VILLEGAS, L. 1967. Changes in volume and turgor pressure in *Valonia* cells. Biochim. Biophys. Acta 136: 590–93.
38. VIRGIN, H. I. 1955. A new method for the determination of the turgor of plant tissues. Physiol. Plantarum 8: 954–62.

Reprinted from Science, March 19, 1965, Vol. 147, No. 3664, pages 1454-1455

Growth and Tissue Formation from Single, Isolated Tobacco Cells in Microculture

Abstract: *Single, isolated cells of tobacco divided and grew to form small colonies of over 50 cells in microcultures with a medium containing fresh liquid coconut milk, but in the absence of neighboring cells or nurse tissue. Subsequently, the mass of cells obtained from the single cell, when transferred to agar coconut milk medium, established itself as a clone of callus tissue. Some of these single-cell clones showed differentiation of many tracheid-like cells and shoots with small leaves.*

Various methods have been employed for growing single plant cells in vitro. These methods can roughly be grouped under four categories: the filter paper, nurse culture method (*1*); the hanging drop technique (*2*); the plating method (*3*); and the microculture method (*4*). Except in a few preliminary trials, in all these methods either a large number of cells in the same environment, a nurse tissue, or a "conditioned" medium (*5*) was used for growing the single cells, so that the single cell was never really free from the influence of the neighboring cells or of the medium from which it was isolated.

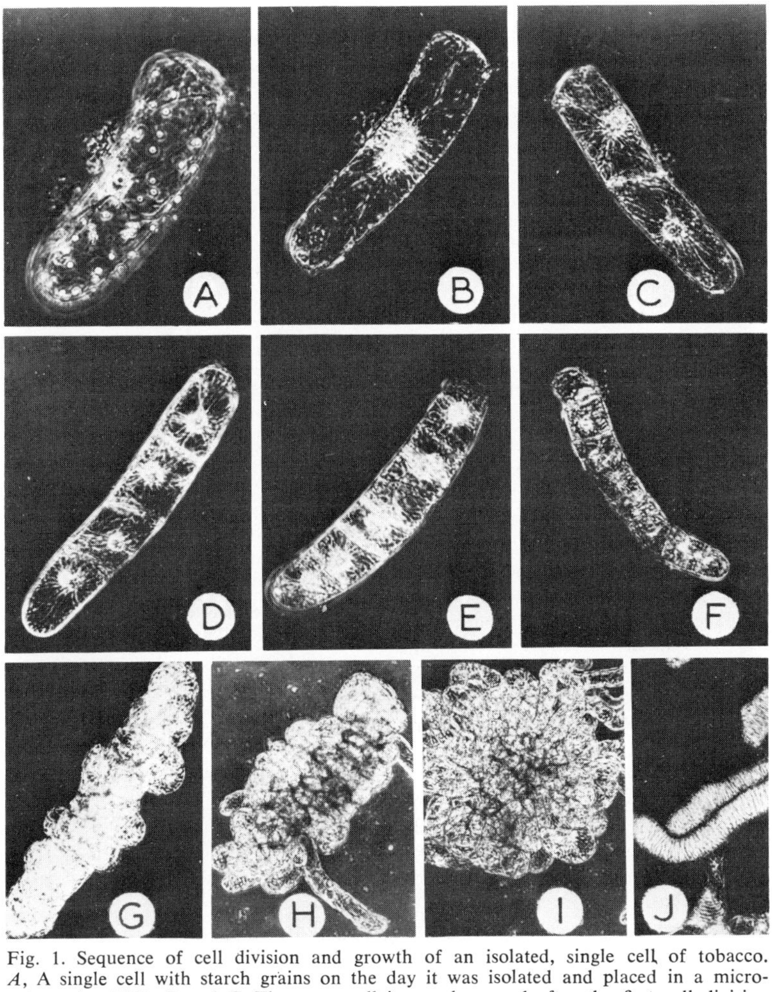

Fig. 1. Sequence of cell division and growth of an isolated, single cell of tobacco. *A*, A single cell with starch grains on the day it was isolated and placed in a microculture (× 700). *B* and *C*, The same cell in prophase and after the first cell division on the 4th day, respectively (*B*, × 560; *C*, × 448). *D–F*, Formation of a linear colony of cells on the 5th, 7th, and 9th days (*D*, × 448; *E*, × 448; *F*, × 350). *G* and *H*, Further stages in the formation of a mass of cells from the original single cell shown in *A* (*G*, × 350; *H*, × 175). *I*, Mass of cells resulting from growth by the single cell shown in *A* (× 175). All stages of development from the single cell shown in *A*. *J*, Portion of tracheid-like cells (× 525).

The plating method has been used (*6*) for studying the growth of colonies from free cells, and the behavior of individual cells of carrot and *Haplopappus*. Pieces of tissue from the secondary phloem of carrot roots in liquid medium produced freely suspended single cells which grew and divided to give rise to groups of cells (*7*). These cell aggregates then formed organized areas or nodules from which roots were produced; when a rooted nodule was transferred to agar medium, it produced shoots. The plants grown in vitro flowered and produced viable seeds when transferred to the greenhouse. In most of these instances, coconut milk was an important constituent of the nutrient medium. Plantlets with shoots and roots were obtained recently in suspension cultures of endive embryo callus tissue in a chemically defined medium (*8*). Entire plants or even continuously growing masses have not been produced from isolated, single cells grown in vitro without nurse tissue, neighboring cells, or "conditioned" medium.

We, therefore, attempted to grow single, completely isolated cells in the absence of any nurse tissue, neighboring cells, or "conditioned" medium and to obtain a continuously growing callus from each cell. These tests were additionally important since the results would throw light on the ability of a single plant cell to form a tissue and to differentiate into a plant.

The plant tissue used was hybrid tobacco (*Nicotiana tabacum* × *N. glutinosa*) secondary single-cell clone H241-18 (*9*) maintained for 8 years on agar *D* medium (inorganic salts, sucrose, coconut milk, vitamins, calcium pantothenate, α-naphthaleneacetic acid, and 2,4-dichlorophenoxyacetic acid) (see *10*). Several media were tried, with or without coconut milk and 2,4-dichlorophenoxyacetic acid (2,4-D). The callus grew and dissociated best in liquid *C* medium (same as *D* medium, but without 2,4-D) on a reciprocating shaker (60 cycles per minute) in the dark. Pieces of callus weighing approximately 200 mg, when transferred to liquid *C* medium, produced the greatest number of single cells between 10 and 12 days after transfer to the shake-culture. Single cells were isolated after they had been in shake-cultures from 1 to 18 days to ascertain the most fa-

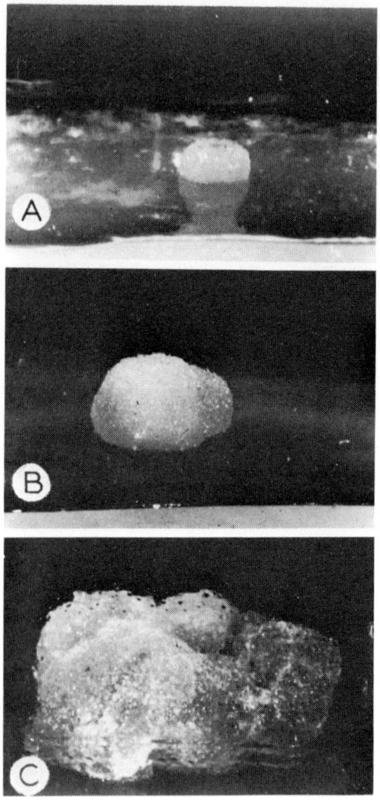

Fig. 2. *A–C*, Calluses, 2, 4, and 6 weeks old, respectively, produced by the cell masses after transfer to agar *C* medium (× 4.8).

vorable age of the single cells for isolation and subsequent growth in microculture. The best age was 10 or 11 days, which coincided with the period for the greatest dissociation of single cells.

Isolated, single cells varied in size and shape. Small cells were generally selected for culturing since a larger number of these divided. Single cells were isolated from a drop (0.5 ml) of the 10- or 11-day-old culture from the shaker and placed in a petri plate to which 15 to 20 ml of fresh liquid *C* medium had been added. Single cells were hand-picked with "Pyrex" glass micropipettes from the petri plate under a dissecting microscope. A microculture (*11*) was prepared in which the isolated cell was placed in a drop of fresh liquid *C* medium in complete isolation from any other cells or cell groups. The microcultures, between observations, were stored in petri plates that were kept in steel cans in the dark at 26° ± 1°C and 55 percent relative humidity. All observations were made under a Zeiss phase-contrast microscope.

Originally, 160 single cells were isolated from cultures grown for 10 to 11 days on the shaker and grown in microcultures in fresh liquid *C* medium. Sixty-five of these single cells divided to form colonies of over 50 cells each. Subsequently, in three separate experiments, 10, 12, and 15 single cells were grown singly in the fresh medium, and all but one single cell grew to form small colonies of more than 50 cells each.

The single cells of tobacco showed, besides the usual cell wall, nucleus, vacuoles, and cytoplasm, a large number of starch grains which were more abundant around the nucleus (Fig. 1*A*). The starch was gradually digested after the single cell was put in the microculture. In most instances, the first division of the single cell took place by the 4th day at right angles to the long axis of the cell (Fig. 1*B*). This lag or preparatory period of 3 days seemed essential for the division of the single cell. Similarly, in freely suspended cells of carrot dispersed in nutrient agar medium in petri plates, growth started after an "induction period" of about 5 days, but some cells remained apparently unchanged for a considerably longer period after inoculation before they started to grow (*6*). Unlike carrot, where 90 percent of the single cells that ultimately grew began to grow continuously after the 9th day, in tobacco, the single cells that failed to divide by the 6th day remained undivided, but continued to live for as long as a month. After the first division, the next few divisions were also in the same plane so that a filament of 8 to 12 cells would form within 7 to 9 days (Fig. 1, *C–F*). In none of the single cell cultures did the pattern of the first four divisions deviate from this plan. Later, divisions started parallel and obliquely to the walls on the long sides of the cells; gradually, the uniseriate filament was converted into a multiseriate cordlike or globose structure (Fig. 1*G*). Further divisions produced a mass of over 50 cells within 15 to 20 days (Fig. 1, *H* and *I*). This small mass of callus, if allowed to remain in the same medium, did not show any increase in cell number. However, starch grains appeared again in abundance in the cells. The small cell colony thus formed was aseptically transferred to fresh agar *C* medium in 6-ounce (170-g) prescription bottles where it continued to grow to form a large mass of callus tissue (Fig. 2, *A–C*). Many such clones of tissue have now been established; some of them showed a large number of tracheid-like cells (Fig. 1*J*). Experiments to differentiate this callus into root and shoot were not successful, probably because the original tissue from which single cells were isolated was several years old. However, clones similarly established from single cells isolated from fresh callus of hybrid tobacco pith tissue, when grown on Murashige and Skoog's chemically defined medium (*12*) with varying concentrations of indoleacetic acid and kinetin, showed shoot formation with several leaves in some instances. The balance of factors encouraging further differentiation of the single cell clones is of interest.

VIMLA VASIL
A. C. HILDEBRANDT
Department of Plant Pathology, University of Wisconsin, Madison

References and Notes

1. W. H. Muir, A. C. Hildebrandt, A. J. Riker, *Am. J. Botany* **45**, 589 (1958).
2. J. G. Torrey, *Proc. Natl. Acad. Sci. U.S.* **43**, 887 (1957).
3. L. Bergmann, *J. Gen. Physiol.* **43**, 941 (1959); J. G. Torrey and J. Reinert, *Plant Physiol.* **36**, 483 (1961); J. Reinert, *Nature* **200**, 90 (1963); J. L. Gibbs and D. K. Dougall, *Science* **141**, 1059 (1963).
4. L. E. Jones, A. C. Hildebrandt, A. J. Riker, J. H. Wu, *Am. J. Botany* **47**, 468 (1960).
5. The term "conditioned" refers to a medium in which tissues or cells have been grown for some time.
6. L. M. Blakely and F. C. Steward, *Am. J. Botany* **49**, 653 (1962); *ibid.* **51**, 780 (1964); L. M. Blakely, *ibid.*, p. 792.
7. F. C. Steward, M. O. Mapes, K. Mears, *ibid.* **45**, 705 (1958).
8. I. K. Vasil, A. C. Hildebrandt, A. J. Riker, *Science* **146**, 76 (1964).
9. R. C. Sievert, A. C. Hildebrandt, A. J. Riker, *Plant Physiol.* **38**, 51 (1963).
10. A. C. Hildebrandt, in *Modernen Methoden der Pflanzenanalyse*, M. V. Tracey and H. F. Linskens, Eds. (Springer, Berlin, 1963), vol. 5, p. 383.
11. A microculture was prepared in the following way: A droplet of paraffin oil (U.S. Pharmacopeia heavy mineral oil) was placed near each end of a standard microscope slide. A 22-mm square No. 1 coverglass was lowered onto each droplet to form "risers" on the slide. A rectangle of mineral oil was then placed on the slide connecting the two coverglass risers and covering their inner ends. Next, the isolated cell along with the drop of fresh liquid *C* medium in the micropipette was placed on the slide between the coverglass risers and a third coverglass was inverted over the microdrop and well surrounded with the mineral oil. The resulting microculture permitted microscopic observations of the isolated cell as it grew and divided to form a colony.
12. T. Murashige and F. Skoog, *Physiol. Plantarum* **15**, 473 (1962).
13. This work was supported by AEC contract AT(11-1)-1304, by the American Cancer Society, by NIH (project CA05860), and by the Research Committee of the Graduate School from funds supplied by the Wisconsin Alumni Research Foundation. We are indebted to E. H. Herrling for assistance in preparing the illustrations.

16 November 1964

Reprinted from Science, February 12, 1965, Vol. 147, No. 3659, pages 756-758

Ontogeny of Adventive Embryos of Wild Carrot

Abstract. *Somatic carrot cells in culture divide to produce undifferentiated preglobular proembryos which exhibit a wide variety of segmentation patterns. One or more globular proembryos, which exhibit normal histological zonation, may develop from single preglobular proembryos. Regeneration of normal embryos from cultured cells grown on media containing only minerals, sucrose, vitamins, and 2,4-dichlorophenoxyacetic acid suggests that the embryo sac may have less of a formative role than currently ascribed to it.*

The recent demonstration that cultured somatic cells of the wild carrot *Daucus carota*, when grown on simple defined media, will regenerate carrot embryos focuses attention on the forces which direct embryo development (*1, 2*). The view that embryo development is under considerable regulation by unique physical and chemical factors in the embryo sac (*3*) gained support from experimental studies which purported to show that coconut milk induced cultured cells to behave as zygotes (*4*). However, the occurrence of embryogenesis in the presence of coconut milk does not establish a causal relationship.

It is now clear that coconut milk is not required for the regeneration of embryos from cultured cells of the wild carrot and is, in fact, inhibitory to the development of carrot proembryos (*2*). It is also significant that in the only well-documented report of embryos from cultured cells of the domestic carrot (*5*), coconut milk was not used in the medium. This point needs to be emphasized since studies of the molecular basis for the regeneration of embryos from cultured cells will be hampered by the use of such complex substances as liquid endosperm.

We now present evidence that embryos which develop from wild carrot cells grown on a basal medium (*2*) containing only minerals, sucrose, vitamins, and 2,4-dichlorophenoxyacetic acid (2,4-D) undergo a series of developmental changes which may be more similar to embryogenesis in the ovule than previously reported, and that variations in the ontogeny of undifferentiated proembryos, caused by culture conditions, have little effect on subsequent embryo development.

We stated previously (*1*) that adventive embryos differed from seed embryos in that they usually did not have a suspensor-like component. This conclusion was erroneous, and attributable to the fact that most observations were made of embryos dissected from callus or embryos which developed in rotating liquid cultures—circumstances which led to the loss of the fragile suspensor (*6*). We have now studied numerous microtome sections of embryos embedded in callus in the original position where they formed, and it is evident that nearly always a distinct suspensor of variable size and shape is present (Fig. 1). In addition, we have grown groups of single cells and observed the entire developmental sequence from single cells to mature embryos. Evidently, the suspensor is nothing more than those cells of the proembryo (*7*) which are not incorporated into the embryo proper, and the size and shape of the preglobular proembryo varies, depending upon the number and orientation of mitoses which occur in random fashion prior to the establishment of polarity. Under the culture conditions used, polarized growth and histological differentiation were inhibited if proembryos were exposed to a concentration of 2,4-D higher than 0.1 mg/liter. When polarity was established in a preglobular proembryo of few cells, subsequent growth and differentiation usually resulted in a globular proembryo with an insignificant suspensor (Fig. 1*C*). If polarity was delayed until a sizable cell mass developed, a single globular proembryo with a massive suspensor (Fig. 1, *A* and *B*) might differentiate, or, commonly, two or more embryos might develop from the same cell mass. In the latter case, the multiple embryos shared a common suspensor and were thus joined at the radicle (Fig. 2).

These variations in ontogeny are illustrated in Fig. 2. Small starch-filled cells divide to form preglobular proembryos which may be spherical, filamentous, or irregular in shape—the particular segmentation pattern being a response to random chemical and physical factors in the callus mass or cell suspension where such proembryos develop. The disorganized spheres (Fig. 2*b*) generally develop into single embryos, as shown by the sequence *b, e, h*, in Fig. 2. Although the well-organized filamentous proembryos appear to have axial polarity and may proliferate at one pole to yield a single embryo with a filamentous suspensor (Fig. 1*A*), unrestricted access to nutrients can also lead to proliferation at several points on the filament rather than just at one end. This type of development is illustrated by the dumbbell-shaped structure in Fig. 2*f*, which consists of two globular proembryos that have developed at either end of a massive filament. Proliferation at the center of such a structure would lead to the triplet shown in Fig. 2*g*. Mature embryos united at their (radicle) ends, as shown in Fig. 2, *i* and *j*, are commonly found in tissue cultures. Cleavage polyembryony, in which multiple embryos develop from an irregularly shaped or branched filamentous proembryo, has been reported as a common occurrence in the embryo sac of several Angiosperm genera (*8*).

Regardless of the segmentation pattern which results from random growth of the preglobular proembryo, the globular proembryo which ultimately differentiates shows normal histological organization. Figure 1 illustrates the normal tissue zonation of three globular proembryos which have obviously had different ontogenetic histories. Proto-

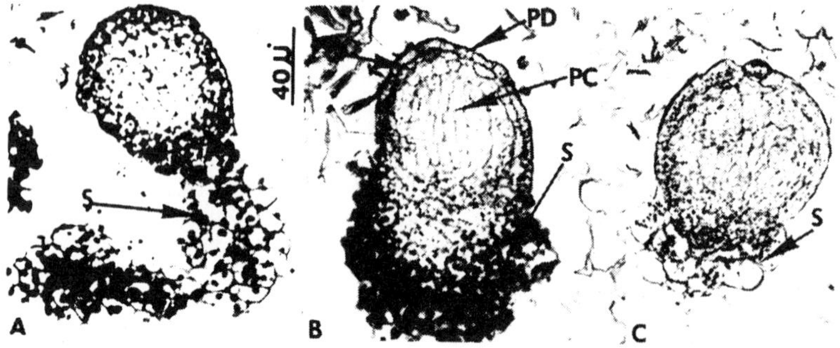

Fig. 1. Sections (15 μ) of globular proembryos embedded in callus. *A*, A proembryo which has differentiated at one end of a massive filament which contains starch grains; *B*, a proembryo which has developed from an undifferentiated spherical mass of cells which are dense with starch; *C*, a proembryo with ephemeral suspensor and relatively little starch (*S*, suspensor; G, ground meristem; *PD*, protoderm; *PC*, procambium; scale in *A* applies to all three embryos).

derm, ground meristem, and procambium are readily discernible. The embryos in Fig. 1, *A* and *B*, show prominent starch grains in the suspensor; the embryo in *A* also shows starch in the region of the ground meristem; that in *C* shows little starch. Starch grains are usually found in the cells of the preglobular proembryo and disappear as the globular embryo differentiates, disappearing first in the region of the procambium. Often, however, numerous grains are present in the cortical region of mature embryos.

Single cells and proembryos containing fewer than five cells were sieved from freshly isolated callus (one to three subcultures) in large numbers, thoroughly washed, and plated on fresh basal media according to the techniques previously described (*2*). Embryogenesis occurred readily under these circumstances. Similar culturing of cells derived from callus which had undergone prolonged subculturing did not yield embryos. Attempts to stimulate embryogenesis in the uniform parenchymatous cells of these older cultures, by adding coconut milk (10 percent by volume), carrot root extracts, or mixtures of natural and synthetic growth regulators to the medium, occasionally resulted in the development of embryos. However, these few embryos are now believed to have been the result of encouraging the growth of a few proembryos carried over from earlier subcultures, rather than the result of inducing some change in "noncompetent" cells.

These findings indicate that cells which are embryologically competent can serve as the starting point in a developmental pathway in which the morphogenetic controls are largely internal. In other words, developing carrot embryos do not appear to require a complex array of exogeneous nutrients or an environment with precise chemical and physical gradients. The factors leading to the "competency" of the zygote or of certain cultured cells are unknown (*9*).

WALTER HALPERIN
DONALD F. WETHERELL
Botany Department,
University of Connecticut, Storrs

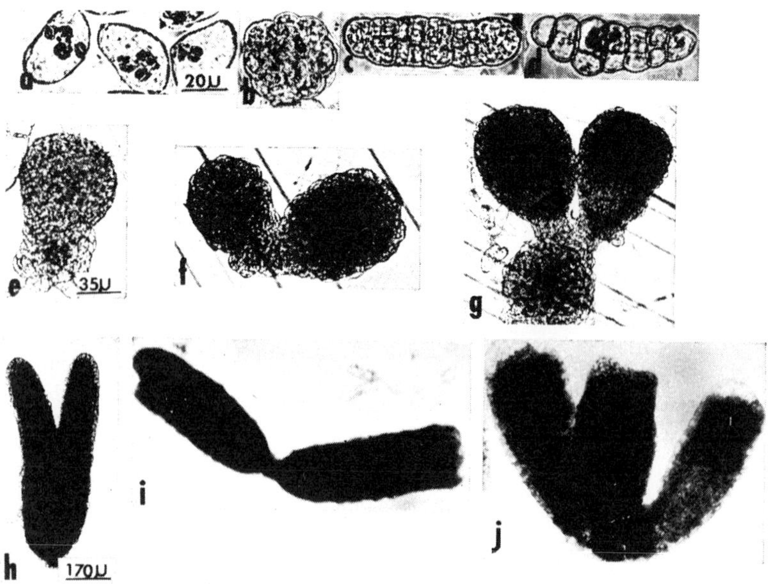

Fig. 2. Various stages in the ontogeny of adventive carrot embryos. Starch-filled embryo initial (*a*) develops into preglobular proembryos which are either spherical (*b*) or filamentous (*c* and *d*); the filamentous pattern is typical of proembryos in the carrot family. Spherical preglobular proembryos (*b*) usually develop into single embryos (*e* and *h*), whereas filamentous proembryos may proliferate to yield either one embryo as shown in (*e*) and (*h*), or multiple embryos as shown in (*f*), (*g*), (*i*), and (*j*). Multiple embryos are commonly found in tissue cultures along with single embryos. The scale, 20 μ in (*a*) applies to (*b*), (*c*), and (*d*); 35 μ in (*e*) applies also to (*f*) and (*g*); 170 μ in (*h*) applies also to (*i*) and (*j*).

References and Notes

1. W. Halperin and D. F. Wetherell, *Am. J. Botany* **51**, 274 (1964).
2. W. Halperin, *Science* **146**, 408 (1964).
3. C. W. Wardlaw, *Embryogenesis in Plants* (Methuen, London, 1955), pp. 224–227.
4. F. C. Steward, M. O. Mapes, A. E. Kent, R. D. Holsten, *Science* **143**, 20 (1964); F. C. Steward, M. O. Mapes, K. Mears, *Am. J. Botany* **45**, 705 (1958).
5. H. Kato and M. Takeuchi, *Plant Cell Physiol.* **4**, 243 (1963).
6. The term suspensor here refers to those cells of the proembryo which are not incorporated into the embryo proper, and which thus have the same topological significance as the suspensor cells of seed embryos. Virtually nothing is known of the function of the suspensor, other than its possible role as a device for pushing the embryo into the endosperm, so no real homology can be established.
7. We have adopted a modification of the terminology of Soueges, in which the term proembryo refers to any developmental stage preceding cotyledon initiation [R. Soueges, *La Differenciation* III. *La Differenciation Organique* (Hermann, Paris, 1936)]. The term preglobular proembryo refers to the histologically undifferentiated condition preceding the initiation of protoderm, and globular proembryo refers to the stage following protoderm initiation and preceding cotyledon development.
8. P. Maheshwari, *An Introduction to the Embryology of Angiosperms* (McGraw-Hill, New York, 1950), p. 343.
9. Since this paper was submitted, we have found that for embryogenesis to proceed in vitro the presence of ammonium ions is essential. Callus will grow indefinitely on a medium containing nitrate alone, but segmentation patterns typical of proembryos will not form. Preliminary experiments indicate that glutamine, aspartate, a mixture of amino acids, or coconut milk will not substitute for the ammonium requirement.
10. This work was done during the tenure of an NIH fellowship (W.H.) and with the aid of NSF grant GB-1936 to D.F.W.

21 December 1964

Reprinted from PLANT PHYSIOLOGY, Vol. 39 No. 5, September, 1964, pages 832–835.
(PRINTED IN U.S.A.)

Dividing Cells as the Prerequisite for Vernalization [1, 2]

S. J. Wellensiek
Laboratorium voor Tuinbouwplantenteelt, Landbouwhogeschool, Wageningen, The Netherlands

Introduction

It is generally accepted that the locus of vernalization is the growing point and preferably the terminal growing tip. For plant vernalization this is based on the work of Curtis and Chang (2) with celery, of Chroboczek (1) with beets, of Schwabe (13) with the chrysant, who all applied a localized cooling of the growing tip. Melchers (8) arrived at the same conclusion for *Hyoscyamus niger* after grafting experiments. He also established that after defoliation vernalization is possible.

For seed vernalization Gregory and Purvis (3, 4) demonstrated that excised rye embryos can be vernalized. Purvis (11) found that even parts of the embryo are susceptible, provided that a growing point is present. Purvis (12) quotes similar results as Drozzin in 1936, Konovalov in 1937, and Bassarskaja in 1941, while she also refers to Lyssenko in 1932.

All the above cited evidence is convincing enough for the concept that growing tips perceive the vernalizing action of low temperatures, but they certainly do not prove that growing tips are the only parts of the plant to do so. In two preliminary notes (15, 16) the present author demonstrated that both isolated leaves and isolated roots of *Lunaria biennis* can be vernalized, as is observed from the flowering of the plants which regenerate from these cuttings. He arrived at the concept that dividing cells are necessary for the action of low temperatures, no matter where in the plant they occur.

The scope of the present paper is to present further details on leaf vernalization and direct evidence that dividing cells are necessary.

Materials and Methods

Plant Material. The experimental plant is *Lunaria biennis* L., a typical cold-requiring biennial. From sowing, the plants pass through a juvenile phase and a transitory phase, lasting approximately 6 and 4 weeks respectively. Then they enter into the vegetative adult phase in which they readily respond to cold treatments. The durations of the juvenile and the transitory phase depend on the light intensity (photosynthesis) during the raising of the plants (6). Also, a relation exists between age and duration of vernalization, older plants requiring less cold. Usually 8 weeks of treatment at 5° is sufficient for a complete and rapid flowering of adult plants (14).

Lunaria biennis is an excellent plant for the present investigations because both leaf cuttings and root cuttings regenerate quite easily into complete plants (5). Leaf cuttings consist of a lamina and part of a petiole. It is advisable to reduce the surface of the lamina to a fixed value by means of a razor blade in order to avoid an eventual influence of the leaf size. In the present experiments the laminae were cut into isosceles triangles of 17.5 cm^2. After planting in sand, about one cm deep, and keeping the cuttings in ordinary greenhouse temperature averaging 20° with a high relative humidity, root formation starts from the cut surface of the petiole within 2 weeks. As soon as the new roots reach a length of about 4 cm, the cuttings are potted in ordinary soil. Some of the roots soon thicken and become typically contractile. On these contractile roots adventitious buds are formed near the cut surface of the petiole and these buds immediately grow into sprouts.

Leaf cuttings were also grown on a nutrient solution. This succeeded rather well, but did not offer any special advantage.

Vernalizing Treatments. Vernalizing cold treatments were given in a cold room of 5° under 12 hours light from fluorescent lamps with an intensity approximating 250 μ W/sec/cm^2. The cold treatments usually started immediately after the cuttings were made. The age of the mother plants, from which the cuttings were taken, and the duration of the cold treatments were varied, but the sowing of the mother plants was timed in such a way that the cold treatments ended simultaneously. Hence the after-treat-

[1] Received May 10, 1963.
[2] Publication 240 of Laboratorium voor Tuinbouwplantenteelt, Landbouwhogeschool, Wageningen, The Netherlands.

ment of all cuttings in one experiment could take place under identical environmental conditions as were found in an ordinary greenhouse.

Experimental Results

Age of Mother Plants and Duration of Cold Treatment (experiment I). Leaf cuttings were taken from plants of 6, 8, 10, 12 or 14 weeks old with average numbers of leaf pairs of 4, 5, 6, 7 or 8 respectively. The cuttings were taken from the third leaf from the top and were exposed to cold treatments during 0, 8, 12, 16 or 20 weeks. Each of the 25 treatments started with 25 cuttings. The percentages of flowering are given in table I.

Table I. *Percentages of Flowering Among Regenerated Leaf Cuttings of* Lunaria biennis, *Taken From Mother Plants of 5 Ages After Cold Treatments During 5 Periods*

Age of mother plants in weeks	Cold treatment in weeks				
	0	8	12	16	20
6	0	0	0	0	3.6
8	0	0	0	0	21.4
10	0	0	0	7.1	25.0
12	0	0	12.5	40.7	40.6
14	0	0	7.5	18.4	40.0

For unknown reasons the percentages of flowering have remained relatively low in this experiment. Nevertheless, some conclusions present themselves. The percentages increase from top to bottom with the exception of the 14 weeks old group. Evidently cuttings from 14 weeks old plants were affected less than those of 12 weeks old. The percentages increase from left to right, so that better results were obtained after cold treatments of longer durations.

The Locus of Vernalization in Leaf Cuttings (experiment II). All the treatments of experiment I were duplicated, but immediately after the cold treatments a portion of 0.5 cm of the petioles was removed. The cuttings re-regenerated rather well, but none of the re-regenerated plants flowered, with one exception which occurred in the 8 weeks old group after a cold treatment during 20 weeks. In absolute figures, 59 out of 299 plants flowered in those treatments of experiment I where flowering occurred. In the comparable treatments of experiment II, 1 out of 195 plants flowered.

We conclude that the vernalization of leaf cuttings is restricted to the lowest part of the petiole in the neighborhood of the cut surface, where the regeneration and hence the cell divisions take place.

Leaf Vernalization on Intact Plants (experiment III). The question can be raised whether leaf vernalization is restricted to isolated leaves or whether also leaves on intact plants can be vernalized. When vernalization is indeed restricted to dividing cells, leaf vernalization on intact plants will also depend on dividing cells during the cold treatment. In other words it will be more successful the younger the leaves are. The experimental approach consisted of periodically making cuttings of a series of leaves with decreasing ages during the cold treatment and hence with increasing numbers of mitoses.

Intact plants of 10, 12 or 14 weeks old were exposed to 5° during 12, 16 or 20 weeks. Ten plants were used in all the 9 treatments. Immediately after the cold treatment, leaf cuttings could be made of the oldest leaf pairs which were already full-grown when the vernalization started, and of the second leaf pairs. These leaf pairs are indicated as I and II respectively. The following leaf pairs, III to VII, were still too small but after 12, 20, 28, 35 and 42 days respectively, leaf cuttings of these leaf pairs could be made. The percentages of flowering of the regenerated leaf cuttings of the leaf pairs I to VII are plotted against the ages of the mother plants, totalized over the 3 cold treatments (fig 1), and are plotted against the durations of the cold treatment, totalized over the 3 ages of the mother plants (fig. 2). In figure 1 we find slightly better flowering in the 12 weeks old group compared with the 10 weeks old group, but the 14 weeks old group is slightly worse than the 12 weeks old group. This is a similar observation as was discussed in experiment I. In figure 2 the flowering percentages increase as the durations of the cold treatment increase. The most important conclusion from both figures 1 and 2 is that the flowering increases as the leaf number increases or as the age of the leaf during the cold treatment decreases. This conclusion is also clear when the percentages of flowering of the cuttings from the leaf

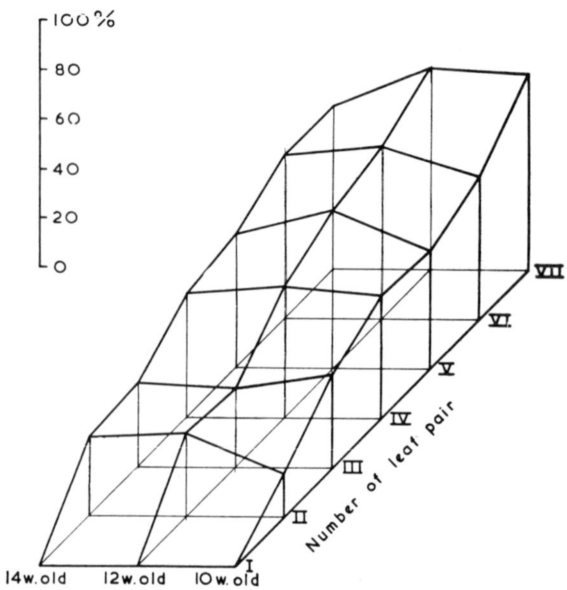

Fig. 1. Percentages of flowering of regenerated leaf cuttings of *Lunaria biennis,* taken from leaf pairs I, II, . . . VII of plants of 10, 12 or 14 weeks old, totalized over periods of 12 + 16 + 20 weeks of vernalization.

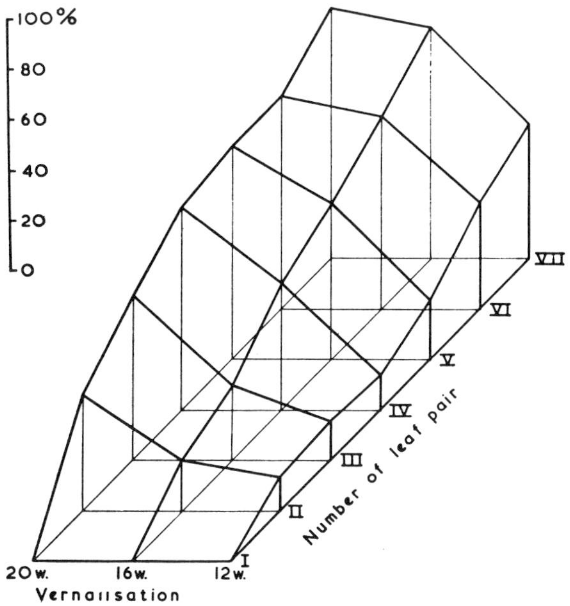

FIG. 2. Percentages of flowering of regenerated leaf cuttings of *Lunaria biennis*, taken from leaf pairs I. II, . . . VII of plants which were exposed to 5° during 12, 16 or 20 weeks, totalized over plant ages of 10 + 12 + 14 weeks.

pairs I to VII are calculated after totalization over the durations of the cold treatment and over the ages of the mother plants. We then find: 0.0, 25.2, 35.6, 51.5, 56.5, 64.7, and 76.1 % respectively. Special attention is drawn to the first 2 figures relating to the leaf pairs I and II, whose cuttings were made simultaneously.

We conclude that leaves on intact plants cannot be vernalized when they are full-grown at the beginning of the cold treatment, but that such a vernalization is possible for younger leaves. In addition, the vernalization increases as the age of the leaves decreases.

An interesting side observation was made in the present material. Among the regenerated leaf cuttings with more than 1 stem, varying numbers of the stems per plant flowered. The percentages of flowering stems per plant clearly increased as the percentages of flowering per plant and per treatment increased. This means that a more optimal vernalization finds its expression in a greater percentage of flowering stems per plant.

Discussion

Germinating seeds, apical meristems, leaf cuttings and root cuttings, which all can be vernalized, have in common the fact that they possess dividing cells. The concept that dividing cells are the prerequisite for vernalization was attractive by indirect evidence, but has now been proven by the direct evidence of the above experiments II and III. Support is found in the prenatal vernalization of rye as studied by Gregory and Purvis (4) who state: ". . . the low temperature is effective only during the period of active growth of the embryo and ceases as the embryo becomes dormant." This points to the necessity of the presence of mitotic cells. Support is also provided in the work of Oehlkers (9, 10) with Streptocarpus, in which plant leaf vernalization is a common occurrence.

Considering the process of vernalization in *Lunaria biennis* as a whole, we come to a picture which is very similar to the one which Zeevaart (18) has drawn up of the photoperiodic reactions in *Perilla crispa*. He distinguishes the induced state, called photoperiodic impression by Lona (7), and the production of the floral stimulus. In *Lunaria biennis* the direct effect of the low temperature would be a vernalized condition which only arises in dividing cells and which only is transmissible from cell to cell by mitosis. Cells which are in the vernalized condition can produce a floral stimulus which can be translocated to cells which themselves need not be vernalized. This was experimentally demonstrated in grafting experiments. Reference is made to Wycherley (17) who studied proliferations in the cold-requiring long day plant *Cynosurus cristatus*, and concluded that the direct effect of vernalization is immobile.

Another feature which *Lunaria biennis* has in common with *Perilla crispa* is the occurrence of physiological or functional chimera, found both by Zeevaart (18) and Lona (7) after grafting. In *Lunaria biennis* the regenerated leaf cuttings with 1 or more flowering branches side by side with 1 or more vegetative branches, as mentioned in the above experiment III, can be considered as physiological chimera. The generative or vegetative condition of a branch will depend on whether or not it regenerated from a cell which was in a mitotic condition during the cold treatment. In this respect, attention should be called to the quantitative action of vernalization, expressed as the ratio of vernalized to unvernalized cells in regenerating tissues. Something like a competition between vernalized and unvernalized cells must occur.

When a transmissible floral stimulus exists, as is clear from the grafting experiments, a comment can be made on the interpretation of experiment III that the flowering of the leaf cuttings is caused by this stimulus. No convincing argument against this comment is known at present for the cuttings of the younger leaves. However, it is quite unlikely for the cuttings of the older leaves, especially for the cuttings of the leaves II which were made simultaneously with those of leaves I at a moment when a floral stimulus could hardly have been formed and translocated. The latter did not flower; the former flowered for 25.2 %.

An interesting observation with regard to juvenility in leaf cuttings is that evidently juvenility is a character of the whole plant, while the condition of the individual leaf is the same as that of the whole plant, no matter what its age.

The concept of dividing cells as the locus of vernalization has several implications which, for the moment, are certainly not completely known. A few will be mentioned as illustrations. Normally the flower formation in *Lunaria biennis* starts at the top and proceeds to lower stem nodes, but not to the lowest nodes. It is suggestive to suppose a relation with leaf vernalization. This would imply a limited supply of the floral stimulus or a limited translocation. After suboptimal vernalization the flower formation frequently starts laterally, while the top of the plant may remain vegetative. This has been established by Higazy (6), but his explanation of leaf vernalization still leaves a number of questions.

In practice it has been observed that interrupted cold treatments often give better flowering results than an uninterrupted cold treatment. This might be explained by increased mitosis during short periods of relatively high temperatures, so that more dividing cells are vernalized.

It is remarkable that seeds of cold-requiring plants yield seedlings which have to be vernalized again. Hence the vernalized condition disappears at a certain stage of the development and it does not seem unlikely that this might be at meiosis, so that during meiosis a devernalization would take place.

The most important implication is doubtlessly that the problem of vernalization can now be approached from a biochemical side as the effect of low temperature on dividing cells.

Summary

Experiments on leaf vernalization in *Lunaria biennis* L. and direct evidence for the necessary presence of dividing cells during the vernalizing cold treatment have been presented. Leaf cuttings are vernalized more completely as the age of the mother plants increases up to a limit and as the cold treatment lasts longer. Flowering of leaf cuttings does not take place when after the cold treatment the lowest 0.5 cm of the petiole is removed. This means that vernalization is restricted to the part of the petiole where mitoses occur during regeneration. Leaves on intact plants can be vernalized, provided they are not full-grown when the cold treatment starts. This vernalization is more complete the younger the leaves are, or the more mitoses they have during the cold treatment. Grafting experiments have demonstrated the existence of a transmissible floral stimulus. These experiments are only successful when donor scions are grafted on receptor stocks.

From the experiments the conclusion is drawn that vernalization only takes place when dividing cells are present during the cold treatment.

It is suggested that as a direct effect of vernalization, a vernalized condition arises which is only transmissible from cell to cell by mitosis. Tissues in the vernalized condition produce a transmissible floral stimulus. Also, in *Lunaria biennis* physiological chimera occur, consisting of both vernalized and unvernalized parts side by side. Some sort of competition between these parts must exist.

Literature Cited

1. Chroboczek, E. 1934. A study of some ecological factors influencing seed-stalk developments in beets (*Beta vulgaris* L.). Cornell Univ. Agr. Exp. Sta. Mem. 154: 84.
2. Curtis, O. F. and H. T. Chang. 1930. The relative effectiveness of the temperature of the crown as contrasted with that of the rest of the plant upon the flowering of celery plants. Am. J. Botany 17: 1047–48.
3. Gregory, F. G. and O. N. Purvis. 1936. Vernalization. Nature 138: 249.
4. Gregory, F. G. and O. N. Purvis. 1938. Studies in vernalization of cereals. II. The vernalization of excised mature embryos, and of developing ears. Ann. Botany 2: 237–51.
5. Hagemann, A. 1932. Untersuchungen an Blattstecklingen. Gartenbauwissenschaft 6: 69–195.
6. Higazy, M. K. M. T. 1962. Shortening the juvenile phase for flowering. Mededel. Landbouwhogeschool 62: No. 8, 53 p.
7. Lona, F. 1959. Results of 12 years of work on the photoperiodic responses of Perilla ocymoides. Proc. Koninkl. Ned. Akad. Wetenschap. C 62: 204–10.
8. Melchers, G. 1936. Versuche zur Genetik und Entwicklungs-physiologie der Blühreife. Biol. Zentralbl. 56: 567–70.
9. Oehlkers, F. 1955. Blattstecklinge als Indikatoren für blütenbildende Substanzen. Z. Naturforsch. 10 b: 158–60.
10. Oehlkers, F. 1956. Veränderungen in der Blühbereitschaft vernalisierter Cotyledonen von Streptocarpus, kenntlich gemacht durch Blattstecklinge. Z. Naturforsch. 11 b: 471–80.
11. Purvis, O. N. 1940. Vernalization of fragments of embryo tissue. Nature 145: 462.
12. Purvis, O. N. 1944. Studies in the vernalization of cereals. VIII. The role of carbohydrate and nitrogen supply in the vernalization of excised embryos of Petkus winter rye. Ann. Botany 8: 285–314.
13. Schwabe, W. W. 1954. Factors controlling flowering in the chrysanthemum. IV. The site of vernalization and translocation of the stimulus. J. Exp. Botany 5: 389–400.
14. Wellensiek, S. J. 1958. Vernalization and age in *Lunaria biennis*. Proc. Koninkl. Ned. Akad. Wetenschap. C 61: 561–71.
15. Wellensiek, S. J. 1961. Leaf vernalization. Nature 192: 1097–98.
16. Wellensiek, S. J. 1962. Dividing cells as the locus for vernalization. Nature 195: 307–08.
17. Wycherley, P. R. 1952. Temperature and photoperiod in relation to flowering in 3 perennial grass species. Mededel. Landbouwhogeschool 52: 75–92.
18. Zeevaart, J. A. D. 1958. Flower formation as studied by grafting. Mededel. Landbouwhogeschool 58: No. 3, 88 p.

Amer. Jour. Bot. 52(8): 805–810. 1965.

THE BEHAVIOR OF TISSUE CULTURES FROM ENGLISH AND ALGERIAN IVY IN DIFFERENT GROWTH PHASES[1]

V. T. Stoutemyer and O. K. Britt

Department of Agricultural Sciences, University of California, Los Angeles, California

ABSTRACT

In both the English and Algerian ivies, *Hedera helix* L. and *Hedera canariensis* Willd., leaf dimorphism of the juvenile and the mature, fruiting growth phases is pronounced, the former being a vine with lobed leaves and the latter a shrubby, upright form with entire leaves. Tissue cultures of English ivy started from stems of the two growth phases on the same plant consistently behaved differently, those from the juvenile stage invariably having the higher proliferation rate and larger cells. These differences were maintained consistently in monthly transfers over a period of two years. No medium was found which would support the growth of tissue subcultures of the adult stage of Algerian ivy, but all growth phases of the English ivy were cultured freely in a modified White's medium with additions of coconut water, naphthaleneacetic acid, enzymatic casein hydrolysate, and inositol. Cultures from individual open-pollinated seedlings of both species varied greatly in proliferation rate but were usually high.

Progressive changes in external morphology accompany the development of plants from the seedling to the mature fruiting stages. In some cases these differences are pronounced and have caused the accidental description of a species under two names. Usually they are relatively inconspicuous.

Wareing (1959) and Robbins (1957, 1960) have attributed the gradual disappearance of juvenile plant characteristics to basic biochemical or physiological changes on the cellular level. Schaffalitsky de Muckadell (1959) used the term "meristematic aging" to describe hypothetical cellular changes which cause external manifestations of morphology or growth phases. Brink (1962) has reviewed the phenomena of growth phase changes from the standpoint of possible relationships to the normal genetic control mechanisms.

The following investigations show that tissue cultures can be used as a means of comparing cellular activity in various growth stages of ivy, thereby demonstrating consistent differences between the juvenile and adult growth phases. A similar attempt to establish cultures of English ivy was made by Privat (1958). He was able to establish tissue subcultures from the hypocotyls of young seedlings but failed to make successful transfers of callus derived from mature fruiting stems. Our studies indicate that the allowable ranges of temperature are different with tissue cultures from the various vegetative growth phases and that the temperatures which he used were adequate for juvenile growth stages but were too low for the mature stage of growth.

Materials and methods—*Sources of the materials*—These experiments were conducted with a non-variegated clone of *Hedera canariensis* Willd. collected in an ornamental planting in Santa Monica, California, and with two selections of *Hedera helix* L. collected in San Diego and in West Los Angeles. Tissue cultures from the two selections of the latter were maintained separately, although no differences in behavior could be observed. The collections were all in the "tree" or mature fruiting phase of growth.

In both of these species the growth phases can be easily distinguished by leaf and stem morphology. The immature stage is a vine with somewhat flattened stems, often producing aerial roots and abundant reddish anthocyanin pigments. The leaves are lobed and have a dorsiventral arrangement. The mature fruiting form has shrubby, stiff, upright stems which are round and rarely have adventitious roots. The leaves are entire and spirally arranged. The ivy seedling normally takes perhaps a decade or more to attain maturity. The mature fruiting stems, when rooted, form upright plants which are known as "tree ivies" and are often grown for ornamental purposes. Typical foliage forms of the various growth phases are shown in Fig. 1.

A large proportion of adult plants can be made to change from the mature to the juvenile type of growth by grafting on juvenile understocks (Doorenbos, 1954; Stoutemyer and Britt, 1961), by treatment with gibberellin (Robbins, 1957; Goodin and Stoutemyer, 1961), and by etiolation (Hackett, personal communication). Heavy pruning has also been observed to favor reversions. Maintenance of adequately high temperatures is an important factor in the success of these treatments.

Tree ivies occasionally produce juvenile shoots at or below the ground level and only such reversions produced naturally were used in these

[1]Received for publication November 10, 1964.
This work was supported in part by cancer research grant funds of the University of California.

Fig. 1. Juvenile, transitional and adult foliage forms of *Hedera helix* L.

experiments in order to avoid possible side effects from the treatments. Reversion shoots were compared with the adult stems produced on the same plant. The seedlings which were included in some of the experiments were grown from open-pollinated seed produced by the various selections of Algerian and English ivy.

Tissue culture media—White's medium (1943) was used in most of these experiments, but it was modified according to a suggestion of Shantz and Steward (1959) which involved replacing 10% of the inorganic nitrate nitrogen with ammonia nitrogen from ammonium chloride. Each liter of the medium also included 1 mg naphthaleneacetic acid, 100 ml of coconut water, 100 mg inositol, and 200 mg enzymatic casein hydrolysate.

Murashige and Skoog's (1962) high-salt formula with the same supplements generally produced a higher growth rate of the tissues, but it was little used in our experiments since the cell masses tended to hold more of the agar, thus decreasing accuracy of weighing of the cultures. The culture medium was autoclaved at 18 lbs pressure for 30 min. The use of coconut water and naphthaleneacetic acid was essential for good growth rates. The value of inositol and casein hydrolysate was doubtful.

These experiments were conducted in incubators with temperature control and in darkness since light was not important in influencing growth rates. Root formation took place most freely in cultures maintained without transfer for 2 or 3 months. Since the cultures used in these experiments were transferred at monthly intervals, root production was ordinarily sparse. Minimum temperatures were critical, especially with cultures of the adult growth phase. A temperature of 27 C or slightly higher should be maintained, although the cultures from seedlings or from juvenile reversion shoots can be grown at slightly lower temperatures.

A source of variability which was occasionally troublesome was in the coconut water. Rarely, coconuts were obtained on the local markets which either had a rancidity of the flesh or an unusual appearance of the fluid. The medium prepared from these gave inferior growth of the cultures.

Extensive experiments were conducted in an effort to eliminate the use of coconut water entirely. Kinetin and sorbitol permitted the continued transfer of cultures without this supplement, but the growth rate was unsatisfactory. Since comparisons of the behavior of the various types of tissue cultures were difficult under unfavorable conditions of growth, coconut water was used for most of our experiments.

Starting tissue cultures—The tissue cultures were started from stems which were washed in tap water after removing the leaves. This was followed by rinsing in sterile distilled water, then soaking 1 min in a 1% solution of sodium hypochlorite. The stems were again rinsed in sterile distilled water, dipped in 95% ethanol, and flamed quickly. The stems were handled with sterile instruments flamed in ethanol. The epidermis was peeled off, and cortical explants about 2 mm square, weighing between 1 and 2 mg, were excised with a sterile scalpel and placed on the culture medium with the cambial side down.

Alternatively, stem cross sections about 1 mm thick were cut with a sharp, single-edge razor blade or scalpel and were placed on the medium with the basal end in contact with the agar.

Subcultures were started from the original explants in one month and subsequent transfers were continued at monthly intervals. The cortical explants often provided material for about 4 or 5 subcultures and the stem cross sections for about 10 subcultures. In order to eliminate the direct influence of the explants, all quantitative experiments were conducted with subcultures.

TRANSFER OF TISSUE AND GROWTH MEASUREMENTS—Subcultures were started by taking the tissues from the vials with sterile forceps and placing on sterile weighing pans. A small double-hook torsion balance with a maximum capacity of 1500 mg was used. These pans were made of heavy aluminum foil of the type used for disposable trays. The mass of tissue was cut to provide 10 portions, each weighing about 15 mg each. The individual weights were determined by subtraction as each piece of inoculum was removed from the pan.

The pan was washed in ethanol and flamed before preparing the next set of subcultures. Contaminations have been rare in our experiments. At the end of the incubation period (4 weeks) the cultures were then weighed individually on a sterile pan of the balance and returned to the culture vial where they were held until subcultured again or discarded. The pan was dipped in ethanol and flamed after 10 weighings.

Six-dram, flint glass vials covered by plastic screw caps with the paper liners removed have been used for most of the cultures grown over the last 3 years. Approximately 10 ml of the medium were placed in each. These vials were easy to handle under sterile conditions and required little space in the incubators. Tissue has been grown for as long as 2–4 months in them; however, the more vigorous tissues exhaust the medium in the shorter period. For holding tissues over longer periods or for the production of tissues for chemical studies, 125-ml flasks were used.

Both the initial weight of the inoculum and the final weight of the culture were determined in a manner which kept the tissue sterile and usable for further experimentation. Data expressing rate of growth were recorded as the initial fresh weight of the inoculum and the final fresh weight of the callus. The latter were ordinarily taken at the end of a month of growth, at which time transfers were made. Averages of both were obtained and the growth increment as defined by Klein (1964) was used, except that this is expressed here as a ratio rather than as a percentage of increase. The formula as used here is:

$$\frac{\text{Final wt} - \text{Initial wt}}{\text{Initial wt}}$$

Thus in our tables a growth increment of 4.0 indicates a fourfold increase in fresh weight during the experimental period.

EXPERIMENTS WITH ALGERIAN IVY—Algerian ivy was not used extensively in our experiments since, in spite of many trials with various supplements to the growing medium, we have not yet been able to devise a method for the successful subculture of adult callus, which is produced abundantly from explants in the adult growth phase. Open-pollinated seedlings of Algerian ivy, as also with English ivy, yield tissue cultures with somewhat variable but consistently maintained growth rates. During a growth period of one month, tissue cultures often yielded a 5-fold or greater increase in fresh weight.

Adult plants of Algerian ivy occasionally produce juvenile reversion shoots, usually near the base. Callus grown from explants of these stems was used in attempts to start tissue cultures. These were only partially successful and failures were numerous. A culture started from one isolate was found to be difficult to maintain, although when transfers were successful, they grew at rates approaching those recorded for seedlings.

As a result of these experimental difficulties, trials with Algerian ivy were abandoned. However, these experiments show, even more clearly than do similar experiments with English ivy, the difference in physiological behavior between the adult shoots and the naturally produced reversion shoots or seedlings.

EXPERIMENTS WITH ENGLISH IVY—Extensive preliminary experiments proved that the modified White's medium would sustain the growth of tissue cultures of all growth phases of English ivy. In these experiments, also, all reversion shoots were those which were occasionally produced without artificial treatments at the base of tree ivies. Thus the adult and reversion shoots represent the same plant but different morphological forms or growth phases.

Table 1 presents the pooled results from the first 3 subcultures made at monthly intervals from 2 different selections and from various open-pollinated seedlings obtained from the same plants. The results of this experiment are typical of those from many trials in our laboratories. Tissue cultures from reversion shoots have invariably had a higher proliferation rate than those from stems exhibiting the adult or tree form. The growth rates from open-pollinated seedlings vary, but they are usually comparable with those of the

TABLE 1. *Growth of tissue cultures of English ivy*

Type of tissue	Plant selection	No. of cultures	Mean initial weight (mg)	Mean final weight (mg)	Growth increment
Adult	P[a]	50	17.3	97.4	4.6
Reversion	P	50	17.5	147.8	7.4
Adult	P	50	16.6	89.4	4.4
Reversion	P	50	16.6	130.0	6.8
Seedling 15 mo.	P	50	16.3	109.4	5.7
Seedling 22 mo.	P	50	17.1	143.8	7.4
Adult	I[b]	50	16.1	93.5	4.8
Reversion	I	50	17.2	159.7	8.3
Seedling 9 mo.	I	50	17.8	159.8	8.0
Seedling 9 mo.	I	50	15.4	83.8	4.4
Seedling 1 mo	I	40	17.2	145.1	7.4
Seedling 1 mo.	I	32	16.0	168.8	9.5

[a] Collected at San Diego.
[b] Collected at Beverly Hills.

TABLE 2. *Growth of tissue cultures of English ivy*[a]

Type of tissue	Mean initial weight (mg)	Mean final weight (mg)	Growth increment
Isolation 1			
Adult	16.72 ± 0.48	106.78 ± 8.36	5.33 ± 0.40
Reversion	16.11 ± 0.78	132.72 ± 10.66	7.26 ± 0.46
Isolation 2			
Adult	15.51 ± 0.55	84.00 ± 6.87	4.40 ± 0.38
Reversion	16.43 ± 0.41	165.48 ± 9.67	9.07 ± 0.61

[a] Standard error of the mean in this table is calculated on the basis of 14 averages (n = 14). However, each average was based on 20 cultures, and each figure in the table above represents 280 cultures.

reversion shoots and tend to exceed the latter. In this instance the growth rates of cultures from the two selections were closely similar on a fresh weight basis.

In the following experiment two separate isolations were made from the adult and reversion shoots on the same plant of English ivy on January 23, 1962 and were transferred in pedigreed cultures, using 20 cultures at each transfer. The pooled data for 14 successive monthly transfers are shown in Table 2. This table shows that variability within the cultures was small in comparison with differences between the two different phases of growth.

In another series of trials pooled data were obtained from 14 successive subcultures. A seedling derived from the experimental plant was likewise included. The results are shown in Table 3 and show essentially the same trends, with the cultures from the seedling and from the reversion shoot having the higher proliferation rates.

An additional experiment which likewise included cultures from a seedling of the same plant is shown in Table 4. These data include the mean growth increments for the 11th to the 20th subcultures and indicate the range of variation encountered in successive transfers. Changes in the composition of the medium involving the omission of inositol made earlier transfers not entirely comparable, and they are not included in this summary.

The results of this experiment are closely similar to those of the others, showing clearly the greater proliferation rate of the callus from the juvenile reversion shoots or from seedlings. In a few cases, however, seedlings were found which had unusually low growth rates. One example may be noticed in Table 1.

Tissue cultures show a considerable range of variability, and many experimenters have reported considerable instability and drift. Variability has existed in our cultures, but this has not obscured the proportional relationships between the tissue cultures from different growth phases which have remained relatively constant.

An additional consistent difference in the cultures, aside from the proliferation rate, was in the number of cells per unit of fresh weight. The

TABLE 3. *Growth of tissue cultures of English ivy*

Type of tissue	Number of cultures	Mean initial weight (mg)	Mean final weight (mg)	Growth increment
Adult	320	14.94	82.23	4.50
Reversion	320	15.23	100.38	5.59
Seedling	310	16.26	221.46	13.62

TABLE 4. *Growth of tissue cultures of English ivy*[a]

Transfer number	Adult			Reversion			Seedling		
	Mean initial weight (mg)	Mean final weight (mg)	Growth increment	Mean initial weight (mg)	Mean final weight (mg)	Growth increment	Mean initial weight (mg)	Mean final weight (mg)	Growth increment
11	13.7	109.3	7.0	13.5	101.9	6.6	14.2	118.6	7.4
12	15.5	105.1	5.8	15.5	116.0	6.5	16.4	149.4	8.1
13	13.0	73.9	4.7	13.3	120.0	8.0	15.9	104.2	5.6
14	18.7	116.8	5.2	19.9	173.8	7.7	20.3	163.0	7.0
15	16.1	85.2	4.3	15.9	129.1	7.1	16.3	125.0	6.6
16	13.2	70.3	4.3	14.0	131.8	8.4	13.9	115.3	7.3
17	16.3	97.4	5.0	18.0	137.2	6.6	17.4	159.2	8.2
18	16.2	78.0	3.8	17.0	118.1	5.9	16.7	143.0	7.6
19	16.0	83.0	4.2	14.6	94.0	5.4	15.4	171.0	10.1
20	16.7	83.3	4.0	17.1	112.0	5.5	17.3	146.0	7.4
Means	15.53 ±0.30[b]	87.87 ±4.69[b]	4.70 ±0.03[b]	15.88 ±0.35[b]	123.29 ±7.00[b]	6.69 ±0.04[b]	16.38 ±0.32[b]	139.49 ±7.51[b]	7.48 ±0.04[b]

[a] 50 cultures in each transfer grown in 5 replicates of 10 cultures each. Data shown for the 11th through the 20th subcultures.

[b] Standard error of the mean calculated from the means for the 50 replicates of 10 cultures each; therefore n = 50.

maceration-haemocytometer method of Brown and Broadbent, as described by Jensen (1962), was used for cell counting. This involved maceration in 5% chromic acid and counting in a Fuchs-Rosenthal haemacytometer grid. Some difficulty with clumping of cells was encountered, but this was virtually eliminated and higher and doubtless more accurate cell counts were obtained by placing the tissues in vials at the end of 1-foot extension arms from a wrist-action shaker. Violent agitation was continued for 8 hours. Cells in the tissue cultures made from shoots in the adult growth phase were smaller and invariably more numerous per unit of fresh weight than in those started from reversion shoots which showed juvenile characters. Pooled means of 80 counts for each type were 23,000 cells per mg of fresh weight of tissue for adult cultures, and 18,800 for those from the reversion shoots. Counts made from cultures of a seedling closely resembled those from the reversion shoots.

DISCUSSION—Theories of juvenility or heteroblastic development in plants have reflected divergent viewpoints. The idea of physiological or meristematic aging has sometimes been advanced, and in recent years investigators and reviewers have generally postulated a correlation between the external morphological changes and biochemical differences on the cellular level. Brink (1962) has also suggested some relationship between phasic development and the genetic mechanism. The present studies use one method of testing these theories.

Instability is frequently encountered in tissue cultures. However, in the present studies the cultures from adult and reversion shoots of various selections of ivy retained their essential differences over periods of time exceeding 2 years. There was considerable variability within the cultures, but in no case was the difference in rate of growth of cultures from the different growth phases obscured. In extensive studies with other culture media than the one used here, over 50,000 tissue cultures were grown and recorded, and in no experiments were the growth rates inconsistent with these results. Naturally, the differences were greatest on the most favorable culture media, and were less pronounced on the unfavorable media.

Roots were occasionally produced, especially in the juvenile cultures, but were never abundant with transfer at monthly intervals. Transferring less frequently, and the use of flasks with larger quantities of culture medium, would have produced significant numbers of roots. Tissue cultures from juvenile types of tissue are characterized by a more rapid proliferation rate, and more, larger parenchymatous cells which are less numerous per weight unit of the tissue culture.

Seedlings, even though derived from a common parent plant, were quite variable in their behavior and, therefore, were difficult to compare with adult forms or with reversion shoots. Occasionally seedlings with abnormally high or low proliferation rates were encountered.

Preliminary trials with some other evergreen trees and shrubs grown outdoors in the climate of Los Angeles, including eucalyptus, aralia and acacias, suggest that similar relationships between the proliferation rates of tissues from juvenile and adult stems exist in these species. However, Trippi (1963) compared proliferation rates of callus from two species of deciduous trees and did not always find the most rapid growth on juvenile tissue. The time of year has long been known to influence growth of tissue cultures from the cambium of deciduous trees. Also, Trippi did not establish transfer cultures of the callus. For these reasons his results are difficult to evaluate in relation to our experiments.

These studies indicate that differences in cellular activity accompany the external morphological changes indicative of the growth phases. These changes can be followed and comparisons made by establishing tissue cultures from the shoots in the various growth stages.

LITERATURE CITED

Brink, R. A. 1962. Phase change in higher plants and somatic cell heredity. Quart. Rev. Biol. 37: 1–22.

Doorenbos, J. 1954. Rejuvenation of *Hedera helix* in graft combinations. Proc. Kon. Ned. Akad. Wetensch. Ser. C, 57: 99–102.

Goodin, J. R., and V. T. Stoutemyer. 1961. Effect of temperature and potassium gibberellate on phases of growth of Algerian ivy. Nature 192: 677–678.

Jensen, W. A. 1962. Botanical histochemistry. W. H. Freeman & Co., San Francisco.

Klein, R. M. 1964. Repression of tissue culture growth by visible and near visible radiation. Plant Physiol. 39: 536–539.

Murashige, T., and F. Skoog. 1962. A revised medium for rapid growth and bioassays with tobacco tissue cultures. Plant Physiol. 15: 473–497.

Privat, G. 1958. La culture *in vitro* des tissues de lierre. Nat. Monspel., Ser. Bot. 10. 91–96.

Robbins, W. J. 1957. Gibberellic acid and the reversal of adult *Hedera* to a juvenile state. Amer. Jour. Bot. 44: 743–746.

———. 1960. Further observations on juvenile and adult *Hedera*. Amer. Jour. Bot. 46: 485–491.

Schaffalitsky de Muckadell, M. 1959. Investigations of aging of apical meristems in woody plants and its importance in silviculture. Kandrup and Wunsch's Bogtrykkeri. Copenhagen.

Shantz, E. M., and F. C. Steward. 1959. Investigations on the growth and metabolism of plant cells. VII. Sources of nitrogen for tissue cultures under optimal conditions for their growth. Ann. Bot. n.s. 23: 371.

Stoutemyer, V. T., and O. K. Britt. 1961. The effect of temperature and grafting on vegetative growth phases of Algerian ivy. Nature 189: 854–855.

Trippi, V. S. 1963. Studies on ontogeny and senility in plants. II. Seasonal variation in proliferative capacity *in vitro* of tissues from branches from juvenile and adult zones of *Aesculus hippocastanum* and *Castanea vulgaris*. Phyton 20: 146–152.

Wareing, P. F. 1959. Problems of juvenility and flowering in trees. Jour. Linn. Soc. London (Bot.) 56 282–289.

White, P. R. 1943. Handbook of plant tissue culture Jacques Cattell Press.

THE USE OF "IN VITRO" CULTURES IN THE INVESTIGATION OF GROWTH AND DIFFERENTIATION IN VASCULAR PLANTS*

Ralph H. Wetmore**

Biological Laboratories, Harvard University, Cambridge, Massachusetts

Vascular plants owe their characteristic architectural form to the possession of meristems at the tips of all roots and shoots. They are in continuous embryogeny, as Bower (1919) has suggested. The studies of Nast (1941), Miller and Wetmore (1945), Allen (1946), and Spurr (1949) indicate that this habit derives ontogenetically from the fact that the multicellular embryo arising from the fertilized egg forms a morula-like stage (Figures 1, 2, 5-13). In contrast to developing animals, however, this plant morula never gastrulates. Instead, its two poles become respectively the apical meristem of the shoot and the root (Figures 14 and 15). The median part of the embryo, between the two meristems, is soon differentiated. The cell aggregates left behind by the continuously advancing shoot and root meristems are added to either end of this differentiated middle. Inevitably, then, the formative regions of root and shoot and their derivative new lateral offshoots, branches of shoot systems and root system, respectively, get farther and farther apart.

A consequence of this unique architectural plan of the vascular plant is that primary growth and development, involving both cell division and cell enlargement, can be found in the distal portions of all axes, shoot and root, throughout the growing season; the mature differentiated parts of the axis are proximal to these growing regions. However, a physiological handicap is imposed upon the vascular plant by this architectural plan. As the plant grows, its absorbing system is year by year farther separated from its photosynthetic system. Water and mineral salts must be transported farther and farther to reach the young regions of the shoot system. Materials elaborated in the green parts must be carried equal distances in the reverse direction to nourish the roots.

Even without any consideration of secondary activity, this working system can be further complicated to the degree that certain metabolic activities are segregated in different parts of the organism. We have much to learn before we are clear about the degree to which such a segregation is carried out. We do know that carbohydrates are synthesized in green parts of the plant; we are equally certain that protein synthesis is not a function of these green tissues only. Moreover, the site or sites of protein synthesis are in no way proof of the place of origin of their constituent amino acids. Nor is the generally accepted statement that

*The investigations reported in this paper are part of a study of growth and differentiation of tissues in vascular plants which has been supported by grants from the American Cancer Society, Inc., upon the recommendation of the Committee on Growth, and the Massachusetts Division of the American Cancer Society, Inc.

**The author wishes to acknowledge his indebtedness to Dr. Georges Morel whose cooperation made the *in vitro* culture techniques a reality in these investigations.

an auxin is produced in the apical regions of plants proof positive that it is produced only there. Our present knowledge does not tell us whether all living cells in the apex or throughout the plant may not be capable of both producing and using auxin. In other words, despite the fact that we know a great deal about the organization of plants, about their physiological processes, about aspects of their biochemistry, an assessment of what we can say as established fact from a morphogenetic point of view is slight indeed.

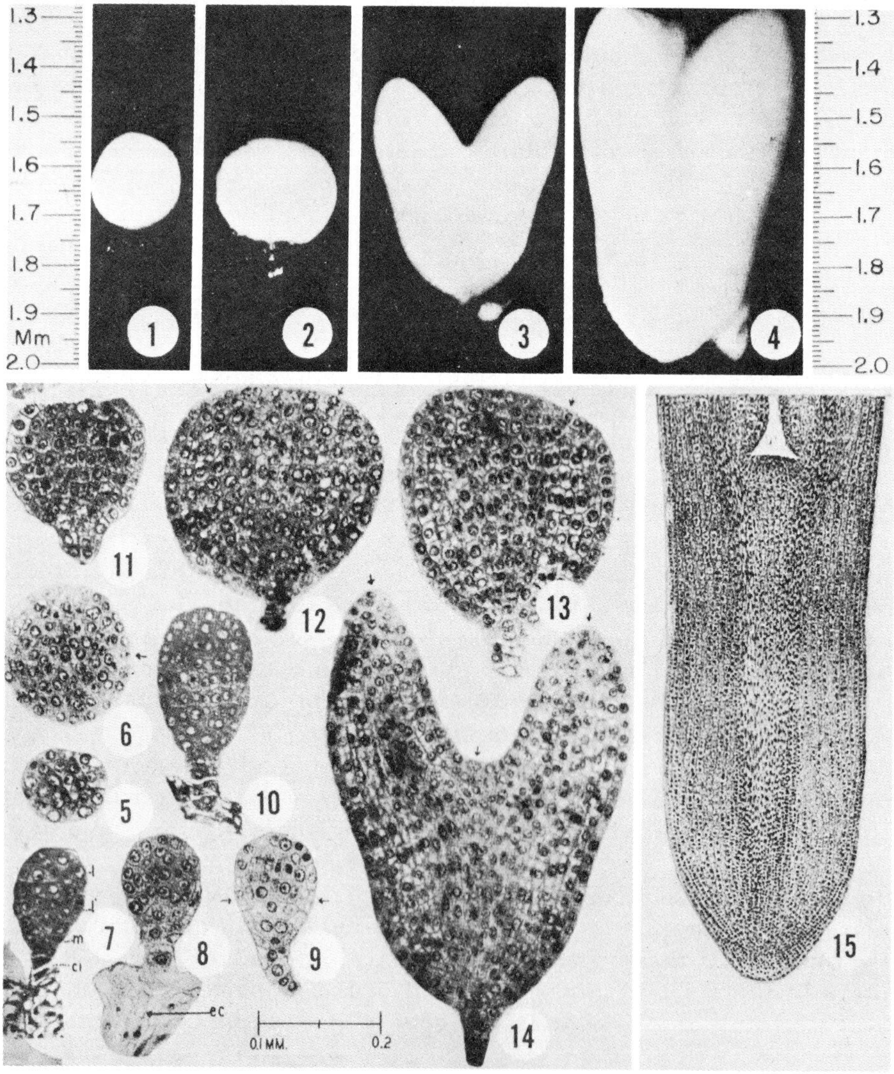

Figures 1 - 15. Proembryos and embryo of *Phlox Drummondii.* (Reprinted with the permission of the American Journal of Botany.)

Figures 1 - 4. Photographs of 5-, 6-, 8- and 9-day proembryos, respectively, after they have been dissected out of the ovules (Magnification x62). Suspensors broken in dissection.

Figures 5 - 14. Sections of early stages of proembryos (Magnification x101). Figures 5, 6. Tranverse sections. Figures 7 and 8, 9 and 10, 11, 12 and 13, and 14 are proembryos, respectively, of 3 days, 4 days, 5 days, 6 days, and 8 days.

Figure 15. 30-day embryo (Magnification x26).

Figures 16 - 21. *In vitro* cultures of rhizome apices of *Adiantum pedatum.*

Figures 16 - 19. Cultures about 5 months old grown on agar medium containing Knop's solution, minor elements, 2 percent sucrose and nothing extra (Figure 16), 0.5 g/liter yeast extract extra (Figure 17), 0.5 g/liter yeast extract and 0.05 mg/liter naphthaleneacetic acid Figure 18), and 0.05 mg/liter NAA (Figure 19) (Magnification x0.75).

Figure 20. Rhizome apex on above medium with yeast extract, 6 months old (x0.75).

Figure 21. *Adiantum* plant, 12.5 cm tall, grown from rhizome apex in about 16 months: 5 months on standard medium above with no additions and about 11 months in soil. Plant fruited the next season (Magnification x0.20).

If any one piece of evidence is clear, it is that the apical meristem of the shoot is an organizer tissue; that is, it not only supplies the increments of cells on the tip of any branch of the shoot system, but it provides the biochemical "know-how" for their continuing organization. This fact has been maintained strongly by Ball (1946, 1948, 1949, 1950, 1951, 1952) and supported by the work of Wardlaw (1944, 1945, 1947, 1948, 1949a,b) and of others (Wetmore and Wardlaw, 1951). Remove the apical meristem, and organization ceases behind the region of excision. No biochemical system is yet known which will regularly initiate the formation of a new apical meristem to supplant that removed, although Skoog (1948) has achieved some success in this direction.

It is important to point out that pieces of the apical meristem, removed from certain plants, with or without foliar primordia - pieces not over 200 to 250 micra in length - will grow into new plants without difficulty if planted *in vitro* on a sterile agar medium containing mineral salts, minor elements, and sucrose. This has been done repeatedly with fern apices, e.g., *Adiantum pedatum* (Figure 26; Wetmore and Morel, 1949), and with apices of *Selaginella willdenovii, Lycopodium cernuum* and *Equisetum hyemale.* For example, Figures 16 to 21 illustrate interesting aspects of the sterile culture *in vitro* of the apices of *Adiantum pedatum.* The first four figures (Figures 16 - 19, marked A, B, C, and D, respectively) show in each illustration three representative cultures of about 5 months' age grown on a standard Knop's solution with a slightly modified Nitsch's (1951)* solution of minor elements, 2 percent sucrose, and 1 percent agar; to A, nothing extra was added; to B, 0.5 g/liter of yeast extract; to C, 0.5 g/liter of yeast extract and 0.05 mg/liter of the auxin, naphthaleneacetic acid; and to D, 0.05 mg/liter of the auxin alone. It will be noticed that the medium supplemented by yeast extract (B) was definitely superior to all the others, even that with the auxin supplementing the yeast extract (C), and that the latter medium with auxin added to that already present in the yeast extract seems consistently of a high enough concentration to be somewhat inhibitory. It should be pointed out too that in five months the cultures with no auxin are not strikingly dissimilar in size to those with 0.05 mg/liter of an auxin, although they have fewer leaves. This similarity did not exist in their early growth, for those with no auxin were much slower in development. In time, however, the leaves on these 0 auxin cultures had acquired adequate size and adequate capacity to produce the missing auxin in amounts necessary for the usual rate of growth. With its own production of auxin, A slowly caught up to D. The continued growth of the whole plant in a larger container with more medium is evident (Figure 20). When transplanted to an alkaline humus-containing soil after five months, the plants grew well in the greenhouses (Figure 21) and bore fruit the second season.

Although it is true that apices of the vascular cryptogams will grow to maturity on a medium containing only Knop's solution, minor elements and sucrose - the last is not needed after a few green leaves are established - it must be stated that they develop faster with nitrogen sources other than nitrates, and with an auxin in low concentration. Thus, the addition of autoclaved coconut milk or yeast extract, supplemented with a low concentration of auxin - we generally used 0.05 mg/liter - definitely speeds up the growth and development. With a constant source of sugar, e.g., 2 percent sucrose, it is noteworthy that characteristic plants develop. When

*Nitsch has since added 25 mg $CoCl_2$ to his stock solution.

transplanted to soil, they go to fruition; in fact, it deserves comment that plants of *Lycopodium cernuum* have produced cones in our culture tubes.

Despite Ball's proven ability to grow apices of *Lupinus* and *Tropaeolum,* we have had little success in obtaining mature plants from the apices of angiosperms. Apices of *Syringa vulgaris, Parthenocissus tricuspidata,* or *Lupinus*

Figures 22 - 25. Apices of *Syringa vulgaris* grown *in vitro* (Magnification x0.75).

Figure 22. Apices having been grown seven weeks as grafts into lilac callus; medium, Knop's solution with 0.05 mg/liter NAA and dextrose.

Figure 23. The same apices but 4-1/2 months after grafting; this is as large as such grafts grow on this medium.

Figure 24. The right and left apices of lilac were grafted into lilac callus growing on medium containing autoclaved coconut milk and a mixture of B-vitamins in addition to above ingredients. The plant on the right was rooted though barely visible in the picture. The middle plant is the control grown on the same medium as the apices in Figures 22 and 23.

Figure 25. Right and left apices were grown on the same medium used in Figure 24, but these were not grafted; the middle apex is the control; note that left apex developed wound callus at base of apex; this plant is rooted.

albus (Figure 27) grew little *in vitro* upon media that had given excellent growth for the vascular cryptogams, unless large pieces, 0.5 mm or more, were taken as explants. However, we found that when apices of lilac were grafted into lilac callus, they grew for a time and became plants 2 cm or so in height which looked normal in all respects except for their lack of roots (Figures 22 and 23). In other words, it seemed that for a time the callus provided a limited reserve of something which was needed by the apex and was not in the medium. It merits comment here that in the callus, below the graft bridge, numerous vascular strands originate which are usually collateral, though the larger ones often have a pith. In our present investigation these strands have in no case been in direct continuity with those of the grafted apex. This is quite unlike the results reported by Camus (1949), who successfully grafted buds of endive upon root callus of the same plant growing *in vitro.* Camus found that in the presence of the grafted bud, vascular tissue was induced in the root fragments across the graft bridge and below the bud. In this case, however, the induced vascular tissue was in essential continuity with that developed in the growing bud.

To understand the contrastive behavior of the apices of vascular cryptogams, which need no other source of nitrogen than nitrates, and those of angiosperms, which we could not grow on this medium, it seemed a possible working concept to assume that enzymes necessary for the synthesis of certain amino acids were missing or in low concentration. On this premise, we tested *Syringa* apices on a medium containing 15 percent of autoclaved coconut milk and 0.5 mg/liter of casein hydrolysate. As is indicated in Figures 24 and 25, they grew, rooted, and gave indication of normal habit, whether grown directly on the medium or grafted into callus of the same species. With these findings before us, we began collections of various parts of *Lupinus* plants (Table 1), namely, apices with no leaf primordia, leaf primordia, partially grown leaves, fully expanded leaves, sections of stems, young roots, older roots and cotyledons, for chromatographic studies of amide and amino acid nitrogen.

These studies were carried out at Cornell University, with the active cooperation of Professor F.C. Steward.* The results for cold 80 percent alcohol-soluble extractions of samples of the several plant parts are given in Table 2. In this table, the nitrogen of each amino acid and amide is expressed as a percentage of the total nitrogen present in the sum of all the compounds to reflect the relative importance of each compound in this soluble extract. Two observations are immediately obvious: 1) The meristematic regions of apex and leaf primordia are well supplied with free amino acids, both in variety and in amount. 2) They are especially rich in the amide asparagine, for even though the shoot meristem and leaf primordia have 63.28 percent and 68.36 percent of their alcohol-soluble nitrogen as asparagine-nitrogen, the concentration is lower in these parts than in any others reported in this study.

It is worthy of comment also that the concentrations of basic amino acids, arginine and lysine - histidine was not demonstrable by the techniques employed - were high in the soluble nitrogenous compounds of the

*The author wishes to express his appreciation of the complete cooperation offered by Professor Steward and his group in following up this problem, both in its technical aspects and in the interpretation of the results (see Steward *et al*, 1954).

Table 1

Description of samples analyzed.

*Lupinus albus**	14-day plants			30-day plants	
	Number	Dry wt. (mg)	Wt. per unit (μg)	Number	Dry wt. (mg)
Apical meristems-shoot	1504	12.0	8.0	472	3.0
Leaf primordia**	3008	43.0	14.0	944	7.2
Unexpanded leaves**	50	740.0	14,800.0	20	320.0
Expanded leaves**	20	2407.0	120,000.0	20	1400.0
Stem sections†	50	322.0	6,450.0	15	380.0
Hypocotyls				8	600.0
Old roots††				12	400.0
Young roots††				14	270.0
Cotyledons*†	20	154.8	77,300.0		

*Plants grown from seed in July and August in greenhouse at Cambridge, Mass., on well manured soil.

**Leaf primordia - herein are included the two youngest leaves; unexpanded leaves - one to two of the youngest leaves which are still growing actively; expanded leaves - the youngest leaves to have obtained full size, though they are probably not completely organized as mature leaves and still may, in fact, show some cell divisions at this time due to plate meristem activity.

†The stem sections were comprised of the first internode above the cotyledons in the 14-day plants and the third internode above the cotyledons in the 30-day plants.

††Collections of roots include whole roots; root apices have not yet been analyzed.

*†Cotyledons collected for analysis 16 days after germination.

apical meristem and leaf primordia. The percentage of these residues drops sharply from the meristem and leaf primordia to the older leaves and stem; the change occurs in a vertical distance of a few millimeters only. The importance of these basic amino acids in high concentration in the meristematic regions may well be looked for in the synthesis of DNA-proteins and in the relationship of the latter to the phenomenon of cell division.

Other points of interest are evident. For example, the contrast between the behavior of glutamine, glutamic acid, and γ-aminobutyric acid on the one hand and asparagine and aspartic acid on the other is at once clear. The two groups seem to be differently related to the metabolic activities of the apical region; otherwise the gradients would be difficult to understand. The evidence here could well support the idea that glutamic acid is more nearly related to the metabolic activity and the protein synthesis of the dividing cells of the apex than is aspartic acid.

Most of the other amino acids recorded in Table 2 show varyingly higher concentrations in the meristem than in the older leaves or stems. That this varying concentration exists suggests that this is not just a relative increase in dry weight; rather the data would permit the suggestion that some of these amino acids may well have been formed at the ultimate expense of asparagine and aspartic acid, for these are the only acid residues which decrease as the meristem is approached.

One amino acid not reported in this table, tryptophan, deserves mention at this time. By a bacteriological technique, Nitsch (Nitsch and Wetmore, 1952) reported a relatively high content in the meristem in contrast to that in the older leaves. This is important in its relation to the naturally occurring auxin, indole-3-acetic acid, to which it is a precursor.

Table 2

Amino acid composition of the non-protein nitrogen fraction. Nitrogen of each amino acid as percentage of total nitrogen estimated.

1	2	3	4	5	6	7	8
Amino acid	*Lupinus* - 14 days old						
	Shoot meristem	Leaf primordia	Unexpanded leaves	Expanded leaves	Stem	Cotyledons	Change from shoot meristem to expanded leaves
Aspartic acid	0.45	0.46	0.64	1.30	0.30	0.18	↗
Glutamic acid	2.12	1.15	0.78	0.35	0.33	0.16	↘
Serine	1.7	1.22	0.45	1.14	0.91	0.85	→
Glycine	---	---	---	0.15	---	0.09	---
Asparagine	63.28	68.36	91.75	92.09	89.97	89.05	↗
Threonine	1.18	2.02	0.39	0.60	0.77	0.81	→
Alanine	2.00	1.44	0.61	0.94	0.77	0.34	→
Glutamine	2.71	2.44	1.80	1.22	0.84	0.13	→
Lysine	3.32	0.58	0.15	0.29	0.20	0.52	↘
Arginine	7.65	7.97	0.56	0.19	0.54	5.18	↘
Proline	2.27	2.44	0.60	---	0.76	---	↘
Valine	5.62	4.88	0.91	0.06	2.19	0.61	↘
Leucine	2.68	1.49	0.57	0.18	1.13	1.04	↘
Phenylalanine	3.36	4.36	0.42	0.16	0.78	0.065	↘
Tyrosine	0.49	---	0.08	0.13	0.13	0.15	→
γ-aminobutyric acid	1.18	1.31	0.30	0.63	0.40	0.83	→

If we turn now to the hydrolyzed proteins for amino acid residues (Table 3), again arranged to give the same comparative percentage relationship between the nitrogen present in each amino acid and the total nitrogen of all the amino acids, certain conclusions can be drawn at once. Perhaps the most important and also the most striking point is the relative constancy of the percentage of nitrogen of most of the amino acid residues in the composition of the apical meristem and of the varying stages of developing leaves. Notable exceptions are the basic amino acids, arginine and lysine, and tyrosine. In all three of these, there is an appreciable drop in percentage between the apical meristem and leaf primordia on the one hand and the mature leaves on the other. By contrast, valine and glutamic acid are the two acid residues which are lower in percentage fraction of the total nitrogen in the apical meristem than elsewhere in the plant. However, it is perhaps important that (columns 2

Table 3

Amino acid composition of the total protein fraction. Nitrogen of each amino acid as percentage of total nitrogen determined.

1	2*	3*	4*	5*	6*	7*	8†	9†	10†
	Lupinus								
Amino acid	Shoot meri-stem	Leaf pri-mordia	Unex-panded leaves	Expanded leaves	Stem	Cotyledon	Hypocotyl	Young roots	Old roots
Aspartic acid	6.3	6.6	8.0	5.2	4.7	8.2	16.9	9.6	8.9
Glutamic acid	8.2	8.4	7.3	10.2	9.4	8.9	9.3	9.1	6.9
Serine	5.2	6.0	5.7	4.3	6.0	4.6	5.7	4.8	5.4
Glycine	11.3	8.4	12.3	10.2	10.4	7.0	6.8	8.6	8.0
Threonine	4.1	3.6	5.0	4.3	3.7	3.8	2.8	3.5	3.7
Alanine	6.8	6.7	8.0	8.6	8.8	8.2	6.9	6.1	7.1
Lysine	11.7	12.6	10.5	10.4	8.2	11.3	10.7	11.5	10.0
Arginine	16.8	17.9	13.9	13.4	13.0	13.7	8.5	10.0	8.1
Methionine	1.4	0.8	1.3	0.6	0.5	0.3	0.3	0.6	0.3
Proline	5.0	4.6	7.6	5.6	4.2	5.9	4.0	5.6	9.3
Valine	4.8	4.7	6.3	7.1	6.7	6.5	6.6	6.5	7.2
Leucine	10.6	11.6	10.1	12.4	12.7	12.6	12.6	14.4	12.0
Phenylalanine	3.7	4.6	1.9	3.7	3.7	4.2	3.3	3.5	3.6
Tyrosine	4.0	2.3	2.3	2.5	2.6	2.5	1.2	2.8	2.3
Hydroxyproline	---	1.3	---	1.6	5.3	2.5	4.4	3.6	6.6

*From *Lupinus albus* plants 14 days old.
†From *Lupinus albus* plants 30 days old, with cotyledons removed at 23 days.

to 7) the sum of the two dicarboxylic acids (aspartic and glutamic acids) is essentially constant, except in the cotyledons, where the total is higher (equals 17.1 percent of the total nitrogen, against 14.1 to 15.4 percent elsewhere).

One is forced to comment on the higher concentration of arginine and lysine in the meristematic regions over that in the alcohol-soluble fraction. Our increasing knowledge of DNA-proteins in relation to cell division and of the high percentage composition of the basic amino acids in these proteins of relatively simple nature and of few amino acid residues gives one a certain feeling of confidence in the data and in the wisdom of extending these data for what they may tell us of the fundamental biochemical background of cell division.

Turning to the significance of these data in the *in vitro* culture of *Lupinus,* I should like to be able to state that when the amino acid and amide mixture was made up according to the concentrations predicated by our findings, the apices (Figure 27) grew as uniformly and as rapidly as did the vascular cryptogams (Figure 26). Had this been so, we could have stated that quite obviously the young, actively dividing cells were limited in enzymes and that fundamental amino acids, at least the basic amino acids, had to be transported to these regions from those parts of the plant in which they are synthesized or stored. Unfortunately, I cannot say that this is

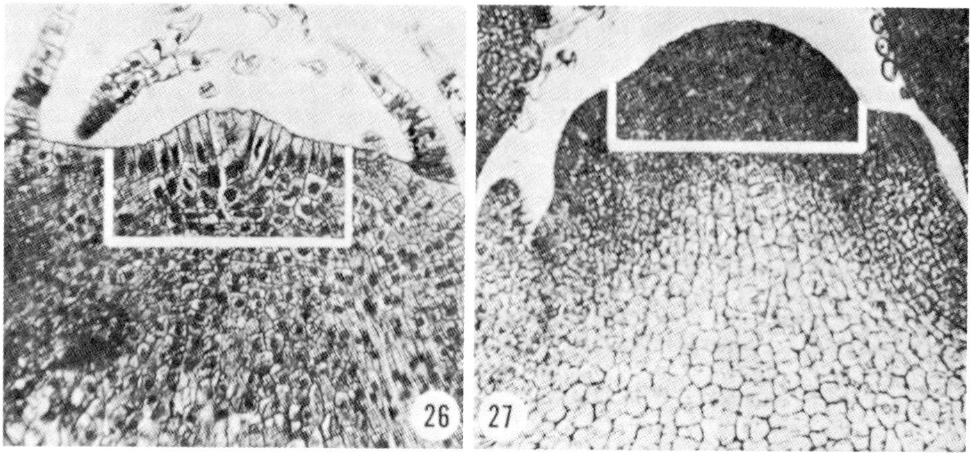

Figures 26 - 27. Apices of *Adiantum pedatum* (Magnification x127.5) and *Lupinus albus* (Magnification x85.5) in longisection. The average size of the pieces excised and the relation of each to the meristem as a whole are indicated.

true. The apices which we have planted since we obtained working results of our chromatographic studies are growing, the minute hairy leaves are normal in appearance, but the plants are not rooted and the largest since June 6 has not been over 1/4 in. in height. In fact, I do not believe we know the complete story at present; there may well be other nitrogen compounds necessary which as yet have not been tried on these minute, leafless explants. But certainly the *in vitro* culture technique has permitted progress in what seems to be a correct direction; by it one can try out various ideas which otherwise seem difficult to test on living material.

As a consequence of these chromatographic findings, we have been forced to re-examine our conclusions on the vascular cryptogams. If dividing cells lack enough enzymes for synthesis of necessary amino acids, then why do apices of vascular cryptogams grow on nitrates as a sole source of nitrogen? It has seemed to us that perhaps we ought to re-examine the studies on the meristems of ferns, horsetails, and *Selaginellas*. It can be seen that in *Adiantum*, and in *Equisetum, Selaginella,* and *Psilotum* as well, the apical cell and the derived prismatic layer are the true meristematic cover of the so-called apical meristem and that little general meristematic activity occurs behind this. In order to test the amino acid content of growing meristems of *Adiantum* we are this summer endeavoring to assemble enough material of essentially the prismatic layer of this species for chromatographic studies. This is a laborious task; moreover, it is inevitable that there will always be retained some of the subjacent, less active tissue. We have planted such pieces *in vitro* with a view to testing their ability to duplicate the growth of the entire so-called meristem. We await the results of this test and those of the chromatographic studies with interest. At the moment, I can only state that even explants of former size from *Adiantum* grow and develop at a faster rate on the *Lupinus* amino acid medium than they did on a nitrate-sugar medium, and at about the rate which they achieved on a medium containing yeast extract or autoclaved coconut milk.

Although no solution to the metabolic activities of the apical meristems of vascular plants is offered by this technique, it seems to us that the leads opened by the preliminary studies are promising and may well suggest lines of investigation important to the morphogeneticist and biochemist alike. If some indication of the paucity of certain enzyme systems in dividing cells becomes apparent, enforcing the notion of contrast between cells in active division, with their high concentration of DNA, RNA, and of their nuclear proteins, and mature cells, with their higher enzyme complement, some direct relation may be suggested between enzyme content and metabolic capacity on the one hand and the length of time between cell divisions on the other. Some evidence points in this direction.

Figures 28 - 31. Sporelings and sporeling apices of *Todea barbara* grown on a good fern medium with different concentrations of sucrose (Magnification x0.75).

Figures 28 and 29. Sporelings grown on 0.1 and 0.5 percent sucrose, respectively. Note two-lobed leaves on sporelings of Figure 28 and three-lobed leaves on those of Figure 29.

Figures 30 and 31 are rooted plants grown from apices of sporelings, transplanted without leaf primordia, on the same medium as above but with 0.1 and 0.5 percent sucrose. Note predominance of two-lobed leaves in Figure 30 and of three-lobed leaves in Figure 31.

May we turn to another aspect of our *in vitro* culture studies? In the course of exploratory studies on the differentiation of vascular tissues, we grew a great many fern sporelings. These sporelings always bear small leaves, are initially protostelic, and only gradually take on the organization of the mature plant. The leaf types which characterize these gradual changes are varied and tend to occur in a somewhat regular sequence (Figure 32). Many workers (e.g., Bower, 1923, 1925; and Orth, 1938) have considered this sequence to be recapitulatory. The first leaves appearing on the sporeling are two-lobed, dichotomous as ordinarily described (Figure 32, 1-6). Some of the early leaves tend to be three-lobed (Figure 32, 7-10), later leaves becoming pinnate in type with few pinnae (Figure 32, 11-12); these are gradually followed by others of a more decided pinnate adult type (Figure 32, 13-14). The growing of these sporelings called for some investigation of the most suitable medium to show the stelar changes. It early became apparent here, as elsewhere, that mineral salts, B-vitamins, auxin, and autoclaved coconut milk in varying amounts in the medium caused variation only in the rate of change of events. However, to our surprise, when concentration of sucrose was varied, there was a definite alteration of pattern.

In order to have sterile fern plants on hand, we have grown large numbers of sporelings of five species: *Todea barbara, Osmunda cinnamomea, Cyrtomium falcatum, Pteridium aquilinum,* var. *latiusculum,* and *Lygodium japonicum.*

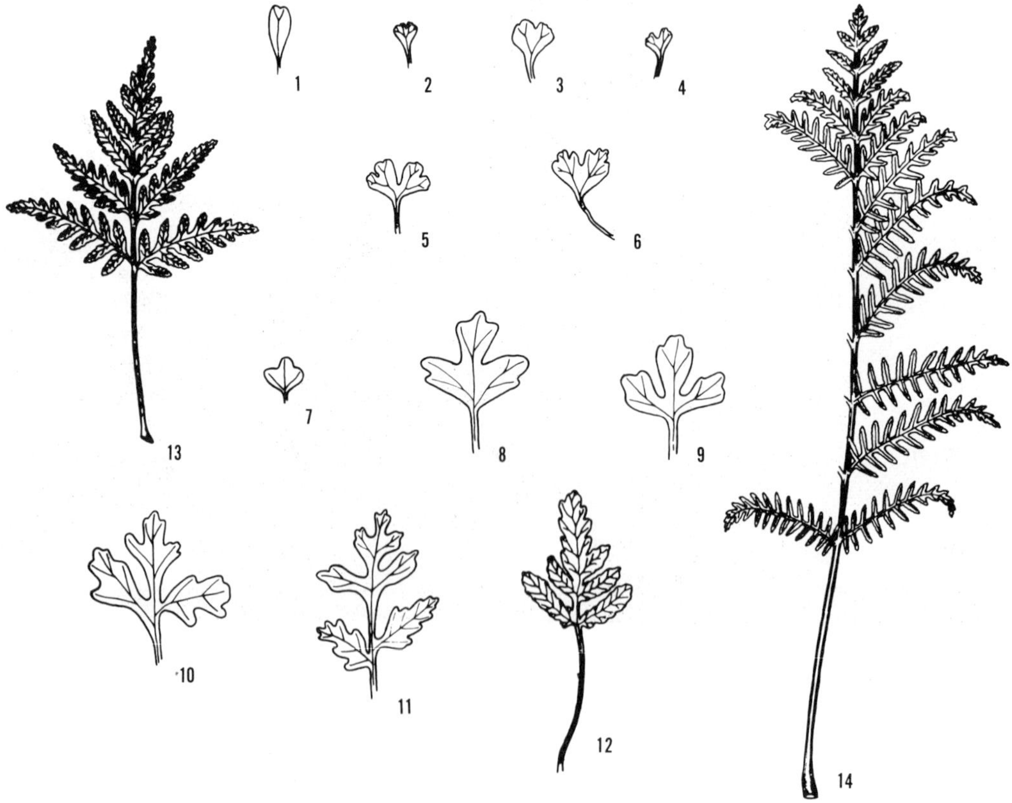

Figure 32. Drawings of varying types of leaf found in fern *Todea barbara.* No. 14 represents a mature leaf; all others are types of sporeling leaves (Magnification: 1-13 x1.2; 14 x0.25).

A favorable medium for these species, which varied only in sucrose concentration - 0.1, 0.25, 0.5, 1, and 2 percent - gave an interesting and consistent story of pattern change. All new leaves on the sporelings grown upon 0.1 percent sucrose tended to be small and two-lobed (Figure 28), and this tendency continued for some time; an occasional short root developed from near a leaf base, before the first three-lobed leaf became evident. On 0.25 percent sucrose, both two-lobed and three-lobed leaves resulted, often in about equal numbers, before the pinnate condition appeared. On 0.5 percent sucrose, only rarely was a two-lobed leaf found, several larger three-lobed types generally arising in advance of the earliest pinnate leaf (Figure 29). Never was a new two-lobed leaf evident on media containing 1 percent and 2 percent sucrose. Here the normal sequence was quickly dispensed with and pinnate leaves of adult nature were the regular type.

When this pattern of change became evident, it was resolved to try the experiment with apices of sporelings of varying ages in which the number and shape of leaves, the number of leaf primordia, and the number of roots already on the sporeling at the initiation of the experiment were known. The results for all five species were generally in agreement and were startlingly sharp. All apices on 0.1 percent sucrose, if they grew at all, grew slowly, developed small leaves, which were almost invariably two-lobed, and only rarely gave rise to a root. Moreover, this continuous sequence of two-lobed leaves went on to many leaves, often 20 or more (Figure 30). This sequence of leaves bore no relationship to the age of the sporeling from which the apex was taken. Old or young, all apices gave two-lobed leaves, except when an already initiated leaf primordium was left on an apex; then a three-lobed leaf or even a short pinnate leaf might precede the normal sequence of two-lobed leaves.

Upon 0.25 percent sucrose, usually an intermixture of two- and three-lobed leaves occurred and an occasional root became evident. Upon 0.5 percent sucrose, only rarely was a two-lobed leaf found. Three-lobed leaves appeared early (Figure 31), and these in turn were replaced by pinnate types with few pinnae. More and more pinnae appeared on each successive and larger leaf. Apices planted upon 1 and 2 percent sucrose never produced two-lobed leaves, gave origin to few or no three-lobed leaves, but early produced pinnate leaves of quite advanced types, with well-developed roots.

In the light of these results, the obvious test was to take apices of mature rhizomes and treat them in the same way. This was done with *Adiantum pedatum.* Examples of our results are illustrated (Figure 33). Two-lobed leaves appeared here on cultures of apices planted on 0.1 percent sucrose, and nowhere else. The significant thing is that they appeared at all, for in the hundreds of plants of *Adiantum* grown from apices upon the standard medium containing 2 percent sucrose or 1 percent glucose, *never* has a two-lobed leaf been seen.

Any contemplation of this set of changes ultimately forces one to a consideration of steps in leaf development and the relation of these steps to carbohydrate metabolism. The early growth of leaves is generally a result of apical division (Foster, 1936). In adult fern leaves, this apical multiplication of cells may continue for some time and the resulting leaf may be several feet long as in some tree ferns; these lengths are duplicated

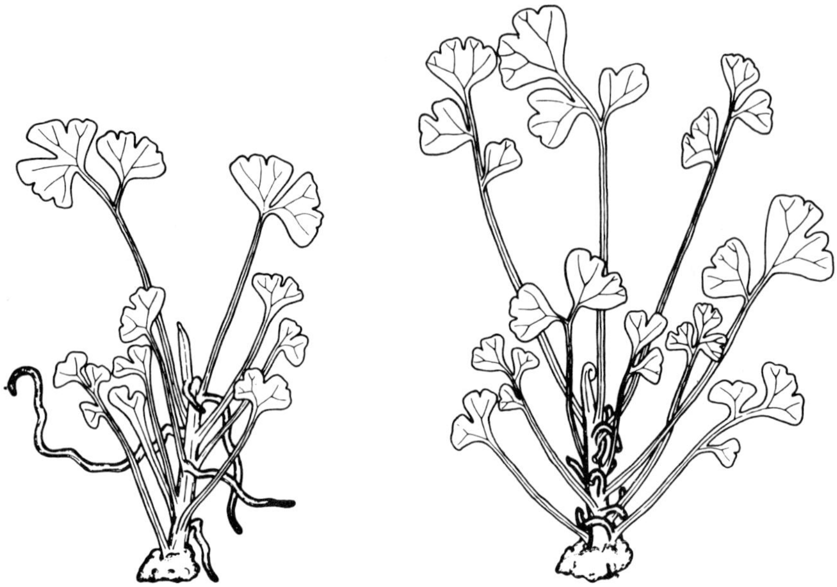

Figure 33. Drawings of plants grown from apices of mature rhizome of *Adiantum pedatum* on what has proved to be a satisfactory medium for *Adiantum* except that the sucrose concentration was dropped from 2 percent to 0.1 percent. Note some two-lobed leaves among the shorter first leaves produced on each (Magnification x1).

by leaves on many plants of varied affinities other than ferns. Generally this period of terminal cell division is followed by the development of a pair of marginal meristems, one along each edge of the leaf, which give the leaf blade its dorsiventrality. As this activity ceases, many cell divisions occur in the various layers of cells comprising the thickness of the leaf. The overlapping effects of these three steps, followed by differential cell enlargement, provide leaves with their varying shapes, sizes, and internal structures.

In the early fern sporeling leaves, terminal growth is essentially nonexistent; marginal growth, therefore, gives a two-lobed structure. At times, however, one may find a kind of bud in the notch between the two lobes of the heart-shaped structure that represents a quiescent potentiality in terminal growth which can often be forced into activity by drastic amputation of the two lobes. Wagner (1952) interprets these heart-shaped leaves as comprised of only the two lowermost pinnae, the terminal midvein region being represented by a "bud" or not at all. In this kind of leaf, terminal growth by cell division is limited directly or indirectly by the carbohydrate supply, it would seem, for when this supply is increased, terminal growth is represented by the three-lobed or multiple-pinnate type, the leaves being larger as well as having more numerous parts. It is not unreasonable to construe this limited growth as being one with limited available energy, as Bullough (1952) has done with his studies on the epidermis taken from the ears of mice. In fact, although no tests have actually been made as yet, the relation of leaf type to sucrose concentration in the medium would suggest that the fundamental mitotic processes of leaf formation are related to the aerobic part of a respiratory cycle. It may prove interesting to substitute the proper concentration of pyruvic acid for sucrose in an effort to test this hypothesis, it being hoped that the substitution will permit growth. In fact, this possible direct relationship between carbohydrate concentration and mitosis is very worth further studies in the attempted extension of our knowledge of meristematic activity in plants.

Morphogenesis involves cell division, cell enlargement, and cell differentiation. It would seem from the studies of Bullough, as well as from our own, that the *in vitro* culture technique might well provide a tool for systematic investigation of energy relations of mitosis and at the same time permit one to observe the modifying effect of carbohydrate-controlled cell division on the pattern of development or on the genesis of form. For a study of cell enlargement as opposed to cell division, it has been suggested by Gautheret (1945) that low concentrations of an auxin favor cell division, whereas higher concentrations foster cell enlargement while inhibiting mitosis. We are not informed in this statement of experience, however, whether the low concentrations favor cell division by stimulating it or by not being of high enough concentration to inhibit it. The ease of administering an auxin to a plant part growing *in vitro* makes it simple to learn the effects of the auxin added over and above that available to the growing part from its own production. One should be able to resolve this problem by *in vitro* methods.

The study of differentiation as a morphogenetic process has hardly begun. Progress by the biochemist is truly evident. The broad need is for concerted studies from all available directions. Certainly the morphogeneticist has a clarion call to provide an understanding of "normal" growth as a background for those interested in the "abnormal." On the other hand, the findings of the student of "abnormal" growth and development may actually provide the clue, the *modus operandi,* for the student of "normal" growth, as they have many times before this. The insect with its gall or the bacterium or virus with its tumor may well open a morphogenetic door for all to enter who see the door open.

Summary

1) It has proven possible to grow mature plants from pieces of stem apices of the vascular cryptogams - *Adiantum pedatum, Selaginella willdenovii, Lycopodium cernuum* and *Equisetum hyemale* - with or without leaf primordia, the pieces being no longer than 0.25 mm. A medium containing only mineral salts (Knop's solution), minor elements (a modified Nitsch's solution), and 2 percent sucrose proves adequate for the growth reported; the rates of growth, however, are speeded up without change in developmental pattern by the addition of yeast extract or autoclaved coconut milk. The increased rates can be attributed to the amino- and/or other nitrogen-containing substances.

2) Correspondingly sized pieces of shoot apices, with or without leaf primordia, of the angiosperms tried - *Lupinus albus, Syringa vulgaris* and *Parthenocissus tricuspidata* - failed to grow at all on the same minimal medium. On this medium, supplemented with coconut milk, a B-vitamin mixture, and an auxin, apices of *Syringa* grew well and rooted, and gave evidence of an acquired capacity for near normal development. When mixtures of amino acids and amides were added to the minimal medium in proportions

suggested by quantitative paper chromatographic studies of stem apices from *Lupinus albus,* some growth occurred, although rooted plants of normal growth and development have not yet been obtained. It appears that some nitrogenous substance or substances may still be lacking or is in inadequate concentration.

3) The contrast in behavior in *in vitro* cultures of shoot apices of vascular cryptogams and angiosperms may well prove helpful in understanding the respective organizations of the apices of these two groups of plants. Work is being done which already suggests the soundness of this premise.

4) The growth *in vitro* of apices of fern sporelings on the above-mentioned minimal salt-sucrose medium, but with sucrose concentrations varying from 0.1 to 2 percent, has provided an explanation of the sequence of leaf shapes of young ferns, a sequence often spoken of as recapitulatory. The explanation is found in a direct relation between the concentration of sucrose as an energy source and the amount of cell division, which determines the shape of the sporeling leaf.

Bibliography

Allen, G.S. 1946. Embryogeny and development of the apical meristems of *Pseudotsuga.* I. Fertilization and early embryogeny. Am. J. Botany 33: 666-677.

Ball, E. 1946. Development in sterile culture of stem tips and subjacent regions of *Tropaeolum majus* L. and of *Lupinus albus* L. Am. J. Botany 33: 301-318.

——— . 1948. Differentiation in the primary shoots of *Lupinus albus* L. and of *Tropaeolum majus* L. Symposia Soc. Exptl. Biol. 2 (Growth): 246-262.

——— . 1949. The shoot apex and normal plant of *Lupinus albus* L., bases for experimental morphology. Am. J. Botany 36: 440-454.

——— . 1950. Regeneration of the shoot apex of *Lupinus albus* after operations upon the central initials. Science 112: 16-17.

——— . 1951. Influence of shoot apex on subjacent shoots of *Lupinus albus* L. Nature 167: 36-38.

——— . 1952. Morphogenesis of shoots after isolation of the shoot apex of *Lupinus albus.* Am. J. Botany 39: 167-191.

Bower, F.O. 1919. "Botany of the Living Plant." Macmillan and Company, Ltd., London.

——— . 1923. The ferns (Filicales). I. Analytical examination and critical comparison. Cambridge University Press, Cambridge.

——— . 1925. Botany. "Evolution in the Light of Modern Knowledge," a collective work. Ch. 5. Blackie and Son, Ltd., London.

Bullough, W.S. 1952. The energy relations of mitotic activity. Biol. Rev. Cambridge Phil. Soc. 27: 133-168.

Camus, G. 1949. Recherches sur le rôle des bourgeons dans les phénomènes de morphogénèse. Rev. cytol. et biol. végétales 11: 1-199.

Foster, A.S. 1936. Leaf differentiation in angiosperms. Botan. Rev. 2: 349-372.

Gautheret, R.J. 1946. Plant tissue cultures. Growth (Suppl.) 10:21-43.

Meiss, A.N. 1952. The formation of asparagine in etiolated seedlings of *Lupinus albus* L. Connecticut Agr. Exptl. Sta. Bull. 553: 1-74.

Miller, H.A., and R.H. Wetmore. 1945. Studies in the developmental anatomy of *Phlox drummondii* Hook. I. The embryo. Am. J. Botany 32: 588-599.

Nast, C.G. 1941. The embryogeny and seedling morphology of *Juglans regia* L. Lilloa 6: 163-205.

Nitsch, J.P. 1951. Growth and development *in vitro* of excised ovaries. Am. J. Botany 38: 566-577.

Nitsch, J.P., and R.H. Wetmore. 1952. The microdetermination of free *L*-tryptophane in *Lupinus albus*. Science 116: 256-257.

Orth, R. 1938. Zur Morphologie des Primärblätter einheimischer Färne. Flora 133: 1-55.

Skoog, F., and C. Tsui. 1948. Chemical control of growth and bud formation in tobacco stem segments and callus cultured *in vitro*. Am. J. Botany 35: 782-787.

Spurr, A.R. 1949. Histogenesis and organization of the embryo in *Pinus strobus* L. Am. J. Botany 36: 629-641.

Steward, F.C., R.H. Wetmore, J.F. Thompson, and J.P. Nitsch. 1954. The nitrogen components of certain shoot apical growing regions. Am. J. Botany 41 (in press).

Wagner, W.H., Jr. 1952. Types of foliar dichotomy in living ferns. Am. J. Botany 39: 578-592.

Wardlaw, C.W. 1944. Experimental and analytical studies of pteridophytes. III. Stelar morphology: The initial differentiation of vascular tissue. Ann. Botany 8: 173-188.

______ . 1945. The shoot apex in pteridophytes. Biol. Rev. Cambridge Phil. Soc. 20: 100-114.

______ . 1947. Experimental investigations of the shoot apex of *Dryopteris aristata* Druce. Trans. Roy. Soc. (London), Ser. B. 232: 343-384.

______ . 1948. Experimental morphology with special reference to pteridophytes. Symposia Soc. Exptl. Biol. 2 (Growth): 276-285.

______ . 1949a. Experiments on organogenesis in ferns. Growth 9: 93-131.

______ . 1949b. Further experimental observations on the shoot apex of *Dryopteris aristata* Druce. Trans. Roy. Soc. (London), Ser. B. 233: 415-451.

Wetmore, R.H., and G. Morel. 1949. Growth and development of *Adiantum pedatum* L. on nutrient agar. Am. J. Botany (Suppl.) 36: 805-806.

______ , and C.W. Wardlaw. 1951. Experimental morphogenesis in vascular plants. Ann. Rev. Plant Physiol. 2: 269-292.

Discussion

BALL: I would like to point out that one fact about this paper of Professor Wetmore's is that it shows a very real approach to our ideal of attempting to grow an entire vegetative plant from a single vegetative cell. I assume that in time that ideal will be achieved. I also wonder if Dr. Wetmore would care to say at this point whether he thinks these amino acids are transported exclusively to the shoot apex or, on the other hand, are some or all of them manufactured in the apex? Another point that interests me is the very extraordinary concentration of asparagine in the alcohol-soluble fraction of the shoot apex, being 69 or 70 percent of the total nitrogen. That appears to me as some evidence that the materials are transported to the apex because one would not expect that much asparagine to be put together in any conceivable sort of protein that we presently know anything about.

WETMORE: In the *Adiantum* material the apices are very small pieces with no leaf primordia. On a medium which contains no added nitrogen other than nitrates, I can definitely state that the apical explants develop faster when not provided with the amino acid mixture than with the spectrum of amino acids which we obtained from the *Lupinus* material. It would seem that the whole *Adiantum* apex must synthesize all the amino acids used. In *Lupinus*, certainly, whole apices without leaf primordia and not including pith rib meristems and pith do not grow in our cultures unless provided with a battery of amino acids. This is perhaps a partial answer but all one can give at the present time. As to the asparagine, of course, *Lupine* is a time-honored plant for acquiring asparagine, as Dr. Vickery and his co-workers have shown; it is really an unfortunate plant from the standpoint of paper chromatography in that the asparagine is so abundant there that it oftentimes obscures the amino acids. One very definitely has to dilute the solution greatly to get the acids to show. *Syringa* is another plant which should be carried through chromatographic study in the same way; this we hope to do this fall. *Syringa* on the one hand and *Lupinus* on the other should give us a little more realistic picture of the amino acids in common as well as the variation between two species of very different families. Also, *Syringa* does not have the same high concentrations of asparagine.

LARUE: I am very much impressed with this work of Dr. Wetmore's. But the fact is that years ago (1936) I did succeed in getting growth of a meristem of *Radicula aquatica,* which is now, I believe, *Nasturtium officinale.* I did succeed in growing just a short typical meristem into a whole plant. Now that seems much more remarkable to me in the view of what Dr. Wetmore has said, than it did before. I would like to call to Dr. Wetmore's attention that this water cress is a most extraordinary plant in its ability to regenerate from almost any part of the leaf or stem, or even root, with the greatest of speed. It may be that this plant is more like your *Adiantum,* which develops, whereas other types of plants cannot. It could well be a plant worthy of your investigation, because it has at least done this one thing of producing growth of a whole plant from a piece of meristem.

WETMORE: I am intrigued by that as a possibility, but it seems to me that one of the things which we have to consider equally are the plants that are difficult to grow. They are the ones that I think are going to give us trouble. Now I believe that in the angiosperms, Dr. White at one time grew *Stellaria media* from an apex and there are a few other instances that are known. In the main, I feel that I can simply sign off a plant which I succeed in growing. The ones that intrigue me very much are the ones I cannot grow because in these lie the problems.

LARUE: Sometimes it is possible to approach the ones that are harder to grow by using the ones that are easier to grow. I believe both approaches are legitimate.

KEHR: Dr. Wetmore, I would like to know what happened to the *Syringa* meristems that were grown on coconut milk. Do they ultimately develop into a fully grown plant?

WETMORE: I grew them *in vitro,* and when they rooted I just assumed that if they were placed in soil they would carry on.

WHITE: I would like to emphasize this matter of controlling the developmental pattern by manipulating the level of a single sugar. I believe it is quite significant and very important that one can get a totally different pattern by the use of a simple substance of that kind. We are running into a good many difficulties with both plant and animal tissues. Some of these things which we can deal with satisfactorily, which we do not ordinarily think of as being the controlling factors, make it of extreme importance. I think it is important to call to mind the relationship that the sugars have in this complex mechanism. We have four different things which are actually related: the auxin, the sugars, the adenine, and the heavy metal elements, of which iron is the particularly significant one. This is one of the places, I believe, where we are going to get more results than by going into all the individual amino acids.

I also wanted to raise a point in regard to the asparagine itself. Of course, Borodin showed 75 years ago that asparagine is the major transport material for the movement of nitrogen through the plant, particularly in the case of the lupines, and also in a great many other of the dicotyledons and is replaced by glutamine in the many monocotyledons, as Archibald has shown. I think the fact that you have a relatively low level of asparagine in the apex and that this is the one level which goes up would lead me to suspect that it is the substance to which we should pay less attention than the amino acids. It seems to me that it is the substance that can be moved away as fast as it is formed. This asparagine, then, is essentially a storage form and its origin will center around other amino acids rather than asparagine itself.

KROTKOV: A comment on the subject of asparagine and glutamine in plant cells. We have been interested for some time in the question of the synthesis of these two amides in living plants. As was mentioned a few minutes earlier, asparagine is found normally in fairly large quantities in lupine plants, but if you take young lupine seedlings and put them on a solution which contains some source of nitrogen, such as ammonium chloride or nitrate, one will find that the asparagine content will be increased even more. And conversely, if you decrease the nitrogen content of the external medium, one will probably find that the asparagine content of the plant will decrease.

We have just finished an experiment along similar lines on glutamine. We have found that a number of plants produce large quantities of glutamine when they are provided with an abundance of either nitrates or ammonia salts. In this connection, a reference can be made to swiss chard plant which we have selected for the synthesis of radioactive glutamine. If one puts a swiss chard leaf on a solution of ammonium nitrate and leaves it illuminated for five days in the presence of an excess of carbon dioxide, one will find that up to 5 percent of the fresh weight of the leaf will be in the form of glutamine. If you use $C^{14}O_2$ and determine the specific activity of glutamine, starch, and sugars present in leaves at the end of this experiment, one will find that the specific activities of all these three substances will be the same. Or to put it in still other terms, the carbon chains of both glutamine and of carbohydrates either come from a common precursor, or glutamine is derived from carbohydrates.

GORDON: The comment on sugar as a morphogenetic determinant has the undesirable implication of direct regulatory functions. After all, developmental differences resulting from increments of added sugar, to what is essentially a starved system, should not be surprising. Sugar is not only a primary energy-yielding substrate, but also supplies the carbon skeleton in the constitution of what are already recognized as more direct morphogenetic determinants. In this connection it is pertinent to mention the experiments reviewed this morning by Dr. Skoog on evocation and individuation through adenine-auxin variations. Also pertinent, with particular reference to the *Adiantum* lobing, is the discussion by Went at the Madison Growth Conferences, in which he described the experimental control of leaf lobe and pinnate form by differential vein and mesophyll growth mediated through variation of adenine-auxin ratios.

Reprinted from the AMERICAN JOURNAL OF BOTANY, Vol. 49, No. 6, 583-589, July, 1962
Printed in U.S.A.

NONESSENTIALITY OF CONCURRENT CELL DIVISIONS FOR DEGREE OF POLARIZATION OF LEAF GROWTH. I. STUDIES WITH RADIATION-INDUCED MITOTIC INHIBITION[1]

ALAN H. HABER
Biology Division, Oak Ridge National Laboratory,[2] Oak Ridge, Tennessee

ABSTRACT

HABER, ALAN H. (Oak Ridge Natl. Lab., Oak Ridge, Tenn.) Nonessentiality of concurrent cell divisions for degree of polarization of leaf growth. I. Studies with radiation-induced mitotic inhibition. Amer. Jour. Bot. 49(6): 583–589. Illus. 1962.—During germination and seedling growth, the first foliage leaf of wheat shows polarized growth with striking changes in shape. During this polarized growth, most of the new cell plates are formed perpendicular to the axis of greatest elongation (longitudinal axis). The relation between leaf length and leaf width is the same in leaves normally growing with cell division as in leaves growing with negligible cell division after gamma-irradiation (800 kr) of the unsown grain. Although cell division is necessary for continued growth and normal development, concurrent cell divisions play no role in determining the degree of polarization, i.e., the ratio of the rates of linear extension in length per mm length to linear extension in width per mm width. The majority of the new cell plates formed *before* the growth period studied were parallel to the axis of greatest absolute and relative elongation during the growth period studied. Irradiated and unirradiated wheat having the same leaf size, and thus the same leaf shape, have quite different cell sizes, shapes, and numbers in both the leaf epidermis and the chlorenchymatous mesophyll. It is concluded that: (1) the orientations of cell divisions, whether concurrent with or prior to the growth period studied, play no role in the basic mechanism which determines the manner in which leaves change shape; and (2) genetic information specifies leaf form independently of genetic influences on sizes and shapes of cells and on the extents and orientations of cell divisions.

THERE is presumably a relation between the form of an organ and the orientations of divisions of its constituent cells. In those instances which have been studied, the new cell plates formed during cytokinesis are mostly perpendicular to the axis undergoing greatest relative elongation (Sinnott, 1960). A fundamental question in plant morphogenesis,[3] however, has remained unanswered: To what extent do oriented cell divisions contribute to the determination of form? Seedlings from heavily gamma-irradiated wheat grain can grow with negligible cell division, but resemble unirradiated seedlings in many aspects of metabolism (Haber, Carrier, and Foard, 1961), growth (Haber and Luippold, 1960; Haber, 1962), certain nuclear functions (unpublished results), and maturation of cells present in the embryo (Foard and Haber, 1961). This paper reports a study of size and shape of the first foliage leaf and of the sizes, shapes, and numbers of its constituent cells in such plants growing without cell division and in unirradiated controls. It will be shown that in this system the oriented cell divisions play no significant role in the basic mechanisms by which growth is polarized. It will also be shown that genetic information does not determine leaf form by specifying the sizes, shapes, and numbers of cells within the leaf.

MATERIALS AND METHODS—Grains of wheat (*Triticum vulgare*, 'Thorne'), from the same batch as in previous studies (Foard and Haber, 1961), were sown in 9-cm Petri dishes containing 1 piece of Whatman's No. 1 filter paper moistened with 5 ml of distilled water under conditions of temperature, light intensity, and photoperiod described previously (Haber, 1962). Some of the grains were exposed to 800 kr of Co^{60} gamma-radiation (at 45 kr/hr) before sowing. Seedlings from these irradiated grain grew with negligible cell division, and are hereafter called "gamma-plantlets." Leaves were cut off at the base, fixed in FAA and mounted and cleared in 75% lactic acid (Simpson, 1929). Measurements of leaf blades less than 30 mm long were made from these mounted sections. Leaf blades longer than 30 mm were fixed in FAA and measured directly.

RESULTS—The rates and final extents of growth of the first foliage leaf are very different in gamma-plantlets and in unirradiated controls (Haber et al., 1961). For any given leaf length, however, the leaf width is the same in both (Fig. 1). The 2 curves (straight lines on the

[1] Received for publication December 10, 1961.

The author is grateful to Mr. William L. Carrier for measurements and cell counts, to Dr. Donald E. Foard for many stimulating discussions, and to Drs. R. F. Kimball, R. B. Setlow, and Mr. D. A. Fuccillo, Jr. for critically reviewing the manuscript.

[2] Operated by Union Carbide Corporation for the U. S. Atomic Energy Commission.

[3] "Morphogenesis" is here used to mean the origin of geometric form (see Sinnott, 1960), and is not intended to be synonymous with all internal and external aspects of development.

double logarithmic plots) represent equations calculated for each of the 2 sets of plants fitted to the general equation for allometric growth, $y = bx^k$, where x and y refer to leaf width and length, respectively, and b and k are constants (Gregory, 1956; Sinnott, 1960).

By 10–12 days a gamma-plantlet reaches the limit of its growth but is comparable in size with a 3-day-old unirradiated control. The data in Table I show that (1) with respect to leaf size and shape, the gamma-plantlet resembles the unirradiated control, and both are quite different from the leaf in the embryo of the unsown grain; and (2) with respect to cell numbers, the gamma-plantlet leaf resembles the leaf in the embryo of the unsown grain, and both are quite different from the unirradiated control. These relations can be more easily visualized from Fig. 2, which shows typical tracings of flattened leaves together with a diagrammatic representation of the cell numbers presented in Table 1. The secondary thickenings and loss of end walls during maturation of the vascular bundle prevent accurate counts of cell number along the longitudinal axis of gamma-plantlet leaves. The absence of significant cell division in gamma-plantlet vascular tissue is indicated by the absence of mitotic figures in these regions during growth and by a constancy of cell number in cross sections of the vascular bundles (D. E. Foard, unpublished results).

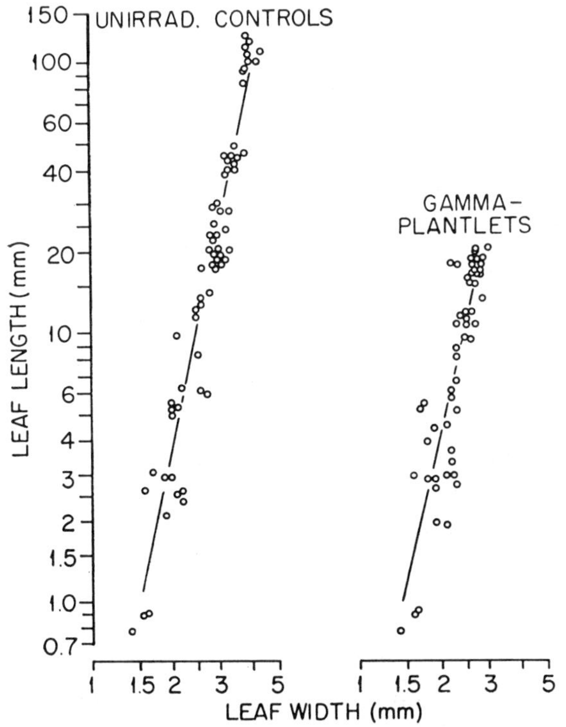

Fig. 1. Shapes of first foliage leaves growing with (unirradiated) and without (gamma-plantlets) cell division. Each point represents an individual leaf blade. Width was measured on the flattened leaf blade halfway between the apex and the base. The ages ranged from 0 to 10 days. The lines represent the data for each of the 2 treatments fitted to the equation for allometric growth, $y = bx^k$, where x and y represent leaf width and length, respectively, and b and k are constants (see text). The lines were fitted to the data by the method of least squares applied to the logarithms of the leaf measurements.

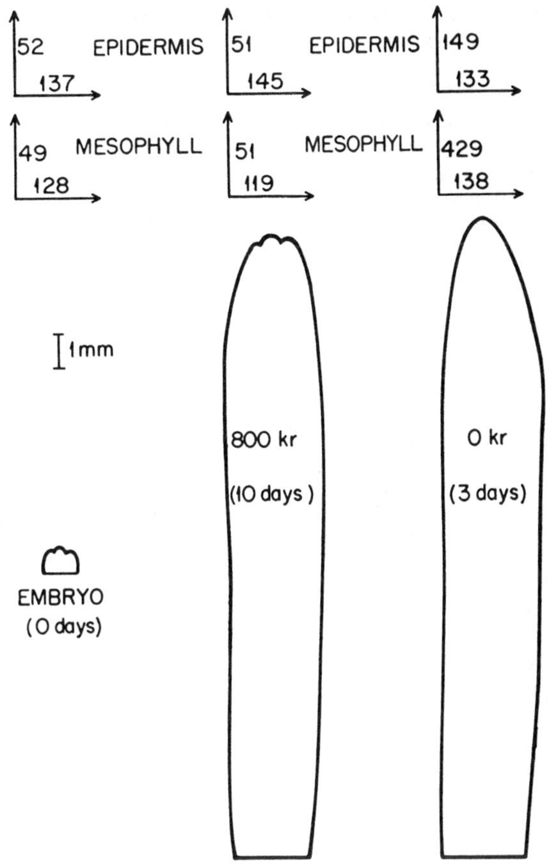

Fig. 2. Outlines of representative first foliage leaves from embryo of unsown grain, from 10-day-old gamma-plantlet, and from 3-day-old unirradiated control. Notches at the apex are produced during preparation of the slides when the leaves are flattened. The diagrams above each leaf outline represent cell numbers taken from Table 1 and are intended to help in visualizing the relations between polarized growth and cell numbers discussed in the text.

Figures 1 and 2 show the greatly polarized growth of these leaves, which change from a shape with width greater than length to shapes with length many times greater than width.[4] The data in Table 1 and Fig. 2 are consistent with our observations that almost all the spindles in Feulgen-stained mounts of seedling leaves are parallel to the longitudinal axis. Exceptions to this rule are the final 3 divisions concerned with the formation of each of the stomatal com-

[4] "Length" of leaves and cells in this paper refers to the apical-basal axis, and not necessarily to the longest dimension.

TABLE 1. *Dimensions and cell numbers in leaves*

Source of leaf	Length		Width[a]	
	mm	No. cells	mm	No. cells
Embryo in unsown grain	0.66 ± 0.04		1.23 ± 0.17	
Epidermis		51.6 ± 1.2[d]		137.0 ± 17.4[d]
Mesophyll[c]		49.1 ± 3.1		~128[e]
10-day-old gamma plantlet	18.1 ± 0.7		2.6 ± 0.1	
Epidermis		50.8 ± 4.5		145.5 ± 5.3
Mesophyll[c]		51.2 ± 2.8		118.9 ± 8.1
3-day-old unirradiated control	18.2 ± 0.7		2.8 ± 0.1	
Epidermis[b]		148.9 ± 12.5		132.8 ± 5.1
Mesophyll[c]		429.2 ± 30.0		137.6 ± 4.5

[a] Width was measured halfway between the apex and base.
[b] Undifferentiated rows in abaxial (smooth) epidermis.
[c] Layer of parenchymatous mesophyll immediately underneath abaxial epidermis.
[d] 95% confidence limits calculated in such a way that one plant contributed only one datum to each entry in the table.
[e] Estimated by dividing average leaf width by average cell width.

plexes occurring in specialized epidermal rows. The great majority of newly formed cell plates are nonetheless perpendicular to the longitudinal axis. The gamma-plantlet leaf has a somewhat less-pointed apex than the unirradiated control (Fig. 2). However, from our measurements of length and width, it is clear that despite the differences between gamma-plantlet leaves and unirradiated controls in the rates and final extents of growth, the degree to which the growth is polarized is the same.

Measurements of cells in the abaxial epidermis of a 10-day-old gamma plantlet and a 3-day-old unirradiated control are given in Table 2. Only cells in undifferentiated rows were measured; consequently, these data overestimate the average cell lengths for the unirradiated leaf epidermis since cells in stomatal rows are shorter. The data nonetheless indicate that the longest cells in the gamma-plantlet leaf are longer, but not wider, than the longest cells in the control leaf of the same length. Figure 3 illustrates the cells in the abaxial epidermis of the embryo within the unsown grain, in the 10-day-old gamma-plantlet, and in the 3-day-old unirradiated control. Absence of cell division prevents the formation of those idioblastic cells that give rise to the stomatal complex and those that bear leaf hairs. In the extreme apical portion of the gamma-plantlets, however, some guard mother cells may have divided to form guard cells, but the absence of cell division in the 2 lateral epidermal cells on either side of the guard mother cell prevents the formation of any subsidiaries in these abnormal stomates. Such uniquely radioresistant cell divisions in the gamma-plantlets would have increased the total cell number by less than 0.1% in the leaf and by less than 0.3% in the leaf epidermal layers.

There are striking differences in cell size of the chlorenchymatous mesophyll tissue between gamma-plantlets and unirradiated controls; this

TABLE 2. *Length and width of cells in abaxial epidermis of seedling leaves*

Distance from apex (mm)	10-day-old gamma plantlet[a]			3-day-old unirradiated control[a]		
	Leaf width (mm)	Cell length (μ)	Cell width (μ)	Leaf width (mm)	Cell length[b] (μ)	Cell width[b] (μ)
3.8	2.8	449 ± 47[c]	31 ± 3[c]	2.6	447 ± 69[c]	29 ± 2[c]
7.5	2.9	650 ± 43	26 ± 2	2.8	412 ± 89	23 ± 1
11.2	2.6	649 ± 62	23	2.8	400 ± 64	23 ± 2
14.9	2.7	352 ± 44	20 ± 1	2.8	171 ± 20	18 ± 2
17.4	2.1	342 ± 68	17 ± 1	2.2	36 ± 6	19 ± 4

[a] The 2 leaves chosen are those whose tracings are shown in Fig. 2. Each is 18.4 mm long.
[b] Chosen from undifferentiated rows.
[c] 95% confidence limits.

difference is evident, although to a much smaller degree, in the data from the abaxial epidermis. Cell sizes and shapes in the layer of chlorenchymatous mesophyll immediately beneath the abaxial epidermis are shown in Fig. 4a–d. These cells, unlike the epidermal cells, are not very dissimilar in size and shape in different portions of a given leaf. The mesophyll cells in gamma-plantlets are greatly elongate compared either to the cells in a 3-day-old unirradiated plant of the same size or to a fully grown 10-day-old unirradiated leaf (Fig. 4b–d). Ten-day-old unirradiated controls have shorter mesophyll cells than the same leaves had when they were 3 days old (and the leaf was less than 1/5 as long as at 10 days). This indicates that cell division occurs throughout this tissue (instead of just near the base) during the entire period of growth. That cell division occurs in the chlorenchymatous mesophyll in the apical regions of the leaf in unirradiated controls was also shown by detection of a few mitotic figures (in other studies) and the fact that 49 chlorenchymatous mesophyll cell lengths from the apex of a 3-day-old plant was about 4.0 mm, whereas 49 cell lengths from the apex was only about 1.4 mm in 10-day-old plants (which had a total length of about 4000 cell lengths in the layer of chlorenchymatous mesophyll immediately underneath the abaxial epidermis).

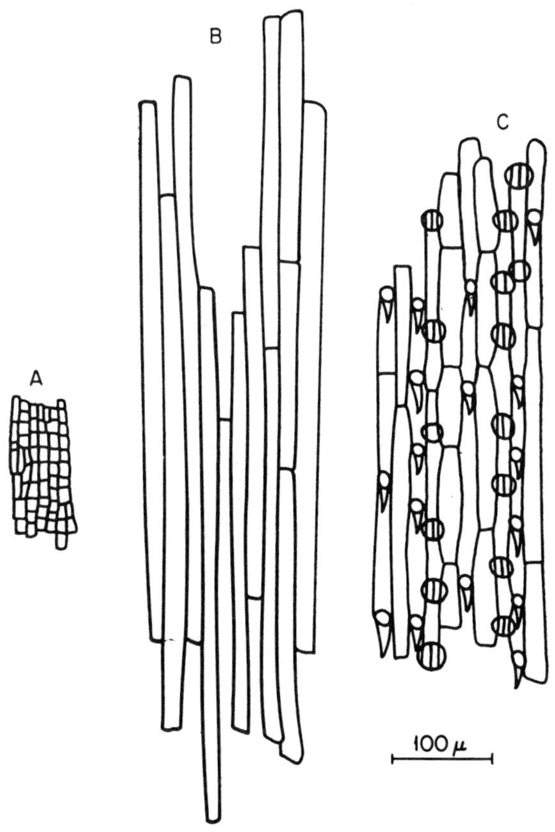

Fig. 3. Camera lucida drawings of cells of abaxial epidermis at a level 1/5 of the distance from leaf base to apex: a, from embryo of unsown grain; b, from 10-day-old gamma-plantlet; c, from 3-day-old unirradiated control. Stomatal complexes and leaf hairs are absent in these portions of the gamma-plantlet leaf.

Discussion—*What is meant by "polar growth"?* —"Polar growth" implies differential extension along different axes. Usually, however, the concept is not precisely explained. We will distinguish between 2 conceptions of "polar growth" in studies in which only length (y) and width (x) of an organ are considered. *Axially unequal relative extension* will designate growth in which the (rate of) increase in length per unit length is unequal to the (rate of) increase in width per unit width, i.e.,

$$\frac{y_2 - y_1}{y_1} \neq \frac{x_2 - x_1}{x_1}$$

Axially unequal absolute extension will designate growth in which the (rate of) increase in length is unequal to the (rate of) increase in width, i.e., $y_2 - y_1 \neq x_2 - x_1$. During growth of the wheat leaves here studied, the longitudinal (Y) axis is the axis of both greater relative extension and greater absolute extension. Axially unequal relative extension does not imply axially unequal absolute extension and vice versa, since

$$\frac{y_2 - y_1}{y_1} \neq \frac{x_2 - x_1}{x_1}$$

does not imply $y_2 - y_1 \neq x_2 - x_1$ and vice versa. The axes of greater relative extension and greater absolute extension may even be different. For example, in certain gourd fruits (Fig. 2 of Sinnott, 1936), the axis of greater relative extension is the transverse axis, whereas the axis of greater absolute extension is the longitudinal axis when $y < 13$ cm. Axially unequal *relative* extension involves different rates of linear extension per mm in length and width, and is therefore the necessary and sufficient condition for growth causing a change in shape.[5] It is thus the type of polarized growth crucial for morphogenesis. It is, moreover, the axis undergoing greatest relative extension, and not necessarily the axis undergoing greatest absolute extension, that is perpendicular to the majority of newly formed cell plates in cucurbit fruits (Sinnott, 1944).

[5] That axially unequal relative extension implies a change in shape can be seen from the following argument. Shape in this instance can be defined as y/x. Then, by the definition of axially unequal relative extension,

$$\frac{y_2 - y_1}{y_1} \neq \frac{x_2 - x_1}{x_1} \therefore \frac{y_2}{y_1} - 1 \neq \frac{x_2}{x_1} - 1 \therefore \frac{y_2}{y_1} \neq \frac{x_2}{x_1} \therefore \frac{y_2}{x_2} \neq \frac{y_1}{x_1} \quad \text{Q.E.D.}$$

By reversing these steps it can be shown that a change in shape implies axially unequal relative extension.

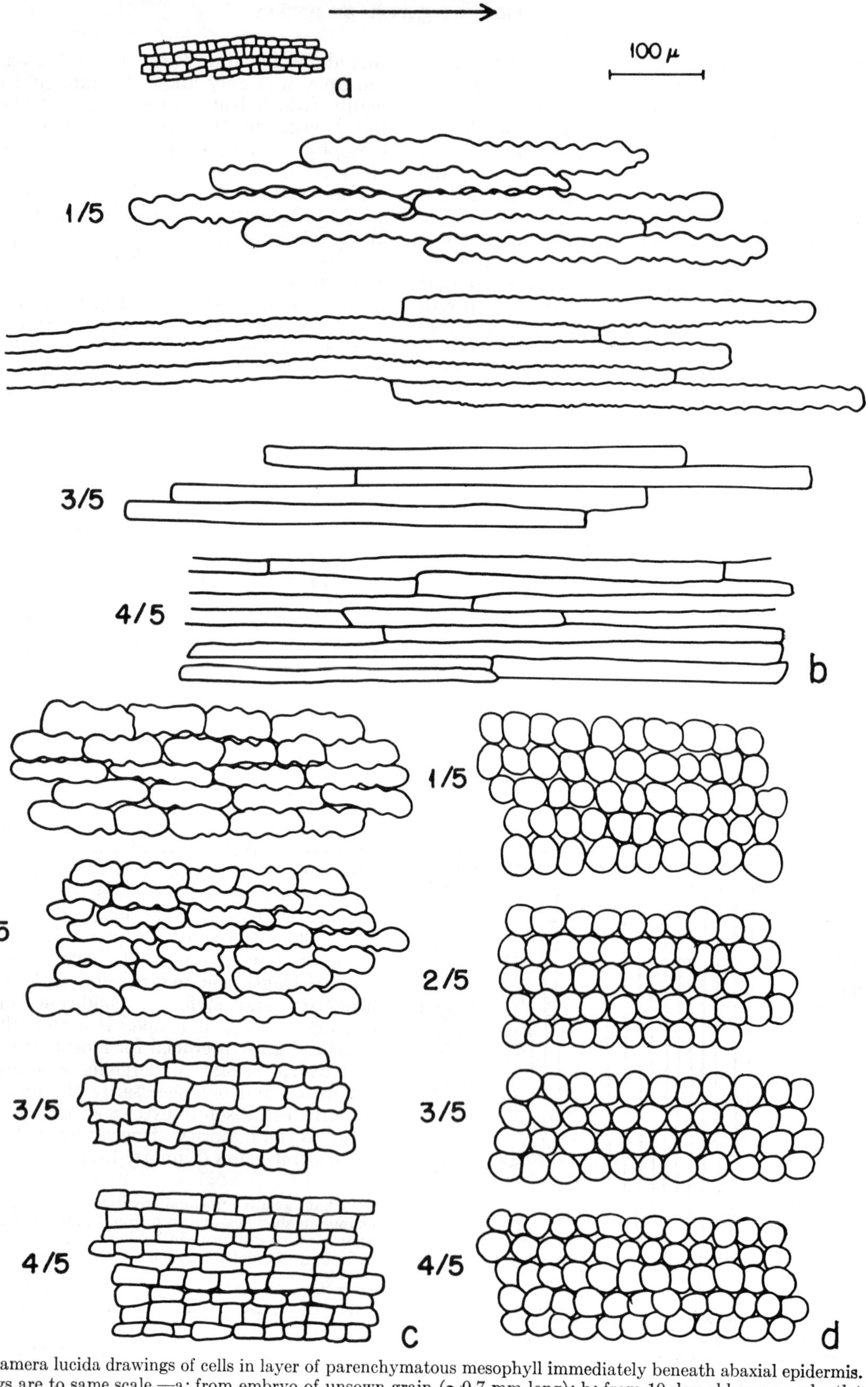

Fig. 4. Camera lucida drawings of cells in layer of parenchymatous mesophyll immediately beneath abaxial epidermis. All drawings are to same scale.—a: from embryo of unsown grain (~0.7 mm long); b: from 10-day-old gamma-plantlet (17 mm long); c: from 3-day-old unirradiated control (18 mm long); d: from 10-day-old unirradiated control (115 mm long). The fractions (i.e., ⅕, ⅖, etc.) beside the drawings in Figs. 4a–4d indicate the fraction of the total leaf blade length from the apex. The arrow indicates the orientation of the leaf longitudinal axis.

Concurrent cell divisions do not contribute to the degree of polarization of growth—The slope of a straight line on a double logarithmic plot of leaf length (y) versus width (x) ("shape curve") is the allometric constant, k, in the equation:

$$y = bx^k$$

(see Gregory, 1956; Sinnott, 1960). It can be shown that this algebraic equation may be written as the differential equation:

$$k = \frac{\frac{1}{y}\left(\frac{dy}{dt}\right)}{\frac{1}{x}\left(\frac{dx}{dt}\right)}.$$

Thus the slope of the straight line in Fig. 1 in each case is equal to the ratio of the rate of linear extension in length per mm length to the rate of linear extension in width per mm width throughout the period of allometric growth. Since the 2 "shape curves" in Fig. 1 are practically identical, it follows that the gamma-irradiation, which inhibits cell division and greatly reduces the subsequent rate and extent of growth, has no effect on the allometric constant. (Perhaps the more vivid term "degree of polarization" might be preferable to the term "allometric constant" in these studies.) Therefore, the oriented cell divisions in the unirradiated leaves play no significant role in determining the degree to which concurrent growth is polarized, i.e., the ratio of the *rate* of increase in length per mm length to the *rate* of increase in width per mm width.

Role of cell division in determining leaf shape—Despite the nonessentiality of concurrent cell division in determining the degree of polarization of the growth of the leaf here studied, cell division does play a role in determining shape of this leaf at maturity. Fully grown gamma-plantlet (10-day-old) and unirradiated control leaves (10-day-old) have quite different shapes (Fig. 1). Since concurrent cell division here plays no role in the degree to which growth is polarized, its contribution to leaf shape must result from its being necessary for sustained growth. The formal reasoning is similar to Sinnott's interpretation concerning the different shapes at maturity of "giant" and "miniature" bottle gourds which follow identical shape curves ($k \neq 1$) during growth: "These 2 races of gourds may thus be regarded as differing primarily in size and only secondarily in shape" (Sinnott, 1936). These results demonstrate the nonessentiality of cell divisions for the degree of polarization of growth of the first foliage leaf of wheat during seedling growth. Subsequent papers in this series will discuss: (1) the general applicability of this finding for growing leaves and other determinate organs; and (2) experimental modifications of the degree of polarization of leaf growth. It must be emphasized, however, that the roles of the earliest cell divisions in polarization of growth during initiation of the leaf primordium are unknown.

Polarization of growth not related to orientations of previous cell divisions—The fact that the polarized growth in leaves growing without cell division is so similar to the polarized growth of normal leaves shows that the oriented cell divisions normally occurring simultaneously with polarized growth are not necessary for this polarization. This finding does not directly answer the question: To what extent is such polarized growth without cell division a consequence of orientations of cell divisions which occurred *before* the growth period studied (i.e., during embryogeny)? The orientations of the previous cell divisions, however, probably were such that most of the new cell plates were *parallel* to the subsequent axis of greater relative and absolute extension, since the leaf in the mature embryo in the unsown grain has more than twice as many cells along a line perpendicular to the longitudinal axis than along the longitudinal axis itself (Table 1, Fig. 2). Some early cell divisions during embryogeny may have been essential for polarization of growth during establishment of the leaf primordium. From the orientations of the majority of cell divisions which occurred during embryogeny, however, one would not have predicted the axis of greater relative extension during seedling growth.

Does "morphogenetic information" specify leaf form or cell forms?—Another fundamental question in plant morphogenesis can be visualized in the diagram below.

Theory No. 1

a—genetic information—
- —specifies → *b*—cell sizes
- —specifies → *c*—cell shapes
- —specifies → *d*—extents of cell division
- —specifies → *e*—orientations of cell divisions

} uniquely determine → *f*—leaf form

Theory No. 2.

a—genetic information—
- specifies → *f*—leaf form
- specifies → *b*, *c*, *d*, *e*—cell sizes, shapes, etc.

There is no question that genetic information, *a*, specifies leaf form, *f*, as well as: cell sizes, *b*; cell shapes, *c*; extents of cell division, *d*; and orientations of cell divisions, *e*, within the various regions of tissues in the leaf. We now ask the following question: Does genetic information, *a*, specify *b*, *c*, *d*, and *e*, which uniquely determine *f*? Or does *a* specify *f* independently of the means by which *a* influences *b*, *c*, *d* and *e*? The experiments here reported show that the shapes of leaves comparable in size, whether growing with or without cell division, are essentially identical, whereas the sizes and shapes of comparable cells in these leaves are quite different. The gamma-irradiation treatments thus greatly alter *b*, *c*, *d*, and *e* without significantly affecting the basic mechanisms which determine *f*. Therefore, *f* is not a dependent variable that is uniquely determined by the independent variables *b*, *c*, *d*, and *e*. Our data disprove theory no. 1 and are consistent with theory no. 2. In other words, genetic information specifies leaf form independently of genetic influences on cell sizes and shapes and on extents and orientations of cell divisions. These results provide direct experimental evidence supporting the view once succinctly expressed by Sinnott (1939): "... the development of an organ proceeds with little relation to the manner in which it is cut up into organized cellular units. The division into cells, though of great physiological advantage since it makes possible the separation of various processes, is apparently of much less significance for development."

LITERATURE CITED

Foard, D. E., and A. H. Haber. 1961. Anatomic studies of gamma-irradiated wheat growing without cell division. Amer. Jour. Bot. 48: 438–446.

Gregory, F. G. 1956. General aspects of leaf growth, p. 3–17. *In* F. L. Milthorpe, [ed.], The growth of leaves. (Proceedings of the Third Easter School in Agricultural Science, University of Nottingham, 1956). Butterworth's Scientific Publications, London.

Haber, A. H. 1962. Effects of indoleacetic acid on growth without mitosis and on mitotic activity in absence of growth by expansion. Plant Physiol. 37: 18–26.

———, W. L. Carrier, and D. E. Foard. 1961. Metabolic studies of gamma-irradiated wheat growing without cell division. Amer. Jour. Bot. 48: 431–438.

———, and Helen J. Luippold. 1960. Effects of gibberellin on gamma-irradiated wheat. Amer. Jour. Bot. 47: 140–144.

Simpson, Jennie L. S. 1929. A short method of clearing plant tissues for anatomical studies. Stain. Technol. 4: 131–132.

Sinnott, E. W. 1936. A developmental analysis of inherited shape differences in cucurbit fruits. Amer. Nat. 70: 245–254.

———. 1939. The cell-organ relationship in plant organization. Growth 3(Suppl.): 77–86.

———. 1944. Cell polarity and the development of form in cucurbit fruits. Amer. Jour. Bot. 31: 388–391.

———. 1960. Plant morphogenesis. McGraw-Hill Book Co., New York.

Geotropism and the Modified Sine Rule; an Interpretation based on the Amyloplast Statolith Theory

By

L. J. AUDUS

Bedford College, London University, London

(Received January 20th, 1964)

I. Introduction

The relationships between the angle of displacement of a gravi-sensitive organ from its normal orientation and its sensitivity to geotropic stimulation in that direction have been much studied and an excellent summary of the results are given in Larsen (1962). It was Sachs (1882) who first suggested that the component of gravity at right angles to the long axis of an orthotropic organ should determine the stimulus intensity which should thus be a linear function of the sine of the angle of displacement. Contrary to this expectation, the maximum response in most cases so far studied is obtained with displacements of the order of 120°. In the days when it was popular to invoke "tonic" effects of the environment, Metzner (1929) called upon the component of the gravitational field parallel to the long axis of the organ to modify appropriately the "tonus" of the responding organ and hence its geotropic sensitivity and derived from such purely theoretical speculations a mathematical expression which could be made to fit the angular displacement-response data available at that time. In this formula the geotropic stimulus G was related to the angle of displacement θ by the formula

$$G = gE\sin\theta - KgE\sin\theta\cos\theta$$

where E is the time of exposure and K is a constant. The first term of this equation represents the direct action on the gravi-sensitive system of the component of gravity *perpendicular* to the main axis of the organ whereas the second term represents the modifying tonic effect on this system of the component acting *along* the main axis (geotonus). Such curve-fitting afforded little insight into the mechanisms of graviperception in plants or of the nature of tonus but Larsen has very recently attempted to take the matter one step further by speculating on these relationships in terms of the statolith bodies responsible. Larsen envisages these statoliths as molecular pendulums, hinged at one end onto some rigid cytoplasmic structure. The movements of re-orientation of these pendulums under gravity would constitute the stimulus leading to geotropic response. Were gravity the only force involved, the sine law would be strictly obeyed, but Larsen provides his pendulum "bobs" with an electrostatic charge which, in the longitudinal electrostatic field of the organ, provides a longitudinal force on the "bob" and this prevents the pendulum from hanging vertical. Larsen worked out from these considerations the expected relationship between the stimulus intensity and this angle of displacement θ. He suggested that the stimulus should be proportional to

$$\frac{\sin \theta}{\sqrt{r^2 + 2r \cos \theta + 1}}$$

where r is the ratio of the longitudinal (electrostatic) force to the gravitational force on the statolith pendulum "bobs". This formula yielded curves which fitted very closely the geotropic curvature-angular displacement relationships he himself recorded for cress (*Lepidium sativum*) roots.

However, there are serious objections to Larsen's pendulum theory on theoretical grounds, the main being that movements of re-orientation of protoplasmic bodies of the order of size postulated by Larsen could not possibly be rapid enough to yield the presentation times normally obtaining. The arguments in support of this contention are set out in full elsewhere (Audus 1962), and will not be repeated here. In the same article I have shown that the starch statolith theory still has the majority of evidence in its favour and for these reasons it would seem logical to ask whether these departures from the simple sine rule can also be explained in terms of the same theory.

It is clear that the sedimenting amyloplasts of the statocyte cells do not move independently. Theoretical consideration of surface tension and viscosity relationships and observations of behaviour under both the light and electron microscopes suggest that these bodies tend to influence each other's sedimentation and also to be influenced by the close proximity of the cell wall. In other words, they do not sediment directly across to another wall of the cell when its orientation is changed but tend to slide or roll in a clump along the wall until a new position of equilibrium is reached. The starch-statolith theory implies that the contact of these amyloplasts with protoplasmic structures on the lateral walls of the statocyte constitutes the immediate geotropic stimulus. This being so the stimulus intensity, and subsequently the degree of geotropic curvature, should be some positive function of the number of amyloplasts reaching the lateral walls during the course of the stimulation period. The type of amyloplast migration movements outlined above strongly suggests that the maximum number of amyloplasts would not come to lie on the lateral wall with displacements of 90°, since they would tend to remain clumped in that corner of the cell made by the lateral (now lowermost) wall and the end wall which had formerly been the lowermost. Displacement beyond 90° would encourage more of this clump to slide or roll or perhaps do both these things and thus spread over the lateral wall, increasing the stimulus.

One way, a difficult and tedious way, of obtaining quantitative evidence to check this would be a series of careful microscope studies on amyloplast behaviour in statocytes after fixation and microtoming. An easier and perhaps more precise way is to make a model of a statocyte cell, and observe its behaviour directly.

II. The Statocyte Model

The basis for this model was the statocyte cell from the root cap of *Vicia faba,* whose dimensions are of the order of 20 μ long and 15 μ wide. To simplify the construction and subsequent measurement, a box of rectangular cross-section was constructed of $^{3}/_{16}''$ transparent perspex on the approximate scale of 1 cm = 2 μ. The internal dimensions of this box were 9 cm × 7 cm × 7 cm and this constituted the model of the statocyte cell. The models of the statoliths were solid and made of white polythene. They were of two kinds. In one series of measurements they were smooth spheres of 1 cm diameter while in another they were cubical and of 0.7 cm side. The cubes were cut from a block with a coarse saw to give them a rough surface. These also were of dimension roughly to scale (*i.e.,* corresponding to amyloplast diameters of 1.4 to 2.0 μ).

The internal surface of one long (9 cm) wall was deeply scored by a series of parallel grooves running transversely across the cell (see Figure 1), the purpose being to increase the friction between the model statoliths and the wall and to allow studies of the effect of this on statolith "sliding" and hence distribution. A number of such model statoliths (either 30 spheres or 36 cubes) were placed in the perspex cell, this representing the order of number of amyloplast in the Vicia statocyte. The cell was then filled completely with very high viscosity medicinal paraffin and sealed tightly with a rubber bung, excluding all air bubbles in the

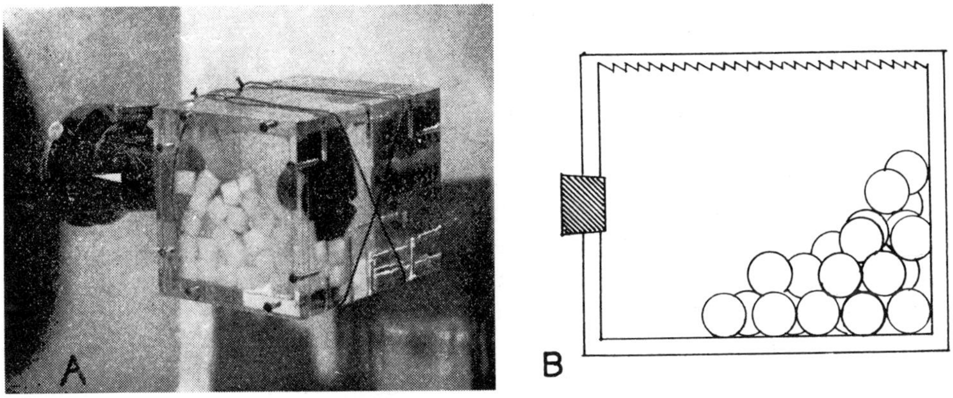

Figure 1. *Photograph* (A) *and diagram* (B) *of model statocyte cell.* The photograph shows the cubes sedimenting on the "rough" wall, while the diagram shows the spherical "statoliths" sedimenting on the "smooth" wall. The rotating support with scale of angles can be seen behind and to the left of the model in the photograph.

process. Polythene has a density only slightly greater than medicinal paraffin and so the model "statoliths" sedimented very slowly in this viscous medium. The cell was fixed to a horizontal rotatable spindle provided with an angular scale, so that it could be quickly and accurately placed in any orientation for sedimentation studies (see Figure 1).

When the cell was orientated with its long axis vertical (corresponding to the normal orientation of an orthotropic plant organ) 40–60 % of the spheres or cubes formed one layer on the bottom wall with about 40–60 % lying on top. Thus some "statoliths" in the bottom layer were in contact with all the lateral walls of the cell. In the sedimentation counts to be described these "contacts" had to be eliminated and to this end a slightly restricted area of lateral wall was used for "contact" measurements. Two transverse lines were scored across the lateral walls at a position 0.6 cm distant from the internal surfaces of the end walls, thus eliminating those areas of the lateral walls where contact with the basal layer of "statoliths" would be likely at equilibrium in the normal or inverted positions (displacement of 0° or 180°). The "sensitive" central area thus represented about 88 per cent of the total lateral walls.

The technique of studying the behaviour of these models was as follows. The model was first inverted and the polythene statoliths allowed to come to rest on one end wall (the "top" end where it was sealed with the bung). It was then turned to its "normal" orientation (rotation through 180°) and the statoliths allowed to sediment randomly on the opposite (bottom) end wall. When they had come to rest, the cell was rotated slowly (*i.e.*, over 1 to 4 seconds depending on the displacement) to a new orientation and the statoliths allowed to sediment to their new positions. Slow rotation was necessary to reduce mass movement of the cell contents. With small displacements ($< 30°$) sedimentation took about 20 seconds but with larger ones (*i.e.*, $> 90°$) equilibrium was reached in about 2–3 minutes. The number of spheres or cubes actually in contact with the "sensitive" area of the lowermost wall was then recorded using an adjustable mirror for ease of viewing. For each angle at least ten independent sedimentations were made. The mean of the ten "contact" counts and its standard error was calculated for each angle studied.

III. Results

To try to reproduce the two possible extremes of sedimentation conditions two types of migration were studied. The first involved smooth spheres sedimenting on to the smooth side of the cell. These spheres very easily slide over and past one another and along the surfaces of the cell and hence in their equilibrium positions they pack themselves fairly snugly into layers, leaving few large spaces. The spheres tend to slide rather than to roll over the cell surfaces. The second involved the rough cubes sedimenting on to the grooved side of the cell. Such cubes tend to roll rather than slide over each other and the grooved surface of the cell. They consequently do not pack snugly as do the spheres but tend to pile up leaving irregular spaces between them. It was felt on a basis of what had been observed with Vicia statocytes under the electron microscope that the cubes sedimenting on the grooved wall were most likely to reproduce the behaviour of the apparently sticky and plastic amyloplasts, whose movements were obviously impeded

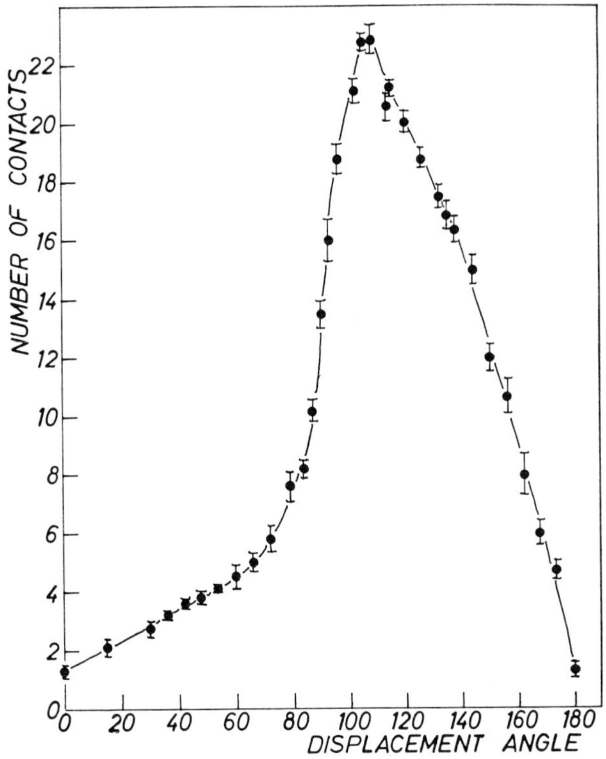

Figure 2. *Relationship between number of smooth spheres in contact with the "sensitive" area of the smooth wall of the model at equilibrium and the angle of displacement of the cell.* The vertical line through a mean represents ± the standard error of that mean.

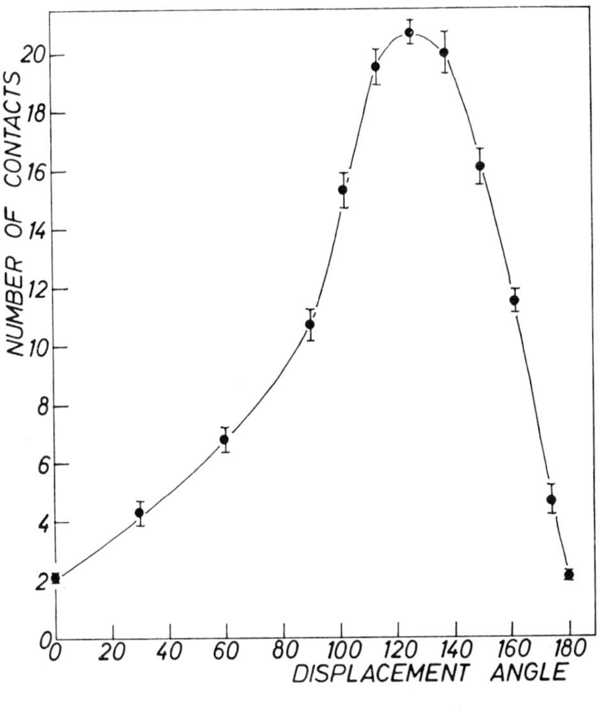

Figure 3. *Relationship between the number of rough cubes in contact with the "sensitive" area of the "rough" wall of the model at equilibrium and the angle of displacement of the cell.* The vertical line through a mean represents ± the standard error of that mean.

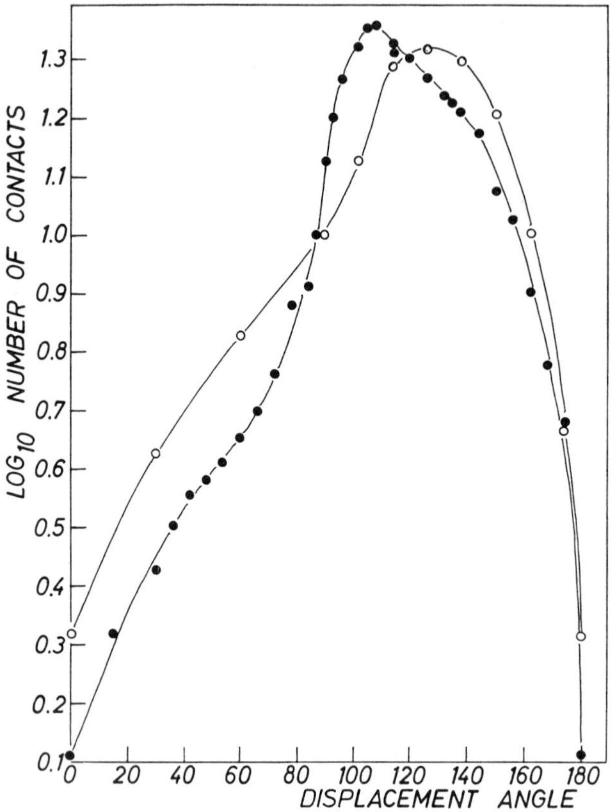

Figure 4. *Relationship between the logarithm of the number of "statoliths" in contact with the sensitive area of lateral wall at equilibrium and the angle of displacement of the cell.* Closed circles — smooth spheres on smooth wall. Open circles — rough cubes on rough wall.

both by their neighbours and possibly by such structures as the endoplasmic reticulum lining the cell wall.

The results of these counts are seen in Figures 2 and 3 where the average number of statolith "contacts" at equilibrium are plotted against orientation for both spheres and cubes. The most obvious feature of both these curves is that the maximum number of contacts occurs at angles considerably exceeding 90°. With the smooth spheres the optimum angle is between 105° and 108° while with rough cubes it is between 125° and 130°. Thus the main contention that starch statoliths can, in principal, explain the departure from the simple sine law is amply supported. However, the shapes of these curves are not very similar to those of the geotropic response curves in that they show a very marked concavity over the range 40° to 90°, whereas the geotropic response curves are much more linear over the wider range 0° to 120°. But it must be remembered that physiological responses are not usually linearly related to the intensity of the stimulus since, according to the Weber-Fechner law, which is of wide application, the response should be linearly related to the logarithms of the stimulus. If then we take the logarithms of the numbers of contacts and plot them against displacement angles we get the curves of Figure 4, where those concavities have virtually been removed and a much closer resemblance to observed response curves is thereby obtained.

The question now arises — how close is the fit of these results to observed

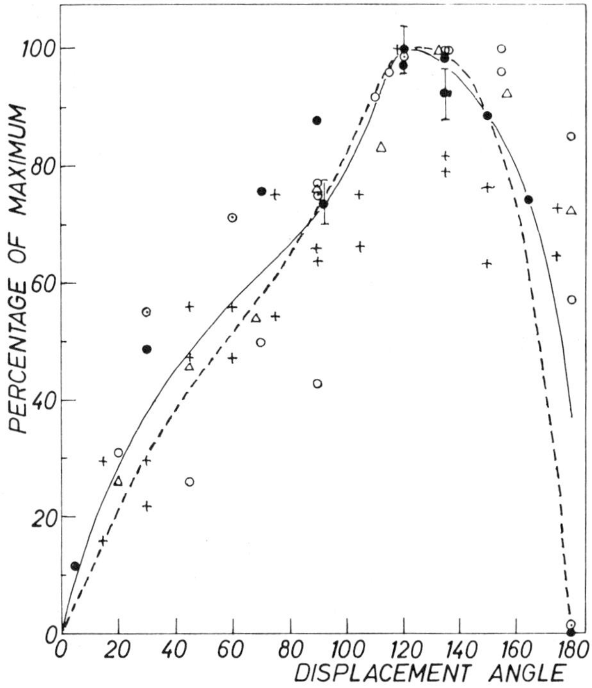

Figure 5. *Comparison of curves from the model statocyte with geotropic curvature results from the literature.* + Ahrens (1933/34) *Vicia faba* roots. ○ Czapek (1895) *Lupinus* roots. ⊙ Czapek (1895) *Secale* node. △ Czapek (1906) *Lupinus* root. ● Larsen (1962) *Lepidium* root. ● Metzner (1929) *Lupinus* root (Mean ± S.E.).

Continuous line = curve of nearest fit to the geotropic curvature data. Dotted line = curve of $\log_{10}N - \log_{10}N_0$ for the rough cubes contacting the rough side of the model cell.

responses? A considerable number of quantitative investigations have been made into the relationship between angular displacement and a number of aspects of geotropic response. The majority have used curvature intensity after a standard exposure in excess of the presentation time or have measured physiological changes resulting from such stimuli. Curvatures arising from saturating standard exposures should be the result of the stimuli arising from the full number of statolith "contacts" *at equilibrium,* a condition represented in my model.

To check this then the results of some major quantitative investigations involving curvatures from this kind of stimulation (Czapek 1895, 1906, Metzner 1929, Ahrens 1933/34, Larsen 1962) have been brought together in standard form for comparison with results from the model. Responses have all been recalculated as a percentage of the maximum for the series and are plotted together in Figure 5. The continuous curve is the curve of nearest fit to all the points and shows that the optimum angle for maximum response falls between 120° and 125°.

Some slight manipulation of the model results has been considered necessary in order to make a proper comparison with these pooled results. Thus, in spite of the precaution of including an "insensitive" wall area at the bottom end of the lateral wall to eliminate "contacts" at zero displacement yet an occasional sphere or cube would make contact with the sensitive area of lateral wall during the preparatory random sedimentations on to the end wall prior to making the displacement. It will thus be seen from Figures 2–4 that, for both types of statolith, zero displacement angle gives a small positive number of "contacts". To eliminate this small residual value and to get a more realistic estimate of the increase in "contacts" on displacement these values for zero displacement have been subtracted from those for final equilibrium. Thus if N represents the final number of contacts and N_0 the number at zero displacement, the values of $N - N_0$ and $\log_{10}N - \log_{10}N_0$ have been calculated to represent the true change. In order to facilitate comparison with the observed curves these values have all been scaled to a maximum of 100 at the optimum angles. Thus four curves have been obtained (two from the "spheres" model and two from the "cubes" model). Of these, one fits the geotropic curvature results of Figure 5 with remarkable closeness; this is the curve of $\log_{10}N - \log_{10}N_0$ for rough cubes and has been plotted as the dotted line in Figure 5. Statistical analysis shows that the fit is almost exact (the difference in shape lies only just on the 5 % probability point) and this is quite remarkable in view of the great diversity of sources from which the observed results of Figure 5 were taken. The fit is at least as good as those provided by the mathematical manipulations of Metzner and Larsen and in addition is much more convincing in that the figures have been obtained by strictly objective observations on a model made to represent as closely as possible a statocyte whose structure and behaviour are very well known. In support of these claims of objectivity it should be noted that this was the first and only model made and that there have been no adjustments of dimensions to get a better fit of the respective curves.

IV. Conclusions

It may therefore be concluded from these relatively simple observations that the well-established departures of geotropic reaction from the simple sine law of Sachs can be more easily, even more convincingly, explained in terms of amyloplast statoliths than in terms of Metzner's "geotonus" effects or Larsen's electrostatically charged pendulums. Although it is indirect, it is nevertheless important, evidence to be added to that great mass already accumulated in support of the starch statolith theory.

Summary

A perspex model has been constructed to simulate the statocyte cell of a higher plant and to study the sedimentation of polythene "statoliths" in a highly viscous "protoplasm" of medicinal paraffin. The relationship between the sedimentation of these "statoliths" on to the lateral walls and the angle of displacement of the model from the vertical has been observed. The logarithm of the number of "statoliths" in contact with the lateral wall at equilibrium bears the same relationship to angular displacement as does the geotropic curvature response in orthotropic organs as shown by results taken from the literature. It is claimed that the results from this model strongly support the amyloplast statolith theory of graviperception and offers the best explanation to date of the modified sine rule for geotropic response.

I wish to acknowledge with gratitude the skilful help of Mr. S. E. White of Bedford College who constructed the perspex cell used in this model.

References

Ahrens, A.: Untersuchungen über die optimale Reizlage bei der geotropischen Reaktion. — Z. Bot. 26: 561–569. 1933/34.

Audus, L. J.: The mechanism of the perception of gravity by plants. — "Biological Receptor Mechanisms", Symposium XVI, Soc. Exp. Biol., pp. 197–226. 1962.

Czapek, F.: Untersuchungen über Geotropismus. — Jahrb. wiss. Bot. 27: 243–339. 1895.

— Die Wirkung verschiedener Neigungslagen auf den Geotropismus parallelotroper Organe. — Jahrb. wiss. Bot. 43: 145–175. 1906.

Larsen, P.: Orthogeotropism in roots. — Encyclopaedia of Plant Physiology (Springer Verlag) 17(2): 153–199. 1962.

Metzner, P.: Über die Wirkung der Längskraft beim Geotropismus. — Jahrb. wiss. Bot. 71: 325–385. 1929.

Sachs, J.: Über orthotrope und plagiotrope Pflanzenteile. — Arb. bot. Inst. Würzburg 2: 226–284. 1882.

The effect of two audible sound frequencies on the germination and growth of a spring and winter wheat[1]

PEARL WEINBERGER AND MARY MEASURES
Department of Biology, University of Ottawa, Ottawa, Canada

Received April 23, 1968

The effect of sound of 5 kc/s and 12 kc/s on the germination and growth of grains and seedlings of vernalized spring (Marquis) and winter (Rideau) wheats was investigated. Untreated controls were simultaneously followed, and all experimental series were replicated at least three times. Growth cabinets were systematically interchanged with the replications to eliminate other variables. Sound at both frequencies significantly stimulated the germination of Marquis grains maintained at 2° and 10 °C. At 25 °C the rate of germination was unchanged. An indication of a slightly elevated germination rate was observed in Rideau grains held at 2° and exposed to 5 kc/s. Neither frequency significantly affected germination at 10° or 25 °C.

Growth of plants derived from fully vernalized grains was followed by measuring total plant height, wet and dry weight of "tops", number of roots, and their fresh and dry weights. Rideau wheat grew more vigorously at both sound frequencies. Continuous exposure to 5 kc/s both during vernalization and subsequent growth resulted in a 200% or greater increase in all parameters followed. The growth response of Marquis wheat was dependent upon treatment. Some conditions of sound exposure accelerated growth, whereas in others growth was depressed.

Canadian Journal of Botany, **46**, 1151 (1968)

Introduction

Stimulation of germination and growth of barley, sunflower, pine, spruce, and Siberian pea tree seedlings by treatment with ultrasonic frequencies has been found to occur by Johnson (1), Obolensky (1, 2), Kozubov (3), Timonin (4), and Lisenkov (5). In all cases, stimulated growth was accompanied by increased activity of the enzymes γ-amylase and peroxidase both in the seed and seedlings. An increase in the respiration rate was also observed. Obolensky and Timonin independently found some species to be inhibited by the same frequency and length of exposure time that stimulated other species. Both found a critical intensity level, below which germination and growth were enhanced, but above which they were inhibited (1, 2, 4). The frequency and amount of energy required to produce these effects appear to vary widely between species.

Very few studies concerning the effects of audible sound have appeared in the literature. Several descriptive papers reporting stimulation of plant growth by exposure to daily periods of music have been noted (6, 7, 8). However, the effect of single-frequency audible sound on plant growth has not, so far, been critically assessed. No information is presently available on the effect of audible sound on the rate of germination of any seeds or grain. The present study on the effect of sound frequencies in the audio range on wheat grain seedlings was therefore initiated.

[1]Supported in part by NRCC Grant A-1737.

Materials and Method

Marquis (spring) and Rideau (winter) wheats, *Triticum aestivum* L., were used throughout the study in parallel experiments. Grain was obtained from the Department of Agriculture, Division of Cereal Crops, Ottawa.

1. *Germination Studies*

This series consisted of a control group, C, not sound treated, and two groups 5K and 12K, exposed to sound of frequency 5 kc/s and 12 kc/s respectively. Each group was held at three temperatures, 25°, 10° and 2 °C, requiring a total of nine cabinets (Sherer-Gillett, model CEL 255-6). In each cabinet were placed five blackened petri dishes, filter paper lined, each containing 10 medium-sized, morphologically comparable grains, to which excess water was added (7.5 ml per 8.5-cm diam petri dish).

In the 5K and 12K groups the sound was transmitted to the cabinets by a 4-in. loud-speaker mounted inside, and in the middle of the rear wall. Each speaker was driven via an impedance matching transformer by the full output of an audio-oscillator (Hewlett-Packard model 200CD) located outside the cabinet.

The sound pressure level was monitored at various places in the cabinet by a meter (Bruel Kjaer).

The sound pressure level at the petri dishes was 95 db re 0.002 dyne/cm^2 at 12 kc/s, and 92 db re 0.002 dyne/cm^2 at 5 kc/s, and varied by less than ±1 db over the various positions of the petri dishes. General background noise (from machinery) was 45 db in all cabinets, but contributions from noise at 5 and 12 kc/s were negligible.

The number of grains germinated was observed at hourly intervals for grains held at 25°, every 4 hours at 10° and every 12 hours on the grains at 2 °C. Germination was assessed to be complete when the radicle just pierced the grain coat.

Replication of the Germination Studies

The set of nine growth conditions described above was repeated at three different times during a 15-month period. In each such repetition cabinets were systematically interchanged, so that control cabinets became 5 kc/s or 12 kc/s cabinets and vice versa. The cabinets assigned to different temperatures were also interchanged during one replication.

2. *Growth Studies*

In this series the grains (10 per blackened petri dish) were first allowed to imbibe at 24 ±1 °C for 6 hours, and were then divided into three groups which were chilled at 2 °C (9) until vernalization was completed (1 week for Marquis, 4 weeks for Rideau wheat). During chilling one group was not exposed to sound, the other two were exposed to 5 kc/s and 12 kc/s respectively. After vernalization, the sound-treated groups were each divided into two sets. With one of the sets the same treatment was continued during the subsequent growth period, whilst the other was transferred to control cabinets and not exposed to sound. Using a two symbol notation to describe the conditions during vernalization and growth respectively, the four resultant sets are represented by 5K,5K; 5K,C; 12K,12K; and 12K,C. The vernalized, but sound-unexposed group was split into three sets, two of which were sound treated during the growth period, giving the sets represented by C,C; C,5K; and C,12K. Each of the final sets consisted of 5 petri dishes with 10 young seedlings per dish.

When the seedling roots reached a length of 3 cm, two seedlings were randomly selected from each petri dish and singly transferred to blackened polyethylene bottles containing a culture medium (Hoagland solution, ¼ strength). The original sample of 50 grains per growth condition, now reduced to 10 seedlings per treatment, was analyzed after the selected growth period with respect to the following parameters: plant height, number of roots, fresh and dry weights of total root system, fresh and dry weight of "tops".

Growth periods chosen were 1, 2, 4, 6, and 8 weeks. For each growth period a complete set of the combinations of vernalization and growth conditions was required.

To reduce the number of cabinets, various samples undergoing the same treatment at a particular time were often housed together.

The culture medium was changed twice weekly initially, and subsequently three or four times weekly as warranted by plant growth. A regime of a 16-hour day and 8-hour night was maintained in all cabinets. The total light intensity during the daytime period was 1500–1750 ft-c provided by four 25-watt incandescent, and eight fluorescent lights (Sylvania F48 T12-cw-Ho).

Dry weights were obtained after the plant material was heated at 95 ±1 °C for 24 hours and then equilibrated with atmospheric humidity.

Replication of Growth Series

Each of the above experiments C,C; C,5K; C,12K; 5K,C; 5K,5K; 12K,C; 12K,12K was repeated three times for Rideau wheat and five times for Marquis wheat, at all the growth periods stated. In addition, systematic cabinet interchanges were made between replications, as described for the germination studies.

Statistical Analysis

The data were subjected to statistical analysis using Scheffé's test at the 5% level.

Results

Germination Studies

The percentage germination was plotted against time. At each temperature the time was deduced at which the divergence between the curves for the sound-treated and control groups was largest. The results are shown in Table I for Marquis and

TABLE I

Percentage germination of Marquis and Rideau wheat as affected by sound and germination temperature (Scheffé's test at 5% level)

Temp., °C	Time	% germination		
		Marquis wheat		
2	6 days	5K	12K	C
		84	82	24
10	3 days	12K	5K	C
		98	64	50
25	14 hours	5K	12K	C
		64	49	45
		Rideau wheat		
2	6 days	5K	C	12K
		94	72	50
10	3 days	12K	C	5K
		100	98	96
25	14 hours	5K	C	12K
		90	90	84

Rideau wheat. Both wheat varieties appear to show the greatest divergence between sound-treated and control samples at the same times, although for Rideau wheat the divergences are not statistically significant.

With Marquis wheat, the effect of sound was temperature dependent. At 2° a larger percentage of Marquis seeds had germinated after 6 days, when exposed to either 5 kc/s or 12 kc/s sound, than when left under control conditions. At 10° only seeds kept under 12 kc/s germinated faster than controls, whereas at 25 °C there were no significant differences between treatments.

Rideau grain, on the other hand, did not show statistically significant differences in germination between treatments at any of the temperatures.

In all cases, 99.8% of the grains germinated when left for longer periods at each of the three temperatures.

TABLE II

Marquis wheat after 4 weeks growth (Scheffé's test at the 5% level)

Height cm						
12K,12K*	C,5K	5K,C	5K,5K	12K,C	C,C	C,12K
48.8	46.9	46.0	45.2	42.1	41.8	41.6
Number of roots						
5K,5K	12K,12K	C,5K	5K,C	C,C	12K,C	C,12K
27.5	27.4	24.8	24.3	22.9	20.1	16.0
Fresh wt. of roots, g						
12K,12K	C,5K	5K,5K	5K,C	C,C	12K,C	C,12K
2.88	2.85	2.70	2.62	2.45	2.28	1.24
Dry wt. of roots, g						
12K,12K	5K,5K	C,5K	C,C	12K,C	5K,C	C,12K
0.21	0.18	0.18	0.17	0.15	0.15	0.08
Fresh wt. tops, g						
12K,12K	5K,5K	C,5K	5K,C	12K,C	C,C	C,12K
6.49	5.61	5.60	4.96	4.43	3.90	3.48
Dry wt. tops, g						
12K,12K	5K,5K	C,5K	5K,C	12K,C	C,C	C,12K
0.76	0.68	0.65	0.59	0.54	0.49	0.41

*Treatment.

TABLE III

Rideau wheat after 4 weeks growth (Scheffé's test at the 5% level)

Height, cm					
5K,5K*	5K,C	12K,12K	C,5K	12K,C	C,C
42.2	4.00	39.7	34.6	33.1	30.2
Number of roots					
5K,5K	5K,C	12K,C	12K,12K	C,5K	C,C
15.9	15.4	15.4	12.8	11.8	10.7
Fresh wt. roots, g					
5K,C	5K,5K	12K,C	12K,12K	C,5K	C,C
1.40	1.27	1.05	0.88	0.69	0.44
Dry wt. roots, g					
5K,C	5K,5K	12K,12K	12K,C	C,5K	C,C
0.10	0.08	0.07	0.07	0.04	0.03
Fresh wt. tops, g					
5K,5K	5K,C	12K,12K	12K,C	C,5K	C,C
3.17	2.80	1.67	1.62	1.38	0.90
Dry wt. tops, g					
5K,5K	5K,C	12K,12K	12K,C	C,5K	C,C
0.41	0.36	0.22	0.22	0.16	0.13

*Treatment.
NOTE: C,12K data not included because of wide spread of values for all parameters.

Growth Studies

Compared with the controls (C,C) the different responses of Marquis and Rideau wheats exposed to sound did not become evident till after 4 weeks of growth, and the data taken at earlier times are therefore not given. Table II shows the parameters for Marquis wheat, grown for 4 weeks. The C,12K treatment evoked a depressed growth response which was clearly significant in the decreased number of roots and depressed levels of fresh and dry weights. Other treatments were barely significant, except 12K,12K where there was a slight stimulation of the fresh and dry weights of tops.

TABLE IV

Marquis wheat after 6 weeks growth (Scheffé's test at the 5% level)

Height, cm						
5K,5K*	C,5K	5K,C	C,C	12K,12K	12K,C	C,12K
60.1	56.8	55.3	51.6	50.3	47.0	46.1
Number of roots						
5K,5K	C,C	C,5K	5K,C	C,12K	12K,12K	12K,C
63.3	50.4	49.4	47.9	47.7	46.6	45.4
Fresh wt. roots, g						
5K,5K	C,C	5K,C	C,5K	C,12K	12K,12K	12K,C
5.20	4.83	4.53	3.91	3.89	3.72	3.62
Dry wt. roots, g						
5K,5K	C,C	5K,C	C,5K	C,12K	12K,C	12K,12K
0.48	0.42	0.34	0.33	0.32	0.29	0.27
Fresh wt. tops, g						
5K 5K	C,C	5K,C	C,5K	C,12K	12K,12K	12K,C
15.75	13.31	13.17	10.05	8.42	8.42	7.75
Dry wt. tops, g						
5K,5K	C,C	5K,C	C,5K	12K,12K	C,12K	12K,C
2.35	1.79	1.56	1.50	1.25	1.22	1.08

*Treatment.

TABLE V

Rideau wheat after 6 weeks growth (Scheffé's test at the 5% level)

Height, cm					
5K,5K*	5K,C	C,5K	C,C	12K,12K	C,12K
54.3	51.5	51.5	47.4	44.5	42.4
Number of roots					
5K,5K	5K,C	12K,12K	C,5K	C,12K	C,C
35.2	32.3	30.4	30.0	22.2	20.6
Fresh wt. roots, g					
5K,5K	5K,C	C,5K	5K,C	C,12K	C,C
3.90	3.32	2.94	2.54	1.69	1.44
Dry wt. roots, g					
5K,5K	5K,C	C,5K	12K,12K	C,12K	C,C
0.28	0.24	0.20	0.20	0.13	0.09
Fresh wt. tops, g					
5K,5K	5K,C	C,5K	12K,12K	C,C	C,12K
12.43	10.71	9.12	7.28	4.90	4.01
Dry wt. tops, g					
5K,5K	5K,C	C,5K	12K,12K	C,C	C,12K
1.51	1.32	1.07	0.92	0.64	0.60

*Treatment.

In contrast, the results with Rideau wheat, shown in Table III, indicate that it was significantly stimulated by all sound treatments.

At the 6-week growth period many treatments adversely affected the growth of Marquis wheat (Table IV). A slight stimulation, significant only in the height of the plants and dry weight of the tops, was noted for the 5K,5K group. All Rideau wheat plants (Table V) grew more vigorously after continuous exposure to 5 kc/s. The weight of both roots and tops were increased by 250–300% over control in this treatment.

After 8 weeks growth, Tables VI and VII, there was an indication that the three series 5K,5K;

TABLE VI

Marquis wheat after 8 weeks growth (Scheffé's test at the 5% level)

Height, cm						
5K,5K*	C,12K	5K,C	C,5K	C,C	12K,12K	12K,C
72.8	70.8	67.7	67.1	62.7	62.0	61.3
Number of roots						
5K,C	12K,C	5K,5K	C,C	12K,12K	C,12K	C,5K
105.6	101.4	98.9	89.8	89.2	85.4	84.7
Fresh wt. roots, g						
12K,C	5K,C	5K,5K	C,C	C,5K	12K,12K	C,12K
14.43	12.67	10.21	8.60	8.04	7.79	7.50
Dry wt. roots, g						
5K,C	12K,C	5K,5K	C,C	C,5K	12K,12K	C,12K
1.27	1.21	1.07	0.98	0.84	0.76	0.57
Fresh wt. tops, g						
5K,5K	5K,C	12K,C	C,5K	C,C	C,12K	12K,12K
31.55	30.08	29.36	25.73	25.18	23.13	18.68
Dry wt. tops, g						
5K,5K	5K,C	12K,C	C,12K	C,5K	C,C	12K,12K
5.90	5.30	4.90	4.37	4.13	4.03	3.45

*Treatment.

TABLE VII

Rideau wheat after 8 weeks growth (Scheffé's test at the 5% level)

Height, cm						
C,5K*	5K,5K	5K,C	C,C	C,12K	12K,12K	12K,C
61.0	60.3	57.2	57.1	55.4	55.1	55.0
Number of roots						
5K,5K	12K,12K	5K,C	C,5K	C,12K	12K,C	C,C
71.0	67.8	66.8	55.1	54.1	53.2	36.0
Fresh wt. roots, g						
C,12K	5K,5K	C,5K	12K,12K	5K,C	12K,C	C,C
9.64	8.31	7.26	6.76	6.60	6.26	4.79
Dry wt. roots, g						
C,12K	5K,5K	12K,12K	5K,C	C,5K	12K,C	C,C
0.96	0.77	0.69	0.64	0.59	0.50	0.35
Fresh wt. tops, g						
5K,5K	C,5K	5K,C	12K,12K	C,12K	12K,C	C,C
29.24	23.79	23.72	21.11	21.04	18.68	15.27
Dry wt. tops, g						
5K,5K	5K,C	12K,12K	C,12K	C,5K	12K,C	C,C
4.50	3.73	3.31	3.14	3.03	2.60	1.85

*Treatment.

12K,C; and 5K,C may have slightly stimulated the growth of Marquis plants. Continuing stimulation was seen in all treated Rideau plants after 8 weeks growth. Highly significant increments in the fresh and dry weights of roots and tops were obtained.

The parameter values together with their variability are also shown in Fig. 1, *a–f*, for the 7 different combinations of vernalization and growth conditions. With respect to most of the parameters, the 5K,5K treatment produced the largest growth stimulation compared to the controls. The photographs in Fig. 2, *a–g*, show the 8-week-old Rideau wheat plants and the stimulating effect of sound is readily visible.

Discussion

Sound appears to have a greater effect on the

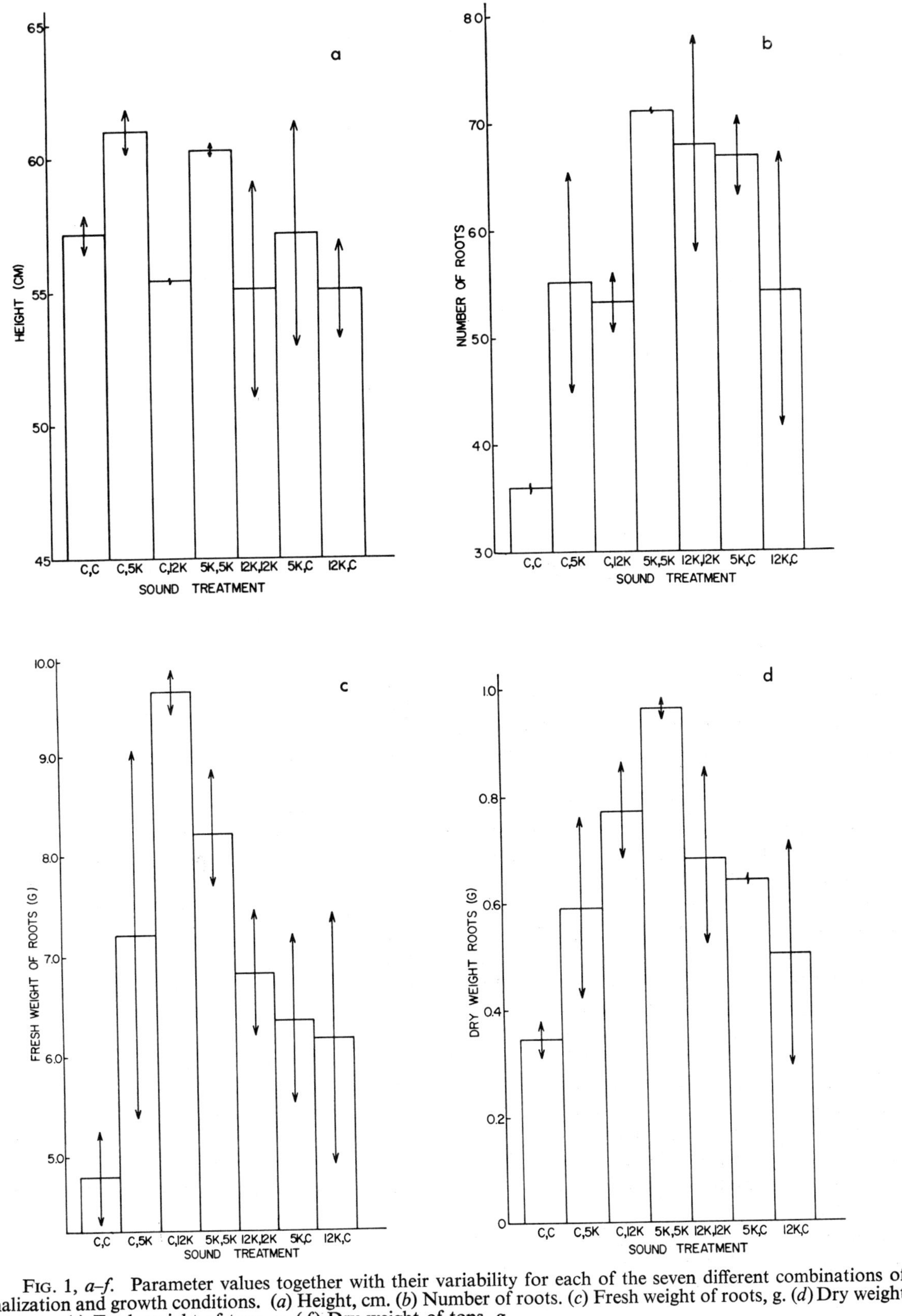

FIG. 1, *a–f*. Parameter values together with their variability for each of the seven different combinations of vernalization and growth conditions. (*a*) Height, cm. (*b*) Number of roots. (*c*) Fresh weight of roots, g. (*d*) Dry weight of roots, g. (*e*) Fresh weight of tops, g. (*f*) Dry weight of tops, g.

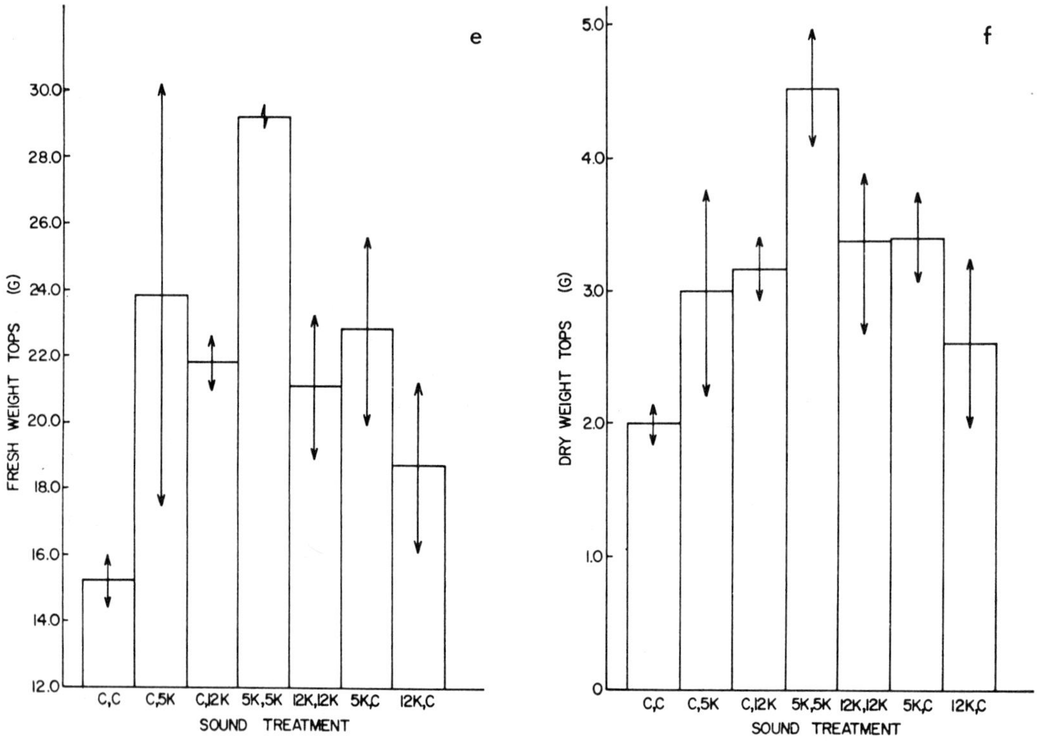

FIG. 1. (*Concluded*)

growth of Rideau wheat plants than on that of Marquis wheat plants, but a greater effect on the germination of the Marquis wheat. This suggests that at least two systems must be affected by these sound frequencies.

Neither the germination response to audible sound nor the different growth responses of the winter and spring wheats to the same sound levels can be explained at this point. It appears, however, that the effect of exposure to sound is transmitted to subsequent cell generations.

The amount of energy added by the sound frequencies in this experiment is very minute, being only one-billionth that required to break a chemical bond and is too low to cause cavitation of water. However, the amount of energy is sufficient to produce a resonance effect in any cellular organelle which has a natural period of vibration corresponding to that of the added sound. In such a case, the effects of individual successive vibrations of the sound waves would be cumulative and might induce larger vibrations of the organelle. This might, in some manner, affect cellular biochemical or biophysical processes which are connected with germination and growth. This then could account for the different responses between the two sound frequencies and between the two wheat varieties.

Another hypothesis involves the effects of sound on water. Since there is no thermal expansion of water at 4 °C, a phase shift of the sound wave may occur at lower temperatures. If the rate of germination is not linear with regards to temperature and pressure, grains may germinate faster in one part of the sound wave than in another. If this is true, then temperature and pressure may re-enforce one another at 2 °C, but not at 25 °C. This then could account for increased germination rates of grains subjected to sound at 2 °C but not at 25 °C. It has also been suggested that the production of ethylene may be involved. It remains to be seen whether the growth-stimulating effect of 5 kc/s and 12 kc/s will be similarly expressed under different environmental conditions.

Obviously, much work remains to be done to elucidate the mechanism involved in reshaping the normal germination and growth processes.

Acknowledgments

The authors thank Mr. F. Gfeller, Department of Agriculture, Ottawa, for kindly supplying the grain used in this study and Miss Andrée Barnaby, who assisted in the early stages of the work.

They are also grateful to Dr. E. Shaw, National Research Council of Canada, Ottawa, for the loan of the sound pressure level meter and for helpful discussions. Dr. E. Whalley, National Research Council of Canada, Ottawa, and Dr. R. D. Jones, Department of Agriculture, Ottawa, made valuable suggestions about the possible mechanisms involved.

1. JOHNSON, L. P. V. and OBOLENSKY, G. 1954. Note on the effect of ultrasonics on the development of barley. Can. J. Agr. Sci. **34**: 651–652.
2. OBOLENSKY, G. 1957. Activation des processus fonctionnels (métaboliques et enzymatiques) de l'orge, par ultra-sons. Materiae Vegetabiles, **2**: 298–335.
3. KOZUBOV, G. M. and GANIUSHKINA, L. G. 1964. Effect of ultra-sound on the seeds of trees and shrubs. (In Russian). Botan. Zh. **49**: 957–965.
4. TIMONIN, M. I. 1966. Effect of ultra-sound on the germination of white pine spruce and jack pine seeds. Can. J. Botany, **44**: 113–115.
5. LISENKOV, A. F. 1966. The effect of sound treated water on seeds of woody plants. Fiziol. Rastenii, **13**: 728–729.
6. SINGH, T. C. N. and GNANAM, S. A. 1965. Studies on the effect of sound waves of Nadeswaram on the growth and yield of Paddy. J. Annamalai Univ. **16**: 78–99.
7. SINGH, T. C. N. and PONNIAH, S. 1954. On the effect of musical sound of violin on the growth of Mimosa pudica L. Rappt. Commun. VIIIe Congr. Intern. Botan. Paris, Sect. II, Physiologie Végétale: 195–196.
8. SMITH, G. E. 1963. Growing corn to music. Popular Mechanics, **119**: 118–121.
9. WEINBERGER, P. and KU, T. 1966. The effect of imbibition and vernalization on some parameters of root growth of Triticum × aestivum (L. Rideau and Marquis) Can. J. Botany, **44**: 633–644.

FIG. 2, *a–g*. Effect of sound treatments on the relative size of Rideau plants after 8 weeks growth. (*a*) C,C). (*b*) C,5K. (*c*) C,12K. (*d*) 5K,C. (*e*) 5K,5K. (*f*) 12K,C. (*g*) 12K,12K.

FIG. 2. (*Continued*)

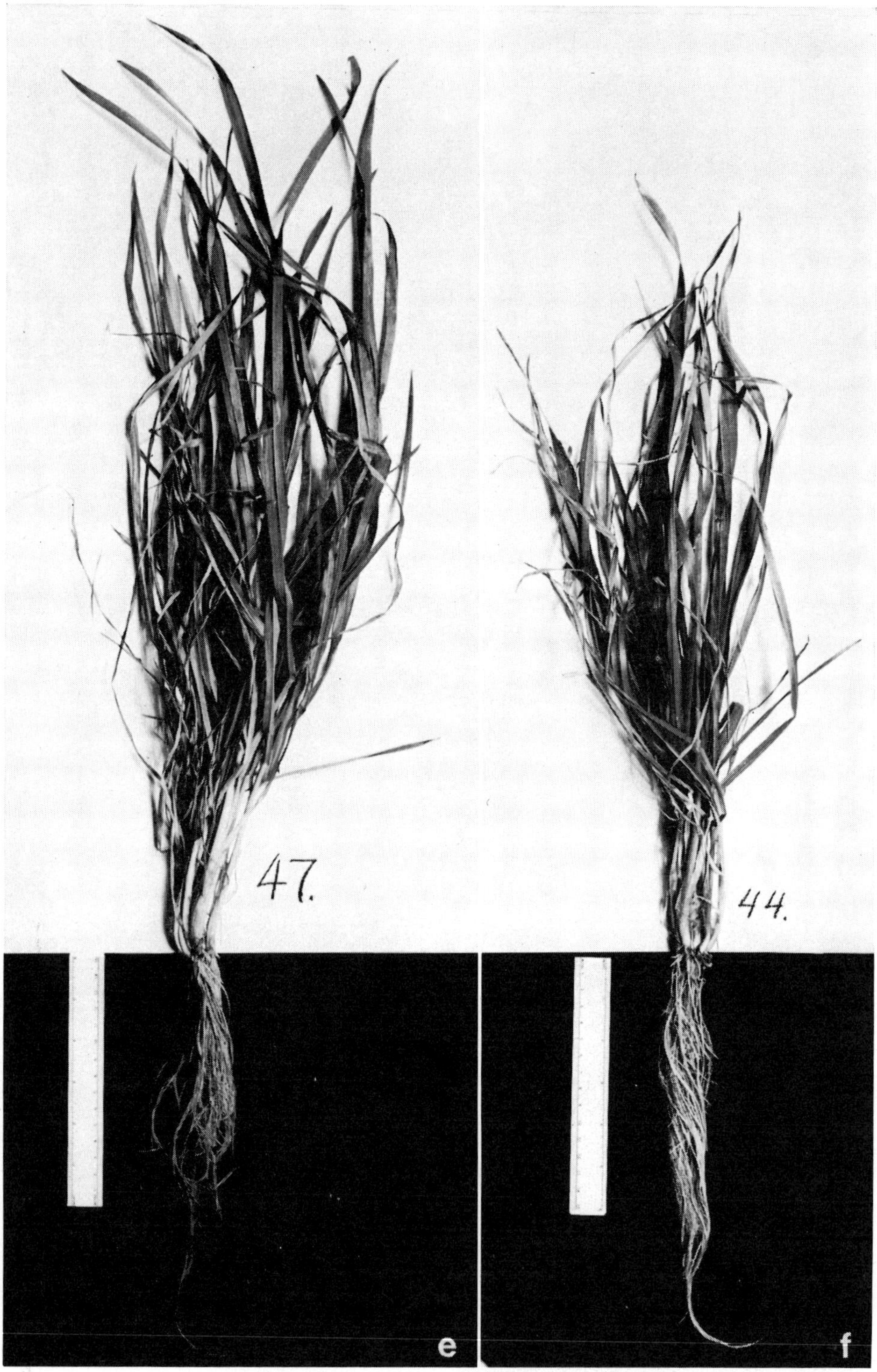

FIG. 2. (*Continued*)

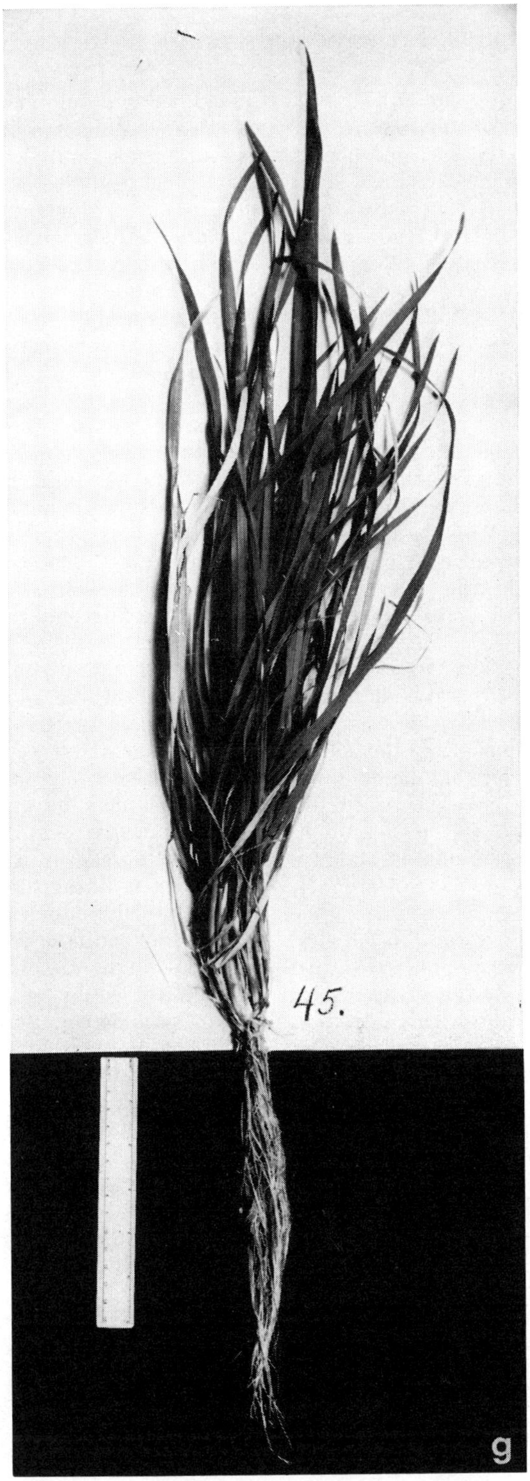

FIG. 2. (*Concluded*)

THE PLASTOCHRON INDEX

Ralph O. Erickson[1] and Francis J. Michelini

IN DEVELOPMENTAL STUDIES one can usually relate only the simplest aspects of the developing organism or organ to time directly. For instance, if one were studying a developing leaf, one could measure its length or width on successive days without injuring it, but many kinds of study which one might wish to make are destructive. Even such a simple measurement as its dry weight cannot be obtained without killing it, and one obviously cannot cut microtome sections of the same leaf more than once. This circumstance applies to almost any study of growth and development, and one must deal with parallel specimens or samples, each of which is sacrificed at a given time to provide the required data. Whether the data are obtained by histological examination, chemical analysis, metabolic study, or other means, they must be related to time if they are to have meaning in developmental terms. In many cases, plotting the observations directly against the chronological age of the plants is of limited value. Unless one is working with genetically uniform plants under exactly controlled conditions, variability may be so great that plants of the same chronological age may have reached quite different stages of development, while plants which are morphologically similar may be of quite different chronological ages. Under these conditions, significant relationships may be obscured, even when the statistical precautions of taking large samples and randomizing the error are observed.

These difficulties can often be minimized by using a numerical *index* of the developmental status of each specimen and relating the observations to time indirectly by way of the index. This developmental index should be based upon measurements of the plant which can be made easily and without injury, and its relationship to time should be worked out by appropriate growth studies. Observational data can then be plotted against the developmental index of each plant used, and the resulting graphs interpreted, on the basis of the growth studies, as a plot of the data in question against time.

The advantages of this procedure are illustrated by studies of developing flower buds and anthers of *Lilium* (Erickson, 1947, 1948; Ogur et al., 1951). The logarithm of length of the flower bud was shown to be linearly related to time, and served as a developmental index for organs and processes within the flower. By relating various measurements and observations to this index, growth curves were constructed and time rates of change were calculated for features of the flower bud and anther which could not have been followed through time by direct means. In addition, this technique served to circumvent the environmental variability of the material to a considerable extent. Although there were large fluctuations in the daily increment of log bud length, the relationships of anther dimensions, and of meiosis and microspore mitosis, for instance, to log bud length were very close indeed.

In the shoot of a higher plant, leaves appear periodically, and the period between initiation of successive leaves has been termed a plastochron. This term was proposed by Askenasy (1880) to designate the interval of time between formation of two successive internode cells in *Nitella flexilis*. He also proposed that the term might be used for all plants having a similar nodal organization, and the concept of the plastochron has gained fairly wide currency, particularly in shoot apex studies of the higher plants (e.g., Esau, 1953). When successive plastochrons are equal in duration, the plastochron may be made to serve as the unit of a developmental scale. A plastochron index for shoot development in *Xanthium* is described below. This index is linearly related to time and serves to specify the developmental status of a shoot with an error of only a few hours. The plastochron index has points in common with procedures which have been used by other authors in physiological and morphological studies of shoot development, as will be pointed out in the discussion.

GROWTH MEASUREMENTS.—The argument presented in this paper is based on growth observations made on *Xanthium* plants. Burs were collected in Delaware County, Pennsylvania, from plants growing as weeds, which should probably be assigned to *X. italicum* Moretti.[2] Seeds were removed from the burs, scarified to insure germination, and planted fortnightly in pots. The plants were kept in a lighted greenhouse and repotted at intervals. Solar illumination was supplemented before dawn and after dusk so as to give the plants 16 hr. of light per day. Since *Xanthium* is a short-day plant, it can be kept in strictly vegetative growth by this regime. The total length, including

[1] Received for publication August 14, 1956.
During a part of the time that this paper was under consideration, the senior author held a fellowship of the John Simon Guggenheim Memorial Foundation and enjoyed the facilities of the Earhart Plant Research Laboratory at the California Institute of Technology.

[2] We are indebted to Dr. Herbert Wahl for identifying our plants. They do not appear to differ in growth characteristics from the strain used by K. C. Hamner and other workers on photoperiodism, which is usually referred to as *X. pennsylvanicum*.

lamina and petiole, of each leaf on each of 20–30 plants was measured daily at approximately 10 A.M. The smaller leaves were measured to the nearest 0.1 mm. with a vernier calipers, and larger ones to the nearest 1 mm. with a rule.

DERIVATION OF THE PLASTOCHRON INDEX.—An inspection of the sigmoid growth curves of the leaves suggests that they may be regarded as exponential in their early portion, as Blackman (1919) and other authors have found for many plants and plant organs. This supposition is supported by plotting leaf length logarithmically against time, as in fig. 1, where the relationship between log leaf length and time for one plant appears to be linear up to about 70 mm. leaf length. Other plants grew in essentially the same manner. Note that we have numbered the leaves serially in order of their appearance in time.

Geometrically considered, the lower portion of fig. 1 may be regarded as a family of straight lines.[3] It is further evident that the lines are approximately parallel, and, with the exception of those for the cotyledons and leaves 1 and 2, that they are approximately equally spaced along the abscissa.

A plastochron is conventionally defined as the time interval between initiation of two successive leaves. It might be more broadly defined as the interval between corresponding stages of development of successive leaves, and one might choose initiation, maturity, or any intermediate stage of development as the stage of reference. For the present purpose, we shall define the reference stage in terms of a measurement of leaf length. A length of 10 mm. is convenient for the *Xanthium* data, since leaves of this order of length are growing exponentially, and yet are large enough to be measured with some accuracy and without injury to the shoot.

In terms of this reference length, the logarithm of which is indicated as a horizontal line in fig. 1, a plastochron may be visualized as the distance between the intersections of the reference line with the growth curves of two successive leaves. The "developmental age" of a plant may be defined arbitrarily in terms of plastochrons conceived in this way. Specifically, we shall regard a plant as being exactly n plastochrons old when leaf n is 10 mm. long, $n + 1$ plastochrons old when leaf $n + 1$ is 10 mm. long, and so on. When, however, no leaf is exactly 10 mm. long, an exact specification of the plastochron age of the shoot requires interpolation. If, for instance, leaf 5, n, were 18 mm. and leaf 6, $n + 1$, 5 mm. long at a given time, it is clear that the plant would be more than 5 but less than 6 plastochrons old. A method for interpolation can be based on the assumptions that the early portions of the logarithmic growth curves are (1) linear, (2) parallel, and (3) equally spaced, as stated above.

Figure 2 is a diagrammatic representation of a portion of fig. 1 in which the diagonal lines represent portions of the leaf growth curves, and the horizontal line is drawn to correspond to the reference length, 10 mm. A vertical line is drawn at a particular time, when the plant is more than n but less than $n + 1$ plastochrons old. Since the plastochrons are assumed to be equal in duration, the plastochron age at this time, by linear interpolation, is $n + (AB/AC)$. This interpolation, however, is dependent on knowing the leaf lengths before and after the time in question. If the plant were sacrificed for study at that time, no growth data would be obtainable thereafter, and, in any case, the collection of growth measurements for a large series of plants is laborious. Since triangles ABD and CBE are similar, it follows that $AB/AC = DB/DE$, and that the plastochron age can equally well be estimated as $n + (DB/DE)$. Since the points D, B, and E represent logarithms of leaf length, the plastochron age can be rewritten

$$P.I. = n + \frac{\log L_n - \log 10}{\log L_n - \log L_{n+1}} \qquad (1)$$

where n is the serial number of that leaf which just exceeds 10 mm. in length, and $\log L_n$ is the logarithm of length in mm. of leaf n. We shall refer to this expression as the "plastochron index" *(P.I.)*, or the plastochron age of a shoot. To calculate it, it is only necessary to count the leaves on the shoot which are longer than 10 mm. and to measure the lengths of the two successive leaves which are respectively just longer and just shorter than 10 mm. at a given time. Logarithms to any base may be used, but the use of natural logarithms leads to an easy method of slide-rule calculation, and has advantages in the growth rate considerations presented below.

A second method of derivation of the plastochron index may be illustrated by replotting the data of fig. 1 as in fig. 3. Here each curve is a graph of log leaf length against node number at a given time. Again, a family of curves appears, which may be regarded as linear at lower values, parallel and equally spaced. However, the lines have negative slopes. In a diagrammatic representation of part of one of these curves, fig. 4, the same relationships shown above may be verified geometrically

$$P.I. = n + \frac{AB}{AC} = n + \frac{DB}{DE} = n + \frac{\log L_n - \log 10}{\log L_n - \log L_{n+1}}$$

[3] Actually, the leaves exhibit a marked diurnal rhythm in growth rate. From a limited series of hourly photographs of leaves growing at a constant temperature of 23°C. with alternating periods of 16 hr. of illumination and 8 hr. of darkness, it is clear that the rate of elongation in the dark may be twice as great as during the light period. In one case, a leaf about 50 mm. long varied in its rate of elongation from about 0.4 mm. · $hr.^{-1}$ in the dark to 0.2 in the light. We have not considered in detail the implications of this periodicity in calculations of the plastochron index, but we feel that it can be ignored in most studies in which interest does not center on direct periodic effects.

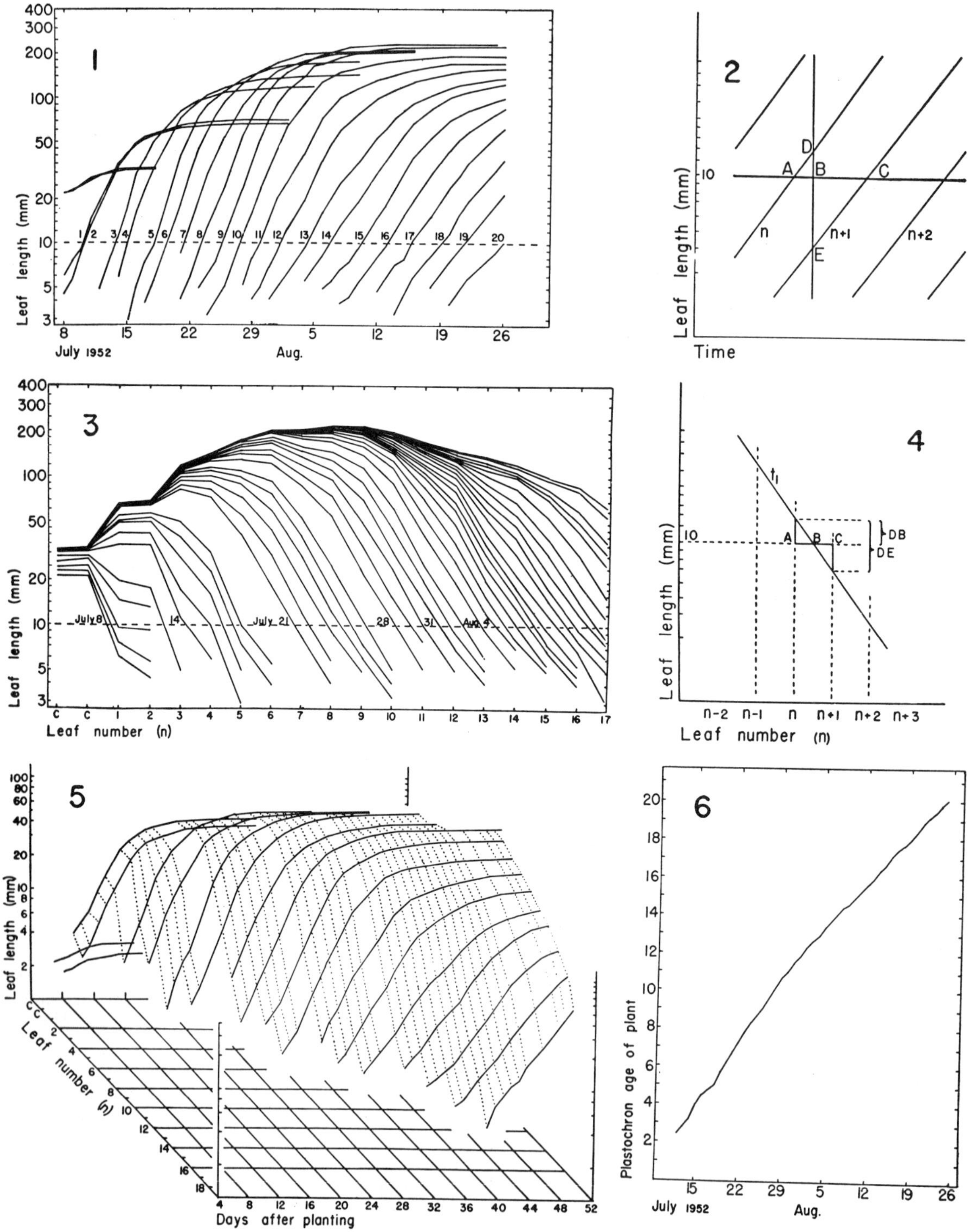

Fig. 1–6.—Fig. 1. Lengths of successive leaves of a *Xanthium* plant plotted logarithmically against time. Each growth curve applies to a single leaf. The unmarked curves at the left apply to the cotyledons.—Fig. 2. Idealized geometric representation of a portion of fig. 1.—Fig. 3. Leaf lengths of fig. 1 plotted logarithmically against node number. Each curve applies to a single day.—Fig. 4. Geometric representation of a portion of fig. 3.—Fig. 5. Leaf lengths of fig. 1 and 3 plotted logarithmically against time and node number.—Fig. 6. Plastochron index, calculated from the data of fig. 1, plotted against time.

These relationships may also be developed from a third point of view, based on growth considerations. If the early portion of the growth curve of a leaf is assumed to be exponential, its equation may be written

$$L = L_o e^{rt} \quad \text{or} \quad \ln L = \ln L_o + rt \qquad (2)$$

where t is time, L is length of a leaf, L_o is length of that leaf at $t = 0$, e is the base of natural logarithms, ln is the natural logarithm, and r is the relative rate of elongation of a leaf, dL/Ldt. If it is assumed that r is constant for a number of successive leaves, equation (2) may be made to describe the early elongation of each of these leaves by changing only the value of L_o for each. On the assumption made above that leaves appear at equal intervals of time, the changes in L_o for successive leaves will be regular. Equation (2) may be written specifically for the two successive leaves n and $n + 1$

$$\ln L_n = \ln L_{no} + rt \text{ and } \ln L_{n+1} = \ln L_{(n+1)} + rt$$

and rt may be eliminated between the two equations, giving

$$\ln L_{no} - \ln L_{(n+1)o} = \ln L_n - \ln L_{n+1} = p$$

This implies that in applying (2) to the second of two successive leaves, the value of $\ln L_o$ used for the first leaf should be decreased by an amount equal to the difference between the natural logarithms of lengths of two successive leaves. For leaf $n + 2$, twice this difference should be substracted, and so on for any number of successive leaves. On the three assumptions we have made (p. 5) this difference is a constant. It is the natural logarithm of the ratio of lengths of two successive leaves and will be symbolized p. A more general equation may now be written, expressing the length of any leaf during its early growth as a function of its serial number and time

$$\ln L = \ln L_o + rt - pn \text{ or } L = L_o e^{rt-pn} \qquad (3)$$

where p is a constant equal to $(\ln L_n - \ln L_{n+1})$ or $\ln(L_n/L_{n+1})$, and n is the serial number of a given leaf, which can be only a positive whole number. Equation (3) suggests an isometric representation of the data of fig. 1. This is drawn in fig. 5, where log leaf length is plotted against time and node number. The solid lines in this figure correspond to the curves drawn in fig. 1, and the dotted lines through the same points to those of fig. 3.

The plastochron index may now be derived from (3). It should first be noted that the plastochron index *(P.I.)* scale corresponds to that of leaf number *(n)*, with the difference that n is discontinuous and *P.I.* is visualized as a continuous variable. *P.I.* and n coincide at the integral values 1, 2, 3 . . . but *P.I.* can take intermediate values as well. In this sense *P.I.* may be substituted for n in (3). The plastochron index can then be formally defined as that value which satisfies (3) for the reference length at a given time. Substituting $L = 10$ into (3) and rearranging

$$P.I. = \frac{\ln L_o + rt - \ln 10}{p} \qquad (4)$$

One may add $(pn - pn)$ to the numerator on the right, substitute $\ln L_n$ from (3) and rearrange to obtain

$$P.I. = n + \frac{\ln L_n - \ln 10}{p}$$

Since $p = \ln L_n - \ln L_{n+1}$

$$P.I. = n + \frac{\ln L_n - \ln 10}{\ln L_n - \ln L_{n+1}}$$

which is identical with (1).

Calculation of the plastochron index may be illustrated with a portion of the data plotted in fig. 1 (table 1). This plant was started from seed on 1 July and the first leaf measurements were made on 8 July. The two leaves used as a basis of calculation each day are italicized. On 24 July, for instance, leaf 8 was longer and leaf 9 shorter than 10 mm. The integral part of the index is therefore 8. The fractional part is most easily evaluated with a log-log slide rule as follows

$$P.I. = 8 + \frac{\ln(12.7/10)}{\ln(12.7/6.1)} = 8 + \frac{\ln 1.27}{\ln 2.08} = 8 + \frac{0.239}{0.733} = 8.33$$

A consideration of the various equations above will suggest other ways of estimating *P.I.* from leaf length measurements.

Rearranging (4) gives

$$P.I. = \frac{\ln L_o - \ln 10}{p} + \frac{r}{p}t \qquad (5)$$

indicating that *P.I.* is a linear function of time, with r/p as the slope and $(\ln L_o - \ln 10)/p$ as the ordinate intercept. In connection with (2) above, it was noted that r is the relative rate of elongation of a leaf, dL/Ldt, and it might also be written $d\ln L/dt$. The term p is analogous, and might be termed the relative plastochron rate of elongation.[4] Since n is discontinuous, however, p cannot properly be represented as a derivative. It might be written $\Delta \ln L/\Delta n$, t being held constant. In exponential growth, r may similarly be written $\Delta \ln L/$

[4] It is of interest to note that Askenasy (1880) calculated values of "*Wachsthumsgeschwindigkeit*" as the natural logarithm of the ratio of lengths of two successive internodes, and symbolized it as a. We feel that this expression, which is equivalent to our p, the relative plastochron rate, may have great morphological interest. In exploratory work with *Xanthium*, which the senior author has done at the Earhart Plant Research Laboratory, it appears that p is relatively invariant with changes of temperature, whereas r, the relative time rate of elongation of the leaves, varies markedly with temperature. Furthermore, it is clear that the value of p for various dicotyledonous species varies widely, and we have preliminary data which suggest that p may have some relation to the phyllotactic divergence of the leaves.

TABLE 1. *Leaf lengths of a Xanthium plant and values of the relative plastochron rate (p), the plastochron index (P.I.), and leaf plastochron index (L.P.I.) on successive days*

Date	Leaf number (n) 3	4	5	6	7	8	9	10	p	P.I.
	Leaf length (mm.)									
July										
21	52	54	40.4	*18.6*	*8.5*	4.1			0.784	6.79
22	55	60	42	30	*12.7*	*5.8*			0.786	7.30
23	57	64	54	45.2	*19.9*	*9.0*	4.9		0.795	7.87
24	57	67	62	47	29.4	*12.7*	*6.1*	3.2	0.733	8.33
25	58	70	67	58	40.3	*17.6*	*7.5*	4.0	0.853	8.66
26	59	70	72	42	42	24.7	*11.1*	*5.1*	0.805	9.13
	Leaf plastochron index									
21	3.79	2.79	1.79	0.79	—0.21	—1.11				
22	4.30	3.30	2.30	1.30	0.30	—0.70				
23	4.87	3.87	2.87	1.87	0.87	—0.13	—1.13			
24	5.33	4.33	3.33	2.33	1.33	0.33	—0.67	—1.67		
25	5.66	4.66	3.66	2.66	1.66	0.66	—0.34	—1.34		
26	6.13	5.13	4.13	3.13	2.13	1.13	0.13	—0.87		

Δt. Taking equal values of the numerators of the two expressions

$$\frac{r}{p} = \frac{\ln L/\Delta t}{\ln L/\Delta n} = \frac{\Delta n}{\Delta t}$$

that is, the rate of change of node number with time, or the rate of appearance of leaves. If Δn is taken as unity, Δt will be the time required for two successive leaves to pass the reference length. That is to say, p/r corresponds to the average duration of a plastochron.

The linearity of the relationship between the plastochron index and time can be tested by plotting *P.I.* as calculated above against days. As shown in fig. 6, a satisfactory straight line results for a considerable period. As in many dicotyledonous seedlings, leaves 1 and 2 are nearly opposite and of nearly equal lengths throughout their development. That is to say, plastochron 1, as we define it, is very short or zero. Hence it is useless to calculate a plastochron index until leaf 2 has exceeded 10 mm., i.e., until the plant has entered the second plastochron. It will also be noted from fig. 1 that plastochron 3 is shorter than either 2 or 4. This is generally true of the plants we have studied and leads to a systematic departure from linearity of the relationship of *P.I.* to time. No attempt has been made to correct for this error. From plastochron 5 on, the leaves seem to appear in a consistent (2 + 3) phyllotactic sequence, and there seem to be no systematic departures from equality of successive plastochrons correlated with the morphology of the shoot. In short, *P.I.* should be a more satisfactory index of development after plastochron 5 than before. After about *P.I.* 13 there is a decrease in slope of the curve. This is undoubtedly to be explained by root-binding of the plant, since in several similar plants the original rate was restored by transferring the plant to a larger pot. A straight line has been fitted by least squares to these data up to about *P.I.* 13 and the standard error of estimated *P.I.* calculated to be 0.018. The slope of the fitted line, r/p, is 0.47 day^{-1}, and the average duration of the plastochron, p/r, is therefore 2.12 days. The standard error, 0.018 plastochron, is therefore equivalent to slightly less than one hour. For other plants standard errors of estimate of 0.15 or less have been calculated, and it is clear that under our conditions the plastochron index serves to specify the developmental status of a shoot with an error of only a few hours.

THE PLASTOCHRON INDEX OF A LEAF.—For studies of leaf development, it is convenient to alter (1) to provide a common scale of plastochrons for any leaf on a shoot. This can be done for any leaf by subtracting its serial number from the plastochron index of the plant on which it is borne. The resulting leaf plastochron index for a given leaf, whose serial number is *a*, is

$$(L.P.I.)_a = (P.I.) - a = n - a + \frac{\ln L_n - \ln 10}{\ln L_n - \ln L_{n+1}} \quad (6)$$

On this scale, a leaf exactly 10 mm. long will have a plastochron age of zero, a leaf shorter than 10 mm. will have a negative plastochron age, and one longer than 10 mm. will have a positive value of the index. *L.P.I.* values have been calculated for each leaf listed in table 1 on each of the six days. Note that on a given date the index increases by one for each successive leaf, counting from the apex toward the base. For a given leaf, it increases approximately linearly with time. This is true even for leaves such as numbers 3 and 4 which have ceased elongating, and from whose length alone no index of age could be obtained.

APPLICATION OF THE PLASTOCHRON INDEX OF THE SHOOT.—The plastochron index as defined by (1) applies to the shoot, and should be useful in developmental studies in which interest centers on characteristics of the shoot or entire plant, such as total fresh or dry weight, content of a biochemical constituent, rate of a metabolic process, etc. While we have no data to illustrate this application, some suggestions might be made. It is usual in such studies to plant a crop of the plant in question in the field or greenhouse, and to dig samples of plants for analysis at weekly or other intervals during the growing period. Because of variability of the plants in their response to environmental fluctuations, fairly large samples are usually used. Average values are then plotted against time. We would propose that in such a study the plastochron index of each plant be found, and that analyses be made on individual plants. The analytical results for each plant could then be plotted against its *P.I.* Plotting *P.I.* against time would provide the information necessary to interpret the *P.I.* scale in temporal terms. It is convenient, though not necessary, that the *P.I.* be a linear function of time. It might be desirable in some cases to plant at frequent intervals until a wide selection of plants in various stages of development are at hand, and at each sampling time to take plants of as wide a range of stages as feasible. This procedure would have the effect of confounding the environmental variability, and should give more precise information with an economy of plants and analytical work. Some limitations of this application of the plastochron index are discussed below.

In experimental studies of the effect of various regimes of light and temperature conditions, nutritional treatments, etc., the rate of progression, r/p, of the plastochron index might be determined for a few plants under each set of conditions. This sort of study would have interest in itself in many cases. Other plants grown under the same sets of conditions could then be taken at various plastochron ages, for analysis. If dry weights, for instance, were determined, rates of change in dry weight per plastochron could be evaluated from a plot of dry weight against *P.I.* These rates multiplied by r/p would yield rates of change of dry weight with time.

The plastochron index should also serve for studies of the development of the shoot apex, such as those of Abbe and Phinney (1951), in which a progressive increase in mass of the apical meristem of *Zea* during the growing season was demonstrated, and Zimmermann (1928) who demonstrated cyclical changes in area of the meristem correlated with leaf initiation in *Hypericum.* Much would be gained by determining the precise plastochron age of each apex, rather than simply grouping them into classes corresponding with integral values of *P.I.*, as Abbe and Phinney have done, or judging the plastochronic phase subjectively as in Zimmermann's work. A reference length of the order of 100μ, which could be measured microscopically, might be suitable in this application.

APPLICATION OF THE LEAF PLASTOCHRON INDEX.—For studies of leaf development, the *L.P.I.* is applicable. This may be illustrated in a simple manner by replotting the leaf lengths of fig. 1 against the *L.P.I.* for each leaf (fig. 7). It is apparent that the growth curves for leaves 3, 5, 7, 9, and 11 are very similar in early stages, and that they differ in the plastochron age at which the relative rate of elongation begins to decrease, as well as in the final lengths attained. A more useful application of the leaf index is reported by Michelini (1955). Fresh weight, chlorophyll content and rate of oxygen uptake were determined for *Xanthium* leaves at various stages of development. The great scatter which resulted from plotting these determinations against days from planting was dramatically reduced by plotting the same values against *L.P.I.* Maksymowych (1956) has also applied the leaf plastochron index to a study of histogenesis of *Xanthium* leaves. A much clearer picture of the temporal relationships of such processes as marginal meristem activity and mesophyll development can be had from this study than from the classical descriptions of leaf development.

DISCUSSION.—Many plant physiologists and morphologists have selected plant material for study or experiment on the basis of measurements or morphological criteria of development. For instance, *Avena* coleoptiles of specified length are used in the standard tests for auxin, and one frequently finds statements that only plants with a specified number of leaves were used for given experiments. In effect, a developmental index is frequently used to provide more or less standardized plant material. We cannot attempt here to review these procedures in any comprehensive way, but we shall discuss a few examples which have come to our attention.

Xanthium is widely used in photoperiodic investigations, and the procedures which have been used for standardizing this material are of interest here. Whole plants which have been grown for a speci-

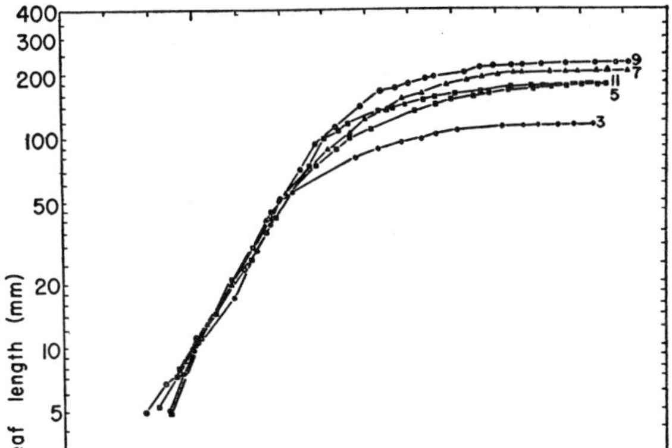

Fig. 7. Lengths of leaves 3, 5, 7, 9, and 11 plotted logarithmically against the leaf plastochron index of each. Same data as fig. 1.

fied time under noninductive conditions have been used in most of the work, and it is sometimes specified that only plants with a given number of expanded leaves have been used. In some cases plants have been defoliated. Parker et al. (1946), for instance, defoliated their plants to a single leaf, which was said to be the youngest "fully-expanded" leaf, and was usually at node 6 or 7. They presumably left an unspecified number of young leaf primordia at the apex. Similar methods of preparing *Xanthium* plants have been used by Bonner and Thurlow (1949), Liverman and Bonner (1953), and others. Khudairi and Hamner (1954) have made a study of the effectiveness of leaves of various developmental stages in photoperiodic induction, using leaf areas and to some extent node numbers as indications of the developmental status of the leaves. Salisbury (1955a, b) has perhaps made the most careful use of such methods of standardization in *Xanthium.* He refers to successive leaves longer than 10 mm. at a given time as the No. 1 leaf, No. 2 leaf, etc., counting from apex to base. These numbers correspond, not to the serial numbers *(n)* which we have assigned to the leaves, but to values of the leaf plastochron index, rounded off to the nearest integer. His fig. 1 is then a plot of leaf length against *L.P.I.* such as our fig. 7. We have calculated from the arbitrary lengths of his No. 1 leaf and No. 2 leaf, 18 and 42 mm. respectively, that the relative plastochron rate, p, is 0.85, in reasonable agreement with our data, and also that his reference leaf length, that is, the length corresponding to $L.P.I. = 0$, is about 7.7 mm. A closer comparison of his methods with ours is not feasible since he has presumably measured leaf blade lengths whereas we have measured total leaf lengths. Incidentally, Salisbury's system of floral stages constitutes a good developmental index for the inflorescence, since he shows (fig. 3) that floral stage increases linearly with time. We feel that the use of the leaf plastochron index to specify leaf age would be appropriate in studies such as these of Khudairi and Hamner, and Salisbury, and would allow more precise standardization.

In his studies of leaf abscission and of xylem regeneration after wounding in *Coleus,* Jacobs (1952) has similarly designated leaf-pairs of increasing age with successive numbers. Similarly, his leaf numbers are integral values of the leaf plastochron index. He states that plants were selected in which leaf-pair no. 1 is 30–50 mm. long and leaf-pair no. 2 is 70–100 mm. long. These values yield a value of the relative plastochron rate, p, of about 0.76 and a reference length of about 18 mm. In preliminary data on the growth of *Coleus* which we have collected, it appears appropriate to use a reference length of 20 mm., since the young leaves are not accessible until about 10–12 mm. long. We suggest that the *L.P.I.* would be quite appropriate for studies of the sort which Jacobs envisages, and would lead to closer correlations with leaf development.

Higgins (1952) has proposed a developmental index for "phenological" observations of cultivated peas *(Pisum)* which has much in common with the plastochron index here described. He proposed counting the number of "fully-developed nodes" on a pea plant, and adding to this integer a decimal fraction representing the stage of development reached by the "terminal bud." The integral part of his index corresponds to the integral part of the plastochron index as here defined. He provides a series of ten drawings and ten descriptive statements to define tenths of a plastochron. We have been told by Dr. F. W. Went that in his experience Higgins' index is not linear with time for fractional parts of a plastochron. We have therefore taken a limited series of photographs of a developing pea shoot with an automatic camera, from which it can be concluded that Higgins' value of 0.5 plastochron corresponds to an actual value of about 0.3. Higgins' index would be a very convenient one provided that accurate descriptive criteria of fractions of a plastochron could be found. In the case of our photographs, however, it is evident that no striking morphological criteria are at hand except during the earlier part of the plastochron as Higgins has defined it. Rather, the changes are quantitative and measurements are, of course, appropriate. The pea leaf, however, is awkward to measure, since it is terminated by a tendril, and in any case is enclosed tightly by the stipules in early stages. From preliminary observations, it seems appropriate in the pea plant to use internode lengths in calculation of a plastochron index. We would suggest assigning to each internode the same serial number as that of the leaf subtending it, and using internode lengths in formula (1) in place of leaf lenths, taking 20 mm. as the reference internode length.

It may be noted that the names orange pekoe, pekoe, souchong, etc., which are used in grading of tea (Hill, 1937), constitute a crude leaf plastochron index.

From the discussion above it may be supposed that the plastochron index as formulated for *Xanthium* would be applicable in studying vegetative shoot growth in a variety of herbaceous dicotyledons, with modifications depending on the morphology of the plant in question. The index fails in *Xanthium* when flowering is induced, since leaf initiation is halted, and this same limitation may apply to many plants. In studying vegetative development, then, it would be desirable, for this and other reasons, to choose plants and growing conditions which will provide strictly vegetative growth. For the study of flowering phenomena, the indices developed by Erickson (1947) for *Lilium* flowers and by Salisbury (1955) and others for *Xanthium* inflorescences are suggestive of ways in which developmental indices may be derived.

In some groups of plants, difficulty would perhaps be encountered in finding an appropriate basis for a plastochron index. In grasses, for instance, young leaves and internodes are enclosed tightly in the sheaths of more mature leaves. On dissection, the young leaves could be measured and a plastochron index calculated for each dissected plant. The difficulty, however, of establishing the relation between this plastochron index and time is one for which we do not see a solution at present. Difficulties might also be encountered in plants with marked seasonal growth. In several tree species which we have observed casually the winter bud contains several leaf primordia of nearly the same length. They expand rapidly and nearly simultaneously after the buds open, and may or may not be followed by later leaves, which follow a plastochronic sequence like that in *Xanthium*. Whether a useful developmental index could be formulated for this process, and the many other growth processes exhibited by plants, would depend in each case on a careful working-out of the growth pattern in time, and some ingenuity in selecting appropriate measurements or criteria which can be related to time in a precise way.

Developmental indices could no doubt be used to advantage in the study of other organisms than the higher plants. Stages of normal development have been defined with greater or lesser formality for many organisms. In vertebrate embryology, for instance, the series of normal stages defined by Harrison and others for a number of amphibians (Hamburger, 1942; Rugh, 1948) and by Hamburger and Hamilton (1951) for the chick embryo are in effect developmental indices, and as we understand, have great utility in experimental embryology. The available data show, however, that the relationship of stage to time is not simple, perhaps because the numbered stages have been chosen primarily for convenience of definition in morphological terms. Furthermore, the possibility of numerical interpolation between the stages does not appear to have been considered seriously. We would suggest that in many organisms numerical stage series, or developmental indices, preferably defined so as to be linear with time, would have great advantages, not only for the standardization and evaluation of experimental material, but also for the calculation of rates of many processes involved in development.

SUMMARY

An index which specifies precisely the developmental status of a vegetative shoot is derived, with particular reference to *Xanthium*. This *plastochron index* has the formula

$$P.I. = n + \frac{\log L_n - \log 10}{\log L_n - \log L_{n+1}},$$

where n is the serial number of the leaf just longer than an arbitrary reference length, 10 mm., and $\log L_n$ is the logarithm of the length of leaf n. *P.I.* is shown to be linearly related to time during vegetative growth, and can therefore be used as a substitute time axis in many studies. The use of the plastochron index in studies of shoot development, and of the similar leaf plastochron index *(L.P.I.)* in studies of leaf development is advocated.

DEPARTMENT OF BOTANY,
UNIVERSITY OF PENNSYLVANIA,
PHILADELPHIA, PENNSYLVANIA

LITERATURE CITED

ABBE, E. C., AND B. O. PHINNEY. 1951. The growth of the shoot apex in maize: external features. Amer. Jour. Bot. 38: 737–744.

ASKENASY, E. 1880. Über eine neue Methode, um die Vertheilung der Wachsthumsintensität in wachsenden Theilen zu bestimmen. Verh. Naturh.-medic. Ver. Heidelberg 2: 70–153.

BLACKMAN, V. H. 1919. The compound interest law and plant growth. Ann. Bot. 33: 353–360.

BONNER, J., AND J. THURLOW. 1949. Inhibition of photoperiodic induction in *Xanthium* by applied auxin. Bot. Gaz. 110: 613–624.

ERICKSON, R. O. 1947. Respiration of developing anthers. Nature 159: 275.

———. 1948. Cytological and growth correlations in the flower bud and anther of *Lilium longiflorum*. Amer. Jour. Bot. 35: 729–739.

ESAU, K. 1953. Plant anatomy. John Wiley and Sons, Inc. New York.

HAMBURGER, V. 1942. A manual of experimental embryology. Univ. of Chicago Press. Chicago.

———, AND H. L. HAMILTON. 1951. A series of normal stages in the development of the chick embryo. Jour. Morph. 88: 49–92.

HIGGINS, J. J. 1952. Instructions for making phenological observations of garden peas. Johns Hopkins Univ. Publ. Climatol. 5(2): 1–11.

HILL, A. F. 1937. Economic botany. McGraw-Hill Book Co. New York.

JACOBS, W. P. 1952. The role of auxin in differentiation of xylem around a wound. Amer. Jour. Bot. 39: 301–309.

KHUDAIRI, A.-K., AND K. C. HAMNER. 1954. The relative sensitivity of *Xanthium* leaves of different ages to photoperiodic induction. Plant Physiol. 29: 251–257.

LIVERMAN, J. L., AND J. BONNER. 1953. Biochemistry of the photoperiodic response: the high-intensity-light reaction. Bot. Gaz. 115: 121–128.

MAKSYMOWYCH, R. 1956. Development of lamina in *Xanthium italicum* Moretti represented by the plastochron index. (M.S. thesis). Univ. Pennsylvania.

MICHELINI, F. J. 1955. The use of the plastochron index in studies of morphological and physiological development in *Xanthium italicum* Moretti. (Ph.D. dissertation). Univ. Pennsylvania.

OGUR, M., R. O. ERICKSON, G. U. ROSEN, K. B. SAX, AND C. HOLDEN. 1951. Nucleic acids in relation to cell division in *Lilium longiflorum.* Exp. Cell Res. 2: 187–203.

Parker, M. W., S. B. Hendricks, H. A. Borthwick, and N. J. Scully. 1946. Action spectrum for the photoperiodic control of floral initiation of short-day plants. Bot. Gaz. 108: 1–26.

Rugh, R. 1948. Experimental embryology. Burgess Publishing Co. Minneapolis.

Salisbury, F. B. 1955a. Kinetic studies on the physiology of flowering. (Ph.D. dissertation). California Inst. Tech.

———. 1955b. The dual role of auxin in flowering. Plant Physiol. 30: 327–334.

Zimmermann, W. A. 1928. Histologische Studien am Vegetationspunkt von *Hypericum uralum.* Jahrb. Wiss. Bot. 68: 289–344.

EXPERIMENTAL MODIFICATION OF PLANT SENESCENCE [1, 2]

A. C. LEOPOLD, ETHEL NIEDERGANG-KAMIEN, AND JULES JANICK

HORTICULTURE DEPARTMENT, PURDUE UNIVERSITY, LAFAYETTE, INDIANA

The termination of the life cycle of annual and biennial plants is sudden and complete. Entire fields of grains and other annual crops die simultaneously, often in the early part of the growing season. Neither the function nor the physiological basis for this senescence is understood.

Plant senescence has received little attention since the publication of a small, perceptive book by Molisch (6) 30 years ago. He noted that plants which were delayed in the completion of the reproductive stages would show delayed senescence. For example, century plant (*Agave americana* L.) has a life span of 8 years in some climates and 100 years in others depending upon how soon it becomes reproductive; whenever it flowers and fruits the plant dies. He carried out experiments with various species of annual plants, and found that removal of flowers effectively extended the life span. From these observations he deduced that ". . . when we hinder flower formation . . . we guard the plant against death . . . but just as soon as it has flowered and fruited, the organic reserves are completely emptied and utilized and a yellowing of the entire plant accompanies this process . . . and for this reason the plant dies after ripening of its seeds."

The present experiments provide a quantitative extension of these observations of Molisch, and provide evidence that the deduced cause of senescence is more complex than a depletion of food reserves.

METHODS

Experiments with soybean (*Glycine max* Merr. var. Lincoln) were carried out in the field with 10-foot rows planted on May 12, 1956. After the first appearance of flowers defloration was carried out 3 times per week on triplicate rows. Flowers were removed as soon as they showed petal color. The experiment was terminated by a killing frost on October 10. Similar experiments were carried out in the summers of 1957 and 1958. More detailed experiments were carried out in the greenhouse in 1958, using various times of flower and fruit removal.

Experiments with spinach (*Spinacia oleracea* L. var. Long Standing Bloomsdale) were carried out in the greenhouse. Plants were selected from large populations for uniformity. In a typical experiment about 60 % of the plants were discarded in order to attain maximal uniformity in the experimental material—an essential if reliable measurements of small differences in senescence rates are to be obtained. At the beginning of the experimental period all plants were decapitated in the internode above the 10th flowering node. At the same time, leaves from the basal rosette were removed in order that the quantitative measurements of senescence should start from zero. As an additional step to increase the precision

FIG. 1. Delay in senescence of soybean plants by defloration (left) as compared to control plants (right). Photographed September 7, 1956.

FIG. 2. Delayed senescence in soybean plants by various treatments: (a) flowers removed as petals show color, (b) fruits removed after pods had elongated but not filled, (c) fruits removed at first sign of yellowing with ripening, and (d) untreated controls.

[1] Received revised manuscript February 17, 1959.

[2] Journal Paper no. 1349, Agricultural Experiment Station, Lafayette, Indiana. This investigation was supported in part by a grant from The Rockefeller Foundation.

of the experiments, only plants at an identical stage in flowering were utilized in comparable groups.

Senescence was followed quantitatively as evidenced by the degeneration of the leaves. Each leaf on each plant was scored by an arbitrary scale as follows: 0) leaves entirely green, 1) leaf tips necrotic, 2) leaf tip and peripheral area dead, 3) half of leaf dead, and 4) entire leaf dead. Scoring of plants was repeated 3 times per week. Each treatment was applied to 6 to 12 plants carefully selected for uniformity. Lateral branches were removed as soon as they appeared. Each test reported was repeated with parallel results.

Experimental

Soybean plants were deflorated as petals showed color or defruited as the pods became nearly filled but before the yellowing color change of the fruit associated with ripening. These treatments completely prevented the senescence of the plants within the growing season. The appearance of the deflorated plants and the untreated controls is illustrated in figure 1, where it will be seen that all plants in the treated group were green and healthy whereas all plants in the fruiting controls were completely dead. This experiment has been repeated for 3 summers. The same complete prevention of senescence is obtained when the fruits are removed, though there is a somewhat more complete leaf retention in the deflorated group than in the defruited group.

The induction of senescence by the reproductive activities of the plant may be related to any of a number of stages in the flowering and fruiting processes. This is readily seen from greenhouse experiments with soybean plants which were (a) deflorated at the time of 1st petal color, (b) defruited after pods had elongated but not filled out, or (c) defruited after the fruits were entirely filled and just starting to show the yellowing color change associated with ripening. Control plants died at an average of 119 days after sowing, and defruited and deflorated plants were increasingly persistent. The appearance of representative plants 119 days after sowing can be seen in figure 2. The deflorated plants are the most vigorous, the early defruited and late defruited were less vigorous, and the untreated plants were entirely dead. The average time of plant death was recorded for the treatments as follows: Deflorated, 179 days; defruited when pod full length, 161 days; defruited when pod filled, 129 days; and control, 119 days.

To permit a more precise analysis of the effects of various stages of reproduction, experiments were carried out with dioecious spinach plants. Defloration and defruiting treatments were carried out on staminate and pistillate plants at different stages of development.

In the spinach it was soon found that once seed stalk development or bolting has begun, senescence will surely follow and the removal of flowers or fruits can only delay this approaching demise. The comparative effects of different treatments on senescence can be estimated by the extent to which senescence is delayed.

The defruiting of pistillate spinach plants was effective in delaying senescence. Effects of this treatment are shown in figure 3. A useful index in comparing the rates of senescence of the various treatments is the time required for the plants to reach an intermediate stage of senescence. For example, pollinated plants reached a senescence score of 2.0 after 25 days, whereas plants from which the fruits were removed or which remained unpollinated reached this value after 31 days. In the latter case the unpollinated ovaries enlarged but never matured into ripe fruits.

In this experiment fruits were removed from some plants on the 5th day after pollination and it can be

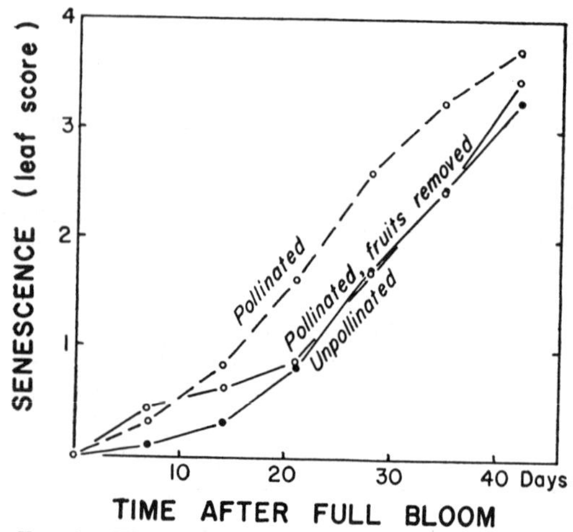

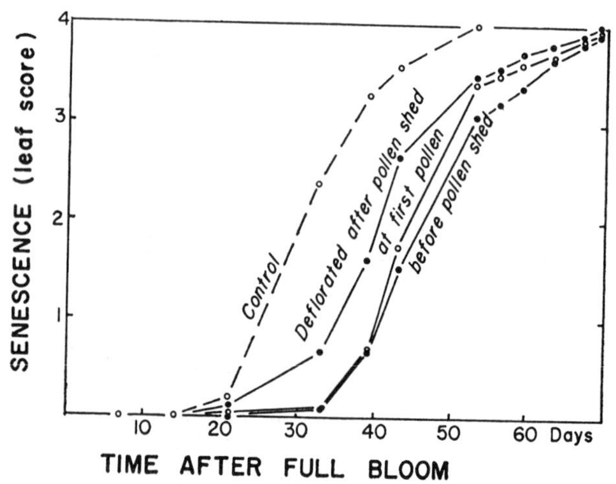

Fig. 3. The development of senescence with time in pistillate spinach plants. Control (pollinated) plants are compared with plants defruited 5 days after pollination and unpollinated plants.

Fig. 4. The development of senescence with time in staminate spinach plants. Control plants are compared with plants deflorated after pollen shed (day 10), at the time of first pollen shed (day 1) or before pollen shed (day 0).

seen that after day 5 the senescence score rapidly became essentially identical with that of the unpollinated controls. In other experiments the removal of fruits was carried out 10 and 20 days after pollination, and in each case the senescence score became essentially the same as that of the unpollinated plants within 10 days after defruiting. Senescence curves of unpollinated plants were not significantly different from those of deflorated plants.

Senescence of staminate spinach plants was likewise delayed by the removal of the flowers. In a sample experiment flowers were removed as they first became identifiable, as dehiscence of the 1st anther occurred, or after the dehiscence of all anthers. The senescence of the plants receiving these 3 treatments is compared in figure 4. It will be seen that the average senescence score for the plants which were not deflorated rose relatively rapidly, and the onset of senescence was delayed increasingly with earlier and earlier defloration. The differences in senescence of plants deflorated when the sex of the flowers became first detectable and when the 1st anther dehisced is probably not real.

The appearance of staminate deflorated plants after 45 days as compared to controls is shown in figure 5. The control plants were dead and the deflorated plants still green and erect.

FIG. 5. Delay in senescence of staminate spinach plants by defloration (left) as compared to control plants (right). Photographed 45 days after full bloom.

Discussion

Senescence appears to be imposed on a plant by an increasing signal, intensified with successive stages in the reproductive development of the plant. In spinach, the induction of flower initiation and bolting appears to start such a signal, for if plants are kept in short photoperiods which maintain a vegetative rosette condition they appear to have a quality approaching immortality, maintaining a healthy green condition for an indefinite period of time. All plants die within a period of 11 weeks after bolting. An intensification of the signal to senescence is associated with the shedding of pollen and, to a larger extent, the degeneration of the staminate flower parts. This is clearly seen in the staminate spinach plants which are delayed in senescence by defloration. A further intensification is associated with the time of fruit enlargement, as indicated by the pistillate spinach plants which are delayed in senescence if the fruits are removed at various times after pollination. And a final increase occurs with the commencement of ripening of the fruits in the bean plants, as indicated by the marked delay in senescence if the fruits are removed shortly before the color changes associated with ripening.

If the concept of Molisch were a complete explanation of the onset of senescence and the organic reserves were emptied from the plant into the developing fruits, then there would be no senescence in the male plants and the delay in senescence in soybeans following the removal of the filled but not yet ripened fruits would not be expected. An interesting side effect of the defloration and early defruiting treatments was an increase in flowering which followed. In fact the defruiting treatment resulted in a considerable increase in the numbers of fruits which were filled and then removed. Thus if the depletion theory were the explanation of senescence then the defruiting treatment might logically have been expected to hasten rather than retard senescence. Surely the depletion effect is involved and contributes to senescence, but we feel that it is only a partial explanation of the onset of death.

It has long been known that flowering and fruiting cause a retardation of growth (7), and it follows that the more vigorous the flowering the more rapid the senescence (11). Many effects have been observed to follow the removal of flowers. In addition to the deferred senescence (6) there is an increase in growth rate (7), reduction in flower abscission (13), an increase in subsequent fruit set (8), and an extension of the duration of flowering (5).

There have been 3 general types of explanations proposed to account for senescence. (a) It has been attributed to excessive water loss associated with the time of flowering (1). This theory has never been seriously considered, though it is clear that wilting of leaves can certainly induce leaf senescence (4). (b) The suggestion has often been made that a decline in available nutrients could cause senescence—the general concept of Molisch. Thus aging leaves show a more rapid utilization of carbohydrates (14, 16), more export of inorganic materials (2, 15), and are less effective in photosynthesis (3, 12). (c) Murneek has proposed that the developing embryos induce senescence in the parent plant. This has been explained as a competition by the embryos for food reserves (9) or, more recently, as a possible production of inhibitors by the developing embryos (10).

While these various physiological events may certainly be contributing to the development of senescence, the evidence provided here cannot be explained on the basis of any or all of them. These experiments suggest an accumulation of influences inducing senescence with each of the developmental stages of reproduction in soybeans and spinach. Establishing a quantitative basis for the measurement of these influences should facilitate the physiological definition of their nature.

Summary

The onset of senescence in soybeans and in spinach plants has been effectively delayed by the removal of either flowers or fruits. The relative effectiveness of such treatments declines progressively as the time of removal is delayed, both in soybeans and in staminate and pistillate plants of spinach. The experiments support a concept of senescence in response to a gradually intensifying signal in reproductive plants, rather than to a depletion of food reserves through mobilization into developing fruits.

Grateful acknowledgement is made to Dr. S. L. Lam for carrying out the greenhouse experiment with soybeans.

Literature Cited

1. Bachuyzen, H. L. S. Physiological phenomena at the time of flowering. Proc. Soc. Exptl. Biol. Med. 24: 143–145. 1926.
2. Chibnall, A. C. Protein Metabolism in the Plant. Yale Univ. Press, New Haven 1939.
3. Egorov, M. A. A propos du role des elements de cendre dans la vie des plantes. Zhur. Opyt. Agron. 16: 270–280. 1915.
4. Gates, C. T. The response of the young tomato plant to a brief period of water shortage I. The whole plant and its principal parts. Australian Jour. Biol. Sci. 8: 196–214. 1955.
5. Gorter, C. J. The flowering response of *Phaseolus vulgaris* to 2,3,5-triiodobenzoic acid and its Cl- and Br- analogues. K. Akad, Wetenschap. Amsterdam. Proc. Sect. Sci. C57: 606–616. 1954.
6. Molisch, H. Der Lebensdauer der Pflanze. G. Fischer Verlag. 1928. Translation of E. H. Fulling, Science Press, Lancaster. 1938.
7. Murneek, A. E. Effects of correlation between vegetative and reproductive functions in the tomato. Plant Physiol. 1: 3–56. 1926.
8. Murneek, A. E. Physiology of reproduction in horticultural plants II. Intermittent sterility of the spider flower. Missouri Agr. Expt. Sta. Res. Bull. 106. 1927.
9. Murneek, A. E. Growth and development as influenced by fruit and seed formation. Plant Physiol. 7: 79–90. 1932.
10. Murneek, A. E. Growth regulators during fertilization and post-fertilization periods. Palestine Jour. Bot. 8: 8–19. 1951.
11. Resende, F. and Viana, M. J. Some effects of synthetic auxins and antiauxins on the floral state. Bol. Soc. Portug. Ci. Nat. 4: 74–78. 1952.
12. Richardson, S. D. The effect of leaf age on the rate of photosynthesis in detached leaves of tree seedlings. Acta Bot. Néerl. 6: 445–457. 1957.
13. Van Stevenincк, R. F. M. Factors affecting the abscission of reproductive organs in yellow Lupins I. Jour. Exptl. Bot. 8: 373–381. 1957.
14. Weatherley, P. E. On the uptake and hydrolysis of sucrose by leaf tissues. New Phytol. 52: 76–79. 1953.
15. Williams, R. F. Redistribution of mineral elements during development. Ann. Rev. Plant Physiol. 6: 25–42. 1955.
16. Yemm, E. W. The metabolism of senescent leaves. In: CIBA Colloquia on Ageing, G.E.W. Wolstenholme, ed. Vol. 2. Pp. 207–214. 1956.

Reprinted from the Proceedings of the NATIONAL ACADEMY OF SCIENCES
Vol. 45, No. 7, pp. 932–938. July, 1959.

*A DEMONSTRATION OF THE RECOVERY OF THE CROWN-GALL TUMOR CELL WITH THE USE OF COMPLEX TUMORS OF SINGLE-CELL ORIGIN**,†

BY ARMIN C. BRAUN

THE ROCKEFELLER INSTITUTE

Communicated by L. O. Kunkel, May 13, 1959

Introduction.—It is a generally accepted belief that the cellular alteration leading to malignant tumor cell types represents a permanent change. Cells that possess low grades of neoplastic change may and frequently do become further altered in the direction of greater malignancy. Once this state has been attained, however, such cells have not been observed to turn back *en masse* toward the benign or normal states. Studies on the origin of a tumor cell (except the virus-induced tumors) have been complicated in the past by an inability to accomplish a controlled recovery of such cells and thus permit an experimental analysis of the several hypotheses that have been advanced to account for the continued abnormal and autonomous growth of a tumor cell.

The typical crown-gall tumor cell of plants, like malignant animal cells, appears to be a permanently altered cell that reproduces true to type and against the growth of which there is no control mechanism in the host. This cell type is characterized both in the host and in culture by a capacity for continued rapid proliferation, by a limited capacity for differentiation, and by the lack of a capacity to organize tissues and organs (Fig. 1, A). Since tumor cells of this type have not in the more than 10

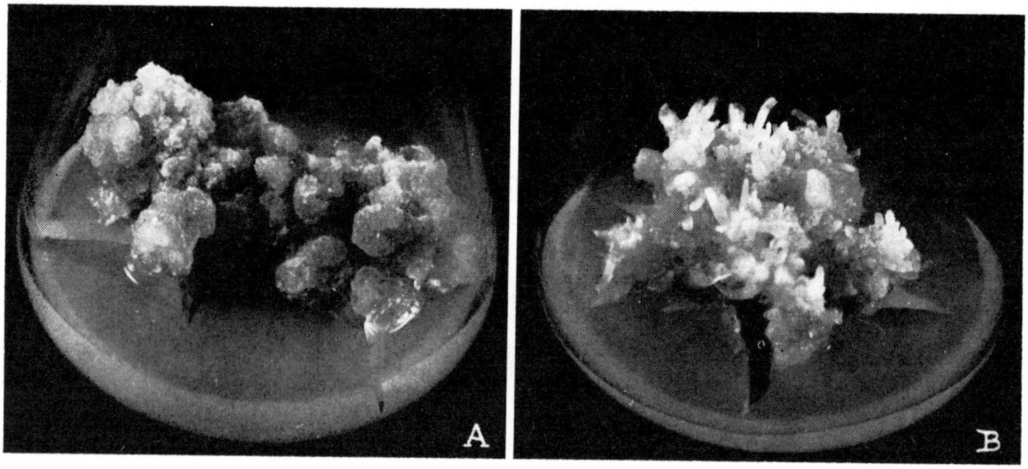

FIG. 1.—A. Typical tobacco crown-gall tumor tissue of the unorganized type. B. Crown-gall teratoma tissue of tobacco derived from a single cell. (Photographs by J. A. Carlile.)

years that they have been kept under observation shown the slightest tendency to become less autonomous, they have generally been considered to be permanently altered cells and hence unsuited for studies dealing with the recovery of a tumor cell.

A second morphologically quite distinct type of crown-gall tumor may be produced in certain plant species the cells of which possess highly developed regenerative capacities. Tumors of this type possess, when they develop in a host, and retain indefinitely when planted in culture, a pronounced capacity to organize highly abnormal leaves and buds (Fig. 1, B). The pluripotent cells of which these teratomata are composed appear to retain, despite their alteration to tumor cells, highly developed regenerative capacities.

The complex tumors, like those of the unorganized type, are transplantable. The growth pattern of the resulting tumor depends, however, upon the position that the implant occupies in a host.[1] Furthermore, cells isolated from teratomata grow indefinitely, as do cells obtained from an unorganized tumor, on a simple inorganic salts-sucrose-containing culture medium that does not support the continued growth of normal cells of the type from which the tumor cells were derived. Since the complex tumors possessed a capacity to organize abnormal tumor buds, they were found to be admirably suited and were used for studies on the origin of the crown-gall tumor cell.[2] An attempt was made in earlier investigations to distinguish between somatic mutation at the nuclear gene level and the presence of a self-perpetuating cytoplasmic entity that had assumed control of the cells and was responsible for the continued abnormal proliferation of the affected cells. It is well known in biology that certain self-duplicating cytoplasmic entities can be eliminated from cells under conditions that favor the increased multiplication of those cells in relation to the multiplication of the self-duplicating factor. Since normal meristematic cells of a rapidly growing bud divide with far greater frequency than do most crown-gall tumor cells, it was hypothesized that, if the tumor buds found to develop from the teratomata could be forced into very rapid growth, recovery of the tumor cells might be accomplished provided the factor responsible for the continued abnormal proliferation of the tumor cells was subject to the effects of dilution in very rapidly dividing cells. The results of that study showed that when tumor shoots derived from tumor buds were forced into rapid growth by a series of graftings to healthy plants, they gradually recovered and became normal in every respect. These results suggested that the cellular alteration in crown gall did not involve a somatic mutation at the nuclear gene level since heritable changes of that type are not generally believed to be lost as a result of rapid growth. They suggested, rather, that some autonomous or partially autonomous entity, which is subject to the effects of dilution in very rapidly dividing cells, is present in and is responsible for the continuity of the tumorous properties from one cell generation to the next. This interpretation of the observed results was subsequently questioned by certain investigators because it was felt that, on the basis of the evidence presented, the possibility that the teratoma tissue was composed of a stable mixture of normal cells and tumor cells had not been eliminated.[3–5] This question could be resolved unequivocally only if clones of teratoma tissue of single-cell origin were used in the investigations. Although methods for the culture of single plant cells had not been developed at the time that the original studies were reported, such methods are now available.[6] This has permitted a re-examination of the question of the recovery of the crown-gall tumor cell. It is with that phase of the crown-gall tumor problem that the present study is concerned.

Experimental Methods.—The clone of tobacco crown-gall teratoma tissue used in this study has now been maintained in culture for somewhat more than 5 years. Throughout this period the tissue has been characterized by a pronounced capacity to organize numerous abnormal-appearing leaves and buds and to grow continuously at a moderately fast rate on a modified White's culture medium containing 1 per cent agar. The teratoma tissue was maintained in this laboratory on a modified medium containing four times the concentration of the basic salts recommended by White.[7]

The single cells used in this study were obtained as follows. The liquid culture medium consisted of White's basic medium supplemented with a 10 per cent extract of juice obtained from rapidly growing *Vinca rosea* crown-gall tumors grown in tissue culture. The tumor extract was found to favor growth of the tumor cells. This medium was further supplemented with naphthalene acetic acid at a final concentration of 0.01 mg/liter. Fifty cc of the medium was placed in each of several 250-ml Erlenmeyer flasks. The flasks were stoppered and sterilized in an autoclave.

Teratomata that had grown on a solid medium for 5–6 weeks were cut into small fragments several millimeters in diameter and a large number of such fragments were placed in the medium described above. The flasks were shaken rapidly on a reciprocal shaker for a period of about 5 weeks. Microscopic examination at the end of that period showed numerous single cells and clumps of cells to be present in the medium (Fig. 2, A, B). A suspension of such cells was either transferred to fresh medium of the type described above and again shaken or it was spread evenly on an agar surface in a Petri dish for isolation of single cells.

The single cells were isolated and cultivated according to the method described by Muir, Hildebrandt, and Riker.[6] The "host" or stock tissue upon which the filter paper (Reeve Angel, crepe surface no. 202) was placed was normal *Nicotiana glutinosa* tissue that had been carried continuously in culture for 10 years on White's basic medium supplemented with 0.5 mg/liter naphthalene acetic acid and fortified with 1 per cent agar. Normal *N. glutinosa* tissue had been found to be a favorable nurse tissue for tobacco teratoma cells. In these studies a cube of the normal tissue was cut in such a way that each side had a diameter of about 7 mm. A square piece of sterile filter paper 8–9 mm in diameter was placed upon the freshly cut upper surface of the tissue fragment. The tissue containing the paper was planted

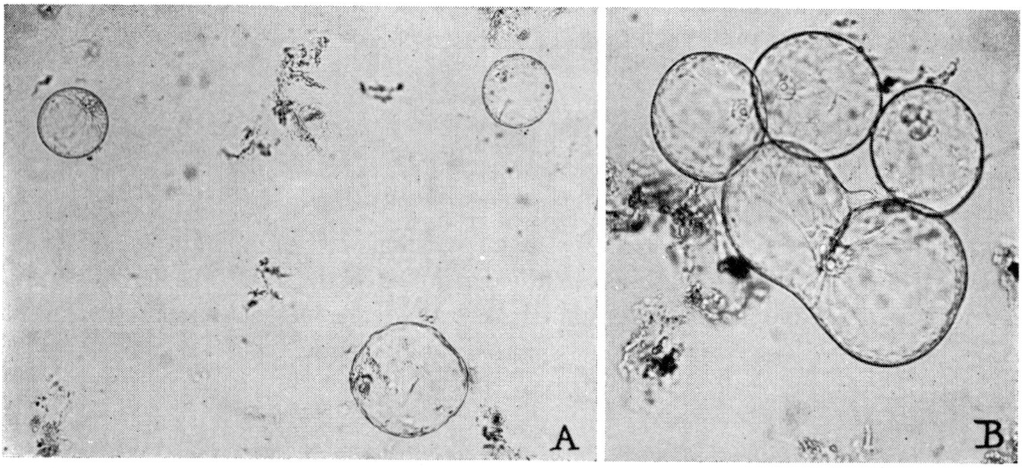

Fig. 2.—A. Single cells of the type from which the teratoma tissue shown in Fig. 1, B was derived. B. Small cluster of cells. Note that one of the cells is undergoing division. (Photographs by J. A. Carlile.)

in 125-ml Erlenmeyer flasks containing White's solid medium supplemented with naphthalene acetic acid at a concentration of 0.05 mg/liter. The tissue was allowed to remain in contact with the agar for 3 days, after which time the tissue and filter paper were removed as a unit and placed in a sterile Petri dish.

Single teratoma cells that had been planted on an agar medium in a Petri dish were isolated under a dissecting microscope with the aid of a platinum microspatula that was flattened and pointed at the tip. The single cells were planted on the filter paper. The tissue containing the filter paper was then again placed on the agar medium from which it had been removed. The entire procedure was carried out as rapidly as possible to prevent drying of the single cell.

The grafting of tissues of single-cell origin and of shoots that developed from such tissues was carried out according to methods previously described.[2,8]

Experimental Results.—Of a total of 267 single teratoma cells isolated and planted on filter paper, 12 or somewhat more than 4 per cent grew. Of these, 1 became contaminated during the course of incubation, 4 multiplied and reached diameters up to

0.3 mm but failed to continue growth, while 7 reached diameters up to 5 mm and were planted on a modified White's basic medium where they continued to grow well independently. The tissue of each of the 7 clones grew in an unorganized manner and showed no macroscopic evidence of organization at the time that they were transplanted to the basic medium from the filter paper. However, after the tissues of 6 of the 7 clones had reached diameters of between 0.7 and 1 cm, they organized morphologically abnormal structures at their surfaces. The tissue of each of the 7 clones was subdivided into 4 pieces and these were again planted on modified White's basic medium. The tissues derived from 6 of the clones grew in a manner comparable in every respect to the teratoma tissue from which the single cells were derived. Tissues of the seventh clone failed to organize during two successive passages in culture. During the third passage, morphologically highly abnormal leaves and buds developed from several of the tissues in this clone. The capacity of tissues of all clones derived from single cells to organize persisted through subsequent subdivisions and plantings on modified White's basic medium.

The results of this study indicate that the capacity of teratoma tissue of single-cell origin to organize is a reflection of the inherent potentialities of pluripotent tumor cells and is not the result, as has been suggested, of a mixture of normal and tumor cells. The possibility exists, of course, that, as the progeny of the single cells grow, some recover, perhaps as a result of unequal cell divisions, from the effects of the tumor-inducing principle associated with this disease and that it is those recovered cells that are responsible for the organizational capacity exhibited by the teratoma tissue. Although this possibility cannot at present be tested experimentally, there is every indication that, if recovery occurs in the manner suggested above, it does not happen with sufficient frequency to account for the observed results.

The finding that clones of teratoma tissue of single-cell origin developed organized structures permitted a re-examination of the question as to whether a controlled recovery of crown-gall tumor cells could be accomplished by forcing organized but morphologically abnormal tumor buds into rapid growth at the shoot apex by means of a series of graftings to healthy tobacco plants. In these studies teratoma tissues of single-cell origin and about 5–7 mm thick were grafted to cut stem ends of tobacco plants from which the axillary buds had been removed. In those instances in which the tumor grafts were successful, they developed into overgrowths of considerable size. The surfaces of these growths were covered with organized structures many of which were highly abnormal in appearance. An examination of free-hand sections of such grafted tumor tissue showed that the organized structures arose from the tumor tissue at the periphery of the expanding tumor mass and not from tissues of the normal plant upon which the graft was made.

The procedures used in an attempt to effect the recovery of crown-gall teratoma cells were similar to those previously described.[2] Shoots that developed from the tumor buds present in the teratoma grafts were removed when they were between 3 and 5 cm long and were grafted to the cut stem ends of healthy tobacco plants. Some of these shoots were very abnormal in appearance and grew slowly and very abnormally in length and in thickness. After they had reached an appropriate length, the tips were again removed and grafted to healthy tobacco plants. These commonly grew more rapidly and became more normal in appearance. The tips of such shoots were again removed and grafted to healthy plants. They developed rapidly and appeared normal in every respect, ultimately flowering and setting seed. Such seed was sown and the resulting plants were found to be normal *Nicotiana tabacum* plants (Fig. 3). In these instances recovery appeared to have been complete. It was a gradual process that progressed in the direction of the normal as the affected shoots were forced into very rapid growth as a result of a series of graftings to healthy plants.

In certain other instances, recovery was achieved more quickly. When, for example, some of the more normal-appearing shoots that developed from the grafted teratomata of single-cell origin were grafted to the cut stem tips of healthy tobacco plants, they grew rapidly after an initial delay and a complete recovery was achieved in a single grafting. The cells of such recovered scions failed to grow on a basic culture medium that supported the continued rapid growth of crown-gall tumor cells. The cells of the recovered plants were also found to be again susceptible to transformation by the tumor-inducing principle associated with this disease.

FIG. 3.—A normal tobacco plant of the type obtained by forcing morphologically abnormal shoots derived from teratoma tissue of single-cell origin into very rapid growth by means of a series of graftings to healthy plants. A complete recovery from the tumorous state has resulted from this procedure. (Photograph by J. A. Carlile.)

Discussion.—Clones of complex crown-gall tumors derived from single cells were used in investigations described above. The results of these studies indicate that when pluripotent plant cells are transformed into crown-gall tumor cells, they may retain indefinitely a capacity to organize morphologically highly abnormal leaves and buds. The organizational capacity exhibited by such cells is an expression of the inherent potentialities of the pluripotent tumor cells and does not in itself appear to affect the recovery of those cells. Only when organized structures such as those found in tumor shoots derived from the morphologically abnormal tumor buds were forced to divide very rapidly at the shoot apex, did they recover and ultimately become normal in every respect. These findings make somatic mutation at the nuclear gene level appear unlikely as a possible explanation for the nature of the cellular alteration in crown gall. They suggest, instead, that some as yet uncharacterized cytoplasmic entity is responsible for the cellular changes that underlie the tumorous state in the crown-gall disease. This does not, of course, preclude the possibility that such postulated cytoplasmic changes may be more or less under control of the nuclear genes. These findings, which are very suggestive of some that have been encountered in microbial genetics, may be interpreted by assuming that the factor responsible for the continued abnormal growth of the crown-gall tumor cell is an autonomous or partially autonomous entity that is subject to the effects of dilution in very rapidly dividing cells.[9] While such a particle hypothesis appears to explain quite satisfactorily the experimental findings reported here, so little is in fact known about the mechanics of cell division, differentiation and organization that other interpretations, such as, for example, Delbrück's[10] suggestion involving steady state chemistry in which alternative chains of metabolic reactions are assumed to compete with one another, may apply.

This study demonstrates unequivocally, furthermore, that the progeny of a single somatic parenchymatous cell of tobacco may possess all of the potentialities necessary to reconstitute an entire tobacco plant.

* This investigation was supported in part by a research grant (PHS C-2944 M & G) from the National Cancer Institute, Public Health Service.

† Some preliminary studies on this problem were carried out in the laboratory of Professor F. C. Steward while the author was Visiting Professor in the Botany Department at Cornell University, Ithaca, New York, in 1957. The author is indebted to Professor Steward and to Miss Kathryn Mears for their kind assistance and interest in this investigation.

1 Braun, A. C., *Botan. Gaz.*, **114,** 363 (1953).

2 Braun, A. C., *Cancer Research,* **11,** 839 (1951).

3 Gautheret, R. J., *Rev. Gén. Bot.*, **62,** 5 (1955).

4 Rack, K., *Phytopathol. Ztschr.*, **21,** 1 (1953).

5 White, P. R., Discussion, p. 125, following article by A. C. Braun in *Brookhaven Symp. Biol., No. 6, Abnormal and Pathological Plant Growth,* 115 (1954).

6 Muir, W. H., A. C. Hildebrandt, and A. J. Riker, *Am. J. Bot.*, **45,** 589 (1958).

7 White, P. R., *The Cultivation of Animal and Plant Cells* (New York: The Ronald Press, 1954).

8 White, P. R., and A. C. Braun, *Cancer Research,* **2,** 597 (1942).

9 Braun, A. C., *Brookhaven Symp. Biol., No.* **6**, *Abnormal and Pathological Plant Growth,* 115 (1954).

10 Delbrück, M., Discussion, p. 33, of article by Sonneborn, T. M., and G. H. Beale in *Unités biologiques douées de continuité génétique, CNRS,* Paris, **7,** 25 (1949), as discussed in: G. H. Beale, *The Genetics of Paramecium aurelia,* (Cambridge Univ. Press, 1954).